Lecture Notes in Computer Science 9665

More information about this series at http://www.springer.com/series/7410

Marc Fischlin · Jean-Sébastien Coron (Eds.)

Advances in Cryptology – EUROCRYPT 2016

35th Annual International Conference
on the Theory and Applications of Cryptographic Techniques
Vienna, Austria, May 8–12, 2016
Proceedings, Part I

Springer

Editors
Marc Fischlin
Technische Universität Darmstadt
Darmstadt
Germany

Jean-Sébastien Coron
University of Luxembourg
Luxembourg
Luxembourg

ISSN 0302-9743 ISSN 1611-3349 (electronic)
Lecture Notes in Computer Science
ISBN 978-3-662-49889-7 ISBN 978-3-662-49890-3 (eBook)
DOI 10.1007/978-3-662-49890-3

Library of Congress Control Number: 2016935585

LNCS Sublibrary: SL4 – Security and Cryptology

Printed on acid-free paper

This Springer imprint is published by Springer Nature
The registered company is Springer-Verlag GmbH Berlin Heidelberg

Preface

Eurocrypt 2016, the 35th annual International Conference on the Theory and Applications of Cryptographic Techniques, was held in Vienna, Austria, during May 8–12, 2016. The conference was sponsored by the International Association for Cryptologic Research (IACR). Krzysztof Pietrzak (IST Austria), together with Joël Alwen, Georg Fuchsbauer, Peter Gaži (all IST Austria), and Eike Kiltz (Ruhr-Universität Bochum), were responsible for the local organization. They were supported by a local organizing team consisting of Hamza Abusalah, Chethan Kamath, and Michal Rybár (all IST Austria). We are indebted to them for their support and smooth collaboration.

The conference program followed the now established parallel track system where the works of the authors were presented in two concurrently running tracks. As in the previous edition of Eurocrypt, one track was labeled $\mathcal{R}$ (for real) and the other one was labeled $\mathcal{I}$ (for ideal). Only the invited talks, the tutorial, the best paper, papers with honorable mentions, and the final session of the conference spanned over both tracks.

The proceedings of Eurocrypt contain 62 papers selected from 274 submissions, which corresponds to a record number of submissions in the history of Eurocrypt. Each submission was anonymized for the reviewing process and was assigned to at least three of the 55 Program Committee members. Submissions co-authored by committee members were assigned to at least four members. Committee members were allowed to submit at most one paper, or two if both were co-authored. The reviewing process included a first-round notification followed by a rebuttal for papers that made it to the second round. After extensive deliberations the Program Committee accepted 62 papers. The revised versions of these papers are included in these two-volume proceedings.

The committee decided to give the Best Paper Award to "Tightly Secure CCA-Secure Encryption Without Pairings" by Romain Gay, Dennis Hofheinz, Eike Kiltz, and Hoeteck Wee. The two runners-up to the award, "Indistinguishability Obfuscation from Constant-Degree Graded Encoding Schemes" by Huijia Lin and "Essentially Optimal Robust Secret Sharing with Maximal Corruptions" by Allison Bishop, Valerio Pastro, Rajmohan Rajaraman, Daniel and Wichs, received honorable mentions. All three papers received invitations for the *Journal of Cryptology*.

The program also included invited talks by Karthikeyan Bhargavan, entitled "Protecting Transport Layer Security from Legacy Vulnerabilities", Bart Preneel, entitled "The Future of Cryptography" (IACR distinguished lecture), and Christian Collberg, entitled "Engineering Code Obfuscation." In addition, Emmanuel Prouff gave a tutorial about "Securing Cryptography Implementations in Embedded Systems." All the speakers were so kind as to also provide a short abstract for the proceedings.

We would like to thank all the authors who submitted papers. We know that the Program Committee's decisions, especially rejections of very good papers that did not find a slot among the sparse number of accepted papers, can be very disappointing. We sincerely hope that the rejected works eventually get the attention they deserve.

We are also indebted to the Program Committee members and all external reviewers for their voluntary work, especially since the newly established and unified page limits and the increasing number of submissions induce quite a workload. It has been an honor to work with everyone. The committee's work was tremendously simplified by Shai Halevi's submission software and his support, including running the service on IACR servers.

Finally, we thank everyone else—speakers, session chairs, and rump session chairs—for their contribution to the program of Eurocrypt 2016.

May 2016

Marc Fischlin
Jean-Sébastien Coron

Eurocrypt 2016

The 35th Annual International Conference on the Theory and Applications of Cryptographic Techniques

Vienna, Austria
May 8–12, 2016

Track $\mathcal{R}$

General Chair

Krzysztof Pietrzak	IST Austria

Program Chairs

Marc Fischlin	Technische Universität Darmstadt, Germany
Jean-Sébastien Coron	University of Luxembourg, Luxembourg

Program Committee

Michel Abdalla	Ecole Normale Superieure and CNRS, France
Shweta Agrawal	IIT Delhi, India
Elette Boyle	IDC Herzliya, Israel
Christina Brzuska	TU Hamburg-Harburg, Germany
Ran Canetti	Tel Aviv University, Israel, and Boston University, USA
David Cash	Rutgers University, USA
Dario Catalano	University of Catania, Italy
Jean-Sébastien Coron	University of Luxembourg, Luxembourg
Cas Cremers	University of Oxford, UK
Yevgeniy Dodis	New York University, USA
Nico Döttling	Aarhus University, Denmark
Pooya Farshim	Queen's University Belfast, UK
Jean-Charles Faugère	Inria Paris-Rocquencourt, France
Sebastian Faust	Ruhr University Bochum, Germany
Dario Fiore	IMDEA Software Institute, Spain
Marc Fischlin	TU Darmstadt, Germany
Georg Fuchsbauer	IST, Austria
Juan A. Garay	Yahoo Labs, USA
Vipul Goyal	Microsoft Research, India
Tim Güneysu	University of Bremen, Germany
Shai Halevi	IBM, USA

Goichiro Hanaoka	AIST, Japan
Martin Hirt	ETH Zurich, Switzerland
Dennis Hofheinz	Karlsruhe KIT, Germany
Tibor Jager	Ruhr University Bochum, Germany
Abhishek Jain	Johns Hopkins University, USA
Aniket Kate	Purdue University, USA
Dmitry Khovratovich	University of Luxembourg, Luxembourg
Vadim Lyubashevsky	Ecole Normale Superieure, France
Sarah Meiklejohn	University College London, UK
Mridul Nandi	Indian Statistical Institute, Kolkata, India
María Naya-Plasencia	Inria, France
Svetla Nikova	KU Leuven, Belgium
Adam O'Neill	Georgetown University, USA
Claudio Orlandi	Aarhus University, Denmark
Josef Pieprzyk	Queensland University of Technology, Australia
Mariana Raykova	Yale University, USA
Thomas Ristenpart	Cornell Tech, USA
Matthieu Rivain	CryptoExperts, France
Arnab Roy	Fujitsu Laboratories of America, USA
Benedikt Schmidt	IMDEA Software Institute, Spain
Thomas Schneider	TU Darmstadt, Germany
Berry Schoenmakers	TU Eindhoven, The Netherlands
Peter Schwabe	Radboud University, The Netherlands
Yannick Seurin	ANSSI, France
Thomas Shrimpton	University of Florida, USA
Nigel P. Smart	University of Bristol, UK
John P. Steinberger	Tsinghua University, China
Ron Steinfeld	Monash University, Australia
Emmanuel Thomé	Inria Nancy, France
Yosuke Todo	NTT, Japan
Dominique Unruh	University of Tartu, Estonia
Daniele Venturi	Sapienza University of Rome, Italy
Ivan Visconti	University of Salerno, Italy
Stefan Wolf	USI Lugano, Switzerland

External Reviewers

Divesh Aggarwal
Shashank Agrawal
Adi Akavia
Martin Albrecht
Joël Alwen
Prabhanjan Ananth
Ewerton Rodrigues Andrade
Elena Andreeva
Kazumaro Aoki
Afonso Arriaga
Gilad Asharov
Gilles Van Assche
Nuttapong Attrapadung
Christian Badertscher
Thomas Baignères
Josep Balasch
Foteini Baldimtsi
Subhadeep Banik
Harry Bartlett
Lejla Batina
Carsten Baum
Aemin Baumeler
Christof Beierle
Sonia Belaïd
Fabrice Benhamouda
David Bernhard

Ritam Bhaumik
Begül Bilgin
Nir Bitansky
Matthieu Bloch
Andrey Bodgnanov
Cecilia Boschini
Vitor Bosshard
Christina Boura
Florian Bourse
Cerys Bradley
Zvika Brakerski
Anne Broadbent
Dan Brown
Seyit Camtepe
Anne Canteaut
Angelo De Caro
Avik Chakraborti
Nishanth Chandran
Melissa Chase
Rahul Chatterjee
Yilei Chen
Jung Hee Cheon
Céline Chevalier
Alessandro Chiesa
Seung Geol Choi
Tom Chothia
Arka Rai Choudhuri
Kai-Min Chung
Yu-Chi Chen
Michele Ciampi
Michael Clear
Aloni Cohen
Ran Cohen
Katriel Cohn-Gordon
Sandro Coretti
Cas Cremers
Dana Dachman-Soled
Yuanxi Dai
Nilanjan Datta
Bernardo Machado David
Gareth T. Davies
Ed Dawson
Jean Paul Degabriele
Martin Dehnel-Wild
Jeroen Delvaux
Grégory Demay
Daniel Demmler
David Derler
Vasil Dimitrov
Yarkin Doroz
Léo Ducas
François Dupressoir
Frederic Dupuis
Avijit Dutta
Stefan Dziembowski
Keita Emura
Antonio Faonio
Serge Fehr
Claus Fieker
Matthieu Finiasz
Viktor Fischer
Jean-Pierre Flori
Pierre-Alain Fouque
Tore Kasper Frederiksen
Tommaso Gagliardoni
Steven Galbraith
David Galindo
Chaya Ganesh
Luke Garratt
Romain Gay
Peter Gaži
Daniel Genkin
Craig Gentry
Hossein Ghodosi
Satrajit Ghosh
Benedikt Gierlichs
Kristian Gjøsteen
Aleksandr Golovnev
Alonso Gonzalez
Dov Gordon
Louis Goubin
Jens Groth
Aurore Guillevic
Sylvain Guilley
Siyao Guo
Divya Gupta
Sourav Sen Gupta
Helene Flyvholm Haagh
Tzipora Halevi
Michael Hamburg
Carmit Hazay
Gottfried Herold
Susan Hohenberger
Justin Holmgren
Pavel Hubacek
Tsung-Hsuan Hung
Christopher Huth
Michael Hutter
Andreas Hülsing
Vincenzo Iovino
Håkon Jacobsen
Aayush Jain
Jérémy Jean
Claude-Pierre Jeannerod
Evan Jeffrey
Ashwin Jha
Daniel Jost
Charanjit Jutla
Ali El Kaafarani
Liang Kaitai
Saqib A. Kakvi
Chethan Kamath
Bhavana Kanukurthi
Pierre Karpman
Elham Kashefi
Tomasz Kazana
Marcel Keller
Dakshita Khurana
Aggelos Kiayias
Paul Kirchner
Elena Kirshanova
Ágnes Kiss
Fuyuki Kitagawa
Ilya Kizhvatov
Thorsten Kleinjung
Vlad Kolesnikov
Venkata Koppala
Luke Kowalczyk
Ranjit Kumaresan
Kaoru Kurosawa
Felipe Lacerda
Virginie Lallemand
Adeline Langlois
Enrique Larraia
Sebastian Lauer
Gregor Leander
Chin Ho Lee
Tancrède Lepoint

Gaëtan Leurent
Benoît Libert
Huijia (Rachel) Lin
Wei-Kai Lin
Bin Liu
Dongxi Liu
Yunwen Liu
Steve Lu
Atul Luykx
Bernardo Magri
Mohammad Mahmoody
Subhamoy Maitra
Hemanta Maji
Giulio Malavolta
Avradip Mandal
Daniel Masny
Takahiro Matsuda
Christian Matt
Willi Meier
Sebastian Meiser
Florian Mendel
Bart Mennink
Eric Miles
Kevin Milner
Ilya Mironov
Arno Mittelbach
Ameer Mohammad
Payman Mohassel
Hart Montgomery
Amir Moradi
François Morain
Paweł Morawiecki
Pedro Moreno-Sanchez
Nicky Mouha
Pratyay Mukherjee
Elke De Mulder
Anderson Nascimento
Muhammad Naveed
Phong Nguyen
Ivica Nikolic
Tobias Nilges
Peter Sebastian Nordholt
Koji Nuida
Maciej Obremski
Frederique Elise Oggier
Emmanuela Orsini
Mohammad Ali Orumiehchi
Elisabeth Oswald
Ekin Ozman
Jiaxin Pan
Giorgos Panagiotakos
Omkant Pandey
Omer Paneth
Dimitris Papadopoulos
Kostas Papagiannopoulos
Bryan Parno
Valerio Pastro
Chris Peikert
Ludovic Perret
Leo Paul Perrin
Christophe Petit
Krzysztof Pietrzak
Benny Pinkas
Oxana Poburinnaya
Bertram Poettering
Joop van de Pol
Antigoni Polychroniadou
Manoj Prabhakaran
Thomas Prest
Emmanuel Prouff
Jörn Müller-Quade
Tal Rabin
Kenneth Radke
Carla Rafols
Mario Di Raimondo
Samuel Ranellucci
Pavel Raykov
Francesco Regazzoni
Omer Reingold
Michał Ren
Guénaël Renault
Oscar Reparaz
Vincent Rijmen
Ben Riva
Tim Ruffing
Ulrich Rührmair
Yusuke Sakai
Amin Sakzad
Benno Salwey
Kai Samelin
Yu Sasaki
Alessandra Scafuro
Christian Schaffner
Tobias Schneider
Peter Scholl
Jacob Schuldt
Gil Segev
Nicolas Sendrier
Abhi Shelat
Leonie Simpson
Shashank Singh
Luisa Siniscalchi
Boris Skoric
Ben Smith
Juraj Somorovsky
John Steinberger
Noah Stephens-Dawidovitz
Björn Tackmann
Vanessa Teague
Sidharth Telang
R. Seth Terashima
Stefano Tessaro
Adrian Thillard
Susan Thomson
Mehdi Tibouchi
Jacques Traoré
Daniel Tschudi
Hoang Viet Tung
Aleksei Udovenko
Margarita Vald
Maria Isabel Gonzalez Vasco
Meilof Veeningen
Vesselin Velichkov
Alexandre Venelli
Muthuramakrishnan Venkitasubramaniam
Frederik Vercauteren
Marion Videau
Vinod Vikuntanathan
Gilles Villard
Damian Vizar
Emmanuel Volte
Christine van Vredendaal
Niels de Vreede
Qingju Wang

Bogdan Warinschi
Hoeteck Wee
Carolyn Whitnall
Daniel Wichs
Alexander Wild
David Wu
Jürg Wullschleger
Masahiro Yagisawa
Shota Yamada
Kan Yasuda
Scott Yilek
Kazuki Yoneyama
Ching-Hua Yu
Samee Zahur
Mark Zhandry
Zongyang Zhang
Vassilis Zikas
Michael Zohner

Invited Talks and Tutorial Papers

Protecting Transport Layer Security from Legacy Vulnerabilities

Karthikeyan Bhargavan

INRIA

Abstract. The Transport Layer Security protocol (TLS) is the most widely-used secure channel protocol on the Web. After 20 years of evolution, TLS has grown to include five protocol versions, dozens of extensions, and hundreds of ciphersuites. The success of TLS as an open standard is at least partially due its *protocol agility*: clients and servers can implement different subsets of protocol features and still interoperate, as long as they can negotiate a common version and ciphersuite. Hence, software vendors can seamlessly deploy newer cryptographic mechanisms while still supporting older algorithms for backwards compatibility.

An undesirable consequence of this agility is that obsolete and broken ciphers can stay enabled in TLS clients and servers for years after cryptographers have explicitly warned against their use. Practitioners consider this relatively safe for two reasons. First, the TLS key exchange protocol incorporates downgrade protection, so if a client and server both support a strong ciphersuite, then they should never negotiate a weaker ciphersuite even if it is enabled. Second, even if a connection uses a cryptographic algorithm with known weaknesses, it is typically hard to exploit the theoretical vulnerability to attack the protocol.

In this talk, we will see that both these assumptions are false. Leaving legacy crypto unattended within TLS configurations has serious consequences, as shown by a recent series of *downgrade attacks* including Logjam [1] and SLOTH [3]. We will show how these attacks expose protocol-level weaknesses in TLS that can be exploited with practical cryptanalysis. We will propose a new notion of *downgrade resilience* for key exchange protocols [2] and use this definition to evaluate the downgrade protections mechanisms built into the upcoming TLS 1.3 protocol.

References

1. Adrian, D., Bhargavan, K., Durumeric, Z., Gaudry, P., Green, M., Halderman, J.A., Heninger, N., Springall, D., Thomé, E., Valenta, L., VanderSloot, B., Wustrow, E., Zanella-Béguelin, S., Zimmermann, P.: Imperfect forward secrecy: how Diffie-Hellman fails in practice. In: ACM Conference on Computer and Communications Security (CCS) (2015)
2. Bhargavan, K., Brzuska, C., Fournet, C., Green, M., Kohlweiss, M., Zanella- Béguelin, S.: Downgrade resilience in key-exchange protocols. In: IEEE Symposium on Security and Privacy (Oakland) (2016)
3. Bhargavan, K., Leurent, G.: Transcript collision attacks: breaking authentication in TLS, IKE, and SSH. In: Network and Distributed System Security Symposium (NDSS) (2016)

The Future of Cryptography

Bart Preneel

KU Leuven and iMinds
Department of Electrical Engineering-ESAT/COSIC
Kasteelpark Arenberg 10 Bus 2452, B-3001 Leuven, Belgium
bart.preneel@esat.kuleuven.be

Abstract. We reflect on the historic role of cryptography. We develop the contrast between its success as an academic discipline and the serious shortcomings of current cryptographic deployments in protecting users against mass surveillance and overreach by corporations. We discuss how the cryptographic research community can contribute towards addressing these challenges.

Since its early days, the goal of cryptography is to protect confidentiality of information, which means that it is used to control who has access to information. A second goal of cryptography is to protect authenticity of data and entities: this allows to protect payment information, transaction records but also configuration files and software. Cryptography also plays a central role in the protection of meta data: in many settings it is important to hide the identities and locations of the communicating parties. In modern cryptography much more complex goals can be achieved beyond protection communications and stored data: cryptographic techniques are used to guarantee the correctness of the execution of a program or to obfuscate programs. Multi-party computation allows parties to compute on data while each one can keep its input private and all can check the correctness of the results, even if some of the parties are malicious. Sophisticated techniques are being developed to compute on encrypted data and to search in the data. Even in a domain as challenging as e-voting progress is being made.

Until the late 1980s, cryptographic devices were expensive, which means that the use of cryptography was limited to military, government, and diplomatic applications as well as a few business contexts such as financial transactions. In the early 1990s the cost of cryptography dropped quickly as the increased power of CPUs made it feasible to implement crypto in software. This resulted in the crypto wars, in which government key escrow schemes were proposed and defeated. One decade later commodity cryptographic hardware started to appear, resulting in a cryptography everywhere. The fast dropping cost of cryptography combined with a rich cryptographic literature leads to the conclusion that today cryptography is widespread.

A quick count shows that there are about 30 billion devices with cryptography. The largest volumes are for mobile communications, the web ecosystem, access cards, bank cards, DRM for media protection, hard disk encryption, and applications such as WhatsApp and Skype. It is remarkable that very few of those mass applications offer end-to-end confidentiality protection; moreover, those that do typically have some key management or governance issue: the specifications or the source code are not public, or the ecosystem is brittle as it relies on trust in hundreds of CAs.

The threat models considered in cryptographic papers can be very strong: we assume powerful opponents who can intercept all communications, corrupt some parties, and perform expensive computations. Since the mid 1990s we take into account opponents who use physics to eavesdrop on signals (side channel attacks) or inject faults in computations. However, the Snowden revelations have shown that our threat models are not sufficiently strong to model intelligence agencies: they undermine the standardization process by injecting stealthily schemes with backdoors, they increase complexity of standards, break supply-chain integrity, undermine end systems using malware, obtain keys using security letters or via malware, and exploit implementation weaknesses, to name just a few.

By combining massive interception with sophisticated search techniques, intelligence agencies have developed mass surveillance systems that are a threat to our values and democracy. In response academic cryptographers have started to publish articles that consider some of these more advanced threat models. Industry has expanded its deployment of cryptography and increased the strength of deployments, e.g., by switching to solutions that offer forward secrecy. However, their efforts are sometimes limited because of the business models that monetize user data and business plans to exploit Big Data at an ever larger scale.

In terms of protection of users, progress is still very slow. The cryptographic literature has plenty of schemes to increase robustness of cryptographic implementations, but few are implemented. The reasons are cost, the lack of open source implementations, and the misalignment with business objectives that are driven by the Big Data gold rush. Moreover, in response to the modest advances made by industry, law enforcement is reviving the early 1990s crypto wars.

Overall, this complex context brings new opportunities for cryptographers: we have the responsibility to help restoring the balance of power between citizens on the one hand, and governments and corporations on the other hand. We can invent new architectures that give users more control and visibility and that avoid single points of failure. We can propose new protocols that are more robust against local compromises by malware, backdoors or security letters. And we can contribute towards developing or analyzing open implementations of these protocols to facilitate their deployment.

Engineering Code Obfuscation

Christian Collberg

University of Arizona, Tucson, AZ, USA
collberg@gmail.com
http://cs.arizona.edu/~collberg

In the *Man-at-the-end* (MATE) security scenario [2] a user (and potential adversary) has physical access to a device and can gain some advantage by extracting or tampering with an asset within that device. Assets can be data (cryptographic keys and media streams) as well as code (security checks and intellectual property). As defenders, our goal is to protect the confidentiality and integrity of these assets.

MATE scenarios are ubiquitous. Consider, for example, the *Advanced Metering Infrastructure* where *smart meters* are installed at house-holds to allow utility companies to connect and disconnect users and to monitor usage. A malicious consumer can tamper with their meter to avoid payment and may even be able to send disconnect commands to other meters [5]. In the mobile *Snapchat* application pictures exchanged between teenagers must be deleted a few seconds after reaching a friend's device. A malicious user can tamper with the application code to save the pictures and use them later for cyber bullying. Finally, in a massive multiplayer online game a malicious player can tamper with the game client to get an unfair advantage over other players [3].

Adversarial Model. In a realistic MATE scenario we must assume that, since the adversary has physical control over his device, in time all assets will be compromised, and, at best, any defenses will be time-limited [4]. Even protection techniques based on tamper-resistant hardware have shown themselves susceptible to attack [1]. In particular, in analogy with *Kerckhoffs's principles*, we must assume an adversary who has complete understanding of our system, including its source code, and who can achieve in-depth understanding of the system through static and dynamic analyses using advanced reverse engineering tools.

Protection Mechanisms and Strategies. MATE protection mechanisms are typically based on the application of obfuscating code transformations that add complexity (for confidentiality) and/or the insertion of tamper-detecting guards (for integrity). Given that individual mechanisms provide limited protection, strategies have to be put in place to extend the in-the-wild survival time.

An important such strategy is *diversity*. *Spatial diversity* (or *defense-in-depth*) means compounding multiple layers of interchangeable primitive protective transformations. *Temporal diversity* (or *renewability*) means to deliver, over time, an infinite and non-repeating sequence of code variants to the user. The basic principle is that every program should be protected with a different combination of transformations, that every user/potential adversary should get a uniquely protected program, and that we will provide an ever-changing attack target to the adversary. In other words, we hope to provide long-term security by overwhelming the adversary's analytical abilities with randomized, unique, and varying code variants.

Evaluation and Benchmarking. MATE protection systems are evaluated on their resilience to attack and their performance, i.e. the increase in size and speed of a protected program over the original. Many real-world applications are interactive (such as the Snapchat and game examples above), and many are running on highly constrained devices (smart meters); performance is thus always of paramount concern.

Finding the combination of primitive transformations and spatial and temporal diversity strategies that achieve the highest level of protection while staying within strict performance bounds is an unsolved engineering problem. Part of the problem is a lack of behavioral models that express the capabilities and limitations of a human adversary, and part of the problem is a lack of universally accepted benchmarks.

Summary. We present an overview of the engineering challenges in providing long-term protection of applications that run under complete control of an adversary. In particular, we discuss the principle of diversity and the need for adversarial modeling and benchmarking.

Acknowledgments. This work was funded in part by NSF grants CNF-1145913 and CNS-1525820 and BSF grant 2008362.

References

1. Anderson, R., Kuhn, M.: Low cost attacks on tamper resistant devices. In: IWSP: International Workshop on Security Protocols (1997)
2. Collberg, C., Nagra, J.: Surreptitious Software: Obfuscation, Watermarking, and Tamperproofing for Software Protection. Addison-Wesley (2009)
3. Hoglund, G., McGraw, G.: Exploiting Online Games: Cheating Massively Distributed Systems. Addison-Wesley (2007)
4. Hohl, F.: Time limited blackbox security: protecting mobile agents from malicious hosts. In: Mobile Agents and Security, pp. 92–113 (1998)
5. Parks, R.C.: Sandia report: advanced metering infrastructure security considerations (2007)

Securing Cryptography Implementations in Embedded Systems

Emmanuel Prouff[1,2]

[1]ANSSI, France
emmanuel.prouff@ssi.gouv.fr
[2] POLSYS, UMR 7606, LIP6
Sorbonne Universities, UPMC University Paris VI

Abstract. Side Channel Analysis is a class of attacks which exploit leakages of information from a cryptographic implementation during execution. To defeat them, various techniques have been introduced during the two last decades, among which masking (aka *implementation sharing*) is a common countermeasure. The principle is to randomly split every sensitive intermediate variable occurring in the computation into several shares and the number of shares, called the order, plays the role of a security parameter. The main issue while applying masking to protect cryptographic implementations is to specify efficient schemes to secure the non-linear steps during the processing. Several solutions, applicable for arbitrary orders, have been recently published. Most of them start from the original concept of *Private Circuits* originally introduced by Ishaï, Sahai and Wagner at Crypto 2003. In parallel, and in order to formally prove the security of the proposed masking schemes, the community has also made important efforts to define leakage models that accurately capture the leakage complexity and simultaneously enable to build accurate security arguments. It is worth noting that there is a tight link between masking/sharing techniques, secure Multi Party Computation, Coding Theory and also Threshold Implementations. During a two hours tutorial, the main classes of countermeasures will be presented, together with models which have been introduced to prove their security. The link with other areas such as secure multi-party computation, error correcting codes and information theory will also be discussed.

Contents – Part I

Best Paper and Honorable Mentions

(Pseudo) Randomness

LPN/LWE

Cryptanalysis I

Masking

Fully Homomorphic Encryption

Cryptanalysis II

Number Theory

Hash Functions

Multilinear Maps

Message Authentication Codes

Attacks on SSL/TLS

Real-World Protocols

Robust Designs

Lattice Reduction

Contents – Part II

Lattice-Based Schemes

Zero-Knowledge I

Pseudorandom Functions

Multi-Party Computation I

Separations

Zero-Knowledge II

Protocols

Round Complexity

Commitments

Lattices

Leakage

Indifferentiability

Multi-Party Computation II

Obfuscation

Automated Analysis, Functional Encryption, and Non-malleable Codes

Tightly CCA-Secure Encryption Without Pairings

Romain Gay[1(✉)], Dennis Hofheinz[3], Eike Kiltz[2], and Hoeteck Wee[1]

[1] ENS, Paris, France
{rgay,wee}@di.ens.fr
[2] Ruhr-Universität Bochum, Bochum, Germany
eike.kiltz@rub.de
[3] Karlsruhe Institute of Technology, Karlsruhe, Germany
Dennis.Hofheinz@kit.edu

Abstract. We present the first CCA-secure public-key encryption scheme based on DDH where the security loss is independent of the number of challenge ciphertexts and the number of decryption queries. Our construction extends also to the standard k-Lin assumption in pairing-free groups, whereas all prior constructions starting with Hofheinz and Jager (Crypto '12) rely on the use of pairings. Moreover, our construction improves upon the concrete efficiency of existing schemes, reducing the ciphertext overhead by about half (to only 3 group elements under DDH), in addition to eliminating the use of pairings.

We also show how to use our techniques in the NIZK setting. Specifically, we construct the first tightly simulation-sound designated-verifier NIZK for linear languages without pairings. Using pairings, we can turn our construction into a highly optimized publicly verifiable NIZK with tight simulation-soundness.

1 Introduction

The most basic security guarantee we require of a public key encryption scheme is that of semantic security against chosen-plaintext attacks (CPA) [14]: it is infeasible to learn anything about the plaintext from the ciphertext. On the other hand, there is a general consensus within the cryptographic research community that in virtually every practical application, we require semantic security against adaptive chosen-ciphertext attacks (CCA) [12,30], wherein an adversary is given access to decryptions of ciphertexts of her choice.

Romain Gay—CNRS, INRIA. Supported by ERC Project aSCEND (639554).
Dennis Hofheinz—Supported by DFG grants HO 4534/2-2, HO 4534/4-1.
Eike Kiltz—Partially supported by DFG grant KI 795/4-1 and ERC Project ERCC (FP7/615074).
Hoeteck Wee—CNRS (UMR 8548), INRIA and Columbia University. Partially supported by the Alexander von Humboldt Foundation, NSF Award CNS-1445424 and ERC Project aSCEND (639554).

M. Fischlin and J.-S. Coron (Eds.): EUROCRYPT 2016, Part I, LNCS 9665, pp. 1–27, 2016.
DOI: 10.1007/978-3-662-49890-3_1

In this work, we focus on the issue of security reduction and security loss in the construction of CPA and CCA-secure public-key encryption from the DDH assumption. Suppose we have such a scheme along with a security reduction showing that attacking the scheme in time t with success probability ϵ implies breaking the DDH assumption in time roughly t with success probability ϵ/L; we refer to L as the security loss. In general, L would depend on the security parameter λ as well as the number of challenge ciphertexts Q_{enc} and the number decryption queries Q_{dec}, and we say that we have a *tight security reduction* if L depends only on the security parameter and is independent of both Q_{enc} and Q_{dec}. Note that for typical settings of parameters (e.g., $\lambda = 80$ and $Q_{\mathsf{enc}}, Q_{\mathsf{dec}} \approx 2^{20}$, or even $Q_{\mathsf{enc}}, Q_{\mathsf{dec}} \approx 2^{30}$ in truly large settings), λ is much smaller than Q_{enc} and Q_{dec}.

In the simpler setting of CPA-secure encryption, the ElGamal encryption scheme already has a tight security reduction to the DDH assumption [6,27], thanks to random self-reducibility of DDH with a tight security reduction. In the case of CCA-secure encryption, the best result is still the seminal Cramer-Shoup encryption scheme [11], which achieves security loss Q_{enc}.[1] This raises the following open problem:

> Does there exist a CCA-secure encryption scheme with a tight security reduction to the DDH assumption?

Hofheinz and Jager [16] gave an affirmative answer to this problem under stronger (and pairing-related) assumptions, notably the 2-Lin assumptions in bilinear groups, albeit with large ciphertexts and secret keys; a series of follow-up works [5,15,22,24] leveraged techniques introduced in the context of tightly-secure IBE [7,10,18] to reduce the size of ciphertext and secret keys to a relatively small constant. However, all of these works rely crucially on the use of pairings, and seem to shed little insight on constructions under the standard DDH assumption; in fact, a pessimist may interpret the recent works as strong indication that the use of pairings is likely to be necessary for tightly CCA-secure encryption.

We may then restate the open problem as eliminating the use of pairings in these prior CCA-secure encryption schemes while still preserving a tight security reduction. From a theoretical stand-point, this is important because an affirmative answer would yield tightly CCA-secure encryption under qualitatively weaker assumptions, and in addition, shed insight into the broader question of whether tight security comes at the cost of qualitative stronger assumptions.

Eliminating the use of pairings is also important in practice as it allows us to instantiate the underlying assumption over a much larger class of groups that admit more efficient group operations and more compact representations, and also avoid the use of expensive pairing operations. Similarly, tight reductions matter in practice because as L increases, we should increase the size of the underlying groups in order to compensate for the security loss, which in turn

[1] We ignore contributions to the security loss that depend only on a statistical security parameter.

increases the running time of the implementation. Note that the impact on performance is quite substantial, as exponentiation in a r-bit group takes time roughly $\mathcal{O}(r^3)$.

1.1 Our Results

We settle the main open problem affirmatively: we construct a tightly CCA-secure encryption scheme from the DDH assumption without pairings. Moreover, our construction improves upon the concrete efficiency of existing schemes, reducing the ciphertext overhead by about half, in addition to eliminating the use of pairings. We refer to Fig. 2 for a comparison with prior works.

Overview of Our Construction. Fix an additively written group $\mathbb{G}$ of order q. We rely on implicit representation notation [13] for group elements: for a fixed generator P of $\mathbb{G}$ and for a matrix $\mathbf{M} \in \mathbb{Z}_q^{n \times t}$, we define $[\mathbf{M}] := \mathbf{M}P \in \mathbb{G}^{n \times t}$ where multiplication is done component-wise. We rely on the $\mathcal{D}_k$-MDDH Assumption [13], which stipulates that given $[\mathbf{M}]$ drawn from a matrix distribution $\mathcal{D}_k$ over $\mathbb{Z}_q^{(k+1)\times k}$, $[\mathbf{Mx}]$ is computationally indistinguishable from a uniform vector in $\mathbb{G}^k$; this is a generalization of the k-Lin Assumption.

We outline the construction under the k-Lin assumption over $\mathbb{G}$, of which the DDH assumption is a special case corresponding to $k = 1$.

In this overview, we will consider a weaker notion of security, namely tag-based KEM security against plaintext check attacks (PCA) [29]. In the PCA security experiment, the adversary gets no decryption oracle (as with CCA security), but a PCA oracle that takes as input a tag and a ciphertext/plaintext pair and checks whether the ciphertext decrypts to the plaintext. Furthermore, we restrict the adversary to only query the PCA oracle on tags different from those used in the challenge ciphertexts. PCA security is strictly weaker than the CCA security we actually strive for, but allows us to present our solution in a clean and simple way. (We show how to obtain full CCA security separately.)

The starting point of our construction is the Cramer-Shoup KEM, in which $\mathsf{Enc}_{\mathsf{KEM}}(\mathsf{pk}, \tau)$ outputs the ciphertext/plaintext pair

$$([\mathbf{y}], [z]) = ([\mathbf{x}^\top \mathbf{M}^\top], [\mathbf{x}^\top \mathbf{M}^\top \mathbf{k}_\tau]), \tag{1}$$

where $\mathbf{k}_\tau = \mathbf{k}_0 + \tau \mathbf{k}_1$ and $\mathsf{pk} := ([\mathbf{M}], [\mathbf{M}^\top \mathbf{k}_0], [\mathbf{M}^\top \mathbf{k}_1])$ for $\mathbf{M} \leftarrow_{\mathrm{R}} \mathbb{Z}_q^{(k+1)\times k}$. The KEM is PCA-secure under k-Lin, with a security loss that depends on the number of ciphertexts Q (via a hybrid argument) but independently of the number of PCA queries [1,11].

Following the "randomized Naor-Reingold" paradigm introduced by Chen and Wee on tightly secure IBE [10], our starting point is (1), where we replace $\mathbf{k}_\tau = \mathbf{k}_0 + \tau \mathbf{k}_1$ with

$$\mathbf{k}_\tau = \sum_{j=1}^{\lambda} \mathbf{k}_{j,\tau_j}$$

and $\mathsf{pk} := ([\mathbf{M}], [\mathbf{M}^\top \mathbf{k}_{j,b}]_{j=1,\ldots,\lambda,b=0,1})$, where $(\tau_1, \ldots, \tau_\lambda)$ denotes the binary representation of the tag $\tau \in \{0,1\}^\lambda$.

Following [10], we want to analyze this construction by a sequence of games in which we first replace $[\mathbf{y}]$ in the challenge ciphertexts by uniformly random group elements via random self-reducibility of MDDH (k-Lin), and then incrementally replace $\mathbf{k}_\tau$ in both the challenge ciphertexts and in the PCA oracle by $\mathbf{k}_\tau + \mathbf{m}^\perp \mathsf{RF}(\tau)$, where RF is a truly random function and $\mathbf{m}^\perp$ is a random element from the kernel of $\mathbf{M}$, i.e., $\mathbf{M}^\top \mathbf{m}^\perp = 0$. Concretely, in Game i, we will replace $\mathbf{k}_\tau$ with $\mathbf{k}_\tau + \mathbf{m}^\perp \mathsf{RF}_i(\tau)$ where RF_i is a random function on $\{0,1\}^i$ applied to the i-bit prefix of τ. We proceed to outline the two main ideas needed to carry out this transition. Looking ahead, note that once we reach Game λ, we would have replaced $\mathbf{k}_\tau$ with $\mathbf{k}_\tau + \mathbf{m}^\perp \mathsf{RF}(\tau)$, upon which security follows from a straight-forward information-theoretic argument (and the fact that ciphertexts and decryption queries carry pairwise different τ).

First Idea. First, we show how to transition from Game i to Game $i+1$, under the restriction that the adversary is only allowed to query the encryption oracle on tags whose $i+1$-st bit is 0; we show how to remove this unreasonable restriction later. Here, we rely on an *information-theoretic* argument similar to that of Cramer and Shoup to increase the entropy from RF_i to RF_{i+1}. This is in contrast to prior works which rely on a computational argument; note that the latter requires encoding secret keys as group elements and thus a pairing to carry out decryption.

More precisely, we pick a random function RF'_i on $\{0,1\}^i$, and implicitly define RF_{i+1} as follows:

$$\mathsf{RF}_{i+1}(\tau) = \begin{cases} \mathsf{RF}_i(\tau) & \text{if } \tau_{i+1} = 0 \\ \mathsf{RF}'_i(\tau) & \text{if } \tau_{i+1} = 1 \end{cases}$$

Observe all of the challenge ciphertexts leak no information about RF'_i or $\mathbf{k}_{i+1,1}$ since they all correspond to tags whose $i+1$-st bit is 0. To handle a PCA query $(\tau, [\mathbf{y}], [z])$, we proceed via a case analysis:

- if $\tau_{i+1} = 0$, then $\mathbf{k}_\tau + \mathsf{RF}_{i+1}(\tau) = \mathbf{k}_\tau + \mathsf{RF}_i(\tau)$ and the PCA oracle returns the same value in both Games i and $i+1$.
- if $\tau_{i+1} = 1$ and $\mathbf{y}$ lies in the span of $\mathbf{M}$, we have

$$\mathbf{y}^\top \mathbf{m}^\perp = 0 \Longrightarrow \mathbf{y}^\top(\mathbf{k}_\tau + \mathbf{m}^\perp \mathsf{RF}_i(\tau)) = \mathbf{y}^\top(\mathbf{k}_\tau + \mathbf{m}^\perp \mathsf{RF}_{i+1}(\tau)),$$

  and again the PCA oracle returns the same value in both Games i and $i+1$.
- if $\tau_{i+1} = 1$ and $\mathbf{y}$ lies outside the span of $\mathbf{M}$, then $\mathbf{y}^\top \mathbf{k}_{i+1,1}$ is uniformly random given $\mathbf{M}, \mathbf{M}^\top \mathbf{k}_{i+1,1}$. (Here, we crucially use that the adversary does not query encryptions with $\tau_{i+1} = 1$, which ensures that the challenge ciphertexts do not leak additional information about $\mathbf{k}_{i+1,1}$.) This means that $\mathbf{y}^\top \mathbf{k}_\tau$ is uniformly random from the adversary's view-point, and therefore the PCA oracle will reject with high probability in both Games i and $i+1$. (At this point, we crucially rely on the fact that the PCA oracle only outputs a *single* check bit and not all of $\mathbf{k}_\tau + \mathsf{RF}(\tau)$.)

Via a hybrid argument, we may deduce that the distinguishing advantage between Games i and $i+1$ is at most Q/q where Q is the number of PCA queries.

Second Idea. Next, we remove the restriction on the encryption queries using an idea of Hofheinz et al. [18] for tightly-secure IBE in the multi-ciphertext setting, and its instantiation in prime-order groups [15]. The idea is to create two "independent copies" of $(\mathbf{m}^\perp, \mathsf{RF}_i)$; we use one to handle encryption queries on tags whose $i+1$-st bit is 0, and the other to handle those whose $i+1$-st bit is 1. We call these two copies $(\mathbf{M}_0^*, \mathsf{RF}_i^{(0)})$ and $(\mathbf{M}_1^*, \mathsf{RF}_i^{(1)})$, where $\mathbf{M}^\top \mathbf{M}_0^* = \mathbf{M}^\top \mathbf{M}_1^* = \mathbf{0}$.

Concretely, we replace $\mathbf{M} \leftarrow_{\mathrm{R}} \mathbb{Z}_q^{(k+1)\times k}$ with $\mathbf{M} \leftarrow_{\mathrm{R}} \mathbb{Z}_q^{3k\times k}$. We decompose $\mathbb{Z}_q^{3k}$ into the span of the respective matrices $\mathbf{M}, \mathbf{M}_0, \mathbf{M}_1$, and we will also decompose the span of $\mathbf{M}^\perp \in \mathbb{Z}_q^{3k\times 2k}$ into that of $\mathbf{M}_0^*, \mathbf{M}_1^*$. Similarly, we decompose $\mathbf{M}^\perp \mathsf{RF}_i(\tau)$ into $\mathbf{M}_0^* \mathsf{RF}_i^{(0)}(\tau) + \mathbf{M}_1^* \mathsf{RF}_i^{(1)}(\tau)$. We then refine the prior transition from Games i to $i+1$ as follows:

- Game $i.0$ (= Game i): pick $\mathbf{y} \leftarrow \mathbb{Z}_q^{3k}$ for ciphertexts, and replace $\mathbf{k}_\tau$ with $\mathbf{k}_\tau + \mathbf{M}_0^* \mathsf{RF}_i^{(0)}(\tau) + \mathbf{M}_1^* \mathsf{RF}_i^{(1)}(\tau)$;
- Game $i.1$: replace $\mathbf{y} \leftarrow_{\mathrm{R}} \mathbb{Z}_q^{3k}$ with $\mathbf{y} \leftarrow_{\mathrm{R}} \mathsf{span}(\mathbf{M}, \mathbf{M}_{\tau_{i+1}})$;
- Game $i.2$: replace $\mathsf{RF}_i^{(0)}(\tau)$ with $\mathsf{RF}_{i+1}^{(0)}(\tau)$;
- Game $i.3$: replace $\mathsf{RF}_i^{(1)}(\tau)$ with $\mathsf{RF}_{i+1}^{(1)}(\tau)$;
- Game $i.4$ (= Game $i+1$): replace $\mathbf{y} \leftarrow_{\mathrm{R}} \mathsf{span}(\mathbf{M}, \mathbf{M}_{\tau_{i+1}})$ with $\mathbf{y} \leftarrow_{\mathrm{R}} \mathbb{Z}_q^{3k}$.

For the transition from Game $i.0$ to Game $i.1$, we rely on the fact that the uniform distributions over $\mathbb{Z}_q^{3k}$ and $\mathsf{span}(\mathbf{M}, \mathbf{M}_{\tau_{i+1}})$ encoded in the group are computationally indistinguishable, even given a random basis for $\mathsf{span}(\mathbf{M}^\perp)$ (in the clear). This extends to the setting with multiple samples, with a tight reduction to the $\mathcal{D}_k$-MDDH Assumption independent of the number of samples.

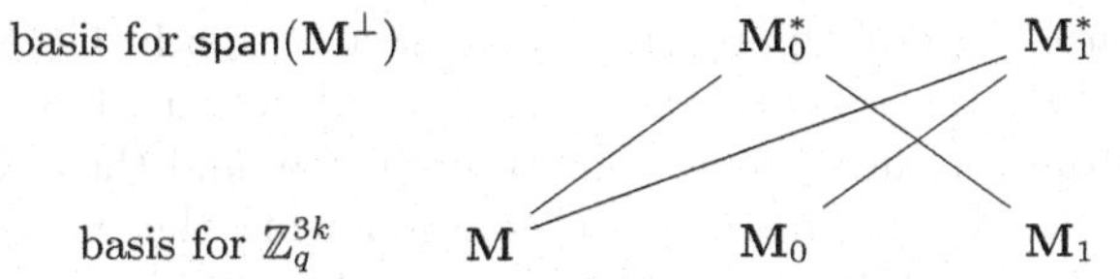

Fig. 1. Solid lines mean orthogonal, that is: $\mathbf{M}^\top \mathbf{M}_0^* = \mathbf{M}_1^\top \mathbf{M}_0^* = \mathbf{0} = \mathbf{M}^\top \mathbf{M}_1^* = \mathbf{M}_0^\top \mathbf{M}_1^*$.

For the transition from Game $i.1$ to $i.2$, we rely on an information-theoretic argument like the one we just outlined, replacing $\mathsf{span}(\mathbf{M})$ with $\mathsf{span}(\mathbf{M}, \mathbf{M}_1)$ and $\mathbf{M}^\perp$ with $\mathbf{M}_0^*$ in the case analysis. In particular, we will exploit the fact that if $\mathbf{y}$ lies outside $\mathsf{span}(\mathbf{M}, \mathbf{M}_1)$, then $\mathbf{y}^\top \mathbf{k}_{i+1,1}$ is uniformly random even given $\mathbf{M}, \mathbf{M}\mathbf{k}_{i+1,1}, \mathbf{M}_1, \mathbf{M}_1\mathbf{k}_{i+1,1}$. The transition from Game $i.2$ to $i.3$ is completely analogous.

From PCA to CCA. Using standard techniques from [4,8,11,19,21], we could transform our basic tag-based PCA-secure scheme into a "full-fledged" CCA-secure encryption scheme by adding another hash proof system (or an authenticated symmetric encryption scheme) and a one-time signature scheme. However, this would incur an additional overhead of several group elements in the ciphertext. Instead, we show how to directly modify our tag-based PCA-secure scheme to obtain a more efficient CCA-secure scheme with the minimal additional overhead of a single symmetric-key authenticated encryption. In particular, the overall ciphertext overhead in our tightly CCA-secure encryption scheme is merely *one* group element more than that for the best known non-tight schemes [17,21].

To encrypt a message M in the CCA-secure encryption scheme, we will (i) pick a random $\mathbf{y}$ as in the tag-based PCA scheme, (ii) derive a tag τ from $\mathbf{y}$, (iii) encrypt M using a one-time authenticated encryption under the KEM key $[\mathbf{y}^\top \mathbf{k}_\tau]$. The naive approach is to derive the tag τ by hashing $[\mathbf{y}] \in \mathbb{G}^{3k}$, as in [21]. However, this creates a circularity in Game $i.1$ where the distribution of $[\mathbf{y}]$ depends on the tag. Instead, we will derive the tag τ by hashing $[\overline{\mathbf{y}}] \in \mathbb{G}^k$, where $\overline{\mathbf{y}} \in \mathbb{Z}_q^k$ are the top k entries of $\mathbf{y} \in \mathbb{Z}_q^{3k}$. We then modify $\mathbf{M}_0, \mathbf{M}_1$ so that the top k rows of both matrices are zero, which avoids the circularity issue. In the proof of security, we will also rely on the fact that for any $\mathbf{y}_0, \mathbf{y}_1 \in \mathbb{Z}_q^{3k}$, if $\overline{\mathbf{y}}_0 = \overline{\mathbf{y}}_1$ and $\mathbf{y}_0 \in \mathsf{span}(\mathbf{M})$, then either $\mathbf{y}_0 = \mathbf{y}_1$ or $\mathbf{y}_1 \notin \mathsf{span}(\mathbf{M})$. This allows us to deduce that if the adversary queries the CCA oracle on a ciphertext which shares the same tag as some challenge ciphertext, then the CCA oracle will reject with overwhelming probability.

Alternative View-Point. Our construction can also be viewed as applying the BCHK IBE→PKE transform [8] to the scheme from [18], and then writing the exponents of the secret keys in the clear, thereby avoiding the pairing. This means that we can no longer apply a computational assumption and the randomized Naor-Reingold argument to the secret key space. Indeed, we replace this with an information-theoretic Cramer-Shoup-like argument as outlined above.

Prior Approaches. Several approaches to construct tightly CCA-secure PKE schemes exist: first, the schemes of [2,3,16,22–24] construct a tightly secure NIZK scheme from a tightly secure signature scheme, and then use the tightly secure NIZK in a CCA-secure PKE scheme following the Naor-Yung double encryption paradigm [12,28]. Since these approaches build on the public verifiability of the used NIZK scheme (in order to faithfully simulate a decryption oracle), their reliance on a pairing seems inherent.

Next, the works of [5,7,10,15,18] used a (Naor-Reingold-based) MAC instead of a signature scheme to design tightly secure IBE schemes. Those IBE schemes can then be converted (using the BCHK transformation [8]) into tightly CCA-secure PKE schemes. However, the derived PKE schemes still rely on pairings, since the original IBE schemes do (and the BCHK does not remove the reliance on pairings).

In contrast, our approach directly fuses a Naor-Reingold-like randomization argument with the encryption process. We are able to do so since we substitute a computational randomization argument (as used in the latter line of works) with an information-theoretic one, as described above. Hence, we can apply that argument to *exponents* rather than group elements. This enables us to trade pairing operations for exponentiations in our scheme.

Efficiency Comparison with Non-tightly Secure Schemes. We finally mention that our DDH-based scheme compares favorably even with the most efficient (non-tightly) CCA-secure DDH-based encryption schemes [17,21]. To make things concrete, assume $\lambda = 80$ and a setting with $Q_{\mathsf{enc}} = Q_{\mathsf{dec}} = 2^{30}$. The best known reductions for the schemes of [17,21] lose a factor of $Q_{\mathsf{enc}} = 2^{30}$, whereas our scheme loses a factor of about $4\lambda \leq 2^9$. Hence, the group size for [17,21] should be at least $2^{2\cdot(80+30)} = 2^{220}$ compared to $2^{2\cdot(80+9)} = 2^{178}$ in our case. Thus, the ciphertext overhead (ignoring the symmetric encryption part) in our scheme is $3 \cdot 178 = 534$ bits, which is close to $2 \cdot 220 = 440$ bits with [17,21].[2]

Perhaps even more interestingly, we can compare computational efficiency of encryption in this scenario. For simplicitly, we only count exponentiations and assume a naive square-and-multiply-based exponentiation with no further multi-exponentiation optimizations.[3] Encryption in [17,21] takes about 3.5 exponentiations (where we count an exponentiation with a $(\lambda + \log_2(Q_{\mathsf{enc}} + Q_{\mathsf{dec}}))$-bit hash value[4] as 0.5 exponentiations). In our scheme, we have about 4.67 exponentiations, where we count the computation of $[\mathbf{M}^\top \mathbf{k}_\tau]$ – which consists of 2λ multiplications – as 0.67 exponentiations.) Since exponentiation (under our assumptions) takes time cubic in the bitlength, we get that encryption with our scheme is actually about 29 % *less expensive* than with [17,21].

However, of course we should also note that public and secret key in our scheme are significantly larger (e.g., $4\lambda + 3 = 323$ group elements in pk) than with [17,21] (4 group elements in pk).

Extension: NIZK Arguments. We also obtain tightly simulation-sound non-interactive zero-knowledge (NIZK) arguments from our encryption scheme in a semi-generic way.

Let us start with any designated-verifier quasi-adaptive NIZK (short: DVQANIZK) argument system Π for a given language. Recall that in a designated-verifier NIZK, proofs can only be verified with a secret verification key, and soundness only holds against adversaries who do not know that key. Furthermore, quasi-adaptivity means that the language has to be fixed at setup time of the scheme. Let Π_{PKE} be the variant of Π in which proofs are encrypted using

[2] In this calculation, we do not consider the symmetric authenticated encryption of the actual plaintext (and a corresponding MAC value), which is the same with [17,21] and our scheme.

[3] Here, optimizations would improve the schemes of [17,21] and ours similarly, since the schemes are very similar.

[4] It is possible to prove the security of [17,21] using a *target*-collision-resistant hash function, such that $|\tau| = \lambda$. However, in the multi-user setting, a hybrid argument is required, such that the output size of the hash function will have to be increased to at least $|\tau| = \lambda + \log_2(Q_{\mathsf{enc}} + Q_{\mathsf{dec}})$.

Reference	$\|\mathsf{pk}\|$	$\|\mathsf{ct}\| - \|m\|$	security loss	assumption	pairing
CS98 [11]	$\mathcal{O}(1)$	3	$\mathcal{O}(Q)$	DDH	no
KD04, HK07 [23, 18]	$\mathcal{O}(1)$	2	$\mathcal{O}(Q)$	DDH	no
HJ12 [17]	$O(1)$	$O(\lambda)$	$\mathcal{O}(1)$	2-Lin	yes
LPJY15 [24, 26]	$\mathcal{O}(\lambda)$	47	$\mathcal{O}(\lambda)$	2-Lin	yes
AHY15 [5]	$\mathcal{O}(\lambda)$	12	$\mathcal{O}(\lambda)$	2-Lin	yes
GCDCT15 [16]	$\mathcal{O}(\lambda)$	10 (resp. $6k+4$)	$\mathcal{O}(\lambda)$	SXDH (resp. k-Lin)	yes
Ours §4	$\mathcal{O}(\lambda)$	3 (resp. $3k$)	$\mathcal{O}(\lambda)$	DDH (resp. k-Lin)	no

Fig. 2. Comparison amongst CCA-secure encryption schemes, where Q is the number of ciphertexts, $|\mathsf{pk}|$ denotes the size (i.e. the number of groups elements, or exponent of group elements) of the public key, and $|\mathsf{ct}| - |m|$ denotes the ciphertext overhead, ignoring smaller contributions from symmetric-key encryption. We omit [18] from this table since we only focus on prime-order groups here.

a CCA-secure PKE scheme PKE. Public and secret key of PKE are of course made part of CRS and verification key, respectively. Observe that Π_{PKE} enjoys simulation-soundness, assuming that simulated proofs are simply encryptions of random plaintexts. Indeed, the CCA security of PKE guarantees that authentic Π_{PKE}-proofs can be substituted with simulated ones, while being able to verify (using a decryption oracle) a purported Π_{PKE}-proof generated by an adversary. Furthermore, if PKE is tightly secure, then so is Π_{PKE}.

When using a hash proof system for Π and our encryption scheme for PKE, this immediately yields a tightly simulation-sound DVQANIZK for linear languages (i.e., languages of the form $\{[\mathbf{Mx}] \mid \mathbf{x} \in \mathbb{Z}_q^t\}$ for some matrix $\mathbf{M} \in \mathbb{Z}_q^{n \times t}$ with $t < n$) that does not require pairings. We stress that our DVQANIZK is tightly secure in a setting with many simulated proofs and many adversarial verification queries.

Using the semi-generic transformation of [20], we can then derive a tightly simulation-sound QANIZK proof system (with public verification), that however relies on pairings. We note that the transformation of [20] only requires a DVQANIZK that is secure against a single adversarial verification query, since the pairing enables the public verifiability of proofs. Hence, we can first optimize and trim down our DVQANIZK (such that only a single adversarial verification query is supported), and then apply the transformation. This yields a QANIZK with particularly compact proofs. See Fig. 3 for a comparison with relevant existing proof systems.

Roadmap. We recall some notation and basic definitions (including those concerning our algebraic setting and for tightly secure encryption) in Sect. 2. Section 3 presents our basic PCA-secure encryption scheme and represents the core of our results. In Sect. 4, we present our optimized CCA-secure PKE scheme. Our NIZK-related applications are presented in the full version of this paper.

Reference	type	$\|\mathsf{crs}\|$	$\|\pi\|$	sec. loss	assumption	pairing
CCS09 [9]	NIZK	$\mathcal{O}(1)$	$2n+6t+52$	$\mathcal{O}(Q_{\mathsf{sim}})$	2-Lin	yes
HJ12 [17]	NIZK	$\mathcal{O}(1)$	$\gg 500$	$\mathcal{O}(1)$	2-Lin	yes
LPJY14 [25]	QANIZK	$\mathcal{O}(n+\lambda)$	20	$\mathcal{O}(Q_{\mathsf{sim}})$	2-Lin	yes
KW15 [22]	QANIZK	$\mathcal{O}(kn)$	$2k+2$	$\mathcal{O}(Q_{\mathsf{sim}})$	k-Lin	yes
LPJY15 [27]	QANIZK	$\mathcal{O}(n+\lambda)$	42	$\mathcal{O}(\lambda)$	2-Lin	yes
Ours (full version)	DVQANIZK	$\mathcal{O}(t+k\lambda)$	$3k+1$	$\mathcal{O}(\lambda)$	k-Lin	no
Ours (full version)	QANIZK	$\mathcal{O}(k^2\lambda+kn)$	$2k+1$	$\mathcal{O}(\lambda)$	k-Lin	yes

Fig. 3. (DV)QANIZK schemes for subspaces of $\mathbb{G}^n$ of dimension $t < n$. $|\mathsf{crs}|$ and $|\pi|$ denote the size (in group elements) of the CRS and of proofs. Q_{sim} is the number of simulated proofs in the simulation-soundness experiment. The scheme from [20] (as well as our own schemes) can also be generalized to matrix assumptions [13], at the cost of a larger CRS.

2 Preliminaries

2.1 Notations

If $\mathbf{x} \in \mathcal{B}^n$, then $|\mathbf{x}|$ denotes the length n of the vector. Further, $x \leftarrow_{\mathrm{R}} \mathcal{B}$ denotes the process of sampling an element x from set $\mathcal{B}$ uniformly at random. For any bit string $\tau \in \{0,1\}^*$, we denote by τ_i the i'th bit of τ. We denote by λ the security parameter, and by $\mathsf{negl}(\cdot)$ any negligible function of λ. For all matrix $\mathbf{A} \in \mathbb{Z}_q^{\ell \times k}$ with $\ell > k$, $\overline{\mathbf{A}} \in \mathbb{Z}_q^{k \times k}$ denotes the upper square matrix of $\mathbf{A}$ and $\underline{\mathbf{A}} \in \mathbb{Z}_q^{\ell-k \times k}$ denotes the lower $\ell - k$ rows of $\mathbf{A}$. With $\mathsf{span}(\mathbf{A}) := \{\mathbf{Ar} \mid \mathbf{r} \in \mathbb{Z}_q^k\} \subset \mathbb{Z}_q^\ell$, we denote the span of $\mathbf{A}$.

2.2 Collision Resistant Hashing

A hash function generator is a PPT algorithm $\mathcal{H}$ that, on input 1^λ , outputs an efficiently computable function $\mathsf{H} : \{0,1\}^* \to \{0,1\}^\lambda$.

Definition 1 (Collision Resistance). *We say that a hash function generator $\mathcal{H}$ outputs collision-resistant functions* H *if for all PPT adversaries $\mathcal{A}$,*

$$\mathbf{Adv}_{\mathcal{H}}^{\mathrm{cr}}(\mathcal{A}) := \Pr\left[x \neq x' \wedge \mathsf{H}(x) = \mathsf{H}(x') \,\middle|\, \begin{array}{l} \mathsf{H} \leftarrow_{\mathrm{R}} \mathcal{H}(1^\lambda), \\ (x, x') \leftarrow \mathcal{A}(1^\lambda, \mathsf{H}) \end{array}\right] = \mathsf{negl}(\lambda).$$

2.3 Prime-Order Groups

Let GGen be a probabilistic polynomial time (PPT) algorithm that on input 1^λ returns a description $\mathcal{G} = (\mathbb{G}, q, P)$ of an additive cyclic group $\mathbb{G}$ of order q for a λ-bit prime q, whose generator is P.

We use implicit representation of group elements as introduced in [13]. For $a \in \mathbb{Z}_q$, define $[a] = aP \in \mathbb{G}$ as the *implicit representation* of a in $\mathbb{G}$. More generally, for a matrix $\mathbf{A} = (a_{ij}) \in \mathbb{Z}_q^{n \times m}$ we define $[\mathbf{A}]$ as the implicit representation of $\mathbf{A}$ in $\mathbb{G}$:

$$[\mathbf{A}] := \begin{pmatrix} a_{11}P \dots a_{1m}P \\ \\ a_{n1}P \dots a_{nm}P \end{pmatrix} \in \mathbb{G}^{n \times m}$$

We will always use this implicit notation of elements in $\mathbb{G}$, i.e., we let $[a] \in \mathbb{G}$ be an element in $\mathbb{G}$. Note that from $[a] \in \mathbb{G}$ it is generally hard to compute the value a (discrete logarithm problem in $\mathbb{G}$). Obviously, given $[a], [b] \in \mathbb{G}$ and a scalar $x \in \mathbb{Z}_q$, one can efficiently compute $[ax] \in \mathbb{G}$ and $[a + b] \in \mathbb{G}$.

2.4 Matrix Diffie-Hellman Assumption

We recall the definitions of the Matrix Decision Diffie-Hellman (MDDH) Assumption [13].

Definition 2 (Matrix Distribution). *Let $k, \ell \in \mathbb{N}$, with $\ell > k$. We call $\mathcal{D}_{\ell,k}$ a matrix distribution if it outputs matrices in $\mathbb{Z}_q^{\ell \times k}$ of full rank k in polynomial time. We write $\mathcal{D}_k := \mathcal{D}_{k+1,k}$.*

Without loss of generality, we assume the first k rows of $\mathbf{A} \leftarrow_{\mathrm{R}} \mathcal{D}_{\ell,k}$ form an invertible matrix. The $\mathcal{D}_{\ell,k}$-Matrix Diffie-Hellman problem is to distinguish the two distributions $([\mathbf{A}], [\mathbf{Aw}])$ and $([\mathbf{A}], [\mathbf{u}])$ where $\mathbf{A} \leftarrow_{\mathrm{R}} \mathcal{D}_{\ell,k}$, $\mathbf{w} \leftarrow_{\mathrm{R}} \mathbb{Z}_q^k$ and $\mathbf{u} \leftarrow_{\mathrm{R}} \mathbb{Z}_q^{\ell}$.

Definition 3 ($\mathcal{D}_{\ell,k}$-Matrix Diffie-Hellman Assumption $\mathcal{D}_{\ell,k}$-MDDH). *Let $\mathcal{D}_{\ell,k}$ be a matrix distribution. We say that the $\mathcal{D}_{\ell,k}$-Matrix Diffie-Hellman ($\mathcal{D}_{\ell,k}$-MDDH) Assumption holds relative to* GGen *if for all PPT adversaries $\mathcal{A}$,*

$$\mathbf{Adv}^{\mathrm{mddh}}_{\mathcal{D}_{\ell,k},\mathsf{GGen}}(\mathcal{A}) := \\ |\Pr[\mathcal{A}(\mathcal{G}, [\mathbf{A}], [\mathbf{Aw}]) = 1] - \Pr[\mathcal{A}(\mathcal{G}, [\mathbf{A}], [\mathbf{u}]) = 1]| = \mathsf{negl}(\lambda),$$

where the probability is over $\mathcal{G} \leftarrow_{\mathrm{R}} \mathsf{GGen}(1^\lambda)$, $\mathbf{A} \leftarrow_{\mathrm{R}} \mathcal{D}_{\ell,k}, \mathbf{w} \leftarrow_{\mathrm{R}} \mathbb{Z}_q^k, \mathbf{u} \leftarrow_{\mathrm{R}} \mathbb{Z}_q^{\ell}$.

For each $k \geq 1$, [13] specifies distributions $\mathcal{L}_k$, $\mathcal{SC}_k$, $\mathcal{C}_k$ (and others) over $\mathbb{Z}_q^{(k+1) \times k}$ such that the corresponding $\mathcal{D}_k$-MDDH assumptions are generically secure in bilinear groups and form a hierarchy of increasingly weaker assumptions. $\mathcal{L}_k$-MDDH is the well known k-Linear Assumption k-Lin with 1-Lin = DDH. In this work we are mostly interested in the uniform matrix distribution $\mathcal{U}_{\ell,k}$.

Definition 4 (Uniform Distribution). *Let $\ell, k \in \mathbb{N}$, with $\ell > k$. We denote by $\mathcal{U}_{\ell,k}$ the uniform distribution over all full-rank $\ell \times k$ matrices over $\mathbb{Z}_q$. Let $\mathcal{U}_k := \mathcal{U}_{k+1,k}$.*

Lemma 1 ($\mathcal{U}_k$-MDDH $\Leftrightarrow$ $\mathcal{U}_{\ell,k}$-MDDH). *Let $\ell, k \in \mathbb{N}$, with $\ell > k$. For any PPT adversary $\mathcal{A}$, there exists an adversary $\mathcal{B}$ (and vice versa) such that $\mathbf{T}(\mathcal{B}) \approx \mathbf{T}(\mathcal{A})$ and $\mathbf{Adv}^{\mathrm{mddh}}_{\mathcal{U}_{\ell,k},\mathsf{GGen}}(\mathcal{A}) = \mathbf{Adv}^{\mathrm{mddh}}_{\mathcal{U}_k,\mathsf{GGen}}(\mathcal{B})$.*

Proof. This follows from the simple fact that a $\mathcal{U}_{\ell,k}$-MDDH instance $([\mathbf{A}],[\mathbf{z}])$ can be transformed into an $\mathcal{U}_k$-MDDH instance $([\mathbf{A}'] = [\mathbf{TA}], [\mathbf{z}'] = [\mathbf{Tz}])$ for a random $(k+1)\times\ell$ matrix $\mathbf{T}$. If $\mathbf{z} = \mathbf{Aw}$, then $\mathbf{z}' = \mathbf{TAw} = \mathbf{A}'\mathbf{w}$; if $\mathbf{z}$ is uniform, so is $\mathbf{z}'$. Similarly, a $\mathcal{U}_k$-MDDH instance $([\mathbf{A}'],[\mathbf{z}'])$ can be transformed into an $\mathcal{U}_{\ell,k}$-MDDH instance $([\mathbf{A}] = [\mathbf{T}'\mathbf{A}'], [\mathbf{z}] = [\mathbf{T}'\mathbf{z}'])$ for a random $\ell \times (k+1)$ matrix $\mathbf{T}'$. □

Among all possible matrix distributions $\mathcal{D}_{\ell,k}$, the uniform matrix distribution $\mathcal{U}_k$ is the hardest possible instance, so in particular k-Lin $\Rightarrow \mathcal{U}_k$-MDDH.

Lemma 2 ($\mathcal{D}_{\ell,k}$-MDDH $\Rightarrow \mathcal{U}_k$-MDDH, [13]). *Let $\mathcal{D}_{\ell,k}$ be a matrix distribution. For any PPT adversary $\mathcal{A}$, there exists an adversary $\mathcal{B}$ such that $\mathbf{T}(\mathcal{B}) \approx \mathbf{T}(\mathcal{A})$ and $\mathbf{Adv}^{\mathrm{mddh}}_{\mathcal{D}_{\ell,k},\mathsf{GGen}}(\mathcal{A}) = \mathbf{Adv}^{\mathrm{mddh}}_{\mathcal{U}_k,\mathsf{GGen}}(\mathcal{B})$.*

Let $Q \geq 1$. For $\mathbf{W} \leftarrow_{\mathrm{R}} \mathbb{Z}_q^{k\times Q}, \mathbf{U} \leftarrow_{\mathrm{R}} \mathbb{Z}_q^{\ell\times Q}$, we consider the Q-fold $\mathcal{D}_{\ell,k}$-MDDH Assumption which consists in distinguishing the distributions $([\mathbf{A}],[\mathbf{AW}])$ from $([\mathbf{A}],[\mathbf{U}])$. That is, a challenge for the Q-fold $\mathcal{D}_{\ell,k}$-MDDH Assumption consists of Q independent challenges of the $\mathcal{D}_{\ell,k}$-MDDH Assumption (with the same $\mathbf{A}$ but different randomness $\mathbf{w}$). In [13] it is shown that the two problems are equivalent, where (for $Q \geq \ell - k$) the reduction loses a factor $\ell - k$. In combination with Lemma 1 we obtain the following tighter version for the special case of $\mathcal{D}_{\ell,k} = \mathcal{U}_{\ell,k}$.

Lemma 3 (Random Self-reducibility of $\mathcal{U}_{\ell,k}$-MDDH, [13]). *Let $\ell, k, Q \in \mathbb{N}$ with $\ell > k$. For any PPT adversary $\mathcal{A}$, there exists an adversary $\mathcal{B}$ such that $\mathbf{T}(\mathcal{B}) \approx \mathbf{T}(\mathcal{A}) + Q \cdot \mathsf{poly}(\lambda)$ with $\mathsf{poly}(\lambda)$ independent of $\mathbf{T}(\mathcal{A})$, and*

$$\mathbf{Adv}^{\mathrm{Q\text{-}mddh}}_{\mathcal{U}_{\ell,k},\mathsf{GGen}}(\mathcal{A}) \leq \mathbf{Adv}^{\mathrm{mddh}}_{\mathcal{U}_{\ell,k},\mathsf{GGen}}(\mathcal{B}) + \frac{1}{q-1}$$

where $\mathbf{Adv}^{\mathrm{Q\text{-}mddh}}_{\mathcal{U}_{\ell,k},\mathsf{GGen}}(\mathcal{B}) := |\Pr[\mathcal{B}(\mathcal{G},[\mathbf{A}],[\mathbf{AW}]) = 1] - \Pr[\mathcal{B}(\mathcal{G},[\mathbf{A}],[\mathbf{U}]) = 1]|$ *and the probability is over* $\mathcal{G} \leftarrow_{\mathrm{R}} \mathsf{GGen}(1^\lambda)$, $\mathbf{A} \leftarrow_{\mathrm{R}} \mathcal{U}_{\ell,k}$, $\mathbf{W} \leftarrow_{\mathrm{R}} \mathbb{Z}_q^{k\times Q}, \mathbf{U} \leftarrow_{\mathrm{R}} \mathbb{Z}_q^{\ell\times Q}$.

2.5 Public-Key Encryption

Definition 5 (PKE). *A* Public-Key Encryption *(PKE) consists of three PPT algorithms* $\mathsf{PKE} = (\mathsf{Param}_{\mathsf{PKE}}, \mathsf{Gen}_{\mathsf{PKE}}, \mathsf{Enc}_{\mathsf{PKE}}, \mathsf{Dec}_{\mathsf{PKE}})$*:*

- *The probabilistic key generation algorithm* $\mathsf{Gen}_{\mathsf{PKE}}(1^\lambda)$ *generates a pair of public and secret keys* $(\mathsf{pk},\mathsf{sk})$.
- *The probabilistic encryption algorithm* $\mathsf{Enc}_{\mathsf{PKE}}(\mathsf{pk}, M)$ *returns a ciphertext* ct.
- *The deterministic decryption algorithm* $\mathsf{Dec}_{\mathsf{PKE}}(\mathsf{pk},\mathsf{sk},\mathsf{ct})$ *returns a message* M *or* $\perp$*, where* $\perp$ *is a special rejection symbol.*

We define the following properties:

Perfect Correctness. *For all* λ*, we have*

$$\Pr\left[\mathsf{Dec}_{\mathsf{PKE}}(\mathsf{pk},\mathsf{sk},\mathsf{ct}) = M \,\middle|\, \begin{array}{l}(\mathsf{pk},\mathsf{sk}) \leftarrow_{\mathrm{R}} \mathsf{Gen}_{\mathsf{PKE}}(1^\lambda);\\ \mathsf{ct} \leftarrow_{\mathrm{R}} \mathsf{Enc}_{\mathsf{PKE}}(\mathsf{pk}, M)\end{array}\right] = 1.$$

Multi-ciphertext CCA Security [6]. *For any adversary* $\mathcal{A}$, *we define*

$$\mathbf{Adv}_{\mathsf{PKE}}^{\mathsf{ind\text{-}cca}}(\mathcal{A}) := \left|\Pr\left[b = b' \,\middle|\, b' \leftarrow \mathcal{A}^{\mathsf{Setup},\mathsf{DecO}(\cdot),\mathsf{EncO}(\cdot,\cdot)}(1^\lambda)\right] - 1/2\right|$$

where:
- Setup *sets* $\mathcal{C}_{\mathsf{enc}} := \emptyset$, *samples* $(\mathsf{pk}, \mathsf{sk}) \leftarrow_R \mathsf{Gen}_{\mathsf{KEM}}(1^\lambda)$ *and* $b \leftarrow_R \{0,1\}$, *and returns* pk. Setup *must be called once at the beginning of the game.*
- $\mathsf{DecO}(\mathsf{ct})$ *returns* $\mathsf{Dec}_{\mathsf{PKE}}(\mathsf{pk}, \mathsf{sk}, \mathsf{ct})$ *if* $\mathsf{ct} \notin \mathcal{C}_{\mathsf{enc}}$, $\perp$ *otherwise.*
- *If* M_0 *and* M_1 *are two messages of equal length,* $\mathsf{EncO}(M_0, M_1)$ *returns* $\mathsf{Enc}_{\mathsf{PKE}}(\mathsf{pk}, M_b)$ *and sets* $\mathcal{C}_{\mathsf{enc}} := \mathcal{C}_{\mathsf{enc}} \cup \{\mathsf{ct}\}$.

We say PKE *is* IND-CCA *secure if for all* PPT *adversaries* $\mathcal{A}$, *the advantage* $\mathbf{Adv}_{\mathsf{PKE}}^{\mathsf{ind\text{-}cca}}(\mathcal{A})$ *is a negligible function of* λ.

2.6 Key-Encapsulation Mechanism

Definition 6 (Tag-based KEM). *A tag-based* Key-Encapsulation Mechanism *(KEM) consists of three PPT algorithms* $\mathsf{KEM} = (\mathsf{Gen}_{\mathsf{KEM}}, \mathsf{Enc}_{\mathsf{KEM}}, \mathsf{Dec}_{\mathsf{KEM}})$:

- *The probabilistic key generation algorithm* $\mathsf{Gen}_{\mathsf{KEM}}(1^\lambda)$ *generates a pair of public and secret keys* $(\mathsf{pk}, \mathsf{sk})$.
- *The probabilistic encryption algorithm* $\mathsf{Enc}_{\mathsf{KEM}}(\mathsf{pk}, \tau)$ *returns a pair* (K, C) *where* K *is a uniformly distributed symmetric key in* $\mathcal{K}$ *and* C *is a ciphertext, with respect to the tag* $\tau \in \mathcal{T}$.
- *The deterministic decryption algorithm* $\mathsf{Dec}_{\mathsf{KEM}}(\mathsf{pk}, \mathsf{sk}, \tau, C)$ *returns a key* $K \in \mathcal{K}$.

We define the following properties:

Perfect Correctness. *For all* λ, *for all tags* $\tau \in \mathcal{T}$, *we have*

$$\Pr\left[\mathsf{Dec}_{\mathsf{KEM}}(\mathsf{pk}, \mathsf{sk}, \tau, C) = K \,\middle|\, \begin{array}{l}(\mathsf{pk}, \mathsf{sk}) \leftarrow_R \mathsf{Gen}_{\mathsf{KEM}}(1^\lambda);\\ (K, C) \leftarrow_R \mathsf{Enc}_{\mathsf{KEM}}(\mathsf{pk}, \tau)\end{array}\right] = 1.$$

Multi-ciphertext PCA Security [29]. *For any adversary* $\mathcal{A}$, *we define*

$$\mathbf{Adv}_{\mathsf{KEM}}^{\mathsf{ind\text{-}pca}}(\mathcal{A}) := \left|\Pr\left[b = b' \,\middle|\, b' \leftarrow \mathcal{A}^{\mathsf{Setup},\mathsf{DecO}(\cdot,\cdot,\cdot),\mathsf{EncO}(\cdot)}(1^\lambda)\right] - 1/2\right|$$

where:
- Setup *sets* $\mathcal{T}_{\mathsf{enc}} = \mathcal{T}_{\mathsf{dec}} := \emptyset$, *samples* $(\mathsf{pk}, \mathsf{sk}) \leftarrow_R \mathsf{Gen}_{\mathsf{KEM}}(1^\lambda)$, *picks* $b \leftarrow_R \{0,1\}$, *and returns* pk. Setup *is called once at the beginning of the game.*
- *The decryption oracle* $\mathsf{DecO}(\tau, C, \widehat{K})$ *computes* $K := \mathsf{Dec}_{\mathsf{KEM}}(\mathsf{pk}, \mathsf{sk}, \tau, C)$. *It returns* 1 *if* $\widehat{K} = K \wedge \tau \notin \mathcal{T}_{\mathsf{enc}}$, 0 *otherwise. Then it sets* $\mathcal{T}_{\mathsf{dec}} := \mathcal{T}_{\mathsf{dec}} \cup \{\tau\}$.
- $\mathsf{EncO}(\tau)$ *computes* $(K, C) \leftarrow_R \mathsf{Enc}_{\mathsf{KEM}}(\mathsf{pk}, \tau)$, *sets* $K_0 := K$ *and* $K_1 \leftarrow_R \mathcal{K}$. *If* $\tau \notin \mathcal{T}_{\mathsf{dec}} \cup \mathcal{T}_{\mathsf{enc}}$, *it returns* (C, K_b), *and sets* $\mathcal{T}_{\mathsf{enc}} := \mathcal{T}_{\mathsf{enc}} \cup \{\tau\}$; *otherwise it returns* $\perp$.

We say KEM *is* IND-PCA *secure if for all* PPT *adversaries* $\mathcal{A}$, *the advantage* $\mathbf{Adv}_{\mathsf{KEM}}^{\mathsf{ind\text{-}pca}}(\mathcal{A})$ *is a negligible function of* λ.

2.7 Authenticated Encryption

Definition 7 (AE [17]). *An* authenticated symmetric encryption *(AE) with message-space $\mathcal{M}$ and key-space $\mathcal{K}$ consists of two polynomial-time deterministic algorithms* $(\mathsf{Enc}_{\mathsf{AE}}, \mathsf{Dec}_{\mathsf{AE}})$*:*

- *The encryption algorithm* $\mathsf{Enc}_{\mathsf{AE}}(K, M)$ *generates* C*, encryption of the message* M *with the secret key* K*.*
- *The decryption algorithm* $\mathsf{Dec}_{\mathsf{AE}}(K, C)$*, returns a message* M *or* $\perp$*.*

We require that the algorithms satisfy the following properties:

Perfect Correctness. *For all* λ*, for all* $K \in \mathcal{K}$ *and* $M \in \mathcal{M}$*, we have*

$$\mathsf{Dec}_{\mathsf{AE}}(K, \mathsf{Enc}_{\mathsf{AE}}(K, M)) = M.$$

One-Time Privacy and Authenticity. *For any PPT adversary* $\mathcal{A}$*,*

$$\mathbf{Adv}_{\mathsf{AE}}^{\mathsf{ae\text{-}ot}}(\mathcal{A}) := \left| \Pr\left[b' = b \,\middle|\, \begin{array}{l} K \leftarrow_{\mathrm{R}} \mathcal{K}; b \leftarrow_{\mathrm{R}} \{0,1\} \\ b' \leftarrow_{\mathrm{R}} \mathcal{A}^{\mathsf{ot\text{-}EncO}(\cdot,\cdot),\mathsf{ot\text{-}DecO}(\cdot)}(1^\lambda, \mathcal{K}) \end{array} \right] - 1/2 \right|$$

is negligible, where $\mathsf{ot\text{-}EncO}(M_0, M_1)$*, on input two messages* M_0 *and* M_1 *of the same length,* $\mathsf{Enc}_{\mathsf{AE}}(K, M_b)$*, and* $\mathsf{ot\text{-}DecO}(\phi)$ *returns* $\mathsf{Dec}_{\mathsf{AE}}(K, \phi)$ *if* $b = 0$*,* $\perp$ *otherwise.* $\mathcal{A}$ *is allowed at most one call to each oracle* $\mathsf{ot\text{-}EncO}$ *and* $\mathsf{ot\text{-}DecO}$*, and the query to* $\mathsf{ot\text{-}DecO}$ *must be different from the output of* $\mathsf{ot\text{-}EncO}$*.* $\mathcal{A}$ *is also given the description of the key-space* $\mathcal{K}$ *as input.*

3 Multi-ciphertext PCA-secure KEM

In this section we describe a tag-based Key Encapsulation Mechanism $\mathsf{KEM}_{\mathsf{PCA}}$ that is IND-PCA-secure (see Definition 6).

For simplicity, we use the matrix distribution $\mathcal{U}_{3k,k}$ in our scheme in Fig. 4, and prove it secure under the $\mathcal{U}_k$-MDDH Assumption ($\Leftrightarrow \mathcal{U}_{3k,k}$-$\mathsf{MDDH}$ Assumption, by Lemma 1), which in turn admits a tight reduction to the standard k-Lin Assumption. However, using a matrix distribution $\mathcal{D}_{3k,k}$ with more compact representation yields a more efficient scheme, secure under the $\mathcal{D}_{3k,k}$-MDDH Assumption (see Remark 1).

3.1 Our Construction

Remark 1 (On the use of the $\mathcal{U}_k$-MDDH *Assumption).* In our scheme, we use a matrix distribution $\mathcal{U}_{3k,k}$ for the matrix $\mathbf{M}$, therefore proving security under the $\mathcal{U}_{3k,k}$-MDDH Assumption $\Leftrightarrow \mathcal{U}_k$-$\mathsf{MDDH}$ Assumption (see Lemma 2). This is for simplicity of presentation. However, for efficiency, one may want to use an assumption with a more compact representation, such as the $\mathcal{CI}_{3k,k}$-MDDH Assumption [26] with representation size $2k$ instead of $3k^2$ for $\mathcal{U}_{3k,k}$.

$\mathsf{Gen}_{\mathsf{KEM}}(1^\lambda)$:
$\mathcal{G} \leftarrow_{\mathrm{R}} \mathsf{GGen}(1^\lambda)$; $\mathbf{M} \leftarrow_{\mathrm{R}} \mathcal{U}_{3k,k}$
$\mathbf{k}_{1,0}, \ldots, \mathbf{k}_{\lambda,1} \leftarrow_{\mathrm{R}} \mathbb{Z}_q^{3k}$
$\mathsf{pk} := \left(\mathcal{G}, [\mathbf{M}], ([\mathbf{M}^\top \mathbf{k}_{j,\beta}])_{1 \leq j \leq \lambda, 0 \leq \beta \leq 1}\right)$
$\mathsf{sk} := (\mathbf{k}_{j,\beta})_{1 \leq j \leq \lambda, 0 \leq \beta \leq 1}$
Return $(\mathsf{pk}, \mathsf{sk})$

$\mathsf{Enc}_{\mathsf{KEM}}(\mathsf{pk}, \tau)$:
$\mathbf{r} \leftarrow_{\mathrm{R}} \mathbb{Z}_q^k$; $C := [\mathbf{r}^\top \mathbf{M}^\top]$
$\mathbf{k}_\tau := \sum_{j=1}^{\lambda} \mathbf{k}_{j,\tau_j}$
$K := [\mathbf{r}^\top \cdot \mathbf{M}^\top \mathbf{k}_\tau]$
Return $(C, K) \in \mathbb{G}^{1 \times 3k} \times \mathbb{G}$

$\mathsf{Dec}_{\mathsf{KEM}}(\mathsf{pk}, \mathsf{sk}, \tau, C)$:
$\mathbf{k}_\tau := \sum_{j=1}^{\lambda} \mathbf{k}_{j,\tau_j}$
Return $K := C \cdot \mathbf{k}_\tau$

Fig. 4. $\mathsf{KEM}_{\mathsf{PCA}}$, an IND-PCA-secure KEM under the $\mathcal{U}_k$-MDDH Assumption, with tag-space $\mathcal{T} = \{0,1\}^\lambda$. Here, GGen is a prime-order group generator (see Sect. 2.3).

3.2 Security Proof

Theorem 1. *The tag-based Key Encapsulation Mechanism* $\mathsf{KEM}_{\mathsf{PCA}}$ *defined in Fig. 4 has perfect correctness. Moreover, if the* $\mathcal{U}_k$-MDDH *Assumption holds in* $\mathbb{G}$, $\mathsf{KEM}_{\mathsf{PCA}}$ *is* IND-PCA *secure. Namely, for any adversary* $\mathcal{A}$, *there exists an adversary* $\mathcal{B}$ *such that* $\mathbf{T}(\mathcal{B}) \approx \mathbf{T}(\mathcal{A}) + (Q_{\mathsf{dec}} + Q_{\mathsf{enc}}) \cdot \mathsf{poly}(\lambda)$ *and*

$$\mathbf{Adv}^{\mathsf{ind\text{-}pca}}_{\mathsf{KEM}_{\mathsf{PCA}}}(\mathcal{A}) \leq (4\lambda + 1) \cdot \mathbf{Adv}^{\mathrm{mddh}}_{\mathcal{U}_k, \mathsf{GGen}}(\mathcal{B}) + (Q_{\mathsf{dec}} + Q_{\mathsf{enc}}) \cdot 2^{-\Omega(\lambda)},$$

where Q_{enc}, Q_{dec} *are the number of times* $\mathcal{A}$ *queries* EncO, DecO, *respectively, and* $\mathsf{poly}(\lambda)$ *is independent of* $\mathbf{T}(\mathcal{A})$.

Proof of Theorem 1. Perfect correctness follows readily from the fact that for all $\mathbf{r} \in \mathbb{Z}_q^k$ and $C = \mathbf{r}^\top \mathbf{M}^\top$, for all $\mathbf{k} \in \mathbb{Z}_q^{3k}$:

$$\mathbf{r}^\top (\mathbf{M}^\top \mathbf{k}) = C \cdot \mathbf{k}.$$

We now prove the IND-PCA security of $\mathsf{KEM}_{\mathsf{PCA}}$. We proceed via a series of games described in Figs. 6 and 7 and we use $\mathbf{Adv}_i$ to denote the advantage of $\mathcal{A}$ in game G_i. We also give a high-level picture of the proof in Fig. 5, summarizing the sequence of games.

Lemma 4 (G_0 *to* G_1)**.** *There exists an adversary* $\mathcal{B}_0$ *such that* $\mathbf{T}(\mathcal{B}_0) \approx \mathbf{T}(\mathcal{A}) + (Q_{\mathsf{enc}} + Q_{\mathsf{dec}}) \cdot \mathsf{poly}(\lambda)$ *and*

$$|\mathbf{Adv}_0 - \mathbf{Adv}_1| \leq \mathbf{Adv}^{\mathrm{mddh}}_{\mathcal{U}_k, \mathsf{GGen}}(\mathcal{B}_0) + \frac{1}{q-1},$$

where Q_{enc}, Q_{dec} *are the number of times* $\mathcal{A}$ *queries* EncO, DecO, *respectively, and* $\mathsf{poly}(\lambda)$ *is independent of* $\mathbf{T}(\mathcal{A})$.

Here, we use the MDDH assumption to "tightly" switch the distribution of all the challenge ciphertexts.

game	$\mathbf{y}$ uniform in:	$\mathbf{k}'_\tau$ used by EncO and DecO	justification/remark
G_0	$\mathsf{span}(\mathbf{M})$	$\mathbf{k}_\tau$	actual scheme
G_1	$\boxed{\mathbb{Z}_q^{3k}}$	$\mathbf{k}_\tau$	$\mathcal{U}_{3k,k}$-MDDH on $[\mathbf{M}]$
$\mathrm{G}_{2.i}$	$\mathbb{Z}_q^{3k}$	$\mathbf{k}_\tau + \boxed{\mathbf{M}^\perp \mathsf{RF}_i(\tau_{\mid i})}$	$\mathrm{G}_1 \equiv \mathrm{G}_{2.0}$
$\mathrm{G}_{2.i.1}$	$\tau_{i+1} = 0 : \boxed{\mathsf{span}(\mathbf{M}, \mathbf{M}_0)}$ $\tau_{i+1} = 1 : \boxed{\mathsf{span}(\mathbf{M}, \mathbf{M}_1)}$	$\mathbf{k}_\tau + \mathbf{M}^\perp \mathsf{RF}_i(\tau_{\mid i})$	$\mathcal{U}_{3k,k}$-MDDH on $[\mathbf{M}_0]$ $\mathcal{U}_{3k,k}$-MDDH on $[\mathbf{M}_1]$
$\mathrm{G}_{2.i.2}$	$\tau_{i+1} = 0 : \mathsf{span}(\mathbf{M}, \mathbf{M}_0)$ $\tau_{i+1} = 1 : \mathsf{span}(\mathbf{M}, \mathbf{M}_1)$	$\mathbf{k}_\tau + \boxed{\mathbf{M}_0^* \mathsf{RF}_{i+1}^{(0)}(\tau_{\mid i+1})} + \mathbf{M}_1^* \mathsf{RF}_i^{(1)}(\tau_{\mid i})$	Cramer-Shoup argument
$\mathrm{G}_{2.i.3}$	$\tau_{i+1} = 0 : \mathsf{span}(\mathbf{M}, \mathbf{M}_0)$ $\tau_{i+1} = 1 : \mathsf{span}(\mathbf{M}, \mathbf{M}_1)$	$\mathbf{k}_\tau + \mathbf{M}_0^* \mathsf{RF}_{i+1}^{(0)}(\tau_{\mid i+1}) + \boxed{\mathbf{M}_1^* \mathsf{RF}_{i+1}^{(1)}(\tau_{\mid i+1})}$	Cramer-Shoup argument
$\mathrm{G}_{2.i+1}$	$\boxed{\mathbb{Z}_q^{3k}}$	$\mathbf{k}_\tau + \mathbf{M}^\perp \mathsf{RF}_{i+1}(\tau_{\mid i+1})$	$\mathcal{U}_{3k,k}$-MDDH on $[\mathbf{M}_0]$ and $[\mathbf{M}_1]$

Fig. 5. Sequence of games for the proof of Theorem 1. Throughout, we have (i) $\mathbf{k}_\tau := \sum_{j=1}^{\lambda} \mathbf{k}_{j,\tau_j}$; (ii) $\mathsf{EncO}(\tau) = ([\mathbf{y}], K_b)$ where $K_0 = [\mathbf{y}^\top \mathbf{k}'_\tau]$ and $K_1 \leftarrow_{\mathrm{R}} \mathbb{G}$; (iii) $\mathsf{DecO}(\tau, [\mathbf{y}], \widehat{K})$ computes the encapsulation key $K := [\mathbf{y}^\top \cdot \mathbf{k}'_\tau]$. Here, $(\mathbf{M}_0^*, \mathbf{M}_1^*)$ is a basis for $\mathsf{span}(\mathbf{M}^\perp)$, so that $\mathbf{M}_1^\perp \mathbf{M}_0^* = \mathbf{M}_0^\perp \mathbf{M}_1^* = \mathbf{0}$, and we write $\mathbf{M}^\perp \mathsf{RF}_i(\tau_{\mid i}) := \mathbf{M}_0^* \mathsf{RF}_i^{(0)}(\tau_{\mid i}) + \mathbf{M}_0^* \mathsf{RF}_i^{(1)}(\tau_{\mid i})$. The second column shows which set $\mathbf{y}$ is uniformly picked from by EncO, the third column shows the value of $\mathbf{k}'_\tau$ used by both EncO and DecO.

Setup: $\mathrm{G}_0, \mathrm{G}_1, \boxed{\mathrm{G}_{2.i}}$

$\mathcal{T}_{\mathsf{enc}} = \mathcal{T}_{\mathsf{dec}} := \emptyset$; $b \leftarrow_{\mathrm{R}} \{0,1\}$
$\mathcal{G} \leftarrow_{\mathrm{R}} \mathsf{GGen}(1^\lambda)$; $\mathbf{M} \leftarrow_{\mathrm{R}} \mathcal{U}_{3k,k}$

$\boxed{\mathbf{M}^\perp \leftarrow_{\mathrm{R}} \mathcal{U}_{3k,2k} \text{ s.t. } \mathbf{M}^\top\mathbf{M}^\perp = \mathbf{0}}$
$\boxed{\text{Pick random } \mathsf{RF}_i : \{0,1\}^i \to \mathbb{Z}_q^{2k}}$

$\mathbf{k}_{1,0}, \ldots, \mathbf{k}_{\lambda,1} \leftarrow_{\mathrm{R}} \mathbb{Z}_q^{3k}$
For all $\tau \in \{0,1\}^\lambda$, $\mathbf{k}_\tau := \sum_{j=1}^{\lambda} \mathbf{k}_{j,\tau_j}$
$\mathbf{k}'_\tau := \mathbf{k}_\tau + \boxed{\mathbf{M}^\perp \mathsf{RF}_i(\tau_{|i})}$
Return
$\mathsf{pk} := \Big(\mathcal{G}, [\mathbf{M}], ([\mathbf{M}^\top \mathbf{k}_{j,\beta}])_{1 \le j \le \lambda, 0 \le \beta \le 1}\Big)$

EncO(τ): G_0, $\mathrm{G}_1, \mathrm{G}_{2.i}$ (dotted frame)

$\mathbf{r} \leftarrow_{\mathrm{R}} \mathbb{Z}_q^k$; $\mathbf{y} := \mathbf{M}\mathbf{r}$; $\mathbf{y} \leftarrow_{\mathrm{R}} \mathbb{Z}_q^{3k}$ (dotted frame)
$K_0 := [\mathbf{y}^\top \cdot \mathbf{k}'_\tau]$; $K_1 \leftarrow_{\mathrm{R}} \mathbb{G}$
If $\tau \notin \mathcal{T}_{\mathsf{dec}} \cup \mathcal{T}_{\mathsf{enc}}$, return $(C := [\mathbf{y}], K_b)$,
and set $\mathcal{T}_{\mathsf{enc}} := \mathcal{T}_{\mathsf{enc}} \cup \{\tau\}$.
Otherwise, return $\perp$.

DecO($\tau, C := [\mathbf{y}], \widehat{K}$): $\mathrm{G}_0, \mathrm{G}_1, \mathrm{G}_{2.i}$

$K := [\mathbf{y}^\top \cdot \mathbf{k}'_\tau]$
Return $\begin{cases} 1 \text{ if } \widehat{K} = K \wedge \tau \notin \mathcal{T}_{\mathsf{enc}} \\ 0 \text{ otherwise} \end{cases}$
$\mathcal{T}_{\mathsf{dec}} := \mathcal{T}_{\mathsf{dec}} \cup \{\tau\}$

Fig. 6. Games $\mathrm{G}_0, \mathrm{G}_1, \mathrm{G}_{2.i}$ (for $1 \le i \le \lambda$) for the proof of multi-ciphertext PCA security of $\mathsf{KEM}_{\mathsf{PCA}}$ in Fig. 4. For all $0 \le i \le \lambda$, $\mathsf{RF}_i : \{0,1\}^i \to \mathbb{Z}_q^{2k}$ is a random function, and for all $\tau \in \mathcal{T}$, $\tau_{|i}$ denotes the i-bit prefix of τ. In each procedure, the components inside a solid (dotted) frame are only present in the games marked by a solid (dotted) frame.

Proof of Lemma 4. To go from G_0 to G_1, we switch the distribution of the vectors $[\mathbf{y}]$ sampled by EncO, using the Q_{enc}-fold $\mathcal{U}_{3k,k}$-MDDH Assumption on $[\mathbf{M}]$ (see Definition 4 and Lemma 3).

We build an adversary $\mathcal{B}'_0$ against the Q_{enc}-fold $\mathcal{U}_{3k,k}$-MDDH Assumption, such that $\mathbf{T}(\mathcal{B}'_0) \approx \mathbf{T}(\mathcal{A}) + (Q_{\mathsf{enc}} + Q_{\mathsf{dec}}) \cdot \mathsf{poly}(\lambda)$ with $\mathsf{poly}(\lambda)$ independent of $\mathbf{T}(\mathcal{A})$, and

$$|\mathbf{Adv}_0 - \mathbf{Adv}_1| \le \mathbf{Adv}_{\mathcal{U}_{3k,k},\mathsf{GGen}}^{Q_{\mathsf{enc}}\text{-mddh}}(\mathcal{B}'_0).$$

This implies the lemma by Lemma 3 (self-reducibility of $\mathcal{U}_{3k,k}$-MDDH), and Lemma 1 ($\mathcal{U}_{3k,k}$-$\mathsf{MDDH} \Leftrightarrow \mathcal{U}_k$-$\mathsf{MDDH}$).

Upon receiving a challenge $(\mathcal{G}, [\mathbf{M}] \in \mathbb{G}^{3k \times k}, [\mathbf{H}] := [\mathbf{h}_1 | \ldots | \mathbf{h}_{Q_{\mathsf{enc}}}] \in \mathbb{G}^{3k \times Q_{\mathsf{enc}}})$ for the Q_{enc}-fold $\mathcal{U}_{3k,k}$-MDDH Assumption, $\mathcal{B}'_0$ picks $b \leftarrow_{\mathrm{R}} \{0,1\}$, $\mathbf{k}_{1,0}, \ldots, \mathbf{k}_{\lambda,1} \leftarrow_{\mathrm{R}} \mathbb{Z}_q^{3k}$, and simulates Setup, DecO as described in Fig. 6. To simulate EncO on its j'th query, for $j = 1, \ldots, Q_{\mathsf{enc}}$, $\mathcal{B}'_0$ sets $[\mathbf{y}] := [\mathbf{h}_j]$, and computes K_b as described in Fig. 6. $\square$

Lemma 5 **(G_1 to $\mathrm{G}_{2.0}$).** $|\mathbf{Adv}_1 - \mathbf{Adv}_{2.0}| = 0$.

Proof of Lemma 5. We show that the two games are statistically equivalent. To go from G_1 to $\mathrm{G}_{2.0}$, we change the distribution of $\mathbf{k}_{1,\beta} \leftarrow_{\mathrm{R}} \mathbb{Z}_q^{3k}$ for $\beta = 0, 1$, to $\mathbf{k}_{1,\beta} + \mathbf{M}^\perp \mathsf{RF}_0(\varepsilon)$, where $\mathbf{k}_{1,\beta} \leftarrow_{\mathrm{R}} \mathbb{Z}_q^{3k}$, $\mathsf{RF}_0(\varepsilon) \leftarrow_{\mathrm{R}} \mathbb{Z}_q^{2k}$, and $\mathbf{M}^\perp \leftarrow_{\mathrm{R}} \mathcal{U}_{3k,2k}$ such that $\mathbf{M}^\top \mathbf{M}^\perp = \mathbf{0}$. Note that the extra term $\mathbf{M}^\perp \mathsf{RF}_0(\varepsilon)$ does not appear in pk, since $\mathbf{M}^\top(\mathbf{k}_{1,\beta} + \mathbf{M}^\perp \mathsf{RF}_0(\varepsilon)) = \mathbf{M}^\top \mathbf{k}_{1,\beta}$. $\square$

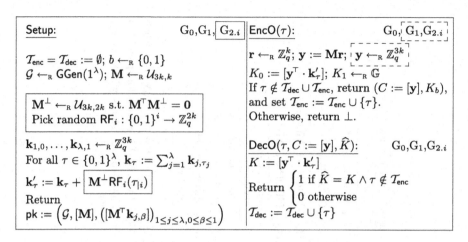

Lemma 6 ($G_{2.i}$ **to** $G_{2.i+1}$). *For all* $0 \le i \le \lambda - 1$, *there exists an adversary* $\mathcal{B}_{2.i}$ *such that* $\mathbf{T}(\mathcal{B}_{2.i}) \approx \mathbf{T}(\mathcal{A}) + (Q_{\mathsf{enc}} + Q_{\mathsf{dec}}) \cdot \mathsf{poly}(\lambda)$ *and*

$$|\mathbf{Adv}_{2.i} - \mathbf{Adv}_{2.i+1}| \le 4 \cdot \mathbf{Adv}^{\mathrm{mddh}}_{\mathcal{U}_k,\mathsf{GGen}}(\mathcal{B}_{2.i}) + \frac{4Q_{\mathsf{dec}} + 2k}{q} + \frac{4}{q-1},$$

where Q_{enc}, Q_{dec} *are the number of times* $\mathcal{A}$ *queries* EncO, DecO, *respectively, and* $\mathsf{poly}(\lambda)$ *is independent of* $\mathbf{T}(\mathcal{A})$.

Proof of Lemma 6. To go from $G_{2.i}$ to $G_{2.i+1}$, we introduce intermediate games $G_{2.i.1}$, $G_{2.i.2}$ and $G_{2.i.3}$, defined in Fig. 7. We prove that these games are indistinguishable in Lemmas 7, 8, 9, and 10.

Setup: $G_{2.i}$, $G_{2.i.1}$, $G_{2.i.2}$, $G_{2.i.3}$

$\mathcal{T}_{\mathsf{enc}} = \mathcal{T}_{\mathsf{dec}} := \emptyset$; $b \leftarrow_R \{0,1\}$
$\mathcal{G} \leftarrow_R \mathsf{GGen}(1^\lambda)$; $\mathbf{M} \leftarrow_R \mathcal{U}_{3k,k}$
$\mathbf{M}^\perp \leftarrow_R \mathcal{U}_{3k,2k}$ s.t. $\mathbf{M}^\top \mathbf{M}^\perp = \mathbf{0}$
$\mathbf{M}_0, \mathbf{M}_1 \leftarrow_R \mathcal{U}_{3k,k}$
$\mathbf{M}_0^*, \mathbf{M}_1^* \leftarrow_R \mathcal{U}_{3k,k}$ s.t.
$\mathsf{span}(\mathbf{M}^\perp) = \mathsf{span}(\mathbf{M}_0^*, \mathbf{M}_1^*)$
$\mathbf{M}^\top \mathbf{M}_0^* = \mathbf{M}_1^\top \mathbf{M}_0^* = \mathbf{0}$
$\mathbf{M}^\top \mathbf{M}_1^* = \mathbf{M}_0^\top \mathbf{M}_1^* = \mathbf{0}$
Pick random $\mathsf{RF}_i : \{0,1\}^i \to \mathbb{Z}_q^{2k}$.
Pick $\mathsf{RF}^{(0)}_{i+1} : \{0,1\}^{i+1} \to \mathbb{Z}_q^k$ and $\mathsf{RF}^{(1)}_i : \{0,1\}^i \to \mathbb{Z}_q^k$
Pick $\mathsf{RF}^{(0)}_{i+1}, \mathsf{RF}^{(1)}_{i+1} : \{0,1\}^{i+1} \to \mathbb{Z}_q^k$.
$\mathbf{k}_{1,0}, \ldots, \mathbf{k}_{\lambda,1} \leftarrow_R \mathbb{Z}_q^{3k}$
For all $\tau \in \{0,1\}^\lambda$, $\mathbf{k}_\tau := \sum_{j=1}^{\lambda} \mathbf{k}_{j,\tau_j}$
$\mathbf{k}'_\tau := \mathbf{k}_\tau + \mathbf{M}^\perp \mathsf{RF}_i(\tau_{|i})$
$\mathbf{k}'_\tau := \mathbf{k}_\tau + \mathbf{M}_0^* \mathsf{RF}^{(0)}_{i+1}(\tau_{|i+1}) + \mathbf{M}_1^* \mathsf{RF}^{(1)}_i(\tau_{|i})$
$\mathbf{k}'_\tau := \mathbf{k}_\tau + \mathbf{M}_0^* \mathsf{RF}^{(0)}_{i+1}(\tau_{|i+1}) + \mathbf{M}_1^* \mathsf{RF}^{(1)}_{i+1}(\tau_{|i+1})$
Return
$\mathsf{pk} := \left(\mathcal{G}, [\mathbf{M}], ([\mathbf{M}^\top \mathbf{k}_{j,\beta}])_{1 \le j \le \lambda, 0 \le \beta \le 1}\right)$

EncO(τ): $G_{2.i}$, $G_{2.i.1}$, $G_{2.i.2}$, $G_{2.i.3}$

$\mathbf{y} \leftarrow_R \mathbb{Z}_q^{3k}$
If $\tau_{i+1} = 0$:
$\mathbf{r} \leftarrow_R \mathbb{Z}_q^k$; $\mathbf{r}_0 \leftarrow_R \mathbb{Z}_q^k$; $\mathbf{y} := \mathbf{M}\mathbf{r} + \mathbf{M}_0 \mathbf{r}_0$
If $\tau_{i+1} = 1$:
$\mathbf{r} \leftarrow_R \mathbb{Z}_q^k$; $\mathbf{r}_1 \leftarrow_R \mathbb{Z}_q^k$; $\mathbf{y} := \mathbf{M}\mathbf{r} + \mathbf{M}_1 \mathbf{r}_1$
$K_0 := [\mathbf{y}^\top \cdot \mathbf{k}'_\tau]$;
$K_1 \leftarrow_R \mathbb{G}$
If $\tau \notin \mathcal{T}_{\mathsf{dec}} \cup \mathcal{T}_{\mathsf{enc}}$, return $([\mathbf{y}], K_b)$ and set $\mathcal{T}_{\mathsf{enc}} := \mathcal{T}_{\mathsf{enc}} \cup \{\tau\}$.
Otherwise, return $\perp$.

DecO($\tau, [\mathbf{y}], \widehat{K}$): $G_{2.i}$, $G_{2.i.1}$, $G_{2.i.2}$, $G_{2.i.3}$

$K := [\mathbf{y}^\top \mathbf{k}'_\tau]$
Return $\begin{cases} 1 \text{ if } \widehat{K} = K \wedge \tau \notin \mathcal{T}_{\mathsf{enc}} \\ 0 \text{ otherwise} \end{cases}$
$\mathcal{T}_{\mathsf{dec}} := \mathcal{T}_{\mathsf{dec}} \cup \{\tau\}$.

Fig. 7. Games $G_{2.i}$ (for $0 \le i \le \lambda$), $G_{2.i.1}$, $G_{2.i.2}$ and $G_{2.i.3}$ (for $0 \le i \le \lambda - 1$) for the proof of Lemma 6. For all $0 \le i \le \lambda$, $\mathsf{RF}_i : \{0,1\}^i \to \mathbb{Z}_q^{2k}$, $\mathsf{RF}_i^{(0)}$, $\mathsf{RF}_i^{(1)} : \{0,1\}^i \to \mathbb{Z}_q^k$ are random functions, and for all $\tau \in \mathcal{T}$, we denote by $\tau_{|i}$ the i-bit prefix of τ. In each procedure, the components inside a solid (dotted, gray) frame are only present in the games marked by a solid (dotted, gray) frame.

Lemma 7 ($G_{2.i}$ **to** $G_{2.i.1}$). *For all* $0 \leq i \leq \lambda - 1$, *there exists an adversary* $\mathcal{B}_{2.i.0}$ *such that* $\mathbf{T}(\mathcal{B}_{2.i.0}) \approx \mathbf{T}(\mathcal{A}) + (Q_{\mathsf{enc}} + Q_{\mathsf{dec}}) \cdot \mathsf{poly}(\lambda)$ *and*

$$|\mathbf{Adv}_{2.i} - \mathbf{Adv}_{2.i.1}| \leq 2 \cdot \mathbf{Adv}^{\mathrm{mddh}}_{\mathcal{U}_k,\mathsf{GGen}}(\mathcal{B}_{2.i.0}) + \frac{2}{q-1},$$

where Q_{enc}, Q_{dec} *are the number of times* $\mathcal{A}$ *queries* EncO, DecO, *respectively, and* $\mathsf{poly}(\lambda)$ *is independent of* $\mathbf{T}(\mathcal{A})$.

Here, we use the MDDH Assumption to "tightly" switch the distribution of all the challenge ciphertexts. We proceed in two steps, first, by changing the distribution of all the ciphertexts with a tag τ such that $\tau_{i+1} = 0$, and then, for those with a tag τ such that $\tau_{i+1} = 1$. We use the MDDH Assumption with respect to an independent matrix for each step.

Proof of Lemma 7. To go from $G_{2.i}$ to $G_{2.i.1}$, we switch the distribution of the vectors $[\mathbf{y}]$ sampled by EncO, using the Q_{enc}-fold $\mathcal{U}_{3k,k}$-MDDH Assumption.

We introduce an intermediate game $G_{2.i.0}$ where $\mathsf{EncO}(\tau)$ is computed as in $G_{2.i.1}$ if $\tau_{i+1} = 0$, and as in $G_{2.i}$ if $\tau_{i+1} = 1$. Setup, DecO are as in $G_{2.i.1}$. We build adversaries $\mathcal{B}'_{2.i.0}$ and $\mathcal{B}''_{2.i.0}$ such that $\mathbf{T}(\mathcal{B}'_{2.i.0}) \approx \mathbf{T}(\mathcal{B}''_{2.i.0}) \approx \mathbf{T}(\mathcal{A}) + (Q_{\mathsf{enc}} + Q_{\mathsf{dec}}) \cdot \mathsf{poly}(\lambda)$ with $\mathsf{poly}(\lambda)$ independent of $\mathbf{T}(\mathcal{A})$, and

Claim 1: $|\mathbf{Adv}_{2.i} - \mathbf{Adv}_{2.i.0}| \leq \mathbf{Adv}^{Q_{\mathsf{enc}}\text{-mddh}}_{\mathcal{U}_{3,k},\mathsf{GGen}}(\mathcal{B}'_{2.i.0})$.
Claim 2: $|\mathbf{Adv}_{2.i.0} - \mathbf{Adv}_{2.i.1}| \leq \mathbf{Adv}^{Q_{\mathsf{enc}}\text{-mddh}}_{\mathcal{U}_{3k,k},\mathsf{GGen}}(\mathcal{B}''_{2.i.0})$.

This implies the lemma by Lemma 3 (self-reducibility of $\mathcal{U}_{3k,k}$-MDDH), and Lemma 1 ($\mathcal{U}_{3k,k}$-MDDH $\Leftrightarrow \mathcal{U}_k$-MDDH).

Let us prove Claim 1. Upon receiving a challenge $(\mathcal{G}, [\mathbf{M}_0] \in \mathbb{G}^{3k \times k}, [\mathbf{H}] := [\mathbf{h}_1|\ldots|\mathbf{h}_{Q_{\mathsf{enc}}}] \in \mathbb{G}^{3k \times Q_{\mathsf{enc}}})$ for the Q_{enc}-fold $\mathcal{U}_{3k,k}$-MDDH Assumption with respect to $\mathbf{M}_0 \leftarrow_{\mathrm{R}} \mathcal{U}_{3k,k}$, $\mathcal{B}'_{2.i.0}$ does as follows:

Setup: $\mathcal{B}'_{2.i.0}$ picks $\mathbf{M} \leftarrow_{\mathrm{R}} \mathcal{U}_{3k,k}$, $\mathbf{k}_{1,0}, \ldots, \mathbf{k}_{\lambda,1} \leftarrow_{\mathrm{R}} \mathbb{Z}_q^{3k}$, and computes pk as described in Fig. 7. For each τ queried to EncO or DecO, it computes on the fly $\mathsf{RF}_i(\tau_{|i})$ and $\mathbf{k}'_\tau := \mathbf{k}_\tau + \mathbf{M}^\perp \mathsf{RF}_i(\tau_{|i})$, where $\mathbf{k}_\tau := \sum_{j=1}^{\lambda} \mathbf{k}_{j,\tau_j}$, $\mathsf{RF}_i : \{0,1\}^i \rightarrow \mathbb{Z}_q^{2k}$ is a random function, and $\tau_{|i}$ denotes the i-bit prefix of τ (see Fig. 7). Note that $\mathcal{B}'_{2.i.0}$ can compute efficiently $\mathbf{M}^\perp$ from $\mathbf{M}$.
EncO: To simulate the oracle $\mathsf{EncO}(\tau)$ on its j'th query, for $j = 1, \ldots, Q_{\mathsf{enc}}$, $\mathcal{B}'_{2.i.0}$ computes $[\mathbf{y}]$ as follows:

$$\begin{aligned}&\text{if } \tau_{i+1} = 0 : \mathbf{r} \leftarrow_{\mathrm{R}} \mathbb{Z}_q^k;\ [\mathbf{y}] := [\mathbf{M}\mathbf{r} + \mathbf{h}_j]\\ &\text{if } \tau_{i+1} = 1 : [\mathbf{y}] \leftarrow_{\mathrm{R}} \mathbb{G}^{3k}\end{aligned}$$

This way, $\mathcal{B}'_{2.i.0}$ simulates EncO as in $G_{2.i.0}$ when $[\mathbf{h}_j] := [\mathbf{M}_0\mathbf{r}_0]$ with $\mathbf{r}_0 \leftarrow_{\mathrm{R}} \mathbb{Z}_q^k$, and as in $G_{2.i}$ when $[\mathbf{h}_j] \leftarrow_{\mathrm{R}} \mathbb{G}^{3k}$.
DecO: Finally, $\mathcal{B}'_{2.i.0}$ simulates DecO as described in Fig. 7.

Therefore, $|\mathbf{Adv}_{2.i} - \mathbf{Adv}_{2.i.0}| \leq \mathbf{Adv}^{Q_{\mathsf{enc}}\text{-mddh}}_{\mathcal{U}_{3k,k},\mathsf{GGen}}(\mathcal{B}'_{2.i.0})$.

To prove Claim 2, we build an adversary $\mathcal{B}''_{2.i.0}$ against the Q_{enc}-fold $\mathcal{U}_{3k,k}$-MDDH Assumption with respect to a matrix $\mathbf{M}_1 \leftarrow_{\mathrm{R}} \mathcal{U}_{3k,k}$, independent from $\mathbf{M}_0$, similarly than $\mathcal{B}'_{2.i.0}$. □

Lemma 8 ($G_{2.i.1}$ **to** $G_{2.i.2}$). *For all* $0 \leq i \leq \lambda - 1$,

$$|\mathbf{Adv}_{2.i.1} - \mathbf{Adv}_{2.i.2}| \leq \frac{2Q_{\mathsf{dec}} + 2k}{q},$$

where Q_{dec} *is the number of times* $\mathcal{A}$ *queries* DecO.

Here, we use a variant of the Cramer-Shoup information-theoretic argument to move from RF_i to RF_{i+1}, thereby increasing the entropy of $\mathbf{k}'_\tau$ computed by Setup. For the sake of readability, we proceed in two steps: in Lemma 8, we move from RF_i to an hybrid between RF_i and RF_{i+1}, and in Lemma 9, we move to RF_{i+1}.

Proof of Lemma 8. In $G_{2.i.2}$, we decompose $\mathsf{span}(\mathbf{M}^\perp)$ into two subspaces $\mathsf{span}(\mathbf{M}_0^*)$ and $\mathsf{span}(\mathbf{M}_1^*)$, and we increase the entropy of the components of $\mathbf{k}'_\tau$ which lie in $\mathsf{span}(\mathbf{M}_0^*)$. To argue that $G_{2.i.1}$ and $G_{2.i.2}$ are statistically close, we use a Cramer-Shoup argument [11].

Let us first explain how the matrices $\mathbf{M}_0^*$ and $\mathbf{M}_1^*$ are sampled. Note that with probability at least $1 - \frac{2k}{q}$ over the random coins of Setup, $(\mathbf{M}\|\mathbf{M}_0\|\mathbf{M}_1)$ forms a basis of $\mathbb{Z}_q^{3k}$. Therefore, we have

$$\mathsf{span}(\mathbf{M}^\perp) = \mathsf{Ker}(\mathbf{M}^\top) = \mathsf{Ker}\big((\mathbf{M}\|\mathbf{M}_1)^\top\big) \oplus \mathsf{Ker}\big((\mathbf{M}\|\mathbf{M}_0)^\top\big).$$

We pick uniformly $\mathbf{M}_0^*$ and $\mathbf{M}_1^*$ in $\mathbb{Z}_q^{3k \times k}$ that generate $\mathsf{Ker}\big((\mathbf{M}\|\mathbf{M}_1)^\top\big)$ and $\mathsf{Ker}\big((\mathbf{M}\|\mathbf{M}_0)^\top\big)$, respectively (see Fig. 1). This way, for all $\tau \in \{0,1\}^\lambda$, we can write

$$\mathbf{M}^\perp \mathsf{RF}_i(\tau_{|i}) := \mathbf{M}_0^* \mathsf{RF}_i^{(0)}(\tau_{|i}) + \mathbf{M}_1^* \mathsf{RF}_i^{(1)}(\tau_{|i}),$$

where $\mathsf{RF}_i^{(0)}, \mathsf{RF}_i^{(1)} : \{0,1\}^i \to \mathbb{Z}_q^k$ are independent random functions.

We define $\mathsf{RF}_{i+1}^{(0)} : \{0,1\}^{i+1} \to \mathbb{Z}_q^k$ as follows:

$$\mathsf{RF}_{i+1}^{(0)}(\tau_{|i+1}) := \begin{cases} \mathsf{RF}_i^{(0)}(\tau_{|i}) & \text{if } \tau_{i+1} = 0 \\ \mathsf{RF}_i^{(0)}(\tau_{|i}) + \mathsf{RF'}_i^{(0)}(\tau_{|i}) & \text{if } \tau_{i+1} = 1 \end{cases}$$

where $\mathsf{RF'}_i^{(0)} : \{0,1\}^i \to \mathbb{Z}_q^k$ is a random function independent from $\mathsf{RF}_i^{(0)}$. This way, $\mathsf{RF}_{i+1}^{(0)}$ is a random function.

We show that the outputs of EncO and DecO are statistically close in $G_{2.i.1}$ and $G_{2.i.2}$. We decompose the proof in two cases (delimited with ■): the queries with a tag $\tau \in \{0,1\}^\lambda$ such that $\tau_{i+1} = 0$, and the queries with a tag τ such that $\tau_{i+1} = 1$.

Queries with $\tau_{i+1} = 0$**:**
The only difference between $G_{2.i.1}$ and $G_{2.i.2}$ is that Setup computes $\mathbf{k}'_\tau$ using the random function $\mathsf{RF}_i^{(0)}$ in $G_{2.i.1}$, whereas it uses the random function $\mathsf{RF}_{i+1}^{(0)}$ in $G_{2.i.2}$ (see Fig. 7). Therefore, by definition of $\mathsf{RF}_{i+1}^{(0)}$, for all $\tau \in \{0,1\}^\lambda$ such that $\tau_{i+1} = 0$, $\mathbf{k}'_\tau$ is the same in $G_{2.i.1}$ and $G_{2.i.2}$, and the outputs of EncO and DecO are identically distributed. ■

Queries with $\tau_{i+1} = 1$:
Observe that for all $\mathbf{y} \in \mathsf{span}(\mathbf{M}, \mathbf{M}_1)$ and all $\tau \in \{0,1\}^\lambda$ such that $\tau_{i+1} = 1$,

$$\overbrace{\mathbf{y}^\top \left(\mathbf{k}_\tau + \mathbf{M}_0^* \mathsf{RF}_i^{(0)}(\tau_{|i}) + \mathbf{M}_1^* \mathsf{RF}_i^{(1)}(\tau_{|i}) + \boxed{\mathbf{M}_0^* \mathsf{RF}'^{(0)}_i(\tau_{|i})}\right)}^{\mathrm{G}_{2.i.2}}$$
$$= \mathbf{y}^\top \left(\mathbf{k}_\tau + \mathbf{M}_0^* \mathsf{RF}_i^{(0)}(\tau_{|i}) + \mathbf{M}_1^* \mathsf{RF}_i^{(1)}(\tau_{|i})\right) + \underbrace{\boxed{\mathbf{y}^\top \mathbf{M}_0^* \mathsf{RF}'^{(0)}_i(\tau_{|i})}}_{=0}$$
$$= \overbrace{\mathbf{y}^\top \cdot \left(\mathbf{k}_\tau + \mathbf{M}_0^* \mathsf{RF}_i^{(0)}(\tau_{|i}) + \mathbf{M}_1^* \mathsf{RF}_i^{(1)}(\tau_{|i})\right)}^{\mathrm{G}_{2.i.1}}$$

where the second equality uses the fact that $\mathbf{M}^\top \mathbf{M}_0^* = \mathbf{M}_1^\top \mathbf{M}_0^* = \mathbf{0}$ and thus $\mathbf{y}^\top \mathbf{M}_0^* = \mathbf{0}$.

This means that:

- the output of EncO on any input τ such that $\tau_{i+1} = 1$ is identically distributed in $\mathrm{G}_{2.i.1}$ and $\mathrm{G}_{2.i.2}$;
- the output of DecO on any input $(\tau, [\mathbf{y}], \widehat{K})$ where $\tau_{i+1} = 1$, and $\mathbf{y} \in \mathsf{span}(\mathbf{M}, \mathbf{M}_1)$ is the same in $\mathrm{G}_{2.i.1}$ and $\mathrm{G}_{2.i.2}$.

Henceforth, we focus on the *ill-formed* queries to DecO, namely those corresponding to $\tau_{i+1} = 1$, and $\mathbf{y} \notin \mathsf{span}(\mathbf{M}, \mathbf{M}_1)$. We introduce intermediate games $\mathrm{G}_{2.i.1.j}$, and $\mathrm{G}'_{2.i.1.j}$ for $j = 0, \ldots, Q_{\mathsf{dec}}$, defined as follows:

- $\mathrm{G}_{2.i.1.j}$: DecO is as in $\mathrm{G}_{2.i.1}$ except that for the first j times it is queried, it outputs 0 to any ill-formed query. EncO is as in $\mathrm{G}_{2.i.2}$.
- $\mathrm{G}'_{2.i.1.j}$: DecO as in $\mathrm{G}_{2.i.2}$ except that for the first j times it is queried, it outputs 0 to any ill-formed query. EncO is as in $\mathrm{G}_{2.i.2}$.

We show that:

$$\mathrm{G}_{2.i.1} \equiv \mathrm{G}_{2.i.1.0} \approx_s \mathrm{G}_{2.i.1.1} \approx_s \ldots \approx_s \mathrm{G}_{2.i.1.Q_{\mathsf{dec}}} \equiv \mathrm{G}'_{2.i.1.Q_{\mathsf{dec}}}$$
$$\mathrm{G}'_{2.i.1.Q_{\mathsf{dec}}} \approx_s \mathrm{G}'_{2.i.1.Q_{\mathsf{dec}}-1} \approx_s \ldots \approx_s \mathrm{G}'_{2.i.1.0} \equiv \mathrm{G}_{2.i.2}$$

where we denote statistical closeness with $\approx_s$ and statistical equality with $\equiv$.

It suffices to show that for all $j = 0, \ldots, Q_{\mathsf{dec}} - 1$:

Claim 1: in $\mathrm{G}_{2.i.1.j}$, if the $j+1$-st query is ill-formed, then DecO outputs 0 with overwhelming probability $1 - 1/q$ (this implies $\mathrm{G}_{2.i.1.j} \approx_s \mathrm{G}_{2.i.1.j+1}$, with statistical difference $1/q$);

Claim 2: in $\mathrm{G}'_{2.i.1.j}$, if the $j+1$-st query is ill-formed, then DecO outputs 0 with overwhelming probability $1 - 1/q$ (this implies $\mathrm{G}'_{2.i.1.j} \approx_s \mathrm{G}'_{2.i.1.j+1}$, with statistical difference $1/q$)

where the probabilities are taken over the random coins of Setup.

Let us prove Claim 1. Recall that in $G_{2.i.1.j}$, on its $j+1$-st query, $\mathsf{DecO}(\tau, [\mathbf{y}], \widehat{K})$ computes $K := [\mathbf{y}^\top \mathbf{k}'_\tau]$, where $\mathbf{k}'_\tau := \big(\mathbf{k}_\tau + \mathbf{M}_0^* \mathsf{RF}_i^{(0)}(\tau_{|i}) + \mathbf{M}_1^* \mathsf{RF}_i^{(1)}(\tau_{|i})\big)$ (see Fig. 7). We prove that if $(\tau, [\mathbf{y}], \widehat{K})$ is ill-formed, then K is completely hidden from $\mathcal{A}$, up to its $j+1$-st query to DecO. The reason is that the vector $\mathbf{k}_{i+1,1}$ in sk contains some entropy that is hidden from $\mathcal{A}$. This entropy is "released" on the $j+1$-st query to DecO if it is ill-formed. More formally, we use the fact that the vector $\mathbf{k}_{i+1,1} \leftarrow_{\mathrm{R}} \mathbb{Z}_q^{3k}$ is identically distributed as $\mathbf{k}_{i+1,1} + \mathbf{M}_0^* \mathbf{w}$, where $\mathbf{k}_{i+1,1} \leftarrow_{\mathrm{R}} \mathbb{Z}_q^{3k}$, and $\mathbf{w} \leftarrow_{\mathrm{R}} \mathbb{Z}_q^{k}$. We show that $\mathbf{w}$ is completely hidden from $\mathcal{A}$, up to its $j+1$-st query to DecO.

- The public key pk does not leak any information about $\mathbf{w}$, since

$$\mathbf{M}^\top(\mathbf{k}_{i+1,1} + \boxed{\mathbf{M}_0^* \mathbf{w}}) = \mathbf{M}^\top \mathbf{k}_{i+1,1}.$$

  This is because $\mathbf{M}^\top \mathbf{M}_0^* = \mathbf{0}$.
- The outputs of EncO also hide $\mathbf{w}$.
  - For τ such that $\tau_{i+1} = 0$, $\mathbf{k}'_\tau$ is independent of $\mathbf{k}_{i+1,1}$, and therefore, so does $\mathsf{EncO}(\tau)$.
  - For τ such that $\tau_{i+1} = 1$, and for any $\mathbf{y} \in \mathsf{span}(\mathbf{M}, \mathbf{M}_1)$, we have:

$$\mathbf{y}^\top(\mathbf{k}'_\tau + \boxed{\mathbf{M}_0^* \mathbf{w}}) = \mathbf{y}^\top \mathbf{k}'_\tau \tag{2}$$

    since $\mathbf{M}^\top \mathbf{M}_0^* = \mathbf{M}_1^\top \mathbf{M}_0^* = \mathbf{0}$, which implies $\mathbf{y}^\top \mathbf{M}_0^* = \mathbf{0}$.
- The first j outputs of DecO also hide $\mathbf{w}$.
  - For τ such that $\tau_{i+1} = 0$, $\mathbf{k}'_\tau$ is independent of $\mathbf{k}_{i+1,1}$, and therefore, so does $\mathsf{DecO}([\mathbf{y}], \tau, \widehat{K})$.
  - For τ such that $\tau_{i+1} = 1$ and $\mathbf{y} \in \mathsf{span}(\mathbf{M}, \mathbf{M}_1)$, the fact that $\mathsf{DecO}(\tau, [\mathbf{y}], \widehat{K})$ is independent of $\mathbf{w}$ follows readily from Equation (2).
  - For τ such that $\tau_{i+1} = 1$ and $\mathbf{y} \notin \mathsf{span}(\mathbf{M}, \mathbf{M}_1)$, that is, for an ill-formed query, DecO outputs 0, independently of $\mathbf{w}$, by definition of $G_{2.i.1.j}$.

This proves that $\mathbf{w}$ is uniformly random from $\mathcal{A}$'s viewpoint.

Finally, because the $j+1$-st query $(\tau, [\mathbf{y}], \widehat{K})$ is ill-formed, we have $\tau_{i+1} = 1$, and $\mathbf{y} \notin \mathsf{span}(\mathbf{M}, \mathbf{M}_1)$, which implies that $\mathbf{y}^\top \mathbf{M}_0^* \neq \mathbf{0}$. Therefore, the value

$$K = [\mathbf{y}^\top(\mathbf{k}'_\tau + \mathbf{M}_0^* \mathbf{w})] = [\mathbf{y}^\top \mathbf{k}'_\tau + \underbrace{\mathbf{y}^\top \mathbf{M}_0^*}_{\neq \mathbf{0}} \mathbf{w}]$$

computed by DecO is uniformly random over $\mathbb{G}$ from $\mathcal{A}$'s viewpoint. Thus, with probability $1 - 1/q$ over $\widehat{K} \leftarrow_{\mathrm{R}} \mathbb{G}$, we have $\widehat{K} \neq K$, and $\mathsf{DecO}(\tau, [\mathbf{y}], \widehat{K}) = 0$.

We prove Claim 2 similarly, arguing than in $G'_{2.i.1.j}$, the value $K := [\mathbf{y}^\top \mathbf{k}'_\tau]$, where $\mathbf{k}'_\tau := \big(\mathbf{k}_\tau + \mathbf{M}_0^* \mathsf{RF}_{i+1}^{(0)}(\tau_{|i+1}) + \mathbf{M}_1^* \mathsf{RF}_i^{(1)}(\tau_{|i})\big)$, computed by $\mathsf{DecO}(\tau, [\mathbf{y}], \widehat{K})$ on its $j+1$-st query, is completely hidden from $\mathcal{A}$, up to its $j+1$-st query to DecO, if $(\tau, [\mathbf{y}], \widehat{K})$ is ill-formed. The argument goes exactly as for Claim 1. ∎

Lemma 9 ($G_{2.i.2}$ **to** $G_{2.i.3}$). *For all* $0 \leq i \leq \lambda - 1$,

$$|\mathbf{Adv}_{2.i.2} - \mathbf{Adv}_{2.i.3}| \leq \frac{2Q_{\mathsf{dec}}}{q},$$

where Q_{dec} *is the number of times* $\mathcal{A}$ *queries* DecO.

Proof of Lemma 9. In $G_{2.i.3}$, we use the same decomposition $\mathsf{span}(\mathbf{M}^{\perp}) = \mathsf{span}(\mathbf{M}_0^*, \mathbf{M}_1^*)$ as that in $G_{2.i.2}$. The entropy of the components of $\mathbf{k}'_\tau$ that lie in $\mathsf{span}(\mathbf{M}_1^*)$ increases from $G_{2.i.2}$ to $G_{2.i.3}$. To argue that these two games are statistically close, we use a Cramer-Shoup argument [11], exactly as for Lemma 8.

We define $\mathsf{RF}^{(1)}_{i+1}\{0,1\}^{i+1} \rightarrow \mathbb{Z}_q^k$ as follows:

$$\mathsf{RF}^{(1)}_{i+1}(\tau_{|i+1}) := \begin{cases} \mathsf{RF}^{(1)}_{i}(\tau_{|i}) + \mathsf{RF}'^{(1)}_{i}(\tau_{|i}) & \text{if } \tau_{i+1} = 0 \\ \mathsf{RF}^{(1)}_{i}(\tau_{|i}) & \text{if } \tau_{i+1} = 1 \end{cases}$$

where $\mathsf{RF}'^{(1)}_{i} : \{0,1\}^i \rightarrow \mathbb{Z}_q^k$ is a random function independent from $\mathsf{RF}^{(1)}_{i}$. This way, $\mathsf{RF}^{(1)}_{i+1}$ is a random function.

We show that the outputs of EncO and DecO are statistically close in $G_{2.i.1}$ and $G_{2.i.2}$. We decompose the proof in two cases (delimited with ■): the queries with a tag $\tau \in \{0,1\}^\lambda$ such that $\tau_{i+1} = 0$, and the queries with tag τ such that $\tau_{i+1} = 1$.

Queries with $\tau_{i+1} = 1$**:**
The only difference between $G_{2.i.2}$ and $G_{2.i.3}$ is that Setup computes $\mathbf{k}'_\tau$ using the random function $\mathsf{RF}^{(1)}_{i}$ in $G_{2.i.2}$, whereas it uses the random function $\mathsf{RF}^{(1)}_{i+1}$ in $G_{2.i.3}$ (see Fig. 7). Therefore, by definition of $\mathsf{RF}^{(1)}_{i+1}$, for all $\tau \in \{0,1\}^\lambda$ such that $\tau_{i+1} = 1$, $\mathbf{k}'_\tau$ is the same in $G_{2.i.2}$ and $G_{2.i.3}$, and the outputs of EncO and DecO are identically distributed. ■

Queries with $\tau_{i+1} = 0$**:**
Observe that for all $\mathbf{y} \in \mathsf{span}(\mathbf{M}, \mathbf{M}_0)$ and all $\tau \in \{0,1\}^\lambda$ such that $\tau_{i+1} = 0$,

$$\begin{aligned} &\overbrace{\mathbf{y}^\top \left(\mathbf{k}_\tau + \mathbf{M}_0^* \mathsf{RF}^{(0)}_{i+1}(\tau_{|i+1}) + \mathbf{M}_1^* \mathsf{RF}^{(1)}_{i}(\tau_{|i}) + \boxed{\mathbf{M}_1^* \mathsf{RF}'^{(1)}_{i}(\tau_{|i})}\right)}^{G_{2.i.3}} \\ &= \mathbf{y}^\top \left(\mathbf{k}_\tau + \mathbf{M}_0^* \mathsf{RF}^{(0)}_{i+1}(\tau_{|i+1}) + \mathbf{M}_1^* \mathsf{RF}^{(1)}_{i}(\tau_{|i})\right) + \underbrace{\boxed{\mathbf{y}^\top \mathbf{M}_1^* \mathsf{RF}'^{(1)}_{i}(\tau_{|i})}}_{=0} \\ &= \overbrace{\mathbf{y}^\top \cdot \left(\mathbf{k}_\tau + \mathbf{M}_0^* \mathsf{RF}^{(0)}_{i+1}(\tau_{|i+1}) + \mathbf{M}_1^* \mathsf{RF}^{(1)}_{i}(\tau_{|i})\right)}^{G_{2.i.2}} \end{aligned}$$

where the second equality uses the fact $\mathbf{M}^\top \mathbf{M}_1^* = \mathbf{M}_0^\top \mathbf{M}_1^* = \mathbf{0}$, which implies $\mathbf{y}^\top \mathbf{M}_1^* = \mathbf{0}$.

This means that:

- the output of EncO on any input τ such that $\tau_{i+1} = 0$ is identically distributed in $\mathrm{G}_{2.i.2}$ and $\mathrm{G}_{2.i.3}$;
- the output of DecO on any input $(\tau, [\mathbf{y}], \widehat{K})$ where $\tau_{i+1} = 0$, and $\mathbf{y} \in \mathsf{span}(\mathbf{M}, \mathbf{M}_0)$ is the same in $\mathrm{G}_{2.i.2}$ and $\mathrm{G}_{2.i.3}$.

Henceforth, we focus on the *ill-formed* queries to DecO, namely those corresponding to $\tau_{i+1} = 0$, and $\mathbf{y} \notin \mathsf{span}(\mathbf{M}, \mathbf{M}_0)$. The rest of the proof goes similarly than the proof of Lemma 8. See the latter for further details. □

Lemma 10 **($\mathrm{G}_{2.i.3}$ to $\mathrm{G}_{2.i+1}$).** *For all $0 \leq i \leq \lambda - 1$, there exists an adversary $\mathcal{B}_{2.i.3}$ such that $\mathbf{T}(\mathcal{B}_{2.i.3}) \approx \mathbf{T}(\mathcal{A}) + (Q_{\mathsf{enc}} + Q_{\mathsf{dec}}) \cdot \mathsf{poly}(\lambda)$ and*

$$|\mathbf{Adv}_{2.i.3} - \mathbf{Adv}_{2.i+1}| \leq 2 \cdot \mathbf{Adv}^{\mathrm{mddh}}_{\mathcal{U}_k, \mathsf{GGen}}(\mathcal{B}_{2.i.3}) + \frac{2}{q-1}$$

where Q_{enc}, Q_{dec} are the number of times $\mathcal{A}$ queries EncO, DecO, respectively, and $\mathsf{poly}(\lambda)$ is independent of $\mathbf{T}(\mathcal{A})$.

Here, we use the MDDH Assumption to "tightly" switch the distribution of all the challenge ciphertexts, as for Lemma 7. We proceed in two steps, first, by changing the distribution of all the ciphertexts with a tag τ such that $\tau_{i+1} = 0$, and then, the distribution of those with a tag τ such that $\tau_{i+1} = 1$, using the MDDH Assumption with respect to an independent matrix for each step.

Proof of Lemma 10. To go from $\mathrm{G}_{2.i.3}$ to $\mathrm{G}_{2.i+1}$, we switch the distribution of the vectors $[\mathbf{y}]$ sampled by EncO, using the Q_{enc}-fold $\mathcal{U}_{3k,k}$-MDDH Assumption. This transition is symmetric to the transition between $\mathrm{G}_{2.i}$ and $\mathrm{G}_{2.i.1}$ (see the proof of Lemma 7 for further details). Finally, we use the fact that for all $\tau \in \{0,1\}^\lambda$, $\mathbf{M}_0^* \mathsf{RF}^{(0)}_{i+1}(\tau_{|i}) + \mathbf{M}_1^* \mathsf{RF}^{(1)}_{i+1}(\tau_{|i+1})$ is identically distributed to $\mathbf{M}^\perp \mathsf{RF}_{i+1}(\tau_{|i+1})$, where $\mathsf{RF}_{i+1} : \{0,1\}^{i+1} \rightarrow \mathbb{Z}_q^{2k}$ is a random function. This is because $(\mathbf{M}_0^*, \mathbf{M}_1^*)$ is a basis of $\mathsf{span}(\mathbf{M}^\perp)$. □

The proof of Lemma 6 follows readily from Lemmas 7, 8, 9, and 10. □

Lemma 11 **($\mathrm{G}_{2.\lambda}$).** $\mathbf{Adv}_{2.\lambda} \leq \frac{Q_{\mathsf{enc}}}{q}$.

Proof of Lemma 11. We show that the joint distribution of all the values K_0 computed by EncO is statistically close to uniform over $\mathbb{G}^{Q_{\mathsf{enc}}}$. Recall that on input τ, $\mathsf{EncO}(\tau)$ computes

$$K_0 := [\mathbf{y}^\top (\mathbf{k}_\tau + \mathbf{M}^\perp \mathsf{RF}_\lambda(\tau))],$$

where $\mathsf{RF}_\lambda : \{0,1\}^\lambda \rightarrow \mathbb{Z}_q^{2k}$ is a random function, and $\mathbf{y} \leftarrow_{\mathrm{R}} \mathbb{Z}_q^{3k}$ (see Fig. 6).

We make use of the following properties:

Property 1: all the tags τ queried to EncO, such that $\mathsf{EncO}(\tau) \neq \perp$, are distinct.

Property 2: the outputs of DecO are independent of $\{\mathsf{RF}(\tau) : \tau \in \mathcal{T}_{\mathsf{enc}}\}$. This is because for all queries $(\tau, [\mathbf{y}], \widehat{K})$ to DecO such that $\tau \in \mathcal{T}_{\mathsf{enc}}$, $\mathsf{DecO}(\tau, [\mathbf{y}], \widehat{K}) = 0$, independently of $\mathsf{RF}_\lambda(\tau)$, by definition of $\mathrm{G}_{2.\lambda}$.

Property 3: with probability at least $1 - \frac{Q_{\mathsf{enc}}}{q}$ over the random coins of EncO, all the vectors $\mathbf{y}$ sampled by EncO are such that $\mathbf{y}^\top \mathbf{M}^\perp \neq \mathbf{0}$.

We deduce that the joint distribution of all the values $\mathsf{RF}_\lambda(\tau)$ computed by EncO is uniformly random over $(\mathbb{Z}_q^{2k})^{Q_{\mathsf{enc}}}$ (from Property 1), independent of the outputs of DecO (from Property 2). Finally, from Property 3, we get that the joint distribution of all the values K_0 computed by EncO is statistically close to uniform over $\mathbb{G}^{Q_{\mathsf{enc}}}$, since:

$$K_0 := [\mathbf{y}^\top(\mathbf{k}_\tau + \mathbf{M}^\perp \mathsf{RF}_\lambda(\tau)) = [\mathbf{y}^\top \mathbf{k}_\tau + \underbrace{\mathbf{y}^\top \mathbf{M}^\perp}_{\neq \mathbf{0} \text{ w.h.p.}} \mathsf{RF}_\lambda(\tau)].$$

This means that the values K_0 and K_1 are statistically close, and therefore, $\mathbf{Adv}_3 \leq \frac{Q_{\mathsf{enc}}}{q}$. □

Finally, Theorem 1 follows readily from Lemmas 4, 5, 6, and 11. □

4 Multi-ciphertext CCA-secure Public Key Encryption Scheme

4.1 Our Construction

We now describe the optimized IND-CCA-secure PKE scheme. Compared to the PCA-secure KEM from Sect. 3, we add an authenticated (symmetric) encryption scheme $(\mathsf{Enc}_{\mathsf{AE}}, \mathsf{Dec}_{\mathsf{AE}})$, and set the KEM tag τ as the hash value of a suitable part of the KEM ciphertext (as explained in the introduction). A formal definition with highlighted differences to our PCA-secure KEM appears in Fig. 8.

We prove the security under the $\mathcal{U}_k$-MDDH Assumption, which admits a tight reduction to the standard k-Lin Assumption.

Theorem 2. *The Public Key Encryption scheme* $\mathsf{PKE}_{\mathsf{CCA}}$ *defined in Fig. 8 has perfect correctness, if the underlying Authenticated Encryption scheme* AE *has perfect correctness. Moreover, if the* $\mathcal{U}_k$-MDDH *Assumption holds in* $\mathbb{G}$, AE *has one-time privacy and authenticity, and* $\mathcal{H}$ *generates collision resistant hash functions, then* $\mathsf{PKE}_{\mathsf{CCA}}$ *is IND-CCA secure. Namely, for any adversary* $\mathcal{A}$*, there exist adversaries* $\mathcal{B}$, $\mathcal{B}'$, $\mathcal{B}''$ *such that* $\mathbf{T}(\mathcal{B}) \approx \mathbf{T}(\mathcal{B}') \approx \mathbf{T}(\mathcal{B}'') \approx \mathbf{T}(\mathcal{A}) + (Q_{\mathsf{dec}} + Q_{\mathsf{enc}}) \cdot \mathsf{poly}(\lambda)$ *and*

$$\begin{aligned}\mathbf{Adv}^{\mathsf{ind\text{-}cca}}_{\mathsf{PKE}_{\mathsf{CCA}}}(\mathcal{A}) &\leq (4\lambda + 1) \cdot \mathbf{Adv}^{\mathrm{mddh}}_{\mathcal{U}_k, \mathsf{GGen}}(\mathcal{B}) \\ &\quad + ((4\lambda + 2)Q_{\mathsf{dec}} + Q_{\mathsf{enc}} + Q_{\mathsf{enc}}Q_{\mathsf{dec}}) \cdot \mathbf{Adv}^{\mathsf{ae\text{-}ot}}_{\mathsf{AE}}(\mathcal{B}') \qquad (3) \\ &\quad + \mathbf{Adv}^{\mathrm{cr}}_{\mathcal{H}}(\mathcal{B}'') + Q_{\mathsf{enc}}(Q_{\mathsf{enc}} + Q_{\mathsf{dec}}) \cdot 2^{-\Omega(\lambda)},\end{aligned}$$

where Q_{enc}, Q_{dec} *are the number of times* $\mathcal{A}$ *queries* EncO, DecO, *respectively, and* $\mathsf{poly}(\lambda)$ *is independent of* $\mathbf{T}(\mathcal{A})$.

$\mathsf{Gen}_{\mathsf{PKE}}(1^\lambda)$:	$\mathsf{Enc}_{\mathsf{PKE}}(\mathsf{pk}, M)$:
$\mathcal{G} \leftarrow_R \mathsf{GGen}(1^\lambda)$; $\mathsf{H} \leftarrow_R \mathcal{H}(1^\lambda)$; $\mathbf{M} \leftarrow_R \mathcal{U}_{3k,k}$ $\mathbf{k}_{1,0}, \ldots, \mathbf{k}_{\lambda,1} \leftarrow_R \mathbb{Z}_q^{3k}$ $\mathsf{pk} := \left(\mathcal{G}, [\mathbf{M}], \mathsf{H}, \left([\mathbf{M}^\top \mathbf{k}_{j,\beta}]\right)_{1 \le j \le \lambda, 0 \le \beta \le 1}\right)$ $\mathsf{sk} := (\mathbf{k}_{j,\beta})_{1 \le j \le \lambda, 0 \le \beta \le 1}$ Return $(\mathsf{pk}, \mathsf{sk})$	$\mathbf{r} \leftarrow_R \mathbb{Z}_q^k$; $\mathbf{y} := \mathbf{M}\mathbf{r}$ $\tau := \mathsf{H}([\overline{\mathbf{y}}])$ $\mathbf{k}_\tau := \sum_{j=1}^{\lambda} \mathbf{k}_{j,\tau_j}$ $K := [\mathbf{r}^\top \cdot \mathbf{M}^\top \mathbf{k}_\tau]$ $\phi := \mathsf{Enc}_{\mathsf{AE}}(K, M)$ Return $([\mathbf{y}], \phi)$ $\mathsf{Dec}_{\mathsf{PKE}}(\mathsf{pk}, \mathsf{sk}, ([\mathbf{y}], \phi))$: $\tau := \mathsf{H}([\overline{\mathbf{y}}])$; $\mathbf{k}_\tau := \sum_{j=1}^{\lambda} \mathbf{k}_{j,\tau_j}$; $K := [\mathbf{y}^\top \mathbf{k}_\tau]$ Return $\mathsf{Dec}_{\mathsf{AE}}(K, \phi)$.

Fig. 8. $\mathsf{PKE}_{\mathsf{CCA}}$, an IND-CCA-secure PKE. We color in blue the differences with $\mathsf{KEM}_{\mathsf{PCA}}$, the IND-PCA-secure KEM in Fig. 4. Here, GGen is a prime-order group generator (see Sect. 2.3) , and $\mathsf{AE} := (\mathsf{Enc}_{\mathsf{AE}}, \mathsf{Dec}_{\mathsf{AE}})$ is an Authenticated Encryption scheme with key-space $\mathcal{K} := \mathcal{G}$ (see Definition 7).

We note that the Q_{enc} and Q_{dec} factors in (3) are only related to AE. Hence, when using a statistically secure (one-time) authenticated encryption scheme, the corresponding terms in (3) become exponentially small.

Remark 2 (Extension to the Multi-user CCA Security). We only provide an analysis in the multi-ciphertext (but single-user) setting. However, we remark (without proof) that our analysis generalizes to the multi-user, multi-ciphertext scenario, similar to [6,16,18]. Indeed, all computational steps (not counting the steps related to the AE scheme) modify all ciphertexts simultaneously, relying for this on the re-randomizability of the $\mathcal{U}_k$-MDDH Assumption relative to a fixed matrix $\mathbf{M}$. The same modifications can be made to many $\mathsf{PKE}_{\mathsf{CCA}}$ simultaneously by using that the $\mathcal{U}_k$-MDDH Assumption is also re-randomizable across many matrices $\mathbf{M}_i$. (A similar property for the DDH, DLIN, and bilinear DDH assumptions is used in [6], [16], and [18], respectively.)

We defer the proof of Theorem 2 to the full version of this paper.

Acknowledgments. We would like to thank Jie Chen for insightful and inspiring discussions, and the reviewers for helpful comments. This work was done in part while the first and last authors were visiting the Simons Institute for the Theory of Computing, supported by the Simons Foundation and NSF grant CNS-1523467.

References

1. Abdalla, M., Benhamouda, F., Pointcheval, D.: Public-key encryption indistinguishable under plaintext-checkable attacks. In: Katz, J. (ed.) PKC 2015. LNCS, vol. 9020, pp. 332–352. Springer, Heidelberg (2015)

2. Abe, M., Chase, M., David, B., Kohlweiss, M., Nishimaki, R., Ohkubo, M.: Constant-size structure-preserving signatures: generic constructions and simple assumptions. In: Wang, X., Sako, K. (eds.) ASIACRYPT 2012. LNCS, vol. 7658, pp. 4–24. Springer, Heidelberg (2012)
3. Abe, M., David, B., Kohlweiss, M., Nishimaki, R., Ohkubo, M.: Tagged one-time signatures: tight security and optimal tag size. In: Kurosawa, K., Hanaoka, G. (eds.) PKC 2013. LNCS, vol. 7778, pp. 312–331. Springer, Heidelberg (2013)
4. Abe, M., Gennaro, R., Kurosawa, K.: Tag-KEM/DEM: a new framework for hybrid encryption. J. Cryptology **21**(1), 97–130 (2008)
5. Attrapadung, N., Hanaoka, G., Yamada, S.: A framework for identity-based encryption with almost tight security. Cryptology ePrint Archive, Report 2015/566 (2015). http://eprint.iacr.org/2015/566
6. Bellare, M., Boldyreva, A., Micali, S.: Public-key encryption in a multi-user setting: security proofs and improvements. In: Preneel, B. (ed.) EUROCRYPT 2000. LNCS, vol. 1807, pp. 259–274. Springer, Heidelberg (2000)
7. Blazy, O., Kiltz, E., Pan, J.: (Hierarchical) Identity-based encryption from affine message authentication. In: Garay, J.A., Gennaro, R. (eds.) CRYPTO 2014, Part I. LNCS, vol. 8616, pp. 408–425. Springer, Heidelberg (2014)
8. Boneh, D., Canetti, R., Halevi, S., Katz, J.: Chosen-ciphertext security from identity-based encryption. SIAM J. Comput. **36**(5), 1301–1328 (2007)
9. Camenisch, J., Chandran, N., Shoup, V.: A public key encryption scheme secure against key dependent chosen plaintext and adaptive chosen ciphertext attacks. In: Joux, A. (ed.) EUROCRYPT 2009. LNCS, vol. 5479, pp. 351–368. Springer, Heidelberg (2009)
10. Chen, J., Wee, H.: Fully, (almost) tightly secure IBE and dual system groups. In: Canetti, R., Garay, J.A. (eds.) CRYPTO 2013, Part II. LNCS, vol. 8043, pp. 435–460. Springer, Heidelberg (2013)
11. Cramer, R., Shoup, V.: Design and analysis of practical public-key encryption schemes secure against adaptive chosen ciphertext attack. SIAM J. Comput. **33**(1), 167–226 (2003)
12. Dolev, D., Dwork, C., Naor, M.: Nonmalleable cryptography. SIAM J. Comput. **30**(2), 391–437 (2000)
13. Escala, A., Herold, G., Kiltz, E., Ràfols, C., Villar, J.: An algebraic framework for Diffie-Hellman assumptions. In: Canetti, R., Garay, J.A. (eds.) CRYPTO 2013, Part II. LNCS, vol. 8043, pp. 129–147. Springer, Heidelberg (2013)
14. Goldwasser, S., Micali, S.: Probabilistic encryption. J. Comput. Syst. Sci. **28**(2), 270–299 (1984)
15. Gong, J., Chen, J., Dong, X., Cao, Z., Tang, S.: Extended nested dual system groups, revisited. Cryptology ePrint Archive, Report 2015/820 (2015). http://eprint.iacr.org/
16. Hofheinz, D., Jager, T.: Tightly secure signatures and public-key encryption. In: Safavi-Naini, R., Canetti, R. (eds.) CRYPTO 2012. LNCS, vol. 7417, pp. 590–607. Springer, Heidelberg (2012)
17. Hofheinz, D., Kiltz, E.: Secure hybrid encryption from weakened key encapsulation. In: Menezes, A. (ed.) CRYPTO 2007. LNCS, vol. 4622, pp. 553–571. Springer, Heidelberg (2007)
18. Hofheinz, D., Koch, J., Striecks, C.: Identity-based encryption with (almost) tight security in the multi-instance, multi-ciphertext setting. In: Katz, J. (ed.) PKC 2015. LNCS, vol. 9020, pp. 799–822. Springer, Heidelberg (2015)

19. Kiltz, E.: Chosen-ciphertext security from tag-based encryption. In: Halevi, S., Rabin, T. (eds.) TCC 2006. LNCS, vol. 3876, pp. 581–600. Springer, Heidelberg (2006)
20. Kiltz, E., Wee, H.: Quasi-adaptive NIZK for linear subspaces revisited. In: Oswald, E., Fischlin, M. (eds.) EUROCRYPT 2015. LNCS, vol. 9057, pp. 101–128. Springer, Heidelberg (2015)
21. Kurosawa, K., Desmedt, Y.G.: A new paradigm of hybrid encryption scheme. In: Franklin, M. (ed.) CRYPTO 2004. LNCS, vol. 3152, pp. 426–442. Springer, Heidelberg (2004)
22. Libert, B., Joye, M., Yung, M., Peters, T.: Concise multi-challenge CCA-secure encryption and signatures with almost tight security. In: Sarkar, P., Iwata, T. (eds.) ASIACRYPT 2014, Part II. LNCS, vol. 8874, pp. 1–21. Springer, Heidelberg (2014)
23. Libert, B., Peters, T., Joye, M., Yung, M.: Non-malleability from malleability: simulation-sound quasi-adaptive NIZK proofs and CCA2-secure encryption from homomorphic signatures. In: Nguyen, P.Q., Oswald, E. (eds.) EUROCRYPT 2014. LNCS, vol. 8441, pp. 514–532. Springer, Heidelberg (2014)
24. Libert, B., Peters, T., Joye, M., Yung, M., Compactly hiding linear spans: Tightly secure constant-size simulation-sound QA-NIZK proofs and applications. Cryptology ePrint Archive, Report 2015/242 (2015). http://eprint.iacr.org/2015/242
25. Libert, B., Peters, T., Joye, M., Yung, M.: Compactly hiding linear spans: tightly secure constant-size simulation-sound QA-NIZK proofs and applications. Cryptology ePrint Archive, Report 2015/242 (2015). http://eprint.iacr.org/
26. Morillo, P., Ràfols, C., Villar, J.L.: Matrix computational assumptions in multilinear groups. IACR Cryptology ePrint Archive, 2015:353 (2015)
27. Naor, M., Reingold, O.: Number-theoretic constructions of efficient pseudo-random functions. J. ACM **51**(2), 231–262 (2004)
28. Naor, M., Yung, M.: Public-key cryptosystems provably secure against chosen ciphertext attacks. In: 22nd ACM STOC, pp. 427–437. ACM Press, May 1990
29. Okamoto, T., Pointcheval, D.: REACT: rapid enhanced-security asymmetric cryptosystem transform. In: Naccache, D. (ed.) CT-RSA 2001. LNCS, vol. 2020, pp. 159–175. Springer, Heidelberg (2001)
30. Rackoff, C., Simon, D.R.: Non-interactive zero-knowledge proof of knowledge and chosen ciphertext attack. In: Feigenbaum, J. (ed.) CRYPTO 1991. LNCS, vol. 576, pp. 433–444. Springer, Heidelberg (1992)

Indistinguishability Obfuscation from Constant-Degree Graded Encoding Schemes

Huijia Lin(✉)

University of California, Santa Barbara, USA
rachel.lin@cs.ucsb.edu

Abstract. We construct an indistinguishability obfuscation (IO) scheme for all polynomial-size circuits from *constant-degree* graded encoding schemes, assuming the existence of a subexponentially secure pseudo-random generator computable by constant-degree arithmetic circuits, and the subexponential hardness of the Learning With Errors (LWE) problems. Previously, all candidate general purpose IO schemes rely on polynomial-degree graded encoding schemes.

1 Introduction

Program obfuscation [13] aims to make computer programs unintelligible while preserving their functionality. Recently, the first candidate general purpose indistinguishability obfuscation (IO) scheme for all polynomial-size circuits was proposed by Garg et al. [36]. Soon after that, an explosion of follow-up works showed the impressive power of IO, not only in obtaining classical cryptographic primitives, from one-way functions [46], trapdoor permutations [15], public-key encryption [60], to fully homomorphic encryption [26], but also in reaching new possibilities, from functional encryption [36], 2-round adaptively secure multi-party computation protocols [24,34,37], succinct garbling in time independent of the computation time [14,25,47,48], to constant-round concurrent ZK protocol [28]. It seems that IO is charting a bigger and more desirable map of cryptography.

However, the Achilles heel of IO research is that it is still unknown whether general purpose IO can be based on standard hardness assumptions. So far, all general purpose IO schemes are constructed in two steps [4,7,12,22,40,59,62]. First, an IO scheme for (polynomial size) NC^1 circuits is built using some candidate graded encoding schemes. The latter is an algebraic structure, introduced by Garg et al. [35], that enables homomorphically evaluating certain polynomials over encoded ring elements and testing whether the output is zero. Next, a bootstrapping theorem transforms an IO scheme for NC^1 into one for $\mathsf{P/poly}$, assuming the LWE assumption [36].

Huijia Lin was partially supported by NSF grants CNS-1528178 and CNS-1514526. This work was done in part while the author was visiting the Simons Institute for the Theory of Computing, supported by the Simons Foundation and by the DIMACS/Simons Collaboration in Cryptography through NSF grant CNS-1523467.

M. Fischlin and J.-S. Coron (Eds.): EUROCRYPT 2016, Part I, LNCS 9665, pp. 28–57, 2016.
DOI: 10.1007/978-3-662-49890-3_2

Tremendous efforts have been spent on basing the first step on more solid foundations. Unfortunately, the state of affairs is that all candidate graded encoding schemes [31,32,35,38] are susceptible to the so called "zeroizing attacks" [20,27,30,35,39] to different degrees.

In this work, we approach the problem from a different, more complexity theoretic, angle.

> *How much can we strengthen the bootstrapping theorem, and hence, simplify the task of building graded encoding schemes?*

We explore answers to this question and obtain the following main result:

Theorem 1 (Informal). *Assuming constant-degree PRG and LWE with subexponential hardness, there is a general purpose IO scheme using only* constant-degree *graded encodings.*

Though our result does not eliminate the need of graded encoding schemes, it weakens the requirement on them to *only supporting evaluation of constant-degree polynomials*; such a scheme is referred to as constant-degree graded encoding schemes. In comparison, previous IO schemes rely on polynomial degree graded encodings, polynomial in the size of the obfuscated circuit. This improvement is established exactly via a stronger bootstrapping theorem.

- *Bootstrapping IO for Constant-Degree Arithmetic Circuits.* We show that there is a class $\mathcal{C}$ of special-purpose constant-degree arithmetic circuits (i.e., corresponding to constant-degree polynomials), such that, special-purpose IO for $\mathcal{C}$ can be bootstrapped to general purpose IO for $\mathsf{P/poly}$, assuming the sub-exponential hardness of LWE, and the existence of a sub-exponentially secure Pseudo-Random Generator (PRG) computable by constant-degree arithmetic circuits. An candidate of such a PRG is Goldreich's PRG in NC^0 [41].
- *Constant-Degree Graded Encodings Suffice.* Then, we show that special purpose IO for $\mathcal{C}$ can be constructed from constant-degree graded encoding schemes.

Relation with Recent Works [16,51,58]. At a first glance, our main theorem is surprising in light of the recent results by [17,51,58]. They showed that any general-purpose IO scheme using *ideal* constant-degree graded encodings can be transformed into an IO scheme in the plain model. Alternatively, their results can be interpreted as: Ideal constant-degree graded encodings do not "help" constructing general-purpose IO schemes. In contrast, our results says that *concrete* constant-degree graded encodings imply general-purpose IO (assuming sub-exponentially secure constant-degree PRG and LWE). The divide stems from the fact that ideal graded encodings can only be used in a black-box manner, whereas our IO scheme crucially makes non-black-box use of the underlying graded encoding scheme. Because of the non-black-box nature of our construction, we actually do not obtain an IO scheme for $\mathsf{P/poly}$ in the ideal constant-degree graded encoding model, and hence we cannot apply the transformation of [17,51,58] to eliminate the use of graded encodings.

Moreover, it is interesting to note that our construction of IO for $\mathsf{P/poly}$ uses as a component the transformation from sub-exponentially secure compact functional encryption to general purpose IO by [3,16]. Their transformation makes non-black-box use to the underlying functional encryption, and is in fact the only non-black-box component in our construction. Therefore, if there were a black-box transformation from sub-exponentially secure compact functional encryption to general purpose IO, we would have obtained a general purpose IO scheme in the ideal constant-degree graded encoding model, and then by [17,51,58], a general purpose IO in the plain model. In summary, the following corollary suggests another avenue towards realizing IO.

Corollary 1. *Assume constant-degree PRG and LWE (with subexponential hardness). If there is a black-box transformation from any (subexponentially secure) compact functional encryption to an IO scheme for* $\mathsf{P/poly}$*, there is an IO scheme for* $\mathsf{P/poly}$ *in the plain model.*

1.1 Overview

Our results contain three parts: First, we establish a stronger bootstrapping theorem from IO for a class $\{\mathcal{C}_\lambda\}$ of special-purpose constant-degree arithmetic circuits to general-purpose IO. Second, we show that thanks to the constant-degree property and the simple structure of the special-purpose circuits, IO for $\{\mathcal{C}_\lambda\}$ can be constructed using only constant-degree graded encodings. The construction of the special-purpose IO scheme makes only black-box use of the constant-degree graded encodings, and is secure in the ideal model; but, the bootstrapping requires using the code of the special-purpose IO scheme. Therefore, to stitch the two first parts together, in the third part, we instantiate the special-purpose IO scheme using semantically secure graded encodings (c.f. [59]), and obtain general-purpose IO via bootstrapping. Below, we explain each part in more detail.

Part 1: Bootstrapping IO for Constant-Degree Arithmetic Circuits. So far, there are only two bootstrapping techniques in the literature, both starting from IO for NC^1. The first technique, proposed by [36], combines fully homomorphic encryption (FHE) and IO for NC^1, where the latter is used to obfuscate a circuit that performs FHE decryption and verifying the correctness of a computation trace, both can be done in logarithmic depth. The second technique by [26] is based on Applebaum's idea of bootstrapping VBB for NC^0 [5], where the underlying IO for NC^1 is used for obfuscating a circuit that computes for each input a randomized encoding (w.r.t. that input and the obfuscated circuit), using independent randomness produced by a Puncturable Pseudo-Random Functions (PPRF) [61] computable in NC^1 [23].

In sum, current bootstrapping techniques require the basic IO scheme to be able to handle complex cryptographic functions. It is an interesting question to ask what is the simplest circuit class — referred to as a "seed class" — such that, IO for it is sufficient for bootstrapping. In this work, we reduce the complexity of "seed classes" for NC^1 to circuits computable in constant degree. More specifically,

Proposition 1 (Informal, Bootstrapping Constant-Degree Computations). *Assume constant-degree PRG and LWE with subexponential hardness. There is a class of special-purpose* constant-degree *circuits* $\{C_\lambda\}$ *with domains* $\{\mathbb{D}_\lambda\}$*, where* $\mathbb{D}_\lambda \subseteq \{0,1\}^{\text{poly}(\lambda)}$*, such that, IO for* $\{C_\lambda\}$ *with* universal efficiency *(explained below) can be bootstrapped into IO for* $\mathsf{P/poly}$.

Let us explain our bootstrapping theorem in more detail.

Model of Constant Degree Computations. Every arithmetic circuit AC naturally corresponds to a polynomial by associating the ith input wire with a formal variable x_i; the degree of AC is exactly the degree of the polynomial. In this work, we consider using arithmetic circuit to compute Boolean functions $f : \{0,1\}^n \rightarrow \{0,1\}^m$, or logic circuits C. A natural model of computation is the following: Fix a ring $\mathcal{R}$ (say, the integers or the reals); a Boolean function f (or logic circuit C) is computed by an arithmetic circuit AC, if $\forall x \in \{0,1\}^n$, $C(x) = AC(x)$ over $\mathcal{R}$ (the 0 and 1 bits are mapped to the additive and multiplicative identities of $\mathcal{R}$ respectively). However, in this work, we consider a even weaker computation model that requires AC to agree with C over any choice of ring $\mathcal{R}$.

- *Constant-Degree Computations:* We say that a Boolean function f (or logic circuit C) is computed by an arithmetic circuit AC, if $\forall x \in \{0,1\}^n$, $C(x) = AC(x)$, over any ring $\mathcal{R}$.

This model of constant-degree computations is quite weak, in fact, so weak that it is equivalent to NC^0. Nisan and Szegedy [56] showed that the degree of the polynomial that computes a Boolean function f over the ring of reals is polynomially related with the decision tree complexity of f. Therefore, if f has constant degree in our model, it has constant decision tree complexity, implying that it is in NC^0.

On the other hand, it is well known that IO for NC^0 can be trivially constructed by searching for canonical representations, which can be done efficiently as every output bit is computed by a constant-size circuit. Though it would be ideal to bootstrap IO for NC^0, we do not achieve this. Instead, we strengthen the above model of computation by considering partial Boolean functions (or logic circuits) defined only over a subset $\mathbb{D} \in \{0,1\}^n$ (i.e., we only care about inputs in $\mathbb{D}$).

- *Constant-Degree Computations with Partial Domains:* We say that a Boolean function f (or logic circuit C) with domain $\mathbb{D} \in \{0,1\}^n$ is computed by an arithmetic circuit AC, if $\forall x \in \mathbb{D}$, $C(x) = AC(x)$, over any ring $\mathcal{R}$.

A concrete constant-degree partial function that is not computable in NC^0 is a variant of the multiplexer function mux that on input (x, e_i), where $x, e_i \in \{0,1\}^n$ and the hamming weight of e_i is 1, outputs x_i. Clearly, the output bit has to depend on all bits of e_i and cannot be computed in NC^0. But, x_i can be computed in degree 2 as the inner product of x and e_i over any ring $\mathcal{R}$.

Nevertheless, our model of constant degree computations (with potentially partial domains) is still weak. In particular, it is separated from AC^0, since we cannot compute unbounded AND in it. In the body of the paper, we put even more constraints and say that a class of circuits (as opposed to a single circuit) is constant degree only if they have universal circuits of constant degrees; we omit this detail in the introduction. As a further evidence on how weak our model of constant degree computations are, we show next that even statistical IO is feasible for such circuits.

Trivial, Statistical IO for Constant Degree Computations. Let C be a logic circuit computable by a degree-d arithmetic circuit AC, which corresponds to a degree-d polynomial. At a high-level, because degree-d polynomials can be learned in poly(n^d) time, we can obfuscate C in the same time with statistical security. More specifically, the degree-d polynomial $p(x)$ can be equivalently represented as a linear function $L(X)$ over $\ell = n^d$ variables, each associated with a degree d monomial over $x_1 \cdots x_n$. To obfuscate C, we simply pick ℓ inputs $x_1, \cdots, x_\ell \in \{0,1\}^n$, such that, their corresponding monomial values $X_1, \cdots, X_\ell$ are linearly independent. Now, the obfuscation $\tilde{C}$ of C is simply the set of input output pairs $(x_1, y_1), \cdots, (x_\ell, y_\ell)$ where $y_i = C(x_i)$.

Given $\tilde{C}$, we can to evaluate C on any input x, since $C(x) = L(x)$ over any ring, and the linear function L can be learned from the set of input output pairs using Gaussian elimination. Moreover, it is easy to see that obfuscation of any two functionally equivalent circuits C and C' are identically distributed, as C and C' have the same truth table and their obfuscations simply reveal a part of their truth tables.

The above construction, though achieve statistical security, is however, trivial: The truth table of a degree-d circuit effectively has only size n^d (by Gaussian elimination), and the above construction simply publishes the effective truth table. As a result, it is not sufficient for our bootstrapping.

Computational IO for Constant Degree Computations, with Universal Efficiency. Instead, we require IO for constant degree computations with better, non-trivial, efficiency. More specifically,

- *Universal Efficiency:* We say that IO for constant degree circuits has universal efficiency, if its run-time is independent of the degree of the computation. That is, there is a universal polynomial p, such that, for every d, obfuscating a degree-d circuit C takes time $p(1^\lambda, |C|)$ for sufficiently large λ.

In fact, our bootstrapping theorem works even if the efficiency of IO for constant degree circuits grows with the degree, as long as it is bounded by $n^{h(d)}$ for a sufficiently small function h, say, $h(d) = \log\log\log(d)$. For simplicity, we consider the above universal efficiency.

General Connection between Complexity of PRG and Complexity of Seed Class. Finally, we note that our bootstrapping theorem can be generalized to establish a connection between the complexity of PRG and the complexity of "seed classes" sufficient for bootstrapping IO. Consider any PRG scheme PRG (not necessarily computable in constant degree). There is a family $\{\mathcal{C}_\lambda\}$ of special-purpose *oracle circuits* that have constant degree (such a circuit can be computed by an arithmetic circuit with oracle gates, and its degree is the degree of the arithmetic circuit when replacing the oracle gates with additions), such that, IO for the class of composed circuits in $\{\mathcal{C}_\lambda^{\mathsf{PRG}}\}$, with again universal efficiency, can be bootstrapped into IO for $\mathsf{P/poly}$.

Proposition 2 (Informal, General Bootstrapping Theorem). *Assume a PRG scheme* PRG *and LWE with sub-exponential hardness. There is a class of special-purpose oracle circuits* $\{\mathcal{C}_\lambda\}$ *that have constant degree, such that, special purpose IO for* $\{\mathcal{C}_\lambda^{\mathsf{PRG}}\}$ *with universal efficiency can be bootstrapped into IO for* $\mathsf{P/poly}$.

In particular, plugging in a constant-degree PRG yields Proposition 1, and plugging in a PRG in AC^0 or TC^0 establishes that IO for AC^0 or TC^0 with universal efficiency suffices for constructing general purpose IO.

Given the general connection between the complexity of PRG and that of the seed class, we summarize the state-of-the-art of low depth PRG at the end of the introduction.

Techniques. Our starting point is two beautiful recent works by Bitansky and Vaikuntanathan [16] and Ananth and Jain [3] showing that sub-exponentially secure (sublinearly) compact FE for NC^1 implies IO for $\mathsf{P/poly}$. Unfortunately, so far, the former is only known from IO for NC^1; thus, our goal is to construct compact FE using IO for the simplest circuits.

Technically, the transformation in the first step is similar to that in [3,49]. However, the former [3] requires IO for a class of special-purpose Turing machines (as opposed circuits). Our transformation uses the same idea as in [49], but requires a much more refined analysis in order to identify and simplify the circuits, whose special structure plays a key role later.

The work of Ananth and Jain [3], and another very recent work by the author, Pass et al. [49] already explored this direction: They show that a compact FE scheme for NC^1 circuits with single output bit (which can be based on LWE [43]) can be transformed into a compact FE for all NC^1 circuits with multiple output bits, using IO for circuits (Turing machines in [3]) with only a logarithmic number $c \log \lambda$ of input wires; such circuits have λ^c-sized truth table. [49] further weakens the efficiency requirement on such IO schemes: As long as the IO scheme outputs obfuscated circuits whose size is sub-linear in the size of the truth table (matching the sub-linear compactness of FE), the transformation goes through.

However, the circuits used in [3,49] are complex, in NC^1. In this work, we significantly reduce the complexity of the circuits using more refined analysis and a number of new techniques. For example, we build a special-purpose PPRF for polynomial sized domain that is computable in constant degree. Interestingly, the polynomial-sized domain is not of the form $\{0,1\}^{c \log \lambda}$, rather is embedded sparsely in a much larger domain $\mathbb{D} \subset \{0,1\}^{\mathrm{poly}(\lambda)}$. This crucially allows us to circumvent lower bounds on the complexity of normal PPRF. Moreover, we design ways to perform comparisons, such as, testing $=, \geq, \leq$ relations, between two integers $i, i' \in [\text{poly}(\lambda)]$ in constant degree; here again, we crucially rely on the fact that we can represent the integers in a different way, embedded sparsely in a much larger domain.

Part 2: Special Purpose IO in Constant-Degree Ideal Graded Encoding Model. Ideal grade encoding model [7,12,22,35,62] captures generic algebraic attacks over graded encodings: In this model, players have black-box access to a ring, and can only perform certain restricted operations over ring elements, and determine whether a "legal" polynomial (one satisfying all restrictions) evaluates to 0 or not—this is referred to as a "zero-test query".

An important parameter, called the *degree* of the graded encodings [51,58], is the *maximum degree* of (legal) polynomials that can be "zero-tested". Clearly, the lower the degree is, the weaker of the graded encodings are. Consider for instance, when the degree is one, the ideal graded encoding model is equivalent to the generic group model, in which operations are restricted to be linear (i.e., degree 1 polynomials), and when degree is two, ideal graded encodings capture idealized groups with bilinear maps. Both special cases have been extensively studied.

So far, general-purpose IO schemes in ideal models all require *high degree* graded encodings (polynomial in the *size* of the circuit being obfuscated) [7,12,22,62]. The dilemma is that such models are so powerful that even general purpose VBB obfuscation is feasible, which is impossible in the plain model [13]. Two recent works [51,58] initiated the study of *low-degree* ideal graded encodings, showing that when the degree is restricted to a *constant*, general purpose VBB obfuscation becomes infeasible. Therefore, constant-degree ideal graded encoding model is qualitatively weaker than its high-degree counterpart, and is much closer to the plain model.

Nevertheless, we show that for the simple class of constant-degree computations, it is sufficient.

Proposition 3 (Informal, Special-Purpose IO in Ideal Model). *There is a universally efficient IO scheme for the class $\{C_\lambda\}$ of constant-degree special-purpose circuits in Proposition* 1, *in the constant-degree ideal graded encoding model.*

Our special-purpose IO scheme crucially exploits the constant degree property of our seed class, as well as the simple structure of circuits in the class.

<u>*Type-Degree Preserving IO Construction.*</u> Our main technique is characterizing a general type of circuits that admit IO schemes with low degree ideal graded encodings. More specifically, we define a new measure, called *type degree*, for arithmetic circuits, which is a quantity no smaller than the actual degree of the circuit, and no larger than the maximum degree of circuits with the same topology (achieved by a circuit with only multiplication gates). We show that if a class of circuits have type degree *td*, then there is an IO scheme for this class using ideal graded encodings of roughly the same degree $O(td)$; we say that such an IO construction is *type-degree preserving.* Our type-degree preserving IO construction is based on the IO scheme of Applebaum and Brakerski [7] in the

composite order ideal graded encoding model; we believe that our construction is of independent interests.

Furthermore, thanks to the simplicity of our special purpose circuits in Proposition 1, we can show that they not only have constant degree, but also have constant type degree, leading to Proposition 4.

Part 3: Instantiation with Concrete Graded Encoding Schemes. The final part combine our bootstrapping theorem (Proposition 1) with our special-purpose IO scheme (Proposition 4) to obtain general-purpose IO, for which we must first instantiate the ideal graded encodings with concrete ones, for the bootstrapping theorem makes non-black-box use of the special-purpose IO. Towards this, the technical question is "under what computational hardness assumption over graded encodings can we prove the security of our special-purpose IO scheme in the plain model?"

So far, in the literature, there are two works that answer questions like the above. Pass et al. [59] proposed the meta-assumption of *semantic security* over prime order graded encoding schemes, from which the security of a general purpose IO scheme follows via an explicit security reduction. Subsequently, Gentry et al. [40] proposed the Multilinear Subgroup Elimination assumption over composite order graded encoding schemes which improves upon semantic security in terms of simplicity and the number of assumptions in the family (albeit requiring a sub-exponential security loss).

Following [59], we show that our special purpose IO schemes in Proposition 4 can be instantiated with any composite order graded encoding schemes satisfying an analogue of semantic security for composite order rings; importantly, the graded encoding scheme only need to support constant-degree computation.[1] Hence, combining with our bootstrapping theorem from Part 1, we obtain a general purpose IO scheme from constant-degree graded encoding schemes.

Proposition 4 (Informal, Special-Purpose IO in the Plain Model). *There is a universally efficient IO scheme for the class $\{C_\lambda\}$ of constant-degree special-purpose circuits in Proposition* 1*, assuming semantically-secure constant-degree graded encodings.*

Finally, applying our bootstrapping theorem (Proposition 1) on the special-purpose IO scheme in the above proposition, gives our main theorem (Theorem 1).

1.2 Low Depth PRG

We survey constructions of low depth PRGs. Some of the texts below are taken verbatim from Applebaum's book [6].

The existence of PRG in TC^0 follows from a variety of hardness assumption including intractability of factoring, discrete logarithm, or lattice problems

[1] We note that the security of (variants of) our IO scheme could potentially be proven from the multilinear subgroup elimination assumption of [40]; we leave this as future work.

(e.g. [11,53–55]). Literature on PRG in AC^0 is limited; more works focus directly on PRG in NC^0. On the negative side, it was shown that there is no PRG in NC^0_4 (with output locality 4) achieving super-linear stretch [33,52]. On the positive side, Applebaum et al. [8] showed that any PRG in NC^1 can be efficiently "compiled" into a PRG in NC^0 using randomized encodings, but with only *sub-linear* stretch. The authors further constructed a *linear-stretch* PRG in NC^0 under a specific intractability assumption related to the hardness of decoding "sparsely generated" linear codes [9], previously conjectured by Alekhnovich [1]. Unfortunately, to the best of our knowledge, there is no construction of PRG in NC^0 (or even AC^0) with polynomial stretch from well-known assumptions. But, candidate construction exists.

Goldreich's Candidate PRGs in NC^0. Goldreich's one-way functions $f : \{0,1\}^n \to \{0,1\}^m$ where each bit of output is a fixed predicate P of a constant number d of input bits, chosen at random or specified by a bipartite expander graph with the right degree, is also a candidate PRG when $m > n$. Several works investigated the (in)security of Goldreich's OWFs and PRGs: So far, there are no successfully attacks when the choice of the predicate P avoids certain degenerating cases [10, 18,29,57]. Notably, O'Donnell and Witmer [57] gave evidence for the security of Goldreich's PRGs with polynomial stretch, showing security against both subexponential-time $\mathbb{F}_2$-linear attacks, as well as subexponential-time attacks using SDP hierarchies such as Sherali-Adams$^+$ and Lasserre/Parrilo.

1.3 Organization

We provide more detailed technical overviews at the beginning of Sects. 3, 4, *and* 5.

In Sect. 2, we formalize our model of constant-degree computations, IO with universal efficiency, and provide basic preliminaries. In Sect. 3, we prove a prelude of our bootstrapping theorem that identifies a class of special purpose circuits, such that IO for this class with universal efficiency can be bootstrapped to general purpose IO. In Sect. 4, we show that the class of special purpose circuits identified in Sect. 3 are computable in constant degree, when the underlying PRG is. Then, we construct a universally efficient IO scheme for these special purpose circuits in constant-degree ideal graded encoding model in Sect. 5. Due to the lack of space, we refer the readers to the full version on how to instantiate our special-purpose IO with semantically secure graded encodings.

2 Preliminaries

Let $\mathbb{Z}$ and $\mathbb{N}$ denote the set of integers, and positive integers, $[n]$ the set $\{1, 2, \ldots, n\}$, $\mathcal{R}$ denote a ring, and $\mathbf{0}$,$\mathbf{1}$ the additive and multiplicative identities.

Due to the lack of space, we omit definitions of standard cryptographic primitives such as, PRG, PPRF, (compact) functional encryption and randomized encodings (see [3,8,16]), and only discuss our models of computation and give definitions of IO and universal efficiency below.

2.1 Models of Computation

Logic Circuits and Partial Domains. In this work, by circuit, we mean logic circuits from $\{0,1\}^*$ to $\{0,1\}^*$, consisting of input gates, output gates, and logical operator gates (AND and OR gates with fan-in 2 and fan-out >0, and NEG gate with fan-in 1).

Any circuit with n-bit input wires and m-bit output wires defines a total Boolean function f mapping $\{0,1\}^n$ to $\{0,1\}^m$. In this work, importantly, we also consider partial functions f defined only over a (partial) domain $\mathbb{D} \subset \{0,1\}^n$. Correspondingly, we associate a circuit C with a domain $\mathbb{D} \subset \{0,1\}^n$, meaning that we only care about evaluating C over inputs in $\mathbb{D}$.

Arithmetic Circuits. We also consider arithmetic circuits AC consisting of input gates, output gates and operator gates for addition, subtraction, and multiplication (with fan-in 2 and fan-out >0). Every arithmetic circuit AC with n input gates defines a n-variate polynomial P over $\mathbb{Z}$, by associating the ith input gate with a formal variable x_i. We say that AC has degree d if P has degree d. An arithmetic circuit AC can also be evaluated over any other ring $\mathcal{R}$ (different from $\mathbb{Z}$), corresponding to computing the polynomial P over $\mathcal{R}$.

Boolean Functions Computable by Arithmetic Circuits. In this work, we, however, do not consider evaluating arithmetic circuits over any specific ring. Rather, we say that a Boolean function f from domain $\mathbb{D} \subseteq \{0,1\}^n$ to range $\{0,1\}^m$, is computed/implemented by an arithmetic circuit AC if for every input $x \in \mathbb{D}$ with output $y = C(x)$, AC evaluated on $\mathbf{x}$ equals to $\mathbf{y}$ over *any* ring $\mathcal{R}$, where $\mathbf{x}$ and $\mathbf{y}$ are vectors of ring elements derived from x and y respectively, by mapping 0 to the additive identity $\mathbf{0}$ and 1 to the multiplicative identity $\mathbf{1}$ of $\mathcal{R}$. We omit explicitly mentioning this conversion in the rest of the paper, and simply write $AC(x) = C(x)$.

We stress again that, in our model, a Boolean function f is computable by an arithmetic circuit only if it produces the correct outputs for all inputs in $\mathbb{D}$, *no matter what underlying ring is used.* This restriction makes this model of computation very weak.

Similarly, we say that a circuit C with domain $\mathbb{D} \subset \{0,1\}^n$ is computable by an arithmetic circuit AC, if the Boolean function $f : \mathbb{D} \to \{0,1\}^m$ defined by C is computable by AC.

Circuit Classes and Families of Circuit Classes. We use the following terminologies and notations:

- A family of circuits $\mathcal{C}$ with domain $\mathbb{D}$ is simply a set of circuits $C \in \mathcal{C}$ with common domain $\mathbb{D}$.
- A class of circuits $\{\mathcal{C}_\lambda\}_{\lambda \in \mathbb{N}}$ with domains $\{\mathbb{D}_\lambda\}_{\lambda \in \mathbb{N}}$ is an ensemble of sets of circuits, where each $\mathcal{C}_\lambda$ is associated with domain $\mathbb{D}_\lambda$. We use the shorthands $\{\mathcal{C}_\lambda\}$ and $\{\mathbb{D}_\lambda\}$.
- A family of circuit classes $\{\{\mathcal{C}^x_\lambda\}\}^{x \in X}$ is a set of circuit classes, where each circuit class $\{\mathcal{C}^x_\lambda\}$ is indexed by an element x in a (countable) index set X.

For convenience, when the index set X is clear in the context, we use shorthand $\{\{\mathcal{C}^x_\lambda\}\}$. A family of circuit classes can also be associated with domains $\{\{\mathbb{D}^x_\lambda\}\}$, meaning that each set $\mathcal{C}^x_\lambda$ is associated with domain $\mathbb{D}^x_\lambda$.
For example, NC^1 circuits can be described as a family of circuit classes $\{\{\mathcal{C}^d_\lambda\}\}^{d \in \mathbb{N}}$, where for every $d \in \mathbb{N}$, the circuit class $\{\mathcal{C}^d_\lambda\}$ contains all circuits of depth $d \log \lambda$.

Universal (Arithmetic) Circuits. Let $\mathcal{C}$ be a family of circuits with domain $\mathbb{D}$, where every $C \in \mathcal{C}$ is described as an ℓ-bit string, and let $\mathcal{U}$ be an (arithmetic) circuit. We say that $\mathcal{U}$ is the universal (arithmetic) circuit of $\mathcal{C}$ if every $C \in \mathcal{C}$ is computed by $\mathcal{U}(\star, C)$ over domain $\mathbb{D}$. Moreover, we say that an ensemble of (arithmetic) circuits $\{\mathcal{U}_\lambda\}$ is the universal (arithmetic) circuits of a circuit class $\{\mathcal{C}_\lambda\}$ with domain $\{\mathbb{D}_\lambda\}$ if for every λ, $\mathcal{U}$ is an (arithmetic) universal circuit of $\mathcal{C}_\lambda$ with domain $\mathbb{D}_\lambda$.

Degree of (Logic) Circuits. Degree is naturally defined for arithmetic circuits as described above, but not so for logic circuits and Boolean functions. In this work, we define the degrees of logic circuits and Boolean functions through the degree of the arithmetic circuits that compute them. Moreover, we also define degrees for families of circuits, circuit classes, and families of circuit classes, through the degrees of the universal arithmetic circuits that compute them.

Degree of a (Logic) Circuit: We say that a circuit C with domain $\mathbb{D}$ has degree d, if it is computable by an arithmetic circuit of degree d.

Degree of a Family of Circuits: We say that a family of circuits $\mathcal{C}$ with domain $\mathbb{D}$ has degree d, if it has a universal arithmetic circuit $\mathcal{U}$ of degree d.

Degree of a Class of Circuits: We say that a class of circuits $\{\mathcal{C}_\lambda\}$ with domain $\mathbb{D}_\lambda$ has degree $d(\lambda)$, if it has universal arithmetic circuits $\{\mathcal{U}_\lambda\}$, with degree $d(\lambda)$. If $d(\lambda)$ is a constant function, then we say $\{\mathcal{C}_\lambda\}$ has constant degree.

Degree of a Family of Circuit Classes: We say that a family of circuit classes $\{\{\mathcal{C}^x_\lambda\}\}$ with domains $\{\{\mathbb{D}^x_\lambda\}\}$ has constant degree, if for every $x \in X$, circuit class $\{\mathcal{C}^x_\lambda\}$ with domains $\{\mathbb{D}^x_\lambda\}$ has constant degree.

It is important to note that we define the degree of a class of circuits via the degree of its *universal* arithmetic circuit, not the degree of individual circuits inside. For example, consider the natural class of circuits containing all (polynomial-sized) circuits with a fixed constant degree d (c.f., the class of poly-sized NC^0 circuits with a fixed constant depth d), under our definition, it is not clear whether this class itself has constant degree, as it is not clear (to us) whether there is a constant degree universal arithmetic circuit that computes all of them. Nevertheless, this more stringent definition only makes our bootstrapping result that it suffices to construct IO for a family of circuit classes with constant degree stronger, and makes the task of constructing IO for such a family easier.

2.2 Indistinguishability Obfuscation

We recall the notion of indistinguishability obfuscation for a class of circuit defined by [13], adding the new dimension that the class of circuits may have restricted domains $\{\mathbb{D}_\lambda\}$.

Definition 1 (Indistinguishability Obfuscator ($i\mathcal{O}$) for a Circuit Class). *A uniform* PPT *machine $i\mathcal{O}$ is a indistinguishability obfuscator for a class of circuits $\{\mathcal{C}_\lambda\}_{\lambda\in\mathbb{N}}$ (with potentially restricted domains $\{\mathbb{D}_\lambda\}_{\lambda\in\mathbb{N}}$), if the following conditions are satisfied:*

Correctness: *For all security parameters $\lambda \in \mathbb{N}$, for every $C \in \mathcal{C}_\lambda$, and every input x (in $\mathbb{D}_\lambda$), we have that*

$$\Pr[C' \leftarrow i\mathcal{O}(1^\lambda, C) \;:\; C'(x) = C(x)] = 1$$

μ-Indistinguishability: *For every ensemble of pairs of circuits $\{C_{0,\lambda}, C_{1,\lambda}\}_\lambda$ satisfying that $C_{b,\lambda} \in \mathcal{C}_\lambda$, $|C_{0,\lambda}| = |C_{1,\lambda}|$, and $C_{0,\lambda}(x) = C_{1,\lambda}(x)$ for every x (in $\mathbb{D}_\lambda$), the following ensembles of distributions are μ-indistinguishable,*

$$\left\{C_{1,\lambda}, C_{2,\lambda}, i\mathcal{O}(1^\lambda, C_{1,\lambda})\right\}_\lambda$$
$$\left\{C_{1,\lambda}, C_{2,\lambda}, i\mathcal{O}(1^\lambda, C_{2,\lambda})\right\}_\lambda$$

In the above definition, μ can be either negligible for standard IO, or subexponentially small for sub-exponentially secure IO.

Definition 2 (IO for $\mathsf{P/poly}$). *A uniform* PPT *machine $i\mathcal{O}_{\mathsf{P/poly}}(\star, \star)$ is an indistinguishability obfuscator for* P/poly *if it is an indistinguishability obfuscator for the class $\{\mathcal{C}_\lambda\}$ of circuits of size at most λ.*

2.3 Indistinguishability Obfuscation for Families of Circuit Classes

In this work, we consider families of circuit classes, and the task of building a family of indistinguishability obfuscators, one for each circuit class.

Definition 3 (IO for Families of Circuit Classes). *Let $\{\{\mathcal{C}^x_\lambda\}\}^{x\in X}$ be a family of circuit classes (with potentially restricted domains $\{\mathbb{D}^x_\lambda\}$). A family of uniform machines $\{i\mathcal{O}^x\}^{x\in X}$ is a family of indistinguishability obfuscators for $\{\{\mathcal{C}^x_\lambda\}\}^{x\in X}$, if for every constant $x \in X$, $i\mathcal{O}^x$ is an indistinguishability obfuscator for the circuit class $\{\mathcal{C}^x_\lambda\}$ (with domains $\{\mathbb{D}^x_\lambda\}$).*

The above definition implicitly requires that for every $x \in X$, $i\mathcal{O}^x$ runs in some polynomial time, potentially depending on x. However, depending on how the run-time of $i\mathcal{O}^x$ vary for different x, qualitatively different types of efficiency could be considered.

We consider the following notion of universal efficiency in this work.

Definition 4 (Universal Efficiency). *A family of indistinguishability obfuscators* $\{i\mathcal{O}^x\}^{x \in X}$ *for a family of circuit class* $\{\{\mathcal{C}_\lambda^x\}\}^{x \in X}$ *(with potentially restricted domains* $\{\{\mathbb{D}_\lambda^x\}\}$*) has* universal efficiency, *if there exists a universal polynomial* p*, such that, for every* $x \in X$*,* $i\mathcal{O}^x(1^\lambda, C)$ *runs in time* $p(\lambda, |C|)$*, for every sufficiently large* λ *(i.e., greater than a constant* c^x *depending on* x*), and circuit* $C \in \mathcal{C}_\lambda^x$*.*

We note that it is without loss of generality to only consider the run-time of $i\mathcal{O}^x$ for sufficiently large λ ($> c^x$), because the security of $i\mathcal{O}^x$ already only holds for sufficiently large λ.

3 Bootstrapping IO for Special-Purpose Circuits

In this section, we identify a family of special-purpose circuit classes and show how to bootstrap IO for this family to all polynomial-sized circuits.

Proposition 5. *Assume the following primitives:*

- *a sub-exponentially secure compact FE scheme* FE *for Boolean* NC^1 *circuits,*
- *a sub-exponentially secure PPRF scheme* PPRF*, and*
- *a sub-exponentially secure RE scheme* RE *in* NC^0*.*

Then, there is a family of special-purpose circuit classes $\{\{\mathcal{P}_\lambda^{T,n}\}\}$ *indexed by two polynomials* $T(\star)$ *and* $n(\star)$ *and defined w.r.t.* FE*,* PPRF *and* RE *as in Fig. 1, such that, the following holds:*

- *If there exists a family* $\{i\mathcal{O}^{T,n}\}$ *of IO schemes for* $\{\{\mathcal{P}_\lambda^{T,n}\}\}$ *with* universal efficiency, *then there are two sufficiently large polynomials* T^* *and* n^**, such that,* $i\mathcal{O}^{T^*,n^*}$ *can be transformed into an IO scheme for* P/poly*.*

We note in Sect. 3.1 that all the underlying primitives of the above Proposition are implied by the sub-exp hardness of LWE.

Overview. Towards the proposition, recall that recent works [2,3,16] show that to construct IO for P/poly, it suffices to construct a compact FE scheme for NC^1 circuits. Formally,

Theorem 2. [2,3,16] *Let* n *be a sufficiently large polynomial. Assume the existence of a sub-exponentially secure, and* $(1-\varepsilon)$*-weakly-compact (single-query, public-key) FE scheme for* NC^1 *circuits, and weak PRF in* NC^1*. There exists an indistinguishability obfuscator for* P/poly*.*

Therefore, the natural direction is constructing compact FE scheme for NC^1 circuits using IO for the special-purpose circuits. We proceed in two steps: For any polynomials T and n, let $\mathsf{NC}^{1,T,n}$ be the subclass of NC^1 circuits with at most size $T(\lambda)$ and at most $n(\lambda)$ input bits.

- Our first step (in Sect. 3.2) constructs an FE scheme $\mathsf{FE}^{T,n}$ for $\mathsf{NC}^{1,T,n}$ from any IO scheme $i\mathcal{O}^{T,n}$ for $\{\mathcal{P}_\lambda^{T,n}\}$ (and the underlying primitives of Proposition 5), for arbitrary T and n. Importantly, the encryption time of the resulting FE scheme is directly proportional to the obfuscation time of the underlying IO scheme:

$$\mathsf{Time}_{i\mathcal{O}^{T,n}}(1^\lambda, C) \leq p^{T,n}(\lambda, |C|)$$
$$\mathsf{Time}_{\mathsf{FE.Enc}}(mpk, m) \leq p^{T,n}(\lambda,\ q(\lambda, n(\lambda), \log T(\lambda)))$$

where q is a universal polynomial independent of T and n. Note that, this does not guarantee that the resulting FE scheme is compact, since the run-time of the IO scheme may depend on T arbitrarily, in particular, it is possible that $p^{T,n}(\lambda, |C|) > T(\lambda)$, while $i\mathcal{O}^{T,n}$ is still a valid polynomial time IO scheme for $\{\mathcal{P}_\lambda^{T,n}\}$.
- To overcome the above issue, our next step (in Sect. 3.3) starts with a stronger premise: The existence of a family $\{i\mathcal{O}^{T,n}\}$ of IO schemes for the family $\{\{\mathcal{P}_\lambda^{T,n}\}\}$ with universal efficiency. This means for any T, n, the obfuscation time of $i\mathcal{O}^{T,n}$ is bounded by a universal polynomial p, and (for sufficiently large λ)

$$\mathsf{Time}_{i\mathcal{O}^{T,n}}(1^\lambda, C) \leq p(\lambda, |C|)$$
$$\mathsf{Time}_{\mathsf{FE.Enc}}(mpk, m) \leq p(\lambda,\ q(\lambda, n(\lambda), \log T(\lambda)))$$

This essentially means that the FE schemes are compact — encryption time is independent of $T(\lambda)$. In particular, for some sufficiently large polynomials T^* and n^*, encryption time of FE^{T^*,n^*} is much smaller than the time of the computation, that is, $p(\lambda,\ q(\lambda, n^*(\lambda), \log T^*(\lambda))) < T^*$. With a closer examination, such an FE scheme FE^{T^*,n^*} is sufficient for the transformation of [2,3,16] to go through. More specifically, the final IO scheme for $\mathsf{P/poly}$ they construct only need to use the underlying FE scheme for NC^1 circuits with some sufficiently large size T^* and sufficiently long input length n^*; the proof goes through, as long as encryption time is sub-linearly $(T^*)^{1-\varepsilon}$ in T^*.

Putting the two steps together, we conclude Proposition 5.

Technically, the transformation in the first step is similar to that in [3,49]. However, the former [3] requires IO for a class of special-purpose Turing machines (as opposed circuits). Our transformation uses the same idea as in [49], but requires a much more refined analysis in order to identify and simplify the circuits, whose special structure plays a key role later.

3.1 Instantiating the Underlying Primitives from LWE

The first primitive of Proposition 5—a compact FE for Boolean NC^1 circuits—can be derived from the work of Goldwasser et al. [43]: Assuming sub-exp LWE, they construct a sub-exp secure FE scheme for the class of polynomial-sized *Boolean* circuits $\{\mathcal{C}_{n,d(n)}\}$ with n input bits and depth $d(n)$. Furthermore, the size of the ciphertexts is $\text{poly}(\lambda, n, d)$ (independent of the size of the circuits); when restricting to Boolean circuits in NC^1 (as needed for Proposition 5), the ciphertexts are compact. Summarizing,

Theorem 3 (Compact FE scheme for Boolean NC^1 Circuits [43]). *Assume sub-exponential hardness of the LWE problem. There exists a sub-exponentially secure compact (single-query, public-key) FE scheme for the class of* Boolean NC^1 *circuits.*

The second primitive—a sub-exp secure PPRF—can be constructed from the necessary assumption of sub-exp secure OWFs [19,21,45]; but, the evaluation algorithms of these PPRF schemes have high depth. Recently, Brakerski and Vaikuntanathan [23] showed that assuming LWE, the depth of the evaluation algorithm can be reduced to logarithmic $O(\log \lambda)$.

Finally, the third primitive—a sub-exp secure RE scheme in NC^0—can be constructed from sub-exp secure low-depth PRG [8,44], which is in turn implied by sub-exp secure LWE.

Circuit $P[T, n, mpk, i^*, K, m_<, \hat{\Pi}, m_>](i)$

Constant: A security parameter $\lambda \in \mathbb{N}$, a time bound $T \in \mathbb{N}$, a threshold $i^* \in \{0, \cdots, T+1\}$, a public key $mpk \in \{0,1\}^{\ell_{mpk}}$ of bFE, a punctured key $K \in \{0,1\}^{\ell_{key}}$ of PPRF, strings $m_<$, $m_>$ of equal length n, and an RE encoding $\hat{\Pi}$.

Input: An index $i \in [T]$.

Procedure:

1. $(R_i || R_i') = \mathsf{F}(K, i)$;
2. If $i < i^*$, set $\hat{\Pi}_i = \mathsf{RE.Enc}\left(1^\lambda,\ \mathsf{bFE.Enc},\ (mpk,\ m_< || i\ ;\ R_i)\ ;\ R_i'\right)$.
3. If $i = i^*$, set $\hat{\Pi}_i = \hat{\Pi}$.
4. If $i > i^*$, set $\hat{\Pi}_i = \mathsf{RE.Enc}\left(1^\lambda,\ \mathsf{bFE.Enc},\ (mpk,\ m_> || i\ ;\ R_i)\ ;\ R_i'\right)$.

Output: Encoding $\hat{\Pi}_i$.

Padding: The hardwired encoding $\hat{\Pi}$ is padded to be of length $\eta'(\lambda, n, \log T)$, and the circuit is padded to be of size $\eta(\lambda, n, \log T)$, for some polynomials η' and η.

Circuit classes $\{\mathcal{P}_\lambda^{T,n}\}$ contains all circuits of form $P[\lambda, T(\lambda), n(\lambda), *, *, *, *, *]$, where all wild-card values satisfy length constraints specified above.

Fig. 1. Special-Purpose circuit P

3.2 FE for $\mathsf{NC}^{1,T,n}$ from IO for $\{\mathcal{P}_\lambda^{T,n}\}$

Fix arbitrary polynomials T and n. We present an FE scheme $\mathsf{FE}^{T,n}$ for $\mathsf{NC}^{1,T,n}$ from IO for $\{\mathcal{P}_\lambda^{T,n}\}$. Our construction starts with a compact FE scheme for Boolean NC^1 circuits $\mathsf{bFE} = (\mathsf{bFE.Setup}, \mathsf{bFE.Enc}, \mathsf{bFE.Dec})$ (as discussed in

Sect. 3.1, such a scheme can be constructed from LWE), and transforms it into $\mathsf{FE}^{T,n}$. The transformation makes uses of the following additional building blocks:

- a puncturable PRF $\mathsf{PPRF} = (\mathsf{PRF.Gen}, \mathsf{PRF.Punc}, \mathsf{F})$ for input domain $\{0,1\}^\lambda$.
- a randomized encoding scheme $\mathsf{RE} = (\mathsf{RE.Enc}, \mathsf{RE.Eval})$ in NC^0, and
- an IO scheme $i\mathcal{O}^{T,n}$ for circuit class $\{\mathcal{P}_\lambda^{T,n}\}$ consisting all circuits of the form $P[\lambda, T, n, mpk, i^*, K, m_1, y, m_0]$ defined in Fig. 1.

Let $\ell_{mpk}(\lambda)$ be the maximal length of master public keys of bFE, and $\ell_{key}(\lambda)$ that of punctured keys of PPRF respectively.

Construction of $\mathsf{FE}^{T,n}$. For any λ, $T = T(\lambda)$ and $n = n(\lambda)$, message m of length n and circuit C with size at most T and input length at most n. The FE scheme $\mathsf{FE}^{T,n} = (\mathsf{FE.Setup}, \mathsf{FE.KeyGen}, \mathsf{FE.Enc}, \mathsf{FE.Dec})$ proceeds as follow:

Setup $\mathsf{FE.Setup}(1^\lambda, T)$**:** Samples $(mpk, msk) \xleftarrow{\$} \mathsf{bFE.Setup}(1^\lambda, T')$, where T' is a time bound for circuits $\bar{C}$ defined below.

Key Generation $\mathsf{FE.KeyGen}(msk, C)$**:** Let $\bar{C}(m, i)$ be a circuit that on input m and $i \in [T]$ outputs the ith bit y_i of the output $y = C(m)$.
Sample $sk_{\bar{C}} \leftarrow \mathsf{bFE.KeyGen}(msk, \bar{C})$; output $sk = sk_{\bar{C}}$.

Encryption $\mathsf{FE.Enc}(mpk, m)$**:**

1. Sample $K \xleftarrow{\$} \mathsf{PRF.Gen}(1^\lambda)$, and puncture it at input 0, $K(-0) = \mathsf{PRF.Punc}(K, 0)$.
2. Sample $\tilde{P} \xleftarrow{\$} i\mathcal{O}^{T,n}(1^\lambda, P)$, where $P = P[\lambda, T, n, mpk, 0, K(-0), 0^\lambda, 0^\kappa, m]$ as defined in Fig. 1.
3. Output ciphertext $\xi = \tilde{P}$.

Decryption $\mathsf{FE.Dec}(sk, \xi)$**:**

1. Parse ξ as an obfuscated program $\tilde{P}$; for $i \in [T]$, compute $\hat{\Pi}_i = \tilde{P}(i)$.
2. For every $i \in [T]$, decode $c_i = \mathsf{RE.Eval}(\hat{\Pi}_i)$.
3. For $i \in [T]$, evaluate c_i with sk to obtain $y_i = \mathsf{bFE.Dec}(sk, c_i)$.
4. Output $y = y_1 || \cdots || y_T$.

It is clear that all algorithms above are PPT. Below, we first analyze the encryption efficiency of $\mathsf{FE}^{T,n}$ in Lemma 1 and then show its correctness and security in Lemma 2.

Lemma 1. *There exists a universal polynomial q, such that,*

$$\begin{aligned} if \quad & \mathsf{Time}_{i\mathcal{O}^{T,n}}(1^\lambda, C) \leq p^{T,n}(\lambda, |C|), \\ then, \quad & \mathsf{Time}_{\mathsf{FE.Enc}}(mpk, m) \leq p^{T,n}(\lambda, q(\lambda, n(\lambda), \log T(\lambda))) \end{aligned}$$

Proof. Towards this, we analyze the efficiency of each step of $\mathsf{FE.Enc}(mpk, m)$:

- It follows from the efficiency of PPRF that Step 1 of $\mathsf{FE.Enc}$ takes a fixed, universal, polynomial time $q_1(\lambda)$.

- It follows from the compactness of bFE that the size of the special purpose circuit P is bounded by and padded to a fixed, universal, polynomial $\eta(\lambda, n, \log T)$ (in Fig. 1).
- It follows from the efficiency of $i\mathcal{O}^{T,n}$ that the second step of encryption takes time $\mathsf{Time}_{i\mathcal{O}^{T,n}}(1^\lambda, P) = p^{T,n}(\lambda, \eta(\lambda, n, \log T))$.

Therefore, there exists a sufficiently large universal polynomial q w.r.t. which the lemma holds.

Lemma 2. *Let* bFE, PPRF, RE, $i\mathcal{O}^{T,n}$ *be defined as above.* FE *is correct and selectively secure for* NC^1 *circuits with* $n(\lambda)$*-bit inputs. Moreover, if all primitives are sub-exponentially secure, so is* FE.

We omit the of this lemma due to the lack of space.

3.3 Obtaining IO for P/poly

By the construction of FE scheme $\mathsf{FE}^{T,n}$ for $\mathsf{NC}^{1,T,n}$ in Sect. 3.2, we immediately have the following lemma:

Lemma 3. *Assume the same underlying primitives as Proposition* 5. *Suppose there is a family of IO schemes* $\{i\mathcal{O}^{T,n}\}$ *for* $\{\{\mathcal{P}_\lambda^{T,n}\}\}$ *with universal efficiency, that is,*

$$\mathsf{Time}_{i\mathcal{O}^{T,n}}(1^\lambda, C) \leq p(\lambda, |C|) \ , \ \textit{where } p \textit{ is a universal polynomial.}$$

Then, there is a family of FE schemes $\{\mathsf{FE}^{T,n}\}$ *for* $\{\mathsf{NC}^{1,T,n}\}$ *with the following encryption efficiency*

$$\mathsf{Time}_{\mathsf{FE.Enc}^{T,n}}(mpk, m) \leq p(\lambda, q(\lambda, n(\lambda), \log T(\lambda))) \ , \ \textit{where } q \textit{ is a universal polynomial.}$$

Clearly, this family of FE schemes $\{\mathsf{FE}^{T,n}\}$ gives a compact FE scheme for $\mathsf{NC}^1 = \{\mathsf{NC}^{1,T,n}\}$, and hence already implies IO for $\mathsf{P/poly}$ by Theorem 2 shown in [2,3,16]. We further examine their results, and observe that for any compact FE scheme, there exist some sufficiently large polynomials T^* and n^*, such that, the resulting IO for $\mathsf{P/poly}$ only uses the FE scheme to generate keys for NC^1 circuits with time bound $T^*(\lambda)$ and input length bound $n^*(\lambda)$. More precisely, we observe the more refined results of [2,3,16].

Theorem 4 (Refined Version of Theorem 2, Implicit in [2,3,16]). *Assume the existence of a sub-exponentially secure weak PRF in* NC^1, *and a (single-query, public-key) FE scheme for* $\mathsf{NC}^{1,T,n}$, *with encryption time bounded by* $T(\lambda)^{1-\varepsilon}$, *for sufficiently large polynomials* n *and* T. *Then, there exists an indistinguishability obfuscator for* $\mathsf{P/poly}$.

Fix any constant ε. By Lemma 3, for any two sufficiently large polynomials T^*, n^* that satisfy the following condition, the FE scheme FE^{T^*,n^*} constructed from $i\mathcal{O}^{T^*,n^*}$ satisfy the premise of Theorem 4, in particular, the encryption time is smaller than $T^*(\lambda)^{1-\varepsilon}$.

$$p(\lambda, q(\lambda, n^*(\lambda), \log T^*(\lambda))) \leq T^*(\lambda)^{1-\varepsilon}$$

Hence, by Theorem 4, $i\mathcal{O}^{T^*,n^*}$ suffices for building IO for $\mathsf{P/poly}$. This concludes Proposition 5.

4 Special-Purpose Circuits in Constant Degree

Assuming a constant-degree PRG, we show how to implement the special-purpose circuits in Fig. 1 using constant-degree arithmetic circuits.

Proposition 6. *Instantiated with a constant-degree PRG, the class of special-purpose circuits $\{\mathcal{P}_\lambda^{T,n}\}$ in Fig. 1 has universal arithmetic circuits $\{U_\lambda\}$ of constant degree* deg *and size $u(\lambda, n, \log T)$ for a universal polynomial u independent of T, n.*

Thus, the family of special-purpose circuit classes $\{\{\mathcal{P}_\lambda^{T,n}\}\}$ has constant-degree.

By Proposition 5, and the fact that all underlying primitives of the Proposition are implied by the hardness of LWE (see the discussion in Sect. 3.1), we obtain the following bootstrapping theorem.

Theorem 5 (Bootstrapping IO for Constant Degree Circuits). *Assume sub-exponential hardness of LWE, and the existence of a sub-exponentially secure constant-degree PRG. There exist a family of circuit classes of constant degree, such that, IO for that family with universal efficiency can be bootstrapped into IO for* $\mathsf{P/poly}$.

Overview. The class $\{\mathcal{P}_\lambda^{T,n}\}$ consists of special purpose circuits of the form $P[\lambda, T, n, \star_1](\star_2)$, where $T = T(\lambda)$ and $n = n(\lambda)$, where $\star_1$ represents the rest of the constants (including $mpk, i^*, K, m_<, \hat{\Pi}, m_>$) and $\star_2$ represents the input i. By viewing the rest of the constants as a description of the circuit, $U(\star_2, \star_1) = P[\lambda, T, n, \star_1](\star_2)$ can be viewed as the universal circuit of family $\mathcal{P}_\lambda^{T,n}$. Hence, towards the proposition, we only need to argue that $P[\lambda, T, n, \star](\star)$ can be implemented by an arithmetic circuit of constant degree and size $\mathrm{poly}(\lambda, n, \log T)$.

The computation of P can be broken down into three parts: (i) Evaluating the PPRF in Step 1, (ii) performing comparison between i and i^*, and (iii) depending on the outcome of comparison, potentially compute a RE encoding in NC^0. By definition of RE in NC^0, part (iii) has constant degree. The challenges lie in implementing Part (i) and (ii) in constant degree. More specifically,

Challenge 1: Let $b_{i,<}$, $b_{i,=}$, $b_{i,>}$ be decision bits indicating whether the input i is smaller than, equal to, or greater than the hardwired threshold i^*. Since $i \in [T]$ and $i^* \in \{0, \cdots, T+1\}$, their binary representation has logarithmic length $l = \lceil \log(T+2) \rceil$. Under binary representation, the straightforward way of computing these decision bits also requires logarithmic $O(l) = O(\log T)$ multiplications. E.g., equality testing can be done as $b_{i,=} = \prod_{j\in[l]}(1-(i_j - i_j^*)^2)$ (over any ring, where i_j and i_j^* are the jth bit of i and i^*).

Challenge 2: The state-of-the-art PPRF scheme [23] has an evaluation algorithm in NC^1 (assuming LWE), far from computable in constant degree. Even without the puncturing functionality, standard PRFs cannot be computed in constant degree, or even AC^0, since such functions are learnable [50].

Towards overcoming above challenges, we rely on the simple, but powerful, observation is that in our special-purpose circuits, the input i and threshold i^* both belong to a polynomial-sized set $\{0, \cdots, T+1\}$ (T by definition is polynomial in λ). This allows us to switch the representation of i and i^* from binary strings of length $O(\log T)$ to *strings of constant length over a polynomial-sized alphabet*, presented below.

New Input Representation. Instead of using binary alphabet, we represent the input $i \in [T]$, as well as the hardwired threshold $i^* \in \{0, \cdots, T+1\}$, using an alphabet Σ consisting of a polynomial number of vectors of length λ,

$$\Sigma = \{\mathbf{e}_0, \cdots, \mathbf{e}_\lambda\} \ , \tag{1}$$

where $\mathbf{e}_j$ for $j \in \{0, \cdots, \lambda\}$ contains 1 at position i and 0 everywhere else (in particular, $\mathbf{e}_0$ is the all 0 vector). Since T is polynomial in λ, there is a minimal constant, c such that, i (as well as i^*) can be divided into c blocks of length $\lfloor \log(\lambda+1) \rfloor$, denoted as $i = \mathsf{i}_1 || \mathsf{i}_2 || \cdots || \mathsf{i}_c$. Therefore, using alphabet Σ,

$$i \stackrel{\Sigma}{=} \mathbf{e}_{\mathsf{i}_1} || \cdots || \mathbf{e}_{\mathsf{i}_c} \ , \ \text{with length } |i|_\Sigma = c\lambda \ ,$$

where $a \stackrel{\Sigma}{=} b$ denote that b is the representation of a using alphabet Σ, and $|a|_\Sigma$ denote the number of bits needed in order to describe the representation over Σ.

We sketch how to resolve the two challenges, using the new representation.

Overcoming the First Challenge: consider the simple task of testing equality of one block, i_k and i_k^*—flag $b_{i,=}^k$ is set to 1 iff $\mathsf{i}_k = \mathsf{i}_k^*$. With the new representation, this equality can be tested by simply computing $b_{i,=}^k \mathbf{e}_{\mathsf{i}_k} \cdot \mathbf{e}_{\mathsf{i}_k^*}$ in degree two. Moreover, after testing equality of all blocks, which can be done in parallel, the equality between i and i^* can be computed as $b_{i,=} = \prod_{k\in[c]} b_{i,=}^k$ in constant degree c. Testing other relations, smaller than and greater than, between i and i^* can be performed similarly. See Sect. 4.1 for details.

Overcoming the Second Challenge: To circumvent the impossibility results, we leverage the fact that we only need to construct a PPRF for a special polynomial-sized domain σ^c. Assume the existence of a constant-degree PRG with polynomial stretch. The most natural idea is to construct a PPRF using the GGM PRF tree [42] as done in previous constructions of PPRF [19,21,45]. Clearly, the degree of the PPRF evaluation grows exponentially with the depth of the tree. Therefore, we can tolerate at most a constant depth. Fortunately, our domain is of polynomial size, and if we use a high-degree GGM tree, where each node has λ children, the depth is constant $O(c)$. However, an issue arises when using high-degree tree. Recall that the evaluation of the GGM PRF requires following the path leading to the leaf indexed by the input; at a particular node, the evaluator needs to choose the appropriate child in the next layer. When the tree has degree λ, choosing a child corresponds to the indexing function called the multiplexer $\mathsf{mux}(\boldsymbol{v}, j) = \boldsymbol{v}_j$, which has at least depth $\Omega(\log |\boldsymbol{v}|)$ when j is represented in binary. But, again thanks to our new input presentation $j \stackrel{\Sigma}{=} \mathbf{e}_j$, mux can be implemented as $\boldsymbol{v} \cdot \mathbf{e}_j$ in degree 2. See Sect. 4.2 for details on the PPRF.

Finally, we put all pieces together in Sect. 4.3. Our final implementation of special purpose circuits had degree of order $\exp(\log_\lambda(Tn))$.

4.1 Performing Comparisons in Constant-Degree

We show how to perform various comparison between i and i^* represented using the new input representation in constant degree. Towards this, we first show how to perform comparison over any single block of i and i^* in degree 2. For any $k \in [c]$, let $b^k_{i,<}$, $b^k_{i,=}$, $b^k_{i,>}$ be flags indicating whether the kth block of i, i_k, is smaller than, or equal to, or greater than the corresponding block of i^*_k, i^*_k; they can be computed as follows:

- $b^k_{i,=}$ can be computed as the inner product $b^k_{i,=} = \mathbf{e}_{\mathsf{i}_k} \cdot \mathbf{e}_{\mathsf{i}^*_k}$.
- $b^k_{i,<}$ can be computed as the inner product $b^k_{i,<} = \mathbf{e}_{\mathsf{i}_k} \cdot \mathbf{e}_{<\mathsf{i}^*_k}$, where $\mathbf{e}_{<\mathsf{i}^*_k}$ denote the vector that contains 1s in the first $\mathsf{i}^*_k - 1$ positions, and 0s in the rest.
- $b^k_{i,>}$ can be similarly computed as the inner product $b^k_{i,>} = \mathbf{e}_{\mathsf{i}_k} \cdot \mathbf{e}_{>\mathsf{i}^*_k}$, where $\mathbf{e}_{>\mathsf{i}^*_k}$ denote the vector that contains 0s in the first i^*_k positions, and 1s in the rest.

Next, performing comparison over entire i and i^* involves congregating the results of comparisons over individual blocks, which can be done using only a constant number $O(c)$ of multiplications as described in Fig. 2.

4.2 PRF Evaluation in Constant-Degree

The special purpose circuits require a PPRF function with input domain $\{0, \cdots, T\}$, key domain $\{0,1\}^\lambda$, and range $\{0,1\}^{L(\lambda)}$ for $L(\lambda)$ long enough to supply the random coins for bFE and RE; hence $L(\lambda) = \text{poly}(\lambda, n, \log T)$. The following lemma provides such a PPRF in constant degree.

Lemma 4. *Assume the existence of a degree-d PRG with $\lambda^{1+\varepsilon}$-stretch for some constant $d \in \mathbb{N}$ and $\varepsilon > 0$. For every polynomial D and L, there is a degree $\deg'$ PPRF scheme with input domain $\{0, \cdots, D(\lambda)\}$, key domain $\{0,1\}^\lambda$, and range $\{0,1\}^{L(\lambda)}$, where $\deg' \in \mathbb{N}$ is some constant depending on d, ε, D and L. Furthermore, if the underlying PRG is subexponentially secure, then so is the PPRF.*

Performing Comparisons $\mathsf{Compare}(i)$

Constants: a threshold $i^* \in \{0, \cdots, T+1\}$ represented as $i^* \overset{\Sigma}{=} (\mathbf{e}_{\mathsf{i}_k^*})_{k\in[c]}$ together with vectors $(\mathbf{e}_{<\mathsf{i}_k^*}, \mathbf{e}_{>\mathsf{i}_k^*})_{k\in[c]}$.

Input: an input $i \in [T]$ represented as $i \overset{\Sigma}{=} (\mathbf{e}_{\mathsf{i}_k^*})_{k\in[c]}$.

Procedure:

1. For every $k \in [c]$, compute $b_{i,=}^k = \mathbf{e}_{\mathsf{i}_k} \cdot \mathbf{e}_{\mathsf{i}_k^*}$, $b_{i,<}^k = \mathbf{e}_{\mathsf{i}_k} \cdot \mathbf{e}_{<\mathsf{i}_k^*}$, and $b_{i,>}^k = \mathbf{e}_{\mathsf{i}_k} \cdot \mathbf{e}_{>\mathsf{i}_k^*}$.
2. Do the following in parallel:

<u>*Testing* $i = i^*$</u> requires checking whether all blocks are equal. Therefore,

$$b_{i,=} = \prod_{k\in[c]} b_{i,=}^k . \tag{2}$$

<u>*Testing* $i < i^*$</u> requires checking whether one of the following cases occur: For some $k \in [c]$, the first $k-1$ blocks of i and i^* are equal, and the kth block of i is smaller than that of i^*. Therefore,

$$b_{i,<} = 1 - \prod_{k\in[c]} \left(1 - \left(\prod_{j<k\in[c]} b_{i,=}^j\right) \times b_{i,<}^k\right) . \tag{3}$$

<u>*Testing* $i > i^*$</u> requires checking whether one of the following cases occur: For some $k \in [c]$, the first $k-1$ blocks of i and i^* are equal, and the kth block of i is larger than that of i^*. Therefore,

$$b_{i,>} = 1 - \prod_{k\in[c]} \left(1 - \left(\prod_{j<k\in[c]} b_{i,=}^j\right) \times b_{i,>}^k\right) . \tag{4}$$

Fig. 2. Performing comparisons between i and i^* in constant degree.

Proof. Let PRG be the PRG in the premise. We first make the observation that it implies a constant-degree PRG scheme qPRG with quadratic stretch: If the stretch of PRG is already more than quadratic, (i.e., $1+\varepsilon \geq 2$) simply truncate the output to length λ^2. Otherwise, iteratively evaluate PRG for a sufficient number $I = \lceil 1/\log(1+\varepsilon)\rceil$ of times to expand the output to length λ^2, that is, $\mathsf{qPRG}(s) = \mathsf{PRG}^I(s)$. The degree of qPRG increases to d^I, still a constant, and the security of qPRG follows from standard argument. Below, we will view the output of qPRG as a vector $\mathbf{v} = \mathbf{v}[1], \cdots, \mathbf{v}[\lambda]$ of λ elements, each $\mathbf{v}[i]$ is a λ-bit binary string.

Furthermore, we observe that to get a PPRF with range $\{0,1\}^{L(\lambda)}$, it suffices to construct one with range $\{0,1\}^{\lambda}$, since one can always apply PRG iteratively to expand the output to $L(\lambda)$ as argued above.

Using qPRG, we now construct a PPRF scheme $\mathsf{PPRF} = (\mathsf{PRF.Gen}, \mathsf{PRF.Punc}, \mathsf{F})$ with λ-bit outputs. Since D is a polynomial, there is a minimal integer c such that for all $\lambda \in \mathbb{N}$, $D(\lambda) < \lambda^c$. Fix any security parameter λ, and $D = D(\lambda)$.

Our scheme PPRF with input domain $\{0, \cdots, D\}$ represents inputs under alphabet Σ (in Eq. (1)), or alternatively, the input domain is Σ^c.

Key Generation $\mathsf{PRF.Gen}(1^\lambda)$ samples a random λ-bit string $K \xleftarrow{\$} \{0,1\}^\lambda$.

Key Puncturing $\mathsf{PRF.Punc}(K, i^*)$ sets $K_0 = K$ and computes the following for every $k \in [c]$:
- $\mathbf{v}_k = \mathsf{qPRG}(K_{k-1})$.
- Let $\mathbf{v}_k[\neq i_k^*]$ be the vector identical to $\mathbf{v}_k$, but with the $i_k^{*\text{th}}$ element replaced with 0.

Set the punctured key as Note that the size of $K(-i^*)$ is bounded by $O(\lambda^2)$.

PRF Evaluation $\mathsf{F}(K(-i^*), i)$ is presented in Fig. 3. It is easy to verify that the algorithm indeed has constant-degree.

Efficiency and security: The only difference between the above scheme and the original constructions of PPRF based on GGM tree [19,21,45] is (i) the tree has degree λ instead of degree 2, and (ii) the inputs i and i^* are represented under Σ. For efficiency, the second difference has no impact, since under Σ, the representation of i and i^* are still of fixed polynomial size; the only effect the first difference has is that the punctured key consists of a λ-sized vector per layer of the tree, as opposed to 1 element per layer, but the size of the punctured key is still bounded by a fixed polynomial. For security, the same proof of [19,21,45] goes through even when the tree has higher degree; we omit details here.

PRF Evaluation $\mathsf{F}(K(-i^*), i)$

Input: A punctured key $K(-i^*) = (\mathbf{e}_{\mathsf{i}_k^*}, \mathbf{v}_k[\neq i_k^*])_{k\in[c]}$, and an input $i \in \{0, \cdots, D\}$ represented as $i \stackrel{\Sigma}{=} (\mathbf{e}_{\mathsf{i}_k})_{k\in[c]}$. By definition $i^* \neq i$.

Procedure:
1. For every $k \in [c]$, compute $b_{i,=}^k = \mathbf{e}_{\mathsf{i}_k^*} \cdot \mathbf{e}_{\mathsf{i}_k^*}$, which indicates whether the kth blocks i_k^* and i_k are equal.
2. For every $k \in [c]$, compute d_i^k indicating whether the following occurs: The first $k-1$ blocks of i and i^* are equal, but the kth block differs.
$$d_i^k = \Big(\prod_{j<k\in[c]} b_{i,=}^j\Big) \times \Big(1 - b_{i,=}^k\Big) .$$
3. For every $k \in [c]$, do:
   - Select the i_k^{th} element in $\mathbf{v}_k[\neq i_k^*]$, $K_k^k = \mathbf{v}_k[\neq i_k^*] \cdot \mathbf{e}_{\mathsf{i}_k}$.
   - For $j = k+1$ to c, compute $\mathbf{w}_j = \mathsf{qPRG}(K_{j-1}^k)$, $K_j^k = \mathbf{w}_j \cdot \mathbf{e}_{\mathsf{i}_j}$.
4. Compute the final output $y = \Sigma_{k\in[c]}(K_c^k \times d_i^k)$.

In the last two steps, multiplication between a string z and bit b yields $0^{|z|}$ if $b = 0$ and z if $b = 1$, and addition between two strings is bit-wise addition. Inner product between a vector of strings and a vector of bits are defined accordingly.

Fig. 3. Constant-degree PRF evaluation

Constant Degree Circuit $P[\lambda, T, n, mpk, i^*, K, m_<, \hat{\Pi}, m_>]$

Constants: $\lambda, T, mpk, m_<, \hat{\Pi}, m_>$ are defined as in Figure 1; $i^* \in \{0, \cdots, T+1\}$ is represented as $i^* \stackrel{\Sigma}{=} (\mathsf{e}_{\mathsf{i}^*_k})_{k\in[c]}$, together with vectors $(\mathsf{e}_{<\mathsf{i}^*_k}, \mathsf{e}_{>\mathsf{i}^*_k})_{k\in[c]}$; K is a punctured key of a constant degree PPRF PPRF.
Input: index $i \in [T]$ represented under Σ, that is, $(\mathsf{e}_{\mathsf{i}_k})_{k\in[c]}$.
Procedure:

1. $(R_i||R'_i) = \mathsf{F}(K, (\mathsf{e}_{\mathsf{i}_k})_{k\in[c]})$. (See Figure 3.)
2. $b_{i,<}, b_{i,=}, b_{i,>} = \mathsf{Compare}[i^*]((\mathsf{e}_{\mathsf{i}_k})_{k\in[c]})$. (See Figure 2.)
3. For $\star \in \{<, >\}$, compute
   $\hat{\Pi}_{i,\star} = \mathsf{RE.Enc}\left(1^\lambda,\ \mathsf{bFE.Enc},\ (mpk,\ m_\star||(\mathsf{e}_{\mathsf{i}_k})_{k\in[c]});\ R_i);\ R'_i\right)$.
4. Output $\hat{\Pi}_i = \hat{\Pi}_{i,<} \times b_{i,<} + \hat{\Pi} \times b_{i,=} + \hat{\Pi}_{i,>} \times b_{i,>}$.

Padding: The hardwired encoding $\hat{\Pi}$ is padded to be of length $\bar{\eta}'(\lambda, n, \log T)$, and the circuit is padded to be of size $\bar{\eta}(\lambda, n, \log T)$, for some polynomials $\bar{\eta}'$ and $\bar{\eta}$ set similarly as in Figure 1.

Fig. 4. Special-purpose circuit P in constant degree

4.3 Putting Pieces Together

Given the sub-routine Compare and a constant-degree PPRF scheme PPRF with domain $\{0, \cdots, T+1\}$ and appropriate output length $L(\lambda) = \text{poly}(\lambda, n, \log T)$, a constant-degree implementation the special-purpose circuits is presented in Fig. 4, where Step 1 and 2 evaluate the new functions Compare and PPRF respectively. The choice of which randomized encoding to output, depending on the outcome of comparisons, is made in Step 4 using simple addition and multiplication. Moreover, since the index i is now represented under Σ, each of its appearance in the special purpose circuit (e.g. in Step 3), as well as in the bootstrapping transformation of Proposition 5 is replaced with $(\mathsf{e}_{\mathsf{i}_1}, \cdots, \mathsf{e}_{\mathsf{i}_c})$. Since this representation also has a fixed polynomial size (bounded by λ^2 for sufficiently large λ), all constructions and proofs remain intact.

It is easy to see that the implementation is correct, and furthermore the circuit size of this implementation is still $u(\lambda, n, \log T)$ for some universal polynomial u independent of T, n: In Step 1, the evaluation of the PPRF takes fixed (universal) polynomial time poly(λ), and so is the evaluation of function Compare in Step 2. The run-time of Step 3 and 4 is determined by that of RE and bFE as before, which again is bounded by a fixed (universal) polynomial poly($\lambda, n, \log T$). Therefore, the worst-case run-time and hence circuit size is bounded by $u(\lambda, n, \log T)$, for some universal polynomial u.

5 IO for Special-Purpose Circuits in Ideal Model

In this section, we construct IO for our special-purpose circuits in ideal graded encoding model. Due to the lack of space, we provide only an overview of our construction. We refer the reader to the full version for more details.

Overview. Our goal is to construct IO for $\{\{\mathcal{P}_\lambda^{T,n}\}\}$ with universal efficiency in constant degree ideal graded encoding model. Constructions of IO for NC^1 in the literature follow two approaches: Either obfuscate the branching programs of circuits [12,22,40,59] or directly obfuscate circuits [4,7,62]. The first approach seems to inherently require high-degree graded encodings, since the evaluation of a branching program has degree proportional to its length. This limitation does not hold for the second approach, but known constructions still require polynomial degree. We base our construction on the construction of IO for NC^1 by Applebaum and Brakerski [7] (shorthand AB-IO) in composite order ideal graded encoding model, and use new ideas to reduce the degree of graded encodings.

Review of Applebaum-Brakerski IO Scheme: Let P be a program with universal arithmetic circuit $U(x, P)$. Consider the following simple idea of encoding every bit of P and both values 0 and 1 for each input bit $i \in [n]$, that is, $\hat{P} = \{[b]_{\mathbf{v}_{i,b}}\}_{i\in[n],b\in\{0,1\}}, \{[P_i]_{\mathbf{v}_{i+n}}\}_{i\in[m]}$. Then, given an input x, an evaluator can simply pick the encodings $\{[x_i]_{\mathbf{v}_{i,x_i}}\}_{i\in[n]}$, and homomorphically evaluate U on encodings of (bits of) x and P to obtain an encoding of $U(x, P)$, which can then be learned by zero-testing. This simple idea does not go far. We mention several key issues and their solutions.

1. To prevent an adversary from using inconsistent values for the same input bit at different steps of the evaluation, AB-IO follows the standard solution of "straddling sets" [12], and uses a set of special levels, so that, if both $Z_{i,0} = [0]_{\mathbf{v}_{i,0}}$ and $Z_{i,1} = [1]_{\mathbf{v}_{i,1}}$ for some input bit i are used, the resulting encoding never reaches the zero testing level $\mathbf{v}_{zt}$. To see this, consider a simplified example: Set $\mathbf{v}_{i,0} = (1, 0, 1)$ and $\mathbf{v}_{i,1} = (0, 1, 1)$, and provide two additional encodings $\hat{Z}_{i,b}$ of random values under levels $\hat{\mathbf{v}}_{i,0} = (0, d, 0)$ and $\mathbf{v}_{i,1} = (d, 0, 0)$; the only way to reach level (d, d, d) is to use $Z_{i,b}$ consistently, followed by multiplication with $\hat{Z}_{i,b}$. Note that doing this for every input already requires degree n multiplication.
2. Graded encodings only support addition in the same levels. Since different input and program bits are encoded under different levels, homomorphic evaluation of U cannot be done. To resolve this, AB-IO uses El-Gamal encoding, under which a value w is represented as $(r, rw) \xleftarrow{\$} \mathsf{EG}(w)$ with a random r. Encodings of El-Gamal encodings of w_1 and w_2, $(R_1 = [r_1]_{\mathbf{v}_1}, Z_1 = [r_1 w_1]_{\mathbf{v}_1})$ and $(R_2 = [r_2]_{\mathbf{v}_2}, Z_2 = [r_2 w_2]_{\mathbf{v}_2})$ can be "added" using an addition gadget $\oplus$ that does $(R_1 R_2 = [r_1 r_2]_{\mathbf{v}_1+\mathbf{v}_2}, Z_1 R_2 + Z_2 R_1 = [r_1 r_2 (w_1 + w_2)]_{\mathbf{v}_1+\mathbf{v}_2})$, even if they are under different levels. Note that the new gadget, however, turns every addition in U into multiplications (and additions) in the homomorphic evaluation, which now has much higher degree, up to 2^{depth}, than U.
3. Point 1 ensures that an adversary must use an input x consistently, but, it can still deviate from evaluating U. AB-IO uses an information theoretic authentication method to prevent this. It samples a random value y_i for each input wire, and computes $\bar{y} = U(y_1, \cdots, y_{n+m})$. The idea is to use the structure of the composite order ring to "bind" the program and input bits with

their corresponding y values, for example, instead of encoding $\mathsf{EG}(P_i)$, encode $\mathsf{EG}(w_{n+i})$ where $w_{n+i} = (P_i, y_{n+i})$. Therefore, whichever computation the adversary performs over x and P, the same is performed over $y_1, \cdots, y_{n+m}$. An honest evaluation yields encodings of $\mathsf{EG}((U(x,P), \bar{y}))$. By additionally releasing encodings of $\mathsf{EG}((1, \bar{y}))$, the output $U(x, P)$ can be learned by first subtracting the encodings and zero-test. Moreover, deviating from computing U leads to encodings of $\mathsf{EG}(Y(x,P), Y(y_1, \cdots, y_{n+m}))$ with some $Y \neq U$, and the value $Y(y_1, \cdots, y_{n+m})$ cannot be eliminated to allow zero-testing $Y(x, P)$, which hence remains hidden.

Due to Point 1 and 2, AB-IO requires the graded encodings to support degree-$(n2^{depth})$ computations.

Towards Using Constant-Degree Graded Encodings, we modify AB-IO as follows:

1. We use the same method as AB-IO to prevent an adversary from using inconsistent input values, but we cannot afford to do that for every input bit. Instead, recall that the domain of our special purpose circuits is Σ^c, where Σ has size λ. We view each symbol $x^1, \cdots, x^c$ (though described as a λ-bit string) as a "single input", and apply the straddling sets of AB-IO for each input symbol. (Ignore the El-Gamal encoding and the y-values temporarily.) For the ith symbol, release for every possible value $s \in \Sigma$, encoding $Z^i_s = [s]_{\mathbf{v}^i_s}$, and $\hat{Z}^i_s$ of a random value under set $\hat{\mathbf{v}}^i_s$. Consider a simplified example: Set $\mathbf{v}^i_s = (0 \cdots 0, 1, 0 \cdots 0, 1)$ with 1 at position s and $\lambda + 1$, and $\hat{\mathbf{v}}^i_s = (d \cdots d, 0, d \cdots d, 0)$ correspondingly. (As in Point 1 above,) the only way to reach $(d, \cdots, d)$ is using Z^i_s for some s consistently followed by a multiplication with $\hat{Z}^i_s$. The actual encoding is more complicated as s is described as a λ-bit string $s_1, \cdots, s_\lambda$, and each bit needs to be encoded separately $\boldsymbol{Z}^i_s = \{[s_j]_{\mathbf{v}^i_s}\}_j$.
2. Informally speaking, the addition gadget $\oplus$ of AB-IO turns addition over encodings under different levels into multiplication; to reduce the degree of homomorphic evaluation, we want to have as many additions under the same levels as possible. In particular, encodings of form $(R_1 = [r]_{\mathbf{v}}, Z_1 = [rw_1]_{\mathbf{v}})$ and $(R_2 = [r]_{\mathbf{v}}, Z_2 = [rw_2]_{\mathbf{v}})$ can be directly "added" $(R_1 = [r]_{\mathbf{v}}, Z_1 + Z_2 = [r(w_1 + w_2)]_{\mathbf{v}})$—we call this the constrained addition gadget $\tilde{\oplus}$. Fortunately, thanks to the special domain Σ^c, encodings for different bits of an input symbol $\boldsymbol{Z}^i_s$ have the same level $\mathbf{v}^i_s$. To allow for using $\tilde{\oplus}$, we further let their El-Gamal encodings share the same randomness r^i_s, that is, $R^i_s = [r^i_s]_{\mathbf{v}^i_s}$ and $\boldsymbol{Z}^i_s = \{[r^i_s s_j]_{\mathbf{v}^i_s}\}_j$. Now addition of different bits in the same input symbol can be performed using only homomorphic addition.

   More generally, we assign "types" to input wires—all wires describing P have one type, and these describing x^i for each i has another. Encodings for input wires of the same type share the same level and El-Gamal randomness, and can be added using $\tilde{\oplus}$ for "free", whereas addition across different types is done using $\oplus$ as in AB-IO, involving homomorphic multiplication. We further assign types to all wires in U recursively: When the incoming wires of an addition gate in U have the same types, $\tilde{\oplus}$ can be applied and its outgoing wire keeps the same type; in all other cases, homomorphic multiplication is required, and the types of the incoming wires add up. Careful examination reveals that the degree of homomorphic evaluation is proportional to the 1-norm of the output wire type, which we call the *type-degree of* U.

Combining the above ideas, we obtain a construction of IO for general circuit class in ideal model where the degree of the graded encodings is $O(td + c)$, proportional to the type degree td and the number of input type c of the circuit class; we say such a construction is *type degree preserving.*

For certain circuits, their type-degrees are much smaller than 2^{depth}. For example, our special purpose circuits, instantiated with a constant-degree PRG, have a constant type degree td, and hence constant degree graded encodings suffice. More generally, when PRG has degree $d(\lambda)$, the type degree of the special purpose circuits is polynomial in $d(\lambda)$.

Our actual IO scheme is more complicated than sketched above due to (1) it is based on the robust obfuscator in [7] as opposed to the simple obfuscator described above; like the robust obfuscator of [7], our IO scheme has the property that a generic attacker can only generate encodings of 0 at the zero-testing level. Such a construction can work with graded encoding schemes with unique encodings and seems to be more secure in face of zeroizing attacks on graded encodings. In particular, [30] showed that a simplified version of the simple obfuscator of [7] can be attacked. (2) Our IO scheme directly obfuscates non-Boolean circuits. Previous constructions of IO for NC^1 considers only Boolean circuits; this is w.l.o.g. as a NC^1 circuit C can be turned into a Boolean one $\bar{C}(x, i) = C(x)_i$, still in NC^1. But, when aiming at type-degree preserving constructions of IO, we cannot use this trick, as $\bar{C}$ may have much higher type degree than C.

Acknowledgements. The author would like to thank Ran Canetti, Shafi Goldwasser, Shai Halevi, Shachar Lovett, Rafael Pass, and Vinod Vaikuntanathan for delightful and insightful discussions. Moreover, the author would like to give special thanks to Benny Applebaum and Stefano Tessaro for many helpful inputs.

References

1. Alekhnovich, M.: More on average case vs approximation complexity. In: Proceedings of the 44th Symposium on Foundations of Computer Science (FOCS 2003), 11–14 October 2003, Cambridge, MA, USA, pp. 298–307 (2003)
2. Ananth, P., Brakerski, Z., Segev, G., Vaikuntanathan, V.: From selective to adaptive security in functional encryption. In: Gennaro, R., Robshaw, M. (eds.) CRYPTO 2015, Part II. LNCS, vol. 9216, pp. 657–677. Springer, Heidelberg (2015)
3. Ananth, P., Jain, A.: Indistinguishability obfuscation from compact functional encryption. In: Gennaro, R., Robshaw, M.J.B. (eds.) CRYPTO 2015, Part I. LNCS, vol. 9215, pp. 308–326. Springer, Heidelberg (2015)
4. Ananth, P.V., Gupta, D., Ishai, Y., Sahai, A.: Optimizing obfuscation: avoiding Barrington's theorem. In: Ahn, G.-J., Yung, M., Li, N. (eds.) ACM CCS 2014, Scottsdale, AZ, USA, 3–7 November 2014, pp. 646–658. ACM Press (2014)

5. Applebaum, B.: Bootstrapping obfuscators via fast pseudorandom functions. In: Sarkar, P., Iwata, T. (eds.) ASIACRYPT 2014, Part II. LNCS, vol. 8874, pp. 162–172. Springer, Heidelberg (2014)
6. Applebaum, B.: Cryptography in Constant Parallel Time. Information Security and Cryptography. Springer, Heidelberg (2014)
7. Applebaum, B., Brakerski, Z.: Obfuscating circuits via composite-order graded encoding. In: Dodis, Y., Nielsen, J.B. (eds.) TCC 2015, Part II. LNCS, vol. 9015, pp. 528–556. Springer, Heidelberg (2015)
8. Applebaum, B., Ishai, Y., Kushilevitz, E.: Cryptography in NC^0. In: FOCS, pp. 166–175 (2004)
9. Applebaum, B., Ishai, Y., Kushilevitz, E.: On pseudorandom generators with linear stretch in NC^0. Comput. Complex. **17**(1), 38–69 (2008)
10. Applebaum, B., Lovett, S.: Algebraic attacks against random local functions and their countermeasures. In: Electronic Colloquium on Computational Complexity (ECCC), vol. 22, p. 172 (2015)
11. Banerjee, A., Peikert, C., Rosen, A.: Pseudorandom functions and lattices. In: Pointcheval, D., Johansson, T. (eds.) EUROCRYPT 2012. LNCS, vol. 7237, pp. 719–737. Springer, Heidelberg (2012)
12. Barak, B., Garg, S., Kalai, Y.T., Paneth, O., Sahai, A.: Protecting obfuscation against algebraic attacks. In: Nguyen, P.Q., Oswald, E. (eds.) EUROCRYPT 2014. LNCS, vol. 8441, pp. 221–238. Springer, Heidelberg (2014)
13. Barak, B., Goldreich, O., Impagliazzo, R., Rudich, S., Sahai, A., Vadhan, S.P., Yang, K.: On the (im)possibility of obfuscating programs. In: Kilian, J. (ed.) CRYPTO 2001. LNCS, vol. 2139, pp. 1–18. Springer, Heidelberg (2001)
14. Bitansky, N., Garg, S., Lin, H., Pass, R., Telang, S.: Succinct randomized encodings and their applications. In: Servedio, R.A., Rubinfeld, R. (eds.) 47th ACM STOC, Portland, OR, USA, June 14–17 2015, pp. 439–448. ACM Press (2015)
15. Bitansky, N., Paneth, O., Wichs, D.: Perfect structure on the edge of chaos - trapdoor permutations from indistinguishability obfuscation. In: Kushilevitz, E., et al. (eds.) TCC 2016-A. LNCS, vol. 9562, pp. 474–502. Springer, Heidelberg (2016). doi:10.1007/978-3-662-49096-9_20
16. Bitansky, N., Vaikuntanathan, V.: Indistinguishability obfuscation from functional encryption. In: IEEE 56th Annual Symposium on Foundations of Computer Science, FOCS 2015, Berkeley, CA, USA, 17–20 October 2015, pp. 171–190 (2015)
17. Bitansky, N., Vaikuntanathan, V.: Indistinguishability obfuscation: from approximate to exact. In: Kushilevitz, E., et al. (eds.) TCC 2016-A. LNCS, vol. 9562, pp. 67–95. Springer, Heidelberg (2016). doi:10.1007/978-3-662-49096-9_4
18. Bogdanov, A., Qiao, Y.: On the security of goldreich's one-way function. Comput. Complex. **21**(1), 83–127 (2012)
19. Boneh, D., Waters, B.: Constrained pseudorandom functions and their applications. In: Sako, K., Sarkar, P. (eds.) ASIACRYPT 2013, Part II. LNCS, vol. 8270, pp. 280–300. Springer, Heidelberg (2013)
20. Boneh, D., Wu, D.J., Zimmerman, J.: Immunizing multilinear maps against zeroizing attacks. Cryptology ePrint Archive, Report 2014/930 (2014). http://eprint.iacr.org/2014/930
21. Boyle, E., Goldwasser, S., Ivan, I.: Functional signatures and pseudorandom functions. In: PKC, pp. 501–519 (2014)
22. Brakerski, Z., Rothblum, G.N.: Virtual black-box obfuscation for all circuits via generic graded encoding. In: Lindell, Y. (ed.) TCC 2014. LNCS, vol. 8349, pp. 1–25. Springer, Heidelberg (2014)

23. Brakerski, Z., Vaikuntanathan, V.: Constrained key-homomorphic PRFs from standard lattice assumptions - or: how to secretly embed a circuit in your PRF. In: Dodis, Y., Nielsen, J.B. (eds.) TCC 2015, Part II. LNCS, vol. 9015, pp. 1–30. Springer, Heidelberg (2015)
24. Canetti, R., Goldwasser, S., Poburinnaya, O.: Adaptively secure two-party computation from indistinguishability obfuscation. In: Dodis, Y., Nielsen, J.B. (eds.) TCC 2015, Part II. LNCS, vol. 9015, pp. 557–585. Springer, Heidelberg (2015)
25. Canetti, R., Holmgren, J., Jain, A., Vaikuntanathan, V.: Succinct garbling and indistinguishability obfuscation for RAM programs. In: Servedio, R.A., Rubinfeld, R. (eds.) 47th ACM STOC, Portland, OR, USA, June 14–17 2015, pp. 429–437. ACM Press (2015)
26. Canetti, R., Lin, H., Tessaro, S., Vaikuntanathan, V.: Obfuscation of probabilistic circuits and applications. In: Dodis, Y., Nielsen, J.B. (eds.) TCC 2015, Part II. LNCS, vol. 9015, pp. 468–497. Springer, Heidelberg (2015)
27. Cheon, J.H., Han, K., Lee, C., Ryu, H., Stehlé, D.: Cryptanalysis of the multilinear map over the integers. In: Oswald, E., Fischlin, M. (eds.) EUROCRYPT 2015. LNCS, vol. 9056, pp. 3–12. Springer, Heidelberg (2015)
28. Chung, K., Lin, H., Pass, R.: Constant-round concurrent zero-knowledge from indistinguishability obfuscation. In: Gennaro, R., Robshaw, M. (eds.) CRYPTO 2015, Part I. LNCS, vol. 9215, pp. 287–307. Springer, Berlin (2015)
29. Cook, J., Etesami, O., Miller, R., Trevisan, L.: Goldreich's one-way function candidate and myopic backtracking algorithms. In: Reingold, O. (ed.) TCC 2009. LNCS, vol. 5444, pp. 521–538. Springer, Heidelberg (2009)
30. Coron, J.-S., et al.: Zeroizing without low-level zeroes: new MMAP attacks and their limitations. In: Gennaro, R., Robshaw, M.J.B. (eds.) CRYPTO 2015, Part I. LNCS, vol. 9215, pp. 247–266. Springer, Heidelberg (2015)
31. Coron, J.-S., Lepoint, T., Tibouchi, M.: Practical multilinear maps over the integers. In: Canetti, R., Garay, J.A. (eds.) CRYPTO 2013, Part I. LNCS, vol. 8042, pp. 476–493. Springer, Heidelberg (2013)
32. Coron, J.-S., Lepoint, T., Tibouchi, M.: New multilinear maps over the integers. In: Gennaro, R., Robshaw, M.J.B. (eds.) CRYPTO 2015, Part I. LNCS, vol. 9215, pp. 267–286. Springer, Heidelberg (2015)
33. Cryan, M., Miltersen, P.B.: On pseudorandom generators in NC0. In: Sgall, J., Pultr, A., Kolman, P. (eds.) MFCS 2001. LNCS, vol. 2136, p. 272. Springer, Heidelberg (2001)
34. Dachman-Soled, D., Katz, J., Rao, V.: Adaptively secure, universally composable, multiparty computation in constant rounds. In: Dodis, Y., Nielsen, J.B. (eds.) TCC 2015, Part II. LNCS, vol. 9015, pp. 586–613. Springer, Heidelberg (2015)
35. Garg, S., Gentry, C., Halevi, S.: Candidate multilinear maps from ideal lattices. In: Johansson, T., Nguyen, P.Q. (eds.) EUROCRYPT 2013. LNCS, vol. 7881, pp. 1–17. Springer, Heidelberg (2013)
36. Garg, S., Gentry, C., Halevi, S., Raykova, M., Sahai, A., Waters, B.: Candidate indistinguishability obfuscation and functional encryption for all circuits. In: 54th FOCS, Berkeley, CA, 26–29 October 2013, pp. 40–49. IEEE Computer Society Press (2013)
37. Garg, S., Polychroniadou, A.: Two-round adaptively secure MPC from indistinguishability obfuscation. In: Dodis, Y., Nielsen, J.B. (eds.) TCC 2015, Part II. LNCS, vol. 9015, pp. 614–637. Springer, Heidelberg (2015)
38. Gentry, C., Gorbunov, S., Halevi, S.: Graph-induced multilinear maps from lattices. In: Dodis, Y., Nielsen, J.B. (eds.) TCC 2015, Part II. LNCS, vol. 9015, pp. 498–527. Springer, Heidelberg (2015)

39. Gentry, C., Halevi, S., Maji, H.K., Sahai, A.: Zeroizing without zeroes: cryptanalyzing multilinear maps without encodings of zero. Cryptology ePrint Archive, Report 2014/929 (2014). http://eprint.iacr.org/2014/929
40. Gentry, C., Lewko, A., Sahai, A., Waters, B.: Indistinguishability obfuscation from the multilinear subgroup elimination assumption. Cryptology ePrint Archive, Report 2014/309 (2014). http://eprint.iacr.org/2014/309
41. Goldreich, O.: Candidate one-way functions based on expander graphs. Cryptology ePrint Archive, Report 2000/063 (2000). http://eprint.iacr.org/2000/063
42. Goldreich, O., Goldwasser, S., Micali, S.: How to construct random functions. J. ACM **33**(4), 792–807 (1986)
43. Goldwasser, S., Kalai, Y.T., Popa, R.A., Vaikuntanathan, V., Zeldovich, N.: Reusable garbled circuits and succinct functional encryption. In: Boneh, D., Roughgarden, T., Feigenbaum, J. (eds.) 45th ACM STOC, Palo Alto, CA, USA, 1–4 June 2013, pp. 555–564. ACM Press (2013)
44. Ishai, Y., Kushilevitz, E.: Perfect constant-round secure computation via perfect randomizing polynomials. In: Widmayer, P., Triguero, F., Morales, R., Hennessy, M., Eidenbenz, S., Conejo, R. (eds.) ICALP 2002. LNCS, vol. 2380, pp. 244–256. Springer, Heidelberg (2002)
45. Kiayias, A., Papadopoulos, S., Triandopoulos, N., Zacharias, T.: Delegatable pseudorandom functions and applications. In: CCS, pp. 669–684 (2013)
46. Komargodski, I., Moran, T., Naor, M., Pass, R., Rosen, A., Yogev, E.: One-way functions and (im)perfect obfuscation (2014)
47. Koppula, V., Lewko, A.B., Waters, B.: Indistinguishability obfuscation for turing machines with unbounded memory. In: Servedio, R.A., Rubinfeld, R. (eds.) 47th ACM STOC, Portland, OR, USA, 14–17 June 2015, pp. 419–428. ACM Press (2015)
48. Lin, H., Pass, R., Seth, K., Telang, S.: Output-compressing randomized encodings and applications. IACR Cryptology ePrint Archive 2015:720 (2015)
49. Lin, H., Pass, R., Seth, K., Telang, S.: Indistinguishability obfuscation with non-trivial efficiency. In: Proceedings of the Public-Key Cryptography - PKC 2016 - 19th IACR International Conference on Practice and Theory in Public-Key Cryptography, Taipei, Taiwan, 6–9 March 2016, Part II, pp. 447–462 (2016)
50. Linial, N., Mansour, Y., Nisan, N.: Constant depth circuits, fourier transform, and learnability. In: 30th FOCS, pp. 574–579 (1989)
51. Mahmoody, M., Mohammed, A., Nematihaji, S.: More on impossibility of virtual black-box obfuscation in idealized models. Cryptology ePrint Archive, Report 2015/632 (2015). http://eprint.iacr.org/2015/632
52. Mossel, E., Shpilka, A., Trevisan, L.: On e-biased generators in NC0. In: Proceedings of the 44th Symposium on Foundations of Computer Science (FOCS 2003), Cambridge, MA, USA, 11–14 October 2003, pp. 136–145 (2003)
53. Naor, M., Reingold, O.: Synthesizers and their application to the parallel construction of pseudo-random functions. In: 36th FOCS, Milwaukee, Wisconsin, 23–25 October 1995, pp. 170–181. IEEE Computer Society Press (1995)
54. Naor, M., Reingold, O.: Number-theoretic constructions of efficient pseudo-random functions. In: 38th FOCS, Miami Beach, Florida, 19–22 October 1997, pp. 458–467. IEEE Computer Society Press (1997)
55. Naor, M., Reingold, O., Rosen, A.: Pseudo-random functions and factoring (extended abstract). In: 32nd ACM STOC, Portland, Oregon, USA, 21–23 May 2000, pp. 11–20. ACM Press (2000)
56. Nisan, N., Szegedy, M.: On the degree of boolean functions as real polynomials. Comput. Complex. **4**, 301–313 (1994)

57. O'Donnell, R., Witmer, D.: Goldreich's PRG: evidence for near-optimal polynomial stretch. In: IEEE 29th Conference on Computational Complexity, CCC 2014, Vancouver, BC, Canada, 11–13 June 2014, pp. 1–12 (2014)
58. Pass, R., abhi shelat.: Impossibility of VBB obfuscation with ideal constant-degree graded encodings. Cryptology ePrint Archive, Report 2015/383 (2015). http://eprint.iacr.org/2015/383
59. Pass, R., Seth, K., Telang, S.: Indistinguishability obfuscation from semantically-secure multilinear encodings. In: Garay, J.A., Gennaro, R. (eds.) CRYPTO 2014, Part I. LNCS, vol. 8616, pp. 500–517. Springer, Heidelberg (2014)
60. Sahai, A., Waters, B.: How to use indistinguishability obfuscation: deniable encryption, and more. In: Shmoys, D.B. (ed.) 46th ACM STOC, New York, NY, USA, 31 May – 3 June 2014, pp. 475–484. ACM Press (2014)
61. Sahai, A., Waters, B.: How to use indistinguishability obfuscation: deniable encryption, and more. In: Proceedings of the STOC 2014 (2014)
62. Zimmerman, J.: How to obfuscate programs directly. In: Oswald, E., Fischlin, M. (eds.) EUROCRYPT 2015. LNCS, vol. 9057, pp. 439–467. Springer, Heidelberg (2015)

Essentially Optimal Robust Secret Sharing with Maximal Corruptions

Allison Bishop[1], Valerio Pastro[1](✉), Rajmohan Rajaraman[2], and Daniel Wichs[2](✉)

[1] Columbia University, New York City, USA
{allison,valerio}@cs.columbia.edu
[2] Northeastern University, Boston, USA
{rraj,wichs}@ccs.neu.edu

Abstract. In a t-out-of-n *robust secret sharing scheme*, a secret message is shared among n parties who can reconstruct the message by combining their shares. An adversary can adaptively corrupt up to t of the parties, get their shares, and modify them arbitrarily. The scheme should satisfy *privacy*, meaning that the adversary cannot learn anything about the shared message, and *robustness*, meaning that the adversary cannot cause the reconstruction procedure to output an incorrect message. Such schemes are only possible in the case of an honest majority, and here we focus on unconditional security in the maximal corruption setting where $n = 2t + 1$.

In this scenario, to share an m-bit message with a reconstruction failure probability of at most 2^{-k}, a known lower-bound shows that the share size must be at least $m + k$ bits. On the other hand, all prior constructions have share size that scales linearly with the number of parties n, and the prior state-of-the-art scheme due to Cevallos et al. (EUROCRYPT '12) achieves $m + \widetilde{O}(k + n)$.

In this work, we construct the first robust secret sharing scheme in the maximal corruption setting with $n = 2t + 1$, that avoids the linear dependence between share size and the number of parties n. In particular, we get a share size of only $m + \widetilde{O}(k)$ bits. Our scheme is computationally efficient and relies on approximation algorithms for the minimum graph bisection problem.

1 Introduction

Secret sharing, originally introduced by Shamir [Sha79] and Blakely [Bla79], is a central cryptographic primitive at the heart of a wide variety of applications, including secure multiparty computation, secure storage, secure message transmission, and threshold cryptography. The functionality of secret sharing allows a dealer to split a secret message into shares that are then distributed to n parties. Any authorized subset of parties can reconstruct the secret reliably from their shares, while unauthorized subsets of parties cannot learn anything about the secret from their joint shares. In particular, a t-out-of-n *threshold* secret sharing

M. Fischlin and J.-S. Coron (Eds.): EUROCRYPT 2016, Part I, LNCS 9665, pp. 58–86, 2016.
DOI: 10.1007/978-3-662-49890-3_3

scheme requires that any t shares reveal no information about the secret, while any subset of $t + 1$ shares can be used to reconstruct the secret.

Many works (e.g. [RB89, CSV93, BS97, CDF01, CFOR12, LP14, CDD+15, Che15]) consider a stronger notion of secret sharing called *robust secret sharing* (RSS). Robustness requires that, even if an adversary can replace shares of corrupted parties by maliciously chosen values, the parties can still reconstruct the true secret. (with high probability). secure storage and message transmission. In particular, we consider a computationally unbounded adversary who maliciously (and adaptively) corrupts t out of n of the parties and learns their shares. After corrupting the t parties, the adversary can adaptively modify their shares and replace them with arbitrary values. The reconstruction algorithm is given the shares of all n parties and we require that it recovers the original secret. Note that robustness requires the reconstruction to work given all n shares of which t contain "errors" while threshold reconstruction is given $t + 1$ correct shares, meaning that $n - t - 1$ shares are "erasures". When $n = 2t + 1$, robustness is therefore a strictly stronger requirement than threshold reconstruction (but in other settings this is not the case).

Known Lower Bounds. It is known that robust secret sharing can only be achieved with an honest majority, meaning $t < n/2$. Moreover, for t in the range $n/3 \leq t < n/2$, we cannot achieve perfect robustness, meaning that we must allow at least a small (negligible) failure probability for reconstruction [Cev11]. Furthermore, in the maximal corruption setting with $n = 2t + 1$ parties, any robust secret sharing scheme for m-bit messages with failure probability 2^{-k} must have a share size that exceeds $m + k$ bits [CSV93, LP14].

Prior Constructions. On the positive side, several prior works show how to construct robust sharing schemes in the maximal corruption setting with $n = 2t + 1$ parties. The first such scheme was described in the work of Rabin and Ben-Or [RB89] with a share size of $m + \widetilde{O}(nk)$ bits. Cramer, Damgård and Fehr [CDF01] showed how to improve this to $m + \widetilde{O}(k + n)$ bits, using what later became known as algebraic-manipulation detection (AMD) codes [CDF+08], but at the cost of having an inefficient reconstruction procedure. Cevallos et al. [CFOR12] then presented an efficient scheme with share size $m + \widetilde{O}(k + n)$.

Other Related Work. Two recent works [CDD+15, Che15] study robust secret sharing in the setting where the number of corruptions is below the maximal threshold by some constant fraction; i.e., $t = (1/2 - \delta)n$ for some constant $\delta > 0$. In this setting, robustness does not necessarily imply threshold reconstructability from $t + 1$ correct shares (but only from $(1/2 + \delta)n$ correct shares). This is often called a *ramp* setting, where there is a gap between the privacy threshold t and the reconstructability threshold. The above works show that it is possible to achieve robustness in this setting with share size of only $O(1)$ bits, when n is sufficiently large in relation to k, m. The work of [Che15] also considers a setting that separately requires robustness *and* threshold reconstruction from $t + 1$ correct shares, and gives a scheme with share size $m + \widetilde{O}(k)$. However, we notice that security in this setting could always be achieved with a *generic construction*, by adding a standard (non-robust) threshold secret sharing scheme on top of a robust scheme – when given exactly $t+1$ shares, the reconstruction uses the threshold scheme, and otherwise it uses the robust scheme. This adds m additional bits to the share size of the robust scheme and, in particular, when n is sufficiently large, the share size would be $m + O(1)$ bits[1].

The main technique used to get robustness in [CDD+15, Che15] is to compose list-decodable codes with a special privacy property together with AMD codes from [CDF+08]. Unfortunately, this technique appears to crucially fail in the setting of $n = 2t+1$ parties. In this setting, the parameters of the list-decodable codes either force a large alphabet or an exponential list size. The latter in turn forces us to use an AMD code with large overhead. In either case the overhead on the share size appears to be at least $O(n)$ bits.

Another related work of [LP14] considers a relaxation of robustness to a setting of *local adversaries*, where each corrupted party's modified share can only depend on its own received share, but not on the shares received by the other corrupted parties. They construct a robust scheme in this setting for $n = 2t+1$ parties with share size of $m+\widetilde{O}(k)$ bits. Unfortunately, the construction is clearly insecure in the traditional robustness setting where a monolithic adversary gets to see the shares received by *all* of the corrupted parties before deciding how to modify each such share.

Finally, the work of [JS13] proposes a robust secret sharing scheme with t corruptions out of $n \geq 2t + 2$. Unfortunately, we found a flaw in the security proof and an attack showing that the proposed scheme is insecure (which we communicated to the authors).

In summary, despite the intense study of robust secret sharing since the late 80 s and early 90 s, in the maximal corruption setting with $n = 2t + 1$ parties there is a large gap between the lower bound of $m+k$ bits and the best previously known upper bound of $m+\widetilde{O}(n+k)$ bits on the share size. In particular, prior to this work, it was not known if the linear dependence between the share size and n is necessary in this setting, or whether there exist (even inefficient) schemes that beat this bound.

Our Result. We present an efficient robust secret sharing scheme in the maximal corruption setting with $n = 2t + 1$ parties, where the share size of only $m+\widetilde{O}(k)$ bits (see Sect. 6 for detailed parameters including poly-logarithmic factors). This is the first such scheme which removes the linear dependence between the share size and the number of parties n. A comparison between our work and previous results is given in Table 1.

[1] Yet another intermediate variant is to separately require robustness with $t = (1/2 - \delta)n$ corruptions and ramp reconstruction from $t + \rho n = (1/2 - \delta + \rho)n$ correct shares for some constants $\delta, \rho > 0$. This could always be achieved by adding a good (non-robust) ramp secret sharing scheme on top of a robust scheme while maintaining the $O(1)$ share size when n is sufficiently large.

Table 1. Comparison of robust secret sharing schemes and lower bounds with n parties, t corruptions, m-bit message, and 2^{-k} reconstruction error. *Robust* reconstruction is given n shares with t errors, while *threshold* reconstruction is given just $t+1$ correct shares ($n-t-1$ erasures). The $\widetilde{O}(\cdot)$ notation hides factors that are poly-logarithmic in k, n and m. This is justified if we think of n, m as some arbitrarily large polynomials in the security parameter k. The † requires that n is sufficiently large in relation to m, k.

Setting: $t = (1/2 - \Omega(1))n$			
Reconstruction	Construction	Share Size	Lower bound
Robust Only	[CDD+15, Che15]	$O(1)$ †	
Robust + Threshold	[Che15]	$(1+o(1))m + O(k)$	m
	Generic construction	$m + O(1)$ †	
Setting: $n = 2t+1$			
Reconstruction	Construction	Share Size	Lower bound
Robust ⇒ Threshold	[RB89]	$m + \widetilde{O}(kn)$	$m + k$
	[CDF01, CDF+08, CFOR12]	$m + \widetilde{O}(k+n)$	
	Our work	$m + \widetilde{O}(k)$	

1.1 Our Techniques

Using MACs. We begin with the same high-level idea as the schemes of [RB89, CFOR12], which use information-theoretic message authentication codes (MACs) to help the reconstruction procedure identify illegitimate shares. The basic premise is to start with a standard (non-robust) t-out-of-n scheme, such as Shamir's scheme, and have parties authenticate each others' Shamir shares using MACs. Intuitively, this should make it more difficult for an adversary to present compelling false shares for corrupted parties as it would have to forge the MACs under unknown keys held by the honest parties.

The original implementation of this idea by Rabin and Ben-Or [RB89] required each party to authenticate the share of every other party with a MAC having unforgeability security 2^{-k} and the reconstruction procedure simply checked that the majority of the tags verified. Therefore, the keys and tags added an extra $\widetilde{O}(nk)$ overhead to the share of each party. The work of Cevallos et al. [CFOR12] showed that one can also make this idea work using a MAC with a weaker unforgeability security of only $\frac{1}{\Omega(n)}$, by relying on a more complex reconstruction procedure. This reduced the overhead to $\widetilde{O}(k+n)$ bits.

Random Authentication Graph. Our core insight is to have each party only authenticate a relatively small but randomly chosen subset of other parties' shares. This will result in a much smaller overhead in the share size, essentially independent of the number of parties n.

More precisely, for each party we choose a random subset of $d = \widetilde{O}(k)$ other parties whose shares to authenticate. We can think of this as a random "authentication graph" $G = ([n], E)$ with out-degree d, having directed edges $(i, j) \in E$ if party i authenticates party j. This graph is stored in a distributed manner where each party i is responsible for storing the information about its d outgoing edges. It is important that this graph is not known to the attacker when choosing which parties to corrupt. In fact, as the attacker adaptively corrupts parties, he should not learn anything about the outgoing edges of uncorrupted parties[2].

Requirements and Inefficient Reconstruction. As a first step, let's start by considering an inefficient reconstruction procedure, as this will already highlight several challenges. The reconstruction procedure does not get to see the original graph G but a possibly modified graph $G' = ([n], E')$ where the corrupted parties can modify their set of outgoing edges. However, the edges that originate from uncorrupted parties are the same in G and G'. The reconstruction procedure labels each edge $e \in E'$ as either good or bad depending on whether the MAC corresponding to that edge verifies.

Let's denote the subset of uncorrupted *honest* parties by $H \subseteq [n]$. Let's also distinguish between corrupted parties where the adversary does not modify the share, which we call *passive corruptions* and denote by $P \subseteq [n]$, and the rest which we call *active corruptions* and denote by $A \subseteq [n]$. Assume that we can ensure that the following requirements are met:

(I) All edges between honest/passive parties, $(i, j) \in E' \ : \ i, j \in H \cup P$, are labeled good.
(II) All edges from honest to active parties, $(i, j) \in E' \ : \ i \in H, j \in A$ are labeled bad.

In this case, the reconstruction procedure can (inefficiently) identify the set $H \cup P$ by simply finding the *maximum self-consistent set* of vertices $C \subseteq [n]$, i.e. the largest subset of vertices such that all of the tags corresponding to edges $(i, j) \in E'$ with $i, j \in C$ are labeled good. We show that $C = H \cup P$ is the unique maximum self-consistent set with overwhelming probability (see Sect. 7). Once we identify the set $H \cup P$ we can simply reconstruct the secret message from the Shamir shares of the parties in $H \cup P$ since these have not been modified.

Implementation: Private MAC and Robust Storage of Tags. Let's now see how to implement the authentication process to satisfy requirements I and II defined above. A naive implementation of this idea would be for each party i to have a MAC key key_i for a d-time MAC (i.e., given the authentication tags

[2] If the graph were chosen at random but known to the attacker in advance, then the attacker could always choose some honest party i and corrupt a set of t parties none of which are being authenticated by i. Then the $t + 1$ shares corresponding to the t corrupted parties along with honest party i would be consistent and the reconstruction would not be able to distinguish it from the set of $t+1$ honest parties. However, with an unknown graph, there is a high probability that every honest party i authenticates many corrupted parties.

of d messages one cannot forge the tag of a new message) and, for each edge $(i, j) \in E$, to create a tag $\sigma_{i \to j} = \mathsf{MAC}_{\mathsf{key}_i}(\widetilde{s}_j)$ where $\widetilde{s}_j$ is the Shamir share of party j. The tags $\sigma_{i \to j}$ would then be stored with party j. In particular, the full share of party i would be

$$s_i = (\widetilde{s}_i, E_i, \mathsf{key}_i, \{\sigma_{j \to i}\}_{(j,i) \in E})$$

where $E_i = \{j \in [n] \ : \ (i, j) \in E\}$ are the outgoing edges for party i.

Unfortunately, there are several problems with this. Firstly, if the adversary corrupts party i, it might modify the values key_i, E_i in the share of party i but keep the Shamir share $\widetilde{s}_i$ intact. This will keep the edges going from honest parties to party i labeled good but some of the edges going from party i to honest parties might now be labeled bad. Therefore we cannot define such party as either passive (this would violate requirement I) or active (this would violate requirement II). Indeed, this would break our reconstruction procedure.

To fix this, when party i authenticates another party j, we compute $\sigma_{i \to j} = \mathsf{MAC}_{\mathsf{key}_i}((\widetilde{s}_j, E_j, \mathsf{key}_j))$ where we authenticate the values E_j, key_j along with the Shamir share $\widetilde{s}_j$. This prevents party j from being able to modify these components without being detected. Therefore we can define a party as active if any of the components $\widetilde{s}_j, E_j, \mathsf{key}_j$ are modified and passive otherwise.

Unfortunately, there is still a problem. An adversary corrupting party j might keep the components $\widetilde{s}_j, E_j, \mathsf{key}_j$ intact but modify some subset of the tags $\sigma_{i \to j}$. This will make some of edges going from honest parties to party j become bad while others remain good, which violates the requirements.

To fix this, we don't store tags $\sigma_{i \to j}$ with party j but rather we store all the tags in a distributed manner among the n parties in a way that guarantees recoverability even if t parties are corrupted. However, we do not provide any privacy guarantees for these tags and the adversary may be able to learn all of them in full. We call this *robust distributed storage* (without privacy), and show that we can use it to store the tags without additional asymptotic overhead. The fact that the tags are not stored privately requires us to use a special type of *private (randomized) MAC* where the tags $\sigma_{i \to j}$ do not reveal anything about the authenticated messages even given the secret key key_i. With this implementation, we can guarantee that requirements I, II are satisfied.

Efficient Reconstruction Using Graph Bisection. To get an efficient reconstruction procedure, we need to solve the following *graph identification problem*. An adversary partitions vertices $V = [n]$ into three sets H, P, A corresponding to honest, active and passive parties respectively. We know that the out-going edges from H are chosen randomly and that the edges are labeled as either good or bad subject to requirements I, II above. The goal is to identify $H \cup P$. We know that, with overwhelming probability, $H \cup P$ is the unique maximum self-consistent set having no bad edges between its vertices, but its not clear how to identify it efficiently.

Let's consider two cases of the above problem depending on whether the size of the passive set P is $|P| \geq \varepsilon n$ or $|P| < \varepsilon n$ for some $\varepsilon = 1/\Theta(\log n)$. If P is larger than εn, then we can distinguish between vertices in A and $H \cup P$

by essentially counting the number of incoming bad edges of each vertex. In particular, the vertices in $H \cup P$ only have incoming bad edges from the set A of size $|A| = (1/2 - \varepsilon)n$ while the vertices in A which have incoming bad edged from H of size $|H| = n/2$. Therefore, at least on average, the vertices in A will have more incoming bad edges than those in $H \cup P$. This turns out to be sufficient to then identify all of $H \cup P$.

On the other hand, if $|P| < \varepsilon n$ then the graph has a "bisection" consisting of H and $A \cup P$ (whose sizes only differ by 1 vertex) with only approximately εnd good edges crossing from H to $A \cup P$, corresponding to the edges from H to P. We then rely on the existence of efficient approximation algorithms for the *minimum graph bisection problem*. This is a classic NP-hard optimization problem [GJS76,FK02], and the best known polynomial-time algorithm is an $O(\log n)$-approximation algorithm due to [Räc08]. In particular, this allows us to bisect the graph it into two components X_0, X_1 with only very few edges crossing from X_0 to X_1. This must mean that one of X_0 or X_1 contains the vast majority of the vertices in H as otherwise, if the H vertices were split more evenly, there would be many more edges crossing. Having such components X_0, X_1 turns out to be sufficient to then identify all of $H \cup P$.

There are many details to fill in for the above high-level description, but one major issue is that we only have efficient approximations for the graph bisection problem in *undirected* graphs. However, in the above scenario, we are only guaranteed that there are few good edges from H to $A \cup P$ but there may be many good edges in the reverse direction. To solve this problem, we need to make sure that our graph problem satisfies one additional requirement (in addition to requirements I, II above):

(III) All edges from active to honest parties, $(i, j) \in E' \ : \ i \in A, j \in H$ are labeled bad.

To ensure that this holds, we need to modify the scheme so that, for any edge $(i, j) \in E$ corresponding to party i using its key to authenticate the share of party j with a tag $\sigma_{i \to j}$, we also add a "reverse-authentication" tag $\sigma_{i \leftarrow j}$ where we authenticate the share of party i under the key of party j. This ensures that edges from active parties to honest parties are labeled bad. Therefore, when P is small, there are very few good edges between H and $A \cup P$ in either direction and we can use an algorithm for the undirected version of the graph bisection problem.

Parallel Repetition and Parameters. A naive instantiation of the above scheme would require a share size of $m + \widetilde{O}(k^2)$ since we need $O(k)$ tags per party and each tag needs to have length $O(k)$. To reduce the share size further, we first instantiate our scheme with much smaller parameters which only provide weak security and ensure that the correct message is recovered with probability 3/4. We then use $O(k)$ parallel copies of this scheme to amplify security, where the reconstruction outputs the majority value. One subtlety is that all of the copies need to use the same underlying Shamir shares since we don't want a multiplicative blowup in the message size m. We show that this does not hurt security. Altogether, this results in a share size of only $m + \widetilde{O}(k)$.

2 Notation and Preliminaries

For $n \in \mathbb{N}$, we let $[n] := \{1, \ldots, n\}$. If X is a distribution or a random variable, we let $x \leftarrow X$ denote the process of sampling a value x according to the distribution X. If A is a set, we let $a \leftarrow A$ denote the process of sampling a uniformly at random from A. If f is a randomized algorithm, we let $f(x; r)$ denote the output of f on input x with randomness r. We let $f(x)$ be a random variable for $f(x; r)$ with random r.

Sub-Vector Notation. For a vector $\boldsymbol{s} = (s_1, \ldots, s_n)$ and a set $I \subseteq [n]$, we let $\boldsymbol{s}_I$ denote the vector consisting only of values in indices $i \in I$; we will represent this as $\boldsymbol{s}_I = (s'_1, \ldots, s'_n)$ with $s'_i = s_i$ for $i \in I$ and $s'_i = \perp$ for $i \notin I$.

Graph Notation. For a (directed) graph $G = (V, E)$, and sets $X, Y \subseteq V$, define $E_{X \to Y}$ as the set of edges from X to Y; i.e. $E_{X \to Y} = \{(v_1, v_2) \in E \mid v_1 \in X, v_2 \in Y\}$.

2.1 Hash Functions, Polynomial Evaluation

Definition 1 (Universal Hashing). *Let $\mathcal{H} = \{H_k : \mathcal{U} \to \mathcal{V}\}_{k \in \mathcal{K}}$ be family of hash functions. We say that $\mathcal{H}$ is ε-universal if for all $x, x' \in \mathcal{U}$ with $x \neq x'$ we have* $\Pr_{k \leftarrow \mathcal{K}}[H_k(x) = H_k(x')] \leq \varepsilon$.

Polynomial Evaluation. Let $\mathbb{F}$ be a finite field. Define the *polynomial evaluation* function $\mathsf{PEval} : \mathbb{F}^d \times \mathbb{F} \to \mathbb{F}$ as $\mathsf{PEval}(\boldsymbol{a}, x) = \sum_{i=1}^{d} a_i x^i$. See the full version [BPRW15] for the properties of the polynomial evaluation we rely on.

2.2 Graph Bisection

Let $G = (V, E)$ be an undirected graph. Let (V_1, V_2) be a partition of its edges. The *cross edges* of (V_1, V_2) are the edges in $E_{V_1 \to V_2}$. Given an undirected graph $G = (V, E)$ with an even number of vertices $|V| = n = 2t$ a *graph bisection* for G is a partition (V_1, V_2) of V such that $|V_1| = t = |V_2|$. We also extend the notion of a graph bisection to graphs with an odd number of vertices $|V| = n = 2t + 1$ by defining a bisection to be a partition with $|V_1| = t, |V_2| = t + 1$.

Definition 2 (Approximate Graph Bisection Algorithm). *Let $G = (V, E)$ be an undirected graph with n vertices. Assume that G has a graph bisection V_1, V_2 with $|E_{V_1 \to V_2}| = m$ cross edges. An algorithm* Bisect *that takes as input G and outputs a bisection U_1, U_2 with at most $|E_{U_1 \to U_2}| \leq \delta m$ cross edges is called a* δ-approximate graph bisection algorithm.

We remark that standard definitions of graphs bisection only consider the case where $n = 2t$ is even. However, given any δ-approximate graph bisection algorithm that works in the even case, we can generically adapt it to also work in the odd case $n = 2t + 1$. In particular, given a graph $G = (V, E)$ with $|V| = 2t + 1$ vertices, we can construct a graph $G' = (V \cup \{\perp\}, E)$ with an added dummy vertex $\perp$ that has no outgoing or incoming edges. We then run the δ-approximate graph bisection algorithm that works for an even number of vertices on G' to get a bisection U_1', U_2' where, without loss of generality, we assume that U_1' contains the vertex $\perp$. By simply taking U_1 to be U_1' with $\perp$ removed and $U_2 = U_2'$ we get a δ-approximate bisection for the graph G.

The work of [FK02] gave an efficient $O(\log^{1.5} n)$-approximate graph bisection algorithm, which was then improved to $O(\log n)$ by [Räc08] (Sect. 3, "Min Bisection").

3 Definition of Robust Secret Sharing

Throughout the rest of the paper, we use the following notation:

- t denotes the number of players that are arbitrarily corrupt.
- $n = 2t + 1$ denotes the number of players in the scheme.
- $\mathcal{M}$ is the message space.

Definition 3 (Robust Secret Sharing). *A t-out-of-n, δ-*robust secret sharing scheme *over a message space* $\mathcal{M}$ *and share space* $\mathcal{S}$ *is a tuple* $(\mathsf{Share}, \mathsf{Rec})$ *of algorithms that run as follows:*

$\mathsf{Share}(\mathsf{msg}) \rightarrow (s_1, \ldots, s_n)$: *This is a randomized algorithm that takes as input a message* $\mathsf{msg} \in \mathcal{M}$ *and outputs a sequence of shares* $s_1, \ldots, s_n \in \mathcal{S}$.

$\mathsf{Rec}(s_1, \ldots, s_n) \rightarrow \mathsf{msg}'$: *This is a deterministic algorithm that takes as input n shares* $(s_1, \ldots, s_n)$ *with* $s_i \in \mathcal{S} \cup \perp$ *and outputs a message* $\mathsf{msg}' \in \mathcal{M}$.

We require perfect correctness, meaning that for all $\mathsf{msg} \in \mathcal{M}$: $\Pr[\mathsf{Rec}(\mathsf{Share}(\mathsf{msg})) = \mathsf{msg}] = 1$. *Moreover, the following properties hold:*

Perfect Privacy: *Any t out of n shares of a secret give no information on the secret itself. More formally, for any* $\mathsf{msg}, \mathsf{msg}' \in \mathcal{M}$, *any* $I \subseteq [n]$ *of size* $|I| = t$, *the distributions* $\mathsf{Share}(\mathsf{msg})_I$ *and* $\mathsf{Share}(\mathsf{msg}')_I$ *are identical.*

Perfect Threshold Reconstruction (with Erasures): *The original secret can be reconstructed from any* $t + 1$ *correct shares. More formally, for any* $\mathsf{msg} \in \mathcal{M}$ *and any* $I \subseteq [n]$ *with* $|I| = t + 1$ *we have* $\Pr[\mathsf{Rec}(\mathsf{Share}(\mathsf{msg})_I) = \mathsf{msg}] = 1$.

Adaptive δ-Robustness: *An adversary that adaptively modifies up to t shares can cause the wrong secret to be recovered with probability at most δ. More formally, we define the experiment* $\boldsymbol{Exp}(\mathsf{msg}, \mathsf{Adv})$ *with some secret* $\mathsf{msg} \in \mathcal{M}$ *and interactive adversary* Adv.

$\boldsymbol{Exp}(\mathsf{msg}, \mathsf{Adv})$: *is defined as follows:*

E.1. *Sample* $\boldsymbol{s} = (s_1, \ldots, s_n) \leftarrow \mathsf{Share}(\mathsf{msg})$.

E.2. *Set* $I := \varnothing$. *Repeat the following while* $|I| \leq t$.
  - Adv *chooses* $i \in [n] \setminus I$.
  - *Update* $I := I \cup \{i\}$ *and give* s_i *to* Adv.

E.3. Adv *outputs modified shares* $s_i' : i \in I$ *and we define* $s_i' := s_i$ *for* $i \notin I$.

E.4. Compute $\mathsf{msg}' = \mathsf{Rec}(s'_1, \ldots, s'_n)$.
E.5. If $\mathsf{msg}' \neq \mathsf{msg}$ *output 1 else 0.*

We require that for any (unbounded) adversary Adv *and any* $\mathsf{msg} \in \mathcal{M}$ *we have*

$$\Pr[\boldsymbol{Exp}(\mathsf{msg}, \mathsf{Adv}) = 1] \leq \delta.$$

Remarks. We note that since privacy and threshold reconstruction are required to hold perfectly (rather than statistically) there is no difference between defining non-adaptive and adaptive variants. In other words, we could also define adaptive privacy where the adversary gets to choose which shares to see adaptively, but this is already implied by our non-adaptive definition of perfect privacy. We also note that when $n = 2t + 1$ then robustness implies a *statistically* secure threshold reconstruction with erasures. However, since we can even achieve *perfect* threshold reconstruction, we define it as a separate property.

Definition 4 (Non-Robust Secret Sharing). *We will say that a scheme is a* non-robust *t-out-of-n secret sharing scheme, if it satisfies the above definition with* $\delta = 1$.

Using Shamir secret sharing, we get a non-robust t-out-of-n secret sharing for any $t < n$ where the share size is the same as the message size.

4 The Building Blocks

In this section we introduce the building blocks of our robust secret sharing scheme: *Robust Distributed Storage*, *Private MACs*, and the *Graph Identification* problem.

4.1 Robust Distributed Storage

A robust distributed storage scheme allows us to store a public value among n parties, t of which may be corrupted. There is no secrecy requirement on the shared value. However, we require robustness: if the adversary adaptively corrupts t of the parties and modifies their shares, the reconstruction procedure should recover the correct value with overwhelming probability. We can think of this primitive as a relaxation of an error-correcting code where shares correspond to codeword symbols. The main difference is that the encoding procedure can be randomized and the adversary only gets to see a set of t (adaptively) corrupted positions of the codeword before deciding on the errors in those positions. These restrictions allow us to achieve better parameters than what is possible with standard error-correcting codes.

Definition 5. *A t-out-of-n,* δ-robust distributed storage *over a message space* $\mathcal{M}$ *is a tuple of algorithms* (Share, Rec) *having the same syntax as robust secret sharing, and satisfying the* δ*-robustness property. However, it need not satisfy the privacy or perfect reconstruction with erasures properties.*

We would like to construct such schemes for $n = 2t + 1$ and for a message of size m so that the share of each party is only $O(m/n)$ bits. These parameters are already beyond the reach of error-correcting codes for worst-case errors. We construct a simple robust distributed storage scheme by combining list-decoding and universal hashing.

List Decoding. In list-decoding, the requirement to decode to a unique codeword is relaxed, and it is only required to obtain a polynomially sized list of potential candidates that is guaranteed to include the correct codeword. We can simply use Reed-Solomon codes and the list-decoding algorithm provided by Sudan [Sud97] (better parameters are known but this suffices for our needs):

Theorem 1 [Sud97]. *A Reed-Solomon code formed by evaluating a degree d polynomial on n points can be efficiently list-decoded to recover from $e < n - \sqrt{2dn}$ errors with a list of size $L \leq \sqrt{2n/d}$.*

Setting $d = \lfloor n/8 \rfloor$, we can then therefore recover from t out of $n = 2t + 1$ errors and obtain a list of size $L \leq \sqrt{2n/d} = O(1)$.

Construction of Robust Distributed Storage. Let t be some parameter, let $n = 2t + 1$, and let $\mathbb{F}$ be a field of size $|\mathbb{F}| = 2^u$ with $|\mathbb{F}| > n$. Let $\mathcal{H} = \{H_k : \mathbb{F}^{d+1} \to \mathbb{F}\}_{k \in \mathbb{F}}$ be an ε-universal hash function. For concreteness, we can use the polynomial evaluation hash $H_k(\boldsymbol{a}) = \mathsf{PEval}(\boldsymbol{a}, k)$, which achieves $\varepsilon = (d+1)/2^u$ (see 'XOR-universality' in the full version [BPRW15]). We use list-decoding for the Reed Solomon code with degree $d = \lfloor n/8 \rfloor = \Omega(n)$ which allows us to recover from t out of n errors with a list size $L = O(1)$. We construct a δ-robust t-out-of-n distributed storage scheme with message space $\mathcal{M} = \mathbb{F}^{d+1}$, meaning that the messages have bit-size $m = u(d+1) = \Omega(un)$, share size $3u = O(u)$, and robustness $\delta = nL\varepsilon = O(n^2)/2^u$. The scheme works as follows:

- $(s_1, \ldots, s_n) \leftarrow \mathsf{Share}(\mathsf{msg})$. Encode $\mathsf{msg} \in \mathbb{F}^{d+1}$ using the Reed-Solomon code by interpreting it as a degree d polynomial and evaluating it on n points to get the Reed-Solomon codeword $(\hat{s}_1, \ldots, \hat{s}_n) \in \mathbb{F}^n$. Choose random values $k_1, \ldots, k_n \leftarrow \mathbb{F}$ and define the shares $s_i = (\hat{s}_i, k_i, H_{k_i}(\mathsf{msg})) \in \mathbb{F}^3$.
- $\mathsf{msg}' \leftarrow \mathsf{Rec}(s'_1, \ldots, s'_n)$. Parse $s'_i = (\hat{s}'_i, k'_i, y'_i)$. Use list-decoding on the modified codeword $(\hat{s}'_1, \ldots, \hat{s}'_n) \in \mathbb{F}^n$ to recover a list of $L = O(1)$ possible candidates $\mathsf{msg}^{(1)}, \ldots, \mathsf{msg}^{(L)} \in \mathbb{F}^{d+1}$ for the message. Output the first value $\mathsf{msg}^{(j)}$ that agrees with the majority of the hashes:

$$|\{i \in [n] \; : \; H_{k'_i}(\mathsf{msg}^{(j)}) = y'_i\}| \geq t + 1.$$

Theorem 2. *For any $n = 2t+1$ and any $u \geq \log n$, the above scheme is a t-out-of-n, δ-robust distributed storage scheme for messages of length $m = \lfloor n/8 \rfloor u = \Omega(nu)$ with shares of length $3u = O(u)$ and robustness $\delta = O(n^2)/2^u$.*

The proof is given in the full version [BPRW15].

4.2 Private Labeled MAC

As a tool in our construction of robust secret sharing schemes, we will use a new notion of an information-theoretic message-authentication code (MAC) that has additional privacy guarantees.

The message authentication code $\sigma = \mathsf{MAC}_{\mathsf{key}}(\mathsf{lab}, \mathsf{msg}, r)$ takes as input a *label* lab, a *message* msg, and some additional *randomness* r. The randomness is there to ensure privacy for the message msg even given key, σ.

Definition 6 (Private Labeled MAC). *An* (ℓ, ε) private MAC *is a family of functions* $\{\mathsf{MAC}_{\mathsf{key}} : \mathcal{L} \times \mathcal{M} \times \mathcal{R} \to \mathcal{T}\}_{\mathsf{key} \in \mathcal{K}}$ *with key-space* $\mathcal{K}$*, message space* $\mathcal{M}$*, label space* $\mathcal{L}$*, randomness space* $\mathcal{R}$*, and tag space* $\mathcal{T}$*. It has the following properties:*

Authentication: *For any* ℓ *values* $(\mathsf{lab}_i, \mathsf{msg}_i, r_i, \sigma_i) \in \mathcal{L} \times \mathcal{M} \times \mathcal{R} \times \mathcal{T} \ : \ i = 1, \ldots, \ell$ *such that the labels* lab_i *are distinct, and for any* $(\mathsf{lab}', \mathsf{msg}', r', \sigma') \in \mathcal{L} \times \mathcal{M} \times \mathcal{R} \times \mathcal{T}$ *such that* $(\mathsf{lab}', \mathsf{msg}', r') \notin \{(\mathsf{lab}_i, \mathsf{msg}_i, r_i)\}_{i \in [\ell]}$ *we have:*

$$\Pr_{\mathsf{key} \leftarrow \mathcal{K}}[\mathsf{MAC}_{\mathsf{key}}(\mathsf{lab}', \mathsf{msg}', r') = \sigma' \mid \{\mathsf{MAC}_{\mathsf{key}}(\mathsf{lab}_i, \mathsf{msg}_i, r_i) = \sigma_i\}_{i \in [\ell]}] \leq \varepsilon.$$

This implies that even after seeing the authentication tags σ_i *for* ℓ *tuples* $(\mathsf{lab}_i, \mathsf{msg}_i, r_i)$ with distinct labels lab_i, an adversary cannot come up with a valid tag σ' for any new tuple $(\mathsf{lab}', \mathsf{msg}', r')$.

Privacy Over Randomness: *For any* ℓ *distinct labels* $\mathsf{lab}_1, \ldots, \mathsf{lab}_\ell$*, any keys* $\mathsf{key}_1, \ldots, \mathsf{key}_\ell \in \mathcal{K}$*, and any* $\mathsf{msg} \in \mathcal{M}$*, the* ℓ *values* $\sigma_1 = \mathsf{MAC}_{\mathsf{key}_1}(\mathsf{lab}_1, \mathsf{msg}, r)$*,* $\ldots, \sigma_\ell = \mathsf{MAC}_{\mathsf{key}_\ell}(\mathsf{lab}_\ell, \mathsf{msg}, r)$ *are uniformly random and independent in* $\mathcal{T}$ *over the choice of* $r \leftarrow \mathcal{R}$.

This says that the tags σ_i do not reveal any information about the message msg, or even about the labels lab_i and the keys key_i, as long as the randomness r is unknown.

Privacy Over Keys: *Let* $(\mathsf{lab}_i, \mathsf{msg}_i, r_i) \in \mathcal{L} \times \mathcal{M} \times \mathcal{R} \ : \ i = 1, \ldots, \ell$ *be* ℓ *values such that the labels* lab_i *are distinct. Then the* ℓ *values* $\sigma_1 = \mathsf{MAC}_{\mathsf{key}}(\mathsf{lab}_1, \mathsf{msg}_1, r_1)$*,* $\ldots, \sigma_\ell = \mathsf{MAC}_{\mathsf{key}}(\mathsf{lab}_\ell, \mathsf{msg}_\ell, r_\ell)$ *are uniformly random and independent in* $\mathcal{T}$ *over a random* $\mathsf{key} \leftarrow \mathcal{K}$.

This says that the tags σ_i do not reveal any information about the values $(\mathsf{lab}_i, \mathsf{msg}_i, r_i)$ as long as key is unknown.

Construction. Let $\mathbb{F}$ and $\mathbb{F}'$ be finite fields such that $|\mathbb{F}'| \geq |\mathcal{L}|$ and $|\mathbb{F}| \geq |\mathbb{F}'| \cdot |\mathcal{L}|$. We assume that we can identify the elements of $\mathcal{L}$ as either a subset of $\mathbb{F}'$ or $\mathbb{F}$ and we can also efficiently identify tuples in $\mathbb{F}' \times \mathcal{L}$ as a subset of $\mathbb{F}$. Let $\mathcal{M} = \mathbb{F}^m$, $\mathcal{R} = \mathbb{F}^\ell$, $\mathcal{K} = \mathbb{F}^{\ell+1} \times (\mathbb{F}')^{\ell+1}$, $\mathcal{T} = \mathbb{F}$. Define $\mathsf{MAC}_{\mathsf{key}}(\mathsf{lab}, \mathsf{msg}, r)$ as follows:

- Parse $\mathsf{key} = (\mathsf{key}_1, \mathsf{key}_2)$ where $\mathsf{key}_1 \in (\mathbb{F}')^{\ell+1}, \mathsf{key}_2 \in \mathbb{F}^{\ell+1}$.
- Compute $\mathsf{key}_1^{\mathsf{lab}} := \mathsf{PEval}(\mathsf{key}_1, \mathsf{lab})$, $\mathsf{key}_2^{\mathsf{lab}} := \mathsf{PEval}(\mathsf{key}_2, \mathsf{lab})$ by identifying $\mathsf{lab} \in \mathcal{L}$ as an element of $\mathbb{F}'$ and $\mathbb{F}$ respectively.

- Output $\sigma := \mathsf{PEval}(\ (r, \mathsf{msg})\ ,\ (\mathsf{lab}, \mathsf{key}_1^{\mathsf{lab}})\ ) + \mathsf{key}_2^{\mathsf{lab}}$. Here we interpret $(r, \mathsf{msg}) \in \mathcal{R} \times \mathcal{M} = \mathbb{F}^{\ell+m}$ as a vector of coefficients in $\mathbb{F}$ and we identify $(\mathsf{lab}, \mathsf{key}_{\mathsf{lab}}^1) \in \mathcal{L} \times \mathbb{F}'$ as an element of $\mathbb{F}$.

Theorem 3. *The above construction is an (ℓ, ε) private MAC, where $\varepsilon = \frac{m+\ell}{|\mathbb{F}'|}$.*

The proof is given in the full version [BPRW15].

4.3 Graph Identification

Here, we define an algorithmic problem called the graph identification problem. This abstracts out the core algorithmic problem that we face in designing our reconstruction algorithm.

Definition 7 (Graph Identification Challenge). *A graph identification challenge $\mathsf{Gen}^{\mathsf{Adv}}(n, t, d)$ is a randomized process that outputs a directed graph $G = (V = [n], E)$, where each vertex $v \in V$ has out-degree d, along with a labeling $L : E \to \{\mathsf{good}, \mathsf{bad}\}$. The process is parameterized by an adversary* Adv *and proceeds as follows.*

Adversarial Components. *The adversary $\mathsf{Adv}(n, t, d)$ does the following:*

1. *It partitions $V = [n]$ into three disjoint sets H, A, P (honest, active and passive) such that $V = H \cup A \cup P$ and $|A \cup P| = t$.*
2. *It chooses the set of edges $E_{A \cup P \to V}$ that originate from $A \cup P$ arbitrarily subject to each $v \in A \cup P$ having out-degree d and no self-loops.*
3. *It chooses the labels $L(e)$ arbitrarily for each edge $e \in E_{A \to (A \cup P)} \cup E_{(A \cup P) \to A}$.*

Honest Components. *The procedure* Gen *chooses the remaining edges and labels as follows:*

1. *It chooses the edges $E_{H \to V}$ that originate from H uniformly at random subject to each vertex having out-degree d and no self-loops. In particular, for each $v \in H$ it randomly selects d outgoing edges (without replacement) to vertices in $V \setminus \{v\}$.*
2. *It sets $L(e) := \mathsf{bad}$ for all $e \in E_{H \to A} \cup E_{A \to H}$.*
3. *It sets $L(e) := \mathsf{good}$ for all $e \in E_{(H \cup P) \to (H \cup P)}$.*

Output. *Output $(G = (V, E), L, H, A, P)$.*

The graph identification challenge is intended to model the authentication graph in our eventual robust secret sharing scheme where the labels will be assigned by verifying MAC tags. It allows us to abstract out a problem about graphs without needing to talk about MACs, secret shares, etc. We will need to show that any adversary on our full robust secret sharing scheme can be translated into an adversary in the graph identification challenge game above.

Note that, in the graph challenge game, we allow the adversary to *choose* the outgoing edges for both active and passive parties. This might seem unnecessary since the adversary in our eventual robust secret sharing scheme cannot *modify* the outgoing edges from passive parties in the authentication graph. However, it can choose which parties are active and which are passive after seeing the outgoing edges from all corrupted parties and therefore the adversary has some (limited) control over the outgoing edges of passive parties. In the definition of the graph challenge game, we therefore simply give the adversary complete control over all such outgoing edges.

As our main result for this problem, we show that there is an efficient algorithm that identifies the set $H \cup P$ given only the graph G and the labeling L. Note that, for our application, we do not crucially need to identify all of $H \cup P$; any subset of $H \cup P$ of size $t+1$ would be sufficient to reconstruct the message from the Shamir shares. However, this does not appear to make the task easier.

Theorem 4. *There exists a polynomial time algorithm* $\mathsf{GraphID}$*, called the* graph identification algorithm*, that takes as input a directed graph* $G = (V = [n], E)$ *a labeling* $L : V \to \{\mathsf{good}, \mathsf{bad}\}$ *and outputs a set* $B \subseteq V$*, such that for any* Adv*, we have:*

$$\Pr\left[B = H \cup P : \begin{array}{c}(G, L, H, A, P) \leftarrow \mathsf{Gen}^{\mathsf{Adv}}(n = 2t+1, t, d),\\ B \leftarrow \mathsf{GraphID}(G, L)\end{array}\right] \geq 1 - 2^{-\Omega(d/\log^2 n - \log n)}$$

In Sect. 7 we give a simple *inefficient* algorithm for the graph identification problem. Then, in Sect. 8, we prove Theorem 4 by providing an efficient algorithm.

5 Construction of Robust Secret Sharing

In this section we construct our robust secret sharing scheme using the tools outlined above. We analyze its security by translating an adversary on the scheme into an adversary in the graph identification game.

5.1 The Construction

Let $t, n = 2t+1$ be parameters that are given to us, and let $\mathcal{M}$ be a message space.

Let d be a graph out-degree parameter and let $\mathsf{GraphID}$ be the graph identification algorithm from Theorem 4 with success probability $1 - \delta_{gi}$ where $\delta_{gi} = 2^{-\Omega(d/\log^2 n)}$.

Let $(\mathsf{Share}_{nr}, \mathsf{Rec}_{nr})$ be a t-out-of-n *non-robust* secret sharing (e.g., Shamir secret sharing) with message space $\mathcal{M}$ and share space $\mathcal{S}_{nr}$.

Let $\{\mathsf{MAC}_{\mathsf{key}} : \mathcal{L} \times \mathcal{M}_{mac} \times \mathcal{R} \to \mathcal{T}\}_{\mathsf{key} \in \mathcal{K}}$ be an $(\ell, \varepsilon_{mac})$ *private MAC* with label space $\mathcal{L} = [n]^2 \times \{0,1\}$ and message space $\mathcal{M}_{mac} = \mathcal{S}_{nr} \times [n]^d \times \mathcal{K}$, where $\ell = 3d$.

Finally, let $(\mathsf{Share}_{rds}, \mathsf{Rec}_{rds})$ be a t-out-of-n robust distributed storage (no privacy) with message space $\mathcal{M}_{rds} = \mathcal{T}^{2dn}$, share space $\mathcal{S}_{rds}$ and with robustness δ_{rds}.

Our robust secret sharing scheme $(\mathsf{Share}, \mathsf{Rec})$ is defined as follows.

$\mathsf{Share}(\mathsf{msg})$. On input a message $\mathsf{msg} \in \mathcal{M}$, the sharing procedure proceeds as follows:

S.1. Choose $(\widetilde{s}_1, \ldots, \widetilde{s}_n) \leftarrow \mathsf{Share}_{nr}(\mathsf{msg})$ to be a non-robust secret sharing of msg.
S.2. Choose a uniformly random directed graph $G = ([n], E)$ with out-degree d, in-degree at most $2d$ and no self-loops as follows:
   (a) For each $i \in [n]$ choose a random set $E_i \subseteq [n] \setminus \{i\}$ of size $|E_i| = d$. Set
   $$E := \{(i, j) \ : \ i \in [n], j \in E_i\}.$$
   (b) Check if there is any vertex in G with in-degree $>2d$. If so, go back to step (a)[3].
S.3. For each $i \in [n]$, sample a random MAC key $\mathsf{key}_i \leftarrow \mathcal{K}$ and MAC randomness $r_i \leftarrow \mathcal{R}$.

   For each $j \in E_i$ define
   $$\sigma_{i \to j} := \mathsf{MAC}_{\mathsf{key}_i}((i, j, 0), (\widetilde{s}_j, E_j, \mathsf{key}_j), r_j),$$
   $$\sigma_{i \leftarrow j} := \mathsf{MAC}_{\mathsf{key}_j}((i, j, 1), (\widetilde{s}_i, E_i, \mathsf{key}_i), r_i).$$
   where we treat $(i, j, 0), (i, j, 1) \in \mathcal{L}$ as a label, and we treat $(\widetilde{s}_j, E_j, \mathsf{key}_j) \in \mathcal{M}_{mac}$ as a message.
S.4. For each $i \in [n]$ define $\mathsf{tags}_i = \{(\sigma_{i \to j}, \sigma_{i \leftarrow j})\}_{j \in E_i} \in \mathcal{T}^{2d}$ and define $\mathsf{tags} = (\mathsf{tags}_1, \ldots, \mathsf{tags}_n) \in \mathcal{T}^{2nd}$. Choose $(p_1, \ldots, p_n) \leftarrow \mathsf{Share}_{rds}(\mathsf{tags})$ using the robust distributed storage scheme.
S.5. For $i \in [n]$, define $s_i = (\widetilde{s}_i, E_i, \mathsf{key}_i, r_i, p_i)$ to be the share of party i. Output $(s_1, \ldots, s_n)$.

$\mathsf{Rec}(s'_1, \ldots, s'_n)$. On input $s'_1, \ldots, s'_n$ with $s'_i = (\widetilde{s}'_i, E'_i, \mathsf{key}'_i, r'_i, p'_i)$ do the following.

R.0. If there is a set of exactly $t+1$ values $W = \{i \in [n] \ : \ s'_i \neq \bot\}$ then output $\mathsf{Rec}_{nr}((\widetilde{s}'_i)_{i \in W})$. Else proceed as follows.

R.1. Reconstruct $\mathsf{tags}' = (\mathsf{tags}'_1, \ldots, \mathsf{tags}'_n) = \mathsf{Rec}_{rds}(p'_1, \ldots, p'_n)$. Parse $\mathsf{tags}'_i = \{(\sigma'_{i \to j}, \sigma'_{i \leftarrow j})\}_{j \in E'_i}$.
R.2. Define a graph $G' = ([n], E')$ by setting $E' := \{(i, j) \ : \ i \in [n], j \in E'_i\}$.
R.3. Assign a label $L(e) \in \{\mathsf{good}, \mathsf{bad}\}$ to each edge $e = (i, j) \in E'$ as follows. If the following holds:
   $$\sigma'_{i \to j} = \mathsf{MAC}_{\mathsf{key}'_i}((i, j, 1), (\widetilde{s}'_j, E'_j, \mathsf{key}'_j), r'_j) \quad \text{and}$$
   $$\sigma'_{i \leftarrow j} = \mathsf{MAC}_{\mathsf{key}'_j}((i, j, 1), (\widetilde{s}'_i, E'_i, \mathsf{key}'_i), r'_i)$$
   then set $L(e) := \mathsf{good}$, else set $L(e) := \mathsf{bad}$.
R.4. Call the graph identification algorithm to compute $B \leftarrow \mathsf{GraphID}(G', L)$.
R.5. Choose a subset $B' \subseteq B$ of size $|B'| = t + 1$ arbitrarily and output $\mathsf{Rec}_{nr}((\widetilde{s}'_i)_{i \in B'})$.

[3] This happens with negligible probability. However, we include it in the description of the scheme in order to get perfect rather than statistical privacy.

5.2 Security Analysis

Theorem 5. *The above scheme* (Share, Rec) *is a t-out-of-n δ-robust secret sharing scheme for $n = 2t + 1$ with robustness $\delta = \delta_{rds} + \delta_{gi} + dn\varepsilon_{mac} + n2^{-d/3}$.*

We prove Theorem 5 by separately proving that the scheme satisfies perfect privacy, perfect threshold reconstruction with erasures and adaptive δ-robustness in the following three lemmas.

Lemma 1. *The scheme* (Share, Rec) *satisfies perfect privacy.*

Proof. Let $I \subseteq [n]$ be of size $|I| = t$ and let $\mathsf{msg}, \mathsf{msg}' \in \mathcal{M}$ be any two values. We define a sequence of hybrids as follows:

Hybrid 0: This is $\mathsf{Share}(\mathsf{msg})_I = (s_i)_{i \in I}$. Each $s_i = (\widetilde{s}_i, E_i, \mathsf{key}_i, r_i, p_i)$.

Hybrid 1: In this hybrid, we change the sharing procedure to simply choose all tags $\sigma_{i \to j}$ and $\sigma_{j \leftarrow i}$ for any $j \notin I$ uniformly and independently at random. This is identically distributed by the "privacy over randomness" property of the MAC. In particular, we rely on the fact that the adversary does not see r_j and that there are at most $\ell = 3d$ tags of the form $\sigma_{i \to j}$ and $\sigma_{j \leftarrow i}$ for any $j \notin I$ corresponding to the total degree of vertex j. These are the only tags that rely on the randomness r_j and they are all created with distinct labels.

Hybrid 2: In this hybrid, we choose $(\widetilde{s}_1, \ldots, \widetilde{s}_n) \leftarrow \mathsf{Share}_{nr}(\mathsf{msg}')$. This is identically distributed by the perfect privacy of the non-robust secret sharing scheme. Note that in this hybrid, the shares $s_i \ : \ i \in I$ observed by the adversary do not contain any information about $\widetilde{s}'_j \ : \ j \notin I$.

Hybrid 3:s This is $\mathsf{Share}(\mathsf{msg}')_I = (s_i)_{i \in I}$. Each $s_i = (\widetilde{s}_i, E_i, \mathsf{key}_i, r_i, p_i)$. This is identically distributed by the "privacy over randomness" property of the MAC, using same argument as going from Hybrid 0 to 1.

Lemma 2. *The scheme* (Share, Rec) *satisfies perfect threshold reconstruction with erasures.*

Proof. This follows directly from the fact that the non-robust scheme (Share_{nr}, Rec_{nr}) satisfies perfect threshold reconstruction with erasures and therefore step R.0 of reconstruction is guaranteed to output the correct answer when there are exactly t erasures.

Lemma 3. *The scheme* (Share, Rec) *is δ-robust for $\delta = \delta_{rds} + \delta_{gi} + dn\varepsilon_{mac} + n2^{-d/3}$.*

Proof Overview. Before giving the formal proof of the lemma, we give a simplified proof intuition. To keep it simple, let's consider non-adaptive robustness experiment where the adversary has to choose the set $I \subseteq [n]$, $|I| = t$ of parties to corrupt at the very beginning of the game (in the full proof, we handle adaptive security). Let $s_i = (\widetilde{s}_i, E_i, \mathsf{key}_i, r_i, p_i)$ be the shares created by the sharing procedure and let $s'_i = (\widetilde{s}'_i, E'_i, \mathsf{key}'_i, r'_i, p'_i)$ be the modified shares submitted by the adversary (for $i \notin I$, we have $s'_i = s_i$). Let us define the set of *actively modified* shares as:

$$A = \{i \in I \;:\; (\widetilde{s}'_i, E'_i, \mathsf{key}'_i, r'_i) \neq (\widetilde{s}_i, E_i, \mathsf{key}_i, r_i)\}.$$

Define $H = [n] \setminus I$ to be the set of *honest* shares, and $P = I \setminus A$ to be the *passive shares.*

To prove robustness, we show that the choice of H, A, P and the labeling L created by the reconstruction procedure follow the same distribution as in the graph identification problem $\mathsf{Gen}^{\mathsf{Adv}'}(n, t, d)$ with some adversary Adv'. Therefore the graph identification procedure outputs $B = H \cup P$ which means that reconstruction outputs the correct message. Intuitively, we rely on the fact that: (1) by the privacy properties of the MAC the adversary does not learn anything about outgoing edges from honest parties and therefore we can think of them as being chosen randomly after the adversarial corruption stage, (2) by the authentication property of the MAC the edges between honest and active parties (in either direction) are labeled bad.

More concretely, we define a sequence of "hybrid" distributions to capture the above intuition as follows:

Hybrid 0. This is the non-adaptive version of the robustness game $\boldsymbol{Exp}(\mathsf{msg}, \mathsf{Adv})$ with a message msg and an adversary Adv as in Definition 3.

Hybrid 1. During reconstruction, instead of recovering $\mathsf{tags}' = \mathsf{Rec}_{rds}(p'_1, \ldots, p'_n)$ we just set $\mathsf{tags}' = \mathsf{tags}$ to be the correct value chosen by the sharing procedure. This is indistinguishable by the security of the robust-distributed storage scheme.

Hybrid 2. During the sharing procedure, we can change all of the tags $\sigma_{i \to j}, \sigma_{j \leftarrow i}$ with $j \in H$ to uniformly random values. This is identically distributed by the "privacy over randomness" property of the MAC since the adversary does not see r_j for any such $j \in H$, and there are at most $\ell = 3d$ such tags corresponding to the total degree of the vertex j. In particular, this means that such tags do not reveal any (additional) information to the adversary about E_j, key_j for $j \in H$.

Hybrid 3. During the reconstruction process, when the labeling L is created, we automatically set $L(e) = \mathsf{bad}$ for any edge $e = (i, j)$ or $e = (j, i)$ in E' such that $i \in H, j \in A$ (i.e., one end-point honest and the other active). The only time this introduces a change is if the adversary manages to forge a MAC tag under some key key_i for $i \in H$. Each such key was used to create at most $\ell = 3d$ tags with distinct labels and therefore, we can rely on the authentication security of the MAC to argue that this change is indistinguishable. Note that, by the definition of the labeling, we are also ensured that $L(e) = \mathsf{good}$ for any edge (i, j) where $i, j \in H \cup P$.

Hybrid 4. During the sharing procedure, we can change all of the tags $\sigma_{i \to j}, \sigma_{j \leftarrow i}$ with $i \in H$ to uniformly random values. This is identically distributed by the "privacy over keys" property of the MAC since the adversary does not see key_i for any such $i \in H$, and there are at most $\ell = 3d$ such tags corresponding to the total degree of the vertex i. In particular, this means that such tags do not reveal anything about the outgoing edges E_i for $i \in H$ and therefore the adversary gets no information about these edges throughout the game.

Hybrid 5. When we choose the graph $G = ([n], E)$ during the sharing procedure, we no longer require that every vertex has in-degree $\leq 2d$. Instead, we just choose each set $E_i \subseteq [n] \setminus \{i\}$, $|E_i| = d$ uniformly at random. Since the expected in-degree of every vertex is d, this change is indistinguishable by a simple Chernoff bound.

Hybrid 6. During reconstruction, instead of computing $B \leftarrow \mathsf{GraphID}(G', L)$ we set $B = H \cup P$. We notice that, in the previous hybrid, the distribution of G', L, H, A, P is exactly that of the graph reconstruction game $\mathsf{Gen}^{\mathsf{Adv}'}(n, t, d)$ with some adversary Adv'. In particular, the out-going edges from the honest set H are chosen uniformly at random and the adversary does not see any information about them throughout the game. Furthermore, the labeling satisfies the properties of the graph identification game. Therefore, the above modification is indistinguishable by the correctness of the graph identification algorithm.

In the last hybrid, the last step of the reconstruction procedure runs $\mathsf{msg}' = \mathsf{Rec}_{nr}((\widetilde{s}'_i)_{i \in B'})$ where $B' \subseteq H \cup P$ is of size $|B'| = t + 1$. Therefore and $\widetilde{s}'_i = \widetilde{s}_i$ for $i \in B'$ and, by the Perfect Reconstruction with Erasures property of the non-robust secret sharing scheme, we have $\mathsf{msg}' = \mathsf{msg}$. For a formal proof, see the full version [BPRW15].

5.3 Parameters of Construction

Let $\mathcal{M} = \{0, 1\}^m$ and $t, n = 2t + 1$ be parameters. Furthermore, let λ be a parameter which we will relate to the security parameter k.

We choose the out-degree parameter $d = \lambda \log^3 n$, which then gives $\delta_{gi} = 2^{-\Omega(d/\log^2 n - \log n)} = 2^{-\Omega(\lambda \log n)}$.

We instantiate the non-robust secret scheme $(\mathsf{Share}_{nr}, \mathsf{Rec}_{nr})$ using t-out-of-n Shamir secret sharing where the share space $\mathcal{S}_{nr}$ is a binary field of size $2^{\max\{m, \lfloor \log n \rfloor + 1\}} = 2^{m + O(\log n)}$.

We instantiate the MAC using the construction from Sect. 4.2. We choose the field $\mathbb{F}'$ to be a binary field of size $|\mathbb{F}'| = 2^{\lceil 5 \log n + \log m + \lambda \rceil}$ which is sufficiently large to encode a label in $\mathcal{L} = [n]^2 \times \{0, 1\}$. We choose the field $\mathbb{F}$ to be of size $|\mathbb{F}| = |\mathbb{F}'| 2^{2\lceil \log n \rceil + 1}$ which is sufficiently large to encode an element of $\mathbb{F}' \times \mathcal{L}$. We set $\ell = 3d = O(\lambda \log^3 n)$. This means that the keys and randomness have length $\log |\mathcal{K}|, \log |\mathcal{R}| = O(d \log |\mathbb{F}|) = O(\lambda \log^3 n(\lambda + \log n + \log m))$ and the tags have length $\log |\mathcal{T}| = \log |\mathbb{F}| = O(\lambda + \log n + \log m)$. We set the message space of the MAC to be $\mathcal{M}_{mac} = \mathbb{F}^{m_{mac}}$ which needs to be sufficiently large to encode the Shamir share, edges, and a key and therefore we set $m_{mac} = \lceil (\max\{m, \lfloor \log n \rfloor + 1\} + \log |\mathcal{K}| + d \log n) / \log |\mathbb{F}| \rceil = O(m + \lambda \log^3 n)$. This gives security $\varepsilon_{mac} = \frac{m_{mac} + \ell}{|\mathbb{F}'|} \leq 2^{\log m + \log \lambda + 3 \log n + O(1) - \log |\mathbb{F}'|} = 2^{-\Omega(\lambda) - 2 \log n}$.

Finally, we instantiate the robust distributed storage scheme using the construction from Sect. 4.1. We need to set the message space $\mathcal{M}_{rds} = \mathcal{T}^{2dn}$ which means that the messages are of length $m_{rds} = 2dn \log |\mathcal{T}| = O(n\lambda \log^3 n(\lambda + \log n + \log m))$. We set $u = \lceil 8m_{rds}/n \rceil = O(\lambda \log^3 n(\lambda + \log n + \log m))$. This results in a robust-distributed storage share length $3u = O(\lambda \log^3 n(\lambda + \log n + \log m))$ and we get security $\delta_{rds} = O(n^2)/2^u \leq 2^{-\Omega(\lambda)}$.

With the above we get security

$$\delta \leq \delta_{rds} + \delta_{gi} + dn\varepsilon_{mac} + n2^{-d/3} = 2^{-\Omega(\lambda)} \tag{5.1}$$

and total share length $\log |\mathcal{S}_{nr}| + d\lceil \log n \rceil + \log |\mathcal{K}| + \log |\mathcal{R}| + 3u$ which is

$$m + O(\lambda \log^3 n(\lambda + \log n + \log m)) \tag{5.2}$$

By choosing a sufficiently large $\lambda = O(k)$ we get security $\delta \leq 2^{-k}$ and share size

$$m + O(k^2 \mathsf{polylog}(n+m)) = m + \widetilde{O}(k^2).$$

6 Improved Parameters via Parallel Repetition

In the previous section, we saw how to achieve robust secret sharing with security $\delta = 2^{-k}$ at the cost of having a share size $m + \widetilde{O}(k^2)$. We now show how to improve this to $m + \widetilde{O}(k)$. We do so by instantiating the scheme from the previous section with smaller parameters that only provide weak robustness $\delta = \frac{1}{4}$ and share size $m + \widetilde{O}(1)$. We then use parallel repetition of $q = O(k)$ independent copies of this weak scheme. The q copies of the recovery procedure recover q candidate messages, and we simply output the majority vote. A naive implementation of this idea, using q completely independent copies of the scheme, would result in share size $O(km) + \widetilde{O}(k)$ sine the (non-robust) Shamir share of length m is repeated q times. However, we notice that we can reuse the same Shamir shares across all q copies. This is because the robustness security held even for a worst-case choice of such shares, only over the randomness of the other components. Therefore, we only get a total share size of $m + \widetilde{O}(k)$.

Construction. In more detail, let (Share, Rec) be our robust secret sharing scheme construction from above. For some random coins coins_{nr} of the non-robust (Shamir) secret sharing scheme, we let $(s_1, \ldots, s_n) \leftarrow \mathsf{Share}(\mathsf{msg}; \mathsf{coins}_{nr})$ denote the execution of the sharing procedure Share(msg) where step S.1 uses the fixed randomness coins_{nr} to select the non-robust shares $(\widetilde{s}_1, \ldots, \widetilde{s}_n) \leftarrow \mathsf{Share}_{nr}(\mathsf{msg}; \mathsf{coins}_{nr})$ but steps S.2 – S.5 use fresh randomness to select the graph G, the keys key_i and the randomness r_i. In particular, $\mathsf{Share}(\mathsf{msg}; \mathsf{coins}_{nr})$ remains a randomized algorithm.

We define the q-wise parallel repetition scheme (Share′, Rec′) as follows:

Share′(msg): The sharing procedure proceeds as follows
- Choose uniformly random coins_{nr} for the non-robust sharing procedure Share_{nr}.

- For $j \in [q]$: sample $(s_1^j, \ldots, s_n^j) \leftarrow \mathsf{Share}(\mathsf{msg}; \mathsf{coins}_{nr})$ where s_i^j equals $(\widetilde{s}_i, E_i^j, \mathsf{key}_i^j, r_i^j, p_i^j)$. Note that the non-robust (Shamir) shares $\widetilde{s}_i$ are the same in all q iterations but the other components are selected with fresh randomness in each iteration.
- For $i \in [n]$, define the party i share as $s_i = (\widetilde{s}_i, \{(E_i^j, \mathsf{key}_i^j, r_i^j, p_i^j)\}_{j=1}^q)$ and output $(s_1, \ldots, s_n)$.

$\mathsf{Rec}'(s_1, \ldots, s_n)$: The reconstruction procedure proceeds as follows

- For $i \in [n]$, parse $s_i = (\widetilde{s}_i, \{(E_i^j, \mathsf{key}_i^j, r_i^j, p_i^j)\}_{j=1}^q)$.
  For $j \in [q]$, define $s_i^j := (\widetilde{s}_i, E_i^j, \mathsf{key}_i^j, r_i^j, p_i^j)$.
- For $j \in [q]$, let $\mathsf{msg}_j := \mathsf{Rec}(s_1^j, \ldots, s_n^j)$. If there is a majority value msg such that $|\{j \in [q] \ : \ \mathsf{msg} = \mathsf{msg}_j\}| > q/2$ then output msg, else output $\perp$.

Analysis. We prove that the parallel repetition scheme satisfies robustness. Assume that the parameters of $(\mathsf{Share}, \mathsf{Rec})$ are chosen such that the scheme is δ-robust.

We first claim that the scheme $(\mathsf{Share}, \mathsf{Rec})$ remains robust even if we fix the random coin coins_{nr} for the non-robust secret sharing scheme (in step S.1 of the Share function) to some worst-case value but use fresh randomness in all the other steps. The fact that coins_{nr} are random was essential for privacy but it does not affect robustness. In particular, let us consider the robustness experiment $\boldsymbol{Exp}(\mathsf{msg}, \mathsf{Adv})$ for the scheme $(\mathsf{Share}, \mathsf{Rec})$ and let us define $\boldsymbol{Exp}(\mathsf{msg}, \mathsf{Adv}; \mathsf{coins}_{nr})$ to be the experiment when using some fixed choice of coins_{nr} but fresh randomness everywhere else. We can strengthen the statement of Lemma 3 which proves the robustness of $(\mathsf{Share}, \mathsf{Rec})$ to show the following.

Lemma 4 (Strengthening of Lemma 3). *The scheme* $(\mathsf{Share}, \mathsf{Rec})$ *remains robust even if* coins_{nr} *is fixed to a worst-case value. In particular, for any* $\mathsf{msg} \in \mathcal{M}$*, any choice of* coins_{nr} *and for all adversaries* Adv *we have*

$$\Pr[\mathbf{Exp}(\mathsf{msg}, \mathsf{Adv}; \mathsf{coins}_{nr}) = 1] \leq \delta.$$

The proof of the above lemma follows the lines of that of Lemma 3. See the full version [BPRW15] for more details.

Theorem 6. *Assume that the parameters of* $(\mathsf{Share}, \mathsf{Rec})$ *are chosen such that the scheme is δ-robust for $\delta \leq \frac{1}{4}$. Then the q-wise parallel repetition scheme* $(\mathsf{Share}', \mathsf{Rec}')$ *is a δ'-robust secret sharing scheme with $\delta' = e^{-\frac{3}{128}q}$.*

The proof of the above theorem is given in the full version [BPRW15].

Parameters. We choose the parameters of the underlying scheme $(\mathsf{Share}, \mathsf{Rec})$ to have security $\delta = \frac{1}{4}$. This corresponds to choosing a sufficiently large $\lambda = O(1)$ and results in a share size of $m + O(\log^4 n + \log^3 n \log m)$ bits (Eq. 5.2). By choosing a sufficiently large $q = O(k)$ and setting $(\mathsf{Share}', \mathsf{Rec}')$ to be the q-wise parallel repetition scheme from above, we get a scheme with robustness $\delta' = 2^{-k}$ and share size

$$m + O(k(\log^4 n + \log^3 n \log m)) = m + \widetilde{O}(k).$$

We note that part of the reason for the large poly-logarithmic factors comes from the parameters of our efficient graph identification algorithm which requires us to set the graph degree to $d = O(\log^3 n)$. If we had instead simply relied on our inefficient graph identification algorithm (Corollary 1) we could set $d = O(\log n)$ and would get an inefficient robust secret sharing scheme with share size $m + O(k(\log^2 n + \log n \log m))$. It remains an interesting challenge to optimize the poly-logarithmic factors.

7 Inefficient Graph Identification via Self-Consistency

We now return to the graph identification problem defined in Sect. 4.3. We begin by showing a simple inefficient algorithm for the graph identification problem. In particular, we show that with overwhelming probability the set $H \cup P$ is the unique maximum *self-consistent* set of vertices, meaning that there are no bad edges between vertices in the set.

Definition 8 (Self-Consistency). *Let $G = (V, E)$ be a directed graph and let $L : V \to \{\mathsf{good}, \mathsf{bad}\}$ be a labeling. We say that a subset of vertices $S \subseteq V$ is* self-consistent *if for all $e \in E_{S \to S}$ we have $L(e) = \mathsf{good}$. A subset $S \subseteq V$ is* max self-consistent *if $|S| \geq |S'|$ for every self-consistent $S' \subseteq V$.*

Note that, in general, there may not be a unique max self-consistent set in G. However, the next lemma shows that if the components are sampled as in the graph identification challenge game $\mathsf{Gen}^{\mathsf{Adv}}(n, t, d)$, then with overwhelming probability there is a unique max self-consistent set in G and it is $H \cup P$.

Lemma 5. *For any Adv, and for the distribution $(G, L, H, A, P) \leftarrow \mathsf{Gen}^{\mathsf{Adv}}(n, t, d)$, the set $H \cup P$ is the unique max self-consistent set in G with probability at least $1 - 2^{-\Omega(d - \log n)}$.*

Proof. We know that the set $H \cup P$ is self-consistent by the definition of the graph identification challenge. Assume that it is not the unique max self-consistent set in G, which we denote by the event BAD. Then there exists some set $S \neq H \cup P$ of size $|S| = |H \cup P|$ such that S is self consistent. This means that S must contain at least $q \geq 1$ elements from A and at least $t + 1 - q$ elements from H. In other words there exists some value $q \in \{1, \ldots, t\}$ and some subsets $A' \subseteq S \cap A \subseteq A \subseteq A \cup P$ of size $|A'| = q$ and $H' \subseteq S \cap H \subseteq H$ of size $t + 1 - q$ such that $E_{H' \to A'} = \varnothing$. This is because, by the definition of the graph challenge game, every edge in $E_{H' \to A'} \subseteq E_{H \to A}$ is labeled bad and so it must be empty if S is consistent. For any fixed q, A', H' as above, if we take the probability over the random choice of d outgoing edges for each $v \in H'$, we get:

$$\Pr[E_{H' \to A'} = \varnothing] = \left(\frac{\binom{n-1-q}{d}}{\binom{n-1}{d}} \right)^{t+1-q} \leq \left(1 - \frac{q}{n-1}\right)^{d(t+1-q)} \leq e^{-\frac{d(t+1-q)q}{n}}.$$

By taking a union bound, we get

$$\Pr[\mathsf{BAD}] \leq \Pr\left[\exists \left\{ \begin{array}{c} q \in \{1,\ldots,t\} \\ A' \subseteq A \cup P : |A'| = q \\ H' \subseteq H, |H'| = t+1-q \end{array} \right\} : E_{H' \to A'} = \varnothing \right]$$

$$\leq \sum_{q=1}^{t} \binom{t+1}{t+1-q} \cdot \binom{t}{q} \cdot e^{-\frac{d(t+1-q)q}{n}} \leq \sum_{q=1}^{t} \binom{t+1}{t+1-q} \cdot \binom{t+1}{q} \cdot e^{-\frac{d(t+1-q)q}{n}}$$

$$\leq 2 \sum_{q=1}^{(t+1)/2} \binom{t+1}{q}^2 \cdot e^{-\frac{d(t+1-q)q}{n}} \qquad \text{(symmetry between } q \text{ and } t+1-q)$$

$$\leq 2 \sum_{q=1}^{(t+1)/2} (t+1)^{2q} \cdot e^{-\frac{d(t+1-q)q}{n}}$$

$$\leq 2 \sum_{q=1}^{(t+1)/2} e^{q\left(2\log_e(t+1) - \frac{d(t+1-q)}{n}\right)}$$

$$\leq 2 \sum_{q=1}^{(t+1)/2} e^{q\left(2\log_e(t+1) - \frac{(t+1)d}{2n}\right)} \qquad \text{(since } q \leq (t+1)/2)$$

$$\leq 2 \sum_{q=1}^{(t+1)/2} e^{q(2\log_e(t+1) - d/4)} \qquad \text{(since } t+1 > n/2)$$

$$\leq (t+1)e^{(2\log_e(t+1)-d/4)} \leq 2^{-\Omega(d-\log n)}$$

As a corollary of the above lemma, we get an inefficient algorithm for the graph identification problem, that simply tries every subset of vertices $S \subseteq V$ and outputs the max self-consistent set.

Corollary 1. *There exists an inefficient algorithm* $\mathsf{GraphID}^{\mathsf{ineff}}$ *such that for any* Adv*:*

$$\Pr\left[B = H \cup P \ : \ \begin{array}{c} (G, L, H, A, P) \leftarrow \mathsf{Gen}^{\mathsf{Adv}}(n = 2t+1, t, d), \\ B \leftarrow \mathsf{GraphID}^{\mathsf{ineff}}(G, L) \end{array}\right] \geq 1 - 2^{-\Omega(d-\log n)}$$

Remark. Note that for the analysis of the inefficient graph reconstruction procedure in Lemma 5 and Corollary 1, we did not rely on the fact that edges from active to honest parties $e = (i,j) \ : \ i \in A, j \in H$ are labeled bad. Therefore, if we only wanted an inefficient graph identification procedure, we could relax the requirements in the graph challenge game and allow the adversary to choose arbitrary labels for such edges e. This would also allow us to simplify our robust secret sharing scheme and omit the "reverse-authentication" tags $\sigma_{i \leftarrow j}$.

8 Efficient Graph Identification

In this section, we prove Theorem 4 and given an efficient graph identification algorithm. We begin with an intuitive overview before giving the technical details.

8.1 Overview and Intuition

A Simpler Problem. We will reduce the problem of identifying the full set $H \cup P$ to the simpler problem of only identifying a small set $Y \subseteq H \cup P$ such that $Y \cap H$ is of size at least εn for some $\varepsilon = 1/\Theta(\log n)$. If we are given such a Y, we can use it to identify a larger set S defined as all vertices in $[n]$ with no bad incoming edge originating in Y. We observe that every vertex in $H \cup P$ is included in S, as there are no bad edges from $H \cup P$ to $H \cup P$. On the other hand, since $Y \cap H$ is big enough, it is unlikely that a vertex in A could be included in S, as every vertex in A likely has an incoming edge from $|Y \cap H|$ that is labeled as bad. Therefore, with high probability $S = H \cup P$. There is a bit of subtlety involved in applying this intuition, as it is potentially complicated by dependencies between the formation of the set Y and the distribution of the edges from Y to A. We avoid dealing with such dependencies by "splitting" the graph into multiple independent graphs and building Y from one of these graphs while constructing S in another.

Now the task becomes obtaining such a set Y in the first place. We consider two cases depending on whether the set P is small ($|P| \leq \varepsilon n$) or large ($|P| > \varepsilon n$).

Small P. In this case, there is only a small number of good edges crossing between H and $A \cup P$ (only edges between H and P). Therefore there exists a bisection of the graph into sets H and $A \cup P$ of size $t+1$ and t respectively, where the number of good edges crossing this bisection is approximately εdn. By using an efficient $O(\log n)$-approximation algorithm for the graph bisection problem (on the good edges in G) we can get a bisection X_0, X_1 with very few edges crossing between X_0 and X_1. This means that, with overwhelming probability, one of X_0 or X_1 contains the vast majority of the honest vertices, say $!.9|H|$, as otherwise if the honest vertices were split more evenly, we'd expect more edges crossing this partition. We can then refine such an X to get a suitable smaller subset Y which is fully contained in $H \cup P$, by taking all vertices that don't have too many incoming bad edges from X.

Large P. In this case, the intuition is that every vertex in A is likely to have at least $d/2$ in-coming bad edges (from the honest vertices), but honest/passive vertices will only have $d(1/2 - \varepsilon)$ in-coming bad edges on average from the active vertices. So we can differentiate the two cases just by counting. This isn't precise since many active vertices can point bad edges at a single honest vertex to make it "look bad". However, intuitively, this cannot happen too often.

To make this work, we first start with the full set of vertices $[n]$ and disqualify any vertices that have more than $d/2$ out-going bad edges (all honest vertices remain since they only have $d(1/2-\varepsilon)$ outgoing bad edges on expectation). This potentially eliminates some active vertices. Let's call the remaining smaller set of vertices X. We then further refine X into a subset Y of vertices that do not have too many incoming bad edges (more than $d(1/2 - \varepsilon/2)$) originating in X. The active vertices are likely to all get kicked out in this step since we expect $d/2$ incoming bad edges from honest vertices. On the other hand, we claim that not too many honest vertices get kicked out. The adversary has at most $(1/2-\varepsilon)dn/2$

out-going bad edges in total under his control in the set $X \cap A$ and has to spend $d(1/2 - \varepsilon/2)$ edges to kick out any honest party. Therefore there is a set of at least $\varepsilon n/2$ of the honest parties that must survive. This means that $Y \subseteq H \cup P$ and that Y contains $\Theta(n/\log n)$ honest parties as we wanted.

Unknown P. Of course, our reconstruction procedure will not know a priori whether P is relatively large or small or, in the case that P is small, which one of the bisection sets X_1 or X_2 to use. So it simply tries all of these possibilities and obtains three candidate sets Y_0, Y_1, Y_2, one of which has the properties we need but we do not know which one. To address this, we construct the corresponding sets S_i for each Y_i as described above, and we know that one of these sets S_i is $H \cup P$. From the previous section (Lemma 5), we also know that $H \cup P$ is the unique max self-consistent set in G. Therefore, we can simply output the largest one of the sets S_0, S_1, S_2 which is self-consistent in G and we are guaranteed that this is $H \cup P$.

8.2 Tool: Graph Splitting

As mentioned above, we will need to split the graph G into three sub-graphs G^1, G^2, G^3 such that the outgoing edges from honest parties are distributed randomly and independently in G^1, G^2, G^3. Different parts of our algorithm will use different sub-graphs and it will be essential that we maintain independence between them for our analysis.

In particular, we describe a procedure $(G^1, G^2, G^3) \leftarrow \mathsf{GraphSplit}(G)$ that takes as input a directed graph $G = (V = [n], E)$ produced by $\mathsf{Gen}^{\mathsf{Adv}}(n, t, d)$ and outputs three directed graphs $(G^i = (V, E^i))_{i=1,2,3}$ such that $E^i \subset E$ and the out-degree of each vertex in each graph is $d' := \lfloor d/3 \rfloor$. Furthermore, we require that the three sets $E^i_{H \to V}$ are random and independent subject to each vertex having out-degree d' and no self-loops. Note that forming the sets E^i by simply partitioning the outgoing edges of each vertex into three sets is *not* a good solution, since in that case the sets will always be disjoint and therefore not random and independent. On the other hand, sub-sampling three random subsets of d' outgoing edges from the set of d outgoing edges in E is also not a good solution since in the case the overlap between the sets is likely to be higher than it would be if we sampled random subsets of d' outgoing edges from all possible edges.

Our algorithm proceeds as follows.

$(G^1, G^2, G^3) \leftarrow \mathsf{GraphSplit}(G)$: On input a directed graph $G = (V, E)$ with out-degree d.

1. Define $d' = \lfloor d/3 \rfloor$.
2. For each vertex, for each $v \in V$:
   (a) Define $N_v := \{w \in V \mid (v, w) \in E\}$, the set of neighbors of v in G
   (b) Sample three uniform and independent sets $\{N^i_v\}_{i=1,2,3}$ with $N^i_v \subseteq V \setminus \{v\}$ and $|N^i_v| = d'$.

   (c) Sample a uniformly random injective function $\pi_v : \bigcup_{i=1,2,3} N_v^i \to N_v$.
   (d) Define $\hat{N}_v^i = \pi_v(N_v^i) \subseteq N_v$.
3. Define $E^i := \{(v, w) \in E \mid w \in \hat{N}_v^i\}$ and output $G^i = (V, E^i)$ for $i = 1, 2, 3$.

Intuitively, for each vertex v, we first sample the three sets of outgoing neighbors N_v^i independently at random from all of $V \setminus \{v\}$, but then we apply an injective function $\pi_v(N_v^i)$ to map them into a the original neighbors N_v. The last step ensures that $E^i \subseteq E$.

Lemma 6. *Let* $(G = (V, E), L, H, A, P) \leftarrow \mathsf{Gen}^{\mathsf{Adv}}(n, t, d)$ *for some adversary* Adv. *Let* $(G^i = (V, E^i))_{i=1,2,3} \leftarrow \mathsf{GraphSplit}(G)$. *Then the joint distribution of* $(E^i_{H\to V})_{i=1,2,3}$ *is identical to choosing each set* $E^i_{H\to V}$ *randomly and independently subject to each vertex having out-degree* d' *and no self-loops; i.e., for each* $i = 1, 2, 3$ *form the set* $E^i_{H\to V}$ *by taking each* $v \in H$ *and choosing a set of* d' *outgoing edges uniformly at random (without replacement) to vertices in* $V \setminus \{v'\}$.

Proof. For each $v \in H$, define $c_{\{1,2\}} = |N_v^1 \cap N_v^2|, c_{\{1,3\}} = |N_v^1 \cap N_v^3|, c_{\{2,3\}} = |N_v^2 \cap N_v^3|$ and $c_{\{1,2,3\}} = |N_v^1 \cap N_v^2 \cap N_v^3|$. We call these numbers the *intersection pattern of* $\{N_v^i\}_{i=1,2,3}$ and denote it by C. Analogously, we define the *intersection pattern of* $\{\hat{N}_v^i\}_{i=1,2,3}$ and denote it by $\hat{C}$.

It's easy to see that, for any fixed choice of $\{N_v^i\}_{i=1,2,3}$ with intersection pattern C, the sets $\{\hat{N}_v^i\}_{i=1,2,3}$ are uniformly random and independent subject to their intersection pattern being $\hat{C} = C$. This follows from the random choice of N_v and the injective function π_v.

Furthermore, since the distribution of the intersection pattern $\hat{C} = C$ is the same for $\{N_v^i\}_{i=1,2,3}$ and for $\{\hat{N}_v^i\}_{i=1,2,3}$, the distribution of $\{\hat{N}_v^i\}_{i=1,2,3}$ is identical to that of $\{N_v^i\}_{i=1,2,3}$. In other words, for each $v \in H$ the three sets of outgoing neighbors of v in G^1, G^2, G^3 are random and independent as we wanted to show.

8.3 The Graph Identification Algorithm

We now define the efficient graph identification algorithm $B \leftarrow \mathsf{GraphID}(G, L)$.

Usage. Our procedure $\mathsf{GraphID}(G, L)$ first runs an initialization phase ***Initialize***, and then runs two procedures ***Small*** P and ***Large*** P sequentially. It uses the data generated in these two procedures to then run the output phase **Output**.

Initialize.

1. Let b be a constant such that there exists a polynomial-time $b \log n$-approximate graph bisection algorithm Bisect, such as the one provided in [Räc08]. Let $c = \frac{800}{9} b$, and let $\varepsilon = 1/(c \cdot \log(n))$.
2. Run $(G^1, G^2, G^3) \leftarrow \mathsf{GraphSplit}(G)$ as defined in Sect. 8.2. This produces three graphs $G^i = (V, E^i)$ such that $E^i \subseteq E$ and the out-degree of each vertex in G^i is $d' = \lfloor d/3 \rfloor$.

Small P.

1. Run $(X_0, X_1) \leftarrow \mathsf{Bisect}(G^*)$, where $G^* = (V, E^*)$ is the undirected graph induced by the good edges of G^1:

$$E^* = \left\{ \{i,j\} \ : \ \begin{array}{l} (e = (i,j) \in E^1 \text{ and } L(e) = \mathsf{good}) \text{ or} \\ (e = (j,i) \in E^1 \text{ and } L(e) = \mathsf{good}) \end{array} \right\}$$

2. For $i = 0, 1$: contract X_i to a set of *candidate good* vertices Y_i that have fewer than $0.4d'$ incoming bad edges in the graph G^2.

$$Y_i := \left\{ v \in X_i \ : \ \left| \{ e \in E^2_{X_i \to \{v\}} \ : \ L(e) = \mathsf{bad} \} \right| < 0.4d' \right\}.$$

Large P.

1. Define a set of *candidate legal* vertices X_2 as the set of vertices having fewer than $d'/2$ outgoing bad edges in G^1.

$$X_2 := \left\{ v \in V \ : \ \left| \{ e \in E^1_{\{v\} \to V} \ : \ L(e) = \mathsf{bad} \} \right| < d'/2 \right\}.$$

2. Contract X_2 to a set of *candidate good* vertices Y_2, defined as the set of vertices in X_2 having fewer than $d'(1/2 - \varepsilon/2)$ incoming bad edges from legal vertices in the graph G^1.

$$Y_2 := \left\{ v \in X_2 \ : \ \left| \{ e \in E^1_{X_2 \to \{v\}} \ : \ L(e) = \mathsf{bad} \} \right| < d'(1/2 - \varepsilon/2) \right\}.$$

Output. This subprocedure takes as input the sets Y_0, Y_1 (generated by **Small** P), and Y_2 (generated by **Large** P) and outputs a single set B, according to the following algorithm.

1. For $i = 0, 1, 2$: define S_i as the set of vertices that only have incoming good edges from Y_i in G^3. Formally,

$$S_i := \{ \ v \in V \ : \ \forall e \in E^3_{Y_i \to v} \ \ L(e) = \mathsf{good} \ \}$$

2. For $i = 0, 1, 2$: if $L(e) = \mathsf{good}$ for all $e \in E_{S_i \to S_i}$, define $B_i := S_i$; otherwise, define $B_i = \varnothing$. This ensures that each B_i is self-consistent (Definition 8) in G.
3. Output a set B defined as any of the largest sets among B_0, B_1, B_2.

8.4 Analysis of Correctness – Overview

In this section, we give an intuition for why the algorithm GraphID outlined above satisfies Theorem 4.

We first fix an arbitrary adversary Adv in the graph challenge game. We consider the distribution induced by running the randomized processes $(G, L, H, A, P) \leftarrow \mathsf{Gen}^{\mathsf{Adv}}(n = 2t+1, t, d)$ and $B \leftarrow \mathsf{GraphID}(G, L)$. Note that without loss of generality we can assume Adv is deterministic and therefore the sets H, A, P are fixed. The only randomness in experiment consists of the choice of edges $E_{H \to V}$ in the execution of $\mathsf{Gen}^{\mathsf{Adv}}(n = 2t + 1, t, d)$ and the randomness of the graph splitting procedure $(G^1, G^2, G^3) \leftarrow \mathsf{GraphSplit}(G)$ during the execution of

GraphID(G, L). By the property of graph splitting (Lemma 6) we can think of this as choosing three independent sets $(E^i_{H\to V})_{i=1,2,3}$. This induces a distribution on the sets X_i, Y_i, S_i, B_i and B defined during the course of the GraphID algorithm and we analyze the probability of various events over this distribution.

We define a *sufficient event*:

$$O := \text{"there exists } i \in \{0,1,2\} \text{ such that } |Y_i \cap H| \geq \varepsilon \cdot n/2 \text{ and } Y_i \subseteq H \cup P\text{"}$$

In the full version [BPRW15], we prove the following technical lemmas that allow us to prove Theorem 4.

Lemma 7. *The conditional probability of $B = H \cup P$, given the occurrence of event O, is $1 - 2^{-\Omega(d/\log n)}$.*

Lemma 8. *If $|P| < \varepsilon \cdot n$, then the probability that O occurs is at least $1 - 2^{-\Omega(d/\log n)}$.*

Lemma 9. *If $|P| \geq \varepsilon \cdot n$, then the probability that O occurs is at least $1 - 2^{-\Omega(d/\log^2 n - \log n)}$.*

In Lemma 7, the probability is over the random edges $E^3_{H\to V}$ in G^3, while in Lemmas 8 and 9, the probability is over the random edges $E^i_{H\to V}$ in G^i for $i = 1, 2$. Therefore, conditioning on the event O in Lemma 7 does not effect the probability distribution.

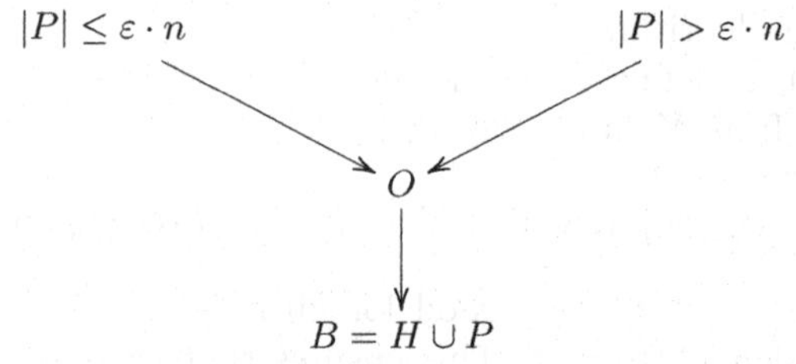

Fig. 1. Structure of our analysis: arrows denote logical implications (happening with high probability).

Our analysis is summarized in Fig. 1. We now complete the proof of Theorem 4.

Proof of Theorem 4. By Lemmas 8 and 9, we obtain that the event O occurs with probability at least $1 - 2^{-\Omega(d/\log^2 n - \log n)}$. Putting together with Lemma 7, we obtain that the probability that the set B returned by the algorithm equals $H \cup P$ is at least $1 - 2^{-\Omega(d/\log^2 n - \log n)}$ completing the proof of the theorem.

9 Conclusion

We constructed an efficient robust secret sharing scheme for the maximal corruption setting with $n = 2t+1$ parties with nearly optimal share size of $m+\widetilde{O}(k)$ bits, where m is the length of the message and 2^{-k} is the failure probability of the reconstruction procedure with adversarial shares.

One open question would be to optimize the poly-logarithmic terms in our construction. It appears to be an interesting question to attempt to go all the way down to $m + O(k)$ or perhaps even just $m + k$ bits for the share size, or to prove a lower bound that (poly)logarithmic factors in n, m are necessary. We leave this as a challenge for future work.

Acknowledgments. Daniel Wichs: Research supported by NSF grants CNS-1347350, CNS-1314722, CNS- 1413964.

Valerio Pastro and Daniel Wichs: This work was done in part while the authors were visiting the Simons Institute for the Theory of Computing, supported by the Simons Foundation and by the DIMACS/Simons Collaboration in Cryptography through NSF grant CNS-1523467.

References

[Bla79] Blakley, G.R.: Safeguarding cryptographic keys. In: International Workshop on Managing Requirements Knowledge, pp. 313–317. IEEE Computer Society (1979)

[BPRW15] Bishop, A., Pastro, V., Rajaraman, R., Wichs, D.: Essentially optimal robust secret sharing with maximal corruptions. IACR Cryptology ePrint Archive, 2015:1032 (2015)

[BS97] Blundo, C., De Santis, A.: Lower bounds for robust secret sharing schemes. Inf. Process. Lett. **63**(6), 317–321 (1997)

[CDD+15] Cramer, R., Damgård, I.B., Döttling, N., Fehr, S., Spini, G.: Linear secret sharing schemes from error correcting codes and universal hash functions. In: Oswald, E., Fischlin, M. (eds.) EUROCRYPT 2015. LNCS, vol. 9057, pp. 313–336. Springer, Heidelberg (2015)

[CDF01] Cramer, R., Damgård, I.B., Fehr, S.: On the cost of reconstructing a secret, or VSS with optimal reconstruction phase. In: Kilian, J. (ed.) CRYPTO 2001. LNCS, vol. 2139, pp. 503–523. Springer, Heidelberg (2001)

[CDF+08] Cramer, R., Dodis, Y., Fehr, S., Padró, C., Wichs, D.: Detection of algebraic manipulation with applications to robust secret sharing and fuzzy extractors. In: Smart, N.P. (ed.) EUROCRYPT 2008. LNCS, vol. 4965, pp. 471–488. Springer, Heidelberg (2008)

[Cev11] Cevallos, A.: Reducing the share size in robust secret sharing (2011). http://www.algant.eu/documents/theses/cevallos.pdf

[CFOR12] Cevallos, A., Fehr, S., Ostrovsky, R., Rabani, Y.: Unconditionally-secure robust secret sharing with compact shares. In: Pointcheval, D., Johansson, T. (eds.) EUROCRYPT 2012. LNCS, vol. 7237, pp. 195–208. Springer, Heidelberg (2012)

[Che15] Cheraghchi, M.: Nearly optimal robust secret sharing. IACR Cryptology ePrint Archive, 2015:951 (2015)

[CSV93] Carpentieri, M., De Santis, A., Vaccaro, U.: Size of shares and probability of cheating in threshold schemes. In: Helleseth, T. (ed.) EUROCRYPT 1993. LNCS, vol. 765, pp. 118–125. Springer, Heidelberg (1994)

[FK02] Feige, U., Krauthgamer, R.: A polylogarithmic approximation of the minimum bisection. SIAM J. Comput. **31**(4), 1090–1118 (2002)

[GJS76] Garey, M.R., Johnson, D.S., Stockmeyer, L.J.: Some simplified np-complete graph problems. Theor. Comput. Sci. **1**(3), 237–267 (1976)

[JS13] Jhanwar, M.P., Safavi-Naini, R.: Unconditionally-secure robust secret sharing with minimum share size. In: Sadeghi, A.-R. (ed.) FC 2013. LNCS, vol. 7859, pp. 96–110. Springer, Heidelberg (2013)

[LP14] Lewko, A.B., Pastro, V.: Robust secret sharing schemes against local adversaries. IACR Cryptology ePrint Archive, 2014:909 (2014)

[Räc08] Räcke, H.: Optimal hierarchical decompositions for congestion minimization in networks. In: Dwork, C. (ed.) Proceedings of the 40th Annual ACM Symposium on Theory of Computing, Victoria, British Columbia, Canada, 17–20 May 2008, pp. 255–264. ACM (2008)

[RB89] Rabin, T., Ben-Or, M.: Verifiable secret sharing and multiparty protocols with honest majority (extended abstract). In: Johnson, D.S. (ed.) Proceedings of the 21st Annual ACM Symposium on Theory of Computing, Seattle, Washigton, USA, 14–17 May 1989, pp. 73–85. ACM (1989)

[Sha79] Shamir, A.: How to share a secret. Commun. ACM **22**(11), 612–613 (1979)

[Sud97] Sudan, M.: Decoding of reed solomon codes beyond the error-correction bound. J. Complex. **13**(1), 180–193 (1997)

Provably Robust Sponge-Based PRNGs and KDFs

Peter Gaži[1(✉)] and Stefano Tessaro[2(✉)]

[1] IST Austria, Klosterneuburg, Austria
peter.gazi@ist.ac.at
[2] UC Santa Barbara, Santa Barbara, USA
tessaro@cs.ucsb.edu

Abstract. We study the problem of devising provably secure PRNGs with input based on the sponge paradigm. Such constructions are very appealing, as efficient software/hardware implementations of SHA-3 can easily be translated into a PRNG in a nearly black-box way. The only existing sponge-based construction, proposed by Bertoni *et al.* (CHES 2010), fails to achieve the security notion of robustness recently considered by Dodis *et al.* (CCS 2013), for two reasons: (1) The construction is deterministic, and thus there are high-entropy input distributions on which the construction fails to extract random bits, and (2) The construction is not forward secure, and presented solutions aiming at restoring forward security have not been rigorously analyzed.

We propose a *seeded* variant of Bertoni *et al.*'s PRNG with input which we prove secure in the sense of robustness, delivering in particular concrete security bounds. On the way, we make what we believe to be an important conceptual contribution, developing a variant of the security framework of Dodis *et al.* tailored at the ideal permutation model that captures PRNG security in settings where the weakly random inputs are provided from a large class of possible adversarial samplers *which are also allowed to query the random permutation.*

As a further application of our techniques, we also present an efficient sponge-based key-derivation function (which can be instantiated from SHA-3 in a black-box fashion), which we also prove secure when fed with samples from permutation-dependent distributions.

Keywords: PRNGs · Sponges · SHA-3 · Key derivation · Weak randomness

1 Introduction

Generating pseudorandom bits is of paramount importance in the design of secure systems – good pseudorandom bits are needed in order for cryptography to be possible. Typically, software-based *pseudorandom number generators* (PRNGs) collect entropy from system events into a so-called *entropy pool*, and then apply cryptographic algorithms (hash functions, block ciphers, PRFs, etc.)

M. Fischlin and J.-S. Coron (Eds.): EUROCRYPT 2016, Part I, LNCS 9665, pp. 87–116, 2016.
DOI: 10.1007/978-3-662-49890-3_4

to extract pseudorandom bits from this pool. These are also often referred to as PRNGs *with input*, as opposed to classical seed-stretching cryptographic PRGs.

There have been significant standardization efforts in the area of PRNGs [1,6,19], and an attack-centric approach [8,18,21,26,30] has mostly driven their evaluation. Indeed, the development of a comprehensive formal framework to *prove* PRNG security has been a slower process, mostly due to the complexity of the desirable security goals. First models [5,13,21] only gave partial coverage of the security desiderata. For instance, Barak and Halevi [5] introduced a strong notion of PRNG robustness, but their model could not capture the ability of a PRNG to collect randomness at a low rate. Two recent works [15,17] considerably improved this state of affairs with a comprehensive security framework for PRNG robustness whose inputs are adversarially generated (under some weak entropy constraints). The framework of [15] was recently applied to the study of the Intel on-chip PRNG by Shrimpton and Terashima [29].

This paper continues the investigation of good candidate constructions for PRNGs with inputs which are both practical and provably secure. In particular, we revisit the question of building PRNGs from permutations, inspired by recent sponge-based designs [10,31]. We provide variants of these designs which are provably robust in the framework of [15]. On the way, we also extend the framework of [15] to properly deal with security proofs in ideal models (e.g. when given a random permutation), in particular considering PRNG inputs sampled by adversaries which can make queries to the permutation.

Overall, this paper contributes to the development of a better understanding of sponge-based constructs when processing weakly random inputs. As a further testament of this, we apply our techniques to analyze key-derivation functions using sponge-based hash functions, like SHA-3.

<u>SPONGE-BASED PRNGS.</u> SHA-3 relies on the elegant sponge paradigm by Bertoni, Daemen, Peeters, and van Assche [9]. Beyond hash functions, sponges have been used to build several cryptographic objects. In particular, in later work [10], the same authors put forward a sponge-based design of a PRNG with input. It uses an efficiently computable (and invertible) permutation π, mapping n-bit strings to n-bit strings, and maintains an n-bit state, which is initially set to $S_0 \leftarrow 0^n$. Then, two types of operations can be alternated (for additional parameters $r \leq n$, and $c = n - r$, the latter being referred to as the *capacity*):

- <u>State refresh.</u> Weakly random material (e.g., resulting from the measurement of system events) can be added r-bit at a time. Concretely, given an r-bit string I_i of weakly random bits, the state is refreshed to

$$S_i \leftarrow \pi(S_{i-1} \oplus (I_i \,\|\, 0^c)) \; .$$

- <u>Random-bit generation.</u> Given the current state S_i, we can extract r bits of randomness by outputting $S_i[1 \ldots r]$, and updating the state as $S_{i+1} \leftarrow \pi(S_i)$. This process can be repeated multiple times to obtain as many bits as necessary.

This construction is very attractive. First off, it is remarkably simple. Second, it resembles the structure of the SHA-3/KECCAK hash function, and thus efficient implementations of this PRNG are possible with the advent of more and more optimized SHA-3 implementations in both software and hardware. In fact, recent work by Van Herrewege and Verbauwhede [31] has already empirically validated the practicality of the design. Also, the permutation π does not need to be the KECCAK permutation – one could for example use AES on a fixed key.

PRNG SECURITY. Of course, we would like the simplicity of this construction to be also backed by strong security guarantees. The minimum security requirement is that whenever a PRNG has accumulated sufficient entropy material, the output bits are indistinguishable from random. The original security analysis of [10] proves this (somewhat indirectly) by showing that the above construction is indifferentiable [23] from a "generalized random oracle" which takes a sequence of inputs $I_1, I_2, \ldots$ through refresh operations, and when asked to produce a certain output after k inputs have been processed, it simply applies a random function to the concatenation of $I_1, I_2, \ldots, I_k$. This definition departs substantially from the literature on PRNG robustness, and only provides minimal security – for example, it does not cover any form of state compromise.

In contrast, here we call for a provably-secure sponge-based PRNG construction which is *robust* in the sense of [15]. However, there are two reasons why the construction, as presented above, is not robust.

(1) NO FORWARD SECRECY. As already recognized in [10], the above PRNG is not forward secure – in particular, learning the state S just after some pseudorandom bits have been output allows to distinguish them from random ones by just computing $\pi^{-1}(S)$. The authors suggest a countermeasure to this: simply zeroing the upper r bits of the input to π before computing the final state, possibly multiple times if r is small. More formally, given the state S'_k obtained after outputting enough pseudorandom bits, and applying π, we compute $S_k, S'_{k+1}, S_{k+1}, \ldots, S'_{k+t}, S_{k+t}$ as

$$S'_{i+1} \leftarrow \pi(S_i) \; ,$$

for $i = k, \ldots, k+t-1$, where S_i is obtained from S'_i by setting the first r bits to 0. This appears to prevent obvious attacks, and to make the construction more secure as t increases, but no formal validation is provided in [10].

In particular, note that the final state S_{k+t} is *not* random, as its first r bits are all 0. Robustness demands that we obtain random bits from S_{k+t} even when no additional entropy is added – unfortunately we cannot just proceed as above, since this will result in outputting r zero bits. (Also note that applying π also does not make the state random, since π is efficiently invertible.) This indicates that a further modification is needed.

(2) LACK OF A SEED. The above sponge-based PRNG is unseeded: This allows for high min-entropy distributions (only short of one bit from maximal entropy) for which the generated bits are not uniform. For example, consider $I = (I_1, \ldots, I_k)$, where each I_j is an r-bit string, and such that I is uniformly

distributed under the sole constraint that the first bit of the state S_k obtained after injecting all k blocks $I_1, \ldots, I_k$ into the state is always 0. Then, we can never expect the construction to provide pseudorandom bits under such inputs.

One could restrict the focus to "special" distributions as done in [5], arguing nothing like the above would arise in practice. As discussed in [15], however, arguing which sources are possible is difficult, and following the traditional cryptographic spirit, it would be highly desirable to reduce assumptions on the input distributions, which ideally should be *adversarially generated*, at the cost of introducing a (short) random seed which is independent of the distribution.

We note that the above distribution would also invalidate the weak security expectation from [10]. However, their treatment bypasses this problem by employing the random permutation model, where effectively the randomly chosen permutation acts as a seed, *independent of the input distribution.* We believe however this approach (which is not unique to [10]) to be problematic – the random permutation model is only used *as a tool in the security proof* due to the lack of standard-model assumptions under which the PRNG can be proved secure. Yet, in instantiations, the permutation is fixed. In contrast, a PRNG seed is an *actual short string which can and should be actually randomly chosen.*

OUR RESULTS. We propose and analyze a new sponge-based seeded construction of a PRNG with input (inspired by the one of [10]) which we prove robust. To this end, we use an extension of the framework of [15] tailored at the ideal-permutation model, and dealing in particular with inputs that are generated by adversarial samplers that can query the permutation. The construction (denoted **SPRG**) uses a seed seed, consisting of s r-bit strings $\mathsf{seed}_0, \ldots, \mathsf{seed}_{s-1}$ (s is not meant to be too large here, not more than 2 or 3 in actual deployment). Then, the construction allows to interleave two operations:

- State refresh. The construction here keeps a state $S_i \in \{0,1\}^n$ and a counter $j \in \{0, 1, \ldots, s-1\}$. Given a string I_i of r weakly random bits, the state is refreshed to

$$S_{i+1} \leftarrow \pi(S_i \oplus (I_i \oplus \mathsf{seed}_j) \,\|\, 0^c) ,$$

  and j is set to $j + 1 \bmod s$.
- Random-bit generation. Given the current state S_i, we can extract r bits of randomness by computing $S_{i+1} \leftarrow \pi(S_i)$, and outputting the first r bits of S_{i+1}. (This process can be repeated multiple times to obtain as many bits as necessary.) When done, we refresh the state by repetitively zeroing its first r bits and applying π, as described above. (How many times we do this is given by a second parameter – t – which ultimately affects the security of the PRNG.)

For a sketch of **SPRG** see Fig. 5. Thus, the main difference over the PRNG of [10] are (1) The use of a seed, (2) The zeroing countermeasure discussed above, and (3) An additional call to π before outputting random bits. In particular, note that

SPRG still follows the sponge principle, and in fact (while this may not be the most efficient implementation), can be realized from a sponge hash function (e.g., SHA-3) in an entirely black-box way[1].

In our proof of security, the permutation is randomly chosen, and both the attacker *and* an adversarial sampler of the PRNG inputs have oracle access to it. In fact, an important contribution of our work is that of introducing a security framework for PRNG security based on [15,29] for the ideal permutation model, and we see this as a step of independent interest towards a proper treatment of ideal-model security for PRNG constructions. As a word of warning, we stress that our proofs consider a *restricted* class of permutation-dependent distribution samplers, where the restriction is in terms of imposing an unpredictability constraint which must hold even under (partial) exposure of some (but not all) of the sampler's queries. While our notion is general enough to generalize previous oracle-free samplers and to encompass non-trivial examples (in particular making seedless extraction impossible, which is what we want for the model to be meaningful), we see potential for future research in relaxing this requirement.

SPONGE-BASED KEY DERIVATION. Our techniques can be used to immediately obtain provable security guarantees for sponge-based key-derivation functions (KDFs). (See Sect. 6.) While the security of sponge-based KDFs already follows from the original proof of [9], our result will be stronger in that it will also hold for larger classes of permutation-dependent sources. We elaborate on this point a bit further down in the last paragraph of the introduction, mentioning further related work.

OUR TECHNIQUES. Our analysis follows from two main results, of independent interest, which we briefly outline here. Both results are obtained using Patarin's H-coefficient method, as reviewed in Sect. 2.

The first result – which we refer to as the *extraction lemma* – deals with the ability of extracting keys from weak sources using sponges. In particular, we consider the seeded construction **Sp** which starting from some initial state $S_0 = \mathsf{IV}$, and obtaining r-bit blocks $I_1, \ldots, I_k$ from a weak random source, and a seed $\mathsf{seed} = (\mathsf{seed}_0, \ldots, \mathsf{seed}_{s-1})$, iteratively computes $S_1, \ldots, S_k$ as

$$S_i \leftarrow \pi(S_{i-1} \oplus (I_i \oplus \mathsf{seed}_j) \,\|\, 0^c) ,$$

where j is incremented modulo s after each iteration. Ideally, we want to prove that if $I_1, \ldots, I_k$ has high min-entropy h, then the output S_k is random, as long as the adversary (who can see the seed and choose the IV) cannot query the permutation more than (roughly) 2^h times[2]. Note that this cannot be true in

[1] Zeroing the upper r bits when refreshing the state after PRNG output can be done by outputting the top r-bit part to be zeroed, and adding it back in.

[2] One may hope to prove a result which is independent of the number of queries, akin to [14], as after all this structure resembles that of CBC. Yet, we will need to restrict the number of queries for the overall security to hold, and given this, we can expect better extraction performance – in particular, the output can be uniform for $h \ll n$, whereas $h \geq n$ would be necessary if we wanted an unrestricted result.

general – take e.g. $k = 1$, and even if I_1 is uniformly random, one single inversion query $\pi^{-1}(S_1)$ is enough to distinguish S_1 from a random string, as in the former case the lower c bits will equal those of the IV. Still, we will be able to prove that this attack is the only way to distinguish – roughly, we will prove that S_k *is uniform* as long as the adversary does not query $\pi^{-1}(S_k)$ when given a random S_k, except with negligible probability. This will be good enough for key-derivation applications, where we will need this result for specific adversaries for which querying $\pi^{-1}(S_k)$ will correspond to querying the *secret key* for an already secure construction. In fact, we believe the approach of showing good extraction properties for restricted adversaries to be novel for ideal-model analyses, and of potential wider appeal. (A moral analogue of this in the standard-model is the work of Barak *et al.* on application-specific entropy-loss reduction for the leftover-hash lemma [4].)

We note that the extraction lemma is even more general – we will consider a generalized extraction game where an adversary can adaptively select a subset of samples from an (also adversarial) distribution sampler with the guarantee of having sufficient min-entropy. We also note that at the technical level this result is inspired by recent analyses of key absorption stages within sponge-based PRFs using key-prepending [3,20]. Nonetheless, these works only considered the case of uniform keys, and not permutation-dependent weakly-random inputs.

Another component of possibly independent interest studies the security of the step generating the actual random bits, when initialized with a state of sufficient pseudorandomness. This result will show that security increases with the number t of zeroing steps applied to the state, i.e., the construction is secure as long as the adversary makes less than 2^{rt} queries.

Related work on oracle dependence. As shown in [28], indifferentiability does not have any implications on multi-stage games such as robustness for permutation-dependent distributions. Indeed, [28] was also the first work (to the best our knowledge) to explicitly consider primitive-dependent samplers, in the context of deterministic and hedged encryption. These results were further extended by a recent notable work of Mittelbach [25], who provided general conditions under which indifferentiability can be used in multi-stage settings.

We note that Mittelbach's techniques can be used to prove that some indifferentiable hash constructions are good extractors. However, this does not help us in proving the extraction lemma, as the construction for which we prove the lemma is not indifferentiable to start with, and thus the result fails. There is hope however that Mittlebach's technique could help us in proving our KDF result of Sect. 6 via the indifferentiability proof for sponges [9] possibly for an even larger class of permutation dependent samplers. We are not sure whether this is the case, and even if possible, what the quantitative implications would be – Mittelbach results are not formulated in the framework of sponges. In contrast, here we obtain our result as a direct corollary of our extraction lemma.

We also note that oracle-dependence was further considered in other multi-stage settings, for instance for related-key security [2]. Also, oracle-dependence can technically be seen as a form of seed-dependence, as considered e.g. in [16], but we are not aware of any of their techniques finding applications in our work.

2 Preliminaries

<u>BASIC NOTATION.</u> We denote $[n] := \{1, \ldots, n\}$. For a finite set $\mathcal{S}$ (e.g., $\mathcal{S} = \{0,1\}$), we let $\mathcal{S}^n$ and $\mathcal{S}^*$ be the sets of sequences of elements of $\mathcal{S}$ of length n and of arbitrary length, respectively. We denote by $S[i]$ the i-th element of $S \in \mathcal{S}^n$ for all $i \in [n]$. Similarly, we denote by $S[i \ldots j]$, for every $1 \leq i \leq j \leq n$, the subsequence consisting of $S[i], S[i+1], \ldots, S[j]$, with the convention that $S[i \ldots i] = S[i]$. $S_1 \parallel S_2$ denotes the concatenation of two sequences $S_1, S_2 \in \mathcal{S}^*$, and if $\mathcal{S}_1, \mathcal{S}_2$ are two subsets of $\mathcal{S}^*$, we denote by $\mathcal{S}_1 \parallel \mathcal{S}_2$ the set $\{S_1 \parallel S_2 : S_1 \in \mathcal{S}_1, S_2 \in \mathcal{S}_2\}$. Moreover, for a single-element set $\mathcal{S}_1 = \{X\}$ we simplify the notation by writing $X \parallel \mathcal{S}_2$ instead of $\{X\} \parallel \mathcal{S}_2$. We let $\mathsf{Perms}(n)$ be the set of all permutations on $\{0,1\}^n$. We denote by $X \xleftarrow{\$} \mathcal{X}$ the process of sampling the value X uniformly at random from a set $\mathcal{X}$. For a bitstring $X \in \{0,1\}^*$, we denote by $X_1, \ldots, X_\ell \xleftarrow{r} X$ parsing it into ℓ r-bit blocks, using some fixed padding method. The distance of two discrete random variables X and Y over a set $\mathcal{X}$ is defined as $\mathbf{SD}(X,Y) = \frac{1}{2}\sum_{x \in \mathcal{X}} |\mathsf{Pr}[X = x] - \mathsf{Pr}[Y = x]|$. Finally, recall that the min-entropy $\mathbf{H}_\infty(X)$ of a random variable X with range $\mathcal{X}$ is defined as $-\log(\max_{x \in \mathcal{X}} \mathsf{Pr}[X = x])$.

<u>GAME-BASED DEFINITIONS.</u> We use the game-playing formalism in the spirit of [7]. For a game G, we denote by $\mathsf{G}(\mathcal{A}) \Rightarrow 1$ the event that after an adversary $\mathcal{A}$ plays this game, the game outputs the bit 1. Similarly, $\mathsf{G}(\mathcal{A}) \rightarrow 1$ denotes the event that the output of the adversary $\mathcal{A}$ itself is 1.

<u>IDEAL PERMUTATION MODEL.</u> We perform our analysis in the *ideal permutation model (IPM)*, where each party has oracle access to a public, uniformly random permutation π selected at the beginning of any security experiment. For any algorithm A, we denote by A^π (or $A[\pi]$) that it has access to an oracle permutation π, which can be queried in *both* the forward and backward direction. In the game descriptions below, we sometimes explicitly mention the availability of π to the adversary as oracles π and π^{-1} for forward and backward queries, respectively. We define a natural distinguishing metric for random variables in the IPM. Given two distributions D_0 and D_1, possibly dependent on the random permutation $\pi \xleftarrow{\$} \mathsf{Perms}(n)$, and an adversary $\mathcal{A}$ querying π, we denote

$$\mathsf{Adv}^{\mathsf{dist}}_{\mathcal{A}}(D_0, D_1) = \mathsf{Pr}\left[X \xleftarrow{\$} D_0^\pi : \mathcal{A}^\pi(X) \Rightarrow 1\right] - \mathsf{Pr}\left[X \xleftarrow{\$} D_1^\pi : \mathcal{A}^\pi(X) \Rightarrow 1\right].$$

We call $\mathcal{A}$ a q_π-adversary if it asks q_π queries to π.

<u>PRNGS WITH INPUT.</u> We use the framework of [15] where a *PRNG with input* is defined as a triple of algorithms $\mathbf{G} = (\mathsf{setup}, \mathsf{refresh}, \mathsf{next})$ parametrized by integers $n, r \in \mathbb{N}$, such that:

- setup is a probabilistic algorithm that outputs a public parameter seed;
- $\mathsf{refresh}$ is a deterministic algorithm that takes seed, a state $S \in \{0,1\}^n$, and an input $I \in \{0,1\}^*$, and outputs a new state $S' \leftarrow \mathsf{refresh}(\mathsf{seed}, S, I) \in \{0,1\}^n$;
- next is a deterministic algorithm that takes seed and a state $S \in \{0,1\}^n$, and outputs a pair $(S', R) \leftarrow \mathsf{next}(\mathsf{seed}, S) \in \{0,1\}^n \times \{0,1\}^r$ where S' is the new state and R is the PRNG output.

The parameters n, r denote the state length and output length, respectively. Note that in contrast to [15], we do not restrict the length of the input I to refresh. In this paper, we repeatedly use the term "PRNG" to denote a PRNG with input in the sense of the above definition.

THE H-COEFFICIENT METHOD. We give the basic theorem underlying the H-Coefficient method [27], as recently revisited by Chen and Steinberger [11].

Let $\mathcal{A}$ be a deterministic, computationally unbounded adversary trying to distinguish two experiments that we call real, respectively ideal, with respective probability measures $\mathsf{Pr}^{\mathsf{real}}$ and $\mathsf{Pr}^{\mathsf{ideal}}$. Let $\mathsf{T}_{\mathsf{real}}$ (resp. $\mathsf{T}_{\mathsf{ideal}}$) denote the random variable of the transcript of the real (resp. ideal) experiment that contains everything that the adversary was able to observe during the experiment. Let $\mathsf{GOOD} \cup \mathsf{BAD}$ be a partition of all valid transcripts into two sets – we refer to the elements of these sets as good and bad transcripts, respectively. Then we have:

Theorem 1 (H-Coefficient Method). *Let $\delta, \varepsilon \in [0, 1]$ be such that:*

(a) $\mathsf{Pr}\left[\mathsf{T}_{\mathsf{ideal}} \in \mathsf{BAD}\right] \le \delta$.
(b) *For all* $\tau \in \mathsf{GOOD}$, $\mathsf{Pr}\left[\mathsf{T}_{\mathsf{real}} = \tau\right] / \mathsf{Pr}\left[\mathsf{T}_{\mathsf{ideal}} = \tau\right] \ge 1 - \varepsilon$.

Then $\left|\mathsf{Pr}^{\mathsf{ideal}}(\mathcal{A} \Rightarrow 1) - \mathsf{Pr}^{\mathsf{real}}(\mathcal{A} \Rightarrow 1)\right| \le \mathbf{SD}(\mathsf{T}_{\mathsf{real}}, \mathsf{T}_{\mathsf{ideal}}) \le \varepsilon + \delta$.

3 PRNG Security in the IPM

In this section, we adapt the notions of robustness, recovering security, and preserving security for PRNGs [15] to the ideal permutation model and to cover sponge-based designs[3]. This will require several extensions.

First, we adjust for the presence of the permutation oracle π available to all parties. In particular, we need a notion of a legitimate distribution sampler that can query the permutation. Second, our definitions take into account that the state of the sponge-based PRNG at some important points (e.g. after extraction) is not required to be close to a uniformly random string, but rather to a uniform element of $0^r \| \{0,1\}^c$ instead. Note that this is an instance of a more general issue raised already in [29], as we discuss below.

We then proceed by proving that these modified notions still maintain the useful property shown in [15,29], namely that the combination of recovering security and preserving security still implies the robustness of the PRNG.

3.1 Oracle-Dependent Randomness and Distribution Samplers

This section discusses the issue of generating randomness in a model where a randomly sampled permutation $\pi \xleftarrow{\$} \mathsf{Perms}(n)$ is available to all parties. We give a formal definition of adversarial distribution samplers to be used within the PRNG security notions formalized further below.

[3] It is straightforward to extend our treatment to any ideal primitive, rather than just a random permutation – we dispense with doing so for ease of notation.

For our purposes, an (oracle-dependent) *source* $\mathcal{S} = \mathcal{S}^{\pi}$ is an input-less randomized oracle algorithm which makes queries to π and outputs a string X. The *range* of $\mathcal{S}$, denoted $[\mathcal{S}]$, is the set of values x output by $\mathcal{S}^{\pi}$ with positive probability, where the probability is taken over the choice of π and the internal random coins of $\mathcal{S}$.

DISTRIBUTION SAMPLERS. We extend the paradigm of (adversarial) *distribution samplers* considered in [15] to allow for oracle queries to a permutation oracle $\pi \xleftarrow{\$} \mathsf{Perms}(n)$[4]. Recall that in the original formalization, a distribution sampler $\mathcal{D}$ is a randomized stateful algorithm which, at every round, outputs a triple (I_i, γ_i, z_i), where z_i is auxiliary information, I_i is a string, and γ_i is an entropy estimate. In order for such sampler to be legitimate, for every i (up to a certain bound $q_{\mathcal{D}}$), given I_j for every $j \neq i$, as well as $(z_1, \gamma_1), \ldots, (z_{q_{\mathcal{D}}}, \gamma_{q_{\mathcal{D}}})$, it must be hard to predict I_i with probability better than $2^{-\gamma_i}$, in a *worst-case sense* over the choice of I_j for $j \neq i$ and $(z_1, \gamma_1), \ldots, (z_{q_{\mathcal{D}}}, \gamma_{q_{\mathcal{D}}})$.

Extending this worst-case requirement will need some care. To facilitiate this, we consider a specific class of oracle-dependent distribution samplers, which explicitly separate the process of sampling the auxiliary information from the processes of sampling the I_i values. Formally, we achieve this by explicitly requiring that $\mathcal{D}$ outputs (the description of) a source $\mathcal{S}_i$, rather than a value I_i, and the actual value I_i is sampled by running this $\mathcal{S}_i$ once with fresh random coins.

Definition 2 (Distribution Samplers). *A* Q-distribution sampler *is a randomized stateful oracle algorithm* $\mathcal{D}$ *which operates as follows:*

- *It takes as input a* state σ_{i-1} *(the initial state is* $\sigma_0 = \perp$*).*
- *On input* σ_{i-1}*,* $\mathcal{D}^{\pi}(\sigma_{i-1})$ *outputs a tuple* $(\sigma_i, \mathcal{S}_i, \gamma_i, z_i)$*, where* σ_i *is a new state,* z_i *is the auxiliary information,* γ_i *is an entropy estimation, and* $\mathcal{S}_i$ *is a source with range* $[\mathcal{S}_i] \subseteq \{0,1\}^{\ell_i}$ *for some* $\ell_i \geq 1$*. Then, we run* $I_i \xleftarrow{\$} \mathcal{S}_i^{\pi}$ *to sample the actual value.*
- *When run for* $q_{\mathcal{D}}$ *times, the overall number of queries made by* $\mathcal{D}$ *and* $\mathcal{S}_1, \ldots, \mathcal{S}_{q_{\mathcal{D}}}$ *is at most* $Q(q_{\mathcal{D}})$*. If* $Q = 0$*, then* $\mathcal{D}$ *is called* oracle independent.

We often abuse notation, and compactly denote by $(\sigma_i, I_i, \gamma_i, z_i) \xleftarrow{\$} \mathcal{D}^{\pi}(\sigma_{i-1})$ *the overall process of running* $\mathcal{D}$ and *the generated source* $\mathcal{S}_i$ *to jointly produce* $(\sigma_i, I_i, \gamma_i, z_i)$*.*

Also we will simply refer to $\mathcal{D}$ as a *distribution sampler*, omitting Q, when the latter is not relevant to the context. Finally, note that in contrast to [15], we consider a relaxed notion where the outputs I_i can be arbitrarily long strings, and are not necessarily fixed length. Still, we assume that the lengths $\ell_1, \ell_2, \ldots$ are a-priori fixed parameters of the samplers, and cannot be chosen dynamically.

We note that this definition appears to exhibit some degree of redundancy. In particular, it seems that without loss of generality one can simply assume

[4] We present the notions here for this specialized case, but needless to say, they extend naturally to other types of randomized oracles, such as random oracles or ideal ciphers.

Game $\mathsf{GLEG}_{q_{\mathcal{D}},i^*}(\mathcal{A}, D)$:

1. Sample $\pi \xleftarrow{\$} \mathsf{Perms}(n)$
2. Run $\mathcal{D}^{\pi}$ $q_{\mathcal{D}}$ rounds, producing outputs $(\gamma_1, z_1), \ldots, (\gamma_{q_{\mathcal{D}}}, z_{q_{\mathcal{D}}})$, as well as $I_1, \ldots, I_{q_{\mathcal{D}}}$. This in particular entails sampling sources $\mathcal{S}_1, \ldots, \mathcal{S}_{q_{\mathcal{D}}}$, and sampling $I_1, \ldots, I_{q_{\mathcal{D}}}$ from them (recall that each $\mathcal{S}_i$ can query π). Let $\mathcal{Q}_{\mathcal{D}}$ be the set of all input-output pairs of permutation queries made by $\mathcal{D}$ and by $\mathcal{S}_j$ (for $j \neq i^*$) in this process. (The queries made by $\mathcal{S}_{i*}$ are omitted from $\mathcal{Q}_{\mathcal{D}}$.)
3. Run $\mathcal{A}^{\pi}$ on input $(\gamma_j, z_j)_{j \in [q_{\mathcal{D}}]}$ and $(I_j)_{j \in [q_{\mathcal{D}}] \setminus \{i^*\}}$, and let $V_{\mathcal{A}}$ be $\mathcal{A}$'s final output.
4. The game then outputs $((I_1, \gamma_1, z_1), \ldots, (I_{q_{\mathcal{D}}}, \gamma_{q_{\mathcal{D}}}, z_{q_{\mathcal{D}}}), V_{\mathcal{A}}, \mathcal{Q}_{\mathcal{D}})$

Fig. 1. Definition of the game $\mathsf{GLEG}_{q_{\mathcal{D}},i^*}(\mathcal{A}, \mathcal{D})$.

that the generated $\mathcal{S}_i$ outputs a fixed value. (Note that $\mathcal{S}_i$ can be chosen itself from a distribution.) However, this separation will be convenient in defining our legitimacy notion for such samplers, as we will distinguish between permutation queries made by $\mathcal{S}_i$, and other permutation queries made by $\mathcal{D}$ (and $\mathcal{S}_j$ for $j \neq i$).

LEGITIMATE DISTRIBUTION SAMPLERS. Intuitively, we want to say that once a source $\mathcal{S}_i$ is output with entropy estimate γ_i, then its output has min-entropy γ_i conditioned on everything we have seen so far. However, due to the availability of the oracle π, which is queried by $\mathcal{D}$, by $\mathcal{S}_i$, and by a potential observer attempting to predict the output of $\mathcal{S}_i$, this is somewhat tricky to formalize.

To this end, let $\mathcal{D}$ be a distribution sampler, $\mathcal{A}$ an adversary, and fix $i^* \in [q_{\mathcal{D}}]$, and consider the game $\mathsf{GLEG}_{q_{\mathcal{D}},i^*}(\mathcal{A}, \mathcal{D})$ given in Fig. 1. Here, the adversary is given I_j for $j \neq i^*$ and $(z_1, \gamma_1), \ldots, (z_{q_{\mathcal{D}}}, \gamma_{q_{\mathcal{D}}})$, and can make some permutation queries. Then, at the end, the game outputs the combination of $(z_1, \gamma_1, I_1), \ldots, (z_{q_{\mathcal{D}}}, \gamma_{q_{\mathcal{D}}}, I_{q_{\mathcal{D}}})$, the adversary's output, and a transcript of all permutation queries made by (1) $\mathcal{D}$, and (2) $\mathcal{S}_j$ for $j \neq i^*$. We ask that in the *worst case*, the value I_{i^*} cannot be predicted with advantage better than $2^{-\gamma_{i^*}}$ given everything else in the output of the game. Formally:

Definition 3 (Legitimate Distribution Sampler). *We say that a distribution sampler $\mathcal{D}$ is $(q_{\mathcal{D}}, q_{\pi})$-legitimate, if for every adversary $\mathcal{A}$ making q_{π} queries and every $i^* \in [q_{\mathcal{D}}]$, and for any possible values $(I_j)_{j \neq i^*}$, $(\gamma_1, z_1), \ldots, (\gamma_{q_{\mathcal{D}}}, z_{q_{\mathcal{D}}})$, $V_{\mathcal{A}}$, $\mathcal{Q}_{\mathcal{D}}$ potentially output by the game $\mathsf{GLEG}_{q_{\mathcal{D}},i^*}(\mathcal{A}, \mathcal{D})$ with positive probability,*

$$\mathsf{Pr}\left[I_{i^*} = x \,\middle|\, (I_j)_{j \neq i^*}, (\gamma_1, z_1), \ldots, (\gamma_{q_{\mathcal{D}}}, z_{q_{\mathcal{D}}}), V_{\mathcal{A}}, \mathcal{Q}_{\mathcal{D}}\right] \leq 2^{-\gamma_{i^*}} \tag{1}$$

for all $x \in \{0,1\}^{\ell_{i^*}}$, where the probability is conditioned on these particular values being output by the game.

Note that the unpredictability of I_{i^*} is due to what is *not* revealed, including the oracle queries made by $\mathcal{S}_{i^*}$, and the internal random coins of $\mathcal{S}_{i^*}$ *and* $\mathcal{D}$. For instance, for oracle-independent distribution samplers (which we can think of as outputting "constant" sources), our notion of legitimacy is equivalent to the definition of [15]. We show a more interesting example next.

AN EXAMPLE: PERMUTATION-BASED RANDOMNESS EXTRACTION. Consider the simple construction $\mathsf{H}^\pi : \{0,1\}^n \to \{0,1\}^{n/2}$ which on input X outputs the first $n/2$ bits of $\pi(X)$. It is not hard to prove that if X is an n-bit random variable with high min-entropy k, i.e., $\Pr[\mathsf{X} = X] \leq 2^{-k}$ for all $X \in \{0,1\}^n$, and $\mathsf{U}_{n/2}$ is uniform over the $(n/2)$-bit strings, then for all adversaries $\mathcal{A}$ making q_π queries,

$$\mathsf{Adv}_{\mathcal{A}}^{\mathsf{dist}}(\mathsf{H}^\pi(\mathsf{X}), \mathsf{U}_{n/2}) \leq \mathcal{O}\left(\frac{q_\pi}{2^{n/2}}\right) + \frac{q_\pi}{2^k} . \tag{2}$$

The proof (which we omit) would simply go by saying that as long as the attacker does not query X to π (on which it has k bit of uncertainty), or queries $\pi(\mathsf{X})$ to π^{-1} (on which it has only $n/2$ bits of uncertainty), the output looks sufficiently close to uniform (with a tiny bias due to the gathered information about π via $\mathcal{A}$'s direct queries).

Now, let us consider a simple distribution sampler $\mathcal{D}$ which does the following – at every round, regardless of this input, it always outputs the following source $\mathcal{S} = \mathcal{S}^\pi$, as well as $\gamma = n - 1$, and $z = \perp$. The source $\mathcal{S}$ does the following: It queries random n-bit strings X_i to π, until the first bit of $\pi(X_i)$ is 0, and then outputs X_i. It is not hard to show that for any $q_\mathcal{D}$ and q_π, this sampler is $(q_\mathcal{D}, q_\pi)$-legitimate. This is because even if $\mathcal{A}$ knows the entire description of π, $\mathcal{S}$ always outputs an independent uniformly distributed n-bit string X conditioned on $\pi(X)$ having the first bit equal 0, and the distribution is uniform over 2^{n-1} possible such X's. Yet, given X sampled from $\mathcal{D}$ (and thus from $\mathcal{S}$), it is very easy to distinguish $\mathsf{H}^\pi(\mathsf{X})$ and $\mathsf{U}_{n/2}$ with advantage $\frac{1}{2}$, by having $\mathcal{A}$ simply output the first bit of its input, and thus *without even making a query to* π!

We stress that this is nothing more than the ideal-model analogue of the classical textbook proof that seedless extractors cannot exist for the class all k-sources, even when k is as large as $n - 1$. Above all, this shows that our class of legitimate samplers is strong enough to encompass such pathological examples, thus allowing to eliminate the odd artificiality of ideal models.

A BRIEF DISCUSSION. The example above shows that our notion is strong enough to include (1) non-trivial distributions forcing us to use seeds and (2) permutation-independent samplers. It is meaningful to ask whether it is possible to weaken the requirement so that the output of $\mathcal{S}_{i^*}$ is only unpredictable when the π queries issued by $\mathcal{S}_j$ for $j \neq i^*$ and by $\mathcal{D}$ are not revealed by the game, and still get meaningful results. We believe this is possible in general, but without restrictions, there are non-trivial dependencies arising (thanks to the auxiliary input) between what the adversary can see and the sampling of I_{i^*} which we cannot handle in our proofs in a *generic* way.

3.2 Robustness, Recovering and Preserving Security in the IPM

ROBUSTNESS. The definition of robustness follows the one from [15], with the aforementioned modifications tailored at our setting.

Procedure initialize:
$\pi \xleftarrow{\$} \mathsf{Perms}(n)$
$\mathsf{seed} \xleftarrow{\$} \mathsf{setup}^{\pi}()$
$S \xleftarrow{\$} 0^r \,\|\, \{0,1\}^c$
$\sigma \leftarrow \bot$
corrupt $\leftarrow$ `false`
$e \leftarrow c$
$b \xleftarrow{\$} \{0,1\}$
return seed

Procedure finalize(b^*):
return $(b = b^*)$

Procedure $\mathcal{D}$-refresh:
$(\sigma, I, \gamma, z) \xleftarrow{\$} \mathcal{D}^{\pi}(\sigma)$
$S \leftarrow \mathsf{refresh}^{\pi}(\mathsf{seed}, S, I)$
$e \leftarrow e + \gamma$
if $e \geq \gamma^*$:
  corrupt $\leftarrow$ `false`
return (γ, z)

Procedure next-ror:
$(S, R_0) \xleftarrow{\$} \mathsf{next}^{\pi}(\mathsf{seed}, S)$
$R_1 \xleftarrow{\$} \{0,1\}^{\ell}$
if corrupt = `true`:
  $e \leftarrow 0$
  return R_0
return R_b

Procedure get-next:
$(S, R) \xleftarrow{\$} \mathsf{next}^{\pi}(\mathsf{seed}, S)$
if corrupt = `true`:
  $e \leftarrow 0$
return R

Procedure get-state:
$e \leftarrow 0$
corrupt $\leftarrow$ `true`
return S

Procedure set-state(S^*):
$e \leftarrow 0$
corrupt $\leftarrow$ `true`
$S \leftarrow S^*$

Fig. 2. Definition of the game $\mathsf{ROB}_{\mathbf{G}}^{\gamma^*}(\mathcal{A}, \mathcal{D})$.

The formal definition of robustness is based on the game ROB given in Fig. 2 and parametrized by a constant γ^*. The game description consists of special procedures initialize and finalize and 5 additional oracles. It is run as follows: first the initialize procedure is run, its output is given to the adversary which is then allowed to query the 5 oracles described, in addition to π and π^{-1}, and once it outputs a bit b^*, this is then given to the finalize procedure, which generates the final output of the game.

For an adversary $\mathcal{A}$ and a distribution sampler $\mathcal{D}$, the advantage against the robustness of a PRNG with input $\mathbf{G}$ is defined as

$$\mathsf{Adv}_{\mathbf{G}}^{\gamma^*\text{-rob}}(\mathcal{A}, \mathcal{D}) = 2 \cdot \Pr\left[\mathsf{ROB}_{\mathbf{G}}^{\gamma^*}(\mathcal{A}, \mathcal{D}) \Rightarrow 1\right] - 1\,.$$

An adversary against robustness that asks q_π queries to its π/π^{-1} oracles, $q_{\mathcal{D}}$ queries to its $\mathcal{D}$-refresh oracle, q_R queries to its next-ror/get-next oracles, and q_S queries to its get-state/set-state oracles, is called a $(q_\pi, q_{\mathcal{D}}, q_R, q_S)$-adversary.

Recovering security. We follow the definition from [15], again adapted to our setting. In particular, we only require that the state resulting from the final next call in the experiment has to be indistinguishable from a c-bit uniformly random string preceded by r zeroes, instead of a random n-bit string.

Recovering security is defined in terms of the game REC parametrized by $q_{\mathcal{D}}$, γ^*, given in Fig. 3. For an adversary $\mathcal{A}$ and a distribution sampler $\mathcal{D}$, the advantage against the recovering security of a PRNG with input $\mathbf{G}$ is defined as

$$\mathsf{Adv}_{\mathbf{G}}^{(\gamma^*, q_{\mathcal{D}})\text{-rec}}(\mathcal{A}, \mathcal{D}) = 2 \cdot \Pr\left[\mathsf{REC}_{\mathbf{G}}^{\gamma^*, q_{\mathcal{D}}}(\mathcal{A}, \mathcal{D}) \Rightarrow 1\right] - 1\,.$$

An adversary against recovering security that asks q_π queries to its π/π^{-1} oracles is called a q_π-adversary.

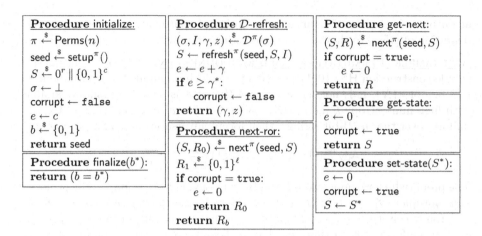

1. The challenger chooses $\pi \xleftarrow{\$} \mathsf{Perms}(n)$, $\mathsf{seed} \xleftarrow{\$} \mathsf{setup}^{\pi}()$, and $b \xleftarrow{\$} \{0,1\}$ and sets $\sigma_0 \leftarrow \perp$. For $k = 1, \ldots, q_{\mathcal{D}}$, the challenger computes $(\sigma_k, I_k, \gamma_k, z_k) \leftarrow \mathcal{D}^{\pi}(\sigma_{k-1})$.
2. The attacker $\mathcal{A}$ gets seed and $\gamma_1, \ldots, \gamma_{q_{\mathcal{D}}}, z_1, \ldots, z_{q_{\mathcal{D}}}$. It also gets access to oracles π/π^{-1}, and to an oracle $\mathsf{get\text{-}refresh}()$ which initially sets $k \leftarrow 0$ and on each invocation increments $k \leftarrow k+1$ and outputs I_k. At some point, $\mathcal{A}$ outputs a value $S_0 \in \{0,1\}^n$ and an integer d such that $k + d \leq q_{\mathcal{D}}$ and $\sum_{j=k+1}^{k+d} \gamma_j \geq \gamma^*$.
3. For $j = 1, \ldots, d$ the challenger computes $S_j \leftarrow \mathsf{refresh}^{\pi}(\mathsf{seed}, S_{j-1}, I_{k+j})$. If $b = 0$ it sets $(S^*, R) \leftarrow \mathsf{next}^{\pi}(\mathsf{seed}, S_d)$, otherwise it sets $(S^*, R) \xleftarrow{\$} (0^r \,\|\, \{0,1\}^c) \times \{0,1\}^r$. The challenger gives $I_{k+d+1}, \ldots, I_{q_{\mathcal{D}}}$ and (S^*, R) to $\mathcal{A}$.
4. The attacker again gets access to π/π^{-1} and outputs a bit b^*. The output of the game is 1 iff $b = b^*$.

Fig. 3. Definition of the game $\mathsf{REC}_{\mathbf{G}}^{\gamma^*, q_{\mathcal{D}}}$.

1. The challenger chooses $\pi \xleftarrow{\$} \mathsf{Perms}(n)$, $\mathsf{seed} \xleftarrow{\$} \mathsf{setup}^{\pi}()$ and $b \xleftarrow{\$} \{0,1\}$ and a state $S_0 \xleftarrow{\$} 0^r \,\|\, \{0,1\}^c$.
2. The attacker $\mathcal{A}$ gets access to oracles π/π^{-1}, and outputs a sequence of values $I_1, \ldots, I_d$ with $I_j \in \{0,1\}^*$ for all $j \in [d]$.
3. The challenger computes $S_j \leftarrow \mathsf{refresh}^{\pi}(\mathsf{seed}, S_{j-1}, I_j)$ for all $j = 1, \ldots, d$. If $b = 0$ it sets $(S^*, R) \leftarrow \mathsf{next}^{\pi}(\mathsf{seed}, S_d)$, otherwise it sets $(S^*, R) \xleftarrow{\$} (0^r \,\|\, \{0,1\}^c) \times \{0,1\}^r$. The challenger gives (S^*, R) to $\mathcal{A}$.
4. The attacker $\mathcal{A}$ again gets access to π/π^{-1} and outputs a bit b^*. The output of the game is 1 iff $b = b^*$.

Fig. 4. Definition of the game $\mathsf{PRES}_{\mathbf{G}}$.

PRESERVING SECURITY. We again follow the definition from [15], with similar modifications as in the case of recovering security above.

The formal definition of preserving security is based on the game PRES given in Fig. 4. For an adversary $\mathcal{A}$, the advantage against the preserving security of a PRNG with input $\mathbf{G}$ is defined as

$$\mathsf{Adv}_{\mathbf{G}}^{\mathsf{pres}}(\mathcal{A}) = 2 \cdot \Pr\left[\mathsf{PRES}_{\mathbf{G}}(\mathcal{A}) \Rightarrow 1\right] - 1\,.$$

An adversary against preserving security that asks q_π queries to its π/π^{-1} oracles is again called a q_π-adversary.

RELATIONSHIP TO [29]. Our need to adapt the notions of [15] confirms that, as observed in [29], assuming that the internal state of a PRNG is pseudorandom is overly restrictive. Indeed, our formalization is a special case of the approach from [29] into the setting of sponge-based constructions, where the so-called *masking function* would be defined as sampling a random $S \in 0^r \,\|\, \{0,1\}^c$ (and preserving the counter j). Our notions would then correspond to the "bootstrapped" notions from [29] and moreover, our results on recovering security below indicate that a naturally-defined procedure setup (for generating the initial state as in [29]) would make this masking function satisfy the *honest-initialization* property.

COMBINING PRESERVING AND RECOVERING SECURITY. This theorem establishes the very useful property that, roughly speaking, the preserving security and the recovering security of a PRNG together imply its robustness. We postpone its proof (following [15]) to the full version.

Theorem 4. *Let* $\mathbf{G}[\pi]$ *be a PRNG with input that issues* q_π^{ref} *(resp.* q_π^{nxt}*)* π*-queries in each invocation of* refresh *(resp.* next*); and let* $\overline{q}_\pi = q_\pi + Q(q_\mathcal{D})$*. For every* $(q_\pi, q_\mathcal{D}, q_R, q_S)$*-adversary* $\mathcal{A}$ *against robustness and for every* Q*-distribution sampler* $\mathcal{D}$*, there exists a family of* $(q_\pi + q_R \cdot q_\pi^{\mathsf{nxt}} + q_\mathcal{D} \cdot q_\pi^{\mathsf{ref}})$*-adversaries* $\mathcal{A}_1^{(i)}$ *against recovering security and a family of* $(\overline{q}_\pi + q_R \cdot q_\pi^{\mathsf{nxt}} + q_\mathcal{D} \cdot q_\pi^{\mathsf{ref}})$*-adversaries* $\mathcal{A}_2^{(i)}$ *against preserving security (for* $i \in \{1, \ldots, q_R\}$*) such that*

$$\mathsf{Adv}_{\mathbf{G}}^{\gamma^*\text{-rob}}(\mathcal{A}, \mathcal{D}) \leq \sum_{i=1}^{q_R} \left(\mathsf{Adv}_{\mathbf{G}}^{(\gamma^*, q_\mathcal{D})\text{-rec}}(\mathcal{A}_1^{(i)}, \mathcal{D}) + \mathsf{Adv}_{\mathbf{G}}^{\mathsf{pres}}(\mathcal{A}_2^{(i)}) \right) .$$

4 Robust Sponge-Based PRNG

We consider the following PRNG construction, using a permutation $\pi \in \mathsf{Perms}(n)$, and depending on parameters s and t. This construction is a seeded variant of the general paradigm introduced by Bertoni *et al.* [10], including countermeasures to prevent attacks against forward secrecy. As we will see in the proof, the parameters s and t are going to enforce increasing degrees of security.

THE CONSTRUCTION. Let $s, t \geq 1$, and $r \leq n$, let $c := n - r$. We define $\mathbf{SPRG}_{s,t,n,r} =$ (setup, refresh, next), where the three algorithms setup, refresh, next make calls to some permutation $\pi \in \mathsf{Perms}(n)$ and operate as follows:

Proc. $\mathsf{setup}^\pi()$:
for $i = 0, \ldots, s-1$ **do**
  $\mathsf{seed}_i \xleftarrow{\$} \{0,1\}^r$
$\mathsf{seed} \leftarrow (\mathsf{seed}_0, \ldots, \mathsf{seed}_{s-1})$
$j \leftarrow 1$
return seed

Proc. $\mathsf{refresh}^\pi(\mathsf{seed}, S, I)$:
$I_1, \ldots, I_\ell \xleftarrow{r} I$
$S_0 \leftarrow S$
for $i = 1, \ldots, \ell$ **do**
  $S_i \leftarrow \pi(S_{i-1} \oplus (I_i \oplus \mathsf{seed}_j \,\|\, 0^c))$
  $j \leftarrow j + 1 \mod s$
return S_ℓ

Proc. $\mathsf{next}^\pi(\mathsf{seed}, S)$:
$S_0 \leftarrow \pi(S)$
$R \leftarrow S_0[1 \ldots r]$
for $i = 1, \ldots, t$ **do**
  $S_i \leftarrow \pi(S_{i-1})$
  $S_i[1 \ldots r] \leftarrow 0^r$
$j \leftarrow 1$
return (S_t, R)

Note that apart from the entropy pool S, the PRNG also keeps a counter j internally as a part of its state. This counter increases (modulo s) as blocks are processed via refresh, and gets resetted whenever next is called. We will often just write **SPRG**, omitting the parameters s, t, n, r whenever the latter are clear from the context. In particular, the parameter s determines the length of the seed in terms of r-bit blocks. The construction **SPRG** is depicted in Fig. 5.

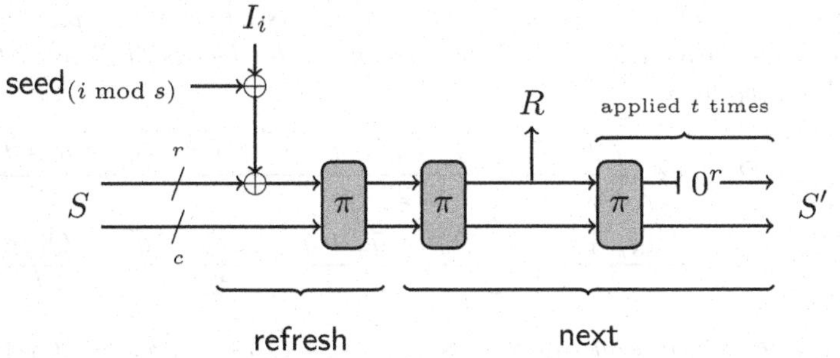

Fig. 5. Procedures refresh (processing a one-block input I_i) and next of the construction $\mathbf{SPRG}_{s,t}[\pi]$.

We also note that it is not hard to modify our treatment to allow for next outputting multiple r-bit blocks at once, instead of just one, and this length could be variable. This could be done by providing an additional input, indicating the number of desired blocks and this would ensure better efficiency. The bounds presented here would only be marginally affected by this, but we decided to keep the presentation simple in this paper.

<u>INSECURITY OF THE UNSEEDED VERSION.</u> We show that seeding is necessary to achieve robustness. A similar argument implies that the original construction of [10] cannot be secure if the distribution sampler is allowed to depend on the public random permutation π.

To this end, we consider the distribution sampler $\mathcal{D}$ which on its first call outputs an $\ell \cdot r$-bit string, for a parameter ℓ such that $(\ell-1)r \geq \gamma^*$. In particular, on its first call $\mathcal{D}$ simply outputs a source $\mathcal{S}_1$ which behaves as follows, given the corresponding entropy estimate $(\ell - 1) \cdot r$:

- It internally samples r-bit strings $I_1, \ldots, I_{\ell-1}$ uniformly at random.
- Then, it samples random $I_\ell^1, I_\ell^2, \ldots$ until it finds one such that $R^j[1] = 0$, where R are the r-bit returned by running next after running refresh, from the some initial state S, with inputs $I_1, \ldots, I_{\ell-1}, I_\ell^j$.

Additionally, consider a robustness adversary $\mathcal{A}$ that first calls set-state(S) and then $\mathcal{D}$-refresh(). Finally, it queries next-ror() obtaining R^*, and checks whether $R^*[1] = 0$. Clearly, $\mathcal{A}$ achieves advantage 1/2 despite $\mathcal{D}$ being legitimate.

5 Security Analysis of SPRG

This section gives a complete security analysis of **SPRG** given in Sect. 4 above, under the assumption that the underlying permutation is a random permutation $\pi \in \mathsf{Perms}(n)$. In particular, we prove the following theorem.

Theorem 5 (Security of SPRG). *Let* $\mathbf{SPRG} = \mathbf{SPRG}_{s,t,n,r}[\pi]$ *denote the PRNG given in Sect. 4. Let* $\gamma^* > 0$*, let* $\mathcal{A}$ *be a* $(q_\pi, q_\mathcal{D}, q_R, q_S)$*-adversary against* *robustness and let $\mathcal{D}$ be a $(q_{\mathcal{D}}, q_\pi)$-legitimate Q-distribution sampler such that the length of its outputs $I_1, \ldots, I_{q_{\mathcal{D}}}$ padded into r-bit blocks is at most $\ell \cdot r$ bits in total. Then we have*

$$\mathsf{Adv}_{\mathbf{SPRG}}^{\gamma^*\text{-rob}}(\mathcal{A}, \mathcal{D}) \leq q_R \cdot \left(\frac{2(2\ell+2)(\bar{q}_\pi + q' + t + \ell) + 4\ell^2}{2^n} + \frac{\bar{q}_\pi + q' + t + 1}{2^{\gamma^*}} + \right.$$

$$\left. + \frac{22(\bar{q}_\pi + q' + t + 1)^2 + \bar{q}_\pi + q'}{2^c} + \frac{2(\bar{q}_\pi + q')}{2^{(r-1)t}} + \frac{Q(q_{\mathcal{D}})}{2^{sr}} \right),$$

where we use the notational shorthands $\bar{q}_\pi = q_\pi + Q(q_{\mathcal{D}})$ and $q' = (t+1)q_R + \ell$.

Note in particular that the construction is secure as long as $q_R \cdot \bar{q}_\pi \cdot \ell < 2^n$, $q_R \cdot \bar{q}_\pi, q_R^2 < 2^c$, $\bar{q}_\pi, q_R^2 \leq 2^{\gamma^*}$, $q_R^2, \bar{q}_\pi q_R \leq 2^{(r-1)t}$. Note that these are more than sufficient margins for SHA-3-like parameters, where $n = 1600$ and $c \geq 1024$ always holds. However, one should assess the bound more carefully for a single-key cipher instantiation, where $n = 128$. In this case, choosing a very small r (note that our construction and bound would support $r \geq 2$) would significantly increase the security margins.

The theorem follows from the bounds on recovering security and preserving security of **SPRG** proven in Lemmas 11 and 12 below, combined using Theorem 4. To establish these two bounds, we first give two underlying lemmas that represent the technical core of our analysis. The first one, Lemma 6, assesses the ability of a seeded sponge construction to act as a randomness extractor on inputs that are coming from a permutation-dependent distribution sampler. The second statement, given in Lemma 10, shows that the procedure next, given a high min-entropy input, produces an output that is very close to random.

5.1   The Sponge Extraction Lemma

The first part of our analysis addresses how the sponge structure can be used to extract (or in fact, condense) randomness. To this end, we first give a general definition of adaptively secure extraction functions.

Let $\mathbf{Ex}[\pi] : \{0,1\}^u \times \{0,1\}^v \times \{0,1\}^* \to \{0,1\}^n$ be an efficiently computable function taking as parameters a u-bit seed seed, a v-bit initialization value IV, together with an input string $X \in \{0,1\}^*$. It makes queries to a permutation $\pi \in \mathsf{Perms}(n)$ to produce the final n-bit output $\mathbf{Ex}^\pi(\mathsf{seed}, \mathsf{IV}, X)$. Then, for every $\gamma^* > 0$ and $q_{\mathcal{D}}$, for such an $\mathbf{Ex}$, an adversary $\mathcal{A}$ and a distribution sampler $\mathcal{D}$, we consider the game $\mathsf{GEXT}_{\mathbf{Ex}}^{\gamma^*, q_{\mathcal{D}}}(\mathcal{A}, \mathcal{D})$ given in Fig. 6. It captures the security of $\mathbf{Ex}$ in producing a random looking output in a setting where an *adaptive* adversary $\mathcal{A}$ can obtain side information and entropy estimates from a sampler $\mathcal{D}$, together with samples $I_1, \ldots, I_k$, until it commits on running $\mathbf{Ex}$ on adaptively chosen IV, as well as $I_{k+1} \ldots I_{k+d}$ for some d such that the guaranteed entropy of these values is $\sum_{i=k+1}^{k+d} \gamma_i \geq \gamma^*$. We define the $(q_{\mathcal{D}}, \gamma^*)$-*extraction advantage* of $\mathcal{A}$ and $\mathcal{D}$ against $\mathbf{Ex}$ as

$$\mathsf{Adv}_{\mathbf{Ex}}^{(\gamma^*, q_{\mathcal{D}})-\mathsf{ext}}(\mathcal{A}, \mathcal{D}) = 2 \cdot \Pr\left[\mathsf{GEXT}_{\mathbf{Ex}}^{\gamma^*, q_{\mathcal{D}}}(\mathcal{A}, \mathcal{D}) \Rightarrow 1\right] - 1 \,.$$

Game $\mathsf{GEXT}^{\gamma^*,q_{\mathcal{D}}}_{\mathbf{Ex}}(\mathcal{A},\mathcal{D})$:

1. The challenger chooses $\mathsf{seed} \xleftarrow{\$} \{0,1\}^u$, $\pi \xleftarrow{\$} \mathsf{Perms}(n)$ and $b \xleftarrow{\$} \{0,1\}$ and sets $\sigma_0 \leftarrow \perp$. For $k = 1, \ldots, q_{\mathcal{D}}$, the challenger computes $(\sigma_k, I_k, \gamma_k, z_k) \leftarrow \mathcal{D}^{\pi}(\sigma_{k-1})$.
2. The attacker $\mathcal{A}$ gets seed and $\gamma_1, \ldots, \gamma_{q_{\mathcal{D}}}, z_1, \ldots, z_{q_{\mathcal{D}}}$. It gets access to oracles π/π^{-1}, and to an oracle get-refresh() which initially sets $k \leftarrow 0$ and on each invocation increments $k \leftarrow k+1$ and outputs I_k. At some point, $\mathcal{A}$ outputs a value IV and an integer d such that $k + d \le q_{\mathcal{D}}$ and $\sum_{j=k+1}^{k+d} \gamma_j \ge \gamma^*$.
3. If $b = 1$, we set $Y^* \xleftarrow{\$} \{0,1\}^n$, and if $b = 0$, we let $Y^* \leftarrow \mathbf{Ex}^{\pi}(\mathsf{seed}, \mathsf{IV}, I_{k+1} \| \cdots \| I_{k+d})$. Then, the challenger gives back Y^* and $I_{k+d+1}, \ldots, I_{q_{\mathcal{D}}}$ to $\mathcal{A}$.
4. The attacker again gets access to π/π^{-1} and outputs a bit b^*. The output of the game is 1 iff $b = b^*$.

Fig. 6. Definition of the game $\mathsf{GEXT}^{\gamma^*,q_{\mathcal{D}}}_{\mathbf{Ex}}(\mathcal{A},\mathcal{D})$.

Also, we denote by $\mathsf{Adv}_n^{(\gamma^*,q_{\mathcal{D}})-\mathsf{hit}}(\mathcal{A},\mathcal{D})$ the probability that $\mathcal{A}$ queries $\pi^{-1}(Y^*)$ conditioned on $b = 1$ in game $\mathsf{GEXT}^{\gamma^*,q_{\mathcal{D}}}_{\mathbf{Ex}}(\mathcal{A},\mathcal{D})$ above, i.e., Y^* is the random n-bit challenge. (The quantity really only depends on n, and not on the actual function $\mathbf{Ex}$, which is dropped from the notation.) Note that in general $\mathsf{Adv}_n^{(\gamma^*,q_{\mathcal{D}})-\mathsf{hit}}(\mathcal{A},\mathcal{D})$ can be one, but we will only consider it for specific adversaries $\mathcal{A}$ for which it can be argued to be small, as we discuss below.

<u>SPONGE-BASED EXTRACTION.</u> We consider a sponge-based instantiation of $\mathbf{Ex}$. That is, for parameters $r \le n$ (recall that we use the shorthand $c = n - r$), we consider the construction $\mathbf{Sp}_{n,r,s}[\pi] : \{0,1\}^{s \cdot r} \times \{0,1\}^n \times \{0,1\}^* \to \{0,1\}^n$ using a permutation $\pi \in \mathsf{Perms}(n)$ which, given seed $\mathsf{seed} = (\mathsf{seed}_0, \ldots, \mathsf{seed}_{s-1})$ (where $\mathsf{seed}_i \in \{0,1\}^r$ for all i), initialization value $\mathsf{IV} \in \{0,1\}^n$, input $X \in \{0,1\}^*$, first encodes X into r-bit blocks $X_1, \ldots, X_\ell$, and then outputs Y_ℓ, where $Y_0 \leftarrow \mathsf{IV}$ and for all $i \in [\ell]$,

$$Y_i \leftarrow \pi(Y_{i-1} \oplus (X_i \oplus \mathsf{seed}_{i \bmod s}) \,\|\, 0^c) .$$

We now turn to the following lemma, which establishes that the above construction $\mathbf{Sp}_{n,r,s}[\pi]$ is indeed a good extractor with respect to the notion defined above, as long as $\mathsf{Adv}_n^{(\gamma^*,q_{\mathcal{D}})-\mathsf{hit}}(\mathcal{A},\mathcal{D})$ is sufficiently small – a condition that will hold in applications of this lemma.

Lemma 6 (Extraction Lemma). *Let r, s be integers, let $q_{\mathcal{D}}, q_\pi$ be arbitrary, and let $\gamma^* > 0$. Also, let $\mathcal{D}$ be a $(q_{\mathcal{D}}, q_\pi)$-legitimate Q-distribution sampler, such that the length of its outputs $I_1, \ldots, I_{q_{\mathcal{D}}}$ padded into r-bit blocks is at most $\ell \cdot r$ bits in total. Then, for any adversary $\mathcal{A}$ making $q_\pi \le 2^{c-2}$ queries,*

$$\mathsf{Adv}^{(\gamma^*,q_\mathcal{D})-\mathsf{ext}}_{\mathbf{Sp}_{n,r,s}}(\mathcal{A},\mathcal{D}) \leq \frac{\overline{q}_\pi}{2^{\gamma^*}} + \frac{Q(q_\mathcal{D})}{2^{sr}} + \frac{14\overline{q}_\pi^2}{2^c} + \frac{2\overline{q}_\pi \ell + 2\ell^2}{2^n} + \mathsf{Adv}^{(\gamma^*,q_\mathcal{D})-\mathsf{hit}}_{n}(\mathcal{A},\mathcal{D}) \,, \quad (3)$$

where $\overline{q}_\pi = q_\pi + Q(q_\mathcal{D})$.

DISCUSSION. Once again, we note that in (3), we cannot *in general* expect the advantage $\mathsf{Adv}^{(\gamma^*,q_\mathcal{D})-\mathsf{hit}}_{n}(\mathcal{A},\mathcal{D})$ to be small – any $\mathcal{A}$ *sees* Y^* and thus *can* query it, and the bound is hence void for such adversaries. The reason why this is not an issue in our context is that the extraction lemma will be applied to *specific* $\mathcal{A}$'s resulting from reductions in scenarios where $\mathbf{Sp}_{n,r,s}$ is used to derive a key for an algorithm which *is already secure* when used with a proper independent random key. In this case, it is usually easy to upper bound $\mathsf{Adv}^{(\gamma^*,q_\mathcal{D})-\mathsf{hit}}_{n}(\mathcal{A},\mathcal{D})$ in terms of the probability of a certain adversary $\mathcal{A}'$ (from which $\mathcal{A}$ is derived) recovering the secret key of a secure construction.

But why is this term necessary? We note that one *can* expect the output to be random even without this restriction on querying $\pi^{-1}(Y)$, if we have the guarantee that the weakly random input fed into $\mathbf{Sp}_{n,r,s}$ is long enough. However, this only yields a weaker result. In particular, if $\mathbf{Sp}_{n,r,s}$ is run on r-bit inputs $I_{k+1},\ldots,I_{k+d}$ to produce an output Y^* (which may be replaced by a random one in the case $b = 1$), it is not hard to see that guessing $I_{k+2},\ldots,I_{k+d}$ is sufficient to distinguish, regardless of I_{k+1}. This is because an adversary $\mathcal{A}$ can simply "invert" the construction starting from computing $S_{k+d-1} \leftarrow \pi^{-1}(Y^*)$, $S_{k+d-2} \leftarrow \pi^{-1}(S_{k+d-1} \oplus (I_{k+d} \oplus \mathsf{seed}_{k+d \bmod s}) \,\|\, 0^c)$, ... until it recovers S_0, and then checks whether $S_0[r+1\ldots n] = \mathsf{IV}[r+1\ldots n]$. This will always succeed (given the right guess) in the $b = 0$ case, but with small probability in the $b = 1$ case. Above all, the crucial point is that I_{k+1} is not necessary to perform this attack. In particular, this would render the result useless for $d = 1$, whereas our statement still makes it useful as long as $q_\pi \leq 2^r$, which is realistic for say $r \geq 80$, and $\mathsf{Adv}^{(\gamma^*,q_\mathcal{D})-\mathsf{hit}}_{n}(\mathcal{A},\mathcal{D})$ is small.

An independent observation is that for oracle-independent distribution samplers (i.e., which do not make any permutation queries), we have $Q(q_\mathcal{D}) = 0$. In this case, the bound becomes independent of s, and indeed one can show that the bound holds even if the seed is contant (i.e., all zero), capturing the common wisdom that seeding is unnecessary for oracle-independent distributions.

PROOF INTUITION. The proof of Lemma 6, which we give in full detail below, is inspired by previous analyses of keyed sponges, which can be seen as a special case where a truly random input is fed into $\mathbf{Sp}_{n,r,s}$[5]. We will show that the advantage of $\mathcal{A}$ and $\mathcal{D}$ is bounded roughly by the probability that they jointly succeed in making all queries necessary to compute $\mathbf{Sp}_{n,r,s}(\mathsf{seed}, \mathsf{IV}, I_{k+1} \,\|\, \cdots \,\|\, I_{k+d})$. Indeed, we show that as long as not all necessary queries are made, then the distinguisher cannot tell apart the case $b = 0$ from the case $b = 1$ with substantial advantage. The core of the proof is bounding the above probability that all queries are issued.

[5] We note that none of these analysis tried to capture a general statement.

To this end, with $X_1, \ldots, X_\ell$ representing the encoding into r-bit blocks of $I_{k+1} \| \cdots \| I_{k+d}$, we consider all possible sequences of ℓ queries to the permutation, each made by $\mathcal{A}$ *or* $\mathcal{D}$, resulting in (not necessarily all distinct) input-output pairs $(\alpha_1, \beta_1), \ldots, (\alpha_\ell, \beta_\ell)$ with the property that

$$\alpha_i[r+1 \ldots n] = \beta_{i-1}[r+1 \ldots n]$$

for every $i \in [\ell]$, where we have set $\beta_0 = \mathsf{IV}$ for notational compactness. We call such sequence of ℓ input-output pairs a *potential chain*. We are interested in the probability that for *some* potential chain we additionally have

$$\alpha_i[1 \ldots r] = \beta_{i-1}[1 \ldots r] \oplus X_i \oplus \mathsf{seed}_{i \bmod s} \tag{4}$$

for all $i \in [\ell]$. Let us see why we can expect the probability that this happens to be small.

Recall that our structural restriction on $\mathcal{D}$ enforces that all of the values $I_{k+1}, \ldots, I_{k+d}$ are explicitly sampled by component sources $\mathcal{S}_{k+1}, \ldots, \mathcal{S}_{k+d}$. One first convenient observation is that as long as the overall number of permutation queries by $\mathcal{D}$ and $\mathcal{A}$, which is denoted by $\bar{q}_\pi$, is smaller than roughly $2^{c/2}$, then every potential chain can have only one of the two following formats:

- *Type A chains.* For $k \in [0 \ldots \ell]$, k input-output pairs $(\alpha_1, \beta_1), \ldots, (\alpha_k, \beta_k)$ resulting from *forward* queries made by $\mathcal{D}$ *outside* the process of sampling $I_{k+1} \ldots I_{k+d}$ by $\mathcal{S}_{k+1}, \ldots, \mathcal{S}_{k+d}$, followed by $\ell - k$ more input-output pairs $(\alpha_{k+1}, \beta_{k+1}), \ldots, (\alpha_\ell, \beta_\ell)$ resulting from queries made by $\mathcal{A}$ directly.
- *Type B chains.* The potential chain is made by some input-output pairs $(\alpha_1, \beta_1), \ldots, (\alpha_\ell, \beta_\ell)$ all resulting from *forward* permutation queries made by $\mathcal{D}$, in particular also possibly by the component sources $\mathcal{S}_{k+1}, \ldots, \mathcal{S}_{k+d}$.

One can also show that for $\overline{q}_\pi < 2^{c/2}$, it is likely that the number of such potential chains (either of Type A or Type B) is at most $\overline{q}_\pi$ and $Q(q_\mathcal{D})$, respectively. Now, we can look at the process of creating Type A and Type B chains *separately*, and note that in the former, the outputs of $\mathcal{S}_{k+1}, \ldots, \mathcal{S}_{k+d}$ have some uncertainty left (roughly, at least γ^* bits of entropy), thus the generated $X_1, \ldots, X_\ell$ end up satisfying (4) for each of the Type A potential chains with probability at most $2^{-\gamma^*}$. Symmetrically, the process of generating Type B chains is totally independent of the seed, and thus once the seed is chosen (which is made of $s \cdot r$ random bits), each one of the at most $Q(q_\mathcal{D})$ potential Type B chains ends up satisfying (4) with probability upper bounded by roughly 2^{-rs}.

We stress that making this high-level intuition formal is quite subtle.

<u>Can we achieve a better bound?</u> The extraction lemma requires $\overline{q}_\pi \leq 2^{c/2}$ for it to be meaningful. One can indeed hope to extend the techniques from [14] and obtain a result (at least for permutation-independent sources) which holds even if π is *completely known* to $\mathcal{A}$, while still being randomly sampled. However, in this regime one can expect the state output by $\mathbf{Sp}_{n,r,s}$ to be random only as long as at least n random bits have been input. In contrast, here we aim at the heuristic expectation (formalized in the ideal model) that as long as the number

of queries is small compared to the entropy of the distribution, then the output looks random.

The restriction $q_\pi \leq 2^{c/2}$ is common for sponges – beyond this, collisions become easy to find, and parameters are set to prevent this. Nonetheless, recent analyses of key absorption (which can be seen as a special case where the inputs are uniform) in sponge-based PRFs [20] trigger hope for security for nearly all $q_\pi \leq 2^c$, as they show that such collisions are by themselves not harmful. Unfortunately, in such high query regimes the number of potential chains as described above effectively explodes, and using the techniques of [20] (which are in turn inspired by [12]) to bound this number results in a fairly weak result.

Proof (of Lemma 6). The proof uses the H-coefficient method, as illustrated in Sect. 2 – indeed, to upper bound $\mathsf{Adv}^{(\gamma^*,q_\mathcal{D})-\mathsf{ext}}_{\mathbf{Sp}_{n,r,s}}(\mathcal{A},\mathcal{D})$, by a standard argument, one needs to upper bound the difference between the probabilities that $\mathcal{A}$ outputs 1 in the $b=1$ and in the $b=0$ cases, respectively. Throughout this proof, we assume that $\mathcal{A}$ is deterministic, and that $\mathcal{D}$ is also deterministic, up to being initialized with a random input R (of sufficient length) consisting of all random coins used by $\mathcal{D}$. In particular, R also contains the random coins used to sample the $I_1, I_2, \ldots, I_{q_\mathcal{D}}$ values by the sources $\mathcal{S}_1, \ldots, \mathcal{S}_{q_\mathcal{D}}$ output by $\mathcal{D}$.

To simplify the proof, we enhance the game $\mathsf{GEXT}^{\gamma^*,q_\mathcal{D}}_{\mathbf{Sp}_{n,r,s}}(\mathcal{A},\mathcal{D})$ so that the adversary $\mathcal{A}$, when done interacting with π, learns some extra information just before outputting the decision bit b'. This extra information includes:

- All strings $I_{k+1}, \ldots, I_{k+d}$ generated by $\mathcal{D}$ and hidden to $\mathcal{A}$ so far.
- The randomness R and all queries to π made by the distribution sampler $\mathcal{D}$ throughout its $q_\mathcal{D}$ calls. This includes all queries made by $\mathcal{S}_1, \ldots, \mathcal{S}_{q_\mathcal{D}}$. Recall that there are at most $Q(q_\mathcal{D})$ such queries by definition.

While this extra information is substantial, note that $\mathcal{A}$ cannot make any further queries to the random permutation after learning it, and, as we will see, this information does not hurt indistinguishability. Introducing it will make reasoning about the proof substantially easier. To start with, note that an execution of $\mathsf{GEXT}^{\gamma^*,q_\mathcal{D}}_{\mathbf{Sp}_{n,r,s}}(\mathcal{A},\mathcal{D})$ defines a *transcript* of the form

$$\tau = ((u_1,v_1),\ldots,(u_{q'},v_{q'}),Y^*,R,\mathsf{seed}=(\mathsf{seed}_1,\ldots,\mathsf{seed}_s),\\ \gamma_1,\ldots,\gamma_{q_\mathcal{D}},I_1,\ldots,I_{q_\mathcal{D}},z_1\ldots z_{q_\mathcal{D}},\mathsf{IV},k,d)\,, \quad (5)$$

where (u_i,v_i) are the input-output pairs resulting from the π-queries by $\mathcal{D}$ and $\mathcal{A}$ (that is, either $\pi(u_i)=v_i$ or $\pi^{-1}(v_i)=u_i$ for each (u_i,v_i) was queried by at least one of $\mathcal{D}$ and $\mathcal{A}$), removing duplicates, and ordered lexicographically. Note in particular that $q' \leq Q(q_\mathcal{D}) + q_\pi = \overline{q}_\pi$, and that the information whether a pair is the result of a forward or a backward query (or both) is omitted from the transcript, as it will not be used explicitly in the following.

We say that a transcript τ as in (5) is *valid* if when running $\mathsf{GEXT}^{\gamma^*,q_\mathcal{D}}_{\mathbf{Sp}_{n,r,s}}(\mathcal{A},\mathcal{D})$ with seed value fixed to seed, feeding Y^* to $\mathcal{A}$, executing $\mathcal{D}$ with randomness R, and answering permutation queries via a partial permutation π' such that $\pi'(u_i)=v_i$ for all $i \in [q']$, then

- The execution terminates, i.e., every permutation query is on a point for which π' is defined. Moreover, *all* queries in $(u_1, v_1), \ldots, (u_{q'}, v_{q'})$ are asked by either $\mathcal{D}$ or $\mathcal{A}$ at some point.
- $\mathcal{D}$ indeed outputs $(I_1, z_1, \gamma_1), \ldots, (I_{q_\mathcal{D}}, z_{q_\mathcal{D}}, \gamma_{q_\mathcal{D}})$.
- $\mathcal{A}$ indeed outputs IV and d, after k calls to $\mathsf{get\text{-}refresh}()$.

Now let T_0 and T_1 be the distributions on valid transcripts resulting from $\mathsf{GEXT}^{\gamma^*, q_\mathcal{D}}_{\mathbf{Sp}_{n,r,s}}(\mathcal{A}, \mathcal{D})$ in the $b = 0$ and $b = 1$ cases, respectively. Then,

$$\mathsf{Adv}^{(\gamma^*, q_\mathcal{D})-\mathsf{ext}}_{\mathbf{Sp}_{n,r,s}}(\mathcal{A}, \mathcal{D}) \leq \mathbf{SD}(\mathsf{T}_0, \mathsf{T}_1) , \tag{6}$$

since the extra information can only help, and a (possibly non-optimal) distinguisher for T_0 and T_1 can still mimic $\mathcal{A}$'s original decision (i.e., output bit), ignoring all additional information contained in the transcripts.

We are now ready to present our partitioning of transcripts into good and bad transcripts. Note first that a transcript explicitly tells us the blocks $I_{k+1}, \ldots, I_{k+d}$ processed by $\mathbf{Sp}_{n,r,s}$, and concretely let $X_1 \ldots X_\ell$ be the encoding into r-bit blocks of $I_{k+1} \parallel \cdots \parallel I_{k+d}$ when processed by $\mathbf{Sp}_{n,r,s}$. In particular we let $\ell = \ell(\tau)$ be the length here (in terms of r-bit blocks) of this encoding.

Definition 7 (Bad Transcript). *We say that a transcript τ as in (5) is* bad *if one of the two following properties is satisfied:*

- Hit. *There exists an (u_i, v_i), for $i \in [q']$, with $v_i = Y$. Note that this may be the result of a forward query $\pi(u_i)$ or a backward query $\pi^{-1}(v_i)$, or both. Which one is the case does not matter here.*
- Chain. *There exist ℓ permutation queries*

$$(\alpha_1, \beta_1), \ldots, (\alpha_\ell, \beta_\ell) \in \{(u_1, v_1), \ldots, (u_{q'}, v_{q'})\}$$

(not necessarily distinct) that constitute a chain, i.e., such that

$$\begin{aligned} \alpha_i[1 \ldots r] &= \beta_{i-1}[1 \ldots r] \oplus X_i \oplus \mathsf{seed}_{i \bmod s} \\ \alpha_i[r+1 \ldots n] &= \beta_{i-1}[r+1 \ldots n] \end{aligned} \tag{7}$$

for every $i \in [\ell]$, where we have set $\beta_0 = \mathsf{IV}$ for notational compactness.

Also, we denote by $\mathcal{B}$ the set of all bad transcripts.

The proof is then concluded by combining the following two lemmas using Theorem 1 in Sect. 2. The proofs of these lemmas are postponed to the full version.

Lemma 8 (Ratio Analysis). *For all good transcripts τ,*

$$\mathsf{Pr}\left[\mathsf{T}_0 = \tau\right] \geq \left(1 - \frac{2q'\ell + 2\ell^2}{2^n}\right) \cdot \mathsf{Pr}\left[\mathsf{T}_1 = \tau\right] .$$

Lemma 9 (Bad Event Analysis). *For $\mathcal{B}$ as defined above,*

$$\Pr[\mathsf{T}_1 \in \mathcal{B}] \leq \frac{Q(q_{\mathcal{D}}) + q_\pi}{2^{\gamma^*}} + \frac{Q(q_{\mathcal{D}})}{2^{rs}} + \frac{14(Q(q_{\mathcal{D}}) + q_\pi)^2}{2^c} + \frac{2Q(q_{\mathcal{D}}) + q_\pi}{2^n} + \mathsf{Adv}_n^{(\gamma^*, q_{\mathcal{D}})-\mathsf{hit}}(\mathcal{A}, \mathcal{D})\,.$$

□

5.2 Analysis of next

We now turn our attention to the procedure next. We are going to prove that if the input state to next has sufficient min-entropy, then the resulting state and the output bits are indistinguishable from a random element from $0^r \parallel \{0,1\}^c$ and $\{0,1\}^r$, respectively. The proof of the following lemma is postponed to the full version of this paper. We give an overview below.

Lemma 10 (Security of next). *Let* S *be a random variable on the n-bit strings. Then, for any* q_π*-adversary* $\mathcal{A}$ *nd all* $t \geq 1$*,*

$$\mathsf{Adv}_{\mathcal{A}}^{\mathsf{dist}}(\mathsf{next}_t^\pi(\mathsf{S}), (0^r \parallel \mathsf{U}_c, \mathsf{U}_r)) \leq \frac{q_\pi}{2^{\mathbf{H}_\infty(\mathsf{S})}} + \frac{q_\pi}{2^{(r-1)t}} + \frac{4(q_\pi + t)^2}{2^c} + \frac{1}{2^n}, \quad (8)$$

where U_r *and* U_c *are uniformly and independently distributed over the r- and c-bit strings, respectively.*

<u>Proof outline.</u> Intuitively, given a value (S_t, R) output by either next(S) or simply by sampling it uniformly as in $0^r \parallel \mathsf{U}_c, \mathsf{U}_r$, the naive attacker would proceed as follows. Starting from S_t, it would try to guess the t r-bit parts in the computation of next (call them $Z_1, \ldots, Z_t$) which have been zeroed out, and repeatedly apply π^{-1} to recover the state S_0 (in the real case) which was used to generate the R-part of the output. Our proof will confirm that this attack is somewhat optimal, but one needs to exercise some care. Indeed, the proof will consist of two steps, which need to be made in the right order:

(1) We first show that if the attacker cannot succeed in doing the above, then it cannot distinguish whether it is given, together with R, the *actual* S_t value output by next on input S, or a value S'_t which is sampled independently of the internal workings of next (while still being given the actual R).
(2) We then show that given S'_t is now sampled independently of next(S), then the adversary will not notice a substantial difference if the *real* R-part of the output of next(S) (which is still given to $\mathcal{A}$) is finally replaced by an independently random one.

While (2) is fairly straightforward, the core of the proof is in (1). Similar to the proof of the extraction lemma, we are going to think here in terms of the adversary attempting to build some potential "chains" of values, which are sequences of queries (α_i, β_i) for $i \in [t]$ where $\beta_{i-1}[r+1 \ldots n] = \alpha_i[r+1 \ldots n]$ for all $i \geq 2$,

$\alpha_i[1\dots r] = 0^r$ for all $i \geq 2$, and $\beta_t[r+1\dots n] = S_t[r+1\dots n]$. The adversary's hope is that one of these chains is such that $\beta_i[1\dots r] = Z_i$ for all $i \in [t]$, and this would allow to distinguish.

It is not hard to show that as long as $q_\pi \leq 2^{c/2}$, there are at most q_π potential chains with high probability. However, it is harder to argue that the probability that one of these potential chains really matchs the Z_i values is small when the adversary is given the real S_t output by $\mathsf{next(S)}$. This is because the values $Z_1, \dots, Z_t$ are already fixed during the execution, and arguing about their conditional distribution is difficult. Rather, our proof (using the H-coefficient technique) shows that it suffices to analyze the probability that the adversary builds such a valid chain in the ideal world, where the adversary is given an independent S'_t. This analysis becomes much easier, as the values $Z_1, \dots, Z_t$ can be sampled lazily after the adversary is done with its permutation queries, and they are essentially *random* and independent of the potential chains they can match.

5.3 Recovering Security

We now use the insights obtained in the previous sections to establish the recovering security of our construction **SPRG**. To slightly simplify the notation, let $\varepsilon_{\text{ext}}(q_\pi, q_\mathcal{D})$ denote the first four terms on the right-hand side of the bound (3) in Lemma 6 as a function of q_π and $q_\mathcal{D}$; and let $\varepsilon_{\text{next}}(q_\pi)$ denote the right-hand side of the bound (8) in Lemma 10 as a function of q_π.

Lemma 11. *Let $\mathbf{SPRG}_{s,t,n,r}$ be the PRNG given in Sect. 4 and let $\varepsilon_{\text{ext}}(\cdot,\cdot)$ and $\varepsilon_{\text{next}}(\cdot)$ be defined as above. Let $\gamma^* > 0$ and $q_\mathcal{D} \geq 0$, let $\mathcal{A}$ be a q_π-adversary against recovering security and $\mathcal{D}$ be a $(q_\mathcal{D}, q_\pi)$-legitimate Q-distribution sampler $\mathcal{D}$ such that the length of its outputs $I_1, \dots, I_{q_\mathcal{D}}$ padded into r-bit blocks is at most $\ell \cdot r$ bits in total. Then we have*

$$\mathsf{Adv}^{(\gamma^*,q_\mathcal{D})\text{-rec}}_{\mathbf{SPRG}[\pi]}(\mathcal{A},\mathcal{D}) \leq \varepsilon_{\text{ext}}(q_\pi + t + 1, q_\mathcal{D}) + 2\varepsilon_{\text{next}}(\overline{q}_\pi) + \frac{q_\pi}{2^n},$$

where $\overline{q}_\pi = q_\pi + Q(q_\mathcal{D})$.

Proof. Intuitively, we argue that due to the extractor properties of $\mathbf{Sp}_{n,r,s}$ shown in Lemma 6, the state S_d in the experiment $\mathsf{REC}^{\gamma^*,q_\mathcal{D}}_{\mathbf{SPRG}}$ (after processing the inputs hidden from the adversary) will be close to random; and due to Lemma 10 the output of next invoked on this state will be close to random as well.

More formally, we start by showing that there exists a $(q_\pi+t+1)$-adversary $\mathcal{A}_1$ and a $\overline{q}_\pi$-adversary $\mathcal{A}_2$ such that

$$\mathsf{Adv}^{(\gamma^*,q_\mathcal{D})\text{-rec}}_{\mathbf{SPRG}[\pi]}(\mathcal{A},\mathcal{D}) \leq \mathsf{Adv}^{(\gamma^*,q_\mathcal{D})-\text{ext}}_{\mathbf{Sp}_{n,r,s},\mathcal{D}}(\mathcal{A}_1) + \mathsf{Adv}^{\text{dist}}_{\mathcal{A}_2}(\mathsf{next}^\pi(\mathsf{U}_n), (0^r \parallel \mathsf{U}_c, \mathsf{U}_r)), \tag{9}$$

where U_ℓ always denotes an independent random ℓ-bit string. Afterwards, we apply Lemmas 6 and 10 to upper-bound the two advantages on the right-hand side of (9).

Let $\mathcal{A}$ be the adversary against recovering security from the statement. Consider an adversary $\mathcal{A}_1$ against extraction that works as follows: Upon receiving seed, $\gamma_1, \ldots, \gamma_{q_\mathcal{D}}, z_1, \ldots, z_{q_\mathcal{D}}$ from the challenger, it runs the adversary $\mathcal{A}$ and provides it with these same values. During its run, $\mathcal{A}$ issues queries to the oracles π/π^{-1} and get-refresh, which are forwarded by $\mathcal{A}_1$ to the equally-named oracles available to it. At some point, $\mathcal{A}$ outputs a pair (S_0, d), $\mathcal{A}_1$ responds by setting $\mathsf{IV} \leftarrow S_0$ and outputting (IV, d) to the challenger. Upon receiving Y^* and $I_{k+d+1}, \ldots, I_{q_\mathcal{D}}$ from the challenger, $\mathcal{A}_1$ computes $(S^*, R^*) \leftarrow \mathsf{next}(Y^*)$ and feeds both (S^*, R^*) and $I_{k+d+1}, \ldots, I_{q_\mathcal{D}}$ to $\mathcal{A}$. Then it responds to the π-queries of $\mathcal{A}$ as before, and upon receiving the final bit b^* from $\mathcal{A}$, $\mathcal{A}_1$ outputs the same bit. It is easy to verify the query complexity of $\mathcal{A}_1$.

For analysis, note that if the bit chosen by the challenger is $b = 0$, for $\mathcal{A}$ this is a perfect simulation of the recovering game $\mathsf{REC}^{\gamma^*, q_\mathcal{D}}_{\mathbf{SPRG}}$ with the challenge bit being also set to 0. On the other hand, if the challenger sets $b = 1$, $\mathcal{A}$ is given $(S^*, R^*) \leftarrow \mathsf{next}(\mathsf{U}_n)$ for an independent random n-bit string U_n, while the game $\mathsf{REC}^{\gamma^*, q_\mathcal{D}}_{\mathbf{SPRG}}$ with challenge bit set to 1 would require randomly chosen $(S^*, R^*) \xleftarrow{\$} (0^r \,\|\, \{0,1\}^c) \times \{0,1\}^r$ instead. The latter term in the bound (9) accounts exactly for this discrepancy – to see this, just consider an adversary $\mathcal{A}_2$ that simulates both $\mathcal{A}_1$ and the game $\mathsf{GEXT}^{\gamma^*, q_\mathcal{D}}_{\mathbf{Sp}_{n,r,s}}(\mathcal{A}_1, \mathcal{D})$ with $b = 1$, and then uses the dist-challenge instead of the challenge for $\mathcal{A}$.

We conclude by upper bounding the advantages on the right-hand side of (9). First, Lemma 6 gives us

$$\mathsf{Adv}^{(\gamma^*, q_\mathcal{D})-\mathsf{ext}}_{\mathbf{Sp}_{n,r,s}, \mathcal{D}}(\mathcal{A}_1) \leq \varepsilon_{\mathrm{ext}}(q_\pi + t + 1, q_\mathcal{D}) + \mathsf{Adv}^{(\gamma^*, q_\mathcal{D})-\mathsf{hit}}_{\mathcal{D}, n}(\mathcal{A}_1) .$$

It hence remains to bound $\mathsf{Adv}^{(\gamma^*, q_\mathcal{D})-\mathsf{hit}}_{\mathcal{D}, n}(\mathcal{A}_1)$, which is the probability that $\mathcal{A}_1$ queries $\pi^{-1}(Y^*)$ in the ideal-case $b = 1$ in $\mathsf{GEXT}^{\gamma^*, q_\mathcal{D}}_{\mathbf{Sp}_{n,r,s}}(\mathcal{A}, \mathcal{D})$. Note that (apart from forwarding $\mathcal{A}$'s π-queries) the only π-queries that $\mathcal{A}_1$ asks "itself" are to evaluate the call $\mathsf{next}(Y^*)$, and these are only forward queries. Therefore, it suffices to bound the probability that $\mathcal{A}$ queries $\pi^{-1}(Y^*)$ and $\mathcal{A}_1$ forwards this query. Since the only information related to Y^* that $\mathcal{A}$ obtains during this experiment is $(S^*, R^*) \leftarrow \mathsf{next}(Y^*)$, if we replace these values by randomly sampled $(S^*, R^*) \xleftarrow{\$} (0^r \,\|\, \{0,1\}^c) \times \{0,1\}^r$, the value Y^* will be completely independent of $\mathcal{A}$'s view. Therefore, again there exists a $\overline{q}_\pi$-adversary $\mathcal{A}_3$ (actually, $\mathcal{A}_3 = \mathcal{A}_2$) such that

$$\mathsf{Adv}^{(\gamma^*, q_\mathcal{D})-\mathsf{hit}}_{\mathcal{D}, n}(\mathcal{A}_1) \leq \frac{q_\pi}{2^n} + \mathsf{Adv}^{\mathsf{dist}}_{\mathcal{A}_3}(\mathsf{next}^\pi(\mathsf{U}_n), (0^r \,\|\, \mathsf{U}_c, \mathsf{U}_r)) .$$

Finally, by Lemma 10 for both $i \in \{2, 3\}$ we have

$$\begin{aligned} &\mathsf{Adv}^{\mathsf{dist}}_{\mathcal{A}_i}(\mathsf{next}^\pi(\mathsf{U}_n), (0^r \,\|\, \mathsf{U}_c, \mathsf{U}_r)) \leq \varepsilon_{\mathrm{next}}(\overline{q}_\pi) \\ &\leq \frac{\overline{q}_\pi}{2^{\mathbf{H}_\infty(\mathsf{U}_n)}} + \frac{\overline{q}_\pi}{2^{(r-1)t}} + \frac{4(\overline{q}_\pi + t)^2}{2^c} + \frac{1}{2^n} = \frac{\overline{q}_\pi + 1}{2^n} + \frac{\overline{q}_\pi}{2^{(r-1)t}} + \frac{4(\overline{q}_\pi + t)^2}{2^c} , \end{aligned}$$

which concludes the proof. □

5.4 Preserving Security

Here we proceed to establish also the preserving security of **SPRG**.

Lemma 12. *Let* **SPRG**$[\pi]$ *be the PRNG given in Sect. 4, and let* $\varepsilon_{\text{next}}(\cdot)$ *be defined as above. For every* q_π*-adversary* $\mathcal{A}$ *against preserving security, we have*

$$\mathsf{Adv}^{\mathsf{pres}}_{\mathbf{SPRG}[\pi]}(\mathcal{A}) \le \varepsilon_{\text{next}}(q_\pi) + \frac{q_\pi}{2^c} + \frac{(2d'+1)(q_\pi+d')}{2^n} \le \\ \le \frac{(2d'+2)(q_\pi+d')}{2^n} + \frac{q_\pi}{2^{(r-1)t}} + \frac{4(q_\pi+t)^2+q_\pi}{2^c},$$

where d' *is the number of* r*-bit blocks resulting from parsing* $\mathcal{A}$*'s output* $I_1,\ldots,I_d$.

Proof. Intuitively, the proof again consists of two steps: showing that (1) since the initial state S_0 is random and hidden from the adversary, the state S_d will most likely look random to it as well; and (2) if S_d is random, we can again rely on Lemma 10 to argue about the pseudorandomness of the outputs of next.

More formally, consider a game PRES' which is defined exactly as the game PRES in Fig. 4, except that instead of computing the value S_d iteratively in Step 3, we sample it freshly at random as $S_d \xleftarrow{\$} \{0,1\}^n$. Moreover, imagine the permutation π as being lazy-sampled in both games.

Let $\mathcal{A}$ be an adversary participating in the game $\mathsf{PRES}_{\mathbf{SPRG}[\pi]}$. Let $\mathcal{QR}^{(1)}_\pi$ denote the set of query-response pairs that the adversary $\mathcal{A}$ asks to π via its oracles π/π^{-1} in its first stage (before submitting $I_1,\ldots,I_d$). More precisely, let $\mathcal{QR}^{(1)}_\pi$ denote the set of pairs $(u,v)\in\{0,1\}^n\times\{0,1\}^n$ such that $\mathcal{A}$ in its first stage either asked the query $\pi(u)$ and received the response v, or asked the query $\pi^{-1}(v)$ and received the response u. Moreover, let us denote by $I'_1,\ldots,I'_{d'}$ the r-bit blocks resulting from parsing the inputs $I_1,\ldots,I_d$ in sequence, using the parsing mechanism from the $\mathsf{refresh}$ procedure. Finally, recall that "$\rightarrow$" denotes the output of the adversary, as opposed to the game output.

We first argue that

$$\Pr\left[\mathsf{PRES}_{\mathbf{SPRG}[\pi]}(\mathcal{A})\rightarrow 0 \,\middle|\, b=0\right] - \Pr\left[\mathsf{PRES}'_{\mathbf{SPRG}[\pi]}(\mathcal{A})\rightarrow 0 \,\middle|\, b=0\right] \\ \le \frac{q_\pi}{2^c} + \frac{(2d'+1)(q_\pi+d')}{2^n}. \qquad (10)$$

To see this, first note that the value S_0 is chosen independently at random from the set $0^r \,\|\, \{0,1\}^c$ and hidden from the adversary. Therefore, we have

$$\Pr\left[\exists (u,v)\in\mathcal{QR}^{(1)}_\pi : S_0 \oplus ((I'_1\oplus \mathsf{seed}_1)\,\|\,0^c) = u\right] \le \frac{\left|\mathcal{QR}^{(1)}_\pi\right|}{2^c} \le \frac{q_\pi}{2^c}.$$

If this does not happen, the first invocation of π during the sequence of evaluations of $\mathsf{refresh}$ on $I_1,\ldots,I_d$ will be on a fresh value and hence its output (call it S'_1) will be chosen uniformly at random from the $2^n - \left|\mathcal{QR}^{(1)}_\pi\right| - 1$ unused values.

Hence, again the probability that the next π-invocation will be on an already defined value is at most $2(q_\pi+1)/2^n$. This same argument can be used iteratively up to the final state S_d: with probability at least $1-q_\pi/2^c-2d'(q_\pi+d')/2^n$ all of the π-invocations used during the sequence of refresh-calls will happen on fresh values, and therefore S_d will be also chosen uniformly at random from the set of at least $2^n-q_\pi-d'$ values. This means that in this case, the statistical distance of S_d in the game $\mathsf{PRES}_{\mathbf{SPRG}[\pi]}$ from S_d in the game $\mathsf{PRES}'_{\mathbf{SPRG}[\pi]}$ where it is chosen at random will be at most $(q_\pi+d')/2^n$. Put together, this proves (10).

Now we observe that there exists a q_π-adversary $\mathcal{A}'$ such that

$$\Pr\left[\mathsf{PRES}'_{\mathbf{SPRG}[\pi]}(\mathcal{A})\to 0 \,\middle|\, b=0\right]-\Pr\left[\mathsf{PRES}'_{\mathbf{SPRG}[\pi]}(\mathcal{A})\to 0 \,\middle|\, b=1\right]\leq$$
$$\leq \mathsf{Adv}^{\mathsf{dist}}_{\mathcal{A}'}(\mathsf{next}^{\pi}_{t}(\mathsf{U}_n),(0^r \,\|\, \mathsf{U}_c,\mathsf{U}_r))\leq \varepsilon_{\mathrm{next}}(q_\pi) \quad (11)$$

where U_ℓ denotes a uniformly random ℓ-bit string. Namely, it suffices to consider $\mathcal{A}'$ that runs the adversary $\mathcal{A}$ and simulates the game PRES' for it (except for the π-queries; also note that $\mathcal{A}'$ does not need to compute the sequence of refresh-calls), then replaces the challenge for $\mathcal{A}$ by its own challenge, and finally outputs the complement of the bit $\mathcal{A}$ outputs.

The proof is finally concluded by combining the bounds (10) and (11) and observing that if $b=1$, the games PRES and PRES' are identical. □

6 Key-Derivation Functions from Sponges

This section applies the sponge extraction lemma (Lemma 6) to key-derivation functions (KDFs). We follow the formalization of Krawczyk [22]. While the fact that sponges can be used as KDFs is widely believed thanks to the existing indifferentiability analysis [9], our treatment allows for a stronger result for adversarial and oracle-dependent distributions.

<u>KDFs AND THEIR SECURITY.</u> A *key derivation function* is an algorithm $\mathsf{KDF}: \{0,1\}^s\times\{0,1\}^*\times\{0,1\}^*\times\mathbb{N}\to\{0,1\}^*$, where the first input is the *seed*, the second is the *source material*, the third is the *context variable*, and the fourth is the *output length*. In particular, for all $\mathsf{seed}\in\{0,1\}^s$, $W,C\in\{0,1\}^*$ and $\mathsf{len}\in\mathbb{N}$, we have $|\mathsf{KDF}(\mathsf{seed},W,C,\mathsf{len})|=\mathsf{len}$, and moreover $\mathsf{KDF}(\mathsf{seed},W,C,\mathsf{len}')$ is a prefix of $\mathsf{KDF}(\mathsf{seed},W,C,\mathsf{len})$ for all $\mathsf{len}'\leq\mathsf{len}$.

We consider KDF constructions making calls to an underlying permutation $\pi\in\mathsf{Perms}(n)$[6]. We define security of KDFs in terms of a security game $\mathsf{GKDF}^{\gamma^*,q_\mathcal{D}}_{\mathsf{KDF}}(\mathcal{A},\mathcal{D})$ which is slightly more general than the one used in [22], and described in Fig. 7. In particular, similar to GEXT above, the game considers an incoming stream of $q_\mathcal{D}$ weakly random values, coming from a legitimate and oracle-dependent distribution sampler, and the attacker can choose a subset of these values with sufficient min-entropy *adaptively* to derive randomness from, as long as these values are guaranteed to have (jointly) min-entropy at

[6] Once again, our treatment easily extends to other ideal models, but we dispense here with a generalization to keep our treatment sufficiently compact.

Game $\mathsf{GKDF}^{\gamma^*,q_\mathcal{D}}_{\mathsf{KDF}}(\mathcal{A},\mathcal{D})$:

1. The challenger chooses $\mathsf{seed} \xleftarrow{\$} \{0,1\}^u$, $\pi \xleftarrow{\$} \mathsf{Perms}(n)$ and $b \xleftarrow{\$} \{0,1\}$ and sets $\sigma_0 \leftarrow \perp$. For $k = 1, \ldots, q_\mathcal{D}$, the challenger computes $(\sigma_k, I_k, \gamma_k, z_k) \leftarrow \mathcal{D}^\pi(\sigma_{k-1})$.
2. The attacker $\mathcal{A}$ gets seed and $\gamma_1, \ldots, \gamma_{q_\mathcal{D}}, z_1, \ldots, z_{q_\mathcal{D}}$. It gets access to oracles π/π^{-1}, and to an oracle get-refresh() which initially sets $k \leftarrow 0$ and on each invocation increments $k \leftarrow k+1$ and outputs I_k. At some point, $\mathcal{A}$ outputs an integer d such that $k + d \leq q_\mathcal{D}$ and $\sum_{j=k+1}^{k+d} \gamma_j \geq \gamma^*$.
3. If $b = 1$, we let $F = \mathbf{RO}(\cdot,\cdot)$, and if $b = 0$, $F = \mathsf{KDF}^\pi(\mathsf{seed}, I_{k+1} \| \ldots \| I_{k+d}, \cdot, \cdot)$. Then, the challenger gives back $I_{k+d+1}, \ldots, I_{q_\mathcal{D}}$ to $\mathcal{A}$.
4. The attacker gets access to π/π^{-1}, and in addition to F, and outputs a bit b^*. The output of the game is 1 iff $b = b^*$.

Fig. 7. Definition of the game $\mathsf{GKDF}^{\gamma^*,q_\mathcal{D}}_{\mathsf{KDF},\mathcal{D}}(\mathcal{A})$. Here, **RO** is an oracle which associates with each string x a potentially infinitely long string $\rho(x)$, and on input (x, len), it returns the first len bits of $\rho(x)$.

least γ^*. The game then requires the attacker $\mathcal{A}$, given seed, to distinguish $\mathsf{KDF}(\mathsf{seed}, I_{k+1} \| \cdots \| I_{k+d}, \cdot, \cdot)$ from $\mathbf{RO}(\cdot,\cdot)$, where the latter returns for every $X \in \{0,1\}^*$ and $\mathsf{len} \in \mathbb{N}$, the first len bits of an infinitely long stream $\rho(X)$ of random bits associated with X.

Then, the kdf advantage of $\mathcal{A}$ is

$$\mathsf{Adv}^{(\gamma^*,q_\mathcal{D})-\mathsf{kdf}}_{\mathsf{KDF}}(\mathcal{A},\mathcal{D}) = 2 \cdot \Pr\left[\mathsf{GKDF}^{\gamma^*,q_\mathcal{D}}_{\mathsf{KDF}}(\mathcal{A},\mathcal{D}) \Rightarrow 1\right] - 1\,.$$

SPONGE-BASED KDF. We present a sponge based KDF construction – denoted $\mathbf{SpKDF}_{n,r,s}$ – that can easily be implemented on top of SHA-3. It depends on three parameters n, r, s, and uses a seed of length $k = r \cdot s$ bits, represented as $\mathsf{seed} = (\mathsf{seed}_0, \ldots, \mathsf{seed}_{s-1})$. It uses a permutation π, and given $W, C \in \{0,1\}^*$, and $\mathsf{len} \in \mathbb{N}$, it operates as follows: It first splits W and C into r-bit blocks $W_1 \ldots W_d$ and $C_1 \ldots C_{d'}$,[7] and then computes, starting with $S_0 = \mathsf{IV}$, the states $S_1, \ldots, S_d, S_{d+1}, \ldots, S_{d+d'}$, where

$$S_i \leftarrow \pi((W_i \oplus \mathsf{seed}_{i \bmod s}) \| 0^c \oplus S_{i-1}) \text{ for all } i \in [d]$$
$$S_i \leftarrow \pi((C_i \| 0^c) \oplus S_{i-1}) \text{ for all } i \in [d+1 \ldots d+d']$$

Then, for $t := \lceil \mathsf{len}/r \rceil$, if $t \geq 2$, it computes the values $S_{d+d'+1}, \ldots, S_{d+d'+t-1}$ as $S_i \leftarrow \pi(S_{i-1})$ for $i \in [d+d'+1 \ldots d+d'+t-1]$. Finally, $\mathbf{SpKDF}^\pi_{n,r,s}(\mathsf{seed}, W, C, \mathsf{len})$ outputs the first len bits of $S_{d+d'}[1 \ldots r] \| \cdots \| S_{d+d'+t-1}[1 \ldots r]$.

[7] As in the original sponge construction, we need to assume that C is always encoded so that every block $C_i \neq 0^r$.

SECURITY OF SPONGE-BASED KDF. The proof of the following theorem (given in the full version) is an application of the sponge extraction lemma (Lemma 6), combined with existing analyses of the PRF security of keyed sponges with variable-output-length [24].

Theorem 13 (Security of SpKDF). *Let r, s be integers, let $q_{\mathcal{D}}, q_\pi$ be arbitrary, and let $\gamma^* > 0$. Also, let $\mathcal{D}$ be a $(q_{\mathcal{D}}, q_\pi)$-legitimate Q-distribution sampler $\mathcal{D}$ for which the overall output length (when invoked $q_{\mathcal{D}}$ times) is at most $\ell \cdot r$ bits after padding. Then, for all adversaries $\mathcal{A}$ making $q_\pi \leq 2^{c-2}$ queries to π, and q queries to F, where every query to the latter results in an input C encoded into at most ℓ' r-bit blocks, and in an output of at most* len *bits, we have*

$$\mathsf{Adv}^{(\gamma^*, q_{\mathcal{D}})-\mathsf{kdf}}_{\mathbf{SpKDF}_{n,r,s}}(\mathcal{A}, \mathcal{D}) \leq \frac{\tilde{q}_\pi}{2^{\gamma^*}} + \frac{Q(q_{\mathcal{D}})}{2^{sr}} + \frac{14\tilde{q}_\pi^2 + 6q^2\bar{\ell} + 3q\bar{\ell}\tilde{q}_\pi}{2^c} + \frac{2\tilde{q}_\pi \ell + 2\ell^2 + 6q^2\bar{\ell}^2 + \tilde{q}_\pi}{2^n},$$

where $\tilde{q}_\pi = (q_\pi + Q(q_{\mathcal{D}}))(1 + 2\lceil \frac{n}{r} \rceil)$ and $\bar{\ell} = \ell + \ell' + \lceil \mathsf{len}/r \rceil$.

Acknowledgments. We thank Mihir Bellare for insightful comments in an earlier stage of this project. Peter Gaži is supported by the European Research Council under an ERC Starting Grant (259668-PSPC). Stefano Tessaro was partially supported by NSF grants CNS-1423566, CNS-1528178, and the Glen and Susanne Culler Chair.

References

1. Information Technology - Security Techniques - Random Bit Generation. ISO/IEC 18031 (2011)
2. Albrecht, M.R., Farshim, P., Paterson, K.G., Watson, G.J.: On cipher-dependent related-key attacks in the ideal-cipher model. In: Joux, A. (ed.) FSE 2011. LNCS, vol. 6733, pp. 128–145. Springer, Heidelberg (2011)
3. Andreeva, E., Daemen, J., Mennink, B., Van Assche, G.: Security of keyed sponge constructions using a modular proof approach. In: Leander, G. (ed.) FSE 2015. LNCS, vol. 9054, pp. 364–384. Springer, Heidelberg (2015)
4. Barak, B., Dodis, Y., Krawczyk, H., Pereira, O., Pietrzak, K., Standaert, F.-X., Yu, Y.: Leftover hash lemma, revisited. In: Rogaway, P. (ed.) CRYPTO 2011. LNCS, vol. 6841, pp. 1–20. Springer, Heidelberg (2011)
5. Barak, B., Halevi, S.: A model and architecture for pseudo-random generation with applications to /dev/random. In: Atluri, V., Meadows, C., Juels, A. (eds.) ACM CCS 2005, pp. 203–212. ACM Press, November 2005
6. Barker, E.B., Kelsey, J.M.: Sp 800–90a. recommendation for random number generation using deterministic random bit generators. Technical report, Gaithersburg, MD, United States (2012)
7. Bellare, M., Rogaway, P.: The security of triple encryption and a framework for code-based game-playing proofs. In: Vaudenay, S. (ed.) EUROCRYPT 2006. LNCS, vol. 4004, pp. 409–426. Springer, Heidelberg (2006)
8. Bello, L.: Dsa-1571-1 openssl-predictable random number generator. Debian Security Advisory (2008)

9. Bertoni, G., Daemen, J., Peeters, M., Van Assche, G.: On the indifferentiability of the sponge construction. In: Smart, N.P. (ed.) EUROCRYPT 2008. LNCS, vol. 4965, pp. 181–197. Springer, Heidelberg (2008)
10. Bertoni, G., Daemen, J., Peeters, M., Van Assche, G.: Sponge-based pseudo-random number generators. In: Mangard, S., Standaert, F.-X. (eds.) CHES 2010. LNCS, vol. 6225, pp. 33–47. Springer, Heidelberg (2010)
11. Chen, S., Steinberger, J.: Tight security bounds for key-alternating ciphers. In: Nguyen, P.Q., Oswald, E. (eds.) EUROCRYPT 2014. LNCS, vol. 8441, pp. 327–350. Springer, Heidelberg (2014)
12. Dai, Y., Steinberger, J.: Tight security bounds for multiple encryption. Cryptology ePrint Archive, Report 2014/096 (2014). http://eprint.iacr.org/2014/096
13. Desai, A., Hevia, A., Yin, Y.L.: A practice-oriented treatment of pseudorandom number generators. In: Knudsen, L.R. (ed.) EUROCRYPT 2002. LNCS, vol. 2332, pp. 368–383. Springer, Heidelberg (2002)
14. Dodis, Y., Gennaro, R., Håstad, J., Krawczyk, H., Rabin, T.: Randomness extraction and key derivation using the CBC, cascade and HMAC modes. In: Franklin, M. (ed.) CRYPTO 2004. LNCS, vol. 3152, pp. 494–510. Springer, Heidelberg (2004)
15. Dodis, Y., Pointcheval, D., Ruhault, S., Vergnaud, D., Wichs, D.: Security analysis of pseudo-random number generators with input: /dev/random is not robust. In: Sadeghi, A.-R., Gligor, V.D., Yung, M. (eds.) ACM CCS 2013, pp. 647–658. ACM Press, November 2013
16. Dodis, Y., Ristenpart, T., Vadhan, S.: Randomness condensers for efficiently samplable, seed-dependent sources. In: Cramer, R. (ed.) TCC 2012. LNCS, vol. 7194, pp. 618–635. Springer, Heidelberg (2012)
17. Dodis, Y., Shamir, A., Stephens-Davidowitz, N., Wichs, D.: How to eat your entropy and have it too – optimal recovery strategies for compromised RNGs. In: Garay, J.A., Gennaro, R. (eds.) CRYPTO 2014, Part II. LNCS, vol. 8617, pp. 37–54. Springer, Heidelberg (2014)
18. Dorrendorf, L., Gutterman, Z., Pinkas, B.: Cryptanalysis of the windows random number generator. In: Ning, P., De Capitani di Vimercati, S., Syverson, P.F. (eds.) ACM CCS 2007, pp. 476–485. ACM Press, October 2007
19. Eastlake, D., Schiller, J., Crocker, S.: Randomness Requirements for Security. RFC 4086 (Best Current Practice), June 2005
20. Gaži, P., Pietrzak, K., Tessaro, S.: The exact PRF security of truncation: tight bounds for keyed sponges and truncated CBC. In: Gennaro, R., Robshaw, M. (eds.) CRYPTO 2015, Part I. LNCS, vol. 9215, pp. 368–387. Springer, Heidelberg (2015)
21. Kelsey, J., Schneier, B., Wagner, D., Hall, C.: Cryptanalytic attacks on pseudorandom number generators. In: Vaudenay, S. (ed.) FSE 1998. LNCS, vol. 1372, pp. 168–188. Springer, Heidelberg (1998)
22. Krawczyk, H.: Cryptographic extraction and key derivation: the HKDF scheme. In: Rabin, T. (ed.) CRYPTO 2010. LNCS, vol. 6223, pp. 631–648. Springer, Heidelberg (2010)
23. Maurer, U.M., Renner, R.S., Holenstein, C.: Indifferentiability, impossibility results on reductions, and applications to the random oracle methodology. In: Naor, M. (ed.) TCC 2004. LNCS, vol. 2951, pp. 21–39. Springer, Heidelberg (2004)
24. Mennink, B., Reyhanitabar, R., Vizár, D.: Security of full-state keyed and duplex sponge: Applications to authenticated encryption. Technical report, Cryptology ePrint Archive, Report 2015/541 (2015). http://eprint.iacr.org. To appear at ASIACRYPT 2015

25. Mittelbach, A.: Salvaging indifferentiability in a multi-stage setting. In: Nguyen, P.Q., Oswald, E. (eds.) EUROCRYPT 2014. LNCS, vol. 8441, pp. 603–621. Springer, Heidelberg (2014)
26. Nohl, K., Evans, D., Starbug, S., Plötz, H.: Reverse-engineering a cryptographic RFID tag. In: USENIX Security Symposium, vol. 28 (2008)
27. Patarin, J.: The "Coefficients H" technique. In: Avanzi, R.M., Keliher, L., Sica, F. (eds.) SAC 2008. LNCS, vol. 5381, pp. 328–345. Springer, Heidelberg (2009)
28. Ristenpart, T., Shacham, H., Shrimpton, T.: Careful with composition: limitations of the indifferentiability framework. In: Paterson, K.G. (ed.) EUROCRYPT 2011. LNCS, vol. 6632, pp. 487–506. Springer, Heidelberg (2011)
29. Shrimpton, T., Seth Terashima, R.: A provable-security analysis of intel's secure key RNG. In: Oswald, E., Fischlin, M. (eds.) EUROCRYPT 2015, Part I. LNCS, vol. 9056, pp. 77–100. Springer, Heidelberg (2015)
30. Shumow, D., Ferguson, N.: On the possibility of a back door in the NIST SP800-90 Dual Ec Prng. CRYPTO 2007 Rump Session, August 2007
31. Van Herrewege, A., Verbauwhede, I.: Software only, extremely compact, keccak-based secure prng on arm cortex-m. In: Proceedings of the 51st Annual Design Automation Conference, DAC 2014, pp. 111: 1–111: 6. ACM, New York (2014)

Reusable Fuzzy Extractors for Low-Entropy Distributions

Ran Canetti[1,2], Benjamin Fuller[1,3(✉)], Omer Paneth[1], Leonid Reyzin[1], and Adam Smith[4]

[1] Boston University, Boston, MA, USA
{canetti,bfuller,paneth,reyzin}@cs.bu.edu
[2] Tel Aviv University, Tel Aviv, Israel
[3] MIT Lincoln Laboratory, Lexington, MA, USA
[4] Pennsylvania State University, University Park, PA, USA
asmith@cse.psu.edu

Abstract. Fuzzy extractors (Dodis et al., Eurocrypt 2004) convert repeated noisy readings of a secret into the same uniformly distributed key. To eliminate noise, they require an initial enrollment phase that takes the first noisy reading of the secret and produces a nonsecret helper string to be used in subsequent readings. *Reusable* fuzzy extractors (Boyen, CCS 2004) remain secure even when this initial enrollment phase is repeated multiple times with noisy versions of the same secret, producing multiple helper strings (for example, when a single person's biometric is enrolled with multiple unrelated organizations).

We construct the first reusable fuzzy extractor that makes no assumptions about how multiple readings of the source are correlated (the only prior construction assumed a very specific, unrealistic class of correlations). The extractor works for binary strings with Hamming noise; it achieves computational security under assumptions on the security of hash functions or in the random oracle model. It is simple and efficient and tolerates near-linear error rates.

Our reusable extractor is secure for source distributions of linear min-entropy rate. The construction is also secure for sources with much lower entropy rates—lower than those supported by prior (nonreusable) constructions—assuming that the distribution has some additional structure, namely, that random subsequences of the source have sufficient minentropy. We show that such structural assumptions are necessary to support low entropy rates.

We then explore further how different structural properties of a noisy source can be used to construct fuzzy extractors when the error rates are high, building a computationally secure and an information-theoretically secure construction for large-alphabet sources.

Keywords: Fuzzy extractors · Reusability · Key derivation · Error-correcting codes · Computational entropy · Digital lockers · Point obfuscation

M. Fischlin and J.-S. Coron (Eds.): EUROCRYPT 2016, Part I, LNCS 9665, pp. 117–146, 2016.
DOI: 10.1007/978-3-662-49890-3_5

1 Introduction

Fuzzy Extractors. Cryptography relies on long-term secrets for key derivation and authentication. However, many sources with sufficient randomness to form long-term secrets provide similar but not identical values of the secret at repeated readings. Prominent examples include biometrics and other human-generated data [11,19,24,44,45,57], physically unclonable functions (PUFs) [26,47,52,54], and quantum information [3]. Turning similar readings into identical values is known as *information reconciliation*; further converting those values into uniformly random secret strings is known as *privacy amplification* [3]. Both of these problems have interactive and non-interactive versions. In this paper, we are interested in the non-interactive case, which is useful for a single user trying to produce the same key from multiple noisy readings of a secret at different times. A *fuzzy extractor* [22] is the primitive that accomplishes both information reconciliation and privacy amplification non-interactively.

Fuzzy extractors consist of a pair of algorithms: Gen (used once, at "enrollment") takes a source value w, and produces a key r and a public helper value p. The second algorithm Rep (used subsequently) takes this helper value p and a close w' to reproduce the original key r. The standard correctness guarantee is that r will be correctly reproduced by Rep as long as w' is no farther than t from w according to some notion of distance (specifically, we work with Hamming distance; our primary focus is on binary strings, although we also consider larger alphabets). The security guarantee is that r produced by Gen is indistinguishable from uniform, even given p. In this work, we consider computational indistinguishability [25] rather than the more traditional information-theoretic notion. (Note that so-called "robust" fuzzy extractors [10,17,21,41,43] additionally protect against active attackers who modify p; we do not consider them here, except to point out that our constructions can be easily made robust by the random-oracle-based transform of [10, Theorem 1].)

Reusability. A fuzzy extractor is *reusable* (Boyen [9]) if it remains secure even when a user enrolls the same or correlated values multiple times. For example, if the source is a biometric reading, the user may enroll the same biometric with different noncooperating organizations. Reusability is particularly important for biometrics which, unlike passwords, cannot be changed or created. It is also useful in other contexts, for example, to permit a user to reuse the same visual password across many services or to make a single physical token (embodying a PUF) usable for many applications.

Each enrollment process will get a slightly different enrollment reading w_i, and will run Gen(w_i) to get a key r_i and a helper value p_i. Security for each r_i should hold even when an adversary is given all the values $p_1, \ldots, p_\rho$ and even the keys r_j for $j \neq i$ (because one organization cannot be sure how other organizations will use the derived keys).

As pointed out by Dodis et al. [20, Sect. 6], reusable extractors for the non-fuzzy case (i.e., without p and Rep) can be constructed using leakage-resilient cryptography. However, adding error-tolerance makes the problem harder. Most constructions of fuzzy extractors are not reusable [6,7,9,50]. In fact, the only known construction of reusable fuzzy extractors [9] requires very particular relationships between w_i values[1], which are unlikely to hold for a practical source.

1.1 Our Contribution

A Reusable Fuzzy Extractor. We construct the first reusable fuzzy extractor whose security holds even if the multiple readings w_i used in Gen are *arbitrarily correlated*, as long as the fuzzy extractor is secure for each w_i *individually*. This construction is the first to provide reusability for a realistic class of correlated readings. Our construction is based on *digital lockers*; in the most efficient instantiation, it requires only evaluation of cryptographic hash functions and is secure in the random oracle model or under strong computational assumptions on the hash functions[2]. The construction can output arbitrarily long r.

Our construction handles a wider class of sources than prior work. It is secure if the bits of w are partially independent. Namely, we require that, for some known parameter k, the substring formed by the bits at k randomly chosen positions in w is unguessable (i.e., has minentropy that is superlogarithmic is the security parameter). We call sources with this feature "sources with high-entropy samples." This requirement is in contrast to most constructions of fuzzy extractors that require w to have sufficient minentropy.

All sources of sufficient minentropy have high-entropy samples (because sampling preserves the entropy rate [55]). However, as we now explain, the family of sources with high-entropy samples also includes some low-entropy sources. (Note that, of course, the entropy of a substring never exceeds the entropy of the entire string; the terms "high" and "low" are relative to the length.)

Low-entropy sources with high-entropy samples are easy to construct artificially: for example, we can build a source of length n whose bits are k-wise independent by multiplying (over GF(2)) a fixed $n \times k$ matrix of rank k by a random k-bit vector. In this source, the entropy rate of any substring of length k is 1, while the entropy rate of the entire string is just k/n.

Such sources also arise naturally whenever w exhibits a lot of redundancy. For example, when the binary string w is obtained via signal processing from some underlying reading (such as an image of an iris or an audio recording of a voice), the signal itself is likely to have a lot of redundancy (for example, nearby pixels of an image are highly correlated). By requiring only high-entropy samples rather than a high entropy rate, we free the signal processing designer from the need to remove redundancy when converting the underlying reading to a string w used in the fuzzy extractor. Thus, we enable the use of oversampled signals.

---

[1] Specifically, Boyen's construction requires that the exclusive or $w_i \oplus w_j$ of any two secrets not leak any information about w_i.

[2] The term "digital lockers" was introduced by Canetti and Dakdouk [13]; the fact that such digital lockers can be built easily out cryptographic hash functions was shown by [38, Sect. 4].

Our construction can tolerate $\frac{n \ln n}{k}$ errors (out of the n bits of w) if we allow the running time of the construction (the number of hash function evaluations) to be linear in n. More generally, we can tolerate $c\frac{n \ln n}{k}$ errors if we allow running time linear in n^c. Note that, since in principle k needs to be only slightly super-logarithmic to ensure the high-entropy condition on the samples, our allowable error rate is only slightly sublinear.

The Advantage of Exploiting the Structure of the Distribution. Following the tradition of extractor literature [16,46], much work on fuzzy extractors has focused on providing constructions that work for *any* source of a given minentropy m. In contrast, our construction exploits more about the structure of the distribution than just its entropy. As a result, it supports not only all sources of a given (sufficiently high) minentropy, but also many sources with an entropy rate much lower than the error rate. We know of no prior constructions with this property. We now explain why, in order to achieve this property, exploiting the structure of the distribution is necessary.

A fuzzy extractor that supports t errors out of a string of n bits and works for *all* sources of minentropy m must have the entropy rate $\frac{m}{n}$ at least as big as the binary entropy[3] of the error rate, $h_2(\frac{t}{n})$ (to be exact, $m \geq nh_2(\frac{t}{n}) - \frac{1}{2}\log n - \frac{1}{2}$). The reason for this requirement is simple: if m too small, then a single ball of radius t, which contains at least $2^{nh_2(\frac{t}{n})-\frac{1}{2}\log n-\frac{1}{2}}$ points [1, Lemma 4.7.2, Eq. 4.7.5, p. 115], may contain the entire distribution of 2^m points inside it. For this distribution, an adversary can run Rep on the center of this ball and always learn the key r. This argument leads to the following proposition, which holds regardless of whether the fuzzy extractor is information-theoretic or computational, and extends even to the interactive setting.

Proposition 1. *If the security guarantee of a fuzzy extractor holds for* any *source of minentropy m and the correctness guarantees holds for any t errors and $m < \log |B_t|$ (where $|B_t|$ denotes the number of points in a ball of radius t), the fuzzy extractor must provide no security. In particular, for the binary Hamming case, m must exceed $nh_2(\frac{t}{n}) - \frac{1}{2}\log n - \frac{1}{2} \approx nh_2(\frac{t}{n}) > t\log_2 \frac{n}{t}$.*

Thus, in order to correct t errors regardless of the structure of the distribution, we would have to assume a high total minentropy m. In contrast, by taking advantage of the specific properties of the distribution, we can handle all distributions of sufficiently high minentropy, but also some distributions whose minentropy that is much less than $t < nh_2(\frac{t}{n})$.

Beating the bound of Proposition 1 is important. For example, the IrisCode [19], which is the state of the art approach to handling what is believed to be the best biometric [49], produces a source where m is less than $nh_2(\frac{t}{n})$ [8, Sect. 5]. PUFs with slightly nonuniform outputs suffer from similar problems [36].

[3] Binary entropy $h_2(\alpha)$ for $0 < \alpha < 1$ is defined as $-\alpha \log_2 \alpha - (1-\alpha)\log_2(1-\alpha)$; it is greater than $\alpha \log_2 \frac{1}{\alpha}$ and, in particular, greater than α for interesting range $\alpha < \frac{1}{2}$.

We emphasize that in applications of fuzzy extractors to physical sources, any constraint on the source—whether minentropy-based or more structured—is always, by necessity, an assumption about the physical world. It is no more possible to verify that a source has high minentropy than it is to verify that it has high-entropy samples[4]. Both statements about the source can be derived only by modeling the source—for example, by modeling the physical processes that generate irises or PUFs.

Some prior work on key agreement from noisy data also made assumptions on the structure of the source (often assuming that it consists of independent identically distributed symbols, e.g. [29,39,40,42,56]). However, we are not aware of any work that beat the bound of Propostion 1, with the exception of the work by Holenstein and Renner [30, Theorem 4]. Their construction supports a uniform length n binary w, with a random selection of $(n - m)$ bits leaked to the adversary and t random bits flipped in w'. They show that it is possible to support any $m > 4t(1 - \frac{t}{n})$, which is lower than $\log |B_t| \approx nh_2(\frac{t}{n})$, but still higher than t.

Constructions Exploiting the Structure of the Distribution for Larger Alphabets. In addition to the binary alphabet construction that supports reuse and low entropy rates, as discussed above, we explore how low entropy rates can be supported when symbols of the string w comes from a large, rather than a binary, alphabet. We obtain two additional constructions, both of which allow for distributions whose total minentropy is lower than the volume of the ball of radius t (in the large-alphabet Hamming space). Unfortunately, neither of them provides reusability, but both can tolerate a linear error rate (of course, over the larger alphabet, where errors may be more likely, because each symbol carries more information).

Our second construction for large alphabets works for *sources with sparse high-entropy marginals*: sources for which sufficiently many symbols have high entropy individually, but no independence among symbols is assumed (thus, the total entropy may be as low as the entropy of a single symbol).

Our third construction for large alphabets provides information-theoretic, rather than computational, security. It works for *sparse block sources*. These are sources in which a sufficient fraction of the symbols have entropy conditioned on previous symbols.

Both constructions should be viewed as evidence that assumptions on the source other than total minentropy may provide new opportunities for increasing the error tolerance of fuzzy extractors.

Our Approach. Our approach in all three constructions is different from most known constructions of fuzzy extractors, which put sufficient information in p

[4] However, standard heuristics for estimating entropy can also be used to indicate whether a source has high-entropy samples. For a corpus of noisy signals, repeat the following a statistically significant number of times: (1) sample k indices (2) run the heuristic entropy test on the corpus which each sample restricted to the k indices.

to recover the original w from a nearby w' during Rep (this procedure is called a *secure sketch*). We deliberately do not recover w, because known techniques for building secure sketches do not work for sources whose entropy rate is lower than its error rate. (This is because they lose at least $\log|B_t|$ bits of entropy regardless of the source. This loss is necessary when the source is uniform [22, Lemma C.1] or when reusability against a sufficiently rich class of correlations is desired [9, Theorem 11]; computational definitions of secure sketches suffer from similar problems [25, Corollary 1].) Instead, in the computational constructions, we lock up a freshly generated random r using parts of w in an error-tolerant way; in the information-theoretic construction, we reduce the alphabet in order to reduce the ball volume while maintaining entropy.

We note that the idea of locking up a random r has appeared in a prior theoretical construction of a computational fuzzy extractor for any source. Namely, Bitansky et al. [5] show how to obfuscate a proximity point program that tests if an input w' is within distance t of the value w hidden inside the obfuscated program and, if so, outputs the secret r (such a program would be output by Gen as p and run by Rep). However, such a construction is based on very strong assumptions (semantically secure graded encodings [48]) and, in contrast to our construction, is highly impractical in terms of efficiency. Moreover, it is not known to provide reusability, because known obfuscation of proximity point programs is not known to be composable.

2 Definitions

For a random variables X_i over some alphabet $\mathcal{Z}$ we denote by $X = X_1, ..., X_n$ the tuple $(X_1, \ldots, X_n)$. For a set of indices J, X_J is the restriction of X to the indices in J. The set J^c is the complement of J. The *minentropy* of X is $\mathrm{H}_\infty(X) = -\log(\max_x \Pr[X = x])$, and the *average (conditional)* minentropy of X given Y is $\tilde{\mathrm{H}}_\infty(X|Y) = -\log(\mathbb{E}_{y\in Y} \max_x \Pr[X = x|Y = y])$ [22, Sect. 2.4]. For a random variable W, let $H_0(W)$ be the logarithm of the size of the support of W, that is $H_0(W) = \log|\{w|\Pr[W = w] > 0\}|$. The *statistical distance* between random variables X and Y with the same domain is $\Delta(X, Y) = \frac{1}{2}\sum_x |\Pr[X = x] - \Pr[Y = x]|$. For a distinguisher D we write the *computational distance* between X and Y as $\delta^D(X, Y) = |\mathbb{E}[D(X)] - \mathbb{E}[D(Y)]|$ (we extend it to a class of distinguishers $\mathcal{D}$ by taking the maximum over all distinguishers $D \in \mathcal{D}$). We denote by $\mathcal{D}_s$ the class of randomized circuits which output a single bit and have size at most s.

For a metric space $(\mathcal{M}, \mathsf{dis})$, the *(closed) ball of radius t around x* is the set of all points within radius t, that is, $B_t(x) = \{y|\mathsf{dis}(x, y) \le t\}$. If the size of a ball in a metric space does not depend on x, we denote by $|B_t|$ the size of a ball of radius t. We consider the Hamming metric over vectors in $\mathcal{Z}^n$, defined via $\mathsf{dis}(x, y) = |\{i|x_i \ne y_i\}|$. For this metric, $|B_t| = \sum_{i=0}^{t}\binom{n}{i}(|\mathcal{Z}| - 1)^i$. U_n denotes the uniformly distributed random variable on $\{0, 1\}^n$. Unless otherwise noted logarithms are base 2. Usually, we use capitalized letters for random variables and corresponding lowercase letters for their samples.

2.1 Fuzzy Extractors

In this section we define computational fuzzy extractors. Similar definitions for information-theoretic fuzzy extractors can be found in the work of Dodis et al. [22, Sects. 2.5–4.1]. The definition of computational fuzzy extractors allows for a small probability of error.

Definition 1 *[25, Definition 4]. Let $\mathcal{W}$ be a family of probability distributions over $\mathcal{M}$. A pair of randomized procedures "generate" (*Gen*) and "reproduce" (*Rep*) is an $(\mathcal{M}, \mathcal{W}, \kappa, t)$-*computational fuzzy extractor *that is $(\epsilon_{sec}, s_{sec})$-hard with error δ if* Gen *and* Rep *satisfy the following properties:*

- *The generate procedure* Gen *on input $w \in \mathcal{M}$ outputs an extracted string $r \in \{0,1\}^\kappa$ and a helper string $p \in \{0,1\}^*$.*
- *The reproduction procedure* Rep *takes an element $w' \in \mathcal{M}$ and a bit string $p \in \{0,1\}^*$ as inputs. The* correctness *property guarantees that if $\mathsf{dis}(w, w') \leq t$ and $(r, p) \leftarrow \mathsf{Gen}(w)$, then $\Pr[\mathsf{Rep}(w', p) = r] \geq 1 - \delta$, where the probability is over the randomness of* (Gen, Rep).
- *The* security *property guarantees that for any distribution $W \in \mathcal{W}$, the string r is pseudorandom conditioned on p, that is $\delta^{\mathcal{D}_{s_{sec}}}((R, P), (U_\kappa, P)) \leq \epsilon_{sec}$.*

In the above definition, the errors are chosen before P: if the error pattern between w and w' depends on the output of Gen, then there is no guarantee about the probability of correctness. In Constructions 1 and 2 it is crucial that w' is chosen independently of the outcome of Gen.

Information-theoretic fuzzy extractors are obtained by replacing computational distance by statistical distance. We do make a second definitional modification. The standard definition of information-theoretic fuzzy extractors considers $\mathcal{W}$ consisting of all distributions of a given entropy. As described in the introduction, we construct fuzzy extractors for parameter regimes where it is impossible to provide security for all distributions with a particular minentropy. In both the computational and information-theoretic settings we consider a family of distributions $\mathcal{W}$.

2.2 Reusable Fuzzy Extractors

A desirable feature of fuzzy extractors is reusability [9]. Intuitively, it is the ability to support multiple independent enrollments of the same value, allowing users to reuse the same biometric or PUF, for example, with multiple noncooperating providers[5]. More precisely, the algorithm Gen may be run multiple times on correlated readings $w_1, ..., w_\rho$ of a given source. Each time, Gen will produce a different pair of values $(r_1, p_1), ..., (r_\rho, p_\rho)$. Security for each extracted string r_i should hold even in the presence of all the helper strings $p_1, \ldots, p_\rho$ (the reproduction procedure Rep at the ith provider still obtains only a single w_i' close to

[5] Reusability and unlinkability are two different properties. Unlinkability prevents an adversary from telling if two enrollments correspond to the same physical source [15, 35]. We do not consider this property in this work.

w_i and uses a single helper string p_i). Because the multiple providers may not trust each other, a stronger security feature (which we satisfy) ensures that each r_i is secure even when all r_j for $j \neq i$ are also given to the adversary.

Our ability to construct reusable fuzzy extractors depends on the types of correlations allowed among $w_1, \ldots, w_\rho$. Boyen [9] showed how to do so when each w_i is a shift of w_1 by a value that is oblivious to the value of w_1 itself (formally, w_i is a result of a transitive isometry applied to w_1). Boyen also showed that even for this weak class of correlations, any secure sketch must lose at least $\log |B_t|$ entropy [9, Theorem 11].

We modify the definition of Boyen [9, Definition 6] for the computational setting. We first present our definition and then compare to the definitions of Boyen.

Definition 2 (Reusable Fuzzy Extractors). *Let $\mathcal{W}$ be a family of distributions over $\mathcal{M}$. Let $(\mathsf{Gen}, \mathsf{Rep})$ be a $(\mathcal{M}, \mathcal{W}, \kappa, t)$-computational fuzzy extractor that is $(\epsilon_{sec}, s_{sec})$-hard with error δ. Let $(W^1, W^2, \ldots, W^\rho)$ be ρ correlated random variables such that each $W^j \in \mathcal{W}$. Let D be an adversary. Define the following game for all $j = 1, ..., \rho$:*

- ***Sampling*** *The challenger samples $w^j \leftarrow W^j$ and $u \leftarrow \{0,1\}^\kappa$.*
- ***Generation*** *The challenger computes $(r^j, p^j) \leftarrow \mathsf{Gen}(w^j)$.*
- ***Distinguishing*** *The advantage of D is*

$$\begin{aligned} Adv(D) \overset{def}{=} &\Pr[D(r^1, ..., r^{j-1}, r^j, r^{j+1}, ..., r^\rho, p^1, ..., p^\rho) = 1] \\ &- \Pr[D(r^1, ..., r^{j-1}, u, r^{j+1}, ..., r^\rho, p^1, ..., p^\rho) = 1]. \end{aligned}$$

$(\mathsf{Gen}, \mathsf{Rep})$ is $(\rho, \epsilon_{sec}, s_{sec})$-reusable if for all $D \in \mathcal{D}_{s_{sec}}$ and for all $j = 1, ..., \rho$, the advantage is at most ϵ_{sec}.

Comparison with the Definition of Boyen. Boyen considers two versions of reusable fuzzy extractors. In the first version (called "outsider security" [9, Definition 6]), the adversary sees $p^1, ..., p^\rho$ and tries to learn about the values $w^1, ..., w^\rho$ or the keys $r^1, ..., r^\rho$. This version is weaker than our version, because the adversary is not given any r^i values. In the second version (called "insider security" [9, Definition 7]), the adversary controls some subset of the servers and can run Rep on arbitrary $\tilde{p}^i$. This definition allows the adversary, in particular, to learn a subset of keys r^i (by performing key generation on the valid p^i), just like in our definition. However, it also handles the case when the p^i values are actively compromised. We do not consider such an active compromise attack. As explained in Sect. 1, protection against such an attack is called "robustness" and can be handled separately—for example, by techniques from [10, Theorem 1].

In Boyen's definitions, the adversary creates a perturbation function f^i after seeing $p^1, ..., p^{i-1}$ (and generated keys in case of insider security) and the challenger generates $w^i = f^i(w^1)$. The definition is parameterized by the class of allowed perturbation functions. Boyen constructs an outsider reusable fuzzy extractor for unbounded ρ when the perturbation family is a family of transitive isometries; Boyen then adds insider security using random oracles.

In contrast, instead of considering perturbation functions to generate w^i, we simply consider all tuples of distributions as long as each distribution is in $\mathcal{W}$, because we support arbitrary correlations among them.

3 Tools: Digital Lockers, Point Functions, and Hash Functions

Our main construction uses digital lockers, which are computationally secure symmetric encryption schemes that retain security even when used multiple times with correlated and weak (i.e., nonuniform) keys [14]. In a digital locker, obtaining any information about the plaintext from the ciphertext is as hard as guessing the key. They have the additional feature that the wrong key can be recognized as such (with high probability). We use notation $\mathsf{c} = \mathsf{lock}(\mathsf{key}, \mathsf{val})$ for the algorithm that performs the locking of the value val using the key key, and $\mathsf{unlock}(\mathsf{key}, \mathsf{c})$ for the algorithm that performs the unlocking (which will output val if key is correct and $\perp$ with high probability otherwise).

The following simple and efficient construction of digital lockers was shown to provide the desired security in the random oracle model of [2] by Lynn, Prabhakaran, and Sahai [38, Sect. 4]. Let H be a cryptographic hash function, modeled as a random oracle. The locking algorithm $\mathsf{lock}(\mathsf{key}, \mathsf{val})$ outputs the pair $\mathsf{nonce}, H(\mathsf{nonce}, \mathsf{key}) \oplus (\mathsf{val}||0^{\mathsf{s}})$, where nonce is a nonce, $||$ denotes concatenation, and s is a security parameter. As long as the entropy of key is superlogarithmic, the adversary has negligible probability of finding the correct key; and if the adversary doesn't find the correct key, then the adversarial knowledge about key and val is not significantly affected by this locker. Concatenation with 0^{s} is used to make sure that unlock can tell (with certainty $1 - 2^{-\mathsf{s}}$) when the correct value is unlocked.

It is seems plausible that in the standard model (without random oracles), specific cryptographic hash functions, if used in this construction, will provide the necessary security [13, Sect. 3.2], [18, Section 8.2.3]. Moreover, Bitansky and Canetti [4], building on the work of [13,14], show how to obtain composable digital lockers based on a strong version of the Decisional Diffie-Hellman assumption without random oracles.

The security of digital lockers is defined via virtual-grey-box simulatability [4], where the simulator is allowed unbounded running time but only a bounded number of queries to the ideal locker. Intuitively, the definition gives the primitive we need: if the keys to the ideal locker are hard to guess, the simulator will not be able to unlock the ideal locker, and so the real adversary will not be able to, either. Formally, let $\mathsf{idealUnlock}(\mathsf{key}, \mathsf{val})$ be the oracle that returns val when given key, and $\perp$ otherwise.

Definition 3. *The pair of algorithm* $(\mathsf{lock}, \mathsf{unlock})$ *with security parameter* λ *is an* ℓ*-composable secure digital locker with error* γ *if the following hold:*

- ***Correctness*** *For all* key *and* val, $\Pr[\mathsf{unlock}(\mathsf{key}, \mathsf{lock}(\mathsf{key}, \mathsf{val})) = \mathsf{val}] \geq 1-\gamma$. *Furthermore, for any* $\mathsf{key}' \neq \mathsf{key}$, $\Pr[\mathsf{unlock}(\mathsf{key}', \mathsf{lock}(\mathsf{key}, \mathsf{val})) = \perp] \geq 1-\gamma$.
- ***Security*** *For every PPT adversary A and every positive polynomial p, there exists a (possibly inefficient) simulator S and a polynomial $q(\lambda)$ such that for any sufficiently large* s, *any polynomially-long sequence of values* $(\mathsf{val}_i, \mathsf{key}_i)$ *for* $i = 1, \ldots, \ell$, *and any auxiliary input* $z \in \{0,1\}^*$,

$$\left|\Pr\left[A\left(z, \{\mathsf{lock}\left(\mathsf{key}_i, \mathsf{val}_i\right)\}_{i=1}^{\ell}\right) = 1\right] - \Pr\left[S\left(z, \{|\mathsf{key}_i|, |\mathsf{val}_i|\}_{i=1}^{\ell}\right) = 1\right]\right| \leq \frac{1}{p(\mathsf{s})}$$

where S is allowed $q(\lambda)$ oracle queries to the oracles

$$\{\mathsf{idealUnlock}(\mathsf{key}_i, \mathsf{val}_i)\}_{i=1}^{\ell}.$$

Point Functions. In one of the constructions for large alphabets, we use a weaker primitive: an obfuscated point function. This primitive can be viewed as a digital locker without the plaintext: it simply outputs 1 if the key is correct and 0 otherwise. Such a function can be easily constructed from the digital locker above with the empty ciphertext, or from a strong version of the Decisional Diffie-Hellman assumption [12]. We use notation $\mathsf{c} = \mathsf{lockPoint}(\mathsf{key})$ and $\mathsf{unlockPoint}(\mathsf{key}, \mathsf{c})$; security is defined the same way as for digital lockers with a fixed plaintext.

4 Main Result: Reusable Construction for Sources with High-Entropy Samples

Sources with High-Entropy Samples. Let the source $W = W_1, \ldots, W_n$ consist of strings of length n over some arbitrary alphabet $\mathcal{Z}$ (the case of greatest interest is that of the binary alphabet $\mathcal{Z} = \{0,1\}$; however, we describe the construction more generally). For some parameters k, α, we say that the source W is a source with *α-entropy k-samples* if

$$\tilde{\mathrm{H}}_\infty(W_{j_1}, \ldots, W_{j_k} \mid j_1, \ldots, j_k) \geq \alpha$$

for uniformly random $1 \leq j_1, \ldots, j_k \leq n$. See Sect. 1 for a discussion of how sources with this property come up naturally.

The Sample-then-Lock Construction. The construction first chooses a random r to be used as the output of the fuzzy extractor. It then samples a random subset of symbols $v_1 = w_{j_1}, \ldots, w_{j_k}$ and creates a digital locker that hides r using v_1[6]. This process is repeated to produce some number ℓ of digital lockers all containing r, each unlockable with $v_1, \ldots, v_\ell$, respectively. The use of the composable

[6] We present and analyze the construction with uniformly random subsets; however, if necessary, it is possible to substantially decrease the required public randomness and the length of p by using more sophisticated samplers. See [27] for an introduction to samplers.

digital lockers allows us to sample multiple times, because we need to argue only about individual entropy of V_i. Composability also allows reusability[7].

Note that the output r can be as long as the digital locker construction can handle (in particular, the constructions discussed in Sect. 3 allow r to be arbitrarily long). Also note that it suffices to have r that is as long as a seed for a pseudorandom generator, because a longer output can be obtained by running this pseudorandom generator on r.

Construction 1 (Sample-then-Lock). *Let $\mathcal{Z}$ be an alphabet, and let $W = W_1, ..., W_n$ be a source with α-entropy k-samples, where each W_j is over $\mathcal{Z}$. Let ℓ be a parameter, to be determined later. Let* $\mathsf{lock}, \mathsf{unlock}$ *be an ℓ-composable secure digital locker with error γ (for κ-bit values and keys over $\mathcal{Z}^k$). Define* $\mathsf{Gen}, \mathsf{Rep}$ *as:*

Gen

1. *Input:* $w = w_1, ..., w_n$
2. *Sample* $r \xleftarrow{\$} \{0,1\}^\kappa$.
3. *For* $i = 1, ..., \ell$:
   (i) *Choose uniformly random* $1 \leq j_{i,1}, ..., j_{i,k} \leq n$
   (ii) *Set* $v_i = w_{j_{i,1}}, ..., w_{j_{i,k}}$.
   (iii) *Set* $\mathsf{c}_i = \mathsf{lock}(v_i, r)$.
   (iv) *Set* $p_i = \mathsf{c}_i, (j_{i,1}, ..., j_{i,k})$.
4. *Output* (r, p), *where* $p = p_1 \dots p_\ell$.

Rep

1. *Input:* $(w' = w'_1, ..., w'_n, p = p_1 \cdots p_\ell)$
2. *For* $i = 1, ..., \ell$:
   (i) *Parse* p_i *as* $\mathsf{c}_i, (j_{i,1}, ..., j_{i,k})$.
   (ii) *Set* $v'_i = w'_{j_{i,1}}, ..., w'_{j_{i,k}}$.
   (iii) *Set* $r_i = \mathsf{unlock}(v'_i, \mathsf{c}_i)$. *If* $r_i \neq \perp$ *output* r_i.
3. *Output* $\perp$.

How to Set Parameters: Correctness vs. Efficiency Tradeoff. To instantiate Construction 1, we need to choose a value for ℓ. Recall we assume that $\mathsf{dis}(w, w') \leq t$. For any given i, the probability that $v'_i = v_i$ is at least $(1 - \frac{t}{n})^k$. Therefore, the probability that no v'_i matches during Rep, causing Rep output to $\perp$, is at most

$$\left(1 - \left(1 - \frac{t}{n}\right)^k\right)^\ell.$$

In addition, Rep may be incorrect due to an error in one of the lockers, which happens with probability at most $\ell \cdot \gamma$. Thus, to make the overall error probability less than fuzzy extractor's allowable error parameter δ we need to set ℓ so that

$$\left(1 - \left(1 - \frac{t}{n}\right)^k\right)^\ell + \ell \cdot \gamma \leq \delta.$$

This provides a way to set ℓ to get a desirable δ, given a digital locker with error γ and source parameters n, t, k.

[7] For the construction to be reusable ρ times the digital locker must be composable $\ell \cdot \rho$ times.

To get a bit more insight, we need to simplify the above expression. We can use the approximation $e^x \approx 1 + x$ to get

$$\left(1 - \left(1 - \frac{t}{n}\right)^k\right)^\ell \approx (1 - e^{-\frac{tk}{n}})^\ell \approx \exp(-\ell e^{-\frac{tk}{n}}).$$

The value γ can be made very small very cheaply in known locker constructions, so let us assume that γ is small enough so that $\ell \cdot \gamma \leq \delta/2$. Then if $tk = cn \ln n$ for some constant c, setting $\ell \approx n^c \log \frac{2}{\delta}$ suffices.

We now provide the formal statement of security for Construction 1; we consider reusability of this construction below, in Theorem 2.

Theorem 1. *Let λ be a security parameter, Let $\mathcal{W}$ be a family of sources over $\mathcal{Z}^n$ with α-entropy k-samples for $\alpha = \omega(\log \lambda)$. Then for any $s_{sec} = \mathtt{poly}(\lambda)$ there exists some $\epsilon_{sec} = \mathtt{ngl}(\lambda)$ such that Construction 1 is a $(\mathcal{Z}^n, \mathcal{W}, \kappa, t)$-computational fuzzy extractor that is $(\epsilon_{sec}, s_{sec})$-hard with error $\delta = (1 - (1 - \frac{t}{n})^k)^\ell + \ell\gamma \approx \exp(-\ell e^{-\frac{tk}{n}}) + \ell\gamma$. (See above for an expression of ℓ as a function the other parameters.)*

Proof. Correctness is already argued above. We now argue security.

Our goal is to show that for all $s_{sec} = \mathtt{poly}(\lambda)$ there exists $\epsilon_{sec} = \mathtt{ngl}(\lambda)$ such that $\delta^{\mathcal{D}_{s_{sec}}}((R, P), (U, P)) \leq \epsilon_{sec}$. Fix some polynomial s_{sec} and let D be a distinguisher of size at most s_{sec}. We want to bound

$$|\mathbb{E}[D(R, P)] - \mathbb{E}[D(U, P)]|$$

by a negligible function.

We proceed by contradiction: suppose this difference is not negligible. That is, suppose that there is some polynomial $p(\cdot)$ such that for all λ_0 there exists some $\lambda > \lambda_0$ such that

$$|\mathbb{E}[D(R, P)] - \mathbb{E}[D(U, P)]| > 1/p(\lambda).$$

We note that λ is a function of λ_0 but we omit this notation for the remainder of the proof for clarity.

By the security of digital lockers (Definition 3), there is a polynomial q and an unbounded time simulator S (making at most $q(\lambda)$ queries to the oracles $\{\mathsf{idealUnlock}(v_i, r)\}_{i=1}^\ell$) such that

$$\left|\mathbb{E}[D(R, P_1, ..., P_\ell)] - \mathbb{E}\left[S^{\{\mathsf{idealUnlock}(v_i,r)\}_{i=1}^\ell}\left(R, \{j_{i,1}, ..., j_{i,k}\}_{i=1}^\ell, k, \kappa\right)\right]\right| \leq \frac{1}{3p(\lambda)}. \quad (1)$$

The same is true if we replaced R above by an independent uniform random variable U over $\{0, 1\}^\kappa$. We now prove the following lemma, which shows that S cannot distinguish between R and U.

Lemma 1. *Let U denote the uniform distribution over $\{0,1\}^\kappa$. Then*

$$\left|\mathbb{E}\left[S^{\{\mathsf{idealUnlock}(v_i,r)\}_{i=1}^{\ell}}\left(R,\{j_{i,1},...,j_{i,k}\}_{i=1}^{\ell},k,\kappa\right)\right]\right.$$
$$\left.-\mathbb{E}\left[S^{\{\mathsf{idealUnlock}(v_i,r)\}_{i=1}^{\ell}}\left(U,\{j_{i,1},...,j_{i,k}\}_{i=1}^{\ell},k,\kappa\right)\right]\right|$$
$$\leq \frac{q(q+1)}{2^\alpha} \leq \frac{1}{3p(\lambda)}, \tag{2}$$

where q is the maximum number of queries S can make.

Proof. Fix any $u \in \{0,1\}^\kappa$ (the lemma will follow by averaging over all u). Let r be the correct value of R. The only information about whether the value is r or u can obtained by S through the query responses. First, modify S slightly to quit immediately if it gets a response not equal to $\perp$ (such S is equally successful at distinguishing between r and u, because the first non-$\perp$ response tells S if its input is equal to the locked value r, and subsequent responses add nothing to this knowledge; formally, it is easy to argue that for any S, there is an S' that quits after the first non-$\perp$ response and is just as successful). There are $q+1$ possible values for the view of S on a given input (q of those views consist of some number of $\perp$ responses followed by the first non-$\perp$ response, and one view has all q responses equal to $\perp$). By [22, Lemma 2.2b], $\tilde{\mathrm{H}}_\infty(V_i|View(S),\{j_{ik}\}) \geq \tilde{\mathrm{H}}_\infty(V_j|\{j_{ik}\}) - \log(q+1) \geq \alpha - \log(q+1)$. Therefore, at each query, the probability that S gets a non-$\perp$ answer (equivalently, guesses V_i) is at most $(q+1)2^{-\alpha}$. Since there are q queries of S, the overall probability is at most $q(q+1)/2^\alpha$. Then since 2^α is $\mathtt{ngl}(\lambda)$, there exists some λ_0 such that for all $\lambda > \lambda_0$, $q(q+1)/2^\alpha \leq 1/(3p(\lambda))$.

Adding together Eqs. 1, 2, and 1 in which R is replaced with U, we obtain that

$$\delta^D((R,P),(U,P)) \leq \frac{1}{p(\lambda)}.$$

This is a contradiction and completes the proof of Theorem 1.

Reusability of Construction 1. The reusability of Construction 1 follows from the security of digital clockers. Consider any ρ number of reuses. For each fixed $i \in \{1,...,\rho\}$, we can treat the keys $r^1,\ldots,r^{i-1},r^{i+1},\ldots,r^\rho$ and the sampled positions as auxiliary input to the digital locker adversary. The result follows by simulatability of this adversary, using the same argument as the proof of Theorem 1 above. Note that this argument now requires the digital locker to be $\rho \cdot \ell$-composable.

Theorem 2. *Fix ρ and let all the variables be as in Theorem 1, except that* $(\mathsf{lock},\mathsf{unlock})$ *is an $\ell \cdot \rho$-composable secure digital locker (for κ-bit values and keys over $\mathcal{Z}^k$). Then for all $s_{sec} = \mathtt{poly}(n)$ there exists some $\epsilon_{sec} = \mathtt{ngl}(n)$ such that Construction 1 is $(\rho, \epsilon_{sec}, s_{sec})$-reusable fuzzy extractor.*

Comparison with work of [51]. The work of Škorić and Tuyls [51] can be viewed as a fuzzy extractor that places the entire string into a single digital locker (in their paper, they use the language of hash functions). Their Rec procedure symbol searches for a nearby value that unlocks the digital locker, limiting Rec to a polynomial number of error patterns. We use a subset of symbols to lock and take multiple samples, greatly increasing the error tolerance.

5 Additional Constructions for the Case of Large Alphabets

In this section we provide additional constructions of fuzzy extractors that exploit the structure of the distribution w (instead of working for all distributions of a particular min-entropy). As stated in the introduction, both constructions work for low entropy rates when w comes from a large source alphabet $\mathcal{Z}$.

5.1 Construction for Sources with Sparse High-Entropy Marginals

In this section, we consider an alternative construction that is suited to sources over large alphabets. Intuitively, we use single symbols of w to lock bits of a secret that we then transform into r; we use error-correcting codes to handle bits of the secret that cannot be retrieved due to errors in w'. Our main technical tool is obfuscated point functions (a weaker primitive than digital lockers; see Sect. 3 for the definition).

This construction requires enough symbols individually to contain sufficient entropy, but does not require independence of symbols, or even "fresh" entropy from them. Unlike the previous construction, it tolerates a linear fraction of errors (but over a larger alphabet, where errors may be more likely.). However, it cannot work for small alphabets, and is not reusable.

Sources with Sparse High-Entropy Marginals. This construction works for distributions $W = W_1, ..., W_n$ over $\mathcal{Z}^n$ in which enough symbols W_j are unpredictable even after adaptive queries to equality oracles for other symbols. This quality of a distribution is captured in the following definition.

Definition 4. *Let* $\mathsf{idealUnlock}(\mathsf{key})$ *be an oracle that returns* 1 *when given* key *and* 0 *otherwise. A source* $W = W_1, ..., W_n$ *has* β*-sparse* α*-entropy* q*-marginals if there exists a set* $J \subset \{1, ..., n\}$ *of size at least* $n - \beta$ *such that for any unbounded adversary* S,

$$\forall j \in J, \tilde{\mathrm{H}}_\infty(W_j | View(S(\cdot)))) \geq \alpha.$$

where S *is allowed* q *queries to the oracles* $\{\mathsf{idealUnlock}(W_i)\}_{i=1}^n$.

We show some examples of such sources in Appendix A.4. In particular, any source W where for all j, $\mathrm{H}_\infty(W_j) \geq \alpha = \omega(\log \lambda)$ (but all symbols may arbitrarily correlated) is a source with sparse high-entropy marginals (Proposition 3).

The Error-Correct-and-Obfuscate Construction. This construction is inspired by the construction of Canetti and Dakdouk [13]. Instead of having large parts of the string w unlock r, we have individual symbols unlock bits of the output.

Before presenting the construction we provide some definitions from error correcting codes. We use error-correct codes over $\{0,1\}^n$ which correct up to t bit flips from 0 to 1 but no bit flips from 1 to 0 (this is the Hamming analog of the Z-channel [53])[8].

Definition 5. *Let $e, c \in \{0,1\}^n$ be vectors. Let $x = \mathsf{Err}(c, e)$ be defined as follows*

$$x_i = \begin{cases} 1 & c_i = 1 \vee e_i = 1 \\ 0 & \text{otherwise.} \end{cases}$$

Definition 6. *A set C (over $\{0,1\}^n$) is a (t, δ_{code})-Z code if there exists an efficient procedure* Decode *such that*

$$\forall e \in \{0,1\}^n | \mathsf{Wgt}(e) \leq t, \Pr_{c \in C}[\mathsf{Decode}(\mathsf{Err}(c, e)) \neq c] \leq \delta_{code}.$$

Construction 2 (Lock-and-Error-Correct). *Let $\mathcal{Z}$ be an alphabet and let $W = W_1, ..., W_n$ be a distribution over $\mathcal{Z}^n$. Let $C \subset \{0,1\}^n$ be (t, δ_{code})-Z code. Let* lockPoint, unlockPoint *be an n-composable secure obfuscated point function with error γ (for keys over $\mathcal{Z}$). Define* Gen, Rep *as:*

Gen	Rep
1. *Input:* $w = w_1, ..., w_n$	1. *Input:* (w', p)
2. *Sample* $c \leftarrow C$.	2. *For* $j = 1, ..., n$:
3. *For* $j = 1, ..., n$:	(i) *If* $\mathsf{unlockPoint}(w'_j, p_j) = 1$: *set* $c'_j = 0$.
(i) *If* $c_j = 0$:	(ii) *Else: set* $c'_j = 1$.
- *Let* $p_j = \mathsf{lockPoint}(w_j)$.	3. *Set* $c = \mathsf{Decode}(c')$.
(ii) *Else:* $r_j \xleftarrow{\$} \mathcal{Z}$.	4. *Output* c.
- *Let* $p_j = \mathsf{lockPoint}(r_j)$.	
4. *Output* (c, p), *where* $p = p_1 \dots p_n$.	

As presented, Construction 2 is not yet a computational fuzzy extractor. The codewords c are not uniformly distributed and it is possible to learn some bits of c (for the symbols of W without much entropy). However, we can show that c looks like it has entropy to a computationally bounded adversary who

[8] Any code that corrects t Hamming errors also corrects t $0 \to 1$ errors, but more efficient codes exist for this type of error [53]. Codes with $2^{\Theta(n)}$ codewords and $t = \Theta(n)$ over the binary alphabet exist for Hamming errors and suffice for our purposes (first constructed by Justensen [32]). These codes also yield a constant error tolerance for $0 \to 1$ bit flips. The class of errors we support in our source (t Hamming errors over a large alphabet) and the class of errors for which we need codes (t $0 \to 1$ errors) are different.

knows p. Applying a randomness extractor with outputs over $\{0,1\}^\kappa$ (technically, an average-case computational randomness extractor) to c, and adding the extractor seed to p, will give us the desired fuzzy extractor. See Appendix A.1 for the formal details.

Construction 2 is secure if no distinguisher can tell whether it is working with r_j or w_j. By the security of point obfuscation, anything learnable from the obfuscation is learnable from oracle access to the function. Therefore, our construction is secure as long as enough symbols are unpredictable even after adaptive queries to equality oracles for individual symbols, which is exactly the property satisfied by sources with sparse high-entropy marginals.

The following theorem formalizes this intuition (proof in Appendix B).

Theorem 3. *Let λ be a security parameter. Let $\mathcal{Z}$ be an alphabet. Let $\mathcal{W}$ be a family of sources with β-sparse $\alpha = \omega(\log \lambda)$-entropy q-marginals over $\mathcal{Z}^n$, for any $q = \mathtt{poly}(n)$. Furthermore, let C be a (t, δ_{code})-Z code over $\mathcal{Z}^n$. Then for any $s_{sec} = \mathtt{poly}(n)$ there exists some $\epsilon_{sec} = \mathtt{ngl}(n)$ such that Construction 2, followed by a κ-bit randomness extractor (whose required input entropy is $\leq H_0(C) - \beta$), is a $(\mathcal{Z}^n, \mathcal{W}, \kappa, t)$-computational fuzzy extractor that is $(\epsilon_{sec}, s_{sec})$-hard with error $\delta_{code} + n(1/|\mathcal{Z}| + \gamma)$.*

Entropy vs. Error Rate. The minimum entropy necessary to satisfy Definition 4 is $\omega(\log \lambda)$ (for example, when all symbols are completely dependent but are all individually unguessable). The construction corrects a constant fraction of errors. When $n = \lambda^{1/c}$ then the entropy is smaller than the number of errors $m = \omega(\log \lambda) < \Theta(n) = \lambda^{1/c}$.

Output Length. The extractor that follows Construction 2 can output $H_0(C) - \beta - 2\log(1/\epsilon_{sec})$ bits using standard information-theoretic techniques (such as the average-case leftover hash lemma [22, Lemma 2.2b, Lemma 2.4]). To get a longer output, Construction 2 can be run multiple (say, μ) times with the same input and independent randomness to get multiple values c, concatenate them, and extract from the concatenation, to obtain an output of sufficient length $\mu(H_0(C) - \beta) - 2\log(1/\epsilon_{sec})$. The goal is to get an output long enough to use as a pseudorandom generator seed: once the seed is obtained, it can be used to generate arbitrary polynomial-length r, just like Construction 1.

Further Improvement. If most codewords have Hamming weight close to 1/2, we can decrease the error tolerance needed from the code from t to about t/2, because roughly half of the mismatches between w and w' occur where $c_j = 1$.

Lack of Reusability. Even though Construction 2 uses composable obfuscated point functions, it is *not* reusable. Definition 4 allows sources with some "weak" symbols that can be completely learned by an adversary observing p. If a source is enrolled multiple times this partial information may add up over time to reveal the original value w_1. In contrast, Construction 1, leaks no partial information for the supported sources, allowing reusability.

5.2 Information-Theoretic Construction for Sparse Block Sources

The construction in this section has information-theoretic security, in contrast to only computational security of the previous two constructions. It uses symbol-by-symbol condensers to reduce the alphabet size while preserving most of the entropy, and then applies a standard fuzzy extractor to the resulting string.

This construction requires less entropy from each symbol than the previous construction; however, it places more stringent independence requirements on the symbols. It tolerates a linear number of errors.

Sparse Block Sources. This construction works for sources $W = W_1, ..., W_n$ over $\mathcal{Z}^n$ in which enough symbols W_j contribute fresh entropy conditioned on previous symbols. We call this such sources *sparse block sources*, weakening the notion of block sources (introduced by Chor and Goldreich [16]), which require every symbol to contribute fresh entropy.

Definition 7. *A distribution $W = W_1, ..., W_n$ is an (α, β)-sparse block source if there exists a set of indices J where $|J| \geq n - \beta$ such that the following holds:*

$$\forall j \in J, \forall w_1, ..., w_{j-1} \in W_1, ..., W_{j-1}, \mathrm{H}_\infty(W_j | W_1 = w_1, ..., W_{j-1} = w_{j-1}) \geq \alpha.$$

The choice of conditioning on the past is arbitrary: a more general sufficient condition is that there exists some ordering of indices where most items have entropy conditioned on all previous items in this ordering (for example, is possible to consider a sparse reverse block source [55]).

The Condense-then-Fuzzy-Extract Construction. The construction first condenses entropy from each symbol of the source and then applies a fuzzy extractor to the condensed symbols. We'll denote the fuzzy extractor on the smaller alphabet as $(\mathsf{Gen}', \mathsf{Rep}')$. A condenser is like a randomness extractor but the output is allowed to be slightly entropy deficient. Condensers are known with smaller entropy loss than possible for randomness extractors (e.g. [23]).

Definition 8. *A function* $\mathtt{cond} : \mathcal{Z} \times \{0,1\}^d \to \mathcal{Y}$ *is a $(m, \tilde{m}, \epsilon)$-randomness condenser if whenever $\mathrm{H}_\infty(W) \geq m$, then there exists a distribution Y with $\mathrm{H}_\infty(Y) \geq \tilde{m}$ and $(\mathtt{cond}(W, seed), seed) \approx_\epsilon (Y, seed)$.*

The main idea of the construction is that errors are "corrected" on the large alphabet (before condensing) while the entropy loss for the error correction is incurred on a smaller alphabet (after condensing).

Construction 3. *Let $\mathcal{Z}$ be an alphabet and let $W = W_1, ..., W_n$ be a distribution over $\mathcal{Z}^n$. We describe* $\mathsf{Gen}, \mathsf{Rep}$ *as follows:*

Gen	Rep
1. *Input:* $w = w_1, ..., w_n$ 2. *For* $j = 1, ..., n$: (i) *Sample* $seed_i \leftarrow \{0,1\}^d$. (ii) *Set* $v_i = \mathsf{cond}(w_i, seed_i)$. 3. *Set* $(r, p') \leftarrow \mathsf{Gen}'(v_1, ..., v_n)$. 4. *Set* $p = (p', seed_1, ..., seed_n)$. 5. *Output* (r, p).	1. *Input:* $(w', p = (p', seed_1, ..., seed_n))$ 2. *For* $j = 1, ..., n$: (i) *Set* $v'_i = \mathsf{cond}(w'_i, seed_i)$. 3. *Output* $r = \mathsf{Rep}'(v', p')$.

The following theorem shows the security of this construction (proof in Appendix B).

Theorem 4. *Let $\mathcal{W}$ be a family of $(\alpha = \Omega(1), \beta \leq n(1 - \Theta(1)))$-sparse block sources over $\mathcal{Z}^n$ and let $\mathsf{cond} : \mathcal{Z} \times \{0,1\}^d \to \mathcal{Y}$ be a $(\alpha, \tilde{\alpha}, \epsilon_{cond})$-randomness conductor. Define $\mathcal{V}$ as the family of all distributions with minentropy at least $\tilde{\alpha}(n - \beta)$ and let $(\mathsf{Gen}', \mathsf{Rep}')$ be $(\mathcal{Y}^n, \mathcal{V}, \kappa, t, \epsilon_{fext})$-fuzzy extractor with error δ*[9]*. Then $(\mathsf{Gen}, \mathsf{Rep})$ is a $(\mathcal{Z}^n, \mathcal{W}, \kappa, t, n\epsilon_{cond} + \epsilon_{fext})$-fuzzy extractor with error δ.*

Overcoming Proposition 1. Proposition 1 shows that no fuzzy extractor can be secure for all sources of a given minentropy $m < \log |B_t|$. Construction 3 supports sparse block sources whose overall entropy is less than $\log |B_t|$. The structure of a sparse block source implies that $\mathrm{H}_\infty(W) \geq \alpha(n - \beta) = \Theta(n)$. We assume that $\mathrm{H}_\infty(W) = \Theta(n)$. Using standard fuzzy extractors (for $\mathsf{Gen}', \mathsf{Rep}'$) it is possible to correct $t = \Theta(n)$ errors, yielding $\log |B_t| > \Theta(n)$ when $|\mathcal{Z}| = \omega(1)$.

Acknowledgements. The authors are grateful to Nishanth Chandran, Nir Bitansky, Sharon Goldberg, Gene Itkis, Bhavana Kanukurthi, and Mayank Varia for helpful discussions, creative ideas, and important references. The authors also thank the anonymous referees for useful feedback on the paper.

The work of A.S. was performed while at Boston University's Hariri Institute for Computing and RISCS Center, and Harvard University's "Privacy Tools" project.

Ran Canetti is supported by the NSF MACS project, an NSF Algorithmic foundations grant 1218461, the Check Point Institute for Information Security, and ISF grant 1523/14. Omer Paneth is additionally supported by the Simons award for graduate students in theoretical computer science. The work of Benjamin Fuller is sponsored in part by US NSF grants 1012910 and 1012798 and the United States Air Force under Air Force Contract FA8721-05-C-0002. Opinions, interpretations, conclusions and recommendations are those of the authors and are not necessarily endorsed by the United States Government. Leonid Reyzin is supported in part by US NSF grants 0831281, 1012910, 1012798, and 1422965. Adam Smith is supported in part by NSF awards 0747294 and 0941553.

[9] We actually need $(\mathsf{Gen}', \mathsf{Rep}')$ to be an average case fuzzy extractor (see [22, Definition 4] and the accompanying discussion). Most known constructions of fuzzy extractors are average-case fuzzy extractors. For simplicity we refer to $\mathsf{Gen}', \mathsf{Rep}'$ as simply a fuzzy extractor.

References

1. Ash, R.: Information Theory. Interscience Publishers, New York (1965)
2. Bellare, M., Rogaway, P.: Random oracles are practical: a paradigm for designing efficientprotocols. In: ACM Conference on Computer and Communications Security, pp. 62–73 (1993)
3. Bennett, C.H., Brassard, G., Robert, J.-M.: Privacy amplification by public discussion. SIAM J. Comput. **17**(2), 210–229 (1988)
4. Bitansky, N., Canetti, R.: On strong simulation and composable point obfuscation. In: Rabin, T. (ed.) CRYPTO 2010. LNCS, vol. 6223, pp. 520–537. Springer, Heidelberg (2010)
5. Bitansky, N., Canetti, R., Kalai, Y.T., Paneth, O.: On virtual grey box obfuscation for general circuits. In: Garay, J.A., Gennaro, R. (eds.) CRYPTO 2014, Part II. LNCS, vol. 8617, pp. 108–125. Springer, Heidelberg (2014)
6. Blanton, M., Aliasgari, M.: On the (non-) reusability of fuzzy sketches and extractors and security improvements in the computational setting. IACR Cryptology ePrint Archive, 2012:608 (2012)
7. Blanton, M., Aliasgari, M.: Analysis of reusability of secure sketches and fuzzy extractors. IEEE Trans. Inf. Forensics Secur. **8**(9–10), 1433–1445 (2013)
8. Blanton, M., Hudelson, W.M.P.: Biometric-based non-transferable anonymous credentials. In: Qing, S., Mitchell, C.J., Wang, G. (eds.) ICICS 2009. LNCS, vol. 5927, pp. 165–180. Springer, Heidelberg (2009)
9. Boyen, X.: Reusable cryptographic fuzzy extractors. In: Proceedings of the 11th ACM Conference on Computer and Communications Security, CCS 2004, pp. 82–91. ACM, New York (2004)
10. Boyen, X., Dodis, Y., Katz, J., Ostrovsky, R., Smith, A.: Secure remote authentication using biometric data. In: Cramer, R. (ed.) EUROCRYPT 2005. LNCS, vol. 3494, pp. 147–163. Springer, Heidelberg (2005)
11. Brostoff, S., Sasse, M.A.: Are passfaces more usable than passwords?: A Field Trial Investigation. People and Computers, pp. 405–424 (2000)
12. Canetti, R.: Towards realizing random oracles: hash functions that hide all partial information. In: Kaliski Jr., B.S. (ed.) CRYPTO 1997. LNCS, vol. 1294, pp. 455–469. Springer, Heidelberg (1997)
13. Canetti, R., Dakdouk, R.R.: Obfuscating point functions with multibit output. In: Smart, N.P. (ed.) EUROCRYPT 2008. LNCS, vol. 4965, pp. 489–508. Springer, Heidelberg (2008)
14. Canetti, R., Tauman Kalai, Y., Varia, M., Wichs, D.: On symmetric encryption and point obfuscation. In: Micciancio, D. (ed.) TCC 2010. LNCS, vol. 5978, pp. 52–71. Springer, Heidelberg (2010)
15. Carter, F., Stoianov, A: Implications of biometric encryption on wide spread use of biometrics. In: EBF Biometric Encryption Seminar, June 2008
16. Chor, B., Goldreich, O.: Unbiased bits from sources of weak randomness and probabilistic communication complexity. SIAM J. Comput. **17**(2), 230–261 (1988)
17. Cramer, R., Dodis, Y., Fehr, S., Padró, C., Wichs, D.: Detection of algebraic manipulation with applications to robust secret sharing and fuzzy extractors. In: Smart, N.P. (ed.) EUROCRYPT 2008. LNCS, vol. 4965, pp. 471–488. Springer, Heidelberg (2008)
18. Dakdouk, R.R.: Theory and Application of Extractable Functions. Ph.D. thesis, Yale University (2009). http://www.cs.yale.edu/homes/jf/Ronny-thesis.pdf

19. Daugman, J.: How iris recognition works. IEEE Trans. Circuits Syst. Video Technol. **14**(1), 21–30 (2004)
20. Dodis, Y., Kalai, Y.T., Lovett, S.: On cryptography with auxiliary input. In: Mitzenmacher, M. (ed.) Proceedings of the 41st Annual ACM Symposium on Theory of Computing, STOC, Bethesda, MD, USA, May 31 - June 2, 2009, pp. 621–630. ACM (2009)
21. Dodis, Y., Kanukurthi, B., Katz, J., Reyzin, L., Smith, A.: Robust fuzzy extractors and authenticated key agreement from close secrets. IEEE Trans. Inf. Theory **58**(9), 6207–6222 (2012)
22. Dodis, Y., Ostrovsky, R., Reyzin, L., Smith, A.: Fuzzy extractors: how to generate strong keys from biometrics and other noisy data. SIAM J. Comput. **38**(1), 97–139 (2008)
23. Dodis, Y., Pietrzak, K., Wichs, D.: Key derivation without entropy waste. In: Nguyen, P.Q., Oswald, E. (eds.) EUROCRYPT 2014. LNCS, vol. 8441, pp. 93–110. Springer, Heidelberg (2014)
24. Ellison, C., Hall, C., Milbert, R., Schneier, B.: Protecting secret keys with personal entropy. Future Gener. Comput. Syst. **16**(4), 311–318 (2000)
25. Fuller, B., Meng, X., Reyzin, L.: Computational fuzzy extractors. In: Sako, K., Sarkar, P. (eds.) ASIACRYPT 2013, Part I. LNCS, vol. 8269, pp. 174–193. Springer, Heidelberg (2013)
26. Gassend, B., Clarke, D., Van Dijk, M., Devadas, S.: Silicon physical random functions. In: Proceedings of the 9th ACM Conference on Computer and Communications Security, pp. 148–160. ACM (2002)
27. Goldreich, O.: A sample of samplers: a computational perspective on sampling. In: Goldreich, O. (ed.) Studies in Complexity and Cryptography. LNCS, vol. 6650, pp. 302–332. Springer, Heidelberg (2011)
28. Håstad, J., Impagliazzo, R., Levin, L.A., Luby, M.: A pseudorandom generator from any one-way function. SIAM J. Comput. **28**(4), 1364–1396 (1999)
29. Hiller, M., Merli, D., Stumpf, F., Sigl, G.: Complementary IBS: application specific error correction for PUFs. In: IEEE International Symposium on Hardware-Oriented Securityand Trust (HOST), pp. 1–6. IEEE (2012)
30. Holenstein, T., Renner, R.S.: One-way secret-key agreement and applications to circuit polarization and immunization of public-key encryption. In: Shoup, V. (ed.) CRYPTO 2005. LNCS, vol. 3621, pp. 478–493. Springer, Heidelberg (2005)
31. Hsiao, C.-Y., Lu, C.-J., Reyzin, L.: Conditional computational entropy, or toward separating pseudoentropy from compressibility. In: Naor, M. (ed.) EUROCRYPT 2007. LNCS, vol. 4515, pp. 169–186. Springer, Heidelberg (2007)
32. Justesen, J.: Class of constructive asymptotically good algebraic codes. IEEE Trans. Inf. Theory **18**(5), 652–656 (1972)
33. Kamp, J., Zuckerman, D.: Deterministic extractors for bit-fixing sources and exposure-resilient cryptography. SIAM J. Comput. **36**(5), 1231–1247 (2007)
34. Kanukurthi, B., Reyzin, L.: Key agreement from close secrets over unsecured channels. In: Joux, A. (ed.) EUROCRYPT 2009. LNCS, vol. 5479, pp. 206–223. Springer, Heidelberg (2009)
35. Kelkboom, E.J.C., Breebaart, J., Kevenaar, T.A.M., Buhan, I., Veldhuis, R.N.J.: Preventing the decodability attack based cross-matching in a fuzzy commitment scheme. IEEE Trans. Inf. Forensics Secur. **6**(1), 107–121 (2011)
36. Koeberl, P., Li, J., Rajan, A., Wu, W.: Entropy loss in PUF-based key generation schemes: the repetition code pitfall. In: 2014 IEEE International Symposium on Hardware-Oriented Security and Trust (HOST), pp. 44–49. IEEE (2014)

37. Krawczyk, H.: Cryptographic extraction and key derivation: the HKDF scheme. In: Rabin, T. (ed.) CRYPTO 2010. LNCS, vol. 6223, pp. 631–648. Springer, Heidelberg (2010)
38. Lynn, B.Y.S., Prabhakaran, M., Sahai, A.: Positive results and techniques for obfuscation. In: Cachin, C., Camenisch, J.L. (eds.) EUROCRYPT 2004. LNCS, vol. 3027, pp. 20–39. Springer, Heidelberg (2004)
39. Maes, R., Tuyls, P., Verbauwhede, I.: Low-overhead implementation of a soft decision helper data algorithm for SRAM PUFs. In: Clavier, C., Gaj, K. (eds.) CHES 2009. LNCS, vol. 5747, pp. 332–347. Springer, Heidelberg (2009)
40. Maurer, U.M.: Secret key agreement by public discussion from common information. IEEE Trans. Inf. Theory **39**(3), 733–742 (1993)
41. Maurer, U.M.: Information-theoretically secure secret-key agreement by NOT authenticated public discussion. In: Fumy, W. (ed.) EUROCRYPT 1997. LNCS, vol. 1233, pp. 209–225. Springer, Heidelberg (1997)
42. Maurer, U.M., Wolf, S.: Towards characterizing when information-theoretic secret key agreement is possible. In: Kim, K., Matsumoto, T. (eds.) ASIACRYPT 1996. LNCS, vol. 1163, pp. 196–209. Springer, Heidelberg (1996)
43. Maurer, U.M., Wolf, S.: Privacy amplification secure against active adversaries. In: Kaliski Jr., B.S. (ed.) CRYPTO 1997. LNCS, vol. 1294, pp. 307–321. Springer, Heidelberg (1997)
44. Mayrhofer, R., Gellersen, H.: Shake well before use: intuitive and secure pairing of mobile devices. IEEE Trans. Mobile Comput. **8**(6), 792–806 (2009)
45. Monrose, F., Reiter, M.K., Wetzel, S.: Password hardening based on keystroke dynamics. Int. J. Inf. Secur. **1**(2), 69–83 (2002)
46. Nisan, N., Zuckerman, D.: Randomness is linear in space. J. Comput. Syst. Sci. **52**, 43–52 (1993)
47. Pappu, R., Recht, B., Taylor, J., Gershenfeld, N.: Physical one-way functions. Science **297**(5589), 2026–2030 (2002)
48. Pass, R., Seth, K., Telang, S.: Obfuscation from semantically-secure multi-linear encodings. Cryptology ePrint Archive, Report 2013/781 (2013). http://eprint.iacr.org/
49. Prabhakar, S., Sharath Pankanti, A.K., Jain, A.K.: Biometric recognition: security and privacy concerns. IEEE Secur. Priv. **1**(2), 33–42 (2003)
50. Simoens, K., Tuyls, P., Preneel, B.: Privacy weaknesses in biometric sketches. In: 2009 30th IEEE Symposium on Security and Privacy, pp. 188–203. IEEE (2009)
51. Škorić, B., Tuyls, P.: An efficient fuzzy extractor for limited noise. Cryptology ePrint Archive, Report 2009/030 (2009). http://eprint.iacr.org/
52. Suh, G.E., Devadas, S.: Physical unclonable functions for device authentication and secret key generation. In: Proceedings of the 44th Annual Design Automation Conference, pp. 9–14. ACM (2007)
53. Tallini, L.G., Al-Bassam, S., Bose, B.: On the capacity and codes for the Z-channel. In: IEEE International Symposium on Information Theory, p. 422 (2002)
54. Tuyls, P., Schrijen, G.-J., Škorić, B., van Geloven, J., Verhaegh, N., Wolters, R.: Read-proof hardware from protective coatings. In: Goubin, L., Matsui, M. (eds.) CHES 2006. LNCS, vol. 4249, pp. 369–383. Springer, Heidelberg (2006)
55. Vadhan, S.P.: On constructing locally computable extractors and cryptosystems in the bounded storage model. In: Boneh, D. (ed.) CRYPTO 2003. LNCS, vol. 2729, pp. 61–77. Springer, Heidelberg (2003)
56. Yu, M.-D.M., Devadas, S.: Secure and robust error correction for physical unclonable functions. IEEE Des. Test **27**(1), 48–65 (2010)

57. Zviran, M., Haga, W.J.: A comparison of password techniques for multilevel authentication mechanisms. Comput. J. **36**(3), 227–237 (1993)

A Analysis of Construction 2

A.1 Computational Fuzzy Conductors and Computational Extractors

In this section we introduce tools necessary to convert Construction 2 to a computation fuzzy extractor. We first define an object weaker than a computational fuzzy extractor: it outputs a key with computational entropy (instead of a pseudorandom key). We call this object a computational fuzzy conductor. It is the computational analogue of a fuzzy conductor (introduced by Kanukurthi and Reyzin [34]). Before defining this object, we define conditional computational "HILL" [28] entropy.

Definition 9. *[31, Definition 3] Let (W, S) be a pair of random variables. W has* HILL entropy *at least m conditioned on S, denoted $H^{\text{HILL}}_{\epsilon_{sec},s_{sec}}(W|S) \geq m$ if there exists a joint distribution (X, S), such that $\tilde{\mathrm{H}}_\infty(X|S) \geq m$ and $\delta^{\mathcal{D}_{s_{sec}}}((W, S), (X, S)) \leq \epsilon_{sec}$.*

Definition 10. *A pair of randomized procedures "generate" (*Gen*) and "reproduce" (*Rep*) is an $(\mathcal{M}, \mathcal{W}, \tilde{m}, t)$-computational fuzzy conductor that is $(\epsilon_{sec}, s_{sec})$-hard with error δ if* Gen *and* Rep *satisfy Definition 1, except the last condition is replaced with the following weaker condition:*

- *for any distribution $W \in \mathcal{W}$, the string r has high HILL entropy conditioned on P. That is $H^{\text{HILL}}_{\epsilon_{sec},s_{sec}}(R|P) \geq \tilde{m}$.*

Computational fuzzy conductors can be converted to computational fuzzy extractors (Definition 1) using standard techniques, as follows. The transformation uses a computational extractor. A computational extractor is the adaption of a randomness extractor to the computational setting. Any information-theoretic randomness extractor is also a computational extractor; however, unlike information-theoretic extractors, computational extractors can expand their output arbitrarily via pseudorandom generators once a long-enough output is obtained. We adapt the definition of Krawczyk [37] to the average case:

Definition 11. *A function* $\texttt{cext} : \{0,1\}^n \times \{0,1\}^d \to \{0,1\}^\kappa$ *a $(m, \epsilon_{sec}, s_{sec})$-*average-case computational extractor *if for all pairs of random variables X, Y (with X over $\{0,1\}^n$) such that $\tilde{H}_\infty(X|Y) \geq m$, we have*

$$\delta^{\mathcal{D}_{s_{sec}}}((\texttt{cext}(X; U_d), U_d, Y), U_\kappa \times U_d \times Y) \leq \epsilon_{sec}.$$

Combining a computational fuzzy conductor and a computational extractor yields a computational fuzzy extractor:

Lemma 2. *Let* $(\mathsf{Gen}', \mathsf{Rep}')$ *be a* $(\mathcal{M}, \mathcal{W}, \tilde{m}, t)$*-computational fuzzy conductor that is* $(\epsilon_{cond}, s_{cond})$*-hard with error* δ *and outputs in* $\{0,1\}^n$*. Let* $\mathtt{cext} : \{0,1\}^n \times \{0,1\}^d \rightarrow \{0,1\}^\kappa$ *be a* $(\tilde{m}, \epsilon_{ext}, s_{ext})$*-average case computational extractor. Define* $(\mathsf{Gen}, \mathsf{Rep})$ *as:*

- $\mathsf{Gen}(w; seed)$ *(where* $seed \in \{0,1\}^d$*): run* $(r', p') = \mathsf{Gen}'(w)$ *and output* $r = \mathtt{cext}(r'; seed)$*,* $p = (p', seed)$*.*
- $\mathsf{Rep}(w', (p', seed))$ *: run* $r' = \mathsf{Rep}'(w'; p')$ *and output* $r = \mathtt{cext}(r'; seed)$*.*

Then $(\mathsf{Gen}, \mathsf{Rep})$ *is a* $(\mathcal{M}, \mathcal{W}, \kappa, t)$*-computational fuzzy extractor that is* $(\epsilon_{cond} + \epsilon_{ext}, s')$*-hard with error* δ *where* $s' = \min\{s_{cond} - |\mathtt{cext}| - d, s_{ext}\}$*.*

Proof. It suffices to show if there is some distinguisher D' of size s' where

$$\delta^{D'}((\mathtt{cext}(X; U_d), U_d, P'), (U_\kappa, U_d, P')) > \epsilon_{cond} + \epsilon_{ext}$$

then there is an distinguisher D of size s_{cond} such that for all Y with $\tilde{\mathrm{H}}_\infty(Y|P') \geq \tilde{m}$,

$$\delta^D((X, P'), (Y, P')) \geq \epsilon_{cond}.$$

Let D' be such a distinguisher. That is,

$$\delta^{D'}(\mathtt{cext}(X, U_d) \times U_d \times P', U_\kappa \times U_d \times P') > \epsilon_{ext} + \epsilon_{cond}.$$

Then define D as follows. On input (y, p') sample $seed \leftarrow U_d$, compute $r \leftarrow \mathtt{cext}(y; seed)$ and output $D(r, seed, p')$. Note that $|D| \approx s' + |\mathtt{cext}| + d = s_{cond}$. Then we have the following:

$$\begin{aligned}\delta^D((X, P'), (Y, P')) &= \delta^{D'}((\mathtt{cext}(X, U_d), U_d, P'), \mathtt{cext}(Y, U_d), U_d, P')\\ &\geq \delta^{D'}((\mathtt{cext}(X, U_d), U_d, P'), (U_\kappa \times U_d \times P'))\\ &- \delta^{D'}((U_\kappa \times U_d \times P'), (\mathtt{cext}(Y, U_d), U_d, P'))\\ &> \epsilon_{cond} + \epsilon_{ext} - \epsilon_{ext} = \epsilon_{cond}.\end{aligned}$$

where the last line follows by noting that D' is of size at most s_{ext}. Thus D distinguishes X from all Y with sufficient conditional minentropy. This is a contradiction.

A.2 Security of Construction 2

It suffices to prove that Construction 2 is a $(\mathcal{Z}^n, \mathcal{W}, \tilde{m} = H_0(C) - \beta, t)$-comp. fuzzy conductor, i.e., that C has HILL entropy $H_0(C) - \beta$ conditioned on P. The final extraction step will convert it to a computational fuzzy extractor (see Lemma 2).

The security proof of Construction 2 is similar to the security proof of Construction 1. However, it is made more complicated by the fact that the definition of sources with sparse high-entropy marginals (Definition 4) allows for certain weak symbols that can easily be guessed. This means we must limit our indistinguishable distribution to symbols that are difficult to guess. Security is proved via the following lemma:

Lemma 3. *Let all variables be as in Theorem 3. For every* $s_{sec} = \mathtt{poly}(n)$ *there exists some* $\epsilon_{sec} = \mathtt{ngl}(n)$ *such that* $H^{\mathtt{HILL}}_{\epsilon_{sec}, s_{sec}}(C|P) \geq H_0(C) - \beta$.

We give a brief outline of the proof, followed by the proof of the new statement. It is sufficient to show that there exists a distribution C' with conditional minentropy and $\delta^{\mathcal{D}_{s_{sec}}}((C, P), (C', P)) \leq \mathtt{ngl}(n)$. Let J be the set of indices that exist according to Definition 4. Define the distribution C' as a uniform codeword conditioned on the values of C and C' being equal on all indices outside of J. We first note that C' has sufficient entropy, because $\tilde{\mathrm{H}}_\infty(C'|P) = \tilde{\mathrm{H}}_\infty(C'|C_{J^c}) \geq \mathrm{H}_\infty(C', C_{J^c}) - H_0(C_{J^c}) = H_0(C) - |J^c|$ (the second step is by [22, Lemma 2.2b]). It is left to show $\delta^{\mathcal{D}_{s_{sec}}}((C, P), (C', P)) \leq \mathtt{ngl}(n)$. The outline for the rest of the proof is as follows:

- Let D be a distinguisher between (C, P) and (C', P). By the security of obfuscated point functions,

$$\left| \mathbb{E}[D(C, P_1, ..., P_n)] - \mathbb{E}\left[S^{\{\mathsf{idealUnlock}(\cdot)\}_{i=1}^n}(C, n \cdot |\mathcal{Z}|)\right] \right|$$

  is small.
- Show that even an unbounded S making a polynomial number of queries to the stored points cannot distinguish between C and C'. That is,

$$\left| \mathbb{E}\left[S^{\{\mathsf{idealUnlock}(\cdot)\}_{i=1}^n}(C, n \cdot |\mathcal{Z}|)\right] - \mathbb{E}\left[S^{\{\mathsf{idealUnlock}(\cdot)\}_{i=1}^n}(C', n \cdot |\mathcal{Z}|)\right] \right|$$

  is small.
- By the security of obfuscated point functions,

$$\left| \mathbb{E}\left[S^{\{\mathsf{idealUnlock}(\cdot)\}_{i=1}^n}(C', n \cdot |\mathcal{Z}|)\right] - \mathbb{E}[D(C', P_1, ..., P_n)] \right|$$

  is small.

Proof (Proof of Lemma 3). The overall approach and the proof of the first and third bullet as in Theorem 1. We only prove the second bullet. Define the distribution X as follows:

$$X_j = \begin{cases} W_j & C_j = 0 \\ R_j & C_j = 1. \end{cases}$$

Lemma 4. $\Delta\left(S^{\{\mathsf{idealUnlock}(X_i)\}_{i=1}^n}(C, n \cdot |\mathcal{Z}|), S^{\{\mathsf{idealUnlock}(X_i)\}_{i=1}^n}(C', n \cdot |\mathcal{Z}|)\right) \leq (n - \beta)2^{-(\alpha+1)}$.

Proof. It suffices to show that for any two codewords that agree on J^c, the statistical distance is at most $(n - \beta)2^{-(\alpha+1)}$.

Lemma 5. *Let c^* be true value encoded in X and let c' a codeword in C'. Then,*

$$\Delta\left(S^{\{\mathsf{idealUnlock}(X_i)\}_{i=1}^n}(c^*, n\cdot|\mathcal{Z}|), S^{\{\mathsf{idealUnlock}(X_i)\}_{i=1}^n}(c', n\cdot|\mathcal{Z}|)\right) \leq (n-\beta)2^{-(\alpha+1)}.$$

Proof. Recall that for all $j \in J$, $\tilde{\mathrm{H}}_\infty(W_j|View(S)) \geq \alpha$. The only information about the correct value of c_j^* is contained in the query responses. When all responses are 0 the view of S is identical when presented with c^* or c'. We now show that for any value of c^* all queries on $j \in J$ return 0 with probability $1-2^{-\alpha+1}$. Suppose not. That is, suppose the probability of at least one nonzero response on index j is $> 2^{-(\alpha+1)}$. Since w, w' are independent of r_j, the probability of this happening when $c_j^* = 1$ is at most $q/\mathcal{Z}$ or equivalently $2^{-\log|\mathcal{Z}|+\log q}$. Thus, it must occur with probability:

$$\begin{aligned} 2^{-\alpha+1} &< \Pr[\text{non zero response location } j] \quad (3)\\ &= \Pr[c_j^*=1]\Pr[\text{non zero response location } j \wedge c_j^*=1]\\ &+ \Pr[c_j^*=0]\Pr[\text{non zero response location } j \wedge c_j^*=0]\\ &\leq 1\times 2^{-\log|\mathcal{Z}|+\log q} + 1\times\Pr[\text{non zero response location } j\wedge c_j^*=0] \end{aligned}$$

We now show that for $\alpha \leq \log|\mathcal{Z}| - \log q$:

Claim. If W is a source with β-sparse α-entropy q-marginals over $\mathcal{Z}$, then $\alpha \leq \log|\mathcal{Z}| - \log q$.

Proof. Let $J \subset \{1, ..., n\}$ the set of good indices. It suffices to show that there exists an S making q queries such that for some

$$j \in J, \tilde{\mathrm{H}}_\infty(W_j|S^{\{\mathsf{idealUnlock}(X_i)\}_{i=1}^n}) \leq \log|\mathcal{Z}| - \log q.$$

Let $j \in J$ be some arbitrary element of J and denote by $w_{j,1}, ..., w_{j,q}$ the q most likely outcomes of W_j (breaking ties arbitrarily). Then $\sum_{i=1}^q \Pr[W_j = w_{j,i}] \geq q/|\mathcal{Z}|$. Suppose not. This means that there is some $w_{j,i}$ with probability $\Pr[W_j = w_{j,i}] < 1/|\mathcal{Z}|$. Since there are $\mathcal{Z} - q$ remaining possible values of W_j for their total probability to be at least $1 - q/|\mathcal{Z}|$ at least of these values has probability at least $1/\mathcal{Z}$. This contradicts the statement $w_{j,1}, ..., w_{j,q}$ are the most likely values. Consider S that queries the jth oracle on $w_{j,1}, .., w_{j,q}$. Denote by Bad the random variable when $W_j \in \{w_{j,1}, .., w_{j,q}\}$ After these queries the remaining minentropy is at most:

$$\begin{aligned} &\tilde{\mathrm{H}}_\infty(W_j|S^{J_W(\cdot,\cdot)})\\ &\quad= -\log\left(\Pr[Bad=1]\times 1 + \Pr[Bad=0]\times\max_w \Pr[W_j = w|Bad=0]\right)\\ &\quad\leq -\log(\Pr[Bad=1]\times 1)\\ &\quad= -\log\left(\frac{q}{|\mathcal{Z}|}\right) = \log|\mathcal{Z}| - \log q \end{aligned}$$

This completes the proof of Claim A.2.

Rearranging terms in Eq. 3, we have:

$$\Pr[\text{non zero response location } j \wedge c_j = 0] > 2^{-\alpha+1} - 2^{-(\log|\mathcal{Z}|-\log q)} = 2^{-\alpha}$$

When there is a 1 response and $c_j = 0$ this means that there is no remaining minentropy. If this occurs with over $2^{-\alpha}$ probability this violates the condition on W (Definition 4). By the union bound over the indices $j \in J$ the total probability of a 1 in J is at most $(n-\beta)2^{-\alpha+1}$. Recall that c^*, c' match on all indices outside of J. Thus, for all c^*, c' the statistical distance is at most $(n-\beta)2^{-\alpha+1}$. This concludes the proof of Lemma 5.

Lemma 4 follows by averaging over all points in C'.

A.3 Correctness of Construction 2

We now argue correctness of Construction 2. We first assume ideal functionality of the obfuscated point functions. Consider a coordinate j for which $c_j = 1$. Since w' is chosen independently of the points r_j, and r_j is uniform, $\Pr[r_j = w'_j] = 1/|\mathcal{Z}|$. Thus, the probability of at least one $1 \to 0$ bit flip (the random choice r_i being the same as w'_i) is $\leq n(1//|\mathcal{Z}|)$. Since there are most t locations for which $w_j \neq w'_j$ there are at most t $0 \to 1$ bit flips in c, which the code will correct with probability $1 - \delta_{code}$, because c was chosen uniformly. Finally, since each obfuscated point function is correct with probability $1 - \gamma$, Construction 2 is correct with error at most $\delta_{code} + n(1/|\mathcal{Z}| + \gamma)$.

A.4 Characterizing Sources with Sparse High-Entropy Marginals

Definition 4 is an inherently adaptive definition and a little unwieldy. In this section, we partially characterize sources that satisfy Definition 4. The majority of the difficulty in characterizing Definition 4 is that different symbols may be dependent, so an equality query on symbol i may reshape the distribution of symbol j. In the examples that follow we denote the adversary by S as the simulator in Definition 3. We first show some sources that have sparse high-entropy marginals and then show sources with high overall entropy that do not have sparse high-entropy marginals

Positive Examples. We begin with the case of independent symbols.

Proposition 2. *Let $W = W_1, ..., W_n$ be a source in which all symbols W_j are mutually independent. Let α be a parameter. Let $J \subset \{1, ..., n\}$ be a set of indices such that for all $j \in J$, $\mathrm{H}_\infty(W_j) \geq \alpha$. Then for any q, W is a source with $(n - |J|)$-sparse $(\alpha - \log(q+1))$-entropy q-marginals. In particular, when $\alpha = \omega(\log n)$ and $q = \mathtt{poly}(n)$, then W is a source with $(n - |J|)$-sparse $\omega(\log n)$-entropy q-marginals.*

Proof. It suffices to show that for all $j \in J, \tilde{\mathrm{H}}_\infty(W_j|View(S(\cdot))) = \alpha - \log(q+1)$ where S is allowed q queries to the oracles $\{\mathsf{idealUnlock}(W_i)\}_{i=1}^n$. We can ignore queries for all symbols but the jth, as the symbols are independent. Furthermore, without loss of generality, we can assume that no duplicate queries are asked, and that the adversary is deterministic (S can calculate the best coins). Let $A_1, A_2, \ldots A_q$ be the random variables representing the oracle answers for an adversary S making q queries about the ith symbol. Each A_k is just a bit, and at most one of them is equal to 1 (because duplicate queries are disallowed). Thus, the total number of possible responses is $q+1$. Thus, we have the following,

$$\begin{aligned}\tilde{\mathrm{H}}_\infty(W_j|View(S(\cdot))) &= \tilde{\mathrm{H}}_\infty(W_j|A_1,\ldots,A_q)\\ &= \mathrm{H}_\infty(W_j) - |A_1,\ldots,A_q|\\ &= \alpha - \log(q+1)\,,\end{aligned}$$

where the second line follows from the first by [22, Lemma 2.2].

In their work on computational fuzzy extractors, Fuller, Meng, and Reyzin [25] show a construction for symbol-fixing sources, where each symbol is either uniform or a fixed symbol (symbol-fixing sources were introduced by Kamp and Zuckerman [33]). Proposition 2 shows that Definition 4 captures, in particular, this class of distributions. However, Definition 4 captures more distributions. We now consider more complicated distributions where symbols are not independent.

Proposition 3. *Let $f : \{0,1\}^e \to \mathcal{Z}^n$ be a function. Furthermore, let f_j denote the restriction of f's output to its jth coordinate. If for all j, f_j is injective then $W = f(U_e)$ is a source with 0-sparse $(e - \log(q+1))$-entropy q-marginals.*

Proof. f is injective on each symbol, so $\tilde{\mathrm{H}}_\infty(W_j|View(S)) = \tilde{\mathrm{H}}_\infty(U_e|View(S))$. Consider a query q_k on symbol j. There are two possibilities: either q_k is not in the image of f_j, or q_k can be considered a query on the preimage $f_j^{-1}(q_k)$. Then (by assuming S knows f) we can eliminate queries which correspond to the same value of U_e. Then the possible responses are strings with Hamming weight at most 1 (like in the proof of Claim 2), and by [22, Lemma 2.2] we have for all j, $\tilde{\mathrm{H}}_\infty(W_j|View(S)) \geq \mathrm{H}_\infty(W_j) - \log(q+1)$.

Note the total entropy of a source in Proposition 3 is e, so there is a family of distributions with total entropy $\omega(\log n)$ for which Construction 2 is secure. For these distributions, all the coordinates are as dependent as possible: one determines all others. We can prove a slightly weaker claim when the correlation between the coordinates W_j is arbitrary:

Proposition 4. *Let $W = W_1, \ldots, W_n$. Suppose that for all j, $\mathrm{H}_\infty(W_j) \geq \alpha$, and that $q \leq 2^\alpha/4$ (this holds asymptotically, in particular, if q is polynomial and α is super-logarithmic). Then W is a source with 0-sparse $(\alpha - 1 - \log(q+1))$-entropy q-marginals.*

Proof. Intuitively, the claim is true because the oracle is not likely to return 1 on any query. Formally, we proceed by induction on oracle queries, using the same notation as in the proof of Proposition 2. Our inductive hypothesis is that

$\Pr[A_1 \neq 0 \vee \cdots \vee A_{i-1} \neq 0] \leq (i-1)2^{1-\alpha}$. If the inductive hypothesis holds, then, for each j,

$$\mathrm{H}_\infty(W_j | A_1 = \cdots = A_{i-1} = 0) \geq \alpha - 1 . \tag{4}$$

This is true for $i = 1$ by the condition of the theorem. It is true for $i > 1$ because, as a consequence of the definition of H_∞, for any random variable X and event E, $\mathrm{H}_\infty(X|E) \geq \mathrm{H}_\infty(X) + \log \Pr[E]$; and $(i-1)2^{1-\alpha} \leq 2q2^{-\alpha} \leq 1/2$.

We now show that $\Pr[A_1 \neq 0 \vee \cdots \vee A_i \neq 0] \leq i2^{1-\alpha}$, assuming that $\Pr[A_1 \neq 0 \vee \cdots \vee A_{i-1} \neq 0] \leq (i-1)2^{1-\alpha}$.

$$\begin{aligned}
&\Pr[A_1 \neq 0 \vee \cdots \vee A_{i-1} \neq 0 \vee A_i \neq 0] \\
&\quad = \Pr[A_1 \neq 0 \vee \cdots \vee A_{i-1} \neq 0] + \Pr[A_1 = \cdots = A_{i-1} = 0 \wedge A_i = 1] \\
&\quad \leq (i-1)2^{1-\alpha} + \Pr[A_i = 1 \mid A_1 = \cdots = A_{i-1} = 0] \\
&\quad \leq (i-1)2^{1-\alpha} + \max_j 2^{-\mathrm{H}_\infty(W_j | A_1 = \cdots = A_{i-1} = 0)} \\
&\quad \leq (i-1)2^{1-\alpha} + 2^{1-\alpha} \\
&\quad = i2^{1-\alpha}
\end{aligned}$$

(where the third line follows by considering that to get $A_i = 1$, the adversary needs to guess some W_j, and the fourth line follows by (4)). Thus, using $i = q+1$ in (4), we know $\mathrm{H}_\infty(W_j | A_1 = \cdots = A_q = 0) \geq \alpha - 1$. Finally this means that

$$\begin{aligned}
\tilde{\mathrm{H}}_\infty(W_j | A_1, \ldots, A_q) &\geq -\log(2^{-\mathrm{H}_\infty(W_j | A_1 = \cdots = A_q = 0)} \Pr[A_1 = \cdots = A_q = 0] \\
&\qquad + 1 \cdot \Pr[A_1 \neq 0 \vee \cdots \vee A_q \neq 0]) \\
&\geq -\log\left(2^{-\mathrm{H}_\infty(W_j | A_1 = \cdots = A_q = 0)} + q2^{1-\alpha}\right) \\
&\geq -\log\left((q+1)2^{1-\alpha}\right) = \alpha - 1 - \log(q+1) .
\end{aligned}$$

Negative Examples. Propositions 2 and 3 rest on there being no easy "entry" point to the distribution. This is not always the case. Indeed it is possible for some symbols to have very high entropy but lose all of it after equality queries.

Proposition 5. *Let* $p = (\mathtt{poly}(\lambda))$ *and let* $f_1, ..., f_n$ *be injective functions where* $f_j : \{0,1\}^{j \times \log p} \rightarrow \mathcal{Z}$[10]. *Then define the distribution* U_n *and consider* $W_1 = f_1(U_{1,...,\log p})$, $W_2 = f_2(U_{1,...,2\log p})$, *....,* $W_n = f_n(U)$. *There is an adversary making* $p \times n$ *queries such that* $\tilde{\mathrm{H}}_\infty(W | View(S(\cdot))) = 0$.

Proof. Let x be the true value for $U_{p \times n}$. We present an adversary S that completely determines x. S computes $y_1^1 = f_1(x_1^1), ..., y_1^p = f(x_1^p)$. Then S queries on $(y_1), ..., (y_p)$ to the first oracle, exactly one answer returns 1. Let this value be y_1^* and its preimage x_1^*. Then S computes $y_2^1 = f_2(x_1^*, x_2^1), ..., y_2^p = f_2(x_1^*, x_2^p)$ and queries $y_2^1, ..., y_2^p$. Again, exactly one of these queries returns 1. This process is repeated until all of x is recovered (and thus w).

[10] Here we assume that $|\mathcal{Z}| \geq n \times \log p$, that is the source has a small number of symbols.

The previous example relies on an adversary's ability to determine a symbol from the previous symbols. We formalize this notion next. We define the entropy jump of a source as the remaining entropy of a symbol when previous symbols are known:

Definition 12. *Let $W = W_1, ..., W_n$ be a source under ordering $i_1, ..., i_n$. The* jump *of a symbol i_j is $\mathtt{Jump}(i_j) = \max_{w_{i_1},...,w_{i_{j-1}}} H_0(W_{i_j} | W_{i_1} = w_{i_1}, ..., W_{i_{j-1}} = w_{i_{j-1}})$.*

An adversary who can learn symbols in succession can eventually recover the entire secret. In order for a source to have sparse high-entropy marginals, the adversary must get "stuck" early enough in this recovery process. This translates to having a super-logarithmic jump early enough.

Proposition 6. *Let W be a distribution and let q be a parameter, if there exists an ordering $i_1, ..., i_n$ such that for all $j \le n - \beta + 1$, $\mathtt{Jump}(i_j) = \log q/(n - \beta + 1)$, then W is not a source with β-sparse high-entropy q-marginals.*

Proof. For convenience relabel the ordering that violates the condition as $1, ..., n$. We describe an unbounded adversary S that determines $W_1, ..., W_{n-\beta+1}$. As before S queries the q/n possible values for W_1 and determines W_1. Then S queries the (at most) $q/(n - \beta + 1)$ possible values for $W_2|W_1$. This process is repeated until $W_{n-\beta+1}$ is learned.

Presenting a sufficient condition for security is more difficult as S may interleave queries to different symbols. It seems like the optimum strategy for S is to focus on a single symbol at a time, but it is unclear how to formalize this intuition.

B Analysis of Construction 3

Proof. Let $W \in \mathcal{W}$. It suffices to argue correctness and security. We first argue correctness.

Correctness: When $w_i = w_i'$, then $\mathsf{cond}(w_i, seed_i) = \mathsf{cond}(w_i', seed_i)$ and thus $v_i = v_i'$. Thus, for all w, w' where $\mathsf{dis}(w, w') \le t$, then $\mathsf{dis}(v, v') \le t$. Then by correctness of $(\mathsf{Gen}', \mathsf{Rep}')$, $\Pr[(r, p) \leftarrow \mathsf{Gen}'(v) \wedge r' \leftarrow \mathsf{Rep}(v', p) \wedge r' = r] \ge 1 - \delta$.

Security: We now argue security. Denote by *seed* the random variable consisting of all n seeds and V the entire string of generated $V_1, ..., V_n$. To show that

$$R|P, seed \approx_{n\epsilon_{cond} + \epsilon_{fext}} U|P, seed,$$

it suffices to show that $\tilde{\mathrm{H}}_\infty(V|seed)$ is $n\epsilon_{cond}$ close to a distribution with average minentropy $\tilde{\alpha}(n - \beta)$. The lemma then follows by the security of $(\mathsf{Gen}', \mathsf{Rep}')$[11].

[11] Note, again, that $(\mathsf{Gen}', \mathsf{Rep}')$ must be an average-case fuzzy extractor. Most known constructions are average-case and we omit this notation.

We now argue that there exists a distribution Y where $\tilde{\mathrm{H}}_\infty(Y|seed) \geq \tilde{\alpha}(n-\beta)$ and $(V, seed_1, ..., seed_n) \approx (Y, seed_1, .., seed_n)$. First note since W is (α, β) sparse block source that there exists a set of indices J where $|J| \geq n-\beta$ such that the following holds:

$$\forall j \in J, \forall w_1, ..., w_{j-1} \in W_1, ..., W_{j-1}, \mathrm{H}_\infty(W_j|W_1 = w_1, ..., W_{j-1} = w_{j-1}) \geq \alpha.$$

Then consider the first element of $j_1 \in J$, $\forall w_1, ..., w_{j_1-1} \in W_1, ..., W_{j_1-1}$,

$$\mathrm{H}_\infty(W_{j_1}|W_1 = w_1, ..., W_{j_1-1} = w_{j_1-1}) \geq \alpha.$$

Thus, there exists a distribution Y_{j_1} with $\tilde{\mathrm{H}}_\infty(Y_{j_1}|seed_{j_1}) \geq \tilde{\alpha}$ such that

$$(\mathtt{cond}(W_{j_1}, seed_{j_1}), seed_{j_1}, W_1, ..., W_{j_1-1}) \approx_{\epsilon_{cond}} (Y_{j_1}, seed_{j_1}, W_1, ..., W_{j_1-1})$$

and since $(seed_1, ..., seed_{j_1})$ are independent of these values

$$(\mathtt{cond}(W_{j_1}, seed_{j_1}), W_{j_1-1}, ..., W_1, seed_{j_1}, ..., seed_1) \approx_{\epsilon_{cond}} (Y_{j_1}, W_{nj_1-1}, ..., W_1, seed_{j_1},, ..., seed_1).$$

Consider the random variable

$$Z_{j_1} = (Y_{j_1}, \mathtt{cond}(W_{j_1-1}, seed_{j_1-1}), ..., \mathtt{cond}(W_1, seed_1))$$

and note that $\tilde{\mathrm{H}}_\infty(Z_{j_1}|seed_1, ..., seed_{j_1}) \geq \alpha'$. Applying a deterministic function does not increase statistical distance and thus,

$$(\mathtt{cond}(W_{j_1}, seed_{j_1}), \mathtt{cond}(W_{j_1-1}, seed_{j_1-1}), .., \mathtt{cond}(W_1, seed_1), seed_{j_1}, ..., seed_1) \approx_{n\epsilon_{cond}} (Z_{j_1}, seed_{j_1}, ..., seed_1)$$

By a hybrid argument there exists a distribution Z with $\tilde{\mathrm{H}}_\infty(Z|seed) \geq \tilde{\alpha}(n-\beta)$ where

$$(\mathtt{cond}(W_n, seed_n), ..., \mathtt{cond}(W_1, seed_1), seed_n, ..., seed_1) \approx_{n\epsilon_{cond}} (Z, seed_n, ..., seed_1).$$

This completes the proof of Theorem 4.

Provably Weak Instances of Ring-LWE Revisited

Wouter Castryck[1,2(✉)], Ilia Iliashenko[1], and Frederik Vercauteren[1]

[1] KU Leuven ESAT/COSIC and iMinds, Kasteelpark Arenberg 10, 3001 Leuven-Heverlee, Belgium
{wouter.castryck,ilia.iliashenko,frederik.vercauteren}@esat.kuleuven.be
[2] Vakgroep Wiskunde, Universiteit Gent, Krijgslaan 281/S22, 9000 Ghent, Belgium

Abstract. In CRYPTO 2015, Elias, Lauter, Ozman and Stange described an attack on the non-dual *decision* version of the ring learning with errors problem (RLWE) for two special families of defining polynomials, whose construction *depends on the modulus* q that is being used. For particularly chosen error parameters, they managed to solve non-dual decision RLWE given 20 samples, with a success rate ranging from 10 % to 80 %. In this paper we show how to solve the *search* version for the same families and error parameters, using only 7 samples with a success rate of 100 %. Moreover our attack works for *every modulus* q' instead of the q that was used to construct the defining polynomial. The attack is based on the observation that the RLWE error distribution for these families of polynomials is very skewed in the directions of the polynomial basis. For the parameters chosen by Elias et al. the smallest errors are negligible and simple linear algebra suffices to recover the secret. But enlarging the error paremeters makes the largest errors wrap around, thereby turning the RLWE problem unsuitable for cryptographic applications. These observations also apply to dual RLWE, but do not contradict the seminal work by Lyubashevsky, Peikert and Regev.

1 Introduction

Hard problems on lattices have become popular building blocks for cryptographic primitives mainly because of two reasons: firstly, lattice based cryptography appears to remain secure even in the presence of quantum computers, and secondly, the security of the primitives can be based on worst-case hardness assumptions. Although it seems appealing to use classical hard lattice problems such as the shortest vector problem or closest vector problem for cryptographic applications, the learning with errors problem (LWE) has proven much more versatile. This problem was introduced by Regev [12,13] who showed that an efficient algorithm for LWE results in efficient quantum algorithms for approximate lattice problems. The *decision* version of LWE can be defined informally as the problem of distinguishing noisy linear equations from truly random ones. More precisely, let $n \geq 1$ be an integer dimension and $q \geq 2$ an integer modulus, then the problem is to distinguish polynomially many pairs of the form $(\mathbf{a}_i, b_i \approx \langle \mathbf{a}_i, \mathbf{s} \rangle)$ from uniformly random and independent pairs. The vectors $\mathbf{a}_i$ are chosen uniformly random in $\mathbb{Z}_q^n$, the vector $\mathbf{s}$ is secret and the same for all pairs, and the element

M. Fischlin and J.-S. Coron (Eds.): EUROCRYPT 2016, Part I, LNCS 9665, pp. 147–167, 2016.
DOI: 10.1007/978-3-662-49890-3_6

b_i is computed as $b_i = \langle \mathbf{a}_i, \mathbf{s} \rangle + e_i$ where e_i is a random error term drawn from an error distribution on $\mathbb{Z}_q$, such as a discretized Gaussian. The *search* version of LWE asks to recover the secret vector $\mathbf{s}$. The hardness of the LWE problem has been analyzed in [3,8,11–13].

The main downside of LWE is that it is not very practical, basically due to the fact that each new $\mathbf{a}_i$ only gives rise to one element b_i (and not a vector of n elements as one could hope). The result is that the public key size and the computation time of LWE-based cryptosystems are typically quadratic in the security parameter. Lyubashevsky, Peikert and Regev [9] solved this issue by introducing the Ring-LWE (RLWE) problem and showing its hardness under wost-case assumptions on ideal lattices. Its flavour is distantly similar to that of NTRU [7]. Informally, the secret key space $\mathbb{Z}_q^n$ is replaced by $R_q = R/qR$ where R is the ring of integers in a number field $K = \mathbb{Q}[x]/(f)$ with f a monic irreducible integral polynomial of degree n and $q \geq 2$ an integer modulus. The inner product on $\mathbb{Z}_q^n$ is replaced by the ring product in R_q. In its *non-dual* form the *decision* version of RLWE is then roughly defined as follows: distinguish polynomially many samples of the form $(\mathbf{a}_i, \mathbf{b}_i \approx \mathbf{a}_i \cdot \mathbf{s})$ from uniformly random and independent pairs. Here the $\mathbf{a}_i \in R_q$ are uniformly random and independent, $\mathbf{s} \in R_q$ is a fixed random secret, and $\mathbf{b}_i$ is computed as $\mathbf{b}_i = \mathbf{a}_i \cdot \mathbf{s} + \mathbf{e}_i$ where $\mathbf{e}_i \in R_q$ is a short random error term that is drawn from a specific error distribution ψ on R_q. The *search* version of the problem is to recover the secret $\mathbf{s}$ from the list of samples. We stress that the actual problem described and analyzed in [9] is the *dual* RLWE problem, in which the secret and the error term are taken from the reduction modulo q of a certain fractional ideal of K, denoted by $R_q^\vee$; see Sect. 2 for more details.

As explained in [9], the search and decision problems are equivalent when K is Galois and q is a prime number that splits into prime ideals with small norm (polynomial in n). In general, no such reduction is known and it is easy to see that search RLWE is at least as hard as decision RLWE.

The definition of the error distribution ψ on R_q (or on $R_q^\vee$) plays a crucial role in RLWE and is obtained by pulling back a near-spherical Gaussian distribution under the canonical embedding of the number field. An alternative problem [5] is called Polynomial-LWE (PLWE) and uses an error distribution on R_q where each coordinate of the error term with respect to the polynomial basis $1, x, x^2, \ldots, x^{n-1}$ is drawn independently from a fixed one-dimensional Gaussian distribution. Again we refer to Sect. 2 for more details.

In [5], Eisentraeger, Hallgren and Lauter presented families of defining polynomials $f \in \mathbb{Z}[x]$ and moduli q such that the *decision* version of PLWE is weak. The attack can be described in a nutshell as follows: assume that $f(1) \equiv 0 \bmod q$, then evaluation at 1 defines a ring homomorphism ϕ from R_q to $\mathbb{Z}_q$. Applying ϕ to the PLWE samples results in equations of the form $\mathbf{a}_i(1) \cdot \mathbf{s}(1) + \mathbf{e}_i(1) = \mathbf{b}_i(1)$. Therefore, if the images $\mathbf{e}_i(1)$ of the error terms can be distinguished from uniform, one can simply loop through all possibilities for $\mathbf{s}(1) \in \mathbb{Z}_q$ and determine if the corresponding $\mathbf{e}_i(1)$ are uniform on $\mathbb{Z}_q$ or not. So as long as q is small enough (such that one can exhaustively run through $\mathbb{Z}_q$), $f(1) \equiv 0 \bmod q$, and the images $\mathbf{e}_i(1)$ do not wrap around too much modulo q, this attack breaks decision PLWE.

In [6], Elias, Lauter, Ozman and Stange extended this attack to the *decision* version of non-dual RLWE, rather than PLWE, by showing that for defining polynomials of the form

$$f_{n,a,b} = x^n + ax + b \in \mathbb{Z}[x]$$

where n, a, b are specifically chosen parameters such that i.a. $f_{n,a,b}(1) \equiv 0 \bmod q$, the distortion introduced by pulling back the Gaussian error terms through the canonical embedding is small enough such that the attack on PLWE still applies. This attack was executed for three parameter sets n, a, b, r where given 20 samples, non-dual decision RLWE could be solved with success rates ranging from 10 % to 80 % depending on the particular family considered [6, Sect. 9]. Here the parameter r determines the width of the Gaussian that is being pulled back, which Elias et al. chose to be spherical.

Our contributions in this paper are as follows. Firstly, we explain how to solve the *search* version of non-dual RLWE, which one might expect to be a harder problem than the decision version (due to the fact that the corresponding number fields are not Galois), for the same parameter sets, using only 7 samples with a success rate of 100 %. The attack invokes simple linear algebra to recover the secret element $\mathbf{s}$ and does not use that $f_{n,a,b}(1) \equiv 0 \bmod q$: in fact, for the same defining polynomial and the same error parameter r our attack works for *every* modulus q'. Secondly, we show that if one tries to adjust r in order to obtain a hard instance of non-dual RLWE, the first few components of the noise wrap around modulo q and become indistinguishable from uniform, thereby obstructing certain cryptographic applications. Thirdly, we show that our observations also apply to the *dual* RLWE problem when set up for the same number fields: either the errors wrap around or linear algebra can be used to reveal the secret. The latter situation only occurs for error widths that are way too small for the hardness results of Lyubashekvsy, Peikert and Regev [9] to be applicable. Therefore neither the results from [6] nor our present attack seem to form a threat on RLWE, at least when set up along the guidelines in [9,10].

Our observations are easiest to explain for $a = 0$, a case which covers two of the three parameter sets. From $f_{n,a,b}(1) = b + 1 \equiv 0 \bmod q$ and $f_{n,a,b}(1) \neq 0$ (by irreducibility) it follows that the roots of $f_{n,a,b}$ lie on a circle with radius

$$\rho \geq \sqrt[n]{q-1} > 1.$$

With respect to the polynomial basis $1, x, x^2, \ldots, x^{n-1}$, the canonical embedding matrix is essentially the Vandermonde matrix generated by these roots, whose column norms grow geometrically as

$$\sqrt{n}, \sqrt{n}\rho, \ldots, \sqrt{n}\rho^{n-1}.$$

This simple observation has major implications for the distortion introduced by the inverse of the canonical embedding: the distribution of the error terms will be extremely stretched at the terms of low degree, whereas they will be squashed at the terms of high degree. For the parameter sets attacked by Elias et al., the latter are so small that after rounding they simply become zero, thereby resulting in exact linear equations in the coefficients of the secret element $\mathbf{s}$. Given enough samples (in the cases considered, between 4 and 7 samples suffice), the secret $\mathbf{s}$ can be recovered using elementary linear algebra. Furthermore, since the ratio between the maximal and the minimal distortion is roughly $\rho^n \geq q-1$, it is impossible to increase the width of the Gaussians used without causing the errors at the terms of low degree to wrap around modulo q.

The remainder of the paper is organized as follows: in Sect. 2 we recall the definition of PLWE and of dual and non-dual RLWE, with particular focus on the error distributions involved. Section 3 reviews the attacks on decision PLWE by Eisentraeger, Hallgren and Lauter and non-dual decision RLWE by Elias, Lauter, Ozman and Stange. Section 4 describes our attack on non-dual search RLWE by analyzing the singular value decomposition of the canonical embedding. We also report on an implementation of our attack in Magma [2], which shows that we can indeed easily break the families considered in [6] using less samples, with a higher success probability, and for every choice of modulus q' (instead of just the q that was used to define $f_{n,a,b}$). We also discuss how switching to dual RLWE affects these observations. In Sect. 5 we study the effect of increasing the error parameter as an attempt to counter our attack, and compare with the hardness results from [9]. Section 6 concludes the paper.

2 Preliminaries

In this section we briefly recall the necessary background on number fields, the canonical embedding and Gaussian distributions to give proper definitions of PLWE and dual and non-dual RLWE.

2.1 Number Fields and the Canonical Embedding

Let $f \in \mathbb{Z}[x]$ be a monic irreducible polynomial of degree n and consider the number field $K = \mathbb{Q}[x]/(f)$ it defines. Let $R \subset K$ denote the ring of integers of K, i.e. the set of all algebraic integers that are contained in K. If f can be taken such that $R = \mathbb{Z}[x]/(f)$, then K is called a *monogenic* number field and f a monogenic polynomial.

The field K has exactly n embeddings into $\mathbb{C}$ denoted by $\sigma_i : K \to \mathbb{C}$ for $i = 1, \ldots, n$. These n embeddings correspond precisely to evaluation in each of the n distinct roots α_i of f, i.e. an element $a(x) \in K$ is mapped to $\sigma_i(a(x)) = a(\alpha_i) \in \mathbb{C}$. Assume that f has s_1 real roots and $n - s_1 = 2s_2$ complex conjugate roots and order the roots such that $\overline{\alpha_{s_1+k}} = \alpha_{s_1+s_2+k}$ for $k = 1, \ldots, s_2$. The *canonical embedding* (also known as the Minkowski embedding) $\sigma : K \to \mathbb{C}^n$ is then defined as:

$$\sigma(a) = (\sigma_1(a), \ldots, \sigma_{s_1}(a), \sigma_{s_1+1}(a), \ldots, \sigma_{s_1+s_2}(a), \overline{\sigma_{s_1+1}}(a), \ldots, \overline{\sigma_{s_1+s_2}}(a)) .$$

It is easy to see that the canonical embedding maps into the space $H \subset \mathbb{R}^{s_1} \times \mathbb{C}^{2s_2}$ given by

$$H = \{(x_1, \dots, x_n) \in \mathbb{R}^{s_1} \times \mathbb{C}^{2s_2} : \overline{x_{s_1+j}} = x_{s_1+s_2+j}\,, \forall j \in [1 \dots s_2]\}\,.$$

The space H is isomorphic to $\mathbb{R}^n$ as an inner product space by considering the orthonormal basis for H given by the columns of

$$B = \begin{pmatrix} I_{s_1 \times s_1} & 0 & 0 \\ 0 & \frac{1}{\sqrt{2}} I_{s_2 \times s_2} & \frac{\mathrm{i}}{\sqrt{2}} I_{s_2 \times s_2} \\ 0 & \frac{1}{\sqrt{2}} I_{s_2 \times s_2} & -\frac{\mathrm{i}}{\sqrt{2}} I_{s_2 \times s_2} \end{pmatrix}.$$

With respect to this basis, the coordinates of $\sigma(a)$ are given by a real vector

$$\begin{aligned}(\tilde{a}_1, \dots, \tilde{a}_n) := (\sigma_1(a), \dots, \sigma_{s_1}(a), \sqrt{2}\,\mathfrak{Re}(\sigma_{s_1+1}(a)), \dots, \sqrt{2}\,\mathfrak{Re}(\sigma_{s_1+s_2}(a)),\\ \sqrt{2}\,\mathfrak{Im}(\sigma_{s_1+1}(a)), \dots, \sqrt{2}\,\mathfrak{Im}(\sigma_{s_1+s_2}(a))).\end{aligned}$$

Note that in [6] the authors did not include the factor $\sqrt{2}$, but we choose to keep it since it makes B unitary.

In summary, an element $a(x) \in K$ can be represented in the polynomial basis as $(a_0, \dots, a_{n-1})$ where $a(x) = \sum_{i=0}^{n-1} a_i x^i$ but also by a real vector $(\tilde{a}_1, \dots, \tilde{a}_n)$ where the canonical embedding of a is given by:

$$\sigma(a) = B \cdot (\tilde{a}_1, \dots, \tilde{a}_n)^t\,.$$

Let M_f denote the Vandermonde matrix $(\alpha_i^{j-1})_{i,j}$ for $i, j = 1, \dots, n$, then the polynomial basis representation is related to the (real) canonical embedding representation by the following transformation

$$(a_0, \dots, a_{n-1})^t = M_f^{-1} \cdot B \cdot (\tilde{a}_1, \dots, \tilde{a}_n)^t\,.$$

Since M_f^{-1} will play a crucial role in the following, we denote it with N_f. Later on, to ease notation we will just write M_f instead of $M_{f_{n,a,b}}$, and similarly for N_f.

2.2 Ideals of the Ring of Integers and Their Dual

An *integral ideal* $I \subseteq R$ is an additive subgroup of R closed under multiplication by elements of R, i.e. $rI \subset I$ for any $r \in R$. A *fractional ideal* $I \subset K$ is a set such that $dI \subseteq R$ is an integral ideal for some $d \in R$. A *principal* (fractional or integral) ideal I is one that is generated by some $u \in K$, i.e. $I = uR$; we denote it as $I = \langle u \rangle$. The sum $I + J$ of two (fractional or integral) ideals is the set of all $x + y$ with $x \in I, y \in J$ and the product $I \cdot J$ is the smallest (fractional or integral) ideal containing all products $x \cdot y$ with $x \in I, y \in J$. The set of non-zero fractional ideals forms a group under multiplication; this is not true for integral ideals. The inverse of a non-zero fractional ideal is denoted by I^{-1}.

Every fractional ideal I is a free $\mathbb{Z}$-module of rank n, and therefore $I \otimes \mathbb{Q} = K$. Its image $\sigma(I)$ under the canonical embedding is a lattice of rank n inside the space H.

The trace $\mathrm{Tr} = \mathrm{Tr}_{K/\mathbb{Q}} : K \to \mathbb{Q}$ maps an element x to the sum of its embeddings $\mathrm{Tr}(x) = \sum_{i=1}^{n} \sigma_i(x)$ and defines an additive homomorphism from R to $\mathbb{Z}$. The norm $\mathrm{No} = \mathrm{No}_{K/\mathbb{Q}} : K \to \mathbb{Q}$ takes the product of all embeddings $\mathrm{No}(x) = \prod_{i=1}^{n} \sigma_i(x)$ and is multiplicative.

For a fractional ideal I, its dual $I^\vee$ is defined as

$$I^\vee = \{x \in K : \mathrm{Tr}(xI) \subseteq \mathbb{Z}\}.$$

It is easy to see that $(I^\vee)^\vee = I$ and that $I^\vee$ is also a fractional ideal. (Under the canonical embedding, this corresponds to the usual notion of dual lattice, modulo complex conjugation.) Furthermore, for any fractional ideal I, its dual is $I^\vee = I^{-1}R^\vee$. The factor $R^\vee$ is a fractional ideal called the *codifferent* and its inverse is called the *different ideal* which is integral. For a monogenic defining polynomial f, i.e. $R = \mathbb{Z}[x]/(f)$ we have that $R^\vee = \langle 1/f'(\alpha)\rangle$ where α is a root of f. Applying this fact to the cyclotomic number field of degree $n = 2^k$ with defining polynomial $f(x) = x^n + 1$, we get that $f'(\xi_{2n}) = n\xi_{2n}^{n-1}$ with ξ_{2n} a primitive $2n$-th root of unity. Thus $R^\vee = \langle n^{-1}\rangle$, since ξ_{2n}^{n-1} is a unit.

2.3 Gaussian Distributions and Discretization

Denote by Γ_r the normal Gaussian distribution on $\mathbb{R}$ with mean 0 and parameter r given by $\Gamma_r(x) = r^{-1}\exp(-\pi x^2/r^2)$. Note that we have $r = \sqrt{2\pi}\rho$ with ρ the standard deviation. We can define an elliptical Gaussian distribution $\Gamma_{\mathbf{r}}$ on H as follows: let $\mathbf{r} = (r_1, \ldots, r_n) \in (\mathbb{R}^+)^n$ be a vector of n positive real numbers, then a sample of $\Gamma_{\mathbf{r}}$ is given by $B\cdot(x_1, \ldots, x_n)^t$ where each x_i is sampled independently from Γ_{r_i} on $\mathbb{R}$. Note that via the inverse of the canonical embedding this also defines a distribution $\Psi_{\mathbf{r}}$ on $K \otimes \mathbb{R}$, in other words

$$N_f \cdot B \cdot (x_1, \ldots, x_n)^t$$

gives us the coordinates of $\Gamma_{\mathbf{r}} \leftarrow (x_1, \ldots, x_n)$ with respect to the polynomial basis $1, x, x^2, \ldots, x^{n-1}$.

In practice we sample from the continuous distribution $\Gamma_{\mathbf{r}}$ modulo some finite but sufficiently high precision (e.g. using the Box-Muller method). In particular our samples live over $\mathbb{Q}$ rather than $\mathbb{R}$, so that an element sampled from $\Psi_{\mathbf{r}}$ can be truly seen as an element of the field K. For use in RLWE one even wants to draw elements from I for some fixed fractional ideal $I \subset K$, where $I = R$ (non-dual RLWE) and $I = R^\vee$ (dual RLWE) are the main examples. In this case one should discretize the Gaussian distribution $\Gamma_{\mathbf{r}}$ to the lattice $\sigma(I)$. There are several ways of doing this, e.g. by rounding coordinates with respect to some given $\mathbb{Z}$-module basis; see [9,10] and the references therein. But for our conclusions this discretization is not relevant, and because it would needlessly complicate things we will just omit it.

2.4 The Polynomial-LWE and Ring-LWE Problem

In this section we provide formal definitions of PLWE [5] and RLWE [4,9], both in its dual and its non-dual version [6]. We stress that it is the dual version of RLWE that was introduced in [9] and for which certain hardness results are available, one of which is recalled in Theorem 1 below.

Let $f \in \mathbb{Z}[x]$ be a monic irreducible polynomial of degree n and let $q \geq 2$ be an integer modulus. Consider the quotient ring $P = \mathbb{Z}[x]/(f)$ and denote with P_q the residue ring P/qP. Denote with Γ_r^n the spherical Gaussian on $\mathbb{R}^n$ with parameter r and interpret this as a distribution on $P \otimes \mathbb{R}$ by mapping the standard basis of $\mathbb{R}^n$ to the polynomial basis $1, x, x^2, \ldots, x^{n-1}$ of P. In particular, elements $\mathbf{e}(x) = \sum_{i=0}^{n-1} e_i x^i \leftarrow \Gamma_r^n$ have each coefficient e_i drawn independently from Γ_r. Let $\mathfrak{U}(P_q)$ denote the uniform distribution on P_q and let $\mathfrak{U}(P_{q,\mathbb{R}})$ be the uniform distribution on the torus $P_{q,\mathbb{R}} = (P \otimes \mathbb{R})/qP$.

With these ingredients we can define the decision and search PLWE problems.

Definition 1 (PLWE Distribution). *For* $\mathbf{s}(x) \in P_q$ *and* $r \in \mathbb{R}^+$, *a sample from the* PLWE *distribution* $A_{\mathbf{s}(x),r}$ *over* $P_q \times P_{q,\mathbb{R}}$ *is generated by choosing* $\mathbf{a}(x) \leftarrow \mathfrak{U}(P_q)$, *choosing* $\mathbf{e}(x) \leftarrow \Gamma_r^n$ *and outputting* $(\mathbf{a}(x), \mathbf{b}(x) = \mathbf{a}(x) \cdot \mathbf{s}(x) + \mathbf{e}(x) \bmod qP)$.

Definition 2 (Decision PLWE). *The decision* PLWE *problem is to distinguish, for a random but fixed choice of* $\mathbf{s}(x) \leftarrow \mathfrak{U}(P_q)$, *with non-negligible advantage between arbitrarily many independent samples from* $A_{\mathbf{s}(x),r}$ *and the same number of independent samples from* $\mathfrak{U}(P_q) \times \mathfrak{U}(P_{q,\mathbb{R}})$.

Definition 3 (Search PLWE). *For a random but fixed choice of* $\mathbf{s}(x) \leftarrow \mathfrak{U}(P_q)$, *the search* PLWE *problem is to recover* $\mathbf{s}(x)$ *with non-negligible probability from arbitrarily many independent samples from* $A_{\mathbf{s}(x),r}$.

To define the dual and non-dual RLWE problems we require a degree n number field K with ring of integers R. We also fix a fractional ideal $I \subset K$, for which two choices are available: in the *dual* RLWE problems we let $I = R^\vee$, while in the *non-dual* RLWE problems we take $I = R$. Note that $I \otimes \mathbb{R} = K \otimes \mathbb{R}$, so we can view the distribution $\Psi_\mathbf{r}$ from the previous section as a distribution on $I \otimes \mathbb{R}$. We let I_q denote I/qI and write $I_{q,\mathbb{R}}$ for the torus $(I \otimes \mathbb{R})/qI$. As before we let $\mathfrak{U}(I_q)$ denote the uniform distribution on I_q and let $\mathfrak{U}(I_{q,\mathbb{R}})$ be the uniform distribution on $I_{q,\mathbb{R}}$.

Definition 4 (RLWE Distribution). *For* $\mathbf{s}(x) \in I_q$ *and* $\mathbf{r} \in (\mathbb{R}^+)^n$, *a sample from the* RLWE *distribution* $A_{\mathbf{s}(x),\mathbf{r}}$ *over* $R_q \times I_{q,\mathbb{R}}$ *is generated by choosing* $\mathbf{a}(x) \leftarrow \mathfrak{U}(R_q)$, *choosing* $\mathbf{e}(x) \leftarrow \Psi_\mathbf{r}$ *and returning* $(\mathbf{a}(x), \mathbf{b}(x) = \mathbf{a}(x) \cdot \mathbf{s}(x) + \mathbf{e}(x) \bmod qI)$.

Definition 5 (Decision RLWE). *The decision* RLWE *problem is to distinguish, for a random but fixed choice of* $\mathbf{s}(x) \leftarrow \mathfrak{U}(I_q)$, *with non-negligible advantage between arbitrarily many independent samples from* $A_{\mathbf{s}(x),\mathbf{r}}$ *and the same number of independent samples from* $\mathfrak{U}(R_q) \times \mathfrak{U}(I_{q,\mathbb{R}})$.

Definition 6 (Search RLWE). *For a random but fixed choice of* $\mathbf{s}(x) \leftarrow \mathfrak{U}(I_q)$, *the search* RLWE *problem is to recover* $\mathbf{s}(x)$ *with non-negligible probability from arbitrarily many independent samples from* $A_{\mathbf{s}(x),\mathbf{r}}$.

A hardness statement on the search RLWE problem in its dual form (i.e. with $I = R^\vee$) was provided by Lyubashevsky, Peikert and Regev. For proof-technical reasons their result actually deals with a slight variant called the search $\text{RLWE}_{\leq r}$ problem, where $r \in \mathbb{R}^+$. In this variant each sample is taken from $A_{\mathbf{s}(x),\mathbf{r}}$ for a new choice of $\mathbf{r}$, chosen uniformly at random from $\{(r_1, \ldots, r_n) \in (\mathbb{R}^+)^n \mid r_i \leq r \text{ for all } i\}$. Think of this parameter r and the modulus $q \geq 2$ as quantities that vary with n, and let ω be a superlinear function. Then Lyubashevsky et al. proved:

Theorem 1 ([9, Theorem 4.1]). *If* $r \geq 2\omega(\sqrt{\log n})$ *then for some negligible* ε *(depending on* n*) there is a probabilistic polynomial-time quantum reduction from* KDGS_γ *to* $\text{RLWE}_{\leq r}$, *where*

$$\gamma : I \mapsto \max\left\{\eta_\varepsilon(I) \cdot (\sqrt{2}q/r) \cdot \omega(\sqrt{\log n}), \sqrt{2n}/\lambda_1(I^\vee)\right\}.$$

Here $\eta_\varepsilon(I)$ *is the smoothing parameter of* $\sigma(I)$ *with threshold* ε, *and* $\lambda_1(I^\vee)$ *is the length of a shortest vector of* $\sigma(I^\vee)$.

In the above statement KDGS_γ refers to the *discrete Gaussian sampling problem*, which is about producing samples from a spherical Gaussian in H with parameter r', discretized to the lattice $\sigma(I)$, for any given non-zero ideal $I \subset R$ and any $r' \geq \gamma(I)$. As discussed in [9] there are easy reductions from certain standard lattice problems to the discrete Gaussian sampling problem.

As an intermediate step in their proof Lyubashevsky et al. obtain a classical (i.e. non-quantum) reduction from an instance of the bounded distance decoding problem in ideal lattices to $\text{RLWE}_{\leq r}$; see [9, Lemma 4.5].

In contrast, Elias, Lauter, Ozman and Stange [6] study RLWE in its non-dual version, and for the sake of comparison our main focus will also be on that setting, i.e. we will mostly take $I = R$. In Sect. 4.3 we will look at the effect of switching to the dual case where $I = R^\vee$, and in Sect. 5 we will include the above hardness result in the discussion. Moreover, again as in [6], the noise parameter $\mathbf{r} = (r_1, \ldots, r_n)$ will usually be taken fixed and spherical, i.e. $r_1 = \cdots = r_n = r$.

3 Provably Weak Instances of Non-dual Decision RLWE

In [5], Eisentraeger, Hallgren and Lauter presented families of defining polynomials $f \in \mathbb{Z}[x]$ such that the *decision* version of PLWE is weak. This attack was later extended to non-dual decision RLWE [6] by Elias, Lauter, Ozman and Stange. In this section we recall the attack, first for PLWE and then how it transfers to non-dual RLWE. We provide a detailed analysis of the singular value decomposition of the matrix N_f for these polynomial families, since this will play an instructive role in our exposition.

3.1 Attack on Decision PLWE

The simplest form of the attack on decision PLWE requires that the defining polynomial f of P and the modulus q satisfy the relation $f(1) \equiv 0 \bmod q$. This implies that evaluation at 1 induces a ring homomorphism $\phi : P_q \to \mathbb{Z}_q : a(x) \mapsto a(1) \bmod q$. By applying ϕ to the PLWE samples $(\mathbf{a}_i, \mathbf{b}_i = \mathbf{a}_i \cdot \mathbf{s} + \mathbf{e}_i)$ we obtain tuples in $\mathbb{Z}_q^2$ namely $(\phi(\mathbf{a}_i), \phi(\mathbf{a}_i) \cdot \phi(\mathbf{s}) + \phi(\lfloor \mathbf{e}_i \rceil))$. Here $\lfloor \mathbf{e}_i \rceil$ denotes the polynomial obtained by rounding each coefficient of $\mathbf{e}_i$ to the nearest integer (with ties broken upward, say).

Assuming that the images of the error terms $\mathbf{e}_i$ under the homomorphism ϕ can be distinguished from uniform with sufficiently high probability, one obtains the following straightforward attack: for each guess $s \in \mathbb{Z}_q$ for the value of $\phi(\mathbf{s}) = \mathbf{s}(1) \bmod q$, compute the corresponding image of the (rounded) error term $\phi(\lfloor \mathbf{e}_i \rceil)$ as $\phi(\lfloor \mathbf{b}_i \rceil) - \phi(\mathbf{a}_i)s$, assuming that the guess is correct. If there exists an s such that the corresponding images $\phi(\lfloor \mathbf{e}_i \rceil)$ are more or less distributed like a discretized Gaussian, rather than uniform, the samples were indeed likely to be actual PLWE samples and the secret $\mathbf{s}$ satisfies $\mathbf{s}(1) = s$. If no such guess is found, the samples were likely to be uniform samples. The attack succeeds if the following three conditions are met:

1. $f(1) \equiv 0 \bmod q$,
2. q is small enough that $\mathbb{Z}_q$ can be enumerated,
3. $\phi(\lfloor \Gamma_r^n \rceil)$ is distinguishable from uniform $\mathfrak{U}(\mathbb{Z}_q)$.

Note that if $\mathbf{e}_i$ is sampled from Γ_r^n, then the $\mathbf{e}_i(1)$ are also Gaussian distributed but with parameter $\sqrt{n} \cdot r$. Therefore, as long as $\sqrt{n} \cdot r$ is sufficiently smaller than q, it should be possible to distinguish $\phi(\lfloor \Gamma_r^n \rceil)$ from uniform.

3.2 Attack on Non-dual Decision RLWE

The attack of Elias et al. on non-dual decision RLWE basically works by interpreting the RLWE samples as PLWE samples and then executing the above attack. For this approach to work, two requirements need to be fulfilled. Firstly, the ring of integers R of the number field K should be a quotient ring of the form $R = \mathbb{Z}[x]/(f)$, i.e. the number field should be monogenic.

The second condition deals with the difference between the error distributions of PLWE and non-dual RLWE. For PLWE one simply uses a spherical Gaussian Γ_r^n on $R \otimes \mathbb{R}$ with respect to the polynomial basis $1, x, x^2, \ldots, x^{n-1}$, whereas the RLWE distribution $\Psi_\mathbf{r}$ is obtained by pulling back a near-spherical Gaussian distribution on H through the canonical embedding σ. With respect to the polynomial basis one can view $\Psi_\mathbf{r}$ as a near-spherical Gaussion that got distorted by $N_f \cdot B$. Since B is a unitary transformation, the only actual distortion comes from N_f.

The maximum distortion of N_f is captured by its spectral norm $s_1(N_f)$, i.e. its largest singular value. The other singular values are denoted by $s_i(N_f)$ ordered by size such that $s_n(N_f)$ denotes its smallest singular value. A spherical Gaussian distribution on H of parameter $\mathbf{r} = (r, r, \ldots, r)$ will therefore be transformed into an elliptical Gaussian distribution on $R \otimes \mathbb{R} = K \otimes \mathbb{R}$ where the maximum parameter will be given by $s_1(N_f) \cdot r$. The attack on non-dual decision RLWE then proceeds by considering the samples with errors coming from $\Psi_{\mathbf{r}}$ as PLWE samples where the error is bounded by a spherical Gaussian with deviation $s_1(N_f) \cdot r$, with $r = \max(\mathbf{r})$.

For the attack to succeed we therefore need the following four conditions:

1. K is monogenic,
2. $f(1) \equiv 0 \bmod q$,
3. q is small enough that $\mathbb{Z}_q$ can be enumerated,
4. $r' = s_1(N_f) \cdot r$ is small enough such that $\phi(\lfloor \Gamma_{r'}^n \rceil)$ can be distinguished from uniform.

Again note that if $\mathbf{e}_i$ is bounded by Γ_r^n then the $\mathbf{e}_i(1)$ are bounded by a Gaussian with parameter $\sqrt{n} \cdot r' = \sqrt{n} \cdot s_1(N_f) \cdot r$. So the requirement is that the latter quantity is sufficiently smaller than q. In fact this is a very rough estimate, and indeed Elias et al. empirically observe in [6, Sect. 9] that their attack works more often than this bound predicts. We will explain this observation in Sect. 4.1.

In [6] the authors remark that given a parameter set (n, q, r) for PLWE, one cannot simply use the same parameter set for non-dual RLWE since the canonical embedding of the ring R into H might be very sparse, i.e. the covolume (volume of a fundamental domain) of $\sigma(R)$ in H might be very large. They therefore propose to scale up the parameter r by a factor of $|\det(M_f B)|^{1/n} = |\det(M_f)|^{1/n}$, which is the n-th root of the covolume. Thus given a PLWE parameter set (n, q, r), their corresponding RLWE parameter set reads $(n, q, \tilde{r})$ with $\tilde{r} = r \cdot |\det(M_f)|^{1/n}$.

3.3 Provably Weak Number Fields for Non-dual Decision RLWE

The first type of polynomials to which the attack of [6] was applied are polynomials of the form $f_{n,a,b}$ with $a = 0$. More precisely they considered

$$f_{n,q} := f_{n,0,q-1} = x^n + q - 1\,,$$

where $n \geq 1$ and q is a prime. Note that the roots of these polynomials are simply the primitive $2n$-th roots of unity scaled up by $(q-1)^{1/n}$. These polynomials satisfy $f_{n,q}(1) \equiv 0 \bmod q$ and are irreducible by Eisenstein's criterion whenever $q - 1$ has a prime factor with exponent one. As shown in [6, Proposition 3], the polynomials $f_{n,q}$ are monogenic whenever $q - 1$ is squarefree, n is a power of a prime ℓ, and $\ell^2 \nmid ((1-q)^n - (1-q))$. In particular it is easy to construct examples for $n = 2^k$.

The final missing ingredient is a bound on the spectral norm $s_1(N_f)$. In [6], a slightly different matrix M_f is used (it is a real matrix containing the real and imaginary parts of the roots of f). For use further down, we adapt the proof of [6, Proposition 4] to derive *all* singular values $s_i(N_f)$. Due to its practical importance we will only deal with the case where n is even, since we are particularly interested in the case where $n = 2^k$.

Proposition 1 (Adapted from [6, Proposition 4]**).** *Assume that $f_{n,q}$ is irreducible and that n is even, then the singular values $s_i(N_f)$ are given by*

$$s_i(N_f) = \frac{1}{\sqrt{n}(q-1)^{(i-1)/n}} .$$

PROOF: The roots of $f_{n,q}$ are given by $a \cdot \xi_{2n}^j$ for $0 < j < 2n$ and j odd, with $a = (q-1)^{1/n} \in \mathbb{R}^+$ and ξ_{2n} a primitive $2n$-th root of unity. To derive the singular values of $N_f = M_f^{-1}$ it suffices to derive the singular values of M_f. Recall that the u-th column of M_f (counting from 0) is given by

$$a^u \cdot (\xi_{2n}^u, \xi_{2n}^{3u}, \ldots, \xi_{2n}^{(2n-1)u})^t .$$

The (Hermitian) inner product of the u-th and v-th column is therefore given by

$$S = a^{u+v} \cdot \sum_{k=0}^{n-1} \xi_{2n}^{(2k+1)(u-v)} .$$

Since $\xi_{2n}^{2n+1} = \xi_{2n}$, we obtain that $\xi_{2n}^{2(u-v)} S = S$. For $u \neq v$ we have that $\xi_{2n}^{2(u-v)} \neq 1$, which implies that $S = 0$. For $u = v$ we obtain $S = na^{2u}$. This shows that the matrix M_f has columns that are orthogonal. The singular values of M_f can be read off from the diagonal of $\overline{M_f}^t \cdot M_f$, in particular $s_i(M_f) = \sqrt{n} a^{n-i}$ for $i = 1, \ldots, n$. This also shows that $s_i(N_f) = 1/(\sqrt{n} a^{i-1})$ for $i = 1, \ldots, n$. One finishes the proof by using that $a^n = q - 1$. □

The above proposition gives $s_1(N_f) = 1/\sqrt{n}$ which is small enough for the attack described in Sect. 3.2 to apply. In [6, Sect. 9], two examples of this family were attacked, giving the following results:

$f_{n,q}$	q	r	$\tilde{r}$	Samples per run	Successful runs	Time per run
$x^{192} + 4092$	4093	8.87	5440.28	20	1 of 10	25 s
$x^{256} + 8190$	8191	8.35	8399.70	20	2 of 10	44 s

Recall that $\tilde{r}$ is simply r scaled up by a factor $|\det(M_f)|^{1/n}$. We remark, as do Elias et al. [6, Sect. 9], that these two examples unfortunately do *not* satisfy that $q - 1$ is squarefree. As a consequence the RLWE problem is not set up in the full ring of integers of the number field $K = \mathbb{Q}[x]/(f)$. We will nevertheless keep using these examples for the sake of comparison; it should be clear from the exposition below that this is not a crucial issue.

As a second instance, the authors of [6] considered polynomials of the form $f_{n,a,b} = x^n + ax + b$ with $a \approx b$, again chosen such that $f_{n,a,b}(1) \equiv 0$ modulo q, which is assumed to be an odd prime. More precisely, they let $a = (q-1)/2 + \Delta$ and $b = (q-1)/2 - \Delta - 1$, or $a = q + \Delta$ and $b = q - \Delta - 1$, for a small value of Δ. Heuristically these polynomials also result in weak instances of non-dual decision RLWE, even though the analysis cannot be made as precise as in the foregoing case. In particular, no explicit formula is known for the spectral norm $s_1(N_f)$, but in [6] a heuristic perturbation argument is given that implies that it is bounded by $\sqrt{\max(a,b)} \cdot \det(N_f)^{1/n}$ infinitely often. They ran their attack for the particular case where $q = 524287$, $\Delta = 1$, $a = q + \Delta$ and $b = q - \Delta - 1$:

$f_{n,a,b}$	q	r	$\tilde{r}$	Samples per run	Successful runs	Time per run
$x^{128} + 524288x + 524285$	524287	8.00	45540	20	8 of 10	24 s

4 A Simple Attack on Search RLWE

We derive a very simple attack on *search* RLWE for the families and parameter sets considered by Elias, Lauter, Ozman and Stange in [6]. The attack is based on two observations.

Firstly, a unit ball in the H-space gets severely deformed when being pulled back to $K \otimes \mathbb{R}$ along the canonical embedding. With respect to the polynomial basis $1, x, x^2, \ldots, x^{n-1}$ we end up with an ellipsoid whose axes have lengths $s_1(N_f), \ldots, s_n(N_f)$. For the first family of polynomials (i.e. where $a = 0$) this is a geometrically decreasing sequence, while for the second family this statement remains almost true. In particular a spherical Gaussian distribution $\Gamma_{\mathbf{r}}$ with $\mathbf{r} = (r, \ldots, r)$ on the H-space will result in a very skew elliptical Gaussian distribution on $K \otimes \mathbb{R}$ with parameters $s_1(N_f) \cdot r, \ldots, s_n(N_f) \cdot r$. For the choices of r (or in fact $\tilde{r}$) made by Elias et al., the errors along the shortest axes of the ellipsoid are so small that after rounding they become zero.

The second observation is that the axes of the error distribution ellipsoid coincide almost perfectly with the polynomial basis. Again for the first family this is exactly the case, while for the second family the distribution is consistent enough, in the sense that the axes do not line up perfectly, but the coordinates of the error samples with respect to $1, x, x^2, \ldots, x^{n-1}$ still tend to go down geometrically. The result is that the directions that get squashed simply correspond to the coefficients of the higher powers of x in the error terms $\mathbf{e}(x)$.

To make these statements precise we will compute the singular value decomposition of the whole transformation matrix $N_f \cdot B$. Recall that the singular value decomposition of an $n \times n$ matrix M is given by

$$M = U\Sigma\overline{V}^t,$$

where U, V are $n \times n$ unitary matrices and Σ is an $n \times n$ matrix with non-negative real numbers on the diagonal, namely the singular values. The image of a unit sphere under M will therefore result in an ellipsoid where the axes are given by the columns of U, with lengths equal to the corresponding singular values.

4.1 Singular Value Decomposition and Error Distribution

For the first family of polynomials $f_{n,q}$ everything can be made totally explicit:

Proposition 2. *The singular value decomposition of $N_f \cdot B$ is*

$$I_{n\times n} \cdot \Sigma \cdot \overline{V}^t, \qquad \textit{where} \quad V = \overline{B}^t \cdot M_f \cdot \Sigma$$

and Σ is the diagonal matrix containing the singular values of N_f.

PROOF: Recall from the proof of Proposition 1 that the Vandermonde matrix M_f has mutually orthogonal columns, where the ith column has norm $\sqrt{n}a^{i-1}$. Thus the normalized matrix

$$M_f \cdot \Sigma \qquad \text{where } \Sigma = \operatorname{diag}\left(1/(\sqrt{n}a^{i-1})\right)_i = \operatorname{diag}\left(s_i(N_f)\right)_i$$

is unitary. But then so is $V = \overline{B}^t \cdot M_f \cdot \Sigma$, and since $\Sigma = \Sigma^2 \cdot \Sigma^{-1} = N_f \overline{N_f}^t \cdot \Sigma^{-1}$, we see that

$$N_f \cdot B = I_{n\times n} \cdot \Sigma \cdot \overline{V}^t$$

is the singular value decomposition of our transformation matrix $N_f \cdot B$. □

The factor $I_{n\times n}$ implies that the axes of our ellipsoid match perfectly with the polynomial basis $1, x, x^2, \ldots, x^{n-1}$. In other words, if we start from a spherical error distribution $\Gamma_{\mathbf{r}}$ on H, $\mathbf{r} = (r, r, \ldots, r)$, then the induced error distribution $\Psi_{\mathbf{r}}$ on $K \otimes \mathbb{R}$ in the ith coordinate (coefficient of x^{i-1}) is a Gaussian with parameter

$$s_i(N_f) \cdot r = \frac{r}{\sqrt{n} \cdot (q-1)^{(i-1)/n}}$$

by Proposition 1. This indeed decreases geometrically with i.

As a side remark, note that this implies that for $\mathbf{e}(x) \leftarrow \Psi_{\mathbf{r}}$ the evaluation $\mathbf{e}(1)$ is sampled from a Gaussian with parameter

$$\left(\sum_{i=1}^{n} s_i(N_f)^2\right)^{1/2} \cdot r = s_1(N_f)\sqrt{\frac{(q-1)^2 - 1}{(q-1)^2 - (q-1)^{2(n-1)/n}}} \cdot r.$$

This is considerably smaller than $\sqrt{n} \cdot s_1(N_f) \cdot r$ and explains why the attack from [6] works better than what their theory predicts [6, Sect. 9].

To illustrate the geometric behavior of the coordinates of the errors $\mathbf{e}(x)$ with respect to the polynomial basis, we have plotted the average and standard deviation of their high order coefficients for the second example $x^{256} + 8190$ from [6] in Fig. 1 (the results for the first example are totally similar), using the error parameter that they used to attack non-dual decision RLWE. The plot shows that for the given parameter set, the highest $\lceil n/7 \rceil$ error coefficients in the polynomial basis of $K \otimes \mathbb{R}$ are all extremely likely to be smaller than $1/2$ (indicated by the dashed line) in absolute value and therefore become zero after rounding.

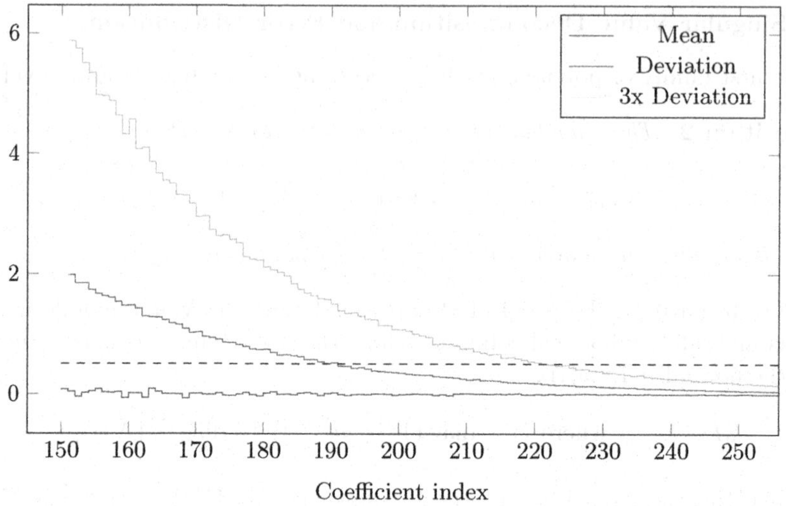

Fig. 1. Distribution of the error terms in the polynomial basis for $f = x^{256} + 8190$

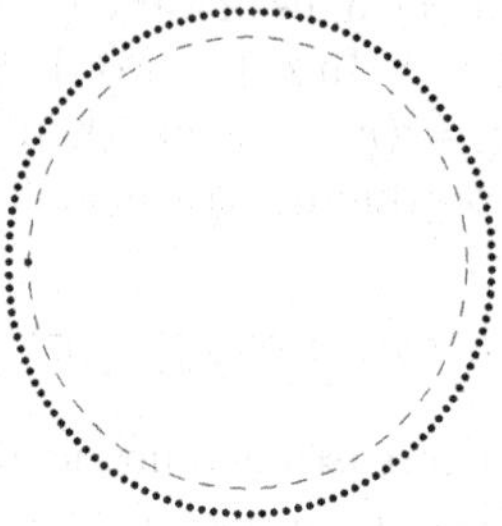

Fig. 2. Zeroes in $\mathbb{C}$ of $x^{128} + 524288x + 524285$, along with the unit circle (dashed)

For the second family of polynomials $f_{n,a,b}$ with $a \neq 0$, we were not able to derive the singular value decomposition in such an explicit form. To get a handle on them, we have computed it explicitly for $f = x^{128} + 524288x + 524285$. For this particular example, the roots of $f_{n,a,b}$ again lie roughly on a circle (except for the real root close to -1): see Fig. 2. So through the Vandermonde matrix we again expect geometric growth of the singular values, as is confirmed by the explicit numerics in Fig. 3, which shows a plot of their logarithms. There is only one outlier, caused by the real root of f close to -1.

The heat map in Fig. 4 plots the norms of the entries in the U-matrix of the singular value decomposition of $N_f \cdot B$ and shows that U is close to being diagonal, implying that the axes of the ellipsoid are indeed lining up almost perfectly with the polynomial basis. Finally Fig. 5 contains a similar plot as Fig. 1, namely, the distribution of the errors terms (highest powers only) for the

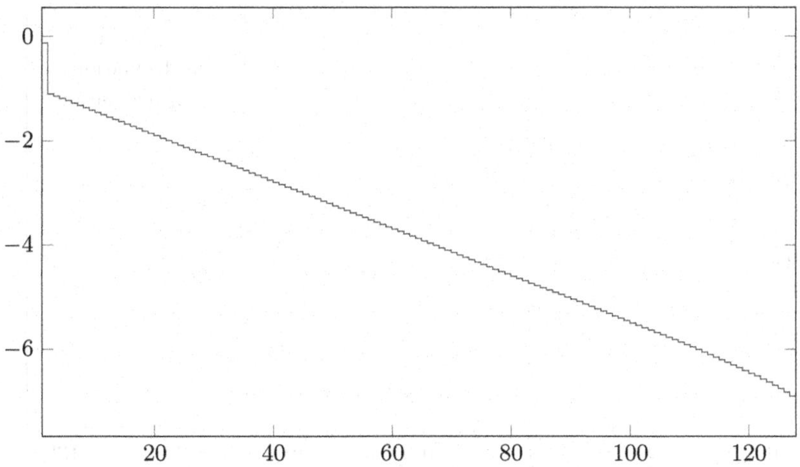

Fig. 3. $\log_{10}$ of the singular values of N_f for $f = x^{128} + 524288x + 524285$

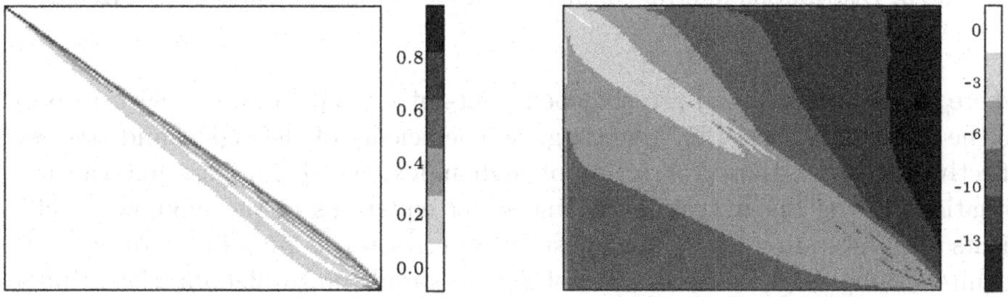

Fig. 4. Heat maps of the norms of the entries of U (left) and $\log_{10}$ of the norms of the entries of $U\Sigma$ (right), where $U\Sigma\overline{V}^t$ is the singular value decomposition of $N_f B$

polynomial $f = x^{128} + 524288x + 524285$. Again we conclude that with very high probability, the last $\lceil n/6 \rceil$ coefficients of the error terms in the polynomial basis will be smaller than $1/2$, and therefore they become zero after rounding.

4.2 Linear Algebra Attack on Non-dual Search RLWE

Turning the above observations into an attack on non-dual search RLWE for these families is straightforward. Recall that the samples are of the form $(\mathbf{a}, \mathbf{b} = \mathbf{a} \cdot \mathbf{s} + \mathbf{e} \bmod q)$ where the errors were sampled from the distribution $\Psi_\mathbf{r}$ on $K \otimes \mathbb{R}$. Since $\mathbf{a}$ is known, we can express multiplication by $\mathbf{a}$ as a linear operation, i.e. we can compute the $n \times n$ matrix $M_\mathbf{a}$ that corresponds to multiplication by $\mathbf{a}$ with respect to the polynomial basis $1, x, x^2, \ldots, x^{n-1}$. Each RLWE sample can therefore be written as a linear algebra problem as follows:

$$M_\mathbf{a} \cdot (s_0, s_1, \ldots, s_{n-1})^t = (b_0, b_1, \ldots, b_{n-1})^t - (e_0, e_1, \ldots, e_{n-1})^t \qquad (1)$$

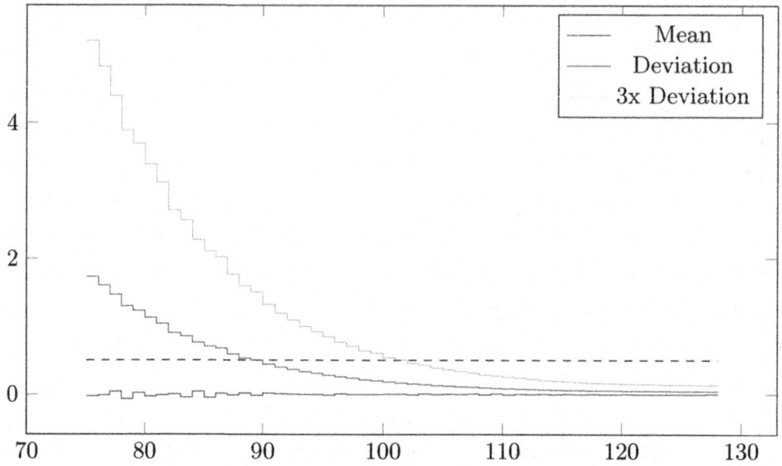

Fig. 5. Distribution of errors of high index for $f = x^{128} + 524288x + 524285$

where the s_i (resp. b_i, e_i) are the coefficients of $\mathbf{s}$ (resp. $\mathbf{b}$ and $\mathbf{e}$) with respect to the polynomial basis. By rounding the coefficients of the right-hand side, we effectively remove the error terms of high index, which implies that the last equations in the linear system become *exact* equations in the unknown coefficients of $\mathbf{s}$. Assuming that the highest $\lceil n/k \rceil$ error terms round to zero, we only require k samples to recover the secret $\mathbf{s}$ using simple linear algebra with a 100 % success rate.

We have implemented this attack in Magma [2] with the following results.

$f_{n,a,b}$	q	r	$\tilde{r}$	Samples per run	Successful runs	Time per run
$x^{192} + 4092$	4093	8.87	5440	7	10 of 10	8.37 s
$x^{256} + 8190$	8191	8.35	8390	6	10 of 10	17.2 s
$x^{128} + 524288x + 524285$	524287	8.00	45540	4	10 of 10	1.96 s

We note that using less samples per run is also possible, but results in a lower than 100 % success rate. A more elaborate strategy would construct several linear systems of equations by discarding some of the equations of lower index (which are most likely to be off by 1) and running exhaustively through the kernel of the resulting underdetermined system of equations. However, we did not implement this strategy since it needlessly complicates the attack.

In fact for errors of the above size one can also use the linearization technique developed by Arora and Ge [1, Theorem 3.1] to retrieve $\mathbf{s}(x)$, but this requires a lot more samples.

We stress that our attack does not use that $f(1) \equiv 0 \bmod q$. For the above defining polynomials our attack works modulo *every* modulus q', as long as the same error parameters are used (or smaller ones).

4.3 Modifications for Dual Search RLWE

In this section we discuss how switching from non-dual RLWE (i.e. from $I = R$) to dual RLWE (where one takes $I = R^\vee$) affects our observations. Recall that in the case of a monogenic defining polynomial f, the codifferent $R^\vee$ is generated as a fractional ideal by $1/f'(\alpha)$ with $\alpha \in \mathbb{C}$ a root of f. We will again work with respect to the polynomial basis $1, x, x^2, \ldots, x^{n-1}$ of $K = \mathbb{Q}[x]/(f)$ over $\mathbb{Q}$, which is also a basis of R over $\mathbb{Z}$, and take $\alpha = x$. For technical reasons we will only do the analysis for the first family of polynomials, namely those of the form

$$f_{n,q} = f_{n,0,q-1} = x^n + q - 1,$$

where one has $f'_{n,q} = nx^{n-1}$. Since

$$1 = \frac{1}{q-1} f_{n,q} - \frac{x}{n(q-1)} f'_{n,q}$$

we find that

$$R^\vee = R \frac{x}{n(q-1)}.$$

Proposition 3. *The elements*

$$\frac{1}{n}, \frac{x}{n(q-1)}, \frac{x^2}{n(q-1)}, \frac{x^3}{n(q-1)}, \ldots, \frac{x^{n-1}}{n(q-1)} \tag{2}$$

form a $\mathbb{Z}$-basis of $R^\vee$.

Proof. It is immediate that

$$\frac{x}{n(q-1)}, \frac{x^2}{n(q-1)}, \frac{x^3}{n(q-1)}, \ldots, \frac{x^{n-1}}{n(q-1)}, \frac{x^n}{n(q-1)}$$

form a $\mathbb{Z}$-basis. But modulo $f_{n,q}$ the last element is just $-1/n$.

Thus we can think of our secret $\mathbf{s}(x) \in R_q^\vee$ as a $\mathbb{Z}$-linear combination of the elements in (2), where the coefficients are considered modulo q. A corresponding RLWE-sample is then of the form $(\mathbf{a}(x), \mathbf{a}(x) \cdot \mathbf{s}(x) + \mathbf{e}(x) \bmod qR^\vee)$ with $\mathbf{e}(x) \in R^\vee \otimes \mathbb{R} = K \otimes \mathbb{R}$ sampled from $\Psi_\mathbf{r}$ for an appropriate choice of $\mathbf{r} \in (\mathbb{R}^+)^n$. To make a comparison with our attack in the non-dual case, involving the parameters from [6], we have to make an honest choice of $\mathbf{r}$, which we again take spherical. Note that the lattice $\sigma(R^\vee)$ is much denser than $\sigma(R)$: the covolume gets scaled down by a factor

$$\left|\mathrm{No}(f'_{n,q}(\alpha))\right| = n^n (q-1)^{n-1}.$$

Therefore, in view of the discussion concluding Sect. 3.2, we scale down our scaled-up error parameter $\tilde{r}$ by a factor

$$\sqrt[n]{n^n(q-1)^{n-1}} \approx n(q-1).$$

Let us denote the result by $\tilde{r}^\vee$.

It follows that the dual setting is essentially just a scaled version of its non-dual counterpart: both the errors and the basis elements become divided by a factor of roughly $n(q-1)$. In particular, for the same choices of r we again find that with near certainty the highest $\lceil n/7 \rceil$ error coefficients are all smaller than

$$\frac{1}{2} \cdot \frac{1}{n(q-1)}$$

in absolute value, and therefore become zero after rounding to the nearest multiple of $1/(n(q-1))$. This then again results in exact equations in the coefficients of the secret $\mathbf{s}(x) \in R_q^\vee$ with respect to the basis (2), that can be solved using linear algebra.

Here too, the attack does not use that $f(1) \equiv 0 \bmod q$ so it works for whatever choice of modulus q' instead of q, as long as the same error parameters are used (or smaller ones).

5 Range of Applicability

One obvious way of countering our attack is by modifying the error parameter. In principle the skewness of $N_f \cdot B$ could be addressed by using an equally distorted elliptical Gaussian rather than a near-spherical one, but that conflicts with the philosophy of RLWE (as opposed to PLWE), namely that the more natural way of viewing a number field is through its canonical embedding. So we will not discuss this option and stick to spherical distributions. Then the only remaining way out is to enlarge the width of the distribution. Again for technical reasons we will restrict our discussion to the first family of polynomials, namely those of the form $f_{n,q} = x^n + q - 1$; the conclusions for the second family should be similar.

In the non-dual case we see that a version of the attack works as long as a sample drawn from a univariate Gaussian with parameter $s_n(N_f) \cdot \tilde{r}$ has absolute value less than $1/2$ with non-negligible probability: then by rounding one obtains at least one exact equation in the unknown secret $\mathbf{s}(x)$. For this one needs that

$$s_n(N_f) \cdot \tilde{r} \leq \frac{C}{2}$$

for some absolute constant $C > 0$ that quantifies what it means to be 'non-negligible'.

Remark 1. In order to recover the *entire* secret, one even wants a non-negligible probability for n consecutive samples to be less than $1/2$, for which one should replace $s_n(N_f)\cdot\tilde{r}$ by $s_n(N_f)\cdot\tilde{r}\cdot\sqrt{\log n}$ (roughly). In fact a slightly better approach is to find the optimal $1 \leq k \leq n$ for which $s_{n-k+1}(N_f)\cdot\tilde{r}$ is likely to be less than $1/2$, thereby yielding at least k exact equations at once, for $\lceil n/k \rceil$ consecutive times.

Let us take $C = 1$ in what follows: for this choice meeting the upper bound corresponds to a chance of about 98.78 % of recovering at least one exact equation. Using Proposition 1 this can be rewritten as

$$\tilde{r} \leq \frac{1}{2}\cdot\sqrt{n}\cdot(q-1)^{1-1/n}. \tag{3}$$

For our two specific polynomials $x^{192} + 4092$ and $x^{256} + 8190$ the right-hand side reads 27148.39 and 63253.95 whereas Elias et al. took $\tilde{r}$ to be 5440.28 and 8399.70, respectively.

Note that the bound in (3) does not depend on the modulus q' that is being used: the q that appears there is just part of the data defining our number field. In other words, whenever $\tilde{r}$ satisfies (3) then for every choice of modulus q' we are very likely to recover at least one exact equation in the coefficients of the secret $\mathbf{s}(x)$.

Unfortunately the bound (3) does not allow for an immediate comparison with the hardness result of Lyubashevsky, Peikert and Regev (see Theorem 1), which was formulated for dual RLWE only. But for dual RLWE one can make a similar analysis. From Sect. 4.3 it follows that we want error coefficients that are smaller than $1/(2n(q-1))$ with a non-negligible probability. The same discussion then leads to the bound

$$\tilde{r}^{\vee} \leq \frac{1}{2}\cdot\frac{1}{\sqrt{n}\cdot(q-1)^{\frac{1}{n}}} \tag{4}$$

which is highly incompatible with the condition $\tilde{r}^{\vee} \geq 2\omega(\sqrt{\log n})$ from Theorem 1. Thus we conclude that it is impossible to enlarge the error parameter up to a range where our attack would form an actual threat to RLWE, as defined in [9, Sect. 3].

Another issue with modifying the error parameter is decodability. In the non-dual case, from (3) we see that $s_n(N_f)\cdot\tilde{r} \gg 1$ is needed to avoid being vulnerable to our skewness attack. But it automatically follows that $s_1(N_f)\cdot\tilde{r} \gg q$. Indeed, this is implied by the fact that the condition number $k(N_f) := s_1(N_f)/s_n(N_f)$ equals

$$(q-1)^{1-1/n} \approx q$$

by Proposition 1. This causes the errors at the terms of low degree to wrap around modulo q. In the dual case the same observation applies, where now the error terms of low degree tend to wrap around modulo multiples of $q \cdot 1/(n(q-1))$. In both cases the effect is that several of these terms become indistinguishable from uniform, requiring more samples for the RLWE problem to become information theoretically solvable. This obstructs, or at least complicates, certain cryptographic applications.

So overall, the conclusion is that the defining polynomials $f_{n,a,b}$ are just not well-suited for use in RLWE: either the error parameter is too small for the RLWE problem to be hard, or the error parameter is too large for the problem to be convenient for use in cryptography. But we stress once more that neither the attack from [6] nor our attack form a genuine threat to RLWE, as it was defined in [9, Sect. 3].

6 Conclusions

In this paper we have shown that non-dual *search* RLWE can be solved efficiently for the families of polynomials and parameter sets from [6] which were shown to be weak for the *decision* version of the problem. The central reason for this weakness lies in the (exponential) skewness of the canonical embedding transformation. We analyzed the singular value decomposition of this transformation and showed that the singular values form an (approximate) geometric series. Furthermore, we also showed that the axes of the error ellipsoid are consistent with the polynomial basis, allowing us to readily identify very small noise coefficients. The attack applies to wider ranges of moduli, and also applies to the dual version, but does not contradict any statement in the work of Lyubashevsky, Peikert and Regev [9].

It is worth remarking that while we used the language of singular value decomposition, for our skewness attack it merely suffices that $N_f \cdot B$ has a very short row, so that the corresponding error coefficient e_i vanishes after rounding and (1) provides an exact equation in the coefficients of the secret. For general number fields this is a strictly weaker condition than having a very small singular value whose corresponding axis lines up perfectly with one of the polynomial basis vectors. But for the particular families of [6] the singular value decomposition turned out to be a convenient tool in proving this, and in visualizing how the RLWE errors are transformed under pull-back along the canonical embedding.

Acknowledgments. This work was supported by the European Commission through the ICT programme under contract H2020-ICT-2014-1 644209 HEAT and contract H2020-ICT-2014-1 645622 PQCRYPTO. We would like to thank Ron Steinfeld and the anonymous referees for their valuable comments.

References

1. Arora, S., Ge, R.: New algorithms for learning in presence of errors. In: Aceto, L., Henzinger, M., Sgall, J. (eds.) ICALP 2011, Part I. LNCS, vol. 6755, pp. 403–415. Springer, Heidelberg (2011)
2. Bosma, W., Cannon, J., Playoust, C.: The Magma algebra system. I. The user language. J. Symbolic Comput. **24**(3–4), 235–265 (1997). Computational algebra and number theory, London (1993)
3. Brakerski, Z., Langlois, A., Peikert, C., Regev, O., Stehlé, D.: Classical hardness of learning with errors. In: Symposium on Theory of Computing Conference, STOC 2013, pp. 575–584. ACM (2013)

4. Ducas, L., Durmus, A.: Ring-LWE in polynomial rings. In: Fischlin, M., Buchmann, J., Manulis, M. (eds.) PKC 2012. LNCS, vol. 7293, pp. 34–51. Springer, Heidelberg (2012)
5. Eisenträger, K., Hallgren, S., Lauter, K.: Weak instances of PLWE. In: Joux, A., Youssef, A. (eds.) SAC 2014. LNCS, vol. 8781, pp. 183–194. Springer, Heidelberg (2014)
6. Elias, Y., Lauter, K.E., Ozman, E., Stange, K.E.: Provably weak instances of ring-LWE. In: Gennaro, R., Robshaw, M. (eds.) CRYPTO, Part I. LNCS, vol. 9215, pp. 63–92. Springer, Heidelberg (2015)
7. Hoffstein, J., Pipher, J., Silverman, J.H.: NTRU: a ring-based public key cryptosystem. In: Buhler, J.P. (ed.) ANTS 1998. LNCS, vol. 1423, pp. 267–288. Springer, Heidelberg (1998)
8. Lindner, R., Peikert, C.: Better key sizes (and attacks) for LWE-based encryption. In: Kiayias, A. (ed.) CT-RSA 2011. LNCS, vol. 6558, pp. 319–339. Springer, Heidelberg (2011)
9. Lyubashevsky, V., Peikert, C., Regev, O.: On ideal lattices and learning with errors over rings. In: Gilbert, H. (ed.) EUROCRYPT 2010. LNCS, vol. 6110, pp. 1–23. Springer, Heidelberg (2010)
10. Lyubashevsky, V., Peikert, C., Regev, O.: A toolkit for ring-LWE cryptography. In: Johansson, T., Nguyen, P.Q. (eds.) EUROCRYPT 2013. LNCS, vol. 7881, pp. 35–54. Springer, Heidelberg (2013)
11. Peikert, C.: Public-key cryptosystems from the worst-case shortest vector problem: extended abstract. In: Symposium on Theory of Computing, STOC 2009, pp. 333–342. ACM (2009)
12. Regev, O.: On lattices, learning with errors, random linear codes, and cryptography. In: Symposium on Theory of Computing, pp. 84–93. ACM (2005)
13. Regev, O.: On lattices, learning with errors, random linear codes, and cryptography. J. ACM **56**(6), 34 (2009)

Faster Algorithms for Solving LPN

Bin Zhang[1,2(✉)], Lin Jiao[1,3(✉)], and Mingsheng Wang[4]

[1] TCA Laboratory, SKLCS, Institute of Software, Chinese Academy of Sciences, Beijing 100190, China
{zhangbin,jiaolin}@tca.iscas.ac.cn
[2] State Key Laboratory of Cryptology, P.O. Box 5159, Beijing 100878, China
[3] University of Chinese Academy of Sciences, Beijing 100049, China
[4] State Key Laboratory of Information Security, Institute of Information Engineering, Chinese Academy of Sciences, Beijing 100093, China

Abstract. The LPN problem, lying at the core of many cryptographic constructions for lightweight and post-quantum cryptography, receives quite a lot attention recently. The best published algorithm for solving it at Asiacrypt 2014 improved the classical BKW algorithm by using covering codes, which claimed to marginally compromise the 80-bit security of HB variants, LPN-C and Lapin. In this paper, we develop faster algorithms for solving LPN based on an optimal precise embedding of cascaded concrete perfect codes, in a similar framework but with many optimizations. Our algorithm outperforms the previous methods for the proposed parameter choices and distinctly break the 80-bit security bound of the instances suggested in cryptographic schemes like HB$^+$, HB$^\#$, LPN-C and Lapin.

Keywords: LPN · BKW · Perfect code · HB · Lapin

1 Introduction

The Learning Parity with Noise (LPN) problem is a fundamental problem in modern cryptography, coding theory and machine learning, whose hardness serves as the security source of many primitives in lightweight and post-quantum cryptography. It is closely related to the problem of decoding random linear codes, which is one of the most important problems in coding theory, and has been extensively studied in the last half century.

In the LPN problem, there is a secret $\mathbf{x} \in \{0,1\}^k$ and the adversary is asked to find $\mathbf{x}$ given many noisy inner products $\langle \mathbf{x}, \mathbf{g} \rangle + e$, where each $\mathbf{g} \in \{0,1\}^k$ is a random vector and the noise e is 1 with some probability η deviating from 1/2. Thus, the problem is how to efficiently restore the secret vector given some amount of noisy queries of the inner products between itself and certain random vectors.

The cryptographic schemes based on LPN are appealing both for theoretical and practical reasons. The earliest proposal dated back to the HB, HB$^+$, HB$^\#$ and AUTH authentication protocols [12,18–20]. While HB is a minimalistic protocol secure in a passive attack model, the modified scheme HB$^+$ with one extra

M. Fischlin and J.-S. Coron (Eds.): EUROCRYPT 2016, Part I, LNCS 9665, pp. 168–195, 2016.
DOI: 10.1007/978-3-662-49890-3_7

round is found to be vulnerable to active attacks, i.e., man-in-the-middle attacks [14]. HB$^{\#}$ was subsequently proposed with a more efficient key representation using a variant called TOEPLITZ-LPN. Besides, there is also a message encryption scheme based on LPN, i.e., the LPN-C scheme in [13] and some message authentication codes (MACs) using LPN in [10,20], allowing for constructions of identification schemes provably secure against active attacks. Another notable scheme, Lapin, was proposed as a two-round identification protocol [16], based on the LPN variant called Ring-LPN where the samples are elements of a polynomial ring. Recently, an LPN-based encryption scheme called Helen was proposed with concrete parameters for different security levels [11].

It is of primordial importance to study the best possible algorithms that can efficiently solve the LPN problem. The seminal work of Blum et al. in [5], known as the BKW algorithm, employs an iterated collision procedure of the queries to reduce the dependency on the information bits with a folded noise level. Levieil and Fouque proposed to exploit the Fast Walsh-Hadamard (FWHT) Transform in the process of searching for the secret in [22]. They also provided different security levels achieved by different instances of LPN, which are referenced by most of the work thereafter. In [21], Kirchner suggested to transform the problem into a systematic form, where each secret bit appears as an observed symbol perturbed by noise. Then Bernstein and Lange demonstrated in [4] the utilization of the ring structure of Ring-LPN in matrix inversion to further reduce the attack complexity, which can be applied to the common LPN instances by a slight modification as well. None of the above algorithms manage to break the 80-bit security of Lapin, nor the parameters suggested in [22] as 80-bit security for LPN-C [13]. At Asiacrypt 2014, a new algorithm for solving LPN was presented in [15] by using covering codes. It was claimed that the 80-bit security bound of the common $(512, 1/8)$-LPN instance can be broken within a complexity of $2^{79.7}$, and so do the previously unbroken parameters of HB variants, Lapin and LPN-C[1]. It shared the same beginning steps of Gaussian elimination and collision procedure as that in [4], followed by the covering code technique to further reduce the dimension of the secret with an increased noise level, also it borrowed the well known Walsh Transform technique from fast correlation attacks on stream ciphers [2,7,23], renamed as subspace hypothesis testing.

In this paper, we propose faster algorithms for solving LPN based on an *optimal* precise embedding of cascaded perfect codes with the parameters found by integer linear programming to efficiently reduce the dimension of the secret information. Our new technique is generic and can be applied to any given (k, η)-LPN instance, while in [15] the code construction methods for covering are missing and only several specific parameters for $(512, 1/8)$-LPN instance were given in their presentation at Asiacrypt 2014. From the explicit covering, we can derive the bias introduced by covering in our construction *accurately*, and derive the attack complexity precisely. It is shown that following some tradeoff techniques,

---

1 The authors of [15] redeclared their results in their presentation at Asiacrypt 2014, for the original results are incorrect due to an insufficient number of samples used to learn an LPN secret via Walsh-Hadamard Transform.

appropriate optimization of the algorithm steps in a similar framework as that in [15] can reduce the overall complexity further. We begin with a theoretical justification of the experimental results on the current existing BKW algorithms, and then propose a general form of the BKW algorithm which exploits tuples in collision procedure with a simulation verification. We also propose a technique to overcome the data restriction efficiently based on goodness-of-fit test using χ^2-statistic both in theory and experiments. In the process, we theoretically analyze the number of queries needed for making a reliable choice for the best candidate and found that the quantity $8l\ln 2/\epsilon_f^2$ is much more appropriate when taking a high success probability into account[2], where ϵ_f is the bias of the final approximation and l is the bit length of the remaining secret information. We also provide the terminal condition of the solving algorithm, correct an error that may otherwise dominate the complexity of the algorithm in [15] and push the upper bound up further, which are omitted in [15]. We present the complexity analysis of the improved algorithm based on three BKW types respectively, and the results show that our algorithm well outperforms the previous ones. Now it is the first time to distinctly break the 80-bit security of both the $(512, 1/8)$- and $(532, 1/8)$- LPN instances, and the complexity for the $(592, 1/8)$-instance just slightly exceeds the bound. A complexity comparison of our algorithm with the previous attacks is shown in Table 1. More tradeoff choices are possible and can be found in Sect. 6.2.

Table 1. Comparison of different algorithms with the instance $(512, 1/8)$

Algorithm	Complexities ($\log_2$)		
	Data	Memory	Time
Levieil-Fouque [22]	75.7	84.8	87.5
Bernstein-Lange [4]	68.6	77.6	85.8
Corrected [15]	63.6	72.6	79.7[1]
This paper	63.5	68.2	72.8

[1] The number of queries we chosen is the twice as that presented for correction in the presentation at Asiacrypt 2014 to assure a success probability of almost 1. Note that if the same success probability is achieved, the complexity of the attack in [15] will exceed the 2^{80} bound.

This paper is organized as follows. We first introduce some preliminaries of the LPN problem in Sect. 2 with a brief review of the BKW algorithm. In Sect. 3, a short description of the algorithm using covering codes in [15] is presented. In

[2] The authors of [15] have chosen $4l\ln 2/\epsilon_f^2$ to correct the original estimate of $1/\epsilon_f^2$ as the number of queries in their presentation at Asiacrypt 2014.

Sect. 4, we present the main improvements and more precise data complexity analysis of the algorithm for solving LPN. Then we propose and analyze certain BKW techniques in Sect. 5. In Sect. 6, we complete the faster algorithm with more specific accelerated techniques at each step, together with the applications to the various LPN-based cryptosystems. Finally, some conclusions are provided in Sect. 7.

2 Preliminaries

In this Section, some basic notations of the LPN problem are introduced with a review of the BKW algorithm that is relevant to our analysis later.

2.1 The LPN Problem

Definition 1 (LPN Problem). *Let Ber_η be the Bernoulli distribution, i.e., if $e \leftarrow Ber_\eta$ then $Pr[e = 1] = \eta$ and $Pr[e = 0] = 1 - \eta$. Let $\langle \mathbf{x}, \mathbf{g} \rangle$ denote the scalar product of the vectors $\mathbf{x}$ and $\mathbf{g}$, i.e., $\mathbf{x} \cdot \mathbf{g}^T$, where $\mathbf{g}^T$ denotes the transpose of $\mathbf{g}$. Then an LPN oracle $\Pi_{LPN}(k, \eta)$ for an unknown random vector $\mathbf{x} \in \{0,1\}^k$ with a noise parameter $\eta \in (0, \frac{1}{2})$ returns independent samples of*

$$(\mathbf{g} \xleftarrow{\$} \{0,1\}^k, e \leftarrow Ber_\eta : \langle \mathbf{x}, \mathbf{g} \rangle + e).$$

The (k, η)-LPN problem consists of recovering the vector $\mathbf{x}$ according to the samples output by the oracle $\Pi_{LPN}(k, \eta)$. An algorithm $\mathcal{S}$ is called (n, t, m, δ)-solver if $Pr[\mathcal{S} = \mathbf{x} : \mathbf{x} \xleftarrow{\$} \{0,1\}^k] \geq \delta$, and runs in time at most t and memory at most m with at most n oracle queries.

This problem can be rewritten in a matrix form as $\mathbf{z} = \mathbf{x}\mathbf{G} + \mathbf{e}$, where $\mathbf{e} = [e_1\ e_2\ \cdots\ e_n]$ and $\mathbf{z} = [z_1\ z_2\ \cdots\ z_n]$, each $z_i = \langle \mathbf{x}, \mathbf{g}_i \rangle + e_i, i = 1, 2, \ldots, n$. The $k \times n$ matrix $\mathbf{G}$ is formed as $\mathbf{G} = [\mathbf{g}_1^T\ \mathbf{g}_2^T\ \cdots\ \mathbf{g}_n^T]$. Note that the cost of solving the first block of the secret vector $\mathbf{x}$ dominates the total cost of recovering $\mathbf{x}$ according to the strategy applied in [1].

Lemma 1 (Piling-up Lemma). *Let $X_1, X_2, \ldots, X_n$ be independent binary random variables where each $Pr[X_i = 0] = \frac{1}{2}(1 + \epsilon_i)$, for $1 \leq i \leq n$. Then,*

$$Pr[X_1 + X_2 + \cdots + X_n = 0] = \frac{1}{2}(1 + \prod_{i=1}^{n} \epsilon_i).$$

2.2 The BKW Algorithm

The BKW algorithm is proposed in the spirit of the generalized birthday algorithm [25], working on the columns of $\mathbf{G}$ as

$$\mathbf{g}_i + \mathbf{g}_j = [*\ *\ \cdots\ *\ \underbrace{0\ 0\ \cdots\ 0}_{b}], \text{ and } (z_i + z_j) = \mathbf{x}(\mathbf{g}_i^T + \mathbf{g}_j^T) + (e_i + e_j),$$

which iteratively reduces the effective dimension of the secret vector. Let the bias ϵ be defined by $\Pr[e = 0] = \frac{1}{2}(1+\epsilon)$, then $\Pr[e_i + e_j = 0] = \frac{1}{2}(1+\epsilon^2)$ according to the pilling-up lemma. Formally, the BKW algorithm works in two phases: reduction and solving. It applies an iterative sort-and-merge procedure to the queries and produces new entries with the decreasing dimension and increasing noise level; finally it solves the secret by exhausting the remaining and test the presence of the expected bias. The framework is as follows.

There are two approaches, called LF1 and LF2 in [22] to fulfill the merging procedure, sharing the same sorting approach with different merging strategies, which is described in the following Algorithms 2 and 3. It is easy to see that LF1 works on pairs with a representative in each partition, which is discarded at last; while LF2 works on any pair. For each iteration in the reduction phase, the noise level is squared, as $e_j^{(i)} = e_{j_1}^{(i-1)} + e_{j_2}^{(i-1)}$ with the superscript (i) being the iteration step. Assume the noises remain independent at each step, we have $\Pr[\sum_{j=1}^{2^t} e_j = 0] = \frac{1}{2}(1+\epsilon^{2^t})$ by the piling-up lemma.

Algorithm 1. Framework of the BKW Algorithm

Input: The $k \times n$ matrix $\mathbf{G}$ and received $\mathbf{z}$, the parameters b, t.

Put the received vector as a first row in the matrix, $\mathbf{G}_0 \leftarrow \begin{bmatrix} \mathbf{z} \\ \mathbf{G} \end{bmatrix}$.

Reduction phase:

for $i = 1$ to t **do**
- **Sorting**: Partition the columns of $\mathbf{G}_{i-1}$ by the last b bits.
- **Merging**: Form pairs of columns in each partition to obtain $\mathbf{G}_i$

end for

Solving phase:

for $\mathbf{x} \in \{0,1\}^{k-bt}$ **do**
- **return** the vector $\mathbf{x}$ that $[1\ \mathbf{x}]\mathbf{G}_t$ has minimal weight.

end for

Algorithm 2. Reduction of LF1

Partition $\mathbf{G}_{i-1} = V_0 \cup V_1 \cup \cdots \cup V_{2^b-1}$ s.t. the columns in V_j have the same last b bits.

for each V_j **do**
- Randomly choose $\mathbf{v}^* \in V_j$ as the representative.
  For $\mathbf{v} \in V_j, \mathbf{v} \neq \mathbf{v}^*$, $\mathbf{G}_i = \mathbf{G}_i \cup (\mathbf{v} + \mathbf{v}^*)$, ignoring the last b entries of 0.

end for

Algorithm 3. Reduction of LF2

Partition $\mathbf{G}_{i-1} = V_0 \cup V_1 \cup \cdots \cup V_{2^b-1}$ s.t. columns in V_j have the same last b bits.

for each V_j **do**
- For each pair $\mathbf{v}, \mathbf{v}' \in V_j, \mathbf{v} \neq \mathbf{v}'$, $\mathbf{G}_i = \mathbf{G}_i \cup (\mathbf{v} + \mathbf{v}')$, ignoring the last b entries of 0.

end for

---

3 The Previous Algorithm Using Covering Codes

In this section, we present a brief review of the algorithm using covering codes in [15], described in the following Algorithm 4.

Algorithm 4 contains five main steps: step 1 transforms the problem into systematic form by Gaussian elimination (Line 2); step 2 performs several BKW steps (Line 3–5); jumping to step 4, it uses a covering code to rearrange the samples (Line 6); step 3 guesses partial secret and Step 5 uses the FWHT to find the best candidate under the guessing, moreover it performs hypothesis testing to determine whether to repeat the algorithm (Line 7–13). Now we take a closer look at each step respectively.

Algorithm 4. The Algorithm using covering codes [15]

Input: n queries $(\mathbf{g}, z)$s of the (k, η)-LPN instance, the parameters b, t, k_2, l, w_1, w_2.
1: **repeat**
2: Pick random column permutation π and perform Gaussian elimination on $\pi(\mathbf{G})$, resulting in $[\mathbf{I}\ \mathbf{L}_0]$;
3: **for** $i = 1$ to t **do**
4: Perform LF1 reduction phase on $\mathbf{L}_{i-1}$ resulting in $\mathbf{L}_i$.
5: **end for**
6: Pick a $[k_2, l]$ linear code and group the columns of $\mathbf{L}_t$ by the last k_2 bits according to their nearest codewords.
7: Set $k_1 = k - tb - k_2$;
8: **for** $\mathbf{x}'_1 \in \{0,1\}^{k_1}$ with $wt(\mathbf{x}'_1) \leq w_1$ **do**
9: Update the observed samples.
10: Use FWHT to compute the numbers of 1s and 0s for each $\mathbf{y} \in \{0,1\}^l$, and pick the best candidate.
11: Perform hypothesis testing with a threshold.
12: **end for**
13: **until**: Acceptable hypothesis is found.

Step 1. Gaussian Elimination. This step systematizes the problem, i.e., change the positions of the secret vector bits without changing the associated noise level [21]. Precisely, from $\mathbf{z} = \mathbf{xG} + \mathbf{e}$, apply a column permutation π to make the first k columns of $\mathbf{G}$ linearly independent. Then form the matrix $\mathbf{D}$ such that $\hat{\mathbf{G}} = \mathbf{DG} = [\mathbf{I}\ \hat{\mathbf{g}}_{k+1}^T\ \hat{\mathbf{g}}_{k+2}^T\ \cdots\ \hat{\mathbf{g}}_n^T]$. Let $\hat{\mathbf{z}} = \mathbf{z} + [z_1\ z_2\ \cdots\ z_k]\hat{\mathbf{G}}$, thus $\hat{\mathbf{z}} = \mathbf{xD}^{-1}\hat{\mathbf{G}} + \mathbf{e} + [z_1\ z_2\ \cdots\ z_k]\hat{\mathbf{G}} = (\mathbf{xD}^{-1} + [z_1\ z_2\ \cdots\ z_k])\hat{\mathbf{G}} + \mathbf{e}$, where $\hat{\mathbf{z}} = [\mathbf{0}\ \hat{z}_{k+1}\ \hat{z}_{k+2}\ \cdots\ \hat{z}_n]$. Let $\hat{\mathbf{x}} = \mathbf{xD}^{-1} + [z_1\ z_2\ \cdots\ z_k]$, then $\hat{\mathbf{z}} = \hat{\mathbf{x}}\hat{\mathbf{G}} + \mathbf{e}$. From the special form of the first k components of $\hat{\mathbf{G}}$ and $\hat{\mathbf{z}}$, it is clear that $\Pr[\hat{x}_i = 1] = \Pr[e_i = 1] = \eta$. The cost of this step is dominated by the computation of $\mathbf{DG}$, which was reduced to $C_1 = (n-k)ka$ bit operations through table look-up in [15], where a is some fixed value.

Step 2. Collision Procedure. This is the BKW part with the sort-and-match technique to reduce the dependency on the information bits [5,22].

From $\hat{\mathbf{G}} = [\mathbf{I}\ \mathbf{L}_0]$, we iteratively process t steps of the BKW reduction on $\mathbf{L}_0$, resulting in a sequence of matrices $\mathbf{L}_i, i = 1, 2, \ldots, t$. Each $\mathbf{L}_i$ has $n - k - i2^b$ columns when adopting the LF1[3] type that discards about 2^b samples at each step. One also needs to update $\hat{\mathbf{z}}$ in the same fashion. Let $m = n - k - t2^b$, this procedure ends with $\mathbf{z}' = \mathbf{x}'\mathbf{G}' + \mathbf{e}'$, where $\mathbf{G}' = [\mathbf{I}\ \mathbf{L}_t]$ and $\mathbf{z}' = [\mathbf{0}\ z'_1\ z'_2 \cdots z'_m]$. The secret vector is reduced to a dimension of $k' = k - tb$, and also remains $\Pr[x'_i = 1] = \eta$ for $1 \le i \le k'$. The noise vector $\mathbf{e}' = [e_1 \cdots e_{k'}\ e'_1 \cdots e'_m]$, where $e'_i = \sum_{j \in \tau_i, |\tau_i| \le 2^t} e_j$ and τ_i contains the positions added up to form the $(k' + i)$-th column. The bias for e'_i is ϵ^{2^t} accordingly, where $\epsilon = 1 - 2\eta$. The complexity of this step is dominated by $C_2 = \sum_{i=1}^{t}(k + 1 - ib)(n - k - i2^b)$.

Step 3. Partial Secret Guessing. Divide $\mathbf{x}'$ into $[\mathbf{x}'_1\ \mathbf{x}'_2]$, accordingly divide $\mathbf{G}' = \begin{bmatrix} \mathbf{G}'_1 \\ \mathbf{G}'_2 \end{bmatrix}$, where $\mathbf{x}'_1$ is of length k_1 and $\mathbf{x}'_2$ is of length k_2 with $k' = k_1 + k_2$. This step guesses all vectors $\mathbf{x}'_1 \in \{0, 1\}^{k_1}$ that $wt(\mathbf{x}'_1) \le w_1$, where $wt(\ )$ is the Hamming weight of vectors. The complexity of this step is determined by updating $\mathbf{z}'$ with $\mathbf{z}' + \mathbf{x}'_1\mathbf{G}'_1$, denoted by $C_3 = m\sum_{i=0}^{w_1}\binom{k_1}{i}i$. The problem becomes $\mathbf{z}' = \mathbf{x}'_2\mathbf{G}'_2 + \mathbf{e}'$.

Step 4. Covering-Code. A linear covering code is used in this step to further decrease the dimension of the secret vector. Use a $[k_2, l]$ linear code $\mathcal{C}$ with covering radius $d_{\mathcal{C}}$ to rewrite any $\mathbf{g}'_i \in \mathbf{G}'_2$ as $\mathbf{g}'_i = \mathbf{c}_i + \tilde{\mathbf{e}}_i$, where $\mathbf{c}_i$ is the nearest codeword in $\mathcal{C}$ and $wt(\tilde{\mathbf{e}}_i) \le d_{\mathcal{C}}$. Let the systematic generator matrix and its parity-check matrix of $\mathcal{C}$ be $\mathbf{F}$ and $\mathbf{H}$, respectively. Then the syndrome decoding technique is applied to select the nearest codeword. The complexity is cost in calculating syndromes $\mathbf{H}\mathbf{g}_i'^T, i = 1, 2, \ldots, m$, which was recursively computed in [15], as $C_4 = (k_2 - l)(2m + 2^l)$. Thus, $z'_i = \mathbf{x}'_2\mathbf{c}_i^T + \mathbf{x}'_2\tilde{\mathbf{e}}_i^T + e'_i, i = 1, 2, \ldots, m$. But if we use a specific concatenated code, the complexity formula of the syndrome decoding step will differ, as we stated later.

In [15], $\epsilon' = (1 - 2\frac{d}{k_2})^{w_2}$ is used to determine the bias introduced by covering, where d is the expected distance bounded by the sphere-covering bound, i.e., d is the smallest integer that $\sum_{i=0}^{d}\binom{k_2}{i} > 2^{k_2 - l}$, and w_2 is an integer that bounds $wt(\mathbf{x}'_2)$. But, we find that it is not proper to consider the components of error vector $\tilde{\mathbf{e}}_i$ as independent variables, which is also pointed out in [6]. Then Bogos et. al. update the bias estimation as follows: when the code has the optimal covering radius, the bias of $\langle \mathbf{x}'_2, \tilde{\mathbf{e}}_i \rangle = 1$ assuming that $\mathbf{x}'_2$ has weight w_2 can be found according to

$$\Pr[\langle \mathbf{x}'_2, \tilde{\mathbf{e}}_i \rangle = 1 | wt(\mathbf{x}'_2) = w_2] = \frac{1}{S(k_2, d)} \sum_{i \le d,\ i \text{ odd}} \binom{c}{i} S(k_2 - w_2, d - i)$$

where $S(k_2, d)$ is the number of k_2-bit strings with weight at most d. Then the bias is computed as $\delta = 1 - 2\Pr[\langle \mathbf{x}'_2, \tilde{\mathbf{e}}_i \rangle = 1 | wt(\mathbf{x}'_2) = w_2]$, and the final

[3] With the corrected number of queries, the algorithm in [15] exceeds the security bound of 80-bit. In order to obtain a complexity smaller than 2^{80}, the LF2 reduction step is actually applied.

complexity is derived by dividing a factor of the sum of covering chunks.[4] Later based on the calculation of the bias in [6], the authors of [15] further require that the Hamming weight bound w_2 is the largest weight of $\mathbf{x}'_2$ that the bias $\tilde{\epsilon}(w_2)$ is not smaller than ϵ_{set}, where ϵ_{set} is a preset bias. Still this holds with probability.

Step 5. Subspace Hypothesis Testing. It is to count the number of equality $z'_i = \mathbf{x}'_2\mathbf{c}_i^T$ in this step. Since $\mathbf{c}_i = \mathbf{u}_i\mathbf{F}$, one can count the number of equality $z'_i = \mathbf{y}\mathbf{u}_i^T$ equivalently, for $\mathbf{y} = \mathbf{x}'_2\mathbf{F}^T$. Group the samples $(\mathbf{g}'_i, z'_i)$ in sets $\mathcal{L}(\mathbf{c}_i)$ according to the nearest codewords and define the function $f(\mathbf{c}_i) = \sum_{(\mathbf{g}'_i,z'_i)\in\mathcal{L}(\mathbf{c}_i)}(-1)^{z'_i}$ on the domain of $\mathcal{C}$. Due to the bijection between $\mathbb{F}_2^l$ and $\mathcal{C}$, define the function $g(\mathbf{u}) = f(\mathbf{c}_i)$ on the domain of $\mathbb{F}_2^l$, where $\mathbf{u}$ represents the first l bits of $\mathbf{c}_i$ for the systematic feature of $\mathbf{F}$. The Walsh transform of g is defined as $\{G(\mathbf{y})\}_{\mathbf{y}\in\mathbb{F}_2^l}$, where $G(\mathbf{y}) = \sum_{\mathbf{u}\in\mathbb{F}_2^l} g(\mathbf{u})(-1)^{\langle\mathbf{y},\mathbf{u}\rangle}$. The authors considered the best candidate as $\mathbf{y}_0 = \arg\max_{\mathbf{y}\in\mathbb{F}_2^l}|G(\mathbf{y})|$. This step calls for the complexity $C_5 = l2^l\sum_{i=0}^{w_1}\binom{k_1}{i}$, which runs for every guess of $\mathbf{x}'_1$ using the FWHT [7]. Note that if some column can be decoded into several codewords, one needs to run this step more times.

Analysis. In [15], it is claimed that it calls for approximately $m \approx 1/(\epsilon^{2^{t+1}}\epsilon'^2)$ samples to distinguish the correct guess from the others, and estimated $n \approx m + k + t2^b$ as the initial queries needed when adopting LF1 in the process. We find that this is highly underestimated. Then they correct it as $4l\ln 2/\epsilon_f^2$ in the presentation at Asiacrypt 2014, and adopt LF2 reduction steps with about $3 \cdot 2^b$ initial queries.

Recall that two assumptions are made regarding to the Hamming weight of secret vector, and it holds with probability $\Pr(w_1, k_1)\cdot \Pr(w_2, k_2)$, where $\Pr(w, k) = \sum_{i=0}^{w}(1-\eta)^{k-i}\eta^i\binom{k}{i}$ since $\Pr[x'_i = 1] = \eta$. If any assumption is invalid, one needs to choose another permutation to run the algorithm again. The authors showed the number of bit operations required for a success run of the algorithm using covering codes as

$$C = \frac{C_1 + C_2 + C_3 + C_4 + C_5}{\Pr(w_1, k_1)\Pr(w_2, k_2)}.$$

[4] We feel that there are some problems in the bias estimation in Bogos et. al. paper. In their work, the bias is computed as $1 - 2\Pr[(x, e) = 1 \mid wt(x) = c]$ with the conditional probability other than the normal probability Pr[(x,e)=1]. Note that the latter can be derived from the total probability formula by traversing all the conditions. Further, the weights of several secret chunks are assumed in a way that facilitates the analysis, which need to be divided at last to assure its occurrence. Here instead of summing up these partial conditional probabilities, they should be multiplied together. The Asiacrypt'14 paper has the similar problem in their analysis. Our theoretical derivation is different and new. We compute $\Pr[(x, e) = 1]$ according to the total probability formula strictly and thus the resultant bias precisely without any assumption, traversing all the conditional probabilities $\Pr[(x, e) = 1 \mid wt(e) = i]$.

4 Our Improvements and Analysis

In this section, we present the core improvement and optimizations of our new algorithm with complexity analysis.

4.1 Embedding Cascaded Perfect Codes

First note that in [15], the explicit code constructions for solving those LPN instances to support the claimed attacks[5] are not provided. Second, it is suspicious whether there will be a good estimation of the bias, with the assumption of Hamming weight restriction, which is crucial for the exact estimate of the complexity. Instead, here we provide a *generic* method to construct the covering codes explicitly and compute the bias accurately.

Covering code is a set of codewords in a space with the property that every element of the space is within a fixed distance to some codeword, while in particular, perfect code is a covering code of minimal size. Let us first look at the perfect codes.

Definition 2 (Perfect code [24]). *A code $\mathcal{C} \subset Q^n$ with a minimum distance $2e+1$ is called a perfect code if every $\mathbf{x} \in Q^n$ has distance $\leq e$ to exactly one codeword, where Q^n is the n-dimensional space.*

From this definition, there exists one and only one decoding codeword in the perfect code for each vector in the space[6]. It is well known that there exists only a limited kinds of the binary perfect codes, shown in Table 2. Here e is indeed the covering radius $d_{\mathcal{C}}$.

Table 2. Types of all binary perfect codes

e	n	l	*Type*
0	n	n	$\{0,1\}^n$
1	2^r-1	2^r-r-1	Hamming code
3	23	12	Golay code
e	$2e+1$	1	Repetition code
e	e	0	$\{\mathbf{0}\}$

Confined to finite types of binary perfect codes and given fixed parameters of $[k_2, l]$, now the challenge is to efficiently find the configuration of some perfect codes that maximize the bias. To solve this problem, we first divide the

[5] There is just a group of particular parameters for $(512, 1/8)$-LPN instance given in their presentation at Asiacrypt 2014, but the other LPN instances are missing.

[6] There exists exactly one decodable code word for each vector in the space, which facilitates the definition of the basic function in the Walsh transform. For other codes, the covering sphere may be overlapped, which may complicate the bias/complexity analysis in an unexpected way. It is our future work to study this problem further.

$\mathbf{G}_2'$ matrix into several chunks by rows partition, and then cover each chunk by a certain perfect code. Thereby each $\mathbf{g}_i' \in \mathbf{G}_2'$ can be uniquely decoded as $\mathbf{c}_i$ chunk by chunk. Precisely, divide $\mathbf{G}_2'$ into h sub-matrices as

$$\mathbf{G}_2' = \begin{bmatrix} \mathbf{G}_{2,1}' \\ \mathbf{G}_{2,2}' \\ \vdots \\ \mathbf{G}_{2,h}' \end{bmatrix}.$$

For each sub-matrix $\mathbf{G}_{2,j}'$, select a $[k_{2,j}, l_j]$ perfect code $\mathcal{C}_j$ with the covering radius $d_{\mathcal{C}_j}$ to regroup its columns, where $j = 1, 2, \ldots, h$. That is, $\mathbf{g}_{i,j}' = \mathbf{c}_{i,j} + \tilde{\mathbf{e}}_{i,j}$, $wt(\tilde{\mathbf{e}}_{i,j}) \leq d_{\mathcal{C}_j}$ for $\mathbf{g}_{i,j}' \in \mathbf{G}_{2,j}'$, where $\mathbf{c}_{i,j}$ is the only decoded codeword in $\mathcal{C}_j$. Then we have

$$\begin{aligned} z_i' &= \mathbf{x}_2' \mathbf{g}_i'^T + e_i' = \sum_{j=1}^{h} \mathbf{x}_{2,j}' \mathbf{g}_{i,j}'^T + e_i' \\ &= \sum_{j=1}^{h} \mathbf{x}_{2,j}' (\mathbf{c}_{i,j} + \tilde{\mathbf{e}}_{i,j})^T + e_i' = \sum_{j=1}^{h} \mathbf{x}_{2,j}' \mathbf{c}_{i,j}^T + \sum_{j=1}^{h} \mathbf{x}_{2,j}' \tilde{\mathbf{e}}_{i,j}^T + e_i', \end{aligned}$$

where $\mathbf{x}_2' = [\mathbf{x}_{2,1}', \mathbf{x}_{2,2}', \ldots, \mathbf{x}_{2,h}']$ is partitioned in the same fashion as that of $\mathbf{G}_2'$. Denote the systematic generator matrix of $\mathcal{C}_j$ by $\mathbf{F}_j$. Since $\mathbf{c}_{i,j} = \mathbf{u}_{i,j} \mathbf{F}_j$, we have

$$\begin{aligned} z_i' &= \sum_{j=1}^{h} \mathbf{x}_{2,j}' \mathbf{F}_j^T \mathbf{u}_{i,j}^T + \sum_{j=1}^{h} \mathbf{x}_{2,j}' \tilde{\mathbf{e}}_{i,j}^T + e_i' \\ &= [\mathbf{x}_{2,1}' \mathbf{F}_1^T, \mathbf{x}_{2,2}' \mathbf{F}_2^T, \ldots, \mathbf{x}_{2,h}' \mathbf{F}_h^T] \cdot \begin{bmatrix} \mathbf{u}_{i,1}^T \\ \mathbf{u}_{i,2}^T \\ \vdots \\ \mathbf{u}_{i,h}^T \end{bmatrix} + \sum_{j=1}^{h} \mathbf{x}_{2,j}' \tilde{\mathbf{e}}_{i,j}^T + e_i'. \end{aligned}$$

Let $\mathbf{y} = [\mathbf{x}_{2,1}' \mathbf{F}_1^T, \mathbf{x}_{2,2}' \mathbf{F}_2^T, \ldots, \mathbf{x}_{2,h}' \mathbf{F}_h^T]$, $\mathbf{u}_i = [\mathbf{u}_{i,1}, \mathbf{u}_{i,1}, \ldots, \mathbf{u}_{i,h}]$, and $\tilde{e}_i = \sum_{j=1}^{h} \tilde{e}_{i,j} = \sum_{j=1}^{h} \mathbf{x}_{2,j}' \tilde{\mathbf{e}}_{i,j}^T$. Then $z_i' = \mathbf{y}\mathbf{u}_i^T + \tilde{e}_i + e_i'$, which conforms to the procedure of Step 5. Actually, we can directly group $(\mathbf{g}_i', z_i')$ in the sets $\mathcal{L}(\mathbf{u})$ and define the function $g(\mathbf{u}) = \sum_{(\mathbf{g}_i', z_i') \in \mathcal{L}(\mathbf{u})} (-1)^{z_i'}$, for each $\mathbf{u}_i$ still can be read from $\mathbf{c}_i$ directly without other redundant bits due to the systematic feature of those generator matrices. According to this grouping method, each $(\mathbf{g}_i', z_i')$ belongs to only one set. Then we examine all the $\mathbf{y} \in \mathbb{F}_2^l$ by the Walsh transform $G(\mathbf{y}) = \sum_{\mathbf{u} \in \mathbb{F}_2^l} g(\mathbf{u})(-1)^{\langle \mathbf{y}, \mathbf{u} \rangle}$ and choose the best candidate.

Next, we consider the bias introduced by such a covering fashion. We find that it is reasonable to treat the error bits $\tilde{e}_{\cdot,j}$ coming from different perfect codes as independent variables, while the error components of $\tilde{\mathbf{e}}_{\cdot,j}$ within one perfect code will have correlations to each other (here we elide the first subscript i for simplicity). Thus, we need an algorithm to estimate the bias introduced by a single $[k, l]$ perfect code with the covering radius $d_{\mathcal{C}}$, denoted by bias$(k, l, d_{\mathcal{C}}, \eta)$.

Equivalently, it has to compute the probability $\Pr[\mathbf{x}\mathbf{e}^T = 1]$ at first, where $\Pr[x_i = 1] = \eta$. In order to ensure $\mathbf{x}\mathbf{e}^T = 1$, within the components equal to 1 in $\mathbf{e}$, there must be an odd number of corresponding components equal to 1 in $\mathbf{x}$, i.e., $|supp(\mathbf{x}) \cap supp(\mathbf{e})|$ is odd. Thereby for $wt(\mathbf{e}) = i, 0 \leq i \leq d_{\mathcal{C}}$, we have

$$\Pr[\mathbf{x}\mathbf{e}^T = 1 | wt(\mathbf{e}) = i] = \sum_{\substack{1 \leq j \leq i \\ j \text{ is odd}}} \eta^j (1-\eta)^{i-j} \binom{i}{j}.$$

Moreover, $\Pr[wt(\mathbf{e}) = i] = 2^l \binom{k}{i} / 2^k$, as the covering spheres are disjoint for perfect codes. We have

$$\Pr[\mathbf{x}\mathbf{e}^T = 1] = \sum_{i=0}^{d_{\mathcal{C}}} \frac{2^l \binom{k}{i}}{2^k} \left(\sum_{\substack{1 \leq j \leq i \\ j \text{ is odd}}} \eta^j (1-\eta)^{i-j} \binom{i}{j} \right).$$

Additionally,

$$\sum_{\substack{1 \leq j \leq i \\ j \text{ is even}}} \eta^j (1-\eta)^{i-j} \binom{i}{j} + \sum_{\substack{1 \leq j \leq i \\ j \text{ is odd}}} \eta^j (1-\eta)^{i-j} \binom{i}{j} = (\eta + 1 - \eta)^i,$$

$$\sum_{\substack{1 \leq j \leq i \\ j \text{ is even}}} \eta^j (1-\eta)^{i-j} \binom{i}{j} - \sum_{\substack{1 \leq j \leq i \\ j \text{ is odd}}} \eta^j (1-\eta)^{i-j} \binom{i}{j} = (1 - \eta - \eta)^i,$$

we can simplify $\sum_{1 \leq j \leq i,\ j \text{ is odd}} \eta^j (1-\eta)^{i-j} \binom{i}{j}$ as $[1 - (1-2\eta)^i]/2$. Then we derive the bias introduced by embedding the cascading as $\tilde{\epsilon} = \prod_{j=1}^{h} \epsilon_j$ according to the pilling-up lemma, where $\epsilon_j = \text{bias}(k_{2,j}, l_j, d_{\mathcal{C}_j}, \eta) = 1 - 2\Pr[\mathbf{x}_{2,j} \tilde{\mathbf{e}}_{\cdot,j}^T = 1]$ for $\mathcal{C}_j$. Note that this is an *accurate* estimation without any assumption on the hamming weights.

Now we turn to the task to search for the optimal cascaded perfect codes $\mathcal{C}_1, \mathcal{C}_2, \ldots, \mathcal{C}_h$ that will maximize the final bias, given a fixed $[k_2, l]$ pair according to the LPN instances.

Denote this process by an algorithm, called construction(k_2, l, η). First, we calculate the bias introduced by each type of perfect code exploiting the above algorithm bias($k, l, d_{\mathcal{C}}, \eta$). In particular, for Hamming code, we compute bias $(2^r - 1, 2^r - r - 1, 1, \eta)$ for $r : 2^r - 1 \leq k_2$ and $2^r - r - 1 \leq l$. For repetition code, we compute bias$(2r+1, 1, r, \eta)$ for $r : 2r + 1 \leq k_2$. We compute bias$(23, 12, 3, \eta)$ for the [23, 12] Golay code, and always have bias$(n, n, 0, \eta)$ equal to 1 for $\{0,1\}^n, n = 1, 2, \ldots$. Also it can be proved that bias$(r, 0, r, \eta) = [\text{bias}(1, 0, 1, \eta)]^r$ for any r. Second, we transform the searching problem into an integer linear programming problem. Let the number of $[2^r - 1, 2^r - r - 1]$ Hamming code be x_r and the number of $[2r+1, 1]$ repetition code be y_r in the cascading. Also let the number of [23, 12] Golay code, $[1, 0]$ code $\{0\}$ and $[1, 1]$ code $\{0, 1\}$ in the cascading be

z, v and w respectively. Then the searching problem converts into the following form.

$$\begin{cases} \sum\limits_{\substack{r:2^r-1\leq k_2 \\ 2^r-r-1\leq l}} (2^r-1)x_r + \sum\limits_{r:2r+1\leq k_2} (2r+1)y_r + 23z + v + w = k_2, \\ \sum\limits_{\substack{r:2^r-1\leq k_2 \\ 2^r-r-1\leq l}} (2^r-r-1)x_r + \sum\limits_{r:2r+1\leq k_2} y_r + 12z + w = l, \end{cases}$$

$$\max \left(\prod_{\substack{r:2^r-1\leq k_2 \\ 2^r-r-1\leq l}} \text{bias}(2^r-1, 2^r-r-1, 1, \eta)^{x_r} \right) \cdot$$

$$\cdot \left(\prod_{r:2r+1\leq k_2} \text{bias}(2r+1, 1, r, \eta)^{y_r} \right) \text{bias}(23, 12, 3, \eta)^z \text{bias}(1, 0, 1, \eta)^v.$$

We perform the logarithm operations on the target function and make it linear as

$$\max \sum_{\substack{r:2^r-1\leq k_2 \\ 2^r-r-1\leq l}} x_r \log[\text{bias}(2^r-1, 2^r-r-1, 1, \eta)] + v\log[\text{bias}(1, 0, 1, \eta)] +$$

$$+ \sum_{r:2r+1\leq k_2} y_r \log[\text{bias}(2r+1, 1, r, \eta)] + z\log[\text{bias}(23, 12, 3, \eta)].$$

Given the concrete value of η, we can provide the optimal cascaded perfect codes with fixed parameters by Maple. We present in Table 3 the optimal cascaded perfect codes with the parameters chosen in Sect. 6.4 for our improved algorithm when adopting various BKW algorithms for different LPN instances. From this table, we can find that the optimal cascaded perfect codes usually select the [23, 12] Golay code and the repetition codes with r at most 4.

It is worth noting that the above mentioned process is a *generic* method that can be applied to any (k, η)-LPN instance, and finds the optimal perfect codes

Table 3. Optimal cascaded perfect codes employed in Sect. 6.4

	LPN instances (k, η)	Parameters $[k_2, l]$	Cascaded perfect codes	h	$\log_2 \tilde{\epsilon}$
LF1	(512, 1/8)	[172, 62]	$y_2 = 9$, $y_3 = 5$, $z = 4$	18	−15.1360
	(532, 1/8)	[182, 64]	$y_2 = 11$, $y_3 = 5$, $z = 4$	20	−16.3859
	(592, 1/8)	[207, 72]	$y_3 = 8$, $y_4 = 4$, $z = 5$	17	−18.8117
LF2	(512, 1/8)	[170, 62]	$y_2 = 10$, $y_3 = 4$, $z = 4$	18	−14.7978
	(532, 1/8)	[178, 64]	$y_2 = 13$, $y_3 = 3$, $z = 4$	20	−15.7096
	(592, 1/8)	[209, 72]	$y_3 = 7$, $y_4 = 5$, $z = 5$	17	−19.1578
LF(4)	(512, 1/8)	[174, 60]	$y_2 = 1$, $y_3 = 11$, $z = 4$	16	−15.9152
	(532, 1/8)	[180, 61]	$y_2 = 2$, $y_3 = 11$, $z = 4$, $v = 1$	18	−16.7328
	(592, 1/8)	[204, 68]	$y_2 = 14$, $y_3 = 6$, $z = 4$	24	−19.2240

combination to be embedded. In the current framework, the results are optimal in the sense that the concrete code construction/the bias is optimally derived from the integer linear programming.

4.2 Data Complexity Analysis

In this section, we present an analysis of the accurate number of queries needed for choosing the best candidate in details. We first point out the distinguisher statistic S between the two distributions corresponding to the correct guess and the others. It is obvious to see that S obeys $Ber_{\frac{1}{2}(1-\epsilon_f)}$ if $\mathbf{y}$ is correct, thus we deduce $\Pr[S_i = 1] = \frac{1}{2}(1-\epsilon_f) = \Pr[z'_i \neq \mathbf{y}\mathbf{u}_i^T]$, where $\epsilon_f = \epsilon^{2^t}\epsilon'$ indicates the bias of the final noise for simplicity. Since $G(\mathbf{y})$ calculates the difference between the number of equalities and inequalities, we have $S = \sum_{i=1}^{m} S_i = \frac{1}{2}(m - G(\mathbf{y}))$. It is clear that the number of inequalities should be minimum if $\mathbf{y}$ is correct. Thus the best candidate is $\mathbf{y}_0 = \arg\min_{\mathbf{y}\in\mathbb{F}_2^l} S = \arg\max_{\mathbf{y}\in\mathbb{F}_2^l} G(\mathbf{y})$, rather than $\arg\max_{\mathbf{y}\in\mathbb{F}_2^l} |G(\mathbf{y})|$ claimed in [15]. Then, let $X_{A=B}$ be the indicator function of equality. Rewrite $S_i = X_{z'_i=\mathbf{y}\mathbf{u}_i^T}$ as usual. Then S_i is drawn from $Ber_{\frac{1}{2}(1+\epsilon_f)}$ if $\mathbf{y}$ is correct and $Ber_{\frac{1}{2}}$ if $\mathbf{y}$ is wrong, which is considered to be random. Take $S = \sum_{i=1}^{m} S_i$, we consider the ranking procedure for each possible $\mathbf{y} \in \mathbb{F}_2^l$ according to the decreasing order of the grade $S_{\mathbf{y}}$.

Let $\mathbf{y}_r$ denote the correct guess, and $\mathbf{y}_w$ otherwise. Given the independency assumption and the central limit theorem, we have

$$\frac{S_{\mathbf{y}_r} - \frac{1}{2}(1+\epsilon_f)m}{\sqrt{\frac{1}{2}(1-\epsilon_f)\frac{1}{2}(1+\epsilon_f)m}} \sim \mathcal{N}(0,1), \text{ and } \frac{S_{\mathbf{y}_w} - \frac{1}{2}m}{\frac{1}{2}\sqrt{m}} \sim \mathcal{N}(0,1),$$

where $\mathcal{N}(\mu, \sigma^2)$ is the normal distribution with the expectation μ and variance σ^2. Thus we can derive $S_{\mathbf{y}_r} \sim \mathcal{N}(\frac{1}{2}(1+\epsilon_f)m, \frac{1}{4}(1-\epsilon_f^2)m)$ and $S_{\mathbf{y}_w} \sim \mathcal{N}(\frac{m}{2}, \frac{m}{4})$. According to the additivity property of normal distributions, $S_{\mathbf{y}_r} - S_{\mathbf{y}_w} \sim \mathcal{N}(\frac{1}{2}\epsilon_f m, \frac{1}{4}(2-\epsilon_f^2)m)$. Therefore, we obtain the probability that a wrong $\mathbf{y}_w$ has a better rank than the right $\mathbf{y}_r$, i.e., $S_{\mathbf{y}_r} < S_{\mathbf{y}_w}$ is approximately $\Phi\left(-\sqrt{\epsilon_f^2 m/(2-\epsilon_f^2)}\right)$, where $\Phi(\cdot)$ is the distribution function of the standard normal distribution. Let $\rho = \epsilon_f^2 m/(2-\epsilon_f^2) \approx \frac{1}{2}\epsilon_f^2 m$, and this probability becomes $\Phi(-\sqrt{\rho}) \approx e^{-\rho/2}/\sqrt{2\pi}$. Since we just select the best candidate, i.e., $S_{\mathbf{y}_r}$ should rank the highest to be chosen. Thus $S_{\mathbf{y}_r}$ gets the highest grade with probability approximatively equal to $(1-\Pr[S_{\mathbf{y}_r} < S_{\mathbf{y}_w}])^{2^l-1} \approx \exp(-2^l e^{-\rho/2}/\sqrt{2\pi})$. It is necessary to have $2^l \leq e^{\rho/2}$, i.e., at least $m \geq 4l\ln 2/\epsilon_f^2$ to make the probability high. So far, we have derived the number of queries used by the authors of [15] in their presentation at Asiacrypt 2014.

Furthermore, we have made extensive experiments to check the real success probability according to different multiples of the queries. The simulations show that $m = 4l\ln 2/\epsilon_f^2$ provides a success probability of about 70 %, while for

$m = 8l\ln 2/\epsilon_f^2$ the success rate is closed to 1[7]. To be consistent with the practical experiments, we finally estimate m as $8l\ln 2/\epsilon_f^2$ hereafter. Updating the complexities for solving different LPN instances in [15] with the $m = 8l\ln 2/\epsilon_f^2$ number of queries, the results reveal that the algorithm in [15] is *not* so valid to break the 80-bit security bound.

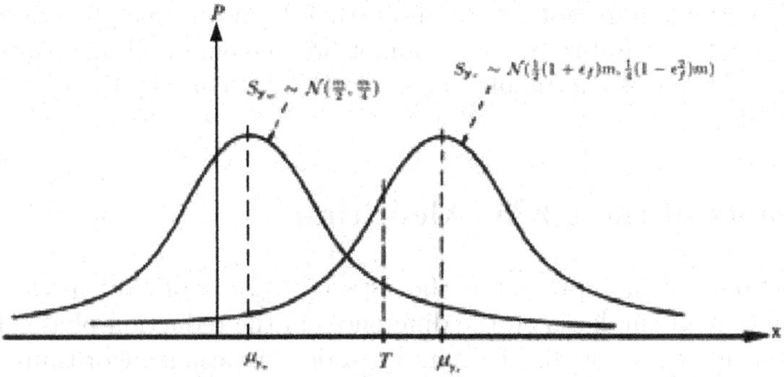

Fig. 1. Distributions according to $\mathbf{y}_w$ and $\mathbf{y}_r$

In addition, in [15] there is a regrettable missing of the concrete terminal condition. It was said that a false alarm can be recognized by hypothesis testing, without the derivation of the specific threshold. To ensure the completeness of our improved algorithm, we solve this problem as follows. Denote the threshold by T, and adopt the selected best candidate $\mathbf{y}$ as correct if $S_{\mathbf{y}} \geq T$. The density functions of the corresponding distributions according to $\mathbf{y}_w$ and $\mathbf{y}_r$ are depicted in Fig. 1, respectively. Then it is clear to see that the probability for a false alarm is $P_f = \Pr[S_{\mathbf{y}} \geq T | \mathbf{y} \text{ is wrong}]$. It is easy to estimate P_f as $1 - \Phi(\lambda)$, where $\lambda = (T - \frac{m}{2})/\sqrt{\frac{m}{4}}$. Following our improvements described in Sect. 4.1, there is no assumption on the weight of $\mathbf{x}_2'$ now. The restricted condition is that the expected number of false alarms over all the $(2^l \sum_{i=0}^{w_1} \binom{k_1}{i})/\Pr(w_1, k_1)$ basic tests is lower than 1. Thus we derive $\lambda = -\Phi^{-1}\left(\frac{\Pr(w_1,k_1)}{2^l \sum_{i=0}^{w_1} \binom{k_1}{i}}\right)$ and the algorithm terminal condition is $S_{\mathbf{y}} \geq T$, i.e., $G(\mathbf{y}) \geq \lambda\sqrt{m}$, for $S_{\mathbf{y}} = \frac{1}{2}(m+G(\mathbf{y}))$ according to the definition above.

4.3 An Vital Flaw in [15] that May Affect the Ultimate Results

As stated in [15], it took an optimized approach to calculate $\mathbf{D}\mathbf{g}^T$ for each column in $\mathbf{G}$. Concretely, for a fixed value s, divide the matrix $\mathbf{D}$ into $a = \lceil k/s \rceil$ parts, i.e., $\mathbf{D} = [\mathbf{D}_1, \mathbf{D}_2, \ldots, \mathbf{D}_a]$, each sub-matrix containing s columns

[7] It is also analyzed that it calls for a number of $8l\ln 2/\epsilon_f^2$ to bound the failure probability in [6]. However, it uses the Hoeffding inequality, which is the different analysis method from ours.

(possibly except the last one). Then store all the possible values of $\mathbf{D}_i\mathbf{x}^T$ for $\mathbf{x} \in \mathbb{F}_2^s$ in tables indexed by $i = 1, 2, \ldots, a$. For a vector $\mathbf{g} = [\mathbf{g}_1, \mathbf{g}_2, \cdots, \mathbf{g}_a]$ partitioned according to $\mathbf{D}$, we have $\mathbf{D}\mathbf{g}^T = \mathbf{D}_1\mathbf{g}_1^T + \mathbf{D}_2\mathbf{g}_2^T + \cdots + \mathbf{D}_a\mathbf{g}_a^T$, where $\mathbf{D}_i\mathbf{g}_i^T$ can be read directly from the stored tables. The complexity of this step is to add those intermediate results together to derive the final result, shown as C_1 in Sect. 3. It was stated that the cost of constructing the tables is about $O(2^s)$, which can be negligible. Since the matrix $\mathbf{D}$ can only be obtained from the online querying and then refreshed for each iteration, this procedure should be reprocessed for each iteration and cannot be pre-computed in advance in the offline phase. Thus it is reasonable to include $PC_1/(\Pr(w_1, k_1)\Pr(w_2, k_2))$ in the overall complexity.

5 Variants of the BKW Algorithm

In this section, we first present a theoretical analysis of the previous BKW algorithms with an emphasis on the differences in the reduction phase. Then we extend the heuristic algorithm LF2 into a series of variant algorithms denoted by LF(κ), as a basis of the improved algorithm proposed in Sect. 6. Furthermore, we verify the performance of these BKW algorithms in experiments

5.1 LF1

LF1 works as follows. Choose a representative in each partition, add it to the rest of samples in the same partition and at last discard the representative, shown in the Algorithm 2 in Sect. 2.2. It is commonly believed that LF1 has no heuristics, and follows a rigorous analysis of its correctness and performance in theory. However, having checked the proof in [22], we find that the authors have overlooked the fact that the noise bits are no more independent after performing the XOR operations among the pairs of queries, which can be easily examined in the small instances. Thus there is no reason in theory to apply the pilling-up lemma for calculating the bias as shown in the proof. Thereby there is no need to treat it superior to other heuristic algorithms for the claimed strict proof.

Fortunately, by implementing LF1 algorithm, we find that the dependency does not affect the performance of the algorithm, shown in the Table 4. That is, the number of queries in theory with the independency assumption supports the corresponding success rate in practice. Thus we keep the independence assumption for the noise bits hereafter.

5.2 LF2

LF2 computes the sum of pairs from the same partition, shown in Algorithm 3 in Sect. 2.2. LF2 is more efficient and allows fewer queries compared to LF1. Let $n[i], i = 1, 2, \ldots, t$ be the excepted number of samples via the i-th BKW step, and $n[0] = n$. We impose a restriction that $n[i]$ is not larger than n for any i to LF2 with the following considerations. One is to control the dependence within certain limits, another is to preserve the number of samples not overgrowing, which will also stress on the complexity otherwise. The simulations done with the parameters under the restriction confirm the performance of LF2 shown in Table 4, and encounter with the statement that the operations of every pair have no visible effect on the success rate of the algorithm in [22].

5.3 Variants: LF(κ)

Here we propose a series of variants of the BKW algorithm, called LF(κ), which not only consider pairs of columns, but also consider κ-tuples that add to 0 in the last b entries. We describe the algorithm as follows. It is easy to see that the number of tuples satisfying the condition has an expectation of $E = \binom{n}{\kappa}2^{-b}$, given the birthday paradox. Similarly, define the excepted number of samples via the i-th BKW step as $n[i], i = 1, 2, \ldots, t$. We have $n[i] = \binom{n[i-1]}{\kappa}2^{-b}$, which also applies to LF2 when $\kappa = 2$. We still impose the restriction that $n[i]$ is not larger than n for any i. The bias introduced by the variant decreases as ϵ^{κ^t}. We have implemented and run the variant algorithm for $\kappa = 3, 4$ under the data restriction, and verified the validity of these algorithms, shown in Table 4.

Algorithm 5. Reduction of LF(κ)

1: Find sufficient κ-tuples from $\mathbf{G}_{i-1}$ that add to 0 in the last b entries.
2: **for** each κ-tuple **do**
3: Calculate the sum of κ-tuple, joint it into $\mathbf{G}_i$ after discarding its last b bits of 0.
4: **end for**

The extra cost of these variant BKW algorithms is to find a number of $n[i]$ such κ-tuples at each step. It is fortunate that this is the same as the κ-sum problem investigated in [25], which stated that the κ-sum problem for a single solution can be solved in $\kappa 2^{b/(1+\lfloor \log_2 \kappa \rfloor)}$ time and space [25]; moreover, one can find $n[i]$ solutions to the κ-sum problem with $n[i]^{1/(1+\lfloor \log_2 \kappa \rfloor)}$ times of the work for a single solution, as long as $n[i] \leq 2^{b/\lfloor \log_2 \kappa \rfloor}$ [25]. Thus this procedure of the variant BKW algorithms adds a complexity of $\kappa(2^b n[i])^{1/(1+\lfloor \log_2 \kappa \rfloor)}$ in time at each step, and $\kappa 2^{b/(1+\lfloor \log_2 \kappa \rfloor)}$ in space. Additionally, it stated that the lower bound of the computational complexity of κ-sum problem is $2^{b/\kappa}$. Thus it is possible to remove the limitation of the extra cost from the variant BKW algorithms if a better algorithm for κ-sum problem is proposed.

In Sect. 6, we present the results of the improved algorithms by embedding optimal cascaded perfect codes, which adopt LF1, LF2 and LF(4) at Step 2 respectively. We choose $\kappa = 4$ when adopting the variant BKW algorithm for the following reasons. If κ increases, the bias ϵ^{κ^t} introduced by LF(κ) falls sharply, and then we cannot find effective attack parameters. Since particularly stressed in [25] it takes a complexity of $2^{b/3}$ in time and space for a single solution when $\kappa = 4$, it calls for an extra cost of $(2^b n[i])^{1/3}$ in time at each step and

$2^{b/3}$ in space when adopting the variant LF(4)[8]. Additionally, since the birthday paradox that $n[i] = (\binom{n[i-1]}{4}) \cdot 2^{-b}$, we derive that $n[i] = \frac{n[i-1]^4}{4!} \cdot 2^{-b}$, then $n[i-1] = \left(4! \cdot 2^b \cdot n[i]\right)^{1/4}$.

5.4 Simulations

We have checked the performance as well as the correctness of the heuristic assumption by extensive simulations shown below. The experimental data adopt the numbers of queries estimated in theory, and obtain the corresponding success rate. In general, the simulations confirmed the validity of our theoretical analysis and the heuristic assumption, though the simulations are just conducted on some small versions of (k, η) for the limitation of computing power.

Table 4. Simulation data

LF1

Problem	Parameters				Results	
η	k	t	b	l	$\log_2 n$	success rate
0.01	60	5	10	10	12.38	98/100
0.02	65	5	12	5	14.35	92/100
0.05	60	4	12	12	14.16	96/100
0.10	70	4	15	10	17.62	88/100
0.15	40	3	10	10	15.27	99/100
0.20	50	3	14	8	18.66	92/100
0.30	65	2	20	25	22.14	84/100

LF2

Problem	Parameters				Results	
η	k	t	b	l	$\log_2 n$	success rate
0.01	60	5	9	15	9.95	96/100
0.02	50	5	9	5	9.96	89/100
0.05	55	4	11	11	11.93	99/100
0.10	70	4	16	6	16.90	91/100
0.15	55	3	15	10	16.75	97/100
0.20	60	3	18	6	19.76	90/100
0.30	60	2	18	24	19.66	80/100

LF(3)

Problem	Parameters				Results	
η	k	t	b	l	$\log_2 n$	success rate
0.01	60	3	14	18	8.30	98/100
0.05	80	3	25	5	13.76	94/100
0.10	55	2	22	11	12.85	90/100
0.15	60	2	27	6	15.34	93/100
0.20	40	1	20	20	12.07	90/100

LF(4)

Problem	Parameters				Results	
η	k	t	b	l	$\log_2 n$	success rate
0.02	65	2	22	21	8.87	95/100
0.05	70	2	29	12	11.38	92/100
0.15	50	1	29	21	12.14	90/100
0.20	45	1	32	13	13.16	91/100

[8] Can LF(4) be equivalent to two consecutive LF2 steps? The answer is no. We can illustrate this in the aspect of the number of vectors remained. If we adopt LF(4) one step, then the number of vectors remained is about $s = \binom{n}{4} \cdot 2^{-b} = \frac{n^4}{4!} \cdot 2^{-b}$. While if we first to reduce the last b_1 positions with LF2, then the number of vectors remained is $t_1 = \binom{n}{2} \cdot 2^{-b_1} = \frac{n^2}{2} \cdot 2^{-b_1}$. Then, we do the second LF2 step regarding to the next $b - b_1$ positions and the number of vectors remained changes into $t_2 = \binom{t_1}{2} \cdot 2^{-(b-b_1)} = \frac{(n^2/2 \cdot 2^{-b_1})^2}{2} \cdot 2^{-(b-b_1)} = \frac{n^4}{8} \cdot 2^{-b-b_1}$. We can see that s is obviously not equal to t_2, so they are not equivalent. Indeed, LF(k) is algorithmically converted into several LF2, the point here is that we need to run the process a suitable number of times to find a sufficient number of samples.

5.5 A Technique to Reduce the Data Complexity

We briefly present a technique that applies to the last reduction step in LF2, aiming to reduce the number of queries. For simplicity, we still denote the samples at last step by $(\mathbf{g}_{t-1}, z)$, which accords with $z = \mathbf{x}\mathbf{g}_{t-1}^T + e$ and consists of k dimensions. Let the number of samples remained before running the last reduction step be n_{t-1} and the aim is to further reduce b dimensions at this step, i.e., $l = k - b$ for solving. This step runs as: (1) Sort the samples according to the last a bits of $\mathbf{g}_{t-1}$ $(a \leq b)$, and merge pairs from the same partition to generate the new samples $(\mathbf{g'}_{t-1}, z')$. (2) Sort $(\mathbf{g'}_{t-1}, z')$ into about $B = 2^{b-a}$ groups according to the middle $b - a$ bits of $\mathbf{g'}_{t-1}$ with each group containing about $m = \binom{n}{2}2^{-b}$ new samples. In each group, the value of $\mathbf{x}_{[l+1,l+a]}\mathbf{g}_{[l+1,l+a]}'^T$ is constant given an assumed value of $\mathbf{x}_{[l+1,l+a]}$, denoted by α. Here $\mathbf{x}_{[a,b]}$ means the interval of components within the vector $\mathbf{x}$. Thus in each group, the bias (denoted by ϵ) for the approximation of $z' = \mathbf{x}_{[1,l]}\mathbf{g}_{[1,l]}'^T$ contains the same sign, which may be different from the real one, for $z' = \mathbf{x}_{[1,l]}\mathbf{g}_{[1,l]}'^T + \alpha + e'$ according to the assumed value of $\mathbf{x}_{[l+1,l+a]}$.

To eliminate the influence of the different signs among the different groups, i.e., no matter what value of $\mathbf{x}_{[l+1,l+a]}$ is assumed, we take χ^2-test to distinguish the correct $\mathbf{x}_{[1,l]}$. For each group, let the number of equalities be m_0. Then we have $S_i = (m_0 - \frac{m}{2})^2/(\frac{m}{2}) + (m - m_0 - \frac{m}{2})^2/(\frac{m}{2}) = W^2/m$, to estimate the distance between the distributions according to the correct $\mathbf{x}_{[1,l]}$ and wrong candidates considered as random, where W is the value of the Walsh transform for each $\mathbf{x}_{[1,l]}$. If $\mathbf{x}_{[1,l]}$ is correct, then $S_i \sim \chi_1^2(m\epsilon^2)$; otherwise, $S_i \sim \chi_1^2$, referred to [9]. Here, χ_M^2 denotes the χ^2-distribution with M degrees of freedom, whose mean is M and the variance is $2M$; and $\chi_M^2(\xi)$ is the non-central χ^2-distribution with the mean of $\xi + M$ and variance of $2(2\xi + M)$. Moreover, If $M > 30$, we have the approximation $\chi_M^2(\xi) \sim \mathcal{N}(\xi + M, 2(2\xi + M))$ [8]. We assume that the S_is are independent, and have the statistic $S = \sum_{j=1}^{B} S_i$. If $\mathbf{x}_{[1,l]}$ is correct, then $S \sim \chi_B^2(Bm\epsilon^2) \sim \mathcal{N}(B(1+\epsilon^2 m), 2B(1+2\epsilon^2 m)$; otherwise, $S \sim \chi_B^2 \sim \mathcal{N}(B, 2B)$, when $B > 30$, according to the additivity property of χ^2-distribution. Hereto, we exploit a similar analysis as Sect. 4.1 to estimate the queries needed to make the correct $\mathbf{x}_{[1,l]}$ rank highest according to the grade S, and result in $m \geq \frac{4l\ln 2}{B\epsilon^2}\left(1 + \sqrt{1 + \frac{B}{2l\ln 2}}\right)$ by solving a quadratic equation in m. Further amplify the result as $m > \frac{4l\ln 2}{\sqrt{B}\epsilon^2}$ and we find that this number decreases to a $\sqrt{B}$-th of the original one with the success rate close to 1. We have simulated this procedure for the independence assumption, and the experimental data verified the pprocess. This technique allows us to overcome the lack of the data queries, while at the expense of some increase of the time complexity, thus we do not adopt it in the following Sect. 6.

6 Faster Algorithm with the Embedded Perfect Codes

In this section, we develop the improved algorithm for solving LPN with the key ideas introduced in the above Sects. 4 and 5. Our improved algorithm for solving

LPN has a similar framework as that in [15], but exploits a precise embedding of cascaded concrete perfect codes at Step 4 and optimizes each step with a series of techniques. The notations here follow those in Sect. 3.

We first formalize our improved algorithm in high level according to the framework, shown in Algorithm 6. We start by including an additional step to select the favorable queries.

Algorithm 6. Improved Algorithm with the embedding cascaded perfect codes

---

Input: (k, η)-LPN instance of N queries, algorithm parameters $c, s, u, t, b, f, l, k_2, w_1$.

Select n samples that the last c bits of $\mathbf{g}$ all equal 0 from the initial queries, and store those selected z and $\mathbf{g}$ without the last c bits of 0.

Run the algorithm construction(k_2, l, η) to generate the optimal cascaded perfect codes and deduce the bias $\tilde{\epsilon}$ introduced by this embedding.

repeat

  Pick a random column permutation π and perform Gaussian elimination on $\pi(\mathbf{G})$ to derive $[\mathbf{I}\ \mathbf{L}_0]$;

  for $i = 1$ to t **do**

    Perform the BKW reduction step (LF1, LF2, LF(4)) on $\mathbf{L}_{i-1}$ resulting in $\mathbf{L}_i$.

  end for

  Based on the cascading, group the columns of $\mathbf{L}_t$ by the last k_2 bits according to their nearest codewords chunk by chunk.

  Set $k_1 = k - c - tb - k_2$;

  for $\mathbf{x}'_1 \in \{0,1\}^{k_1}$ with $wt(\mathbf{x}'_1) \leq w_1$ **do**

    Update the observed samples.

    Use FWHT to compute the numbers of 1s and 0s for each $\mathbf{y} \in \{0,1\}^l$, and pick the best candidate.

    Perform hypothesis testing with a threshold.

  end for

until: Acceptable hypothesis is found.

Step 0. Sample Selection. We select the data queries in order to transfer the (k, η)-LPN problem into a $(k - c, \eta)$-LPN problem compulsively. It works as follows. Just take the samples that the last c entries of $\mathbf{g}$ all equal 0 from the initial queries. Thus we reduce the dimension of the secret vector $\mathbf{x}$ by c-bit accordingly, while the number of initial queries needed may increase by a factor of 2^c. We only store these selected z and $\mathbf{g}$ without the last c bits of 0 to form the new queries of the reduced $(k - c, \eta)$-LPN instance, and for simplicity still denoted by $(\mathbf{g}, z)$. Note that the parameter k used hereafter is actually $k - c$, and we do not substitute it for a clear comparison with [15].

As just stated, the number of initial queries is $N = 2^c n$. To search for the desirable samples that the last c entries of $\mathbf{g}$ being 0, we can just search for the samples that the Hamming weight of its last c entries equals to 0 with a complexity of $\log_2 c$ [17]. But, this can be sped up by pre-computing a small table of Hamming weight, and look up the table within a unit constant time $O(1)$.

The pre-computation time and space can be ignored compared to those of the other procedures, for usually c is quite small. Thus, the complexity of this step is about $C_0 = N$.

Step 1. Gaussian Elimination. This step is still to change the position distribution of the coordinates of the secret vector and is similar as that described in Sect. 3. Here we present several improvements of the dominant calculation of the matrix multiplication $\mathbf{DG}$.

In [15], this multiplication is optimized by table look-up as $\mathbf{D}\mathbf{g}^T = \mathbf{D}_1\mathbf{g}_1^T + \mathbf{D}_2\mathbf{g}_2^T + \cdots + \mathbf{D}_a\mathbf{g}_a^T$, now we further improve the procedure of constructing each table $\mathbf{D}_i\mathbf{x}^T$ for $\mathbf{x} \in \mathbb{F}_2^s$ by computing the products in a certain order. It is easy to see that the product $\mathbf{D}_i\mathbf{x}^T$ is a linear combination of some columns in $\mathbf{D}_i$. We partition $\mathbf{x} \in \mathbb{F}_2^s$ according to its weight. For $\mathbf{x}$ with $wt(\mathbf{x}) = 0$, $\mathbf{D}_i\mathbf{x}^T = 0$. For all $\mathbf{x}$ with $wt(\mathbf{x}) = 1$, we can directly read the corresponding column. Then for any $\mathbf{x}$ with $wt(\mathbf{x}) = 2$, there must be a $\mathbf{x}'$ with $wt(\mathbf{x}) = 1$ such that $wt(\mathbf{x} + \mathbf{x}') = 1$. Namely, we just need to add one column to the already obtained product $\mathbf{D}_i\mathbf{x}'^T$, which is within k bit operations. Inductively, for any $\mathbf{x}$ with $wt(\mathbf{x}) = w$, there must be a $\mathbf{x}'$ with $wt(\mathbf{x}') = w - 1$ such that $wt(\mathbf{x} + \mathbf{x}') = 1$. Thus, the cost of constructing one table $\mathbf{D}_i\mathbf{x}^T, \mathbf{x} \in \mathbb{F}_2^s$ can be reduced to $k\sum_{i=2}^{s}\binom{s}{i} = (2^s - s - 1)k$. The total complexity is $PC_{11} = (2^s - s - 1)ka$ for a tables, which is much lower than the original $k^2 2^s$. We also analyze the memory needed for storing the tables $[\mathbf{x}, \mathbf{D}_i\mathbf{x}^T]_{\mathbf{x}\in\mathbb{F}_2^s}, i = 1, 2, \ldots, a$, which is $M_{11} = 2^s(s + k)a$.

Next, we present an optimization of the second table look-up to sum up the a columns, each of which has k dimensions. Based on the direct reading from the above tables, i.e., $\mathbf{D}_1\mathbf{g}_1^T + \mathbf{D}_2\mathbf{g}_2^T + \cdots + \mathbf{D}_a\mathbf{g}_a^T$, this addition can be divided into $\lceil k/u \rceil$ additions of a u-dimensional columns, depicted in the following schematic (Fig. 2).

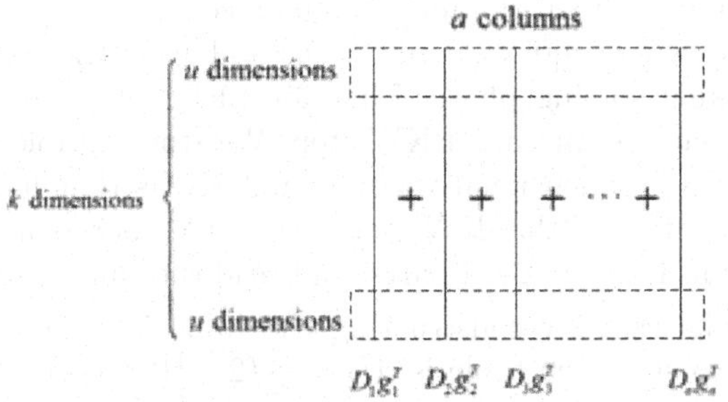

Fig. 2. Schematic of addition

We store a table of all the possible additions of a u-dimensional vectors and read it $\lceil k/u \rceil$ times to compose the sum of a k-dimensional vectors required. Thus the complexity of Step 1 can be reduced to $C_1 = (n - k)(a + \lceil k/u \rceil)$ from $(n - k)ak$.

Now we consider the cost for constructing the tables of all the possible additions of a u-dimensional vectors. It will cost $2^{ua}u(a - 1)$ bit operations by the simple exhaustive enumeration. We optimize this procedure as follows. It is true that any addition of a u-dimensional repeatable vectors can be reduced to the addition of less than or equal to a u-dimensional distinct nonzero vectors, for the sum of even number of same vectors equals $\mathbf{0}$. Thus, the problem transforms into the one that enumerates all the additions of i distinct nonzero u-dimensional vectors, where $i = 2, \ldots, a$. We find that every nonzero vector appears $\binom{2^u-1-1}{i-1}$ times in the enumeration of all the additions of the i distinct nonzero vectors. Then the total number of nonzero components of the vectors in the enumeration for i can be the upper bound of the bit operations for the addition, i.e., $\leq \binom{2^u-2}{i-1} \cdot \sum_{j=1}^{u} \binom{u}{j} j$ bit operations. Moreover, each nonzero vector appears the same number of times in the sums of the enumeration, which can be bounded by $\binom{2^u-1}{i}/(2^u - 1)$. We store the vectors as storing sparse matrix expertly and the memory required for this table is confined to $\left[\binom{2^u-2}{i-1} + \binom{2^u-1}{i}/(2^u-1)\right] \cdot \left[\sum_{j=1}^{u} \binom{u}{j} j\right]$. We can simply $\sum_{j=1}^{u} \binom{u}{j} j$ as $\sum_{j=0}^{u-1} \binom{u-1}{j-1} u = u2^{u-1}$. Thus the total complexity for constructing this table is $PC_{12} = \sum_{i=2}^{a} u2^{u-1}\binom{2^u-2}{i-1}$ in time and $M_{12} = \sum_{i=2}^{a} \frac{i+1}{i} u2^{u-1}\binom{2^u-2}{i-1}$ in memory. For each addition of a u-dimensional columns derived from the original a k-dimensional columns, we discard all the even number of reduplicative columns and read the sum from the table. Moreover, this table can be pre-computed in the off-line phase and applied to each iteration. Since the table here is in the similar size to that used at Step 1 in [15], we still consider the complexity of the first table look-up as $O(1)$, which is the same as that in [15]. This step has another improved method in [3]

Step 2. Collision Procedure. This step still exploits the BKW reduction to make the length of the secret vector shorter. As stated in Sect. 5, we adopt the reduction mode of LF1, LF2 and LF(4) to this step, respectively. Similarly, denote the expected number of samples remained via the i-th BKW step by $n[i], i = 0, 1, \ldots, t$, where $n[0] = n, n[t] = m$ and m is the number of queries required for the final solving phase. First for LF1, $n[i] = n - k - i2^b$, as it discards about 2^b samples at each BKW step. We store a table of all the possible additions of two f-dimensional vectors similarly as that in Step 1. For the merging procedure at the i-th BKW step, we divide each pair of $(k+1-ib)$-dimensional columns into $\lceil \frac{k+1-ib}{f} \rceil$ parts, and read the sum of each part directly from the table. The cost for constructing the table is $PC_2 = f2^{f-1}(2^f - 2)$ and the memory to store the table is $M_2 = \frac{3}{2} f2^{f-1}(2^f - 2)$. Then the cost of this step is $C_2 = \sum_{i=1}^{t} \lceil \frac{k+1-ib}{f} \rceil (n - k - i2^b)$ for LF1, for the samples remained indicated the pairs found.

Second for LF2, we have $n[i] = \binom{n[i-1]}{2} 2^{-b}$ following the birthday paradox. We still do the merging as LF1 above and it calls for a cost of $C_2 = \sum_{i=1}^{t} \lceil \frac{k+1-ib}{f} \rceil n[i]$, and also a pre-computation of $PC_2 = f2^{f-1}(2^f - 2)$ in time and $M_2 = \frac{3}{2} f2^{f-1}(2^f - 2)$ in memory.

Third for LF(4), we similarly have $n[i] = \binom{n[i-1]}{4} 2^{-b}$. We need to store a table of all the possible additions of four f-dimensional vectors. For the merging procedure at the i-th BKW step, we divide each 4-tuple of $(k+1-ib)$-dimensional columns into $\lceil \frac{k+1-ib}{f} \rceil$ parts, and read the sum of each part directly from the table. The cost of constructing the table is $PC_2 = \sum_{i=2}^{4} f 2^{f-1} \binom{2^f-2}{i-1}$ and the memory to store the table is $\sum_{i=2}^{4} \frac{i+1}{i} f 2^{f-1} \binom{2^f-2}{i-1}$. Additionally, there is still one more cost for finding the 4-tuples that add to 0 in the last b entries. This procedure has a cost of $(2^b n[i])^{1/3}$ as stated in Sect. 5.3. Hence, Step 2 has the complexity of $C_2 = \sum_{i=1}^{t} \left(\lceil \frac{k+1-ib}{f} \rceil n[i] + (2^b n[i])^{1/3} \right)$ for LF(4). Moreover, it needs another memory of $2^{b/3}$ to search for the tuples, i.e., $M_2 = \sum_{i=2}^{4} \frac{i+1}{i} f 2^{f-1} \binom{2^f-2}{i-1} + 2^{b/3}$ for LF(4).

Step 3. Partial Secret Guessing. It still guesses all the vectors $\mathbf{x}_1' \in \{0,1\}^{k_1}$ that $wt(\mathbf{x}_1') \leq w_1$ and updates $\mathbf{z}'$ with $\mathbf{z}' + \mathbf{x}_1'\mathbf{G}_1'$ at this step. We can optimize this step by the same technique used at Step 1 for multiplication. Concretely, the product $\mathbf{x}_1'\mathbf{G}_1'$ is a linear combination of some rows in $\mathbf{G}_1'$. We calculate these linear combinations in the increasing order of $wt(\mathbf{x}_1')$. For $wt(\mathbf{x}_1') = 0$, $\mathbf{x}_1'\mathbf{G}_1' = 0$ and $\mathbf{z}'$ does not change. For all the $\mathbf{x}_1'$ with $wt(\mathbf{x}_1') = 1$, we calculate the sum of $\mathbf{z}'$ and the corresponding row in $\mathbf{G}_1'$, which costs a complexity of m. Inductively, for any $\mathbf{x}_1'$ with $wt(\mathbf{x}_1') = w$, there must be another $\mathbf{x}_1'$ of weight $w-1$ such that the weight of their sum equals 1, and the cost for calculating this sum based on the former result is m. Thus the cost of this step can be reduced from $m\sum_{i=0}^{w_1} \binom{k_1}{i} i$ to $C_3 = m\sum_{i=1}^{w_1} \binom{k_1}{i}$.

Step 4. Covering-Coding. This step still works as covering each $\mathbf{g}_i' \in \mathbf{G}_2'$ with some fixed code to reduce the dimension of the secret vector. The difference is that we propose the explicit code constructions of the optimal cascaded perfect codes. According to the analysis in Sect. 6.1, we have already known the specific constructing method, the exact bias introduced by this constructing method and covering fashion, thus we do not repeat it here. One point to illustrate is how to optimize the process of selecting the codeword for each $\mathbf{g}_i' \in \mathbf{G}_2'$ online. From Table 3, we find that the perfect code of the longest length chosen for those LPN instances is the [23, 12] Golay code. Thus we can pre-compute and store the nearest codewords for all the vectors in the space corresponding to each perfect code with a small complexity, and read it for each part of $\mathbf{g}_i'$ online directly. Here we take the [23, 12] Golay code as an example, and the complexity for the other cascaded perfect codes can be ignored by taking into consideration their small scale of code length. Let $\mathbf{H}$ be the parity-check matrix of the [23, 12] Golay code corresponding to its systematic generator matrix. We calculate all the syndromes $\mathbf{H}\mathbf{g}^T$ for $\mathbf{g} \in \mathbb{F}_2^{23}$ in a complexity of $PC_4 = 11 \cdot 23 \cdot 2^{23}$. Similarly, it can be further reduced by calculating them in an order of the increasing weight and also for the reason that the last 11 columns of $\mathbf{H}$ construct an identity matrix. Based on the syndrome decoding table of the [23, 12] Golay code, we find the corresponding error vector $\mathbf{e}$ to $\mathbf{H}\mathbf{g}^T$ since $\mathbf{H}\mathbf{g}^T = \mathbf{H}(\mathbf{c}^T + \mathbf{e}^T) = \mathbf{H}\mathbf{e}^T$, and derive the codeword $\mathbf{c} = \mathbf{g} + \mathbf{e}$. We also obtain $\mathbf{u}$, which is the first 12 bits of $\mathbf{c}$. We store the pairs of $(\mathbf{g}, \mathbf{u})$ in the table and it has a cost of $M_4 = (23+12) \cdot 2^{23}$ in space. For the online phase, we just need to read from the pre-computed tables chunk by chunk, and the complexity of this step is reduced to $C_4 = mh$.

Step 5. Subspace Hypothesis Testing. At this step, it still follows the solving phase but with a little difference according to the analysis in Sect. 6.1. The complexity for this step is still $C_5 = l2^l \sum_{i=0}^{w_1} \binom{k_1}{i}$. Here we have to point that the best candidate chosen is $\mathbf{y}_0 = \arg\max_{\mathbf{y}\in\mathbb{F}_2^l} G(\mathbf{y})$, rather than $\arg\max_{\mathbf{y}\in\mathbb{F}_2^l} |G(\mathbf{y})|$. According to concrete data shown in the following tables, we estimate the success probability as $1 - \Pr[S_\mathbf{y} < T | \mathbf{y} \text{ is correct}]$ considering the missing events, and the results will be close to 1.

6.1 Complexity Analysis

First, we consider the final bias of the approximation $z_i' = \mathbf{y}\mathbf{u}_i^T$ in our improved algorithm. As the analysis in Sect. 4.1, we derive the bias introduced by embedding optimal cascaded perfect codes, denoted by $\tilde{\epsilon}$. The bias introduced by the reduction of the BKW steps is ϵ^{2^t} for adopting LF1 or LF2 at Step 2, while ϵ^{4^t} for adopting LF(4). Thus the final bias is $\epsilon_f = \epsilon^{2^t}\tilde{\epsilon}$ for adopting LF1 or LF2, and $\epsilon_f = \epsilon^{4^t}\tilde{\epsilon}$ for adopting LF(4).

Second, we estimate the number of queries needed. As stated in Sect. 4.1, it needs $m = 8l\ln 2/\epsilon_f^2$ queries to distinguish the correct guess from the others in the final solving phase. Then the number of queries for adopting LF1 at Step 2 is $n = m + k + t2^b$. For adopting LF2, the number of queries is computed as follows, $n[i] \approx \lceil (2^{b+1} n[i+1])^{1/2} \rceil, i = t-1, t-2, \ldots, 0$, where $n[t] = m$ and $n = n[0]$. Similarly for adopting LF(4), the number of queries is computed as $n[i] \approx \lceil (4!2^b n[i+1])^{1/4} \rceil$. Note that the number of initial queries needed in our improved algorithm should be $N = n2^c$ for the selection at Step 0, whatever the reduction mode is adopted at Step 2.

Finally, we present the complexity of our improved algorithm. The tables for vector additions at Step 1 and Step 2 can be constructed offline, as there is no need of the online querying data. The table for decoding at Step 4 can be calculated offline as well. Then the complexity of pre-computation is $Pre = PC_{12} + PC_2 + PC_4$. The memory complexity is about $M = nk + M_{11} + M_{12} + M_2 + M_4$ for storing those tables and the queries data selected. There remains one assumption regarding to the Hamming weight of the secret vector $\mathbf{x}_1'$, and it holds with the probability $\Pr(w_1, k_1)$, where $\Pr(w, k) = \sum_{i=0}^{w} (1-\eta)^{k-i} \eta^i \binom{k}{i}$ for $\Pr[x_i' = 1] = \eta$. Similarly, we need to choose another permutation to run the algorithm again if the assumption is invalid. Thus we are expected to meet this assumption within $1/\Pr(w_1, k_1)$ times iterations. The complexity of each iteration is $PC_{11} + C_1 + C_2 + C_3 + C_4 + C_5$. Hence, the overall time complexity of our improved algorithm online is

$$C = C_0 + \frac{PC_{11} + C_1 + C_2 + C_3 + C_4 + C_5}{\Pr(w_1, k_1)}.$$

6.2 Complexity Results

Now we present the complexity results for solving the three core LPN instances by our improved algorithm when adopting LF1, LF2 and LF(4) at Step 2, respectively.

Note that all of the three LPN instances aim to achieve a security level of 80-bit, and this is indeed the *first* time to distinctly break the first two instances. Although we do not break the third instance, the complexity of our algorithm is quite close to the security level and the remained security margin is quite thin. More significantly, our improved algorithm can provide security evaluations to *any* LPN instances with the proposed parameters, which may be a basis of some cryptosystems (Table 5).

Table 5. The complexity for solving the three LPN instances by our improved algorithm when adopting LF1

LPN instance	Parameters									Selected data $\log_2 n$
	c	s	u	t	b	f	l	k_2	w_1	
(512, 1/8)	5	51	8	5	63	31	62	172	1	66.291
(532, 1/8)	5	53	8	5	65	32	64	182	1	68.584
(592, 1/8)	4	59	9	5	73	36	72	207	1	75.557

LPN instance	Complexities			
	Time $\log_2 C$	Initial data $\log_2 N$	Memory $\log_2 M$	Pre-computation $\log_2 Pre$
(512, 1/8)	75.897	71.291	75.281	66.164
(532, 1/8)	78.182	73.584	77.629	68.053
(592, 1/8)	84.715	79.557	84.764	76.391

We briefly illustrate the following tables. Here for each algorithm adopting a different BKW reduction type, we provide a corresponding table respectively; each contains a sheet of parameters chosen in the order of appearance and a sheet of overall complexity of our improved algorithm (may a sheet of queries numbers via each BKW step as well). There can be several choices when choosing the parameters, and we make a tradeoff in the aspects of time, data and memory. From Tables 6 and 7, we can see the parameters chosen strictly follow the restriction that $n[i] \leq n$ for $i = 1, 2, \ldots, t$ for LF2 and LF(4), as stated in Sect. 5. We also present an attack adopting LF(4) to $(592, 1/8)$-instance without the covering method but directly solving. The parameters are $c = 15$, $s = 59$, $u = 8$, $t = 3$, $b = 178$, $f = 17$, $l = k_2 = 35$, $w_1 = 1$, and the queries data via each BKW step is $n = 2^{60.859}$, $n[1] = 2^{60.853}$, $n[2] = 2^{60.827}$, $n[3] = 2^{60.725}$. The overall complexity is $C = 2^{81.655}$ in time, $N = 2^{75.859}$ for initial data, $M = 2^{72.196}$ in memory and $Pre = 2^{68.540}$ for the pre-computation.

Remark 1. All the results above strictly obey that $m = 8l\ln 2/\epsilon_f^2$, which is much more appropriate for evaluating the success probability, rather than the

Table 6. The complexity for solving three LPN instances by our improved algorithm when adopting LF2

LPN instance	Parameters									Selected data $\log_2 n$
	c	s	u	t	b	f	l	k_2	w_1	
(512, 1/8)	5	51	8	5	64	31	62	170	1	64.987
(532, 1/8)	7	53	8	5	66	32	64	178	1	66.983
(592, 1/8)	4	59	9	5	73	36	72	209	1	73.985

LPN instance	Data via each BKW step				
	$\log_2 n[1]$	$\log_2 n[2]$	$\log_2 n[3]$	$\log_2 n[4]$	$\log_2 n[5]$
(512, 1/8)	64.974	64.948	64.896	64.792	64.583
(532, 1/8)	66.966	66.932	66.863	66.726	66.453
(592, 1/8)	73.970	73.940	73.880	73.759	73.519

LPN instance	Complexities			
	Time $\log_2 C$	Initial data $\log_2 N$	Memory $\log_2 M$	Pre-computation $\log_2 Pre$
(512, 1/8)	74.732	69.987	73.983	66.164
(532, 1/8)	76.902	73.983	76.028	68.053
(592, 1/8)	83.843	77.985	83.204	76.391

Table 7. The complexity for solving three LPN instances by our improved algorithm when adopting LF(4)

LPN instance	Parameters									Selected data	Data via each BKW step	
	c	s	u	t	b	f	l	k_2	w_1	$\log_2 n$	$\log_2 n[1]$	$\log_2 n[2]$
(512, 1/8)	10	47	8	2	156	16	60	174	1	53.526	53.519	53.490
(532, 1/8)	15	47	8	2	162	17	61	180	1	55.504	55.433	55.149
(592, 1/8)	18	53	8	2	177	17	68	204	1	60.513	60.468	60.288

LPN instance	Complexities			
	Time $\log_2 C$	Initial data $\log_2 N$	Memory $\log_2 M$	Pre-computation $\log_2 Pre$
(512, 1/8)	72.844	63.526	68.197	68.020
(532, 1/8)	74.709	70.504	69.528	69.231
(592, 1/8)	81.963	78.513	70.806	69.231

$m = 4l\ln 2/\epsilon_f^2$ which is chosen by the authors of [15] in their presentation at Asiacrypt 2014. If we choose the data as theirs for comparison, our complexity can be reduced to around 2^{71}, which is about 2^8 time lower than that in [15].

6.3 Concrete Attacks

Now we briefly introduce these three key LPN instances and the protocols based on them. The first one with parameter of (512, 1/8) is widely accepted in various LPN-based cryptosystems, e.g., HB^+ [19], $HB^\#$ [12] and LPN-C [13]. The 80-bit security of HB^+ is directly based on that of (512, 1/8)-LPN instance. Thus we can yield an active attack to break HB^+ authentication protocol straight forwardly. $HB^\#$ and LPN-C are two cryptosystems with the similar structures for authentication and encryption. There exist an active attack on $HB^\#$ and a chosen-plaintext attack on LPN-C. The typical parameter settings of the columns number are 441 for $HB^\#$, and 80 (or 160) for LPN-C. These two cryptosystems both consist of two version: secure version as RANDOM- $HB^\#$ and LPN-C, efficient version as TOEPLITZ- $HB^\#$ and LPN-C. For the particularity of Toeplitz Matrices, if we attack its first column successively, then the cost for determining the remaining vectors can be bounded by 2^{40}. Thus we break the 80-bit security of these efficient versions employing Toeplitz matrices, i.e., TOEPLITZ-$HB^\#$ and LPN-C. For the random matrix case, the most common method is to attack it column by column. Then the complexity becomes a columns number multiple of the complexity attacking one $(512, 1/8)$-LPN instance[9]. That is, it has a cost of $441 \times 2^{72.177} \approx 2^{80.962}$ to attack RANDOM-$HB^\#$, which slightly exceeds the security level, and may be improved by some advanced method when conducting different columns. Similarly, the 80-bit security of RANDOM- LPN-C can be broke with a complexity of at most $160 \times 2^{72.177} \approx 2^{79.499}$.

The second LPN instance with the increased length (532, 1/8) is adopted as the parameter of an irreducible Ring-LPN instance employed in Lapin to achieve 80-bit security [16]. Since the Ring-LPN problem is believed to be not harder than the standard LPN problem, the security level can be break easily. It is urgent and necessary to increase the size of the employed irreducible polynomial in Lapin for 80-bit security. The last LPN instance with $(592, 1/8)$ is a new design parameter recommended to use in the future. However, we do not suggest to use it, for the security margin between our attack complexity and the security level is too small.

7 Conclusions

In this paper, we have proposed faster algorithms for solving the LPN problem based on an optimal precise embedding of cascaded concrete perfect codes, in the similar framework to that in [15], but with more careful analysis and optimized procedures. We have also proposed variants of BKW algorithms using tuples for collision and a technique to reduce the requirement of queries. The results beat all the previous approaches, and present efficient attacks against the LPN instances suggested in various cryptographic primitives. Our new approach is *generic* and is the best known algorithm for solving the LPN problem so far, which is practical to provide concrete security evaluations to the LPN instances with any parameters in the future designs. It is our further work to study the problem how to cut down the limitation of candidates, and meanwhile employ other type of good codes, such as nearly perfect codes.

[9] Here, we adjust the parameter of $(512, 1/8)$-LPN instance in Table 6 that c changes into 16. Then the complexity of time, initial data, memory and pre-computation are respectively $C = 2^{72.177}$, $N = 2^{69.526}$, $M = 2^{68.196}$ and $Pre = 2^{68.020}$.

Acknowledgements. The authors would like to thank one of the anonymous reviewers for very helpful comments. This work is supported by the program of the National Natural Science Foundation of China (Grant No. 61572482), National Grand Fundamental Research 973 Programs of China (Grant No. 2013CB-338002 and 2013CB834203) and the program of the National Natural Science Foundation of China (Grant No. 61379142).

References

1. Albrecht, M., Cid, C., Faugère, J.C., Fitzpatrick, R., Perret, L.: On the complexity of the BKW algorithm on LWE. Des. Codes Crypt. **74**(2), 325–354 (2015)
2. Berbain, C., Gilbert, H., Maximov, A.: Cryptanalysis of grain. In: Robshaw, M. (ed.) FSE 2006. LNCS, vol. 4047, pp. 15–29. Springer, Heidelberg (2006)
3. Bernstein, D.: Optimizing linear maps modulo 2. http://binary.cr.yp.to/linearmod2-20090830.pdf
4. Bernstein, D.J., Lange, T.: Never trust a bunny. In: Hoepman, J.-H., Verbauwhede, I. (eds.) RFIDSec 2012. LNCS, vol. 7739, pp. 137–148. Springer, Heidelberg (2013)
5. Blum, A., Kalai, A., Wasserman, H.: Noise-tolerant learning, the parity problem, and the statistical query model. J. ACM **50**(4), 506–519 (2003)
6. Bogos, S., Tramer, F., Vaudenay, S.: On solving LPN using BKW and variants. https://eprint.iacr.org/2015/049.pdf
7. Chose, P., Joux, A., Mitton, M.: Fast correlation attacks: an algorithmic point of view. In: Knudsen, L.R. (ed.) EUROCRYPT 2002. LNCS, vol. 2332, p. 209. Springer, Heidelberg (2002)
8. Cramér, H.: Mathematical Methods of Statistics, vol. 9. Princeton University Press, Princeton (1999)
9. Drost, F., Kallenberg, W., Moore, D., Oosterhoff, J.: Power approximations to multinomial tests of fit. J. Am. Stat. Assoc. **84**(405), 130–141 (1989)
10. Dodis, Y., Kiltz, E., Pietrzak, K., Wichs, D.: Message authentication, revisited. In: Pointcheval, D., Johansson, T. (eds.) EUROCRYPT 2012. LNCS, vol. 7237, pp. 355–374. Springer, Heidelberg (2012)
11. Duc, A., Vaudenay, S.: HELEN: a public-key cryptosystem based on the LPN and the decisional minimal distance problems. In: Youssef, A., Nitaj, A., Hassanien, A.E. (eds.) AFRICACRYPT 2013. LNCS, vol. 7918, pp. 107–126. Springer, Heidelberg (2013)
12. Gilbert, H., Robshaw, M., Seurin, Y.: $HB^{\#}$: increasing the security and efficiency of HB^{+}. In: Smart, N.P. (ed.) EUROCRYPT 2008. LNCS, vol. 4965, pp. 361–378. Springer, Heidelberg (2008)
13. Gilbert, H., Robshaw, M., Seurin, Y.: How to encrypt with the LPN problem. In: Aceto, L., Damgård, I., Goldberg, L.A., Halldórsson, M.M., Ingólfsdóttir, A., Walukiewicz, I. (eds.) ICALP 2008, Part II. LNCS, vol. 5126, pp. 679–690. Springer, Heidelberg (2008)
14. Gilbert, H., Robshaw, M., Sibert, H.: Active attack against HB^{+}: a provably secure lightweight authentication protocol. Electron. Lett. **41**(21), 1169–1170 (2005)
15. Guo, Q., Johansson, T., Löndahl, C.: Solving LPN using covering codes. In: Sarkar, P., Iwata, T. (eds.) ASIACRYPT 2014. LNCS, vol. 8873, pp. 1–20. Springer, Heidelberg (2014)

16. Heyse, S., Kiltz, E., Lyubashevsky, V., Paar, C., Pietrzak, K.: Lapin: an efficient authentication protocol based on ring-LPN. In: Canteaut, A. (ed.) FSE 2012. LNCS, vol. 7549, pp. 346–365. Springer, Heidelberg (2012)
17. Lipmaa, H., Moriai, S.: Efficient algorithms for computing differential properties of addition. In: Matsui, M. (ed.) FSE 2001. LNCS, vol. 2355, p. 336. Springer, Heidelberg (2002)
18. Hopper, N.J., Blum, M.: Secure human identification protocols. In: Boyd, C. (ed.) ASIACRYPT 2001. LNCS, vol. 2248, pp. 52–66. Springer, Heidelberg (2001)
19. Juels, A., Weis, S.A.: Authenticating pervasive devices with human protocols. In: Shoup, V. (ed.) CRYPTO 2005. LNCS, vol. 3621, pp. 293–308. Springer, Heidelberg (2005)
20. Kiltz, E., Pietrzak, K., Cash, D., Jain, A., Venturi, D.: Efficient authentication from hard learning problems. In: Paterson, K.G. (ed.) EUROCRYPT 2011. LNCS, vol. 6632, pp. 7–26. Springer, Heidelberg (2011)
21. Kirchner, P.: Improved generalized birthday attack. http://eprint.iacr.org/2011/377.pdf
22. Levieil, É., Fouque, P.-A.: An improved LPN algorithm. In: De Prisco, R., Yung, M. (eds.) SCN 2006. LNCS, vol. 4116, pp. 348–359. Springer, Heidelberg (2006)
23. Lu, Y., Vaudenay, S.: Faster correlation attack on bluetooth keystream generator E0. In: Franklin, M. (ed.) CRYPTO 2004. LNCS, vol. 3152, pp. 407–425. Springer, Heidelberg (2004)
24. Van Lint, J.H.: Introduction to Coding Theory, vol. 86. Springer Science+Business Media, Berlin (1999)
25. Wagner, D.: A generalized birthday problem. In: Yung, M. (ed.) CRYPTO 2002. LNCS, vol. 2442, pp. 288–304. Springer, Heidelberg (2002)

Provable Security Evaluation of Structures Against Impossible Differential and Zero Correlation Linear Cryptanalysis

Bing Sun[1,2,4(✉)], Meicheng Liu[2,3], Jian Guo[2], Vincent Rijmen[5], and Ruilin Li[6]

[1] College of Science, National University of Defense Technology, Changsha 410073, Hunan, People's Republic of China
happy_come@163.com
[2] Nanyang Technological University, Singapore, Singapore
meicheng.liu@gmail.com, ntu.guo@gmail.com
[3] State Key Laboratory of Information Security, Institute of Information Engineering, Chinese Academy of Sciences, Beijing 100093, People's Republic of China
[4] State Key Laboratory of Cryptology, P.O. Box 5159, Beijing 100878, People's Republic of China
[5] Department of Electrical Engineering (ESAT), KU Leuven and iMinds, Leuven, Belgium
vincent.rijmen@esat.kuleuven.be
[6] College of Electronic Science and Engineering, National University of Defense Technology, Changsha 410073, Hunan, People's Republic of China
securitylrl@163.com

Abstract. Impossible differential and zero correlation linear cryptanalysis are two of the most important cryptanalytic vectors. To characterize the impossible differentials and zero correlation linear hulls which are independent of the choices of the non-linear components, Sun *et al.* proposed the structure deduced by a block cipher at CRYPTO 2015. Based on that, we concentrate in this paper on the security of the SPN structure and Feistel structure with SP-type round functions. Firstly, we prove that for an SPN structure, if $\alpha_1 \rightarrow \beta_1$ and $\alpha_2 \rightarrow \beta_2$ are possible differentials, $\alpha_1|\alpha_2 \rightarrow \beta_1|\beta_2$ is also a possible differential, i.e., the OR "|" operation preserves differentials. Secondly, we show that for an SPN structure, there exists an r-round impossible differential if and only if there exists an r-round impossible differential $\alpha \not\rightarrow \beta$ where the Hamming weights of both α and β are 1. Thus for an SPN structure operating on m bytes, the computation complexity for deciding whether there exists an impossible

The work in this paper is supported by the National Natural Science Foundation of China (No: 61303258, 61379139, 61402515, 61572026, 11526215), National Basic Research Program of China (973 Program) (2013CB338002), the Strategic Priority Research Program of the Chinese Academy of Science under Grant No. XDA06010701 and the Research Fund KU Leuven, OT/13/071. Part of the work was done while the first author was visiting Nanyang Technological University in Singapore.

M. Fischlin and J.-S. Coron (Eds.): EUROCRYPT 2016, Part I, LNCS 9665, pp. 196–213, 2016.
DOI: 10.1007/978-3-662-49890-3_8

differential can be reduced from $\mathcal{O}(2^{2m})$ to $\mathcal{O}(m^2)$. Thirdly, we associate a primitive index with the linear layers of SPN structures. Based on the matrices theory over integer rings, we prove that the length of impossible differentials of an SPN structure is upper bounded by the primitive index of the linear layers. As a result we show that, unless the details of the S-boxes are considered, there do not exist 5-round impossible differentials for the AES and ARIA. Lastly, based on the links between impossible differential and zero correlation linear hull, we projected these results on impossible differentials to zero correlation linear hulls. It is interesting to note some of our results also apply to the Feistel structures with SP-type round functions.

Keywords: Impossible differential · Zero correlation linear · SPN structure · Feistel structure · AES · Camellia · ARIA

1 Introduction

Block ciphers are the vital elements in constructing many symmetric cryptographic schemes and the core security of these schemes depends on the resistance of the underlying block ciphers to known cryptanalytic techniques. Differential cryptanalysis [4] and linear cryptanalysis [20] are among the most famous cryptanalytic tools. Nowadays, most block ciphers are designed to be resilient to these two attacks. To prove the security of a block cipher against differential/linear attack, a common way is to give an upper bound on the rounds of the differential characteristics/linear trails that can distinguish a round-reduced cipher from a random permutation. Or equivalently, one can show when the number of the rounds of a block cipher is more than a certain r, there do not exist any useful differential characteristics or linear trails. However, the security margin of the ciphers against extended differential and linear cryptanalysis, such as impossible differential [3,13] and zero correlation linear cryptanalysis [6], may not be yet well studied and formulated. To some extend, the success of such attacks relies mainly on the attackers' intensive analysis of the structures used in each individual designs.

In differential cryptanalysis, one usually finds differential characteristics with high probability and then uses statistical methods to sieve the right keys. However, the main idea of impossible differential cryptanalysis, which was independently proposed by Knudsen [13] and Biham *et al.* [3], is to use differentials that hold with probability zero to discard the wrong keys. So far, impossible differential cryptanalysis has received lots of attention and been used to attack a variety of well-known block ciphers [5,7,16,22].

The first step in impossible differential cryptanalysis is to construct some impossible differentials that cover as many rounds as possible. For any function $F : \mathbb{F}_{2^b} \to \mathbb{F}_{2^b}$, we can always find some α and β such that $\alpha \to \beta$ is an impossible differential of F. However, when b is large and we know little about the algebraic structure of F, it is hard to determine whether $\alpha \to \beta$ is a possible differential or an impossible one. A block cipher $E(\cdot, k)$ may exhibit a differential $\alpha \to \beta$ that is a possible differential for some key k while it is impossible for the rest. In practice, such differentials are difficult to determine in most of the cases. Generally, in a search for impossible differentials it is difficult to guarantee completeness. Therefore, from the practical point of view, we are more interested in the impossible differentials that are independent of the secret keys. Since in most cases the non-linear transformations applied to x can be written as $S(x \oplus k)$, we always employ impossible differentials that are independent of the S-boxes, which are called *truncated impossible differentials*, i.e., we only detect whether there are differences on some bytes and we do not care about the values of the differences. Usually, an impossible differential is constructed by the miss-in-the-middle technique, i.e., trace the properties of input and output differences from the encryption and decryption directions, respectively, if there are some contradictions in the middle, an impossible differential is then found. Several automatic approaches have been proposed to derive truncated impossible differentials of a block cipher effectively such as the $\mathcal{U}$-method [12], *UID*-method [18] and the extended tool of the former two methods generalized by Wu and Wang [24] (WW-method). It has been proved in [21] that the WW-method can find all impossible differentials of a structure, or equivalently, it can find all impossible differentials of a block cipher which are independent of the choices of the non-linear components. Similar ideas have found applications in cryptanalysis against hash functions BMW [10] and BLAKE [2].

In linear cryptanalysis, one uses linear characteristics with high correlations. Zero correlation cryptanalysis is a novel technique for cryptanalysis of block ciphers [6]. The distinguishing property used in zero correlation cryptanalysis is the zero correlation linear approximations, i.e., those linear approximations that hold with a probability $p = 1/2$, that is, strictly unbiased approximations having a correlation $c = 2p - 1$ equal to 0. As in impossible differential cryptanalysis, we are more interested in the zero correlation linear hulls that are independent of the choices of the non-linear layers.

In CRYPTO 2015, Sun *et al.* proposed the concept of *structure* to characterize what "being independent of the choices of the S-boxes" means, and proposed *dual structure* to study the link between impossible differentials and zero correlation linear hulls [21]. One of the basic statements in [21] is that constructing impossible differentials of a structure is equivalent to constructing zero correlation linear hulls of the dual structure. Therefore, all the known methods to construct impossible differentials of structures can also be used to construct zero correlation linear hulls.

Despite the known 4-/4-/8-round impossible differentials for the AES, ARIA and Camellia without FL/FL^{-1} layers [1,9,14,17,19,25], effort to find new impossible differentials of these ciphers that cover more rounds has never stopped. On the other hand, we already have some novel techniques such as the wide trail strategy [8] and the decorrelation theory [23] to prove that a cipher is resilient to differential and linear attacks. However, the provable security of block ciphers against impossible differential and zero correlation linear cryptanalysis is still missing. Noting that for a dedicated iterated block cipher, there always exist impossible differentials for any rounds with some keys, we wonder that if we consider the impossible differentials that are independent of the choices of the S-boxes, there may exist an integer R such that there does not exist any impossible differentials that cover more than R rounds, which can give some insights on provable security of block ciphers against impossible differential and zero correlation linear cryptanalysis, i.e., R is the upper bound of such attacks. Furthermore, since the WW-method can only determine whether a given differential/mask is an impossible differential/zero correlation linear hull or not, though it can theoretically find all impossible differentials/zero correlation linear hulls of a structure, it is impractical to exhaust all the differentials/masks to determine whether there exist r-round impossible differentials/zero correlation linear hulls or not. Therefore, finding new techniques to solve these problems in a practical way remains as an open problem.

Our Contributions. Inspired by the provable security of differential and linear cryptanalysis, this paper mainly concentrates on the provable security of block ciphers against impossible differential/zero correlation linear cryptanalysis and we aim at determining an upper bound for the longest rounds of impossible differentials/zero correlation linear hulls of SPN structures and Feistel structures with SPN round functions. The main results of this paper are as follows:

(1) For SPN structures, we prove that if $\alpha_1 \rightarrow \beta_1$ and $\alpha_2 \rightarrow \beta_2$ are possible differentials, then $\alpha_1|\alpha_2 \rightarrow \beta_1|\beta_2$ is also a possible differential, based on which we conclude that there exists an r-round impossible differential if and only if there exists an impossible differential $\alpha \rightarrow \beta$ where the Hamming weight of both α and β is 1. Therefore, for an SPN structure with m bytes, the complexity of testing whether there exist r-round impossible differentials is reduced significantly from $\mathcal{O}(2^{2m})$ to $\mathcal{O}(m^2)$.

(2) For Feistel structures with SP-type round functions, we prove that if $\alpha_1 \rightarrow \beta_1$ and $\alpha_2 \rightarrow \beta_2$ are *independent possible differentials* (we will define it later), then $\alpha_1|\alpha_2 \rightarrow \beta_1|\beta_2$ is also an independent possible differential, then similar result as in (1) applies.

(3) For any matrix over finite fields, we can always define two polynomials to calculate an upper bound on the highest possible rounds of impossible differentials of SPN structures and independent impossible differentials of Feistel structures with SP-type round functions. Our results show that, unless we take the details of the S-boxes into consideration, there do not exist 5-round impossible differentials of the AES and ARIA, and 9-round independent impossible differentials of Camellia without FL/FL^{-1} layers.

(4) Since the zero correlation linear hull of a structure is equivalent to the impossible differential of its dual structure, our results on impossible differentials cryptanalysis also apply to zero correlation linear cryptanalysis.

From the theoretical point of view, our results demonstrate some direct insight to the longest possible rounds of truncated impossible differentials and zero correlation linear hulls. And from the practical point of view, our results could reduce the work effort to find impossible differentials and zero correlation linear hulls of a structure.

Organization. The rest of this paper is organized as follows. Section 2 will introduce some definitions that will be used throughout this paper. In Sect. 3, we give some new features of the structures. We investigate on the SPN structures and Feistel structures with SP-type round functions in Sects. 4 and 5, respectively. Section 6 concludes the paper.

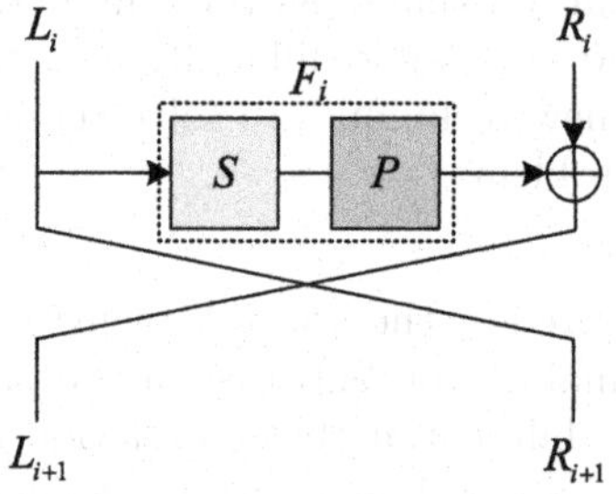

Fig. 1. Feistel structure with SP-type round functions

2 Preliminaries

2.1 Block Ciphers

SPN Ciphers. The SPN structure is widely used in constructing cryptographic primitives. It iterates some SP-type round functions to achieve confusion and diffusion. Specifically, the SP-type function $f : \mathbb{F}_{2^b}^m \to \mathbb{F}_{2^b}^m$ used in this paper is defined as follows where $\mathbb{F}_{2^b}$ is the finite field with 2^b elements.

Assume the input x is divided into m pieces $x = (x_0, \ldots, x_{m-1})$, where x_i is a b-bit byte. First, apply the non-linear transformation s_i to x_i,

$$y = S(x) \triangleq (s_0(x_0), \ldots, s_{m-1}(x_{m-1})) \in \mathbb{F}_{2^b}^m.$$

Then, apply a linear transformation $P : \mathbb{F}_{2^b}^m \to \mathbb{F}_{2^b}^m$ to y, and Py is the output of f. Notice that the linear transformation in the last round of an r-round SPN structure is omitted, i.e., an r-round SPN cipher is simply denoted as $(SP)^{r-1}S$.

Feistel Ciphers. An r-round Feistel cipher E is defined as follows: Let $(L_0, R_0) \in \mathbb{F}_2^{2m}$ be the input of E. Iterate the following transformation r times:

$$\begin{cases} L_{i+1} = F_i(L_i) \oplus R_i \\ R_{i+1} = L_i \end{cases} \quad 0 \le i \le r-1,$$

where $L_i, R_i \in \mathbb{F}_2^m$, see Fig. 1. The output of E is defined as the output of the r-th iteration. In this paper, we will focus on the case that F_i's are defined as SP-type functions.

2.2 Vectors and Matrices

Assume $\alpha, \beta \in \mathbb{F}_{2^b}^m$, where $\mathbb{F}_{2^b}^m$ is the vector space over $\mathbb{F}_{2^b}$ with dimension m. Then $\alpha|\beta$ is defined as the bit-wise OR operation of α and β. Let $\theta : \mathbb{F}_{2^b} \rightarrow \mathbb{F}_2$ be defined as

$$\theta(x) = \begin{cases} 0 & x = 0, \\ 1 & x \neq 0. \end{cases}$$

Then, for $X = (x_0, \ldots, x_{m-1}) \in \mathbb{F}_{2^b}^m$, the *truncated characteristic* of X is defined as

$$\chi(X) \triangleq (\theta(x_0), \ldots, \theta(x_{m-1})) \in \mathbb{F}_2^m.$$

The *Hamming weight* of X is defined as the number of non-zero elements of the vector, i.e. $H(X) = \#\{i|x_i \neq 0, i = 0, 1, \ldots, m-1\}$.

For $P = (p_{ij}) \in \mathbb{F}_{2^b}^{m \times m}$, denote by $\mathbb{Z}$ the integer ring, the *characteristic matrix* of P is defined as $P^* = (p_{ij}^*) \in \mathbb{Z}^{m \times m}$, where $p_{ij}^* = 0$ if $p_{ij} = 0$ and $p_{ij}^* = 1$ otherwise. A matrix $M \in \mathbb{Z}^{m \times m}$ is *non-negative* if all elements of M are non-negative, and *positive* if all elements of M are positive. Therefore, the characteristic matrix is always non-negative.

Definition 1. *Let $P \in \mathbb{F}_{2^b}^{m \times m}$, P^* be the characteristic matrix of P, and*

$$f_t(x) = x^t,$$
$$g_t(x) = \begin{cases} \sum_{i=0}^{h} x^{2i} & t = 2h, \\ \sum_{i=1}^{h} x^{2i-1} & t = 2h-1. \end{cases}$$

Then the minimal integer t such that $f_t(P^*)$ is a positive matrix is called type 1 primitive index of P, and the minimal integer t such that $g_t(P^*)$ is positive is called type 2 primitive index of P.

If the input X to the linear layer P is viewed as a column vector, then the output Y can also be viewed as a column vector which is computed as $Y = PX$. According to the definition of characteristic matrix, $p_{ij}^* = 0$ means the i-th output byte of the first round is independent of the j-th input byte. Generally, let $f_t(P^*) = (P^*)^t = (q_{ij})$, then $q_{ij} = 0$ means the i-th output byte of the t-round SPN cipher is independent of the j-th input byte. Furthermore, let $(P^*)^{t_1} + (P^*)^{t_2} = (u_{ij})$, then $u_{ij} = 0$ means the i-th output bytes of both the t_1-round and t_2-round SPN cipher are independent of j-th input byte. Similarly, let $g_t(P^*) = (w_{ij})$, then $w_{ij} = 0$ means the i-th output byte of the t-round Feistel cipher is independent of the j-th input byte.

2.3 Impossible Differentials and Zero Correlation Linear Hulls

Given a function $G : \mathbb{F}_2^n \rightarrow \mathbb{F}_2$, the correlation c of G is defined by

$$c(G(x)) \triangleq \frac{1}{2^n} \sum_{x \in \mathbb{F}_2^n} (-1)^{G(x)}.$$

Given a function $G : \mathbb{F}_2^n \rightarrow \mathbb{F}_2^k$, the correlation c of the linear approximation for a k-bit output mask b and an n-bit input mask a is defined by

$$c(ax \oplus bG(x)) = \frac{1}{2^n} \sum_{x \in \mathbb{F}_2^n} (-1)^{ax \oplus bG(x)}.$$

If $c(ax \oplus bG(x)) = 0$, then $(a \rightarrow b)$ is called a zero correlation linear hull of G. This definition can be extended as follows: let $A \subseteq \mathbb{F}_2^n$, $B \subseteq \mathbb{F}_2^k$, if for all $a \in A$ and $b \in B$, $c(ax \oplus bG(x)) = 0$, then $(A \rightarrow B)$ is also called a zero correlation linear hull of G.

Let $\delta \in \mathbb{F}_2^n$ and $\Delta \in \mathbb{F}_2^k$. The differential probability of $\delta \rightarrow \Delta$ is defined as

$$p(\delta \rightarrow \Delta) \triangleq \frac{\#\{x \in \mathbb{F}_2^n | G(x) \oplus G(x \oplus \delta) = \Delta\}}{2^n}.$$

If $p(\delta \rightarrow \Delta) = 0$, then $\delta \rightarrow \Delta$ is called an *impossible differential* of G, this definition follows that in [3,13]. Let $A \subseteq \mathbb{F}_2^n$, $B \subseteq \mathbb{F}_2^k$. If for all $a \in A$ and $b \in B$, $p(a \rightarrow b) = 0$, $A \rightarrow B$ is called an *impossible differential* of G.

3 Differential Properties of Structures

In many cases, when constructing impossible differentials and zero correlation linear hulls, we are only interested in detecting whether there is a difference (mask) in an S-box or not, regardless of the actual value of the difference (mask) which leads to the following definition:

Definition 2 [21]. *Let $E : \mathbb{F}_2^n \rightarrow \mathbb{F}_2^n$ be a block cipher with bijective S-boxes as the basic non-linear components.*

(1) *A* structure *$\mathcal{E}^E$ on $\mathbb{F}_2^n$ is defined as a set of block ciphers E' which is exactly the same as E except that the S-boxes can take all possible bijective transformations on the corresponding domains.*
(2) *Let $\alpha, \beta \in \mathbb{F}_2^n$. If for any $E' \in \mathcal{E}^E$, $\alpha \nrightarrow \beta$ is an impossible differential (zero correlation linear hull) of E', $\alpha \nrightarrow \beta$ is called an impossible differential (zero correlation linear hull) of $\mathcal{E}^E$.*

Thus the structure deduced by a single S layer can be written as $\mathcal{E}^S$; the structure deduced by a single S layer followed by a P layer can be written as $\mathcal{E}^{SP}$. If $\alpha \rightarrow \beta$ is not an impossible differential of $\mathcal{E}^E$, i.e., there exist some x and $E' \in \mathcal{E}^E$ such that $E'(x) \oplus E'(x \oplus \alpha) = \beta$, we call it a *possible differential* of $\mathcal{E}^E$.

Definition 3. *Let $\mathcal{E}$ be a structure and $\alpha \nrightarrow \beta$ an impossible differential of $\mathcal{E}$. If for all α^* and β^* satisfying $\chi(\alpha^*) = \chi(\alpha)$ and $\chi(\beta^*) = \chi(\beta)$, $\alpha^* \nrightarrow \beta^*$ are impossible differentials, we call $\alpha \nrightarrow \beta$ an* independent impossible differential *of $\mathcal{E}$. Otherwise, we call it a* dependent impossible differential *of $\mathcal{E}$.*

As shown in [25], for any $\alpha \neq 0$ and $\beta \neq 0$,

$$(0|0|0|0|0|0|0|0, \alpha|0|0|0|0|0|0|0) \not\rightarrow (\beta|0|0|0|0|0|0|0, 0|0|0|0|0|0|0|0)$$

is an 8-round impossible differential of Camellia without FL/FL^{-1} layers. According to the definition, such an impossible differential is an independent impossible differential of Camellia without FL/FL^{-1} layers.

A dependent impossible differential means that there are some constraints on actual differences of both the input and output bytes. For example, for any given α, $(0, \alpha) \not\rightarrow (0, \alpha)$ is a 5-round impossible differential of Feistel structures with bijective round functions. However, we cannot determine that $(0, \alpha) \not\rightarrow (0, \beta)$ is an impossible differential for any $\alpha \neq \beta$. Thus, $(0, \alpha) \not\rightarrow (0, \alpha)$ is a dependent impossible differential of 5-round Feistel structure with bijective round functions.

Usually, we have many different ways to define a linear transformation, which means we have many different ways to express the matrix of the linear transformation. However, no matter which one we use, the transformation is always linear over $\mathbb{F}_2$, thus the bit-wise matrix representation of a linear transformation is call the *primitive representation.* The definition of dual structure is proposed to study the link between impossible differential and zero correlation linear hulls:

Definition 4 [21]. *Let $\mathcal{F}_{SP}$ be a* Feistel structure *with SP-type round function, and let the primitive representation of the linear transformation be P. Let σ be the operation that exchanges the left and right halves of a state. Then the* dual structure *$\mathcal{F}_{SP}^{\perp}$ of $\mathcal{F}_{SP}$ is defined as $\sigma \circ \mathcal{F}_{P^T S} \circ \sigma$.*

Let $\mathcal{E}_{SP}$ be an SPN structure *with primitive representation of the linear transformation being P. Then the* dual structure *$\mathcal{E}_{SP}^{\perp}$ of $\mathcal{E}_{SP}$ is defined as $\mathcal{E}_{S(P^{-1})^T}$.*

Next, we are going to give some statements on the differential properties of structures while they may not hold for dedicated block ciphers.

Let $\mathcal{E}^{(r)}$ be an r-round iterated structure. If $\alpha \rightarrow \beta$ is a possible differential of $\mathcal{E}^{(r_1)}$, then for any x, there always exists $E_1 \in \mathcal{E}^{(r_1)}$ such that $E_1(x) \oplus E_1(x \oplus \alpha) = \beta$. If $\beta \rightarrow \gamma$ is a possible differential of $\mathcal{E}^{(r_2)}$, for $y = E_2(x)$, there always exists $E_2 \in \mathcal{E}^{(r_2)}$ such that $E_2(y) \oplus E_2(y \oplus \beta) = \gamma$. Let $E = E_2 \circ E_1$, we have $E(x) \oplus E(x \oplus \alpha) = \gamma$ which means $\alpha \rightarrow \gamma$ is a possible differential $\mathcal{E}^{(r_1+r_2)}$. See (1) for the procedures. Accordingly, for a structure $\mathcal{E}$, if there do not exist r-round impossible differentials, there do not exist R-round impossible differentials for any $R \geq r$.

$$E: \begin{array}{ccccc} x & \xrightarrow{E_1} & y & \xrightarrow{E_2} & z \\ | & & | & & | \\ x \oplus \alpha & \xrightarrow{E_1} & y \oplus \beta & \xrightarrow{E_2} & z \oplus \gamma \end{array} \tag{1}$$

Next we show that $\alpha \rightarrow \beta$ is a possible differential of a single S layer $\mathcal{E}^S$ if and only if $\chi(\alpha) = \chi(\beta)$. Firstly, we cannot construct a bijective S-box such that a zero difference causes a non-zero difference. Secondly, let $\alpha = (\alpha_0, \ldots, \alpha_{m-1}), \beta = (\beta_0, \ldots, \beta_{m-1}) \in \mathbb{F}_{2^b}^m$. If $\chi(\alpha) = \chi(\beta)$, for any $x = (x_0, \ldots, x_{m-1}) \in \mathbb{F}_{2^b}^m$, we can always construct an $S = (s_0, \ldots, s_{m-1})$ where $s_i : \mathbb{F}_{2^b} \rightarrow \mathbb{F}_{2^b}$, such that $S(x) \oplus S(x \oplus \alpha) = \beta$, i.e., $s_i(x_i) \oplus s_i(x_i \oplus \alpha_i) = \beta_i$, $i = 0, \ldots, m-1$.

4 Cryptanalysis of SPN Structures

In this section, we will simply use $\mathcal{E}_{SP}^{(r)}$ to denote an r-round SPN structure.

4.1 How to Check Whether a Differential is Impossible or Not

Assume $\alpha \rightarrow \beta$ is a possible differential of $\mathcal{E}_{SP}^{(r)}$. Then, there always exist some α' and β' such that

$$\alpha \xrightarrow{\mathcal{E}^S} \alpha' \xrightarrow{\mathcal{E}^{PS\cdots SP}} \beta' \xrightarrow{\mathcal{E}^S} \beta$$

is a possible differential of $\mathcal{E}_{SP}^{(r)}$. Thus for any α^* and β^* such that $\chi(\alpha^*) = \chi(\alpha)$, $\chi(\beta^*) = \chi(\beta)$,

$$\alpha^* \xrightarrow{\mathcal{E}^S} \alpha' \xrightarrow{\mathcal{E}^{PS\cdots SP}} \beta' \xrightarrow{\mathcal{E}^S} \beta^*$$

is still a possible differential. In other words, impossible differentials of SPN structures are independent impossible differentials.

Therefore, for an SPN structure, to check whether there exists an r-round impossible differential or not, one needs to test $(2^m - 1) \times (2^m - 1) \approx 2^{2m}$ candidates. However, this complexity could be further reduced as illustrated in the following.

Lemma 1. *Assume $m \leq 2^{b-1} - 1$. If $\alpha_1 \rightarrow \beta_1$ and $\alpha_2 \rightarrow \beta_2$ are possible differentials of $\mathcal{E}^{SP}$, then there always exist α and β such that*

$$\begin{cases} \chi(\alpha) = \chi(\alpha_1)|\chi(\alpha_2), \\ \chi(\beta) = \chi(\beta_1)|\chi(\beta_2), \end{cases}$$

and $\alpha \rightarrow \beta$ is a possible differential of $\mathcal{E}^{SP}$.

The proof of this lemma is shown in Appendix A. In the following, we always assume $m \leq 2^{b-1} - 1$ which fits well with most cases. Furthermore, since the last round only has the S layer, we have:

Corollary 1. *If $\alpha_1 \rightarrow \beta_1$ and $\alpha_2 \rightarrow \beta_2$ are possible differentials of $\mathcal{E}_{SP}^{(r)}$, $\alpha_1|\alpha_2 \rightarrow \beta_1|\beta_2$ is also a possible differential of $\mathcal{E}_{SP}^{(r)}$.*

Assume $(x_0, 0, \ldots, 0) \rightarrow (y_0, 0, \ldots, 0)$ and $(0, x_1, 0, \ldots, 0) \rightarrow (0, y_1, 0, \ldots, 0)$ are possible differentials of $\mathcal{E}_{SP}$, where x_0, x_1, y_0, y_1 are non-zero. Then according to Corollary 1, $(x_0, x_1, 0, \ldots, 0) \rightarrow (y_0, y_1, 0, \ldots, 0)$ is a possible differential. In other words, if $(x_0, x_1, 0, \ldots, 0) \rightarrow (y_0, y_1, 0, \ldots, 0)$ is an impossible differential of $\mathcal{E}_{SP}$, either $(x_0, 0, \ldots, 0) \rightarrow (y_0, 0, \ldots, 0)$ or $(0, x_1, 0, \ldots, 0) \rightarrow (0, y_1, 0, \ldots, 0)$ is an impossible differential. Generally, we have the following theorem:

Theorem 1. *There exists an impossible differential of $\mathcal{E}_{SP}^{(r)}$ if and only if there exists an impossible differential $\alpha \not\rightarrow \beta$ of $\mathcal{E}_{SP}^{(r)}$, where $H(\alpha) = H(\beta) = 1$, with $H(x)$ denoting the Hamming weight of x.*

Thus with the help of Theorem 1, for every SPN structure, and any (α, β) where $H(\alpha) = H(\beta) = 1$, we can use the WW-method to check whether $\alpha \to \beta$ is a possible differential or not. Therefore, we could reduce the complexities of checking whether there exists an impossible differential of an SPN structure from $\mathcal{O}(2^{2m})$ to $\mathcal{O}(m^2)$.

Since the zero correlation linear hull of $\mathcal{E}_{SP}$ is the impossible differential of $\mathcal{E}_{S(P^{-1})^T}$ which is also an SPN structure, we have the following:

Corollary 2. *There exists a zero correlation linear hull of $\mathcal{E}_{SP}^{(r)}$ if and only if there exists a zero correlation linear hull $\alpha \not\to \beta$ of $\mathcal{E}_{SP}^{(r)}$ where $H(\alpha) = H(\beta) = 1$.*

4.2 An Upper Bound for the Rounds of Impossible Differentials

As discussed above, we can use the WW-method to determine the maximal length of impossible differentials for an SPN structure. In the following, we are going to show an upper bound for the length of impossible differentials for an SPN structure, which only uses the property of the P layer. To characterize the longest impossible differential of an SPN cipher, we first recall that if $\beta = P\alpha$, then there always exist α_0 and β_0 such that $\chi(\alpha_0) = \chi(\alpha)$, $\chi(\beta_0) = \chi(\beta)$ and $\alpha_0 \to \beta_0$ is a possible differential of a single round of SPN structure. Then according to Corollary 1, the following theorem holds.

Theorem 2. *Let $R_1(P)$ and $R_{-1}(P)$ be the type 1 primitive indexes of P and P^{-1} respectively. Then there does not exist any impossible differential or zero correlation linear hull of $\mathcal{E}_{SP}^{(r)}$ for $r \geq R_1(P) + R_{-1}(P) + 1$.*

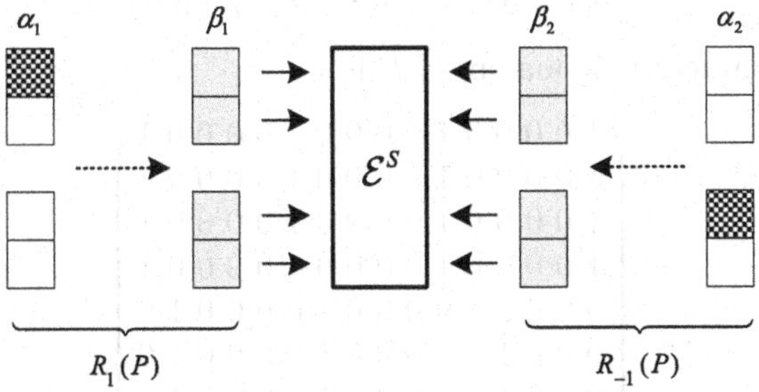

Fig. 2. Constructing $(R_1(P) + R_{-1}(P) + 1)$-round differential for $\mathcal{E}_{SP}$

Proof. See Fig. 2. Firstly, for any $\alpha_1 \neq 0$, $H(\alpha_1) = 1$, according to Lemma 1, there always exist some β_1 where $H(\beta_1) = m$ such that $\alpha_1 \to \beta_1$ is a possible differential of $R_1(P)$-round $\mathcal{E}_{SP}$. Secondly, for any $\alpha_2 \neq 0$, $H(\alpha_2) = 1$, according to Lemma 1, there always exist some β_2 where $H(\beta_2) = m$ such that $\alpha_2 \to \beta_2$ is a possible differential of $R_{-1}(P)$-round decryption of $\mathcal{E}_{SP}$.

Since $\chi(\beta_1) = \chi(\beta_2)$, $\beta_1 \to \beta_2$ is a possible differential of the single S layer $\mathcal{E}^S$, we conclude that $\alpha_1 \to \alpha_2$ is a possible differential of $(R_1(P)+R_{-1}(P)+1)$-round $\mathcal{E}_{SP}$. By Theorem 1, there does not exist any impossible differential or zero correlation linear hull of $\mathcal{E}_{SP}^{(r)}$ for $r \geq R_1(P) + R_{-1}(P) + 1$. □

4.3 Applications

The Advanced Encryption Standard (AES) is one of the most popular SPN ciphers up to date. Firstly, if we consider the 4×4 state as a vector in $\mathbb{F}_{2^8}^{16}$, the composition of the ShiftRows and MixColumns can be written as the following 16×16 matrix over $\mathbb{F}_{2^8}$:

$$P = \begin{pmatrix}
2&0&0&0&0&3&0&0&0&0&1&0&0&0&0&1\\
1&0&0&0&0&2&0&0&0&0&3&0&0&0&0&1\\
1&0&0&0&0&1&0&0&0&0&2&0&0&0&0&3\\
3&0&0&0&0&1&0&0&0&0&1&0&0&0&0&2\\
0&0&0&1&2&0&0&0&0&3&0&0&0&0&1&0\\
0&0&0&1&1&0&0&0&0&2&0&0&0&0&3&0\\
0&0&0&3&1&0&0&0&0&1&0&0&0&0&2&0\\
0&0&0&2&3&0&0&0&0&1&0&0&0&0&1&0\\
0&0&1&0&0&0&0&1&2&0&0&0&0&3&0&0\\
0&0&3&0&0&0&0&1&1&0&0&0&0&2&0&0\\
0&0&2&0&0&0&0&3&1&0&0&0&0&1&0&0\\
0&0&1&0&0&0&0&2&3&0&0&0&0&1&0&0\\
0&3&0&0&0&0&1&0&0&0&0&1&2&0&0&0\\
0&2&0&0&0&0&3&0&0&0&0&1&1&0&0&0\\
0&1&0&0&0&0&2&0&0&0&0&3&1&0&0&0\\
0&1&0&0&0&0&1&0&0&0&0&2&3&0&0&0
\end{pmatrix}.$$

Therefore, the characteristic matrix of P is

$$P^* = \begin{pmatrix}
1&0&0&0&0&1&0&0&0&0&1&0&0&0&0&1\\
1&0&0&0&0&1&0&0&0&0&1&0&0&0&0&1\\
1&0&0&0&0&1&0&0&0&0&1&0&0&0&0&1\\
1&0&0&0&0&1&0&0&0&0&1&0&0&0&0&1\\
0&0&0&1&1&0&0&0&0&1&0&0&0&0&1&0\\
0&0&0&1&1&0&0&0&0&1&0&0&0&0&1&0\\
0&0&0&1&1&0&0&0&0&1&0&0&0&0&1&0\\
0&0&0&1&1&0&0&0&0&1&0&0&0&0&1&0\\
0&0&1&0&0&0&0&1&1&0&0&0&0&1&0&0\\
0&0&1&0&0&0&0&1&1&0&0&0&0&1&0&0\\
0&0&1&0&0&0&0&1&1&0&0&0&0&1&0&0\\
0&0&1&0&0&0&0&1&1&0&0&0&0&1&0&0\\
0&1&0&0&0&0&1&0&0&0&0&1&1&0&0&0\\
0&1&0&0&0&0&1&0&0&0&0&1&1&0&0&0\\
0&1&0&0&0&0&1&0&0&0&0&1&1&0&0&0\\
0&1&0&0&0&0&1&0&0&0&0&1&1&0&0&0
\end{pmatrix}.$$

Since

$$(P^*)^2 = \begin{pmatrix} 1&1&1&1&1&1&1&1&1&1&1&1&1&1&1&1 \\ 1&1&1&1&1&1&1&1&1&1&1&1&1&1&1&1 \\ 1&1&1&1&1&1&1&1&1&1&1&1&1&1&1&1 \\ 1&1&1&1&1&1&1&1&1&1&1&1&1&1&1&1 \\ 1&1&1&1&1&1&1&1&1&1&1&1&1&1&1&1 \\ 1&1&1&1&1&1&1&1&1&1&1&1&1&1&1&1 \\ 1&1&1&1&1&1&1&1&1&1&1&1&1&1&1&1 \\ 1&1&1&1&1&1&1&1&1&1&1&1&1&1&1&1 \\ 1&1&1&1&1&1&1&1&1&1&1&1&1&1&1&1 \\ 1&1&1&1&1&1&1&1&1&1&1&1&1&1&1&1 \\ 1&1&1&1&1&1&1&1&1&1&1&1&1&1&1&1 \\ 1&1&1&1&1&1&1&1&1&1&1&1&1&1&1&1 \\ 1&1&1&1&1&1&1&1&1&1&1&1&1&1&1&1 \\ 1&1&1&1&1&1&1&1&1&1&1&1&1&1&1&1 \\ 1&1&1&1&1&1&1&1&1&1&1&1&1&1&1&1 \\ 1&1&1&1&1&1&1&1&1&1&1&1&1&1&1&1 \end{pmatrix},$$

we get $R_1(P) = 2$. Similarly, we can get $R_{-1}(P) = 2$. Therefore, we have

Proposition 1. *There does not exist any impossible differential or zero correlation linear hull of* $\mathcal{E}^{AES}$ *which covers* $r \geq 5$ *rounds. Or equivalently, there does not exist any* 5*-round impossible differential or zero correlation linear hull of the AES unless the details of the S-boxes are considered.*

ARIA is another famous SPN cipher which uses a linear transformation P such that $P = P^{-1}$. Since

$$(P^*)^2 = \begin{pmatrix} 7&2&2&2&2&4&2&4&2&2&4&4&2&4&4&2 \\ 2&7&2&2&4&2&4&2&2&2&4&4&4&2&2&4 \\ 2&2&7&2&2&4&2&4&4&4&2&2&4&2&2&4 \\ 2&2&2&7&4&2&4&2&4&4&2&2&2&4&4&2 \\ 2&4&2&4&7&2&2&2&2&4&4&2&2&2&4&4 \\ 4&2&4&2&2&7&2&2&4&2&2&4&2&2&4&4 \\ 2&4&2&4&2&2&7&2&4&2&2&4&4&4&2&2 \\ 4&2&4&2&2&2&2&7&2&4&4&2&4&4&2&2 \\ 2&2&4&4&2&4&4&2&7&2&2&2&2&4&2&4 \\ 2&2&4&4&4&2&2&4&2&7&2&2&4&2&4&2 \\ 4&4&2&2&4&2&2&4&2&2&7&2&2&4&2&4 \\ 4&4&2&2&2&4&4&2&2&2&2&7&4&2&4&2 \\ 2&4&4&2&2&2&4&4&2&4&2&4&7&2&2&2 \\ 4&2&2&4&2&2&4&4&4&2&4&2&2&7&2&2 \\ 4&2&2&4&4&4&2&2&2&4&2&4&2&2&7&2 \\ 2&4&4&2&4&4&2&2&4&2&4&2&2&2&2&7 \end{pmatrix}$$

we have $R_1(P) = R_{-1}(P) = 2$. Therefore, we have

Proposition 2. *There does not exist any impossible differential or zero correlation linear hull of $\mathcal{E}^{ARIA}$ which covers $r \geq 5$ rounds. Or equivalently, there does not exist any 5-round impossible differential or zero correlation linear hull of the ARIA unless the details of the S-boxes are considered.*

Since we already have 4-round impossible differential and 4-round zero correlation linear hull of $\mathcal{E}^{AES}$ and $\mathcal{E}^{ARIA}$, unless we investigate on the details of the S-boxes, with respect to the rounds, we cannot find neither better impossible differentials nor zero correlation linear hulls for the AES and ARIA.

5 Cryptanalysis of Feistel Structures with SP-Type Round Functions

In the following, we simply use $\mathcal{F}_{SP}^{(r)}$ to denote an r-round Feistel structure with SP-type round functions. Since the techniques to study the Feistel structure with SPN round functions are almost the same, we only give the results as follows.

Lemma 2. *Assume $m \leq 2^{b-1}-1$. If $(\alpha_1, \beta_1) \to (\gamma_1, \alpha_1)$ and $(\alpha_2, \beta_2) \to (\gamma_2, \alpha_2)$ are possible differentials of $\mathcal{F}_{SP}^{(1)}$. Then, there always exist α, β and γ, such that $\chi(\alpha) = \chi(\alpha_1)|\chi(\alpha_2)$, $\chi(\beta) = \chi(\beta_1)|\chi(\beta_2)$, $\chi(\gamma) = \chi(\gamma_1)|\chi(\gamma_2)$, and $(\alpha, \beta) \to (\gamma, \alpha)$ is a possible differential of $\mathcal{F}_{SP}^{(1)}$.*

We have shown that all impossible differentials of an SPN structure are independent impossible differentials. However, this does not hold for the Feistel structure. In the following, we only consider the independent impossible differentials of a Feistel structure which fits well with most of the practical cases.

Lemma 3. *If $\alpha_1 \to \beta_1$ and $\alpha_2 \to \beta_2$ are independent possible differentials of $\mathcal{F}_{SP}^{(r)}$, $(\alpha_1|\alpha_2) \to (\beta_1|\beta_2)$ is also an independent possible differential.*

Theorem 3. *There exists an independent impossible differential of $\mathcal{F}_{SP}^{(r)}$ if and only if there exists an impossible differential $\alpha \not\to \beta$ of $\mathcal{F}_{SP}^{(r)}$ where $H(\alpha) = H(\beta) = 1$.*

Therefore, checking whether there exists an r-round independent impossible differential of a Feistel structure with SP-type round functions can also be reduced to checking whether there exists an r-round independent impossible differential with the Hamming weights of both the input and output difference being 1. Since the dual structure of $\mathcal{F}_{SP}$ is $\sigma \circ \mathcal{F}_{P^TS} \circ \sigma$, the results on impossible differentials cannot be applied to zero correlation linear hulls directly. However, in case P is invertible, we always have

$$\mathcal{F}_{P^TS} = \left((P^T)^{-1}, (P^T)^{-1}\right) \circ \mathcal{F}_{SP^T} \circ \left(P^T, P^T\right) \triangleq P_{\text{in}} \circ \mathcal{F}_{SP^T} \circ P_{\text{out}},$$

which indicates that despite some linear transformations applied to the input and output masks, respectively, both $\mathcal{F}_{SP}$ and $\mathcal{F}_{SP}^{\perp}$ are Feistel structures with SPN round functions. We use the following definition of independent zero correlation linear hulls for $\mathcal{F}_{SP}$.

Definition 5. *Let $\alpha \not\rightarrow \beta$ be a zero correlation linear hull of $\mathcal{F}_{SP}$. If for all α^* and β^* satisfying $\chi(P_{\text{in}}\alpha^*) = \chi(P_{\text{in}}\alpha)$ and $\chi(P_{\text{out}}\beta^*) = \chi(P_{\text{out}}\beta)$, $\alpha^* \not\rightarrow \beta^*$ are zero correlation linear hulls, we call $\alpha \not\rightarrow \beta$ an* independent zero correlation linear hull *of $\mathcal{F}_{SP}$. Otherwise, we call it a* dependent zero correlation linear hull *of $\mathcal{F}_{SP}$.*

Then based on the links between impossible differentials and zero correlation linear hulls, we have:

Corollary 3. *There exists an independent zero correlation linear hull of $\mathcal{F}_{SP}^{(r)}$ if and only if there exists an independent zero correlation linear hull $\alpha \not\rightarrow \beta$ of $\mathcal{F}_{SP}^{(r)}$ where $H(P_{\text{in}}\alpha) = H(P_{\text{out}}\beta) = 1$.*

Theorem 4. *Let $R_2(P)$ be the type 2 primitive indexes of P. Then, there does not exist any independent impossible differential or zero correlation linear hull of $\mathcal{F}_{SP}^{(r)}$ for $r \geq 2R_2(P) + 5$.*

The proof is similar with the SPN structures. The key point is that, as in the proof of Lemma 1, we can always choose $\beta_1, \beta_2, \gamma_1, \gamma_2$ and φ, where $H(\beta_1) = H(\beta_2) = H(\varphi) = m$ such that the differential shown in Fig. 3 is a possible one.

To avoid some potential attack, an FL/FL^{-1} layer is inserted to the Feistel structure every 6 rounds in Camellia. Denote by $\mathcal{E}^{Camellia*}$ the structure deduced by Camellia without the FL/FL^{-1} layer. Since

$$(P^*)^2 + I = \begin{pmatrix} 4&3&5&4&5&5&4&4 \\ 4&4&3&5&4&5&5&4 \\ 5&4&4&3&4&4&5&5 \\ 3&5&4&4&5&4&4&5 \\ 3&2&3&4&5&3&4&4 \\ 4&3&2&3&4&5&3&4 \\ 3&4&3&2&4&4&5&3 \\ 2&3&4&3&3&4&4&5 \end{pmatrix},$$

where I is the identity matrix, we have $R_2(P) = 2$. Therefore, we obtain the following proposition:

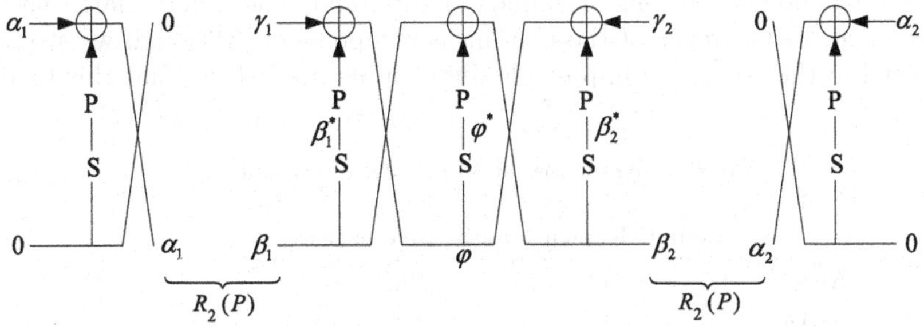

Fig. 3. Constructing $(2R_2(P) + 5)$-round differential for $\mathcal{F}_{SP}$

Proposition 3. *There does not exist any independent impossible differential of* $\mathcal{E}^{Camellia^*}$ *which covers* $r \geq 9$ *rounds. Or equivalently, there does not exist any* 9-*round independent impossible differential of Camellia without* FL/FL^{-1} *unless the details of the S-boxes are considered.*

In other words, unless we investigate the details of the S-boxes, the known independent impossible differentials of Camellia without FL/FL^{-1} cannot be improved with respect to the rounds.

Zodiac is another Feistel cipher with SP-type round function. Please refer to [11,15] for more details of Zodiac. Since we have $R_2(P) = 6$, if we do not exploit the details of the S-boxes, there does not exist any $2 \times 6 + 5 = 17$ independent impossible differential of Zodiac, while the longest impossible differential of Zodiac is 16 rounds [22].

Although there may exist some dependent impossible differentials of Feistel structures with SP-type round functions, we believe that the bound given above is also applicable to all impossible differentials.

6 Conclusion

In this paper, we mainly investigated the security of structures against impossible differential and zero correlation linear cryptanalysis. Our approach is to determine an upper bound for the longest impossible differentials for a structure. We first reduced the problem whether there exists an r-round impossible differential to the problem whether there exists an r-round impossible differential where the Hamming weights of the input and output differentials are 1. Therefore, we reduced the time complexity of checking whether there exists an impossible differential of an SPN structure or an independent impossible differential of a Feistel structure with SP-type round functions from $\mathcal{O}(2^{2m})$ to $\mathcal{O}(m^2)$. Then, by using the structures and dual structures, as well as the matrices theory, we have given an upper bound for the rounds of impossible differentials and zero correlation linear hulls for both SPN structures and Feistel structures with SP-type round functions.

As in the provable security of differential and linear cryptanalysis, we gave an upper bound on the longest rounds of the impossible differentials that are independent of the choice of the non-linear components. Although we are only interested in the truncated impossible differentials, we believe that this kind of

Table 1. Known results for some block ciphers

	Bound	Known rounds	References	
AES	4	4	[19]	
ARIA	4	4	[25]	
Camellia	8	8	[25]	Independent ID
Zodiac	16	16	[22]	Independent ID

impossible differentials cover most of the known cases. Therefore, they not only have theoretical significance, but also have practical significance. As a result, see Table 1, we show that unless the details of the non-linear layer are considered, there does not exist any 5-round impossible differentials of the AES or ARIA, and there does not exist any 9-round independent impossible differentials of the Camellia without FL/FL^{-1} layer.

Acknowledgment. The authors would like to thank the anonymous reviewers for their useful comments, and Shaojing Fu, Lei Cheng and Xuan Shen for fruitful discussions.

References

1. Aoki, K., Ichikawa, T., Kanda, M., Matsui, M., Moriai, S., Nakajima, J., Tokita, T.: *Camellia*: a 128-bit block cipher suitable for multiple platforms - design and analysis. In: Stinson, D.R., Tavares, S. (eds.) SAC 2000. LNCS, vol. 2012, pp. 39–56. Springer, Heidelberg (2001)
2. Aumasson, J.-P., Guo, J., Knellwolf, S., Matusiewicz, K., Meier, W.: Differential and invertibility properties of BLAKE. In: Hong, S., Iwata, T. (eds.) FSE 2010. LNCS, vol. 6147, pp. 318–332. Springer, Heidelberg (2010)
3. Biham, E., Biryukov, A., Shamir, A.: Cryptanalysis of Skipjack reduced to 31 rounds using impossible differentials. In: Stern, J. (ed.) EUROCRYPT 1999. LNCS, vol. 1592, pp. 12–23. Springer, Heidelberg (1999)
4. Biham, E., Shamir, A.: Differential Cryptanalysis of the Data Encryption Standard. Springer, New York (1993)
5. Blondeau, C.: Impossible differential attack on 13-round Camellia-192. Inf. Process. Lett. **115**(9), 660–666 (2015)
6. Bogdanov, A., Rijmen, V.: Linear hulls with correlation zero and linear cryptanalysis of block ciphers. Des. Codes Crypt. **70**(3), 369–383 (2014)
7. Boura, C., Naya-Plasencia, M., Suder, V.: Scrutinizing and improving impossible differential attacks: applications to CLEFIA, Camellia, LBlock and SIMON. In: Sarkar, P., Iwata, T. (eds.) ASIACRYPT 2014. LNCS, vol. 8873, pp. 179–199. Springer, Heidelberg (2014)
8. Daemen, J., Rijmen, V.: AES and the wide trail design strategy. In: Knudsen, L.R. (ed.) EUROCRYPT 2002. LNCS, vol. 2332, pp. 108–109. Springer, Heidelberg (2002)
9. Daemen, J., Rijmen, V.: The Design of Rijndael: AES - The Advanced Encryption Standard. Information Security and Cryptography. Springer, Berlin (2002)
10. Guo, J., Thomsen, S.S.: Deterministic differential properties of the compression function of BMW. In: Biryukov, A., Gong, G., Stinson, D.R. (eds.) SAC 2010. LNCS, vol. 6544, pp. 338–350. Springer, Heidelberg (2011)
11. Hong, D., Sung, J., Moriai, S., Lee, S.-J., Lim, J.-I.: Impossible differential cryptanalysis of Zodiac. In: Matsui, M. (ed.) FSE 2001. LNCS, vol. 2355, pp. 300–311. Springer, Heidelberg (2002)
12. Kim, J., Hong, S., Lim, J.: Impossible differential cryptanalysis using matrix method. Discrete Math. **310**(5), 988–1002 (2010)
13. Knudsen, L.R.: DEAL - A 128-bit Block Cipher. Technical report, Department of Informatics, University of Bergen, Norway (1998)

14. Kwon, D., Kim, J., Park, S., Sung, S.H., Sohn, Y., Song, J.H., Yeom, Y., Yoon, E., et al.: New block cipher: ARIA. In: Lim, J.-I., Lee, D.-H. (eds.) ICISC 2003. LNCS, vol. 2971, pp. 432–445. Springer, Heidelberg (2004)
15. Lee, C., Jun, K., Jung, M.S., Park, S., Kim, J.: Zodiac version 1.0 (revised) architecture and specification. In: Standardization Workshop on Information Security Technology 2000, Korean Contribution on MP18033, ISO/IEC JTC1/SC27 N2563 (2000). http://www.kisa.or.kr/seed/index.html
16. Li, R., Sun, B., Li, C.: Impossible differential cryptanalysis of SPN ciphers. IET Inf. Secur. **5**(2), 111–120 (2011)
17. Lu, J., Dunkelman, O., Keller, N., Kim, J.-S.: New impossible differential attacks on AES. In: Chowdhury, D.R., Rijmen, V., Das, A. (eds.) INDOCRYPT 2008. LNCS, vol. 5365, pp. 279–293. Springer, Heidelberg (2008)
18. Luo, Y., Lai, X., Wu, Z., Gong, G.: A unified method for finding impossible differentials of block cipher structures. Inf. Sci. **263**, 211–220 (2014)
19. Mala, H., Dakhilalian, M., Rijmen, V., Modarres-Hashemi, M.: Improved impossible differential cryptanalysis of 7-round AES-128. In: Gong, G., Gupta, K.C. (eds.) INDOCRYPT 2010. LNCS, vol. 6498, pp. 282–291. Springer, Heidelberg (2010)
20. Matsui, M.: Linear cryptanalysis method for DES cipher. In: Helleseth, T. (ed.) EUROCRYPT 1993. LNCS, vol. 765, pp. 386–397. Springer, Heidelberg (1994)
21. Sun, B., Liu, Z., Rijmen, V., Li, R., Cheng, L., Wang, Q., AlKhzaimi, H., Li, C.: Links among impossible differential, integral and zero correlation linear cryptanalysis. In: Gennaro, R., Robshaw, M. (eds.) CRYPTO 2015. LNCS, vol. 9215, pp. 95–115. Springer, Berlin (2015)
22. Sun, B., Zhang, P., Li, C.: Impossible differential and integral cryptanalysis of Zodiac. J. Softw. **22**(8), 1911–1917 (2011)
23. Vaudenay, S.: Provable security for block ciphers by decorrelation. In: Meinel, C., Morvan, M., Krob, D. (eds.) STACS 1998. LNCS, vol. 1373, pp. 249–275. Springer, Heidelberg (1998)
24. Wu, S., Wang, M.: Automatic search of truncated impossible differentials for word-oriented block ciphers. In: Galbraith, S., Nandi, M. (eds.) INDOCRYPT 2012. LNCS, vol. 7668, pp. 283–302. Springer, Heidelberg (2012)
25. Wu, W., Zhang, W., Feng, D.: Impossible differential cryptanalysis of reduced-round ARIA and Camellia. J. Comput. Sci. Technol. **22**(3), 449–456 (2007)

A Proof of Lemma 1

Firstly, $\alpha_1 \rightarrow \beta_1$ and $\alpha_2 \rightarrow \beta_2$ are possible differentials of $\mathcal{E}^{SP}$ implies that there exist some α_1^*, α_2^*, $\chi(\alpha_1^*) = \chi(\alpha_1)$, $\chi(\alpha_2^*) = \chi(\alpha_2)$, such that the following differentials hold:

$$\begin{cases} \alpha_1 \xrightarrow{S} \alpha_1^* \xrightarrow{P} \beta_1, \\ \alpha_2 \xrightarrow{S} \alpha_2^* \xrightarrow{P} \beta_2. \end{cases}$$

For any $\lambda \in \mathbb{F}_{2^b}^*$, since $\chi(\lambda\alpha_2^*) = \chi(\alpha_2)$, $\alpha_2 \xrightarrow{S} \lambda\alpha_2^* \xrightarrow{P} \lambda\beta_2$ is also a possible differential of $\mathcal{E}^{SP}$.

Without loss of generality, let

$$\begin{cases} \alpha_1^* = (x_{w_1}^{(1)}, x_{r_1}^{(1)}, 0_{m-r_1-w_1}) \\ \alpha_2^* = (x_{w_1}^{(2)}, 0_{r_1}, x_{m-r_1-w_1}^{(2)}) \\ \beta_1 = (y_{w_2}^{(1)}, y_{r_2}^{(1)}, 0_{m-r_2-w_2}) \\ \beta_2 = (y_{w_2}^{(2)}, 0_{r_2}, y_{m-r_2-w_2}^{(2)}) \end{cases}$$

where $0_t = \underbrace{0 \cdots 0}_{t}$, $x_r^{(i)}, y_r^{(i)} \in (\mathbb{F}_{2^b}^*)^r$. Let

$$\begin{cases} x_{w_1}^{(1)} = (a_0^{(1)}, \ldots, a_{w_1-1}^{(1)}) \\ x_{w_1}^{(2)} = (a_0^{(2)}, \ldots, a_{w_1-1}^{(2)}) \\ y_{w_2}^{(1)} = (b_0^{(1)}, \ldots, b_{w_2-1}^{(1)}) \\ y_{w_2}^{(2)} = (b_0^{(2)}, \ldots, b_{w_2-1}^{(2)}) \end{cases}$$

and let

$$\Lambda = \left\{ \frac{a_0^{(1)}}{a_0^{(2)}}, \ldots, \frac{a_{w_1-1}^{(1)}}{a_{w_1-1}^{(2)}}, \frac{b_0^{(1)}}{b_0^{(2)}}, \ldots, \frac{b_{w_2-1}^{(1)}}{b_{w_2-1}^{(2)}} \right\}.$$

Since $\#\Lambda \le w_1 + w_2 \le m + m = 2m \le 2 \times (2^{b-1} - 1) = 2^b - 2$, $\mathbb{F}_{2^b}^* \setminus \Lambda$ is a non-empty set. Therefore, for $\lambda \in \mathbb{F}_{2^b}^* \setminus \Lambda$, we always have

$$\begin{cases} \chi(\alpha_1^* \oplus \lambda\alpha_2^*) = \chi(\alpha_1^* | \alpha_2^*) \\ \chi(\beta_1 \oplus \lambda\beta_2) = \chi(\beta_1 | \beta_2), \end{cases}$$

which implies that

$$\alpha_1 | \alpha_2 \xrightarrow{\mathcal{S}} \alpha_1^* \oplus \lambda\alpha_2^* \xrightarrow{P} \beta_1 \oplus \lambda\beta_2$$

is a possible differential of $\mathcal{E}^{SP}$.

Polytopic Cryptanalysis

Tyge Tiessen(✉)

DTU Compute, Technical University of Denmark, Kongens Lyngby, Denmark
tyti@dtu.dk

Abstract. Standard differential cryptanalysis uses statistical dependencies between the difference of two plaintexts and the difference of the respective two ciphertexts to attack a cipher. Here we introduce polytopic cryptanalysis which considers interdependencies between larger sets of texts as they traverse through the cipher. We prove that the methodology of standard differential cryptanalysis can unambiguously be extended and transferred to the polytopic case including impossible differentials. We show that impossible polytopic transitions have generic advantages over impossible differentials. To demonstrate the practical relevance of the generalization, we present new low-data attacks on round-reduced DES and AES using impossible polytopic transitions that are able to compete with existing attacks, partially outperforming these.

1 Introduction

Without doubt is differential cryptanalysis one of the most important tools that the cryptanalyst has at hand when trying to evaluate the security of a block cipher. Since its conception by Biham and Shamir [2] in their effort to break the Data Encryption Standard [26], it has been successfully applied to many block ciphers such that any modern block cipher is expected to have strong security arguments against this attack.

The methodology of differential cryptanalysis has been extended several times with a number of attack vectors, most importantly truncated differentials [19], impossible differentials [1,20], and higher-order differentials [19,22]. Further attacks include the boomerang attack [29], which bears some resemblance of second-order differential attacks, and differential-linear attacks [24].

Nonetheless many open problems remain in the field of differential cryptanalysis. Although the concept of higher-order differentials is almost 20 years old, it has not seen many good use cases. One reason has been the difficulty of determining the probability of higher-order differentials accurately without evaluating Boolean functions with prohibitively many terms. Thus the common use case remains probability 1 higher-order differentials where we know that a derivative of a certain order has to evaluate to zero because of a limit in the degree of the function.

M. Fischlin and J.-S. Coron (Eds.): EUROCRYPT 2016, Part I, LNCS 9665, pp. 214–239, 2016.
DOI: 10.1007/978-3-662-49890-3_9

Another open problem is the exact determination of the success probability of boomerang attacks and their extensions. It has correctly been observed that the correlation between differentials must be taken into account to accurately determine the success probability [25]. The true probability can otherwise deviate arbitrarily from the estimated one.

Starting with Chabaud and Vaudenay [12], considerable effort has gone into shedding light on the relation and interdependencies of various cryptographic attacks (see for example [5,6,30]). With this paper, we offer a generalized view on the various types of differential attacks that might help to understand both the interrelation between the attacks as well as the probabilities of the attacks better.

Our Contribution

In this paper we introduce polytopic cryptanalysis. It can be viewed as a generalization of standard differential cryptanalysis which it embeds as a special case. We prove that the definitions and methodology of differential cryptanalysis can unambiguously be extended to polytopic cryptanalysis, including the concept of impossible differentials. Polytopic cryptanalysis is general enough to even encompass attacks such as higher-order differentials and might thus be valuable as a reference framework.

For impossible polytopic transitions, we show that they exhibit properties that allow them to be very effective in scenarios where ordinary impossible differentials fail. This is mostly due to a generic limit in the diffusion of any block cipher that guarantees that only a negligible number of all polytopic transitions is possible for a sufficiently high choice of dimension. This also makes impossible polytopic transitions ideal for low-data attacks where standard impossible differentials usually have a high data complexity.

Finally we prove that polytopic cryptanalysis is not only theoretically intriguing but indeed relevant for practical cryptanalysis by demonstrating competitive impossible polytopic attacks on round-reduced DES and AES that partly outperform existing low-data attacks and offer different trade-offs between time and data complexity.

In the appendix, we further prove that higher-order differentials can be expressed as truncated polytopic transitions and are hence a special case of these. Thus higher-order differentials can be expressed in terms of a collection of polytopic trails just as differentials can be expressed as a collection of differential trails. A consequence of this is that it is principally possible to determine lower bounds for the probability of a higher-order differential by summing over the probabilities of a subset of the polytopic trails which it contains.

Related Work

To our knowledge, the concept of polytopic transitions is new and has not been used in cryptanalysis before. Nonetheless there is other work that shares some similarities with polytopic cryptanalysis.

Higher-order differentials [22] can in some sense also be seen as a higher-dimensional version of a differential. However, most concepts of ordinary differentials do not seem to extend to higher-order differentials, such as characteristics or iterated differentials.

The idea of using several differentials simultaneously in an attack is not new (see for example [4]). However as opposed to assuming independence of the differentials, which does not hold in general (see [25]), we explicitly take their correlation into account and use it in our framework.

Another type of cryptanalysis that uses a larger set of texts instead of a single pair is integral cryptanalysis (see for example [3,14]), in which structural properties of the cipher are used to elegantly determine a higher-order derivative to be zero without relying on bounds in the degree. These attacks can be considered a particular form of higher-order differentials.

Finally decorrelation theory [28] also considers relations between multiple plaintext-ciphertext pairs but takes a different direction by considering security proofs based on a lack of correlation between the texts.

Organization of the Paper

In Sect. 2, notation and concepts necessary for polytopic cryptanalysis are introduced. It is demonstrated how the concepts of differential cryptanalysis naturally extend to polytopic cryptanalysis. We also take a closer look at the probability of polytopic transitions and applicability of simple polytopic cryptanalysis.

In Sect. 3, we introduce impossible polytopic transitions. We show that impossible polytopic transitions offer some inherent advantages over impossible differentials and are particularly interesting for low-data attacks. We show that, given an efficient method to determine the possibility of a polytopic transition, generic impossible polytopic attack always exist.

In Sect. 4, we demonstrate the practicability of impossible polytopic transition attacks. We present some attacks on DES and AES that are able to compete with existing attacks with low-data complexity, partially outperforming these.

Furthermore, in Appendix B truncated polytopic transitions are introduced. We then give a proof that higher-order differentials are a special case of these. The cryptanalytic ramifications of the fact that higher-order differentials consist of polytopic trails are then discussed.

Notation

We use $\mathbb{F}_2^n$ to denote the n-dimensional binary vector space. To identify numbers in hexadecimal notation we use a typewriter font as in `3af179`. Random variables are denoted with bold capital letters ($\mathbf{X}$). We will denote d-difference (introduced later) by bold Greek letters ($\boldsymbol{\alpha}$) and standard differences by Roman (i.e., non-bold) Greek letters (α).

2 Polytopes and Polytopic Transitions

Classical differential cryptanalysis utilizes the statistical interdependency of two texts as they traverse through the cipher. When we are not interested in the absolute position of the two texts in the state space, the difference between the two texts completely determines their relative positioning.

But there is no inherent reason that forces us to be restricted to only using a pair of texts. Let us instead consider an ordered set of texts as they traverse through the cipher.

Definition 1 (s-polytope). *An s-polytope in $\mathbb{F}_2^n$ is an s-tuple of values in $\mathbb{F}_2^n$.*

Similar to differential cryptanalysis, we are not so much interested in the absolute position of these texts but the relations between the texts. If we choose one of the texts as the point of reference, the relations between all texts are already uniquely determined by only considering their differences with respect to the reference text. If we thus have $d+1$ texts, we can describe their relative positioning by a tuple of d differences (see also Fig. 1).

Definition 2 (d-difference). *A d-difference over $\mathbb{F}_2^n$ is a d-tuple of values in $\mathbb{F}_2^n$ describing the relative position of the texts of a $(d+1)$-polytope from one point of reference.*

When we reduce a $(d+1)$-polytope to a corresponding d-difference, we loose the information of the absolute position of the polytope. A d-difference thus corresponds to an equivalence class of $(d+1)$-polytopes where polytopes are equivalent if and only if they can be transformed into each other by simple shifting in state space. We will mostly be dealing with these equivalence classes.

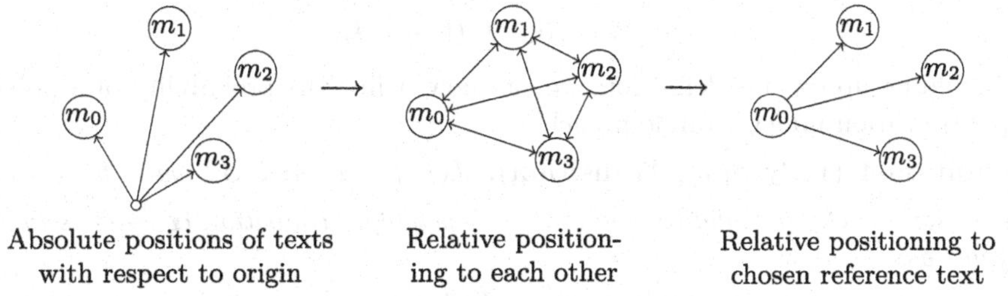

Fig. 1. Depiction of three views of a polytope with four vertices

In principal there are many d-differences that correspond to one $(d+1)$-polytope depending on the choice of reference text and the order of the differences. As a convention we will construct a d-difference from a $(d+1)$-polytope as follows:

Convention. For a $(d+1)$-polytope $(m_0, m_1, \ldots, m_d)$, the corresponding d-difference is created as $(m_0 \oplus m_1, m_0 \oplus m_2, \ldots, m_0 \oplus m_d)$.

This means, we use the first text of the polytope as the reference text and write the differences in the same order as the remaining texts of the polytope. We will call the reference text the *anchor* of the d-difference. Hence if we are given a d-difference and the value of the anchor, we can reconstruct the corresponding $(d+1)$-polytope uniquely.

Example. Let (m_0, m_1, m_2, m_3) be a 4-polytope in $\mathbb{F}_2^n$. Then $(m_0 \oplus m_1, m_0 \oplus m_2, m_0 \oplus m_3)$ is the corresponding 3-difference with m_0 as the anchor.

In the following, we will now show that we can build a theory of polytopic cryptanalysis in which the same methodology as in standard differential cryptanalysis applies. Standard differential cryptanalysis is contained in this framework as a special case.

A short note regarding possible definitions of difference: in this paper we restrict ourselves to XOR-differences as the most common choice. Most, if not all, statements in this paper naturally extend to other definitions of difference, e.g., in modular arithmetic.

The equivalent of a differential in polytopic cryptanalysis is the polytopic transition. We use d-differences for the definition.

Definition 3 (Polytopic Transition with Fixed Anchor). *Let $f : \mathbb{F}_2^n \to \mathbb{F}_2^q$. Let $\boldsymbol{\alpha}$ be a d-difference $(\alpha_1, \alpha_2, \ldots, \alpha_d)$ over $\mathbb{F}_2^n$ and let $\boldsymbol{\beta}$ be the d-difference $(\beta_1, \beta_2, \ldots, \beta_d)$ over $\mathbb{F}_2^q$. By the $(d+1)$-polytopic transition $\boldsymbol{\alpha} \xrightarrow[x]{f} \boldsymbol{\beta}$ we denote that f maps the polytope corresponding to $\boldsymbol{\alpha}$ with anchor x to a polytope corresponding to $\boldsymbol{\beta}$. More precisely, we have $\boldsymbol{\alpha} \xrightarrow[x]{f} \boldsymbol{\beta}$ if and only if*

$$\begin{aligned} f(x \oplus \alpha_1) \oplus f(x) &= \beta_1 \\ \textit{and } f(x \oplus \alpha_2) \oplus f(x) &= \beta_2 \\ &\cdots \\ \textit{and } f(x \oplus \alpha_d) \oplus f(x) &= \beta_d. \end{aligned}$$

Building up on this definition, we can now define the probability of a polytopic transition under a random anchor.

Definition 4 (Polytopic Transition). *Let f, $\boldsymbol{\alpha}$, and $\boldsymbol{\beta}$ again be as in Definition 3. The probability of the $(d+1)$-polytopic transition $\boldsymbol{\alpha} \xrightarrow{f} \boldsymbol{\beta}$ is then defined as:*

$$\Pr\left(\boldsymbol{\alpha} \xrightarrow{f} \boldsymbol{\beta}\right) := \Pr_{\mathbf{X}}\left(\boldsymbol{\alpha} \xrightarrow[\mathbf{X}]{f} \boldsymbol{\beta}\right) \tag{1}$$

where $\mathbf{X}$ is a random variable, uniformly distributed on $\mathbb{F}_2^n$. We will at times also write $\boldsymbol{\alpha} \longrightarrow \boldsymbol{\beta}$ if the function is clear from the context or not important.

Note that this definition coincides with the definition of the differential probability when differences between only two texts (2-polytopes) are considered.

Let $f : \mathbb{F}_2^n \to \mathbb{F}_2^n$ now be a function that is the repeated composition of round functions $f_i : \mathbb{F}_2^n \to \mathbb{F}_2^n$:

$$f := f_r \circ \cdots \circ f_2 \circ f_1. \tag{2}$$

Similarly to differential cryptanalysis, we can now define trails of polytopes:

Definition 5 (Polytopic Trail). *Let f be as in Eq. (2). A polytopic trail on f is an $(r+1)$-tuple of d-differences $(\boldsymbol{\alpha}_0, \boldsymbol{\alpha}_1, \ldots, \boldsymbol{\alpha}_r)$ written as*

$$\boldsymbol{\alpha}_0 \xrightarrow{f_1} \boldsymbol{\alpha}_1 \xrightarrow{f_2} \cdots \xrightarrow{f_r} \boldsymbol{\alpha}_r. \tag{3}$$

The probability of such a polytopic trail is defined as

$$\Pr_{\mathbf{X}}\left(\boldsymbol{\alpha}_0 \xrightarrow[\mathbf{X}]{f_1} \boldsymbol{\alpha}_1 \text{ and } \boldsymbol{\alpha}_1 \xrightarrow[f_1(\mathbf{X})]{f_2} \boldsymbol{\alpha}_2 \text{ and } \cdots \text{ and } \boldsymbol{\alpha}_{r-1} \xrightarrow[f_{r-1}\circ\cdots\circ f_1(\mathbf{X})]{f_r} \boldsymbol{\alpha}_r\right) \tag{4}$$

where $\mathbf{X}$ is a random variable, distributed uniformly on $\mathbb{F}_2^n$.

Similarly to differentials, it is possible to partition a polytopic transition over a composed function into all polytopic trails that feature the same input and output differences as the polytopic transition.

Proposition 1. *The probability of a polytopic transition $\boldsymbol{\alpha}_0 \xrightarrow{f} \boldsymbol{\alpha}_r$ over a function $f : \mathbb{F}_2^n \to \mathbb{F}_2^n, f = f_r \circ \cdots \circ f_2 \circ f_1$ is the sum of the probabilities of all polytopic trails $(\boldsymbol{\alpha}_0, \boldsymbol{\alpha}_1, \ldots, \boldsymbol{\alpha}_r)$ which it contains:*

$$\Pr\left(\boldsymbol{\alpha}_0 \xrightarrow{f} \boldsymbol{\alpha}_r\right) = \sum_{\boldsymbol{\alpha}_1,\ldots,\boldsymbol{\alpha}_{r-1}} \Pr\left(\boldsymbol{\alpha}_0 \xrightarrow{f_1} \boldsymbol{\alpha}_1 \xrightarrow{f_2} \cdots \xrightarrow{f_{r-1}} \boldsymbol{\alpha}_{r-1} \xrightarrow{f_r} \boldsymbol{\alpha}_r\right) \tag{5}$$

where $\boldsymbol{\alpha}_0, \ldots, \boldsymbol{\alpha}_r$ are d-differences and as such lie in $\mathbb{F}_2^{dn}$.

Proof. If we fix the initial value of the anchor, we also fix the trail that the polytope has to take. The set of polytopic trails gives us thus a partition of the possible anchor values and in particular a partition of the anchors for which the output polytope is of type $\boldsymbol{\alpha}_r$. Using the above definitions we thus get:

$$\begin{aligned}
\Pr\left(\boldsymbol{\alpha}_0 \xrightarrow{f} \boldsymbol{\alpha}_r\right) &= \Pr_{\mathbf{X}}\left(\boldsymbol{\alpha}_0 \xrightarrow[\mathbf{X}]{f} \boldsymbol{\alpha}_r\right) \\
&= 2^{-n} \cdot \left|\left\{x \in \mathbb{F}_2^n \;\middle|\; \boldsymbol{\alpha}_0 \xrightarrow[x]{f} \boldsymbol{\alpha}_r\right\}\right| \\
&= 2^{-n} \cdot \sum_{\boldsymbol{\alpha}_1,\ldots,\boldsymbol{\alpha}_{r-1}} \left|\left\{x \in \mathbb{F}_2^n \;\middle|\; \boldsymbol{\alpha}_0 \xrightarrow[x]{f_1} \boldsymbol{\alpha}_1,\ \boldsymbol{\alpha}_1 \xrightarrow[f_1(x)]{f_2} \boldsymbol{\alpha}_2, \ldots \right.\right. \\
&\qquad\qquad \left.\left. \ldots,\ \boldsymbol{\alpha}_{r-1} \xrightarrow[f_{r-1}\circ\cdots\circ f_1(x)]{f_r} \boldsymbol{\alpha}_r\right\}\right| \\
&= \sum_{\boldsymbol{\alpha}_1,\ldots,\boldsymbol{\alpha}_{r-1}} \Pr_{\mathbf{X}}\left(\boldsymbol{\alpha}_0 \xrightarrow[\mathbf{X}]{f_1} \boldsymbol{\alpha}_1 \text{ and } \boldsymbol{\alpha}_1 \xrightarrow[f_1(\mathbf{X})]{f_2} \boldsymbol{\alpha}_2 \text{ and } \ldots\right. \\
&\qquad\qquad \left.\ldots \text{ and } \boldsymbol{\alpha}_{r-1} \xrightarrow[f_{r-1}\circ\cdots\circ f_1(\mathbf{X})]{f_r} \boldsymbol{\alpha}_r\right) \\
&= \sum_{\boldsymbol{\alpha}_1,\ldots,\boldsymbol{\alpha}_{r-1}} \Pr\left(\boldsymbol{\alpha}_0 \xrightarrow{f_1} \boldsymbol{\alpha}_1 \xrightarrow{f_2} \cdots \xrightarrow{f_{r-1}} \boldsymbol{\alpha}_{r-1} \xrightarrow{f_r} \boldsymbol{\alpha}_r\right)
\end{aligned}$$

which proves the proposition. □

To be able to calculate the probability of a differential trail, it is common in differential cryptanalysis to make an assumption on the independence of the round transitions. This is usually justified by showing that the cipher is a Markov cipher and by assuming the stochastic equivalence hypothesis (see [23]). As we will mostly be working with impossible trails where these assumptions are not needed, we will assume for now that this independence holds and refer the interested reader to Appendix A where the Markov model is adapted to polytopic cryptanalysis.

Under the assumption that the single round transitions are independent, we can work with polytopic transitions just as with differentials:

1. The probability of a polytopic transition is the sum of the probabilities of all polytopic trails with the same input and output d-difference.
2. The probability of a polytopic trail is the product of the probabilities of the 1-round polytopic transitions that constitute the trail.

We are thus principally able to calculate the probability of a polytopic transition over many rounds by knowing how to calculate the polytopic transition over single rounds.

Now to calculate the probability of a 1-round polytopic transition, we can use the following observations:

3. A linear function maps a d-difference with probability 1 to the d-difference that is the result of applying the linear function to each single difference in the d-difference.
4. Addition of a constant to the anchor leaves the d-difference unchanged.
5. The probability of a polytopic transition over an S-box layer is the product of the polytopic transitions for each S-box.

We are thus able to determine probabilities of polytopic transitions and polytopic trails just as we are used to from standard differential cryptanalysis.

A Note on Correlation, Diffusion and the Difference Distribution Table

When estimating the probability of a polytopic transition a first guess might be that it is just the product of the individual 1-dimensional differentials. For a 3-polytopic transition we might for example expect:

$$\Pr\Big((\alpha_0,\alpha_1) \longrightarrow (\beta_0,\beta_1)\Big) \overset{?}{=} \Pr\Big(\alpha_0 \longrightarrow \beta_0\Big) \cdot \Pr\Big(\alpha_1 \longrightarrow \beta_1\Big).$$

That this is generally *not* the case is a consequence of the following lemma.

Lemma 1. *Let $f : \mathbb{F}_2^n \rightarrow \mathbb{F}_2^n$. For a given input d-difference $\boldsymbol{\alpha}$ the number of output d-differences to which $\boldsymbol{\alpha}$ is mapped with non-zero probability is upper bounded by 2^n.*

Proof. This is just a result of the fact that the number of anchors for the transition is limited to 2^n:

$$\left|\left\{\beta \in \mathbb{F}_2^{dn} \;\middle|\; \Pr\left(\alpha \xrightarrow{f} \beta\right) > 0\right\}\right| = \left|\left\{\beta \in \mathbb{F}_2^{dn} \;\middle|\; \exists x \in \mathbb{F}_2^n : \alpha \xrightarrow[x]{f} \beta\right\}\right| \leq 2^n$$

□

One implication of this limitation of possible output d-differences is a correlation between differentials: the closer the distribution of differences of a function is to a uniform distribution, the stronger is the correlation of differentials over that function.

Example. Let us take the AES 8-bit S-box (denoted by S here) which is differentially 4-uniform. Consider the three differentials, $7 \xrightarrow{S} 166$, $25 \xrightarrow{S} 183$, and $25 \xrightarrow{S} 1$ which have probabilities 2^{-6}, 2^{-6}, and 2^{-7} respectively. The probabilities of the polytopic transitions of the combined differentials deviate strongly from the product of the single probabilities:

$$\Pr\left((7,25) \xrightarrow{S} (166,183)\right) = 2^{-6} > \Pr\left(7 \xrightarrow{S} 166\right) \cdot \Pr\left(25 \xrightarrow{S} 183\right) = 2^{-12}$$

$$\Pr\left((7,25) \xrightarrow{S} (166,1)\right) = 0 < \Pr\left(7 \xrightarrow{S} 166\right) \cdot \Pr\left(25 \xrightarrow{S} 1\right) = 2^{-13}.$$

Another consequence of Lemma 1 is that it sets an inherent limit to the maximal diffusion possible over one round. A one d-difference can at most be mapped to 2^n possible d-differences over one round, the number of possible d-differences reachable can only increase by a factor of 2^n over each round. Thus when starting from one d-difference, after one round at most 2^n d-differences are possible, after two rounds at most 2^{2n} differences are possible, after three rounds at most 2^{3n} are possible and generally after round r at most 2^{rn} d-differences are possible.

In standard differential cryptanalysis, the number of possible output differences for a given input difference is limited by the state size of the function. This is no longer true for d-differences: if the state space is $\mathbb{F}_2^n$, the space of d-differences is $\mathbb{F}_2^{dn}$. The number of possible d-differences thus increases exponentially with the dimension d. This has a consequence for the size of the difference distribution table (DDT). For an 8-bit S-box, a classical DDT has a size of 2^{16} entries, i.e., 64 kilobytes. But already the DDT for 3-differences has a size of 2^{48}, i.e., 256 terabytes. Fortunately though, a third consequence of Lemma 1 is that the DDT table is sparse for $d > 1$. As a matter of fact, we can calculate any row of the DDT with a time complexity of 2^n by trying out all possible values for the anchor.

Relation to Decorrelation Theory. Decorrelation theory [28] is a framework that can be used to design ciphers which are provably secure against a range of attacks including differential and linear cryptanalysis. A cipher is called perfectly decorrelated of order d when the image of any d-tuple of distinct plaintexts is uniformly distributed on all d-tuples of ciphertexts with distinct values under a uniformly distributed random key. It can for example be proved that a cipher which is perfectly decorrelated of order 2 is secure against standard differential and linear cryptanalysis.

When we consider $(d+1)$-polytopes in polytopic cryptanalysis, we can naturally circumvent security proofs for order-d perfectly decorrelated ciphers. The boomerang attack [29] for example – invented to break an order-2 perfectly decorrelated cipher – can be described as a 4-polytopic attack.

Limitations of Simple Polytopic Cryptanalysis

Can simple polytopic cryptanalysis, i.e., using a single polytopic transition, outperform standard differential cryptanalysis? Unfortunately this is generally not the case as is shown in the following.

Definition 6. *Let $\boldsymbol{\alpha} \longrightarrow \boldsymbol{\beta}$ be a $(d+1)$-polytopic transition with d-differences $\boldsymbol{\alpha}$ and $\boldsymbol{\beta}$. Let $\boldsymbol{\alpha}' \longrightarrow \boldsymbol{\beta}'$ be a d'-difference with $d' \leq d$. We then write $(\boldsymbol{\alpha}', \boldsymbol{\beta}') \sqsubseteq (\boldsymbol{\alpha}, \boldsymbol{\beta})$ if and only if for each $i \in [1, d']$ there exists $j \in [1, d]$ such that the ith differences in $\boldsymbol{\alpha}'$ and $\boldsymbol{\beta}'$ correspond to the jth differences in $\boldsymbol{\alpha}$ and $\boldsymbol{\beta}$.*

Using this notation, we have the following lemma:

Lemma 2. *Let $\boldsymbol{\alpha} \longrightarrow \boldsymbol{\beta}$ be a $(d+1)$-polytopic transition and let $\boldsymbol{\alpha}' \longrightarrow \boldsymbol{\beta}'$ be a $(d'+1)$-polytopic transition with $d' \leq d$ and $(\boldsymbol{\alpha}', \boldsymbol{\beta}') \sqsubseteq (\boldsymbol{\alpha}, \boldsymbol{\beta})$. Then*

$$\Pr(\boldsymbol{\alpha} \longrightarrow \boldsymbol{\beta}) \leq \Pr(\boldsymbol{\alpha}' \longrightarrow \boldsymbol{\beta}'). \tag{6}$$

Proof. This follows directly from the fact that $\boldsymbol{\alpha} \xrightarrow[x]{f} \boldsymbol{\beta}$ implies $\boldsymbol{\alpha}' \xrightarrow[x]{f} \boldsymbol{\beta}'$. □

In words, the probability of a polytopic transition is always at most as high as the probability of any lower dimensional polytopic transition that it contains. In particular, it can never have a higher probability than any standard differential that it contains.

It can in some instances still be profitable to use a single polytopic transition instead of a standard differential that it contains. This is the case when the probability of the polytopic transition is the same as (or close to) the probability of the best standard differential it contains. Due to the fact that the space of d-differences is much larger than that of standard differentials (2^{dn} vs. 2^n), one set of texts that follows the polytopic transition is usually enough to distinguish the biased distribution from a uniform distribution as opposed to standard differentials where at least two are needed. Nonetheless the cryptanalytic advantages of polytopic cryptanalysis lie elsewhere as we will see in the next sections.

3 Impossible Polytopic Cryptanalysis

Impossible differential cryptanalysis makes use of differentials with probability zero to distinguish a cipher from an ideal cipher. In this section, we extend the definition to encompass polytopic transitions.

Impossible polytopic cryptanalysis offers distinct advantages over standard impossible differential cryptanalysis that are a result of the exponential increase in the size of the space of d-differences with increasing dimension d. This not only allows impossible $(d+1)$-polytopic attacks using just a single set of $d+1$ chosen plaintexts, it also allows generic distinguishing attacks on $(d-1)$-round block ciphers whenever it is computationally easy to determine whether a $(d+1)$-polytopic transition is possible or not. We elaborate this in more detail later in this section.

Definition 7. *An impossible $(d+1)$-polytopic transition is a $(d+1)$-polytopic transition that occurs with probability zero.*

In impossible differential attacks, we use knowledge of an impossible differential over $r-1$ rounds to filter out wrong round key guesses for the last round: any round key that decrypts a text pair such that their difference adheres to the impossible differential has to be wrong. The large disadvantage of this attack is that it always requires a large number of text pairs to sufficiently reduce the number of possible keys. This is due to the fact that the filtering probability corresponds to the fraction of the impossible differentials among all differentials. Unfortunately for the attacker, most ciphers are designed to provide good diffusion, such that this ratio is usually low after a few rounds.

This is exactly where the advantage of impossible polytopic transitions lies. Due to the exponential increase in the size of the space of d-differences (from $\mathbb{F}_2^n$ to $\mathbb{F}_2^{dn}$) and the limitation of the diffusion to maximally a factor of 2^n (see Lemma 1), the ratio of possible $(d+1)$-polytopic transitions to impossible $(d+1)$-polytopic transitions will be low for many more rounds than possible for standard differentials. In fact, by increasing the dimension d of the polytopic transition, it can be assured that the ratio of possible to impossible polytopic transitions is close to zero for an almost arbitrary number of rounds.

An impossible $(d+1)$-polytopic attack could then proceed as follows. Let n be the block size of the cipher and let l be the number of bits in the last round key.

1. Choose a d and a d-difference such that the ratio of possible to impossible $(d+1)$-polytopic transitions is lower than 2^{-l-1}.
2. Get the r-round encryption of $d+1$ plaintexts chosen such that they adhere to the input d-difference.
3. For each guess of the round key k_r decrypt the last round. If the obtained d-difference after the $(r-1)$th round is possible, keep the key as a candidate. Otherwise discard it.

Clearly this should leave us on average with one round key candidate which is bound to be the correct one. In practice, an attack would likely be more complex, e.g., with only partially guessed round keys and tradeoffs in the filtering probability and the data/time complexities.

While the data complexity is limited to $d+1$ chosen plaintexts (and thus very low), the time complexity is harder to determine and depends on the difficulty of determining whether an obtained $(r-1)$-round $(d+1)$-polytopic transition is possible or not. The straightforward approach is to precompute a list of possible d-differences after round $r - 1$. Both the exponentially increasing memory requirements and the time of the precomputation limit this approach though. In spite of this, attacks using this approach are competitive with existing low-data attacks as we show in Sect. 4.

One possibility to reduce the memory complexity is to use a meet-in-the-middle approach where one searches for a collision in the possible d-differences reachable from the input d-difference and the calculated d-difference after round $(r-1)$ at a round somewhere in the middle of the cipher. This however requires to repeat the computation for every newly calculated d-difference and thus limits its use in the scenario where we calculated a new d-difference after round $(r-1)$ for each key guess (not in a distinguishing attack though).

Clearly any method that could efficiently determine the impossibility of most impossible polytopic transitions would prove extremely useful in an attack. Intuitively it might seem that it is generally a hard problem to determine the possibility of a polytopic transition. As a matter of fact though, there already exists a cryptographic technique that provides a very efficient distinguisher for certain types of polytopic transitions, namely higher-order differentials which are shown in Appendix B to correspond to truncated polytopic transitions. This raises the hope that better distinguishing techniques could still be discovered.

There is one important further effect of the increase in the size of the difference space: it allows us to restrict ourselves to impossible d-differences on only a part of the state. It is even possible to restrict the d-difference to a d-difference in one bit and still use it for efficient filtering[1]. In Sect. 4 we will use these techniques in impossible polytopic attacks to demonstrate the validity of the attacks and provide a usage scenario.

Wrong Keys and Random Permutations

Note that while impossible polytopic attacks – just like impossible differential attacks – do not require the stochastic equivalence hypothesis, practical attacks require another hypothesis: the wrong-key randomization hypothesis. This hypothesis states that when decrypting one or several rounds with a wrong key guess creates a function that behaves like a random function. For our setting, we formulate it is as following:

Wrong-Key Randomization Hypothesis. When decrypting one or multiple rounds of a block cipher with a wrong key guess, the resulting polytopic transition probability will be close to the transition probability over a random permutation for almost all key guesses.

Let us therefore take a look at the polytopic transition probabilities over random functions and random permutation. To simplify the treatment, we make the following definition:

[1] In standard differential cryptanalysis, this would require a probability 1 truncated differential.

Definition 8 (Degenerate d-difference). *Let $\boldsymbol{\alpha}$ be a d-difference over $\mathbb{F}_2^n$: $\boldsymbol{\alpha} = (\alpha_1, \ldots, \alpha_d)$. We call $\boldsymbol{\alpha}$* degenerate *if there exists an i with $1 \le i \le d$ with $\alpha_i = 0$ or if there exists a pair i, j with $1 \le i < j \le d$ and $\alpha_i = \alpha_j$. Otherwise we call $\boldsymbol{\alpha}$* non-degenerate.

Clearly if and only if a d-difference $\boldsymbol{\alpha}$ is degenerate, there exist two texts in the underlying $(d+1)$-polytope that are identical. To understand the transition probability of a degenerate d-difference it is thus sufficient to evaluate the transition probability of a non-degenerate d'-difference ($d' < d$) that contains the same set of texts. For the following two propositions, we will thus restrict ourselves to non-degenerate d-differences.

Proposition 2 (Distribution over Random Function). *Let $\boldsymbol{\alpha}$ be a non-degenerate d-difference over $\mathbb{F}_2^n$. Let $\mathbf{F}$ be a uniformly distributed random function from $\mathbb{F}_2^n$ to $\mathbb{F}_2^m$. The image of $\boldsymbol{\alpha}$ is then uniformly distributed over all d-difference over $\mathbb{F}_2^m$. In particular* $\Pr\left(\boldsymbol{\alpha} \xrightarrow{F} \boldsymbol{\beta}\right) = 2^{-md}$ *for any d-difference $\boldsymbol{\beta} \in (\mathbb{F}_2^m)^d$.*

Proof. Let $(m_0, m_1, \ldots, m_d)$ be a $(d+1)$-polytope that adheres to $\boldsymbol{\alpha}$. Then the polytope $(\mathbf{F}(m_0), \mathbf{F}(m_1), \ldots, \mathbf{F}(m_d))$ is clearly uniformly randomly distributed on $(\mathbb{F}_2^m)^{d+1}$ and accordingly $\boldsymbol{\beta}$ with $\boldsymbol{\alpha} \xrightarrow{\mathbf{F}} \boldsymbol{\beta}$ is distributed uniformly randomly on $(\mathbb{F}_2^m)^d$. □

For the image of a d-difference over a random permutation, we have a similar result:

Proposition 3 (Distribution over Random Permutation). *Let $\boldsymbol{\alpha}$ be a non-degenerate d-difference over $\mathbb{F}_2^n$. Let $\mathbf{F}$ be a uniformly distributed random permutation on $\mathbb{F}_2^n$. The image of $\boldsymbol{\alpha}$ is then uniformly distributed over all non-degenerate d-difference over $\mathbb{F}_2^n$.*

Proof. Let $(m_0, m_1, \ldots, m_d)$ be a $(d+1)$-polytope that adheres to $\boldsymbol{\alpha}$. As $\boldsymbol{\alpha}$ is non-degenerate, all m_i are distinct. Thus the polytope $(\mathbf{F}(m_0), \mathbf{F}(m_1), \ldots, \mathbf{F}(m_d))$ is clearly uniformly randomly distributed on all $(d+1)$-polytopes in $(\mathbb{F}_2^m)^{d+1}$ with distinct values. Accordingly $\boldsymbol{\beta}$ with $\boldsymbol{\alpha} \xrightarrow{F} \boldsymbol{\beta}$ is distributed uniformly randomly on all non-degenerate d-differences over $\mathbb{F}_2^n$. □

As long as $d \ll 2^n$, we can thus well approximate the probability $\Pr\left(\boldsymbol{\alpha} \xrightarrow{\mathbf{F}} \boldsymbol{\beta}\right)$ by 2^{-dn} when $\boldsymbol{\beta}$ is non-degenerate.

In the following, these proposition will be useful when we try to estimate the probability that a partial decryption with a wrong key guess will still give us a possible intermediate d-difference. We will then always assume that the wrong-key randomization hypothesis holds and that the probability of getting a particular d-difference on m bits is the same as if we had used a random permutation, i.e., it is 2^{-dm} (as our d is always small).

4 Impossible Polytopic Attacks on DES and AES

Without much doubt are the Data Encryption Standard (DES) [26] and the Advanced Encryption Standard (AES) [15] the most studied and best cryptanalyzed block ciphers. Any cryptanalytic improvement on these ciphers should thus be a good indicator of the novelty and quality of a new cryptanalytic attack. We believe that these ciphers thus pose ideal candidates to demonstrate that the generalization of differential cryptanalysis to polytopic cryptanalysis is not a mere intellectual exercise but useful for practical cryptanalysis.

In the following, we demonstrate several impossible polytopic attacks on reduced-round versions of DES and AES that make only use of a very small set of chosen plaintexts. The natural reference frame for these attacks are hence low-data attacks. In Tables 1 and 2 we compare our attacks to other low-data attacks on round-reduced versions of DES and AES respectively. We should mention here that [11] only states attacks on 7 and 8 rounds of DES. It is not clear whether the techniques therein could also be used to improve complexities of meet-in-the-middle attacks for 5- and 6-round versions of that cipher.

We stress here that in contrast to at least some of the other low-data attacks, our attacks make no assumption on the key schedule and work equally well with independent round keys. In fact, all of our attacks are straight-forward applications of the ideas developed in this paper. There is likely still room for improvement of these attacks using details of the ciphers and more finely controlled trade-offs.

In all of the following attacks, we determine the possibility or impossibility of a polytopic transition by deterministically generating a list of all d-differences that are reachable from the starting d-difference, i.e., we generate and keep a list of all possible d-differences. The determination of these lists is straightforward using the rules described in Sect. 2. The sizes of these lists are the limiting factors of the attacks both for the time and the memory complexities.

4.1 Attacks on the DES

For a good reference for the DES, we refer to [21]. A summary of the results for DES can be found in Table 1.

A 5-Round Attack. Let us start with an impossible 4-polytopic attack on 5-round DES. We split our input 3-difference into two parts, one for the left 32 state bits and one for the right 32 state bits. Let us denote the left 3-difference as (α, β, γ). For the right half we choose the 3-difference $(0, 0, 0)$. This allows us to pass the first round for free (as can be seen in Fig. 2).

The number of possible 3-differences after the second round depends now on our choice of α, β, and γ. To keep this number low, clearly it is good to choose the differences to activate as few S-boxes as possible. We experimentally tried out different natural choices and chose the values

$$(\alpha, \beta, \gamma) = (\mathtt{02000000}, \mathtt{04000000}, \mathtt{06000000}).$$

Table 1. Comparison table of low-data attacks on round-reduced DES. Data complexity is measured in number of required known plaintexts (KP) or chosen plaintexts (CP). Time complexity is measured in round-reduced DES encryption equivalents. Memory complexity is measured in plaintexts (8 bytes). For the other attacks no memory requirements were explicitly specified in the publications. They should be low though. The attacks of this paper are in bold.

Rounds	Attack type	Time	Data	Memory	Source
5	Differential	$> 2^{11.7}$	64 CP	-	As in [18]
	Linear	$> 2^{13.8}$	72 CP	-	As in [18]
	MitM	$2^{45.5}$	1 KP	-	From [13]
	MitM	$2^{37.9}$	28 KP	-	From [18]
	MitM	2^{30}	8 CP	-	From [18]
	Imp. polytopic	$\mathbf{2^{13.2}}$	**4 CP**	$\mathbf{2^{9}}$	**This paper**
6	Differential	$2^{13.7}$	256 CP	-	As in [18]
	Linear	$2^{13.9}$	>104 KP	-	As in [18]
	MitM	$2^{51.8}$	1 KP	-	From [18]
	Truncated diff	2^{48}	7 CP	-	From [19]
	Truncated diff	$2^{11.8}$	46 CP	-	From [19]
	Imp. polytopic	$\mathbf{2^{32.2}}$	**4 CP**	$\mathbf{2^{10}}$	**This paper**
	Imp. polytopic	$\mathbf{2^{18.4}}$	**48 CP**	$\mathbf{2^{9}}$	**This paper**
7	MitM Sieve	2^{53}	1 KP	-	From [11]
	Imp. polytopic	$\mathbf{2^{43}}$	**16 CP**	$\mathbf{2^{43}}$	**This paper**
	Imp. polytopic	$\mathbf{2^{37.8}}$	**48 CP**	$\mathbf{2^{10}}$	**This paper**
8	MitM Sieve	2^{53}	16 KP	-	From [11]

All of these three differences only activate S-box 2 in round 2. With this choice we get 35 possible 3-differences after round 2. Note that the left 3-difference is still (α, β, γ) after round 2 while the 35 variations only appear in the right half.

As discussed earlier, the maximal number of output d-differences for a fixed input d-difference is inherently limited by the size of the domain of the function. A consequence of this is that for any of the 35 3-differences after round 2 the possible number of output 3-differences of any S-box in round 3 is limited to 2^6 as shown in Fig. 2. But by guessing the 6 bits of round key 5 that go into the corresponding S-box in round 5, we can determine the 3-difference in the same four output bits of round 3 now coming from the ciphertexts. For the right guess of the 6 key bits, the determined 3-difference will be possible. For a wrong key guess though, we expect the 3-difference to take a random value in the set of all 3-differences on 4 bits.

But the size of the space of 3-differences in these four output bits is now $2^{4 \cdot 3} = 2^{12}$. Thus when fixing one of the 35 possible 3-differences after round 2, we expect on average to get one suggestion for the 6 key bits in that S-box.

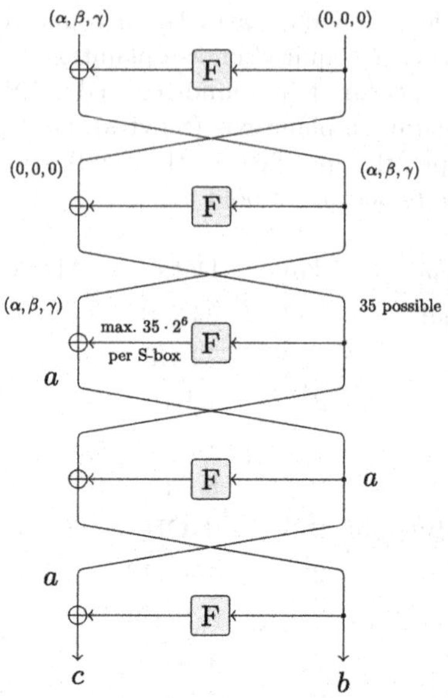

Fig. 2. Outline of the 5-round attack on DES.

Repeating this for every S-box, we get on average one suggestion for the last round key for each of the 35 possible 3-differences after round 2, leaving us with an average of 35 key candidates for the last round key.

What are the complexities of the attack? Clearly we only need 4 chosen plaintexts. For the time complexity we get the following: For each of the 35 possible 3-differences after round 2, we have to determine the 2^6 possible output 3-differences and for each of these, we have to see in the list of possible 3-differences obtained from the key guesses whether there is a guess of the 6 key bits that gives us exactly that 3-difference. Thus we have a total of $35 \cdot 8 \cdot 2^6 = 2^{14.2}$ steps each of which should be easier than one round of DES encryption. This leaves us with a time complexity of $\approx 2^{12}$ 5-round DES encryptions equivalents. But to completely determine the DES key we need 8 additional bits that are not present in the last round key. As we expect on average maximally 35 round keys, we are left with trying out the $35 \cdot 2^8 = 2^{13.2}$ full key candidates, setting the time complexity of this attack to that value.

The only memory requirement in this attack is storing the list of possible 3-differences for each key guess in each S-box. This should roughly be no more than 2^{12} bytes.

A 6-Round Attack. The 6-round attack proceeds exactly as the 5-round attack, with the only difference being that instead of determining the possible 3-difference output of each S-box in round 3, we do the same in round 4 and thus have to repeat the attack for every possible 3-difference after round 3.

Experimental testing revealed that it is beneficial for this attack to choose a different choice of α, β, and γ, namely

$$(\alpha, \beta, \gamma) = (\texttt{20000000}, \texttt{40000000}, \texttt{60000000}),$$

which now activates S-box 1 instead of S-box 2 as it gives us the lowest number of 3-differences after round 3. For this choice, we get a number of 48 possible 3-differences after round 2 and $2^{24.12}$ possible 3-differences after round 3. Now substituting 35 with this number in the previous attack, gives us the time complexity for this 6-round attack.

A note regarding the memory requirement of this attack: As we loop over the $2^{24.12}$ possible 3-differences after round 3, we are not required to store all of them at any time. By doing the attack while creating these possible 3-differences we can keep the memory complexity nearly as low as before, namely to roughly 2^{13} bytes.

A 7-Round Attack. Unfortunately extending from 6 to 7 rounds as done when going from 5 to 6 rounds is not possible, due to the prohibitively large number of possible 3-differences after round 4. Instead we use a different angle.

It is well known that when attacking r-round DES, guessing the appropriate 36 round key bits of the last round key and the appropriate 6 bits of the round key in round $r-1$ allows us to determine the output state bits of an S-box of our choice after round $r-3$. We will thus restrict ourselves to looking for impossible d-differences in only one S-box. We choose S-box 1 here.

In order to have a sufficiently high success rate, we need to increase the dimension of our polytopic transitions to increase the size of the d-difference space of the four output bits of the S-box of our choice. For this attack we choose $d = 15$ giving us a 15-difference space size of 2^{60} in four bits.

For our choice of input 15-difference, we again leave all differences in the right side to 0, while choosing for the 15-difference on the left side:

$$\begin{aligned}(&\texttt{00000002}, \texttt{00000004}, \texttt{00000006}, \texttt{02000000}, \texttt{02000002}, \texttt{02000004},\\ &\texttt{02000006}, \texttt{04000000}, \texttt{04000002}, \texttt{04000004}, \texttt{04000006}, \texttt{06000000},\\ &\texttt{06000002}, \texttt{06000004}, \texttt{06000006})\end{aligned}$$

which only activates S-boxes 2 and 8. For this choice of input 15-difference we get 1470 possible 15-differences after round 2 and $2^{36.43}$ possible 15-differences after round 3.

For each of these $2^{36.43}$ possible 15-differences after round 3, we calculate the 2^6 possible output 15-differences of S-box 1 after round 4. Now having precomputed a list of possible 15-differences in the output bits of S-box 1 after round 4 for each of the 2^{42} guessed key bits of round 7 and 6, we can easily test whether we get a collision. What is the probability of this? The 15-difference space size in the four bits is 2^{60} and, we get maximally 2^{42} possible 15-differences from the key guesses. This leaves us with a chance of 2^{-18} that we find a 15-difference in that list. Thus for each of the $2^{36.43}$ possible 15-differences after round 3, we expect on average at most 2^{-12} suggestions for the guessed 42 key bits, a total of $2^{24.43}$ suggestions.

What are the complexities for this attack? Clearly again, the data complexity is 16 chosen plaintexts. For the time complexity, for each of the $2^{42.42}$ possible 4-bit 15-differences obtained after round 4, we have to see whether it is contained in the list of 2^{42} 3-differences which we obtained from the key guesses. To do this efficiently, we first have to sort the list which should take $2^{42} \cdot 42 = 2^{47.4}$ elementary steps. Assuming that a 7-round DES encryption takes at least 42 elementary steps, we can upperbound this complexity with 2^{42} DES encryption equivalents. As finding an entry in a list of 2^{42} entries also takes approximately 42 elementary steps, this leaves us with a total time complexity of at most 2^{43} 7-round DES encryption equivalents. As each suggestion gives us 42 DES key bits and as the list of suggestions has a size of $2^{24.23}$, we can find the correct full key with $2^{38.23}$ 7-round DES trial encryptions which is lower than then the previously mentioned time complexity and can thus be disregarded.

The data complexity is determined by the size of the list of 4-bit 15-differences generated from the key guesses. This gives us a memory requirement of $2^{42}(15 \cdot 4 + 42)$ bits $\approx 2^{46}$ bytes.

Extension of the Attacks Using More Plaintexts. The attacks for 5 and 6 rounds can be extended by one round at the cost of a higher data complexity. The extension can be made at the beginning of the cipher in the following way.

Let us suppose we start with a 3-difference $(\delta_1, \delta_2, \delta_3)$ in the left half and the 3-difference (α, β, γ) in the right half. If we knew the output 3-difference of the round function in the first round, we could choose $(\delta_1, \delta_2, \delta_3)$ accordingly to make sure that we end up at the starting position of the original attack. Thus by guessing this value and repeating the attack for each guessed value of this 3-difference we can make sure we still retrieve the key.

Fortunately the values of (α, β, γ) are already chosen to give a minimal number of possible 3-difference in the round function. Thus the time complexity only increases by this value, i.e., 35 and 48. The data complexity increases even less. As it turns out, 12 different values for the left half of the input text are enough to generate all of the 35 resp. 48 3-differences. Thus the data complexity only increases to 48 chosen plaintexts.

We should mention that the same technique can be used to extend the 7-round attack to an 8-round attack. But this leaves us with the same time complexity as the 8-round attack in [11], albeit at a much higher data cost.

Experimental Results. To verify the correctness of the above attacks and their complexities, we implemented the 5-round and 6-round attacks that use 4 chosen plaintexts. We ran the attacks on a single core of an Intel Core i5-4300U processor. We ran the 5-round attack 100000 times which took about 140 s. The average number of suggested round keys was 47 which is slightly higher than the expected number of 35. The suggested number of round keys was below 35 though in 84 percent of the cases and below 100 in 95 percent of the cases. In fact, the raised average is created by a few outliers in the distribution: taking the average on all but the 0.02 percent worst cases, we get 33.1 round key suggestions per case. While this shows that the estimated probability is generally good, it also demonstrates that the wrong-key randomization hypothesis has to be handled with care.

Running the six-round attack 10 times, an attack ran an average time of 10 min and produced an average of $2^{22.3}$ candidates for the last round key. As expected, the correct round key was always in the list of candidate round keys for both the 5-round and 6-round attacks.

4.2 Attacks on the AES.

For a good reference for the AES, we refer to [15]. A summary of the results for AES can be found in Table 2.

Table 2. Comparison table of low-data attacks on round-reduced AES. Data complexity is measured in number of required chosen plaintexts (CP). Time complexity is measured in round-reduced AES encryption equivalents. Memory complexity is measured in plaintexts (16 bytes). The column 'keyschedule' denotes whether the attacks use the AES key schedule. All attacks that rely on the keyschedule are attacks on AES-128. The attacks of this paper are in bold.

Rounds	Attack type	Time	Data	Memory	Keyschedule	Source
4	Guess & Det.	2^{120}	1 KP	2^{120}	Yes	As in [10]
	Diff. MitM	2^{104}	3 CP	1	Yes	As in [8,9]
	Guess & Det.	2^{80}	2 CP	2^{80}	Yes	As in [10]
	Guess & Det.	2^{32}	4 CP	2^{24}	Yes	As in [10]
	Imp. polytopic	$\mathbf{2^{38}}$	**8 CP**	$\mathbf{2^{15}}$	**No**	**This paper**
5	MitM	2^{64}	8 CP	2^{56}	Yes	As in [17], Sec. 7.5.1
	Imp. polytopic	$\mathbf{2^{70}}$	**15 CP**	$\mathbf{2^{41}}$	**No**	**This paper**

A 4-Round Attack. We first present here an impossible 8-polytopic attack on 4-round AES. For the input 7-difference, we choose a 7-difference that activates only the first byte, i.e., that is all-zero in all other bytes. Such a 7-difference can be mapped after round 1 to at most 2^8 different 7-differences. If we restrict ourselves to the 7-differences in the first column after round 2, we can then at most have 2^{16} different 7-differences in this column. In particular, we can have at most have 2^{16} different 7-differences in the first byte. For a depiction of this, see Fig. 3.

If we now request the encryptions of 8 plaintexts that adhere to our chosen start 7-difference, we can now determine the corresponding 7-difference after round 2 in the first byte by guessing 40 round key bits of round keys 3 and 4. If this 7-difference does not belong to the set of 2^{16} possible ones, we can discard the key guess as wrong.

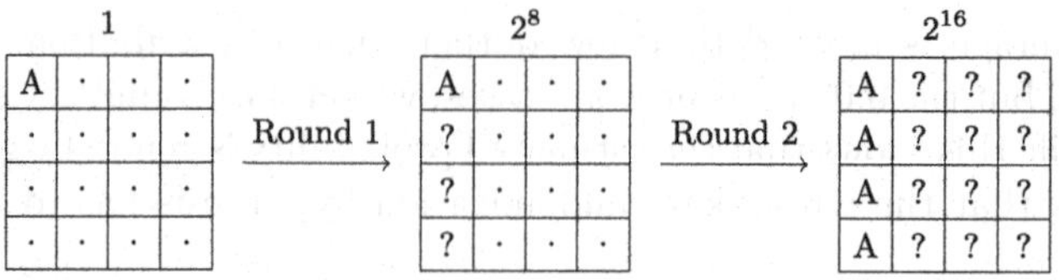

Fig. 3. Diffusion of the starting 7-difference for the 4-round attack on AES. The letter A shows a byte position in which a possible 7-difference is non-zero and known. A dot indicates a byte position where the 7-difference is known to be zero. A question mark indicates a byte position where arbitrary values for the 7-differences are allowed. In total there are 2^{16} different 7-differences possible in the first column after the second round.

How many guesses of the 40 key bits, do we expect to survive the filtering? There are 2^{56} possible 7-difference on a byte and only 2^{16} possible ones coming from our chosen input 7-difference. This leaves a chance of 2^{-40} for a wrong key guess to produce a correct 7-difference. We thus expect on average 2 suggestions for the 40 key bits, among them the right one. To determine the remaining round key bits, we can use the same texts, only restricting ourselves to different columns.

The data complexity of the attack is limited to 8 chosen plaintexts. The time complexity is dominated by determining the 7-difference in the first byte after round 2 for each guess of the 40 key bits and checking whether it is among the 2^{16} possible ones. This can be done in less than 16 table lookups on average for each key guess. Thus the time complexity corresponds to $2^{40} \cdot 2^{-2} = 2^{38}$ 4-round AES encryption equivalents, assuming one 4-round encryption corresponds to $4 \cdot 16$ table lookups. The memory complexity is limited to a table of the 2^{16} allowed 7-difference in one byte, corresponding to 2^{19} bytes or 2^{15} plaintext equivalents.

A 5-Round Attack. In this attack, we are working with 15-polytopes and trace the possible 14-differences one round further than in the 4-round attack. Again we choose our starting 14-difference such that it only activates the first byte. After two rounds we then have maximally 2^{40} different 14-differences on the whole state. If we restrict ourselves to only the first column of the state after round 3, we then get an additional 2^{32} possible 14-differences in this column for each of the 2^{40} possible 14-differences after round 2, resulting in a total of 2^{72} possible 14-differences in the first column after round 3. This is depicted in Fig. 4. In particular again, we can have at most have 2^{72} different 14-differences in the first byte.

Let us suppose now we have the encrypted values of a 15-polytope that adheres to our starting 14-difference. We can then again find the respective 14-difference in the first byte after the third round by guessing 40 key bits in round keys 4 and 5. There are in total 2^{112} different 14-differences in one byte. The chance of a wrong key guess to produce one of the possible 2^{72} 14-differences is thus 2^{-40}. We thus expect on average 2 suggestions for the 40 key bits, among them the right one. To determine the remaining round key bits, we can again use the same texts but restricting ourselves to a different column.

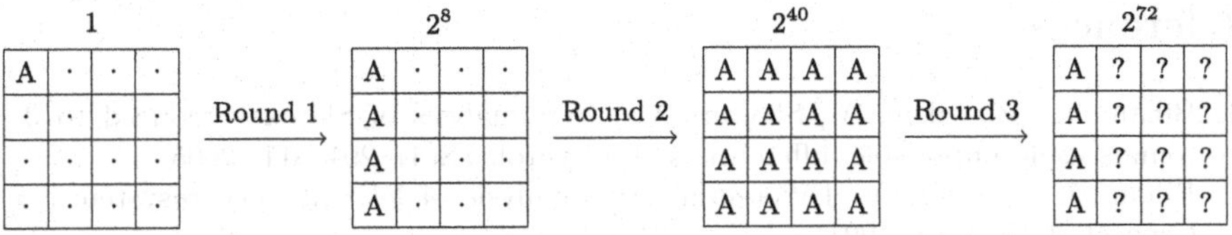

Fig. 4. Diffusion of the starting 14-difference for the 5-round attack on AES. The letter A shows a byte position in which a possible 14-difference is non-zero and known. A dot indicates a byte position where the 14-difference is known to be zero. A question mark indicates a byte position where arbitrary values for the 14-differences are allowed. In total there are 2^{72} different 14-differences possible in the first column after the third round.

To lower the memory complexity of this attack it is advantageous to not store the 2^{72} possible 14-differences but store for each of the 2^{40} key guesses the obtained 14-difference. This gives a memory complexity of $2^{40} \cdot (14 + 5)$ bytes corresponding to 2^{41} plaintext equivalents. The time complexity is then dominated by constructing the 2^{72} possible 14-differences and testing whether they correspond to one of the key guesses. This should not take more than the equivalent of $2^{72} \cdot 16$ table lookups resulting in a time complexity of 2^{70} 5-round AES encryption equivalents. The data complexity is restricted to the 15 chosen plaintexts needed to construct one 15-polytope corresponding to the starting 14-difference.

5 Conclusion

In this paper, we developed and studied polytopic cryptanalysis. We were able to show that the methodology and notation of standard cryptanalysis can be unambiguously extended to polytopic cryptanalysis, including the concept of impossible differentials. Standard differential cryptanalysis remains as a special case of polytopic cryptanalysis.

For impossible polytopic transitions, we demonstrated that both the increase in the size of the space of d-differences and the inherent limit in the diffusion of d-differences in a cipher allow them to be very effective in settings where ordinary impossible differentials fail. This is the case when the number of rounds is so high that impossible differentials do no longer exist or when the allowed data complexity is too low.

Finally we showed the practical relevance of this framework by demonstrating novel low-data attacks on DES and AES that are able to compete with existing attacks.

Acknowledgements. The author thanks Christian Rechberger, Stefan Kölbl, and Martin M. Lauridsen for fruitful discussions. The author also thanks Dmitry Khovratovich and the anonymous reviewers for comments that helped to considerably improve the quality of the paper.

References

1. Biham, E., Biryukov, A., Shamir, A.: Cryptanalysis of Skipjack reduced to 31 rounds using impossible differentials. J. Cryptol. **18**(4), 291–311 (2005)
2. Biham, E., Shamir, A.: Differential cryptanalysis of DES-like cryptosystems. J. Cryptol. **4**(1), 3–72 (1991)
3. Biryukov, A., Shamir, A.: Structural cryptanalysis of SASAS. J. Cryptol. **23**(4), 505–518 (2010)
4. Blondeau, C., Gérard, B.: Multiple differential cryptanalysis: theory and practice. In: Joux, A. (ed.) FSE 2011. LNCS, vol. 6733, pp. 35–54. Springer, Heidelberg (2011)
5. Blondeau, C., Leander, G., Nyberg, K.: Differential-linear cryptanalysis revisited. In: Cid, C., Rechberger, C. (eds.) FSE 2014. LNCS, vol. 8540, pp. 411–430. Springer, Heidelberg (2015)
6. Blondeau, C., Nyberg, K.: Links between truncated differential and multidimensional linear properties of block ciphers and underlying attack complexities. In: Nguyen, P.Q., Oswald, E. (eds.) EUROCRYPT 2014. LNCS, vol. 8441, pp. 165–182. Springer, Heidelberg (2014)
7. Bogdanov, A., Knudsen, L.R., Leander, G., Paar, C., Poschmann, A., Robshaw, M., Seurin, Y., Vikkelsoe, C.: PRESENT: an ultra-lightweight block cipher. In: Paillier, P., Verbauwhede, I. (eds.) CHES 2007. LNCS, vol. 4727, pp. 450–466. Springer, Heidelberg (2007)
8. Bouillaguet, C., Derbez, P., Dunkelman, O., Fouque, P., Keller, N., Rijmen, V.: Low-data complexity attacks on AES. IEEE Trans. Inf. Theor. **58**(11), 7002–7017 (2012)
9. Bouillaguet, C., Derbez, P., Dunkelman, O., Keller, N., Rijmen, V., Fouque, P.: Low data complexity attacks on AES. Cryptology ePrint Archive, Report 2010/633 (2010). http://eprint.iacr.org/
10. Bouillaguet, C., Derbez, P., Fouque, P.-A.: Automatic search of attacks on round-reduced AES and applications. In: Rogaway, P. (ed.) CRYPTO 2011. LNCS, vol. 6841, pp. 169–187. Springer, Heidelberg (2011)
11. Canteaut, A., Naya-Plasencia, M., Vayssière, B.: Sieve-in-the-middle: improved MITM attacks. In: Canetti, R., Garay, J.A. (eds.) CRYPTO 2013, Part I. LNCS, vol. 8042, pp. 222–240. Springer, Heidelberg (2013)
12. Chabaud, F., Vaudenay, S.: Links between differential and linear cryptanalysis. In: De Santis, A. (ed.) EUROCRYPT 1994. LNCS, vol. 950, pp. 356–365. Springer, Heidelberg (1995)
13. Chaum, D., Evertse, J.-H.: Cryptanalysis of DES with a reduced number of rounds. In: Williams, H.C. (ed.) CRYPTO 1985. LNCS, vol. 218, pp. 192–211. Springer, Heidelberg (1986)
14. Daemen, J., Knudsen, L.R., Rijmen, V.: The block cipher SQUARE. In: Biham, E. (ed.) FSE 1997. LNCS, vol. 1267, pp. 149–165. Springer, Heidelberg (1997)
15. Daemen, J., Rijmen, V.: The Design of Rijndael: AES - The Advanced Encryption Standard. Springer, Information Security and Cryptography, Heidelberg (2002)
16. De Cannière, C., Dunkelman, O., Knežević, M.: KATAN and KTANTAN — a family of small and efficient hardware-oriented block ciphers. In: Clavier, C., Gaj, K. (eds.) CHES 2009. LNCS, vol. 5747, pp. 272–288. Springer, Heidelberg (2009)
17. Derbez, P.: Meet-in-the-middle attacks on AES. Ph.D. thesis, Ecole Normale Supérieure de Paris - ENS Paris, December 2013. https://tel.archives-ouvertes.fr/tel-00918146

18. Dunkelman, O., Sekar, G., Preneel, B.: Improved meet-in-the-middle attacks on reduced-round DES. In: Srinathan, K., Rangan, C.P., Yung, M. (eds.) INDOCRYPT 2007. LNCS, vol. 4859, pp. 86–100. Springer, Heidelberg (2007)
19. Knudsen, L.R.: Truncated and higher order differentials. In: Preneel, B. (ed.) FSE 1994. LNCS, vol. 1008, pp. 196–211. Springer, Heidelberg (1995)
20. Knudsen, L.R.: DEAL - a 128-bit block cipher. Technical report 151, Department of Informatics, University of Bergen, Norway, submitted as an AES candidate by Richard Outerbridge, February 1998
21. Knudsen, L.R., Robshaw, M.: The Block Cipher Companion. Springer, Information Security and Cryptography, Heidelberg (2011)
22. Lai, X.: Higher order derivatives and differential cryptanalysis. In: Blahut, R.E., Costello Jr., D.J., Maurer, U., Mittelholzer, T. (eds.) Communications and Cryptography, Two Sides of One Tapestry, pp. 227–233. Kluwer Academic Publishers, Berlin (1994)
23. Lai, X., Massey, J.L.: Markov ciphers and differential cryptanalysis. In: Davies, D.W. (ed.) EUROCRYPT 1991. LNCS, vol. 547, pp. 17–38. Springer, Heidelberg (1991)
24. Langford, S.K., Hellman, M.E.: Differential-linear cryptanalysis. In: Desmedt, Y.G. (ed.) CRYPTO 1994. LNCS, vol. 839, pp. 17–25. Springer, Heidelberg (1994)
25. Murphy, S.: The return of the cryptographic boomerang. IEEE Trans. Inf. Theor. **57**(4), 2517–2521 (2011)
26. National Institute of Standards and Technology: Data Encryption Standard. Federal Information Processing Standard (FIPS), Publication 46, U.S. Department of Commerce, Washington D.C., January 1977
27. Shirai, T., Shibutani, K., Akishita, T., Moriai, S., Iwata, T.: The 128-bit blockcipher CLEFIA (extended abstract). In: Biryukov, A. (ed.) FSE 2007. LNCS, vol. 4593, pp. 181–195. Springer, Heidelberg (2007)
28. Vaudenay, S.: Decorrelation: a theory for block cipher security. J. Cryptol. **16**(4), 249–286 (2003)
29. Wagner, D.: The boomerang attack. In: Knudsen, L.R. (ed.) FSE 1999. LNCS, vol. 1636, pp. 156–170. Springer, Heidelberg (1999)
30. Wagner, D.: Towards a unifying view of block cipher cryptanalysis. In: Roy, B., Meier, W. (eds.) FSE 2004. LNCS, vol. 3017, pp. 16–33. Springer, Heidelberg (2004)

A Markov Model in Polytopic Cryptanalysis

To develop the Markov model, we first need to introduce keys in the function over which the transitions take place. We will thus restrict our discussion to product ciphers i.e., block ciphers that are constructed through repeated composition of round functions. In contrast to Eq. (2), each round function f^i is now keyed with its own round key k_i which itself is derived from the key k of the cipher via a key schedule[2]. We can then write the block cipher f_k as:

$$f_k := f^r_{k_r} \circ \cdots \circ f^2_{k_2} \circ f^1_{k_1}. \tag{7}$$

[2] For a clearer notation, we moved the index from subscript to superscript.

The first assumption that we now need to make, is that the round keys are independent. The second assumption is that the product cipher is a Markov cipher. Here we adopt the notion of a Markov cipher from [23] to polytopic cryptanalysis:

Definition 9. *A product cipher is a $(d+1)$-polytopic Markov cipher if and only if for all round functions f^i, for any $(d+1)$-polytopic transition $\boldsymbol{\alpha} \longrightarrow \boldsymbol{\beta}$ for that round function and any fixed inputs $x, y \in \mathbb{F}_2^n$, we have*

$$\Pr_{\mathbf{K}}\left(\boldsymbol{\alpha} \xrightarrow[x]{f^i_{\mathbf{K}}} \boldsymbol{\beta}\right) = \Pr_{\mathbf{K}}\left(\boldsymbol{\alpha} \xrightarrow[y]{f^i_{\mathbf{K}}} \boldsymbol{\beta}\right) \tag{8}$$

where $\mathbf{K}$ is a random variable distributed uniformly over the spaces of round keys.

In words, a cipher is a $(d+1)$-polytopic Markov cipher if and only if the probabilities of 1-round $(d+1)$-polytopic transitions do not depend on the specific anchor as long as the round key is distributed uniformly at random. For $d = 1$, the definition coincides with the classical definition.

Just as with the standard definition of Markov ciphers, most block ciphers are $(d+1)$-polytopic Markov ciphers for any d as the round keys are usually added to any part of the state that enters the non-linear part of the round function (for a counterexample, see [16]). Examples of $(d+1)$-polytopic Markov ciphers are SPN ciphers such as AES [15] or PRESENT [7], and Feistel ciphers such as DES [26] or CLEFIA [27]. We are not aware of any cipher that is Markov in the classical definition but not $(d+1)$-polytopic Markov.

In the following, we extend the central theorem from [23] (Theorem 2) to the case of $(d+1)$-polytopes.

Theorem 1. *Let $f_k = f^r_{k_r} \circ \cdots \circ f^1_{k_1}$ be a $(d+1)$-polytopic Markov cipher with independent round keys, chosen uniformly at random and let $\boldsymbol{\delta}_0, \boldsymbol{\delta}_1, \ldots, \boldsymbol{\delta}_r$ be a series of d-differences such that $\boldsymbol{\delta}_0$ is the input d-difference of round 1 and $\boldsymbol{\delta}_i$ is the output d-difference of round i of some fixed input $(d+1)$-polytope. The series $\boldsymbol{\delta}_0, \boldsymbol{\delta}_1, \ldots, \boldsymbol{\delta}_r$ then forms a Markov chain.*

The following proof follows the lines of the original proof from [23].

Proof. We limit ourselves here to showing that

$$\Pr_{\mathbf{K}_1,\mathbf{K}_2}\left(\boldsymbol{\delta}_1 \xrightarrow[f^1_{\mathbf{K}_1}(x)]{f^2_{\mathbf{K}_2}} \boldsymbol{\delta}_2 \,\middle|\, \boldsymbol{\delta}_0 \xrightarrow[x]{f^1_{\mathbf{K}_1}} \boldsymbol{\delta}_1\right) = \Pr_{\mathbf{K}_2}\left(\boldsymbol{\delta}_1 \xrightarrow[z]{f^2_{\mathbf{K}_2}} \boldsymbol{\delta}_2\right) \tag{9}$$

where x and z are any elements from $\mathbb{F}_2^n$ and $\mathbf{K}_1$ and $\mathbf{K}_2$ are distributed uniformly at random over their respective round key spaces and the conditioned event has positive probability. The theorem then follows easily by induction and application of the same arguments to the other rounds.

For any $x, z \in \mathbb{F}_2^n$, we now have

$$\begin{aligned}
&\Pr_{\mathbf{K}_1,\mathbf{K}_2}\left(\boldsymbol{\delta}_1 \xrightarrow[f^1_{\mathbf{K}_1}(x)]{f^2_{\mathbf{K}_2}} \boldsymbol{\delta}_2 \text{ and } \boldsymbol{\delta}_0 \xrightarrow[x]{f^1_{\mathbf{K}_1}} \boldsymbol{\delta}_1\right)\\
&= \sum_{y\in\mathbb{F}_2^n} \Pr_{\mathbf{K}_1,\mathbf{K}_2}\left(\boldsymbol{\delta}_1 \xrightarrow[y]{f^2_{\mathbf{K}_2}} \boldsymbol{\delta}_2 \text{ and } \boldsymbol{\delta}_0 \xrightarrow[x]{f^1_{\mathbf{K}_1}} \boldsymbol{\delta}_1 \text{ and } f^1_{\mathbf{K}_1}(x) = y\right)\\
&= \sum_{y\in\mathbb{F}_2^n} \Pr_{\mathbf{K}_2}\left(\boldsymbol{\delta}_1 \xrightarrow[y]{f^2_{\mathbf{K}_2}} \boldsymbol{\delta}_2\right) \cdot \Pr_{\mathbf{K}_1}\left(\boldsymbol{\delta}_0 \xrightarrow[x]{f^1_{\mathbf{K}_1}} \boldsymbol{\delta}_1 \text{ and } f^1_{\mathbf{K}_1}(x) = y\right)\\
&= \Pr_{\mathbf{K}_2}\left(\boldsymbol{\delta}_1 \xrightarrow[z]{f^2_{\mathbf{K}_2}} \boldsymbol{\delta}_2\right) \cdot \sum_{y\in\mathbb{F}_2^n} \Pr_{\mathbf{K}_1}\left(\boldsymbol{\delta}_0 \xrightarrow[x]{f^1_{\mathbf{K}_1}} \boldsymbol{\delta}_1 \text{ and } f^1_{\mathbf{K}_1}(x) = y\right)\\
&= \Pr_{\mathbf{K}_2}\left(\boldsymbol{\delta}_1 \xrightarrow[z]{f^2_{\mathbf{K}_2}} \boldsymbol{\delta}_2\right) \cdot \Pr_{\mathbf{K}_1}\left(\boldsymbol{\delta}_0 \xrightarrow[x]{f^1_{\mathbf{K}_1}} \boldsymbol{\delta}_1\right)
\end{aligned}$$

where the second equality comes from the independence of keys K_1 and K_2 and the third equality comes from the Markov property of the cipher. From this, Eq. (9) follows directly. □

The important consequence of the fact that the sequence of d-differences forms a Markov chain is that, just as in standard differential cryptanalysis, the average probability of a particular polytopic trail with respect to independent random round keys is the product of the single polytopic 1-round transitions of which it consists. We then have the following result:

Corollary 1. *Let f_k, $f^i_{k_i}, 1 \leq i \leq r$ be as before. Let $\boldsymbol{\alpha}_0 \xrightarrow{f_1} \boldsymbol{\alpha}_1 \xrightarrow{f_2} \cdots \xrightarrow{f_r} \boldsymbol{\alpha}_r$ be an r-round $(d+1)$-polytopic trail. Then*

$$\Pr\left(\boldsymbol{\alpha}_0 \xrightarrow{f^1_{\mathbf{K}_1}} \boldsymbol{\alpha}_1 \xrightarrow{f^2_{\mathbf{K}_2}} \cdots \xrightarrow{f^r_{\mathbf{K}_r}} \boldsymbol{\alpha}_r\right) = \prod_{i=1}^{r} \Pr\left(\boldsymbol{\alpha}_{i-1} \xrightarrow{f^i_{\mathbf{K}_i}} \boldsymbol{\alpha}_i\right) \qquad (10)$$

where $x \in \mathbb{F}_2^n$ and the $\mathbf{K}_i$ are uniformly randomly distributed on their respective spaces.

Proof. This is a direct consequence of the fact that d-differences form a Markov chain. □

In most attacks though, we are attacking one fixed key and can not average the attack over all keys. Thus the following assumption is necessary:

Hypothesis of Stochastic Equivalence. *Let f be as above. The hypothesis of stochastic equivalence then refers to the assumption that the probability of any polytopic trail $\boldsymbol{\alpha}_0 \xrightarrow{f_1} \boldsymbol{\alpha}_1 \xrightarrow{f_2} \cdots \xrightarrow{f_r} \boldsymbol{\alpha}_r$ is roughly the same for the large majority of keys:*

$$\Pr\left(\boldsymbol{\alpha}_0 \xrightarrow{f^1_{\mathbf{K}_1}} \boldsymbol{\alpha}_1 \xrightarrow{f^2_{\mathbf{K}_2}} \cdots \xrightarrow{f^r_{\mathbf{K}_r}} \boldsymbol{\alpha}_r\right) \approx \Pr\left(\boldsymbol{\alpha}_0 \xrightarrow{f^1_{k_1}} \boldsymbol{\alpha}_1 \xrightarrow{f^2_{k_2}} \cdots \xrightarrow{f^r_{k_r}} \boldsymbol{\alpha}_r\right) \quad (11)$$

for almost all tuples of round keys $(k_1, k_2, \ldots, k_r)$.

B Truncated Polytopic Transitions and Higher-Order Differentials

In this section, we extend the definition of truncated differentials to polytopic transitions and prove that higher-order differentials are a special case of these. We then gauge the cryptographic ramifications of this.

In accordance with usual definitions for standard truncated differentials (see for example [6], we define:

Definition 10. *A truncated d-difference is an affine subspace of the space of d-differences. A truncated $(d+1)$-polytopic transition is a pair (A, B) of truncated d-differences, mostly denoted as $A \xrightarrow{f} B$. The probability of a truncated $(d+1)$-polytopic transition (A, B) is defined as the probability that an input d-difference chosen uniformly randomly from A maps to a d-difference in B:*

$$\Pr\left(A \xrightarrow{f} B\right) := |A|^{-1} \sum_{\substack{\boldsymbol{\alpha} \in A \\ \boldsymbol{\beta} \in B}} \Pr\left(\boldsymbol{\alpha} \xrightarrow{f} \boldsymbol{\beta}\right) \tag{12}$$

As the truncated input d-difference is usually just a single d-difference, the probability of a truncated differential is then just the probability that this input d-difference maps to any of the output d-differences in the output truncated d-difference. With a slight abuse of notation, we will denote the truncated polytopic transition then also as $\boldsymbol{\alpha} \xrightarrow{f} B$ where $\boldsymbol{\alpha}$ is the single d-difference of the input truncated d-difference.

A particular case of a truncated d-difference is the case where the individual differences of the d-differences always add up to the same value. This is in fact just the kind of d-differences one is interested in when working with higher-order derivatives. We refer here to Lai's original paper on higher-order derivatives [22] and Knudsen's paper on higher-order differentials [19] for reference and notation.

Theorem 2. *A differential of order t is a special case of a truncated 2^t-polytopic transition. In particular, its probability is the sum of the probabilities of all 2^t-polytopic trails that adhere to the truncated 2^t-polytopic transition.*

Proof. Let $f : \mathbb{F}_2^n \to \mathbb{F}_2^n$. Let $(\alpha_1, \ldots, \alpha_t)$ be the set of linearly independent differences that are used as the base for our derivative. Let $L(\alpha_1, \ldots, \alpha_t)$ denote the linear space spanned by these differences. Let furthermore β be the output difference we are interested in. The probability of the t-th order differential $\Delta_{\alpha_1,\ldots,\alpha_t} f(\mathbf{X}) = \beta$ is then defined as the probability that

$$\sum_{\gamma \in L(\alpha_1,\ldots,\alpha_t)} f(\mathbf{X} \oplus \gamma) = \beta \tag{13}$$

holds with $\mathbf{X}$ being a random variable, uniformly distributed on $\mathbb{F}_2^n$.

Let B now be the truncated $(2^t - 1)$-difference defined as

$$B := \left\{ (\delta_1, \ldots, \delta_{2^t-1}) \;\middle|\; \sum_{i=1}^{2^t-1} \delta_i = \beta \right\}. \tag{14}$$

Let $\gamma_1, \gamma_2, \ldots, \gamma_{2^t-1}$ be an arbitrary ordering of the non-zero elements of the linear space $L(\alpha_1, \ldots, \alpha_t)$ and let $\boldsymbol{\alpha} = (\gamma_1, \ldots, \gamma_{2^t-1})$ be the $(2^t - 1)$-difference consisting of these. We will then show that the probability of the t-th order differential $(\alpha_1, \ldots, \alpha_t, \beta)$ is equal to the the probability of the truncated 2^t-polytopic transition $\boldsymbol{\alpha} \xrightarrow{f} B$. With $\mathbf{X}$ being a random variable, uniformly distributed on $\mathbb{F}_2^n$, we have

$$\begin{aligned}
&\Pr\left(\boldsymbol{\alpha} \xrightarrow{f} B\right) \\
&= \Pr_{\mathbf{X}}\left(\sum_{i=1}^{2^t-1} \Big(f(\mathbf{X} \oplus \gamma_i) \oplus f(\mathbf{X}) \Big) = \beta \right) \\
&= \Pr_{\mathbf{X}}\left(\sum_{i=1}^{2^t-1} \Big(f(\mathbf{X} \oplus \gamma_i) \Big) \oplus f(\mathbf{X}) = \beta \right) \\
&= \Pr_{\mathbf{X}}\left(\sum_{\gamma \in L(\alpha_1, \ldots, \alpha_t)} \Big(f(\mathbf{X} \oplus \gamma) \Big) = \beta \right)
\end{aligned}$$

which proves the theorem. □

Example. Let α_1 and α_2 be two differences with respect to which we want to take the second order derivative and let β be the output value we are interested in. The probability that $\Delta_{\alpha_1, \alpha_2} f(\mathbf{X}) = \beta$ for uniformly randomly chosen $\mathbf{X}$ is then nothing else than the probability that the 3-difference $(\alpha_1, \alpha_2, \alpha_1 \oplus \alpha_2)$ is mapped to a 3-difference $(\beta_1, \beta_2, \beta_3)$ with $\beta_1 \oplus \beta_2 \oplus \beta_3 = \beta$.

This theoretical connection between truncated and higher-order differentials has an interesting consequence: a higher-order differentials can be regarded as the union of polytopic trails. This principally allows us to determine lower bounds for the probability of higher-order differentials by summing over the probabilities of a subset of all polytopic trails that it contains – just as we are used to from standard differentials.

As shown in Lemma 2, the probability of a $(d + 1)$-polytopic trail is always at most as high as the probability of the worst standard differential trail that it contains. A situation in which the probability of a higher-order differential at the same time is dominated by a single polytopic trail and has a higher probability than any ordinary differential can thus never occur. To find a higher-order differential with a higher probability than any ordinary differential for a given cipher, we are thus always forced to sum over many polytopic trails. Whether this number can remain manageable for a large number of rounds will require further research and is beyond the scope of this paper.

From Improved Leakage Detection to the Detection of Points of Interests in Leakage Traces

François Durvaux([✉]) and François-Xavier Standaert

ICTEAM/ELEN/Crypto Group, Université catholique de Louvain, Louvain-la-neuve, Belgium
francois.durvaux@gmail.com, fstandae@uclouvain.be

Abstract. Leakage detection usually refers to the task of identifying data-dependent information in side-channel measurements, independent of whether this information can be exploited. Detecting Points-Of-Interest (POIs) in leakage traces is a complementary task that is a necessary first step in most side-channel attacks, where the adversary wants to turn this information into (e.g.) a key recovery. In this paper, we discuss the differences between these tasks, by investigating a popular solution to leakage detection based on a t-test, and an alternative method exploiting Pearson's correlation coefficient. We first show that the simpler t-test has better sampling complexity, and that its gain over the correlation-based test can be predicted by looking at the Signal-to-Noise Ratio (SNR) of the leakage partitions used in these tests. This implies that the sampling complexity of both tests relates more to their implicit leakage assumptions than to the actual statistics exploited. We also put forward that this gain comes at the cost of some intuition loss regarding the localization of the exploitable leakage samples in the traces, and their informativeness. Next, and more importantly, we highlight that our reasoning based on the SNR allows defining an improved t-test with significantly faster detection speed (with approximately 5 times less measurements in our experiments), which is therefore highly relevant for evaluation laboratories. We finally conclude that whereas t-tests are the method of choice for leakage detection only, correlation-based tests exploiting larger partitions are preferable for detecting POIs. We confirm this intuition by improving automated tools for the detection of POIs in the leakage measurements of a masked implementation, in a black box manner and without key knowledge, thanks to a correlation-based leakage detection test.

1 Introduction

Leakage detection tests have recently emerged as a convenient solution to perform preliminary (black box) evaluations of resistance against side-channel analysis. Cryptography Research (CRI)'s non specific (fixed vs. random) t-test is a popular example of this trend [4,10]. It works by comparing the leakages of a cryptographic (e.g. block cipher) implementation with fixed plaintexts (and key)

M. Fischlin and J.-S. Coron (Eds.): EUROCRYPT 2016, Part I, LNCS 9665, pp. 240–262, 2016.
DOI: 10.1007/978-3-662-49890-3_10

to the leakages of the same implementation with random plaintexts (and fixed key)[1], thanks to Welch's t-test [38]. Besides their conceptual simplicity, the main advantage of such tests, that were carefully discussed in [18], is their low *sampling complexity*. That is, by comparing only two (fixed vs. random) classes of leakages, one reduces the detection problem to a simpler estimation task. In this paper, we want to push the understanding of leakage detection one step further, by underlining more precisely its pros and cons, and clarifying its difference with the problem of detecting Points-Of-Interest (POIs) in leakage traces. As clear from [9], those two problems are indeed related, and one can also exploit t-tests for the detection of POIs in leakage traces. So as for any side-channel analysis, the main factor influencing the intuitions that one can extract from leakage detection is the implicit assumptions that we make about the partitioning of the leakages (aka leakage model). Our contributions in this respect are threefold.

First, we notice that CRI's fixed vs. random t-test is one extreme in this direction (since it relies on a partitioning in two classes), which is reminiscent of Kocher's single-bit Differential Power Analysis (DPA) [14]. For comparison purposes, we therefore start by specifying an alternative leakage detection test based on the popular Correlation Power Analysis (CPA) distinguisher [3]. The resulting ρ-test directly derives from the hypothesis tests for CPA provided in [16], and relies on a partitioning into 2^s classes, where s is the bitsize of the fixed portion of plaintext in the test. We then compare the t-test and ρ-test approaches, both in terms of sampling complexity and based on their *exploitability*.[2] That is, does a positive answer to leakage detection imply exploitable leakage, and does a negative answer to leakage detection imply no exploitable leakage? Our experimental analysis based on real and simulated data leads to the following observations:

- First, the *sampling complexity* of the t-test is (on average) lower than the one of the ρ-test, as previously hinted [10,18]. Interestingly, we show that the sampling complexity ratio between the two tests can be simply approximated as a function of a Signal-to-Noise Ratio (SNR) for the leakage partition used in these tests. This underlines that the difference between the tests is mainly due to their different leakage assumptions, i.e. is somewhat independent of statistics used (backing up the conclusions of [17] for "standard DPA attacks").
- Second, the *exploitability* of the tests is quite different. On the one hand, leakages that are informative (and therefore can be detected with the ρ-test) but cannot be detected with the t-test are easy to produce (resp. can be observed in practice). Take for example a fixed class of which the mean leakage is the same as (resp. close to) the mean leakage of the random class. On the

[1] The Test Vector Leakage Assessment methodology in [4,10] includes other options, e.g. non specific semi-fixed vs. random tests and specific tests – we focus on this non specific fixed vs. random test that is the most popular in the literature [2,18].

[2] One could also compare the computational complexity of the tests. Since they are based on simple statistics, we will assume that both the t-test and ρ-test can be implemented efficiently. Besides, a minor advantage of the ρ-test is that it can be implemented in a known-plaintexts scenario (vs. chosen-plaintext for the t-test).

other hand, the fixed vs. random t-test leads to the detection of many time samples spread around the complete leakage traces. Hence, not all of these samples can be exploited in a standard DPA (because of the diffusion within the cipher).

Concretely, these observations refine the analysis in [18], where it was argued that leakage detection is a useful preliminary to white box (worst-case) security evaluations such as advertized in [34]. This is indeed the case. Yet, certain leakage detection tests are more connected with the actual security level of a leaking implementation. In this respect, the fixed vs. random t-test is a more efficient way to perform leakage detection only. And the minor drawback regarding its unability to detect certain leakages (e.g. our example with identical means) is easily mitigated in practice, by running the test on large enough traces, or for a couple of keys (as suggested in [4,10]). By contrast, the main price to pay for this efficiency is a loss of intuition regarding (*i*) the localisation of the leakage samples that are exploitable by standard DPA, and (*ii*) the complexity of a side-channel attack taking advantage of the leakage samples for which the detection test is positive. As a result, the ρ-test can be viewed as a perfect complement, since it provides these intuitions (at the cost of higher sampling complexity).

Second, we show that our reasoning based on the SNR not only allows a better statistical understanding of leakage detection, but can also lead to more efficient t-tests. Namely, it directly suggests that if the evaluator's goal is to minimize the number of samples needed to detect data-dependent information in side-channel measurements, considering a partitioning based on two fixed plaintexts (rather than one fixed and one random plaintext) leads to significantly faster detection speeds. This is both due to an improved signal (since when integrated over large execution times, samples with large differences between the two fixed classes will inevitably occur) and a reduced noise (since the random class in CRI's t-test implies a larger algorithmic noise that is cancelled in our proposal). We also confirm these intuitions experimentally, with two representative AES implementations: an 8-bit software one and a 128-bit hardware one. In both cases, we exhibit detections with roughly 5 times less measurements than when using the previous fixed vs. random non specific t-test. We believe these results are highly relevant to evaluation laboratories since (*i*) they lead to reductions of the measurement cost of a leakage detection by a large factor (whereas improvements of a couple of percents are usually considered as significant in the side-channel literature), and (*ii*) they imply that a device for which no leakages have been detected with one million measurements using a fixed vs. random t-test could in fact have detectable leakages with 200,000 (or even less) measurements.

These observations lead to the last contribution of the paper. That is, when extending leakage detection towards the detection of POIs, the ρ-test naturally gains additional interest, since it provides more intuitions regarding the exploitable samples in side-channel traces. More precisely, it allows a better selection of POIs based on the criteria that these POIs depend on an enumerable part of the key. It also maximizes the SNR metric that can be easily connected to the worst-case complexity of standard DPA attacks [5]. Therefore, and more concretely, our results directly imply that the automated tools for the detection of POIs recently proposed in [7] are also applicable in a fully black box setting, without any key knowledge, by simply adapting the objective function used in their optimization (i.e. replacing it by the ρ-test in this paper). We finally confirm this claim with an experimental evaluation, in the context of first-order secure masked implementations. Doing so, we put forward that the detection of a threshold for which an improvement of the objective function is considered as significative in the optimizations of [7] is made easier when using the ρ-test. We also improve the latter methods by adapting the objective function to the multivariate case and taking advantage of cross-validation to evaluating it.

2 Background

2.1 Measurement Setups

Most of our experiments are based on measurements of an AES Furious implementation (http://point-at-infinity.org/avraes/) run by an 8-bit Atmel ATMega644P microcontroller, at a 20 MHz clock frequency. We monitored the voltage variations across a 22 Ω resistor introduced in the supply circuit of our target chip. Acquisitions were performed using a Lecroy HRO66ZI oscilloscope running at 200 MHz and providing 8-bit samples. In each of our evaluations, the 128-bit AES master key remains the same for all the measurements and is denoted as $\kappa = s_0||s_1||\ldots||s_{15}$, where the s_i's represent the 16 key bytes. When evaluating the fixed vs. random t-test, we built sets of 2000 traces divided in two subsets of 1000 traces each, one corresponding to a fixed plaintext and key, the other corresponding to random plaintexts and a fixed key, next denoted as $\mathcal{L}_f$ and $\mathcal{L}_r$ respectively. When evaluating the correlation-based test, we built a single set of 2000 traces $\mathcal{L}$, corresponding to random plaintexts and a fixed key. In the following, we denote the encryption traces obtained from a plaintext p including the target byte x under a key κ including the subkey s as: $\mathsf{AES}_{\kappa_s}(p_x) \rightsquigarrow l_y$ (with $y = x \oplus s$). Whenever accessing the points of these traces, we use the notation $l_y(\tau)$ (with $\tau \in [1; 20\,000]$, typically). These different subscripts and indexes will be omitted when not necessary. In Sect. 5, we additionally consider a hardware implementation of the AES of which the design is described in [13]. The same amount of measurement as for the previous Atmel case were taken, based on a prototype chip embedding an AES core with a 128-bit architecture requiring 11 cycles per encryption, implemented in a 65-nanometer low power technology, running at 60 MHz and sampled at 2 GHz. Eventually, Sect. 6 considered masked implementation of the AES in our Atmel microcontroller, based on precomputed table lookups [25,31]. For every pair of input/output masks (m, q), it-precomutes an S-box S^* such that $\mathsf{S}^*(x \oplus s \oplus m) = \mathsf{S}(x \oplus s) \oplus q$. This pre-computation is part of the adversary's measurements, which leads to quite large traces with $30,000$ samples. In this last case, we used an evaluation set with 256×50 traces in total, i.e. 50 per fixed value of the target key byte.

2.2 CPA Distinguisher

Our correlation-based leakage detection test will be based on the Correlation Power Analysis (CPA) distinguisher [3], extended to a profiled setting. In this case, and for each time sample τ, the evaluator starts by estimating a model for his target intermediate variable Y from N_p profiling traces: $\hat{\mathsf{model}}_\tau(Y) \leftarrow \mathcal{L}_p$. This model corresponds to the mean leakages associated with the different values of Y. He then estimates the correlation between measured leakages and modeled leakages. In our AES example, it would lead to $\hat{\rho}(L_Y(\tau), \hat{\mathsf{model}}_\tau(Y))$. In practice, this estimation is performed by sampling (i.e. measuring) a set of N_t test traces from the leakage distribution, that we denote as $\mathcal{L}_t$ (with $\mathcal{L}_p \perp\!\!\!\perp \mathcal{L}_t$).

2.3 Fixed vs. Random Leakage Detection Test

CRI's fixed vs. random t-test essentially works by comparing the leakages corresponding to the fixed and random sets of traces defined in Sect. 2.1. For this purpose, and for each sample, one simply has to estimate and compare two mean values. The first one, denoted as $\hat{\mu}_f(\tau)$, corresponds to the samples in the fixed set of traces $\mathcal{L}_f$. The second one, denoted as $\hat{\mu}_r(\tau)$, corresponds to the samples in the random set of traces $\mathcal{L}_f$. Intuitively, being able to distinguish these two mean values indicates the presence of data-dependencies in the leakages. For this purpose, and in order to determine whether some difference observed in practice is meaningful, Welch's t-test is applied (which is a variant of Student's t-test that considers different variances and sample size for the sets $\mathcal{L}_f$ and $\mathcal{L}_r$). The statistic to be tested is defined as:

$$\Delta(\tau) = \frac{\hat{\mu}_f(\tau) - \hat{\mu}_r(\tau)}{\sqrt{\frac{\hat{\sigma}_f^2(\tau)}{N_f} + \frac{\hat{\sigma}_r^2(\tau)}{N_r}}},$$

where $\hat{\sigma}_f^2(\tau)$ (resp. $\hat{\sigma}_r^2(\tau)$) is the estimated variance over the N_f (resp. N_r) samples of $\mathcal{L}_f$ (resp. $\mathcal{L}_r$). Its p-value, i.e. the probability of the null hypothesis which assumes $\Delta(\tau) = 0$, can be computed as follows:

$$p = 2 \times (1 - \mathsf{CDF_t}(|\Delta(\tau)|, \nu)),$$

where $\mathsf{CDF_t}$ is the cumulative function of a Student's t distribution, and ν is its number of freedom degrees, which is derived from the previous means and variances as: $\nu = (\hat{\sigma}_f^2/N_f + \hat{\sigma}_r^2/N_r)/[(\hat{\sigma}_f^2/N_f)/(N_f - 1) + (\hat{\sigma}_r^2/N_r)/(N_r - 1)]$. Intuitively, the value of ν is proportional to the number of samples N_f and N_r. When increasing, Student's t distribution gets closer to a normal distribution $\mathcal{N}(0,1)$.

3 A Correlation-Based Leakage Detection Test

We start by describing an alternative leakage detection test based on the CPA distinguisher, inspired from the hypothesis test described in [16], and further taking advantage of the cross-validation techniques recently introduced in [6]. For k-fold cross–validation, the set of acquired traces $\mathcal{L}$ is first split into k (non overlapping) sets $\mathcal{L}^{(i)}$ of approximately the same size. We then define the profiling sets $\mathcal{L}_p^{(j)} = \bigcup_{i \neq j} \mathcal{L}^{(i)}$ and the test sets $\mathcal{L}_t^{(j)} = \mathcal{L} \setminus \mathcal{L}_p^{(j)}$. Based on these notations, our ρ-test is defined as follows, for a target plaintext byte variable X. First, and for each cross-validation set j with $1 \leq j \leq k$, a model is estimated: $\hat{\mathsf{model}}_\tau^{(j)}(X) \leftarrow \mathcal{L}_p^{(j)}$. For s-bit plaintext bytes, this model corresponds to the sample means of the leakage sample τ corresponding to each value of the plaintext byte, i.e. $\hat{\mu}_x^{(j)}(\tau)$.[3] Next, the correlation between this model and the leakage samples in the test sets $\mathcal{L}_t^{(j)}$ is computed as follows:

$$\hat{r}^{(j)}(\tau) = \hat{\rho}(L_X^{(j)}(\tau), \hat{\mathsf{model}}_\tau^{(j)}(X)).$$

The k cross-validation results $\hat{r}^{(j)}(\tau)$ can then be averaged in order to get a single (unbiased) result $\hat{r}(\tau)$ obtained from the full measurement set $\mathcal{L}$. Following, and as in [16], Fisher's z-transformation is applied to obtain:

$$\hat{r}_z(\tau) = \frac{1}{2} \times \ln\left(\frac{1 + \hat{r}(\tau)}{1 - \hat{r}(\tau)}\right).$$

By normalizing this value with the standard deviation $\frac{1}{\sqrt{N-3}}$, where N is the size of the evaluation set $\mathcal{L}$, we obtain a sample that can be (approximately) interpreted according to a normal distribution $\mathcal{N}(0, 1)$. This allows us to compute the following p-value for a null hypothesis assuming no correlation:

$$p = 2 \times (1 - \mathsf{CDF}_{\mathcal{N}(0,1)}(|\hat{r}_z(\tau)|)),$$

where $\mathsf{CDF}_{\mathcal{N}(0,1)}$ is the cumulative function of a standard normal distribution. Besides exploiting cross-validation (which allows us to obtain unbiased estimates for Pearson's correlation coefficient), the main difference between this test and the hypothesis test in [16] is that our model is built based on a plaintext byte rather than a key-dependent intermediate value. This allows us to implement it in a black box manner and without key knowledge, just as the previous t-test.

4 Experimental Results

In order to discuss the pros and cons of the two previous leakage detection test, we now consider various experimental results. We start with a simulated setting which allows us to control all the parameters of the leakages to detect, in order to discuss the sampling complexity of both methods. Next, we analyze actual leakage traces obtained from the measurement setup described in Sect. 2.1, which allows us to put forward the intuitions provided by the t-test and ρ-test regarding the time localization of the informative samples in our traces.

[3] If there is no available trace for a given value of x, which happens when the evaluation set is small, the model takes the mean leakage taken over all the traces in $\mathcal{L}_p^{(j)}$.

4.1 Simulated Experiments

We define a standard simulated setting for the leakages of a block cipher, where an intermediate computation $z = \mathsf{S}(y = x \oplus s)$ is performed, with S an 8-bit S-box. It gives rise to a (multivariate) leakage variable of the form:

$$L_X = [\mathsf{HW}(X) + R_1,\ \mathsf{HW}(Y) + R_2,\ \mathsf{HW}(Z) + R_3],$$

where HW is the Hamming weight function, R_1, R_2 and R_3 are Gaussian distributed random noises with mean 0 and variance σ_n^2, and the index X recalls that in our detection setup, the evaluator only varies the plaintext. For t-tests, the set $\mathcal{L}_f$ contains leakages corresponding to fixed values of x, y or z, denoted as x_f, y_f, z_f, while the set $\mathcal{L}_r$ corresponds uniformly random x's, y's or z's. For ρ-tests, the leakages all correspond to uniformly random x's, y's or z's.

Concretely, we analyzed the t-test based on the third sample of L_X (which corresponds to the target intermediate value z), and for different fixed values of this intermediate value. This choice is naturally motivated by the counter-example given in introduction. That is, since the average leakage of the random class equals 4 in our simulation setting, a fixed class such that $\mathsf{HW}(z_f) = 4$ should not lead to any detection. And extending this example, the bigger the difference between $\mathsf{HW}(z_f)$ and 4, the easier the detection should be.

In parallel, we investigated the ρ-test for the same sample in two cases. First the realistic case, where the model estimation using k-fold cross-validation described in Sect. 3 is applied (using a standard $k = 10$). Second, a theoretical simplification where we assume that the evaluator knows the perfect (Hamming weight) model, which implies that all the samples in the set $\mathcal{L}$ are directly used to compute a single estimate for the correlation $\hat{r}(\tau) = \hat{\rho}(L_X(\tau), \mathsf{model}_\tau(X))$.

The results of our experiments are given in Fig. 1, where the upper part corresponds to a noise variance $\sigma_n^2 = 50$ and the lower part to a noise variance $\sigma_n^2 = 100$. In both cases, we set the detection threshold to 5, which is the value suggested in [2]. They allow the following relevant observations.

(1) *On the impact of the noise.* As doubling the noise variance generally doubles the measurement complexity of a side-channel attack, it has the same impact on the sample complexity of a leakage detection test. For example, detecting a difference between a fixed class such that $\mathsf{HW}(z_f) = 2$ and a random class with the t-test requires ≈ 1300 traces in the upper part of the figure and ≈ 2600 traces in its lower part. Similar observations hold for all the tests.
(2) *On the impact of the fixed value for the t-test.* As expected, for both σ_n^2, a fixed class such that $\mathsf{HW}(z_f) = 4$ cannot be distinguished at all from the random class (since they have the same mean). By contrast, a fixed class such that $\mathsf{HW}(z_f) = 0$ is extremely fast to distinguish from the random class.
(3) *The ρ-test can have (much) larger sampling complexity.* This naturally depends on the fixed value for the t-test. But assuming that several samples from a trace are used in a the leakage detection (which is usually the case, as will be shown in our following measured experiments), there should be some of them that lead to faster leakage detection with the t-test than with the ρ-test.

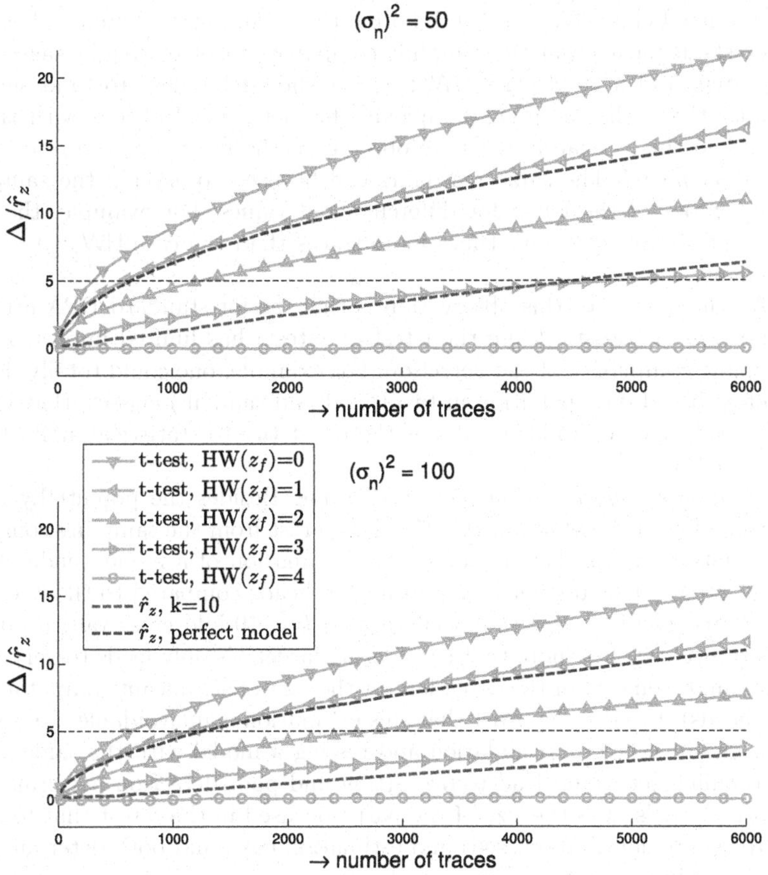

Fig. 1. Leakage detection on simulated traces, Hamming weight leakage function.

(4) *It's all in the SNR.* Most importantly, and just as in standard DPA, the sampling complexity of a detection test essentially depends on the SNR of its leakage partitioning. For the ρ-test, we can directly exploit Mangard's definition from CT-RSA 2004 for this purpose [15]. That is, the signal corresponds to the variance of the random variable $\mathsf{HW}(Z)$ with Z uniform, which equals 2 for 8-bit values, and the noise variance equals to σ_n^2. As for the t-test, we need to define an binary random variable B that is worth $\mathsf{HW}(z_f)$ with probability 1/2 and $\mathsf{HW} = 4$ with probability 1/2. For each value of the fixed z_f, the signal then corresponds to the variance of B, and the noise variance equals to σ_n^2 for the fixed class, and $\sigma_n^2 + 2$ for the random class (since in this case, the noise comes both from the variable Z and from the noise R). For example, this means a signal 0 for the fixed class $\mathsf{HW}(z_f) = 4$, a signal 0.25 for the fixed class $\mathsf{HW}(z_f) = 3$, a signal 1 for the fixed class $\mathsf{HW}(z_f) = 2$, a signal 2.25 for the fixed class $\mathsf{HW}(z_f) = 1$, and a signal 4

for the fixed class $\mathsf{HW}(z_f) = 0$. Ignoring the small noise differences between the tests, it means that the sampling complexity for detecting leakages with the t-test and a fixed class $\mathsf{HW}(z_f) = 1$ should be close to (and slightly smaller than) the sampling complexity for detecting leakages with the ρ-test. And this is exactly what we observe on the figure, for the ρ-test *with a perfect model.* The same reasoning can be used to explain the sampling complexities of the t-test for different fixed values. For example, the case $\mathsf{HW}(z_f) = 3$ requires four times more traces than the case $\mathsf{HW}(z_f) = 2$ on the figure.

A consequence of this observation is that, as for standard DPA attacks, the choice of statistic (here the t-test or ρ-test) has limited impact on the sampling complexity of the detection. For example, one could totally design a ρ-test based on a partition in two (fixed and random) classes, that would then lead to very similar results as the t-test (up to statistical artifacts, as discussed in [17]).

(5) *Estimating a model can only make it worse.* Besides the potentially lower signal, another drawback of the 256-class ρ-test from the sampling complexity point-of-view is that it requires the estimation of a model made of 256 mean values. This further increases its overheads compared to the t-test, as illustrated in Fig. 1 (see the $\hat{r}_z$ curve with $k = 10$-fold cross-validation). In this respect, we first note that considering larger k's only leads to very marginal improvements of the detection (at the cost of significant computational overheads). Besides, we insist that this estimation is unavoidable. For example, ignoring the cross-validation and testing a model with the same set as its profiling set would lead to overfitting and poor detection performances. In other words, it is the size of the partition used in the ρ-test that fixes its SNR (as previously discussed) and estimation cost, and both determine the final sampling complexity of the test.

Note that the above conclusions are independent of the leakage function considered (we repeated experiments with identity rather than Hamming weight leakages, and reached the same conclusions). Therefore, these simulated results confirm our introduction claim that for leakage detection only, a non specific t-test is the method of choice, and that its gains over a ρ-test can be easily predicted from a leakage function/partition and its resulting SNR metric.

4.2 Measured Experiments

We now extend the previous simulated analysis to the practically-relevant case of actual AES measurements, obtained from the setup described in Sect. 2.1. We will divide our investigations in two parts. First, a global analysis will consider the leakage traces of the full AES executions, in order to discuss the sampling complexity and intuitions regarding the POIs for our two detection tests. Next, a local analysis will be used in order to discuss possible false negatives in the t-test, and intuitions regarding the informativeness of the detected samples.

Global Analysis. The results of a fixed vs. random t-test and a ρ-test for leakage traces corresponding to an entire AES Furious execution are provided in Fig. 2, from which two main observations can be extracted.

(1) *The t-test has lower sampling complexity on average.* This is essentially the concrete counterpart of observation (3) in the previous section. That is, we already know that for some fixed values of the plaintext, the t-test should have a lower sampling complexity. Figure 2 confirms that when looking at complete AES traces, those "easy-to-detect" fixed values are indeed observed (which is natural since the AES Furious implementation accounts for a bit more than 3000 clock cycles, and the intermediate values within such a block cipher execution should be uniformly distributed after a couple of rounds). Concretely, this means that the sampling complexity for detecting leakages with a similar confidence increases from ≈ 200 traces for the t-test to ≈ 2000 traces for the ρ-test, i.e. a factor ≈ 10 which is consistent with the previous simulations. Note that even in the context of a hardware implementation with a reduced cycle count (e.g. 11 cycles per AES execution), finding fixed values that are easy-to-detect for the t-test is feasible by trying a couple of fixed plaintexts and keys.

(2) *The ρ-test (resp. t-test) does (resp. not) provide intuitions regarding exploitable leakage samples.* This is easily seen from the figure as well. Whereas the t-test detects information leakage everywhere in the trace, the ρ-test is much more localized, and points towards the samples that depend on the single plaintext byte of which the leakage is considered as signal (here corresponding to the first round and first S-box). Since the key is fixed in leakage detection, it implies that peaks are observed whenever this (useless)

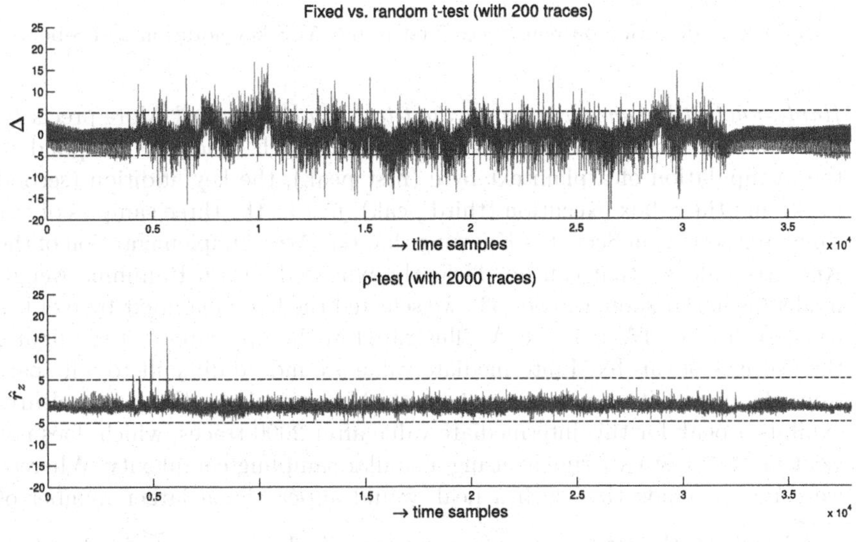

Fig. 2. Leakage detection on real traces, entire AES execution.

plaintext byte and the (useful) intermediate values that bijectively depend on it are manipulated, e.g. the key addition and S-box outputs in Fig. 2. In other words, the ρ-test is mostly relevant for the detection of POIs that are exploitable in a standard DPA attack (i.e. excluding the false positives corresponding to plaintext manipulations).

Local Analysis. The results of a fixed vs. random t-test and a ρ-test for leakage traces corresponding to the beginning of the first AES round execution are provided in Fig. 3, from which two main observations can be extracted.[4]

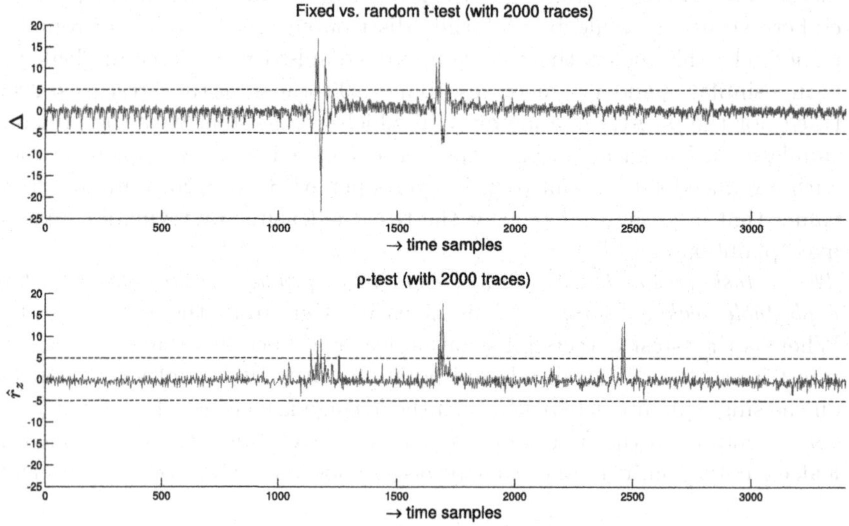

Fig. 3. Leakage detection on real traces, first-round AES key addition and S-box.

(1) *Hard-to-detect leakage samples for the t-test can be observed.* More precisely, the lower part of Fig. 3 exhibits three peaks which exactly correspond to the manipulation of a plaintext byte (first peak), the key addition (second peak) and the S-box execution (third peak), just as the three samples of our simulated setting in Sect. 4.1. Knowing that our Atmel implementation of the AES has leakages that can be efficiently exploited with a Hamming weight model (as in our simulations) [33], we selected the fixed plaintext byte of the t-test such that $\mathsf{HW}(z_f) = 4$. As illustrated in the upper part of the figure, the leakages of this fixed intermediate value are indeed difficult to tell apart from the one of its random counterpart. More precisely, the ρ-test clearly exhibits a peak for this intermediate value after 2000 traces, which does not exist in the t-test experiment using a similar sampling complexity. Whereas we cannot exclude that such a peak would appear for a larger number of

[4] Exceptionally for this experiment, we considered a single varying byte for the t-test, in order to better exhibit intuitions regarding the detected samples for a single S-box.

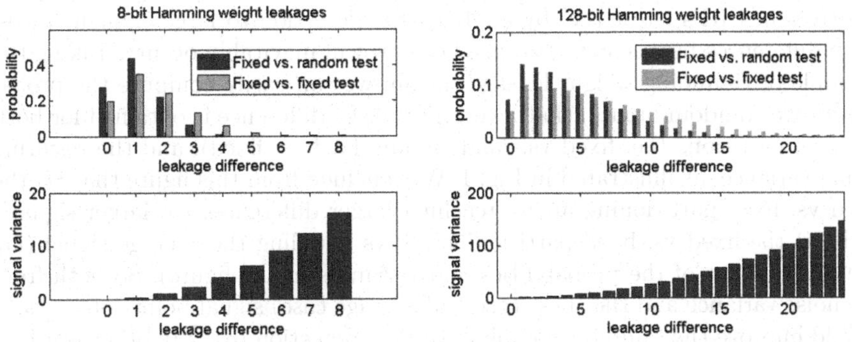

Fig. 4. Fixed vs. random and fixed vs. fixed leakage detection signal.

traces (since the chip does not exactly follow the Hamming weight leakage model), this confirms that not all leakage samples are easier to detect with the t-test than with the ρ-test.

(2) *The ρ-test does provide intuitions regarding the informativeness of the leakage samples.* Eventually, a straightforward advantage of the ρ-test is that the value of its correlation coefficient estimates brings some intuition regarding the complexity of a side-channel attack exploiting this sample, which is only provided up to a limited extent by the t-test. Indeed, a side-channel attack exploiting an s-bit intermediate value is most efficient if it relies on an s-bit model, as considered by the ρ-test (otherwise $s-1$ bits out of s will produce "algorithmic noise"). In this context, we can take advantage of the connection between Pearson's correlation coefficient and the information theoretic metrics in [34] (see [17]), themselves related to the worst-case complexity of standard DPA attacks [5].

5 Improved Leakage Detection Test

One central conclusion of the previous section is that the sampling complexity of leakage detection tests highly depends on the SNR of the leakage partition on which they are based. Interestingly, this observation directly suggests a natural improvement of CRI's non specific (fixed vs. random) t-test. Namely, rather than performing the test based on a fixed and a random class, a more efficient solution is to perform a similar test based on two fixed classes (i.e. two fixed plaintexts). On the one hand, this directly reduces the detection noise from $2\sigma_n^2 + \sigma_{\text{alg}}^2$ to $2\sigma_n^2$, since it cancels the algorithmic noise due to the variations of the random class. Taking the example of Hamming weight leakages, this algorithmic noise corresponds to $\sigma_{\text{alg}}^2 = 2$ for 8-bit values, but it increases for larger parallel implementations (e.g. it is worth $\sigma_{\text{alg}}^2 = 32$ for 128-bit implementations). On the other hand, and when applied to large traces, such a partitioning also increases the signal with high probability, for the same argument as used to avoid false positives in CRI's t-test (i.e. by applying the detection to large enough traces, large differences between the two fixed classes will inevitably occur). Taking the example of Hamming weight leakages again, we can easily compute the probability (over random inputs) that a certain leakage difference is obtained for both types of partitions (i.e. fixed vs. random and fixed vs. fixed), and the resulting signal variance, as illustrated in Fig. 4. We conclude from this figure that (*i*) the fixed vs. fixed partitioning allows reaching larger differences (so larger signals) and (*ii*) the fixed vs. fixed partitioning allows doubling the average signal (i.e. the dot product of the probabilities and variances in the figure). So both from the noise variance and the (best-case and average case) signal points-of-views, it should improve the sampling complexity of the detection test.[5] In other words, a leakage detection based on a fixed vs. fixed leakage partition should theoretically have better sampling complexity than with a fixed vs. random one.

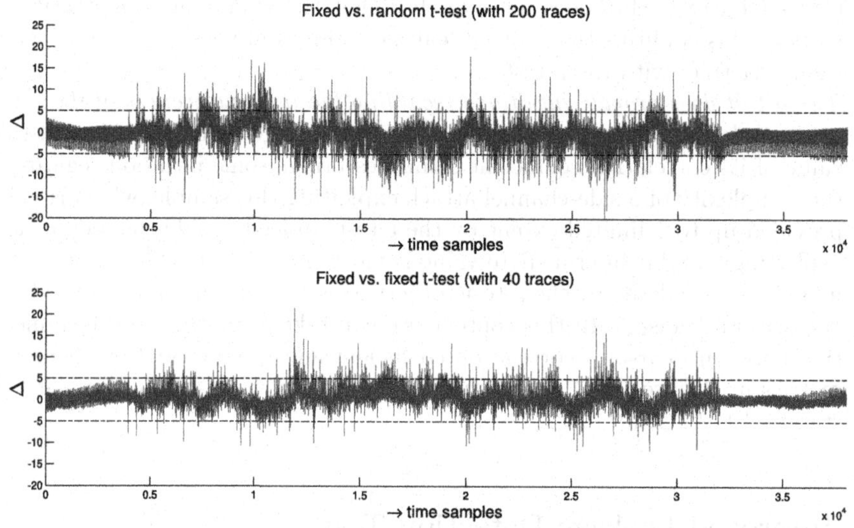

Fig. 5. Improved leakage detection on real traces (Atmel implementation).

Quite naturally, the exact gains of this new detection test depend on the actual leakages. So as in the previous section, we confirmed our expectations with two case studies. First, we compared the fixed vs. random and fixed vs. fixed t-tests based on our software AES implementation. The results of this experiment are in Fig. 5 where we observe that data-dependent leakages are detected with similar confidence with approximately 5 times less traces thanks to our new partitioning. Next, we investigated the context of the hardware implementation of the AES described in Sect. 2.1. As illustrated in Fig. 6, similar gains are obtained.

[5] A similar conclusion can be obtained for other leakage functions, though the binomial distribution of the Hamming weight leakages naturally make computations easier.

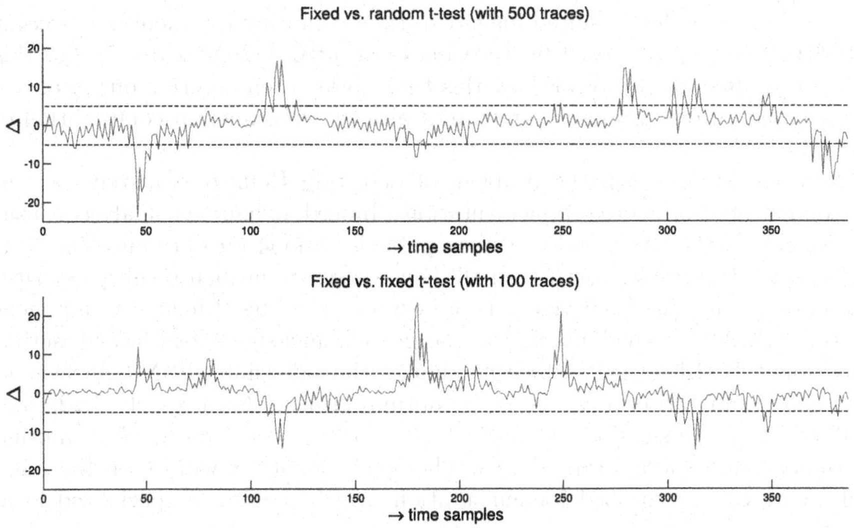

Fig. 6. Improved leakage detection on real traces (ASIC implementation).

Note however that despite we gain an approximate factor 5 in both cases, the reasons of this gain are different. Indeed, the software implementation case is dominated by an increase of signal (due to its large cycle count) and has limited algorithmic noise. By contrast, the hardware implementation has larger algorithmic noise (corresponding to 128-bit random values) but less improvements of the signal (because its traces are only 11-cycle long). Even larger gains could be obtained by combining both the signal and noise effects (e.g. by considering multiple keys for the hardware implementation). Based on these theoretical arguments and experimental confirmation, we expect our fixed vs. fixed partitioning to lead to faster leakage detections in most practical scenarios.

Remark. The fixed vs. fixed test can only be successful if the vectors used in the test exhibit significant differences for the target algorithm's intermediate computations (which is the counterpart of having fixed leakages different from average leakages in CRI's fixed vs. random test). This is easily obtained with block cipher implementations for which these intermediates are pseudorandom.

6 From Leakage Detection to POI Detection

The previous sections lead to the natural conclusion that non specific tests are a method of choice for leakage detection. In particular, their application to full leakage traces (or multiple keys) allows overcoming the issue of false positives mentioned in Sect. 4.1. By contrast, the correlation-based test is better suited to the detection of POIs because it provides useful intuitions regarding the exploitable samples in side-channel traces and their informativeness. As a result, it is a good candidate for the more specific task of detecting POIs for mounting an attack. In this section, we conclude the paper by putting forward that a

ρ-test is in fact perfectly suited for integration in (an improvement of) a recent POI detection tool proposed by Durvaux et al. at COSADE 2015 [7]. For this purpose, we first briefly recall how this tool works, then describe our improvements based on our proposed ρ-test, and provide experimental confirmation of our claims.

Note that in general, the problem of detecting POIs is relatively easy in the context of unprotected implementations. Indeed, exhaustive analysis is usually feasible in this case, and it is even possible to look for optimal transforms that project the samples towards small (hence easier-to-evaluate) subspaces such that most of their informativeness is preserved, e.g. using Principal Component Analysis (PCA) [1], which maximizes the side-channel signal, or Linear Discriminant Analysis (LDA) [32], which maximizes the side-channel SNR. In fact, in this context, any criteria can be easily optimized using local search [7,22], and most criteria are essentially equivalent anyway (i.e. correlation, SNR, mutual information and success rate [5,17]). Therefore, our focus will be on the more challenging case of masked implementation, which requires a specialized local search.

6.1 The COSADE 2015 POI Detection Tool

The COSADE 2015 POI detection aims at finding a projection vector $\boldsymbol{\alpha}$ that converts the N_s samples $l_y(\tau)$ of a leakage trace into a single projected sample λ_y:

$$\lambda_y = \sum_{\tau=1}^{N_s} \alpha(\tau) \cdot l_y(\tau).$$

In the case of unprotected implementations, and as previously mentioned, it is possible to find projections $\boldsymbol{\alpha}$ that optimize the informativeness of the projected sample (where the $\alpha(\tau)$ coefficients are real numbers, typically). By contrast, in the context of masking, the task is arguably more difficult since (*i*) single samples may not contain information (e.g. in the context of software implementations where the different shares are manipulated at different time samples), and (*ii*) the information about the target intermediate variables lies in higher-order moments of the leakage distribution. Therefore, Durvaux et al. focused on the simplified problem of finding a projection such that $\alpha(\tau) = 1$ if the time sample $l_y(\tau)$ contains some information about a share, and $\alpha(\tau) = 0$ otherwise.

In this context, a naive solution would be to consider each possible combination of time samples, but this scales badly (i.e. exponentially) with the number of shares to detect, and is rapidly prohibitive in practice, even for two shares (since masked implementations generally imply traces with many samples). In order to avoid this drawback, the algorithm proposed in [7] works by considering d non-overlapping windows of length W_{len} that set the covered weights to 1 (and leaves to others stuck at 0). Algorithm 1 provides a succinct description of this method. Besides the previously mentioned window length, it mainly requires defining an objective function f_{obj} and a detection threshold T_{det}, and works in two steps. First, the find_solution phase places the windows randomly at different

Algorithm 1. Local search algorithm for finding POIs in masked traces.

```
Local_Search(d, W_len, T_det, @f_obj)
   α = find_solution(d, W_len, T_det, @f_obj);
   if(α ≠ null)
      return improve_solution(α, @f_obj);
   end
end
```

locations of the trace, until the returned value of the objective function crosses the threshold. Then, the improve_solution modifies the windows' size in order to best fit the informative time samples. As a result, we obtain the position and the size of each window that maximizes f_{obj}. By changing the number of windows and objective function, this approach can easily be extended to masking schemes of any order and number of shares. Intuitively, the W_{len} parameter leads to a natural tradeoff between the time complexity and sampling complexity of the algorithm. Namely, small window lengths are more time intensive[6], and large ones more rapidly cover POIs, but imply an estimation of the objective function for samples projected according to larger windows, which are potentially more noisy. Eventually, the objective function proposed in the COSADE paper is the Moments-Correlating Profiled DPA (MCP-DPA) introduced in [21], which can be viewed as a classical higher-order DPA based on the CPA distinguisher given in Sect. 2.2, where one correlates the leakages samples raised to a power d with a model corresponding to the dth-order statistical moment of the leakage distribution. We refer to the previous papers for the details on these tools.

6.2 Our Contribution

We first recall that the COSADE 2015 POI detection tool is *black box* in the sense that it does not require any knowledge of the target implementation. By contrast, it does require *key profiling*, since the MCP-DPA distinguisher is a profiled one. In this respect, our first contribution is the simple but useful observation that one can easiliy apply such a black box POI detection without key profiling, by simply profiling the MCP-DPA objective function based on plaintext knowledge, just as the ρ-test in this paper. Indeed, when detecting POIs, it is sufficient to know the leakage model up to a permutation corresponding to key knowledge (a quite similar idea has been exploited in [28] for similar purposes). As previously discussed, this solution will suffer from the (minor) risk of detecting plaintext samples, but as will be detailed next, this can be easily mitigated in practice.

Based on these premises, our second contribution starts from the equally simple observation that the ρ-test of this paper can be used identically with the MCP-DPA distinguisher. So it is theoretically eligible for detecting leakages and POIs of any order. Therefore, by replacing the MCP-DPA objective function in [7] by the ρ-test in this paper (based on CPA or MCP-DPA), we obtain

[6] For example, $W_{len} = 1$ is equivalent to testing all combinations of time samples.

a very simple and rigorous way to set the detection threshold in Algorithm 1. That is, one just has to use the same "five sigma rule" as used in the leakage detections of Figs. 2 and 3. Note by changing the objective function and selection of a detection threshold in this way, we benefit from the additional advantage of (more efficiently) estimating the objective function with cross-validation, which is another improvement over the method described at COSADE 2015.

Third and maybe most importantly, we notice that the COSADE 2015 objective function is based on the estimation of central (second-order) statistical moments. That is, given the (average) leakage of the two windows, it first sums them and then computes a variance. But looking at the discussion in [35], Sect. 4, it is clear that whenever the leakages are relatively noisy – which will always happen in our context of which the goal is to exploit the largest possible windows in order to reduce the time needed to detect POIs – considering mixed statistical moments is a better choice. In other words, by exploiting the multivariate MCP-DPA mentioned at the end of [21], we should be able to detect POIs with larger windows. In our following experiments based on a first-order (2-shares) masking scheme, this just means using the normalized product between the (average) leakage of the two windows as objective function, which has been shown optimal in the context of Hamming weight leakages in [26].

6.3 Experimental Validation

In order to confirm the previous claims, we tested Algorithm 1 using exactly the previously described modifications, based on a target implementation and measurement setup very similar to the one in [7]. That is, we first analyzed the leakages of the masked implementation described in Sect. 2.1 which leads to large traces with $N_s = 30,000$ samples (for which an exhaustive analysis of all the pairs of samples is out of reach). As in the COSADE 2015 paper, we verified that our implementation does not lead to any first-order leakages (this time with the ρ-based test from Sect. 3). We further set the window length to 25 samples, which corresponds to a bit more than two clock cycles at our clock frequency and sampling rate. With these parameters, the local search was able to return a solution within the same number of objective function calls as [7], namely $\approx 12\,000$ on average. An example of leakage trace together with windows obtained thanks to Algorithm 1 is given in Fig. 7. As clear from the zoomed plots at the bottom of the figure (where we represent the sum of the projection vectors obtained after 100 experiments), the selection of POIs corresponds to leakage samples that combine the precomputation and masked S-box computation. Interestingly, we could expect some false positives due to the detection of plaintext bytes that is possible in our non-profiled scenario. However, the improve_solution of Algorithm 1 (where the window size is adapted to be most informative) combined with the fact that the most informative leakage samples in our traces correspond to memory accesses (i.e. the S-box computations) prevented these to happen. Note that even if the leakage of the plaintext manipulations was more informative, we could easily "mark" the cycles that correspond to plaintext knowledge only, and exclude them from our optimization. Since the number of POIs corresponding

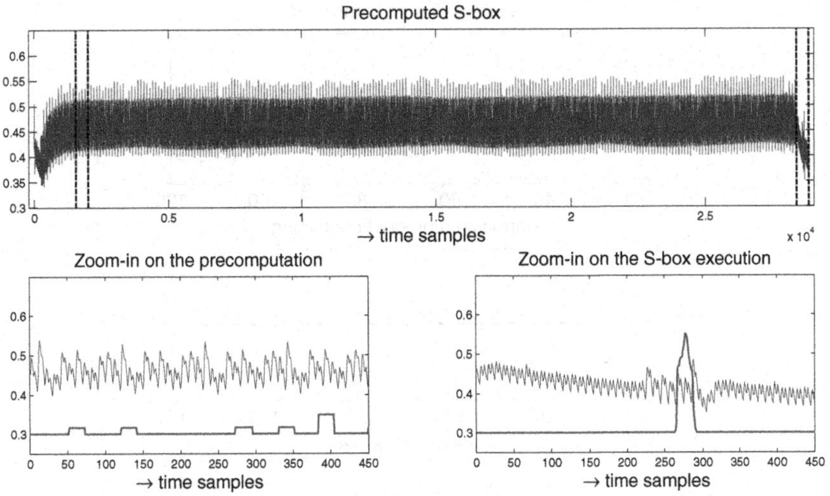

Fig. 7. Non-profiled detection of POIs based on our ρ-test.

to a single plaintext byte is usually limited, this would lead to the detection of a valid pair of POIs after a couple of iterations of Algorithm 1. Besides, we note that Simple Power Analysis, or DPA against the plaintext (before it is XORed with the masks) are other simple ways to gain the minimum intuition about the time localization of the POIs, in order to avoid false positives when running a non-profiled local search.

Next, we analyzed the impact of our modified objective function on the largest window lengths for which we could detect POIs. As illustrated in Fig. 8, we can clearly observe a (significant) gain of an approximate factor > 3 when using a normalized product combining as objective function rather than the previously used square of sum (for which the figure also suggests that the window of length 25 was close to optimal). It means that for exactly the same amount of traces in an evaluation set, we are able to detect POIs with > 3 times larger windows with our improved objective function and detection threshold. Concretely, this corresponds to a reduction of the time complexity by a factor > 3 compared to the COSADE 2015 results (and by a factor ≈ 90 compared to a naive combinatorial search). Interestingly, we also see that increasing the window length is not always detrimental, which corresponds to the fact that larger windows do not only contain more noise, but also more informative samples.

To conclude, we believe the connections made in this section are important to raise awareness that up to the selection of POIs in the leakage traces, side-channel security evaluations can essentially be performed in a black box way, and without any key profiling. In this respect, leakage detection and the detection of POIs are indeed related tasks, with the significant difference that the latter has to take the exploitability of the detected samples into account. And this is exactly the difference between simple t-tests and more measurement-intensive ρ-tests based on larger leakage partitions. Note that the non-profiled detection in

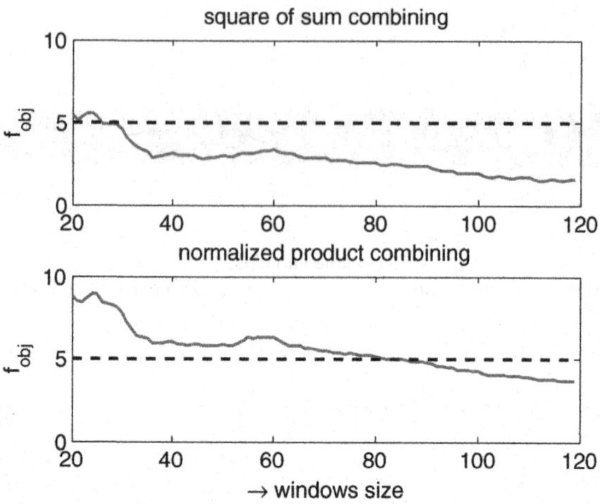

Fig. 8. Estimation of objective functions (with cross-validation) based on central and mixed statistical moments, in front of a detection threshold of five sigma.

this section only applies to the first/last block cipher rounds (i.e. before diffusion is complete), which captures many relevant practical scenarios but could be an issue, e.g. in contexts where these extreme rounds are better protected than the central ones. Besides, and more generally, we recall that as soon as the POIs are detected and the evaluator has to build a model for these samples, key profiling becomes strictly necessary to evaluate a worst-case security level [39].

7 Summary and Open Problems

The discussion in this paper highlights that there are significant differences between current approaches to side-channel security evaluation. On the one hand, CRI's Test Vector Assessment Methodology (TVLA) aims at minimizing the evaluator's efforts. Very concretely, non specific t-tests as proposed in [4,10] are indeed good to detect univariate and first-order leakages. As we observed in Sect. 5, slightly tweaking the selection of the classes (from fixed vs. random to fixed vs. fixed) allows significantly improving the detection speed in this case. We can expect these gains to be even more significant in the context of masked implementations (for which the impact of noise is exponentially amplified). The fixed vs. fixed test also has good potential for evaluating the implementations of asymmetric cryptographic primitives. So despite minor theoretical caveats (i.e. the possibility of false positives and negatives), the application of such 2-class t-tests turns out to be extremely efficient. On the other side of the spectrum, complete (ideally worst-case) security evaluations such as discussed in [34] rather aim at a precise rating of the security level, possibly considering the adversary's computing power [36], which is an arguably more expensive task. In this case, the selection of POIs is a usually a necessary first step. As also discussed in this paper, and when restriced to univariate and first-order leakages, the main reason for the additional cost of this approach (including the selection of POIs) is the larger number of classes for which the leakage distribution has to be well estimated. In this context as well, our investigations focused on non-profiled POI detection (which can be performed efficiently for the first/last cipher rounds). But similar conclusions hold in the profiled evaluation setting, which allows finding POIs in all the cipher rounds, and is necessary for worst-case analysis.

These different methodologies naturally raise the question of which one to use in which context, and whether they can be connected to some extent, leading to the following open problems. First, how to generalize (simple) detection tests to capture more types of leakages? Moving from univariate first-order leakages to univariate higher-order leakages is already reachable with existing tools. One option, already described in Sect. 6, is to work "by moments". This implied to implement a Moments-Correlating DPA in our multi-class context, but could naturally be specialized to simpler t-tests, F-tests,..., if only 2 classes were considered: see [30] for a recent discussion that is complementary to our results. Another option is to exploit more general statistical tests, e.g. Mutual Information Analysis [8], as already applied in the context of leakage detection by Mather et al. [18]. Moving to multivariate leakage detection appears much more difficult. At least, testing all pairs/triples/... of samples in a trace rapidly turns out to be unfeasible as the size of the traces increase, which usually leads current evaluations to be based on heuristics (e.g. the ones discussed in Sect. 6). Note that the gap between univariate and multivariate attacks is probably among the most important remaining challenge in side-channel security evaluations, where significant risks of false negatives remain. A typical example is the case of static leakages that may only be revealed in the context of (highly) multivariate analyses [20,24]. More generally, limiting an evaluation to univariate leakages typically ignores the significant gains that can be obtained with dimensionality reductions techniques (aka projections), and multi-target attacks [12,19,37].

Second, can we extrapolate or bound the worst-case security level of an implementation based on simple statistical tests? For example, the recent work in [5] shows that one can (in certain well-defined conditions) bound the security level of an implementation, measured with a success rate and in function of the number of measurements and computing power of the adversary, based on information theoretic metrics (such as the mutual information in general, and the SNR if we only consider univariate attacks). But as discussed in this paper, evaluating an SNR is still significantly more expensive than detecting leakages with non specific tests. So of course, it would be interesting to investigate whether it is possible to bound the security level based on simpler leakage detection tests. In case of negative answer, it anyway remains that such leakage detection tests can always be used as a prelimininary to more expensive approaches (detecting POIs, security evaluations), e.g. to reduce the dimensionality of the traces.

Acknowledgements. F.-X. Standaert is a research associate of the Belgian Fund for Scientific Research (FNRS-F.R.S.). This work has been funded in parts by the European Commission through the ERC project 280141 (CRASH).

References

1. Archambeau, C., Peeters, E., Standaert, F.-X., Quisquater, J.-J.: Template attacks in principal subspaces. In: Goubin and Matsui [11], pp. 1–14
2. Balasch, J., Gierlichs, B., Grosso, V., Reparaz, O., Standaert, F.-X.: On the cost of lazy engineering for masked software implementations. In: Joye, M., Moradi, A. (eds.) CARDIS 2014. LNCS, vol. 8968, pp. 64–81. Springer, Heidelberg (2015)
3. Brier, E., Clavier, C., Olivier, F.: Correlation power analysis with a leakage model. In: Joye, M., Quisquater, J.-J. (eds.) CHES 2004. LNCS, vol. 3156, pp. 16–29. Springer, Heidelberg (2004)
4. Cooper, J., De Mulder, E., Goodwill, G., Jaffe, J., Kenworthy, G., Rohatgi, P.: Test vector leakage assessment (tvla) methodology in practice (extended abstract). ICMC 2013. http://icmc-2013.org/wp/wp-content/uploads/2013/09/goodwillkenworthtestvector.pdf
5. Duc, A., Faust, S., Standaert, F.-X.: Making masking security proofs concrete. In: Oswald, E., Fischlin, M. (eds.) EUROCRYPT 2015. LNCS, vol. 9056, pp. 401–429. Springer, Heidelberg (2015)
6. Durvaux, F., Standaert, F.-X., Veyrat-Charvillon, N.: How to certify the leakage of a chip? In: Nguyen, P.Q., Oswald, E. (eds.) EUROCRYPT 2014. LNCS, vol. 8441, pp. 459–476. Springer, Heidelberg (2014)
7. Durvaux, F., Standaert, F.-X., Veyrat-Charvillon, N., Mairy, J.-B., Deville, Y.: Efficient selection of time samples for higher-order DPA with projection pursuits. In: Mangard, S., Poschmann, A.Y. (eds.) COSADE 2015. LNCS, vol. 9064, pp. 34–50. Springer, Heidelberg (2015)
8. Gierlichs, B., Batina, L., Tuyls, P., Preneel, B.: Mutual information analysis. In: Oswald and Rohatgi [23], pp. 426–442
9. Gierlichs, B., Lemke-Rust, K., Paar, C.: Templates vs. stochastic methods. In: Goubin and Matsui [11], pp. 15–29
10. Goodwill, G., Jun, B., Jaffe, J., Rohatgi, P.: A testing methodology for side channel resistance validation. NIST non-invasive attack testing workshopp (2011). http://csrc.nist.gov/news_events/non-invasive-attack-testing-workshop/papers/08_Goodwill.pdf
11. Goubin, L., Matsui, M. (eds.): CHES 2006. LNCS, vol. 4249. Springer, Heidelberg (2006)
12. Grosso, V., Standaert, F.-X.: ASCA, SASCA and DPA with enumeration: which one beats the other and when? In: Iwata, T., et al. (eds.) ASIACRYPT 2015. LNCS, vol. 9453, pp. 291–312. Springer, Heidelberg (2015). doi:10.1007/978-3-662-48800-3_12
13. Kerckhof, S., Durvaux, F., Hocquet, C., Bol, D., Standaert, F.-X.: Towards green cryptography: A comparison of lightweight ciphers from the energy viewpoint. In: Prouff and Schaumont [27], pp. 390–407
14. Kocher, P.C., Jaffe, J., Jun, B.: Differential power analysis. In: Wiener, M. (ed.) CRYPTO 1999. LNCS, vol. 1666, pp. 388–397. Springer, Heidelberg (1999)
15. Mangard, S.: Hardware countermeasures against DPA – a statistical analysis of their effectiveness. In: Okamoto, T. (ed.) CT-RSA 2004. LNCS, vol. 2964, pp. 222–235. Springer, Heidelberg (2004)

16. Mangard, S., Oswald, E., Popp, T.: Power Analysis Attacks - Revealing the Secrets of Smart Cards. Springer, New York (2007)
17. Mangard, S., Oswald, E., Standaert, F.-X.: One for all - all for one: unifying standard differential power analysis attacks. IET Inf. Secur. **5**(2), 100–110 (2011)
18. Mather, L., Oswald, E., Bandenburg, J., Wójcik, M.: Does my device leak information? an *a priori* statistical power analysis of leakage detection tests. In: Sako, K., Sarkar, P. (eds.) ASIACRYPT 2013, Part I. LNCS, vol. 8269, pp. 486–505. Springer, Heidelberg (2013)
19. Mather, L., Oswald, E., Whitnall, C.: Multi-target DPA attacks: Pushing DPA beyond the limits of a desktop computer. In: Sarkar and Iwata [29], pp. 243–261
20. Moradi, A.: Side-channel leakage through static power. In: Batina, L., Robshaw, M. (eds.) CHES 2014. LNCS, vol. 8731, pp. 562–579. Springer, Heidelberg (2014)
21. Moradi, A., Standaert, F.-X.: Moments-correlating DPA. IACR Cryptology ePrint Archive, 2014:409 (2014)
22. Oswald, D., Paar, C.: Improving side-channel analysis with optimal linear transforms. In: Mangard, S. (ed.) CARDIS 2012. LNCS, vol. 7771, pp. 219–233. Springer, Heidelberg (2013)
23. Oswald, E., Rohatgi, P. (eds.): CHES 2008. LNCS, vol. 5154. Springer, Heidelberg (2008)
24. Pozo, S.M., Del Standaert, F.-X., Kamel, D., Moradi, A.: Side-channel attacks from static power: when should we care? In: Nebel, W., Atienza, D. (eds.) DATE 2015, Grenoble, France, March 9–13, 2015, pp. 145–150. ACM (2015)
25. Prouff, E., Rivain, M.: A generic method for secure sbox implementation. In: Kim, S., Yung, M., Lee, H.-W. (eds.) WISA 2007. LNCS, vol. 4867, pp. 227–244. Springer, Heidelberg (2008)
26. Prouff, E., Rivain, M., Bevan, R.: Statistical analysis of second order differential power analysis. IEEE Trans. Comput. **58**(6), 799–811 (2009)
27. Prouff, E., Schaumont, P. (eds.): CHES 2012. LNCS, vol. 7428. Springer, Heidelberg (2012)
28. Reparaz, O., Gierlichs, B., Verbauwhede, I.: Selecting time samples for multivariate DPA attacks. In: Prouff and Schaumont [27], pp. 155–174
29. Sarkar, P., Iwata, T. (eds.): ASIACRYPT 2014. LNCS, vol. 8873. Springer, Heidelberg (2014)
30. Schneider, T., Moradi, A.: Leakage assessment methodology. In: Güneysu, T., Handschuh, H. (eds.) CHES 2015. LNCS, vol. 9293, pp. 495–513. Springer, Heidelberg (2015)
31. Schramm, K., Paar, C.: Higher order masking of the AES. In: Pointcheval, D. (ed.) CT-RSA 2006. LNCS, vol. 3860, pp. 208–225. Springer, Heidelberg (2006)
32. Standaert, F.-X., Archambeau, C.: Using subspace-based template attacks to compare and combine power and electromagnetic information leakages. In: Oswald and Rohatgi [23], pp. 411–425
33. Standaert, F.-X., Gierlichs, B., Verbauwhede, I.: Partition *vs.* comparison side-channel distinguishers: an empirical evaluation of statistical tests for univariate side-channel attacks against two unprotected CMOS devices. In: Lee, P.J., Cheon, J.H. (eds.) ICISC 2008. LNCS, vol. 5461, pp. 253–267. Springer, Heidelberg (2009)
34. Standaert, F.-X., Malkin, T.G., Yung, M.: A unified framework for the analysis of side-channel key recovery attacks. In: Joux, A. (ed.) EUROCRYPT 2009. LNCS, vol. 5479, pp. 443–461. Springer, Heidelberg (2009)

35. Standaert, F.-X., Veyrat-Charvillon, N., Oswald, E., Gierlichs, B., Medwed, M., Kasper, M., Mangard, S.: The world is not enough: another look on second-order DPA. In: Abe, M. (ed.) ASIACRYPT 2010. LNCS, vol. 6477, pp. 112–129. Springer, Heidelberg (2010)
36. Veyrat-Charvillon, N., Gérard, B., Standaert, F.-X.: Security evaluations beyond computing power. In: Johansson, T., Nguyen, P.Q. (eds.) EUROCRYPT 2013. LNCS, vol. 7881, pp. 126–141. Springer, Heidelberg (2013)
37. Veyrat-Charvillon, N., Gérard, B., Standaert, F.-X.: Soft analytical side-channel attacks. In: Sarkar and Iwata [29], pp. 282–296
38. Welch, B.L.: The generalization of Student's problem when several different population variances are involved. Biometrika, pp. 28–35 (1947)
39. Whitnall, C., Oswald, E., Standaert, F.-X.: The myth of generic DPA. . .and the magic of learning. In: Benaloh, J. (ed.) CT-RSA 2014. LNCS, vol. 8366, pp. 183–205. Springer, Heidelberg (2014)

Improved Masking for Tweakable Blockciphers with Applications to Authenticated Encryption

Robert Granger[1], Philipp Jovanovic[2(✉)], Bart Mennink[3], and Samuel Neves[4]

[1] Laboratory for Cryptologic Algorithms,
École polytechnique fédérale de Lausanne, Lausanne, Switzerland
robert.granger@epfl.ch

[2] Decentralized and Distributed Systems Lab,
École polytechnique fédérale de Lausanne, Lausanne, Switzerland
philipp.jovanovic@epfl.ch

[3] Department of Electrical Engineering, ESAT/COSIC,
KU Leuven, and iMinds, Leuven, Belgium
bart.mennink@esat.kuleuven.be

[4] CISUC, Department of Informatics Engineering,
University of Coimbra, Coimbra, Portugal
sneves@dei.uc.pt

Abstract. A popular approach to tweakable blockcipher design is via masking, where a certain primitive (a blockcipher or a permutation) is preceded and followed by an easy-to-compute tweak-dependent mask. In this work, we revisit the principle of masking. We do so alongside the introduction of the tweakable Even-Mansour construction MEM. Its masking function combines the advantages of word-oriented LFSR- and powering-up-based methods. We show in particular how recent advancements in computing discrete logarithms over finite fields of characteristic 2 can be exploited in a constructive way to realize highly efficient, constant-time masking functions. If the masking satisfies a set of simple conditions, then MEM is a secure tweakable blockcipher up to the birthday bound. The strengths of MEM are exhibited by the design of fully parallelizable authenticated encryption schemes OPP (nonce-respecting) and MRO (misuse-resistant). If instantiated with a reduced-round BLAKE2b permutation, OPP and MRO achieve speeds up to 0.55 and 1.06 cycles per byte on the Intel Haswell microarchitecture, and are able to significantly outperform their closest competitors.

Keywords: Tweakable Even-Mansour · Masking · Optimization · Discrete logarithms · Authenticated encryption · BLAKE2

1 Introduction

Authenticated encryption (AE) has faced significant attention in light of the ongoing CAESAR competition [15]. An AE scheme aims to provide both confidentiality and integrity of processed data. While the classical approach is predominantly blockcipher-based, where an underlying blockcipher is used to encrypt,

M. Fischlin and J.-S. Coron (Eds.): EUROCRYPT 2016, Part I, LNCS 9665, pp. 263–293, 2016.
DOI: 10.1007/978-3-662-49890-3_11

novel approaches start from a permutation and either rely on Sponge-based principles or on the fact that the Even-Mansour construction $E(K, M) = P(K \oplus M) \oplus K$ is a blockcipher.

Characteristic for the majority of blockcipher-based AE schemes is that they rely on a tweakable blockcipher where changes in the tweak can be realized efficiently. The most prominent example of this is the OCB2 mode which internally uses the XEX tweakable blockcipher [71]:

$$\mathrm{XEX}(K, (X, i_0, i_1, i_2), M) = E(K, \delta \oplus M) \oplus \delta,$$

where $\delta = \mathbf{2}^{i_0}\mathbf{3}^{i_1}\mathbf{7}^{i_2}E(K, X)$. The idea is that every associated data or message block is transformed using a different tweak, where increasing i_0, i_1, or i_2 can be done efficiently. This approach is furthermore used in second-round CAESAR candidates AEZ, COPA, ELmD, OTR, POET, and SHELL. Other approaches to masking include Gray code ordering (used in OCB1 and OCB3 [55,72] and OMD) and the word-oriented LFSR-based approach where $\delta = \varphi^i(E(K, X))$ for some LFSR φ (suggested by Chakraborty and Sarkar [18]).

The same masking techniques can also be used for permutation-based tweakable blockciphers. For instance, Minalpher uses the Tweakable Even-Mansour (TEM) construction [75] with XEX-like masking, and similar for Prøst. This TEM construction has faced generalizations by Cogliati et al. [23,24] and Mennink [62], but none of them considers efficiency improvements of the masking.

1.1 Masked Even-Mansour (MEM) Tweakable Cipher

As a first contribution, we revisit the state of the art in masking with the introduction of the "Masked Even-Mansour" tweakable blockcipher in Sect. 3. At a high level, MEM is a Tweakable Even-Mansour construction, where the masking combines ideas from both word-oriented LFSR- and powering-up-based masking. As such, MEM combines "the best of both" masking approaches, leading to significant improvements in simplicity, error-proneness, and efficiency.

In more detail, let P be a b-bit permutation. MEM's encryption function is defined as

$$\widetilde{E}(K, X, \bar{i}, M) = P(\delta(K, X, \bar{i}) \oplus M) \oplus \delta(K, X, \bar{i}),$$

where $\bar{i} = (i_0, \ldots, i_{u-1})$ and where the masking function is of the form

$$\delta(K, X, \bar{i}) = \varphi_{u-1}^{i_{u-1}} \circ \cdots \circ \varphi_0^{i_0}(P(X \parallel K)),$$

for a certain set of LFSRs $(\varphi_0, \ldots, \varphi_{u-1})$. MEM's decryption function $\widetilde{D}$ is specified analogously but using P^{-1} instead of P.

The tweak space and the list of LFSRs are clearly required to satisfy some randomness condition. Indeed, if a distinguisher can choose a list of tweaks $\bar{i}$ such that $\varphi_{u-1}^{i_{u-1}} \circ \cdots \circ \varphi_0^{i_0}(L)$ for a uniformly random L offers no or limited entropy, it can easily distinguish MEM from a random primitive. A similar case applies if the distinguisher can make two different maskings collide with high probability. Denote by ϵ the minimal amount of entropy offered by the functions $\varphi_{u-1}^{i_{u-1}} \circ \cdots \circ \varphi_0^{i_0}$ and $\varphi_{u-1}^{i_{u-1}} \circ \cdots \circ \varphi_0^{i_0} \oplus \varphi_{u-1}^{i'_{u-1}} \circ \cdots \circ \varphi_0^{i'_0}$ for any two maskings $\bar{i}, \bar{i}'$ (see Definition 1 for the formal definition). Then, we prove that MEM is a secure tweakable blockcipher in the ideal permutation model up to $\frac{4.5q^2+3qp}{2^\epsilon} + \frac{p}{2^k}$, where q is the number of construction queries, p the number of primitive queries, and k the key length. The security proof follows Patarin's H-coefficient technique, which has shown its use to Even-Mansour security proofs before in, among others, [3,20,21,23,25,63].

To guarantee that the maskings offer enough randomness, it is of pivotal importance to define a proper domain of the masking. At the least, the functions $\varphi_{u-1}^{i_{u-1}} \circ \cdots \circ \varphi_0^{i_0}$ should be different for all possible choices of $\bar{i}$, or more formally, such that there do not exist $\bar{i}, \bar{i}'$ such that

$$\varphi_{u-1}^{i_{u-1}} \circ \cdots \circ \varphi_0^{i_0} = \varphi_{u-1}^{i'_{u-1}} \circ \cdots \circ \varphi_0^{i'_0}.$$

Guaranteeing this requires the computation of discrete logarithms. For small cases, such as $b = 64$ and $b = 128$, we can inherit the computations from Rogaway for XEX [71]. For instance, for $b = 128$, it is known that $u = 3$, $(\varphi_0, \varphi_1, \varphi_2) = (\mathbf{2}, \mathbf{3}, \mathbf{7})$, and $(i_0, i_1, i_2) \in \{-2^{108}, \ldots, 2^{108}\} \times \{-2^7, \ldots, 2^7\} \times \{-2^7, \ldots, 2^7\}$ does the job.

We extend the XEX approach to much larger block sizes by taking advantage of the recent breakthroughs in the computation of discrete logs in small characteristic fields, beginning with [30], followed by [45]. Computation of individual discrete logarithms for the 1024-bit block used in our MEM instantiation takes about 8 h on a single core of a standard desktop computer, after an initial precomputation, applicable to all logarithms, of 33.3 h. Larger blocks are also attainable, rendering workarounds such as subgroups [77] or different modes [74] largely unnecessary.

Peculiarly, there have been uses of XEX for state sizes larger than $b = 128$ bits, even though it has been unclear what restrictions on the indices are due. For instance, Prøst [50] defines a COPA and OTR instance for a 256- and 512-bit blockcipher; both use maskings of the form $\mathbf{2}^{i_0}\mathbf{3}^{i_1}\mathbf{7}^{i_2}$ for i_0 ranging between 0 and the maximal message length. For COPA, it has $(i_1, i_2) \in \{0, \ldots, 5\} \times \{0, 1\}$ and for OTR it has $(i_1, i_2) \in \{0, 1\} \times \{0\}$. The security proof of Prøst never formally computes conditions on the indices, and simply inherits the conditions for $b = 128$. By computing the discrete logarithms in the respective fields—a computationally easy task, demonstrated in Sect. 3.6—we can confirm that the tweaks are unique for $i_0 \in \{0, \ldots, 2^{246} - 1\}$ in the 256-bit block case, and $i_0 \in \{0, \ldots, 2^{505} - 1\}$ in the 512-bit block case.

1.2 Application to Nonce-Based AE

As first application, we present the Offset Public Permutation (OPP) mode in Sect. 4, a parallelizable nonce-based AE based on MEM. It can be considered as a permutation-based generalization of OCB3 [55] to arbitrary block sizes using permutations and using the improved masking from MEM. Particularly, assuming security of MEM, the proof of [55] mostly carries over, and we obtain that OPP behaves like a random AE up to attack complexity dominated by $\min\{2^{b/2}, 2^k\}$, where b is the size of the permutation and k is the key length. OPP also shows similarities with Kurosawa's adaptation of IAPM and OCB to the permutation-based setting [56].

Using the masking techniques described later in this paper, OPP has excellent performance when compared to contemporary permutation-based schemes, such as first-round CAESAR [15] submissions Artemia, Ascon, CBEAM, ICEPOLE, Keyak, NORX, π-Cipher, PRIMATEs, and STRIBOB, or SpongeWrap schemes in general [9,63]. OPP improves upon these by being inherently parallel and rate-1; the total overhead of the mode reduces to 2 extra permutation calls and the aforementioned efficient masking.

In particular, when instantiated with a reduced-round BLAKE2b permutation [5], OPP achieves a peak speed of 0.55 cycles per byte on an Intel Haswell processor (see Sect. 8). This is faster than any other permutation-based CAESAR submission. In fact, there are only a few CAESAR ciphers, such as Tiaoxin (0.28 cpb) or AEGIS (0.35 cpb), which are faster than the above instantiation of OPP. However, both require AES-NI to reach their best performance and neither of them is arbitrarily parallelizable.

1.3 Application to Nonce-Misuse Resistant AE

We also consider permutation-based authenticated encryption schemes that are resistant against nonce-reuse. We consider "full" nonce-misuse resistance, where the output is completely random for different inputs, but we remark that similarly schemes can be designed to achieve "online" nonce-misuse resistance [26,41], for instance starting from COPA [2]. It is a well-known result that nonce-misuse resistant schemes are inherently offline, meaning that two passes over the data must be made in order to perform the authenticated encryption.

The first misuse-resistant AE we consider is the parallelizable Misuse-Resistant Offset (MRO) mode (Sect. 5). It starts from OPP, but with the absorption performed on the entire data and with encryption done in counter mode instead[1]. As the underlying MEM is used by the absorption and encryption parts for different maskings, we can view the absorption and encryption as two independent functions and a classical MAC-then-Encrypt security proof shows that MRO is secure up to complexity dominated by $\min\{2^{b/2}, 2^k, 2^{\tau/2}\}$, where b and k are as before and τ denotes the tag length.

Next, we consider Misuse-Resistant Sponge (MRS) in Sect. 6. It is not directly based on MEM; it can merely be seen as a cascaded evaluation of the Full-state Keyed Duplex of Mennink et al. [63], a generalization of the Duplex of Bertoni et al. [9]: a first evaluation computes the tag on input of all data, the second evaluation encrypts the message with the tag functioning as the nonce.

[1] MRO's structure is comparable with the independently introduced Synthetic Counter in Tweak [43,44,70].

MRS is mostly presented to suit the introduction of the Misuse-Resistant Sponge-Offset hybrid (MRSO) in Sect. 7, which absorbs like MRS and encrypts like MRO. (It is also possible to consider the complementary Offset-Sponge hybrid, but we see no potential applications of this construction.) The schemes MRS and MRSO are proven secure up to complexity of about $\min\{2^{c/2}, 2^{k/2}, 2^{\tau/2}\}$ and $\min\{2^{(b-\tau)/2}, 2^k, 2^{\tau/2}\}$, respectively, where c denotes the capacity of the Sponge.

While various blockcipher-based fully misuse-resistant AE schemes exist (such as SIV [73], GCM-SIV [37], HS1-SIV [54], AEZ [40], Deoxys$^=$ and Joltik$^=$ [43,44] (using Synthetic Counter in Tweak mode [70]), and DAEAD [19]), the state of the art for permutation-based schemes is rather scarce. In particular, the only misuse-resistant AE schemes known in literature are Haddoc and Mr. Monster Burrito by Bertoni et al. [11]. Haddoc lacks a proper formalization but appears to be similar to MRSO, and the security and efficiency bounds mostly carry over. Mr. Monster Burrito is a proof of concept to design a permutation-based robust AE comparable with AEZ [40], but it is four-pass and thus not very practical[2].

When instantiated with a reduced-round BLAKE2b permutation, MRO achieves a peak speed of 1.06 cycles per byte on the Intel Haswell platform (see Sect. 8). This puts MRO on the same level as AES-GCM-SIV [37] (1.17 cpb), which, however, requires AES-NI to reach its best performance. We further remark that MRO is also more efficient than MRSO, and thus the Haddoc mode.

2 Notation

Denote by $\mathbb{F}_{2^n}$ the finite field of order 2^n with $n \geq 1$. A b-bit string X is an element of $\{0,1\}^b$ or equivalently of the $\mathbb{F}_2$-vector space $\mathbb{F}_2^b$. The length of a bit string X in bits is denoted by $|X|$ $(= b)$ and in r-bit blocks by $|X|_r$. For example, the size of X in bytes is $|X|_8$. The bit string of length 0 is identified with ε. The *concatenation* of two bit strings X and Y is denoted by $X \parallel Y$. The encoding of an integer x as an n-bit string is denoted by $\langle x \rangle_n$. The symbols $\neg$, $\vee$, $\wedge$, $\oplus$, $\ll$, $\gg$, $\lll$, and $\ggg$, denote bit-wise *NOT*, *OR*, *AND*, *XOR*, *left-shift*, *right-shift*, *left-rotation*, and *right-rotation*, respectively.

Given a b-bit string $X = x_0 \parallel \cdots \parallel x_{b-1}$ we define $\mathsf{left}_l(X) = x_0 \parallel \cdots \parallel x_{l-1}$ to be the l left-most and $\mathsf{right}_r(X) = x_{b-r} \parallel \cdots \parallel x_{b-1}$ to be the r right-most bits of X, respectively, where $1 \leq l, r \leq b$. In particular, note that $X = \mathsf{left}_l(X) \parallel \mathsf{right}_{b-l}(X) = \mathsf{left}_{b-r}(X) \parallel \mathsf{right}_r(X)$. We define the following mapping functions which extend a given input string X to a multiple of the block size b and cut it into chunks of b bits:

2 We remark that the state of the art on online misuse-resistant permutation-based AE is a bit more advanced. For instance, APE [1] is online misuse-resistant, and achieves security against the release of unverified plaintext, but satisfies the undesirable property of backwards decryption. Also Minalpher and Prøst-COPA are online misuse-resistant.

$$\mathsf{pad}_b^0 : \{0,1\}^* \to (\{0,1\}^b)^+, X \mapsto X \parallel 0^{(b-|X|) \bmod b},$$
$$\mathsf{pad}_b^{10} : \{0,1\}^* \to (\{0,1\}^b)^+, X \mapsto X \parallel 1 \parallel 0^{(b-|X|-1) \bmod b}.$$

The set of all permutations of *width* $b \geq 0$ bits is denoted by $\mathsf{Perm}(b)$. The parameters $k, n, \tau \geq 0$ conventionally define the size of the key, nonce, and tag, respectively, for which we require that $n \leq b - k - 1$. In the context of Sponge functions $r \geq 0$ and $c \geq 0$ denote *rate* and *capacity* such that $b = r + c$, and we require $k \leq c$. When writing $X \xleftarrow{\$} \mathcal{X}$ for some finite set $\mathcal{X}$, we mean that X gets sampled uniformly at random from $\mathcal{X}$.

2.1 Distinguishers

A distinguisher $\mathbf{D}$ is a computationally unbounded probabilistic algorithm. By $\mathbf{D}^{\mathcal{O}}$ we denote the setting that $\mathbf{D}$ is given query access to an oracle $\mathcal{O}$: it can make queries to $\mathcal{O}$ adaptively, and after this, the distinguisher outputs 0 or 1. If we consider two different oracles $\mathcal{O}$ and $\mathcal{P}$ with the same interface, we define the distinguishing advantage of $\mathbf{D}$ by

$$\Delta_{\mathbf{D}}(\mathcal{O} ; \mathcal{P}) = \left|\mathbf{Pr}\left(\mathbf{D}^{\mathcal{O}} = 1\right) - \mathbf{Pr}\left(\mathbf{D}^{\mathcal{P}} = 1\right)\right|. \tag{1}$$

Here, the probabilities are taken over the randomness from $\mathcal{O}$ and $\mathcal{P}$. The distinguisher is usually bounded by a limited set of resources, e.g., it is allowed to make at most q queries to its oracle. We will use the definition of Δ for our formalization of the security (tweakable) blockciphers and authenticated encryption. Later in the paper, Δ is used to measure the security of PRFs, etc.

2.2 Tweakable Blockciphers

Let $\mathcal{T}$ be a set of "tweaks." A tweakable blockcipher $\widetilde{E} : \{0,1\}^k \times \mathcal{T} \times \{0,1\}^b \to \{0,1\}^b$ is a function such that for every key $K \in \{0,1\}^k$ and tweak $T \in \mathcal{T}$, $\widetilde{E}(K, T, \cdot)$ is a permutation in $\mathsf{Perm}(b)$. We denote its inverse by $\widetilde{E}^{-1}(K, T, \cdot)$. Denote by $\widetilde{\mathsf{Perm}}(\mathcal{T}, b)$ the set of families of tweakable permutations $\widetilde{\pi}$ such that $\widetilde{\pi}(T, \cdot) \in \mathsf{Perm}(b)$ for every $T \in \mathcal{T}$.

The conventional security definitions for tweakable blockciphers are tweakable pseudorandom permutation (TPRP) security and *strong* TPRP (STPRP) security: in the former, the distinguisher can only make forward construction queries, while in the latter it is additionally allowed to make inverse construction queries. We will consider a mixed security notion, where the distinguisher may only make forward queries for a subset of tweaks. It is inspired by earlier definitions from Rogaway [71] and Andreeva et al. [2].

Let $P \xleftarrow{\$} \mathsf{Perm}(b)$ be a b-bit permutation, and consider a tweakable blockcipher $\widetilde{E}$ based on permutation P. Consider a partition $\mathcal{T}_0 \cup \mathcal{T}_1 = \mathcal{T}$ of the tweak space into *forward-only* tweaks $\mathcal{T}_0$ and *forward-and-inverse* tweaks $\mathcal{T}_1$.

We define the *mixed* tweakable pseudorandom permutation (MTPRP) security of $\widetilde{E}$ against a distinguisher $\mathbf{D}$ as

$$\mathbf{Adv}^{\widetilde{\text{mprp}}}_{\widetilde{E},P}(\mathbf{D}) = \Delta_{\mathbf{D}}(\widetilde{E}^{\pm}_K, P^{\pm}\ ;\ \widetilde{\pi}^{\pm}, P^{\pm}), \tag{2}$$

where the probabilities are taken over the random choices of K, $\widetilde{\pi}$, and P. The distinguisher is not allowed to query $\widetilde{E}_K^{-1}$ for tweaks from $\mathcal{T}_0$. By $\mathbf{Adv}^{\widetilde{\text{mprp}}}_{\widetilde{E},P}(q,p)$ we denote the maximum advantage over all distinguishers that make at most q construction queries and at most p queries to $P^{\pm}$.

Note that the definition of MTPRP matches TPRP if $(\mathcal{T}_0, \mathcal{T}_1) = (\mathcal{T}, \emptyset)$ and STPRP if $(\mathcal{T}_0, \mathcal{T}_1) = (\emptyset, \mathcal{T})$. It is a straightforward observation that if a tweakable cipher $\widetilde{E}$ is MTPRP for two sets $(\mathcal{T}_0, \mathcal{T}_1)$, then it is MTPRP for $(\mathcal{T}_0 \cup \{T\}, \mathcal{T}_1 \backslash \{T\})$ for any $T \in \mathcal{T}_1$. Ultimately, this observation implies that an STPRP is a TPRP.

2.3 Authenticated Encryption

Let $\Pi = (\mathcal{E}, \mathcal{D})$ be a deterministic authenticated encryption (AE) scheme which is keyed via a secret key $K \in \{0,1\}^k$ and operates as follows:

$$\mathcal{E}_K(N, H, M) = (C, T),$$
$$\mathcal{D}_K(N, H, C, T) = M/\bot.$$

Here, N is the nonce, H the associated data, M the message, C the ciphertext, and T the tag. In our analysis, we always have $|M| = |C|$, and we require that

$$\mathcal{D}_K(N, H, \mathcal{E}_K(N, H, M)) = M$$

for all N, H, M. By $\$_{\mathcal{E}}$ we define the idealized version of $\mathcal{E}_K$, which returns $(C, T) \xleftarrow{\$} \{0,1\}^{|M|+\tau}$ for every input. Finally, we denote by $\bot$ a function that returns $\bot$ upon every query.

Our AE schemes are based on a b-bit permutation P, and we will analyze the security of them in the setting where P is a random permutation: $P \xleftarrow{\$} \mathsf{Perm}(b)$. Following, Rogaway and Shrimpton [73], Namprempre et al. [65], and Gueron and Lindell [37], we define the AE security of Π against a distinguisher $\mathbf{D}$ as

$$\mathbf{Adv}^{\text{ae}}_{\Pi,P}(\mathbf{D}) = \Delta_{\mathbf{D}}(\mathcal{E}_K, \mathcal{D}_K, P^{\pm}\ ;\ \$_{\mathcal{E}}, \bot, P^{\pm}), \tag{3}$$

where the probabilities are taken over the random choices of K, $\$_{\mathcal{E}}$, and P. The distinguisher is not allowed (i) to repeat any query and (ii) to relay the output of $\mathcal{E}_K$ to the input of $\mathcal{D}_K$. Note that we do *not* a priori require the distinguisher to be nonce-respecting: depending on the setting, it may repeat nonces at its own discretion. We will always mention whether we consider nonce-respecting or nonce-reusing distinguishers. By $\mathbf{Adv}^{\text{ae}}_{\Pi,P}(q_{\mathcal{E}}, q_{\mathcal{D}}, \sigma, p)$ we denote the maximum advantage over all (nonce-respecting/reusing) distinguishers that make at most $q_{\mathcal{E}}$ queries to the encryption oracle and at most $q_{\mathcal{D}}$ to the decryption oracle, of total length at most σ padded blocks, and that make at most p queries to $P^{\pm}$.

3 Tweakable Even-Mansour with General Masking

We present the tweakable Even-Mansour construction MEM. Earlier appearances of tweakable Even-Mansour constructions include Sasaki et al. [75], Cogliati et al. [23], and Mennink [62], but these constructions target different settings, do not easily capture the improved maskings as introduced below, and are therefore not applicable in this work.

Our specification can be seen as a generalization of both the XE(X) construction of Rogaway [71] and the tweakable blockcipher from Chakraborty and Sarkar [18] to the permutation-based setting. While Rogaway limited himself to 128-bit fields, we realize our approach to fields well beyond the reach of Pohlig-Hellman: historically the large block size would have been a severe obstruction, as observed in works by Yasuda and Sarkar [74,77], and some schemes simply ignored the issue [50]. The breakthroughs in computing discrete logarithms in small characteristic fields [7,30,34,45] allow to easily pass the 128-bit barrier. In particular, for blocks of 2^n bits, it is eminently practical to compute discrete logarithms for $n \leq 13$. Further details of our solution of discrete logarithms over $\mathbb{F}_{2^{512}}$ and $\mathbb{F}_{2^{1024}}$ are described in Sect. 3.6.

3.1 Definition

Let $b \geq 0$ and $P \in \mathsf{Perm}(b)$. In the following we specify MEM, a *tweakable Even-Mansour block cipher with general masking* $(\widetilde{E}, \widetilde{D})$ where $\widetilde{E}$ and $\widetilde{D}$ denote encryption and decryption functions, respectively. Let $u \geq 1$, and let $\Phi = \{\varphi_0, \ldots, \varphi_{u-1}\}$ be a set of functions $\varphi_j : \{0,1\}^b \to \{0,1\}^b$. Consider a *tweak space* $\mathcal{T}$ of the form

$$\mathcal{T} \subseteq \{0,1\}^{b-k} \times \mathbb{N}^u \tag{4}$$

and specify the *general masking function* $\delta : \{0,1\}^k \times \mathcal{T} \to \{0,1\}^b$ as

$$\delta : (K, X, i_0, \ldots, i_{u-1}) \mapsto \varphi_{u-1}^{i_{u-1}} \circ \cdots \circ \varphi_0^{i_0}(P(X \parallel K)).$$

By convention, we set $\varphi_j^{i_j} = id$ for $i_j = 0$, for each $0 \leq j \leq u-1$. For brevity of notation we write $\bar{i} = (i_0, \ldots, i_{u-1})$, and set

$$\mathcal{T}_{\bar{i}} = \{\bar{i} \mid \exists X \text{ such that } (X, \bar{i}) \in \mathcal{T}\}.$$

The encryption function $\widetilde{E} : \{0,1\}^k \times \mathcal{T} \times \{0,1\}^b \to \{0,1\}^b$ is now defined as

$$\widetilde{E} : (K, X, \bar{i}, M) \mapsto P(\delta(K, X, \bar{i}) \oplus M) \oplus \delta(K, X, \bar{i}),$$

where M denotes the to be encrypted message. The decryption function $\widetilde{D} : \{0,1\}^k \times \mathcal{T} \times \{0,1\}^b \to \{0,1\}^b$ is defined analogously as

$$\widetilde{D} : (K, X, \bar{i}, C) \mapsto P^{-1}(\delta(K, X, \bar{i}) \oplus C) \oplus \delta(K, X, \bar{i}),$$

where C denotes the to be decrypted ciphertext. Note that the usual block cipher property $\widetilde{D}(K, X, \bar{i}, \widetilde{E}(K, X, \bar{i}, M)) = M$ is obviously satisfied. Throughout the document, we will often use the following shorthand notation for $\widetilde{E}_{K,X}^{\bar{i}}(M) = \widetilde{E}(K, X, \bar{i}, M)$, $\widetilde{D}_{K,X}^{\bar{i}}(C) = \widetilde{D}(K, X, \bar{i}, C)$, and $\delta_{K,X}^{\bar{i}} = \delta(K, X, \bar{i})$.

3.2 Security

Equation (4) already reveals that we require some kind of restriction on $\mathcal{T}$. Informally, we require the masking functions $\varphi_{u-1}^{i_{u-1}} \circ \cdots \circ \varphi_0^{i_0}$ to generate pairwise independent values for different tweaks. More formally, we define proper tweak spaces in Definition 1. This definition is related to earlier observations in Rogaway [71] and Chakraborty and Sarkar [18,74].

Definition 1. *Let $b \geq 0$, $u \geq 1$, and $\Phi = \{\varphi_0, \ldots, \varphi_{u-1}\}$ be a set of functions. The tweak space $\mathcal{T}$ is* ϵ-proper *relative to the function set Φ if the following two properties are satisfied.*

1. *For any $y \in \{0,1\}^b$, $(i_0, \ldots, i_{u-1}) \in \mathcal{T}_{\bar{i}}$, and uniformly random $L \xleftarrow{\$} \{0,1\}^b$:*
$$\Pr\left[\varphi_{u-1}^{i_{u-1}} \circ \cdots \circ \varphi_0^{i_0}(L) = y\right] = 2^{-\epsilon}.$$
2. *For any $y \in \{0,1\}^b$, distinct $(i_0, \ldots, i_{u-1}), (i'_0, \ldots, i'_{u-1}) \in \mathcal{T}_{\bar{i}}$, and uniformly random $L \xleftarrow{\$} \{0,1\}^b$:*
$$\Pr\left[\varphi_{u-1}^{i_{u-1}} \circ \cdots \circ \varphi_0^{i_0}(L) \oplus \varphi_{u-1}^{i'_{u-1}} \circ \cdots \circ \varphi_0^{i'_0}(L) = y\right] = 2^{-\epsilon}.$$

The definition is reminiscent of the definition of universal hash functions (as also noted in [18]), but we will stick to the convention. We are now ready to prove the security of MEM.

Theorem 2. *Let $b \geq 0$, $u \geq 1$, and $\Phi = \{\varphi_0, \ldots, \varphi_{u-1}\}$ be a set of functions. Let $P \xleftarrow{\$} \mathsf{Perm}(b)$. Assume that the tweak space $\mathcal{T}$ is ϵ-proper relative to Φ. Let $\mathcal{T}_0 \cup \mathcal{T}_1 = \mathcal{T}$ be a partition such that $(0, \ldots, 0) \notin \mathcal{T}_{1\bar{i}}$. Then,*

$$\mathbf{Adv}_{\widetilde{E},P}^{\widetilde{\mathrm{mprp}}}(q,p) \leq \frac{4.5q^2}{2^\epsilon} + \frac{3qp}{2^\epsilon} + \frac{p}{2^k}.$$

The proof can be found in the full version of this work. It is based on Patarin's H-coefficient technique [21,68], and borrows ideas from [18,62,71,74].

3.3 History of Masking

Originally, IAPM [49] proposed the masking to be a subset sum of c encrypted blocks derived from the nonce, where 2^c is the maximum number of blocks a message can have. In the same document Jutla also suggested masking the jth block with $(j+1)K + IV \bmod p$, for some prime p near the block size. XCBC [27] used a similar masking function, but replaced arithmetic modulo p by arithmetic modulo 2^b, at the cost of some tightness in security reductions.

OCB [55,71,72] and PMAC [12] used the field $\mathbb{F}_{2^b}$ for their masking. There are two different masking functions used in variants of OCB:

- The powering-up method of OCB2 [71] computes $\varphi^i(L) = x^i \cdot L$, where $\cdot$ is multiplication in $\mathbb{F}_{2^b}$, and x is a generator of the field.

- The Gray code masking of OCB1 [72] and OCB3 [55] computes $\varphi^i(L) = \gamma_i \cdot L$, where $\gamma_i = i \oplus (i \gg 1)$. This method requires one XOR to compute $\varphi^{i+1}(L)$ given $\varphi^i(L)$, provided a precomputation of $\log_2 |M|$ multiples of L is carried out in advance. Otherwise, up to $\log_2 i$ field doublings are required to obtain $\gamma_i \cdot L$. This Gray code trick was also applicable to IAPM's subset-sum masking.

Another family of masking functions, word-oriented LFSRs, was suggested by Chakraborty and Sarkar [18]. Instead of working directly with the polynomial representation $\mathbb{F}_2[x]/f$ for some primitive polynomial f, word-oriented LFSRs treat the block as the field $\mathbb{F}_{2^{wn}}$, where w is the native word size. Thus, the block can be represented as a polynomial of degree n over F_{2^w}, which makes the arithmetic more software-friendly. A further generalized variant of this family of generators is described (and rejected) in [55, Appendix B], who also attribute the same technique to [78]. Instead of working with explicitly-constructed field representations, one starts by trying to find a $b \times b$ matrix $M \in \mathrm{GL}(b, \mathbb{F}_2)$ that is very efficient to compute. Then, if this matrix has a primitive minimal polynomial of degree b, this transformation is in fact isomorphic to $\mathbb{F}_{2^b}$ and has desirable masking properties. The masking function is then $\varphi^i(L) = M^i \cdot L$.

Although the above maximal-period matrix recursions have only recently been suggested for use in efficient masking, the approach has been long studied by designers of non-cryptographic pseudorandom generators. For example, Niederreiter [67, Sect. 4] proposed a pseudorandom generator design based on a matrix recursion. Later methods, like the Mersenne Twister family [60] and the Xorshift [59] generator, improved the efficiency significantly by cleverly choosing the matrix shape to be CPU-friendly.

More recently, Minematsu [64] suggested a different approach to masking based on data-dependent rotation. In particular,

$$\varphi^i(L) = \bigoplus_{0 \le j < b} \begin{cases} (L \lll j) & \text{if } \lfloor i/2^j \rfloor \bmod 2 = 1, \\ 0 & \text{otherwise.} \end{cases}$$

where the block size b is prime. With Gray code ordering, one only needs one rotation and XOR per sequential mask *without* storing previous masks. That being said, the prime block size is inconvenient, and data-dependent rotation is a relatively expensive operation compared to some of the previous techniques.

3.4 Proposed Masking for $u = 1$

We loosely follow the Xorshift [59] design approach for our masking procedure. Let $b = nw$ be the block size, interpreted as n words of w bits. We begin with fast linear operations available in most current CPUs and encode them as $w \times w$ matrices. More precisely, we denote by 0 the all-zero matrix, by I the identity matrix, by SHL_c and SHR_c matrices corresponding to left- and right-shift by c bits, by ROT_c the matrix realizing left-rotation by c bits, and by AND_c the matrix corresponding to bit-wise AND with a constant c. Then, we construct block matrices using those operations in a way that minimizes computational effort. To maximize efficiency we consider $b \times b$ matrices over $\mathbb{F}_2$ of the form

$$M = \begin{pmatrix} 0 & I & \cdots & 0 \\ \vdots & \vdots & \ddots & \vdots \\ 0 & 0 & \cdots & I \\ X_0 & X_1 & \cdots & X_{n-1} \end{pmatrix} \tag{5}$$

with $X_i \in \{0, I, \mathsf{SHL}_c, \mathsf{SHR}_c, \mathsf{ROT}_c, \mathsf{AND}_c\}$ where $\dim(X_i) = w$ for $0 \le i \le n-1$. We favor matrices where only a minimal amount of X_i are nonzero. For a concrete selection of $X_0, \ldots, X_{n-1}$ we check if the matrix order is maximal, that is, if the smallest integer $t > 0$ such that $M^t = I$ equals $2^b - 1$; if so, this matrix is suitable for a masking function that respects the conditions listed above.

Testing candidate masks for maximal order may be efficiently performed without any explicit matrix operations. Given a candidate linear map corresponding to a matrix M of the form Eq. (5),

$$(x_0, \ldots, x_{n-1}) \mapsto (x_1, \ldots, x_{n-1}, f(x_0, \ldots, x_{n-1})),$$

one can simply select $x_0, \ldots, x_{n-1}$ at random, define $x_{i+n} = f(x_i, \ldots, x_{i+n-1})$, and obtain the connection polynomial $p(x)$ from the sequence of least significant bits of $x_0, \ldots, x_{2b}$ using, e.g., Berlekamp-Massey. If $p(x)$ is a primitive polynomial of degree b, $p(x)$ is also the minimal polynomial of the associated matrix M.

This approach yields a number of simple and efficient masking functions. In particular, the 3-operation primitives $(x_0 \lll r_0) \oplus (x_i \gg r_1)$ and $(x_0 \lll r_0) \oplus (x_i \ll r_1)$ are found for several useful block and word sizes, as Table 1 illustrates. Some block sizes do not yield such small generators so easily; in particular, 128-bit blocks require at least 4 operations, which is consistent—albeit somewhat better—with the results of [55, Appendix B]. Using an extra basic instruction, double-word shift, another noteworthy class of maskings appears: $(x_0 \lll r_0) \oplus (x_i \ll r_1) \oplus (x_j \gg (w - r_1))$, or in other words $(x_0 \lll r_0) \oplus ((x_i \parallel x_j) \ll r_1)$. This leads to more block sizes with 3-operation masks, e.g., $(x_1, x_2, x_3, (x_0 \lll 15) \oplus ((x_1 \parallel x_0) \gg 11))$ for 128-bit blocks. Lemma 3 shows that this approach yields proper masking functions according to Definition 1.

Lemma 3. *Let M be an $b \times b$ matrix over $\mathbb{F}_2$ of the form shown in Eq.* (5). *Furthermore, let M's minimal polynomial be primitive and of degree b. Then given the function $\varphi_0^i(L) = M^i \cdot L$, any tweak set with $\mathcal{T}_{\bar{i}} \subseteq \{0, \ldots, 2^b - 2\}$ is a b-proper tweak space by Definition 1.*

Proof. [18, Proposition 1] directly applies.

One may wonder whether there is any significant advantage of the above technique over, say, the Gray code sequence with the standard polynomial representation. We argue that our approach improves on it in several ways:

Simplicity. OCB (especially OCB2) requires implementers to be aware of Galois field arithmetic. Our approach requires no user knowledge—even implicitly—of field or polynomial arithmetic, but only *unconditional* shifts and XOR operations. Even Sarkar's word-based LFSRs [74] do not hide the finite field structure from implementers, thus making it easier to make mistakes.

Table 1. Sample masking functions for various state sizes b and respective decompositions into n words of w bits

b	w	n	φ
128	8	16	$(x_1, \ldots, x_{15}, \ (x_0 \lll 1) \oplus (x_9 \gg 1) \oplus (x_{10} \ll 1))$
128	32	4	$(x_1, \ldots, x_3, \ (x_0 \lll 1) \oplus (x_1 \wedge 31) \oplus (x_2 \wedge 127))$
128	32	4	$(x_1, \ldots, x_3, \ (x_0 \lll 5) \oplus x_1 \oplus (x_1 \ll 13))$
128	64	2	$(x_1, \ (x_0 \lll 11) \oplus x_1 \oplus (x_1 \ll 13))$
256	32	8	$(x_1, \ldots, x_7, \ (x_0 \lll 17) \oplus x_5 \oplus (x_5 \gg 13))$
256	64	4	$(x_1, \ldots, x_3, \ (x_0 \lll 3) \oplus (x_3 \gg 5))$
512	32	16	$(x_1, \ldots, x_{15}, \ (x_0 \lll 5) \oplus (x_3 \gg 7))$
512	64	8	$(x_1, \ldots, x_7, \ (x_0 \lll 29) \oplus (x_1 \ll 9))$
800	32	25	$(x_1, \ldots, x_{15}, \ (x_0 \lll 25) \oplus x_{21} \oplus (x_{21} \gg 13))$
1024	8	128	$(x_1, \ldots, x_{127}, \ (x_0 \lll 1) \oplus x_{125} \oplus (x_{125} \gg 5))$
1024	64	16	$(x_1, \ldots, x_{15}, \ (x_0 \lll 53) \oplus (x_5 \ll 13))$
1600	32	50	$(x_1, \ldots, x_{49}, \ (x_0 \lll 3) \oplus (x_{23} \gg 3))$
1600	64	25	$(x_1, \ldots, x_{24}, \ (x_0 \lll 55) \oplus x_{21} \oplus (x_{21} \ll 21))$
1600	64	25	$(x_1, \ldots, x_{24}, \ (x_0 \lll 15) \oplus x_{23} \oplus (x_{23} \ll 23))$

Constant-Time. Both OCB masking schemes require potentially variable-time operations to compute each mask—be it conditional XOR, number of trailing zeroes, or memory accesses indexed by $\mathsf{ntz}(i+1)$. This is easily avoidable by clever implementers, but it is also a pitfall avoidable by our design choice. Even in specifications aimed at developers [53], $\mathsf{double}(S)$ is defined as a variable-time operation.

Efficiency. Word-based masking has the best space-time efficiency tradeoff of all considered masking schemes. It requires only minimal space usage—one block—while also involving a very small number of operations beyond the XOR with the block (as low as 3, cf. Table 1). It is also SIMD-friendly, allowing the generation of several consecutive masks with a single short SIMD instruction sequence.

In particular, for permutations that can take advantage of a CPU's vector units via "word-slicing"—which is the case for Salsa20, ChaCha, Threefish, and many other ARX designs—it is possible to compute a few consecutive masks at virtually the same cost as computing a single mask transition. It is also efficient to add the mask to the plaintext both in transposed order (word-sliced) and regular order.

For concreteness, consider the mask sequence $(x_1, \ldots, x_{15}, (x_0 \lll 5) \oplus (x_3 \gg 7))$ and a permutation using 512-bit blocks of 32-bit words. Suppose further that we are working with a CPU with 8-wide vectors, e.g., AVX2. Given 8 additional words of storage, it is possible to compute $L = (x_1, \ldots, x_{15}, (x_0 \lll 5) \oplus (x_3 \gg 7), \ldots, (x_7 \lll 5) \oplus (x_{10} \gg 7))$ entirely in parallel.

Consider now the transposed set of 8 blocks $m_0, \ldots, m_7$; adding the mask consists of $m_0 \oplus L_{0-15}, m_1 \oplus L_{1-16}, \ldots$. On the other hand, when the blocks are word-sliced—with m'_0 being the first 32-bit word of m_i, m'_1 being the second, and so on—adding the mask is still efficient, as $m'_0 \oplus L_{0-7}, m'_1 \oplus L_{1-8}, \ldots$. This would be impossible with the standard masking schemes used in, e.g., OCB.

There is also an advantage at the low-end—φ can easily be implemented as a circular array, which implies that only an index increment and the logical operations must be executed for each mask update. This improves on both the typical Gray code and powering-up approach, in that shifting by one requires moving *every word* of the mask, instead of only one of them. Additionally, storage is often a precious resource in low-end systems, and the Gray code method requires significantly more than one block to achieve its best performance.

3.5 Proposed Masking for $u = 2$ and $u = 3$

Modes often require the tweak space to have multiple dimensions. In particular, the modes of Sects. 4 and 5 require the tweak space to have 2 and 3 "coordinates." To extend the masking function from Sect. 3.4 to a tweak space divided into disjunct sets, we have several options. We can simply split the range $[0, 2^b - 1]$ into equivalence classes, e.g., $i_0 = 4k + 0, i_1 = 4k + 1, \ldots$ for at most 4 different tweak indexes. Some constructions instead store a few fixed tweak values that are used later as "extra" finalization tweaks.

The approach we follow takes a cue from XEX [71]. Before introducing the scheme itself, we need a deeper understanding of the masking function φ introduced in Sect. 3.4. At its core, φ is a linear map representable by a matrix M with primitive minimal polynomial $p(x)$. In fact, φ can be interpreted as the matrix representation [58, § 2.52] of $\mathbb{F}_{2^b}$, where M is, up to a change of basis, the companion matrix of $p(x)$. This property may be exploited to quickly jump ahead to an arbitrary state $\varphi^i(L)$: since $\varphi^i(L) = M^i \cdot L$ and additionally $p(M) = 0$, then $(x^i \bmod p(x))(M) = (x^i)(M) + (p(x)q(x))(M) = (x^i)(M) = M^i$. Therefore we can implement arbitrarily large "jumps" in the tweak space by evaluating the right polynomials over M. This property—like fast word-oriented shift registers—has had its first uses in the pseudorandom number generation literature [39].

Since we may control the polynomials here, we choose the very same polynomials as Rogaway for the best performance: $x + 1$, and $x^2 + x + 1$, denoted in [71] as **3** and **7**. Putting everything together, our masking for $u = 3$ becomes

$$\begin{aligned}\delta(K, X, i_0, i_1, i_2) &= ((x)(M))^{i_0}((x+1)(M))^{i_1}((x^2+x+1)(M))^{i_2} \cdot P(K \parallel X) \\ &= M^{i_0}(M+I)^{i_1}(M^2+M+I)^{i_2} \cdot P(K \parallel X).\end{aligned}$$

To ensure that the tweak space is b-proper we need one extra detail: we need to ensure that the logarithms $\log_x(x+1)$ and $\log_x(x^2+x+1)$ are sufficiently apart. While for $\mathbb{F}_{2^{128}}$ Rogaway already computed the corresponding discrete logarithms [71] using generic methods, larger blocks make it nontrivial to show b-properness. The following lemma shows that one particular function satisfies Definition 1. The lemma uses the discrete logarithms whose computation is described in Sect. 3.6.

Lemma 4. *Let $\varphi(x) : \{0,1\}^{1024} \mapsto \{0,1\}^{1024}$ be the linear map $(x_0, \ldots, x_{15}) \mapsto (x_1, \ldots, x_{15}, (x_0 \lll 53) \oplus (x_5 \ll 13))$. Further, let M be the 1024×1024 matrix associated with φ such that $\varphi(L) = M \cdot L$. Let $\Phi = \{\varphi_0^{i_0}, \varphi_1^{i_1}, \varphi_2^{i_2}\}$ be the set of functions used in the masking, with $\varphi_0^{i_0}(L) = M^{i_0} \cdot L$, $\varphi_1^{i_1}(L) = (M+I)^{i_1} \cdot L$, and $\varphi_2^{i_2}(L) = (M^2+M+I)^{i_2} \cdot L$. The tweak space*

$$\mathcal{T} = \mathcal{T}_0 \times \mathcal{T}_1 \times \mathcal{T}_2 = \{0, 1, \ldots, 2^{1020}-1\} \times \{0,1,2,3\} \times \{0,1\}$$

is b-proper relative to the function set Φ.

Proof. The proof closely follows [71, Proposition 5]. Let $i_0 \in \mathcal{T}_0$, $i_1 \in \mathcal{T}_1$, and $i_2 \in \mathcal{T}_2$. We first show that $\varphi_0^{i_0} \circ \varphi_1^{i_1} \circ \varphi_2^{i_2}$ is unique for any distinct set of tweaks.

An easy computation shows that $p(x) = x^{1024} + x^{901} + x^{695} + x^{572} + x^{409} + x^{366} + x^{203} + x^{163} + 1$ is the minimal polynomial of M. This polynomial is both irreducible and primitive, which implies that the order of M is $2^{1024}-1$. We begin by determining the logarithms of $M+I$ and M^2+M+I relatively to M. This may be accomplished by computing $l_1 = \log_x(x+1)$ and $l_2 = \log_x(x^2+x+1)$ in the field $\mathbb{F}_2[x]/p(x)$, see Sect. 3.6.

The values l_1 and l_2 let us represent $M^{i_0}M^{i_1}M^{i_2}$ as $M^{i_0}M^{l_1 i_1}M^{l_2 i_2}$. Given a second distinct pair (i'_0, i'_1, i'_2), we have that $M^{i_0}M^{l_1 i_1}M^{l_2 i_2} = M^{i'_0}M^{l_1 i'_1}M^{l_2 i'_2}$ iff $i_0 + l_1 i_1 + l_2 i_2 = i'_0 + l_1 i'_1 + l_2 i'_2 \pmod{2^{1024}-1}$. Equivalently, $i_0 - i'_0 = (i_1 - i'_1)l_1 + (i_2 - i'_2)l_2 \pmod{2^{1024}-1}$. By a simple exhaustive search through the valid ranges of i_1 and i_2 we are able to see that the smallest absolute difference $(i_1 - i'_1)l_1 + (i_2 - i'_2)l_2$ occurs when $i_1 - i'_1 = -1$ and $i_2 - i'_2 = -1$, and is $\approx 2^{1020.58}$. Since $i_0 - i'_0$ is at most $\pm(2^{1020}-1)$, collisions cannot happen. Since each mask is unique, the fact that $\mathcal{T}$ is b-proper follows from Lemma 3. □

Remark. Nontrivial bounds for $\mathcal{T}$, such as in the case where one desires $\mathcal{T}_0$, $\mathcal{T}_1$, and $\mathcal{T}_2$ to be balanced, cannot be easily found by exhaustive search. Such bounds can be found, however, with lattice reduction. Consider the lattice spanned by the rows

$$\begin{pmatrix} K \cdot 1 & w_0 & 0 & 0 \\ K \cdot l_1 & 0 & w_1 & 0 \\ K \cdot l_2 & 0 & 0 & w_2 \\ K \cdot m & 0 & 0 & 0 \end{pmatrix},$$

for a suitable integer K, $m = 2^b - 1$, and weights w_i. A shortest vector for low-dimensional lattices such as this can be computed exactly in polynomial time [66]. A shortest vector for this lattice has the form $(\Delta i_0 + \Delta i_1 l_1 + \Delta i_2 l_2 + km, \Delta i_0 w_0, \Delta i_1 w_1, \Delta i_2 w_2)$, and will be shortest when $\Delta i_0 + \Delta i_1 l_1 + \Delta i_2 l_2 \equiv 0 \pmod{2^n - 1}$. This yields concrete bounds on i_0, i_1, and i_2. The constant K needs to be large enough to avoid trivial shortest vectors such as $(K, 1, 0, 0)$. The weights w_i can be used to manipulate the relative size of each domain; for example, using the weights 1, 2^{1019}, and 2^{1022} results in a similar bound as Lemma 4, with $\mathcal{T}_0$ dominating the tweak space.

3.6 Computing Discrete Logarithms in $\mathbb{F}_{2^{512}}$ and $\mathbb{F}_{2^{1024}}$

While the classical incarnation of the Function Field Sieve (FFS) with $\mathbb{F}_2$ as the base field could no doubt solve logarithms in $\mathbb{F}_{2^{512}}$ with relatively modest computational resources—see for example [46, 76]—the larger field would require a significant amount of work [6]. One could instead use subfields other than $\mathbb{F}_2$ and apply the medium-base-field method of Joux and Lercier [47], which would be relatively quick for $\mathbb{F}_{2^{512}}$, but not so easy for $\mathbb{F}_{2^{1024}}$.

However, with the advent of the more sophisticated modern incarnation of the FFS, development of which began in early 2013 [7, 30–34, 45, 48], the target fields are now regarded as small, even tiny, at least relative to the largest such example computation where a DLP in $\mathbb{F}_{2^{9234}}$ was solved [35]. Since these developments have effectively rendered small characteristic DLPs useless for public key cryptography, (despite perhaps some potential doubters [17, Appendix D]) it is edifying that there is a constructive application in cryptography[3] for what is generally regarded as a purely cryptanalytic pursuit.

Due to the many subfields present in the fields in question, there is a large parameter space to explore with regard to the application of the modern techniques, and it becomes an interesting optimization exercise to find the most efficient approach. Moreover, such is the size of these fields that coding time rather than computing time is the dominant term in the overall cost. We therefore solved the relevant DLPs using MAGMA V2.19-1 [14], which allowed us to develop rapidly. All computations were executed on a standard desktop computer with a 2.0 GHz AMD Opteron processor.

3.6.1 Fields Setup. For reasons of both efficiency and convenience we use $\mathbb{F}_{2^{16}}$ as base field for both target fields, given by the following extensions:

$$\mathbb{F}_{2^4} = \mathbb{F}_2[U]/(U^4 + U + 1) = \mathbb{F}_2(u),$$
$$\mathbb{F}_{2^{16}} = \mathbb{F}_{2^4}[V]/(V^4 + V^3 + V + u) = \mathbb{F}_{2^4}(v).$$

We represent $\mathbb{F}_{2^{512}}$ as $\mathbb{F}_{2^{16}}[X]/(I_{32}(X)) = \mathbb{F}_{2^{16}}(x)$, where I_{32} is the degree 32 irreducible factor of $H_{32}(X) = h_1(X^{16})X + h_0(X^{16})$, where $h_1 = (X + u^9 + u^5v + u^{13}v^2 + u^3v^3)^3$ and $h_0 = X^3 + u^2 + u^9v^2 + u^{13}v^3$. The other irreducible factors of $H_{32}(X)$ have degrees 6 and 11.

We represent $\mathbb{F}_{2^{1024}}$ as $\mathbb{F}_{2^{16}}[X]/(I_{64}(X)) = \mathbb{F}_{2^{16}}(x)$, where I_{64} is the degree 64 irreducible factor of $H_{64}(X) = h_1(X^{16})X + h_0(X^{16})$, where $h_1 = (X + u + u^7v + u^4v^2 + u^7c^3)^5$ and $h_0 = X^5 + u^9 + u^4v + u^6v^2 + v^3$. The other irreducible factors of $H_{64}(X)$ have degrees 7 and 10. Transforming from the original representations of Sect. 3.6.3 to these is a simple matter [57].

Note that ideally one would only have to use h_i's of degree 2 and 4 to obtain degree 32 and 64 irreducibles, respectively. However, no such h_i's exist and so we

[3] Beyond cryptography, examples abound in computational mathematics: in finite geometry; representation theory; matrix problems; group theory; and Lie algebras in the modular case; to name but a few.

are forced to use h_i's of degree 3 and 5. The penalty for doing so incurs during the relation generation, see Sect. 3.6.2, and during the descent, in particular for degree 2 elimination, see Sect. 3.6.3.

Remark. The degrees of the irreducible cofactors of I_{32} in H_{32} and of I_{64} in H_{64} is an essential consideration in the set up of the two fields. In particular, if the degree d_f of a cofactor f has a non-trivial GCD with the degree of the main irreducible, then it should be considered as a 'trap' for the computation of the logarithms of the factor base elements, modulo all primes dividing $2^{16\cdot\gcd(d_f,32i)}-1$ for $i=1,2$, for $\mathbb{F}_{2^{512}}$ and $\mathbb{F}_{2^{1024}}$, respectively [22,42]. This is because $\mathbb{F}_{2^{16}}[X]/(H_{32i}(X))$ will contain another copy of $\mathbb{F}_{2^{16\cdot\gcd(d_f,32i)}}$ which arises from f, and hence the solution space modulo primes dividing $2^{16\cdot\gcd(d_f,32i)}-1$ has rank >1. Our choice of h_0 and h_1 in each case limits the effect of this problem to prime factors of $2^{32}-1$, namely subgroups of tiny order within which we solve the DLPs using a linear search. The irreducible cofactors are also traps for the descent phase [32], but are easily avoided.

3.6.2 Relation Generation and Logarithms of Linear Elements.

The factor base is defined to be $\mathcal{F}=\{x+d \mid d\in\mathbb{F}_{2^{16}}\}$. To generate relations over $\mathcal{F}$, we use the technique from [30], described most simply in [32]. In particular, for both target fields let $y=x^{16}$; by the definitions of I_{32} and I_{64} it follows in both cases that $x=h_0(y)/h_1(y)$. Using these field isomorphisms, for any $a,b,c\in\mathbb{F}_{2^{16}}$ we have the field equality

$$x^{17}+ax^{16}+bx+c=\frac{1}{h_1(y)}(yh_0(y)+ayh_1(y)+bh_0(y)+ch_1(y) \qquad (6)$$

One can easily generate (a,b,c) triples such that the left hand side of Eq. (6) always splits completely over $\mathcal{F}$. Indeed, one first computes the set $\mathcal{B}$ of 16 values $B\in\mathbb{F}_{2^{16}}$ such that the polynomial $f_B(X)=X^{17}+BX+B$ splits completely over $\mathbb{F}_{2^{16}}$ [13]. Assuming $c\neq ab$ and $b\neq a^{16}$, the left hand side of Eq. (6) can be transformed (up to a scalar factor) into f_B, where $B=\frac{(b+a^{16})^{17}}{(c+ab)^{16}}$. Hence if this B is in $\mathcal{B}$ then the left hand side also splits. In order to generate relations, one repeatedly chooses random $B\in\mathcal{B}$ and random $a,b\neq a^{16}\in\mathbb{F}_{2^{16}}$, computes $c=((b+a^{16})^{17})^{1/16}+ab$, and tests whether the right hand side of Eq. (6) also splits over $\mathbb{F}_{2^{16}}$. If it does then one has a relation, since $(y+d)=(x+d^{1/16})^{16}$, and each h_1 is a power of a factor base element.

The probability that the right hand side of Eq. (6) splits completely is heuristically $1/4!$ and $1/6!$ for $\mathbb{F}_{2^{512}}$ and $\mathbb{F}_{2^{1024}}$ respectively. In both cases we obtain $2^{16}+200$ relations, which took about 0.3 h and 8.8 h, respectively. To compute the logarithms of the factor base elements, we used MAGMA's `ModularSolution` function, with its `Lanczos` option set, modulo the 9th to 13th largest prime factors of $2^{512}-1$ for the smaller field and modulo the 10th to 16th largest prime factors of $2^{1024}-1$ for the larger field. These took about 13.5 h and 24.5 h, respectively.

3.6.3 Individual Logarithms. The original representations of the target fields are:

$$\begin{aligned}\mathbb{F}_{2^{512}} &= \mathbb{F}_2[T]/(T^{512}+T^{335}+T^{201}+T^{67}+1) = \mathbb{F}_2(t),\\ \mathbb{F}_{2^{1024}} &= \mathbb{F}_2[T]/(T^{1024}+T^{901}+T^{695}+T^{572}+T^{409}+T^{366}+T^{203}+T^{163}+1)\\ &= \mathbb{F}_2(t).\end{aligned}$$

In order to solve the two relevant DLPs in each original field, we need to compute three logarithms in each of our preferred field representations, namely the logarithms of the images of t, $t+1$ and t^2+t+1—which we denote by t_0, t_1 and t_2—relative to some generator. We use the generator x in both cases.

For $\mathbb{F}_{2^{512}}$, we multiply the targets t_i by random powers of x and apply a continued fraction initial split so that $x^k t_i \equiv n/d \pmod{I_{32}}$, with n of degree 16 and d of degree 15, until both n and d are 4-smooth. One then just needs to eliminate irreducible elements of degree 2, 3, 4 into elements of smaller degree. For degree 4 elements, we apply the building block for the quasi-polynomial algorithm due to Granger, Kleinjung, and Zumbrägel [33,34], which is just degree 2 elimination but over a degree 2 extended base field. This results in each degree 4 element being expressed as a product of powers of at most 19 degree 2 elements, and possibly some linear elements. For degree 3 elimination we use Joux's bilinear quadratic system approach [45], which expresses each degree 3 element as a product of powers of again at most 19 degree 2 elements and at least one linear element. For degree 2 elimination, we use the on-the-fly technique from [30], but with the quadratic system approach from [31], which works for an expected proportion $1-(1-1/2!)^{16} = 255/256$ of degree 2's, since the cofactor in each case has degree 2. On average each descent takes about 10 s, and if it fails due to a degree 2 being ineliminable, we simply rerun it with a different random seed. Computing logarithms modulo the remaining primes only takes a few seconds with a linear search, which completes the following results:

$$\begin{aligned}\log_t(t+1) = {}& 5016323028665706705636609709550289619036901979668873\\ & 4872643788516514405882411611155920582686309266723854\\ & 5122357792870542653280226105514939849018182092980 2,\\ \log_t(t^2+t+1) = {}& 7789795054597035122960933502653082209865724780784381\\ & 2166626513019333878034142500477941950081303675633401\\ & 1185966465812007766565485320190254829936577378946 2.\end{aligned}$$

The total computation time for these logarithms is less than 14 h.

For $\mathbb{F}_{2^{1024}}$, we use the same continued fraction initial split, but now with n and d of degree 32 and 31, until each is 4-smooth, but also allowing a number of degree 8 elements. Finding such an expression takes on average 7 h, which, while not optimal, means that the classical special-Q elimination method could be obviated, i.e., not coded. For degree 8 elimination, we again use the building block for the quasi-polynomial algorithm of Granger et al., which expresses such a degree 8 element as a product of powers of at most 21 degree 4 elements, and possibly some degree 2 and 1 elements. Degree 4 and 3 elimination proceed as before, but with a larger cofactor of the element to be eliminated on the r.h.s. due to the larger degrees of h_0 and h_1. Degree 2 elimination is significantly harder in this case, since the larger degrees of the h_i's mean that the elimination probability for a random degree 2 element was only $1 - (1 - 1/4!)^{16} \approx 0.494$. However, using the recursive method from the DLP computation in $\mathbb{F}_{2^{4404}}$ [32] allows this to be performed with near certainty. If any of the eliminations fails, then as before we simply rerun the eliminations with a different random seed. In total, after the initial rewrite of the target elements into a product of degree 1, 2, 3, 4, and 8 elements, each descent takes just under an hour. Again, computing logarithms modulo the remaining primes takes less than a minute with a linear search resulting in:

$$\begin{aligned}
\log_t(t+1) = {} & 356031381070238016894189506806176884676865287991652 4\\
& 27967534565655098427076557554137531006209790218857 20\\
& 19667853514803076973117094568313720185984991744411 96\\
& 14703326022161615833783625836575707566310249359279 84\\
& 24982722386995285762306852428057639389511554481264 95\\
& 512475014867387149681903876406067502645471152193,\\
\log_t(t^2+t+1) = {} & 16100564391890287934521444613155584470201173764326 42\\
& 55248594862381613746542797178003007061367496076306 01\\
& 49673626737775471400899387001441124240813887118712 90\\
& 79733192516296283613982673518809480691614597930522 57\\
& 19071179482911643233555281698543543964820295077819 47\\
& 253417131307693777579790915978887936187609988 8834.
\end{aligned}$$

The total computation time for these logarithms is about 57 h.

Note that it is possible to avoid the computations in $\mathbb{F}_{2^{512}}$ altogether by embedding the relevant DLPs into $\mathbb{F}_{2^{1024}}$. However, the descent time would take longer than the total time, at least with the non-optimal descent that we used. We considered the possibility of using "jokers" [32], which permit one to halve the degree of even degree irreducibles when they are elements of a subfield of index 2. However, it seems to only be possible when one uses compositums, which is not possible in the context of the fields $\mathbb{F}_{2^{2^n}}$. In any case, such optimizations are academic when the total computation time is as modest as those recorded here, and our approach has the bonus of demonstrating the easiness of computing logarithms in $\mathbb{F}_{2^{512}}$, as well as in $\mathbb{F}_{2^{1024}}$.

With regard to larger n, it would certainly be possible to extend the approach of Kleinjung [52] to solve logarithms in the fields $\mathbb{F}_{2^{2^n}}$ for $n = 11, 12$ and 13, should this be needed for applications, without too much additional effort.

4 Offset Public Permutation Mode (OPP)

We present the *Offset Public Permutation Mode* (OPP), a nonce-respecting authenticated encryption mode with support for associated data which uses the techniques presented in Sect. 3. It can be seen as a generalization of OCB3 [55] to arbitrary block sizes using permutations and using improved masking techniques from Sect. 3.

4.1 Specification of OPP

Let b, k, n, τ as outlined in Sect. 2. OPP uses MEM of Sect. 3.1 for $u = 3$ and $\Phi = \{\alpha, \beta, \gamma\}$ with $\alpha(x) = \varphi(x)$, $\beta(x) = \varphi(x) \oplus x$ and $\gamma(x) = \varphi(x)^2 \oplus \varphi(x) \oplus x$, employing φ as introduced in Sect. 3.4. Furthermore, the general masking function is specified as

$$\delta : (K, X, i_0, i_1, i_2) \mapsto \gamma^{i_2} \circ \beta^{i_1} \circ \alpha^{i_0}(P(X \parallel K)).$$

We require that the tweak space of MEM used in OPP is b-proper with respect to Φ as introduced in Definition 1 and proven in Lemma 4.

The formal specification of OPP is given in Fig. 1. We refer to the authentication part of OPP as OPPAbs and to the encryption part as OPPEnc. The OPPAbs mode requires only the encryption function $\widetilde{E}$, while the OPPEnc mode uses both $\widetilde{E}$ and $\widetilde{D}$ of MEM.

Let H_i and M_j denote b-bit header and message blocks with $0 \le i \le h-1$ and $0 \le j \le m-1$ where $h = |H|_b$ and $m = |M|_b$. Note that the size of the last blocks H_{h-1} and M_{m-1} is potentially smaller than b bits. To realize proper domain separation between full and partial data blocks, and different data types, OPP uses the following setup:

OPPAbs			OPPEnc		
Data block	Condition	(i_0, i_1, i_2)	Data block	Condition	(i_0, i_1, i_2)
H_i	$0 \le i < h-1$	$(i, 0, 0)$	M_j	$0 \le j < m-1$	$(j, 0, 1)$
H_{h-1}	$\lvert H\rvert \bmod b = 0$	$(h-1, 0, 0)$	M_{m-1}	$\lvert M\rvert \bmod b = 0$	$(m-1, 0, 1)$
H_{h-1}	$\lvert H\rvert \bmod b \neq 0$	$(h-1, 1, 0)$	M_{m-1}	$\lvert M\rvert \bmod b \neq 0$	$(m-1, 1, 1)$
$\bigoplus_{j=0}^{m-1} M_j$	$\lvert M\rvert \bmod b = 0$	$(h-1, 2, 0)$			
$\bigoplus_{j=0}^{m-1} M_j$	$\lvert M\rvert \bmod b \neq 0$	$(h-1, 3, 0)$			

4.2 Security of OPP

Theorem 5. *Let b, k, n, τ as outlined in Sect. 2. Let $P \xleftarrow{\$} \mathsf{Perm}(b)$. Then, in the* nonce-respecting *setting,*

$$\mathbf{Adv}^{\mathrm{ae}}_{\mathsf{OPP},P}(q_{\mathcal{E}}, q_{\mathcal{D}}, \sigma, p) \le \frac{4.5\sigma^2}{2^b} + \frac{3\sigma p}{2^b} + \frac{p}{2^k} + \frac{2^{n-\tau}}{2^n - 1}.$$

Algorithm: $\mathsf{OPPEnc}(K, X, M)$

$M_0 \,\|\, \cdots \,\|\, M_{m-1} \leftarrow M$, s.t. $|M_i| = b, 0 \leq |M_{m-1}| < b$
$C \leftarrow \varepsilon$
$S \leftarrow 0^b$
for $i \in \{0, \ldots, m-2\}$ **do**
  $C_i \leftarrow \widetilde{E}_{K,X}^{i,0,1}(M_i)$
  $C \leftarrow C \,\|\, C_i$
  $S \leftarrow S \oplus M_i$
end
if $|M_{m-1}| > 0$ **then**
  $Z \leftarrow \widetilde{E}_{K,X}^{m-1,1,1}(0)$
  $C_{m-1} \leftarrow \mathsf{left}_{|M_{m-1}|}(\mathsf{pad}_b^0(M_{m-1}) \oplus Z)$
  $C \leftarrow C \,\|\, C_{m-1}$
  $S \leftarrow S \oplus \mathsf{pad}_b^{10}(M_{m-1})$
end
return C, S

Algorithm: $\mathsf{OPPDec}(K, X, C)$

$C_0 \,\|\, \cdots \,\|\, C_{m-1} \leftarrow C$, s.t. $|C_i| = b, 0 \leq |C_{m-1}| < b$
$M \leftarrow \varepsilon$
$S \leftarrow 0^b$
for $i \in \{0, \ldots, m-2\}$ **do**
  $M_i \leftarrow \widetilde{D}_{K,X}^{i,0,1}(C_i)$
  $M \leftarrow M \,\|\, M_i$
  $S \leftarrow S \oplus M_i$
end
if $|C_{m-1}| > 0$ **then**
  $Z \leftarrow \widetilde{E}_{K,X}^{m-1,1,1}(0)$
  $M_{m-1} \leftarrow \mathsf{left}_{|C_{m-1}|}(\mathsf{pad}_b^0(C_{m-1}) \oplus Z)$
  $M \leftarrow M \,\|\, M_{m-1}$
  $S \leftarrow S \oplus \mathsf{pad}_b^{10}(M_{m-1})$
end
return M, S

Algorithm: $\mathsf{OPPAbs}(K, X, H, S, l)$

$H_0 \,\|\, \cdots \,\|\, H_{h-1} \leftarrow H$, s.t. $|H_i| = b, 0 \leq |H_{h-1}| < b$
$S' \leftarrow 0^b$
for $i \in \{0, \ldots, h-2\}$ **do**
  $S' \leftarrow S' \oplus \widetilde{E}_{K,X}^{i,0,0}(H_i)$
end
if $|H_{h-1}| > 0$ **then**
  $S' \leftarrow S' \oplus \widetilde{E}_{K,X}^{h-1,1,0}(\mathsf{pad}_b^{10}(H_{h-1}))$
end
$j \leftarrow \lceil l/b \rceil + 2$
return $\mathsf{left}_\tau(S' \oplus \widetilde{E}_{K,X}^{h-1,j,0}(S))$

Algorithm: $\mathsf{OPP}\mathcal{E}(K, N, H, M)$

$X \leftarrow \mathsf{pad}_{b-n-k}^0(N)$
$C, S \leftarrow \mathsf{OPPEnc}(K, X, M)$
$T \leftarrow \mathsf{OPPAbs}(K, X, H, S, |M| \bmod b)$
return C, T

Algorithm: $\mathsf{OPP}\mathcal{D}(K, N, H, C, T)$

$X \leftarrow \mathsf{pad}_{b-n-k}^0(N)$
$M, S \leftarrow \mathsf{OPPDec}(K, X, C)$
$T' \leftarrow \mathsf{OPPAbs}(K, X, H, S, |M| \bmod b)$
if $T = T'$ **then return** M **else return** $\perp$ **end**

Fig. 1. Offset Public Permutation Mode (OPP)

The proof is given in the full version of this work. Note that OPP shares its structure with OCB3 of Krovetz and Rogaway [55]. In more detail, we will show that once MEM gets replaced by a random tweakable permutation $\widetilde{\pi}$, OPP becomes exactly the ΘCB3 construction [55]. The proof follows by combining the security of MEM and the security of ΘCB3. The first three terms of Theorem 5 come from the security of MEM and the b-properness of the masking.

5 Misuse-Resistant Offset Mode (MRO)

We present the *Misuse-Resistant Offset Mode* (MRO), a MAC-then-Encrypt AE mode with support for associated data which fully tolerates nonce re-usage. In some sense, MRO is the misuse-resistant variant of OPP and also uses the techniques presented in Sect. 3. It can be seen as a permutation-based variation of PMAC [12] followed by a permutation-based variation of CTR mode, and shares ideas with the Synthetic Counter in Tweak (SCT) mode [70] used in Deoxys v1.3 and Joltik v1.3 [43,44], though MRO is permutation-based and employs the improved masking schedule of Sect. 3.

5.1 Specification of MRO

Let b, k, n, τ as outlined in Sect. 2. The formal specification of MRO is given in Fig. 2. Similar to OPP, we refer to the authentication part of MRO as MROAbs and to the encryption part as MROEnc. In contrast to OPP, MRO only requires the encryption function $\widetilde{E}$ of MEM. Using notation as in the OPP mode, MRO uses the following setup for masking:

MROAbs			MROEnc		
Data block	Condition	(i_0, i_1, i_2)	Data block	Condition	(i_0, i_1, i_2)
H_i	$0 \le i \le h-1$	$(i, 0, 0)$	M_j	$0 \le j \le m-1$	$(0, 0, 1)$
M_j	$0 \le j \le m-1$	$(j, 1, 0)$			
$\lvert H\rvert \parallel \lvert M\rvert$	n.a	$(0, 2, 0)$			

Algorithm: $\mathsf{Absorb}(K, X, S, A, j)$

if $|A| > 0$ **then**
  $A_0 \parallel \cdots \parallel A_{a-1} \leftarrow \mathsf{pad}_b^0(A)$
  for $i \in \{0, \ldots, a-1\}$ **do**
    $S \leftarrow S \oplus \widetilde{E}_{K,X}^{i,j,0}(A_i)$
  end
end
return S

Algorithm: $\mathsf{MROAbs}(K, X, H, M)$

$S \leftarrow 0^b$
$S \leftarrow \mathsf{Absorb}(K, X, S, H, 0)$
$S \leftarrow \mathsf{Absorb}(K, X, S, M, 1)$
$S \leftarrow \widetilde{E}_{K,X}^{0,2,0}(S \oplus |H| \parallel |M|)$
return $\mathsf{left}_\tau(S)$

Algorithm: $\mathsf{MROEnc}(K, X, T, M)$

$C \leftarrow \varepsilon$
if $|M| > 0$ **then**
  $M_0 \parallel \cdots \parallel M_{m-1} \leftarrow \mathsf{pad}_b^0(M)$
  for $i \in \{0, \ldots, m-1\}$ **do**
    $C_i \leftarrow M_i \oplus \widetilde{E}_{K,X}^{0,0,1}(T \parallel i)$
    $C \leftarrow C \parallel C_i$
  end
end
return $\mathsf{left}_{|M|}(C)$

Algorithm: $\mathsf{MRODec}(K, X, T, C)$

$M \leftarrow \varepsilon$
if $|C| > 0$ **then**
  $C_0 \parallel \cdots \parallel C_{m-1} \leftarrow \mathsf{pad}_b^0(C)$
  for $i \in \{0, \ldots, m-1\}$ **do**
    $M_i \leftarrow C_i \oplus \widetilde{E}_{K,X}^{0,0,1}(T \parallel i)$
    $M \leftarrow M \parallel M_i$
  end
end
return $\mathsf{left}_{|C|}(M)$

Algorithm: $\mathsf{MRO}\mathcal{E}(K, N, H, M)$

$X \leftarrow \mathsf{pad}_{b-n-k}^0(N)$
$T \leftarrow \mathsf{MROAbs}(K, X, H, M)$
$C \leftarrow \mathsf{MROEnc}(K, X, T, M)$
return C, T

Algorithm: $\mathsf{MRO}\mathcal{D}(K, N, H, C, T)$

$X \leftarrow \mathsf{pad}_{b-n-k}^0(N)$
$M \leftarrow \mathsf{MRODec}(K, X, T, C)$
$T' \leftarrow \mathsf{MROAbs}(K, X, H, M)$
if $T = T'$ **then return** M **else return** $\bot$ **end**

Fig. 2. Misuse-Resistant Offset Mode (MRO)

5.2 Security of MRO

Theorem 6. *Let b, k, n, τ as outlined in Sect. 2. Let $P \xleftarrow{\$} \mathsf{Perm}(b)$. Then, in the* nonce-reuse *setting,*

$$\mathbf{Adv}_{\mathsf{MRO},P}^{\mathrm{ae}}(q_\mathcal{E}, q_\mathcal{D}, \sigma, p) \le \frac{6.5\sigma^2}{2^b} + \frac{3\sigma p}{2^b} + \frac{p}{2^k} + \frac{q_\mathcal{E}^2/2 + q_\mathcal{D}}{2^\tau}.$$

The proof is given in the full version of this work. The proof is in fact a standard-model proof where the scheme is considered to be based on MEM. It is a modular proof that, at a high level, consists of the following steps:

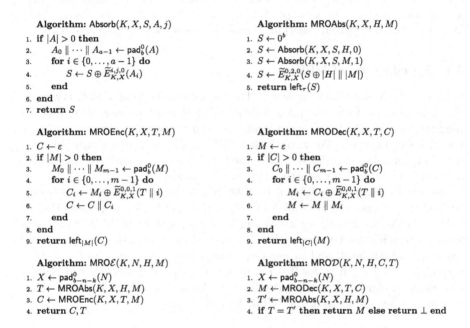

(i) The first step in the analysis is to replace MEM with a random secret tweakable permutation. It costs the MTPRP security of MEM, $\frac{4.5\sigma^2}{2^b} + \frac{3\sigma p}{2^b} + \frac{p}{2^k}$, using that the masking is b-proper.
(ii) The absorption function and encryption function call the tweakable cipher for *distinct* tweaks. Hence, using an adaption of the MAC-then-Encrypt paradigm to misuse resistance [37,65] allows us to analyze the MAC parts and the encryption parts separately.

6 Misuse-Resistant Sponge (MRS)

We introduce the *Misuse-Resistant Sponge Mode* (MRS), a MAC-then-Encrypt Sponge-based AE mode with support for associated data which fully tolerates nonce re-usage. The absorption function is a full-state keyed Sponge MAC [3,10,63]. The encryption function follows the SpongeWrap approach [9,63].

6.1 Specification of MRS

Let b, k, n, τ, r, c as outlined in Sect. 2. The formal specification of MRS is given in Fig. 3. It consists of an absorption function MRSAbs and an encryption function MRSEnc, in a MAC-then-Encrypt mode, but using the same primitive and same key in both functions. We remark that MRS as given in Fig. 3 only does one round of squeezing in order to obtain the tag. This can be easily generalized to multiple rounds, without affecting the security proofs.

We briefly discuss the differences of MRS with Haddoc, the misuse-resistant AE scheme presented by Bertoni et al. [11] at the 2014 SHA-3 workshop. Haddoc follows the MAC-then-Encrypt paradigm as well, where the MAC function is identical to MRSAbs. For encryption, however, Haddoc uses the Sponge in CTR mode. At a high level, and in our terminology, this boils down to $C_i = M_i \oplus \mathsf{left}_r(P(T \parallel \langle i \rangle \parallel 1 \parallel K))$, for $0 \le i \le m-1$. In other words, MRS and Haddoc structurally differ in the way encryption is performed, and in fact, Haddoc more closely matches the ideas of the MRSO hybrid of Sect. 7.

6.2 Security of MRS

Theorem 7. *Let b, k, n, τ, r, c as outlined in Sect. 2. Let $P \xleftarrow{\$} \mathsf{Perm}(b)$. Then, in the* nonce-reuse *setting,*

$$\mathbf{Adv}^{\mathsf{ae}}_{\mathsf{MRS},P}(q_{\mathcal{E}}, q_{\mathcal{D}}, \sigma, p) \le \frac{4\sigma^2}{2^b} + \frac{4\sigma^2}{2^c} + \frac{2\sigma p}{2^k} + \frac{q_{\mathcal{E}}^2/2 + q_{\mathcal{D}}q_{\mathcal{E}} + q_{\mathcal{D}}}{2^\tau}.$$

The proof is given in the full version of this work. It is different from the proofs for OPP and MRO, although it is also effectively a standard-model proof. It relies on the observation that both the absorption and the encryption phase are in fact evaluations of the Full-state Keyed Duplex [9,63]. This construction has been proven to behave like a random functionality, with the property that

Algorithm: $\mathsf{Absorb}(S, A)$

if $|A| > 0$ **then**
  $A_0 \parallel \cdots \parallel A_{a-1} \leftarrow \mathsf{pad}_b^0(A)$
  for $i \in \{0, \dots, a-1\}$ **do**
    $S \leftarrow P(S)$
    $S \leftarrow S \oplus A_i$
  end
end
return S

Algorithm: $\mathsf{MRSAbs}(K, N, H, M)$

$S \leftarrow N \parallel 0^* \parallel 0 \parallel K$
$S \leftarrow \mathsf{Absorb}(S, H)$
$S \leftarrow \mathsf{Absorb}(S, M)$
$S \leftarrow P(S)$
$S \leftarrow S \oplus |H| \parallel |M|$
$S \leftarrow P(S)$
$T \leftarrow \mathsf{left}_\tau(S)$
return T

Algorithm: $\mathsf{MRSEnc}(K, T, M)$

$C \leftarrow \varepsilon$
if $|M| > 0$ **then**
  $S \leftarrow T \parallel 0^* \parallel 1 \parallel K$
  $M_0 \parallel \cdots \parallel M_{m-1} \leftarrow \mathsf{pad}_r^0(M)$
  for $i \in \{0, \dots, m-1\}$ **do**
    $S \leftarrow P(S)$
    $S \leftarrow S \oplus (M_i \parallel 0^c)$
    $C \leftarrow C \parallel \mathsf{left}_r(S)$
  end
end
return $\mathsf{left}_{|M|}(C)$

Algorithm: $\mathsf{MRSDec}(K, T, C)$

$M \leftarrow \varepsilon$
if $|C| > 0$ **then**
  $S \leftarrow T \parallel 0^* \parallel 1 \parallel K$
  $C_0 \parallel \cdots \parallel C_{m-1} \leftarrow \mathsf{pad}_r^0(C)$
  for $i \in \{0, \dots, m-1\}$ **do**
    $S \leftarrow P(S)$
    $M \leftarrow M \parallel \mathsf{left}_r(S \oplus (C_i \parallel 0^c))$
    $S \leftarrow C_i \parallel \mathsf{right}_c(S)$
  end
end
return $\mathsf{left}_{|C|}(M)$

Algorithm: $\mathsf{MRS}\mathcal{E}(K, N, H, M)$

$T \leftarrow \mathsf{MRSAbs}(K, N, H, M)$
$C \leftarrow \mathsf{MRSEnc}(K, T, M)$
return C, T

Algorithm: $\mathsf{MRS}\mathcal{D}(K, N, H, C, T)$

$M \leftarrow \mathsf{MRSDec}(K, T, C)$
$T' \leftarrow \mathsf{MRSAbs}(K, N, H, M)$
if $T = T'$ **then return** M **else return** $\perp$ **end**

Fig. 3. Misuse-Resistant Sponge (MRS)

it always outputs uniformly random data, up to common prefix in the input. Assuming that the distinguisher never makes duplicate queries, MRSAbs never has common prefixes; assuming tags never collide, MRSEnc never has common prefixes; and finally, the initial inputs to MRSAbs versus MRSEnc are always different due to the 0/1 domain separation. The proof then easily follows.

7 Misuse-Resistant Sponge-Offset (MRSO)

The constructions of Sects. 5 and 6 can be combined in a straightforward way to obtain two hybrids: the *Misuse-Resistant Sponge-then-Offset Mode* (MRSO) and the *Misuse-Resistant Offset-then-Sponge Mode* (MROS). While we cannot think of any practical use-case for MROS, we do think MRSO is useful. As suggested in Sect. 6, MRSO is comparable with—and in fact improves over—Haddoc.

7.1 Specification of MRSO

Let b, k, n, τ as outlined in Sect. 2. The formal specification of the MRSO AE scheme is formalized in Fig. 4. It MACs the data using MRSAbs and encrypts using MROEnc. MRSO uses MEM as specified for OPP but requires only a very limited selection of tweaks and has $i_1 = i_2 = 0$ fixed. Thus, the general masking function can be simplified to

$$\delta : (K, X, i_0) \mapsto \alpha^{i_0}(P(X \parallel K)).$$

For the encryption part MROEnc this is clear (cf. Sect. 5). For the absorption part MRSAbs, this is less clear: informally, it is based on the idea of setting $L = P(N \parallel 0^* \parallel K)$, and of XORing this value everywhere in-between two consecutive evaluations of P. Because at the end of MRSAbs, a part of the rate is extracted, this "trick" only works if performed with the rightmost $b - \tau$ bits of L. Therefore, MRSO is based on a slight adjustment of MEM with $b - \tau$-bit maskings only. Let $h = |H|_b$ and $m = |M|_b$ denote the number of b-bit header and message blocks, respectively. We use the following setup for masking:

MRSAbs			MROEnc		
Data block	Condition	i_0	Data block	Condition	i_0
H_i	$0 \le i \le h-1$	0	M_j	$0 \le j \le m-1$	1
M_j	$0 \le j \le m-1$	0			
$\|H\| \parallel \|M\|$	n.a	0			

Algorithm: $\mathsf{MRSO}\mathcal{E}(K, N, H, M)$

1. $X \leftarrow \mathsf{pad}^0_{b-n-k}(N)$
2. $T \leftarrow \mathsf{MRSAbs}(K, N, H, M)$
3. $C \leftarrow \mathsf{MROEnc}(K, X, T, M)$
4. **return** C, T

Algorithm: $\mathsf{MRSO}\mathcal{D}(K, N, H, C, T)$

1. $X \leftarrow \mathsf{pad}^0_{b-n-k}(N)$
2. $M \leftarrow \mathsf{MRODec}(K, X, T, C)$
3. $T' \leftarrow \mathsf{MRSAbs}(K, N, H, M)$
4. **if** $T = T'$ **then return** M **else return** $\perp$ **end**

Fig. 4. Sponge-Offset mode MRSO. Refer to Figs. 2 and 3 for the sub-algorithms

7.2 Security of MRSO

Theorem 8. *Let b, k, n, τ as outlined in Sect. 2. Let $P \xleftarrow{\$} \mathsf{Perm}(b)$. Then, in the* nonce-reuse *setting,*

$$\mathbf{Adv}^{\mathrm{ae}}_{\mathsf{MRSO},P}(q_\mathcal{E}, q_\mathcal{D}, \sigma, p) \le \frac{2\sigma^2}{2^b} + \frac{5.5\sigma^2}{2^{b-\tau}} + \frac{3\sigma p}{2^{b-\tau}} + \frac{p}{2^k} + \frac{q_\mathcal{E}^2/2 + q_\mathcal{D}}{2^\tau}.$$

The proof is similar to the proof of MRO, with the difference that now we use $(b - \tau)$-properness of the masking. It is given in the full version of the work.

8 Implementation

In this section we discuss our results on the implementations of concrete instantiations of OPP, MRO, and MRS. For all three schemes we use state, key, tag, and nonce sizes of $b = 1024$, $k = \tau = 256$, and $n = 128$ bits. For P we employ the BLAKE2b [5] permutation with $l \in \{4, 6\}$ rounds. For OPP and MRO we use $\varphi(x_0, \ldots, x_{15}) = (x_1, ..., x_{15}, (x_0 \lll 53) \oplus (x_5 \ll 13))$ and for MRSEnc we set rate and capacity to $r = 768$ and $c = 256$ bits. To remain self-contained, we now recall the BLAKE2b permutation. It operates on a state $S = (s_0, \ldots, s_{15})$ with 64-bit words s_i. A single round $F(S)$ consists of the sequence of operations

$$\begin{array}{llll} G(s_0, s_4, s_8, s_{12}); & G(s_1, s_5, s_9, s_{13}); & G(s_2, s_6, s_{10}, s_{14}); & G(s_3, s_7, s_{11}, s_{15}); \\ G(s_0, s_5, s_{10}, s_{15}); & G(s_1, s_6, s_{11}, s_{12}); & G(s_2, s_7, s_8, s_{13}); & G(s_3, s_4, s_9, s_{14}); \end{array}$$

where

$$G(a, b, c, d) = \begin{cases} a = a + b; \; d = (d \oplus a) \ggg 32; \; c = c + d; \; b = (b \oplus c) \ggg 24; \\ a = a + b; \; d = (d \oplus a) \ggg 16; \; c = c + d; \; b = (b \oplus c) \ggg 63; \end{cases}$$

BLAKE2 and its predecessors have been heavily analyzed, e.g., [38,51]. These results are mostly of theoretical interest though since the complexity of the attacks vastly outweigh our targeted security level. Nevertheless, the BLAKE2 permutation family has some evident and well-known non-random characteristics [4]: for any $l > 0$, it holds that $F^l(0) = 0$ and $F^l(a, a, a, a, b, b, b, b, c, c, c, c, d, d, d, d) = (w, w, w, w, x, x, x, x, y, y, y, y, z, z, z, z)$ for arbitrary values a, b, c, and d. These symmetric states can be easily avoided with a careful design, so that they cannot be exploited as a distinguisher. Thus, we use slightly modified variants of the schemes from Sects. 4, 5, 6 and 7. Instead of initializing the masks with $P(N \parallel 0^{640} \parallel K)$ in OPP and MRO, we encode the round number l and tag size τ as 64-bit strings and use $P(N \parallel 0^{512} \parallel \langle l \rangle_{64} \parallel \langle \tau \rangle_{64} \parallel K)$. Analogously, MRSAbs and MRSEnc are initialized with $N \parallel 0^{448} \parallel \langle l \rangle_{64} \parallel \langle \tau \rangle_{64} \parallel \langle 0 \rangle_{64} \parallel K$ and $T \parallel 0^{320} \parallel \langle l \rangle_{64} \parallel \langle \tau \rangle_{64} \parallel \langle 1 \rangle_{64} \parallel K$, respectively.

We wrote reference implementations of all schemes in plain C and optimized variants using the AVX, AVX2, and NEON instruction sets[4]. Performance was measured on the Intel Sandy Bridge and Haswell and on the ARM Cortex-A8 and compared to some reference AEAD schemes, see Tables 2 and 3. All values are given for "long messages" ($\geq$ 4 KiB) with cycles per byte (cpb) as unit.

In the nonce-respecting scenario our fastest proposal is OPP with 4 BLAKE2b rounds. Our 4-fold word-sliced AVX2-implementation achieves 0.55 cpb on

Table 2. Performance of OPP, MRO, and MRS instantiated with the BLAKE2b permutation

		$l = 4$			$l = 6$		
Platform	Impl.	OPP	MRO	MRS	OPP	MRO	MRS
Cortex-A8	NEON	4.26	8.07	8.50	5.91	11.32	12.21
Sandy Bridge	AVX	1.24	2.41	2.55	1.91	3.58	3.87
Haswell	AVX2	0.55	1.06	2.40	0.75	1.39	3.58

4 The source code of our schemes is freely available at [61] under a CC0 license.

Table 3. Performance of some reference AEAD modes

	Nonce-respecting					Misuse-resistant	
Platform	AES-GCM	OCB3	ChaCha20-Poly1305	Salsa20-Poly1305	Deoxys$^{\neq}$-128-128	GCM-SIV	Deoxys$^{=}$ – 128 – 128
Cortex-A8	38.6	28.9	-	5.60+2.60	-	-	-
Sandy Bridge	2.55	0.98	-	-	1.29	-	≈ 2.58
Haswell	1.03	0.69	1.43+0.66	-	0.96	1.17	≈ 1.92
References	[16,36]	[36,55]	[28,29]	[8]	[43,69]	[37]	[43,69]

Haswell, amounting to a throughput of 6.36 GiBps and assuming a CPU frequency of 3.5 GHz. Compared to its competitors AES-GCM, OCB3, ChaCha20-Poly1305 and Deoxys$^{\neq}$ (v1.3)[5], this instantiation of OPP is faster by factors of about 1.87, 1.25, 3.80, and 1.74 respectively. Even the 6-round variant of OPP is able to maintain high speeds at 0.75 cpb (4.67 GiBps) reducing the distance to the above competitors to factors of 1.37, 0.92, 2.78, and 1.28. On ARM platforms, without AES-NI, OPP's advantage is even more significant. The NEON-variant outperforms the AES-based ciphers OCB3 and AES-GCM by factors of about 6.78 and 9.06. The highly optimized Salsa20-Poly1305 implementation of [8] is slower by a factor of around 1.92.

In the misuse-resistant scenario our fastest proposal is MRO with 4 BLAKE2b rounds. Our 4-fold word-sliced AVX2-implementation achieves 1.06 cpb on Haswell which is equivalent to a throughput of 3.30 GiBps at a frequency of 3.5 GHz. In comparison to schemes such as AES-GCM-SIV and Deoxys$^{=}$ (v.1.3), the above instantiation of MRO is faster by factors of about 1.10 and 1.81. For the 6-round version with 1.39 cpb these factors are reduced to 0.79 and 1.38, respectively. Unfortunately, there is not enough published data on performance of misuse-resistant AE schemes on ARM. As for OPP in the nonce-respecting scenario, one can expect similar performance gaps between the misuse-resistant AES-based schemes and MRO.

Due to the inherently sequential Sponge-construction used in MRS, advanced implementation techniques like 4-fold word-slicing are not possible. In general, MRS performs therefore worse than MRO. On Haswell MRS achieves 2.40 cpb ($l = 4$) and 3.58 cpb ($l = 6$) which translate to throughputs of 1.45 GiBps and 0.97 GiBps, respectively. Thus, MRS is still competitive to other misuse-resistant AE schemes on Intel platforms. On ARM it shows good performance as well, almost on the level of MRO. We have not written any implementations for MRSO but it is to be expected that its performance lies between MRO and MRS.

Acknowledgements. Robert Granger is supported by the Swiss National Science Foundation via grant number 200021-156420. Bart Mennink is a Postdoctoral Fellow of the Research Foundation – Flanders (FWO). He is supported in part by the Research Council KU Leuven: GOA TENSE (GOA/11/007).

[5] We point out that Deoxys$^{\neq}$, unlike the other considered modes, aims for security beyond the birthday bound up to the full block size.

We kindly thank Thorsten Kleinjung for the observations on irreducible cofactors presented in the remark at the end of Sect. 3.6.1 and for further helpful discussions during our work. We also thank Antoine Joux for interesting discussions on using lattice reduction to compute different separations of the masking domains (as discussed in the remark at the end of Sect. 3.5). We also thank the anonymous reviewers of EUROCRYPT 2016 for their useful comments and suggestions.

References

1. Andreeva, E., Bilgin, B., Bogdanov, A., Luykx, A., Mennink, B., Mouha, N., Yasuda, K.: APE: authenticated permutation-based encryption for lightweight cryptography. In: Cid, C., Rechberger, C. (eds.) FSE 2014. LNCS, vol. 8540, pp. 168–186. Springer, Heidelberg (2015)
2. Andreeva, E., Bogdanov, A., Luykx, A., Mennink, B., Tischhauser, E., Yasuda, K.: Parallelizable and authenticated online ciphers. In: Sako, K., Sarkar, P. (eds.) ASIACRYPT 2013, Part I. LNCS, vol. 8269, pp. 424–443. Springer, Heidelberg (2013)
3. Andreeva, E., Daemen, J., Mennink, B., Van Assche, G.: Security of keyed sponge constructions using a modular proof approach. In: Leander, G. (ed.) FSE 2015. LNCS, vol. 9054, pp. 364–384. Springer, Heidelberg (2015)
4. Aumasson, J.P., Jovanovic, P., Neves, S.: Analysis of NORX: investigating differential and rotational properties. In: Aranha, D.F., Menezes, A. (eds.) LATINCRYPT 2014. LNCS, vol. 8895, pp. 306–324. Springer, Heidelberg (2015)
5. Aumasson, J.P., Neves, S., Wilcox-O'Hearn, Z., Winnerlein, C.: BLAKE2: simpler, smaller, fast as MD5. In: Jacobson Jr., M.J., Locasto, M.E., Mohassel, P., Safavi-Naini, R. (eds.) ACNS 13. LNCS, vol. 7954, pp. 119–135. Springer, Heidelberg (2013)
6. Barbulescu, R., Bouvier, C., Detrey, J., Gaudry, P., Jeljeli, H., Thomé, E., Videau, M., Zimmermann, P.: Discrete logarithm in GF(2^{809}) with FFS. In: Krawczyk, H. (ed.) PKC 2014. LNCS, vol. 8383, pp. 221–238. Springer, Heidelberg (2014)
7. Barbulescu, R., Gaudry, P., Joux, A., Thomé, E.: A heuristic quasi-polynomial algorithm for discrete logarithm in finite fields of small characteristic. In: Nguyen, P.Q., Oswald, E. (eds.) EUROCRYPT 2014. LNCS, vol. 8441, pp. 1–16. Springer, Heidelberg (2014)
8. Bernstein, D.J., Schwabe, P.: NEON crypto. In: Prouff, E., Schaumont, P. (eds.) CHES 2012. LNCS, vol. 7428, pp. 320–339. Springer, Heidelberg (2012)
9. Bertoni, G., Daemen, J., Peeters, M., Van Assche, G.: Duplexing the sponge: single-pass authenticated encryption and other applications. In: Miri, A., Vaudenay, S. (eds.) SAC 2011. LNCS, vol. 7118, pp. 320–337. Springer, Heidelberg (2011)
10. Bertoni, G., Daemen, J., Peeters, M., Van Assche, G.: On the security of the keyed sponge construction. In: SKEW 2011 (2011)
11. Bertoni, G., Daemen, J., Peeters, M., Van Assche, G., Van Keer, R.: Using Keccak technology for AE: Ketje, Keyak and more. In: SHA-3 2014 Workshop (2014)
12. Black, J., Rogaway, P.: A block-cipher mode of operation for parallelizable message authentication. In: Knudsen, L.R. (ed.) EUROCRYPT 2002. LNCS, vol. 2332, pp. 384–397. Springer, Heidelberg (2002)
13. Bluher, A.W.: On $x^{q+1} + ax + b$. Finite Fields Appl. **10**(3), 285–305 (2004)
14. Bosma, W., Cannon, J., Playoust, C.: The Magma algebra system. I. The user language. J. Symbolic Comput. **24**(3–4), 235–265 (1997). Computational algebra and number theory (London, 1993)

15. CAESAR – Competition for Authenticated Encryption: Security, Applicability, and Robustness (2014)
16. Câmara, D.F., Gouvêa, C.P.L., López, J., Dahab, R.: Fast software polynomial multiplication on ARM processors using the NEON engine. In: Cuzzocrea, A., Kittl, C., Simos, D.E., Weippl, E.R., Xu, L. (eds.) Security Engineering and Intelligence Informatics. LNCS, vol. 8128, pp. 137–154. Springer, Heidelberg (2013)
17. Canteaut, A., Carpov, S., Fontaine, C., Lepoint, T., Naya-Plasencia, M., Paillier, P., Sirdey, R.: Stream ciphers: a practical solution for efficient homomorphic-ciphertext compression. In: FSE 2016. LNCS. Springer, Heidelberg, March 2016 (to appear)
18. Chakraborty, D., Sarkar, P.: A general construction of tweakable block ciphers and different modes of operations. IEEE Trans. Inf. Theory **54**(5), 1991–2006 (2008)
19. Chakraborty, D., Sarkar, P.: On modes of operations of a block cipher for authentication and authenticated encryption. Cryptology ePrint Archive, Report 2014/627 (2014)
20. Chen, S., Lampe, R., Lee, J., Seurin, Y., Steinberger, J.P.: Minimizing the two-round Even-Mansour cipher. In: Garay, J.A., Gennaro, R. (eds.) CRYPTO 2014, Part I. LNCS, vol. 8616, pp. 39–56. Springer, Heidelberg (2014)
21. Chen, S., Steinberger, J.P.: Tight security bounds for key-alternating ciphers. In: Nguyen, P.Q., Oswald, E. (eds.) EUROCRYPT 2014. LNCS, vol. 8441, pp. 327–350. Springer, Heidelberg (2014)
22. Cheng, Q., Wan, D., Zhuang, J.: Traps to the BGJT-algorithm for discrete logarithms. LMS J. Comput. Math. **17**, 218–229 (2014)
23. Cogliati, B., Lampe, R., Seurin, Y.: Tweaking Even-Mansour ciphers. In: Gennaro, R., Robshaw, M.J.B. (eds.) CRYPTO 2015, Part I. LNCS, vol. 9215, pp. 189–208. Springer, Heidelberg (2015)
24. Cogliati, B., Seurin, Y.: Beyond-birthday-bound security for tweakable Even-Mansour ciphers with linear tweak and key mixing. In: Iwata, T., Cheon, J.H. (eds.) ASIACRYPT 2015, Part II. LNCS, vol. 9453, pp. 134–158. Springer, Heidelberg (2015)
25. Cogliati, B., Seurin, Y.: On the provable security of the iterated Even-Mansour cipher against related-key and chosen-key attacks. In: Oswald, E., Fischlin, M. (eds.) EUROCRYPT 2015, Part I. LNCS, vol. 9056, pp. 584–613. Springer, Heidelberg (2015)
26. Fleischmann, E., Forler, C., Lucks, S.: McOE: a family of almost foolproof on-line authenticated encryption schemes. In: Canteaut, A. (ed.) FSE 2012. LNCS, vol. 7549, pp. 196–215. Springer, Heidelberg (2012)
27. Gligor, V.D., Donescu, P.: Fast encryption and authentication: XCBC encryption and XECB authentication modes. In: Matsui, M. (ed.) FSE 2001. LNCS, vol. 2355, pp. 92–108. Springer, Heidelberg (2002)
28. Goll, M., Gueron, S.: Vectorization on ChaCha stream cipher. In: Latifi, S. (ed.) ITNG 2014, pp. 612–615. IEEE Computer Society (2014)
29. Goll, M., Gueron, S.: Vectorization of Poly1305 message authentication code. In: Latifi, S. (ed.) ITNG 2015, pp. 612–615. IEEE Computer Society (2015)
30. Göloglu, F., Granger, R., McGuire, G., Zumbrägel, J.: On the function field sieve and the impact of higher splitting probabilities – application to discrete logarithms in $\mathbb{F}_{2^{1971}}$ and $\mathbb{F}_{2^{3164}}$. In: Canetti, R., Garay, J.A. (eds.) CRYPTO 2013, Part II. LNCS, vol. 8043, pp. 109–128. Springer, Heidelberg (2013)
31. Göloglu, F., Granger, R., McGuire, G., Zumbrägel, J.: Solving a 6120-bit DLP on a desktop computer. In: Lange, T., Lauter, K., Lisonek, P. (eds.) SAC 2013. LNCS, vol. 8282, pp. 136–152. Springer, Heidelberg (2014)

32. Granger, R., Kleinjung, T., Zumbrägel, J.: Breaking '128-bit secure' supersingular binary curves – (or how to solve discrete logarithms in $F_{2^{4 \cdot 1223}}$ and $F_{2^{12 \cdot 367}}$). In: Garay, J.A., Gennaro, R. (eds.) CRYPTO 2014, Part II. LNCS, vol. 8617, pp. 126–145. Springer, Heidelberg (2014)
33. Granger, R., Kleinjung, T., Zumbrägel, J.: On the powers of 2. Cryptology ePrint Archive, Report 2014/300 (2014)
34. Granger, R., Kleinjung, T., Zumbrägel, J.: On the discrete logarithm problem in finite fields of fixed characteristic. Cryptology ePrint Archive, Report 2015/685 (2015)
35. Granger, R., Kleinjung, T., Zumbrägel, J.: Discrete Logarithms in $GF(2^{9234})$. NMBRTHRY list, 31 January 2014
36. Gueron, S.: AES-GCM software performance on the current high end CPUs as a performance baseline for CAESAR competition. In: DIAC 2013 (2013)
37. Gueron, S., Lindell, Y.: GCM-SIV: full nonce misuse-resistant authenticated encryption at under one cycle per byte. In: Ray, I., Li, N., Kruegel: C. (eds.) ACM CCS 2015, pp. 109–119. ACM Press, October 2015
38. Guo, J., Karpman, P., Nikolic, I., Wang, L., Wu, S.: Analysis of BLAKE2. In: Benaloh, J. (ed.) CT-RSA 2014. LNCS, vol. 8366, pp. 402–423. Springer, Heidelberg (2014)
39. Haramoto, H., Matsumoto, M., Nishimura, T., Panneton, F., L'Ecuyer, P.: Efficient jump ahead for $\mathbb{F}_2$-linear random number generators. INFORMS J. Comput. **20**(3), 385–390 (2008)
40. Hoang, V.T., Krovetz, T., Rogaway, P.: Robust authenticated-encryption AEZ and the problem that it solves. In: Oswald, E., Fischlin, M. (eds.) EUROCRYPT 2015, Part I. LNCS, vol. 9056, pp. 15–44. Springer, Heidelberg (2015)
41. Hoang, V.T., Reyhanitabar, R., Rogaway, P., Vizár, D.: Online authenticated-encryption and its nonce-reuse misuse-resistance. In: Gennaro, R., Robshaw, M.J.B. (eds.) CRYPTO 2015, Part I. LNCS, vol. 9215, pp. 493–517. Springer, Heidelberg (2015)
42. Huang, M., Narayanan, A.K.: On the relation generation method of Joux for computing discrete logarithms. CoRR abs/1312.1674 (2013)
43. Jean, J., Nikolić, I., Peyrin, T.: Deoxys v1.3. CAESAR Round 2 submission (2015)
44. Jean, J., Nikolić, I., Peyrin, T.: Joltik v1.3. CAESAR Round 2 submission (2015)
45. Joux, A.: A new index calculus algorithm with complexity $L(1/4 + o(1))$ in small characteristic. In: Lange, T., Lauter, K., Lisonek, P. (eds.) SAC 2013. LNCS, vol. 8282, pp. 355–379. Springer, Heidelberg (2014)
46. Joux, A., Lercier, R.: The function field sieve is quite special. In: Fieker, C., Kohel, D.R. (eds.) Algorithmic Number Theory. LNCS, vol. 2369, pp. 431–445. Springer, Heidelberg (2002)
47. Joux, A., Lercier, R.: The function field sieve in the medium prime case. In: Vaudenay, S. (ed.) EUROCRYPT 2006. LNCS, vol. 4004, pp. 254–270. Springer, Heidelberg (2006)
48. Joux, A., Pierrot, C.: Improving the polynomial time precomputation of Frobenius representation discrete logarithm algorithms – simplified setting for small characteristic finite fields. In: Sarkar, P., Iwata, T. (eds.) ASIACRYPT 2014, Part I. LNCS, vol. 8873, pp. 378–397. Springer, Heidelberg (2014)
49. Jutla, C.S.: Encryption modes with almost free message integrity. J. Cryptology **21**(4), 547–578 (2008)
50. Kavun, E.B., Lauridsen, M.M., Leander, G., Rechberger, C., Schwabe, P., Yalçın, T.: Prøst v1. CAESAR Round 1 submission (2014)

51. Khovratovich, D., Nikolic, I., Pieprzyk, J., Sokolowski, P., Steinfeld, R.: Rotational cryptanalysis of ARX revisited. In: Leander, G. (ed.) FSE 2015. LNCS, vol. 9054, pp. 519–536. Springer, Heidelberg (2015)
52. Kleinjung, T.: Discrete logarithms in GF(2^{1279}). NMBRTHRY list, 17 October 2014
53. Krovetz, T., Rogaway, P.: The OCB authenticated-encryption algorithm. RFC 7253 (Informational) (2014)
54. Krovetz, T.: HS1-SIV v1. CAESAR Round 1 submission (2014)
55. Krovetz, T., Rogaway, P.: The software performance of authenticated-encryption modes. In: Joux, A. (ed.) FSE 2011. LNCS, vol. 6733, pp. 306–327. Springer, Heidelberg (2011)
56. Kurosawa, K.: Power of a public random permutation and its application to authenticated encryption. IEEE Trans. Inf. Theory **56**(10), 5366–5374 (2010)
57. Lenstra Jr., H.W.: Finding isomorphisms between finite fields. Math. Comput. **56**(193), 329–347 (1991)
58. Lidl, R., Niederreiter, H.: Finite Fields, Encyclopedia of Mathematics and its Applications, vol. 20, 2nd edn. Cambridge University Press, Cambridge, United Kingdom (1997)
59. Marsaglia, G.: Xorshift RNGs. J. Stat. Softw. **8**(14), 1–6 (2003)
60. Matsumoto, M., Nishimura, T.: Mersenne Twister: a 623-dimensionally equidistributed uniform pseudo-random number generator. ACM Trans. Model. Comput. Simul. **8**(1), 3–30 (1998)
61. MEM Family of AEAD Schemes (2015). https://github.com/MEM-AEAD
62. Mennink, B.: XPX: Generalized Tweakable Even-Mansour with Improved Security Guarantees. Cryptology ePrint Archive, Report 2015/476 (2015)
63. Mennink, B., Reyhanitabar, R., Vizár, D.: Security of full-state keyed sponge and duplex: applications to authenticated encryption. In: Iwata, T., Cheon, J.H. (eds.) ASIACRYPT 2015, Part II. LNCS, vol. 9453, pp. 465–489. Springer, Heidelberg (2015)
64. Minematsu, K.: A short universal hash function from bit rotation, and applications to blockcipher modes. In: Susilo, W., Reyhanitabar, R. (eds.) ProvSec 2013. LNCS, vol. 8209, pp. 221–238. Springer, Heidelberg (2013)
65. Namprempre, C., Rogaway, P., Shrimpton, T.: Reconsidering generic composition. In: Nguyen, P.Q., Oswald, E. (eds.) EUROCRYPT 2014. LNCS, vol. 8441, pp. 257–274. Springer, Heidelberg (2014)
66. Nguyen, P.Q., Stehlé, D.: Low-dimensional lattice basis reduction revisited. ACM Trans. Algorithms **5**(4), 46 (2009)
67. Niederreiter, H.: Factorization of polynomials and some linear-algebra problems over finite fields. Linear Algebra Appl. **192**, 301–328 (1993)
68. Patarin, J.: The "coefficients H" technique (invited talk). In: Avanzi, R.M., Keliher, L., Sica, F. (eds.) SAC 2008. LNCS, vol. 5381, pp. 328–345. Springer, Heidelberg (2009)
69. Peyrin, T.: Personal communication, February 2016
70. Peyrin, T., Seurin, Y.: Counter-in-tweak: authenticated encryption modes for tweakable block ciphers. Cryptology ePrint Archive, Report 2015/1049 (2015)
71. Rogaway, P.: Efficient instantiations of tweakable blockciphers and refinements to modes OCB and PMAC. In: Lee, P.J. (ed.) ASIACRYPT 2004. LNCS, vol. 3329, pp. 16–31. Springer, Heidelberg (2004)
72. Rogaway, P., Bellare, M., Black, J., Krovetz, T.: OCB: a block-cipher mode of operation for efficient authenticated encryption. In: ACM CCS 2001, pp. 196–205. ACM Press (2001)

73. Rogaway, P., Shrimpton, T.: A provable-security treatment of the key-wrap problem. In: Vaudenay, S. (ed.) EUROCRYPT 2006. LNCS, vol. 4004, pp. 373–390. Springer, Heidelberg (2006)
74. Sarkar, P.: Pseudo-random functions and parallelizable modes of operations of a block cipher. IEEE Trans. Inf. Theory **56**(8), 4025–4037 (2010)
75. Sasaki, Y., Todo, Y., Aoki, K., Naito, Y., Sugawara, T., Murakami, Y., Matsui, M., Hirose, S.: Minalpher v1. CAESAR Round 1 submission (2014)
76. Thomé, E.: Computation of discrete logarithms in $F_{2^{607}}$. In: Boyd, C. (ed.) ASIACRYPT 2001. LNCS, vol. 2248, pp. 107–124. Springer, Heidelberg (2001)
77. Yasuda, K.: A one-pass mode of operation for deterministic message authentication- security beyond the birthday barrier. In: Nyberg, K. (ed.) FSE 2008. LNCS, vol. 5086, pp. 316–333. Springer, Heidelberg (2008)
78. Zeng, G., Han, W., He, K.: High efficiency feedback shift register: $\sigma-$LFSR. Cryptology ePrint Archive, Report 2007/114 (2007)

Sanitization of FHE Ciphertexts

Léo Ducas[1(✉)] and Damien Stehlé[2]

[1] Cryptology Group, CWI, Amsterdam, The Netherlands
ducas@cwi.nl
[2] Laboratoire LIP (U. Lyon, CNRS, ENSL, INRIA, UCBL), ENS de Lyon, Lyon, France
damien.stehle@ens-lyon.fr

Abstract. By definition, fully homomorphic encryption (FHE) schemes support homomorphic decryption, and all known FHE constructions are bootstrapped from a Somewhat Homomorphic Encryption (SHE) scheme via this technique. Additionally, when a public key is provided, ciphertexts are also re-randomizable, e.g., by adding to them fresh encryptions of 0. From those two operations we devise an algorithm to sanitize a ciphertext, by making its distribution canonical. In particular, the distribution of the ciphertext does not depend on the circuit that led to it via homomorphic evaluation, thus providing circuit privacy in the honest-but-curious model. Unlike the previous approach based on noise flooding, our approach does not degrade much the security/efficiency trade-off of the underlying FHE. The technique can be applied to all lattice-based FHE proposed so far, without substantially affecting their concrete parameters.

1 Introduction

A fully homomorphic encryption (FHE) scheme enables the efficient and compact public transformation of ciphertexts decrypting to plaintexts $\mu_1, \ldots, \mu_k$, into a ciphertext decrypting to $\mathcal{C}(\mu_1, \ldots, \mu_k)$, for any circuit $\mathcal{C}$ with any number k of input wires. Since Gentry's first proposal of a candidate FHE scheme [Gen09a, Gen09b], plenty of FHE schemes have been proposed (see [SV10, DGHV10, BV11a, BV11b, Bra12, GHS12, GSW13], to name just a few).

A typical application of FHE is to offshore heavy computations on privacy-sensitive data: a computationally limited user encrypts its data, sends it to a distant powerful server, tells the server which operations to perform on the encrypted data, retrieves the result and decrypts. For this mainstream application, confidentiality, malleability and compactness seem sufficient. However, for other invaluable applications of FHE, another property, which we will call *ciphertext sanitizability*, has proved central. Statistical (resp. computational) ciphertext sanitizability requires that there exists a probabilistic polynomial time algorithm $\mathsf{Sanitize}$ taking as inputs a public key pk and a ciphertext c decrypting to a plaintext μ under the secret key sk associated to pk, such that the distributions $\mathsf{Sanitize}(pk, c)$ and $\mathsf{Sanitize}(pk, \mathsf{Enc}(pk, \mu))$ are statistically (resp. computationally) indistinguishable, given pk *and* sk (here Enc refers to the encryption

M. Fischlin and J.-S. Coron (Eds.): EUROCRYPT 2016, Part I, LNCS 9665, pp. 294–310, 2016.
DOI: 10.1007/978-3-662-49890-3_12

algorithm). For all applications we are aware of, computational ciphertext sanitizability suffices. Nevertheless, all known approaches (including ours) provide statistical ciphertext sanitizability.

IMPORTANCE OF CIPHERTEXT SANITIZABILITY. The ciphertext sanitizability property is closely related to the concept of (honest-but-curious) *circuit privacy.* The latter was introduced in the context of FHE by Gentry (see [Gen09a, Chapter 2]). Ciphertext sanitizability implies that if C_0 and C_1 are respectively obtained by the homomorphic evaluation of circuits $\mathcal{C}_0$ and $\mathcal{C}_1$ on honestly formed public key and ciphertexts, and if they decrypt to the same plaintext, then their distributions should be indistinguishable. This property is convenient in the following context: a first user wants a second user to apply a circuit on its plaintexts, but the first user wants to retain privacy of its plaintexts, while the second user wants to retain privacy of its circuit. A circuit private FHE with compact ciphertexts leads to a 2-flow protocol with communication cost bounded independently of the circuit size (this is not the case when directly using Yao's garbled circuit). The communication cost is proportional to the ciphertext bit-size and the number of data bits owned by the first user.

Two other potential applications of ciphertext sanitizability are mentioned in Sect. 5.

FLOODING-BASED CIPHERTEXT SANITIZABILITY. The only known approach to realize ciphertext sanitizability, already described in [Gen09a, Chapter 21], is via the *noise flooding* technique (also called noise smudging and noise drowning). The ciphertexts of existing FHE schemes all contain a noise component, which grows (with respect to the Euclidean norm) and whose distribution gets skewed with homomorphic evaluations. Assume that at the end of the computation, its norm is below some bound B. The noise flooding technique consists in adding a statistically independant noise with much larger standard deviation. This may be done publicly by adding an encryption of plaintext 0 with large noise. The mathematical property that is used to prove ciphertext sanitizability is that the statistical distance between the uniform distribution over $[-B', B']$ and the uniform distribution over $[-B' + c, B' + c]$ for c such that $|c| \leq B$ is $\leq B/B'$ (see [AJL+12]). In the context of noise flooding, the parameter B' is taken of the order of $B \cdot 2^\lambda$, where λ refers to the security parameter, so that the statistical distance is exponentially small[1].

The noise flooding technique results in impractical schemes. To enable correct decryption, the scheme must tolerate much larger noise components: up to magnitude $B \cdot 2^\lambda$ instead of B, where B can be as small as $\lambda^{O(1)}$. In the case of schemes based on the Learning With Errors problem (LWE) [Reg09], the encryption noise rate α must be set exponentially small as a function of λ, to guarantee decryption correctness. Then, to ensure IND-CPA security against all known attacks costing $2^{o(\lambda)}$ operations, the LWE dimension n and modulus q

[1] Note that in some works, it is only required that $\sigma \geq B \cdot \lambda^{\omega(1)}$. These works consider resistance only against polynomial-time attackers. Here we consider the more realistic setting where attackers can have up to sub-exponential run-time $2^{o(\lambda)}$.

must satisfy the condition $n \log q \geq \lambda^3$ up to poly-logarithmic factors in λ (lattice reduction algorithms [Sch87] may be used to solve LWE with parameters n, q and α in time $2^{n \log q / \log^2 \alpha}$ up to polylogarithmic factors in the exponent). This impacts key sizes, ciphertext expansion, and efficiency of encryption, decryption and homomorphic evaluation. For example, a ciphertext from the Brakerski-Vaikuntanathan FHE [BV11a] would have bit-size $O(n \log q) = \widetilde{O}(\lambda)$ if there is no need to support noise flooding, and $O(n \log q) = \widetilde{O}(\lambda^3)$ if it is to support noise flooding. A related impact is that the weakest hardness assumption on lattice problems allowing to get ciphertext sanitizability via noise flooding is the quantum hardness of standard worst case lattice problems such as SVP with approximation factors of the order of $2^{\sqrt{n}}$ in dimension n (this is obtained via the quantum reduction of [Reg09]).

CONTRIBUTION. We propose a novel approach to realize the ciphertext sanitizability property, based on successive iterations of bootstrapping. In short, we replace the *flooding* strategy by a *soak-spin-repeat* strategy. It allows to take much smaller parameters (both in practice and in theory) and to rely on less aggressive hardness assumptions. In the case of LWE-based FHE schemes such as [BV11a,BV11b,Bra12,GSW13], the proposed scheme modification to realize ciphertext sanitizability allows to keep the same underlying hardness assumption (up to a small constant factor in the lattice approximation parameter) as for basic FHE without ciphertext sanitizability, and the same parameters (up to a small constant factor). On the downside, sanitizing a ciphertext requires successive iterations of bootstrapping. Note that the cost of bootstrapping has been recently decreased [AP14,DM15].

FHE bootstrapping consists in encrypting an FHE ciphertext under a second encryption layer, and removing the inner encryption layer by homomorphically evaluating the decryption circuit. If a ciphertext c decrypts to a plaintext μ, bootstrapping produces a ciphertext c' that also decrypts to μ, as if c was decrypted to μ and then μ re-encrypted to c'. The latter simplification is misleading, as one may think that c' is a fresh encryption of μ and hence that its distribution is canonical. This is incorrect. Homomorphic evaluation results in a ciphertext whose distribution may depend on the plaintexts underlying the input ciphertexts. In the context of bootstrapping, the input plaintexts are the bits of the decryption key and the bits of c. The distribution of ciphertext c' output by bootstrapping depends on the distribution of c.

Rather, we propose to bootstrap several times and inject some entropy in the ciphertext between each bootstrapping step. Suppose we start with two ciphertexts c_0 and c_1 decrypting to the same plaintext μ. We randomize them by adding a fresh encryption of 0. After a bootstrapping step, we obtain ciphertexts $c_0^{(1)}$ and $c_1^{(1)}$ decrypting to μ. By the data processing inequality, the statistical distance between them is no greater than before the bootstrapping. We then inject entropy in $c_0^{(1)}$ and $c_1^{(1)}$ to decrease their statistical distance by a constant factor, e.g., by a factor 2: this is achieved by adding a fresh encryption of 0. This process is iterated λ times, resulting in a pair of ciphertexts decrypting to μ

and whose statistical distance is $\leq 2^{-\lambda}$. The process is akin to a dynamical system, approaching to a fixed point, canonical, distribution. This technique almost provides a solution to a problem suggested by Gentry in [Gen09a, page 30], stating that bootstrapping could imply circuit privacy.

It remains to explain how to realize the entropy injection step, whose aim is to decrease the statistical distance between the two ciphertexts by a constant factor. In the case of FHEs with a noise component, we use a *tiny flooding*. We add a fresh independent noise to the noise component, by publicly adding a fresh encryption of plaintext 0 to the ciphertext. As opposed to traditional flooding, this noise term is not required to be huge, as we do not aim at statistical closeness in one go. Both noise terms (the polluted one and the fresh one) may be of the same orders of magnitude.

COMPARISON WITH OTHER APPROACHES. We have already mentioned that in the case of FHE schemes based on LWE, the flooding based approach requires assuming that LWE with noise rate $\alpha = 2^{-\lambda}$ is hard, and hence setting $n \log q \geq \lambda^3$ (up to poly-logarithmic factors in λ). The inefficacy impact can be mitigated by performing the homomorphic evaluation of the circuit using small LWE parameters, bootstrapping the resulting ciphertext to large LWE parameters, flooding with noise and then bootstrapping to small parameters (or, when it is feasible, switching modulus) before transmitting the result. This still involves one bootstrapping with resulting LWE parameters satisfying $n \log q \geq \lambda^3$. Our approach compares favorably in terms of sanitization efficiency, as it involves λ bootstrapping with parameters satisfying $n \log q \geq \lambda$ (still up to polylogarithmic factors).

In the context of (honest-but-curious) circuit privacy with communication bounded independently of the circuit size, van Dijk *et al.* [DGHV10, Appendix C] suggested using an FHE scheme and, instead of sending back the resulting ciphertext c, sending a garbling of a circuit taking as input the secret key and decrypting c. Using Yao's garbled circuit results in a communication cost that is at least λ times larger than the decryption circuit, which is itself at least linear in the ciphertext bit-length. Therefore, our approach compares favorably in terms of communication.

RELATED WORKS. In [OPP14], Ostrovsky *et al.* study circuit privacy in the malicious setting: circuit privacy (or ciphertext sanitizability) must hold even if the public key and ciphertexts are not properly generated. This is a stronger property than the one we study in the present work. Ostrovsky *et al.* combine a compact FHE and a (possibly non-compact) homomorphic encryption scheme that enjoys circuit privacy in the malicious setting, to obtain a compact FHE which is maliciously circuit private. Their construction proceeds in two steps, and our work can be used as an alternative to the first step.

Noise flooding is a powerful technique to obtain new functionalities and security properties in lattice-based cryptography. As explained above, however, it leads to impractical schemes. It is hence desirable to find alternatives that allow for more efficient realizations of the same functionalities. For example, Lyubashesvky [Lyu09] used rejection sampling in the context of signatures (see also [Lyu12, DDLL13]). Alwen *et al.* [AKPW13] used the lossy mode of LWE to prove hardness of the Learning With Rounding problem (LWR) for smaller parameters than [BPR12]. LWR is for example used to designing pseudo-random functions [BPR12, BLMR13, BP14]. Langlois *et al.* [LSS14] used the Rényi divergence as an alternative to the statistical distance to circumvent noise flooding in encoding re-randomization for the Garg *et al.* cryptographic multi-linear map candidate [GGH13].[2] Further, in [BLP+13], Brakerski *et al.* introduced the first-is-errorless LWE problem to prove hardness of the Extended LWE problem without noise flooding, hence improving over a result from [OPW11]. They also gave a flooding-free hardness proof for binary LWE based on the hardness of Extended LWE, hence improving a hardness result from [GKPV10]. LWE with binary secrets was introduced to construct a leakage resilient encryption scheme [GKPV10]. Extended LWE was introduced to design a bi-deniable encryption scheme [OPW11], and was also used in the context of encryption with key-dependent message security [AP12]. The tools developed to circumvent noise flooding seem quite diverse, and it is unclear whether a general approach could be used.

ROADMAP. In Sect. 2, we provide some necessary reminders. In Sect. 3, we describe our ciphertext sanitation procedure. We instantiate our approach to LWE-based FHE schemes in Sect. 4.

2 Preliminaries

We give some background definitions and properties on Fully Homomorphic Encryption and probability distributions.

2.1 Fully Homomorphic Encryption

We let S denote the set of secret keys, P the set of public keys (which, in our convention includes what is usually referred to as the evaluation key), C the ciphertext space and M the message space. For simplicity, we set $M = \{0, 1\}$. Additionally, we let C_μ denote the set of all ciphertexts that decrypt to $\mu \in M$ (under an implicitly fixed secret key $sk \in S$). We also assume that every ciphertext decrypts to a message: $C = \bigcup_{\mu \in M} C_\mu$ (i.e., decryption never fails). All those sets implicitly depend on a security parameter λ.

An FHE scheme (for S, P, M, C) is given by four polynomial time algorithms:

- a (randomized) key generation algorithm $\mathsf{KeyGen} : \{1^\lambda\} \to P \times S$,
- a (randomized) encryption algorithm $\mathsf{Enc} : P \times M \to C$,
- a (deterministic) decryption algorithm $\mathsf{Dec} : S \times C \to M$,
- a (deterministic) homomorphic evaluation function $\mathsf{Eval} : \forall k, P \times (M^k \to M) \times C^k \to C$.

[2] Note that the Garg *et al.* and hence its Langlois *et al.* improvement have recently been cryptanalysed [HJ15].

Correctness requires that for any input circuit $\mathcal{C}$ with any number of input wires k, and for any $\mu_1, \ldots, \mu_k \in \{0,1\}$, we have (with overwhelming probability $1 - \lambda^{-\omega(1)}$ over the random coins used by the algorithms):

$$\mathsf{Dec}\left(sk, \mathsf{Eval}(pk, \mathcal{C}, (c_1, \ldots, c_k))\right) = \mathcal{C}(\mu_1, \ldots, \mu_k),$$

where $(pk, sk) = \mathsf{KeyGen}(1^\lambda)$ and $c_i = \mathsf{Enc}(pk, \mu_i)$ for all $i \leq k$.

Compactness requires that elements in C can be stored on $\lambda^{O(1)}$ bits.

Indistinguishability under chosen plaintext attacks (IND-CPA) requires that given pk (where $(pk, sk) = \mathsf{KeyGen}(1^\lambda)$), the distributions of $\mathsf{Enc}(pk, 0)$ and $\mathsf{Enc}(pk, 1)$ are computationally indistinguishable.

In addition to the above four algorithms, we define the function

$$\mathsf{Refresh}(pk, c) = \mathsf{Eval}\left(pk, \mathcal{C}_{\mathsf{Dec}}, (bk_1, \ldots, bk_k, c_1', \ldots, c_\ell')\right),$$

where $\mathcal{C}_{\mathsf{Dec}}$ refers to a polynomial-size circuit implementing Dec, $bk_i = \mathsf{Enc}(pk, sk_i)$ for all k bits sk_i of secret key sk, and $c_i' = \mathsf{Enc}(pk, c_i)$ for all ℓ bits c_i of ciphertext c. Note that $\mathsf{Refresh}$ is the typical bootstrapping step of current FHE constructions.

We assume that the bk_i's are given as part of pk, and do not impact IND-CPA security of the FHE scheme. This circular security assumption is standard in the context of FHE. We may circumvent it by using a sequence of key pairs (pk_j, sk_j) and encrypting the bits of sk_j under pk_{j+1} for all j. This drastically increases the bit-size of pk and does not provide FHE per say, but only homomorphic encryption for circuits of size bounded by any a priori known polynomial.

2.2 Properties of the Statistical Distance

For a probability distribution $\mathcal{D}$ over a countable set $\mathcal{S}$, we let $\mathcal{D}(x)$ denote the weight of $\mathcal{D}$ at x, i.e., $\mathcal{D}(x) = \Pr[\tilde{x} = x | \tilde{x} \leftarrow \mathcal{D}]$.

Let X and X' be two random variables taking values in a countable set $\mathcal{S}$. Let $\mathcal{D}$ and $\mathcal{D}'$ be the probability distributions of X and X'. The statistical distance $\Delta(X, X')$ is defined by

$$\Delta(X, X') = \frac{1}{2} \sum_{x \in \mathcal{S}} |\mathcal{D}(x) - \mathcal{D}'(x)|.$$

By abuse of notation, we aso write $\Delta(\mathcal{D}, \mathcal{D}')$. Note that $0 \leq \Delta(X, X') \leq 1$ always holds.

Assuming that $\delta = \Delta(X, X') < 1$, the intersection distribution $\mathcal{C} = \mathcal{D} \cap \mathcal{D}'$ is defined over $\mathcal{S}$ by $\mathcal{C}(x) = \frac{1}{1-\delta} \min(\mathcal{D}(x), \mathcal{D}'(x))$. It may be checked that $\mathcal{C}$ is indeed a distribution (i.e., $\sum_{x \in \mathcal{S}} \mathcal{C}(x) = 1$), by using the following identity, holding for any reals a and b: $2\min(a, b) = a + b - |a - b|$. We also define the mixture of two distributions $\mathcal{B} = \alpha \cdot \mathcal{D} + (1 - \alpha) \cdot \mathcal{D}'$ for $0 \leq \alpha \leq 1$ by $\mathcal{B}(x) = \alpha \cdot \mathcal{D}(x) + (1 - \alpha) \cdot \mathcal{D}'(x)$. If X and X' are random variables with distributions $\mathcal{D}$ and $\mathcal{D}'$ respectively, then $\mathcal{B}$ is the density function of the random variable obtained with the following experiment: sample a bit from the Bernoulli distribution giving probability α to 0; if the bit is 0, then return a sample from X; if the bit is 1, then return a sample from X'.

We will use the following two lemmas.

Lemma 2.1. *For any $\delta \in [0,1)$ and any distributions $\mathcal{B}, \mathcal{B}'$ such that $\delta \geq \Delta(\mathcal{B}, \mathcal{B}')$, there exist two distributions $\mathcal{D}$ and $\mathcal{D}'$ such that:*

$$\mathcal{B} = (1-\delta)\cdot \mathcal{B}\cap\mathcal{B}' + \delta\cdot\mathcal{D} \quad \textit{and} \quad \mathcal{B}' = (1-\delta)\cdot \mathcal{B}\cap\mathcal{B}' + \delta\cdot\mathcal{D}'.$$

Proof. Let $\mathcal{C} = \mathcal{B}\cap\mathcal{B}'$. One builds $\mathcal{D}$ as the renormalization to sum 1 of the non-negative function $\mathcal{B}(x) - (1-\delta)\cdot\mathcal{C}(x)$, and proceeds similarly for $\mathcal{D}'$. □

Lemma 2.2. *For any $\alpha \in [0,1]$ and any distributions $\mathcal{C}, \mathcal{D}, \mathcal{D}'$, we have*

$$\Delta\left((1-\alpha)\cdot\mathcal{C} + \alpha\cdot\mathcal{D}, (1-\alpha)\cdot\mathcal{C} + \alpha\cdot\mathcal{D}'\right) \;=\; \alpha\cdot\Delta(\mathcal{D},\mathcal{D}').$$

Proof. Let $\mathcal{B} = (1-\alpha)\mathcal{C} + \alpha\mathcal{D}$ and $\mathcal{B}' = (1-\alpha)\mathcal{C} + \alpha\mathcal{D}'$. We derive

$$\begin{aligned} 2\cdot\Delta(\mathcal{B},\mathcal{B}') &= \sum |((1-\alpha)\mathcal{C}(x) + \alpha\mathcal{D}(x)) - ((1-\alpha)\mathcal{C}(x) + \alpha\mathcal{D}'(x))| \\ &= \sum |\alpha\mathcal{D}(x) - \alpha\mathcal{D}'(x)| \\ &= 2\alpha\cdot\Delta(\mathcal{D},\mathcal{D}'). \end{aligned}$$

This completes the proof. □

The following lemma is at the core of our main result. It states that if applying a randomized function f to any two inputs $a, b \in \mathcal{S}$ leads to two somewhat close-by distributions, then iterating f several times provides extremely close distributions.

Lemma 2.3. *Let $\delta \in [0,1]$ and $f : \mathcal{S} \to \mathcal{S}$ be a randomized function such that $\Delta(f(a), f(b)) \leq \delta$ holds for all $a, b \in \mathcal{S}$. Then:*

$$\forall k \geq 0, \forall a, b \in \mathcal{S},\; \Delta(f^k(a), f^k(b)) \leq \delta^k.$$

Proof. We prove the result by induction on $k \geq 0$. It trivially holds for $k = 0$. We now assume that $\Delta(f^k(a), f^k(b)) \leq \delta^k$ holds for all $a, b \in \mathcal{S}$ and some $k \geq 0$, and aim at showing that $\Delta(f^{k+1}(a), f^{k+1}(b)) \leq \delta^{k+1}$.

By Lemma 2.1, there exist two distributions $\mathcal{D}$ and $\mathcal{D}'$ such that:

$$\begin{aligned} f^k(a) &= (1-\delta^k)\cdot f^k(a)\cap f^k(b) + \delta^k\cdot\mathcal{D}, \\ f^k(b) &= (1-\delta^k)\cdot f^k(a)\cap f^k(b) + \delta^k\cdot\mathcal{D}'. \end{aligned}$$

By composing with f, we obtain that:

$$\begin{aligned} f^{k+1}(a) &= (1-\delta^k)\cdot f(f^k(a)\cap f^k(b)) + \delta^k\cdot f(\mathcal{D}), \\ f^{k+1}(b) &= (1-\delta^k)\cdot f(f^k(a)\cap f^k(b)) + \delta^k\cdot f(\mathcal{D}'). \end{aligned}$$

Now, Lemma 2.2 implies that

$$\Delta(f^{k+1}(a), f^{k+1}(b)) = \delta^k \cdot \Delta(f(\mathcal{D}), f(\mathcal{D}')).$$

To complete the proof, note that

$$\begin{aligned}\Delta(f(\mathcal{D}), f(\mathcal{D}')) &= \sum_{x\in\mathcal{S}} \Big| \sum_{a'\in\mathcal{S}} \mathcal{D}(a') \Pr_f[f(a') = x] - \sum_{b'\in\mathcal{S}} \mathcal{D}'(b') \Pr_f[f(b') = x]\Big| \\ &= \sum_{x\in\mathcal{S}} \Big| \sum_{a',b'\in\mathcal{S}} \mathcal{D}(a')\mathcal{D}'(b') \big[\Pr_f[f(a') = x] - \Pr_f[f(b') = x]\big]\Big| \\ &\le \sum_{a',b'\in\mathcal{S}} \mathcal{D}(a')\mathcal{D}'(b') \Big| \sum_{x\in\mathcal{S}} \big[\Pr_f[f(a') = x] - \Pr_f[f(b') = x]\big]\Big|.\end{aligned}$$

The latter quantity is $\le \delta$, by assumption. □

3 Sanitization of Ciphertexts

We first formally state the correctness and security requirements of a sanitization algorithm for an encryption scheme (KeyGen, Enc, Dec) with secret key space S, public key space P, message space M and ciphertext space C.

Definition 3.1 (Sanitization Algorithm). *A polynomial-time (randomized) algorithm* Sanitize : $P \times C \to C$ *is said to be message-preserving if the following holds with probability* $\ge 1 - \lambda^{-\omega(1)}$ *over the choice of* $(pk, sk) = \mathsf{KeyGen}(1^\lambda)$*:*

$$\forall c \in C : \mathsf{Dec}(sk, \mathsf{Sanitize}(pk, c)) = \mathsf{Dec}(sk, c).$$

It is said (statistically) sanitizing if the following holds with probability $\ge 1-2^{-\lambda}$ *over the choice of* $(pk, sk) = \mathsf{KeyGen}(1^\lambda)$*: for all* $c, c' \in C$ *such that* $\mathsf{Dec}(sk, c) = \mathsf{Dec}(sk, c')$*, we have*

$$\Delta\big(\mathsf{Sanitize}(pk, c)|(pk, sk), \mathsf{Sanitize}(pk, c')|(pk, sk)\big) \le 2^{-\lambda}.$$

In what follows, we fix the key pair $(pk, sk) = \mathsf{KeyGen}(1^\lambda)$ and assume it is given. To simplify notations, we will omit the conditioning of distributions $\mathsf{Sanitize}(pk, c)$ and $\mathsf{Sanitize}(pk, c')$ by (pk, sk).

3.1 Generic Algorithm

For each $\mu \in M$, we let C^*_μ denote $\mathsf{Refresh}(pk, C_\mu)$.[3] We assume that one may build an efficient randomized algorithm Rerand : $P \times C \mapsto C$ such that

$$c \in C^*_\mu \;\Rightarrow\; \mathsf{Rerand}(pk, c) \in C_\mu. \qquad (1)$$

[3] To give intuition, note that in our LWE instantiation, the set C^*_μ will correspond to low-noise ciphertexts decrypting to μ.

We choose a cycle parameter $\kappa > 0$ as an implicit function of λ. We now define

$$\mathsf{Wash} : (pk, c) \mapsto \mathsf{Rerand}(pk, \mathsf{Refresh}(pk, c)),$$

and $\mathsf{Sanitize}(pk, c)$ as the κ-th iteration of $(pk, c) \mapsto \mathsf{Wash}(pk, c)$. The following statement follows from the definitions.

Lemma 3.2 (Sanitize is Message-Preserving). *Under assumption (1), algorithms* Wash *and* Sanitize *are message-preserving.*

In practical FHEs, implication (1) would typically only hold with overwhelming probability $1 - \lambda^{-\omega(1)}$ over the random coins used during key generation, encryption and execution of Rerand: guaranteeing that those bounds always hold requires larger parameters, leading to slightly worse practical performance. If so, the membership $\mathsf{Sanitize}(pk, c) \in C_\mu$ of Lemma 3.2 holds only with overwhelming probability. This impacts our main result, Theorem 3.3 below, as follows: the statistical distance bound becomes

$$\Delta\left(\mathsf{Sanitize}(pk, c), \mathsf{Sanitize}(pk, c')\right) \leq \delta^\kappa + \kappa \cdot \lambda^{-\omega(1)}.$$

Such a bound does not allow to prove that all sub-exponential attacks can be prevented. To obtain this, one can increase the scheme parameters a little to enable correct decryption with probability $\geq 1 - 2^{-\Omega(\lambda)}$. Then the statistical distance bound of Theorem 3.3 becomes

$$\Delta\left(\mathsf{Sanitize}(pk, c), \mathsf{Sanitize}(pk, c')\right) \leq \delta^\kappa + \kappa \cdot 2^{-\Omega(\lambda)},$$

hence providing security against all sub-exponential attackers.

3.2 Security

Note that the trivial case $C_\mu = C^*_\mu$ and $\mathsf{Rerand}(pk, \cdot) = \mathrm{Id}$ with Refresh replaced by the identity map fits our assumptions, but is exactly the possibly non-sanitized initial scheme. For security, we require that $\mathsf{Rerand}(pk, c)$ does introduce some ambiguity about c, but unlike the previous flooding-based techniques, we do not require that it completely updates the distribution of c. More precisely:

Theorem 3.3 (Sanitization Security). *Assume that (1) holds, and that*

$$\forall \mu \in M, \forall c, c' \in C^*_\mu,\ \Delta\left(\mathsf{Rerand}(pk, c), \mathsf{Rerand}(pk, c')\right) \leq \delta$$

for some constant $\delta \in [0, 1]$. *Then*

$$\Delta\left(\mathsf{Sanitize}(pk, c), \mathsf{Sanitize}(pk, c')\right) \leq \delta^\kappa.$$

In particular if $\delta^\kappa \leq 2^{-\lambda}$, *then* Sanitize *is statistically sanitizing.*

Proof. The result is obtained by applying Lemma 2.3, with $S = C^*_\mu$, $k = \kappa$ and f set to $c \mapsto \mathsf{Rerand}(pk, c)$. □

4 Application to Some FHE Schemes

We now apply our technique to LWE-based schemes built upon Regev's encryption scheme [Reg09]. These include the schemes following the designs of [BV11a] and [GSW13]. We comment practical aspects for HElib [HS] and FHEW [DM].

Our technique can also be applied to Gentry's original scheme and its variants [Gen09a, Gen09b, Gen10, SV10, SS10]. It may also be applied to the FHE scheme "based on the integers" of van Dijk *et al.* [DGHV10] and its improvements (see [CS15] and references therein).

4.1 Rerandomizing a Regev Ciphertext

We let $\mathrm{LWE}_{\boldsymbol{s}}^q(\mu, \eta)$ denote the set of LWE-encryptions of $\mu \in M$ under key $sk = \boldsymbol{s} \in \mathbb{Z}_q^n$ with modulus q and error rate less than η, i.e., the set

$$\mathrm{LWE}_{\boldsymbol{s}}^q(\mu, \eta) = \left\{(\boldsymbol{a}, \langle \boldsymbol{a}, \boldsymbol{s}\rangle + \mu \cdot \lfloor q/2 \rfloor + e) \in \mathbb{Z}_q^{n+1} \text{ such that } |e| < \eta q\right\}.$$

One may recover μ from an element $(\boldsymbol{c}_1, c_2)$ from $\mathrm{LWE}_{\boldsymbol{s}}^q(\mu, \eta)$ by looking at the most significant bit of $c_2 - \langle \boldsymbol{c}_1, \boldsymbol{s}\rangle \mod q$. Correctness of decryption is ensured up to $\eta < 1/4$.

We assume that the public key pk contains $\ell = O(n \log q)$ encryptions of 0, called rerandomizers:

$$\forall i \leq \ell,\ r_i = (\boldsymbol{a}_i, b_i = \langle \boldsymbol{a}, \boldsymbol{s}\rangle + e_i) \in \mathrm{LWE}_{\boldsymbol{s}}^q(0, \eta).$$

We also assume that the $\boldsymbol{a}_i$'s are uniform and independent (they have been freshly sampled).

For a ciphertext $c \in \mathrm{LWE}_{\boldsymbol{s}}^q(\mu, \eta)$, we may now define

$$\mathsf{Rerand}(pk, c) = c + \sum_i \varepsilon_i r_i + (\boldsymbol{0}, f),$$

where the ε_i's are uniformly and independently sampled from $\{0, \pm 1\}$, and f is sampled uniformly in an interval $[-B, B]$ for some B to be chosen below. By an appropriate version of the leftover hash lemma (see, e.g., [Reg09, Section 5]), writing

$$c' = c + \sum_i \varepsilon_i r_i = (\boldsymbol{a}', \langle \boldsymbol{a}', \boldsymbol{s}\rangle + \mu \lfloor q/2 \rfloor + e'),$$

we know that $\boldsymbol{a}'$ is (within exponentially small distance from) uniform in $\mathbb{Z}_q^n$, independently of c. That is, the only information about c contained in c' is carried by e' (and plaintext μ). Additionally, we have that $|e'| < (\ell + 1) \cdot \eta \cdot q$.

To conclude, it remains to remark that for any $x, y \in [-(\ell+1)\eta q, (\ell+1)\eta q]$, we have:

$$\Delta(x + U([-B, B]), y + U([-B, B])) \leq \frac{(\ell+1)\eta q}{B} =: \delta.$$

Therefore, for any $c_0, c_1 \in \mathrm{LWE}_s^q(\mu, \eta)$, it holds that

$$\Delta(\mathsf{Rerand}(pk, c_0), \mathsf{Rerand}(pk, c_1)) \leq \delta,$$

and that

$$\mathsf{Rerand}(pk, c_0), \mathsf{Rerand}(pk, c_1) \in \mathrm{LWE}_s^q(\mu, \frac{(\delta+1)B}{q}).$$

To ensure the correctness of decryption after rerandomization, we may set the parameters so that $(\delta+1)B/q < 1/4$.

4.2 Application to FHE à la [BV11a]

For simplicity, we only present the case of the (non-ring) LWE-based FHE scheme of [BV11a].

Let us first recall how an FHE scheme is bootstrapped from a given SHE scheme. Assume the SHE scheme supports the homomorphic evaluation of any (binary) circuit of multiplicative depth f, and that the decryption operation can be implemented by a circuit of multiplicative depth $g < f$. The SHE scheme is bootstrapped to an FHE scheme using the Refresh function, and evaluates sub-circuit of depth $f - g \geq 1$ between each refreshing procedure.

The construction of the SHE from [BV11a] is made more efficient by the use of modulus switching. This induces a leveled ciphertext-space: for each $i \leq f$, the ciphertext space C^i is $\mathrm{LWE}_s^{q_i}(\cdot, \eta)$ for a sequence of $q_0 > q_1 > \cdots > q_f$ and a fixed $\eta < 1/4$.

The modulus switching technique allows, without any key material, to map $\mathrm{LWE}_s^q(\mu, \eta)$ to $\mathrm{LWE}_s^{q'}(\mu, \eta')$ where $\eta' = \eta + n \cdot (\log\, n)^{O(1)}/q'$ (or even $\eta + \sqrt{n} \cdot (\log\, n)^{O(1)}/q'$ allowing up to negligible probability of incorrect computation).

By sequentially applying so-called ciphertext tensoring, key switching and modulus switching steps, one may compute—given appropriate key material—a ciphertext $c'' \in \mathrm{LWE}_s^{q_{i+1}}(\mu\mu', \eta)$ from two ciphertexts $c \in \mathrm{LWE}_s^{q_i}(\mu, \eta)$ and $c' \in \mathrm{LWE}_s^{q_i}(\mu', \eta)$, on the condition that $q_{i+1}/q_i \geq n \cdot (\log\, n)^{O(1)}$.

Technically, the Refresh function may only be applied to ciphertext $c \in C^f$, as the naive decryption of ciphertexts with a large modulus $q_i > q_f$ could require larger multiplicative depth. To extend Refresh over the whole ciphertext space, one can switch the modulus to the last level beforehand, which, for appropriate parameters q_i's does not affect the error bound.

Instantiating Rerand. Let $C_\mu^g = \mathrm{LWE}_s^{q_g}(\mu, \eta)$. Then, according to the description above, we have $C_\mu^* = \mathsf{Refresh}(pk, C_\mu) \subseteq C_\mu^g$. We use the Rerand function described in Sect. 4.1, with $q = q_g$.

To ensure the correctness of the whole scheme, it suffices that

$$(\eta(\ell+1) + B/q_g) + n(\log n)^{O(1)}/q_f < 1/4.$$

Setting $B \geq 2\eta(\ell+1)q_g$, $\eta < 1/(8(\ell+1))$ and $q_f \geq 8n^{1+o(1)}$ allows to fulfill the conditions of Theorem 3.3 for some $\delta \leq 1/2$.

A larger gap $q_f/q_g > n^{f-g}$ is beneficial to our sanitizing technique, as it allows one to choose $\delta \approx 1/n^{f-g-1}$, and therefore decrease the length κ of the washing program: soaking in a large bucket makes the soak-spin-repeat program shorter. A striking example is given below.

Application to HElib. It turns out that the parameters given in the bootstrappable version of HElib [HS15] lead to $\kappa = 1$ or 2, which means that, in this setting, the flooding strategy is, or almost is, already applicable. Indeed, choosing for example the set of parameters corresponding to $n = \phi(m) = 16384$, we have $f = 22$ and $f - g = 10$. The parameters q_f and q_g are not given, yet it is typical to have $q_{i+1}/q_i = \sqrt{n} \cdot (\log\ n)^{O(1)}$ (guaranteeing correctness only with probability $1 - n^{-\omega(1)}$). We can therefore assume that a single soaking step may achieve $\delta \leq n/\sqrt{n}^{f-g} \approx 2^{-14 \cdot 10/2+14} = 2^{-56}$. This gives, according to [HS15] a batch sanitization procedure of 720 ciphertexts in 500 to 1000 s with the current software [HS15,HS] (on an IntelX5570 processor at 2.93 GHz, with a single thread).

4.3 Application to FHEW

Because the constructions à la [BV11a] rely on the hardness of LWE with inverse noise rate $2^{(\log n)^c}$ for some $c > 1$ in theory (and necessarily larger than $\sqrt{n}^f \approx 2^{14 \cdot 22/2} = 2^{154}$ in practice), it is not so surprising that the implementations allow to straightforwardly apply the flooding strategy in practice (which theoretically requires assuming the hardness of LWE with inverse noise rate $2^{\sqrt{n}}$). It is therefore more interesting to study our sanitization strategy for FHE schemes based on the hardness of LWE with inverse polynomial noise rates [GSW13,BV14,AP14], in particular the concrete instantiation FHEW proposed in [DM15]. For comparison, the security of this scheme is based on a (Ring)-LWE problem [LPR10] with inverse noise rate $\approx 2^{32}$.

Warning. The following analysis is only given as an attempt to estimate the practical cost of our technique, yet the application with the original parameters of FHEW is not to be considered secure. Indeed, for efficiency purposes, the authors [DM15] have chosen to guarentee correctness only *heuristically*, and with a rather large failure probability $\approx 2^{-45}$. Because decryption correctness is essential in our argument (see remark at the end of Sect. 3.1), a serious implementation should first revise the parameters to *provably* ensure decryption correctness with *higher probability*.

Sanitizing FHEW. We proceed to modify the original scheme recalled in Fig. 1 to implement the sanitizing strategy, as described in Fig. 2. This scheme uses two plaintext moduli $t = 2, 4$, and this extends the definition of LWE ciphertexts as follows.

$$\mathrm{LWE}_{\boldsymbol{s}}^{t:q}(\mu, \eta) = \left\{ (\boldsymbol{a}, \langle \boldsymbol{a}, \boldsymbol{s} \rangle + \mu \cdot \lfloor q/t \rfloor + e) \in \mathbb{Z}_q^{n+1} \text{ such that} |e| < \eta q \right\}.$$

Correct decryption now requires $\eta < q/(2t)$. The scheme uses two LWE dimensions: dimension $n = 500$ for a first secret vector $\boldsymbol{s}$, and dimension $N = 1024$

for a second secret vector $\boldsymbol{z}$. It also switches between two ciphertext moduli $q = 2^9$ and $Q = 2^{32}$. According to the analysis from [DM15], the parameters allow to securely encrypt in dimension N and modulus Q, with a (discrete) Gaussian error of standard deviation $\varsigma = 1.4$.

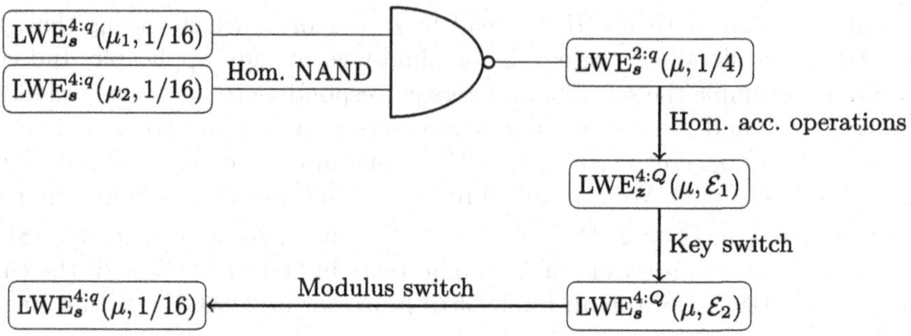

Fig. 1. Original cycle of FHEW

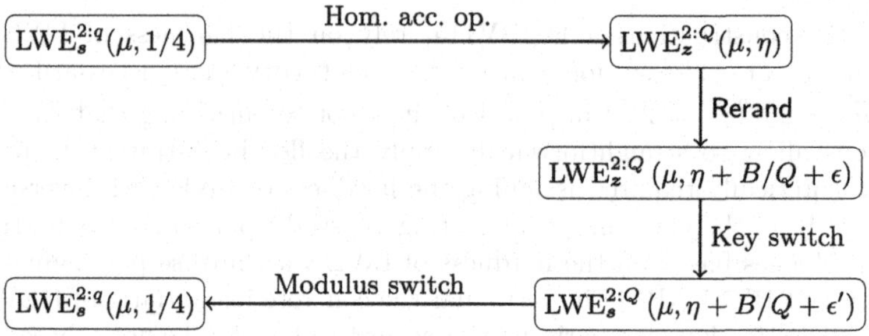

Fig. 2. Washing cycle for FHEW. The only internal modification required is setting $u = Q/4 + 1$ instead of $Q/8 + 1$. See [DM15] for more details.

Following the heuristic central-limit estimate of [DM15], the first step of Fig. 2 (i.e., the homomorphic accumulator operations) returns a ciphertext with a Gaussian-like error of standard deviation $\approx 2^{18}$, so that error is of magnitude less than $Q\eta = 2^{21}$ (with probability $\geq 1 - 2^{-45}$). Also, the choice $\varsigma = 1.4$ makes the error introduced by the key switch negligible. Similarly, the re-randomization of the $\boldsymbol{a}$ part of the ciphertext $c = (\boldsymbol{a}, b)$ using fresh encryption of 0 with error parameter ς given in the public-key ensure that (with notation similar than in the previous section) $b = Q\eta + Q\varepsilon$ where ε is small compared to η.

Not having to compute any NAND also improves the error tolerance from 1/16 to 1/4. We may, in return, introduce a soaking noise of parameter B such that $Bq/Q \approx 3q/16$, that is $B \approx 2^{29}$. This results in $\delta = b/B \approx 2^{-8}$.

In conclusion, setting κ between 8 and 16 (depending on the desired security level) should suffice to achieve appropriate statistical sanitation. This gives sanitization of a single ciphertext in 5 to 10 s with the current software [DM] (on an unspecified Intel processor at 3 Ghz, with a single thread).

5 Conclusion and Open Problems

We have shown that both in theory and in practice, the sanitization of FHE ciphertexts can be performed at a reasonable cost and without substantial modification of current schemes. It remains that FHE is too slow for many real world scenarios and SHE is often much preferable. In a credible scenario where the circuit to evaluate is shallow, with potentially many inputs but few outputs, the best strategy may be to use HElib in SHE mode for the main computation, and sanitizing the final result using FHEW.

When applied to circuit privacy, our approach only provides passive (honest-but-curious) security. Standard (interactive or not) zero-knowledge proofs help prevent malicious attackers using fake public keys and/or fake ciphertexts. Yet ad-hoc techniques surely need to be developed: with public key size of several gigabytes, the statement to be proved is gigantic.

A worthy remark toward this goal, is that malicious ciphertexts are easily tackled once the honest generation of the public key has been established. Indeed, a single Refresh operation on each input ciphertexts will ensure that they are in the subset of valid ciphertexts (formally proving such statement using, e.g., the circuit privacy definition of [OPP14] is rather direct). This strategy may effectively reduce interactivity in secure multi-party computation (MPC) protocols based on FHE, and offer amortization of an initial zero-knowledge proof on the public key.

Ciphertext sanitizability may have further applications in MPC based on FHE, or, more precisely, based on Threshold FHE. Threshold FHE is a variant of FHE in which 1- several parties can execute a key generation protocol to generate a common public key and secret key shares, and 2- to decrypt, the parties execute a decryption protocol using their secret key shares. It is theoretically possible to generically convert any FHE into a Threshold FHE, but this is too cumbersome for practical use: in particular, it results in a significant number of communication rounds. Instead, Threshold FHE schemes have been designed directly by modifying existing FHE schemes [AJL+12, LTV12, CLO+13, CM15, MW15], eventually allowing for MPC in two communication rounds [MW15]. A crucial security property of Threshold FHE, called simulatability of partial decryptions, is that the partial decryptions obtained by individual users do not reveal anything about the confidential data of the other users. Ciphertext sanitization may help enforce this property without resorting to noise flooding.

Acknowledgments. The authors thank Lisa Kohl, Ron Steinfeld and Daniel Wichs for helpful discussions. This work has been supported by an NWO Free Competition Grant and by ERC Starting Grant ERC-2013-StG-335086-LATTAC.

References

[AJL+12] Asharov, G., Jain, A., López-Alt, A., Tromer, E., Vaikuntanathan, V., Wichs, D.: Multiparty computation with low communication, computation and interaction via threshold FHE. In: Pointcheval, D., Johansson, T. (eds.) EUROCRYPT 2012. LNCS, vol. 7237, pp. 483–501. Springer, Heidelberg (2012)

[AKPW13] Alwen, J., Krenn, S., Pietrzak, K., Wichs, D.: Learning with rounding, revisited. In: Canetti, R., Garay, J.A. (eds.) CRYPTO 2013, Part I. LNCS, vol. 8042, pp. 57–74. Springer, Heidelberg (2013)

[AP12] Alperin-Sheriff, J., Peikert, C.: Circular and KDM security for identity-based encryption. In: Fischlin, M., Buchmann, J., Manulis, M. (eds.) PKC 2012. LNCS, vol. 7293, pp. 334–352. Springer, Heidelberg (2012)

[AP14] Alperin-Sheriff, J., Peikert, C.: Faster bootstrapping with polynomial error. In: Garay, J.A., Gennaro, R. (eds.) CRYPTO 2014, Part I. LNCS, vol. 8616, pp. 297–314. Springer, Heidelberg (2014)

[BLMR13] Boneh, D., Lewi, K., Montgomery, H., Raghunathan, A.: Key homomorphic PRFs and their applications. In: Canetti, R., Garay, J.A. (eds.) CRYPTO 2013, Part I. LNCS, vol. 8042, pp. 410–428. Springer, Heidelberg (2013)

[BLP+13] Brakerski, Z., Langlois, A., Peikert, C., Regev, O., Stehlé, D.: Classical hardness of learning with errors. In: Proceedings of STOC, pp. 575–584. ACM (2013)

[BP14] Banerjee, A., Peikert, C.: New and improved key-homomorphic pseudorandom functions. In: Garay, J.A., Gennaro, R. (eds.) CRYPTO 2014, Part I. LNCS, vol. 8616, pp. 353–370. Springer, Heidelberg (2014)

[BPR12] Banerjee, A., Peikert, C., Rosen, A.: Pseudorandom functions and lattices. In: Pointcheval, D., Johansson, T. (eds.) EUROCRYPT 2012. LNCS, vol. 7237, pp. 719–737. Springer, Heidelberg (2012)

[Bra12] Brakerski, Z.: Fully homomorphic encryption without modulus switching from classical GapSVP. In: Safavi-Naini, R., Canetti, R. (eds.) CRYPTO 2012. LNCS, vol. 7417, pp. 868–886. Springer, Heidelberg (2012)

[BV11a] Brakerski, Z., Vaikuntanathan, V.: Efficient fully homomorphic encryption from (standard) LWE. In: Proceedings of FOCS, pp. 97–106. IEEE Computer Society Press (2011)

[BV11b] Brakerski, Z., Vaikuntanathan, V.: Fully homomorphic encryption from ring-LWE and security for key dependent messages. In: Rogaway, P. (ed.) CRYPTO 2011. LNCS, vol. 6841, pp. 505–524. Springer, Heidelberg (2011)

[BV14] Brakerski, Z., Vaikuntanathan, V.: Lattice-based FHE as secure as PKE. In: Proceedings of ITCS, pp. 1–12. ACM (2014)

[CLO+13] Choudhury, A., Loftus, J., Orsini, E., Patra, A., Smart, Nigel P.: Between a rock and a hard place: interpolating between MPC and FHE. In: Sako, K., Sarkar, P. (eds.) ASIACRYPT 2013, Part II. LNCS, vol. 8270, pp. 221–240. Springer, Heidelberg (2013)

[CM15] Clear, M., McGoldrick, C.: Multi-identity and multi-key leveled FHE from learning with errors. In: Gennaro, R., Robshaw, M. (eds.) CRYPTO 2015. LNCS, vol. 9216, pp. 630–656. Springer, Heidelberg (2015)

[CS15] Cheon, J.H., Stehlé, D.: Fully homomophic encryption over the integers revisited. In: Oswald, E., Fischlin, M. (eds.) EUROCRYPT 2015. LNCS, vol. 9056, pp. 513–536. Springer, Heidelberg (2015)

[DDLL13] Ducas, L., Durmus, A., Lepoint, T., Lyubashevsky, V.: Lattice signatures and bimodal gaussians. In: Canetti, R., Garay, J.A. (eds.) CRYPTO 2013, Part I. LNCS, vol. 8042, pp. 40–56. Springer, Heidelberg (2013)

[DGHV10] van Dijk, M., Gentry, C., Halevi, S., Vaikuntanathan, V.: Fully homomorphic encryption over the integers. In: Gilbert, H. (ed.) EUROCRYPT 2010. LNCS, vol. 6110, pp. 24–43. Springer, Heidelberg (2010)

[DM] Ducas, L., Micciancio, D.: Fhew. https://github.com/lducas/FHEW

[DM15] Ducas, L., Micciancio, D.: FHEW: bootstrapping homomorphic encryption in less than a second. In: Oswald, E., Fischlin, M. (eds.) EUROCRYPT 2015. LNCS, vol. 9056, pp. 617–640. Springer, Heidelberg (2015)

[Gen09a] Gentry, C.: A fully homomorphic encryption scheme. PhD thesis, Stanford University (2009) Manuscript available at. http://crypto.stanford.edu/craig

[Gen09b] Gentry, C.: Fully homomorphic encryption using ideal lattices. In: Proceedings of STOC, pp. 169–178. ACM (2009)

[Gen10] Gentry, C.: Toward basing fully homomorphic encryption on worst-case hardness. In: Rabin, T. (ed.) CRYPTO 2010. LNCS, vol. 6223, pp. 116–137. Springer, Heidelberg (2010)

[GGH13] Garg, S., Gentry, C., Halevi, S.: Candidate multilinear maps from ideal lattices. In: Johansson, T., Nguyen, P.Q. (eds.) EUROCRYPT 2013. LNCS, vol. 7881, pp. 1–17. Springer, Heidelberg (2013)

[GHS12] Gentry, C., Halevi, S., Smart, N.P.: Homomorphic evaluation of the AES circuit. In: Safavi-Naini, R., Canetti, R. (eds.) CRYPTO 2012. LNCS, vol. 7417, pp. 850–867. Springer, Heidelberg (2012)

[GKPV10] Goldwasser, S., Kalai, Y.T., Peikert, C., Vaikuntanathan, V.: Robustness of the learning with errors assumption. In: Proceedings of Innovations in Computer Science - ICS, pp. 230–240. Tsinghua University Press (2010)

[GSW13] Gentry, C., Sahai, A., Waters, B.: Homomorphic encryption from learning with errors: conceptually-simpler, asymptotically-faster, attribute-based. In: Canetti, R., Garay, J.A. (eds.) CRYPTO 2013, Part I. LNCS, vol. 8042, pp. 75–92. Springer, Heidelberg (2013)

[HJ15] Hu, Y., Jia, H.: Cryptanalysis of GGH map. IACR Cryptology ePrint Archive **2015**, 301 (2015)

[HS] Halevi, S., Shoup, V.: Helib. https://github.com/shaih/HElib

[HS15] Halevi, S., Shoup, V.: Bootstrapping for HElib. In: Oswald, E., Fischlin, M. (eds.) EUROCRYPT 2015. LNCS, vol. 9056, pp. 641–670. Springer, Heidelberg (2015)

[LPR10] Lyubashevsky, V., Peikert, C., Regev, O.: On ideal lattices and learning with errors over rings. In: Gilbert, H. (ed.) EUROCRYPT 2010. LNCS, vol. 6110, pp. 1–23. Springer, Heidelberg (2010)

[LSS14] Langlois, A., Stehlé, D., Steinfeld, R.: GGHLite: more efficient multilinear maps from ideal lattices. In: Nguyen, P.Q., Oswald, E. (eds.) EUROCRYPT 2014. LNCS, vol. 8441, pp. 239–256. Springer, Heidelberg (2014)

[LTV12] López-Alt, A., Tromer, E., Vaikuntanathan, V.: On-the-fly multiparty computation on the cloud via multikey fully homomorphic encryption. In: Proceedings of STOC, pp. 1219–1234. ACM (2012)

[Lyu09] Lyubashevsky, V.: Fiat-shamir with aborts: applications to lattice and factoring-based signatures. In: Matsui, M. (ed.) ASIACRYPT 2009. LNCS, vol. 5912, pp. 598–616. Springer, Heidelberg (2009)

[Lyu12] Lyubashevsky, V.: Lattice signatures without trapdoors. In: Pointcheval, D., Johansson, T. (eds.) EUROCRYPT 2012. LNCS, vol. 7237, pp. 738–755. Springer, Heidelberg (2012)

[MW15] Mukherjee, P., Wichs, D.: Two round MPC from LWE via multi-key FHE. IACR Cryptology ePrint Archive **2015**, 345 (2015)

[OPP14] Ostrovsky, R., Paskin-Cherniavsky, A., Paskin-Cherniavsky, B.: Maliciously circuit-private FHE. In: Garay, J.A., Gennaro, R. (eds.) CRYPTO 2014, Part I. LNCS, vol. 8616, pp. 536–553. Springer, Heidelberg (2014)

[OPW11] O'Neill, A., Peikert, C., Waters, B.: Bi-deniable public-key encryption. In: Rogaway, P. (ed.) CRYPTO 2011. LNCS, vol. 6841, pp. 525–542. Springer, Heidelberg (2011)

[Reg09] Regev, O.: On lattices, learning with errors, random linear codes, and cryptography. J. ACM **56**(6), 1–40 (2009)

[Sch87] Schnorr, C.P.: A hierarchy of polynomial lattice basis reduction algorithms. Theor. Comput. Sci. **53**, 201–224 (1987)

[SS10] Stehlé, D., Steinfeld, R.: Faster fully homomorphic encryption. In: Abe, M. (ed.) ASIACRYPT 2010. LNCS, vol. 6477, pp. 377–394. Springer, Heidelberg (2010)

[SV10] Smart, N.P., Vercauteren, F.: Fully homomorphic encryption with relatively small key and ciphertext sizes. In: Nguyen, P.Q., Pointcheval, D. (eds.) PKC 2010. LNCS, vol. 6056, pp. 420–443. Springer, Heidelberg (2010)

Towards Stream Ciphers for Efficient FHE with Low-Noise Ciphertexts

Pierrick Méaux[1(✉)], Anthony Journault[2], François-Xavier Standaert[2], and Claude Carlet[3]

[1] INRIA, CNRS, ENS and PSL Research University, Paris, France
Pierrick.Meaux@ens.fr
[2] ICTEAM/ELEN/Crypto Group, Université catholique de Louvain, Louvain-la-Neuve, Belgium
{anthony.journault,fstandae}@uclouvain.be
[3] LAGA, Department of Mathematics, University of Paris VIII and University of Paris XIII, Paris, France
claude.carlet@gmail.com

Abstract. Symmetric ciphers purposed for Fully Homomorphic Encryption (FHE) have recently been proposed for two main reasons. First, minimizing the implementation (time and memory) overheads that are inherent to current FHE schemes. Second, improving the homomorphic capacity, *i.e.* the amount of operations that one can perform on homomorphic ciphertexts before bootstrapping, which amounts to limit their level of noise. Existing solutions for this purpose suggest a gap between block ciphers and stream ciphers. The first ones typically allow a constant but small homomorphic capacity, due to the iteration of rounds eventually leading to complex Boolean functions (hence large noise). The second ones typically allow a larger homomorphic capacity for the first ciphertext blocks, that decreases with the number of ciphertext blocks (due to the increasing Boolean complexity of the stream ciphers' output). In this paper, we aim to combine the best of these two worlds, and propose a new stream cipher construction that allows constant and small(er) noise. Its main idea is to apply a Boolean (filter) function to a public bit permutation of a constant key register, so that the Boolean complexity of the stream cipher outputs is constant. We also propose an instantiation of the filter function designed to exploit recent (3rd-generation) FHE schemes, where the error growth is quasi-additive when adequately multiplying ciphertexts with the same amount of noise. In order to stimulate further investigation, we then specify a few instances of this stream cipher, for which we provide a preliminary security analysis. We finally highlight the good properties of our stream cipher regarding the other goal of minimizing the time and memory complexity of calculus delegation (for 2nd-generation FHE schemes). We conclude the paper with open problems related to the large design space opened by these new constructions.

1 Introduction

Purpose: Calculus Delegation. Recent years have witnessed massive changes in communication technologies, that can be summarized as a combination of

M. Fischlin and J.-S. Coron (Eds.): EUROCRYPT 2016, Part I, LNCS 9665, pp. 311–343, 2016.
DOI: 10.1007/978-3-662-49890-3_13

two trends: (1) the proliferation of small embedded devices with limited storage and computing facilities, and (2) the apparition of cloud services with extensive storage and computing facilities. In this context, the outsourcing of data and the delegation of data processing gains more and more interest. Yet, such new opportunities also raise new security and privacy concerns. Namely, users typically want to prevent the server from learning about their data and processing. For this purpose, Gentry's breakthrough Fully Homomorphic Encryption (FHE) scheme [30] brought a perfect conceptual answer. Namely, it allows applying processing on ciphertexts in a homomorphic way so that after decryption, plaintexts have undergone the same operations as ciphertexts, but the server has not learned anything about these plaintexts.[1]

Application Scenario. Cloud services can be exploited in a plethora of applications, some of them surveyed in [51]. In general, they are always characterized by the aforementioned asymmetry between the communication parties. For illustration, we start by providing a simple example where data outsourcing and data processing delegation require security and privacy. Let us say that a patient, Alice, has undergone a surgery and is coming back home. The hospital gave her a monitoring watch (with limited storage) to measure her metabolic data on a regular basis. And this metabolic data should be made available to the doctor Bob, to follow the evolution of the post-surgery treatment. Quite naturally, Bob has numerous patients and no advanced computing facilities to store and process the data of all his patients. So this is a typical case where sending the data to a cloud service would be very convenient. That is, Alice's data could be sent to and stored on the cloud, and associated to both her and the doctor Bob. And the cloud would provide Bob with processed information in a number of situations such as when the metabolic data of Alice is abnormal (in which case an error message should be sent to Bob), or during an appointment between Alice and Bob, so that Bob can follow the evolution of Alice's data (possibly after some processing). Bob could in fact even be interested by accessing some other patient's data, in order to compare the effect of different medications. And of course, we would like to avoid the cloud to know anything about the (private) data it is manipulating.

Typical Framework. More technically, the previous exemplary application can be integrated in a quite general cloud service application framework, that can be seen as a combination of 5 steps, combining a symmetric encryption scheme and an asymmetric homomorphic encryption scheme, as summarised in Fig. 1 and described next:

1. *Initialization.* Alice runs the key generation algorithms $H.\mathsf{KeyGen}$ and $S.\mathsf{KeyGen}$ of the two schemes, and sends her homomorphic public key pk^H and the homomorphic ciphertext of her symmetric key $\mathbf{C}^H(\mathsf{sk}_i^S)$.

[1] In the remaining of the paper, and when not specified otherwise, the term FHE will also be used for related schemes such as Leveled HE, SomeWhat HE, Scalable HE, *etc.*.

2. *Storage.* Alice encrypts her data m_i with the symmetric encryption scheme, and sends $\mathbf{C}^S(m_i)$ to Claude.
3. *Evaluation.* Claude homomorphically evaluates, with the $H.\mathsf{Eval}$ algorithm, the decryption $\mathbf{C}^H(m_i)$ of the symmetric scheme on Alice's data $\mathbf{C}^S(m_i)$.
4. *Computation.* Claude homomorphically executes the treatment f on Alice's encrypted data.
5. *Result.* Claude sends a compressed encrypted result of the data treatment $\mathbf{c}^H(f(m_i))$, obtained with the $H.\mathsf{Comp}$ algorithm, and Alice decrypts it.

Note that if we assume the existence of a trusted third party active only during the initialization step, Alice can avoid Step 1, which needs a significant computational and memory storage effort. Note also that this framework is versatile: computation can be done in parallel (in a batch setting) or can be turned into a secret key FHE.

	Alice		Claude
1: Initialization	$(\mathsf{sk}^H, \mathsf{pk}^H) \leftarrow H.\mathsf{KeyGen}(\lambda)$ $\mathsf{sk}^S \leftarrow S.\mathsf{KeyGen}(\lambda)$ $\mathbf{C}^H(\mathsf{sk}_i^S) = H.\mathsf{Enc}(\mathsf{sk}_i^S, \mathsf{pk}^H)$	$\xrightarrow{\mathbf{C}^H(\mathsf{sk}_i^S), \mathsf{pk}^H}$	$\mathbf{C}^H(\mathsf{sk}_i^S), \mathsf{pk}^H$
2: Storage	$\mathbf{C}^S(m_i) = S.\mathsf{Enc}(m_i, \mathsf{sk}^S)$	$\xrightarrow{\mathbf{C}^S(m_i)}$	$\mathbf{C}^S(m_i)$
3: Evaluation			$\mathbf{C}^H(m_i)$ $=$ $H.\mathsf{Eval}(S.\mathsf{Dec}(\mathbf{C}^S(m_i), \mathbf{C}^H(\mathsf{sk}_i^S), \mathsf{pk}^H)$
4: Computation	f	$\xrightarrow{f}$	$\mathbf{C}^H(f(m_i)) = H.\mathsf{Eval}(f(\mathbf{C}^H(m_i))$
5: Result	$\mathbf{c}^H(f(m_i))$ $f(m_i) = H.\mathsf{Dec}(\mathbf{c}^H(f(m_i)), \mathsf{sk}^H)$	$\xleftarrow{\mathbf{c}^H(f(m_i))}$	$\mathbf{c}^H(f(m_i)) = H.\mathsf{Comp}(\mathbf{C}^H(f(m_i)))$

Fig. 1. Homomorphic Encryption - Symmetric Encryption framework. H and S respectively refer to homomorphic and symmetric encryption schemes, for algorithms (*e.g.* $H.\mathsf{KeyGen}$) or scheme components (*e.g.* sk^S).

FHE Bottlenecks. The main limitation for the deployment of cloud services based on such FHE frameworks relates to its important overheads, that can be related to two main concerns: computational and memory costs (especially on the client side) and limited homomorphic capacity (*i.e.* noise increase). More precisely:

- The computational and memory costs for the client depend overwhelmingly on the homomorphic encryption and decryption algorithms during the steps 1 and 5. The memory cost is mostly influenced by the homomorphic ciphertexts and public key sizes. Solving these two problems consists in building size-efficient FHE schemes with low computational cost [35,38]. On the server side, this computational cost further depends on the symmetric encryption scheme and function to evaluate.
- The homomorphic capacity relates to the fact that FHE constructions are built on noise-based cryptography, where the unbounded amount of homomorphic operations is guaranteed by an expensive bootstrapping technique. The homomorphic capacity corresponds to the amount of operations doable before the noise grows too much forcing to use bootstrapping. Therefore, and in order to reduce the time and computational cost of the framework, it is important to manage the error growth during the homomorphic operations (*i.e.* steps 3 and 4). Furthermore, since the 4th step is the most important one from the application point-of-view (since this is where the useful operations are performed by the cloud), there is strong incentive to minimize the cost of the homomorphic decryption in the 3rd step.

Previous Works. In order to mitigate these bottlenecks, several works tried to reduce more and more the homomorphic cost of evaluating a symmetric decryption algorithm. First attempts in this direction, which were also used as benchmark for FHE implementations, used the AES for this purpose [15,31]. Various alternative schemes were also considered, all with error and sizes depending on the multiplicative depth of the symmetric encryption scheme, such as BGV [9] and FV [26]. Additional optimizations exploited batching and bitslicing, leading to the best results of performing 120 AES decryptions in 4 minutes [31].

Since the multiplicative depth of the AES decryption evaluation was a restrictive bound in these works, other symmetric encryption schemes were then considered. The most representative attempts in this direction are the family of block ciphers LowMC [1] and the stream cipher Kreyvium [11]. These constructions led to reduced and more suitable multiplicative depths. Yet, and intuitively, these attempts were still limited by complementary drawbacks. First for LowMC, the remaining multiplicative depth remains large enough to significantly reduce the homomorphic capacity (*i.e.* increase the noise). Such a drawback seems to be inherent in block cipher structures where the iteration of rounds eventually leads to Boolean functions with large algebraic degree, which inevitably imply a constant per block but high noise after homomorphic evaluation. For example, ciphers dedicated to efficient masking against side-channel attacks [33,34,52], which share the goal of minimizing the multiplicative complexity, suffer from similar issues and it seems hard to break the barrier of one multiplication per round (and therefore of 12 to 16 multiplications for 128-bit ciphers). Second for Kreyvium, the error actually grows with the number of evaluated ciphertexts, which implies that at some point, the output ciphertexts are too noisy, and cannot be decrypted (which requires either to bootstrap or to re-initialize the stream cipher).

Our Contribution. In view of this state-of-the-art, a natural direction would be to try combining the best of these two previous works. That is, to design a cipher inheriting from the constant noise property offered by block ciphers, and the lower noise levels of stream ciphers (due to the lower algebraic degree of their outputs), leading to the following contributions.

First, we introduce a new stream cipher construction, next denoted as a filter permutator (by analogy with filter generators). Its main design principle is to filter a constant key register with a variable (public) bit permutation. More precisely, at each cycle, the key register is (bit) permuted with a pseudorandomly generated permutation, and we apply a non-linear filtering function to the output of this permuted key register. The main advantage of this construction is to always apply the non-linear filtering directly on the key bits, which allows maintaining the noise level of our outputs constant. Conceptually, this type of construction seems appealing for any FHE scheme.

Second, and going deeper in the specification of a concrete scheme, we discuss the optimization of the components in a filter permutator, with a focus on the filtering function (which determines the output noise after homomorphic evaluation). For this purpose, we first notice that existing FHE schemes can be split in (roughly) two main categories. On one hand the so-called *2nd-generation* FHE (such as [9,15]) where the metric for the noise growth is essentially the multiplicative depth of the circuit to homomorphically evaluate. On the other hand, the so-called *3rd-generation* FHE (such as [2,32]) where the error growth is asymmetric, and in particular quasi-additive when considering a multiplicative chain. From these observations, we formalize a *comb* structure which can be represented as a (possibly long) multiplicative chain, in order to take the best advantage of 3rd-generation FHE schemes. We then design a filtering function based on this comb structure (combined with other technical ingredients in order to prevent various classes of possible attacks against stream ciphers) and specify a family of filter permutators (called FLIP).

Third, and in order to stimulate further investigations, we instantiate a few version of FLIP designs, for 80-bit and 128-bit security. We then provide a preliminary evaluation of their security against some of the prevailing cryptanalysis from the open literature – such as (fast) algebraic attacks, (fast) correlation attacks, BKW-like attacks [6], guess and determine attacks, *etc.* – based on state-of-the-art tools. We also analyze the noise brought by their filtering functions in the context of 3rd-generation FHE. In this respect, our main result is that we can limit the noise after the homomorphic evaluation of a decryption to a level of the same order of magnitude as for a single homomorphic multiplication - hence essentially making the impact of the symmetric encryption scheme as small as possible.

We finally observe that our FLIP designs have a very reduced multiplicative depth, which makes them suitable for 2nd-generation FHE schemes as well, and provide preliminary results of prototype implementations using HElib that confirm their good behavior compared to state-of-the-art block and stream ciphers designed for efficient FHE.

Overall, filter permutators in general and FLIP instances in particular open a large design space of new symmetric constructions to investigate. Hence, we conclude the paper with a list of open problems regarding these algorithms, their best cryptanalysis, the Boolean functions used in their filter and their efficient implementation in concrete applications.

2 Background

2.1 Boolean Functions

In this section, we recall the cryptographic properties of Boolean functions that we will need in the rest of the paper (mostly taken from [12]).

Definition 1 (Boolean Function). *A Boolean function f with n variables is a function from $\mathbb{F}_2^n$ to $\mathbb{F}_2$. The set of all Boolean functions in n variables is denoted by $\mathcal{B}_n$.*

Definition 2 (Walsh Transform). *Let $f \in \mathcal{B}_n$ a Boolean function. Its Walsh Transform $\mathsf{W_f}$ at $a \in \mathbb{F}_2^n$ is defined as:*

$$\mathsf{W_f}(a) = \sum_{x \in \mathbb{F}_2^n} (-1)^{f(x)+\langle a,x \rangle},$$

where $\langle a, x \rangle$ denotes the inner product in $\mathbb{F}_2^n$.

Definition 3 (Balancedness). *A Boolean function $f \in \mathcal{B}_n$ is said to be balanced if its outputs are uniformly distributed over $\{0,1\}$.*

Definition 4 (Non-linearity). *The non-linearity NL of a Boolean function $f \in \mathcal{B}_n$, where n is a positive integer, is the minimum Hamming distance between f and all the affine functions g:*

$$\mathsf{NL}(f) = \min_g \{d_H(f,g)\},$$

with $d_H(f,g) = \#\{x \in \mathbb{F}_2^n \mid f(x) \neq g(x)\}$ the Hamming distance between f and g. The non-linearity of a Boolean function can also be defined by its Walsh Transform:

$$\mathsf{NL}(f) = 2^{n-1} - \frac{1}{2} \max_{a \in \mathbb{F}_2^n} |\mathsf{W_f}(a)|.$$

Definition 5 (Resiliency). *A Boolean function $f \in \mathcal{B}_n$ is said m-resilient if any of its restrictions obtained by fixing at most m of its coordinates is balanced. We will denote by $\mathsf{res}(f)$ the resiliency m of f and set $\mathsf{res}(f) = -1$ if f is unbalanced.*

Definition 6 (Algebraic Immunity). *The algebraic immunity of a Boolean function $f \in \mathcal{B}_n$, denoted as* $\mathsf{AI}(f)$, *is defined as:*

$$\mathsf{AI}(f) = \min_{g \neq 0}\{\mathsf{deg}(g) \mid fg = 0 \text{ or } (f \oplus 1)g = 0\},$$

where $\mathsf{deg}(g)$ *is the degree of g. The function g is called an annihilator of f (or $(f \oplus 1)$).*

Definition 7 (Fast Algebraic Immunity). *The fast algebraic immunity of a Boolean function $f \in \mathcal{B}_n$, denoted as* $\mathsf{FAI}(f)$, *is defined as:*

$$\mathsf{FAI}(f) = \min\{2\mathsf{AI}(f), \min_{1 \leq \mathsf{deg}(g) < \mathsf{AI}(f)} (\max[\mathsf{deg}(g) + \mathsf{deg}(fg), 3\mathsf{deg}(g)])\}.$$

Summarizing, the good balancedness, non-linearity and resiliency properties have to be ensured to widthstand correlation attacks [56] and fast correlation attacks [48]. The high algebraic immunity and fast algebraic immunity have to be ensured to widthstand algebraic attacks [13].

2.2 (Ring) Learning with Errors

In this section, we recall useful notations and definitions needed about the decisional LWE problem and its ring variation. For an integer modulus q, we denote by $\mathbb{Z}_q$ the quotient ring of integers modulo q. We denote vectors with bold letters $\mathbf{e}$ and matrices with bold capital letters $\mathbf{A}$. The notation $s \leftarrow_\$ S$ (*resp.* $s \leftarrow_\$ \chi$) denotes that s is picked uniformly at random from a finite set S (*resp.* from a distribution χ).

The decisional Learning With Error problem (dLWE) was introduced by Regev [53].

Definition 8 (dLWE). *For an integer $q = q(n) \geq 2$, an adversary $\mathcal{A}$ and an error distribution $\chi = \chi(n)$ over $\mathbb{Z}_q$, we define the following advantage function:*

$$\mathsf{Adv}_{\mathcal{A}}^{\mathsf{dLWE}_{n,m,q,\chi}} := |\Pr[\mathcal{A}(\mathbf{A}, \mathbf{z}_0) = 1] - \Pr[\mathcal{A}(\mathbf{A}, \mathbf{z}_1) = 1]|,$$

where

$$\mathbf{A} \leftarrow_\$ \mathbb{Z}_q^{n \times m}, \mathbf{s} \leftarrow_\$ \mathbb{Z}_q^n, \mathbf{e} \leftarrow_\$ \chi^m, \mathbf{z}_0 := \mathbf{s}^\top \mathbf{A} + \mathbf{e}^\top \quad \text{and} \quad \mathbf{z}_1 \leftarrow_\$ \mathbb{Z}_q^m.$$

The $\mathsf{dLWE}_{n,m,q,\chi}$ *assumption asserts that for all* PPT *adversaries $\mathcal{A}$, the advantage* $\mathsf{Adv}_{\mathcal{A}}^{\mathsf{dLWE}_{n,m,q,\chi}}$ *is a negligible function in n.*

The ring variant was introduced by Lyubashevsky, Peikert and Regev in [46].

Definition 9 (dR-LWE). *For a polynomial ring $R = \mathbb{Z}[X]/f(X)$ with f of degree n, an integer $q \geq 2$, an adversary $\mathcal{A}$ and an error distribution χ over $R_q = R/qR$, $R^\vee$ being R dual fractional ideal, we define the following advantage function:*

$$\mathsf{Adv}_{\mathcal{A}}^{\mathsf{d}R\mathsf{LWE}_{R,q,\chi}} := |\Pr[\mathcal{A}(a, z_0) = 1] - \Pr[\mathcal{A}(a, z_1) = 1]|,$$

where

$$a \leftarrow_{\$} R_q,\ s \leftarrow_{\$} R_q^{\vee},\ e \leftarrow_{\$} \chi,\ z_0 := a \cdot s + e \quad and \quad z_1 \leftarrow_{\$} R.$$

With $f(X)$ a cyclotomic polynomial, the $\mathsf{d}R\mathsf{LWE}_{R,q,\chi}$ *assumption asserts that for all* PPT *adversaries $\mathcal{A}$, the advantage* $\mathsf{Adv}_{\mathcal{A}}^{\mathsf{d}R\mathsf{LWE}_{R,q,\chi}}$ *is a negligible function in n.*

For our constructions, we need to take the distribution χ as a subgaussian random variable which we define hereafter. More details about the subgaussian distribution and the lemmas' proof can be found in [2,58].

Definition 10 (Subgaussian Random Variables). *Let X be a random variable. We say X is subgaussian with parameter σ if there exists σ such that:*

$$\forall t \in \mathbb{R}, \mathbb{E}[e^{tX}] \leq e^{\sigma^2 t^2/2},$$

where $\mathbb{E}[e^{tX}]$ is the moment generating function of X.

Lemma 1 (Subgaussian Random Variables Properties). *Let X, X' be independent subgaussian random variables of parameter σ and σ' respectively. Assuming $\mathbb{E}(X) = \mathbb{E}(X') = 0$ we have the following properties:*

- *Tails: $\forall t \geq 0$ we have $Pr[|X| \geq t] \leq 2e^{-\pi t^2/\sigma^2}$.*
- *Homogeneity: $\forall c \in \mathbb{R}$, cX is subgaussian with parameter $|c|\sigma$.*
- *Pythagorean additivity: $X + X'$ is subgaussian with parameter $\sqrt{\sigma^2 + \sigma'^2}$.*

We extend the notion of subgaussianity to vectors and polynomials. Since the coefficients of a polynomial are seen as a vector, we call subgaussian vector of parameter σ a vector where each coefficient follows an independent subgaussian distribution with parameter σ.

Lemma 2 (Subgaussian Vector Norm, Adapted from [2], Lemma 2.1). *Let $\mathbf{x} \in \mathbb{R}^n$ be a random vector where each coordinate follows an independent subgaussian distribution of parameter σ. Then for some universal constant $C > 0$ we have $Pr[||\mathbf{x}||_2 > C\sigma\sqrt{n}] \leq 2^{-\Omega(n)}$ and therefore $||\mathbf{x}||_2 = \mathcal{O}(\sigma\sqrt{n})$.*

2.3 Fully Homomorphic Encryption

In this section we recall the definition of (Fully) Homomorphic Encryption and present the Homomorphic Encryption schemes we will use, both based on GSW [32].

Definition 11 (Homomorphic Encryption Scheme). *Let $\mathcal{M}$ be the plaintext space, C the ciphertext space and λ the security parameter. A homomorphic encryption scheme consists of four algorithms:*

- $H.\mathsf{KeyGen}(1^\lambda)$. *Output* pk^H *and* sk^H *the public and secret keys of the scheme.*
- $H.\mathsf{Enc}(m, \mathsf{pk}^H)$. *From the plaintext* $m \in \mathcal{M}$ *and the public key, output a ciphertext* $c \in \mathcal{C}$.
- $H.\mathsf{Dec}(c, \mathsf{sk}^H)$. *From the ciphertext* $c \in \mathcal{C}$ *and the secret key, output* $m' \in \mathcal{M}$.
- $H.\mathsf{Eval}(f, c_1, \cdots, c_k, \mathsf{pk}^H)$. *With* $c_i = H.\mathsf{Enc}(m_i, \mathsf{pk}^H)$ *for* $1 \leq i \leq k$, *output a ciphertext* $c_f \in \mathcal{C}$ *such that* $H.\mathsf{Dec}(c_f) = f(m_1, \cdots, m_k)$.

A homomorphic encryption scheme is called a Fully Homomorphic Encryption (FHE) scheme when f can be any function and $|\mathcal{C}|$ is finite. A simpler primitive to consider is the SomeWhat Homomorphic Encryption (SWHE) scheme, where f is restricted to be any univariate polynomial of finite degree.

Since the breakthrough work of Gentry [30], the only known way to obtain FHE consists in adding a bootstrapping technique to a SWHE. As bootstrapping computational cost is still expensive in comparison to the other FHE algorithms, in the following part of the article we will only consider SWHE for our applications.

GSW Homomorphic Encryption Scheme. In 2013, Gentry, Sahai and Waters [32] introduced a Homomorphic Encryption scheme based on LWE using a new technique stemming from the *approximate eigenvector problem*. This new technique led to a new family of FHE, called 3rd-generation FHE, consisting in Homomorphic Encryption schemes such that the multiplicative error growth is quasi-additive. Hereafter, we present two schemes belonging to this generation, the first one with security based on dLWE and the second one based on dRLWE. We first set some useful notations considering the different schemes.

For a matrix $\mathbf{E}$ we refer to the i-th row as $\mathbf{e}_i^\top$ and to the j-th column as $\mathbf{e}_j$. The $\log q$ notation refers to the logarithm in base 2 of q. The notation $[a]_q$ is for $a \bmod q$ and $\lfloor [a]_q \rceil_2 \in \{0,1\}$ is a function in $a \in \mathbb{Z}_q$ giving 1 if $\lfloor \frac{q}{4} \rfloor \leq a \leq \lfloor \frac{3q}{4} \rfloor \bmod q$ and 0 otherwise. We denote by $[n]$ the set of integers $\{1, \cdots, n\}$. We finally use $|x|$ and $||x||_2$ for the standard norms 1 and 2 on vectors $x \in \mathbb{R}^n$.

Batched GSW. This scheme is a batched version of GSW presented in [36], enabling to pack independently r plaintexts in one ciphertext. From the security parameter λ and the considered applications, we can derive the parameters n, q, r, χ of the scheme described below.

$H.\mathsf{KeyGen}(n, q, r, \chi)$. On inputs the lattice dimension n, the modulus q, the number of bits by ciphertext r and the error distribution χ do:

- Set $\ell = \lceil \log q \rceil$, $m = \mathcal{O}(n\ell)$, $N = (r+n)\ell$, $\mathcal{M} = \{0,1\}^r$ and $\mathcal{C} = \mathbb{Z}_q^{(r+n)\times N}$.
- Pick $\mathbf{A} \leftarrow_\$ \mathbb{Z}_q^{n\times m}$, $\mathbf{S}' \leftarrow_\$ \chi^{r\times n}$ and $\mathbf{E} \leftarrow_\$ \chi^{r\times m}$.
- Set $\mathbf{S} = [\mathbf{I}| - \mathbf{S}'] \in \mathbb{Z}_q^{r\times(r+n)}$ and $\mathbf{B} = \left[\frac{\mathbf{S}'\mathbf{A} + \mathbf{E}}{\mathbf{A}}\right]_q \in \mathbb{Z}_q^{(r+n)\times m}$.
- For all $\mathbf{m} \in \{0,1\}^r$:
  - Pick $\mathbf{R_m} \leftarrow_\$ \{0,1\}^{m\times N}$.

- Set $\mathbf{P_m} = \left[\mathbf{BR_m} + \begin{pmatrix} m_1 \cdot \mathbf{s}_1^\top \\ \vdots \\ m_r \cdot \mathbf{s}_r^\top \\ \mathbf{0} \end{pmatrix} \mathbf{G}\right]_q \in \mathbb{Z}_q^{(r+n)\times N}$.

  with $\mathbf{s}_i^\top$ the i-th row of $\mathbf{S}$ and $\mathbf{G} = (2^0, \cdots, 2^{\ell-1})^\top \otimes \mathbf{I} \in \mathbb{Z}_q^{(r+n)\times N}$.
- Output $\mathsf{pk}^H := (\{\mathbf{P_m}\}, \mathbf{B})$ and $\mathsf{sk}^H := \mathbf{S}$.

$H.\mathsf{Enc}(\mathsf{pk}^H, \mathbf{m})$. On input pk^H, and $\mathbf{m} \in \{0,1\}^r$, do:

- Pick $\mathbf{R} \leftarrow_\$ \{0,1\}^{m\times N}$, and output $\mathbf{C} = [\mathbf{BR} + \mathbf{P_m}]_q \in \mathbb{Z}_q^{(r+n)\times N}$.

$H.\mathsf{Dec}(\mathbf{C}, \mathsf{sk}^H)$. On input the secret key sk^H, and a ciphertext $\mathbf{C}$, do:

- For all $i \in [r] : m_i' = \lfloor [\langle \mathbf{s}_i^\top, \mathbf{c}_{i\ell}\rangle]_q \rceil_2$ where $\mathbf{c}_{i\ell}$ is the column $i\ell$ of $\mathbf{C}$.
- Output $m_1', \cdots, m_r' \in \{0,1\}^r$.

Note that $\mathbf{SC} = \mathbf{SBR}+\mathbf{SP_m} = \mathbf{ER}+\mathbf{ER_m}+\begin{pmatrix} m_1 \cdot \mathbf{s}_1^\top \\ \vdots \\ m_r \cdot \mathbf{s}_r^\top \end{pmatrix}\mathbf{G} = \mathbf{E}'+\begin{pmatrix} m_1 \cdot \mathbf{s}_1^\top \\ \vdots \\ m_r \cdot \mathbf{s}_r^\top \end{pmatrix}\mathbf{G}$.

The $H.\mathsf{Eval}$ algorithm finally consists in iterating, following a circuit f, the homomorphic operations $H.\mathsf{Add}$ and $H.\mathsf{Mul}$:

- $H.\mathsf{Add}(\mathbf{C}_1, \mathbf{C}_2) : \mathbf{C}_+ = \mathbf{C}_1 + \mathbf{C}_2$.
- $H.\mathsf{Mul}(\mathbf{C}_1, \mathbf{C}_2) : \mathbf{C}_\times = \mathbf{C}_1 \times \mathbf{G}^{-1}\mathbf{C}_2$ with $\mathbf{G}^{-1}$ a function such that $\forall \mathbf{C} \in \mathbb{Z}_q^{(r+n)\times N}, \mathbf{G}\mathbf{G}^{-1}(\mathbf{C}) = \mathbf{C}$ and the values of $\mathbf{G}^{-1}(\mathbf{C})$ follow a subgaussian distribution with parameter $\mathcal{O}(1)$ (see [49] for the existence and proof of $\mathbf{G}^{-1}$).

The correctness and security of this scheme are proven in the extended version of this paper.

Remark 1. For practical use, we only need to store $r+1$ matrices $\mathbf{P}_m$, namely the $r+1$ ones with $\mathbf{m}$ of hamming weight equal to 0 or 1 are sufficient to generate correct encryption of all $\mathbf{m} \in \{0,1\}^r$ with at most r additions of the corresponding $\mathbf{P}_m$ matrices.

Ring-GSW. This scheme is a ring version of GSW presented in [38], transposing the *approximate eigenvector problem* into the ring setting. From λ the security parameter and the considered applications, we can derive the parameters n, q and $\mathcal{M}$ of the scheme described below.

$H.\mathsf{KeyGen}(n, q, \chi, \mathcal{M})$. On inputs the lattice dimension n, which is set to a power of 2, the modulus q, the error distribution χ and the plaintext space $\mathcal{M}$ do:

- Set $R = \mathbb{Z}[X]/(X^n+1)$, $R_q = R/qR$, $\ell = \lceil \log q \rceil$, $N = 2\ell$ and $\mathcal{C} = R_q^{2\times N}$.
- Set $R_{0,1} = \{P \in R_q, p_i \in \{0,1\}, 0 \leq i < n\}$.
- Pick $a \leftarrow_\$ R_q$, $s' \leftarrow_\$ \chi$ and $e \leftarrow_\$ \chi$.

- Set $\mathbf{s} = [1| - s']^\top \in R_q^{1\times 2}$ and $\mathbf{b} = \begin{pmatrix} s'a + e \\ a \end{pmatrix} \in R_q^{2\times 1}$.
- Output $\mathsf{pk}^H := \mathbf{b}$ and $\mathsf{sk}^H := \mathbf{s}$.

$H.\mathsf{Enc}(\mathsf{pk}^H, m)$. On input pk^H, and $m \in \mathcal{M}$, do:

- Pick $\mathbf{E} \leftarrow_\$ \chi^{2\times N}$.
- Pick $\mathbf{r} \leftarrow_\$ R_{0,1}^N$, and output $\mathbf{C} = [\mathbf{b}\mathbf{r}^\top + m\mathbf{G} + \mathbf{E}]_q \in R_q^{2\times N}$.

$H.\mathsf{Dec}(\mathbf{C}, \mathsf{sk}^H)$. On input the secret key sk^H, and a ciphertext $\mathbf{C}$, do:

- Compute $m' = \lfloor [< \mathbf{s}, \mathbf{c}_l >]_q \rceil_2$.
- Output $m' \in R_q$.

The $H.\mathsf{Eval}$ algorithm finally consists in iterating $H.\mathsf{Add}$ and $H.\mathsf{Mul}$:

- $H.\mathsf{Add}(\mathbf{C}_1, \mathbf{C}_2) : \mathbf{C}_+ = \mathbf{C}_1 + \mathbf{C}_2$.
- $H.\mathsf{Mul}(\mathbf{C}_1, \mathbf{C}_2) : \mathbf{C}_\times = \mathbf{C}_1 \times \mathbf{G}^{-1}\mathbf{C}_2$.

The correctness and security of this scheme are proven in the extended version of this paper.

Remark 2. The plaintext space $\mathcal{M}$ has a major influence on the considered application in terms of quantity of information contained in a single ciphertext and error growth. For our application we choose $\mathcal{M}$ as the set of polynomials with all coefficients of degree greater than 0 being zero, and the constant coefficient being bounded.

3 New Stream Cipher Constructions

In this section, we introduce our new stream cipher construction. We first describe the general filter permutator structure. Next we list a number of Boolean building blocks together with their necessary cryptographic properties. Third, we specify a family of filter permutators (denoted as FLIP) and analyze its security based on state-of-the art cryptanalysis and design tools. Finally, we propose a couple of parameters to fully instantiate a few examples of FLIP designs.

3.1 Filter Permutators

The general structure of filter permutators is depicted in Fig. 2. It is composed of three parts: a register where the key is stored, a (bit) permutation generator parametrised by a Pseudo Random Number Generator (PRNG) [7,37] (which is initialized with a public IV), and a filtering function which generates a keystream. The filter permutator can be compared to a filter generator, in which the LFSR is replaced by a permuted key register. In other words, the register is

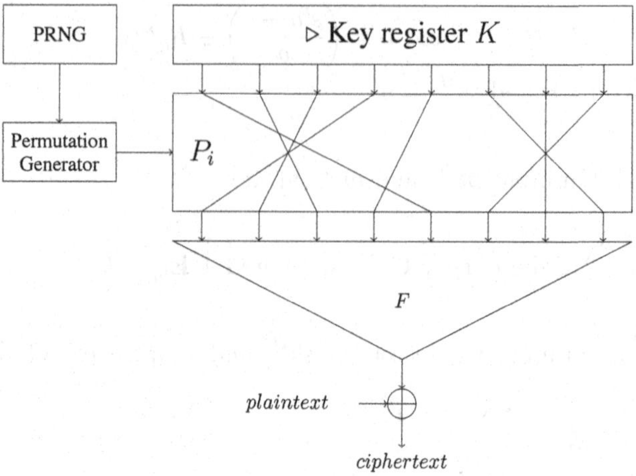

Fig. 2. Filter permutator construction.

no longer updated by means of the LFSR, but with pseudorandom bit permutations. More precisely, at each cycle (*i.e.* each time the filtering function outputs a bit), a pseudo-random permutation is applied to the register and the permuted key register is filtered. Eventually, the encryption (*resp.* decryption) with a filter permutator simply consists in XORing the bits output by the filtering function with those of the plaintext (*resp.* ciphertext).

3.2 Boolean Building Blocks for the Filter Permutator

We will first exploit direct sums of Boolean functions defined as follows:

Definition 12 (Direct Sum). *Let $f_1(x_0, \cdots, x_{n_1-1})$ and $f_2(x_{n_1}, \cdots, x_{n_1+n_2-1})$ be two Boolean functions in respectively n_1 and n_2 variables. The direct sum of f_1 and f_2 is defined as $f = f_1 \oplus f_2$, which is a Boolean function in $n_1 + n_2$ variables such that:*

$$f(x_0, \cdots, x_{n_1+n_2-1}) = f_1(x_0, \cdots, x_{n_1-1}) \oplus f_2(x_{n_1}, \cdots, x_{n_1+n_2-1}).$$

They inherit from the following set of properties, proven in the extended version of this paper.

Lemma 3 (Direct Sum Properties). *Let f be the direct sum of f_1 and f_2 with n_1 and n_2 variables respectively. Then f has the following cryptographic properties:*

1. *Non Linearity:* $\mathsf{NL}(f) = 2^{n_2}\mathsf{NL}(f_1) + 2^{n_1}\mathsf{NL}(f_2) - 2\mathsf{NL}(f_1)\mathsf{NL}(f_2)$.
2. *Resiliency:* $\mathsf{res}(f) = \mathsf{res}(f_1) + \mathsf{res}(f_2) + 1$.
3. *Algebraic Immunity:* $\mathsf{AI}(f_1) + \mathsf{AI}(f_2) \geq \mathsf{AI}(f) \geq \max(\mathsf{AI}(f_1), \mathsf{AI}(f_2))$.
4. *Fast Algebraic Immunity:* $\mathsf{FAI}(f) \geq \max(\mathsf{FAI}(f_1), \mathsf{FAI}(f_2))$.

Our direct sums will then be based on three parts: a linear function, a quadratic function and triangular functions, defined as follows.

Definition 13 (Linear Functions). *Let $n > 0$ be a positive integer, the L_n linear function is a n-variable Boolean function defined as:*

$$L_n(x_0, \cdots, x_{n-1}) = \sum_{i=0}^{n-1} x_i.$$

Definition 14 (Quadratic Functions). *Let $n > 0$ be a positive integer, the Q_n linear function is a $2n$-variable Boolean function defined as:*

$$Q_n(x_0, \cdots, x_{2n-1}) = \sum_{i=0}^{n-1} x_{2i}x_{2i+1}.$$

Definition 15 (Triangular Functions). *Let $k > 0$ be a positive integer. The k-th triangular function T_k is a $\frac{k(k+1)}{2}$-variable Boolean function defined as:*

$$T_k(x_0, \cdots, x_{\frac{k(k+1)}{2}-1}) = \Sigma_{i=1}^{k} \Pi_{j=0}^{i-1} x_{j+\Sigma_{\ell=0}^{i-1}\ell}.$$

For example, the 4th triangular function T_4 is:

$$T_4 = x_0 \oplus x_1x_2 \oplus x_3x_4x_5 \oplus x_6x_7x_8x_9.$$

These three types of functions allow us to guarantee the following properties.

Lemma 4 (Linear Functions Properties). *Let L_n be a linear function in n variables, then L_n has the following cryptographic properties:*

1. *Non Linearity:* $\mathsf{NL}(L_n) = 0$.
2. *Resiliency:* $\mathsf{res}(L_n) = n - 1$.
3. *Algebraic Immunity:* $\mathsf{AI}(L_n) = 1$.
4. *Fast Algebraic Immunity:* $\mathsf{FAI}(L_n) = 2$.

Lemma 5 (Quadratic Functions Properties). *Let Q_n be a linear function in $2n$ variables, then Q_n has the following cryptographic properties:*

1. *Non Linearity:* $\mathsf{NL}(Q_n) = 2^{2n-1} - 2^{n-1}$.
2. *Resiliency:* $\mathsf{res}(Q_n) = -1$.
3. *Algebraic Immunity:* $\mathsf{AI}(Q_1) = 1$ *and* $\forall n > 1, \mathsf{AI}(Q_n) = 2$.
4. *Fast Algebraic Immunity:* $\mathsf{FAI}(Q_1) = 2$ *and* $\forall n > 1,\ \mathsf{FAI}(Q_n) = 4$.

Lemma 6 (Triangular Functions Properties). *Let k a positive integer and let T_k the k-th triangular function. Then the following properties hold:*

1. *Non Linearity follows the recursive formula defined as:*
   (i) $\mathsf{NL}(T_1 = 0)$,
   (ii) $\mathsf{NL}(T_{k+1}) = (2^{k+1} - 2)\mathsf{NL}(T_k) + 2^{k(k+1)/2}$.
2. *Resiliency:* $\mathsf{res}(T_k) = 0$.
3. *Algebraic Immunity:* $\mathsf{AI}(T_k) = k$.
4. *Fast Algebraic Immunity:* $\mathsf{FAI}(T_k) = k + 1$.

The proof of Lemma 6 can be found in the extended version of this paper.

3.3 The FLIP Family of Stream Ciphers

Based on the previous definitions, we specify the FLIP family of stream ciphers as a filter permutator using a forward secure PRNG [5] based on the AES-128 (*e.g.* as instantiated in the context of leakage-resilient cryptography [57]), the Knuth shuffle (see below) as bit permutation generator and such that the filter F is the N-variable Boolean function defined by the direct sum of three Boolean functions f_1, f_2 and f_3 of respectively n_1, n_2 and n_3 variables, such that:

- $f_1(x_0, \cdots, x_{n_1-1}) = L_{n_1}$,
- $f_2(x_{n_1}, \cdots, x_{n_1+n_2-1}) = Q_{n_2/2}$,
- $f_3(x_{n_1+n_2}, \cdots, x_{n_1+n_2+n_3-1})$ is the direct sum of nb triangular functions T_k, *i.e.* such that each T_k acts on different and independent variables, that we denote as ${}^{nb}\Delta^k$.

That is, we have $F : \mathbb{F}_2^{n_1+n_2+n_3} \to \mathbb{F}_2$ the Boolean function such that:

$$F(x_0, \cdots, x_{n_1+n_2+n_3-1}) = L_{n_1} \oplus Q_{n_2/2} \oplus \bigoplus_{i=1}^{nb} T_k.$$

In the following section, we provide a preliminary security analysis of the FLIP filter permutators against a couple of standard attacks against stream ciphers, based on state-of-the-art tools. For this purpose, we will assume that no additional weaknesses arise from its PRNG and bit permutation generator. In this respect, we note that our forward secure PRNG does not allow malleability, so it should be hard to obtain a collision in the chosen IV model better than with birthday probability. This should prevent collisions on the generated permutations. Besides, the Knuth shuffle [41] (or Fisher-Yates shuffle) is an algorithm allowing to generate a random permutation on a finite set. This algorithm has the interesting property of giving the same probability to all permutations if used with a random number generator. As a result, we expect that any deviation between a bit permutation based on a Knuth shuffle fed with the PRNG will be hard to exploit by an adversary. Our motivation for this assumption is twofold. First, it allows us to focus on whether the filter permutator construction is theoretically sound. Second, if such a choice was leading to an exploitable weakness, it remains possible to build a pseudorandom permutation based on standard cryptographic constructions [45].

Remark 3. Since the permutation generation part of FLIP has only birthday security (with respect to the size of the PRNG), it implies that it is only secure up to 2^{64} PRNG outputs when implemented with the AES-128. Generating more keystream using larger block ciphers should be feasible. However, in view of the novelty of the FLIP instances, our claims are only made for this limited (birthday) data complexity so far, which should not be limiting for the intended FHE applications. We leave the investigation of their security against attacks using larger data complexities as a scope for further research. Besides, we note that using a PRNG based on a tweakable block cipher [44] (where a part of the larger IV would be used as tweak) could be an interesting track to reduce the impact of a collision on the PRNG output in the known IV model, which we also leave as an open research direction.

3.4 Security Analysis

Since the filter permutator shares similarities with a filter generator, it is natural to start our investigations with the typical attacks considered against such types of stream ciphers. For this purpose, we next study the applicability of algebraic attacks and correlation attacks, together with more specialized cryptanalyses that have been considered against stream ciphers. Note that the attacks considered in the rest of this section frequently require to solve systems of equations and to implement a Gaussian reduction. Our complexity estimations will consider Strassen's algorithm for this purpose and assume $\omega = \log 7$ to be the exponent in a Gaussian reduction. Admittedly, approaches based on Gröbner bases [27] or taking advantage of the sparsity of the matrices [59] could lead to even faster algorithms. We ignore them for simplicity in these preliminary investigations. Note also that we only claim security in the single-key setting.

Algebraic Attacks were first introduced by Courtois and Meier in [18] and applied to the stream cipher Toyocrypt. Their main idea is to build an overdefined system of equations with the initial state of the LFSR as unknown, and to solve this system with Gaussian elimination. More precisely, by using a nonzero function g such that both g and $h = gF$ have low algebraic degree, an adversary is able to obtain T equations with monomials of degree at most $AI(f)$. It is easily shown that g can be taken equal to the annihilator of F or of $F \oplus 1$, *i.e.* such that $gF = 0$ or $g(F \oplus 1) = 0$. After a linearisation step, the adversary obtains a system of T equations in $D = \sum_{i=0}^{\mathsf{AI}(F)} \binom{N}{i}$ variables. Therefore, the time complexity of the algebraic attack is $\mathcal{O}(D^{\omega})$, that is, $\mathcal{O}(N^{\omega AI(f)})$.

Fast Algebraic Attacks are a variation of the previous algebraic attacks introduced by Courtois at Crypto 2003 [17]. Considering the relation $gF = h$, their goal is to find and use functions g of low algebraic degree e, possibly smaller than $AI(f)$, and h of low but possibly larger degree d, and to lower the degree of the resulting equations by an off-line elimination of the monomials of degrees larger than e (several equations being needed to obtain each one with degree at most e). Following [4], this attack can be decomposed into four steps:

1. The search of the polynomials g and h generating a system of $D+E$ equations in $D+E$ unknowns, where $D = \sum_{i=0}^{d} \binom{N}{i}$ and $E = \sum_{i=0}^{e} \binom{N}{i}$. This step has a time complexity in $\mathcal{O}(\sum_{i=0}^{d} \binom{n}{i} + \sum_{i=0}^{e} \binom{n}{i})^{\omega}$.
2. The search of linear relations which allow the suppression of the monomials of degree more than e. This step has a time complexity in $\mathcal{O}(D \log^2(D))$.
3. The elimination of monomials of degree larger than e using the Berlekamp-Massey algorithm. This step has a time complexity in $\mathcal{O}(ED \log(D))$.
4. The resolution of the system. This step has a time complexity in $\mathcal{O}(E^{\omega})$.

Given the FAI of F, the time complexity of this attack is in $\mathcal{O}(N^{\mathsf{FAI}})$, or more precisely $O(D \log^2 D + E^2 D + E^{\omega})$ (ignoring Step 1 which is trivial for our choice of F).

Correlation Attacks. In their basic versions, correlation attacks try to distinguish the output sequence of a stream cipher from a random one, by exploiting the bias δ of the filtering function. We can easily rule out such attacks by considering a (much) simplified version of filter permutator where the bit permutations P_i's would be made on two independent permutations P_i^1 and $P_i^{2,3}$ (respectively acting on the $n_1 + 1$ bits of the linear part and the $n_2 + n_3 - 1$ bits of the non-linear part of F). Suppose for simplicity that P_i^1 is kept constant t times, then the output distribution of F has a bias δ and it can be distinguished for the right choice of the $n_1 + 1 = \mathsf{res} + 1$ bits of the linear part. In this case, a correlation attack would have a data complexity of $\mathcal{O}(\delta^{-2})$ and a time complexity of $\mathcal{O}(2^{\mathsf{res}(F)+1}\delta^{-2})$, with $\delta = \frac{1}{2} - \left(\frac{\mathsf{NL}(F)}{2^N}\right)$. For simplicity, we will consider this conservative estimation in our following selection of security parameters. Yet, we note that since the permutation P_i of a filter permutator is acting on all the N bits of the filter F, the probability that the linear part of F is kept invariant by the permutations t times is in fact considerably smaller than what is predicted by the resilience.

BKW-like Attack. The BKW algorithm was introduced in [6] as a solution to solve the LPN problem using smart combinations of well chosen vectors and their associated bias. Intuitively, our stream cipher construction simplified as just explained (with two independent permutations P_i^1 and $P_i^{2,3}$ rather than a single one P_i) also shares similarities with this problem. Indeed, we could see the linear part as the parity of an LPN problem and the non-linear one (with a small bias) as a (large) noise. Adapting the BKW algorithm to our setting amounts to XOR some linear parts of F in order to obtain vectors of low Hamming weight, and then to consider a distinguishing attack with the associated bias. Denoting h the target Hamming weight, x the log of the number of XORs and δ the bias, the resulting attack (which can be viewed as an extension of the previous correlation attack) has data complexity $\mathcal{O}(2^h\delta^{-2(x+1)})$ (more details are given in the extended version of this paper).

Higher-Order Correlation Attacks were introduced by Courtois [16] and exploit the so-called XL algorithm. They look for good correlations between F and an approximation g of degree $d > 1$, in order to solve a linearised system based on the values of this approximation. The value ε is defined such that g is equal to F with probability greater than $1-\varepsilon$. Such attacks have a (conservative) time complexity estimate:

$$\mathcal{O}\left(\binom{N}{D}^{\omega}(1-\varepsilon)^{-m}\right), \text{ where } D \geq d \text{ and } m \geq \frac{\binom{N}{D}}{\binom{N}{D-d}}.$$

Guess and Determine Attacks. Note that this section has been motivated by a private communication from Sébastien Duval, Virginie Lallemand and Yann Rotella, of which the details will be available in an upcoming ePrint report [25]. Guess and determine attacks are generic attacks which consist in guessing ℓ bits of the key in order to cancel some monomials. In our context, it allows an adversary to focus on a filtering function restricted to a subset of variables. This weaker function can then be cryptanalyzed, *e.g.* analyzed with the four aforementioned attacks, *i.e.* the algebraic attack, the fast algebraic attack, the correlation/BKW-like attacks and the higher-order correlation attack. The complexity of a guess and determine attack against a function F of N variables is $\min_\ell\{2^\ell C(F[\ell])\}$ where $F[\ell]$ is a function of $N[\ell]$ variables obtained by fixing ℓ variables of F, $C(F)$ is the complexity of the best of the four attacks considered on the filtering function F and the minimum is taken over all ℓ's. The case $\ell = 0$ corresponds to attacking the scheme without guess and determine. We next bound the minimal complexity over these four attacks considering the weakest functions obtained by guessing. To do so, we introduce some notations and criteria allowing us to specify the cryptographic properties of Boolean functions obtained by guessing ℓ variables of Boolean functions being direct sums of monomials. As the impact of guessing is most relevant for fast algebraic attacks and CA/BKW-like attacks, we defer the other part of the analysis and extra lemmas to the extended version of this paper.

Definition 16 (Direct Sum Vector). *For a boolean function F of N variables obtained as a direct sum of monomials we associate its* ***direct sum vector*** *: $\mathbf{m}_F$ of length $k = \deg(F)$: $[m_1, m_2, \cdots, m_k]$ such that m_i is the number of monomials of degree i of F and $N = \sum_{i=1}^{k} i m_i$. We define two quantities related to this vector:*

- $\mathbf{m}_F^*$ *is the number of nonzero values of* $\mathbf{m}_F$.
- $\delta_{\mathbf{m}_F} = \frac{1}{2} - \frac{\mathsf{NL}(F)}{2^N}$.

These notations will be useful to quantify the impact of guessing some bits on the cryptographic properties of a Boolean function obtained by direct sums. $\mathbf{m}_F$, $\mathbf{m}_F^*$ and $\delta_{\mathbf{m}_F}$ are easily computable from the description of F, the latter can be computed recursively using Lemma 3.

Lemma 7 (Guessing and Direct Sum Vector). *For all guessing of $0 \leq \ell \leq N$ variables of a Boolean function F in N variables obtained by direct sums associated with $\mathbf{m}_F$, we obtain a function $F[\ell]$ in $N[\ell]$ variables associated with $\mathbf{m}_{F[\ell]}$ such that:*

1. $\sum_{i=1}^{\deg(F[\ell])} m_i[\ell] \geq (\sum_{i=1}^{\deg(F)} m_i) - \ell$.
2. $\mathbf{m}_{F[\ell]}^* \geq \mathbf{m}_F^* - \lfloor \frac{\ell}{\min_{1\leq i\leq \deg(F)} m_i} \rfloor$.
3. $\delta_{\mathbf{m}_{F[\ell]}} \leq \delta_{\mathbf{m}_F} 2^\ell$.

Hereafter we describe the bounds we have used in order to assess the security of our instances.

Lemma 8 (Guess And Determine & Fast Algebraic Attacks). *Let F be a boolean function in N variables and $C_{GDFAA}(F)$ be the minimum complexity of the Guess And Determine with Fast Algebraic Attacks on F, then:*

$$C_{GDFAA}(F) \geq \min_{0 \leq \ell \leq N} \left[2^{\ell} \binom{\min N[\ell]}{\mathbf{m}^*_{F[\ell]}} \log^2 \binom{\min N[\ell]}{\mathbf{m}^*_{F[\ell]}} + (\min N[\ell])^2 \binom{\min N[\ell]}{\mathbf{m}^*_{F[\ell]}} + (\min N[\ell])^{\omega} \right],$$

where $\mathbf{m}^*_{F[\ell]} = \mathbf{m}^*_F - \lfloor \frac{\ell}{\min_{i \in [\deg(F)]} m_i} \rfloor$.

Lemma 9 (Guess and Determine & CA/BKW-like Attacks). *Let F be a boolean function in N variables and $C_{GDCA/BKW}(F)$ be the minimum complexity of the Guess And Determine with Correlation/BKW Attacks on F, then:*

$$C_{GDCA/BKW}(F) \geq \min_{0 \leq \ell \leq N} \{2^{-\ell} \delta^{-2}_{\mathbf{m}_F}\}.$$

Other Attacks. Besides the previous attacks that will be taken into account quantitatively when selecting our concrete instances of FLIP designs, we also investigated the following other cryptanalyses. First, *fast correlation attacks* were introduced by Meier and Staffelbach at Eurocrypt 1988 [48]. A recent survey can be found in [47]. The attack is divided into two phases. The first one aims at looking for relations between the output bits a_i of the LFSR to generate a system of parity-check equations. The second one uses a fast decoding algorithm (*e.g.* the belief propagation algorithm) in order to decode the words of the code $z_i = F(a_i)$ satisfying the previous relations, where the channel has an error probability $p = \frac{\mathsf{NL}(F)}{2^N}$. The working principles of this attack are quite similar to the ones of the previously mentioned correlation attacks and BKW-like attacks. So we assume that the previous (conservative) complexity estimates rule out this variation as well. Besides, note that intuitively, the belief propagation algorithm is best suited to the decoding of low-density parities, which is what our construction (and the LPN problem) typically avoids.

Second, *weak keys* (*i.e.* keys of low or high hamming weights) can produce a keystream sufficiently biased to determine this hamming weight, and then to recover the key among the small amount of possible ones. The complexity of such attacks can be computed from the resiliency of F. However, since our N parameter will typically be significantly larger than the bit-security of our filter permutator instances, we suggest to restrict the key space to keys of Hamming weight $N/2$ to rule out this concern. For this purpose, master keys can simply be generated by applying a first (secret) random permutation to any stream with Hamming weight $N/2$.

Third, *augmented function attacks* are attacks focusing on multiple outputs of the function rather than one. The goal is to find coefficients $j_1, \cdots, j_r$ such that a relation between the key and the outputs $s_{i+j_1}, \cdots, s_{i+j_r}$ can be exploited. This relation can be a correlation (as explained in [3]) or simply algebraic [28]. In both cases, a prerequisite is that the relation holds on a sufficient number of i. As each bit output by FLIP depends on a different permutation, we believe that there is no exploitable relation between different outputs.

Eventually, *cube attacks* were introduced by Dinur and Shamir at Eurocrypt 2009 [20] as a variant of algebraic attacks taking advantage of the public parameters of a cryptographic protocol (plaintext in block ciphers, IV in stream cipher) in order to generate a system of equations of low degree. However in filter permutator constructions, the only such public parameter is the seed of the PRNG allowing to generate the pseudo-random bit permutations P_i. Since controlling this seed hardly allows any control of the F function's inputs, such attacks do not seem applicable. A similar observation holds for conditional differential cryptanalysis [39] and for integral/zero-sum distinguishers [8,40].

3.5 Cautionary Note and Design Tweaks

As already mentioned, all the previous analyzes are based on standard cryptanalysis and design tools. In particular, the security of our FLIP designs is based on properties of Boolean functions that are generally computed assuming a uniform input distribution. Yet, for filter permutators this condition is not strictly respected since the Hamming weight of the key register is fixed (we decided to set it to $N/2$ in order to avoid weak keys, but even without this condition, it would be fixed to an unknown value). This means the input distribution of our linear, quadratic and triangle functions is not uniform. We verified experimentally that the output of FLIP is sufficiently balanced despite this non-uniformity. More precisely, we could not detect biases larger than $2^{\frac{q}{2}}$ when generating 2^q bits of keystream (based on small-scale experiments with $q = 32$). But we did not study the impact of this non-uniformity for other attacks, which we leave as an important research scope, both from the cryptanalysis and the Boolean functions points-of-view.

Note that in case the filter permutator of Sect. 3.1 turns out to have weaknesses specifically due to the imbalanced F function's inputs, there are tweaks that could be used to mitigate their impact. The simplest one is to apply a public whitening to the input bits of the non-linear parts of F (using additional public PRNG outputs), which has no impact on the homomorphic capacity. The adversary could then bias the F function's inputs based on his knowledge of the whitening bits, but to a lower extent than with our fixed Hamming weight keys. Alternatively, one could add a (more or less complex) linear layer before the non-linear part of F, which would then make the filter permutator conceptually more similar to filter generators, and (at least for certain layers) only imply moderate cost from the FHE point-of-view.

3.6 80- & 128-bit Security Instances

We selected a few instances aiming at 80- and 128-bit security based on the previous bounds, leading to the attack complexities listed in Table 1, where $\mathsf{FLIP}(n_1, n_2, {}^{nb}\Delta^k)$ denotes the instantiation of FLIP with linear part of n_1 bits, quadratic part of n_2 bits and nb triangular functions of degree k. These instances are naturally contrasted. On the one hand, the bounds taken are conservative with respect to the attacks considered: if these attacks were the best ones, more aggressive instances could be proposed (*e.g.* in order to reduce the key size). On the other hand, filter permutators are based on non-standard design principles, and our security analysis is only preliminary, which naturally suggests the need of security margins. Overall, we believe the proposed instances are a reasonable trade-off between efficiency and security based on our current understanding of filter permutators, and therefore are a good target for further investigations.

Table 1. Attack complexities in function of n_1, n_2 and ${}^{nb}\Delta^k$. AA stands for algebraic attacks, FAA stands for fast algebraic attacks, CA/BKW stands for correlation or BKW-like attacks, HOC stands for higher-order correlation attacks and ℓ stands for the number of bits guessed leading to the best complexity for guess and determine attacks. For the CA/BKW column, we reported the minimum complexity between the correlation and BKW-like attack. Eventually, λ stands for the security parameter of F and is simply taken as the minimum between AA, FAA,CA/BKW and HOC.

Instance	N	AA	ℓ	FAA	l	CA/BKW	ℓ	HOC	ℓ	λ
FLIP$(42, 128, {}^{8}\Delta^{9})$	530	95	56	81	0	86	72	94	55	81
FLIP$(46, 136, {}^{4}\Delta^{15})$	662	91	52	81	52	80	72	90	48	80
FLIP$(82, 224, {}^{8}\Delta^{16})$	1394	156	112	140	40	134	120	155	109	134
FLIP$(86, 238, {}^{5}\Delta^{23})$	1704	149	105	137	105	133	124	128	74	128

3.7 Indirect Sums

Before analyzing the FHE properties of filter permutators, we finally suggest FLIP designs based on indirect sums as another interesting topic for evaluation, since they lead to quite different challenges. Namely, the main motivation to use direct sums in the previous sections was the possibility to assess their cryptographic properties based on existing tools. By contrast, filter permutator designs based on indirect sums seem harder to analyze (both for designers and cryptanalysts). This is mainly because in this case, not only the inputs of the Boolean functions vary, but also the Boolean functions themselves. For illustration, we can specify "multi-FLIP" designs, next denoted as b-FLIP designs, such that we compute b instances of FLIP in parallel, each with the same filtering function but with different permutations, and then to XOR the b computed bits in order to produce a keystream bit. We conjecture that such b-FLIP designs could lead to secure stream ciphers with smaller states, and suggest 10-FLIP$(10, 20, {}^{1}\Delta^{20})$ and 15-FLIP$(15, 30, {}^{1}\Delta^{30})$ as exemplary instances for 80- and 128-bit security.

4 Application to FHE

4.1 80- & 128-bit Security Parameters

For the security parameters choices, we follow the analysis of Lindner and Peikert [43] for the hardness of LWE and RLWE, considering distinguishing and decoding attacks using BKZ [14,55]. We assume that the distribution χ in the considered LWE instances is the discrete Gaussian distribution with mean 0 and standard deviation σ. First we compute the best root Hermite factor δ of a basis (see [29]) computable with complexity 2^λ from the conservative lower bound of [43]:

$$\log(\delta) = 1.8/(110 + \lambda). \tag{1}$$

The distinguishing attack described in [43,50,54] is successful with advantage ε by finding vectors of length $\alpha\frac{q}{\sigma}$ with $\alpha = \sqrt{\ln(1/\varepsilon)/\pi}$. The length of the shortest vector that can be computed is $2^{2\sqrt{n \log q \log \delta}}$, leading to the inequality:

$$\alpha\frac{q}{\sigma} < 2^{2\sqrt{n \log q \log \delta}}. \tag{2}$$

Given $\sigma \geq 2\sqrt{n}$ from Regev's reduction [53], we can choose parameters for n and q matching Eq. (2) for the considered security parameter λ. The parameters we select for our application are summarized in Table 2.

Table 2. (R)LWE parameters used in our applications.

Security λ	n	$\log q$
80	256	80
128	512	120

4.2 Noise Analysis

Considering our framework of Fig. 1, Claude has at its disposal the homomorphic encryption of the symmetric key $\mathbf{C}^H(\mathsf{sk}_i^S)$, the homomorphic public key pk^H and the symmetric encrypted messages $\mathbf{C}^S(m_i)$. He has to perform the homomorphic evaluation of the symmetric decryption circuit, *i.e.* to perform homomorphic operations on the ciphertexts $\mathbf{C}^H(\mathsf{sk}_i^S)$ in order to get $\mathbf{C}^H(m_i)$, the homomorphic encryption of m_i. In this section, we study the error growth in these ciphertexts after the application of the homomorphic operations. As we are considering SWHE, we need to control the magnitude of the error and keep it below a critical level to ensure the correctness of a final ciphertext. This noise management is crucial for the applications, it is directly linked with the quantity of computation that the server can do for the client. We now study the error growth stemming from the homomorphic evaluation of FLIP. In this case, all the ciphertexts used by the server in the computation step will have a same starting error. The knowledge of this starting error (defined by some parameter) and its growth for additions and multiplications (in a chosen order) is enough to determine the amount of computation that can be performed correctly by the server.

In the remainder of this section we proceed in three steps. First we recall the error growth implied by the H.Add and H.Mul operations: for GSW-like HE it has already been done in [2,10,24,32,36]. As our homomorphic encryption schemes are slightly differently written to fit our applications (batched version to perform in parallel the same computations, generic notations for various frameworks), we give these error growth with our notations for completeness and consistency of the paper. Then we analyse the error for a sub-case of homomorphic product, namely H.Comb, which gives a practical tool to study the error growth in FLIP. As the asymmetric property of GSW multiplication and plaintext norm have been pointed out relatively to the error growth, we consider important to focus on both when analysing this error metric. Considering H.Comb types of operations is therefore suited to be consistent with this metric and is very important for practical purpose (in term of real life applications). Finally we analyse the error in a ciphertext output by FLIP and study some optimizations to reduce the noise growth further.

Error Growth in H.Add and H.Mul. We first need to evaluate the error growth of the basic homomorphic operations, the addition and the multiplication of ciphertexts. We use the analysis of [2] based on subgaussian distributions to study the error growth in these homomorphic operations. From a coefficient or a vector following a subgaussian distribution of parameter σ, we can bound its norm with overwhelming probability and then study the evolution of this parameter while performing the homomorphic operations. Hence we can bound the final error to ensure correctness.

For simplicity we use two notations arising from the error growth depending on the arithmetic of the underlying ring of the two schemes, γ the expansion factor (see [9]) and $Norm(m_j)$ such that:

- Batched GSW: $\gamma = 1$ and $Norm(m_j) = |m_j|$ (arithmetic in $\mathbb{Z}$).
- Ring GSW: $\gamma = n$ and $Norm(m_j) = ||m_j||_2$ (arithmetic in R).

Lemma 10 (*H*.Add **Error Growth**). *Suppose $\mathbf{C}_i$ for $1 \le i \le k$ are ciphertexts of a GSW based Homomorphic Encryption scheme with error components $\mathbf{e}_i$ of coefficients following a distribution of parameter σ_i. Let $\mathbf{C}_f = H.\mathsf{Add}(\mathbf{C}_i,$ for $1 \le i \le k)$ and $\mathbf{e}_f$ the related error with subgaussian parameter σ' such that:*

$$\sigma' = \sqrt{\sum_{i=1}^{k} \sigma_i^2} \quad or \quad \sigma' = \sigma\sqrt{k} \ if \ \sigma_i = \sigma, \forall i \in [k].$$

Lemma 11 (*H*.Mul **Error Growth**). *Suppose $\mathbf{C}_i$ for $1 \le i \le k$ are ciphertexts of a GSW based Homomorphic Encryption scheme with error components $\mathbf{e}_i$, of coefficients following a subgaussian distribution of parameter σ_i, and plaintext m_i. $\mathbf{C}_f$ is the result of a multiplicative homomorphic chain such that:*

$$\mathbf{C}_f = H.\mathsf{Mul}(\mathbf{C}_1, H.\mathsf{Mul}(\mathbf{C}_2, H.\mathsf{Mul}(\cdots, H.\mathsf{Mul}(\mathbf{C}_k, \mathbf{G})))),$$

and $\mathbf{e}_f$ the corresponding error with subgaussian parameter σ' such that:

$$\sigma' = \mathcal{O}\left(\sqrt{N\gamma}\sqrt{\sigma_1^2 + \sum_{i=2}^{k} \left(\sigma_i \Pi_{j=1}^{i-1} Norm(m_j)\right)^2}\right).$$

Lemmas 10 and 11 are proven in the extended version of this paper.

Error Growth in $H.\mathsf{Comb}$. For the sake of clarity, we formalize hereafter the comb homomorphic product $H.\mathsf{Comb}$ and the quantity σ_{comb} which stands for the subgaussian parameter. We study the error growth of $H.\mathsf{Comb}$ as we will use it as a tool for the error growth analysis of FLIP.

Definition 17 (Homomorphic Comb $H.\mathsf{Comb}$). *Let $\mathbf{C}_1, \cdots, \mathbf{C}_k$ be k ciphertexts of a GSW based Homomorphic Encryption scheme with error coefficients from independent distributions with same subgaussian parameter σ. We define $H.\mathsf{Comb}(y, \sigma, c, k) = H.\mathsf{Mul}(\mathbf{C}_1, \cdots, \mathbf{C}_k, \mathbf{G})$ where:*

- *$y = \sqrt{N\gamma}$ is a constant depending on the ring,*
- *$c = \max_{1 \le i \le k}(Norm(m_i))$ is a constant which depends on the plaintexts,*

and $\mathbf{C}_{comb} = H.\mathsf{Comb}(y, \sigma, c, k)$ as error components following a subgaussian distribution of parameter $\mathcal{O}(\sigma_{comb})$.

Lemma 12 (σ_{comb} Quantity). *Let $\mathbf{C}_1, \cdots, \mathbf{C}_k$ be k ciphertexts of a GSW based Homomorphic Encryption scheme with same error parameter σ and $\mathbf{C}_{comb} = H.\mathsf{Comb}(y, \sigma, c, k)$. Then we have:*

$$\sigma_{comb}(y, \sigma, c, k) = y\sigma c_k, \quad where\, c_k = \sqrt{\sum_{i=0}^{k-1} c^{2i}}.$$

Proof. Thanks to Lemma 11 we obtain:

$$\begin{aligned}
\sigma_{comb} &= \sqrt{N\gamma}\sqrt{\sigma^2 + \textstyle\sum_{i=2}^{k}(\sigma \Pi_{j=1}^{i-1} Norm(m_j))^2}, \\
\sigma_{comb} &= y\sqrt{\sigma^2 + \textstyle\sum_{i=2}^{k}(\sigma c^{i-1})^2}, \\
\sigma_{comb} &= y\sigma\sqrt{\textstyle\sum_{i=1}^{k}(c^{i-1})^2}, \\
\sigma_{comb} &= y\sigma c_k.
\end{aligned}$$

□

The compatibility of this comb structure with the asymmetric multiplicative error growth property of GSW enables us to easily quantify the error in our construction, with a better accuracy than computing the multiplicative depth. In order to minimize the quantity σ_{comb}, we choose the plaintext space such that $c = 1$ for freshly generated ciphertexts. The resulting $\sigma_{comb}(y, \sigma, 1, k)$ quantity is therefore $y\sigma\sqrt{k}$, growing less than linearly in the number of ciphertexts. Fixing the constant c to be 1 is usual with FHE. As we mostly consider Boolean circuits, it is usual to use plaintexts in $\{-1, 0, 1\}$ to encrypt bits, leading to $c = 1$ and therefore $c_k = \sqrt{k}$.

Error Growth in FLIP. In the previous paragraphs, we have evaluated the error growth in the basic homomorphic operations $H.\mathsf{Add}$, $H.\mathsf{Mul}$ and $H.\mathsf{Comb}$. We will use them as building blocks in order to evaluate the error growth in the homomorphic evaluation of FLIP. Coming back to the framework of Fig. 1, the error in the ciphertexts $\mathbf{C}^H(m_i)$ is of major importance as it will determine the possible number of homomorphic computations f that Claude is able to perform.

The main feature of the filter permutator model, considering FHE settings, is that it allows to handle ciphertexts having the same error level, whatever the number of output bits. Consequently all ciphertexts obtained by FLIP evaluation will have the same constant and small amount of noise and will be considered as fresh start for more computation.

Evaluating homomorphically the FLIP decryption (*resp.* encryption) algorithm consists in applying three steps of homomorphic operations on the ciphertexts $\mathbf{C}^H(\mathsf{sk}_i^S)$ in our application framework, each one encoding one bit of the key register. For each ciphertext bit, these steps are: a (bit) permutation, the application of the filtering function F and a XOR with the ciphertext (*resp.* plaintext). The (bit) permutation consists only in a public rearrangement of the key ciphertexts, leading to a noise-free operation. The last XOR is done with a freshly encrypted bit. Hence the error growth depends mostly on the homomorphic evaluation of F.

As $H.\mathsf{Dec}$ outputs quantities modulus 2, we can evaluate the XORs of F by $H.\mathsf{Add}$ and the ANDs by $H.\mathsf{Mul}$. We then determine the subgaussian parameter of the error of a ciphertext from the homomorphic evaluation of F. For a given encrypted key, this parameter will be the same for every homomorphic evaluation of FLIP and is computed from σ_{comb}.

Lemma 13 (Error Growth Evaluating F). *Let F be the FLIP filtering function in N variables defined in Sect. 3.3. Assume that $\mathbf{C}_i$ for $0 \leq i \leq N-1$ are ciphertexts of a GSW HE scheme with same subgaussian parameter σ and $c = 1$. We define $\mathbf{C}_F = H.\mathsf{Eval}(F, \mathbf{C}_i)$ the output of the homomorphic evaluation of the ciphertexts $\mathbf{C}_i$'s along the circuit F. Then the error parameter σ' is:*

$$\sigma' = \mathcal{O}\left(\sigma\sqrt{n_1 + y^2(n_2+n_3)}\right) \approx \mathcal{O}\left(\sigma y \sqrt{N}\right).$$

Proof. We first evaluate the noise brought by F for each of its components $L_{n_1}, Q_{n_2}, {}^{nb}\Delta^k$, defining the respective ciphertexts $\mathbf{C}_{L_{n_1}}, \mathbf{C}_{Q_{n_2}}, \mathbf{C}_{T_k}$ (the last one standing for one triangle only) and the subgaussian parameter of the respective error distributions (of the components of the error vectors) $\sigma_{L_{n_1}}, \sigma_{Q_{n_2}}, \sigma_{T_k}$:

- L_{n_1}: $\mathbf{C}_{L_{n_1}} = H.\mathsf{Eval}(L_{n_1}, \mathbf{C}_0, \cdots, \mathbf{C}_{n_1-1}) = H.\mathsf{Add}(\mathbf{C}_0, \cdots, \mathbf{C}_{n_1-1})$ then $\sigma_{L_{n_1}} = \sigma\sqrt{n_1}$.
- Q_{n_2}: $\mathbf{C}_{Q_{n_2}} = H.\mathsf{Add}(H.\mathsf{Mul}(\mathbf{C}_{n_1+2j}, \mathbf{C}_{n_1+2j+1}, \mathbf{G}))$ for $0 \leq j \leq n_2$. $H.\mathsf{Mul}(\mathbf{C}_{n_1+2j}, \mathbf{C}_{n_1+2j+1}, \mathbf{G}) = H.\mathsf{Comb}(y, \sigma, 1, 2)$ has subgaussian parameter $\mathcal{O}(\sigma_{comb}(y, \sigma, 1, 2)) = \mathcal{O}(y\sigma\sqrt{2})$ for $0 \leq j \leq n_2$.
  Then $\sigma_{Q_{n_2}} = \mathcal{O}(y\sigma\sqrt{2}\sqrt{\frac{n_2}{2}}) = \mathcal{O}(y\sigma\sqrt{n_2})$.
- T_k: $\mathbf{C}_{T_k} = H.\mathsf{Add}(H.\mathsf{Mul}(\mathbf{C}_{n_1+n_2+j+(i-1)(i-2)/2}; 1 \leq j \leq i); 1 \leq i \leq k)$.
  $\mathbf{C}_{T_k} = H.\mathsf{Add}(H.\mathsf{Comb}(y, \sigma, 1, i), 1 \leq i \leq k)$.
  then $\sigma_{T_k} = \mathcal{O}(\sqrt{\sum_{i=1}^k (y\sigma\sqrt{i})^2}) = \mathcal{O}(y\sigma\sqrt{\frac{k(k+1)}{2}})$.
  As ${}^{nb}\Delta^k$ is obtained by adding nb independent triangles, we get:
  $\mathbf{C}_{{}^{nb}\Delta^k} = H.\mathsf{Add}(\mathbf{C}_{T_k,i}, 1 \leq i \leq nb)$,
  and $\sigma_{{}^{nb}\Delta^k} = \mathcal{O}(y\sigma\sqrt{nb}\sqrt{\frac{k(k+1)}{2}}) = \mathcal{O}(y\sigma\sqrt{n_3})$.

By Pythagorean additivity the subgaussian parameter of $\mathbf{C}_F$ is finally:

$$\sigma' = \mathcal{O}(\sqrt{(\sigma\sqrt{n_1})^2 + (y\sigma\sqrt{n_2})^2 + (y\sigma\sqrt{n_3})^2}) = \mathcal{O}(\sigma\sqrt{n_1 + y^2(n_2 + n_3)}). \quad \square$$

Optimizations. The particular error growth in GSW Homomorphic Encryption enables to use more optimizations to reduce the error norm and perform more operations without increasing the parameter's size. The error growth in $H.\mathsf{Comb}$ depends on the quantity c_k derived from bounds on norms of the plaintexts; these quantities can be reduced using negative numbers. A typical example is in the LWE-based scheme to use $m \in \{-1, 0, 1\}$ rather than $\{0, 1\}$; the c_k quantity is the same and in average the sums in $\mathbb{Z}$ are smaller. Then the norm $|\sum m_i|$ is smaller which is important when multiplying. Conserving this norm as low as possible gives better bounds and c_k coefficients, leading to smaller noise when performing distinct levels of operations. An equivalent way of minimizing the error growth is to still use $\mathcal{M} = \{0, 1\}$ but with $H.\mathsf{Add}(\mathbf{C}_1, \mathbf{C}_2) = \mathbf{C}_1 \pm \mathbf{C}_2$. This homomorphic addition is still correct because:

$$\mathbf{S} - \mathbf{C}_2 = -\mathbf{E}_2' - \begin{pmatrix} m_{2,1} \cdot \mathbf{s}_1^\top \\ \hline \vdots \\ \hline m_{2,r} \cdot \mathbf{s}_r^\top \end{pmatrix} \mathbf{G} = \mathbf{E}_2'' + \begin{pmatrix} -m_{2,1} \cdot \mathbf{s}_1^\top \\ \hline \vdots \\ \hline -m_{2,r} \cdot \mathbf{s}_r^\top \end{pmatrix},$$

where the coefficients in $\mathbf{E}_2''$ rows follow distribution of same subgaussian parameter as the one in $\mathbf{E}_2'$ by homogeneity and $-m = m \mod 2$.

4.3 Concrete Results

Contrary to other works published in the context of symmetric encryption schemes for efficient FHE [1,11,31], our primary focus is not on the performances (see SHIELD [38] for efficient implementation of Ring-GSW) but rather on the error growth. As pointed out in [11], in most of these previous works, after the decryption process the noise inside the ciphertexts was too high to perform any other operation on them, whereas it is the main motivation for a practical use of FHE.

In this section, we consequently provide experimental results about this error growth in the ciphertexts after different operations evaluated on the Ring GSW scheme. As the link between subgaussian parameter, ciphertext error and homomorphic computation is not direct, we make some choices for representing these results focusing on giving intuition on how the error behaves.

The choice of the Ring GSW setting rather than Batched GSW is for convenience. It allows to deal with smaller matrices and faster evaluations, providing the same confirmation on the heuristic error growth. We give the parameters n and ℓ defining the polynomial ring and fix $\sigma = 2\lceil\sqrt{n}\rceil$ for the error distribution.

An efficient way of measuring the error growth within the ciphertexts is to compute the difference by applying the rounding $\lfloor\cdot\rceil_2$ in $H.\mathsf{Dec}$ between various ciphertexts with known plaintext. This difference (for each polynomial coefficient or vector component) corresponds to the amount of noise contained in this ciphertext. The correctness requires this quantity to be inferior to $2^{\ell-2}$. Then, considering its logarithm in base 2, it enables to have an intuitive and practical measure of the ciphertext noise: this quantity grows with the homomorphic operations until this log equals $\ell - 2$. Concretely, in our experiments we encrypt polynomials being $m = 0$ or $m = 1$, compute on the constant coefficient the quantity $e = |(\langle \mathbf{s}, \mathbf{c}_\ell \rangle - m2^{\ell-1}) \mod q|$, and give its logarithm. We give another quantity in order to provide intuition about the homomorphic computation possibilities over the ciphertexts, by simply computing a percentage of the actual level of noise relatively to the maximal level $\ell - 2$.

Remark 4. The quantity exhibited by our measures is roughly the subgaussian parameter of the distribution of the error contained in the ciphertexts. Considering the simpler case of a real Gaussian distribution $\mathcal{N}(0, \sigma^2)$, the difference that we compute then follows a half normal distribution with mean $\sigma\frac{\sqrt{2}}{\sqrt{\pi}}$.

We based our prototype implementation on the NTL library combined with GMP and the discrete gaussian sampler of BLISS [23]. We report in Table 3 experimental results on the error growth for different RLWE and FLIP parameters, based on an average over a hundred of samples.

The results confirm the quasi-additive error growth when considering the specific metric of GSW given by the asymptotic bounds. The main conclusion of these results is that the error inside the ciphertexts after a homomorphic evaluation of FLIP is of the same order of magnitude as the one after a multiplication. The only difference between these noise increases is a term provided by the root of the symmetric key register size, that is linear in λ. Therefore, with the FLIP construction the error growth is roughly the basic multiplicative error growth of two ciphertexts. Hence, we conclude that filter permutators as FLIP release the bottleneck of evaluating symmetric decryption, and lead the further improvement of the calculus delegation framework to depend overwhelmingly on improvements of the homomorphic operations.

Table 3. Experimental error growth of Ring-GSW. Fresh, H.Add, H.Mul and H.Eval(FLIP) respectively stands for the noise e measure after a fresh homomorphic encryption, the homomorphic addition of two fresh ciphertexts, the homomorphic multiplication of two fresh ciphertexts and the homomorphic evaluation of FLIP on fresh ciphertexts. The first value is the log of the error e inside the corresponding ciphertexts and the percentage represents the proportion of the noise with respect to the capacity of decryption (*i.e.* $\ell - 2$).

Ring (n, ℓ)		FLIP	Fresh		H.Add		H.Mul		H.Eval(FLIP)	
			$\log e$	%	$\log e$	%	$\log e$	%	$\log e$	%
256	80	$(42, 128, {}^8\Delta^9)$	13,07	17 %	13,96	18 %	19,82	25 %	24,71	31 %
512	120	$(82, 224, {}^8\Delta^{16})$	14,68	12 %	15,14	13 %	23,27	20 %	28,77	24 %

4.4 Performances for 2nd-Generation Schemes

Despite our new constructions are primarily designed for 3rd-generation FHE, a look at Table 4 suggests that also from the multiplicative depth point of view, FLIP instances bring good results compared to their natural competitors such as LowMC [1] and Trivium/Kreyvium [11]. In Trivium/Kreyvium, the multiplicative depth of the decryption circuit is at most 13, while the LowMC family has a record multiplicative depth of 11 which is still larger than our FLIP instances. For completeness, we finally investigated the performances of some instances of FLIP for 2nd-generation FHE schemes using HElib, as reported in Table 5, where the latency is the amount of time (in seconds) needed to homomorphically decrypt (Nb * Number of Slots) bits, and the throughput is calculated as (Nb * Number of Slots * 60)/latency. As in [11], we have considered two noise levels: a first one that does not allow any other operations on the ciphertexts, and a second one where we allow operations of multiplicative depth up to 7. Note that the (max) parenthesis in the Nb column recalls that for Trivium/Kreyvium, the homomorphic capacity decreases with the number of keystream bits generated, which therefore bounds the number of such bits before re-keying. We observe that for 80-bit security, our instances outperform the ones based on Trivium. As for 128-bit security, the gap between our instances and Kreyvium is limited (despite the larger state of FLIP), and LowMC has better throughput in this context. Note also that our results correspond to the evaluation of the F function of FLIP (we verified that the time needed to generate the permutations only marginally affects the overall performances of homomorphic FLIP evaluations). We finally mention that these results should certainly not be viewed as strict comparisons, since obtained on different computers and for relatively new ciphers for which we have limited understanding of the security margins (especially

Table 4. Multiplicative depth of different symmetric ciphers.

Algorithm	Reference	Multiplicative depth	Security
SIMON-32/64	[42]	32	64
Trivium-12	[11]	12	80
Trivium-13	[11]	13	80
LowMc-80	[1]	11	80
FLIP$(42, 128, {}^{8}\Delta^{9})$	This work	$\lceil \log 9 \rceil = 4$	80
AES-128	[15,31]	40	128
SIMON-64/128	[42]	44	128
Prince	[22]	24	128
Kreyvium-12	[11]	12	128
Kreyvium-13	[11]	13	128
LowMc-128	[1]	12	128
FLIP$(82, 224, {}^{8}\Delta^{16})$	This work	$\lceil \log 16 \rceil = 4$	128

Table 5. Timings of the homomorphic evaluation of several instances of the Boolean function of FLIP using HElib on an Intel Core i7-3770. The other results are taken from [11]. L and Number of Slots are HElib parameters which stand respectively for the level of noise and the number of bits packed in one ciphertext. (Nb * Number of Slots) corresponds to the number of decrypted bits.

Algorithm	Security	Nb	L	Number of Slots	Latency (sec)	Throughput (bits/min)
Trivium-12	80	45 (max)	12	600	1417.4	1143.0
	80	45 (max)	19	720	4420.3	439.8
Trivium-13	80	136 (max)	13	600	3650.3	1341.3
	80	136 (max)	20	720	11379.7	516.3
Kreyvium-12	128	42 (max)	12	504	1715.0	740.5
	128	42 (max)	19	756	4956.0	384.4
Kreyvium-13	128	124 (max)	13	682	3987.2	1272.6
	128	124 (max)	20	420	12450.8	286.8
LowMC-128	$? \leq 128$	256	13	682	3368.8	3109.6
	$? \leq 128$	256	20	480	9977.1	739.0
$\mathsf{FLIP}(42, 128, {}^{8}\Delta^{9})$	80	1	5	378	4.72	4805.08
	80	1	12	600	17.39	2070.16
$\mathsf{FLIP}(82, 224, {}^{8}\Delta^{16})$	128	1	6	630	14.53	2601.51
	128	1	13	720	102.51	421.42

for LowMC [19,21] and FLIP). So they should mainly be seen as an indication that besides their excellent features from the FHE capacity point-of-view, filter permutators inherently have good properties for efficient 2nd-generation FHE implementations as well.

5 Conclusions and Open Problems

In the context of our Homomorphic Encryption - Symmetric Encryption framework, where most of the computations are delegated to a server, we have designed a symmetric encryption scheme which fits the FHE settings, with as main goal to get the homomorphic evaluation of the symmetric decryption circuit as cheap as possible, with respect to the error growth. In particular the error growth obtained by our construction, only one level of multiplication considering the metric of third generation FHE, achieves the lowest bound we can get with a secure symmetric encryption scheme. The use of zero-noise operations as permutations enables us to combine the advantages of block ciphers and stream ciphers evaluation, namely constant noise on the one hand and starting low noise on the other hand. As a result, the homomorphic evaluation of filter permutators can be made insignificant relatively to a complete FHE framework.

The general construction of our encryption scheme – *i.e.* the filter permutator – and its FLIP instances are admittedly provocative. As a result, we believe an important contribution of this paper is to open a wide design space of symmetric constructions to investigate, ranging from the very efficient solutions we suggest to more classical stream ciphers such as filter generators. Such a design space leads to various interesting directions for further research. Overall, the main question raised by filter permutators is whether it is possible to build a secure symmetric encryption scheme with aggressively reduced algebraic degree. Such a question naturally relates to several more concrete problems. First, and probably most importantly, additional cryptanalysis is needed in view of the non-standard design principles exploited in filter permutators. It typically includes algebraic attack taking advantage of the sparsity of their systems of equations, attacks exploiting the imbalances at the input of the filter, and the possibility to exploit chosen IVs to improve those attacks. Second, our analyses also raise interesting problems in the field of Boolean functions, *e.g.* the analysis of such functions with non-uniform input distributions and the investigation of the best fixed degree approximations of a Boolean function (which is needed in our study of higher-order correlation attacks). More directly related to the FLIP instances, it would also be interesting to refine our security analyses, with a stronger focus on the attacks data complexity, and to evaluate whether instances with smaller key register could be sufficiently secure. In case of new cryptanalysis results, the design tweaks we suggest in the paper are yet another interesting research path. Eventually, and from the FHE application point-of-view, optimizing the implementations of filter permutators, *e.g.* by taking advantage of parallel computing clusters that we did not exploit so far, would be useful in order to evaluate their applicability to real-world scenarii.

Acknowledgements. We are highly grateful to Sébastien Duval, Virginie Lallemand and Yann Rotella for sharing their ideas about guess and determine attacks before the publication of this paper, which allowed us to modify the instances of FLIP accordingly. We are also indebted to Anne Canteaut for numerous useful suggestions about the design of filter permutators, and for putting forward some important open problems they raise. Finally, we would like to thank Thierry Berger, Sergiu Carpov, Rafaël Delpino, Malika Izabachene, Nicky Mouha, Thomas Prest and Renaud Sirdey for their feedback about early (and less early) versions of this paper. This work was funded in parts by the H2020 ICT COST CryptoAction, by the H2020 ICT Project SAFECrypto, by the H2020 ERC Staring Grant CRASH and by the INNOVIRIS SCAUT project. François-Xavier Standaert is a research associate of the Belgian Fund for Scientific Research (F.R.S.-FNRS).

References

1. Albrecht, M.R., Rechberger, C., Schneider, T., Tiessen, T., Zohner, M.: Ciphers for MPC and FHE. In: Advances in Cryptology - EUROCRYPT 2015–34th Annual International Conference on the Theory and Applications of Cryptographic Techniques, Sofia, Bulgaria, April 26–30, 2015, Proceedings, Part I. pp. 430–454 (2015)
2. Alperin-Sheriff, J., Peikert, C.: Faster bootstrapping with polynomial error. In: Advances in Cryptology - CRYPTO 2014 - 34th Annual Cryptology Conference, Santa Barbara, CA, USA, August 17-21, 2014, Proceedings, Part I, pp. 297–314 (2014)
3. Anderson, R.J.: Searching for the optimum correlation attack. In: Fast Software Encryption: Second International Workshop. Leuven, Belgium, 14–16 December 1994, Proceedings, pp. 137–143 (1994)
4. Armknecht, F., Carlet, C., Gaborit, P., Künzli, S., Meier, W., Ruatta, O.: Efficient computation of algebraic immunity for algebraic and fast algebraic attacks. In: Vaudenay, S. (ed.) EUROCRYPT 2006. LNCS, vol. 4004, pp. 147–164. Springer, Heidelberg (2006)
5. Bellare, M., Yee, B.S.: Forward-security in private-key cryptography. IACR Cryptology ePrint Archive 2001, 35 (2001)
6. Blum, A., Kalai, A., Wasserman, H.: Noise-tolerant learning, the parity problem, and the statistical query model. J. ACM **50**(4), 506–519 (2003)
7. Blum, M., Micali, S.: How to generate cryptographically strong sequences of pseudo random bits. SIAM J. Comput. **13**(4), 850–864 (1984)
8. Boura, C., Canteaut, A.: Zero-sum distinguishers for iterated permutations and application to keccak-f and hamsi-256. In: Biryukov, A., Gong, G., Stinson, D.R. (eds.) Selected Areas in Cryptography. LNCS, vol. 6544, pp. 1–17. Springer, Heidelberg (2011)
9. Brakerski, Z., Gentry, C., Vaikuntanathan, V.: (leveled) fully homomorphic encryption without bootstrapping. In: Innovations in Theoretical Computer Science 2012, Cambridge, MA, USA, January 8–10, 2012, pp. 309–325 (2012)
10. Brakerski, Z., Vaikuntanathan, V.: Lattice-based FHE as secure as PKE. In: Innovations in Theoretical Computer Science, ITCS 2014, Princeton, NJ, USA, January 12–14, 2014, pp. 1–12 (2014)
11. Canteaut, A., Carpov, S., Fontaine, C., Lepoint, T., Naya-Plasencia, M., Paillier, P., Sirdey, R.: Stream ciphers: A practical solution for efficient homomorphic-ciphertext. IACR Cryptology ePrint Archive 2015, 113 (2015)
12. Carlet, C.: Boolean Models and Methods in Mathematics, Computer Science, and Engineering, chap. Boolean Functions for Cryptography and Error Correcting Codes, pp. 257–397 (2010)
13. Carlet, C., Tang, D.: Enhanced Boolean functions suitable for the filter model of pseudo-random generator. Des. Codes Crypt. **76**(3), 571–587 (2015)
14. Chen, Y., Nguyen, P.Q.: BKZ 2.0: Better lattice security estimates. In: Wang, X., Lee, D.H. (eds.) ASIACRYPT 2011. LNCS, vol. 7073, pp. 1–20. Springer, Heidelberg (2011)
15. Coron, J.-S., Lepoint, T., Tibouchi, M.: Scale-invariant fully homomorphic encryption over the integers. In: Krawczyk, H. (ed.) PKC 2014. LNCS, vol. 8383, pp. 311–328. Springer, Heidelberg (2014)
16. Courtois, N.T.: Higher order correlation attacks, XL algorithm and cryptanalysis of toyocrypt. In: Lee, P.J., Lim, C.H. (eds.) ICISC 2002. LNCS, vol. 2587, pp. 182–199. Springer, Heidelberg (2003)

17. Courtois, N.T.: Fast algebraic attacks on stream ciphers with linear feedback. In: Boneh, D. (ed.) CRYPTO 2003. LNCS, vol. 2729, pp. 176–194. Springer, Heidelberg (2003)
18. Courtois, N.T., Meier, W.: Algebraic attacks on stream ciphers with linear feedback. In: Biham, E. (ed.) Advances in Cryptology – EUROCRYPT 2003. LNCS, vol. 2656, pp. 345–359. Springer, Heidelberg (2003)
19. Dinur, I., Liu, Y., Meier, W., Wang, Q.: Optimized interpolation attacks on lowmc. IACR Cryptology ePrint Archive 2015, 418 (2015)
20. Dinur, I., Shamir, A.: Cube attacks on tweakable black box polynomials. In: Joux, A. (ed.) EUROCRYPT 2009. LNCS, vol. 5479, pp. 278–299. Springer, Heidelberg (2009)
21. Dobraunig, C., Eichlseder, M., Mendel, F.: Higher-order cryptanalysis of lowmc. IACR Cryptology ePrint Archive 2015, 407 (2015)
22. Doröz, Y., Shahverdi, A., Eisenbarth, T., Sunar, B.: Toward practical homomorphic evaluation of block ciphers using prince. In: Böhme, R., Brenner, M., Moore, T., Smith, M. (eds.) FC 2014 Workshops. LNCS, vol. 8438, pp. 208–220. Springer, Heidelberg (2014)
23. Ducas, L., Durmus, A., Lepoint, T., Lyubashevsky, V.: Lattice signatures and bimodal gaussians. In: Canetti, R., Garay, J.A. (eds.) CRYPTO 2013, Part I. LNCS, vol. 8042, pp. 40–56. Springer, Heidelberg (2013)
24. Ducas, L., Micciancio, D.: FHEW: Bootstrapping homomorphic encryption in less than a second. In: Oswald, E., Fischlin, M. (eds.) Advances in Cryptology – EUROCRYPT 2015. LNCS, vol. 9056, pp. 617–640. Springer, Heidelberg (2015)
25. Duval, S., Lallemand, V., Rotella, Y.: Cryptanalysis of the FLIP family of stream ciphers. Cryptology ePrint Archive, Report 2016/271 (2016). http://eprint.iacr.org/
26. Fan, J., Vercauteren, F.: Somewhat practical fully homomorphic encryption. IACR Cryptology ePrint Archive 2012, 144 (2012)
27. Faugère, J.C.: A new efficient algorithm for computing grobner bases (f4). J. Pure Appl. Algebra **139**(13), 61–88 (1999)
28. Fischer, S.: Algebraic immunity of S-boxes and augmented functions. In: Fischer, S., Meier, W. (eds.) Fast Software Encryption. LNCS, vol. 4593, pp. 366–381. Springer, Heidelberg (2007)
29. Gama, N., Nguyen, P.Q.: Predicting lattice reduction. In: Smart, N. (ed.) Advances in Cryptology – EUROCRYPT 2008. LNCS, vol. 4965, pp. 31–51. Springer, Heidelberg (2008)
30. Gentry, C.: Fully homomorphic encryption using ideal lattices. In: Proceedings of the 41st Annual ACM Symposium on Theory of Computing, STOC 2009, Bethesda, MD, USA, May 31 - June 2, 2009, pp. 169–178 (2009)
31. Gentry, C., Halevi, S., Smart, N.P.: Homomorphic evaluation of the AES circuit. In: Safavi-Naini, R., Canetti, R. (eds.) Advances in Cryptology – CRYPTO 2012. LNCS, vol. 7417, pp. 850–867. Springer, Heidelberg (2012)
32. Gentry, C., Sahai, A., Waters, B.: Homomorphic encryption from learning with errors: conceptually-simpler, asymptotically-faster, attribute-based. In: Canetti, R., Garay, J.A. (eds.) Advances in Cryptology – CRYPTO 2013. LNCS, vol. 8042, pp. 75–92. Springer, Heidelberg (2013)
33. Gérard, B., Grosso, V., Naya-Plasencia, M., Standaert, F.-X.: Block ciphers that are easier to mask: How far can we go? In: Bertoni, G., Coron, J.-S. (eds.) CHES 2013. LNCS, vol. 8086, pp. 383–399. Springer, Heidelberg (2013)

34. Grosso, V., Leurent, G., Standaert, F.-X., Varici, K.: LS-Designs: Bitslice encryption for efficient masked software implementations. In: Cid, C., Rechberger, C. (eds.) FSE 2014. LNCS, vol. 8540, pp. 18–37. Springer, Heidelberg (2015)
35. Halevi, S., Shoup, V.: Algorithms in HElib. In: Garay, J.A., Gennaro, R. (eds.) CRYPTO 2014, Part I. LNCS, vol. 8616, pp. 554–571. Springer, Heidelberg (2014)
36. Hiromasa, R., Abe, M., Okamoto, T.: Packing messages and optimizing bootstrapping in GSW-FHE. In: Katz, J. (ed.) PKC 2015. LNCS, vol. 9020, pp. 699–715. Springer, Heidelberg (2015)
37. Katz, J., Lindell, Y.: Introduction to Modern Cryptography. Chapman and Hall/CRC Press, Boca Raton (2007)
38. Khedr, A., Gulak, P.G., Vaikuntanathan, V.: SHIELD: Scalable homomorphic implementation of encrypted data-classifiers. IACR Cryptology ePrint Archive 2014, 838 (2014)
39. Knellwolf, S., Meier, W., Naya-Plasencia, M.: Conditional differential cryptanalysis of NLFSR-based cryptosystems. In: Abe, M. (ed.) Advances in Cryptology - ASIACRYPT 2010. LNCS, vol. 6477, pp. 130–145. Springer, Heidelberg (2010)
40. Knudsen, L.R., Wagner, D.: Integral cryptanalysis. In: Daemen, J., Rijmen, V. (eds.) FSE 2002. LNCS, vol. 2365, pp. 112–127. Springer, Heidelberg (2002)
41. Knuth, D.E.: The Art of Computer Programming. Seminumerical Algorithms. Addison-Wesley, Boston (1969)
42. Lepoint, T., Naehrig, M.: A comparison of the homomorphic encryption schemes FV and YASHE. In: Pointcheval, D., Vergnaud, D. (eds.) AFRICACRYPT 2014. LNCS, vol. 8469, pp. 318–335. Springer, Heidelberg (2014)
43. Lindner, R., Peikert, C.: Better key sizes (and attacks) for lwe-based encryption. In: Kiayias, A. (ed.) Topics in Cryptology – CT-RSA 2011. LNCS, vol. 6558, pp. 319–339. Springer, Heidelberg (2011)
44. Liskov, M., Rivest, R.L., Wagner, D.: Tweakable block ciphers. J. Cryptology **24**(3), 588–613 (2011)
45. Luby, M., Rackoff, C.: How to construct pseudorandom permutations from pseudorandom functions. SIAM J. Comput. **17**(2), 373–386 (1988)
46. Lyubashevsky, V., Peikert, C., Regev, O.: On ideal lattices and learning with errors over rings. In: Gilbert, H. (ed.) EUROCRYPT 2010. LNCS, vol. 6110, pp. 1–23. Springer, Heidelberg (2010)
47. Meier, W.: Fast correlation attacks: Methods and countermeasures. In: Joux, A. (ed.) Fast Software Encryption. LNCS, vol. 6733, pp. 55–67. Springer, Heidelberg (2011)
48. Meier, W., Staffelbach, O.: Fast correlation attacks on stream ciphers. In: Günther, C.G. (ed.) EUROCRYPT 1988. LNCS, vol. 330, pp. 301–314. Springer, Heidelberg (1988)
49. Micciancio, D., Peikert, C.: Trapdoors for lattices: simpler, tighter, faster, smaller. In: Pointcheval, D., Johansson, T. (eds.) EUROCRYPT 2012. LNCS, vol. 7237, pp. 700–718. Springer, Heidelberg (2012)
50. Micciancio, D., Regev, O.: Lattice-based cryptography. Springer, Heidelberg (2009)
51. Naehrig, M., Lauter, K.E., Vaikuntanathan, V.: Can homomorphic encryption be practical? In: Proceedings of the 3rd ACM Cloud Computing Security Workshop, CCSW 2011, Chicago, IL, USA, October 21, 2011, pp. 113–124 (2011)
52. Piret, G., Roche, T., Carlet, C.: PICARO - A block cipher allowing efficient higher-order side-channel resistance. In: Bao, F., Samarati, P., Zhou, J. (eds.) Applied Cryptography and Network Security. LNCS, vol. 7341, pp. 311–328. Springer, Heidelberg (2012)

53. Regev, O.: On lattices, learning with errors, random linear codes, and cryptography. In: Proceedings of the 37th Annual ACM Symposium on Theory of Computing, Baltimore, MD, USA, May 22–24, 2005, pp. 84–93 (2005)
54. Rückert, M., Schneider, M.: Estimating the security of lattice-based cryptosystems. IACR Cryptology ePrint Archive 2010, 137 (2010)
55. Schnorr, C., Euchner, M.: Lattice basis reduction: Improved practical algorithms and solving subset sum problems. Math. Program. **66**, 181–199 (1994)
56. Siegenthaler, T.: Decrypting a class of stream ciphers using ciphertext only. IEEE Trans. Comput. **34**(1), 81–85 (1985)
57. Standaert, F.-X., Pereira, O., Yu, Y.: Leakage-resilient symmetric cryptography under empirically verifiable assumptions. In: Canetti, R., Garay, J.A. (eds.) CRYPTO 2013, Part I. LNCS, vol. 8042, pp. 335–352. Springer, Heidelberg (2013)
58. Vershynin, R.: Introduction to the non-asymptotic analysis of random matrices. CoRR abs/1011.3027 (2010)
59. Wiedemann, D.H.: Solving sparse linear equations over finite fields. IEEE Trans. Inf. Theor. **32**(1), 54–62 (1986)

Improved Differential-Linear Cryptanalysis of 7-Round Chaskey with Partitioning

Gaëtan Leurent(✉)

Inria, Paris, France
Gaetan.Leurent@inria.fr

Abstract. In this work we study the security of Chaskey, a recent lightweight MAC designed by Mouha *et al.*, currently being considered for standardization by ISO/IEC and ITU-T. Chaskey uses an ARX structure very similar to SipHash. We present the first cryptanalysis of Chaskey in the single user setting, with a differential-linear attack against 6 and 7 rounds, hinting that the full version of Chaskey with 8 rounds has a rather small security margin. In response to these attacks, a 12-round version has been proposed by the designers.

To improve the complexity of the differential-linear cryptanalysis, we refine a partitioning technique recently proposed by Biham and Carmeli to improve the linear cryptanalysis of addition operations. We also propose an analogue improvement of differential cryptanalysis of addition operations. Roughly speaking, these techniques reduce the data complexity of linear and differential attacks, at the cost of more processing time per data. It can be seen as the analogue for ARX ciphers of partial key guess and partial decryption for SBox-based ciphers.

When applied to the differential-linear attack against Chaskey, this partitioning technique greatly reduces the data complexity, and this also results in a reduced time complexity. While a basic differential-linear attack on 7 round takes 2^{78} data and time (respectively 2^{35} for 6 rounds), the improved attack requires only 2^{48} data and 2^{67} time (respectively 2^{25} data and 2^{29} time for 6 rounds). We also show an application of the partitioning technique to FEAL-8X, and we hope that this technique will lead to a better understanding of the security of ARX designs.

Keywords: Differential cryptanalysis · Linear cryptanalysis · ARX · Addition · Partitioning · Chaskey · FEAL

1 Introduction

Linear cryptanalysis and differential cryptanalysis are the two major cryptanalysis techniques in symmetric cryptography. Differential cryptanalysis was introduced by Biham and Shamir in 1990 [6], by studying the propagation of differences in a cipher. Linear cryptanalysis was discovered in 1992 by Matsui [25,26], using a linear approximation of the non-linear round function.

In order to apply differential cryptanalysis (respectively, linear cryptanalysis), the cryptanalyst has to build differentials (resp. linear approximations)

M. Fischlin and J.-S. Coron (Eds.): EUROCRYPT 2016, Part I, LNCS 9665, pp. 344–371, 2016.
DOI: 10.1007/978-3-662-49890-3_14

for each round of a cipher, such the output difference of a round matches the input difference of the next round (resp. linear masks). The probability of the full differential or the imbalance of the full linear approximation is computed by multiplying the probabilities (respectively imbalances) of each round. This yields a statistical distinguisher for several rounds:

- A differential distinguisher is given by a plaintext difference δ_P and a ciphertext difference δ_C, so that the corresponding probability p is non-negligible:
$$p = \Pr\left[E(P \oplus \delta_P) = E(P) \oplus \delta_C\right] \gg 2^{-n}.$$
The attacker collects $D = \mathcal{O}(1/p)$ pairs of plaintexts (P_i, P_i') with $P_i' = P_i \oplus \delta_P$, and checks whether a pair of corresponding ciphertexts satisfies $C_i' = C_i \oplus \delta_C$. This happens with high probability for the cipher, but with low probability for a random permutation.
- A linear distinguisher is given by a plaintext mask χ_P and a ciphertext mask χ_C, so that the corresponding imbalance[1] ε is non-negligible:
$$\varepsilon = \left|2 \cdot \Pr\left[P[\chi_P] = C[\chi_C]\right] - 1\right| \gg 2^{-n/2}.$$
The attacker collects $D = \mathcal{O}(1/\varepsilon^2)$ known plaintexts P_i and the corresponding ciphertexts C_i, and computes the observed imbalance $\hat{\varepsilon}$:
$$\hat{\varepsilon} = \left|2 \cdot \#\left\{i : P_i[\chi_P] = C_i[\chi_C]\right\}/D - 1\right|.$$
The observed imbalance is close to ε for the attacked cipher, and smaller than $1/\sqrt{D}$ (with high probability) for a random function.

Last Round Attacks. The distinguishers are usually extended to a key-recovery attack on a few more rounds using partial decryption. The main idea is to guess the subkeys of the last rounds, and to compute an intermediate state value from the ciphertext and the subkeys. This allows to apply the distinguisher on the intermediate value: if the subkey guess was correct the distinguisher should succeed, but it is expected to fail for wrong key guesses. In a Feistel cipher, the subkey for one round is usually much shorter than the master key, so that this attack recovers a partial key without considering the remaining bits. This allows a divide and conquer strategy were the remaining key bits are recovered by exhaustive search. For an SBox-based cipher, this technique can be applied if the difference δ_C or the linear mask χ_C only affect a small number of SBoxes, because guessing the key bits affecting those SBoxes is sufficient to invert the last round.

ARX Ciphers. In this paper we study the application of differential and linear cryptanalysis to ARX ciphers. ARX ciphers are a popular category of ciphers built using only additions ($x \boxplus y$), bit rotations ($x \lll n$), and bitwise xors ($x \oplus y$). These simple operations are very efficient in software and in hardware,

[1] The imbalance is also called correlation.

but they interact in complex ways that make analysis difficult and is expected to provide security. ARX constructions have been used for block ciphers (*e.g.* TEA, XTEA, FEAL, Speck), stream ciphers (*e.g.* Salsa20, ChaCha), hash functions (*e.g.* Skein, BLAKE), and for MAC algorithms (*e.g.* SipHash, Chaskey).

The only non-linear operation in ARX ciphers is the modular addition. Its linear and differential properties are well understood [14,24,29,32,33,37,39], and differential and linear cryptanalysis have been use to analyze many ARX designs (see for instance the following papers: [4,8,16,21,22,26,40,41]).

However, there is no simple way to extend differential or linear distinguishers to last-round attack for ARX ciphers. The problem is that they typically have 32-bit or 64-bit words, but differential and linear characteristics have a few active bits in each word[2]. Therefore a large portion of the key has to be guessed in order to perform partial decryption, and this doesn't give efficient attacks.

Besides, differential and linear cryptanalysis usually reach a limited number of rounds in ARX designs because the trails diverge quickly and we don't have good techniques to keep a low number of active bits. This should be contrasted with SBox-based designs where it is sometimes possible to build iterative trails, or trails with only a few active SBoxes per round. For instance, this is case for differential characteristics in DES [7] and linear trails in PRESENT [13].

Because of this, cryptanalysis methods that allow to divide a cipher E into two sub-ciphers $E = E_\perp \circ E_\top$ are particularly interesting for the analysis of ARX designs. In particular this is the case with boomerang attacks [38] and differential-linear cryptanalysis [5,20]. A boomerang attack uses differentials with probabilities $p_\top$ and $p_\perp$ in $E_\top$ and $E_\perp$, to build a distinguisher with complexity $\mathcal{O}(1/p_\top^2 p_\perp^2)$. A differential-linear attack uses a differential with probability p for $E_\top$ and a linear approximation with imbalance ε for $E_\perp$ to build a distinguisher with complexity about $\mathcal{O}(1/p^2\varepsilon^4)$ (using a heuristic analysis).

Our Results. In this paper, we consider improved techniques to attack ARX ciphers, with application to Chaskey. Since Chaskey has a strong diffusion, we start with differential-linear cryptanalysis, and we study in detail how to build a good differential-linear distinguisher, and how to improve the attack with partial key guesses.

Our main technique follows a recent paper by Biham and Carmeli [3], by partitioning the available data according to some plaintext and ciphertext bits. In each subset, some data bits have a fixed value and we can combine this information with key bit guesses to deduce bits after the key addition. These known bits result in improved probabilities for differential and linear cryptanalysis. While Biham and Carmeli considered partitioning with a single control bit (*i.e.* two partitions), and only for linear cryptanalysis, we extend this analysis to multiple control bits, and also apply it to differential cryptanalysis.

When applied to differential and linear cryptanalysis, this results in a significant reduction of the data complexity. Alternatively, we can extend the attack to a larger number of rounds with the same data complexity. Those results are

[2] A notable counterexample is FEAL, which uses only 8-bit additions.

Table 1. Key-recovery attacks on Chaskey

Rounds	Data	Time	Gain	
6	2^{35}	2^{35}	1 bit	Differential-Linear
6	2^{25}	$2^{28.6}$	6 bits	Differential-Linear with partitioning
7	2^{78}	2^{78}	1 bit	Differential-Linear
7	2^{48}	2^{67}	6 bits	Differential-Linear with partitioning

very similar to the effect of partial key guess and partial decryption in a last-round attack: we turn a distinguisher into a key recovery attack, and we can add some rounds to the distinguisher. While this can increase the time complexity in some cases, we show that the reduced data complexity usually leads to a reduced time complexity. In particular, we adapt a convolution technique used for linear cryptanalysis with partial key guesses [15] in the context of partitioning.

These techniques result in significant improvements over the basic differential-linear technique: for 7 rounds of Chaskey (respectively 6 rounds), the differential-linear distinguisher requires 2^{78} data and time (respectively 2^{35}), but this can be reduced to 2^{48} data and 2^{67} time (respectively 2^{25} data and 2^{29} time) (see Table 1). The full version of Chaskey has 8 rounds, and is claimed to be secure against attacks with 2^{48} data and 2^{80} time.

The paper is organized as follows: we first explain the partitioning technique for linear cryptanalysis in Sect. 2 and for differential cryptanalysis in Sect. 3. We discuss the time complexity of the attacks in Sect. 4. Then we demonstrate the application of this technique to the differential-linear cryptanalysis of Chaskey in Sect. 5. Finally, we show how to apply the partitioning technique to reduce the data complexity of linear cryptanalysis against FEAL-8X in Appendix A.

2 Linear Analysis of Addition

We first discuss linear cryptanalysis applied to addition operations, and the improvement using partitioning. We describe the linear approximations using linear masks; for instance an approximation for E is written as $\Pr\left[E(x)[\chi'] = x[\chi]\right] = 1/2 \pm \varepsilon/2$ where χ and χ' are the input and output linear masks ($x[\chi]$ denotes $x[\chi_1] \oplus x[\chi_2] \oplus \cdots x[\chi_\ell]$, where $\chi = (\chi_1, \dots \chi_\ell)$ and $x[\chi_i]$ is bit χ_i of x), and $\varepsilon \geq 0$ is the imbalance. We also denote the imbalance of a random variable x as $\mathcal{I}(x) = 2 \cdot \Pr[x = 0] - 1$, and $\varepsilon(x) = |\mathcal{I}(x)|$. We will sometimes identify a mask with the integer with the same binary representation, and use an hexadecimal notation.

We first study linear properties of the addition operation, and use an ARX cipher E as example. We denote the word size as w. We assume that the cipher starts with an xor key addition, and a modular addition of two state variables[3].

[3] This setting is quite general, because any operation before a key addition can be removed, as well as any linear operation after the key addition. Ciphers where the key addition is made with a modular addition do not fit this model, but the technique can easily be adapted.

We denote the remaining operations as E', and we assume that we know a linear approximation $(\alpha, \beta, \gamma) \xrightarrow{E'} (\alpha', \beta', \gamma')$ with imbalance ε for E'. We further assume that the masks are sparse, and don't have adjacent active bits. Following previous works, the easier way to extend the linear approximation is to use the following masks for the addition:

$$(\alpha \oplus \alpha \gg 1, \alpha) \xrightarrow{\boxplus} \alpha. \tag{1}$$

As shown in Fig. 1, this gives the following linear approximation for E:

$$(\alpha \oplus \alpha \gg 1, \beta \oplus \alpha, \gamma) \xrightarrow{E} (\alpha', \beta', \gamma'). \tag{2}$$

In order to explain our technique, we initially assume that α has a single active bit, *i.e.* $\alpha = 2^i$. We explain how to deal with several active bits in Sect. 2.3. If $i = 0$, the approximation of the linear addition has imbalance 1, but for other values of i, it is only 1/2 [39]. In the following we study the case $i > 0$, where the linear approximation (2) for E has imbalance $\varepsilon/2$.

2.1 Improved Analysis with Partitioning

We now explain the improved analysis of Biham and Carmeli [3]. A simple way to understand their idea is to look at the carry bits in the addition. More precisely, we study an addition operation $s = a \boxplus b$, and we are interested in the value $s[\alpha]$. We assume that $\alpha = 2^i, i > 0$, and that we have some amount of input/output pairs. We denote individual bits of a as $a_0, a_1, \ldots a_{n-1}$, where a_0 is the LSB (respectively, b_i for b and s_i for s). In addition, we consider the

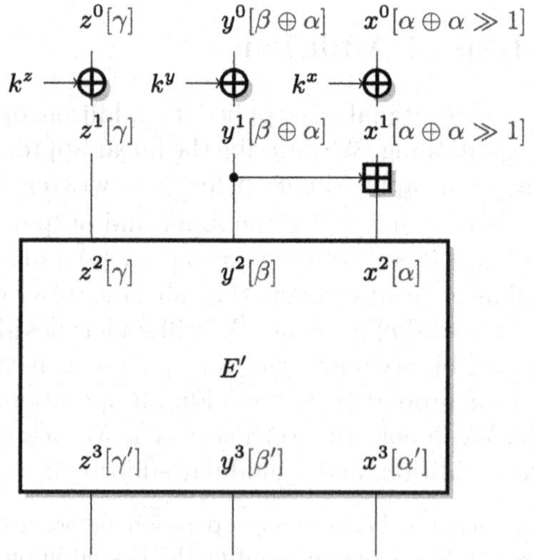

Fig. 1. Linear attack against the first addition

carry bits c_i, defined as $c_0 = 0$, $c_{i+1} = \mathrm{MAJ}(a_i, b_i, c_i)$ (where $\mathrm{MAJ}(a, b, c) = (a \wedge b) \vee (b \wedge c) \vee (c \wedge a)$). Therefore, we have $s_i = a_i \oplus b_i \oplus c_i$.

Note that the classical approximation $s_i = a_i \oplus a_{i-1} \oplus b_i$ holds with probability 3/4 because $c_i = a_{i-1}$ with probability 3/4. In order to improve this approximation, Biham and Carmeli partition the data according to the value of bits a_{i-1} and b_{i-1}. This gives four subsets:

00 If $(a_{i-1}, b_{i-1}) = (0, 0)$, then $c_i = 0$ and $s_i = a_i \oplus b_i$.
01 If $(a_{i-1}, b_{i-1}) = (0, 1)$, then $\varepsilon(c_i) = 0$ and $\varepsilon(s_i \oplus a_i \oplus a_{i-1}) = 0$.
10 If $(a_{i-1}, b_{i-1}) = (1, 0)$, then $\varepsilon(c_i) = 0$ and $\varepsilon(s_i \oplus a_i \oplus a_{i-1}) = 0$.
11 If $(a_{i-1}, b_{i-1}) = (1, 1)$, then $c_i = 1$ and $s_i = a_i \oplus b_i \oplus 1$.

If bits of a and b are known, filtering the data in subsets 00 and 11 gives a trail for the addition with imbalance 1 over one half of the data, rather than imbalance 1/2 over the full data-set. This can be further simplified to the following:

$$s_i = a_i \oplus b_i \oplus a_{i-1} \qquad \text{if } a_{i-1} = b_{i-1} \tag{3}$$

In order to apply this analysis to the setting of Fig. 1, we guess the key bits k^x_{i-1} and k^y_{i-1}, so that we can compute the values of x^1_{i-1} and y^1_{i-1} from x^0 and y^0. More precisely, an attack on E can be performed with a single (logical) key bit guess, using Eq. (3):

$$\begin{aligned} x^2_i &= x^1_i \oplus y^1_i \oplus x^1_{i-1} && \text{if } x^1_{i-1} = y^1_{i-1} \\ x^2_i &= x^0_i \oplus y^0_i \oplus x^0_{i-1} \oplus k^x_i \oplus k^y_i \oplus k^x_{i-1} && \text{if } x^0_{i-1} \oplus y^0_{i-1} = k^x_{i-1} \oplus k^y_{i-1} \end{aligned}$$

If we guess the key bit $k^x_{i-1} \oplus k^y_{i-1}$, we can filter the data satisfying $x^0_{i-1} \oplus y^0_{i-1} = k^x_{i-1} \oplus k^y_{i-1}$, and we have $\varepsilon(x^2_i \oplus x^0_i \oplus y^0_i \oplus x^0_{i-1}) = 1$. Therefore the linear approximation (2) has imbalance ε. We need $1/\varepsilon^2$ data after the filtering for the attack to succeed, *i.e.* $2/\varepsilon^2$ in total. The time complexity is also $2/\varepsilon^2$ because we run the analysis with $1/\varepsilon^2$ data for each key guess. This is an improvement over a simple linear attack using (2) with imbalance $\varepsilon/2$, with $4/\varepsilon^2$ data.

Complexity. In general this partitioning technique multiply the data and time complexity by the following ratio:

$$R^D_{\mathrm{lin}} = \frac{\mu^{-1}/\tilde{\varepsilon}^2}{1/\varepsilon^2} = \varepsilon^2/\mu\tilde{\varepsilon}^2 \qquad R^T_{\mathrm{lin}} = \frac{2^\kappa/\tilde{\varepsilon}^2}{1/\varepsilon^2} = 2^\kappa\varepsilon^2/\tilde{\varepsilon}^2 \tag{4}$$

where μ is the fraction of data used in the attack, κ is the number of guessed key bits, ε is the initial imbalance, and $\tilde{\varepsilon}$ is the improved imbalance for the selected subset. For Biham and Carmeli's attack, we have $\mu = 1/2$, $\kappa = 1$ and $\tilde{\varepsilon} = 2\varepsilon$, hence $R^D_{\mathrm{lin}} = 1/2$ and $R^T_{\mathrm{lin}} = 1/2$.

2.2 Generalized Partitioning

We now refine the technique of Biham and Carmeli using several control bits. In particular, we analyze cases 01 and 10 with extra control bits a_{i-2} and b_{i-2} (some of the cases of shown in Fig. 2):

0	1	0 0	1 1	? ?
? a_i 0 ? ?	? a_i 1 ? ?	? a_i 0 0 ?	? a_i 0 1 ?	? a_i 1 0 ?
+ ? b_i 0 ? ?	+ ? b_i 1 ? ?	+ ? b_i 1 0 ?	+ ? b_i 1 1 ?	+ ? b_i 0 1 ?
? s_i ? ? ?	? s_i ? ? ?	? s_i ? ? ?	? s_i ? ? ?	? s_i ? ? ?

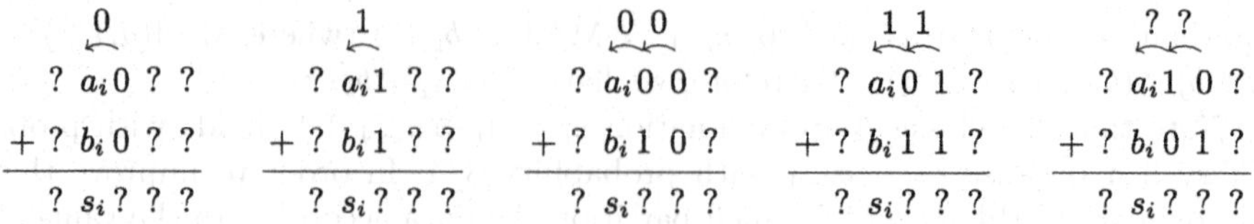

Fig. 2. Some cases of partitioning for linear cryptanalysis of an addition

01.00 If $(a_{i-1}, b_{i-1}, a_{i-2}, b_{i-2}) = (0,1,0,0)$,
    then $c_{i-1} = 0$, $c_i = 0$ and $s_i = a_i \oplus b_i$.
01.01 If $(a_{i-1}, b_{i-1}, a_{i-2}, b_{i-2}) = (0,1,0,1)$,
    then $\varepsilon(c_{i-1}) = 0$, $\varepsilon(c_i) = 0$, and $\varepsilon(s_i \oplus a_i \oplus a_{i-1}) = 0$.
01.10 If $(a_{i-1}, b_{i-1}, a_{i-2}, b_{i-2}) = (0,1,1,0)$,
    then $\varepsilon(c_{i-1}) = 0$, $\varepsilon(c_i) = 0$, and $\varepsilon(s_i \oplus a_i \oplus a_{i-1}) = 0$.
01.11 If $(a_{i-1}, b_{i-1}, a_{i-2}, b_{i-2}) = (0,1,1,1)$,
    then $c_{i-1} = 1$, $c_i = 1$ and $s_i = a_i \oplus b_i \oplus 1$.
10.00 If $(a_{i-1}, b_{i-1}, a_{i-2}, b_{i-2}) = (1,0,0,0)$,
    then $c_{i-1} = 0$, $c_i = 0$ and $s_i = a_i \oplus b_i$.
10.01 If $(a_{i-1}, b_{i-1}, a_{i-2}, b_{i-2}) = (1,0,0,1)$,
    then $\varepsilon(c_{i-1}) = 0$, $\varepsilon(c_i) = 0$, and $\varepsilon(s_i \oplus a_i \oplus a_{i-1}) = 0$.
10.10 If $(a_{i-1}, b_{i-1}, a_{i-2}, b_{i-2}) = (1,0,1,0)$,
    then $\varepsilon(c_{i-1}) = 0$, $\varepsilon(c_i) = 0$, and $\varepsilon(s_i \oplus a_i \oplus a_{i-1}) = 0$.
10.11 If $(a_{i-1}, b_{i-1}, a_{i-2}, b_{i-2}) = (1,0,1,1)$,
    then $c_{i-1} = 1$, $c_i = 1$ and $s_i = a_i \oplus b_i \oplus 1$.

This yields an improved partitioning because we now have a trail for the addition with imbalance 1 in 12 out of 16 subsets: 00.00, 00.01, 00.10, 00.11, 01.00, 01.11, 10.00, 10.11, 11.00, 11.01, 11.10, 11.11. We can also simplify this case analysis:

$$s_i = \begin{cases} a_i \oplus b_i \oplus a_{i-1} & \text{if } a_{i-1} = b_{i-1} \\ a_i \oplus b_i \oplus a_{i-2} & \text{if } a_{i-1} \neq b_{i-1} \text{ and } a_{i-2} = b_{i-2} \end{cases} \qquad (5)$$

This gives an improved analysis of E by guessing more key bits. More precisely we need $k_{i-1}^x \oplus k_{i-1}^y$ and $k_{i-2}^x \oplus k_{i-2}^y$, as shown below:

$$x_i^2 = \begin{cases} x_i^1 \oplus y_i^1 \oplus x_{i-1}^1 & \text{if } x_{i-1}^1 = y_{i-1}^1 \\ x_i^1 \oplus y_i^1 \oplus x_{i-2}^1 & \text{if } x_{i-1}^1 \neq y_{i-1}^1 \text{ and } x_{i-2}^1 = y_{i-2}^1 \end{cases}$$

$$x_i^2 = \begin{cases} x_i^0 \oplus y_i^0 \oplus x_{i-1}^0 \oplus k_i^x \oplus k_i^y \oplus k_{i-1}^x & \text{if } x_{i-1}^0 \oplus y_{i-1}^0 = k_{i-1}^x \oplus k_{i-1}^y \\ x_i^0 \oplus y_i^0 \oplus x_{i-2}^0 \oplus k_i^x \oplus k_i^y \oplus k_{i-2}^x & \text{if } x_{i-1}^0 \oplus y_{i-1}^0 \neq k_{i-1}^x \oplus k_{i-1}^y \\ & \text{and } x_{i-2}^0 \oplus y_{i-2}^0 = k_{i-2}^x \oplus k_{i-2}^y \end{cases}$$

$$\varepsilon(x_i^2 \oplus x_i^0 \oplus y_i^0 \oplus x_{i-1}^0) = 1 \qquad \text{if } x_{i-1}^0 \oplus y_{i-1}^0 = k_{i-1}^x \oplus k_{i-1}^y$$

$$\varepsilon(x_i^2 \oplus x_i^0 \oplus y_i^0 \oplus x_{i-2}^0) = 1 \qquad \begin{aligned} &\text{if } x_{i-1}^0 \oplus y_{i-1}^0 \neq k_{i-1}^x \oplus k_{i-1}^y \\ &\text{and } x_{i-2}^0 \oplus y_{i-2}^0 = k_{i-2}^x \oplus k_{i-2}^y \end{aligned}$$

Since this analysis yields different input masks for different subsets of the data, we use an analysis following multiple linear cryptanalysis [9]. We first divide the data into four subsets, depending on the value of $x^0_{i-1} \oplus y^0_{i-1}$ and $x^0_{i-2} \oplus y^0_{i-2}$, and we compute the measured (signed) imbalance $\hat{\mathcal{I}}[s]$ of each subset. Then, for each guess of the key bits $k^x_{i-1} \oplus k^y_{i-1}$, and $k^x_{i-2} \oplus k^y_{i-2}$, we deduce the expected imbalance $\mathcal{I}_k[s]$ of each subset, and we compute the distance to the observed imbalance as $\sum_s (\hat{\mathcal{I}}[s] - \mathcal{I}_k[s])^2$. According to the analysis of Biryukov, De Cannière and Quisquater, the correct key is ranked first (with minimal distance) with high probability when using $\mathcal{O}(1/c^2)$ samples, where $c^2 = \sum_i \mathcal{I}_i^2 = \sum_i \varepsilon_i^2$ is the capacity of the system of linear approximations. Since we use three approximations with imbalance ε, the capacity of the full system is $3\varepsilon^2$, and we need $1/3 \cdot 1/\varepsilon^2$ data in each subset after partitioning, *i.e.* $4/3 \cdot 1/\varepsilon^2$ in total.

Again, the complexity ratio of this analysis can be computed as $R^D_{\text{lin}} = \varepsilon^2 / \mu\widetilde{\varepsilon}^2$ $R^T_{\text{lin}} = 2^\kappa \varepsilon^2 / \widetilde{\varepsilon}^2$ With $\mu = 3/4$ and $\widetilde{\varepsilon} = 2\varepsilon$, we find:

$$R^D_{\text{lin}} = 1/3 \qquad\qquad R^T_{\text{lin}} = 1.$$

The same technique can be used to refine the partitioning further, and give a complexity ratio of $R^D_{\text{lin}} = 1/4 \times 2^\kappa / (2^\kappa - 1)$ when guessing κ bits.

Time complexity. In general, the time complexity of this improved partitioning technique is the same as the time complexity as the basic attack ($R^T_{\text{lin}} = 1$), because we have to repeat the analysis 4 times (for each key of the key bits) with one fourth of the amount of data. We describe some techniques to reduce the time complexity in Sect. 4.

2.3 Combining Partitions

Finally, we can combine several partitions to analyze an addition with several active bits. If we use k_1 partitions for the first bit, and k_2 for the second bit, this yields a combined partition with $k_1 \cdot k_2$ cases. If the bits are not close to each other, the gains of each bit are multiplied. This can lead to significant improvements even though R_{lin} is small for a single active bit.

For more complex scenarios, we select the filtering bits assuming that the active bits don't interact, and we evaluate experimentally the probability in each subset. We can further study the matrix of probabilities to detect (logical) bits with no or little effect on the total capacity in order to improve the complexity of the attack. This will be used for our applications in Sect. 5 and Appendix A.

3 Differential Analysis of Addition

We now study differential properties of the addition. We perform our analysis in the same way as the analysis of Sect. 2, following Fig. 3. We consider the first addition operation separately, and we assume that we know a differential $(\alpha, \beta, \gamma) \to (\alpha', \beta', \gamma')$ with probability p for the remaining of the cipher. Following previous works, a simple way to extend the differential is to linearize the first addition, yielding the following differences for the addition:

$$\alpha \oplus \beta, \beta \xrightarrow{\boxplus} \alpha.$$

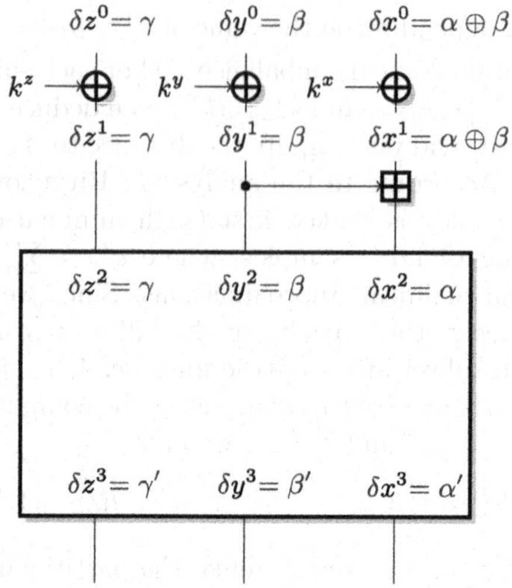

Fig. 3. Differential attack against the first addition

Similarly to our analysis of linear cryptanalysis, we consider a single addition $s = a \boxplus b$, and we first assume that a single bit is active through the addition. However, we have to consider several cases, depending on how many input/output bits are active. The cases are mostly symmetric, but there are important differences in the partitioning.

3.1 Analysis of $(\alpha = 0, \beta = 2^i)$

With $i < w - 1$, the probability for the addition is $\Pr[(2^i, 2^i) \xrightarrow{\boxplus} 0] = 1/2$.

Improved Analysis with Structures. We first discuss a technique using multiple differentials and structures. More precisely, we use the following differentials for the addition:[4]

$$\mathcal{D}_1 : (2^i, 2^i) \xrightarrow{\boxplus} 0 \qquad \Pr\left[(2^i, 2^i) \xrightarrow{\boxplus} 0\right] = 1/2$$

$$\mathcal{D}_2 : (2^i \oplus 2^{i+1}, 2^i) \xrightarrow{\boxplus} 0 \qquad \Pr\left[(2^i \oplus 2^{i+1}, 2^i) \xrightarrow{\boxplus} 0\right] = 1/4$$

We can improve the probability of $\mathcal{D}_2$ using a partitioning according to (a_i, a_{i+1}):

00 If $(a_i, a_{i+1}) = (0, 0)$, then $a' = a \boxplus 2^i \boxplus 2^{i+1}$ and $s \neq s'$.
01 If $(a_i, a_{i+1}) = (0, 1)$, then $a' = a \boxminus 2^i$ and $\Pr[s = s'] = 1/2$.
10 If $(a_i, a_{i+1}) = (1, 0)$, then $a' = a \boxplus 2^i$ and $\Pr[s = s'] = 1/2$.
11 If $(a_i, a_{i+1}) = (1, 1)$, then $a' = a \boxminus 2^i \boxminus 2^{i+1}$ and $s \neq s'$.

[4] Note that in the application to E, we can modify the difference in x^1 but not in y^1.

This can be written as:

$$\Pr\left[(2^i, 2^i) \xrightarrow{\boxplus} 0\right] = 1/2$$

$$\Pr\left[(2^i \oplus 2^{i+1}, 2^i) \xrightarrow{\boxplus} 0\right] = 1/2 \qquad \text{if} a_i \neq a_{i+1}$$

The use of structures allows to build pairs of data for both differentials from the same data set. More precisely, we consider the following inputs:

$$p = (x^0, y^0, z^0) \qquad q = (x^0 \oplus 2^i, y^0 \oplus 2^i, z^0)$$
$$r = (x^0 \oplus 2^{i+1}, y^0, z^0) \qquad s = (x^0 \oplus 2^{i+1} \oplus 2^i, y^0 \oplus 2^i, z^0)$$

We see that (p, q) and (r, s) follow the input difference of $\mathcal{D}_1$, while (p, s) and (r, q) follow the input difference of $\mathcal{D}_2$. Moreover, we have from the partitioning:

$$\Pr[E(p) \oplus E(q) = (\alpha', \beta', \gamma')] = 1/2 \cdot p$$
$$\Pr[E(r) \oplus E(s) = (\alpha', \beta', \gamma')] = 1/2 \cdot p$$
$$\Pr[E(p) \oplus E(s) = (\alpha', \beta', \gamma')] = 1/2 \cdot p \quad \text{if } x_i^0 \oplus x_{i+1}^0 \neq k_i^x \oplus k_{i+1}^x$$
$$\Pr[E(r) \oplus E(q) = (\alpha', \beta', \gamma')] = 1/2 \cdot p \quad \text{if } x_i^0 \oplus x_{i+1}^0 = k_i^x \oplus k_{i+1}^x$$

For each key guess, we select three candidate pair out of a structure of four plaintexts, and every pair follows a differential for E with probability $p/2$. Therefore we need $2/p$ pairs, with a data complexity of $8/3 \cdot 1/p$ rather than $4 \cdot 1/p$.

In general this partitioning technique multiply the data and time complexity by the following ratio:

$$R^D_{\text{diff}} = \frac{\widetilde{p}^{-1}T/(\mu T^2/4)}{p^{-1}T/(T/2)} = \frac{2p}{\mu T \widetilde{p}} \qquad R^T_{\text{diff}} = 2^\kappa \mu R^D_{\text{diff}} = \frac{2^{\kappa+1}p}{T\widetilde{p}}, \tag{6}$$

where μ is the fraction of data used in the attack, κ is the number of guessed key bits, T is the number of plaintexts in a structure (we consider $T^2/4$ pairs rather than $T/2$ without structures) p is the initial probability, and $\widetilde{p}$ is the improved probability for the selected subset. Here we have $\mu = 3/4$, $\kappa = 1$, $T = 4$, and $\widetilde{p} = p$, hence

$$R^D_{\text{diff}} = 2/3 \qquad R^T_{\text{diff}} = 1$$

Moreover, if the differential trail is used in a boomerang attack, or in a differential-linear attack, it impacts the complexity twice, but the involved key bits are the same, and we only need to use the structure once. Therefore, the complexity ratio should be evaluated as:

$$R^D_{\text{diff-2}} = \frac{\widetilde{p}^{-2}T/(\mu T^2/4)}{p^{-2}T/(T/2)} = \frac{2p^2}{\mu T \widetilde{p}^2} \qquad R^T_{\text{diff-2}} = 2^\kappa \mu R^D_{\text{diff-2}} = \frac{2^{\kappa+1}p^2}{T\widetilde{p}^2}, \tag{7}$$

In this scenario, we have the same ratios:

$$R^D_{\text{diff-2}} = 2/3 \qquad R^T_{\text{diff-2}} = 1$$

Generalized Partitioning. We can refine the analysis of the addition by partitioning according to (b_i). This gives the following:

$$\Pr\left[(2^i, 2^i) \to 0\right] = 1 \qquad \text{if } a_i \neq b_i$$
$$\Pr\left[(2^i \oplus 2^{i+1}, 2^i) \to 0\right] = 1 \qquad \text{if } a_i = b_i \text{ and } a_i \neq a_{i+1}$$

This gives an attack with $T = 4$, $\mu = 3/8$, $\kappa = 2$ and $\widetilde{p} = 2p$, which yield the same ratio in a simple differential setting, but a better ratio for a boomerang or differential-linear attack:

$$R^D_{\text{diff}} = 2/3 \qquad R^T_{\text{diff}} = 1$$
$$R^D_{\text{diff-2}} = 1/3 \qquad R^T_{\text{diff-2}} = 1/2$$

In addition, this analysis allows to recover an extra key bit, which can be useful for further steps of an attack.

Larger Structure. Alternatively, we can use a larger structure to reduce the complexity: with a structure of size 2^t, we have an attack with a ratio $R^D_{\text{diff}} = 1/2 \times 2^\kappa/(2^\kappa - 1)$, by guessing $\kappa - 1$ key bits.

3.2 Analysis of $(\alpha = 2^i, \beta = 0)$

With $i < w - 1$, the probability for the addition is $\Pr[(2^i, 0) \xrightarrow{\boxplus} 2^i] = 1/2$.

Improved Analysis with Structures. As in the previous section, we consider multiple differentials, and use partitioning to improve the probability:

$$\mathcal{D}_1 : \qquad \Pr\left[(2^i, 0) \xrightarrow{\boxplus} 2^i\right] = 1/2$$
$$\mathcal{D}_2 : \qquad \Pr\left[(2^i \oplus 2^{i+1}, 0) \xrightarrow{\boxplus} 2^i\right] = 1/2 \qquad \text{if } a_i \neq a_{i+1}$$

We also use structures in order to build pairs of data for both differentials from the same data set. More precisely, we consider the following inputs:

$$p = (x^0, y^0, z^0) \qquad q = (x^0 \oplus 2^i, y^0, z^0)$$
$$r = (x^0 \oplus 2^{i+1}, y^0, z^0) \qquad s = (x^0 \oplus 2^{i+1} \oplus 2^i, y^0, z^0)$$

We see that (p, q) and (r, s) follow the input difference of $\mathcal{D}_1$, while (p, s) and (r, q) follow the input difference of $\mathcal{D}_2$. Moreover, we have from the partitioning:

$$\Pr[E(p) \oplus E(q) = (\alpha', \beta', \gamma')] = 1/2 \cdot p$$
$$\Pr[E(r) \oplus E(s) = (\alpha', \beta', \gamma')] = 1/2 \cdot p$$
$$\Pr[E(p) \oplus E(s) = (\alpha', \beta', \gamma')] = 1/2 \cdot p \quad \text{if } x^0_i \oplus x^0_{i+1} \neq k^x_i \oplus k^x_{i+1}$$
$$\Pr[E(r) \oplus E(q) = (\alpha', \beta', \gamma')] = 1/2 \cdot p \quad \text{if } x^0_i \oplus x^0_{i+1} = k^x_i \oplus k^x_{i+1}$$

In this case, we also have $\mu = 3/4$, $T = 4$, and $\widetilde{p} = p$, hence

$$R^D_{\text{diff}} = 2/3 \qquad R^T_{\text{diff}} = 1$$
$$R^D_{\text{diff-2}} = 2/3 \qquad R^T_{\text{diff-2}} = 1$$

Generalized Partitioning. Again, we can refine the analysis of the addition by partitioning according to (s_i). This gives the following:

$$\Pr\left[(2^i, 0) \to 2^i\right] = 1 \qquad \text{if } a_i = s_i$$
$$\Pr\left[(2^i \oplus 2^{i+1}, 0) \to 2^i\right] = 1 \qquad \text{if } a_i \neq s_i \text{ and } a_i \neq a_{i+1}$$

Since we can not readily filter according to bits of s, we use the results of Sect. 2:

$$a_i \oplus b_i \oplus a_{i-1} = s_i \qquad \text{if } a_{i-1} = b_{i-1}$$

This gives:

$$\Pr\left[(2^i, 0) \to 2^i\right] = 1 \quad \text{if } b_i = a_{i-1} \text{ and } a_{i-1} = b_{i-1}$$
$$\Pr\left[(2^i \oplus 2^{i+1}, 0) \to 2^i\right] = 1 \quad \text{if } b_i \neq a_{i-1} \text{ and } a_{i-1} = b_{i-1} \text{ and } a_i \neq a_{i+1}$$

Unfortunately, we can only use a small fraction of the pairs $\mu = 3/16$. With $T = 4$ and $\widetilde{p} = 2p$, this yields, an increase of the data complexity for a simple differential attack:

$$R^D_{\text{diff}} = 4/3 \qquad R^T_{\text{diff}} = 1/2$$
$$R^D_{\text{diff-2}} = 2/3 \qquad R^T_{\text{diff-2}} = 1/4$$

3.3 Analysis of ($\alpha = 2^i, \beta = 2^i$)

With $i < w - 1$, the probability for the addition is $\Pr[(0, 2^i) \xrightarrow{\boxplus} 2^i] = 1/2$.

The results in this section will be the same as in the previous section, but we have to use a different structure. Indeed when this analysis is applied to E, we can freely modify the difference in x^0 but not in y^0, because it would affect the differential in E'.

More precisely, we use the following differentials:

$$\mathcal{D}_1: \qquad \Pr\left[(0, 2^i) \xrightarrow{\boxplus} 2^i\right] = 1/2$$
$$\mathcal{D}_2: \qquad \Pr\left[(2^{i+1}, 2^i) \xrightarrow{\boxplus} 2^i\right] = 1/2 \qquad \text{if } a_{i+1} \neq b_i$$

and the following structure:

$$p = (x^0, y^0, z^0) \qquad q = (x^0, y^0 \oplus 2^i, z^0)$$
$$r = (x^0 \oplus 2^{i+1}, y^0, z^0) \qquad s = (x^0 \oplus 2^{i+1}, y^0 \oplus 2^i, z^0)$$

This yields:

$$\begin{aligned}
&\Pr[E(p) \oplus E(q) = (\alpha', \beta', \gamma')] = 1/2 \cdot p \\
&\Pr[E(r) \oplus E(s) = (\alpha', \beta', \gamma')] = 1/2 \cdot p \\
&\Pr[E(p) \oplus E(s) = (\alpha', \beta', \gamma')] = 1/2 \cdot p \quad \text{if } x_i^1 \oplus x_{i+1}^0 \neq k_i^y \oplus k_{i+1}^x \\
&\Pr[E(r) \oplus E(q) = (\alpha', \beta', \gamma')] = 1/2 \cdot p \quad \text{if } x_i^1 \oplus x_{i+1}^0 = k_i^y \oplus k_{i+1}^x
\end{aligned}$$

4 Improving the Time Complexity

The analysis of the previous sections assume that we repeat the distinguisher for each key guess, so that the data complexity is reduced in a very generic way. When this is applied to differential or linear cryptanalysis, it usually result in an increased time complexity ($R^T > 1$). However, when the distinguisher is a simple linear of differential distinguisher, we can perform the analysis in a more efficient way, using the same techniques that are used in attacks with partial key guess against SBox-based ciphers. For linear cryptanalysis, we use a variant of Matsui's Algorithm 2 [25], and the improvement using convolution algorithm [15]; for differential cryptanalysis we filter out pairs that can not be a right pair for any key. In the best cases, the time complexity of the attacks can be reduced to essentially the data complexity.

4.1 Linear Analysis

We follow the analysis of Matsui's Algorithm 2, with a distillation phase using counters to keep track of the important features of the data, and an analysis phase for every key that requires only the counters rather than the full dataset.

More precisely, let us explain this idea within the setting of Sect. 2.2 and Fig. 1. For each key guess, the attacker computes the observed imbalance over a subset $\mathcal{S}_k$ corresponding to the data with $x_{i-1}^0 \oplus y_{i-1}^0 = k_{i-1}^x \oplus k_{i-1}^y$, or $(x_{i-1}^0 \oplus y_{i-1}^0 \neq k_{i-1}^x \oplus k_{i-1}^y$ and $x_{i-2}^0 \oplus y_{i-2}^0 = k_{i-2}^x \oplus k_{i-2}^y)$:

$$\begin{aligned}
\hat{\mathcal{I}} &= \mathcal{I}_{\mathcal{S}_k}(P[\chi_P] \oplus C[\chi_C]) \\
&= 1/\left|\mathcal{S}_k\right| \times \sum_{\mathcal{S}_k} (-1)^{P[\chi_P] \oplus C[\chi_C]}
\end{aligned}$$

where (using $\alpha = 2^i$)

$$\begin{aligned}
P[\chi_P] \oplus C[\chi_C] &= x_i^2 \oplus y^2[\beta] \oplus z^2[\gamma] \oplus x^3[\alpha'] \oplus y^3[\beta'] \oplus z^3[\gamma'] \\
&= \begin{cases} x_i^0 \oplus y_i^0 \oplus x_{i-1}^0 \oplus y^0[\beta] \oplus z^0[\gamma] \oplus x^3[\alpha'] \oplus y^3[\beta'] \oplus z^3[\gamma'] \\ \quad \text{if } x_{i-1}^0 \oplus y_{i-1}^0 = k_{i-1}^x \oplus k_{i-1}^y \\ x_i^0 \oplus y_i^0 \oplus x_{i-2}^0 \oplus y^0[\beta] \oplus z^0[\gamma] \oplus x^3[\alpha'] \oplus y^3[\beta'] \oplus z^3[\gamma'] \\ \quad \text{if } x_{i-1}^0 \oplus y_{i-1}^0 \neq k_{i-1}^x \oplus k_{i-1}^y \\ \quad \text{and } x_{i-2}^0 \oplus y_{i-2}^0 = k_{i-2}^x \oplus k_{i-2}^y \end{cases}
\end{aligned}$$

Therefore, the imbalance can be efficiently reconstructed from a series of 2^4 counters keeping track of the amount of data satisfying every possible value of the following bits:

$$x_i^0 \oplus y_i^0 \oplus x_{i-1}^0 \oplus y^0[\beta] \oplus z^0[\gamma] \oplus x^3[\alpha'] \oplus y^3[\beta'] \oplus z^3[\gamma'],$$
$$x_{i-1}^0 \oplus x_{i-2}^0, \quad x_{i-1}^0 \oplus y_{i-1}^0, \quad x_{i-2}^0 \oplus y_{i-2}^0$$

This results in an attack where the time complexity is equal to the data complexity, plus a small cost to compute the imbalance. The analysis phase require only about 2^6 operations in this case (adding 2^4 counters for 2^2 key guesses). When the amount of data required is larger than 2^6, the analysis step is negligible.

When several partitions are combined (with several active bits in the first additions), the number of counters increases to 2^b, where b is the number of control bits. To reduce the complexity of the analysis phase, we can use a convolution algorithm (following [15]), so that the cost of the distillation is only $\mathcal{O}(b \cdot 2^b)$ rather than $\mathcal{O}(2^\kappa \cdot 2^b)$. This will be explained in more details with the application to Chaskey in Sect. 5.

In general, there is a trade-off between the number of partitioning bits, and the complexity. A more precise partitioning allows to reduce the data complexity, but this implies a larger set of counters, hence a larger memory complexity. When the number of partitioning bits reaches the data complexity, the analysis phase becomes the dominant phase, and the time complexity is larger than the data complexity.

4.2 Differential Analysis

For a differential attack with partitioning, we can also reduce the time complexity, by filtering pairs before the analysis phase. In the following, we assume that we use a simple differential distinguisher with output difference δ', following Sect. 3 (where $\delta' = (\alpha', \beta', \gamma')$)

We first define a linear function L with rank $n-1$ (where n is the block size), so that $L(\delta') = 0$. In particular, any pair $x, x' = x \oplus \delta'$ satisfies $L(x) = L(x')$. This allows to detect collisions by looking at all *values* in a structure, rather than all *pairs* in a structure. We just compute $L(E(x))$ for all x's in a structure, and we look for collisions.

5 Application to Chaskey

Chaskey is a recent MAC proposal designed jointly by researchers from COSIC and Hitachi [31]. The mode of operation of Chaskey is based on CBC-MAC with an Even-Mansour cipher; but it can also be described as a permutation-based design as seen in Fig. 4. Chaskey is designed to be extremely fast on 32-bit microcontrollers, and the internal permutation follows an ARX construction with 4 32-bit words based on SipHash; it is depicted in Fig. 5. Since the security of Chaskey is based on an Even-Mansour cipher, the security bound has a birthday

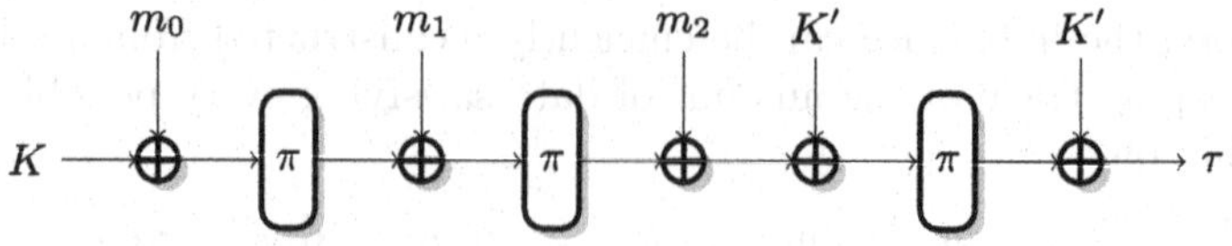

Fig. 4. Chaskey mode of operation (full block message)

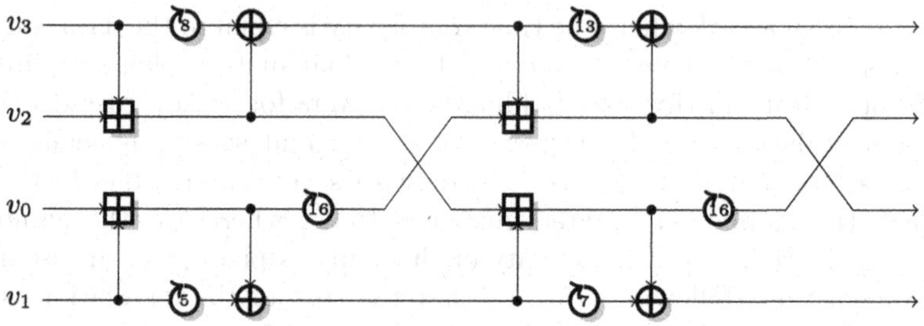

Fig. 5. One round of the Chaskey permutation. The full permutation has 8 rounds.

term $\mathcal{O}(TD \cdot 2^{-128})$. More precisely, the designers claim that it should be secure up to 2^{48} queries, and 2^{80} computations.

So far, the only external cryptanalysis results on Chaskey are generic attacks in the multi-user setting [27]. The only analysis of the permutation is in the submission document; the best result is a 4 round bias, that can probably be extended into a 5 round attack following the method of attacks against the Salsa family [1]. It is important to try more advanced techniques in order to understand the security of Chaskey, in particular because it is being considered for standardization.

Table 2. Probabilities of the best differential characteristics of Chaskey reported by the designers [31]

Rounds:	1	2	3	4	5	6	7	8
Probability:	1	2^{-4}	2^{-16}	2^{-37}	$2^{-73.1}$	$2^{-132.8}$	$2^{-205.6}$	$2^{-289.9}$

5.1 Differential-Linear Cryptanalysis

The best differential characteristics found by the designers of Chaskey quickly become unusable when the number of rounds increase (See Table 2). The designers also report that those characteristics have an "hourglass structure": there is a position in the middle where a single bit is active, and this small difference is expanded by the avalanche effect when propagating in both direction. This is typical of ARX designs: short characteristics have a high probability, but after a few rounds the differences cannot be controlled and the probability decrease very fast. The same observation typically holds also for linear trails.

Because of these properties, attacks that can divide the cipher E in two parts $E = E_\perp \circ E_\top$ and build characteristics or trail for both half independently – such as the boomerang attack or differential-linear cryptanalysis – are particularly interesting. In particular, many attacks on ARX designs are based on the boomerang attack [10,19,23,28,35,42] or differential-linear cryptanalysis [18]. Since Chaskey never uses the inverse permutation, we cannot apply a boomerang attack, and we focus on differential-linear cryptanalysis.

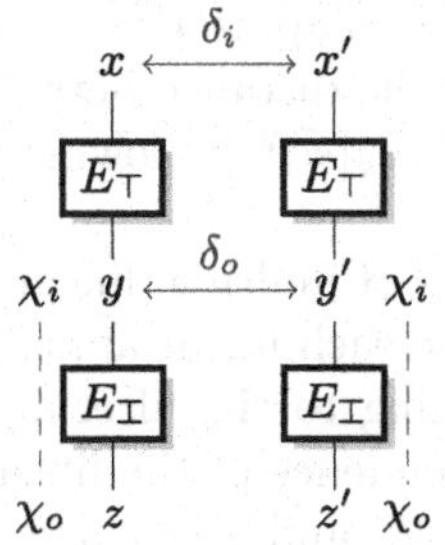

Fig. 6. Differential-linear cryptanalysis

Differential-linear cryptanalysis uses a differential $\delta_i \xrightarrow{E_\top} \delta_o$ with probability p for $E_\top$, and a linear approximation $\chi_i \xrightarrow{E_\perp} \chi_o$ with imbalance ε for $E_\perp$ (see Fig. 6). The attacker uses pairs of plaintexts (P_i, P_i') with $P_i' = P_i \oplus \delta_i$, and computes the observed imbalance $\hat{\varepsilon} = |2 \cdot \#\{i : C_i[\chi_o] = C_i'[\chi_o]\}/D - 1|$. Following the heuristic analysis of [5], the expected imbalance is about $p\varepsilon^2$, which gives an attack complexity of $\mathcal{O}(2/p^2\varepsilon^4)$:

- A pair of plaintext satisfies $E_\top(P) \oplus E_\top(P') = \delta_o$ with probability p. In this case, we have $E_\top(P)[\chi_i] \oplus E_\top(P')[\chi_i] = \delta_o[\chi_i]$. Without loss of generality, we assume that $\delta_o[\chi_i] = 0$.
- Otherwise, we expect that $E_\top(P)[\chi_i] \oplus E_\top(P')[\chi_i]$ is not biased. This gives the following:

$$\Pr\left[E_\top(P)[\chi_i] \oplus E_\top(P')[\chi_i] = 0\right] = p + (1-p)\cdot 1/2 = 1/2 + p/2 \quad (8)$$

$$\varepsilon(E_\top(P)[\chi_i] \oplus E_\top(P')[\chi_i]) = p \quad (9)$$

- We also have $\varepsilon(E_\top(P)[\chi_i] \oplus C[\chi_o]) = \varepsilon(E_\top(P')[\chi_i] \oplus C'[\chi_o]) = \varepsilon$ from the linear approximations. Combining with (9), we get $\varepsilon(C[\chi_o] \oplus C'[\chi_o]) = p\varepsilon^2$.

A more rigorous analysis has been recently provided by Blondeau et al. [12], but since we use experimental values to evaluate the complexity of our attacks, this heuristic explanation will be sufficient.

5.2 Using Partitioning

A differential-linear distinguisher can easily be improved using the results of Sects. 2 and 3. We can improve the differential and linear part separately, and combine the improvements on the differential-linear attack. More precisely, we have to consider structures of plaintexts, and to guess some key bits in the differential and linear parts. We partition all the potential pairs in the structures according to the input difference, and to the filtering bits in the differential and linear part; then we evaluate the observed imbalance $\hat{\mathcal{I}}[s]$ in every subset s. Finally, for each key guess k, we compute the expected imbalance $\mathcal{I}_k[s]$ for each subset s, and then we evaluate the distance between the observed and expected imbalances as $L(k) = \sum_s (\hat{\mathcal{I}}[s] - \mathcal{I}_k[s])^2$ (following the analysis of multiple linear cryptanalysis [9]).

While we follow the analysis of multiple linear cryptanalysis to evaluate the complexity of our attack, we use each linear approximation on a different subset of the data, partitioned according to the filtering bits. In particular, we don't have to worry about the independence of the linear approximations.

If we use structures of size T, and select a fraction μ_{diff} of the input pairs with an improved differential probability $\widetilde{p}$, and a fraction μ_{lin} of the output pairs with an improved linear imbalance $\widetilde{\varepsilon}$, the data complexity of the attack is $\mathcal{O}(\mu_{\mathsf{lin}}\mu_{\mathsf{diff}}^2 T/2 \times 2/\widetilde{p}^2\widetilde{\varepsilon}^4)$. This corresponds to a complexity ratio of $R^D_{\mathsf{diff}\text{-}2} {R^D_{\mathsf{lin}}}^2$.

More precisely, using differential filtering bits p_{diff} and linear filtering bits c_{lin}, the subsets are defined by the input difference Δ, the plaintext bits $P[p_{\mathsf{diff}}]$ and the cipher text bits $C[c_{\mathsf{lin}}]$ and $C'[c_{\mathsf{lin}}]$, with $C = E(P)$ and $C' = E(P \oplus \Delta)$. In practice, for every P, P' in a structure, we update the value of $\hat{\mathcal{I}}[P \oplus P', P[p_{\mathsf{lin}}], C[c_{\mathsf{diff}}], C'[c_{\mathsf{diff}}]]$.

We also take advantage of the Even-Mansour construction of Chaskey, without keys inside the permutation. Indeed, the filtering bits used to define the subsets s correspond to the key bits used in the attack. Therefore, we only need to compute the expected imbalance for the zero key, and we can deduce the expected imbalance for an arbitrary key as $\mathcal{I}_{k_{\mathsf{diff}},k_{\mathsf{lin}}}[\Delta, p, c, c'] = \mathcal{I}_0[\Delta, p \oplus k_{\mathsf{lin}}, c \oplus k_{\mathsf{diff}}, c' \oplus k_{\mathsf{diff}}]$.

Time Complexity. This description lead to an attack with low time complexity using an FFT algorithm, as described previously for linear cryptanalysis [15] and multiple linear cryptanalysis [17]. Indeed, the distance between the observed and expected imbalance can be written as:

$$\begin{aligned} L(k) &= \sum_s (\hat{\mathcal{I}}[s] - \mathcal{I}_k[s])^2 \\ &= \sum_s (\hat{\mathcal{I}}[s] - \mathcal{I}_0[s \oplus \phi(k)])^2, \quad \text{where } \phi(k_{\mathsf{diff}}, k_{\mathsf{lin}}) = (0, k_{\mathsf{lin}}, k_{\mathsf{diff}}, k_{\mathsf{diff}}) \\ &= \sum_s \hat{\mathcal{I}}[s]^2 + \sum_s \mathcal{I}_0[s \oplus \phi(k)]^2 - 2\sum_s \hat{\mathcal{I}}[s]\mathcal{I}_0[s \oplus \phi(k)], \end{aligned}$$

where only the last term depend on the key. Moreover, this term can be seem as the $\phi(k)$-th component of the convolution $\mathcal{I}_0 * \hat{\mathcal{I}}$. Using the convolution theorem, we can compute the convolution efficiently with an FFT algorithm.

This gives the following fast analysis:

1. Compute the expected imbalance $\mathcal{I}_0[s]$ of the differential-linear distinguisher for the zero key, for every subset s.
2. Collect D plaintext-ciphertext pairs, and compute the observed imbalance $\hat{\mathcal{I}}[s]$ of each subset.
3. Compute the convolution $\mathcal{I} * \hat{\mathcal{I}}$, and find k that maximizes coefficient $\phi(k)$.

5.3 Differential-Linear Cryptanalysis of Chaskey

In order to find good differential-linear distinguishers for Chaskey, we use a heuristic approach. We know that most good differential characteristics and good linear trails have an "hourglass structure", with a single active bit in the middle. If a good differential-linear characteristics is given with this "hourglass structure", we can divide E in three parts $E = E_\perp \circ E_I \circ E_\top$, so that the single active bit in the differential characteristic falls between $E_\top$ and E_I, and the single active bit in the linear trail falls between E_I and $E_\perp$. We use this decomposition to look for good differential-linear characteristics: we first divide E in three parts, and we look for a differential characteristic $\delta_i \xrightarrow{E_\top} \delta_o$ in $E_\top$ (with probability p), a differential-linear characteristic $\delta_o \xrightarrow{E_I} \chi_i$ in E_I (with imbalance b), and a linear characteristic $\chi_i \xrightarrow{E_\perp} \chi_o$ in $E_\perp$ (with imbalance ε), where δ_o and χ_i have a single active bit. This gives a differential-linear distinguisher with imbalance close to $bp\varepsilon^2$:

- We consider a pair of plaintext (P, P') with $P' = P \oplus \delta_i$, and we denote $X = E_\top(P)$, $Y = E_I(X)$, $C = E_\perp(Y)$.
- We have $X \oplus X' = \delta_o$ with probability p. In this case, $\varepsilon(Y[\chi_i] \oplus Y'[\chi_i]) = b$
- Otherwise, we expect that $Y[\chi_i] \oplus Y'[\chi_i]$ is not biased. This gives the following:

$$\Pr\left[Y[\chi_i] \oplus Y'[\chi_i] = 0\right] = (1-p) \cdot 1/2 + p \cdot (1/2 + b/2) = 1/2 + bp/2 \quad (10)$$

$$\varepsilon(Y[\chi_i] \oplus Y'[\chi_i]) = bp \quad (11)$$

- We also have $\varepsilon(Y[\chi_i] \oplus C[\chi_o]) = \varepsilon(Y'[\chi_i] \oplus C'[\chi_o]) = \varepsilon$ from the linear approximations. Combining with (11), we get $\varepsilon(C[\chi_o] \oplus C'[\chi_o]) = bp\varepsilon^2$.

In the E_I section, we can see the characteristic as a small differential-linear characteristic with a single active input bit and a single active output bit, or as a truncated differential where the input difference has a single active bit and the output value is truncated to a single bit. In other words, we use pairs of values with a single bit difference, and we look for a biased output bit difference.

We ran an exhaustive search over all possible decompositions $E = E_\perp \circ E_I \circ E_\top$ (varying the number of rounds), and all possible positions for the active bits i at the input of $E_{\top\!\!\!\perp}$ and the biased bit[5] j at the output of $E_{\top\!\!\!\perp}$. For each candidate, we evaluate experimentally the imbalance $\varepsilon(E_{\top\!\!\!\perp}(x)[j] \oplus E_{\top\!\!\!\perp}(x \oplus 2^i)[j])$, and we study the best differential and linear trails to build the full differential-linear distinguisher. This method is similar to the analysis of the Salsa family by Aumasson *et al.* [1]: they decompose the cipher in two parts $E = E_\perp \circ E_{\top\!\!\!\perp}$, in order to combine a biased bit in $E_{\top\!\!\!\perp}$ with an approximation of $E_\perp$.

This approach allows to identify good differential-linear distinguisher more easily than by building full differential and linear trails. In particular, we avoid most of the heuristic problems in the analysis of differential-linear distinguishers (such as the presence of multiple good trails in the middle) by evaluating experimentally $\varepsilon(E_{\top\!\!\!\perp}(x)[j] \oplus E_{\top\!\!\!\perp}(x \oplus 2^i)[j])$ without looking for explicit trails in the middle. In particular, the transition between $E_\top$ and $E_{\top\!\!\!\perp}$ is a transition between two differential characteristics, while the transition between $E_{\top\!\!\!\perp}$ and $E_\perp$ is a transition between two linear characteristics.

5.4 Attack Against 6-Round Chaskey

The best distinguisher we identified for an attack against 6-round Chaskey uses 1 round in $E_\top$, 4 rounds in $E_{\top\!\!\!\perp}$, and 1 round in $E_\perp$. The optimal differences and masks are:

- Differential for $E_\top$ with probability $p_\top \approx 2^{-5}$:

$$v_0[26], v_1[26], v_2[6, 23, 30], v_3[23, 30] \xrightarrow{E_\top} v_2[22]$$

- Biased bit for $E_{\top\!\!\!\perp}$ with imbalance $\varepsilon_{\top\!\!\!\perp} = 2^{-6.05}$:

$$v_2[22] \xrightarrow{E_{\top\!\!\!\perp}} v_2[16]$$

- Linear approximations for $E_\perp$ with imbalance $\varepsilon_\perp = 2^{-2.6}$:

$$v_2[16] \xrightarrow{E_\perp} v_0[5], v_1[23, 31], v_2[0, 8, 15], v_3[5]$$

The differential and linear trails are shown in Fig. 7. The expected imbalance is $p_\top \cdot \varepsilon_{\top\!\!\!\perp} \cdot \varepsilon_\perp^2 = 2^{-16.25}$. This gives a differential-linear distinguisher with expected complexity in the order of $D = 2/p_\top^2 \varepsilon_{\top\!\!\!\perp}^2 \varepsilon_\perp^4 \approx 2^{33.5}$.

We can estimate the data complexity more accurately using [11, Eq. (11)]: we need about $2^{34.1}$ pairs of samples in order to reach a false positive rate of 2^{-4}. Experimentally, with 2^{34} pairs of samples (*i.e.* 2^{35} data), the measured imbalance is larger than $2^{-16.25}$ with probability 0.5; with random data, it is larger than $2^{-16.25}$ with probability 0.1. This matches the predictions of [11], and confirms the validity of our differential-linear analysis.

This simple differential-linear attack is more efficient than generic attacks against the Even-Mansour construction of Chaskey. It follows the usage limit of Chaskey, and reaches more rounds than the analysis of the designers. Moreover, it we can be improved significantly using the results of Sects. 2 and 3.

[5] We also consider pairs of adjacent bits, following the analysis of [14].

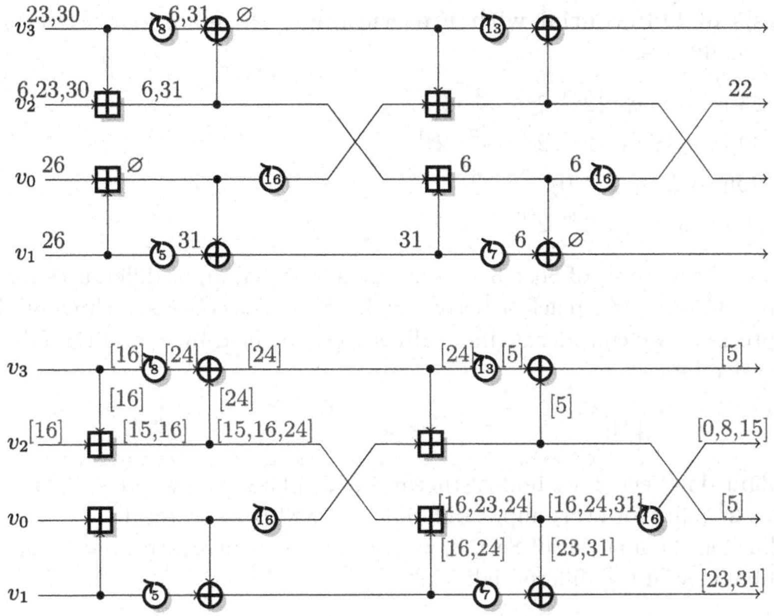

Fig. 7. 6-round attack: differential characteristic, and linear trail.

Analysis of Linear Approximations with Partitioning. To make the description easier, we remove the linear operations at the end, so that the linear trail becomes:

$$v_2[16] \xrightarrow{E_\perp} v_1[16,24], v_2[16,23,24], v_3[24]$$

We select control bits to improve the probability of the addition between v_1 and v_2 on active bits 16 and 24. Following the analysis of Sect. 2.2, we need $v_1[14] \oplus v_2[14]$ and $v_1[15] \oplus v_2[15]$ as control bits for active bit 16. To identify more complex control bits, we consider $v_1[14,15,22,23]$, $v_2[14,15,22,23]$ as potential control bits, as well as $v_3[23]$ because it can affect the addition on the previous half-round. Then, we evaluate the bias experimentally (using the round function as a black box) in order to remove redundant bits. This leads to the following 8 control bits:

$v_1[14] \oplus v_2[14]$	$v_1[14] \oplus v_1[15]$	$v_1[22]$	$v_1[23]$
$v_1[15] \oplus v_2[15]$	$v_1[15] \oplus v_3[23]$	$v_2[22]$	$v_2[23]$

This defines 2^8 partitions of the ciphertexts, after guessing 8 key bits. We evaluated the bias in each partition, and we found that the combined capacity is $c^2 = 2^{6.84}$. This means that we have the following complexity ratio

$$R^D_{\text{lin}} = 2^{-2 \cdot 2.6}/2^{-8}2^{6.84} \approx 2^{-4} \tag{12}$$

Analysis of Differential with Partitioning. There are four active bits in the first additions:

- Bit 23 in $v_2 \boxplus v_3$: $(2^{23}, 2^{23}) \xrightarrow{\boxplus} 0$
- Bit 30 in $v_2 \boxplus v_3$: $(2^{30}, 2^{30}) \xrightarrow{\boxplus} 2^{31}$
- Bit 6 in $v_2 \boxplus v_3$: $(2^{6}, 0) \xrightarrow{\boxplus} 2^{6}$
- Bit 26 in $v_0 \boxplus v_1$: $(2^{26}, 2^{26}) \xrightarrow{\boxplus} 0$

Following the analysis of Sect. 3, we can use additional input differences for each of them. However, we reach a better trade-off by selected only three of them. More precisely, we consider 2^3 input differences, defined by δ_i and the following extra active bits:

$$v_2[24] \qquad\qquad v_2[31] \qquad\qquad v_0[27]$$

As explained in Sect. 2, we build structures of 2^4 plaintexts, where each structure provides 2^3 pairs for every input difference, *i.e.* 2^6 pairs in total.

Following the analysis of Sect. 3, we use the following control bits to improve the probability of the differential:

$$\begin{array}{ccc} v_2[23] \oplus v_2[24] & v_2[30] \oplus v_3[30] & v_0[26] \oplus v_0[27] \\ v_2[24] \oplus v_3[23] & & v_0[27] \oplus v_1[26] \end{array}$$

This divides each set of pairs into 2^5 subsets, after guessing 5 key bits. In total we have 2^8 subsets to analyze, according to the control bits and the multiple differentials. We found that, for 18 of those subsets, there is a probability 2^{-2} to reach δ_o (the probability is 0 for the remaining subsets). This leads to a complexity ratio:

$$R^D_{\text{diff}} = \frac{2 \cdot 2^{-5}}{18/2^8 \times 2^4 \times 2^{-2}} = 2/9$$

$$R^D_{\text{diff-2}} = \frac{2 \cdot 2^{2\times-5}}{18/2^8 \times 2^4 \times 2^{2\times-2}} = 1/36$$

This corresponds to the analysis of Sect. 3: we have a ratio of 2/3 for bits $v_2[23]$ and $v_0[27]$ (Sect. 3.1), and a ratio of 1/2 for $v_2[31]$ in the simple linear case. In the differential-linear case, we have respectively ratios of 1/3 and 1/4.

Finally, the improved attack requires a data complexity in the order of:

$${R^D_{\text{lin}}}^2 R^D_{\text{diff-2}} D \approx 2^{20.3}.$$

We can estimate the data complexity more accurately using the analysis of Biryukov et al. [9]. First, we give an alternate description of the attack similar the multiple linear attack framework. Starting from D chosen plaintexts, we build $2^2 D$ pairs using structures, and we keep $N = 18 \cdot 2^{-8} \cdot 2^{-14} \cdot 2^2 D$ samples per approximation after partitioning the differential and linear parts. The imbalance of the distinguisher is $2^{-2} \cdot 2^{-6.05} \cdot 2^{6.84} = 2^{-1.21}$. Following [9, Corollary 1], the gain of the attack with $D = 2^{24}$ is estimated as 6.6 bits, *i.e.* the average key rank should be about 42 (for the 13-bit subkey).

Using the FFT method of Sect. 5.2, we perform the attack with 2^{24} counters $\hat{\mathcal{I}}[s]$. Each structure of 2^4 plaintexts provides 2^6 pairs, so that we need $2^2 D$ operations to update the counters. Finally, the FFT computation require $24 \times 2^{24} \approx 2^{28.6}$ operations.

We have implemented this analysis, and it runs in about 10 s on a single core of a desktop PC[6]. Experimentally, we have a gain of about 6 bits (average key rank of 64 with 128 experiments); this validates our theoretical analysis. We also notice some key bits don't affect the distinguisher and cannot be recovered. On the other hand, the gain of the attack can be improved using more data, and further trade-offs are possible using larger or smaller partitions.

5.5 Attack Against 7-Round Chaskey

The best distinguisher we identified for an attack against 7-round Chaskey uses 1.5 round in $E_\top$, 4 rounds in $E_\perp$, and 1.5 round in $E_\perp$. The optimal differences and masks are:

- Differential for $E_\top$ with probability $p_\top = 2^{-17}$:
  $v_0[8,18,21,30], v_1[8,13,21,26,30], v_2[3,21,26], v_3[21,26,27] \xrightarrow{E_\top} v_0[31]$
- Biased bit for $E_\perp$ with imbalance $\varepsilon_\perp = 2^{-6.1}$:
  $v_0[31] \xrightarrow{E_\perp} v_2[20]$
- Linear approximations for $E_\perp$ with imbalance $\varepsilon_\perp = 2^{-7.6}$:
  $v_2[20] \xrightarrow{E_\perp} v_0[0,15,16,25,29], v_1[7,11,19,26], v_2[2,10,19,20,23,28], v_3[0,25,29]$

This gives a differential-linear distinguisher with expected complexity in the order of $D = 2/p_\top^2 \varepsilon_\perp^2 \varepsilon_\perp^4 \approx 2^{77.6}$. This attack is more expensive than generic attacks against on the Even-Mansour cipher, but we now improve it using the results of Sects. 2 and 3.

Analysis of Linear Approximations with Partitioning. We use an automatic search to identify good control bits, starting from the bits suggested by the result of Sect. 2. We identified the following control bits:

$v_1[3] \oplus v_1[11] \oplus v_3[10]$	$v_1[3] \oplus v_1[11] \oplus v_3[11]$	$v_0[15] \oplus v_3[14]$
$v_0[15] \oplus v_3[15]$	$v_1[11] \oplus v_1[18] \oplus v_3[17]$	$v_1[11] \oplus v_1[18] \oplus v_3[18]$
$v_1[3] \oplus v_2[2]$	$v_1[3] \oplus v_2[3]$	$v_1[11] \oplus v_2[9]$
$v_1[11] \oplus v_2[10]$	$v_1[11] \oplus v_2[11]$	$v_1[18] \oplus v_2[17]$
$v_1[18] \oplus v_2[18]$	$v_1[2] \oplus v_1[3]$	$v_1[9] \oplus v_1[11]$
$v_1[10] \oplus v_1[11]$	$v_1[17] \oplus v_1[18]$	$v_0[14] \oplus v_0[15]$
$v_0[15] \oplus v_1[3] \oplus v_1[11] \oplus v_1[18]$		

[6] Haswell microarchitecture running at 3.4 GHz.

Note that the control bits identified in Sect. 2 appear as linear combinations of those control bits.

This defines 2^{19} partitions of the ciphertexts, after guessing 19 key bits. We evaluated the bias in each partition, and we found that the combined capacity is $c^2 = 2^{14.38}$. This means that we gain the following factor:

$$R^D_{\text{lin}} = 2^{-2 \cdot 7.6}/2^{-19}2^{14.38} \approx 2^{-10.5} \tag{13}$$

This example clearly shows the power of the partitioning technique: using a few key guesses, we essentially avoid the cost of the last layer of additions.

Analysis of Differential with Partitioning. We consider 2^9 input differences, defined by δ_i and the following extra active bits:

$v_0[9]$	$v_0[22]$	$v_0[31]$	$v_0[19]$	
$v_0[14]$	$v_0[27]$	$v_2[22]$	$v_2[27]$	$v_2[4]$

As explained in Sect. 2, we build structures of 2^{10} plaintexts, where each structure provides 2^9 pairs for every input difference, *i.e.* 2^{18} pairs in total.

Again, we use an automatic search to identify good control bits, starting from the bits suggested in Sect. 3. We use the following control bits to improve the probability of the differential:

$v_0[4] \oplus v_2[3]$	$v_2[22] \oplus v_3[21]$	$v_2[27] \oplus v_3[26]$	$v_2[27] \oplus v_3[27]$
$v_2[3] \oplus v_2[4]$	$v_2[21] \oplus v_2[22]$	$v_2[26] \oplus v_2[27]$	$v_0[9] \oplus v_1[8]$
$v_0[14] \oplus v_1[13]$	$v_0[27] \oplus v_1[26]$	$v_0[30] \oplus v_1[30]$	$v_0[8] \oplus v_0[9]$
$v_0[18] \oplus v_0[19]$	$v_0[21] \oplus v_0[22]$		

This divides each set of pairs into 2^{14} subsets, after guessing 14 key bits. In total we have 2^{23} subsets to analyze, according to the control bits and the multiple differentials. We found that, for 17496 of those subsets, there is a probability 2^{-11} to reach δ_o (the probability is 0 for the remaining subsets). This leads to a ratio:

$$R^D_{\text{diff-2}} = \frac{2 \cdot 2^{-2 \cdot 17}}{17496/2^{23} \times 2^{10} \times 2^{-2 \cdot 11}} = 1/4374 \approx 2^{-12.1}$$

Finally, the improved attack requires a data complexity of:

$$R^{D}_{\text{lin}}{}^{2} R^D_{\text{diff-2}} D \approx 2^{44.5}.$$

Again, we can estimate the data complexity more accurately using [9]. In this attack, starting from N_0 chosen plaintexts, we build $2^8 N_0$ pairs using structures, and we keep $N = 17496 \cdot 2^{-23} \cdot 2^{-38} \cdot 2^8 N_0$ samples per approximation after partitioning the differential and linear parts. The imbalance of the distinguisher is $2^{-11} \cdot 2^{-6.1} \cdot 2^{14.38} = 2^{-2.72}$. Following [9, Corollary 1], the gain of the attack with $N_0 = 2^{48}$ is estimated as 6.3 bits, *i.e.* the average rank of the 33-bit subkey should be about $2^{25.7}$. Following the experimental results of Sect. 5.4, we expect this to estimation to be close to the real gain (the gain can also be increased if more than 2^{48} data is available).

Using the FFT method of Sect. 5.2, we perform the attack with 2^{61} counters $\hat{\mathcal{I}}[s]$. Each structure of 2^{10} plaintexts provides 2^{18} pairs, so that we need $2^8 D$ operations to update the counters. Finally, the FFT computation require $61 \times 2^{61} \approx 2^{67}$ operations.

This attack recovers only a few bits of a 33-bit subkey, but an attacker can run the attack again with a different differential-linear distinguisher to recover other key bits. For instance, a rotated version of the distinguisher will have a complexity close to the optimal one, and the already known key bits can help reduce the complexity.

Conclusion

In this paper, we have described a partitioning technique inspired by Biham and Carmeli's work. While Biham and Carmeli consider only two partitions and a linear approximation for a single subset, we use a large number of partitions, and linear approximations for every subset to take advantage of all the data. We also introduce a technique combining multiple differentials, structures, and partitioning for differential cryptanalysis. This allows a significant reduction of the data complexity of attacks against ARX ciphers, and is particularly efficient with boomerang and differential-linear attacks.

Our main application is a differential-linear attack against Chaskey, that reaches 7 rounds out of 8. In this application, the partitioning technique allows to go through the first and last additions almost for free. This is very similar to the use of partial key guess and partial decryption for SBox-based ciphers. This is an important result because standard bodies (ISO/IEC JTC1 SC27 and ITU-T SG17) are currently considering Chaskey for standardization, but little external cryptanalysis has been published so far. After the first publications of these results, the designers of Chaskey have proposed to standardize a new version with 12 rounds [30].

Acknowledgement. We would like to thank Nicky Mouha for enriching discussions about those results, and the anonymous reviewers for their suggestions to improve the presentation of the paper.

The author is partially supported by the French Agence Nationale de la Recherche through the BRUTUS project under Contract ANR-14-CE28-0015.

A Appendix: Application to FEAL-8X

We now present application of our techniques to reduce the data complexity of differential and linear attacks.

FEAL is an early block cipher proposed by Shimizu and Miyaguchi in 1987 [36]. FEAL uses only addition, rotation and xor operations, which makes it much more efficient than DES in software. FEAL has inspired the development of many cryptanalytic techniques, in particular linear cryptanalysis.

At the rump session of CRYPTO 2012, Matsui announced a challenge for low data complexity attacks on FEAL-8X using only known plaintexts. At the time, the best practical attack required 2^{24} known plaintexts [2] (Matsui and Yamagishi had non-practical attacks with as little as 2^{14} known plaintext [26]), but Biham and Carmeli won the challenge with a new linear attack using 2^{15} known plaintexts, and introduced the partitioning technique to reduce the data complexity to 2^{14} [3]. Later Sakikoyama *et al.* improved this result using multiple linear cryptanalysis, with a data complexity of only 2^{12} [34].

We now explain how to apply the generalized partitioning to attack FEAL-8X. Our attack follows the attack of Biham and Carmeli [3], and uses the generalized partitioning technique to reduce the data complexity further. The attack by Biham and Carmeli requires 2^{14} data and about 2^{45} time, while our attack needs only 2^{12} data, and 2^{45} time. While the attack of Sakikoyama *et al.* is more efficient with the same data complexity, this shows a simple example of application of the generalized partitioning technique.

The attacks are based on a 6-round linear approximation with imbalance 2^{-5}, using partial encryption for the first round (with a 15 bit key guess), and partial decryption for the last round (with a 22 bit key guess). This allows to compute enough bits of the state after the first round and before the last round, respectively, to compute the linear approximation. For more details of the attack, we refer the reader to the description of Biham and Carmeli [3].

In order to improve the attack, we focus on the round function of the second-to-last round. The corresponding linear approximation is $x[\texttt{10115554}] \to y[\texttt{04031004}]$ with imbalance of approximately 2^{-3}.

We partition the data according to the following 4 bits[7] (note that all those bits can be computed in the input of round 6 with the 22-bit key guess of $DK7$):

$$b_0 = f_{0,3} \oplus f_{1,3} \oplus f_{2,2} \oplus f_{3,2} \qquad b_1 = f_{0,3} \oplus f_{1,3} \oplus f_{2,3} \oplus f_{3,3}$$
$$b_2 = f_{0,3} \oplus f_{1,3} \oplus f_{2,5} \oplus f_{3,5} \qquad b_3 = f_{0,2} \oplus f_{1,2} \oplus f_{0,3} \oplus f_{1,3}$$

The probability of the linear approximation in each subset is as follows (indexed by the value of b_3, b_2, b_1, b_0):

$$\begin{array}{llll} p_{0000} = 0.250 & p_{0001} = 0.270 & p_{0010} = 0.531 & p_{0011} = 0.746 \\ p_{0100} = 0.406 & p_{0101} = 0.699 & p_{0110} = 0.750 & p_{0111} = 0.652 \\ p_{1000} = 0.254 & p_{1001} = 0.469 & p_{1010} = 0.730 & p_{1011} = 0.750 \\ p_{1100} = 0.652 & p_{1101} = 0.750 & p_{1110} = 0.699 & p_{1111} = 0.406 \end{array}$$

[7] We use Biham and Carmeli's notation $f_{i,j}$ for bit j of input word i.

This gives a total capacity $c^2 = \sum_i (2 \cdot p_i - 1)^2 = 2.49$, using subsets of 1/16 of the data. For reference, a linear attack without partitioning has a capacity $(2^{-3})^2$, therefore the complexity ratio can be computed as:

$$R_{\text{lin}}^{D} = 2^{-6}/(1/16 \times 2.49) \approx 1/10$$

This can be compared to Biham and Carmeli's partitioning, where they use a single linear approximation with capacity 0.1 for 1/2 of the data, this gives a ratio of only:

$$R_{\text{lin}}^{D} = 2^{-6}/(1/2 \times 0.1) \approx 1/3.2$$

With a naive implementation of this attack, we have to repeat the analysis 16 times, for each guess of 4 key bits. Since the data is reduced by a factor 4, the total time complexity increases by a factor 4 compared to the attack on Biham and Carmeli. This result in an attack with 2^{12} data and 2^{47} time.

However, the time complexity can also be reduced using counters, because the 4 extra key bits only affect the choice of the partitions. This leads to an attack with 2^{12} data and 2^{43} time.

References

1. Aumasson, J.-P., Fischer, S., Khazaei, S., Meier, W., Rechberger, C.: New features of latin dances: analysis of Salsa, ChaCha, and Rumba. In: Nyberg, K. (ed.) FSE 2008. LNCS, vol. 5086, pp. 470–488. Springer, Heidelberg (2008)
2. Biham, E.: On Matsui's linear cryptanalysis. In: De Santis, A. (ed.) EUROCRYPT 1994. LNCS, vol. 950, pp. 341–355. Springer, Heidelberg (1995)
3. Biham, E., Carmeli, Y.: An improvement of linear cryptanalysis with addition operations with applications to FEAL-8X. In: Joux, A., Youssef, A. (eds.) SAC 2014. LNCS, vol. 8781, pp. 59–76. Springer, Heidelberg (2014)
4. Biham, E., Chen, R., Joux, A.: Cryptanalysis of SHA-0 and reduced SHA-1. J. Cryptology **28**(1), 110–160 (2015)
5. Biham, E., Dunkelman, O., Keller, N.: Enhancing differential-linear cryptanalysis. In: Zheng, Y. (ed.) ASIACRYPT 2002. LNCS, vol. 2501, pp. 254–266. Springer, Heidelberg (2002)
6. Biham, E., Shamir, A.: Differential cryptanalysis of DES-like cryptosystems. In: Menezes, A., Vanstone, S.A. (eds.) CRYPTO 1990. LNCS, vol. 537, pp. 2–21. Springer, Heidelberg (1991)
7. Biham, E., Shamir, A.: Differential cryptanalysis of DES-like cryptosystems. J. Cryptology **4**(1), 3–72 (1991)
8. Biham, E., Shamir, A.: Differential cryptanalysis of feal and N-Hash. In: Davies, D.W. (ed.) EUROCRYPT 1991. LNCS, vol. 547, pp. 1–16. Springer, Heidelberg (1991)
9. Biryukov, A., De Cannière, C., Quisquater, M.: On multiple linear approximations. In: Franklin, M. (ed.) CRYPTO 2004. LNCS, vol. 3152, pp. 1–22. Springer, Heidelberg (2004)
10. Biryukov, A., Nikolić, I., Roy, A.: Boomerang attacks on BLAKE-32. In: Joux, A. (ed.) FSE 2011. LNCS, vol. 6733, pp. 218–237. Springer, Heidelberg (2011)

11. Blondeau, C., Gérard, B., Tillich, J.P.: Accurate estimates of the data complexity and success probability for various cryptanalyses. Des. Codes Crypt. **59**(1–3), 3–34 (2011)
12. Blondeau, C., Leander, G., Nyberg, K.: Differential-linear cryptanalysis revisited. In: Cid, C., Rechberger, C. (eds.) FSE 2014. LNCS, vol. 8540, pp. 411–430. Springer, Heidelberg (2015)
13. Bogdanov, A.A., Knudsen, L.R., Leander, G., Paar, C., Poschmann, A., Robshaw, M., Seurin, Y., Vikkelsoe, C.: PRESENT: an ultra-lightweight block cipher. In: Paillier, P., Verbauwhede, I. (eds.) CHES 2007. LNCS, vol. 4727, pp. 450–466. Springer, Heidelberg (2007)
14. Cho, J.Y., Pieprzyk, J.: Crossword puzzle attack on NLS. In: Biham, E., Youssef, A.M. (eds.) SAC 2006. LNCS, vol. 4356, pp. 249–265. Springer, Heidelberg (2007)
15. Collard, B., Standaert, F.-X., Quisquater, J.-J.: Improving the time complexity of Matsui's linear cryptanalysis. In: Nam, K.-H., Rhee, G. (eds.) ICISC 2007. LNCS, vol. 4817, pp. 77–88. Springer, Heidelberg (2007)
16. Gilbert, H., Chassé, G.: A statistical attack of the FEAL-8 cryptosystem. In: Menezes, A., Vanstone, S.A. (eds.) CRYPTO 1990. LNCS, vol. 537, pp. 22–33. Springer, Heidelberg (1991)
17. Hermelin, M., Nyberg, K.: Dependent linear approximations: the algorithm of Biryukov and others revisited. In: Pieprzyk, J. (ed.) CT-RSA 2010. LNCS, vol. 5985, pp. 318–333. Springer, Heidelberg (2010)
18. Huang, T., Tjuawinata, I., Wu, H.: Differential-linear cryptanalysis of ICEPOLE. In: Leander, G. (ed.) FSE 2015. LNCS, vol. 9054, pp. 243–263. Springer, Heidelberg (2015)
19. Lamberger, M., Mendel, F.: Higher-order differential attack on reduced SHA-256. In: IACR Cryptology ePrint Archive, report 2011/37 (2011)
20. Langford, S.K., Hellman, M.E.: Differential-linear cryptanalysis. In: Desmedt, Y.G. (ed.) CRYPTO 1994. LNCS, vol. 839, pp. 17–25. Springer, Heidelberg (1994)
21. Leurent, G.: Analysis of differential attacks in ARX constructions. In: Wang, X., Sako, K. (eds.) ASIACRYPT 2012. LNCS, vol. 7658, pp. 226–243. Springer, Heidelberg (2012)
22. Leurent, G.: Construction of differential characteristics in ARX designs application to Skein. In: Canetti, R., Garay, J.A. (eds.) CRYPTO 2013, Part I. LNCS, vol. 8042, pp. 241–258. Springer, Heidelberg (2013)
23. Leurent, G., Roy, A.: Boomerang attacks on hash function using auxiliary differentials. In: Dunkelman, O. (ed.) CT-RSA 2012. LNCS, vol. 7178, pp. 215–230. Springer, Heidelberg (2012)
24. Lipmaa, H., Moriai, S.: Efficient algorithms for computing differential properties of addition. In: Matsui, M. (ed.) FSE 2001. LNCS, vol. 2355, pp. 336–350. Springer, Heidelberg (2002)
25. Matsui, M.: Linear cryptanalysis method for DES cipher. In: Helleseth, T. (ed.) EUROCRYPT 1993. LNCS, vol. 765, pp. 386–397. Springer, Heidelberg (1994)
26. Matsui, M., Yamagishi, A.: A new method for known plaintext attack of FEAL cipher. In: Rueppel, R.A. (ed.) EUROCRYPT 1992. LNCS, vol. 658, pp. 81–91. Springer, Heidelberg (1993)
27. Mavromati, C.: Key-recovery attacks against the mac algorithm chaskey. In: SAC 2015 (2015)
28. Mendel, F., Nad, T.: Boomerang distinguisher for the SIMD-512 compression function. In: Bernstein, D.J., Chatterjee, S. (eds.) INDOCRYPT 2011. LNCS, vol. 7107, pp. 255–269. Springer, Heidelberg (2011)

29. Miyano, H.: Addend dependency of differential/linear probability of addition. IEICE Trans. Fundam. Electron. Commun. Comput. Sci. **81**(1), 106–109 (1998)
30. Mouha, N.: Chaskey: a MAC algorithm for microcontrollers - status update and proposal of Chaskey-12 -. In: IACR Cryptology ePrint Archive, report 2015/1182 (2015)
31. Mouha, N., Mennink, B., Van Herrewege, A., Watanabe, D., Preneel, B., Verbauwhede, I.: Chaskey: an efficient MAC algorithm for 32-bit microcontrollers. In: Joux, A., Youssef, A. (eds.) SAC 2014. LNCS, vol. 8781, pp. 306–323. Springer, Heidelberg (2014)
32. Mouha, N., Velichkov, V., De Cannière, C., Preneel, B.: The differential analysis of S-Functions. In: Biryukov, A., Gong, G., Stinson, D.R. (eds.) SAC 2010. LNCS, vol. 6544, pp. 36–56. Springer, Heidelberg (2011)
33. Nyberg, K., Wallén, J.: Improved linear distinguishers for SNOW 2.0. In: Robshaw, M. (ed.) FSE 2006. LNCS, vol. 4047, pp. 144–162. Springer, Heidelberg (2006)
34. Sakikoyama, S., Todo, Y., Aoki, K., Morii, M.: How much can complexity of linear cryptanalysis be reduced? In: Lee, J., Kim, J. (eds.) ICISC 2014. LNCS, vol. 8949, pp. 117–131. Springer, Heidelberg (2014)
35. Sasaki, Y.: Boomerang distinguishers on MD4-Family: first practical results on full 5-Pass HAVAL. In: Miri, A., Vaudenay, S. (eds.) SAC 2011. LNCS, vol. 7118, pp. 1–18. Springer, Heidelberg (2012)
36. Shimizu, A., Miyaguchi, S.: Fast data encipherment algorithm FEAL. In: Price, W.L., Chaum, D. (eds.) EUROCRYPT 1987. LNCS, vol. 304, pp. 267–278. Springer, Heidelberg (1988)
37. Tardy-Corfdir, A., Gilbert, H.: A known plaintext attack of FEAL-4 and FEAL-6. In: Feigenbaum, J. (ed.) CRYPTO 1991. LNCS, vol. 576, pp. 172–182. Springer, Heidelberg (1992)
38. Wagner, D.: The boomerang attack. In: Knudsen, L.R. (ed.) FSE 1999. LNCS, vol. 1636, pp. 156–170. Springer, Heidelberg (1999)
39. Wallén, J.: Linear approximations of addition modulo 2^n. In: Johansson, T. (ed.) FSE 2003. LNCS, vol. 2887, pp. 261–273. Springer, Heidelberg (2003)
40. Wang, X., Yin, Y.L., Yu, H.: Finding collisions in the full SHA-1. In: Shoup, V. (ed.) CRYPTO 2005. LNCS, vol. 3621, pp. 17–36. Springer, Heidelberg (2005)
41. Wang, X., Yu, H.: How to break MD5 and other hash functions. In: Cramer, R. (ed.) EUROCRYPT 2005. LNCS, vol. 3494, pp. 19–35. Springer, Heidelberg (2005)
42. Yu, H., Chen, J., Wang, X.: The boomerang attacks on the round-reduced Skein-512. In: Knudsen, L.R., Wu, H. (eds.) SAC 2012. LNCS, vol. 7707, pp. 287–303. Springer, Heidelberg (2013)

Reverse-Engineering the S-Box of Streebog, Kuznyechik and STRIBOBr1

Alex Biryukov[1,2](✉), Léo Perrin[2](✉), and Aleksei Udovenko[2](✉)

[1] University of Luxembourg, Luxembourg City, Luxembourg
alex.biryukov@uni.lu

[2] SnT, University of Luxembourg, Luxembourg City, Luxembourg
{leo.perrin,aleksei.udovenko}@uni.lu

Abstract. The Russian Federation's standardization agency has recently published a hash function called Streebog and a 128-bit block cipher called Kuznyechik. Both of these algorithms use the same 8-bit S-Box but its design rationale was never made public.

In this paper, we reverse-engineer this S-Box and reveal its hidden structure. It is based on a sort of 2-round Feistel Network where exclusive-or is replaced by a finite field multiplication. This structure is hidden by two different linear layers applied before and after. In total, five different 4-bit S-Boxes, a multiplexer, two 8-bit linear permutations and two finite field multiplications in a field of size 2^4 are needed to compute the S-Box.

The knowledge of this decomposition allows a much more efficient hardware implementation by dividing the area and the delay by 2.5 and 8 respectively. However, the small 4-bit S-Boxes do not have very good cryptographic properties. In fact, one of them has a probability 1 differential.

We then generalize the method we used to partially recover the linear layers used to whiten the core of this S-Box and illustrate it with a generic decomposition attack against 4-round Feistel Networks whitened with unknown linear layers. Our attack exploits a particular pattern arising in the Linear Approximations Table of such functions.

Keywords: Reverse-Engineering · S-Box · Streebog · Kuznyechik · STRIBOBr1 · White-Box · Linear Approximation Table · Feistel Network

1 Introduction

S-Boxes are key components of many symmetric cryptographic primitives including block ciphers and hash functions. Their use allows elegant security arguments

The work of Léo Perrin is supported by the CORE ACRYPT project (ID C12-15-4009992) funded by the *Fonds National de la Recherche* (Luxembourg). The work of Aleksei Udovenko is supported by the *Fonds National de la Recherche*, Luxembourg (project reference 9037104).

M. Fischlin and J.-S. Coron (Eds.): EUROCRYPT 2016, Part I, LNCS 9665, pp. 372–402, 2016.
DOI: 10.1007/978-3-662-49890-3_15

based on the so-called wide-trail strategy [1] to justify that the primitive is secure against some of the best known attacks, e.g. differential [2] and linear [3,4] cryptanalysis.

Given the importance of their role, S-Boxes are carefully chosen and the criteria or algorithm used to build them are explained and justified by the designers of new algorithms. For example, since the seminal work of Nyberg on this topic [5], the inverse function in the finite field of size 2^n is often used (see the Advanced Encryption Standard [1], TWINE [6]...).

However, some algorithms are designed secretly and, thus, do not justify their design choices. Notable such instances are the primitives designed by the American National Security Agency (NSA) or standardized by the Russian Federal Agency on Technical Regulation and Metrology (FATRM). While the NSA eventually released some information about the design of the S-Boxes of the Data Encryption Standard [7,8], the criteria they used to pick the S-Box of Skipjack [9] remain mostly unknown despite some recent advances on the topic [10]. Similarly, recent algorithms standardized by FATRM share the same function π, an unexplained 8-bit S-Box. These algorithms are:

Streebog (officially called "GOST R 34.11-2012", sometimes spelled Stribog) is the new standard hash function for the Russian Federation [11]. Several cryptanalyses against this algorithm have been published. A second preimage attack requiring 2^{266} calls to the compression function instead of the expected 2^{512} has been found by Guo et al. [12]. Another attack [13] targets a modified version of the algorithm where only the round constants are modified: for some new round constants, it is actually possible to find collisions for the hash function. To show that the constants were not chosen with malicious intentions, the designers published a note [14] describing how they were derived from a modified version of the hash function. While puzzling at a first glance, the seeds actually correspond to Russian names written backward (see the full version of this paper [15]).

Kuznyechik (officially called "GOST R 34.12-2015"; sometimes the spelling "Kuznechik" is used instead) is the new standard block cipher for the Russian Federation. It was first mentioned in [16] and is now available at [17]. It is a 128-bit block cipher with a 256-bit key consisting of 9 rounds of a Substitution-Permutation Network where the linear layer is a matrix multiplication in $(\mathbb{F}_{2^8})^{16}$ and the S-Box layer consists in the parallel application of an 8-bit S-Box. The best attack so far is a Meet-in-the-Middle attack covering 5 rounds [18]. It should not be mistaken with GOST 28147-89 [19], a 64-bit block cipher standardized in 1989 and which is sometimes referred to as "the GOST cipher" in the literature and "Magma" in the latest Russian documents.

STRIBOB [20] is a CAESAR candidate which made it to the second round of the competition. The designer of this algorithm is not related to the Russian agencies. Still, the submission for the first round (STRIBOBr1) is based on Streebog.[1]

The Russian agency acting among other things as a counterpart of the American NSA is the Federal Security Service (FSB). It was officially involved in the design of Streebog. Interestingly, in a presentation given at RusCrypto'13 [24] by Shishkin on behalf of the FSB, some information about the design process of the S-Box is given: it is supposed not to have an analytic structure — even if that means not having optimal cryptographic properties unlike e.g. the S-Box of the AES [1] — and to minimize the number of operations necessary to compute it so as to optimize hardware and vectorized software implementations. However, the designers did not publish any more details about the rationale behind their choice for π and, as a consequence, very little is known about it apart from its look-up table, which we give in Table 1. In [21], Saarinen et al. summarize a discussion they had with some of the designers of the GOST algorithms at a conference in Moscow:

> We had brief informal discussions with some members of the Streebog and Kuznyechik design team at the CTCrypt'14 workshop (05-06 June 2014, Moscow RU). Their recollection was that the aim was to choose a "randomized" S-Box that meets the basic differential, linear, and algebraic requirements. Randomization using various building blocks was simply iterated until a "good enough" permutation was found. This was seen as an effective countermeasure against yet-unknown attacks [as well as algebraic attacks].

Since we know little to nothing about the design of this S-Box, it is natural to try and gather as much information as we can from its look-up table. In fact, the reverse-engineering of algorithms with unknown design criteria is not a new research area. We can mention for example the work of the community on the American National Security Agency's block cipher Skipjack [9] both before and after its release [25–27]. More recently, Biryukov et al. proved that its S-Box was not selected from a collection of random S-Boxes and was actually the product of an algorithm that optimized its linear properties [10].

Another recent example of reverse-engineering actually deals with Streebog. The linear layer of the permutation used to build its compression function was originally given as a binary matrix. However, it was shown in [28] that it corresponds to a matrix multiplication in $(\mathbb{F}_{2^8})^8$.

More generally, the task of reverse-engineering S-Boxes is related to finding generic attacks against high-level constructions. For instance, the cryptanalysis

[1] The version submitted to the next round, referred to as "STRIBOBr2" and "WHIRLBOB" [21], uses the S-Box of the Whirlpool hash function [22] whose design criteria and structure are public. In fact, the secrecy surrounding the S-Box of Streebog was part of the motivation behind this change [23].

Table 1. The S-Box π in hexadecimal. For example, $\pi(\texttt{0x7a}) = \texttt{0xc6}$.

	.0	.1	.2	.3	.4	.5	.6	.7	.8	.9	.a	.b	.c	.d	.e	.f
0.	fc	ee	dd	11	cf	6e	31	16	fb	c4	fa	da	23	c5	04	4d
1.	e9	77	f0	db	93	2e	99	ba	17	36	f1	bb	14	cd	5f	c1
2.	f9	18	65	5a	e2	5c	ef	21	81	1c	3c	42	8b	01	8e	4f
3.	05	84	02	ae	e3	6a	8f	a0	06	0b	ed	98	7f	d4	d3	1f
4.	eb	34	2c	51	ea	c8	48	ab	f2	2a	68	a2	fd	3a	ce	cc
5.	b5	70	0e	56	08	0c	76	12	bf	72	13	47	9c	b7	5d	87
6.	15	a1	96	29	10	7b	9a	c7	f3	91	78	6f	9d	9e	b2	b1
7.	32	75	19	3d	ff	35	8a	7e	6d	54	c6	80	c3	bd	0d	57
8.	df	f5	24	a9	3e	a8	43	c9	d7	79	d6	f6	7c	22	b9	03
9.	e0	0f	ec	de	7a	94	b0	bc	dc	e8	28	50	4e	33	0a	4a
a.	a7	97	60	73	1e	00	62	44	1a	b8	38	82	64	9f	26	41
b.	ad	45	46	92	27	5e	55	2f	8c	a3	a5	7d	69	d5	95	3b
c.	07	58	b3	40	86	ac	1d	f7	30	37	6b	e4	88	d9	e7	89
d.	e1	1b	83	49	4c	3f	f8	fe	8d	53	aa	90	ca	d8	85	61
e.	20	71	67	a4	2d	2b	09	5b	cb	9b	25	d0	be	e5	6c	52
f.	59	a6	74	d2	e6	f4	b4	c0	d1	66	af	c2	39	4b	63	b6

of SASAS [29], the recent attacks against the ASASA scheme [30,31] and the recovery of the secret Feistel functions for 5-, 6- and 7-round Feistel proposed in [32] can also be interpreted as methods to reverse-engineer S-Boxes built using such structures.

Our Contribution. We managed to reverse-engineer the hidden structure of this S-Box. A simplified high level view is given in Fig. 1. It relies on two rounds reminiscent of a Feistel or Misty-like structure where the output of the Feistel function is combined with the other branch using a finite field multiplication. In each round, a different heuristic is used to prevent issues caused by multiplication by 0. This structure is hidden by two different whitening linear layers applied before and after it.

With the exception of the inverse function which is used once, none of the components of this decomposition exhibits particularly good cryptographic properties. In fact, one of the non-linear 4-bit permutations used has a probability 1 differential.

Our recovery of the structure of π relies on spotting visual patterns in its LAT and exploiting those. We generalize this method and show how visual patterns in the LAT of 4-round Feistel Networks can be exploited to decompose a so-called $\mathsf{AF}^4\mathsf{A}$ structure consisting in a 4-round Feistel Network whitened with affine layers.

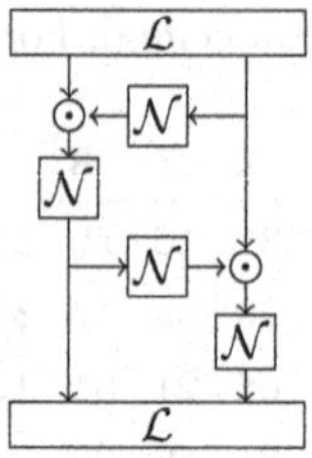

Fig. 1. A simplified view of our decomposition of π. Linear (resp. non linear) functions are denoted $\mathcal{L}$ (resp. $\mathcal{N}$) and $\odot$ is a finite field multiplication.

Outline. Section 2 introduces the definitions we need and describes how a statistical analysis of π rules out its randomness. Then, Sect. 3 explains the steps we used to reverse-engineer the S-Box starting from a picture derived from its linear properties and ending with a full specification of its secret structure. Section 4 is our analysis of the components used by GOST to build this S-Box. Finally, Sect. 5 describes a generic recovery attack against permutations affine-equivalent to 4-round Feistel Networks with secret components.

2 Boolean Functions and Randomness

2.1 Definitions and Notations

Definition 1. *We denote as* $\mathbb{F}_p$ *the finite field of size* p. *A* vectorial Boolean function *is a function mapping* $\mathbb{F}_2^n$ *to* $\mathbb{F}_2^m$. *We call* Boolean permutation *a permutation of* $\mathbb{F}_2^n$.

In what follows, we shall use the following operations and notations:

- exclusive-OR (or XOR) is denoted $\oplus$,
- logical AND is denoted $\wedge$,
- the scalar product of two elements $x = (x_{n-1}, ..., x_0)$ and $y = (y_{n-1}, ..., y_0)$ of $\mathbb{F}_2^n$ is denoted "$\cdot$" and is equal to to $x \cdot y = \bigoplus_{i=0}^{n-1} x_i \wedge y_i$, and
- finite field multiplication is denoted $\odot$.

The following tables are key tools to predict the resilience of an S-Box against linear and differential attacks.

DDT the *Difference Distribution Table* of a function f mapping n bits to m is a $2^n \times 2^m$ matrix T where $T[\delta, \Delta] = \#\{x \in \mathbb{F}_2^n, f(x \oplus \delta) \oplus f(x) = \Delta\}$,

LAT the *Linear Approximation Table* of a function f mapping n bits to m is a $2^n \times 2^m$ matrix $\mathcal{L}$ where $\mathcal{L}[a, b] = \#\{x \in \mathbb{F}_2^n, a \cdot x = b \cdot f(x)\} - 2^{n-1}$. We note that coefficient $\mathcal{L}[a, b]$ can equivalently be expressed as follows:

$$\mathcal{L}[a, b] = \frac{-1}{2} \sum_{x \in \mathbb{F}_2^n} (-1)^{a \cdot x \oplus b \cdot f(x)},$$

where the sum corresponds to the so-called *Walsh transform* of $x \mapsto (-1)^{b \cdot f(x)}$.

Furthermore, the coefficient $\mathcal{T}[\delta, \Delta]$ of a DDT $\mathcal{T}$ is called *cardinal of the differential* $(\delta \rightsquigarrow \Delta)$ and the coefficient $\mathcal{L}[a, b]$ of a LAT $\mathcal{L}$ is called *bias of the approximation* $(a \rightsquigarrow b)$.

From a designer's perspective, it is better to keep both the differential cardinals and the approximation biases low. For instance, the maximum cardinal of a differential is called the *differential uniformity* [5] and is chosen to be small in many primitives including the AES [1]. Such a strategy decreases the individual probability of all differential and linear trails.

Our analysis also requires some specific notations regarding linear functions mapping $\mathbb{F}_2^n$ to $\mathbb{F}_2^m$. Any such linear function can be represented by a matrix of elements in $\mathbb{F}_2$. For the sake of simplicity, we denote M^t the transpose of a matrix M and f^t the linear function obtained from the transpose of the matrix representation of the linear function f.

Finally, we recall the following definition.

Definition 2 (Affine-Equivalence). *Two vectorial Boolean functions f and g are affine-equivalent if there exist two affine mappings μ and η such that $f = \eta \circ g \circ \mu$.*

2.2 Quantifying Non-Randomness

In [10], Biryukov et al. proposed a general approach to try to reverse-engineer an S-Box suspected of having a hidden structure or of being built using secret design criteria. Part of their method allows a cryptanalyst to find out whether or not the S-Box could have been generated at random. It consists in checking if the distributions of the coefficients in both the DDT and the LAT are as it would be expected in the case of a random permutation.

The probability that all coefficients in the DDT of a random 8-bit permutation are at most equal to 8 and that this value occurs at most 25 times (as is the case for Streebog) is given by:

$$P[\max(d) = 8 \text{ and } N(8) \leq 25] = \sum_{\ell=0}^{25} \binom{255^2}{\ell} \cdot \Big[\sum_{d=0}^{3} \mathcal{D}(2d) \Big]^{255^2 - \ell} \cdot \mathcal{D}(8)^{\ell},$$

where $\mathcal{D}(d)$ is the probability that a coefficient of the DDT of a random permutation of $\mathbb{F}_2^8$ is equal to d. It is given in [33] and is equal to

$$\mathcal{D}(d) = \frac{e^{-1/2}}{2^{d/2}(d/2)!}.$$

We find that $P[\max(d) = 8 \text{ and } N(8) \leq 25] \approx 2^{-82.69}$. Therefore, we claim for π what Biryukov and Perrin claimed for the "F-Table" of Skipjack, namely that:

1. this S-Box was not picked uniformly at random from the set of the permutations of $\mathbb{F}_2^8$,
2. this S-Box was not generated by first picking many S-Boxes uniformly at random and then keeping the best according to some criteria, and
3. whatever algorithm was used to build it optimized the differential properties of the result.

3 Reverse-Engineering π

We used the algorithm described in [10] to try and recover possible structures for the S-Box. It has an even signature, meaning that it could be a Substitution-Permutation Network with simple bit permutations or a Feistel Network. However, the SASAS [29] attack and the SAT-based recovery attack against 3- ,4- and 5-round Feistel (both using exclusive-or and modular addition) from [10] failed. We also discarded the idea that π is affine-equivalent to a monomial of $\mathbb{F}_{2^8}$ using the following remark.

Remark 1. If f is affine-equivalent to a monomial, then every line of its DDT corresponding to a non-zero input difference contains the same coefficients (although usually in a different order).

This observation is an immediate consequence of the definition of the *differential spectrum* of monomials in [34]. For example, every line of the DDT of the S-Box of the AES, which is affine-equivalent to $x \mapsto x^{-1}$, contains exactly 129 zeroes, 126 twos and 1 four.

3.1 From a Vague Pattern to a Highly Structured Affine-Equivalent S-Box

It is also suggested in [10] to look at the so-called "Jackson Pollock representation" of the DDT and LAT of an unknown S-Box. These are obtained by assigning a color to each possible coefficient and drawing the table using one pixel per coefficient. The result for the absolute value of the coefficients of the LAT of π is given in Fig. 2. While it may be hard to see on paper, blurry vertical lines appear when looking at a large enough version of this picture. In order to better see this pattern, we introduce the so-called $\oplus$-*texture*. It is a kind of auto-correlation.

Definition 3. *We call* $\oplus$-texture *of the LAT* $\mathcal{L}$ *of an S-Box the matrix* $\mathcal{T}^{\oplus}$ *with coefficients* $\mathcal{T}^{\oplus}[i,j]$ *defined as:*

$$\mathcal{T}^{\oplus}[i,j] \;=\; \#\{(x,y), |\mathcal{L}[x \oplus i, y \oplus j]| = |\mathcal{L}[x,y]|\}.$$

The Jackson Pollock representation of the $\oplus$-texture of the LAT $\mathcal{L}_\pi$ of π is given in Fig. 3. The lines are now much more obvious and, furthermore, we observe dark dots in the very first column. The indices of both the rows containing the black dots and the columns containing the lines are the same and correspond to a binary vector space $\mathcal{V}$ defined, using hexadecimal notations, as:

$$\mathcal{V} = \{00, 1a, 20, 3a, 44, 5e, 64, 7e, 8a, 90, aa, b0, ce, d4, ee, f4\}.$$

In order to cluster the columns together to the left of the picture and the dark dots to the top of it, we can apply a linear mapping L to obtain a new table $\mathcal{L}'_\pi$ where $\mathcal{L}'_\pi[i,j] = \mathcal{L}_\pi[L(i), L(j)]$. We define L so that it maps $i \in \mathbb{F}_2^4$ to the i-th element of $\mathcal{V}$ and then complete it in a natural way to obtain a linear permutation of $\mathbb{F}_2^8$. It maps each bit as described below in hexadecimal notations:

$$L(01) = 1a, L(02) = 20, L(04) = 44, L(08) = 8a,$$

$$L(10) = 01, L(20) = 02, L(40) = 04, L(80) = 08.$$

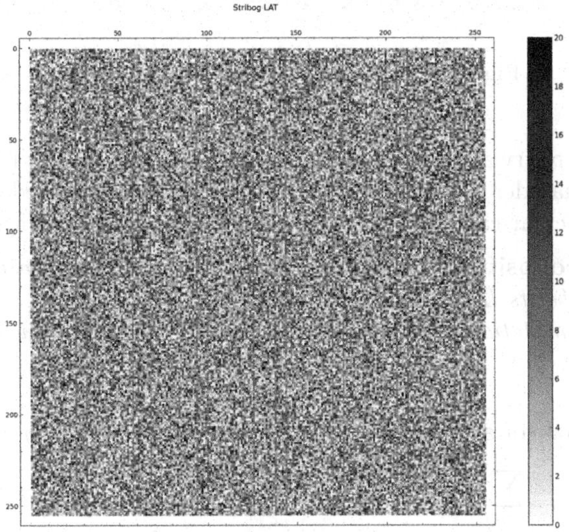

Fig. 2. The Jackson Pollock representation of the LAT of π (absolute value)

The Jackson Pollock representation of $\mathcal{L}'_\pi$ is given in Fig. 4. As we can see, it is highly structured: there is a 16×16 square containing[2] only coefficients equal to 0 in the top left corner. Furthermore, the left-most 15 bits to the right of column 0, exhibit a strange pattern: each of the coefficients in it has an absolute value in $[4, 12]$ although the maximum coefficient in the table is equal to 28. This forms a sort of low-contrast "stripe". The low number of different values it contains implies a low number of colour in the corresponding columns in $\mathcal{L}_\pi$, which in turn correspond to the lines we were able to distinguish in Fig. 2.

[2] Except of course in position $(0, 0)$ where the bias is equal to the maximum of 128.

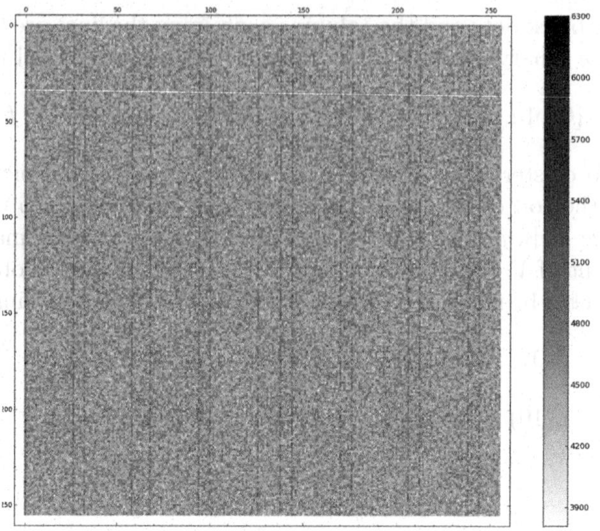

Fig. 3. The $\oplus$-texture of the LAT $\mathcal{L}_\pi$ of π.

It is natural to try and build another S-Box from π such that its LAT is equal to $\mathcal{L}'_\pi$. The remainder of this section describes how we achieved this. First, we describe a particular case[3] of Proposition 8.3 of [35] in the following lemma.

Lemma 1 (Proposition 8.3 of [35]**).** *Let f be a permutation mapping n bits to n and let $\mathcal{L}$ be its LAT. Let $\mathcal{L}'$ be a table defined by $\mathcal{L}'[u, v] = \mathcal{L}[\mu(u), v]$ for some linear permutation μ. Then the function f' has LAT $\mathcal{L}'$, where*

$$f' = f \circ (\mu^{-1})^t.$$

We also note that for a permutation f, the change of variable $y = f(x)$ implies:

$$\sum_{x \in \mathbb{F}_2^n} (-1)^{a \cdot x \oplus b \cdot f(x)} = \sum_{y \in \mathbb{F}_2^n} (-1)^{a \cdot f^{-1}(y) \oplus b \cdot y},$$

which in turn implies the following observation regarding the LAT of a permutation and its inverse.

Remark 2. Let f be a permutation mapping n bits to n and let $\mathcal{L}$ be its LAT. Then the LAT of its inverse f^{-1} is $\mathcal{L}^t$.

We can prove the following theorem using this remark and Lemma 1.

Theorem 1. *Let f be a permutation mapping n bits to n and let $\mathcal{L}$ be its LAT. Let $\mathcal{L}'$ be a table defined by $\mathcal{L}'[u, v] = \mathcal{L}[\mu(u), \eta(v)]$ for some linear permutations μ and η. Then the function f' has LAT $\mathcal{L}'$, where*

$$f' = \eta^t \circ f \circ (\mu^{-1})^t.$$

[3] It is obtained by setting $b = b_0 = a = 0$ in the statement of the original proposition and renaming the functions used.

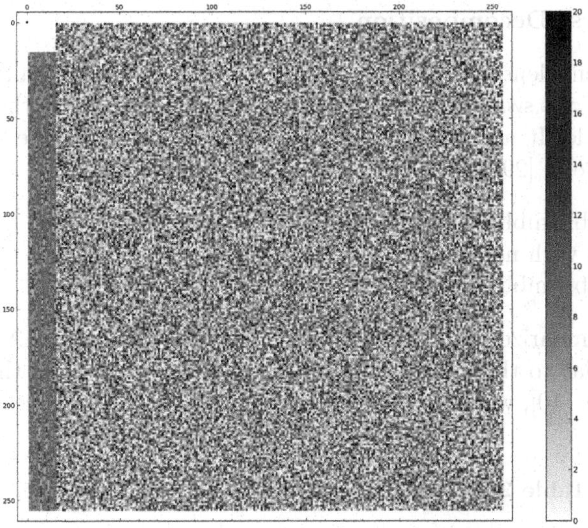

Fig. 4. The Jackson Pollock representation of $\mathcal{L}'_\pi$, where $\mathcal{L}'_\pi[i,j] = \mathcal{L}_\pi[L(i), L(j)]$.

Proof. Let f be a permutation of n bits and let $\mathcal{L}$ be its LAT. We first build $f_\mu = f \circ (\mu^{-1})^t$ using Lemma 1 so that the LAT of f_μ is $\mathcal{L}_\mu$ with $\mathcal{L}_\mu[u,v] = \mathcal{L}[\mu(u), v]$. We then use Remark 2 to note that the inverse of f_μ has LAT $\mathcal{L}_\mu^{\text{inv}}$ with $\mathcal{L}_\mu^{\text{inv}}[u,v] = \mathcal{L}_\mu[v,u] = \mathcal{L}[v,\mu(u)]$. Thus, $f_\eta = f_\mu^{-1} \circ (\eta^{-1})^t$ has LAT $\mathcal{L}_\eta$ with $\mathcal{L}_\eta[u,v] = \mathcal{L}[\eta(v), \mu(u)]$. Using again Remark 2, we obtain that $f_\eta^{-1} = \eta^t \circ f \circ (\mu^{-1})^t$ has LAT $\mathcal{L}'$. □

As a consequence of Theorem 1, the S-Box $L^t \circ \pi \circ (L^t)^{-1}$ has $\mathcal{L}'_\pi$ as its LAT. The mapping L^t consists in a linear Feistel round followed by a permutation of the left and right 4-bit nibbles (which we denote swapNibbles). To simplify the modifications we make, we remove the nibble permutation and define

$$\pi' = L^* \circ \pi \circ L^*$$

where L^* is the Feistel round in L^t and is described in Fig. 5.

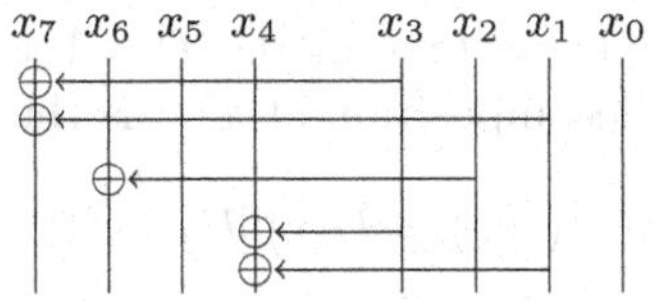

Fig. 5. A circuit computing L^* where its input is given in binary.

3.2 The First Decomposition

This affine-equivalent S-Box π' is highly structured. First of all, the LAT of $(\mathsf{swapNibbles} \circ \pi' \circ \mathsf{swapNibbles})$ is $\mathcal{L}'_\pi$, with its white square in the top left and strange left side. It also has interesting multiset properties. We use notations similar to those in [29], i.e.:

C denotes a 4-bit nibble which is constant,
? denotes a set with no particular structure, and
P denotes a 4-bit nibble taking all 16 values.

Table 2 summarizes the multiset properties of π' and π'^{-1}. As we can see, these are similar to those of a 3-rounds Feistel. However, using the SAT-based algorithm from [10], we ruled out this possibility.

Table 2. The multiset properties of π' and its inverse.

π' input	output
(P, C)	$(?, ?)$
(C, P)	$(P, ?)$

π'^{-1} input	output
(P, C)	$(?, ?)$
(C, P)	$(P, ?)$

When looking at the inverse of π', we notice that the multiset property is actually even stronger. Indeed, for any constant ℓ, the set $\mathcal{S}_\ell = \{\pi'^{-1}(\ell||r), \forall r \in [0, 15]\}$ is almost a vector space. If we replace the unique element of $\mathcal{S}_\ell$ of the form $(?||0)$ by $(0||0)$, the set obtained is a vector space V_ℓ where the right nibble is a linear function of the left nibble. As stated before, the left nibble takes all possible values. If we put aside the outputs of the form $(?||0)$ then π'^{-1} can be seen as

$$\pi'^{-1}(\ell||r) = T_\ell(r)||V_\ell(T_\ell(r)),$$

In this decomposition, T is a 4-bit block cipher with a 4-bit key where the left input of π'^{-1} acts as a key. On the other hand, V is a keyed linear function: for all ℓ, V_ℓ is a linear function mapping 4 bits to 4 bits.

We then complete this alternative representation by replacing $V_\ell(0)$, which should be equal to 0, by the left side of $\pi'^{-1}(\ell||T_\ell^{-1}(0))$. This allows to find a high level decomposition of π'^{-1}.

Finally, we define a new keyed function $U_r(\ell) = V_\ell(r)$ and notice that, for all r, U_r is a permutation. A decomposition of π'^{-1} is thus:

$$\pi'^{-1}(\ell||r) = T_\ell(r)||U_{T_\ell(r)}(\ell),$$

where the full codebooks of both mini-block ciphers T and U are given in Table 3a and b respectively. This structure is summarized in Fig. 6.

We decompose the mini-block ciphers T and U themselves in Sects. 3.3 and 3.4 respectively.

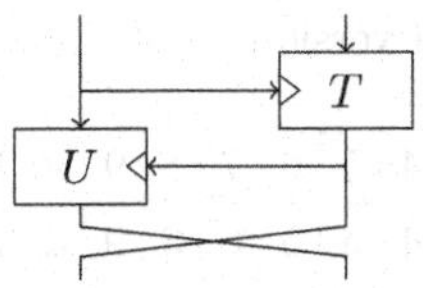

Fig. 6. The high level structure of π'^{-1}.

3.3 Reverse-Engineering T

We note that the mini-block cipher T' defined as $T'_k : x \mapsto T_k(x \oplus t_{\text{in}}(k) \oplus 0xC)$ for $t_{\text{in}}(k) = 0||k_2||k_3||0$ (see Table 4) is such that $T'_k(0) = 0$ for all k.

Furthermore, T' is such that all lines of T'_k can be obtained through a linear combination of T'_6, T'_7, T'_8 and T'_9 as follows:

$$\begin{array}{lll} T'_0 = T'_7 \oplus T'_9 & T'_1 = T'_8 \oplus T'_9 & T'_2 = T'_7 \oplus T'_9 \\ T'_3 = T'_6 \oplus T'_7 & T'_4 = T'_7 & T'_5 = T'_7 \oplus T'_8 \\ T'_a = T'_6 \oplus T'_7 \oplus T'_8 \oplus T'_9 & T'_b = T'_6 \oplus T'_7 & T'_c = T'_6 \oplus T'_7 \oplus T'_8 \\ T'_d = T'_9 & T'_e = T'_8 & T'_f = T'_7 \oplus T'_9. \end{array} \quad (1)$$

We also notice that T'_6, T'_7, T'_8 and T'_9 are all affine equivalent. Indeed, the linear mapping A defined by $A : 1 \mapsto 4, 2 \mapsto 1, 4 \mapsto 8, 8 \mapsto a$ (see Fig. 7a) is such that:

$$\begin{aligned} T'_7 &= A \circ T'_6 \\ T'_8 &= A^2 \circ T'_6 \\ T'_9 &= A^3 \circ T'_6. \end{aligned}$$

Table 3. The mini-block ciphers used to decompose π'^{-1}.

	0	1	2	3	4	5	6	7	8	9	a	b	c	d	e	f
T_0	e	f	2	5	7	b	8	1	3	c	d	a	0	9	4	6
T_1	2	9	a	4	e	6	7	b	1	8	3	d	0	c	f	5
T_2	e	f	2	5	7	b	8	1	3	c	d	a	0	9	4	6
T_3	5	d	4	2	6	7	b	8	c	1	9	f	0	3	a	e
T_4	5	e	6	7	4	3	f	a	0	1	d	2	8	b	c	9
T_5	9	d	f	a	c	6	8	1	0	5	b	3	2	4	e	7
T_6	3	9	d	f	1	e	b	8	0	2	7	c	4	a	5	6
T_7	5	e	6	7	4	3	f	a	0	1	d	2	8	b	c	9
T_8	7	b	8	5	9	d	c	3	2	e	a	f	6	1	0	4
T_9	d	f	a	c	e	6	2	5	1	3	b	7	9	4	0	8
T_a	e	6	7	4	c	3	8	1	a	2	d	9	5	b	0	f
T_b	4	2	5	d	b	8	6	7	9	f	c	1	a	e	0	3
T_c	2	5	a	4	3	9	d	8	c	f	0	7	b	1	6	e
T_d	e	6	2	5	d	f	a	c	9	4	0	8	1	3	b	7
T_e	9	d	c	3	7	b	8	5	6	1	0	4	2	e	a	f
T_f	8	1	7	b	2	5	e	f	4	6	0	9	d	a	3	c

(a) T.

	0	1	2	3	4	5	6	7	8	9	a	b	c	d	e	f
U_0	8	f	0	2	d	5	6	9	e	3	1	7	c	b	4	a
U_1	8	c	7	3	d	f	2	0	e	4	1	b	6	5	9	a
U_2	3	4	e	9	d	8	0	5	1	2	c	f	7	b	a	6
U_3	b	8	9	a	0	7	2	5	f	6	d	4	1	e	3	c
U_4	c	2	5	b	e	8	7	1	4	f	d	6	9	3	0	a
U_5	4	e	2	8	3	7	5	1	a	b	c	d	f	6	9	0
U_6	f	6	b	2	3	0	7	4	5	d	1	9	e	8	a	c
U_7	7	a	c	1	e	f	5	4	b	9	0	2	8	d	3	6
U_8	a	f	b	e	c	4	d	5	7	0	6	1	8	3	9	2
U_9	2	3	c	d	1	b	f	5	9	4	7	a	e	6	0	8
U_a	9	b	5	7	1	c	d	0	6	2	a	e	f	8	3	4
U_b	1	7	2	4	c	3	f	0	8	6	b	5	9	d	a	e
U_c	6	d	e	5	2	c	a	4	3	f	b	7	1	0	9	8
U_d	e	1	9	6	f	3	8	4	d	b	a	c	7	5	0	2
U_e	5	9	0	c	f	4	a	1	2	d	7	8	6	b	3	e
U_f	d	5	7	f	2	b	8	1	c	9	6	3	0	e	a	4

(b) U.

Table 4. A modified version T' of the mini-block cipher T.

	0	1	2	3	4	5	6	7	8	9	a	b	c	d	e	f
6	0	2	7	c	4	a	5	6	3	9	d	f	1	e	b	8
7	0	1	d	2	8	b	c	9	5	e	6	7	4	3	f	a
8	0	4	6	1	a	f	2	e	c	3	9	d	8	5	7	b
9	0	8	9	4	b	7	1	3	2	5	e	6	a	c	d	f
0	0	9	4	6	3	c	d	a	7	b	8	1	e	f	2	5
1	0	c	f	5	1	8	3	d	e	6	7	b	2	9	a	4
2	0	9	4	6	3	c	d	a	7	b	8	1	e	f	2	5
3	0	3	a	e	c	1	9	f	6	7	b	8	5	d	4	2
4	0	1	d	2	8	b	c	9	5	e	6	7	4	3	f	a
5	0	5	b	3	2	4	e	7	9	d	f	a	c	6	8	1
a	0	f	5	b	d	9	a	2	8	1	c	3	7	4	e	6
b	0	3	a	e	c	1	9	f	6	7	b	8	5	d	4	2
c	0	7	c	f	6	e	b	1	a	4	2	5	d	8	3	9
d	0	8	9	4	b	7	1	3	2	5	e	6	a	c	d	f
e	0	4	6	1	a	f	2	e	c	3	9	d	8	5	7	b
f	0	9	4	6	3	c	d	a	7	b	8	1	e	f	2	5

If we swap the two least significant bits (an operation we denote swap2lsb) before and after applying A we see a clear LFSR structure (see Fig. 7b).

(a) Definition of A. (b) Swapping the least significant bits.

Fig. 7. The mapping used to generate T'_7, T'_8 and T'_9 from T'_6.

We deduce the LFSR polynomial to be $X^4 + X^3 + 1$. This points towards finite field multiplication and, indeed, the mapping $\hat{A} = \mathsf{swap2lsb} \circ A \circ \mathsf{swap2lsb}$ can be viewed as a multiplication by X in $\mathbb{F}_{2^4} = \mathbb{F}_2[X]/(X^4 + X^3 + 1)$. To fit the swap into the original scheme we modify T'_6 and the bottom linear layer. Indeed, note that

$$A^i = (\mathsf{swap2lsb} \circ \hat{A} \circ \mathsf{swap2lsb})^i = \mathsf{swap2lsb} \circ \hat{A}^i \circ \mathsf{swap2lsb} \quad \text{for } i = 0, 1, \ldots,$$

so that we can merge one swap2lsb into T'_6 and move the other swap2lsb through XOR's outside T'. Let $t = \mathsf{swap2lsb} \circ T'_6$. Then $\mathsf{swap2lsb} \circ T'_k(x)$ is a linear

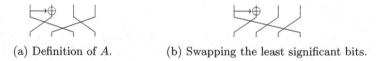

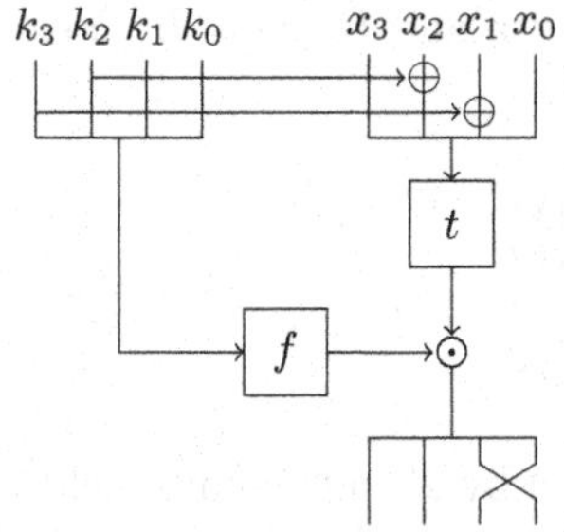

Fig. 8. The mini-block cipher T.

Table 5. Mappings f and t.

	0	1	2	3	4	5	6	7	8	9	a	b	c	d	e	f
f	a	c	a	3	2	6	1	2	4	8	f	3	7	8	4	a
t	2	d	b	8	3	a	e	f	4	9	6	5	0	1	7	c

combination of $X^i \odot t(x)$, where $i \in \{0, 1, 2, 3\}$ and $\odot$ is multiplication in the specified field. Thus, T can be computed as follows:

$$T_k(x) = \mathsf{swap2lsb}\left(f(k) \odot t\big(x \oplus t_{\text{in}}(k) \oplus 0xC\big)\right),$$

where f captures the linear relations from Eq. (1). Both f and t are given in Table 5 and a picture representing the structure of T is given in Fig. 8.

Note that $f(x)$ is never equal to 0: if it were the case then the function would not be invertible. On the other hand, the inverse of T_k is easy to compute: f must be replaced by $1/f$ where the inversion is done in the finite field $\mathbb{F}_{2^4}$, t by its inverse t^{-1} and the order of the operations must be reversed.

3.4 Reverse-Engineering U

Since $U_k(x) = V_x(k)$ and V_x is a linear function when $x \neq 0$, we have

$$U_k(x) = \big(k_3 \times U_8(x)\big) \oplus \big(k_2 \times U_4(x)\big) \oplus \big(k_1 \times U_2(x)\big) \oplus \big(k_0 \times U_1(x)\big)$$

where $k = \sum_{i \leq 3} k_i 2^i$ and $k \neq 0$. We furthermore notice that the permutations U_2, U_4 and U_8 can all be derived from U_1 using some affine functions B_k so that $U_k = B_k \circ U_1$. The values of $B_k(x)$ are given in Table 6.

Table 6. Affine functions such that $U_k = B_k \circ U_1$.

	0	1	2	3	4	5	6	7	8	9	a	b	c	d	e	f
B_2	5	c	0	9	2	b	7	e	3	a	6	f	4	d	1	8
B_4	1	d	7	b	f	3	9	5	c	0	a	6	2	e	4	8
B_8	5	6	d	e	0	3	8	b	a	9	2	1	f	c	7	4

If we let $B(x) = B_4(x) \oplus 1$ then $B_2(x) = B^{-1}(x) \oplus 5$ and $B_8(x) = B^2(x) \oplus 5$. Thus, we can define a linear function u_{out} such that

$$\begin{aligned} U_1(x) &= B^0 \circ U_1(x) \oplus u_{\text{out}}(1) \\ U_2(x) &= B^{-1} \circ U_1(x) \oplus u_{\text{out}}(2) \\ U_4(x) &= B^1 \circ U_1(x) \oplus u_{\text{out}}(4) \\ U_8(x) &= B^2 \circ U_1(x) \oplus u_{\text{out}}(8). \end{aligned} \tag{2}$$

Let M_2 be the matrix representation of multiplication by X in the finite field we used to decompose T, namely $\mathbb{F}_{2^4} = \mathbb{F}_2[X]/(X^4 + X^3 + 1)$. The linear mapping u_f defined by $u_f : 1 \mapsto 5, 2 \mapsto 2, 4 \mapsto 6, 8 \mapsto 8$ is such that $B = u_f \circ M_2 \circ u_f^{-1}$ is so that Eq. (2) can be re-written as

$$\begin{aligned} U_1(x) &= u_f \circ M_2^0 \circ u_f^{-1} \circ U_1(x) \oplus u_{\text{out}}(1) \\ U_2(x) &= u_f \circ M_2^{-1} \circ u_f^{-1} \circ U_1(x) \oplus u_{\text{out}}(2) \\ U_4(x) &= u_f \circ M_2^1 \circ u_f^{-1} \circ U_1(x) \oplus u_{\text{out}}(4) \\ U_8(x) &= u_f \circ M_2^2 \circ u_f^{-1} \circ U_1(x) \oplus u_{\text{out}}(8). \end{aligned} \tag{3}$$

If we swap the two least significant bits of k, then the exponents of matrix M_2 will go in ascending order: $(-1, 0, 1, 2)$. Let $u_1 = M_2^{-1} \circ u_f^{-1} \circ U_1(x)$. Since M_2 is multiplication by X in the finite field, we can write the following expression for U_k (when $k \neq 0$):

$$U_k(x) = u_f\big(u_1(x) \odot \mathsf{swap2lsb}(k)\big) \oplus u_{out}(k). \tag{4}$$

The complete decomposition of U is presented in Fig. 9. It uses the 4-bit permutations u_0 and u_1 specified in Table 7. We could not find a relation between u_1 and $u_0 = u_f^{-1} \circ U_0(x)$ so there has to be a conditional branching: U selects the result of Eq. (4) if $k \neq 0$ and the result of $u_0(x)$ otherwise before applying u_f. This is achieved using a multiplexer which returns the output of u_0 if $k_3 = k_2 = k_1 = k_0 = 0$, and returns the output of u_1 if it is not the case. In other words, U can be computed as follows:

$$U_k(x) = \begin{cases} u_f\big(u_1(x) \odot \mathsf{swap2lsb}(k)\big) \oplus u_{\text{out}}(k), & \text{if } k \neq 0 \\ u_f\big(u_0(x)\big) & \text{if } k = 0. \end{cases}$$

3.5 The Structure of π

In Sects. 3.3 and 3.4, we decomposed the two mini-block ciphers T and U which can be used to build π'^{-1}, the inverse of $L^* \circ \pi \circ L^*$. These mini-block ciphers are based on the non-linear 4-bit functions f, t, u_0, u_1, two finite field multiplications, a "trick" to bypass the non-invertibility of multiplication by 0 and simple linear functions. Let us now use the expressions we identified to express π itself.

First, we associate the linear functions and L^* into α and ω, two linear permutations applied respectively at the beginning and the end of the computation.

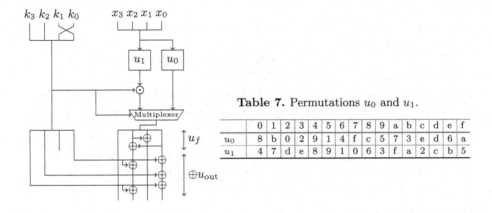

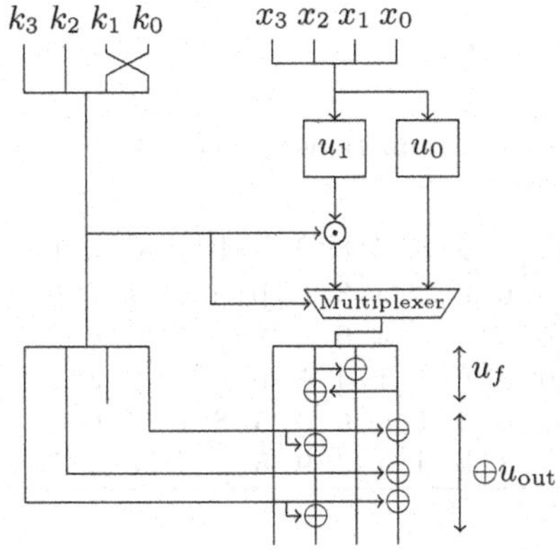

Fig. 9. The mini-block cipher U.

Table 7. Permutations u_0 and u_1.

	0	1	2	3	4	5	6	7	8	9	a	b	c	d	e	f
u_0	8	b	0	2	9	1	4	f	c	5	7	3	e	d	6	a
u_1	4	7	d	e	8	9	1	0	6	3	f	a	2	c	b	5

α First of all, we need to apply L^* as well as the the swap of the left and right branches (swapNibbles) present in the high level decomposition of π'^{-1} (see Fig. 6). Then, we note that the key in U needs a swap of its 2 bits of lowest weight (swap2lsb) and that the ciphertext of T needs the same swap. Thus, we simply apply swap2lsb. Then, we apply the addition of u_{out} and the inverse of u_f.

ω This function is simpler: it is the composition of the addition of t_{in} and of L^*.

The matrix representations of these layers are

$$\alpha = \begin{bmatrix} 0\,0\,0\,0\,1\,0\,0\,0 \\ 0\,1\,0\,0\,0\,0\,0\,1 \\ 0\,1\,0\,0\,0\,0\,1\,1 \\ 1\,1\,1\,0\,1\,1\,1\,1 \\ 1\,0\,0\,0\,1\,0\,1\,0 \\ 0\,1\,0\,0\,0\,1\,0\,0 \\ 0\,0\,0\,1\,1\,0\,1\,0 \\ 0\,0\,1\,0\,0\,0\,0\,0 \end{bmatrix}, \ \omega = \begin{bmatrix} 0\,0\,0\,0\,1\,0\,1\,0 \\ 0\,0\,0\,0\,0\,1\,0\,0 \\ 0\,0\,1\,0\,0\,0\,0\,0 \\ 1\,0\,0\,1\,1\,0\,1\,0 \\ 0\,0\,0\,0\,1\,0\,0\,0 \\ 0\,1\,0\,0\,0\,1\,0\,0 \\ 1\,0\,0\,0\,0\,0\,1\,0 \\ 0\,0\,0\,0\,0\,0\,0\,1 \end{bmatrix}.$$

In order to invert U, we define $\nu_0 = u_0^{-1}$ and $\nu_1 = u_1^{-1}$. If $\ell = 0$, then the output of the inverse of U is $\nu_0(r)$, otherwise it is $\nu_1\big(r \odot \mathcal{I}(\ell)\big)$, where $\mathcal{I} : x \mapsto x^{14}$ is the multiplicative inverse in $\mathbb{F}_{2^4}$. To invert T, we define $\sigma = t^{-1}$ and $\phi = \mathcal{I} \circ f$ and compute $\sigma\big(\phi(\ell) \odot r\big)$.

Figure 10 summarizes how to compute π using these components. The non-linear functions are all given in Table 8. A Sage [36] script performing those computations can be downloaded on Github.[4] The evaluation of $\pi(\ell||r)$ can be done as follows:

[4] https://github.com/picarresursix/GOST-pi

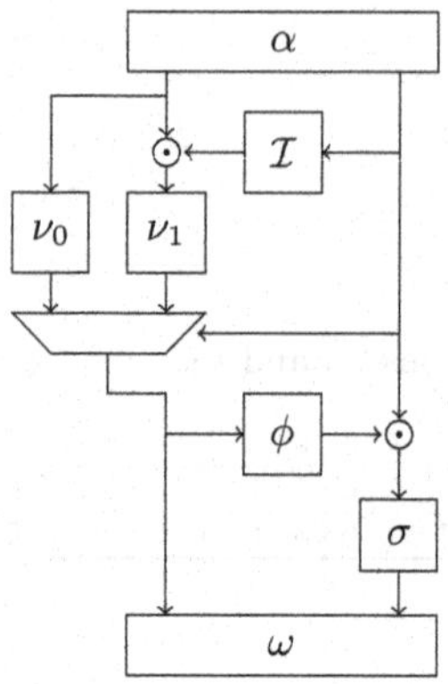

Fig. 10. Our decomposition of π.

Table 8. The non-linear functions needed to compute π.

	0	1	2	3	4	5	6	7	8	9	a	b	c	d	e	f
$\mathcal{I}$	0	1	c	8	6	f	4	e	3	d	b	a	2	9	7	5
ν_0	2	5	3	b	6	9	e	a	0	4	f	1	8	d	c	7
ν_1	7	6	c	9	0	f	8	1	4	5	b	e	d	2	3	a
ϕ	b	2	b	8	c	4	1	c	6	3	5	8	e	3	6	b
σ	c	d	0	4	8	b	a	e	3	9	5	2	f	1	6	7

1. $(\ell||r) := \alpha(\ell||r)$
2. if $r = 0$ then $\ell := \nu_0(\ell)$, else $\ell := \nu_1(\ell \odot \mathcal{I}(r))$
3. $r := \sigma(r \odot \phi(l))$
4. return $\omega(\ell||r)$.

4 Studying the Decomposition of π

4.1 Analyzing the Components

Table 9 summarizes the properties of the non-linear components of our decomposition. While it is not hard to find 4-bit permutations with a differential uniformity of 4, we see that none of the components chosen do except for the inverse function. We can thus discard the idea that the strength of π against differential and linear attacks relies on the individual resilience of each of its components.

As can be seen in Table 9, there is a probability 1 differential in ν_1: $9 \rightsquigarrow 2$. Furthermore, a difference equal to 2 on the left branch corresponds to a 1 bit difference on bit 5 of the input of ω, a bit which is left unchanged by ω.

Table 9. Linear and differential properties of the components of π.

	1-to-1	Best differentials and their probabilities	Best linear approximations and their probabilities
ϕ	No	$1 \rightsquigarrow$ d (8/16)	$3 \rightsquigarrow 8$ (2/16), $7 \rightsquigarrow$ d (2/16)
σ	Yes	f $\rightsquigarrow$ b (6/16)	$1 \rightsquigarrow$ f (14/16)
ν_0	Yes	$6 \rightsquigarrow$ c (6/16), e $\rightsquigarrow$ e (6/16)	30 approximations $(8 \pm 4)/16$
ν_1	Yes	$9 \rightsquigarrow 2$ (16/16)	8 approximations $(8 \pm 6)/16$

The structure itself also implies the existence of a truncated differential with high probability. Indeed, if the value on the left branch is equal to 0 for two different inputs, then the output difference on the left branch will remain equal to 0 with probability 1. This explains why the probability that a difference in $\Delta_{\text{in}} = \{\alpha^{-1}(\ell || 0), \ell \in \mathbb{F}_2^4, x \neq 0\}$ is mapped to a difference in $\Delta_{\text{out}} = \{\omega(\ell || 0), \ell \in \mathbb{F}_2^4, x \neq 0\}$ is higher than the expected 2^{-4}:

$$\frac{1}{2^4 - 1} \sum_{\delta \in \Delta_{\text{in}}} P[\pi(x \oplus \delta) \oplus \pi(x) \in \Delta_{\text{out}}] = \frac{450}{(2^4 - 1) \times 2^8} \approx 2^{-3}.$$

4.2 Comments on the Structure Used

We define $\hat{\pi}$ as $\omega^{-1} \circ \pi \circ \alpha^{-1}$, i.e. π minus its whitening linear layers.

The structure of $\hat{\pi}$ is similar to a 2-round combination of a Misty-like and Feistel structure where the XORs have been replaced by finite field multiplications. To the best of our knowledge, this is the first time such a structure has been used in cryptography. Sophisticated lightweight decompositions of the S-Box of the AES rely on finite field multiplication in $\mathbb{F}_{2^4}$, for instance in [37]. However, the high level structure used in this case is quite different. If π corresponds to such a decomposition then we could not find what it corresponds to. Recall in particular that π cannot be affine-equivalent to a monomial.

The use of finite field multiplication in such a structure yields a problem: if the output of the "Feistel function" is equal to 0 then the structure is not invertible. This issue is solved in a different way in each round. During the first round, a different data-path is used in the case which should correspond to a multiplication by zero. In the second round, the "Feistel function" is not bijective and, in particular, has no pre-image for 0.

Our decomposition also explains the pattern in the LAT[5] of π and π' that we used in Sect. 3.1 to partially recover the linear layers permutations α and ω. This pattern is made of two parts: the white square appearing at the top-left of $\mathcal{L}'_\pi$ and the "stripe" covering the 16 left-most columns of this table (see Fig. 4). In what follows, we explain why the white square and the stripe are present in $\mathcal{L}'_\pi$. We also present an alternative representation of $\hat{\pi}$ which highlights the role of the multiplexer.

On the White Square. We first define a *balanced function* and the concept of *integral distinguishers.* Using those, we can rephrase a result from [38] as Lemma 2.

Definition 4 (Balanced Function). *Let $f : \mathbb{F}_2^n \to \mathbb{F}_2^m$ be a Boolean function. We say that f is* balanced *if the size of the preimage $\{x \in \mathbb{F}_2^n, f(x) = y\}$ of y is the same for all y in $\mathbb{F}_2^m$.*

[5] Note that the LAT of $\hat{\pi}$ is not exactly the same as $\mathcal{L}'_\pi$ which is given in Fig. 4 because e.g. of a nibble swap.

Definition 5 (Integral Distinguisher). *Let f be a Boolean function mapping n bits to m. An integral distinguisher consists in two subsets $\mathcal{C}_{in} \subseteq [0, n-1]$ and $\mathcal{B}_{out} \subseteq [0, m-1]$ of input and output bit indices with the following property: for all c, the sum $\bigoplus_{x \in \mathcal{X}} f(x)$ restricted to the bits with indices in $\mathcal{B}_{out}$ is balanced; where $\mathcal{X}$ is the set containing all x such that the bits with indices in $\mathcal{C}_{in}$ are fixed to the corresponding value in the binary expansion of c (so that $|\mathcal{X}| = 2^{n-|\mathcal{C}_{in}|}$).*

Lemma 2 ([38]). *Let f be a Boolean function mapping n bits to m and with LAT $\mathcal{L}_f$ for which there exists an integral distinguisher $(\mathcal{C}_{in}, \mathcal{B}_{out})$. Then, for all (a, b) such that the 1 in the binary expansion of a all have indices in $\mathcal{C}_{in}$ and the 1 in the binary expansion of b have indices in $\mathcal{B}_{out}$, it holds that $\mathcal{L}_f[a, b] = 0$.*

This theorem explains the presence of the white square in $\mathcal{L}'_{\pi}$. Indeed, fixing the input of the right branch of $\hat{\pi}$ leads to a permutation of the left branch in the plaintext becoming a permutation of the left branch in the ciphertext; hence the existence of an integral distinguisher for $\hat{\pi}$ which in turn explains the white square.

On the Stripe. Biases in the stripe correspond to approximations $(a_L||a_R \rightsquigarrow b_L||0)$ in $\hat{\pi}$, where $b_L > 0$. The computation of the corresponding biases can be found in the full version of this paper [15]. It turns out that the expression of $\mathcal{L}[a_L||a_R, b_L||0]$ is

$$\mathcal{L}[a_L||a_R, b_L||0] = \mathcal{L}_0[a_L, b_L] + 8 \times \left((-1)^{b_L \cdot y_0} - \hat{\delta}(b_L)\right),$$

where $\mathcal{L}_0$ is the LAT of ν_0, y_0 depends on a_R, a_L and the LAT of ν_1, and $\hat{\delta}(b_L)$ is equal to 1 if $b_L = 0$ and to 0 otherwise. Very roughly, ν_1 is responsible for the sign of the biases in the stripe and ν_0 for their values.

Since the minimum and maximum biases in $\mathcal{L}^0$ are -4 and $+4$, the absolute value of $\mathcal{L}[a_L||a_R, b_L||0]$ is indeed in $[4, 12]$. As we deduce from our computation of these biases, the stripe is caused by the conjunction of three elements:

- the use of a multiplexer,
- the use of finite field inversion, and
- the fact that ν_0 has good non-linearity.

Ironically, the only "unsurprising" sub-component of π, namely the inverse function, is one of the reasons why we were able to reverse-engineer this S-Box in the first place. Had $\mathcal{I}$ been replaced by a different (and possibly weaker!) S-Box, there would not have been any of the lines in the LAT which got our reverse-engineering started. Note however that the algorithm based on identifying linear subspaces of zeroes in the LAT of a permutation described in Sect. 5 would still work because of the white-square.

Alternative Representation. Because of the multiplexer, we can deduce an alternative representation of $\hat{\pi}$. If the right nibble of the input is not equal to

0 then $\hat{\pi}$ can be represented as shown in Fig. 11b using a Feistel-like structure. Otherwise, it is essentially equivalent to one call to the 4-bit S-Box ν_0, as shown in Fig. 11a. We also have some freedom in the placement of the branch bearing ϕ. Indeed, as shown in Fig. 11c, we can move it before the call to ν_1 provided we replace ϕ by $\psi = \phi \circ \nu_1$.

Note also that the decomposition we found is not unique. In fact, we can create many equivalent decompositions by e.g. adding multiplication and division by constants around the two finite field multiplications. We can also change the finite field in which the operations are made at the cost of appropriate linear isomorphisms modifying the 4-bit S-Boxes and the whitening linear layers. However the presented decomposition is the most structured that we have found.

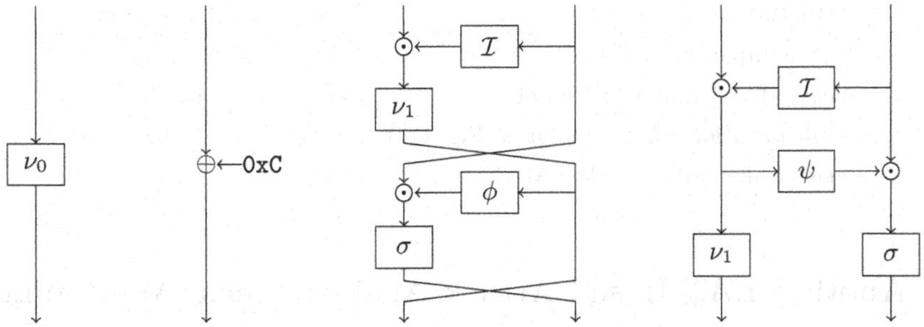

(a) $\hat{\pi}$ when $r = 0$. (b) $\hat{\pi}$ when $r \neq 0$. (c) $\hat{\pi}$ when $r \neq 0$.

Fig. 11. Alternative representations of $\hat{\pi}$ where $\pi = \omega \circ \hat{\pi} \circ \alpha$.

4.3 Hardware Implementation

It is not uncommon for cryptographers to build an S-Box from smaller ones, typically an 8-bit S-Box from several 4-bit S-Boxes. For example, S-Boxes used in Whirlpool [22], Zorro [39], Iceberg [40], Khazad [41], CLEFIA [42], and Robin and Fantomas [43] are permutations of 8 bits based on smaller S-Boxes. In many cases, such a structure is used to allow an efficient implementation of the S-Box in hardware or using a bit-sliced approach. In fact, a recent work by Canteaut et al. [44] focused on how to build efficient 8-bit S-Boxes from 3-round Feistel and Misty-like structures. Another possible reason behind such a choice is given by the designers e.g. of CLEFIA: it is to prevent attacks based on the algebraic properties of the S-Box, especially if it is based on the inverse in $\mathbb{F}_{2^8}$ like in the AES. Finally, a special structure can help achieve special properties. For instance, the S-Box of Iceberg is an involution obtained using smaller 4-bit involutions and a Substitution-Permutation Network.

As stated in the introduction, hardware optimization was supposed to be one of the design criteria used by the designers of π. Thus, it is reasonable to assume that one of the aims of the decomposition we found was to decrease the hardware footprint of the S-Box.

To test this hypothesis, we simulated the implementation of π in hardware.[6] We used four different definitions of π: the look up table given by the designers, our decomposition, a tweaked decomposition where the multiplexer is moved lower[7] and, finally, the alternative decomposition presented in Fig. 11c. Table 10 contains both the area taken by our implementations and the delay, i.e. the time taken to compute the output of the S-Box. For both quantities, the lower is the better. As we can see, the area is divided by up to 2.5 and the delay by 8, meaning that an implementer knowing the decomposition has a significant advantage over one that does not.

Table 10. Results on the hardware implementation of π.

Structure	Area (μm^2)	Delay (ns)
Naive implementation	3889.6	362.52
Feistel-like (similar to Fig. 11b)	1534.7	61.53
Multiplications-first (similar to Fig. 11c)	1530.3	54.01
Feistel-like (with tweaked MUX)	1530.1	46.11

5 Another LAT-Based Attack Against Linear Whitening

Our attack against π worked by identifying patterns in a visual representation of its LAT and exploiting them to recover parts of the whitening linear layers surrounding the core of the permutation.

It is possible to exploit other sophisticated patterns in the LAT of a permutation. In the remainder of this section, we describe a specific pattern in the LAT of a 4-round Feistel Network using bijective Feistel functions. We then use this pattern in conjunction with Theorem 1 to attack the $\mathsf{AF}^4\mathsf{A}$ structure corresponding to a 4-round Feistel Network with whitening linear layers. Note that more generic patterns such as white rectangles caused by integral distinguishers (see Sect. 4.2) could be used in a similar fashion to attack other generic constructions, as we illustrate in Sect. 5.3. The attack principle is always the same:

1. identify patterns in the LAT,
2. deduce partial whitening linear layers,
3. recover the core of the permutation with an *ad hoc* attack.

We also remark that Feistel Networks with affine masking exist in the literature. Indeed, the so-called FL-layers of MISTY [45] can be interpreted as such affine masks. Furthermore, one of the S-Box of the stream cipher ZUC is a 3-round Feistel Network composed with a bit rotation [46] — an affine operation.

6 We used `Synopsys design_compiler` (version J-2014.09-SP2) along with digital library `SAED_EDK90_CORE` (version 1.11).

7 More precisely, the multiplexer is moved after the left side is input to ϕ. This does not change the output: when the output of ν_0 is selected, the right branch is equal to 0 and the input of σ is thus 0 regardless of the left side.

5.1 Patterns in the LAT of a 4-Round Feistel Network

Let $F_0, ..., F_3$ be four n-bit Boolean permutations. Figure 12a represents the 4-round Feistel Network f built using F_i as its i-th Feistel function. Figure 12b is the Pollock representation of the LAT of a 8-bit Feistel Network f_{exp} built using four 4-bit permutations picked uniformly at random.

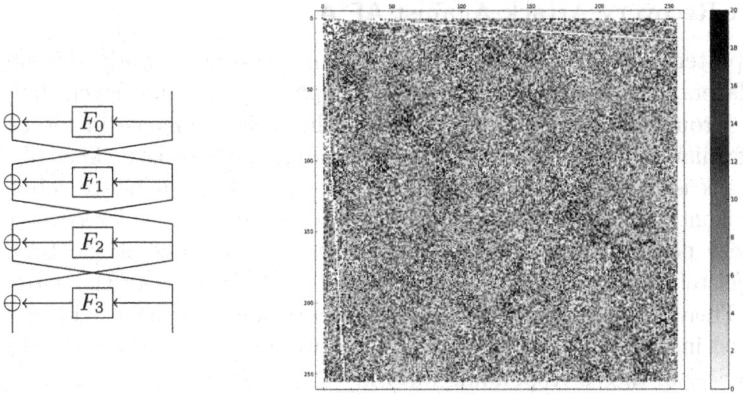

(a) Definition of f. (b) The LAT (Pollock repr.) of f_{exp}.

Fig. 12. A 4-round Feistel Network and its LAT.

In Fig. 12b, we note that the LAT $\mathcal{L}_{\text{exp}}$ of f_{exp} contains both vertical and horizontal segments of length 16 which are made only of zeroes. These segments form two lines starting at (0,0), one ending at (15,255) and another one ending in (255,15), where (0,0) is the top left corner. The vertical segments are in columns 0 to 15 and correspond to entries $\mathcal{L}_{\text{exp}}[a_L||a_R, 0||a_L]$ for any (a_L, a_R). The horizontal ones are in lines 0 to 15 and correspond to entries $\mathcal{L}_{\text{exp}}[0||a_R, a_R||b_L]$ for any (a_L, a_R).

Let us compute the coefficients which correspond to such vertical segments for any 4-round Feistel Network f with LAT $\mathcal{L}$. These are equal to

$$\begin{aligned}\mathcal{L}[a_L||a_R, 0||a_L] &= \sum_{x \in \mathbb{F}_2^{2n}} (-1)^{(a_L||a_R)\cdot x \oplus a_L \cdot f(x)} \\ &= \sum_{r \in \mathbb{F}_2^n} (-1)^{a_R \cdot r} \sum_{\ell \in \mathbb{F}_2^n} (-1)^{a_L \cdot \left(\ell \oplus f_R(\ell||r)\right)},\end{aligned}$$

where $f_R(x)$ is the right word of $f(x)$. This quantity is equal to $\ell \oplus F_0(r) \oplus F_2\big(r \oplus F_1(\ell \oplus F_0(r))\big)$, so that $\mathcal{L}[a_L||a_R, 0||a_L]$ can be re-written as:

$$\mathcal{L}[a_L||a_R, 0||a_L] = \sum_{r \in \mathbb{F}_2^n} (-1)^{a_R \cdot r} \sum_{\ell \in \mathbb{F}_2^n} (-1)^{a_L \cdot \left(F_0(r) \oplus F_2\left(r \oplus F_1(\ell \oplus F_0(r))\right)\right)}.$$

Since $\ell \mapsto F_2(r \oplus F_1(\ell \oplus F_0(r)))$ is balanced for all r, the sum over ℓ is equal to 0 for all r (unless $a_L = 0$). This explains[8] the existence of the vertical "white segments". The existence of the horizontal ones is a simple consequence of Remark 2: as the inverse of f is also a 4-round Feistel, its LAT must contain white vertical segments. Since the LAT of f is the transpose of the LAT of f^{-1}, these vertical white segments become the horizontal ones.

5.2 A Recovery Attack Against AF^{4}A

These patterns can be used to attack a 4-round Feistel Network whitened using affine layers, a structure we call AF^{4}A. Applying the affine layers before and after a 4-round Feistel Network scrambles the white segments in the LAT in a linear fashion - each such segment becomes an affine subspace. The core idea of our attack is to compute the LAT of the target and then try to rebuild both the horizontal and vertical segments. In the process, we will recover parts of the linear permutations applied to the rows and columns of the LAT of the inner Feistel Network and, using Theorem 1, recover parts of the actual linear layers. Then the resulting 4-round Feistel Network can be attacked using results presented in [32]. By parts of a linear layer we understand the four $(n \times n)$-bit submatrices of the corresponding matrix.

First Step: Using the LAT. Let $f : \mathbb{F}_2^{2n} \to \mathbb{F}_2^{2n}$ be a 4-round Feistel Network and let $g = \eta \circ f \circ \mu$ be its composition with some whitening linear layers η and μ. The structure of g is presented in Fig. 13a using $(n \times n)$-bit matrices $\mu_{\ell,\ell}, \mu_{r,\ell}, \mu_{\ell,r}, \mu_{r,r}$ for μ and $\eta_{\ell,\ell}, \eta_{r,\ell}, \eta_{\ell,r}, \eta_{r,r}$ for η.

Assume that we have the full codebook of g and therefore that we can compute the LAT $\mathcal{L}_g$ of g. By Theorem 1, it holds that $\mathcal{L}_g[u,v] = \mathcal{L}_f[(\mu^{-1})^t(u), \eta^t(v)]$ and, equivalently, that $\mathcal{L}_g[\mu^t(u), (\eta^{-1})^t(v)] = \mathcal{L}_f[u,v]$.

We use Algorithm 1 (see next section) to find a linear subspace $\mathcal{S}$ of $\mathbb{F}_2^{2n}$ such that $|\mathcal{S}| = 2^n$ and such that it has the following property: there exists 2^n distinct values c and some c dependent u_c such that $\mathcal{L}_g[u_c \oplus s, c] = 0$ for all s in $\mathcal{S}$. Such a vector space exists because the row indices $(\ell || r)$ for r in $R = \{(0||r) \mid r \in \mathbb{F}_2^n\}$ and a fixed ℓ of each vertical segment in the LAT of f becomes an affine space $\mu^t(\ell||0) \oplus \mu^t(R)$ in the LAT of g, so that the image of the row indices of each of the 2^n vertical segments has an identical linear part. We thus assume that $\mathcal{S} = \mu^t(R)$.

Then, we choose an arbitrary bijective linear mapping[9] $\overline{\mu}^t$ such that $\overline{\mu}^t(\mathcal{S}) = R$. Let a^t, b^t, c^t, d^t be $n \times n$-bit matrices such that

$$\mu'^t = \overline{\mu}^t \circ \mu^t = \begin{pmatrix} a^t & c^t \\ b^t & d^t \end{pmatrix}.$$

Note that $\mu'^t(R) = \overline{\mu}^t(\mu^t(R)) = \overline{\mu}^t(\mathcal{S}) = R$, which implies $c^t = 0$.

[8] Note that our proof actually only requires F_1 and F_2 to be permutations. The pattern would still be present if the first and/or last Feistel functions had inner-collisions.

[9] We make some definitions with transpose to simplify later notations.

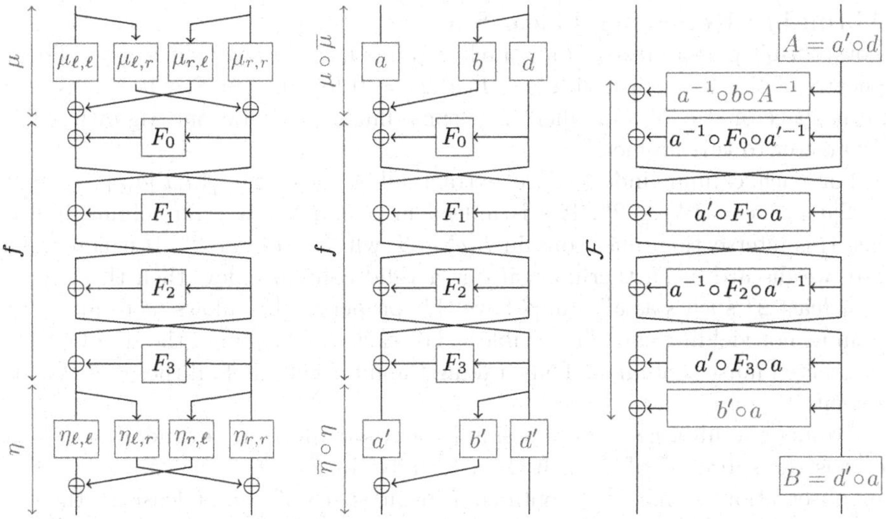

(a) $g = \eta \circ f \circ \mu$. (b) $g'' = \overline{\eta} \circ g \circ \overline{\mu}$. (c) g'' (alt. representation).

Fig. 13. The effect of $\overline{\mu}$ and $\overline{\eta}$ on g. Linear layers are in red and inner Feistel Networks in blue (Color figure online).

We then apply $\overline{\mu}^t$ to columns of $\mathcal{L}_g$ to obtain a new LAT $\mathcal{L}'_g$ such that $\mathcal{L}'_g[\overline{\mu}^t(u), v] = \mathcal{L}_g[u, v]$. Using Theorem 1, we define $g' = g \circ \overline{\mu}$ so that the LAT of g' is $\mathcal{L}'_g$. Note that g' can also be expressed using f and μ':

$$g' = \eta \circ f \circ \mu \circ \overline{\mu} = \eta \circ f \circ \mu', \text{ with } \mu' = \begin{pmatrix} a & b \\ 0 & d \end{pmatrix}.$$

The function g' we obtained is such that there is no branch from the left side to the right side in the input linear layer as the corresponding element of the μ' matrix is equal to zero. We can apply the same method to the inverse of g (thus working with the transpose of the LAT) and find a linear mapping $\overline{\eta}$ allowing us to define a new permutation g'' such that:

$$g'' = \overline{\eta} \circ g \circ \overline{\mu} \text{ where } \overline{\eta} \circ \eta = \begin{pmatrix} a' & b' \\ 0 & d' \end{pmatrix}.$$

The resulting affine-equivalent structure is shown in Fig. 13b. Note that the LAT of g'' exhibits the patterns described in Sect. 5.1. This can be explained using an alternative representation of g'' where the Feistel functions are replaced by some other affine equivalent functions as shown in Fig. 13c. It also implies that we can decompose g'', as described in the next sections.

The dominating step in terms of complexity is Algorithm 1, meaning that building g'' from g takes time $O(2^{6n})$ (see next section).

Subroutine: Recovering Linear Subpaces. Suppose we are given the LAT $\mathcal{L}$ of a $2n$-bit permutation. Our attack requires us to recover efficiently a linear space $\mathcal{S}$ of size 2^n such that $\mathcal{L}[s \oplus L(u), u] = 0$ for all s in $\mathcal{S}$, where u is in a linear space of size 2^n and where L is some linear permutation. Algorithm 1 is our answer to this problem.

For each column index c, we extract all s such that $|\{a, \mathcal{L}[a,c] = 0\} \cap \{a, \mathcal{L}[a, c \oplus s] = 0\}| \geq 2^n$. If s is indeed in $\mathcal{S}$ and c is a valid column index, then this intersection must contain $L(c) \oplus \mathcal{S}$, which is of size 2^n. If it is not the case, we discard s. Furthermore, if c is a valid column index, then there must be at least 2^n such s as all s in $\mathcal{S}$ have this property: this allows us to filter out columns not yielding enough possible s. For each valid column, the set of offsets s extracted must contain $\mathcal{S}$. Thus, taking the intersection of all these sets yields $\mathcal{S}$ itself.

To increase filtering, we use a simple heuristic function `refine`$(\mathcal{Z}, n)$ which returns the subset $\overline{\mathcal{Z}}$ of $\mathcal{Z}$ such that, for all z in $\overline{\mathcal{Z}}$, $|(z \oplus \mathcal{Z}) \cap \mathcal{Z}| \geq 2^n$. The key observation is that if $\mathcal{Z}$ contains a linear space of size at least 2^n then $\overline{\mathcal{Z}}$ contains it too. This subroutine runs in time $O(|\mathcal{Z}|^2)$.

The dominating step in the time complexity of this algorithm is the computation of $|(s \oplus \mathcal{Z}) \cap \mathcal{Z}|$ for every c and s. The complexity of this step is $O(2^{2n} \times 2^{2n} \times |\mathcal{Z}|)$. A (loose) upper bound on the time complexity of this algorithm is thus $O(2^{6n})$ where n is the branch size of the inner Feistel Network, i.e. half of the block size.

Second Step: Using a Yoyo Game. The decomposition of $\mathsf{AF^4A}$ has now been reduced to attacking g'', a $2n$-bit 4-round Feistel Network composed with two n-bit linear permutations A and B. The next step is to recover these linear permutations. To achieve this, we use a simple observation inspired by the so-called *yoyo game* used in [32] to attack a 5-round FN.

Consider the differential trail parametrized by $\gamma \neq 0$ described in Fig. 14. If the pair of plaintexts $(x_L||x_R, x'_L||x'_R)$ follows the trail (i.e. is connected in γ), then so does $(x_L \oplus \gamma||x_R, x'_L \oplus \gamma||x'_R)$. Furthermore, if $(y_L||y_R) = g''(x_L||x_R)$ and $(y'_L||y'_R) = g''(x'_L||x'_R)$, then swapping the right output words and decrypting the results leads to a pair of plaintexts $\left(g''^{-1}(y_L||y'_R), g''^{-1}(y'_L||y_R)\right)$ that still follows the trail. It is thus possible to iterate the addition of γ and the swapping of the right output word to generate many right pairs. More importantly, if x and x' do follow the trail, iterating these transformation must lead to the difference on the right output word being constant and equal to $B(\gamma)$: if it is not the case, we can abort and start again from another pair x, x'.

We can thus recover B fully by trying to iterate the game described above for random pairs (x, x') and different values of γ. Once a good pair has been found, we deduce that $B(\gamma)$ is the difference in the right output words of the ciphertext pairs obtained. We thus perform this step for $\gamma = 1, 2, 4, 8$, etc., until the image by B of all bits has been found. Wrong starting pairs are identified quickly so this step takes time $O(n2^{2n})$. Indeed, we need to recover n linearly independent n-bit vectors; for each vector we try all 2^n candidates and for each

Algorithm 1. Linear subspace extraction
Inputs LAT $\mathcal{L}$, branch size n — **Output** Linear space $\mathcal{S}$

```
C := ∅
for all c ∈ [1, 2^{2n} − 1] do
    Z := {i ∈ [1, 2^{2n} − 1], L[i, c] = 0}
    if Z ≥ 2^n then
        S_c := ∅                                  ▷ S_c is the candidate for S for c.
        for all s ∈ [1, 2^{2n} − 1] do
            if |(s ⊕ Z) ∩ Z| ≥ 2^n then
                S_c := S_c ∪ {s}
            end if
        end for
        S_c := refine(S_c, 2^n)
        if |S_c| ≥ 2^n then
            Store S_c ; C := C ∪ {c}
        end if
    end if
end for
C := refine(C, 2^n) ; S := [0, 2^{2n} − 1]
for all c ∈ C do
    S := S ∩ S_c
end for
return S
```

guess we run a Yoyo game in time 2^n to check the guess. The other linear part, A, is recovered identically by running the same attack while swapping the roles of encryption and decryption.

Final Step: A Full Decomposition. As stated before, we can recover all 4 Feistel functions in g'' minus its linear layers in time $O(2^{3n/2})$ using the guess and determine approach described in [32]. This gives us a 4-round FN, denoted $\mathcal{F}$, which we can use to decompose g like so (where I denotes the identity matrix):

$$g = \overline{\eta}^{-1} \circ \begin{pmatrix} I & 0 \\ 0 & B \end{pmatrix} \circ \mathcal{F} \circ \begin{pmatrix} I & 0 \\ 0 & A \end{pmatrix} \circ \overline{\mu}^{-1}.$$

5.3 Outline of an Attack Against $\mathsf{AF^3A}$

A structure having one less Feistel round could be attacked in a similar fashion. The main modifications would be as follows.

1. The pattern in the LAT we try to rebuild would not be the one described in Sect. 5.1 but a white square similar to the one observed in the LAT of π'. Indeed, an integral distinguisher similar to the one existing in π' exists for any 3-round FN when the second Feistel function is a bijection.

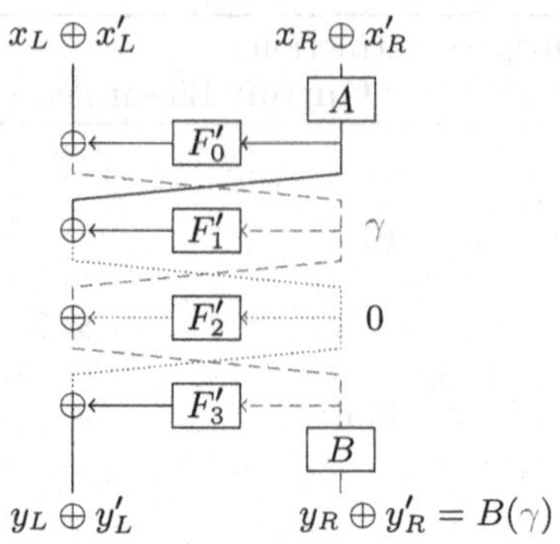

Fig. 14. The differential trail used to recover B.

2. Recovering the remaining $(n \times n)$ mappings with a yoyo game would be much more efficient since there would not be any need to guess that a pair follows the trail.

The complexity of such an attack would be dominated as before by the recovery of the linear subspace embedded in the LAT so that it would take time $O(2^{6n})$.

5.4 Comments on Affine-Whitened Feistel Networks

Table 11 contains a comparison of the complexities of the attack recovering the Feistel functions of Feistel Networks along with, possibly, the linear layers used to whiten it.

Table 11. Complexity of recovery attacks against (possibly linearly whitened) Feistel Networks with n-bit branches and secret bijective Feistel functions.

Target	Type	Time complexity	Ref.
$\mathsf{AF}^3\mathsf{A}$	LAT-based	2^{6n}	Sect. 5.3
F^4	Guess &Det.	$2^{3n/2}$	[32]
$\mathsf{AF}^4\mathsf{A}$	LAT-based	2^{6n}	Sect. 5.2
F^5	Yoyo cryptanalysis	2^{2n}	[32]

The complexities of our attacks against $\mathsf{AF}^k\mathsf{A}$ are dominated by the linear subspace recovery which is much slower than an attack against as much as 5 Feistel Network rounds. It seems like using affine whitening rather than a simpler XOR-based whitening increases the complexity of a cryptanalysis significantly. This observation can be linked with the recent attacks against the ASASA scheme [31]: while attacking SAS is trivial, the cryptanalysis of ASASA requires sophisticated algorithms.

We note that a better algorithm for linear subspace extraction will straightforwardly lead to a lower attack complexity. However, the complexity is lower bounded by LAT computation which is $O(n2^{4n})$ in our case.

For the sake of completeness, we tried this attack against the "F-Table" of Skipjack [9]. It failed, meaning that it has neither an $\mathsf{AF}^3\mathsf{A}$ nor an $\mathsf{AF}^4\mathsf{A}$ structure. Note also that, due to the presence of the white square in the LAT of $\hat{\pi}$, running the linear subspace recovery described in Algorithm 1 on π returns the vector space $\mathcal{V}$ which allowed to start our decomposition of this S-Box.

We implemented the first step of our attack (including Algorithm 1) using SAGE [36] and ran it on a computer with 16 Intel Xeon CPU (E5-2637) v3 clocked at 3.50 GHz. It recovers correct linear layers $\overline{\eta}$ and $\overline{\mu}$ in about 3 seconds for $n = 4$, 14 seconds for $n = 5$, 4 min for $n = 6$ and 1 hour for $n = 7$. Since the first step is the costliest, we expect a complete attack to take a similar time.

6 Conclusion

The S-Box used by the standard hash function Streebog, the standard block cipher Kuznyechik and the CAESAR first round candidate STRIBOBr1 has a hidden structure. Using the three non-linear 4-bit permutations ν_0, ν_1 and σ, the non-linear 4-bit function ϕ and the 8-bit linear permutations α and ω, the computation of $\pi(\ell||r)$ can be made as follows:

1. $(\ell||r) := \alpha(\ell||r)$
2. If $r = 0$ then $\ell := \nu_0(\ell)$, else $\ell := \nu_1(\ell \odot r^{14})$
3. $r := \sigma\big(r \odot \phi(l)\big)$
4. Return $\omega(\ell||r)$

How and why those components were chosen remains an open question. Indeed, their individual cryptographic properties are at best not impressive and, at worst, downright bad. However, knowing this decomposition allows a much more efficient hardware implementation of the S-Box.

We also described a decomposition attack against $\mathsf{AF}^4\mathsf{A}$ which uses the same high level principles as our attack against π: first spot patterns in the LAT, then deduce the whitening linear layers and finally break the core.

Acknowledgment. We thank Yann Le Corre for studying the hardware implementation of the S-Box. We also thank Oleksandr Kazymyrov for suggesting this target and the anonymous reviewers for their helpful comments. The work of Léo Perrin is supported by the CORE ACRYPT project (ID C12-15-4009992) funded by the *Fonds National de la Recherche* (Luxembourg). The work of Aleksei Udovenko is supported by the *Fonds National de la Recherche*, Luxembourg (project reference 9037104).

References

1. Daemen, J., Rijmen, V.: The Design of Rijndael: AES-The Advanced Encryption Standard. Information Security and Cryptography. Springer, Heidelberg (2002)
2. Biham, E., Shamir, A.: Differential cryptanalysis of DES-like cryptosystems. J. Crypt. **4**(1), 3–72 (1991)
3. Tardy-Corfdir, A., Gilbert, H.: A known plaintext attack of FEAL-4 and FEAL-6. In: Feigenbaum, J. (ed.) Advances in Cryptology - CRYPTO 1992. LNCS, vol. 576, pp. 172–182. Springer, Berlin Heidelberg (1992)
4. Matsui, M.: Linear cryptanalysis method for DES cipher. In: Helleseth, T. (ed.) EUROCRYPT 1993. LNCS, vol. 765, pp. 386–397. Springer, Heidelberg (1994)
5. Nyberg, K.: Differentially uniform mappings for cryptography. In: Helleseth, T. (ed.) EUROCRYPT 1993. LNCS, vol. 765, pp. 55–64. Springer, Heidelberg (1994)
6. Suzaki, T., Minematsu, K., Morioka, S., Kobayashi, E.: TWINE : A lightweight block cipher for multiple platforms. In: Knudsen, L.R., Wu, H. (eds.) SAC 2012. LNCS, vol. 7707, pp. 339–354. Springer, Heidelberg (2013)
7. U.S. Department: OF COMMERCE/National Institute of Standards and Technology: Data encryption standard. Publication, Federal Information Processing Standards (1999)
8. Coppersmith, D.: The data encryption standard (DES) and its strength against attacks. IBM J. Res. Develop. **38**(3), 243–250 (1994)
9. National Security Agency, N.S.A.: SKIPJACK and KEA AlgorithmSpecifications (1998)
10. Biryukov, A., Perrin, L.: On reverse-engineering s-boxes with hidden design criteria or structure. In: Gennaro, R., Robshaw, M. (eds.) Advances in Cryptology - CRYPTO 2015. LNCS, vol. 9215, pp. 116–140. Springer, Berlin, Heidelberg (2015)
11. Federal Agency on Technical Regulation and Metrology: GOST R34.11-2012: Streebog hash function (2012). https://www.streebog.net/
12. Guo, J., Jean, J., Leurent, G., Peyrin, T., Wang, L.: The usage of counter revisited: second-preimage attack on new russian standardized hash function. In: Joux, A., Youssef, A. (eds.) SAC 2014. LNCS, vol. 8781, pp. 195–211. Springer International Publishing, Switzerland (2014)
13. AlTawy, R., Youssef, A.M.: Watch your constants: malicious streebog. IET Inf. Secur. **9**(6), 328–333 (2015)
14. Rudskoy, V.: Note on Streebog constants origin (2015). http://www.tc26.ru/en/ISO_IEC/streebog/streebog_constants_eng.pdf
15. Biryukov, A., Perrin, L., Udovenko, A.: Reverse-Engineering the S-Box of Streebog, Kuznyechik and STRIBOBr 1. Cryptology ePrint Archive, report 2016/071 (2016). http://eprint.iacr.org/
16. Shishkin, V., Dygin, D., Lavrikov, I., Marshalko, G., Rudskoy, V., Trifonov, D.: Low-weight and hi-end: draft russian encryption standard. In: Preproceedings of CTCrypt 2014, 05–06 June 2014, Moscow. Russia, pp. 183–188 (2014)
17. Federal Agency on Technical Regulation and Metrology: Block ciphers (2015). http://www.tc26.ru/en/standard/draft/ENG_GOST_R_bsh.pdf
18. AlTawy, R., Youssef, A.M.: A meet in the middle attack on reduced round Kuznyechik. Cryptology ePrint Archive, report 2015/096 (2015). http://eprint.iacr.org/
19. Dolmatov, V.: GOST 28147-89: Encryption, decryption, and message authentication code (MAC) algorithms, RFC 5830, March 2010. http://www.rfc-editor.org/rfc/rfc5830.txt

20. Saarinen, M.J.O.: STRIBOB: Authenticated encryption from GOST R 34.11-2012 LPS permutation. In: Математические вопросы криптографии [Mathematical Aspects of Cryptography]. vol.6(2), pp. 67–78. Steklov Mathematical Institute ofRussian Academy of Sciences (2015)
21. Saarinen, M.J.O., Brumley, B.B.: WHIRLBOB, the whirlpool based variant of STRIBOB. In: Buchegger, S., Dam, M. (eds.) NordSec 2015. LNCS, vol. 9417, pp. 106–122. Springer International Publishing, Cham (2015)
22. Barreto, P., Rijmen, V.: The whirlpool hashing function. In: First open NESSIE Workshop, Leuven, Belgium. vol. 13, p. 14 (2000)
23. Saarinen, M.J.O.: STRIBOBr 2 availability. Mail to the CAESAR mailing list. https://groups.google.com/forum/#!topic/crypto-competitions/_zgi54-NEFM
24. Shishkin, V.: Принципы синтеза перспективного алгоритма блочного шифрования с длиной блока 128 бит (2013). http://www.ruscrypto.ru/resource/summary/rc2013/ruscrypto_2013_066.zip
25. Knudsen, L.R., Robshaw, M.J., Wagner, D.: Truncated differentials and skipjack. In: Wiener, M. (ed.) Advances in Cryptology-CRYPTO 1999. LNCS, vol. 1666, pp. 165–180. Springer, Heidelberg (1999)
26. Biham, E., Biryukov, A., Dunkelman, O., Richardson, E., Shamir, A.: Initial observations on skipjack: cryptanalysis of skipjack-3XOR. In: Tavares, S., Meijer, H. (eds.) SAC 1998. LNCS, vol. 1556, p. 362. Springer, Heidelberg (1999)
27. Knudsen, L., Wagner, D.: On the structure of Skipjack. Discrete Appl. Math. **111**(1), 103–116 (2001)
28. Kazymyrov, O., Kazymyrova, V.: Algebraic aspects of the russian hash standard GOST R 34.11-2012. In: IACR Cryptology ePrint Archive 2013 556 (2013)
29. Biryukov, A., Shamir, A.: Structural cryptanalysis of SASAS. In: Pfitzmann, B. (ed.) Advances in Cryptology - EUROCRYPT 2001. LNCS, vol. 2045, pp. 395–405. Springer, Berlin Heidelberg (2001)
30. Dinur, I., Dunkelman, O., Kranz, T., Leander, G.: Decomposing the ASASA block cipher construction. In: Cryptology ePrint Archive, report 2015/507 (2015). http://eprint.iacr.org/
31. Minaud, B., Derbez, P., Fouque, P.A., Karpman, P.: Key-Recovery attacks on ASASA. In: Iwata, T., Cheon, J.H. (eds.) Advances in Cryptology – ASIACRYPT 2015. LNCS, vol. 9453, pp. 3–27. Springer, Heidelberg (2015)
32. Biryukov, A., Leurent, G., Perrin, L.: Cryptanalysis of Feistel networks with secret round functions. In: Dunkelman, O., Keliher, L. (eds.) SAC 2015. LNCS. Springer International Publishing, Heidelberg (2015)
33. Daemen, J., Rijmen, V.: Probability distributions of correlation and differentials in block ciphers. J. Math. Crypt. JMC **1**(3), 221–242 (2007)
34. Blondeau, C., Canteaut, A., Charpin, P.: Differential properties of power functions. Int. J. Inf. Coding Theory **1**(2), 149–170 (2010)
35. Preneel, B.: Analysis and design of cryptographic hash functions. Ph.D. thesis, Katholieke Universiteit Leuven (1993)
36. The Sage Developers: Sage Mathematics Software (Version 6.8) (2015). http://www.sagemath.org
37. Canright, D.: A very compact S-Box for AES. In: Rao, J., Sunar, B. (eds.) Cryptographic Hardware and Embedded Systems - CHES 2005. LNCS, vol. 3659, pp. 441–455. Springer, Berlin Heidelberg (2005)

38. Bogdanov, A., Leander, G., Nyberg, K., Wang, M.: Integral and multidimensional linear distinguishers with correlation zero. In: Wang, X., Sako, K. (eds.) Advances in Cryptology - ASIACRYPT 2012. LNCS, vol. 7658, pp. 244–261. Springer, Berlin Heidelberg (2012)
39. Gérard, B., Grosso, V., Naya-Plasencia, M., Standaert, F.-X.: Block ciphers that are easier to mask: how far can we go? In: Bertoni, G., Coron, J.-S. (eds.) CHES 2013. LNCS, vol. 8086, pp. 383–399. Springer, Heidelberg (2013)
40. Standaert, F.X., Piret, G., Rouvroy, G., Quisquater, J.J., Legat, J.D.: ICEBERG : An involutional cipher efficient for block encryption in reconfigurable hardware. In: Roy, B., Meier, W. (eds.) Fast Software Encryption. LNCS, vol. 3017, pp. 279–298. Springer, Berlin Heidelberg (2004)
41. Barreto, P., Rijmen, V.: The Khazad legacy-level block cipher. In: Primitive submitted to NESSIE 97 (2000)
42. Shirai, T., Shibutani, K., Akishita, T., Moriai, S., Iwata, T.: The 128-Bit blockcipher CLEFIA (Extended Abstract). In: Biryukov, A. (ed.) FSE 2007. LNCS, vol. 4593, pp. 181–195. Springer, Heidelberg (2007)
43. Grosso, V., Leurent, G., Standaert, F.X., Varıcı, K.: LS-designs: Bitslice encryption for efficient masked software implementations. In: Fast Software Encryption (2014)
44. Canteaut, A., Duval, S., Leurent, G.: Construction of lightweight s-boxes using feistel and MISTY structures. In: Dunkelman, O., Keliher, L. (eds.) Selected Areas in Cryptography - SAC 2015. LNCS, vol. 8731. Springer International Publishing, Heidelberg (2015)
45. Matsui, M.: New block encryption algorithm MISTY. In: Biham, E. (ed.) FSE 1997. LNCS, vol. 1267, pp. 54–68. Springer, Berlin, Heidelberg (1997)
46. Specification of the 3GPP Confidentiality and Integrity Algorithms 128-EEA3 & 128-EIA3. Document 4 : Design and Evaluation Report, Technical report, ETSI/Sage, September 2011. http://www.gsma.com/aboutus/wp-content/uploads/2014/12/EEA3_EIA3_Design_Evaluation_v2_0.pdf

Complete Addition Formulas for Prime Order Elliptic Curves

Joost Renes[1(✉)], Craig Costello[2], and Lejla Batina[1]

[1] Digital Security, Radboud University, Nijmegen, The Netherlands
{j.renes,lejla}@cs.ru.nl
[2] Microsoft Research, Redmond, USA
craigco@microsoft.com

Abstract. An elliptic curve addition law is said to be *complete* if it correctly computes the sum of *any* two points in the elliptic curve group. One of the main reasons for the increased popularity of Edwards curves in the ECC community is that they can allow a complete group law that is also relatively efficient (e.g., when compared to all known addition laws on Edwards curves). Such complete addition formulas can simplify the task of an ECC implementer and, at the same time, can greatly reduce the potential vulnerabilities of a cryptosystem. Unfortunately, until now, complete addition laws that are relatively efficient have only been proposed on curves of composite order and have thus been incompatible with all of the currently standardized prime order curves.

In this paper we present optimized addition formulas that are complete on *every* prime order short Weierstrass curve defined over a field k with $\text{char}(k) \neq 2, 3$. Compared to their incomplete counterparts, these formulas require a larger number of field additions, but interestingly require fewer field multiplications. We discuss how these formulas can be used to achieve secure, exception-free implementations on *all* of the prime order curves in the NIST (and many other) standards.

1 Introduction

Extending the works of Lange–Ruppert [48] and Bosma–Lenstra [19], Arène, Kohel and Ritzenthaler [4] showed that, under any projective embedding of an elliptic curve E/k, *every* addition law has pairs of exceptional points in $(E \times E)(\bar{k})$. That is, over the algebraic closure of k, there are always pairs of points for which a given elliptic curve addition law does not work.

Fortunately, in elliptic curve cryptography (ECC), we are most often only concerned with the k-rational points on E. In this case it is possible to have a single addition law that is well-defined on all pairs of k-rational points, because its exceptional pairs are found in $(E \times E)(\bar{k})$, but not in $(E \times E)(k)$. A celebrated example of this is the Edwards model [31]; when suitably chosen [12], an Edwards curve has a simple addition law that works for all pairs of k-rational points.

This work was supported in part by the Technology Foundation STW (project 13499 - TYPHOON), from the Dutch government.

M. Fischlin and J.-S. Coron (Eds.): EUROCRYPT 2016, Part I, LNCS 9665, pp. 403–428, 2016.
DOI: 10.1007/978-3-662-49890-3_16

This phenomenon was characterized more generally over elliptic curves by Kohel [47], and further generalized to arbitrary abelian varieties in [4]. For our purposes it suffices to state a special case of the more general results in [4,47]: namely, that every elliptic curve E over a finite field $\mathbb{F}_q$ (with $q \geq 5$) has an $\mathbb{F}_q$-complete addition law corresponding to the short Weierstrass model in $\mathbb{P}^2(\mathbb{F}_q)$.

Addition laws that are $\mathbb{F}_q$-complete are highly desirable in ECC. They can significantly simplify the task of an implementer and greatly reduce the potential vulnerabilities of a cryptosystem. We elaborate on this below.

Our Contributions. In Algorithm 1 we present optimized point addition formulas that correctly compute the sum of *any* two points on *any* odd order elliptic curve $E/\mathbb{F}_q\colon y^2 = x^3 + ax + b$ with $q \geq 5$. We do not claim credit for the complete formulas themselves, as these are exactly the formulas given by Bosma and Lenstra two decades ago [19]. What is novel in this paper is optimizing the explicit computation of these formulas for cryptographic application. In particular, Table 1 shows that the computation of the Bosma–Lenstra complete additions can be performed using fewer general field multiplications than the best known (incomplete!) addition formulas on short Weierstrass curves: excluding multiplications by curve constants and field additions, the explicit formulas in this paper compute additions in 12 field multiplications (12**M**), while the fastest known addition formulas in homogeneous coordinates require 14 field multiplications (12**M** + 2**S**) and the fastest known addition formulas in Jacobian coordinates require 16 field multiplications (11**M** + 5**S**). We immediately note, however, that our explicit formulas incur a much larger number of field additions than their incomplete counterparts. Thus, as is discussed at length below, the relative performance of the complete additions will be highly dependent on the platform and/or scenario. However, we stress that outperforming the incomplete addition formulas is not the point of this paper: our aim is to provide the fastest possible complete formulas for prime order curves.

Wide Applicability. While the existence of an $\mathbb{F}_q$-complete addition law for prime order Weierstrass curves is not news to mathematicians (or to anyone that has read, e.g., [4,19]), we hope it might be a pleasant surprise to ECC practitioners. In particular, the benefits of completeness are now accessible to anyone whose task it is to securely implement the prime order curves in the standards. These include:

- The example curves originally specified in the working drafts of the American National Standards Institute (ANSI), versions X9.62 and X9.63 [1,2].
- The five NIST prime curves specified in the current USA digital signature standard (DSS), i.e., FIPS 186-4 – see [55,56]. This includes Curve P-384, which is the National Security Agency (NSA) recommended curve in the most recent Suite B fact sheet for both key exchange and digital signatures [28, 60]; Curve P-256, which is the most widely supported curve in the Secure Shell (SSH) and Transport Layer Security (TLS) protocol [17, Sect. 3.2-3.3];

Table 1. Summary of explicit formulas for the addition law on prime order short Weierstrass elliptic curves $E/k\colon y^2 = x^3 + ax + b$ in either homogeneous (homog.) coordinates or Jacobian coordinates, and the corresponding exceptions (excep.) in both points doublings (DBL) and point additions (ADD). Here the operation counts include multiplications ($\mathbf{M}$), squarings ($\mathbf{S}$), multiplications by a ($\mathbf{m_a}$), multiplications by (small multiples of) b ($\mathbf{m_b}$), and additions ($\mathbf{a}$), all in the ground field k. We note that various trade-offs exist with several of the above formulas, in particular for point doublings in Jacobian coordinates – see [15].

addition formulas	a	excep. Q in ADD(P, Q)	ADD(P,Q)					excep. in DBL(P)	DBL(P)					ref
			$\mathbf{M}$	$\mathbf{S}$	$\mathbf{m_a}$	$\mathbf{m_b}$	$\mathbf{a}$		$\mathbf{M}$	$\mathbf{S}$	$\mathbf{m_a}$	$\mathbf{m_b}$	$\mathbf{a}$	
complete homog.	any	none	12	0	3	2	23	none	8	3	3	2	15	**this work**
	-3		12	0	0	2	29		8	3	0	2	21	
	0		12	0	0	2	19		6	2	0	1	9	
incomplete homog.	any	$\pm P$, $\mathcal{O}$	12	2	0	0	7	$\mathcal{O}$	5	6	1	0	12	[15,27]
	-3		12	2	0	0	7		7	3	0	0	11	[15,27]
	0		-						-					-
incomplete Jacobian	any	$\pm P$, $\mathcal{O}$	12	4	0	0	7	none	3	6	1	0	13	[27]
	-3		12	4	0	0	7		4	4	0	0	8	[27,51]
	0		12	4	0	0	7		3	4	0	0	7	[27,42]

and Curve P-192, which is the most common elliptic curve used in Austria's national e-ID cards [17, Sect. 3.4].

- The seven curves specified in the German brainpool standard [30], i.e., `brainpoolPXXXr1`, where `XXX` $\in$ {`160`, `192`, `224`, `256`, `320`, `384`, `512`}.
- The eight curves specified by the UK-based company Certivox [23], i.e., `ssc-XXX`, where `XXX` $\in$ {`160`, `192`, `224`, `256`, `288`, `320`, `384`, `512`}.
- The curve `FRP256v1` recommended by the French Agence nationale de la sécurité des systèmes d'information (ANSSI) [3].
- The three curves specified (in addition to the above NIST prime curves) in the Certicom SEC 2 standard [22]. This includes `secp256k1`, which is the curve used in the Bitcoin protocol.
- The recommended curve in the Chinese SM2 digital signature algorithm [25].
- The example curve in the Russian GOST R 34.10 standard [35].

In particular, implementers can now write secure, exception-free code that supports all of the above curves without ever having to look further than Algorithm 1 for curve arithmetic. Moreover, in Sect. 5.2 we show how Algorithm 1 can easily be used to securely implement the two composite order curves, Curve25519 [7] and Ed448-Goldilocks [39], recently recommended for inclusion in future versions of TLS by the Internet Research Task Force Crypto Forum Research Group (IRTF CFRG).

Side-Channel Protection. Real-world implementations of ECC have a number of potential side-channel vulnerabilities that can fall victim to simple timing attacks [46] or exceptional point attacks [32,43]. One of the main reasons these attacks pose a threat is the *branching* that is inherent in the schoolbook short Weierstrass elliptic curve addition operation. For example, among the dozens of `if` statements in OpenSSL's[1] standard addition function "`ec_GFp_simple_add`", the initial three that check whether the input points are equal, opposite, or at infinity can cause timing variability (and therefore leak secret data) in ECDH or ECDSA. The complete formulas in this paper remove these vulnerabilities and significantly decrease the attack surface of a cryptosystem. As Bernstein and Lange point out [13], completeness "eases implementations" and "avoids simple side-channel attacks".

Although it is possible to use incomplete formulas safely, e.g., by carefully deriving uniform scalar multiplication algorithms that avoid exceptional pairs of inputs, implementing these routines in *constant-time* and in a provably correct way can be a cumbersome and painstaking process [16, Sect. 4]. Constant-time ECC implementations typically recode scalars from their binary encoding to some other form that allows a uniform execution path (cf. Okeya-Tagaki [57] and Joye-Tunstall [44]), and these recodings can complicate the analysis of exceptional inputs to the point addition functions. For example, it can be difficult to prove that the running value in a scalar multiplication is never equal to (or the inverse of) elements in the lookup table; if this equality occurs before an addition, the incomplete addition function is likely to fail. Furthermore, guaranteeing exception-free, constant-time implementations of more exotic scalar multiplication routines, e.g., multiscalar multiplication for ECDSA verification, fixed-base scalar multiplications [49], scalar multiplications exploiting endomorphisms [34], or scalar multiplications using common power analysis countermeasures [29,33], is even more difficult; that is, unless the routine can call complete addition formulas.

Performance Considerations. While the wide applicability and correctness of Algorithm 1 is at the heart of this paper, we have also aimed to cater to implementers that do not want to sacrifice free performance gains, particularly those concerned with supporting a special curve or special family of curves. To that end, Algorithms 2–9 give faster complete addition formulas in the special (and standardized) cases that the Weierstrass curve constant a is $a = -3$ or $a = 0$, and in the special cases of point doublings; Table 1 summarizes the operation counts for all of these scenarios.

As we mentioned above, outperforming the (previously deployed) incomplete addition formulas is not the point of this paper. Indeed, the high number of field additions present in our complete addition algorithms are likely to introduce an overall slowdown in many scenarios. To give an idea of this *performance hit* in a common software scenario, we plugged our complete addition algorithms into OpenSSL's implementation of the five NIST prime curves. Using the `openssl speed` function to benchmark the performance of the existing incomplete formulas and the new complete formulas shows that the latter incurs between a

[1] See `ec_smpl.c` in `crypto/ec/` in the latest release at http://openssl.org/source/.

1.34x and 1.44x slowdown in an average run of the elliptic curve Diffie-Hellman (ECDH) protocol (see Table 2 for the full details). As we discuss below, and in detail in Sect. 5.3, this factor slowdown should be considered an upper bound on the difference in performance between the fastest incomplete algorithms and our complete ones.

On the contrary, there are example scenarios where plugging in the complete formulas will result in an unnoticeable performance difference, or possibly even a speedup. For example, compared to the incomplete addition function `secp256k1_gej_add_var` used in the Bitcoin code[2], our complete addition function in Algorithm 7 saves **4S** at the cost of **8a**+`1mul_int`[3]; compared to Bitcoin's incomplete mixed addition function `secp256k1_gej_add_ge_var`, our complete mixed addition saves **3S** at the cost of **3M** + **2a** + `1mul_int`; and, compared to Bitcoin's doubling function `secp256k1_gej_double_var`, our formulas save **2S** + `5mul_int` at the cost of **3M** + **3a**. In this case it is unclear which set of formulas would perform faster, but it is likely to be relatively close and to depend on the underlying field arithmetic and/or target platform. Furthermore, the overall speed is not just dependent on the formulas: the `if` statements present in the Bitcoin code also hamper performance. On the contrary, the complete algorithms in this paper have no `if` statements.

There are a number of additional real-world scenarios where the performance gap between the incomplete and the complete formulas will not be as drastic as the OpenSSL example above. The operation counts in Tables 1 and 3 suggest that this will occur when the cost of field multiplications and squarings heavily outweighs the cost of field additions. The benchmarks above were obtained on a 64-bit processor, where the **M**/**a** ratio tends to be much lower than that of low-end (e.g., 8-, 16-, and 32-bit) architectures. For example, field multiplications on wireless sensor nodes commonly require over 10 times more clock cycles than a field addition (e.g., see [50, Table 1] and [59, Table 1]), and in those cases the complete formulas in this paper are likely to be very competitive in terms of raw performance.

In any case, we believe that many practitioners will agree that a small performance difference is a worthwhile cost to pay for branch-free point addition formulas that culminate in much simpler and more compact code, which *guarantees* correctness of the outputs and eliminates several side-channel vulnerabilities. We also note that the Bitcoin curve is not an isolated example of the more favorable formula comparison above: all of the most popular pairing-friendly curves, including Barreto-Naehrig (BN) curves [5] which have appeared in recent IETF drafts[4], also have $a = 0$. In those cases, our specialized, exception-free formulas give implementers an easy way to correctly implement curve arithmetic in both $\mathbb{G}_1$ and $\mathbb{G}_2$ in the setting of cryptographic pairings. On a related note, we point that the word "prime" in our title can be relaxed to "odd"; the completeness

[2] See https://github.com/bitcoin/bitcoin/tree/master/src/secp256k1.

[3] `mul_int` denotes the cost of Bitcoin's specialized function that multiplies field elements by small integers.

[4] See http://datatracker.ietf.org/doc/draft-kasamatsu-bncurves-01.

of the Bosma–Lenstra formulas only requires the non-existence of rational two-torsion points (see Sects. 2 and 3), i.e., that the group order $\#E(\mathbb{F}_q)$ is not even. BN curves define $\mathbb{G}_2$ as (being isomorphic to) a proper subgroup of a curve $E'/\mathbb{F}_{p^2}$, whose group order $\#E'(\mathbb{F}_{p^2})$ is the product of a large prime with odd integers [5, Sect. 3], meaning that our explicit formulas are not only complete in $\mathbb{G}_2 \subset E'(\mathbb{F}_{p^2})$, but also in $E'(\mathbb{F}_{p^2})$.

Related Work. Complete addition laws have been found and studied on non-Weierstrass models of elliptic curves, e.g., on the (twisted) Edwards [8,12] and (twisted) Hessian models [9]. Unfortunately, in all of those scenarios, the models are not compatible with prime order curves and therefore all of the standardized curves mentioned above.

In terms of obtaining a complete and computationally efficient addition algorithm for prime order curves, there has been little success to date. Bernstein and Lange [13] found complete formulas on a non-Weierstrass model that would be compatible with, e.g., the NIST curves, reporting explicit formulas that (ignoring additions and multiplications by curve constants) cost $26\mathbf{M} + 8\mathbf{S}$. Bos *et al.* [16] considered applying the set of two Bosma–Lenstra addition laws to certain prime order Weierstrass curves, missing the observation (cf. [4, Remark 4.4]) that one of the addition laws is enough, and abandoning the high cost of computing both addition laws for an alternative but more complicated approach towards side-channel protection [16, Appendix C]. Brier and Joye [20] developed *unified* formulas[5] for general Weierstrass curves, but these formulas still have exceptions and (again, ignoring additions and multiplications by curve constants) require $11\mathbf{M} + 6\mathbf{S}$, which is significantly slower than our complete algorithms.

Prime Order Curves Can Be Safe. Several of the standardized prime order curves mentioned above have recently been critiqued in [14], where they were deemed not to meet (some or all of) the four "ECC security" requirements: (i) Ladder, (ii) Twists, (iii) Completeness, and (iv) Indistinguishability.

On the contrary, this paper shows that prime order curves have complete formulas that are comparably efficient. In addition, Brier and Joye [20, Sect. 4] extended the Montgomery ladder to all short Weierstrass curves. In particular, when $E/\mathbb{F}_q \colon y^2 = x^3 + ax + b$ is a prime order curve, their formulas give rise to a function `ladder` that computes $x([m]P) = \texttt{ladder}(x(P), m, a, b)$ for the points $P \in E(\mathbb{F}_{q^2})$ with $(x, y) \in \mathbb{F}_q \times \mathbb{F}_{q^2}$, that is, a function that works for all $x \in \mathbb{F}_q$ and that does not distinguish whether x corresponds to a point on the curve E, or to a point on its quadratic twist $E' \colon dy^2 = x^3 + ax + b$, where d is non-square in $\mathbb{F}_q$. If E is chosen to be twist-secure (this presents no problem in the prime order setting), then for all $x \in \mathbb{F}_q$, the function $\texttt{ladder}(x, m, a, b)$ returns an instance of the discrete logarithm problem (whose solution is m)

[5] These are addition formulas that also work for point doublings.

on a cryptographically strong curve, just like the analogous function on twist-secure Montgomery curves [7]. Finally, we note that Tibouchi [61] presented a prime-order analogue of the encoding given for certain composite-order curves in [11], showing that the indistinguishability property can also be achieved on prime order curves.

As is discussed in [14], adopting the Brier-Joye ladder (or, in our case, the complete formulas) in place of the fastest formulas presents implementers with a trade-off between "simplicity, security and speed". However, these same trade-offs also exist on certain choices of Edwards curves, where, for example, the fastest explicit formulas are also not complete: the Curve41417 implementation chooses to sacrifice the fastest coordinate system for the sake of completeness [10, Section 3.1], while the Goldilocks implementation goes to more complicated lengths to use the fastest formulas [37–39]. Furthermore, there is an additional category that is not considered in [14], i.e., the non-trivial security issues related to having a cofactor h greater than 1 [38, Sect. 1.1].

Given the complete explicit formulas in this paper, it is our opinion that well-chosen prime order curves can be considered safe choices for elliptic curve cryptography. It is well-known that curves with cofactors offer efficiency benefits in certain scenarios, but to our knowledge, efficiency and/or bandwidth issues are the only valid justifications for choosing a curve with a cofactor $h > 1$.

Organization. Section 2 briefly gives some preliminaries and notation. Section 3 presents the complete addition algorithms. In Sect. 4 we give intuition as to why these explicit formulas are optimal, or close to optimal, for prime order curves in short Weierstrass form. In Sect. 5 we discuss how these formulas can be used in practice. For Magma [18] scripts that can be used to verify our explicit algorithms and operation counts, we point the reader to the full version of this paper [58].

2 Preliminaries

Let k be a field of characteristic not two or three, and $\mathbb{P}^2(k)$ be the homogeneous projective plane of dimension two. Two points $(X_1 : Y_1 : Z_1)$ and $(X_2 : Y_2 : Z_2)$ in $\mathbb{P}^2(k)$ are equal if and only if there exist $\lambda \in k^\times$ such that $(X_1, Y_1, Z_1) = (\lambda X_2, \lambda Y_2, \lambda Z_2)$.

Let E/k be an elliptic curve embedded in $\mathbb{P}^2(k)$ as a Weierstrass model $E/k \colon Y^2Z = X^3 + aXZ^2 + bZ^3$. The points on E form an abelian group with identity $\mathcal{O} = (0 : 1 : 0)$. An *addition law* on E is a triple of polynomials (X_3, Y_3, Z_3) such that the map

$$P, Q \mapsto (X_3(P,Q) : Y_3(P,Q) : Z_3(P,Q))$$

determines the group law $+$ on an open subset of $(E \times E)(\bar{k})$, where $\bar{k}$ is the algebraic closure of k. For an extension K of k, a set of such addition laws is said to be K-complete if, for any pair of K-rational pair of points (P, Q), at least one addition law in the set is defined at (P, Q).

Lange and Ruppert [48] proved that the space of all addition laws on E has dimension 3, and Bosma and Lenstra [19] proved that a $\bar{k}$-complete set must contain (at least) two addition laws. In other words, Bosma and Lenstra proved that every addition law on E has at least one exceptional pair of inputs over the algebraic closure. More recent work by Arène, Kohel and Ritzenthaler [4] showed that this is true without assuming a Weierstrass embedding of E. That is, they showed that every elliptic curve addition law has exceptional pairs over the algebraic closure, irrespective of the projective embedding.

Following [19], for positive integers μ and ν, we define an *addition law of bidegree* (μ, ν) to be a triple of polynomials

$$X_3, Y_3, Z_3 \in k[X_1, Y_1, Z_1, X_2, Y_2, Z_2]$$

that satisfy the following two properties:

1. The polynomials X_3, Y_3 and Z_3 are homogeneous of degree μ in X_1, Y_1 and Z_1, and are homogeneous of degree ν in X_2 Y_2 and Z_2;
2. Let $P = (X_1 : Y_1 : Z_1)$ and $Q = (X_2 : Y_2 : Z_2)$ be in $E(K)$, where K is an extension of k. Then either X_3, Y_3 and Z_3 are all zero at P and Q, or else $(X_3 : Y_3 : Z_3)$ is an element of $E(K)$ and is equal to $P + Q$. If the first holds, we say that the pair (P, Q) is *exceptional* for the addition law. If there are no exceptional pairs of points, we say that the addition law is K-*complete* (note that this is in line with the definition of a K-complete set of addition laws).

Hereafter, if a single addition law is k-complete, we simply call it *complete*.

3 Complete Addition Formulas

Let $E/k\colon Y^2Z = X^3 + aXZ^2 + bZ^3 \subset \mathbb{P}^2$ with $\mathrm{char}(k) \neq 2, 3$. The complete addition formulas optimized in this section follow from the theorem of Bosma and Lenstra [19, Theorem 2], which states that, for any extension field K/k, there exists a 1-to-1 correspondence between lines in $\mathbb{P}^2(K)$ and addition laws of bidegree $(2, 2)$ on $E(K)$. Two points P and Q in $E(K)$ are then exceptional for an addition law if and only if $P - Q$ lies on the corresponding line. When $K = \bar{k}$, the algebraic closure of k, *every* line intersects $E(K)$; thus, one consequence of this theorem is that every addition law of bidegree $(2, 2)$ has an exceptional pair over the algebraic closure.

The addition law considered in this paper is the addition law corresponding to the line $Y = 0$ in $\mathbb{P}^2$ in [19], specialized to the short Weierstrass embedding of E above. For two points $P = (X_1 : Y_1 : Z_1)$, $Q = (X_2 : Y_2 : Z_2)$ on E, the sum $(X_3 : Y_3 : Z_3) = P + Q$ is given by

$$\begin{aligned}
X_3 &= Y_1Y_2(X_1Y_2 + X_2Y_1) - aX_1X_2(Y_1Z_2 + Y_2Z_1)\\
&\quad - a(X_1Y_2 + X_2Y_1)(X_1Z_2 + X_2Z_1) - 3b(X_1Y_2 + X_2Y_1)Z_1Z_2\\
&\quad - 3b(X_1Z_2 + X_2Z_1)(Y_1Z_2 + Y_2Z_1) + a^2(Y_1Z_2 + Y_2Z_1)Z_1Z_2,\\
Y_3 &= Y_1^2Y_2^2 + 3aX_1^2X_2^2 + 9bX_1X_2(X_1Z_2 + X_2Z_1)\\
&\quad - 2a^2X_1Z_2(X_1Z_2 + 2X_2Z_1) + a^2(X_1Z_2 + X_2Z_1)(X_1Z_2 - X_2Z_1)\\
&\quad - 3abX_1Z_1Z_2^2 - 3abX_2Z_1^2Z_2 - (a^3 + 9b^2)Z_1^2Z_2^2,\\
Z_3 &= 3X_1X_2(X_1Y_2 + X_2Y_1) + Y_1Y_2(Y_1Z_2 + Y_2Z_1)\\
&\quad + a(X_1Y_2 + X_2Y_1)Z_1Z_2 + a(X_1Z_2 + X_2Z_1)(Y_1Z_2 + Y_2Z_1)\\
&\quad + 3b(Y_1Z_2 + Y_2Z_1)Z_1Z_2.
\end{aligned}$$

Bosma and Lenstra prove that a pair of points (P, Q) is exceptional for this addition law if and only if $P - Q$ is a point of order two.

Exceptions. Throughout this paper, we fix $q \geq 5$ and assume throughout that $E(\mathbb{F}_q)$ has prime order to exclude $\mathbb{F}_q$-rational points of order two, so that the above formulas are complete. However, we note that the explicit algorithms that are derived in Sect. 3 will, firstly, be complete for any short Weierstrass curves of odd order, and secondly, also be exception-free for all pairs of points inside odd order subgroups on any short Weierstrass curve. In particular, this means that they can also be used to compute exception-free additions and scalar multiplications on certain curves with an even order. We come back to this in Sect. 5.2.

3.1 The General Case

Despite the attractive properties that come with completeness, this addition law seems to have been overlooked due to its apparent inefficiency. We now begin to show that these formulas are not as inefficient as they seem, to the point where the performance will be competitive with the fastest, incomplete addition laws in current implementations of prime order curves.

We start by rewriting the above formulas as

$$\begin{aligned}
X_3 &= (X_1Y_2 + X_2Y_1)(Y_1Y_2 - a(X_1Z_2 + X_2Z_1) - 3bZ_1Z_2)\\
&\quad - (Y_1Z_2 + Y_2Z_1)(aX_1X_2 + 3b(X_1Z_2 + X_2Z_1) - a^2Z_1Z_2),\\
Y_3 &= (3X_1X_2 + aZ_1Z_2)(aX_1X_2 + 3b(X_1Z_2 + X_2Z_1) - a^2Z_1Z_2)+\\
&(Y_1Y_2 + a(X_1Z_2 + X_2Z_1) + 3bZ_1Z_2)(Y_1Y_2 - a(X_1Z_2 + X_2Z_1) - 3bZ_1Z_2),\\
Z_3 &= (Y_1Z_2 + Y_2Z_1)(Y_1Y_2 + a(X_1Z_2 + X_2Z_1) + 3bZ_1Z_2)\\
&\quad + (X_1Y_2 + X_2Y_1)(3X_1X_2 + aZ_1Z_2).
\end{aligned} \tag{1}$$

Algorithm 1. Complete, projective point addition for arbitrary prime order short Weierstrass curves $E/\mathbb{F}_q\colon y^2 = x^3 + ax + b$.

Require: $P = (X_1 : Y_1 : Z_1)$, $Q = (X_2 : Y_2 : Z_2)$, $E\colon Y^2Z = X^3 + aXZ^2 + bZ^3$, and $b_3 = 3 \cdot b$.

Ensure: $(X_3 : Y_3 : Z_3) = P + Q$.

1. $t_0 \leftarrow X_1 \cdot X_2$	2. $t_1 \leftarrow Y_1 \cdot Y_2$	3. $t_2 \leftarrow Z_1 \cdot Z_2$
4. $t_3 \leftarrow X_1 + Y_1$	5. $t_4 \leftarrow X_2 + Y_2$	6. $t_3 \leftarrow t_3 \cdot t_4$
7. $t_4 \leftarrow t_0 + t_1$	8. $t_3 \leftarrow t_3 - t_4$	9. $t_4 \leftarrow X_1 + Z_1$
10. $t_5 \leftarrow X_2 + Z_2$	11. $t_4 \leftarrow t_4 \cdot t_5$	12. $t_5 \leftarrow t_0 + t_2$
13. $t_4 \leftarrow t_4 - t_5$	14. $t_5 \leftarrow Y_1 + Z_1$	15. $X_3 \leftarrow Y_2 + Z_2$
16. $t_5 \leftarrow t_5 \cdot X_3$	17. $X_3 \leftarrow t_1 + t_2$	18. $t_5 \leftarrow t_5 - X_3$
19. $Z_3 \leftarrow a \cdot t_4$	20. $X_3 \leftarrow b_3 \cdot t_2$	21. $Z_3 \leftarrow X_3 + Z_3$
22. $X_3 \leftarrow t_1 - Z_3$	23. $Z_3 \leftarrow t_1 + Z_3$	24. $Y_3 \leftarrow X_3 \cdot Z_3$
25. $t_1 \leftarrow t_0 + t_0$	26. $t_1 \leftarrow t_1 + t_0$	27. $t_2 \leftarrow a \cdot t_2$
28. $t_4 \leftarrow b_3 \cdot t_4$	29. $t_1 \leftarrow t_1 + t_2$	30. $t_2 \leftarrow t_0 - t_2$
31. $t_2 \leftarrow a \cdot t_2$	32. $t_4 \leftarrow t_4 + t_2$	33. $t_0 \leftarrow t_1 \cdot t_4$
34. $Y_3 \leftarrow Y_3 + t_0$	35. $t_0 \leftarrow t_5 \cdot t_4$	36. $X_3 \leftarrow t_3 \cdot X_3$
37. $X_3 \leftarrow X_3 - t_0$	38. $t_0 \leftarrow t_3 \cdot t_1$	39. $Z_3 \leftarrow t_5 \cdot Z_3$
40. $Z_3 \leftarrow Z_3 + t_0$		

Algorithm 2. Complete, mixed point addition for arbitrary prime order short Weierstrass curves $E/\mathbb{F}_q\colon y^2 = x^3 + ax + b$.

Require: $P = (X_1 : Y_1 : Z_1)$, $Q = (X_2 : Y_2 : 1)$, $E\colon Y^2Z = X^3 + aXZ^2 + bZ^3$, and $b_3 = 3 \cdot b$.

Ensure: $(X_3 : Y_3 : Z_3) = P + Q$.

1. $t_0 \leftarrow X_1 \cdot X_2$	2. $t_1 \leftarrow Y_1 \cdot Y_2$	3. $t_3 \leftarrow X_2 + Y_2$
4. $t_4 \leftarrow X_1 + Y_1$	5. $t_3 \leftarrow t_3 \cdot t_4$	6. $t_4 \leftarrow t_0 + t_1$
7. $t_3 \leftarrow t_3 - t_4$	8. $t_4 \leftarrow X_2 \cdot Z_1$	9. $t_4 \leftarrow t_4 + X_1$
10. $t_5 \leftarrow Y_2 \cdot Z_1$	11. $t_5 \leftarrow t_5 + Y_1$	12. $Z_3 \leftarrow a \cdot t_4$
13. $X_3 \leftarrow b_3 \cdot Z_1$	14. $Z_3 \leftarrow X_3 + Z_3$	15. $X_3 \leftarrow t_1 - Z_3$
16. $Z_3 \leftarrow t_1 + Z_3$	17. $Y_3 \leftarrow X_3 \cdot Z_3$	18. $t_1 \leftarrow t_0 + t_0$
19. $t_1 \leftarrow t_1 + t_0$	20. $t_2 \leftarrow a \cdot Z_1$	21. $t_4 \leftarrow b_3 \cdot t_4$
22. $t_1 \leftarrow t_1 + t_2$	23. $t_2 \leftarrow t_0 - t_2$	24. $t_2 \leftarrow a \cdot t_2$
25. $t_4 \leftarrow t_4 + t_2$	26. $t_0 \leftarrow t_1 \cdot t_4$	27. $Y_3 \leftarrow Y_3 + t_0$
28. $t_0 \leftarrow t_5 \cdot t_4$	29. $X_3 \leftarrow t_3 \cdot X_3$	30. $X_3 \leftarrow X_3 - t_0$
31. $t_0 \leftarrow t_3 \cdot t_1$	32. $Z_3 \leftarrow t_5 \cdot Z_3$	33. $Z_3 \leftarrow Z_3 + t_0$

The rewritten formulas still appear somewhat cumbersome, but a closer inspection of (1) reveals that several terms are repeated. In Algorithm 1, we show that this can in fact be computed[6] using $12\mathbf{M} + 3\mathbf{m_a} + 2\mathbf{m_{3b}} + 23\mathbf{a}$[7].

[6] Notation here is the same as in Table 1, except for $\mathbf{m_{3b}}$ which denotes multiplication by the curve constant $3b$.

Although Algorithm 1 is sufficient for cryptographic implementations, performance gains can be obtained by specializing the point additions to the useful scenarios of mixed additions[8] (i.e., where $Z_2 = 1$) and/or point doublings (i.e., where $P = Q$). The mixed addition follows the same formulas as for point addition; Algorithm 2 shows this can be done in $11\mathbf{M} + 3\mathbf{m_a} + 2\mathbf{m_{3b}} + 17\mathbf{a}$.

For a point $P = (X : Y : Z)$, doubling is computed as

$$\begin{aligned} X_3 &= 2XY(Y^2 - 2aXZ - 3bZ^2) \\ &\quad - 2YZ(aX^2 + 6bXZ - a^2Z^2), \\ Y_3 &= (Y^2 + 2aXZ + 3bZ^2)(Y^2 - 2aXZ - 3bZ^2) \\ &\quad + (3X^2 + aZ^2)(aX^2 + 6bXZ - a^2Z^2), \\ Z_3 &= 8Y^3Z. \end{aligned}$$

Algorithm 3 shows that this can be computed in $8\mathbf{M} + 3\mathbf{S} + 3\mathbf{m_a} + 2\mathbf{m_{3b}} + 15\mathbf{a}$.

Algorithm 3. Exception-free point doubling for arbitrary prime order short Weierstrass curves $E/\mathbb{F}_q\colon y^2 = x^3 + ax + b$.

Require: $P = (X : Y : Z)$ on $E\colon Y^2Z = X^3 + aXZ^2 + bZ^3$, and $b_3 = 3 \cdot b$.
Ensure: $(X_3 : Y_3 : Z_3) = 2P$.

1. $t_0 \leftarrow X \cdot X$	2. $t_1 \leftarrow Y \cdot Y$	3. $t_2 \leftarrow Z \cdot Z$
4. $t_3 \leftarrow X \cdot Y$	5. $t_3 \leftarrow t_3 + t_3$	6. $Z_3 \leftarrow X \cdot Z$
7. $Z_3 \leftarrow Z_3 + Z_3$	8. $X_3 \leftarrow a \cdot Z_3$	9. $Y_3 \leftarrow b_3 \cdot t_2$
10. $Y_3 \leftarrow X_3 + Y_3$	11. $X_3 \leftarrow t_1 - Y_3$	12. $Y_3 \leftarrow t_1 + Y_3$
13. $Y_3 \leftarrow X_3 \cdot Y_3$	14. $X_3 \leftarrow t_3 \cdot X_3$	15. $Z_3 \leftarrow b_3 \cdot Z_3$
16. $t_2 \leftarrow a \cdot t_2$	17. $t_3 \leftarrow t_0 - t_2$	18. $t_3 \leftarrow a \cdot t_3$
19. $t_3 \leftarrow t_3 + Z_3$	20. $Z_3 \leftarrow t_0 + t_0$	21. $t_0 \leftarrow Z_3 + t_0$
22. $t_0 \leftarrow t_0 + t_2$	23. $t_0 \leftarrow t_0 \cdot t_3$	24. $Y_3 \leftarrow Y_3 + t_0$
25. $t_2 \leftarrow Y \cdot Z$	26. $t_2 \leftarrow t_2 + t_2$	27. $t_0 \leftarrow t_2 \cdot t_3$
28. $X_3 \leftarrow X_3 - t_0$	29. $Z_3 \leftarrow t_2 \cdot t_1$	30. $Z_3 \leftarrow Z_3 + Z_3$
31. $Z_3 \leftarrow Z_3 + Z_3$		

3.2 Special Cases of Interest

$\mathbf{a = -3}$. Several standards (e.g., [3,22,23,30,56,60]) adopt short Weierstrass curves with the constant a being $a = -3$, which gives rise to faster explicit formulas for point doubling[9].

[7] We thank Emmanuel Thomé whose careful read-through resulted in a $1\mathbf{m}_a$ saving in all three of the explicit formulas for the general case.

[8] We note that it is not technically correct to call "mixed" additions complete, since $Z_2 = 1$ precludes the second point being the point at infinity. However, this is not a problem in practice as the second point is typically taken from a precomputed lookup table consisting of small multiples of the input point $P \neq \mathcal{O}$. For prime order curves, these small multiples can never be at infinity.

[9] When $\mathbb{F}_q$ is a large prime field, $a = -3$ covers 1/2 (resp. 1/4) of the isomorphism classes for $q \equiv 3 \mod 4$ (resp. $q \equiv 1 \mod 4$) – see [21, Sect. 3].

In this case, the complete formulas in (1) specialize to

$$\begin{aligned}
X_3 &= (X_1Y_2 + X_2Y_1)(Y_1Y_2 + 3(X_1Z_2 + X_2Z_1 - bZ_1Z_2)) \\
&\quad - 3(Y_1Z_2 + Y_2Z_1)(b(X_1Z_2 + X_2Z_1) - X_1X_2 - 3Z_1Z_2), \\
Y_3 &= 3(3X_1X_2 - 3Z_1Z_2)(b(X_1Z_2 + X_2Z_1) - X_1X_2 - 3Z_1Z_2) + \\
&(Y_1Y_2 - 3(X_1Z_2 + X_2Z_1 - bZ_1Z_2))(Y_1Y_2 + 3(X_1Z_2 + X_2Z_1 - bZ_1Z_2)), \\
Z_3 &= (Y_1Z_2 + Y_2Z_1)(Y_1Y_2 - 3(X_1Z_2 + X_2Z_1 - bZ_1Z_2)) \\
&\quad + (X_1Y_2 + X_2Y_1)(3X_1X_2 - 3Z_1Z_2).
\end{aligned}$$

These can be computed at a cost of $12\mathbf{M} + 2\mathbf{m_b} + 29\mathbf{a}$ using Algorithm 4. The mixed addition can be done at a cost of $11\mathbf{M} + 2\mathbf{m_b} + 23\mathbf{a}$, as shown in Algorithm 5.

Algorithm 4. Complete, projective point addition for prime order short Weierstrass curves $E/\mathbb{F}_q \colon y^2 = x^3 + ax + b$ with $a = -3$.

Require: $P = (X_1 : Y_1 : Z_1)$, $Q = (X_2 : Y_2 : Z_2)$, $E \colon Y^2Z = X^3 - 3XZ^2 + bZ^3$
Ensure: $(X_3 : Y_3 : Z_3) = P + Q$;

1. $t_0 \leftarrow X_1 \cdot X_2$	2. $t_1 \leftarrow Y_1 \cdot Y_2$	3. $t_2 \leftarrow Z_1 \cdot Z_2$
4. $t_3 \leftarrow X_1 + Y_1$	5. $t_4 \leftarrow X_2 + Y_2$	6. $t_3 \leftarrow t_3 \cdot t_4$
7. $t_4 \leftarrow t_0 + t_1$	8. $t_3 \leftarrow t_3 - t_4$	9. $t_4 \leftarrow Y_1 + Z_1$
10. $X_3 \leftarrow Y_2 + Z_2$	11. $t_4 \leftarrow t_4 \cdot X_3$	12. $X_3 \leftarrow t_1 + t_2$
13. $t_4 \leftarrow t_4 - X_3$	14. $X_3 \leftarrow X_1 + Z_1$	15. $Y_3 \leftarrow X_2 + Z_2$
16. $X_3 \leftarrow X_3 \cdot Y_3$	17. $Y_3 \leftarrow t_0 + t_2$	18. $Y_3 \leftarrow X_3 - Y_3$
19. $Z_3 \leftarrow b \cdot t_2$	20. $X_3 \leftarrow Y_3 - Z_3$	21. $Z_3 \leftarrow X_3 + X_3$
22. $X_3 \leftarrow X_3 + Z_3$	23. $Z_3 \leftarrow t_1 - X_3$	24. $X_3 \leftarrow t_1 + X_3$
25. $Y_3 \leftarrow b \cdot Y_3$	26. $t_1 \leftarrow t_2 + t_2$	27. $t_2 \leftarrow t_1 + t_2$
28. $Y_3 \leftarrow Y_3 - t_2$	29. $Y_3 \leftarrow Y_3 - t_0$	30. $t_1 \leftarrow Y_3 + Y_3$
31. $Y_3 \leftarrow t_1 + Y_3$	32. $t_1 \leftarrow t_0 + t_0$	33. $t_0 \leftarrow t_1 + t_0$
34. $t_0 \leftarrow t_0 - t_2$	35. $t_1 \leftarrow t_4 \cdot Y_3$	36. $t_2 \leftarrow t_0 \cdot Y_3$
37. $Y_3 \leftarrow X_3 \cdot Z_3$	38. $Y_3 \leftarrow Y_3 + t_2$	39. $X_3 \leftarrow t_3 \cdot X_3$
40. $X_3 \leftarrow X_3 - t_1$	41. $Z_3 \leftarrow t_4 \cdot Z_3$	42. $t_1 \leftarrow t_3 \cdot t_0$
43. $Z_3 \leftarrow Z_3 + t_1$		

In this case, the doubling formulas become

$$\begin{aligned}
X_3 &= 2XY(Y^2 + 3(2XZ - bZ^2)) \\
&\quad - 6YZ(2bXZ - X^2 - 3Z^2), \\
Y_3 &= (Y^2 - 3(2XZ - bZ^2))(Y^2 + 3(2XZ - bZ^2)) \\
&\quad + 3(3X^2 - 3Z^2)(2bXZ - X^2 - 3Z^2), \\
Z_3 &= 8Y^3Z,
\end{aligned}$$

which can be computed at a cost of $8\mathbf{M} + 3\mathbf{S} + 2\mathbf{m_b} + 21\mathbf{a}$ using Algorithm 6.

Algorithm 5. Complete, mixed point addition for prime order short Weierstrass curves $E/\mathbb{F}_q\colon y^2 = x^3 + ax + b$ with $a = -3$.

Require: $P=(X_1 : Y_1 : Z_1)$, $Q=(X_2 : Y_2 : 1)$, $E\colon Y^2Z=X^3 - 3XZ^2 + bZ^3$

Ensure: $(X_3 : Y_3 : Z_3) = P + Q$;

1. $t_0 \leftarrow X_1 \cdot X_2$	2. $t_1 \leftarrow Y_1 \cdot Y_2$	3. $t_3 \leftarrow X_2 + Y_2$
4. $t_4 \leftarrow X_1 + Y_1$	5. $t_3 \leftarrow t_3 \cdot t_4$	6. $t_4 \leftarrow t_0 + t_1$
7. $t_3 \leftarrow t_3 - t_4$	8. $t_4 \leftarrow Y_2 \cdot Z_1$	9. $t_4 \leftarrow t_4 + Y_1$
10. $Y_3 \leftarrow X_2 \cdot Z_1$	11. $Y_3 \leftarrow Y_3 + X_1$	12. $Z_3 \leftarrow b \cdot Z_1$
13. $X_3 \leftarrow Y_3 - Z_3$	14. $Z_3 \leftarrow X_3 + X_3$	15. $X_3 \leftarrow X_3 + Z_3$
16. $Z_3 \leftarrow t_1 - X_3$	17. $X_3 \leftarrow t_1 + X_3$	18. $Y_3 \leftarrow b \cdot Y_3$
19. $t_1 \leftarrow Z_1 + Z_1$	20. $t_2 \leftarrow t_1 + Z_1$	21. $Y_3 \leftarrow Y_3 - t_2$
22. $Y_3 \leftarrow Y_3 - t_0$	23. $t_1 \leftarrow Y_3 + Y_3$	24. $Y_3 \leftarrow t_1 + Y_3$
25. $t_1 \leftarrow t_0 + t_0$	26. $t_0 \leftarrow t_1 + t_0$	27. $t_0 \leftarrow t_0 - t_2$
28. $t_1 \leftarrow t_4 \cdot Y_3$	29. $t_2 \leftarrow t_0 \cdot Y_3$	30. $Y_3 \leftarrow X_3 \cdot Z_3$
31. $Y_3 \leftarrow Y_3 + t_2$	32. $X_3 \leftarrow t_3 \cdot X_3$	33. $X_3 \leftarrow X_3 - t_1$
34. $Z_3 \leftarrow t_4 \cdot Z_3$	35. $t_1 \leftarrow t_3 \cdot t_0$	36. $Z_3 \leftarrow Z_3 + t_1$

Algorithm 6. Exception-free point doubling for prime order short Weierstrass curves $E/\mathbb{F}_q\colon y^2 = x^3 + ax + b$ with $a = -3$.

Require: $P = (X : Y : Z)$ on $E\colon Y^2Z = X^3 - 3XZ^2 + bZ^3$.

Ensure: $(X_3 : Y_3 : Z_3) = 2P$.

1. $t_0 \leftarrow X \cdot X$	2. $t_1 \leftarrow Y \cdot Y$	3. $t_2 \leftarrow Z \cdot Z$
4. $t_3 \leftarrow X \cdot Y$	5. $t_3 \leftarrow t_3 + t_3$	6. $Z_3 \leftarrow X \cdot Z$
7. $Z_3 \leftarrow Z_3 + Z_3$	8. $Y_3 \leftarrow b \cdot t_2$	9. $Y_3 \leftarrow Y_3 - Z_3$
10. $X_3 \leftarrow Y_3 + Y_3$	11. $Y_3 \leftarrow X_3 + Y_3$	12. $X_3 \leftarrow t_1 - Y_3$
13. $Y_3 \leftarrow t_1 + Y_3$	14. $Y_3 \leftarrow X_3 \cdot Y_3$	15. $X_3 \leftarrow X_3 \cdot t_3$
16. $t_3 \leftarrow t_2 + t_2$	17. $t_2 \leftarrow t_2 + t_3$	18. $Z_3 \leftarrow b \cdot Z_3$
19. $Z_3 \leftarrow Z_3 - t_2$	20. $Z_3 \leftarrow Z_3 - t_0$	21. $t_3 \leftarrow Z_3 + Z_3$
22. $Z_3 \leftarrow Z_3 + t_3$	23. $t_3 \leftarrow t_0 + t_0$	24. $t_0 \leftarrow t_3 + t_0$
25. $t_0 \leftarrow t_0 - t_2$	26. $t_0 \leftarrow t_0 \cdot Z_3$	27. $Y_3 \leftarrow Y_3 + t_0$
28. $t_0 \leftarrow Y \cdot Z$	29. $t_0 \leftarrow t_0 + t_0$	30. $Z_3 \leftarrow t_0 \cdot Z_3$
31. $X_3 \leftarrow X_3 - Z_3$	32. $Z_3 \leftarrow t_0 \cdot t_1$	33. $Z_3 \leftarrow Z_3 + Z_3$
34. $Z_3 \leftarrow Z_3 + Z_3$		

a = 0. Short Weierstrass curves with $a = 0$, i.e., with j-invariant 0, have also appeared in the standards. For example, Certicom's SEC-2 standard [22] specifies three such curves; one of these is `secp256k1`, which is the curve used in the Bitcoin protocol. In addition, in the case that pairing-based cryptography becomes standardized, it is most likely that the curve choices will be short Weierstrass curves with $a = 0$, e.g., BN curves [5].

Algorithm 7. Complete, projective point addition for prime order j-invariant 0 short Weierstrass curves $E/\mathbb{F}_q\colon y^2 = x^3 + b$.

Require: $P = (X_1 : Y_1 : Z_1)$, $Q = (X_2 : Y_2 : Z_2)$ on $E\colon Y^2Z = X^3 + bZ^3$ and $b_3 = 3 \cdot b$.

Ensure: $(X_3 : Y_3 : Z_3) = P + Q$;

1. $t_0 \leftarrow X_1 \cdot X_2$	2. $t_1 \leftarrow Y_1 \cdot Y_2$	3. $t_2 \leftarrow Z_1 \cdot Z_2$
4. $t_3 \leftarrow X_1 + Y_1$	5. $t_4 \leftarrow X_2 + Y_2$	6. $t_3 \leftarrow t_3 \cdot t_4$
7. $t_4 \leftarrow t_0 + t_1$	8. $t_3 \leftarrow t_3 - t_4$	9. $t_4 \leftarrow Y_1 + Z_1$
10. $X_3 \leftarrow Y_2 + Z_2$	11. $t_4 \leftarrow t_4 \cdot X_3$	12. $X_3 \leftarrow t_1 + t_2$
13. $t_4 \leftarrow t_4 - X_3$	14. $X_3 \leftarrow X_1 + Z_1$	15. $Y_3 \leftarrow X_2 + Z_2$
16. $X_3 \leftarrow X_3 \cdot Y_3$	17. $Y_3 \leftarrow t_0 + t_2$	18. $Y_3 \leftarrow X_3 - Y_3$
19. $X_3 \leftarrow t_0 + t_0$	20. $t_0 \leftarrow X_3 + t_0$	21. $t_2 \leftarrow b_3 \cdot t_2$
22. $Z_3 \leftarrow t_1 + t_2$	23. $t_1 \leftarrow t_1 - t_2$	24. $Y_3 \leftarrow b_3 \cdot Y_3$
25. $X_3 \leftarrow t_4 \cdot Y_3$	26. $t_2 \leftarrow t_3 \cdot t_1$	27. $X_3 \leftarrow t_2 - X_3$
28. $Y_3 \leftarrow Y_3 \cdot t_0$	29. $t_1 \leftarrow t_1 \cdot Z_3$	30. $Y_3 \leftarrow t_1 + Y_3$
31. $t_0 \leftarrow t_0 \cdot t_3$	32. $Z_3 \leftarrow Z_3 \cdot t_4$	33. $Z_3 \leftarrow Z_3 + t_0$

Algorithm 8. Complete, mixed point addition for prime order j-invariant 0 short Weierstrass curves $E/\mathbb{F}_q\colon y^2 = x^3 + b$.

Require: $P = (X_1 : Y_1 : Z_1)$, $Q = (X_2 : Y_2 : 1)$ on $E\colon Y^2Z = X^3 + bZ^3$ and $b_3 = 3 \cdot b$.

Ensure: $(X_3 : Y_3 : Z_3) = P + Q$;

1. $t_0 \leftarrow X_1 \cdot X_2$	2. $t_1 \leftarrow Y_1 \cdot Y_2$	3. $t_3 \leftarrow X_2 + Y_2$
4. $t_4 \leftarrow X_1 + Y_1$	5. $t_3 \leftarrow t_3 \cdot t_4$	6. $t_4 \leftarrow t_0 + t_1$
7. $t_3 \leftarrow t_3 - t_4$	8. $t_4 \leftarrow Y_2 \cdot Z_1$	9. $t_4 \leftarrow t_4 + Y_1$
10. $Y_3 \leftarrow X_2 \cdot Z_1$	11. $Y_3 \leftarrow Y_3 + X_1$	12. $X_3 \leftarrow t_0 + t_0$
13. $t_0 \leftarrow X_3 + t_0$	14. $t_2 \leftarrow b_3 \cdot Z_1$	15. $Z_3 \leftarrow t_1 + t_2$
16. $t_1 \leftarrow t_1 - t_2$	17. $Y_3 \leftarrow b_3 \cdot Y_3$	18. $X_3 \leftarrow t_4 \cdot Y_3$
19. $t_2 \leftarrow t_3 \cdot t_1$	20. $X_3 \leftarrow t_2 - X_3$	21. $Y_3 \leftarrow Y_3 \cdot t_0$
22. $t_1 \leftarrow t_1 \cdot Z_3$	23. $Y_3 \leftarrow t_1 + Y_3$	24. $t_0 \leftarrow t_0 \cdot t_3$
25. $Z_3 \leftarrow Z_3 \cdot t_4$	26. $Z_3 \leftarrow Z_3 + t_0$	

In this case, the complete additions simplify to

$$\begin{aligned}
X_3 &= (X_1Y_2 + X_2Y_1)(Y_1Y_2 - 3bZ_1Z_2) \\
&\quad - 3b(Y_1Z_2 + Y_2Z_1)(X_1Z_2 + X_2Z_1), \\
Y_3 &= (Y_1Y_2 + 3bZ_1Z_2)(Y_1Y_2 - 3bZ_1Z_2) + 9bX_1X_2(X_1Z_2 + X_2Z_1), \\
Z_3 &= (Y_1Z_2 + Y_2Z_1)(Y_1Y_2 + 3bZ_1Z_2) + 3X_1X_2(X_1Y_2 + X_2Y_1),
\end{aligned}$$

which can be computed in $12\mathbf{M} + 2\mathbf{m_{3b}} + 19\mathbf{a}$ via Algorithm 7. The mixed addition is computed in $11\mathbf{M} + 2\mathbf{m_{3b}} + 13\mathbf{a}$ via Algorithm 8.

The doubling formulas in this case are

$$X_3 = 2XY(Y^2 - 9bZ^2),$$
$$Y_3 = (Y^2 - 9bZ^2)(Y^2 + 3bZ^2) + 24bY^2Z^2,$$
$$Z_3 = 8Y^3Z.$$

These can be computed in $6\mathbf{M} + 2\mathbf{S} + 1\mathbf{m}_{\mathbf{3b}} + 9\mathbf{a}$ via Algorithm 9.

Algorithm 9. Exception-free point doubling for prime order j-invariant 0 short Weierstrass curves $E/\mathbb{F}_q \colon y^2 = x^3 + b$.

Require: $P = (X : Y : Z)$ on $E\colon Y^2Z = X^3 + bZ^3$ and $b_3 = 3 \cdot b$.
Ensure: $(X_3 : Y_3 : Z_3) = 2P$.

1. $t_0 \leftarrow Y \cdot Y$	2. $Z_3 \leftarrow t_0 + t_0$	3. $Z_3 \leftarrow Z_3 + Z_3$
4. $Z_3 \leftarrow Z_3 + Z_3$	5. $t_1 \leftarrow Y \cdot Z$	6. $t_2 \leftarrow Z \cdot Z$
7. $t_2 \leftarrow b_3 \cdot t_2$	8. $X_3 \leftarrow t_2 \cdot Z_3$	9. $Y_3 \leftarrow t_0 + t_2$
10. $Z_3 \leftarrow t_1 \cdot Z_3$	11. $t_1 \leftarrow t_2 + t_2$	12. $t_2 \leftarrow t_1 + t_2$
13. $t_0 \leftarrow t_0 - t_2$	14. $Y_3 \leftarrow t_0 \cdot Y_3$	15. $Y_3 \leftarrow X_3 + Y_3$
16. $t_1 \leftarrow X \cdot Y$	17. $X_3 \leftarrow t_0 \cdot t_1$	18. $X_3 \leftarrow X_3 + X_3$

4 Some Intuition Towards Optimality

In this section we motivate the choice of the complete formulas in (1) that were taken from Bosma and Lenstra [19], by providing reasoning as to why, among the many possible complete addition laws on prime order curves, we chose the set corresponding to the line $Y = 0$ in $\mathbb{P}^2(k)$ under the straightforward homogeneous projection.

We do not claim that this choice is truly optimal, since proving that a certain choice of projective embedding and/or complete addition law for any particular prime order curve is faster than *all* of the other choices for that curve seems extremely difficult, if not impossible. We merely explain why, when aiming to write down explicit algorithms that will simultaneously be complete on all prime order short Weierstrass curves, choosing the Bosma–Lenstra formulas makes sense.

Furthermore, we also do not claim that our explicit algorithms to compute the addition law in (1) are computationally optimal. It is likely that trade-offs can be advantageously exploited on some platforms (cf. [41, Sect. 3.6]) or that alternative operation scheduling could reduce the number of field additions[10].

[10] Our experimentation did suggest that computing (1) in any reasonable way with fewer than 12 generic multiplications appears to be difficult.

4.1 Choice of the Line $Y = 0$ for Bidegree $(2, 2)$ Addition Laws

Let $L_{(\alpha,\beta,\gamma)}$ denote the line given by $\alpha X + \beta Y + \gamma Z = 0$ inside $\mathbb{P}^2(\mathbb{F}_q)$, and, under the necessary assumption that $L_{(\alpha,\beta,\gamma)}$ does not intersect the curve $E\colon Y^2Z = X^3 + aXZ^2 + bZ^3 \subset \mathbb{P}^2(\mathbb{F}_q)$, let $A_{(\alpha,\beta,\gamma)}$ denote the complete addition law of bidegree $(2, 2)$ corresponding to $L_{(\alpha,\beta,\gamma)}$ given by [19, Theorem 2]. So far we have given optimizations for $A_{(0,1,0)}$, but the question remains as to whether there are other lines $L_{(\alpha,\beta,\gamma)}$ which give rise to even faster addition laws $A_{(\alpha,\beta,\gamma)}$.

We first point out that $L_{(0,1,0)}$ is the only line that does not intersect E independently of a, b and q. It is easy to show that any other line in $\mathbb{P}^2(\mathbb{F}_q)$ that does not intersect E will have a dependency on at least one of a, b and q, and the resulting addition law will therefore only be complete on a subset of prime order curves.

Nevertheless, it is possible that there is a better choice than $A_{(0,1,0)}$ for a given short Weierstrass curve, or that there are special choices of prime order curves that give rise to more efficient complete group laws. We now sketch some intuition as to why this is unlikely. For $A_{(\alpha,\beta,\gamma)}$ to be complete, it is necessary that, in particular, $L_{(\alpha,\beta,\gamma)}$ does not intersect E at the point at infinity $(0 : 1 : 0)$. This implies that $\beta \neq 0$. From [19,48], we know that the space of all addition laws has dimension 3 and that

$$A_{(\alpha,\beta,\gamma)} = \alpha A_{(1,0,0)} + \beta A_{(0,1,0)} + \gamma A_{(0,0,1)},$$

where $A_{(1,0,0)}$, $A_{(0,1,0)}$ and $A_{(0,0,1)}$ are the three addition laws given in [19, pp. 236–239], specialized to short Weierstrass curves. Given that $\beta \neq 0$, our only hope of finding a more simple addition law than $A_{(0,1,0)}$ is by choosing α and/or γ in a way that causes an advantageous cross-cancellation of terms. Close inspection of the formulas in [19] strongly suggests that no such cancellation exists.

Remark 1. Interestingly, both $A_{(1,0,0)}$ and $A_{(0,0,1)}$ vanish to zero when specialized to doubling. This means that any doubling formula in bidegree $(2, 2)$ that is not exceptional at the point at infinity is a scalar multiple of $A_{(0,1,0)}$, i.e., the formulas used in this paper.

Remark 2. Although a more efficient addition law might exist for larger bidegrees, it is worth reporting that our experiments to find higher bidegree analogues of the Bosma and Lenstra formulas suggest that this, too, is unlikely. The complexity (and computational cost) of the explicit formulas grows rapidly as the bidegree increases, which is most commonly the case across all models of elliptic curves and projective embeddings (cf. [41]). We could hope for an addition law of bidegree lower than $(2, 2)$, but in [19, Sect. 3] Bosma and Lenstra prove that this is not possible under the short Weierstrass embedding[11] of E.

[11] Lower bidegree addition laws are possible for other embeddings (i.e., models) of E in the case where E has a k-rational torsion structure – see [47].

4.2 Jacobian Coordinates

Since first suggested for short Weierstrass curves by Miller in his seminal paper [52, p. 424], Jacobian coordinates have proven to offer significant performance advantages over other coordinate systems. Given their ubiquity in real-world ECC code, and the fact that their most commonly used sets of efficient point doubling formulas turn out to be exception-free on prime order curves (see Table 1), it is highly desirable to go searching for a Jacobian coordinate analogue of the Bosma–Lenstra (homogeneous coordinates) addition law. Unfortunately, we now show that such addition formulas in Jacobian coordinates must have a higher bidegree, intuitively making them slower to compute.

For the remainder of this section only, let $E \subset \mathbb{P}(2,3,1)(k)$ have odd order. If an addition law $f = (f_X, f_Y, f_Z)$ has f_Z of bidegree (μ, ν), then the bidegrees of f_X and f_Y are $(2\mu, 2\nu)$ and $(3\mu, 3\nu)$, respectively. Below we show that any complete formulas must have $\mu, \nu \geq 3$.

Consider the addition of two points $P = (X_1 : Y_1 : Z_1)$ and $Q = (X_2 : Y_2 : Z_2)$, using the addition law

$$f(P,Q) = (f_X(P,Q) : f_Y(P,Q) : f_Z(P,Q)),$$

with f_Z of bidegree (μ, ν). Suppose that f is complete, and that $\mu < 3$. Then f_Z, viewed as a polynomial in X_1, Y_1, Z_1, has degree $\mu < 3$, and in particular cannot contain Y_1. Now, since $-P = (X_1 : -Y_1 : Z_1)$ on E, it follows that $f_Z(P,Q) = f_Z(-P,Q)$ for all possible Q, and in particular when $Q = P$. But in this case, and given that P cannot have order 2, we have $f_Z(P,Q) \neq 0$ and $f_Z(-P,Q) = 0$, a contradiction. We conclude that $\mu \geq 3$, and (by symmetry) that $\nu \geq 3$. It follows that f_X and f_Y have bidegrees at least $(6,6)$ and $(9,9)$, respectively, which destroys any hope of comparable efficiency to the homogeneous Bosma–Lenstra formulas.

5 Using These Formulas in Practice

In this section we discuss the practical application of the complete algorithms in this paper. We discuss how they can be used for both the prime order curves (Sect. 5.1) and composite order curves (Sect. 5.2) in the standards. In Sect. 5.3, we give performance numbers that shed light on the expected cost of completeness in certain software scenarios, before discussing why this cost is likely to be significantly reduced in many other scenarios, e.g., in hardware.

5.1 Application to Prime Order Curves (or, Secure ECC for Noobs)

Using Algorithm 1 as a black-box point addition routine, non-experts now have a straightforward way to implement the standardized prime order elliptic curves. So long as scalars are *recoded* correctly, the subsequent scalar multiplication routine will always compute the correct result.

Given the vulnerabilities exposed in already-deployed ECC implementations (see Sect. 1), we now provide some implementation recommendations, e.g., for an implementer whose task it is to (re)write a simple and timing-resistant scalar multiplication routine for prime order curves from scratch. The main point is that branches (e.g., `if` statements) inside the elliptic curve point addition algorithms can now be avoided entirely. Our main recommendation is that more streamlined versions of Algorithm 1 should only be introduced to an implementation if they are guaranteed to be exception-free; subsequently, we stress that branching should never be introduced into any point addition algorithms.

Assuming access to branch-free, constant-time field arithmetic in $\mathbb{F}_q$, a first step is to implement Algorithm 1 to be used for *all* point (doubling and addition) operations, working entirely in homogeneous projective space. The natural next step is to implement a basic scalar recoding (e.g., [44,57]) that gives rise to a fixed, uniform, scalar-independent main loop. This typically means that the main loop repeats the same pattern of a fixed number of doublings followed by a single table lookup/extraction and, subsequently, an addition. The important points are that this table lookup must be done in a cache-timing resistant manner (cf. [45, Sect. 3.4]), and that the basic scalar recoding must itself be performed in a uniform manner.

Once the above routine is running correctly, an implementer that is seeking further performance gains can start by viewing stages of the routine where Algorithm 1 can safely be replaced by its specialized, more efficient variants. If the code is intended to support only short Weierstrass curves with either $a = -3$ or $a = 0$, then Algorithm 1 should be replaced by (the faster and more compact) Algorithm 4 or Algorithm 7, respectively. If the performance gains warrant the additional code, then at all stages where the addition function is called to add a point to itself (i.e., the point doubling stages), the respective exception-free point doubling routine(s) in Algorithms 3, 6 and 9 should be implemented and called there instead.

Incomplete short Weierstrass addition routines (e.g., the prior works summarized in Table 1) should only be introduced for further performance gains if the implementer can guarantee that exceptional pairs of points can never be input into the algorithms, and subsequently can implement them without introducing any branches. For example, Bos *et al.* [16, Sect. 4.1] proved that, under their particular choice of scalar multiplication algorithm, all-but-one of the point additions in a variable-base scalar multiplication can be performed without exception using an incomplete addition algorithm. The high-level argument used there was that such additions almost always took place between elements of the lookup table and a running value that had just been output from a point doubling, the former being small odd multiples of the input point (e.g., P, $[3]P$, $[5]P$, etc.) and the latter being some even multiple. Subsequently, they showed that the only possible time when the input points to the addition algorithm could coincide with (or be inverses of) each other is in the final addition, ruling out the exceptional points in all prior additions. On the other hand, as we mentioned in Sect. 1 and as was encountered in [16, Sect. 4.1], it can be significantly more complicated to rule out exceptional input points in more exotic scalar multiplication scenarios like fixed-base scalar multiplications, multiscalar multiplications, or those that exploit endomorphisms. In those cases, it could be that the only option to rule out any exceptional points is to *always* call complete addition algorithms.

Remark 3 (The best of both worlds?). We conclude this subsection by mentioning one more option that may be of interest to implementers who want to combine the fastest complete point addition algorithms with the fastest exception-free point doubling algorithms. Recall from Table 1 that the fastest doubling algorithms for short Weierstrass curves work in Jacobian coordinates and happen to be exception-free in the prime order setting, but recall from Sect. 4.2 that there is little hope of obtaining relatively efficient complete addition algorithms in Jacobian coordinates. This prompts the question as to whether the doubling algorithms that take place in $\mathbb{P}(2,3,1)(k)$ can be combined with our complete addition algorithms that take place in $\mathbb{P}^2(k)$. Generically, we can map the elliptic curve point $(X : Y : Z) \in \mathbb{P}(2,3,1)(k)$ to $(XZ : Y : Z^3) \in \mathbb{P}^2(k)$, and conversely, we can map the point $(X : Y : Z) \in \mathbb{P}^2(k)$ to $(XZ : YZ^2 : Z) \in \mathbb{P}(2,3,1)(k)$; both maps cost $2\mathbf{M} + 1\mathbf{S}$. We note that in the first direction there are no exceptions: in particular, the point at infinity $(1 : 1 : 0) \in \mathbb{P}(2,3,1)(k)$ correctly maps to $(0 : 1 : 0) \in \mathbb{P}^2(k)$. However, in the other direction, the point at infinity $(0 : 1 : 0) \in \mathbb{P}^2(k)$ does not correctly map to $(1 : 1 : 0) \in \mathbb{P}(2,3,1)(k)$, but rather to the point $(0 : 0 : 0) \notin \mathbb{P}(2,3,1)(k)$.

For a variable-base scalar multiplication using a fixed window of width w, one option would be to store the precomputed lookup table in $\mathbb{P}^2(k)$ (or in $\mathbb{A}^2(k)$ if normalizing for the sake of complete mixed additions is preferred), and to compute the main loop as follows. After computing each of the w consecutive doublings in $\mathbb{P}(2,3,1)(k)$, the running value is converted to $\mathbb{P}^2(k)$ at a cost of $2\mathbf{M} + 1\mathbf{S}$, then the result of a complete addition (between the running value and a lookup table element) is converted back to $\mathbb{P}(2,3,1)(k)$ at a cost of $2\mathbf{M} + 1\mathbf{S}$. Even for small window sizes that result in additions (and thus the back-and-forth conversions) occurring relatively often, the operation counts in Table 1 suggest that this trade-off will be favorable; and, for larger window sizes, the resulting scalar multiplication will be significantly faster than one that works entirely in $\mathbb{P}^2(k)$.

The only possible exception that could occur in the above routine is when the result of an addition is the point at infinity $(0 : 1 : 0) \in \mathbb{P}^2(k)$, since the conversion back to $\mathbb{P}(2,3,1)(k)$ fails here. Thus, this strategy should only be used if the scalar multiplication routine is such that the running value is never the inverse of any element in the lookup table, or if the conversion from $\mathbb{P}^2(k)$ to $\mathbb{P}(2,3,1)(k)$ is written to handle this possible exception in a constant-time fashion. In the former case, if (as in [16, Sect. 4.1]) this can only happen in the final addition, then the workaround is easy: either guarantee that the scalars cannot be a multiple of the group order (which rules out this possibility), or else do not apply the conversion back to $\mathbb{P}(2,3,1)(k)$ after the final addition.

5.2 Interoperability with Composite Order Curves

The IRTF CFRG recently selected two composite order curves as a recommendation to the TLS working group for inclusion in upcoming versions of TLS: Bernstein's Curve25519 [7] and Hamburg's Goldilocks [39]. The current IETF internet draft[12] specifies the wire format for these curves to be the u-coordinate corresponding to a point (u, v) on the Montgomery model of these curves $E_{\mathrm{M}}/\mathbb{F}_q\colon v^2 = u^3+Au^2+u$. Curve25519 has $q = 2^{255}-19$ with $A = 486662$ and Goldilocks has $q = 2^{448} - 2^{224} - 1$ with $A = 156326$.

Since our complete formulas are likely to be of interest to practitioners concerned with global interoperability, e.g., those investing a significant budget into one implementation that may be intended to support as many standardized curves as possible, we now show that Algorithm 1 can be adapted to interoperate with the composite order curves in upcoming TLS ciphersuites. We make no attempt to disguise the fact that this will come with a significant performance penalty over the Montgomery ladder, but in this case we are assuming that top performance is not the priority.

A trivial map from the Montgomery curve to a short Weierstrass curve is $\kappa\colon E_{\mathrm{M}} \to E$, $(u, v) \mapsto (x, y) = (u - A/3, v)$; here the short Weierstrass curve is $E\colon y^2 = x^3 + ax + b$, with $a = 1 - A^2/3$ and $b = A(2A^2 - 9)/27$.

Thus, a dedicated short Weierstrass implementation can interoperate with Curve25519 (resp. Goldilocks) as follows. After receiving the u-coordinate on the wire, set $x = u - A/3$ (i.e., add a fixed, global constant), and decompress to compute the corresponding y-coordinate on E via the square root $y = \sqrt{x^3 + ax + b}$ as usual; the choice of square root here does not matter. Setting $P = (x, y)$ and validating that $P \in E$, we can then call Algorithm 1 to compute 3 (resp. 2) successive doublings to get Q. This is in accordance with the scalars being defined with 3 (resp. 2) fixed zero bits to clear the cofactor [7]. The point Q is then multiplied by the secret part of the scalar (using, e.g., the methods we just described in Sect. 5.1), then normalized to give $Q = (x', y')$, and the Montgomery u-coordinate of the result is output as $u' = x' + A/3$.

Note that the above routine is exception free: Algorithm 1 only fails to add the points P_1 and P_2 when $P_1 - P_2$ is a point of exact order 2. Thus, it can be used for point doublings on all short Weierstrass curves (including those of even order). Furthermore, the point Q is in the prime order subgroup, so the subsequent scalar multiplication (which only encounters multiples of Q) cannot find a pair of points that are exceptional to Algorithm 1.

Finally, we note that although neither Curve25519 or Goldilocks are isomorphic to a Weierstrass curve with $a = -3$, both curves have simple isomorphisms to Weierstrass curves with small a values, e.g., $a = 2$ and $a = 1$, respectively. Making use of this would noticeably decrease the overhead of our complete formulas.

[12] See https://datatracker.ietf.org/doc/draft-irtf-cfrg-curves/.

5.3 The Cost of Completeness

In Table 2 we report the factor slowdown obtained when substituting the complete formulas in Algorithms 4–6 for OpenSSL's "`ec_GFp_simple_add`" and "`ec_GFp_simple_dbl`" functions inside the OpenSSL scalar multiplication routine for the five NIST prime curves (which all have $a = -3$).

Table 2. Number of ECDH operations in 10 s for the OpenSSL implementation of the five NIST prime curves, using complete and incomplete addition formulas. Timings were obtained by running the "`openssl speed ecdhpXXX`" command on an Intel Core i5-5300 CPU @ 2.30 GHz, averaged over 100 trials of 10 s each.

NIST curve	no. of ECDH operations (per 10 s) complete	incomplete	factor slowdown
P-192	35274	47431	1.34x
P-224	24810	34313	1.38x
P-256	21853	30158	1.38x
P-384	10109	14252	1.41x
P-521	4580	6634	1.44x

We intentionally left OpenSSL's scalar multiplication routines unaltered in order to provide an unbiased upper bound on the performance penalty that our complete algorithms will introduce. For the remainder of this subsection, we discuss why the performance difference is unlikely to be this large in many practical scenarios.

Referring to Table 3 (which, as well as the counts given in Table 1, includes the operation counts for mixed additions), we see that the mixed addition formulas in Jacobian coordinates are $4\mathbf{M} + 1\mathbf{S}$ faster than full additions, while for our complete formulas the difference is only $1\mathbf{M} + 6\mathbf{a}$. Thus, in Jacobian coordinates, it is often advantageous to normalize the lookup table (using one shared inversion [54]) in order to save $4\mathbf{M}+1\mathbf{S}$ per addition. On the other hand, in the case of the complete formulas, this will not be a favorable trade-off and (assuming there is ample cache space) it is likely to be better to leave all of the lookup elements in $\mathbb{P}^2$. The numbers reported in Table 2 use OpenSSL's scalar multiplication which does normalize the lookup table to use mixed additions, putting the complete formulas at a disadvantage.

As we mentioned in Sect. 1, the slowdowns reported in Table 2 (which were obtained on a 64-bit machine) are likely to be significantly less on low-end architectures where the relative cost of field additions drops. Furthermore, in embedded scenarios where implementations must be protected against more than just timing attacks, a common countermeasure is to randomize the projective coordinates of intermediate points [29]. In these cases, normalized lookup table elements could also give rise to side-channel vulnerabilities [33, Sect. 3.4–3.6], which would take mixed additions out of the equation. As Table 3 suggests, when full

Table 3. Operation counts for the prior incomplete addition algorithms and our complete ones, with the inclusion of mixed addition formulas. Credits for the incomplete formulas are the same as in Table 1, except for the additional mixed formulas which are, in homogeneous coordinates, due to Cohen, Miyaji and Ono [27], and in Jacobian coordinates, due to Hankerson, Menezes and Vanstone [40, p. 91].

addition formulas	a	ADD(P,Q)					mADD(P,Q)					DBL(P)				
		M	S	$\mathbf{m_a}$	$\mathbf{m_b}$	a	M	S	$\mathbf{m_a}$	$\mathbf{m_b}$	a	M	S	$\mathbf{m_a}$	$\mathbf{m_b}$	a
complete homogeneous	any	12	0	3	2	23	11	0	3	2	17	8	3	3	2	15
	-3	12	0	0	2	29	11	0	0	2	23	8	3	0	2	21
	0	12	0	0	2	19	11	0	0	2	13	6	2	0	1	9
incomplete homogeneous	any	12	2	0	0	7	9	2	0	0	7	5	6	1	0	12
	-3	12	2	0	0	7	9	2	0	0	7	7	3	0	0	11
	0	-					-					-				
incomplete Jacobian	any	12	4	0	0	7	8	3	0	0	7	3	6	1	0	13
	-3	12	4	0	0	7	8	3	0	0	7	4	4	0	0	8
	0	12	4	0	0	7	8	3	0	0	7	3	4	0	0	7

additions are used throughout, our complete algorithms will give much better performance relative to their incomplete counterparts.

Hardware implementations of ECC typically rely on using general field hardware multipliers that are often based on the algorithm of Montgomery [53]. These types of hardware modules use a multiplier for both multiplications and squarings [24,36], meaning that the squarings our addition algorithms save (over the prior formulas) are full multiplications. Moreover, hardware architectures that are based on Montgomery multiplication can benefit from modular additions/subtractions computed as non-modular operations. The concept is explained in [6], which is a typical ECC hardware architecture using the "relaxed" Montgomery parameter such that the conditional subtraction (from the original algorithm of Montgomery) can be omitted. In this way, the modular addition/subtraction is implemented not just very efficiently, but also as a time-constant operation. Using this approach implies the only cost to be taken into account is the one of modular multiplication, i.e., modular additions come almost for free. Similar conclusions apply for multiplications with a constant as they can be implemented very efficiently in hardware, assuming a constant is predefined and hence "hardwired". Again, viewing the operation counts in Table 3 suggests that such scenarios are, relatively speaking, likely to give a greater benefit to our complete algorithms.

Finally, we remark that runtime is not the only metric of concern to ECC practitioners; in fact, there was wide consensus (among both speakers and panelists) at the recent NIST workshop[13] that security and simplicity are far more important in real-world ECC than raw performance. While our complete algorithms are likely to be slower in some scenarios, we reiterate that complete formulas reign supreme in all other aspects, including total code size, ease of implementation, and issues relating to side-channel resistance.

[13] See http://www.nist.gov/itl/csd/ct/ecc-workshop.cfm.

Acknowledgements. Special thanks to Emmanuel Thomé who managed to save us $1\mathbf{m}_a$ in the explicit formulas in Sect. 3.1. We thank Joppe Bos and Patrick Longa for their feedback on an earlier version of this paper, and the anonymous Eurocrypt reviewers for their valuable comments.

References

1. Accredited Standards Committee X9. American National Standard X9.62-1999, Public Key Cryptography forthe Financial Services Industry: The Elliptic Curve Digital Signaturealgorithm (ECDSA) (1999) Draft at. http://grouper.ieee.org/groups/1363/Research/Other.html
2. Accredited Standards Committee X9. American National Standard X9.63-2001, Public key cryptography forthe financial services industry: key agreement and key transport usingelliptic curve cryptography (1999) Draft at. http://grouper.ieee.org/groups/1363/Research/Other.html
3. Agence nationale de la sécurité des sysèmes d'information(ANSSI).Mécanismes cryptographiques: Règles et recommandationsconcernant le choix et le dimensionnement des mécanismescryptographiques (2014). http://www.ssi.gouv.fr/uploads/2015/01/RGS_v-2-0_B1.pdf
4. Arène, C., Kohel, D., Ritzenthaler, C.: Complete addition laws on abelian varieties. LMS J. Comput. Math. **15**, 308–316 (2012)
5. Barreto, P.S.L.M., Naehrig, M.: Pairing-friendly elliptic curves of prime order. In: Preneel, B., Tavares, S. (eds.) SAC 2005. LNCS, vol. 3897, pp. 319–331. Springer, Heidelberg (2006)
6. Batina, L., Bruin-Muurling, G., Örs, S.: Flexible hardware design for RSA and elliptic curve cryptosystems. In: Okamoto, T. (ed.) CT-RSA 2004. LNCS, vol. 2964, pp. 250–263. Springer, Heidelberg (2004)
7. Bernstein, D.J.: Curve25519: new diffie-hellman speed records. In: Yung, M., Dodis, Y., Kiayias, A., Malkin, T. (eds.) PKC 2006. LNCS, vol. 3958, pp. 207–228. Springer, Heidelberg (2006)
8. Bernstein, D.J., Birkner, P., Joye, M., Lange, T., Peters, C.: Twisted edwards curves. In: Vaudenay, S. (ed.) AFRICACRYPT 2008. LNCS, vol. 5023, pp. 389–405. Springer, Heidelberg (2008)
9. Bernstein, D.J., Chuengsatiansup, C., Kohel, D., Lange, T.: Twisted hessian curves. In: Lauter, K., Rodríguez-Henríquez, F. (eds.) LatinCrypt 2015. LNCS, vol. 9230, pp. 269–294. Springer, Heidelberg (2015)
10. Bernstein, D.J., Chuengsatiansup, C., Lange, T.: Curve41417: karatsuba revisited. In: Batina, L., Robshaw, M. (eds.) CHES 2014. LNCS, vol. 8731, pp. 316–334. Springer, Heidelberg (2014)
11. Bernstein, D.J., Hamburg, M., Krasnova, A., Lange, T.: Elligator: elliptic-curve points indistinguishable from uniform random strings. In: Sadeghi, A., Gligor, V.D., Yung, M., (eds.) ACM SIGSAC Conference on Computer and Communications Security, CCS 2013, Berlin, Germany, 4–8 November 2013, pp. 967–980. ACM (2013)
12. Bernstein, D.J., Lange, T.: Faster addition and doubling on elliptic curves. In: Kurosawa, K. (ed.) ASIACRYPT 2007. LNCS, vol. 4833, pp. 29–50. Springer, Heidelberg (2007)
13. Bernstein, D.J., Lange, T.: Complete addition laws for elliptic curves. Talk at Algebra and Number Theory Seminar (Universidad Autonomo de Madrid) (2009) Slides at. http://cr.yp.to/talks/2009.04.17/slides.pdf

14. Bernstein, D.J., Lange, T.: Safecurves: choosing safe curves for elliptic-curve cryptography. http://safecurves.cr.yp.to/. Accessed 5 October 2015
15. Bernstein, D.J., Lange, T.: Explicit-Formulas Database. http://hyperelliptic.org/EFD/index.html. Accessed 3 October 2015
16. Bos, J.W., Costello, C., Longa, P., Naehrig, M.: Selecting elliptic curves for cryptography: An efficiency andsecurity analysis. J. Crypt. Eng. 1–28 (2015). http://dx.doi.org/10.1007/s13389-015-0097-y
17. Bos, J.W., Halderman, J.A., Heninger, N., Moore, J., Naehrig, M., Wustrow, E.: Elliptic curve cryptography in practice. In: Christin and Safavi-Naini [26], pp. 157–175
18. Bosma, W., Cannon, J.J., Playoust, C.: The magma algebra system I: the user language. J. Symb. Comput. **24**(3/4), 235–265 (1997)
19. Bosma, W., Lenstra, H.W.: Complete systems of two addition laws for elliptic curves. J. Number Theory **53**(2), 229–240 (1995)
20. Brier, E., Joye, M.: Weierstraß elliptic curves and side-channel attacks. In: Naccache, D., Paillier, P. (eds.) PKC 2002. LNCS, vol. 2274, pp. 335–345. Springer, Heidelberg (2002)
21. Brier, E., Joye, M.: Fast point multiplication on elliptic curves through isogenies. In: Fossorier, M.P.C., Høholdt, T., Poli, A. (eds.) AAECC 2003. LNCS, vol. 2643, pp. 43–50. Springer, Heidelberg (2003)
22. Certicom Research. SEC 2: Recommended Elliptic Curve Domain Parameters, Version 2.0 (2010). http://www.secg.org/sec2-v2.pdf
23. Certivox UK, Ltd. CertiVox Standard Curves. http://docs.certivox.com/docs/miracl/certivox-standard-curves. Accessed 9 September 2015
24. Chen, G., Bai, G., Chen, H.: A high-performance elliptic curve cryptographic processor for general curves over GF(p) based on a systolic arithmetic unit. IEEE Trans. Circ. Syst. II Express Briefs **54**(5), 412–416 (2007)
25. Chinese Commerical Cryptography Administration Office.SM2 Digital Signature Algorithm. See (2010). http://www.oscca.gov.cn/UpFile/2010122214836668.pdf and http://tools.ietf.org/html/draft-shen-sm2-ecdsa-02
26. Christin, N., Safavi-Naini, R. (eds.): FC 2014. LNCS, vol. 8437. Springer, Heidelberg (2014)
27. Cohen, H., Miyaji, A., Ono, T.: Efficient elliptic curve exponentiation using mixed coordinates. In: Ohta, K., Pei, D. (eds.) ASIACRYPT 1998. LNCS, vol. 1514, pp. 51–65. Springer, Heidelberg (1998)
28. Committee on National Security Systems (CNSS). Advisory Memorandum: Use of Public Standards for the Secure Sharingof Information Among National Security Systems (2015). https://www.cnss.gov/CNSS/openDoc.cfm?Q5ww0Xu+7kg/OpTB/R2/MQ==
29. Coron, J.-S.: Resistance against differential power analysis for elliptic curve cryptosystems. In: Koç, Ç.K., Paar, C. (eds.) CHES 1999. LNCS, vol. 1717, p. 292. Springer, Heidelberg (1999)
30. ECC Brainpool. ECC Brainpool Standard Curves and Curve Generation (2005). http://www.ecc-brainpool.org/download/Domain-parameters.pdf
31. Edwards, H.: A normal form for elliptic curves. Bull. Am. Math. Soc. **44**(3), 393–422 (2007)
32. Fan, J., Gierlichs, B., Vercauteren, F.: To infinity and beyond: combined attack on ECC using points of low order. In: Preneel, B., Takagi, T. (eds.) CHES 2011. LNCS, vol. 6917, pp. 143–159. Springer, Heidelberg (2011)

33. Fan, J., Verbauwhede, I.: An updated survey on secure ECC implementations: attacks, countermeasures and cost. In: Naccache, D. (ed.) Cryphtography and Security: From Theory to Applications. LNCS, vol. 6805, pp. 265–282. Springer, Heidelberg (2012)
34. Gallant, R.P., Lambert, R.J., Vanstone, S.A.: Faster point multiplication on elliptic curves with efficient endomorphisms. In: Kilian, J. (ed.) CRYPTO 2001. LNCS, vol. 2139, pp. 190–200. Springer, Heidelberg (2001)
35. Government Committee of Russia for Standards.Information technology. Cryptographic data security. Signature andverification processes of [electronic] digital signature (2001). See. https://tools.ietf.org/html/rfc5832
36. Güneysu, T., Paar, C.: Ultra high performance ECC over NIST primes on commercial FPGAs. In: Oswald, E., Rohatgi, P. (eds.) CHES 2008. LNCS, vol. 5154, pp. 62–78. Springer, Heidelberg (2008)
37. Hamburg, M.: Twisting Edwards curves with isogenies. Cryptology ePrint Archive, report 2014/027 (2014). http://eprint.iacr.org/
38. Hamburg, M.: Decaf: eliminating cofactors through point compression. In: Gennaro, R., Robshaw, M. (eds.) Advances in Cryptology – CRYPTO 2015. LNCS, vol. 9215, pp. 705–723. Springer, Heidelberg (2015)
39. Hamburg, M.: Ed448-Goldilocks, a new elliptic curve. Cryptology ePrint Archive, report 2015/625 (2015). http://eprint.iacr.org/
40. Hankerson, D., Menezes, A.J., Vanstone, S.: Guide to Elliptic Curve Cryptography. Springer Science and Business Media, Heidelberg (2006)
41. Hisil, H.: Elliptic curves, group law, and efficient computation, Ph.D. thesis, Queensland University of Technology (2010). http://eprints.qut.edu.au/33233/
42. Hu, Z., Longa, P., Xu, M.: Implementing the 4-dimensional GLV method on GLS elliptic curves with j-invariant 0. Des. Codes Crypt. **63**(3), 331–343 (2012)
43. Izu, T., Takagi, T.: Exceptional procedure attack on elliptic curve cryptosystems. In: Desmedt, Y. (ed.) Public Key Cryptography - PKC 2003. LNCS, vol. 2567, pp. 224–239. Springer, Heidelberg (2003)
44. Joye, M., Tunstall, M.: Exponent recoding and regular exponentiation algorithms. In: Preneel, B. (ed.) AFRICACRYPT 2009. LNCS, vol. 5580, pp. 334–349. Springer, Heidelberg (2009)
45. Käsper, E.: Fast elliptic curve cryptography in openSSL. In: Danezis, G., Dietrich, S., Sako, K. (eds.) FC 2011 Workshops 2011. LNCS, vol. 7126, pp. 27–39. Springer, Heidelberg (2012)
46. Kocher, P.C.: Timing attacks on implementations of diffie-hellman, RSA, DSS, and other systems. In: Koblitz, N. (ed.) CRYPTO 1996. LNCS, vol. 1109, pp. 104–113. Springer, Heidelberg (1996)
47. Kohel, D.: Addition law structure of elliptic curves. J. Number Theory **131**(5), 894–919 (2011)
48. Lange, H., Ruppert, W.: Complete systems of addition laws on abelian varieties. Inventiones mathematicae **79**(3), 603–610 (1985)
49. Lim, C.H., Lee, P.J.: More flexible exponentiation with precomputation. In: Desmedt, Y.G. (ed.) CRYPTO 1994. LNCS, vol. 839, pp. 95–107. Springer, Heidelberg (1994)
50. Liu, Z., Seo, H., Großschädl, J., Kim, H.: Efficient implementation of NIST-compliant elliptic curve cryptography for sensor nodes. In: Qing, S., Zhou, J., Liu, D. (eds.) ICICS 2013. LNCS, vol. 8233, pp. 302–317. Springer, Heidelberg (2013)
51. Longa, P., Gebotys, C.: Efficient techniques for high-speed elliptic curve cryptography. In: Mangard, S., Standaert, F. (eds.) CHES 2010. LNCS, vol. 6225, pp. 80–94. Springer, Heidelberg (2010)

52. Miller, V.S.: Use of elliptic curves in cryptography. In: Williams, H.C. (ed.) CRYPTO 1985. LNCS, vol. 218, pp. 417–426. Springer, Heidelberg (1986)
53. Montgomery, P.L.: Modular multiplication without trial division. Math. Comput. **44**(170), 519–521 (1985)
54. Montgomery, P.L.: Speeding the pollard and elliptic curve methods of factorization. Math. Comput. **48**(177), 243–264 (1987)
55. National Institute for Standards and Technology (NIST). Digital Signature Standard. Federal Information Processing Standards Publication 186-2 (2000). http://csrc.nist.gov/publications/fips/archive/fips186-2/fips186-2.pdf
56. National Institute for Standards and Technology (NIST). Digital Signature Standard. Federal Information Processing Standards Publication 186-4 (2013). http://nvlpubs.nist.gov/nistpubs/FIPS/NIST.FIPS.186-4.pdf
57. Okeya, K., Takagi, T.: The width-w NAF method provides small memory and fast elliptic scalar multiplications secure against side channel attacks. In: Joye, M. (ed.) CT-RSA 2003. LNCS, vol. 2612, pp. 328–342. Springer, Heidelberg (2003)
58. Renes, J., Costello, C., Batina, L.: Complete addition formulas for prime order elliptic curves. Cryptology ePrint Archive, report 2015/1060 (2015). http://eprint.iacr.org/
59. Szczechowiak, P., Oliveira, L.B., Scott, M., Collier, M., Dahab, R.: NanoECC: testing the limits of elliptic curve cryptography in sensor networks. In: Verdone, R. (ed.) EWSN 2008. LNCS, vol. 4913, pp. 305–320. Springer, Heidelberg (2008)
60. Renes, J., Costello, C., Batina, L.: Complete addition formulas for prime order elliptic curves. Cryptology ePrint Archive, report 2015/1060 (2015). http://eprint.iacr.org/
61. Tibouchi, M.: Elligator squared: Uniform points on elliptic curves of prime order as uniform random strings. In: Christin and Safavi-Naini [26], pp. 139–156

New Complexity Trade-Offs for the (Multiple) Number Field Sieve Algorithm in Non-Prime Fields

Palash Sarkar and Shashank Singh(✉)

Applied Statistics Unit, Indian Statistical Institute, Kolkata, India
palash@isical.ac.in, sha2nk.singh@gmail.com

Abstract. The selection of polynomials to represent number fields crucially determines the efficiency of the Number Field Sieve (NFS) algorithm for solving the discrete logarithm in a finite field. An important recent work due to Barbulescu et al. builds upon existing works to propose two new methods for polynomial selection when the target field is a non-prime field. These methods are called the generalised Joux-Lercier (GJL) and the Conjugation methods. In this work, we propose a new method (which we denote as $\mathcal{A}$) for polynomial selection for the NFS algorithm in fields $\mathbb{F}_Q$, with $Q = p^n$ and $n > 1$. The new method both subsumes and generalises the GJL and the Conjugation methods and provides new trade-offs for both n composite and n prime. Let us denote the variant of the (multiple) NFS algorithm using the polynomial selection method "X" by (M)NFS-X. Asymptotic analysis is performed for both the NFS-$\mathcal{A}$ and the MNFS-$\mathcal{A}$ algorithms. In particular, when $p = L_Q(2/3, c_p)$, for $c_p \in [3.39, 20.91]$, the complexity of NFS-$\mathcal{A}$ is better than the complexities of all previous algorithms whether classical or MNFS. The MNFS-$\mathcal{A}$ algorithm provides lower complexity compared to NFS-$\mathcal{A}$ algorithm; for $c_p \in (0, 1.12] \cup [1.45, 3.15]$, the complexity of MNFS-$\mathcal{A}$ is the same as that of the MNFS-Conjugation and for $c_p \notin (0, 1.12] \cup [1.45, 3.15]$, the complexity of MNFS-$\mathcal{A}$ is lower than that of all previous methods.

1 Introduction

Let $\mathfrak{G} = \langle \mathfrak{g} \rangle$ be a finite cyclic group. The discrete log problem (DLP) in $\mathfrak{G}$ is the following. Given $(\mathfrak{g}, \mathfrak{h})$, compute the minimum non-negative integer $\mathfrak{e}$ such that $\mathfrak{h} = \mathfrak{g}^{\mathfrak{e}}$. For appropriately chosen groups $\mathfrak{G}$, the DLP in $\mathfrak{G}$ is believed to be computationally hard. This forms the basis of security of many important cryptographic protocols.

Studying the hardness of the DLP on subgroups of the multiplicative group of a finite field is an important problem. There are two general algorithms for tackling the DLP on such groups. These are the function field sieve (FFS) [1,2, 16,18] algorithm and the number field sieve (NFS) [11,17,19] algorithm. Both these algorithms follow the framework of index calculus algorithms which is currently the standard approach for attacking the DLP in various groups.

M. Fischlin and J.-S. Coron (Eds.): EUROCRYPT 2016, Part I, LNCS 9665, pp. 429–458, 2016.
DOI: 10.1007/978-3-662-49890-3_17

For small characteristic fields, the FFS algorithm leads to a quasi-polynomial running time [6]. Using the FFS algorithm outlined in [6,15], Granger et al. [12] reported a record computation of discrete log in the binary extension field $\mathbb{F}_{2^{9234}}$. FFS also applies to the medium characteristic fields. Some relevant works along this line are reported in [14,18,25].

For medium to large characteristic finite fields, the NFS algorithm is the state-of-the-art. In the context of the DLP, the NFS was first proposed by Gordon [11] for prime order fields. The algorithm proceeded via number fields and one of the main difficulties in applying the NFS was in the handling of units in the corresponding ring of algebraic integers. Schirokauer [26,28] proposed a method to bypass the problems caused by units. Further, Schirokauer [27] showed the application of the NFS algorithm to composite order fields. Joux and Lercier [17] presented important improvements to the NFS algorithm as applicable to prime order fields.

Joux, Lercier, Smart and Vercauteren [19] later showed that the NFS algorithm is applicable to all finite fields. Since then, several works [5,13,20,24] have gradually improved the NFS in the context of medium to large characteristic finite fields.

The efficiency of the NFS algorithm is crucially dependent on the properties of the polynomials used to construct the number fields. Consequently, polynomial selection is an important step in the NFS algorithm and is an active area of research. The recent work [5] by Barbulescu et al. extends a previous method [17] for polynomial selection and also presents a new method. The extension of [17] is called the generalised Joux-Lercier (GJL) method while the new method proposed in [5] is called the Conjugation method. The paper also provides a comprehensive comparison of the trade-offs in the complexity of the NFS algorithm offered by the various polynomial selection methods.

The NFS based algorithm has been extended to multiple number field sieve algorithm (MNFS). The work [8] showed the application of the MNFS to medium to high characteristic finite fields. Pierrot [24] proposed MNFS variants of the GJL and the Conjugation methods. For more recent works on NFS we refer to [4,7,22].

Our contributions: In this work, we build on the works of [5,17] to propose a new method of polynomial selection for NFS over $\mathbb{F}_{p^n}$. The new method both subsumes and generalises the GJL and the Conjugation methods. There are two parameters to the method, namely a divisor d of the extension degree n and a parameter $r \geq k$ where $k = n/d$.

For $d = 1$, the new method becomes the same as the GJL method. For $d = n$ and $r = k = 1$, the new method becomes the same as the Conjugation method. For $d = n$ and $r > 1$; or, for $1 < d < n$, the new method provides polynomials which leads to different trade-offs than what was previously known. Note that the case $1 < d < n$ can arise only when n is composite, though the case $d = n$ and $r > 1$ arises even when n is prime. So, the new method provides new trade-offs for both n composite and n prime.

Following the works of [5,24] we carry out an asymptotic analysis of new method for the classical NFS as well as for MNFS. For the medium and the large characteristic cases, the results for the new method are exactly the same as those obtained for existing methods in [5,24]. For the boundary case, however, we obtain some interesting asymptotic results. Letting $Q = p^n$, the subexponential expression $L_Q(a, c)$ is defined to be the following:

$$L_Q(a, c) = \exp\left((c + o(1))(\ln Q)^a (\ln \ln Q)^{1-a}\right). \tag{1}$$

Write $p = L_Q(2/3, c_p)$ and let θ_0 and θ_1 be such that the complexity of the MNFS-Conjugation method is $L_Q(1/3, \theta_0)$ and the complexity of the MNFS-GJL method is $L_Q(1/3, \theta_1)$. As shown in [24], $L_Q(1/3, \theta_0)$ is the minimum complexity of MNFS[1] while for $c_p > 4.1$, complexity of new method (MNFS-$\mathcal{A}$) is lower than the complexity $L_Q(1/3, \theta_1)$ of MNFS-GJL method.

The classical variant of the new method, (i.e., NFS-$\mathcal{A}$) itself is powerful enough to provide better complexity than all previously known methods, whether classical or MNFS, for $c_p \in [3.39, 20.91]$. The MNFS variant of the new method provides lower complexity compared to the classical variant of the new method for all c_p.

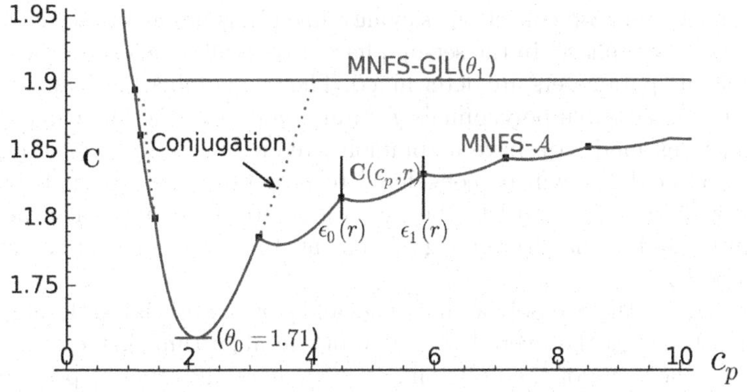

Fig. 1. Complexity plot for MNFS boundary case

The complexity of MNFS-$\mathcal{A}$ with $k = 1$ and using linear sieving polynomials can be written as $L_Q(1/3, \mathbf{C}(c_p, r))$, where $\mathbf{C}(c_p, r)$ is a function of c_p and a parameter r. For every integer $r \geq 1$, there is an interval $[\epsilon_0(r), \epsilon_1(r)]$ such that for $c_p \in [\epsilon_0(r), \epsilon_1(r)]$, $\mathbf{C}(c_p, r) < \mathbf{C}(c_p, r')$ for $r \neq r'$. Further, for a fixed r, let $C(r)$ be the minimum value of $\mathbf{C}(c_p, r)$ over all c_p. We show that $C(r)$ is monotone increasing for $r \geq 1$; $C(1) = \theta_0$; and that $C(r)$ is bounded above by θ_1 which is its limit as r goes to infinity. So, for the new method the minimum complexity is the same as MNFS-Conjugation method. On the other hand, as r

[1] The value of θ_0 obtained in [24] is incorrect.

increases, the complexity of MNFS-$\mathcal{A}$ remains lower than the complexities of all the prior known methods. In particular, the complexity of MNFS-$\mathcal{A}$ interpolates nicely between the complexity of the MNFS-GJL and the minimum possible complexity of the MNFS-Conjugation method. This is depicted in Fig. 1. In Fig. 4 of Sect. 8.1, we provide a more detailed plot of the complexity of MNFS-$\mathcal{A}$ in the boundary case.

The complete statement regarding the complexity of MNFS-$\mathcal{A}$ in the boundary case is the following. For $c_p \in (0, 1.12] \cup [1.45, 3.15]$, the complexity of MNFS-$\mathcal{A}$ is the same as that of MNFS-Conjugation; for $c_p \notin (0, 1.12] \cup [1.45, 3.15]$, the complexity of MNFS-$\mathcal{A}$ is lower than that of all previous methods. In particular, the improvements for c_p in the range $(1.12, 1.45)$ is obtained using $k = 2$ and 3; while the improvements for $c_p > 3.15$ is obtained using $k = 1$ and $r > 1$. In all cases, the minimum complexity is obtained using linear sieving olynomials.

2 Background on NFS for Non-Prime Fields

We provide a brief sketch of the background on the variant of the NFS algorithm that is applicable to the extension fields $\mathbb{F}_Q$, where $Q = p^n$, p is a prime and $n > 1$. More detailed discussions can be found in [5,17].

Following the structure of index calculus algorithms, NFS has three main phases, namely, relation collection (sieving), linear algebra and descent. Prior to these, is the set-up phase. In the set-up phase, two number fields are constructed and the sieving parameters are determined. The two number fields are set up by choosing two irreducible polynomials $f(x)$ and $g(x)$ over the integers such that their reductions modulo p have a common irreducible factor $\varphi(x)$ of degree n over $\mathbb{F}_p$. The field $\mathbb{F}_{p^n}$ will be considered to be represented by $\varphi(x)$. Let $\mathfrak{g}$ be a generator of $\mathfrak{G} = \mathbb{F}_{p^n}^{\star}$ and let q be the largest prime dividing the order of $\mathfrak{G}$. We are interested in the discrete log of elements of $\mathfrak{G}$ to the base $\mathfrak{g}$ modulo this largest prime q.

The choices of the two polynomials $f(x)$ and $g(x)$ are crucial to the algorithm. These greatly affect the overall run time of the algorithm. Let $\alpha, \beta \in \mathbb{C}$ and $m \in \mathbb{F}_{p^n}$ be the roots of the polynomials $f(x)$, $g(x)$ and $\varphi(x)$ respectively. We further let $l(f)$ and $l(g)$ denote the leading coefficient of the polynomials $f(x)$ and $g(x)$ respectively. The two number fields and the finite field are given as follows.

$$\mathbb{K}_1 = \mathbb{Q}(\alpha) = \frac{\mathbb{Q}[x]}{\langle f(x)\rangle}, \mathbb{K}_2 = \mathbb{Q}(\beta) = \frac{Q[x]}{\langle g(x)\rangle} \text{ and } \mathbb{F}_{p^n} = \mathbb{F}_p(m) = \frac{\mathbb{F}_p[x]}{\langle \varphi(x)\rangle}.$$

Thus, we have the following commutative diagram shown in Fig. 2, where we represent the image of $\xi \in \mathbb{Z}(\alpha)$ or $\xi \in \mathbb{Z}(\beta)$ in the finite field $\mathbb{F}_{p^n}$ by $\overline{\xi}$. Actual computations are carried out over these number fields and are then transformed to the finite field via these homomorphisms. In fact, instead of doing the computations over the whole number field $\mathbb{K}_i$, one works over its ring of algebraic integers $\mathcal{O}_i$. These integer rings provide a nice way of constructing a factor basis and moreover, unique factorisation of ideals holds over these rings.

The factor basis $\mathcal{F} = \mathcal{F}_1 \cup \mathcal{F}_2$ is chosen as follows.

$$\mathcal{F}_1 = \left\{ \begin{array}{c} \text{prime ideals } \mathfrak{q}_{1,j} \text{ in } \mathcal{O}_1, \text{ either having norm less than } B \\ \text{or lying above the prime factors of } l(f) \end{array} \right\}$$

$$\mathcal{F}_2 = \left\{ \begin{array}{c} \text{prime ideals } \mathfrak{q}_{2,j} \text{ in } \mathcal{O}_2, \text{ either having norm less than } B \\ \text{or lying above the prime factors of } l(g) \end{array} \right\}$$

where B is the smoothness bound and is to be chosen appropriately. An algebraic integer is said to be B-smooth if the principal ideal generated by it factors into the prime ideals of norms less than B. As mentioned in the paper [5], independently of choice of f and g, the size of the factor basis is $B^{1+o(1)}$. For asymptotic computations, this is simply taken to be B. The work flow of NFS can be understood by the diagram in Fig. 2.

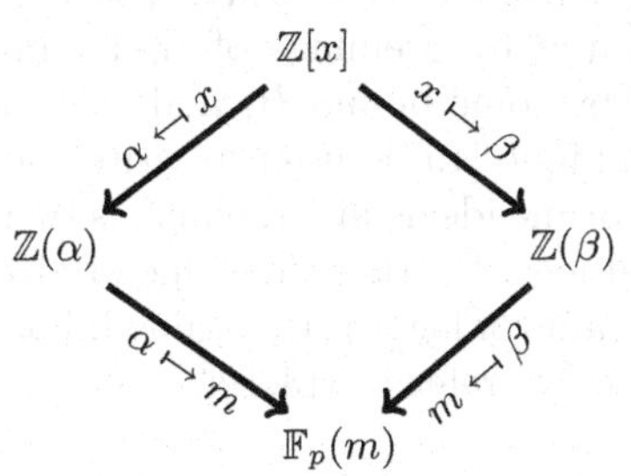

Fig. 2. A work-flow of NFS.

A polynomial $\phi(x) \in \mathbb{Z}[x]$ of degree at most $t-1$ (i.e. having t coefficients) is chosen and the principal ideals generated by its images in the two number fields are checked for smoothness. If both of them are smooth, then

$$\phi(\alpha)\mathcal{O}_1 = \prod_j \mathfrak{q}_{1,j}{}^{e_j} \text{ and } \phi(\beta)\mathcal{O}_2 = \prod_j \mathfrak{q}_{2,j}{}^{e'_j} \tag{2}$$

where $\mathfrak{q}_{1,j}$ and $\mathfrak{q}_{2,j}$ are prime ideals in $\mathcal{F}_1$ and $\mathcal{F}_2$ respectively. For $i = 1, 2$, let h_i denote the class number of $\mathcal{O}_i$ and r_i denote the torsion-free rank of $\mathcal{O}_i^\star$. Then, for some $\varepsilon_{i,j} \in \mathfrak{q}_{i,j}$ and units $u_{i,j} \in \mathcal{O}_i^\star$, we have

$$\log_g \overline{\phi(\alpha)} \equiv \sum_{j=1}^{r_1} \lambda_{1,j}(\phi(\alpha))\Lambda_{1,j} + \sum_j e_j X_{1,j} \pmod{q}, \tag{3}$$

$$\log_g \overline{\phi(\beta)} \equiv \sum_{j=1}^{r_2} \lambda_{2,j}(\phi(\beta))\Lambda_{2,j} + \sum_j e'_j X_{2,j} \pmod{q}, \tag{4}$$

where for $i = 1, 2$ and $j = 1 \ldots r_i$, $\Lambda_{i,j} = \log_g \overline{u_{i,j}}$ is an unknown **virtual logarithm** of the unit $u_{i,j}$, $X_{i,j} = h_i^{-1} \log_g \overline{\varepsilon_{i,j}}$ is an unknown **virtual logarithm** of prime ideal $\mathfrak{q}_{i,j}$ and $\lambda_{i,j} : \mathcal{O}_i \mapsto \mathbb{Z}/q\mathbb{Z}$ is Schirokauer map [19,26,28]. We skip the details of virtual logarithms and Schirokauer maps, as these details will not affect the polynomial selection problem considered in this work.

Since $\overline{\phi(\alpha)} = \overline{\phi(\beta)}$, we have

$$\sum_{j=1}^{r_1} \lambda_{1,j}\left(\phi(\alpha)\right)\Lambda_{1,j} + \sum_j e_j X_{1,j} \equiv \sum_{j=1}^{r_2} \lambda_{2,j}\left(\phi(\beta)\right)\Lambda_{2,j} + \sum_j e_j' X_{2,j} (\bmod\ q) \quad (5)$$

The relation given by (5) is a linear equation modulo q in the unknown virtual logs. More than $(\#\mathcal{F}_1 + \#\mathcal{F}_2 + r_1 + r_2)$ such relations are collected by sieving over suitable $\phi(x)$. The linear algebra step solves the resulting system of linear equations using either the Lanczos or the block Wiedemann algorithms to obtain the virtual logs of factor basis elements.

After the linear algebra phase is over, the descent phase is used to compute the discrete logs of the given elements of the field $\mathbb{F}_{p^n}$. For a given element $\mathfrak{y}$ of $\mathbb{F}_{p^n}$, one looks for an element of the form $\mathfrak{y}^i\mathfrak{g}^j$, for some $i, j \in \mathbb{N}$, such that the principal ideal generated by preimage of $(\mathfrak{y}^i\mathfrak{g}^j)$ in $\mathcal{O}_1$, factors into prime ideals of norms bounded by some bound B_1 and of degree at most $t-1$. Then the special-$\mathfrak{q}$ descent technique [19] is used to write the ideal generated by the preimage as a product of prime ideals in $\mathcal{F}_1$, which is then converted into a linear equation involving virtual logs. Putting the value of virtual logs, obtained after linear algebra phase, the value of $\log_{\mathfrak{g}}(\mathfrak{y})$ is obtained. For more details and recent work on the descent phase, we refer to [13,19].

3 Polynomial Selection and Sizes of Norms

It is evident from the description of NFS that the relation collection phase requires polynomials $\phi(x) \in \mathbb{Z}[x]$ whose images in the two number fields are simultaneously smooth. For ensuring the smoothness of $\phi(\alpha)$ and $\phi(\beta)$, it is enough to ensure that their norms viz, $\text{Res}(f, \phi)$ and $\text{Res}(g, \phi)$ are B-smooth. We refer to [5] for further explanations.

Using the Corollary 2 of Kalkbrener's work [21], we have the following upper bound for the absolute value of the norm.

$$|\text{Res}(f, \phi)| \le \kappa\left(\deg f, \deg \phi\right) \|f\|_\infty^{\deg \phi} \|\phi\|_\infty^{\deg f} \quad (6)$$

where $\kappa(a, b) = \binom{a+b}{a}\binom{a+b-1}{a}$ and $\|f\|_\infty$ is maximum of the absolute values of the coefficients of f.

Following [5], let E be such that the coefficients of ϕ are in $\left[-\frac{1}{2}E^{2/t}, \frac{1}{2}E^{2/t}\right]$. So, $\|\phi\|_\infty \approx E^{2/t}$ and the number of polynomials $\phi(x)$ that is considered for the sieving is E^2. Whenever $p = L_Q(a, c_p)$ with $a > \frac{1}{3}$, we have the following bound on the $\text{Res}(f, \phi) \times \text{Res}(g, \phi)$ (for details we refer to [5]).

$$|\text{Res}(f, \phi) \times \text{Res}(g, \phi)| \approx \left(\|f\|_\infty \|g\|_\infty\right)^{t-1} E^{(\deg f + \deg g)2/t}. \quad (7)$$

For small values of n, the sieving polynomial $\phi(x)$ is taken to be linear, i.e., $t = 2$ and then the norm bound becomes approximately $\|f\|_\infty \|g\|_\infty E^{(\deg f + \deg g)}$.

The methods for choosing f and g result in the coefficients of one or both of these polynomials to depend on Q. So, the right hand side of (7) is determined by Q and E. All polynomial selection algorithms try to minimize the RHS of (7). From the bound in (7), it is evident that during polynomial selection, the goal should be to try and keep the degrees and the coefficients of both f and g to be small. Ensuring both degrees and coefficients to be small is a nontrivial task and leads to a trade-off. Previous methods for polynomial selections provide different trade-offs between the degrees and the coefficients. Estimates of Q-E trade-off values have been provided in [5] and is based on the CADO factoring software [3]. Table 1 reproduces these values where $Q(dd)$ represents the number of decimal digits in Q.

Table 1. Estimate of Q-E values [5].

Q(dd)	100	120	140	160	180	200	220	240	260	280	300
Q(bits)	333	399	466	532	598	665	731	798	864	931	997
E(bits)	20.9	22.7	24.3	25.8	27.2	28.5	29.7	30.9	31.9	33.0	34.0

As mentioned in [5,13], presently the following three polynomial selection methods provide competitive trade-offs.

1. **JLSV1:** Joux, Lercier, Smart, Vercauteren method [19].
2. **GJL:** Generalised Joux Lercier method [5,23].
3. **Conjugation method** [5].

Brief descriptions of these methods are given below.

JLSV1. Repeat the following steps until f and g are obtained to be irreducible over $\mathbb{Z}$ and φ is irreducible over $\mathbb{F}_p$.

1. Randomly choose polynomials $f_0(x)$ and $f_1(x)$ having small coefficients with $\deg(f_1) < \deg(f_0) = n$.
2. Randomly choose an integer ℓ to be slightly greater than $\lceil \sqrt{p} \rceil$.
3. Let (u, v) be the rational reconstruction of ℓ in $\mathbb{F}_p$, i.e., $\ell \equiv u/v \mod p$.
4. Define $f(x) = f_0(x) + \ell f_1(x)$ and $g(x) = v f_0(x) + u f_1(x)$ and $\varphi(x) = f(x) \mod p$.

Note that $\deg(f) = \deg(g) = n$ and both $\|f\|_\infty$ and $\|g\|_\infty$ are $O\left(p^{1/2}\right) = O\left(Q^{1/(2n)}\right)$ and so (7) becomes $E^{4n/t} Q^{(t-1)/n}$ which is $E^{2n} Q^{1/n}$ for $t = 2$.

GJL. The basic Joux-Lercier method [17] works for prime fields. The generalised Joux-Lercier method extends the basic Joux-Lercier method to work over composite fields $\mathbb{F}_{p^n}$.

The heart of the GJL method is the following idea. Let $\varphi(x)$ be a monic polynomial $\varphi(x) = x^n + \varphi_{n-1} x^{n-1} + \cdots + \varphi_1 x + \varphi_0$ and $r \geq \deg(\varphi)$ be an integer. Let $n = \deg(\varphi)$. Given $\varphi(x)$ and r, define an $(r+1) \times (r+1)$ matrix $M_{\varphi,r}$ in the following manner.

$$M_{\varphi,r} = \begin{bmatrix} p & & & & & & \\ & \ddots & & & & & \\ & & \ddots & & & & \\ & & & p & & & \\ \varphi_0 & \varphi_1 & \cdots & \varphi_{n-1} & 1 & & \\ & \ddots & \ddots & & \ddots & & \\ & & \varphi_0 & \varphi_1 & \cdots & \varphi_{n-1} & 1 \end{bmatrix} \tag{8}$$

The first $n \times n$ principal sub-matrix of $M_{\varphi,r}$ is diag$[p, p, \ldots, p]$ corresponding to the polynomials $p, px, \ldots, px^{n-1}$. The last $r - n + 1$ rows correspond to the polynomials $\varphi(x), x\varphi(x), \ldots, x^{r-n}\varphi(x)$.

Apply the LLL algorithm to $M_{\varphi,r}$ and let the first row of the resulting LLL-reduced matrix be $[g_0, g_1, \ldots, g_{r-1}, g_r]$. Define

$$g(x) = g_0 + g_1 x + \cdots + g_{r-1}x^{r-1} + g_r x^r. \tag{9}$$

The notation

$$g = \text{LLL}\left(M_{\varphi,r}\right) \tag{10}$$

will be used to denote the polynomial $g(x)$ given by (9). By construction, $\varphi(x)$ is a factor of $g(x)$ modulo p.

The GJL procedure for polynomial selection is the following. Choose an $r \geq n$ and repeat the following steps until f and g are irreducible over $\mathbb{Z}$ and φ is irreducible over $\mathbb{F}_p$.

1. Randomly choose a degree $(r+1)$-polynomial $f(x)$ which is irreducible over $\mathbb{Z}$ and having coefficients of size $O(\ln(p))$ such that $f(x)$ has a factor $\varphi(x)$ of degree n modulo p which is both monic and irreducible.
2. Let $\varphi(x) = x^n + \varphi_{n-1}x^{n-1} + \cdots + \varphi_1 x + \varphi_0$ and $M_{\varphi,r}$ be the $(r+1) \times (r+1)$ matrix given by (8).
3. Let $g(x) = \text{LLL}\left(M_{\varphi,r}\right)$.

The polynomial $f(x)$ has degree $r+1$ and $g(x)$ has degree r. The procedure is parameterised by the integer r.

The determinant of M is p^n and so from the properties of the LLL-reduced basis, the coefficients of $g(x)$ are of the order $O\left(p^{n/(r+1)}\right) = O\left(Q^{1/(r+1)}\right)$. The coefficients of $f(x)$ are $O(\ln p)$.

The bound on the norm given by (7) in this case is $E^{2(2r+1)/t}Q^{(t-1)/(r+1)}$ which becomes $E^{2r+1}Q^{1/(r+1)}$ for $t = 2$. Increasing r reduces the size of the coefficients of $g(x)$ at the cost of increasing the degrees of f and g. In the concrete example considered in [5] and also in [24], r has been taken to be n and so M is an $(n+1) \times (n+1)$ matrix.

Conjugation. Repeat the following steps until f and g are irreducible over $\mathbb{Z}$ and φ is irreducible over $\mathbb{F}_p$.

1. Choose a quadratic monic polynomial $\mu(x)$, having coefficients of size $O(\ln p)$, which is irreducible over $\mathbb{Z}$ and has a root $\mathfrak{t}$ in $\mathbb{F}_p$.
2. Choose two polynomials $g_0(x)$ and $g_1(x)$ with small coefficients such that $\deg g_1 < \deg g_0 = n$.
3. Let (u, v) be a rational reconstruction of $\mathfrak{t}$ modulo p, i.e., $\mathfrak{t} \equiv u/v \mod p$.
4. Define $g(x) = vg_0(x) + ug_1(x)$ and $f(x) = \mathrm{Res}_y\big(\mu(y), g_0(x) + y\, g_1(x)\big)$.

Note that $\deg(f) = 2n$, $\deg(g) = n$, $\|f\|_\infty = O(\ln p)$ and $\|g\|_\infty = O(p^{1/2}) = O(Q^{1/(2n)})$. In this case, the bound on the norm given by (7) is $E^{6n/t}Q^{(t-1)/(2n)}$ which becomes $E^{3n}Q^{1/(2n)}$ for $t = 2$.

4 A Simple Observation

For the GJL method, while constructing the matrix M, the coefficients of the polynomial $\varphi(x)$ are used. If, however, some of these coefficients are zero, then these may be ignored. The idea is given by the following result.

Proposition 1. *Let n be an integer, d a divisor of n and $k = n/d$. Suppose $A(x)$ is a monic polynomial of degree k. Let $r \geq k$ be an integer and set $\psi(x) = \mathrm{LLL}(M_{A,r})$. Define $g(x) = \psi(x^d)$ and $\varphi(x) = A(x^d)$. Then*

1. $\deg(\varphi) = n$ and $\deg(g) = rd$;
2. $\varphi(x)$ is a factor of $g(x)$ modulo p;
3. $\|g\|_\infty = p^{n/(d(r+1))}$.

Proof. The first point is straightforward. Note that by construction $A(x)$ is a factor of $\psi(x)$ modulo p. So, $A(x^d)$ is a factor of $\psi(x^d) = g(x)$ modulo p. This shows the second point. The coefficients of $g(x)$ are the coefficients of $\psi(x)$. Following the GJL method, $\|\psi\|_\infty = p^{k/(r+1)} = p^{n/(d(r+1))}$ and so the same holds for $\|g\|_\infty$. This shows the third point. □

Note that if we had defined $g(x) = \mathrm{LLL}(M_{\varphi,rd})$, then $\|g\|_\infty$ would have been $p^{n/(rd+1)}$. For $d > 1$, the value of $\|g\|_\infty$ given by Proposition 1 is smaller.

A Variant. The above idea shows how to avoid the zero coefficients of $\varphi(x)$. A similar idea can be used to avoid the coefficients of $\varphi(x)$ which are small. Suppose that the polynomial $\varphi(x)$ can be written in the following form.

$$\varphi(x) = \varphi_{i_1}x^{i_1} + \cdots + \varphi_{i_k}x^{i_k} + x^n + \sum_{j \notin \{i_1,\ldots,i_k\}} \varphi_j x^j \tag{11}$$

where $i_1, \ldots, i_k$ are from the set $\{0, \ldots, n-1\}$ and for $j \in \{0, \ldots, n-1\} \setminus \{i_1, \ldots, i_k\}$, the coefficients φ_j are all $O(1)$. Some or even all of these φ_j's could be zero. A $(k+1) \times (k+1)$ matrix M is constructed in the following manner.

$$M = \begin{bmatrix} p & & & & \\ & \ddots & & & \\ & & \ddots & & \\ & & & p & \\ \varphi_{i_1} & \varphi_{i_2} & \cdots & \varphi_{i_k} & 1 \end{bmatrix} \tag{12}$$

The matrix M has only one row obtained from $\varphi(x)$ and it is difficult to use more than one row. Apply the LLL algorithm to M and write the first row of the resulting LLL-reduced matrix as $[g_{i_1}, \ldots, g_{i_k}, g_n]$. Define

$$g(x) = (g_{i_1}x^{i_1} + \cdots + g_{i_k}x^{i_k} + g_n x^n) + \sum_{j \notin \{i_1,\ldots,i_k,n\}} \varphi_j x^j. \tag{13}$$

The degree of $g(x)$ is n and the bound on the coefficients of $g(x)$ is determined as follows. The determinant of M is p^k and by the LLL-reduced property, each of the coefficients $g_{i_1}, \ldots, g_{i_k}, g_n$ is $O(p^{k/(k+1)}) = O(Q^{k/(n(k+1))})$. Since φ_j for $j \notin \{i_1, \ldots, i_k\}$ are all $O(1)$, it follows from (13) that all the coefficients of $g(x)$ are $O(Q^{k/(n(k+1))})$ and so $\|g\|_\infty = O(Q^{k/(n(k+1))})$.

5 A New Polynomial Selection Method

In the simple observation made in the earlier section, the non-zero terms of the polynomial $g(x)$ are powers of x^d. This creates a restriction and does not turn out to be necessary to apply the main idea of the previous section. Once the polynomial $\psi(x)$ is obtained using the LLL method, it is possible to substitute any degree d polynomial with small coefficients for x and still the norm bound will hold. In fact, the idea can be expressed more generally in terms of resultants. Algorithm $\mathcal{A}$ describes the new general method for polynomial selection.

The following result states the basic properties of Algorithm $\mathcal{A}$.

Algorithm. $\mathcal{A}$: A new method of polynomial selection.

Input: p, n, d (a factor of n) and $r \geq n/d$.
Output: $f(x)$, $g(x)$ and $\varphi(x)$.

Let $k = n/d$;
repeat

Randomly choose a monic irreducible polynomial $A_1(x)$ having the following properties: $\deg A_1(x) = r + 1$; $A_1(x)$ is irreducible over the integers; $A_1(x)$ has coefficients of size $O(\ln(p))$; modulo p, $A_1(x)$ has an irreducible factor $A_2(x)$ of degree k.
Randomly choose monic polynomials $C_0(x)$ and $C_1(x)$ with small coefficients such that $\deg C_0(x) = d$ and $\deg C_1(x) < d$.
Define

$$\begin{aligned} f(x) &= \mathrm{Res}_y\left(A_1(y), C_0(x) + y\,C_1(x)\right); \\ \varphi(x) &= \mathrm{Res}_y\left(A_2(y), C_0(x) + y\,C_1(x)\right) \mod p; \\ \psi(x) &= \mathrm{LLL}(M_{A_2,r}); \\ g(x) &= \mathrm{Res}_y\left(\psi(y), C_0(x) + y\,C_1(x)\right). \end{aligned}$$

until *$f(x)$ and $g(x)$ are irreducible over $\mathbb{Z}$ and $\varphi(x)$ is irreducible over $\mathbb{F}_p$.*
return $f(x)$, $g(x)$ and $\varphi(x)$.

Proposition 2. *The outputs $f(x)$, $g(x)$ and $\varphi(x)$ of Algorithm $\mathcal{A}$ satisfy the following.*

1. $\deg(f) = d(r+1)$; $\deg(g) = rd$ *and* $\deg(\varphi) = n$;
2. *both* $f(x)$ *and* $g(x)$ *have* $\varphi(x)$ *as a factor modulo* p;
3. $\|f\|_\infty = O(\ln(p))$ *and* $\|g\|_\infty = O(Q^{1/(d(r+1))})$.

Consequently,

$$\begin{aligned}|\text{Res}(f, \phi) \times \text{Res}(g, \phi)| &\approx (\|f\|_\infty \|g\|_\infty)^{t-1} \times E^{2(\deg f + \deg g)/t} \\ &= O\left(E^{2d(2r+1)/t} \times Q^{(t-1)/(d(r+1))}\right). \qquad (14)\end{aligned}$$

Proof. By definition $f(x) = \text{Res}_y(A_1(y), C_0(x) + y\,C_1(x))$ where $A_1(x)$ has degree $r+1$, $C_0(x)$ has degree d and $C_1(x)$ has degree $d-1$, so the degree of $f(x)$ is $d(r+1)$. Similarly, one obtains the degree of $\varphi(x)$ to be n. Since $\psi(x)$ is obtained from $A_2(x)$ as $\text{LLL}(M_{A_2,r})$ it follows that the degree of $\psi(x)$ is r and so the degree of $g(x)$ is rd.

Since $A_2(x)$ divides $A_1(x)$ modulo p, it follows from the definition of $f(x)$ and $\varphi(x)$ that modulo p, $\varphi(x)$ divides $f(x)$. Since $\psi(x)$ is a linear combination of the rows of $M_{A_2,r}$, it follows that modulo p, $\psi(x)$ is a multiple of $A_2(x)$. So, $g(x) = \text{Res}_y(\psi(y), C_0(x) + y\,C_1(x))$ is a multiple of $\varphi(x) = \text{Res}_y(A_2(y), C_0(x) + y\,C_1(x))$ modulo p.

Since the coefficients of $C_0(x)$ and $C_1(x)$ are $O(1)$ and the coefficients of $A_1(x)$ are $O(\ln p)$, it follows that $\|f\|_\infty = O(\ln p)$. The coefficients of $g(x)$ are $O(1)$ multiples of the coefficients of $\psi(x)$. By third point of Proposition 1, the coefficients of $\psi(x)$ are $O(p^{n/(d(r+1))}) = Q^{1/(d(r+1))}$ which shows that $\|g\|_\infty = O(Q^{1/(d(r+1))})$. □

Proposition 2 provides the relevant bound on the product of the norms of a sieving polynomial $\phi(x)$ in the two number fields defined by $f(x)$ and $g(x)$. We note the following points.

1. If $d = 1$, then the norm bound is $E^{2(2r+1)/t}Q^{(t-1)/(r+1)}$ which is the same as that obtained using the GJL method.
2. If $d = n$, then the norm bound is $E^{2n(2r+1)/t}Q^{(t-1)/(n(r+1))}$. Further, if $r = k = 1$, then the norm bound is the same as that obtained using the Conjugation method. So, for $d = n$, Algorithm $\mathcal{A}$ is a generalisation of the Conjugation method. Later, we show that choosing $r > 1$ provides asymptotic improvements.
3. If n is a prime, then the only values of d are either 1 or n. The norm bounds in these two cases are covered by the above two points.
4. If n is composite, then there are non-trivial values for d and it is possible to obtain new trade-offs in the norm bound. For concrete situations, this can be of interest. Further, for composite n, as value of d increases from $d = 1$ to $d = n$, the norm bound nicely interpolates between the norm bounds of the GJL method and the Conjugation method.

Existence of $\mathbb{Q}$-automorphisms: The existence of $\mathbb{Q}$-automorphism in the number fields speeds up the NFS algorithm in the non-asymptotic sense [19]. Similar to the existence of $\mathbb{Q}$-automorphism in GJL method, as discussed in [5], the first polynomial generated by the new method, can have a $\mathbb{Q}$-automorphism. In general, it is difficult to get an automorphism for the second polynomial as it is generated by the LLL algorithm. On the other hand, we can have a $\mathbb{Q}$-automorphism for the second polynomial also in the specific cases. Some of the examples are reported in [10].

6 Non-asymptotic Comparisons and Examples

We compare the norm bounds for $t = 2$, i.e., when the sieving polynomial is linear. In this case, Table 2 lists the degrees and norm bounds of polynomials for various methods. Table 3 compares the new method with the JLSV1 and the GJL method for concrete values of n, r and d. This shows that the new method provides different trade-offs which were not known earlier.

As an example, we can see from Table 3 that the new method compares well with GJL and JLSV1 methods for $n = 4$ and Q of 300 dd (refer to Table 1). As mentioned in [5], when the differences between the methods are small, it is not possible to decide by looking only at the size of the norm product. Keeping this in view, we see that the new method is competitive for $n = 6$ as well. These observations are clearly visible in the plots given in the Fig. 3. From the Q-E pairs given in Table 1, it is clear that the increase of E is slower than that of Q. This suggests that the new method will become competitive when Q is sufficiently large.

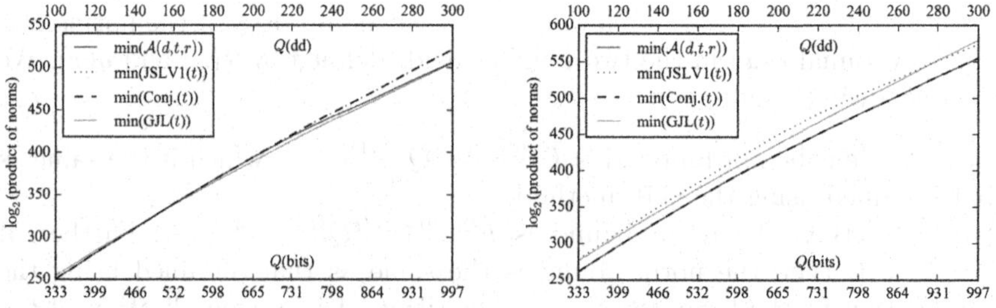

(a) Polynomials for $\mathbb{F}_{p^4}$ (b) Polynomials for $\mathbb{F}_{p^6}$

Fig. 3. Product of norms for various polynomial selection methods

Next we provide some concrete examples of polynomials $f(x), g(x)$ and $\varphi(x)$ obtained using the new method. The examples are for $n = 6$ and $n = 4$. For $n = 6$, we have taken $d = 1, 2, 3$ and 6 and in each case r was chosen to be $r = k = n/d$. For $n = 4$, we consider $d = 2$ with $r = k = n/d$ and $r = k + 1$; and $d = 4$ with $r = k$. These examples are to illustrate that the method works as

Table 2. Parameterised efficiency estimates for NFS obtained from the different polynomial selection methods.

Methods	$\deg f$	$\deg g$	$\|f\|_\infty$	$\|g\|_\infty$	$\|f\|_\infty \|g\|_\infty E^{(\deg f + \deg g)}$
JLSV1	n	n	$Q^{\frac{1}{2n}}$	$Q^{\frac{1}{2n}}$	$E^{2n} Q^{\frac{1}{n}}$
GJL ($r \geq n$)	$r+1$	r	$O(\ln p)$	$Q^{\frac{1}{r+1}}$	$E^{2r+1} Q^{\frac{1}{r+1}}$
Conjugation	$2n$	n	$O(\ln p)$	$Q^{\frac{1}{2n}}$	$E^{3n} Q^{\frac{1}{2n}}$
$\mathcal{A}$ ($d\|n$, $r \geq n/d$)	$d(r+1)$	dr	$O(\ln p)$	$Q^{\frac{1}{d(r+1)}}$	$E^{d(2r+1)} Q^{1/(d(r+1))}$

Table 3. Comparison of efficiency estimates for composite n with $d = 2$ and $r = n/2$.

$\mathbb{F}_Q$	method	$(\deg f, \deg g)$	$\|f\|_\infty$	$\|g\|_\infty$	$\|f\|_\infty \|g\|_\infty E^{(\deg f + \deg g)}$
$\mathbb{F}_{p^4}$	GJL	$(5, 4)$	$O(\ln p)$	$Q^{\frac{1}{5}}$	$E^9 Q^{\frac{1}{5}}$
	JLSV1	$(4, 4)$	$Q^{\frac{1}{8}}$	$Q^{\frac{1}{8}}$	$E^8 Q^{\frac{1}{4}}$
	$\mathcal{A}$	$(6, 4)$	$O(\ln p)$	$Q^{\frac{1}{6}}$	$E^{10} Q^{\frac{1}{6}}$
$\mathbb{F}_{p^6}$	GJL	$(7, 6)$	$O(\ln p)$	$Q^{\frac{1}{7}}$	$E^{13} Q^{\frac{1}{7}}$
	JLSV1	$(6, 6)$	$Q^{\frac{1}{12}}$	$Q^{\frac{1}{12}}$	$E^{12} Q^{\frac{1}{6}}$
	$\mathcal{A}$	$(8, 6)$	$O(\ln p)$	$Q^{\frac{1}{8}}$	$E^{14} Q^{\frac{1}{8}}$
$\mathbb{F}_{p^8}$	GJL	$(9, 8)$	$O(\ln p)$	$Q^{\frac{1}{9}}$	$E^{17} Q^{\frac{1}{9}}$
	JLSV1	$(8, 8)$	$Q^{\frac{1}{16}}$	$Q^{\frac{1}{16}}$	$E^{16} Q^{\frac{1}{8}}$
	$\mathcal{A}$	$(10, 8)$	$O(\ln p)$	$Q^{\frac{1}{10}}$	$E^{18} Q^{\frac{1}{10}}$
$\mathbb{F}_{p^9}$	GJL	$(10, 9)$	$O(\ln p)$	$Q^{\frac{1}{10}}$	$E^{19} Q^{\frac{1}{10}}$
	JLSV1	$(9, 9)$	$Q^{\frac{1}{18}}$	$Q^{\frac{1}{18}}$	$E^{18} Q^{\frac{1}{9}}$
	$\mathcal{A}$	$(12, 9)$	$O(\ln p)$	$Q^{\frac{1}{12}}$	$E^{21} Q^{\frac{1}{12}}$

predicted and returns the desired polynomials very fast. We have used Sage [29] and MAGMA computer algebra system [9] for all the computations done in this work.

Example 1. Let $n = 6$, and p is a 201-bit prime given below.

$$p = 1606938044258990275541962092341162602522202993782792835361211$$

Taking $d = 1$ and $r = n/d$, we get

$$f(x) = x^7 + 18\,x^6 + 99\,x^5 - 107\,x^4 - 3470\,x^3 - 15630\,x^2 - 30664\,x - 23239$$

$$\begin{aligned} g(x) = {} & 712965136783466122384156554261504665235609243446869\,x^6 + 16048203858903 \\ & 2606917662167026525754352818075442477 12\,x^5 + 14867720774814154920358989 \\ & 0852868028274077107624860184\,x^4 + 72408538453914392579556483572292625 61 \\ & 7192085298666037 2\,x^3 + 194693204195493982969795038496468458378024972218 \\ & 5345772\,x^2 + 27189717972702351712342597931428514169233315191786758 74\,x \\ & +1517248296800681060244076172658712224507653769252953211 \end{aligned}$$

$$\begin{aligned}\varphi(x) = {} & x^6 + 6715600759360122754018289503697292868061440059396953492907 60\, x^5 + \\ & 7747058346245540667371991605555115020882703234812683373405 14\, x^4 + 1100 \\ & 6464475526715804379638610850204311451261510579373184797 17\, x^3 + 27131646 \\ & 3864123658232870095113273120009266491174096472632727\, x^2 + 4101717389506 \\ & 73951225351009256251353058695601874372080573092\, x + 132663280496102776 7 \\ & 27233466269357885584536385439823152439060 7\end{aligned}$$

Note that $\|g\|_\infty \approx 2^{180}$. Taking $d = 2$ and $r = n/d$, we get

$$f(x) = x^8 - x^7 - 5\, x^6 - 50\, x^5 - 181\, x^4 - 442\, x^3 - 801\, x^2 - 633\, x - 787$$

$$\begin{aligned}g(x) = {} & 83348093250051649250593583918500819369645778 7\, x^6 + 209259361664128765 5 \\ & 06574003289698634358069861 5\, x^5 + 129854089956895226179153774346833519 4 \\ & 318853332 0\, x^4 + 21869741590966357897620167461539967141532970622\, x^3 + 6 \\ & 44030972246342626772738034719926717478609685 64\, x^2 + 55864711695281584 2 \\ & 83909455665521092749502793807\, x + 92177835405907782725278435670487132 7 \\ & 10722661831\end{aligned}$$

$$\begin{aligned}\varphi(x) = {} & x^6 + 22557756689804128540553922618322150822628658922554614271405 7\, x^5 + \\ & 7261566737238890828953514517397335453283947205232462729551 73\, x^4 + 10214 \\ & 7813205469472157888899400173076493445466063054368834805 6\, x^3 + 674978102 \\ & 5562087428820180277199513084540786093481181587839 1\, x^2 + 632426210761786 \\ & 62210549419431493781792743937291802904271884 3\, x + 104093530686601670252 \\ & 64551437254153796047423390654217938440 38\end{aligned}$$

Note that $\|g\|_\infty \approx 2^{156}$. Taking $d = 3$ and $r = n/d$, we get

$$f(x) = x^9 - 4\, x^8 - 54\, x^7 - 174\, x^6 - 252\, x^5 - 174\, x^4 - 76\, x^3 - 86\, x^2 - 96\, x - 42$$

$$\begin{aligned}g(x) = {} & 2889742364508381557593312392497801006712\, x^6 + 836336953706463060856 10 \\ & 87765146274738509\, x^5 + 1082807880652408570550641278340877294187 7\, x^4 + \\ & 4181282488973040016900039741726719770117 9\, x^3 + 149742134777753247621 3 \\ & 31508897969482387354\, x^2 + 24094671698944321029344296555261130559219 4\, x \\ & +1516964556551047444030737433339404265988 33\end{aligned}$$

$$\begin{aligned}\varphi(x) = {} & x^6 + 265074577705978624915342871970538348132010154368109244143774\, x^5 \\ & +21159801273629654486978970226092134077566675973129512551886\, x^4 + 10 \\ & 6344561144568426694128954082794719939741627633418805583789 2\, x^3 + 1459 \\ & 587283058054365639950761731919998074021438242745336103973\, x^2 + 145654 \\ & 343780057164332563864820718837111792353916826321052299 5\, x + 378129170 \\ & 96051021149160030362367447146841414479717884697700 7\end{aligned}$$

Note that $\|g\|_\infty \approx 2^{137}$. Taking $d = 6$ and $r = n/d$, we get

$$\begin{aligned} f(x) = {} & x^{12} + 3\,x^{10} + 10\,x^9 + 53\,x^8 + 112\,x^7 + 163\,x^6 \\ & +184\,x^5 + 177\,x^4 + 166\,x^3 + 103\,x^2 + 72\,x + 48 \end{aligned}$$

$$\begin{aligned} g(x) = {} & -6668781384023531954988326698 48\,x^6 - 1867253271074924746011849188889\,x^5 \\ & -5601759813224774238035547566667\,x^4 - 6668753801765210948063915265053\,x^3 \\ & -4268003536420067847037882226971\,x^2 - 6935516090029480629033212906363\,x \\ & -7469013084299698984047396755556 \end{aligned}$$

$$\begin{aligned} \varphi(x) = {} & x^6 + 35648533684707409192094459718781128441184904799133426 6185684\,x^5 + \\ & 10694560105412222757628337915634338532355471439740027985570 52\,x^4 + 175 \\ & 4886399763801840627608935978938195370422461738784955672 05\,x^3 + 1069456 \\ & 0105412222757628337915634338532355471439740027985570 50\,x^2 + 1069456010 \\ & 5412222757628337915634338532355471439740027985570 54\,x + 14259413473882 \\ & 96367683778388751245137647396191965337064742736 \end{aligned}$$

In this case we get $\|g\|_\infty \approx 2^{102}$.

Example 2. Let $n = 4$, and p is a 301-bit prime given below.

$$\begin{aligned} p = {} & 2037035976334486086268445688409378161051468393665936250636140449354 38 \\ & 12997633367061834936 07 \end{aligned}$$

Taking $d = 2$ and $r = n/d$, we get

$$f(x) = x^6 + 2\,x^5 + 10\,x^4 + 11\,x^3 + 8\,x^2 + 3\,x + 5$$

$$\begin{aligned} g(x) = {} & 11084862440235762086893604101763003731322206545909767864821 34\,x^4 + 20 \\ & 5076293814498228936009608370556396593557366710355499452804 4\,x^3 + 5523 \\ & 4675803770219347530917862076484798670362096791517930153 19\,x^2 + 456222 \\ & 7246514756745388645848004531501269616133890841445574 058\,x + 441498133 \\ & 63534457260637313760313481067348155550881750065331 85 \end{aligned}$$

$$\begin{aligned} \varphi(x) = {} & x^4 + 13056233606982846851755992777073434575762791461882425862452101 99 \\ & 34477785613829304916553629 2\,x^3 + 16306637647132427224267721755759453 19 \\ & 6406656557949629326536345456905706772528539726899970 48\,x^2 + 1955704168 \\ & 72820075967794507344454718170505216540168327906205889203636349836741 48 \\ & 96214457800\,x + 1630663764713242722426772175575945319640665655794962 93 \\ & 26536345456905706772528539726899970 47 \end{aligned}$$

In this case we have $\|g\|_\infty \approx 2^{201}$. If we take $r = n/d + 1$, we get

$$f(x) = x^8 + 16\,x^7 + 108\,x^6 + 398\,x^5 + 865\,x^4 + 1106\,x^3 + 820\,x^2 + 328\,x + 55$$

$$\begin{aligned} g(x) = {} & 34848214784208386538088134778439992533572855 7\,x^6 + 5536103979982210590 \\ & 1860164454592897730290456 18\,x^5 + 33812545050706664774530525723335145 80 \\ & 1290667783\,x^4 + 9606217195726112476342859064895874518873544533 0\,x^3 + 1 \\ & 24085795781307363759935898131887563792535489069\,x^2 + 73090839973729169 \\ & 96696406142840231613191113 0808\,x + 1609381078327430905535048197202884 1 \\ & 649178007790 \end{aligned}$$

$$\begin{aligned} \varphi(x) = {} & x^4 + 51286909645979432465019623589986762370339308461689674479903342 44 \\ & 5569631918567326276559942 8\,x^3 + 18024087969327494874449747905760220 81 \\ & 70834465922920791127184582765003571338326842766241644 4\,x^2 + 1553341208 \\ & 02632167628916463755257366860311697999082884334755795747728615002384 38 \\ & 04262435184\,x + 26380150755336651349438608287641921059816540537851767 6 \\ & 874745554282946755826248639365618168 \end{aligned}$$

In this case we have $\|g\|_\infty \approx 2^{156}$. If we take $d = 4$ and $r = d/n$, we have

$$f(x) = x^8 - 3\,x^7 - 33\,x^6 - 97\,x^5 - 101\,x^4 + 3\,x^3 + 73\,x^2 - 35\,x - 8$$

$$\begin{aligned} g(x) = {} & 68486288602412597391139186719841584143687727 8\,x^4 + 1925808392957060519 \\ & 248933705295588974777491073 1\,x^3 + 16682478627264257142784499126962718 75 \\ & 703468525\,x^2 + 409615604475389614851823857001230937582717 63\,x + 124094 \\ & 5506932934545337541838097173133338033453 \end{aligned}$$

$$\begin{aligned} \varphi(x) = {} & x^4 + 30012929912905666581877080461131623268227469635765762480590133 80 \\ & 72170670924524605598965 54\,x^3 + 90038789738716999745631241383394869804 6 \\ & 82408907297287441770401421651201277357381679689656\,x^2 + 15006464956452 \\ & 83329093854023056581163411373481788288124029506690360853354622623027 99 \\ & 482756\,x + 30012929912905666581877080461131623268227469635765762480590 \\ & 1338072170670924524605598965 53 \end{aligned}$$

In this case also we have $\|g\|_\infty \approx 2^{150}$.

7 Asymptotic Complexity Analysis

The goal of the asymptotic complexity analysis is to express the runtime of the NFS algorithm using the L-notation and at the same time obtain bounds on p for which the analysis is valid. Our description of the analysis is based on prior works predominantly those in [5, 17, 19, 24].

For $0 < a < 1$, write

$$p = L_Q(a, c_p), \text{ where } c_p = \frac{1}{n}\left(\frac{\ln Q}{\ln\ln Q}\right)^{1-a} \text{ and so } n = \frac{1}{c_p}\left(\frac{\ln Q}{\ln\ln Q}\right)^{1-a}. \tag{15}$$

The value of a will be determined later. Also, for each c_p, the runtime of the NFS algorithm is the same for the family of finite fields $\mathbb{F}_{p^n}$ where p is given by (15).

From Sect. 3, we recall the following.

1. The number of polynomials to be considered for sieving is E^2.
2. The factor base is of size B.

Sparse linear algebra using the Lanczos or the block Wiedemann algorithm takes time $O(B^2)$. For some $0 < b < 1$, let

$$B = L_Q(b, c_b). \tag{16}$$

The value of b will be determined later. Set

$$E = B \tag{17}$$

so that asymptotically, the number of sieving polynomials is equal to the time for the linear algebra step.

Let $\pi = \Psi(\Gamma, B)$ be the probability that a random positive integer which is at most Γ is B-smooth. Let $\Gamma = L_Q(z, \zeta)$ and $B = L_Q(b, c_b)$. Using the L-notation version of the Canfield-Erdös-Pomerance theorem,

$$(\Psi(\Gamma, B))^{-1} = L_Q\left(z - b, (z - b)\frac{\zeta}{c_b}\right). \tag{18}$$

The bound on the product of the norms given by Proposition 2 is

$$\Gamma = E^{\frac{2}{t}d(2r+1)} \times Q^{\frac{t-1}{d(r+1)}}. \tag{19}$$

Note that in (19), $t - 1$ is the degree of the sieving polynomial. Following the usual convention, we assume that the same smoothness probability π holds for the event that a random sieving polynomial $\phi(x)$ is smooth over the factor base.

The expected number of polynomials to consider for obtaining one relation is π^{-1}. Since B relations are required, obtaining this number of relations requires trying $B\pi^{-1}$ trials. Balancing the cost of sieving and the linear algebra steps requires $B\pi^{-1} = B^2$ and so

$$\pi^{-1} = B. \tag{20}$$

Obtaining π^{-1} from (18) and setting it to be equal to B allows solving for c_b. Balancing the costs of the sieving and the linear algebra phases leads to the runtime of the NFS algorithm to be $B^2 = L_Q(b, 2c_b)$. So, to determine the runtime, we need to determine b and c_b. The value of b will turn out to be $1/3$ and the only real issue is the value of c_b.

Lemma 1. *Let $n = kd$ for positive integers k and d. Using the expressions for p and $E(= B)$ given by (15) and (16), we obtain the following.*

$$\left.\begin{aligned} E^{\frac{2}{t}d(2r+1)} &= L_Q\left(1-a+b, \tfrac{2c_b(2r+1)}{c_p kt}\right); \\ Q^{\frac{t-1}{d(r+1)}} &= L_Q\left(a, \tfrac{kc_p(t-1)}{(r+1)}\right). \end{aligned}\right\} \tag{21}$$

Proof. The second expression follows directly from $Q = p^n$, $p = L_Q(a, c_p)$ and $n = kd$. The computation for obtaining the first expression is the following.

$$\begin{aligned} E^{\frac{2}{t}d(2r+1)} &= L_Q\left(b, c_b\frac{2}{t}d(2r+1)\right) \\ &= \exp\left(c_b\frac{2}{t}(2r+1)\frac{n}{k}(\ln Q)^b(\ln\ln Q)^{1-b}\right) \\ &= \exp\left(c_b\frac{2}{c_pkt}(2r+1)\left(\frac{\ln Q}{\ln\ln Q}\right)^{1-a}(\ln Q)^b(\ln\ln Q)^{1-b}\right) \\ &= L_Q\left(1-a+b, \frac{2c_b(2r+1)}{c_pkt}\right). \end{aligned}$$

□

Theorem 1 (Boundary Case). *Let k divide n, $r \geq k$, $t \geq 2$ and $p = L_Q(2/3, c_p)$ for some $0 < c_p < 1$. It is possible to ensure that the runtime of the NFS algorithm with polynomials chosen by Algorithm $\mathcal{A}$ is $L_Q(1/3, 2c_b)$ where*

$$c_b = \frac{2r+1}{3c_pkt} + \sqrt{\left(\frac{2r+1}{3c_pkt}\right)^2 + \frac{kc_p(t-1)}{3(r+1)}}. \tag{22}$$

Proof. Setting $2a = 1 + b$, the two L-expressions given by (21) have the same first component and so the product of the norms is

$$\Gamma = L_Q\left(a, \frac{2c_b(2r+1)}{c_pkt} + \frac{kc_p(t-1)}{(r+1)}\right).$$

Then π^{-1} given by (18) is

$$L_Q\left(a-b, (a-b)\left(\frac{2(2r+1)}{c_pkt} + \frac{kc_p(t-1)}{c_b(r+1)}\right)\right).$$

From the condition $\pi^{-1} = B$, we get $b = a - b$ and

$$c_b = (a-b)\left(\frac{2(2r+1)}{c_pkt} + \frac{kc_p(t-1)}{c_b(r+1)}\right).$$

The conditions $a - b = b$ and $2a = 1 + b$ show that $b = 1/3$ and $a = 2/3$. The second equation then becomes

$$c_b = \frac{1}{3}\left(\frac{2(2r+1)}{c_pkt} + \frac{kc_p(t-1)}{c_b(r+1)}\right). \tag{23}$$

Solving the quadratic for c_b and choosing the positive root gives

$$c_b = \frac{2r+1}{3c_pkt} + \sqrt{\left(\frac{2r+1}{3c_pkt}\right)^2 + \frac{kc_p(t-1)}{3(r+1)}}.$$

□

Corollary 1 (Boundary Case of the Conjugation Method [5]). *Let $r = k = 1$. Then for $p = L_Q(2/3, c_p)$, the runtime of the NFS algorithm is $L_Q(1/3, 2c_b)$ with*

$$c_b = \frac{1}{c_pt} + \sqrt{\left(\frac{1}{c_pt}\right)^2 + \frac{c_p(t-1)}{6}}.$$

Allowing r to be greater than k leads to improved asymptotic complexity. We do not perform this analysis. Instead, we perform the analysis in the similar situation which arises for the multiple number field sieve algorithm.

Theorem 2 (Medium Characteristic Case). *Let $p = L_Q(a, c_p)$ with $a > 1/3$. It is possible to ensure that the runtime of the NFS algorithm with the polynomials produced by Algorithm $\mathcal{A}$ is $L_Q(1/3, (32/3)^{1/3})$.*

Proof. Since $a > 1/3$, the bound Γ on the product of the norms can be taken to be the expression given by (7). The parameter t is chosen as follows [5]. For $0 < c < 1$, let $t = c_t n((\ln Q)/(\ln\ln Q))^{-c}$. For the asymptotic analysis, $t-1$ is also assumed to be given by the same expression for t. Then the expressions given by (21) become the following.

$$E^{\frac{2}{t}d(2r+1)} = L_Q\left(b+c, \tfrac{2c_b(2r+1)}{kc_t}\right); \; Q^{\frac{t-1}{d(r+1)}} = L_Q\left(1-c, \tfrac{kc_t}{r+1}\right). \qquad (24)$$

This can be seen by substituting the expression for t in (21) and further by using the expression for n given in (15).

Setting $2c = 1-b$, the first components of the two expressions in (24) become equal and so

$$\Gamma = L_Q\left(b+c, \frac{2c_b(2r+1)}{kc_t} + \frac{kc_t}{r+1}\right).$$

Using this Γ, the expression for π^{-1} is

$$\pi^{-1} = L_Q\left(c, c\left(\frac{2(2r+1)}{kc_t} + \frac{kc_t}{c_b(r+1)}\right)\right).$$

We wish to choose c_t so as to maximise the probability π and hence to minimise π^{-1}. This is done by setting $2(2r+1)/(kc_t) = (kc_t)/(c_b(r+1))$ whence $kc_t = \sqrt{2c_b(r+1)(2r+1)}$. With this value of kc_t,

$$\pi^{-1} = L_Q\left(c, \frac{2c\sqrt{2c_b(r+1)(2r+1)}}{c_b(r+1)}\right).$$

Setting π^{-1} to be equal to $B = L_Q(b, c_b)$ yields $b = c$ and

$$c_b = \left(\frac{2c\sqrt{2c_b(r+1)(2r+1)}}{c_b(r+1)} \right).$$

From $b = c$ and $2c = 1 - b$ we obtain $c = b = 1/3$. Using this value of c in the equation for c_b, we obtain $c_b = (2/3)^{2/3} \times ((2(2r+1))/(r+1))^{1/3}$. The value of c_b is the minimum for $r = 1$ and this value is $c_b = (4/3)^{1/3}$. □

Note that the parameter a which determines the size of p is not involved in any of the computation. The assumption $a > 1/3$ is required to ensure that the bound on the product of the norms can be taken to be the expression given by (7).

Theorem 3 (Large Characteristic). *It is possible to ensure that the run-time of the NFS algorithm with the polynomials produced by Algorithm $\mathcal{A}$ is $L_Q(1/3, (64/9)^{1/3})$ for $p \geq L_Q(2/3, (8/3)^{1/3})$.*

Proof. Following [5], for $0 < e < 1$, let $r = c_r/2((\ln Q)/(\ln \ln Q))^e$. For the asymptotic analysis, the expression for $2r + 1$ is taken to be two times this expression. Substituting this expression for r in (21), we obtain

$$\left.\begin{aligned} E^{\frac{2}{t}d(2r+1)} &= L_Q\left(1 - a + b + e, \tfrac{2c_b c_r}{c_p k t}\right); \\ Q^{\frac{t-1}{d(r+1)}} &= L_Q\left(a - e, \tfrac{2kc_p(t-1)}{c_r}\right). \end{aligned}\right\} \tag{25}$$

Setting $1 + b = 2(a - e)$, we obtain $\Gamma = L_Q\left(\frac{1+b}{2}, \frac{2c_b c_r}{c_p k t} + \frac{2kc_p(t-1)}{c_r}\right)$ and so the probability π^{-1} is given by

$$L_Q\left(\frac{1-b}{2}, \frac{1-b}{2} \times \left(\frac{2c_r}{c_p k t} + \frac{2kc_p(t-1)}{c_r c_b}\right)\right).$$

The choice of c_r for which the probability π is maximised (and hence π^{-1} is minimised) is obtained by setting $c_r/(c_p k) = \sqrt{(t(t-1))/c_b}$ and the minimum value of π^{-1} is

$$L_Q\left(\frac{1-b}{2}, \frac{1-b}{2} \times \left(4\sqrt{\frac{t-1}{tc_b}}\right)\right).$$

Setting this value of π^{-1} to be equal to B, we obtain

$$b = (1-b)/2; \; c_b = \frac{1-b}{2} \times \left(4\sqrt{\frac{t-1}{tc_b}}\right).$$

The first equation shows $b = 1/3$ and using this in the second equation, we obtain $c_b = (4/3)^{2/3}((t-1)/t)^{1/3}$. This value of c_b is minimised for the minimum value of t which is $t = 2$. This gives $c_b = (8/9)^{1/3}$.

Using $2(a - e) = 1 + b$ and $b = 1/3$ we get $a - e = 2/3$. Note that $r \geq k$ and so $p \geq p^{k/r} = L_Q(a, (c_p k)/r) = L_Q(a - e, (2c_p k)/c_r)$. With $t = 2$, the value of $(c_p k)/c_r$ is equal to $(1/3)^{1/3}$ and so $p \geq L_Q(2/3, (8/3)^{1/3})$. □

Theorems 2 and 3 show that the generality introduced by k and r do not affect the overall asymptotic complexity for the medium and large prime case and the attained complexities in these cases are the same as those obtained for previous methods in [5].

8 Multiple Number Field Sieve Variant

As the name indicates, the multiple number field sieve variant uses several number fields. The discussion and the analysis will follow the works [8,24].

There are two variants of multiple number field sieve algorithm. In the first variant, the image of $\phi(x)$ needs to be smooth in at least any two of the number fields. In the second variant, the image of $\phi(x)$ needs to be smooth in the first number field and at least one of the other number fields.

We have analysed both the variants of multiple number field sieve algorithm and found that the second variant turns out to be better than the first one. So we discuss the second variant of MNFS only. In contrast to the number field sieve algorithm, the right number field is replaced by a collection of V number fields in the second variant of MNFS. The sieving polynomial $\phi(x)$ has to satisfy the smoothness condition on the left number field as before. On the right side, it is sufficient for $\phi(x)$ to satisfy a smoothness condition on at least one of the V number fields.

Recall that Algorithm $\mathcal{A}$ produces two polynomials $f(x)$ and $g(x)$ of degrees $d(r+1)$ and dr respectively. The polynomial $g(x)$ is defined as $\mathrm{Res}_y(\psi(y), C_0(x)+yC_1(x))$ where $\psi(x) = \mathrm{LLL}(M_{A_2,r})$, i.e., $\psi(x)$ is defined from the first row of the matrix obtained after applying the LLL-algorithm to $M_{A_2,r}$.

Methods for obtaining the collection of number fields on the right have been mentioned in [24]. We adapt one of these methods to our setting. Consider Algorithm $\mathcal{A}$. Let $\psi_1(x)$ be $\psi(x)$ as above and let $\psi_2(x)$ be the polynomial defined from the second row of the matrix $M_{A_2,r}$. Define $g_1(x) = \mathrm{Res}_y(\psi_1(y), C_0(x) + yC_1(x))$ and $g_2(x) = \mathrm{Res}_y(\psi_2(y), C_0(x) + yC_1(x))$. Then choose $V-2$ linear combinations $g_i(x) = s_i g_1(x) + t_i g_2(x)$, for $i = 3, \ldots, V$. Note that the coefficients s_i and t_i are of the size of $\sqrt{V}$. All the g_i's have degree dr. Asymptotically, $\|\psi_2\|_\infty = \|\psi_1\|_\infty = Q^{1/(d(r+1))}$. Since we take $V = L_Q(1/3)$, all the g_i's have their infinity norms to be the same as that of $g(x)$ given by Proposition 2.

For the left number field, as before, let B be the bound on the norms of the ideals which are in the factor basis defined by f. For each of the right number fields, let B' be the bound on the norms of the ideals which are in the factor basis defined by each of the g_i's. So, the size of the entire factor basis is $B + VB'$. The following condition balances the left portion and the right portion of the factor basis.

$$B = VB'. \tag{26}$$

With this condition, the size of the factor basis is $B^{1+o(1)}$ as in the classical NFS and so asymptotically, the linear algebra step takes time B^2. As before, the number of sieving polynomials is $E^2 = B^2$ and the coefficients of $\phi(x)$ can take $E^{2/t}$ distinct values.

Let π be the probability that a random sieving polynomial $\phi(x)$ gives rise to a relation. Let π_1 be the probability that $\phi(x)$ is smooth over the left factor basis and π_2 be the probability that $\phi(x)$ is smooth over *at least* one of the right factor bases. Further, let $\Gamma_1 = \mathrm{Res}_x(f(x), \phi(x))$ be the bound on the norm corresponding to the left number field and $\Gamma_2 = \mathrm{Res}_x(g_i(x), \phi(x))$ be the bound on the norm for any of the right number fields. Note that Γ_2 is determined only by the degree and the L_∞-norm of $g_i(x)$ and hence is the same for all $g_i(x)$'s. Heuristically, we have

$$\begin{aligned} \pi_1 &= \Psi(\Gamma_1, B); \\ \pi_2 &= V\Psi(\Gamma_2, B'); \\ \pi &= \pi_1 \times \pi_2. \end{aligned} \tag{27}$$

As before, one relation is obtained in about π^{-1} trials and so B relations are obtained in about $B\pi^{-1}$ trials. Balancing the cost of linear algebra and sieving, we have as before $B = \pi^{-1}$.

The following choices of B and V are made.

$$\begin{aligned} E = B &= L_Q\left(\tfrac{1}{3}, c_b\right); \\ V &= L_Q\left(\tfrac{1}{3}, c_v\right); \text{ and so} \\ B' = B/V &= L_Q\left(\tfrac{1}{3}, c_b - c_v\right). \end{aligned} \tag{28}$$

With these choices of B and V, it is possible to analyse the MNFS variant for Algorithm $\mathcal{A}$ for three cases, namely, the medium prime case, the boundary case and the large characteristic case. Below we present the details of the boundary case. This presents a new asymptotic result.

Theorem 4 (MNFS-Boundary Case). *Let k divide n, $r \geq k$, $t \geq 2$ and*

$$p = L_Q\left(\frac{2}{3}, c_p\right) \text{ where } c_p = \frac{1}{n}\left(\frac{\ln Q}{\ln \ln Q}\right)^{1/3}.$$

It is possible to ensure that the runtime of the MNFS algorithm is $L_Q(1/3, 2c_b)$ where

$$c_b = \frac{4r+2}{6ktc_p} + \sqrt{\frac{r(3r+2)}{(3ktc_p)^2} + \frac{c_p k(t-1)}{3(r+1)}}. \tag{29}$$

Proof. Note the following computations.

$$\begin{aligned}
\Gamma_1 &= \|\phi\|_\infty^{\deg(f)} = E^{2\deg(f)/t} = E^{(2d(r+1))/t} = E^{(2n(r+1))/(kt)} \\
&= L_Q\left(\frac{2}{3}, \frac{2(r+1)c_b}{ktc_p}\right); \\
\pi_1^{-1} &= L_Q\left(\frac{1}{3}, \frac{2(r+1)}{3ktc_p}\right); \\
\Gamma_2 &= \|\phi\|_\infty^{\deg(g)} \times \|g\|_\infty^{\deg(\phi)} = E^{2\deg(g)/t} \times Q^{(t-1)/(d(r+1))} \\
&= E^{(2rd)/t} \times Q^{(t-1)/(d(r+1))} = E^{(2rn)/(kt)} \times Q^{k(t-1)/(n(r+1))} \\
&= L_Q\left(\frac{2}{3}, \frac{2rc_b}{c_pkt} + \frac{kc_p(t-1)}{r+1}\right); \\
\pi_2^{-1} &= L_Q\left(\frac{1}{3}, -c_v + \frac{1}{3(c_b - c_v)}\left(\frac{2rc_b}{c_pkt} + \frac{kc_p(t-1)}{r+1}\right)\right); \\
\pi^{-1} &= L_Q\left(\frac{1}{3}, \frac{2(r+1)}{3ktc_p} - c_v + \frac{1}{3(c_b - c_v)}\left(\frac{2rc_b}{c_pkt} + \frac{kc_p(t-1)}{r+1}\right)\right);
\end{aligned}$$

From the condition $\pi^{-1} = B$, we obtain the following equation.

$$c_b = \frac{2(r+1)}{3ktc_p} - c_v + \frac{1}{3(c_b - c_v)}\left(\frac{2rc_b}{c_pkt} + \frac{kc_p(t-1)}{r+1}\right). \tag{30}$$

We wish to find c_v such that c_b is minimised subject to the constraint (30). Using the method of Lagrange multipliers, the partial derivative of (30) with respect to c_v gives

$$c_v = \frac{r+1}{3ktc_p}.$$

Using this value of c_v in (30) provides the following quadratic in c_b.

$$(3ktc_p)c_b^2 - (4r+2)c_b + \frac{(r+1)^2}{3ktc_p} - \frac{(c_pk)^2 t(t-1)}{r+1} = 0.$$

Solving this and taking the positive square root, we obtain

$$c_b = \frac{4r+2}{6ktc_p} + \sqrt{\frac{r(3r+2)}{(3ktc_p)^2} + \frac{c_pk(t-1)}{3(r+1)}}. \tag{31}$$

Hence the overall complexity of MNFS for the boundary case is $L_Q\left(\frac{1}{3}, 2c_b\right)$. □

8.1 Further Analysis of the Boundary Case

Theorem 4 expresses $2c_b$ as a function of c_p, t, k and r. Let us write this as $2c_b = \mathbf{C}(c_p, t, k, r)$. It turns out that fixing the values of (t, k, r) gives a set $S(t, k, r)$ such that for $c_p \in S(t, k, r)$, $\mathbf{C}(c_p, t, k, r) \leq \mathbf{C}(c_p, t', k', r')$ for any

$(t', k', r') \neq (t, k, r)$. In other words, for a choice of (t, k, r), there is a set of values for c_p where the minimum complexity of MNFS-$\mathcal{A}$ is attained. The set $S(t, k, r)$ could be empty implying that the particular choice of (t, k, r) is sub-optimal.

For $1.12 \leq c_p \leq 4.5$, the appropriate intervals are given in Table 4. Further, the interval $(0, 1.12]$ is the union of $S(t, 1, 1)$ for $t \geq 3$. Note that the choice $(t, k, r) = (t, 1, 1)$ specialises MNFS-$\mathcal{A}$ to MNFS-Conjugation. So, for $c_p \in (0, 1.12] \cup [1.45, 3.15]$ the complexity of MNFS-$\mathcal{A}$ is the same as that of MNFS-Conjugation.

Table 4. Choices of (t, k, r) and the corresponding $S(t, k, r)$.

(t, k, r)	$S(t, k, r)$
$(t, 1, 1)$, $t \geq 3$	$\bigcup_{t \geq 3} S(t, 1, 1) \approx (0, 1.12]$
$(2, 3, 3)$	$[(1/3)(4\sqrt{21} + 20)^{1/3}, (\sqrt{78}/9 + 29/36)^{1/3}] \approx [1.12, 1.21]$
$(2, 2, 2)$	$[(\sqrt{78}/9 + 29/36)^{1/3}, (1/2)(4\sqrt{11} + 11)^{1/3}] \approx [1.21, 1.45]$
$(2, 1, 1)$	$[(1/2)(4\sqrt{11} + 11)^{1/3}, (2\sqrt{62} + 31/2)^{1/3}] \approx [1.45, 3.15]$
$(2, 1, 2)$	$[(2\sqrt{62} + 31/2)^{1/3}, (8\sqrt{33} + 45)^{1/3}] \approx [3.15, 4.5]$

In Fig. 4, we have plotted $2c_b$ given by Theorem 4 against c_p for some values of t, k and r where the minimum complexity of MNFS-$\mathcal{A}$ is attained. The plot is labelled MNFS-$\mathcal{A}$. The sets $S(t, k, r)$ are clearly identifiable from the plot. The figure also shows a similar plot for NFS-$\mathcal{A}$ which shows the complexity in the boundary case given by Theorem 1. For comparison, we have plotted the complexities of the GJL and the Conjugation methods from [5] and the MNFS-GJL and the MNFS-Conjugation methods from [24].

Based on the plots given in Fig. 4, we have the following observations.

1. Complexities of NFS-$\mathcal{A}$ are never worse than the complexities of NFS-GJL and NFS-Conjugation. Similarly, complexities of MNFS-$\mathcal{A}$ are never worse than the complexities of MNFS-GJL and MNFS-Conjugation.
2. For both the NFS-$\mathcal{A}$ and the MNFS-$\mathcal{A}$ methods, increasing the value of r provides new complexity trade-offs.
3. There is a value of c_p for which the minimum complexity is achieved. This corresponds to the MNFS-Conjugation. Let $L_Q(1/3, \theta_0)$ be this complexity. The value of θ_0 is determined later.
4. Let the complexity of the MNFS-GJL be $L_Q(1/3, \theta_1)$. The value of θ_1 was determined in [24]. The plot for MNFS-$\mathcal{A}$ approaches the plot for MNFS-GJL from below.
5. For smaller values of c_p, it is advantageous to choose $t > 2$ or $k > 1$. On the other hand, for larger values of c_p, the minimum complexity is attained for $t = 2$ and $k = 1$.

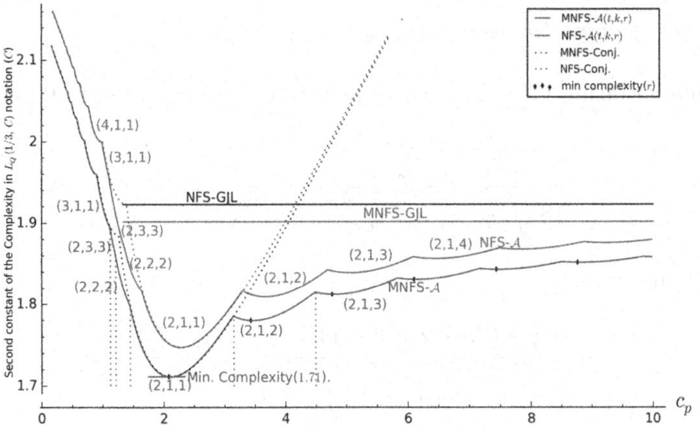

Fig. 4. Complexity plot for boundary case

From the plot, it can be seen that for larger values of c_p, the minimum value of c_b is attained for $t = 2$ and $k = 1$. So, we decided to perform further analysis using these values of t and k.

8.2 Analysis for $t = 2$ and $k = 1$

Fix $t = 2$ and $k = 1$ and let us denote $\mathbf{C}(c_p, 2, 1, r)$ as simply $\mathbf{C}(c_p, r)$. Then from Theorem 4 the complexity of MNFS-$\mathcal{A}$ for $p = L_Q(2/3, c_p)$ is $L_Q(1/3, \mathbf{C}(c_p, r))$ where

$$\mathbf{C}(c_p, r) = 2c_b = 2\sqrt{\frac{c_p}{3\,(r+1)} + \frac{(3\,r+2)r}{36\,c_p^2} + \frac{2\,r+1}{3\,c_p}}. \tag{32}$$

Figure 4 shows that for each $r \geq 1$, there is an interval $[\epsilon_0(r), \epsilon_1(r)]$ such that for $c_p \in [\epsilon_0(r), \epsilon_1(r)]$, $\mathbf{C}(c_p, r) < \mathbf{C}(c_p, r')$ for $r \neq r'$. For $r = 1$, we have

$$\epsilon_0(1) = \frac{1}{2}\left(4\sqrt{11} + 11\right)^{\frac{1}{3}} \approx 1.45;\ \epsilon_1(1) = \left(2\sqrt{62} + \frac{31}{2}\right)^{\frac{1}{3}} \approx 3.15.$$

For $p = L_Q(2/3, c_p)$, the complexity of MNFS-$\mathcal{A}$ is same as the complexity of MNFS-Conj. for c_p in $[1.45, 3.15]$; for $c_p > 3.15$, the complexity of MNFS-$\mathcal{A}$ is lower than the complexity of all prior methods. The following result shows that the minimum complexity attainable by MNFS-$\mathcal{A}$ approaches the complexity of MNFS-GJL from below.

Theorem 5. *For $r \geq 1$, let $C(r) = \min_{c_p > 0} \mathbf{C}(c_p, r)$. Then*

1. $C(1) = \theta_0 = \left(\frac{146}{261}\sqrt{22} + \frac{208}{87}\right)^{1/3}$.
2. *For $r \geq 1$, $C(r)$ is monotone increasing and bounded above.*

3. The limiting upper bound of $C(r)$ is $\theta_1 = \left(\frac{2\times(13\sqrt{13}+46)}{27}\right)^{1/3}$.

Proof. Differentiating $\mathbf{C}(c_p, r)$ with respect to c_p and equating to 0 gives

$$\frac{\frac{6}{r+1} - \frac{(3\,r+2)r}{c_p^3}}{18\sqrt{\frac{c_p}{3\,(r+1)} + \frac{(3\,r+2)r}{36\,c_p^2}}} - \frac{2\,r+1}{3\,c_p^2} = 0 \tag{33}$$

On simplifying we get,

$$\frac{6c_p^3 - (3r+2)r(r+1)}{\sqrt{\left(12c_p^3 + (r+1)(3r+2)r\right)(r+1)}} - \frac{2\,r+1}{1} = 0 \tag{34}$$

Equation (34) is quadratic in c_p^3. On solving we get the following value of c_p.

$$\begin{aligned} c_p &= \left(\frac{7}{6}r^3 + 2\,r^2 + \frac{1}{6}\sqrt{13\,r^2+8\,r+1}(2\,r^2+3\,r+1) + r + \frac{1}{6}\right)^{1/3} \\ &= \rho(r)\ (\text{say}). \end{aligned} \tag{35}$$

Putting the value of c_p back in (32), we get the minimum value of C (in terms of r) as

$$C(r) = 2\sqrt{\frac{\rho(r)}{3\,(r+1)} + \frac{(3\,r+2)r}{36\,\rho(r)^2}} + \frac{2\,r+1}{3\,\rho(r)}. \tag{36}$$

All the three sequences in the expression for $C(r)$, viz, $\frac{\rho(r)}{3\,(r+1)}$, $\frac{(3\,r+2)r}{36\,\rho(r)^2}$ and $\frac{2\,r+1}{3\,\rho(r)}$ are monotonic increasing. This can be verified through computation (with a symbolic algebra package) as follows. Let s_r be any one of these sequences. Then computing s_{r+1}/s_r gives a ratio of polynomial expressions from which it is possible to directly argue that s_{r+1}/s_r is greater than one. We have done these computations but, do not present the details since they are uninteresting and quite messy. Since all the three sequences $\frac{\rho(r)}{3\,(r+1)}$, $\frac{(3\,r+2)r}{36\,\rho(r)^2}$ and $\frac{2\,r+1}{3\,\rho(r)}$ are monotonic increasing so is $C(r)$.

Note that for $r \geq 1$, $\rho(r) > (7/6)^{1/3}r > 1.05r$. So, for $r > 1$,

$$\begin{aligned} \frac{(3r+2)r}{\rho(r)^2} &= 3\left(\frac{r}{\rho(r)}\right)^2 + 2\frac{r}{\rho(r)^2} < 3\times\left(\frac{1}{1.05}\right)^2 + 2\times\frac{1}{1.05}. \\ \frac{(2r+1)}{\rho(r)} &= 2\frac{r}{\rho(r)} + \frac{1}{\rho(r)} < 2\times\frac{1}{1.05} + \frac{1}{1.05}. \end{aligned}$$

This shows that the sequences $\frac{(3r+2)r}{\rho(r)^2}$ and $\frac{(2r+1)}{\rho(r)}$ are bounded above. For $r > 8$, we have $(3r+1) < (8r+1) < r^2$ and $(2r^2+r+1/6) < r^3/3$ which implies that for $r > 8$, $\rho(r) < (7/6+1/6\times\sqrt{14}\times 3+1/3)^{1/3}r < 1.5\,r$. Using $\rho(r) < 1.5r$ for $r > 8$, it can be shown that the sequence $\left(\frac{\rho(r)}{r+1}\right)_{r>8}$ is bounded above by 1.5. Since the three constituent sequences $\frac{\rho(r)}{(r+1)}$, $\frac{(3r+2)r}{\rho(r)^2}$ and $\frac{2r+1}{\rho(r)}$ are bounded above, it follows that $C(r)$ is also bounded above. Being monotone increasing and bounded above $C(r)$ is convergent. We claim that

$$\lim_{r\to\infty} C(r) = \left(\frac{2\times(13\sqrt{13}+46)}{27}\right)^{1/3}.$$

The proof of the claim is the following. Using the expression for $\rho(r)$ from (35) we have $\lim_{r\to\infty}\frac{\rho(r)}{r} = \left(\frac{2}{6}\sqrt{13}+\frac{7}{6}\right)^{\frac{1}{3}}$. Now,

$$C(r) = 2\sqrt{\frac{\rho(r)/r}{3\,(1+1/r)} + \frac{(3+2/r)}{36\,\rho(r)^2/r^2} + \frac{2+1/r}{3\,\rho(r)/r}}. \tag{37}$$

Hence,

$$\lim_{r\to\infty} C(r) = 2\sqrt{\frac{(2\sqrt{13}+7)^{1/3}}{3\times 6^{1/3}} + \frac{3\times 6^{2/3}}{36\,(2\sqrt{13}+7)^{2/3}} + \frac{2\times 6^{1/3}}{3\,(2\sqrt{13}+7)^{1/3}}}$$

After further simplification, we get

$$\lim_{r\to\infty} C(r) = \left(\frac{2\times(13\sqrt{13}+46)}{27}\right)^{1/3}.$$

The limit of $C(r)$ as r goes to infinity is the value of θ_1 where $L_Q(1/3, \theta_1)$ is the complexity of MNFS-GJL as determined in [24]. This shows that as r goes to infinity, the complexity of MNFS-$\mathcal{A}$ approaches the complexity of MNFS-GJL from below.

We have already seen that $C(r)$ is monotone increasing for $r \geq 1$. So, the minimum value of $C(r)$ is obtained for $r = 1$. After simplifying $C(1)$, we get the minimum complexity of MNFS-$\mathcal{A}$ to be

$$L_Q\left(1/3, \left(\frac{146}{261}\sqrt{22}+\frac{208}{87}\right)^{1/3}\right) = L\,(1/3, 1.7116)\,. \tag{38}$$

This minimum complexity is obtained at $c_p = \rho(1) = \left(\sqrt{22}+\frac{13}{3}\right)^{1/3} = 2.0819$. □

Note 1. As mentioned earlier, for $r = k = 1$, the new method of polynomial selection becomes the Conjugation method. So, the minimum complexity of MNFS-$\mathcal{A}$ is the same as the minimum complexity for MNFS-Conjugation. Here we note that the value of the minimum complexity given by (38), is not same as the one reported by Pierrot in [24]. This is due to an error in the calculation in [24][2].

Complexity of NFS-$\mathcal{A}$: From Fig. 4, it can be seen that there is an interval for c_p for which the complexity of NFS-$\mathcal{A}$ is better than both MNFS-Conjugation and MNFS-GJL. An analysis along the lines as done above can be carried out to formally show this. We skip the details since these are very similar to (actually a bit simpler than) the analysis done for MNFS-$\mathcal{A}$. Here we simply mention the following two results:

1. For $c_p \geq \left(2\sqrt{89}+20\right)^{\frac{1}{3}} \approx 3.39$, the complexity of NFS-$\mathcal{A}$ is better than that of MNFS-Conjugation.
2. For $c_p \leq \frac{1}{8}\sqrt{390}\sqrt{(5\sqrt{13}-18)\left(\frac{26}{27}\sqrt{13}+\frac{92}{27}\right)^{\frac{1}{3}}+\frac{45}{8}\left(\frac{26}{27}\sqrt{13}+\frac{92}{27}\right)^{\frac{2}{3}}} \approx 20.91$, the complexity of NFS-$\mathcal{A}$ is better than that of MNFS-GJL.
3. So, for $c_p \in [3.39, 20.91]$, the complexity of NFS-$\mathcal{A}$ is better than the complexity of all previous method including the MNFS variants.

Current state-of-the-art: The complexity of MNFS-$\mathcal{A}$ is lower than that of NFS-$\mathcal{A}$. As mentioned earlier (before Table 4) the interval $(0, 1.12]$ is the union of $\cup_{t\geq 3} S(t,1,1)$. This fact combined with Theorem 5 and Table 4 show the following. For $p = L_Q(2/3, c_p)$, when $c_p \in (0, 1.12] \cup [1.45, 3.15]$, the complexity of MNFS-$\mathcal{A}$ is the same as that of MNFS-Conjugation; for $c_p \notin (0, 1.12] \cup [1.45, 3.15]$ and $c_p > 0$, the complexity of MNFS-$\mathcal{A}$ is smaller than all previous methods. Hence, MNFS-$\mathcal{A}$ should be considered to provide the current state-of-the-art asymptotic complexity in the boundary case.

8.3 Medium and Large Characteristic Cases

In a manner similar to that used to prove Theorem 4, it is possible to work out the complexities for the medium and large characteristic cases of the MNFS corresponding to the new polynomial selection method. To tackle the medium prime case, the value of t is taken to be $t = c_t n\left((\ln Q)(\ln\ln Q)\right)^{-1/3}$ and to tackle the large prime case, the value of r is taken to be $r = c_r/2\left((\ln Q)(\ln\ln Q)\right)^{1/3}$. This will provide a relation between c_b, c_v and r (for the medium prime case) or t (for the large prime case). The method of Lagrange multipliers is then used to find the minimum value of c_b. We have carried out these computations and the complexities turn out to be the same as those obtained in [24] for the MNFS-GJL (for large characteristic) and the MNFS-Conjugation (for medium characteristic) methods. Hence, we do not present these details.

[2] The error is the following. The solution for c_b to the quadratic $(18t^2c_p^2)c_b^2 - (36tc_p)c_b + 8 - 3t^2(t-1)c_p^3 = 0$ is $c_b = 1/(tc_p) + \sqrt{5/(9(c_pt)^2) + (c_p(t-1))/6}$ with the positive sign of the radical. In [24], the solution is erroneously taken to be $1/(tc_p) + \sqrt{5/((9c_pt)^2) + (c_p(t-1))/6}$.

9 Conclusion

In this work, we have proposed a new method for polynomial selection for the NFS algorithm for fields $\mathbb{F}_{p^n}$ with $n > 1$. Asymptotic analysis of the complexity has been carried out both for the classical NFS and the MNFS algorithms for polynomials obtained using the new method. For the boundary case with $p = L_Q(2/3, c_p)$ for c_p outside a small set, the new method provides complexity which is lower than all previously known methods.

References

1. Adleman, L.M.: The function field sieve. In: Adleman, L.M., Huang, M.-D. (eds.) ANTS 1994. LNCS, vol. 877, pp. 108–121. Springer, Heidelberg (1994)
2. Adleman, L.M., Huang, M.-D.A.: Function field sieve method for discrete logarithms over finite fields. Inf. Comput. **151**(1–2), 5–16 (1999)
3. Bai, S., Bouvier, C., Filbois, A., Gaudry, P., Imbert, L., Kruppa, A., Morain, F., Thomé, E., Zimmermann, P.: CADO-NFS, an implementation of the number field sieve algorithm. CADO-NFS, Release 2.1.1 (2014). http://cado-nfs.gforge.inria.fr/
4. Barbulescu, R.: An appendix for a recent paper of Kim. IACR Cryptology ePrint Archive 2015:1076 (2015)
5. Barbulescu, R., Gaudry, P., Guillevic, A., Morain, F.: Improving NFS for the discrete logarithm problem in non-prime finite fields. In: Oswald, E., Fischlin, M. (eds.) EUROCRYPT 2015. LNCS, vol. 9056, pp. 129–155. Springer, Heidelberg (2015)
6. Barbulescu, R., Gaudry, P., Joux, A., Thomé, E.: A heuristic quasi-polynomial algorithm for discrete logarithm in finite fields of small characteristic. In: Nguyen, P.Q., Oswald, E. (eds.) EUROCRYPT 2014. LNCS, vol. 8441, pp. 1–16. Springer, Heidelberg (2014)
7. Barbulescu, R., Gaudry, P., Kleinjung, T.: The tower number field sieve. In: Iwata, T., et al. (eds.) ASIACRYPT 2015. LNCS, vol. 9453, pp. 31–55. Springer, Heidelberg (2015). doi:10.1007/978-3-662-48800-3_2
8. Barbulescu, R., Pierrot, C.: The multiple number field sieve for medium and high characteristic finite fields. LMS J. Comput. Math. **17**, 230–246 (2014)
9. Bosma, W., Cannon, J., Playoust, C.: The Magma algebra system. I. The user language. J. Symbolic Comput. **24**(3–4), 235–265 (1997). Computational algebra and number theory (London, 1993)
10. Gaudry, P., Grmy, L., Videau, M.: Collecting relations for the number field sieve in GF(p^6). Cryptology ePrint Archive, Report 2016/124 (2016). http://eprint.iacr.org/
11. Gordon, D.M.: Discrete logarithms in GF(p) using the number field sieve. SIAM J. Discrete Math. **6**, 124–138 (1993)
12. Granger, R., Kleinjung, T., Zumbrägel, J.: Discrete logarithms in GF(2^{9234}). NMBRTHRY list, January 2014
13. Guillevic, A.: Computing individual discrete logarithms faster in GF(p^n). Cryptology ePrint Archive, Report 2015/513, (2015). http://eprint.iacr.org/
14. Joux, A.: Faster index calculus for the medium prime case application to 1175-bit and 1425-bit finite fields. In: Johansson, T., Nguyen, P.Q. (eds.) EUROCRYPT 2013. LNCS, vol. 7881, pp. 177–193. Springer, Heidelberg (2013)

15. Joux, A.: A new index calculus algorithm with complexity L(1/4 + o(1)) in small characteristic. In: Lange, T., Lauter, K., Lisoněk, P. (eds.) SAC 2013. LNCS, vol. 8282, pp. 355–379. Springer, Heidelberg (2014)
16. Joux, A., Lercier, R.: The function field sieve is quite special. In: Fieker, C., Kohel, D.R. (eds.) ANTS 2002. LNCS, vol. 2369, pp. 431–445. Springer, Heidelberg (2002)
17. Joux, A., Lercier, R.: Improvements to the general number field sieve for discrete logarithms in prime fields. A comparison with the gaussian integer method. Math. Comput. **72**(242), 953–967 (2003)
18. Joux, A., Lercier, R.: The function field sieve in the medium prime case. In: Vaudenay, S. (ed.) EUROCRYPT 2006. LNCS, vol. 4004, pp. 254–270. Springer, Heidelberg (2006)
19. Joux, A., Lercier, R., Smart, N.P., Vercauteren, F.: The number field sieve in the medium prime case. In: Dwork, C. (ed.) CRYPTO 2006. LNCS, vol. 4117, pp. 326–344. Springer, Heidelberg (2006)
20. Joux, A., Pierrot, C.: The special number field sieve in $\mathbb{F}_{p^n}$. In: Cao, Z., Zhang, F. (eds.) Pairing 2013. LNCS, vol. 8365, pp. 45–61. Springer, Heidelberg (2014)
21. Kalkbrener, M.: An upper bound on the number of monomials in determinants of sparse matrices with symbolic entries. Math. Pannonica **8**(1), 73–82 (1997)
22. Kim, T.: Extended tower number field sieve: a new complexity for medium prime case. IACR Cryptology ePrint Archive, 2015:1027 (2015)
23. Matyukhin, D.: Effective version of the number field sieve for discrete logarithm in a field GF(p^k). Trudy po Discretnoi Matematike **9**, 121–151 (2006). (in Russian), 2006. http://m.mathnet.ru/php/archive.phtml?wshow=paper&jrnid=tdm&paperid=144&option_lang=eng
24. Pierrot, C.: The multiple number field sieve with conjugation and generalized joux-lercier methods. In: Oswald, E., Fischlin, M. (eds.) EUROCRYPT 2015. LNCS, vol. 9056, pp. 156–170. Springer, Heidelberg (2015)
25. Sarkar, P., Singh, S.: Fine tuning the function field sieve algorithm for the medium prime case. IEEE Transactions on Information Theory, 99: 1–1 (2016)
26. Schirokauer, O.: Discrete logarithms and local units. Philosophical Transactions: Physical Sciences and Engineering **345**, 409–423 (1993)
27. Schirokauer, O.: Using number fields to compute logarithms in finite fields. Math. Comp. **69**(231), 1267–1283 (2000)
28. Schirokauer, O.: Virtual logarithms. J. Algorithms **57**(2), 140–147 (2005)
29. Stein, W.A., et al.: Sage Mathematics Software. The Sage Development Team (2013). http://www.sagemath.org

Freestart Collision for Full SHA-1

Marc Stevens[1](✉), Pierre Karpman[2,3,4], and Thomas Peyrin[4]

[1] Centrum Wiskunde & Informatica, Amsterdam, The Netherlands
marc.stevens@cwi.nl
[2] Inria, Saclay, France
pierre.karpman@inria.fr
[3] École Polytechnique, Palaiseau, France
[4] Nanyang Technological University, Singapore, Singapore
thomas.peyrin@ntu.edu.sg

Abstract. This article presents an explicit freestart colliding pair for SHA-1, *i.e.* a collision for its internal compression function. This is the first practical break of the full SHA-1, reaching all 80 out of 80 steps. Only 10 days of computation on a 64-GPU cluster were necessary to perform this attack, for a runtime cost equivalent to approximately $2^{57.5}$ calls to the compression function of SHA-1 on GPU. This work builds on a continuous series of cryptanalytic advancements on SHA-1 since the theoretical collision attack breakthrough of 2005. In particular, we reuse the recent work on 76-step SHA-1 of Karpman *et al.* from CRYPTO 2015 that introduced an efficient framework to implement (freestart) collisions on GPUs; we extend it by incorporating more sophisticated accelerating techniques such as boomerangs. We also rely on the results of Stevens from EUROCRYPT 2013 to obtain optimal attack conditions; using these techniques required further refinements for this work.

Freestart collisions do not directly imply a collision for the full hash function. However, this work is an important milestone towards an actual SHA-1 collision and it further shows how GPUs can be used very efficiently for this kind of attack. Based on the state-of-the-art collision attack on SHA-1 by Stevens from EUROCRYPT 2013, we are able to present new projections on the computational and financial cost required for a SHA-1 collision computation. These projections are significantly lower than what was previously anticipated by the industry, due to the use of the more cost efficient GPUs compared to regular CPUs.

We therefore recommend the industry, in particular Internet browser vendors and Certification Authorities, to retract SHA-1 quickly. We hope the industry has learned from the events surrounding the cryptanalytic breaks of MD5 and will retract SHA-1 before concrete attacks such as signature forgeries appear in the near future.

M. Stevens—Supported by the Netherlands Organization for Scientific Research Veni Grant 2014

P. Karpman—Partially supported by the Direction Générale de l'Armement and by the Singapore National Research Foundation Fellowship 2012 (NRF-NRFF2012-06)

T. Peyrin—Supported by the Singapore National Research Foundation Fellowship 2012 (NRF-NRFF2012-06).

M. Fischlin and J.-S. Coron (Eds.): EUROCRYPT 2016, Part I, LNCS 9665, pp. 459–483, 2016.
DOI: 10.1007/978-3-662-49890-3_18

Keywords: SHA-1 · Hash function · Cryptanalysis · Freestart collision · GPU implementation

1 Introduction

A cryptographic hash function H is a function that takes an arbitrarily long message M as input and outputs a fixed-length hash value of size n bits. It is a versatile primitive useful in many applications, such as building digital signature schemes, message authentication codes or password hashing functions. One key security feature expected from a cryptographic hash function is collision resistance: it should not be feasible for an adversary to find two distinct messages M, $\hat{M}$ that hash to the same value $H(M) = H(\hat{M})$ faster than with a generic algorithm, *i.e.* with significantly less than $2^{\frac{n}{2}}$ calls to the hash function.

A widely used hash function construction is the Merkle-Damgård paradigm [6,24]: H is built by iterating a compression function h that updates a fixed-size internal state (also called chaining value) with fixed-size message blocks; the initial chaining value (IV) is a fixed constant of the hash function. This construction is useful in particular for the simple security reduction it allows to make: if the compression function is collision-resistant, then so is the corresponding hash function. This leads to defining variants of collision attacks which allow the attacker to choose the IV: a *freestart* collision is a pair of different message and IV (C, M), $(\hat{C}, \hat{M})$ such that $H_C(M) = H_{\hat{C}}(\hat{M})$; *semi-freestart* collisions are similar but impose $C = \hat{C}$. It is noteworthy that the Merkle-Damgård security reduction assumes that any type of collision (freestart or semi-freestart) must be intractable by the adversary. Thus, a collision attack on the compression function should be taken very seriously as it invalidates the security reduction coming from the operating mode of the hash function.

The most famous hash function family, basis for most hash function industry standards, is undoubtedly the MD-SHA family, which includes notable functions such as MD4, MD5, SHA-1 and SHA-2. This family first originated with MD4 [35] and continued with MD5 [36] (due to serious security weaknesses [9,11] found on MD4 soon after its publication). Even though collision attacks on the compression function were quickly identified [10], the industry widely deployed MD5 in applications where hash functions were required. Yet, in 2005, a team of researchers led by Wang [47] completely broke the collision resistance of MD5, which allowed to efficiently compute colliding messages for the full hash function. This groundbreaking work inspired much further research on the topic; in a major development, Stevens *et al.* [44] showed that a more powerful type of attack (the so-called *chosen-prefix collision attack*) could be performed against MD5. This eventually led to the forgery of a Rogue Certification Authority that in principle completely undermined HTTPS security [45]. This past history of cryptanalysis on MD5 is yet another argument for a very careful treatment of collision cryptanalysis progress: the industry should move away from weak cryptographic hash functions or hash functions built on weak inner components (compression functions that are not collision resistant) before the seemingly theoretic attacks prove to be a direct threat to security (*counter-cryptanalysis* [42] could be used to mitigate some of the risks during the migration).

While lessons should be learned from the case of MD5, it is interesting to observe that the industry is again facing a similar challenge. SHA-1 [28], designed by the NSA and a NIST standard, is one of the main hash functions of today, and it is facing important attacks since 2005. Based on previous successful cryptanalysis works [1,2,4] on SHA-0 [27] (SHA-1's predecessor, that only differs by a single rotation in the message expansion function), a team led again by Wang *et al.* [46] showed in 2005 the very first theoretical collision attack on SHA-1. Unlike the case of MD5, this attack, while groundbreaking, remains mostly theoretical as its expected cost was evaluated to be equivalent to 2^{69} calls to the SHA-1 compression function.

Therefore, as a proof of concept, many teams considered generating real collisions for reduced versions of SHA-1: 64 steps [8] (with a cost of 2^{35} SHA-1 calls), 70 steps [7] (cost 2^{44} SHA-1), 73 steps [14] (cost $2^{50.7}$ SHA-1) and the latest advances for the hash function reached 75 steps [15] (cost $2^{57.7}$ SHA-1) using extensive GPU computation power.

In 2013, building on these advances and a novel rigorous framework for analyzing SHA-1, the current best collision attack on full SHA-1 was presented by Stevens [43] with an estimated cost of 2^{61} calls to the SHA-1 compression function. Nevertheless, a publicly known collision still remains out of reach.

Very recently, collisions on the compression function of SHA-1 reduced to 76 steps (out of 80) were obtained by using a start-from-the-middle approach and a highly efficient GPU framework [19]. This required only a reasonable amount of GPU computation power (less than a week on a single card, equivalent to about $2^{50.3}$ calls to SHA-1 on GPU, whereas the runtime cost equivalent on regular CPUs is about $2^{49.1}$ SHA-1).

Because of these worrisome cryptanalysis advances on SHA-1, one is advised to use *e.g.* SHA-2 [29] or the new hash functions standard SHA-3 [31] when secure hashing is needed. While NIST recommended that SHA-1-based certificates should not be trusted beyond 2014 [30] (by 2010 for governmental use), the industry actors only recently started to move away from SHA-1, about a decade after the first theoretical collision attacks. For example, Microsoft, Google and Mozilla have all announced that their respective browsers will stop accepting SHA-1 SSL certificates by 2017 (and that SHA-1-based certificates should not be issued after 2015). These deadlines are motivated by a simple evaluation of the computational and financial cost required to generate a collision for SHA-1: in 2012, Bruce Schneier (using calculations by Jesse Walker based on a 2^{61} attack cost [43], Amazon EC2 spotprices and Moore's Law) estimated the cost of running one SHA-1 collision attack to be around 700,000 US$ in 2015, down to about 173,000 US$ in 2018, which he deemed to be within the resources of criminals [37]. We observe that while a majority of industry actors already chose to migrate to more secure hashing algorithms, surveys show that in September 2015 SHA-1 remained the hashing primitive for about 28.2 % of certificate signatures [40].

Table 1. A freestart collision for SHA-1. A test program for this colliding pair is available at https://sites.google.com/site/itstheshappening/tester.cpp

	Message 1
IV_1	50 6b 01 78 ff 6d 18 90 20 22 91 fd 3a de 38 71 b2 c6 65 ea
M_1	9d 44 38 28 a5 ea 3d f0 86 ea a0 fa 77 83 a7 36
	33 24 48 4d af 70 2a aa a3 da b6 79 d8 a6 9e 2d
	54 38 20 ed a7 ff fb 52 d3 ff 49 3f c3 ff 55 1e
	fb ff d9 7f 55 fe ee f2 08 5a f3 12 08 86 88 a9
Compr(IV_1,M_1)	f0 20 48 6f 07 1b f1 10 53 54 7a 86 f4 a7 15 3b 3c 95 0f 4b

	Message 2
IV_2	50 6b 01 78 ff 6d 18 91 a0 22 91 fd 3a de 38 71 b2 c6 65 ea
M_2	3f 44 38 38 81 ea 3d ec a0 ea a0 ee 51 83 a7 2c
	33 24 48 5d ab 70 2a b6 6f da b6 6d d4 a6 9e 2f
	94 38 20 fd 13 ff fb 4e ef ff 49 3b 7f ff 55 04
	db ff d9 6f 71 fe ee ee e4 5a f3 06 04 86 88 ab
Compr(IV_2,M_2)	f0 20 48 6f 07 1b f1 10 53 54 7a 86 f4 a7 15 3b 3c 95 0f 4b

1.1 Our Contributions

In this article, we give the first colliding pair for the full SHA-1 compression function (see Table 1), which amounts to a freestart collision for the full hash function. This was obtained at a GPU runtime cost approximately equivalent to $2^{57.5}$ evaluations of SHA-1.[1]

The starting point for this attack is the start-from-the-middle approach and the GPU framework of CRYPTO 2015, which was used to compute freestart collisions on the 76-step reduced SHA-1 [19]. We improve this by incorporating the auxiliary paths (or boomerangs) speed-up technique from Joux and Peyrin [17]. We also rely on the cryptanalytic techniques by Stevens [43] to obtain optimal attack conditions, which required further refinements for this work.

As was mentioned above, previous recommendations on retracting SHA-1 were based on estimations of the resources needed to find SHA-1 collisions. These consist both in the time necessary to mount an attack, for a given computational power, as well as the cost of building and maintaining this capability, or of renting the equipment directly on a platform such as Amazon EC2 [38]. In that respect, our freestart collision attack can be run in about 9 to 10 days on average on a cluster with 64 GeForce GTX970 GPUs, or by renting GPU time on Amazon

[1] Which from previous experience is about a factor 2 higher than the runtime cost in equivalent number of SHA-1 evaluations on regular CPUs.

EC2 for about 2K US$.[2] Based on this experimental data and the 2013 state-of-the-art collision attack, we can project that a complete `SHA-1` collision would take between 49 and 78 days on a 512 GPU cluster, and renting the equivalent GPU time on EC2 would cost between 75K US$ and 120K US$ and would plausibly take at most a few months.

Although freestart collisions do not directly translate to collisions for the hash function, they directly invalidate the security reduction of the hash function to the one of the compression function. Hence, obtaining a concrete example of such a collision further highlights the weaknesses of `SHA-1` and existing users should quickly stop using this hash function. In particular, we believe that our work shows that the industry's plan to move away from `SHA-1` in 2017 might not be soon enough.

Outline. In Sect. 2, we provide our analysis and recommendations regarding the timeline of migration from `SHA-1` to a secure hash function. In Sect. 3 we give a short description of the `SHA-1` hash function and our notations. In Sect. 4, we explain the structure of our cryptanalysis and the various techniques used from a high level point of view, and we later provide in Sect. 5 all the details of our attack for the interested readers.

2 Recommendations for the Swift Removal of SHA-1

Our work allowed to generate a freestart collision for the full `SHA-1`, but a collision for the entire hash algorithm is still unknown. There is no known generic and efficient algorithm that can turn a freestart collision into a plain collision for the hash function. However, the advances we have made do allow us to precisely estimate and update the computational and financial cost to generate such a collision with latest cryptanalysis advances [43] (the computational cost required to generate such a collision was actually a recurrent debate in the academic community since the first theoretical attack from Wang *et al.* [46]).

Schneier's projections [37] on the cost of `SHA-1` collisions in 2012 (on EC2: ≈700K US$ by 2015, ≈173K US$ by 2018 and ≈43K US$ by 2021) were based on (an early announcement of) [43]. As mentioned earlier, these projections have been used to establish the timeline of migrating away from `SHA-1`-based signatures for secure Internet websites, resulting in a migration by January 2017 — one year before Schneier estimated that a `SHA-1` collision would be within the resources of criminal syndicates.

However, as remarked in [19] and now further improved in this article thanks to the use of boomerang speed-up techniques [17], GPUs are much faster for this

[2] This is based on the spot price for Amazon EC2 GPU Instance Type 'g2.8xlarge', featuring 4 GPUs, which is about 0.50 US$ per hour as of October 2015. These four GPU cards are comparable to NVidia Tesla cards and actually contain 2 physical GPU chips each. But due to their lower clock speed and slightly lower performance we estimate that each card is comparable to about one GTX970s.

type of attacks (compared to CPUs) and we now precisely estimate that a full SHA-1 collision should not cost more than between 75 K and 120K US$ by renting Amazon EC2 cloud over a few months at the time of writing, in early autumn 2015. Our new GPU-based projections are now more accurate and they are significantly below Schneier's estimations. More worrying, they are theoretically already within Schneier's estimated resources of criminal syndicates as of today, almost two years earlier than previously expected, and one year before SHA-1 being marked as unsafe in modern Internet browsers. Therefore, we believe that migration from SHA-1 to the secure SHA-2 or SHA-3 hash algorithms should be done sooner than previously planned.

Note that it has previously been shown that a more advanced so-called chosen-prefix collision attack on MD5 allowed the creation of a rogue Certification Authority undermining the security of all secure websites [45]. Collisions on SHA-1 can result in *e.g.* signature forgeries, but do not directly undermine the security of the Internet at large. More advanced so-called chosen-prefix collisions [45] are significantly more threatening, but currently much costlier to mount. Yet, given the lessons learned with the MD5 full collision break, it is not advisable to wait until these become practically possible.

At the time of the submission of this article in October 2015, we learned that in an ironic turn of events the CA/Browser Forum[3] was planning to hold a ballot to decide whether to extend issuance of SHA-1 certificates through the year 2016 [12]. With our new cost projections in mind, we strongly recommended against this extension and the ballot was subsequently withdrawn [13]. Further action is also being considered by major browser providers such as Microsoft [25] and Mozilla [26] in speeding up the removal of SHA-1 certificates.

3 Preliminaries

3.1 Description of SHA-1

We start this section with a brief description of the SHA-1 hash function. We refer to the NIST specification document [28] for a more thorough presentation. SHA-1 is a hash function from the MD-SHA family which produces digests of 160 bits. It is based on the popular Merkle-Damgård paradigm [6,24], where the (padded) message input to the function is divided into k blocks of a fixed size (512 bits in the case of SHA-1). Each block is fed to a compression function h which then updates a 160-bit chaining value cv_i using the message block m_{i+1}, *i.e.* $cv_{i+1} = h(cv_i, m_{i+1})$. The initial value $cv_0 = IV$ is a predefined constant and cv_k is the output of the hash function.

Similarly to other members of the MD-SHA family, the compression function h is built around an *ad hoc* block cipher E used in a Davies-Meyer construction: $cv_{i+1} = E(m_{i+1}, cv_i) + cv_i$, where $E(x, y)$ is the encryption of the plaintext y with the key x and "+" denotes word-wise addition in $\mathbf{Z}/2^{32}\,\mathbf{Z}$. The block cipher itself

[3] The CA/Browser Forum is the main association of industries regulating the use of digital certificates on the Internet.

is an 80-step (4 rounds of 20 steps each) five-branch generalized Feistel network using an Add-Rotate-Xor "ARX" step function. The internal state consists in five 32-bit registers $(A_i, B_i, C_i, D_i, E_i)$; at each step, a 32-bit extended message word W_i is used to update the five registers:

$$\begin{cases} A_{i+1} = (A_i \lll 5) + f_i(B_i, C_i, D_i) + E_i + K_i + W_i \\ B_{i+1} = A_i \\ C_{i+1} = B_i \ggg 2 \\ D_{i+1} = C_i \\ E_{i+1} = D_i \end{cases}$$

where K_i are predetermined constants and f_i are Boolean functions (see Table 2 for their specifications). As all updated registers but A_{i+1} are just rotated copies of another, it is possible to equivalently express the step function in a recursive way using only the register A:

$$A_{i+1} = (A_i \lll 5) + f_i(A_{i-1}, A_{i-2} \ggg 2, A_{i-3} \ggg 2) + (A_{i-4} \ggg 2) + K_i + W_i.$$

Table 2. Boolean functions and constants of SHA-1

round	step i	$f_i(B,C,D)$	K_i
1	$0 \le i < 20$	$f_{IF} = (B \wedge C) \oplus (\overline{B} \wedge D)$	0x5a827999
2	$20 \le i < 40$	$f_{XOR} = B \oplus C \oplus D$	0x6ed6eba1
3	$40 \le i < 60$	$f_{MAJ} = (B \wedge C) \oplus (B \wedge D) \oplus (C \wedge D)$	0x8fabbcdc
4	$60 \le i < 80$	$f_{XOR} = B \oplus C \oplus D$	0xca62c1d6

Finally, the extended message words W_i are computed from the 512-bit message block, which is split into sixteen 32-bit words $M_0, \ldots, M_{15}$. These sixteen words are then expanded linearly into the eighty 32-bit words W_i as follows:

$$W_i = \begin{cases} M_i, & \text{for } 0 \le i \le 15 \\ (W_{i-3} \oplus W_{i-8} \oplus W_{i-14} \oplus W_{i-16}) \lll 1, & \text{for } 16 \le i \le 79 \end{cases}$$

The step function and the message expansion can both easily be inverted.

3.2 Differential Collision Attacks on SHA-1

We now introduce the main notions used in a collision attack on SHA-1 (and more generally on members of the MD-SHA family).

Background. In a differential collision attack on a (Merkle-Damgård) hash function, the goal of the attacker is to find a high-probability differential path (the differences being on the message, and also on the IV in the case of a freestart attack) which entails a zero difference on the final state of the function (*i.e.* the hash value). A pair of messages (and optionally IVs) following such a path indeed leads to a collision.

In the case of SHA-1 (and more generally ARX primitives), the way of expressing differences between messages is less obvious than for *e.g.* bit or byte-oriented primitives. It is indeed natural to consider both "XOR differences" (over $\mathbf{F}_2^n$) and "modular differences" (over $\mathbf{Z}/2^n\mathbf{Z}$) as both operations are used in the function. In practice, the literature on SHA-1 uses several hybrid representations of differences based on *signed XOR differences.* In its most basic form, such a difference is similar to an XOR difference with the additional information of the value of the differing bits for each message (and also of some bits equal in the two messages), which is a "sign" for the difference. This is an important information when one works with modular addition as the sign impacts the (absence of) propagation of carries in the addition of two differences. Let us for instance consider the two pairs of words $a = 11011000001_b$, $\hat{a} = 11011000000_b$ and $b = 10100111000_b$, $\hat{b} = 10100111001_b$; the XOR differences $(a \oplus \hat{a})$ and $(b \oplus \hat{b})$ are both 00000000001_b (which may be written ....x), meaning that $(a \oplus b) = (\hat{a} \oplus \hat{b})$. On the other hand, the signed XOR difference between a and $\hat{a}$ may be written ....- to convey the fact that they are different on their lowest bit *and* that the value of this bit is 1 for a (and thence 0 for $\hat{a}$); similarly, the signed difference between b and $\hat{b}$ may be written ....+, which is a difference in the same position but of a different sign. From these differences, we can deduce that $(a + b) = (\hat{a} + \hat{b})$ because differences of different signs cancel; if we were to swap the values b and $\hat{b}$, both differences on a and b would have the same sign and indeed we would have $(a + b) \neq (\hat{a} + \hat{b})$ (though $(a \oplus b)$ and $(\hat{a} \oplus \hat{b})$ would still be equal). It is possible to extend signed differences to account for more generic combinations of possible values for each message bit; this was for instance done by De Cannière and Rechberger to aid in the automatic search of differential paths [8]. Another possible extension is to consider relations between various bits of the (possibly rotated) state words; this allows to efficiently keep track of the propagation of differences through the step function. Such differences are for instance used by Stevens [43], and also in this work (see Fig. 2).

The structure of differential attacks on SHA-1 evolved to become quite specific. At a high level, they consist of: 1. a *non-linear* differential path of low probability; 2. a *linear* differential path of high probability; 3. accelerating techniques.

The terms *non-linear* and *linear* refer to how the paths were obtained: the latter is derived from a linear (over $\mathbf{F}_2^{32}$) modelling of the step function. This kind of path is used in the probabilistic phase of the attack, where one simply tries many message pairs in order to find one that indeed "behaves" linearly. Computing the exact probability of this event is however not easy, although it is not too hard to find reasonable estimates. This probability is the main factor determining the final complexity of the attack.

The role of a non-linear path is to bootstrap the attack by bridging a state with no differences (the IV) with the start of the linear differential path[4]. In a nutshell, this is necessary because these paths do not typically lie in the kernel of the linearized SHA-1; hence it is impossible to obtain a collision between two messages following a fully linear path. This remains true in the present case of a freestart attack, even if the non-linear path now connects the start of the linear path with an IV containing some differences. Unlike the linear path, the non-linear one has a very low probability of being followed by random messages. However, the attacker can fully choose the messages to guarantee that they do follow the path, as he is free to set the 512 bits of the message. Hence finding conforming message pairs for this path effectively costs nothing in the attack.

Finally, the role of accelerating techniques is to find efficient ways of using the freedom degrees remaining after a pair following the non-linear path has been found, in order to delay the effective moment where the probabilistic phase of the attack starts.

We conclude this section with a short discussion of how to construct these three main parts of a (freestart) collision attack on SHA-1.

Linear Path; Local Collisions. The linear differential paths used in collision attacks are built around the concept of *local collision*, introduced by Chabaud and Joux in 1998 to attack SHA-0. The idea underlying a local collision is first to introduce a difference in one of the intermediate state words of the function, say A_i, through a difference in the message word W_{i-1}. For an internal state made of j words ($j = 5$ in the case of SHA-0 or SHA-1), the attacker then uses subsequent differences in (possibly only some of) the message words $W_{i\ldots i+(j-1)}$ in order to cancel any contribution of the difference in A_i in the computation of a new internal state $A_{i+1\ldots i+j}$, which will therefore have no differences. The positions of these "correcting" differences are dictated by the step function, and there may be different options depending on the used Boolean function, though originally (and in most subsequent cases) these were chosen according to a linearized model (over $\mathbf{F}_2^{32}$) of the step functions.

Local collisions are a fit basis to generate differential paths of good probability. The main obstacle to do this is that the attacker does not control all of the message words, as some are generated by the message expansion. Chabaud and Joux showed how this could be solved by chaining local collisions along a *disturbance vector* (DV) in such a way that the final state of the function contains no difference and that the pattern of the local collisions is compatible with the message expansion. The disturbance vector just consists of a sparse message (of

[4] For the sake of simplicity, we ignore here the fact that a collision attack on SHA-1 usually uses two blocks, with the second one having differences in its chaining value. The general picture is actually the following: once a pair of messages following the linear path $\mathcal{P}^+$ is found, the first block ends with a signed difference $+\Delta$; the sign of the linear path is then switched for the second block to become $\mathcal{P}^-$ and following this path results in a difference $-\Delta$; the feedforward then cancels both differences and yields a collision.

sixteen 32-bit words) that has been expanded with the linear message expansion of SHA-1. Every "one" bit of this expanded message then marks the start of a local collision (and expanding all the local collisions thus produces a complete linear path).

Each local collision in the probabilistic phase of the attack (roughly corresponding to the last three rounds) increases the overall complexity of the attack, hence one should use disturbance vectors that are sparse over these rounds. Initially, the evaluation of the probability of disturbance vector candidates was done mostly heuristically, using *e.g.* the Hamming weight of the vector [1,18,22,33,34], the sum of bit conditions for each local collision independently (not allowing carries) [48,49], and the product of independent local collision probabilities (allowing carries) [21,23]. Manuel [20,21] noticed that all disturbance vectors used in the literature belong to two classes $\mathrm{I}(K,b)$ and $\mathrm{II}(K,b)$. Within each class all disturbance vectors are forward or backward shifts in the step index (controlled by K) and/or bitwise cyclic rotations (controlled by b) of the same expanded message. We will use this notation through the remainder of this article.

Manuel also showed that success probabilities of local collisions are not always independent, causing biases in the above mentioned heuristic cost functions. This was later resolved by Stevens using a technique called joint local-collision analysis (JLCA) [41,43], which allows to analyze entire sets of differential paths over the last three rounds that conform to the (linear path entailed by the) disturbance vector. This is essentially an exhaustive analysis taking into account all local collisions together, using which one can determine the highest possible success probability. This analysis also produces a minimal set of sufficient conditions which, when all fulfilled, ensure that a pair of messages follows the linear path; the conditions are minimal in the sense that meeting all of them happens with this highest probability that was computed by the analysis. Although a direct approach is clearly unfeasible (as it would require dealing with an exponentially growing amount of possible differential paths), JLCA can be done practically by exploiting the large amount of redundancy between all the differential paths to a very large extent.

Non-linear Differential Path. The construction of non-linear differential paths was initially done by hand by Wang, Yin and Yu in their first attack on the full SHA-1 [46]. Efficient algorithmic construction of such differential paths was later proposed in 2006 by De Cannière and Rechberger, who introduced a guess-and-determine approach [8]. A different approach based on a meet-in-the-middle method was also proposed by Stevens *et al.* [16,41].

Accelerating Techniques. For a given differential path, one can derive explicit conditions on state and message bits which are sufficient to ensure that a pair of messages follows the path. This lets the collision search to be entirely defined over a single compression function computation. Furthermore, they also allow detection of "bad" message pairs a few steps earlier compared to computing the state and verifying differences, allowing to abort computations earlier in this case.

An important contribution of Wang, Yin and Yu was the introduction of powerful *message modification* techniques, which followed an earlier work of Biham and Chen who introduced *neutral bits* to produce better attacks on SHA-0 [1]. The goal of both techniques is for the attacker to make a better use of the available freedom in the message words in order to decrease the complexity of the attack. Message modifications try to correct bad message pairs that only slightly deviate from the differential path, and neutral bits try to generate several good message pairs out of a single one (by changing the value of a bit which does not invalidate nearby sufficient conditions with good probability). In essence, both techniques allow to amortize part of the computations, which effectively delays the beginning of the purely probabilistic phase of the attack.

Finally, Joux and Peyrin showed how to construct powerful neutral bits and message modifications by using auxiliary differential paths akin to *boomerangs* [17], which allow more efficient attacks. In a nutshell, a boomerang (in collision attacks) is a small set of bits that together form a local collision. Hence flipping these bits together ensures that the difference introduced by the first bit of the local collision does not propagate to the rest of the state; if the initial difference does not invalidate a sufficient condition, this local collision is indeed a neutral bit. Yet, because the boomerang uses a single (or sometimes a few) local collision, more differences will actually be introduced when it goes through the message expansion. The essence of boomerangs is thus to properly choose where to locate the local collisions so that no differences are introduced for the most steps possible.

4 Attack Overview

In this section we provide an overview of how our attack was constructed. At a high level, it consists of the following steps:

1. *Disturbance Vector Selection*: We need to select the best disturbance vector for our attack. This choice is based on results provided by joint local collision analysis (JLCA), taking into account constraints on the number and position of sufficient conditions on the IV implied by the disturbance vector. We explain this in Sect. 4.1.
2. *Finding Optimal Attack Conditions*: Having selected a disturbance vector, we need to determine a set of attack conditions over all steps consisting of sufficient conditions for state bits up to some step, augmented by message bit relations. We use non-linear differential path construction methods to determine conditions within the first round. Using JLCA we derive an optimal complete set of attack conditions that given the first round path leads to the highest possible success probability over all steps, yet minimizes the number of conditions within this model. We detail this in Sect. 4.2.
3. *Finding and Analyzing Boomerangs and Neutral Bits*: To speed up the freestart collision attack, we exploit advanced message modification techniques such as (multiple) neutral bits and boomerangs. In order to find suitable candidates, we sample partial solutions fulfilling the above attack

conditions up to an early step. The samples are used to test many potential boomerangs and neutral bits, and only ones of good quality and that do not introduce contradictions will be used. In particular, no boomerang or neutral bit may invalidate the attack conditions of the so-called base solution (see below, includes all message bit relations) with non-negligible probability. We also use sampling to estimate the probability of interaction between boomerang and neutral bits with particular sufficient conditions, in the forward and backward direction. Although we do not allow significant interaction in the backward direction, we use these probabilities to determine at which step the boomerang or neutral bit are used. This is explained in Sect. 4.3.

4. *Base Solution Generation*: Before we can apply neutral bits and boomerangs, we first need to compute a partial solution over 16 consecutive steps. Only this partial solution can then be extended to cover more steps by using neutral bits and boomerangs. We call such a solution a *base solution*; it consists of state words $A_{-3}, \ldots, A_{17}$ and message words $W_1, \ldots, W_{16}$. The cost for generating base solutions is relatively low compared to the overall attack cost, therefore it is not heavily optimized and the search is run on regular CPUs. This is further explained in Sect. 4.4.
5. *Application of Neutral Bits and Boomerangs on GPU*: We extend each base solution into solutions over a larger number of steps by successively applying neutral bits and boomerangs and verifying sufficient conditions. Once all neutral bits and boomerangs have been exploited, the remainder of the steps have to be fulfilled probabilistically.

   This is computationally the most intensive part, and it is therefore implemented on GPUs that are significantly more cost-efficient than CPUs, using the highly efficient framework introduced by Karpman, Peyrin and Stevens [19]. More details are provided in Sect. 4.5.

All these steps strongly build upon the continuous series of papers that have advanced the state-of-the-art in SHA-1 cryptanalysis, yet there are still small adaptions and improvements used for this work. We now describe all these points in more details.

4.1 Disturbance Vector Selection

It is possible to compute exactly the highest success probability over the linear part by using joint-local collision analysis [43]. By further using the improvements described in [19], one can restrict carries for the steps where sufficient conditions are used and obtain the sufficient conditions for those steps immediately.

The number of sufficient conditions at the beginning of round 2 and the associated highest success probability for the remaining steps provide insight into the attack complexity under different circumstances. In Table 3 we give our analysis results for various DVs, listing the negative $\log_2$ of the success probability over steps $[24, 80)$ assuming that all sufficient conditions up to A_{24} have been satisfied; we also include the number of conditions on A_{24} and A_{23}. The

Table 3. Disturbance vector analysis. For each DV, under $c_{[24,80)}$, we list the negative $\log_2$ of the success probability over steps $[24, 80)$ assuming that all sufficient conditions up to A_{24} have been satisfied. The columns c_{23} and c_{22} list the number of conditions on A_{24} (in step 23) and A_{23} (in step 22), respectively. The final column represents an estimated runtime in days on a cluster consisting of 64 GTX970s based on $c_{[24,80)}$.

DV	Cost $c_{[24,80)}$	Cost c_{23}	Cost c_{22}	Days on 64 GPUs
I(48,0)	61.6	1	3	39.1
I(49,0)	60.5	3	2	18.3
I(50,0)	61.7	2	1	41.8
I(51,0)	62.1	1	2	55.7
I(48,2)	64.4	1	2	281.9
I(49,2)	62.8	2	3	90.4
II(46,0)	64.8	1	0	369.5
II(50,0)	59.6	1	2	9.9
II(51,0)	57.5	3	3	2.2
II(52,0)	58.3	3	3	4.1
II(53,0)	59.9	3	2	11.8
II(54,0)	61.3	2	1	31.4
II(55,0)	60.7	1	3	21.0
II(56,0)	58.9	3	2	6.3
II(57,0)	59.3	2	3	7.9
II(58,0)	59.7	3	2	10.5
II(59,0)	61.0	3	2	26.2
II(49,2)	61.0	2	3	26.1
II(50,2)	59.4	3	2	8.7
II(51,2)	59.4	2	3	8.5

final column represents an estimated runtime in days on a cluster consisting of 64 GTX970s based on $c_{[24,80)}$, by multiplying the runtime of the 76-step freestart GPU attack [19] with the difference between the costs $c_{[24,80)}$ for the 76-step attack and the DVs in the table.

Considering Table 3, the obvious choice of DV to mount a full collision attack on SHA-1 would be II(51,0). However in the present case of a freestart attack additional constraints need to be taken into account. In particular the, choice of the DV determines the possible differences in the IV, as these have to cancel the differences of the final state $(A_{80}, B_{80}, C_{80}, D_{80}, E_{80})$. This impacts the estimated runtime as follows: if there are sufficient conditions present on A_0 then the estimated runtime in the last column should be multiplied by $2^{c_{23}}$. Indeed, the 76-step freestart attack did not have any sufficient conditions on A_0 and could thus ignore step 0, leading to an offset of one in the probabilistic phase. Moreover, if the IV differences are denser or ill-located (compared to the 76-step attack), then more neutral bits and boomerangs are likely to interact badly with the sufficient conditions on the IV, when propagated backwards. If only few neutral bits and boomerangs can be used for a given DV, the cost of the attack would rise significantly. The number of conditions c_{23} and c_{22} for each DV allow to estimate how much more expensive a vector will be in the case where the probabilistic phase effectively starts sooner than in step A_{25} as for the 76-step attack.

Taking the freestart setting into account, and with a preliminary analysis of available neutral bits and boomerangs, the best option eventually seemed to be II(59,0). This DV is actually a downward shift by four of II(55,0), which was used in the 76-step attack. Consequently, this choice leads to the same IV sufficient conditions as in the latter.

4.2 Finding Optimal Attack Conditions

Using joint local collision analysis, we could obtain sufficient conditions for the beginning of the second round and IV differences that are optimal (*i.e.*, with the highest probability of cancelling differences in the final state). What remains to do is to construct a non-linear differential path for the first round. For this, we used the meet-in-the-middle method using the public HashClash implementation [16]. Although we tried both non-linear differential path construction methods, *i.e.* guess-and-determine, using our own implementation, and the meet-in-the-middle method, we have found that the meet-in-the-middle approach generally resulted in fewer conditions. Furthermore, the control on the position of these conditions was greater with the meet-in-the-middle approach.

This differential path for the first round was then used as input for another run of joint local collision analysis. In this case the run was over all 80 steps, also replacing the differences assumed from the disturbance vector with the differences in the state words coming from the non-linear path of the first round; switching the sign of a difference was also allowed when it resulted in a sparser overall difference. In this manner joint local collision analysis is able to provide a complete set of attack conditions (*i.e.*, sufficient conditions for the state words and linear relations on message bits) that is optimized for the highest success probability over the last three rounds, all the while minimizing the amount of conditions needed.

In fact, JLCA outputs many complete sets that only vary slightly in the signing of the differences. For our selected disturbance vector II(59,0) it turned out that this direct approach is far too costly and far too memory-consuming, as the amount of complete sets grows exponentially with the number of steps for which we desired sufficient conditions sets. We were able to improve this by introducing attack condition classes, where two sets of sufficient conditions belong to the same class if their sufficient conditions over the last five state words are identical. By expressing the attack condition classes over steps $[0, i]$ as extensions of attack conditions classes over steps $[0, i-1]$, we only have to work with a very small number of class representatives at each step, making it very practical.

Note that we do not exploit this to obtain additional freedom for the attack yet, However, it allows us to automatically circumvent random unpredictable contradictions between the attack conditions in the densest part, by randomly sampling complete sets until a usable one is found. We previously used the guess-and-determine approach to resolve such contradictions by changing signs, however this still required some manual interaction.

The resulting sufficient conditions on the state are given in the form of the differential path in Fig. 1 (using the symbols of Fig. 2) and the message bit relations are given in Fig. 3 through Fig. 5.

4.3 Finding and Analyzing Neutral Bits and Boomerangs

Generating and analyzing the boomerangs and neutral bits used in the attack was done entirely automatically as described below. This process depends on a parameter called the *main block offset* (specific to a freestart attack) that determines the offset of the message freedom window used during the attack. We have selected a main block offset of 5 as this led to the best distribution of usable neutral bits and boomerangs. This means that all the neutral bits and boomerangs directly lead to changes in the state from steps 5 up to 20, and that these changes propagate to steps 4 down to 0 backwards and steps 21 up to 79 forwards.

Because the dense area of the attack conditions may implicitly force certain other bits to specific values (resulting in hidden conditions), we use more than 4000 sampled solutions for the given attack conditions (over steps 1 up to 16) in the analysis. The 16 steps fully determine the message block, and also verify the sufficient conditions in the IV and in the dense non-linear differential path of the first round. It should be noted that for this step it is important to generate every sample independently. Indeed using *e.g.* message modification techniques to generate many samples from a single one would result in a biased distribution where many samples would only differ in the last few steps.

Boomerang Analysis. We analyze potential boomerangs that flip a single state bit together with 3 or more message bits. Each boomerang should be orthogonal to the attack conditions, *i.e.*, the state bit should be free of sufficient conditions, while flipping the message bits should not break any of the message bit relations (either directly or through the message propagation). Let $t \in [6, 16], b \in [0, 31]$ be such that the state bit $A_t[b]$ has no sufficient condition.

First, we determine the best usable boomerang on $A_t[b]$ as follows. For every sampled solution, we flip that state bit and compute the signed bit differences between the resulting and the unaltered message words $W_5, \ldots, W_{20}$. We verify that the boomerang is usable by checking that flipping its constituting bits breaks none of the message bit relations. We normalize these signed bit differences by negating them all when the state bit is flipped from 1 to 0. In this manner we obtain a set of usable boomerangs for $A_t[b]$. We determine the auxiliary conditions on message bits and state bits and only keep the best usable boomerang that has the fewest auxiliary conditions.

Secondly, we analyze the behaviour of the boomerang over the backwards steps. For every sampled solution, we simulate the application of the boomerang by flipping the bits of the boomerang. We then recompute steps 4 to 0 backwards and verify if any sufficient condition on these steps is broken. Any boomerang that breaks any sufficient conditions on the early steps with probability higher than 0.1 is dismissed.

Thirdly, we analyze the behaviour of the boomerang over the forward steps. For every sampled solution, we simulate the application of the boomerang by flipping its constituting bits. We then recompute steps 21 up to 79 forwards and keep track of any sufficient condition for the differential path that becomes violated. A boomerang will be used at step i in our attack if it does not break any sufficient condition up to step $i-1$ with probability more than 0.1.

Neutral Bits Analysis. The neutral bit analysis uses the same overall approach as the boomerangs, with the following changes. After boomerangs are determined, their conditions are added to the previous attack conditions and used to generate a new set of solution samples. Usable neutral bits consist of a set of one or more message bits that are flipped simultaneously. However, unlike for boomerangs, the reason for flipping more than one bit is to preserve message bit relations, and not to control the propagation of a state difference. Let $t \in [5, 20], b \in [0, 31]$ be a candidate neutral bit; flipping $W_t[b]$ may possibly break certain message bit relations. We express each message bit relation over $W_5, \ldots, W_{20}$ using linear algebra, and use Gaussian elimination to ensure that each of them has a unique last message bit $W_i[j]$ (*i.e.* where $i*32+j$ is maximal). For each relation involving $W_t[b]$, let $W_i[j]$ be its last message bit. If (i, j) equals (t, b) then this neutral bit is not usable (indeed, it would mean that its value is fully determined by earlier message bits). Otherwise we add bit $W_i[j]$ to be flipped together with $W_t[b]$ as part of the neutral bit. Similarly to boomerangs, we dismiss any neutral bit that breaks sufficient conditions backwards with probability higher than 0.1. The step i in which the neutral bit is used is determined in the same way as for the boomerangs.

The boomerangs we have selected are given in Fig. 7 and the neutral bits are listed in Fig. 6. In the case of the latter, only the first neutral bit is given and not the potential corrections for the message bit relations.

4.4 Base Solution Generation on CPU

We are now equipped with a set of attack conditions, including some that were added by the selected boomerangs and neutral bits. However, before these can be applied, we first need to compute partial solutions over 16 consecutive steps. Since the selected neutral bits and boomerangs cannot be used to correct the sufficient conditions on the IV, these have to be pre-satisfied as well. Therefore, we compute what we call *base solutions* over steps $1, \ldots, 16$ that fulfill all state conditions on $A_{-4}, \ldots, A_{17}$ and all message bit relations. A base solution itself consists of state words $A_{-3}, \ldots, A_{17}$ and message words $W_1, \ldots, W_{16}$ (although the implementation of the GPU step implies that the message words are translated to the equivalent message words $W_5, \ldots, W_{20}$ with the main block offset of 5).

The C++ code generating base solutions is directly compiled from the attack and auxiliary conditions. In this manner, all intermediate steps and condition tables can be hard-coded, and we can apply some static local optimizations eliminating unnecessary computations where possible. However, we do not exploit more advanced message modification techniques within these first 16 steps yet.

Generating the base solutions only represents a small part of the cost of the overall attack, and it is run entirely on CPU. Although theoretically we need only a few thousand base solutions to be successful given the total success probability over the remaining steps and the remaining freedom degrees yet to be used, in practice we need to generate a small factor more to ensure that all GPUs have enough work.

4.5 Applying Neutral Bits and Boomerangs on GPU

We now describe the final phase of the attack, which is also the most computationally intensive; as such, it was entirely implemented on GPUs. In particular, we used 65 recent Nvidia GTX970 [32] GPUs that feature 1664 small cores operating at a clock speed of about 1.2GHz; each card cost about 350 US$ in 2015.[5] In [19], the authors evaluate a single GTX970 to be worth 322 CPU cores[6] for raw SHA-1 operations, and about 140 cores for their SHA-1 attack.

We make use of the same efficient framework for Nvidia GPUs [19]. This makes use of the CUDA toolkit that provides programming extensions to C and C++ for convenient programming. For each step of SHA-1 wherein we use neutral bits and boomerangs, there will be a separate GPU-specific C++ function. Each function will load solutions up to that step from a global cyclic buffer; extend those solutions using the freedom for that step by triggering the available neutral bits or boomerangs; verify the sufficient conditions; and finally save the resulting partial solution extended by one step in the next global cyclic buffer. The smallest unit that can act independently on Nvidia GPUs is the *warp*, which consists of 32 threads that can operate on different data, but should execute the same instruction for best performance. When threads within a warp diverge onto different execution paths, these paths are executed serially, not in parallel. In the framework, the threads within each warp will agree on which function (thus which step) to execute together, resulting in reads, computations, and conditional writes that are coherent between all threads of the warp. We refer the reader to the original paper introducing this framework for a more detailed description [19].

The exact details of which neutral bits and which boomerangs are used for each step are given in Sect. 5.

In the probabilistic phase, after all freedom degrees have been exhausted, we can verify internal state collisions that should happen after steps 39 and 59 (for a message pair that follows the differential path), as these are steps with no active differences in the disturbance vector. These checks are still done on the GPU. Solutions up to A_{60} are passed back to the CPU for further verification to determine if a complete freestart collision has been found.

[5] With the right motherboard one can place up to 4 such GPUs on a single machine.
[6] Intel Haswell Core-i5 3.2GHz CPU.

We would like to note that in comparison with the attack on 76 steps, this work introduces boomerangs and has a slightly bigger count of neutral bits (60 v. 51). As a result, this required to use more intermediate buffers, and consequently a slightly more careful management of the memory. Additionally, in this work there is a relatively high proportion of the neutral bits that need additional message bits to be flipped to ensure no message bit relation is broken, whereas this only happens twice in the 76-step attack. These two factors result in an attack that is slightly more complex to implement, although neither point is a serious issue.

5 Attack Details

5.1 Sufficient Conditions

We give a graphical representation of the differential path used in our attack up to step 28 in Fig. 1, consisting of sufficient conditions for the state, and the associated message signed bit differences. The meaning of the bit condition

```
A-4: ........ ........ ........ ........
A-3: ........ ........ ........ ........
A-2: ........ ........ ........ .....^-.
A-1: 1...1... ........ ........ .0.....+
A0 : 01..0... ........ ........ .1......  W0 : x.+...+. ........ ........ ...+....
A1 : 11+^..+. ........ ....^... ...+....  W1 : ..-..-.. ........ ........ ...-++..
A2 : ..-11-1. 1......^ .....1+1 10.1.0..  W2 : ..+..--. ........ ........ ...-.+..
A3 : .0.0-001 1.^.10.. .+01.011 11^0.1.1  W3 : ..-..--. ........ ........ ...-+.-.
A4 : .1.11+-1 +^^^+1^^ ^011^^.- +++++-.+  W4 : ........ ........ ........ ...+....
A5 : .+.+.-++ ++++++++ ++++++++ .+0-1111  W5 : .....-.. ........ ........ ...+++..
A6 : .0.0.1.0 11.111.1 1110-010 0-1.10-+  W6 : x+..++.. ........ ........ ...-.+..
A7 : 1-.+.1.0 10100010 00000011 1+.-.0.+  W7 : ....-+.. ........ ........ ......+.
A8 : 0+.0.0.. ........ ......0. .+.-.0.1  W8 : x-...... ........ ........ ...+....
A9 : .+.0.0.. ........ ........ .0.+...^  W9 : x.-+.-.. ........ ........ ...-++..
A10: .+...... ........ ........ ...+.0..  W10: ..-+++.. ........ ........ .....-..
A11: ...-.... ........ ........ ........  W11: x.++++.. ........ ........ ...-+.+.
A12: ...0.1.. ........ ........ .....1..  W12: ..-..... ........ ........ ...-....
A13: .1...0.. ........ ........ ......!^  W13: ..+..+.. ........ ........ ...-++..
A14: +-...... ........ ........ ........  W14: x++.+-.. ........ ........ ...-.+..
A15: 1.1-.... ........ ........ ......!.  W15: ....+-.. ........ ........ ......+.
A16: +.10.1.. ........ ........ ........  W16: x+...... ........ ........ ...-....
A17: 1.-..0.. ........ ........ .......^  W17: x.++.+.. ........ ........ ...+--..
A18: .+-.0... ........ ........ .......!  W18: ..+.--.. ........ ........ .....-..
A19: .+.s.... ........ ........ ........  W19: x.+---.. ........ ........ ...-+...
A20: -...R... ........ ........ ........  W20: x.++.... ........ ........ ...+....
A21: -.+R.... ........ ........ ........  W21: ........ ........ ........ ....++..
A22: -...S... ........ ........ .......^  W22: x.---... ........ ........ ...+....
A23: .-..R... ........ ........ ........  W23: ....-... ........ ........ ...+-...
A24: -.rs.... ........ ........ ........  W24: .-+--... ........ ........ ...+....
A25: -.-r.... ........ ........ ........  W25: ....+... ........ ........ ...+.+..
A26: -...s... ........ ........ ........  W26: .+--.... ........ ........ ...+....
A27: -.-.r... ........ ........ ........  W27: x.+-+... ........ ........ ...++-..
A28: ........ ........ ........ ........  W28: x+-.-... ........ ........ ........
A29: ..-..... ........ ........ ........
```

Fig. 1. The differential path used in the attack up to step 28. The meaning of the different symbols is given in Fig. 2

Symbol	*Condition on* $(\mathbf{A_t[i]}, \mathbf{A'_t[i]})$
.	$A_t[i] = A'_t[i]$
x	$A_t[i] \neq A'_t[i]$
+	$A_t[i] = 0, \quad A'_t[i] = 1$
-	$A_t[i] = 1, \quad A'_t[i] = 0$
0	$A_t[i] = A'_t[i] = 0$
1	$A_t[i] = A'_t[i] = 1$
^	$A_t[i] = A'_t[i] = A_{t-1}[i]$
!	$A_t[i] = A'_t[i] \neq A_{t-1}[i]$
r	$A_t[i] = A'_t[i] = (A_{t-1} \ggg 2)[i]$
R	$A_t[i] = A'_t[i] \neq (A_{t-1} \ggg 2)[i]$
s	$A_t[i] = A'_t[i] = (A_{t-2} \ggg 2)[i]$
S	$A_t[i] = A'_t[i] \neq (A_{t-2} \ggg 2)[i]$

Fig. 2. Bit conditions

symbols are defined in Fig. 2. Note that the signs of message bit differences are enforced through message bit relations. All message bit relations used in our attack are given in Fig. 3 through Fig. 5. The remainder of the path can easily be determined by linearization of the step function given the differences in the message.

5.2 The Neutral Bits

We give here the list of the neutral bits used in our attack. There are 60 of them over the 7 message words W_{14} to W_{20}, distributed as follows:

- W_{14}: 6 neutral bits at bit positions (starting with the least significant bit (LSB) at zero) 5,7,8,9,10,11
- W_{15}: 11 neutral bits at positions 4,7,8,9,10,11,12,13,14,15,16
- W_{16}: 9 neutral bits at positions 8,9,10,11,12,13,14,15,16
- W_{17}: 10 neutral bits at positions 10,11,12,13,14,15,16,17,18,19
- W_{18}: 11 neutral bits at positions 4,6,7,8,9,10,11,12,13,14,15
- W_{19}: 8 neutral bits at positions 6,7,8,9,10,11,12,14
- W_{20}: 5 neutral bits at positions 6,11,12,13,15

We give a graphical representation of the position of these neutral bits in Fig. 6.

Not all of the neutral bits of the same word (say W_{14}) are used at the same step during the attack. Their repartition in that respect is as follows

- Bits neutral up to step 18 (excluded): W_{14}[8,9,10,11], W_{15}[13,14,15,16]
- Bits neutral up to step 19 (excluded): W_{14}[5,7], W_{15}[8,9,10,11,12], W_{16}[12,13,14,15,16]

```
- W0[4] = 0
- W0[25] = 0
- W0[29] = 0
- W1[2] = 0
- W1[3] = 0
- W1[4] = 1
- W1[26] = 1
- W1[29] = 1
- W2[2] = 0
- W2[4] = 1
- W2[25] = 1
- W2[26] = 1
- W2[29] = 0
- W3[1] = 1
- W3[3] = 0
- W3[4] = 1
- W3[25] = 1
- W3[26] = 1
- W3[29] = 1
- W4[4] = 0
- W5[2] = 0
- W5[3] = 0
- W5[4] = 0
- W5[26] = 1
- W6[2] = 0
- W6[4] = 1
- W6[26] = 0
- W6[27] = 0
- W6[30] = 0
- W7[1] = 0
- W7[26] = 0
- W7[27] = 1
- W8[4] = 0
- W8[30] = 1
- W9[2] = 0
- W9[3] = 0
- W9[4] = 1
- W9[26] = 1
- W9[28] = 0
- W9[29] = 1
- W10[2] = 1
- W10[26] = 0
- W10[27] = 0
- W10[28] = 0
- W10[29] = 1
- W11[1] = 0
- W11[3] = 0
- W11[4] = 1
- W11[26] = 0
- W11[27] = 0
- W11[28] = 0
- W11[29] = 0
- W12[4] = 1
- W12[29] = 1
- W13[2] = 0
- W13[3] = 0
- W13[4] = 1
- W13[26] = 0
- W13[29] = 0
- W14[2] = 0
- W14[4] = 1
- W14[26] = 1
- W14[27] = 0
- W14[29] = 0
- W14[30] = 0
- W15[1] = 0
- W15[26] = 1
- W15[27] = 0
- W16[4] = 1
- W16[30] = 0
- W17[2] = 1
- W17[3] = 1
- W17[4] = 0
- W17[26] = 0
- W17[28] = 0
- W17[29] = 0
- W18[2] = 1
- W18[26] = 1
- W18[27] = 1
- W18[29] = 0
- W19[3] = 0
- W19[4] = 1
- W19[26] = 1
- W19[27] = 1
- W19[28] = 1
- W19[29] = 0
- W20[4] = 0
- W20[28] = 0
- W20[29] = 0
- W21[2] = 0
- W21[3] = 0
- W22[4] = 0
- W22[27] = 1
- W22[28] = 1
- W22[29] = 1
- W23[3] = 1
- W23[4] = 0
- W23[27] = 1
- W24[4] = 0
- W24[27] = 1
- W24[28] = 1
- W24[29] = 0
- W24[30] = 1
- W26[4] = 0
- W26[28] = 1
- W26[29] = 1
- W26[30] = 0
- W27[2] = 1
- W27[3] = 0
- W27[4] = 0
- W27[27] = 0
- W27[28] = 1
- W27[29] = 0
- W28[27] = 0
- W28[29] = 1
- W28[30] = 0
- W29[2] = 0
- W29[28] = 0
- W29[29] = 0
- W30[27] ^ W30[28] = 1
- W30[30] = 1
- W31[2] = 0
- W31[3] = 0
- W31[28] = 0
- W31[29] = 0
- W33[28] ^ W33[29] = 1
- W30[4] ^ W34[29] = 0
- W35[27] = 0
- W35[28] = 0
- W35[4] ^ W39[29] = 0
- W58[29] ^ W59[29] = 0
- W57[29] ^ W59[29] = 0
- W55[4] ^ W59[29] = 0
- W53[29] ^ W54[29] = 0
- W52[29] ^ W54[29] = 0
- W51[28] ^ W51[29] = 1
- W50[4] ^ W54[29] = 0
- W50[28] ^ W51[28] = 0
- W50[29] ^ W51[28] = 1
- W49[28] ^ W51[28] = 0
- W48[29] ^ W48[30] = 0
- W47[3] ^ W51[28] = 0
- W47[4] ^ W51[28] = 1
- W46[29] ^ W51[28] = 1
- W45[4] ^ W51[28] = 0
- W44[29] ^ W51[28] = 0
- W43[4] ^ W51[28] = 1
- W43[29] ^ W51[28] = 0
- W41[4] ^ W51[28] = 0
- W63[4] ^ W67[29] = 0
- W79[5] = 0
- W78[0] = 1
- W77[1] ^ W78[6] = 1
- W75[5] ^ W79[30] = 0
- W74[0] ^ W79[30] = 1
```

Fig. 3. The message bit-relations used in the attack.

- Bits neutral up to step 20 (excluded): $W_{15}[4,7,8,9]$, $W_{16}[8,9,10,11,12]$, $W_{17}[12,13,14,15,16]$
- Bits neutral up to step 21 (excluded): $W_{17}[10,11,12,13]$, $W_{18}[15]$
- Bits neutral up to step 22 (excluded): $W_{18}[9,10,11,12,13,14]$, $W_{19}[10,14]$
- Bits neutral up to step 23 (excluded): $W_{18}[4,6,7,8]$, $W_{19}[9,11,12]$, $W_{20}[15]$
- Bits neutral up to step 24 (excluded): $W_{19}[6,7,8]$, $W_{20}[11,12,13]$
- Bit neutral up to step 25 (excluded): $W_{20}[7]$

One should note that this list only includes a single bit per neutral bit group. As we mentioned in the previous section, some additional flips may be needed in order to preserve message bit relations.

```
W0 :  . . 0 . . . 0 . . . . . . . . . . . . . . . . . . . . 0 . . . .
W1 :  . . 1 . . 1 . . . . . . . . . . . . . . . . . . . . . 1 0 0 . .
W2 :  . . 0 . . 1 1 . . . . . . . . . . . . . . . . . . . . 1 . 0 . .
W3 :  . . 1 . . 1 1 . . . . . . . . . . . . . . . . . . . . 1 0 . 1 .
W4 :  . . . . . . . . . . . . . . . . . . . . . . . . . . . 0 . . . .
W5 :  . . . . . 1 . . . . . . . . . . . . . . . . . . . . . 0 0 0 . .
W6 :  . 0 . . 0 0 . . . . . . . . . . . . . . . . . . . . . 1 . 0 . .
W7 :  . . . . 1 0 . . . . . . . . . . . . . . . . . . . . . . . . 0 .
W8 :  . 1 . . . . . . . . . . . . . . . . . . . . . . . . . 0 . . . .
W9 :  . . 1 0 . 1 . . . . . . . . . . . . . . . . . . . . . 1 0 0 . .
W10:  . . 1 0 0 0 . . . . . . . . . . . . . . . . . . . . . . . 1 . .
W11:  . . 0 0 0 0 . . . . . . . . . . . . . . . . . . . . . 1 0 . 0 .
W12:  . . 1 . . . . . . . . . . . . . . . . . . . . . . . . 1 . . . .
W13:  . . 0 . . 0 . . . . . . . . . . . . . . . . . . . . . 1 0 0 . .
W14:  . 0 0 . 0 1 . . . . . . . . . . . . . . . . . . . . . 1 . 0 . .
W15:  . . . . 0 1 . . . . . . . . . . . . . . . . . . . . . . . . 0 .
W16:  . 0 . . . . . . . . . . . . . . . . . . . . . . . . . 1 . . . .
W17:  . . 0 0 . 0 . . . . . . . . . . . . . . . . . . . . . 0 1 1 . .
W18:  . . 0 . 1 1 . . . . . . . . . . . . . . . . . . . . . . . 1 . .
W19:  . . 0 1 1 1 . . . . . . . . . . . . . . . . . . . . . 1 0 . . .
W20:  . . 0 0 . . . . . . . . . . . . . . . . . . . . . . . 0 . . . .
W21:  . . . . . . . . . . . . . . . . . . . . . . . . . . . . 0 0 . .
W22:  . . 1 1 1 . . . . . . . . . . . . . . . . . . . . . . 0 . . . .
W23:  . . . . 1 . . . . . . . . . . . . . . . . . . . . . . 0 1 . . .
W24:  . 1 0 1 1 . . . . . . . . . . . . . . . . . . . . . . 0 . . . .
W25:  . . . . . . . . . . . . . . . . . . . . . . . . . . . . . . . .
W26:  . 0 1 1 . . . . . . . . . . . . . . . . . . . . . . . 0 . . . .
W27:  . . 0 1 0 . . . . . . . . . . . . . . . . . . . . . . 0 0 1 . .
W28:  . 0 1 . 0 . . . . . . . . . . . . . . . . . . . . . . . . . . .
W29:  . . 0 0 . . . . . . . . . . . . . . . . . . . . . . . . . 0 . .
W30:  . 1 . A a . . . . . . . . . . . . . . . . . . . . . . c . . . .
W31:  . . 0 0 . . . . . . . . . . . . . . . . . . . . . . . . 0 0 . .
W32:  . . . . . . . . . . . . . . . . . . . . . . . . . . . . . . . .
W33:  . . B b . . . . . . . . . . . . . . . . . . . . . . . . . . . .
W34:  . . c . . . . . . . . . . . . . . . . . . . . . . . . . . . . .
W35:  . . . 0 0 . . . . . . . . . . . . . . . . . . . . . . d . . . .
W36:  . . . . . . . . . . . . . . . . . . . . . . . . . . . . . . . .
W37:  . . . . . . . . . . . . . . . . . . . . . . . . . . . . . . . .
W38:  . . . . . . . . . . . . . . . . . . . . . . . . . . . . . . . .
W39:  . . d . . . . . . . . . . . . . . . . . . . . . . . . . . . . .
```

Fig. 4. The message bit-relations used in the attack for words W_0 to W_{39} (graphical representation). A "0" or "1" character represents a bit unconditionally set to 0 or 1. A pair of two letters x means that the two bits have the same value. A pair of two letters x and X means that the two bits have different values.

5.3 The Boomerangs

We finally give the boomerangs used in the attack, which are regrouped in two sets of two. The first one first introduces a difference in the message on word W_{10}; as it does not significantly impact conditions up to step 27, it is used to increase the number of partial solutions A_{28} that are generated. The second set first introduces a difference on word W_{11}, and is used to generate partial solutions at A_{30}. More precisely, the four boomerangs have their first differences

```
W40: . . . . . . . . . . . . . . . . . . . . . . . . . . . . . . . .
W41: . . . . . . . . . . . . . . . . . . . . . . . . . . . w . . . .
W42: . . . . . . . . . . . . . . . . . . . . . . . . . . . . . . . .
W43: . . v . . . . . . . . . . . . . . . . . . . . . . . . u . . . .
W44: . . t . . . . . . . . . . . . . . . . . . . . . . . . . . . . .
W45: . . . . . . . . . . . . . . . . . . . . . . . . . . . s . . . .
W46: . . r . . . . . . . . . . . . . . . . . . . . . . . . . . . . .
W47: . . . . . . . . . . . . . . . . . . . . . . . . . . . q p . . .
W48: . o o . . . . . . . . . . . . . . . . . . . . . . . . . . . . .
W49: . . . n . . . . . . . . . . . . . . . . . . . . . . . . . . . .
W50: . . m l . . . . . . . . . . . . . . . . . . . . . . . k . . . .
W51: . . J @ . . . . . . . . . . . . . . . . . . . . . . . . . . . .
W52: . . i . . . . . . . . . . . . . . . . . . . . . . . . . . . . .
W53: . . h . . . . . . . . . . . . . . . . . . . . . . . . . . . . .
W54: . . * . . . . . . . . . . . . . . . . . . . . . . . . . . . . .
W55: . . . . . . . . . . . . . . . . . . . . . . . . . . . g . . . .
W56: . . . . . . . . . . . . . . . . . . . . . . . . . . . . . . . .
W57: . . f . . . . . . . . . . . . . . . . . . . . . . . . . . . . .
W58: . . e . . . . . . . . . . . . . . . . . . . . . . . . . . . . .
W59: . . $ . . . . . . . . . . . . . . . . . . . . . . . . . . . . .
W60: . . . . . . . . . . . . . . . . . . . . . . . . . . . . . . . .
W61: . . . . . . . . . . . . . . . . . . . . . . . . . . . . . . . .
W62: . . . . . . . . . . . . . . . . . . . . . . . . . . . . . . . .
W63: . . . . . . . . . . . . . . . . . . . . . . . . . . . x . . . .
W64: . . . . . . . . . . . . . . . . . . . . . . . . . . . . . . . .
W65: . . . . . . . . . . . . . . . . . . . . . . . . . . . . . . . .
W66: . . . . . . . . . . . . . . . . . . . . . . . . . . . . . . . .
W67: . . x . . . . . . . . . . . . . . . . . . . . . . . . . . . . .
W68: . . . . . . . . . . . . . . . . . . . . . . . . . . . . . . . .
W69: . . . . . . . . . . . . . . . . . . . . . . . . . . . . . . . .
W70: . . . . . . . . . . . . . . . . . . . . . . . . . . . . . . . .
W71: . . . . . . . . . . . . . . . . . . . . . . . . . . . . . . . .
W72: . . . . . . . . . . . . . . . . . . . . . . . . . . . . . . . .
W73: . . . . . . . . . . . . . . . . . . . . . . . . . . . . . . . .
W74: . . . . . . . . . . . . . . . . . . . . . . . . . . . . . . . a
W75: . . . . . . . . . . . . . . . . . . . . . . . . . . z . . . . .
W76: . . . . . . . . . . . . . . . . . . . . . . . . . . . . . . . .
W77: . . . . . . . . . . . . . . . . . . . . . . . . . . . . . . y .
W78: . . . . . . . . . . . . . . . . . . . . . . . . . Y . . . . . 1
W79: . % . . . . . . . . . . . . . . . . . . . . . . . . 0 . . . . .
@ = jlMnpQRstUvw
* = hik
$ = efg
% = zA
```

Fig. 5. The message bit-relations used in the attack for words W_{40} to W_{79} (graphical representation, continued). Non-alphanumeric symbols are used as shorthand for bit positions with more than one relation.

at bits 7,8 of W_{10} and 8,9 of W_{11}. In Fig. 7, we give a graphical representation of the complete set of message bits to be flipped for each boomerang. One can see that these indeed follow the pattern of a local collisions.

Software Disclosure Policy. To allow verification and improve understanding of our new results, we intend to release our engineered freestart attack code for graphic cards at https://sites.google.com/site/itstheshappening/.

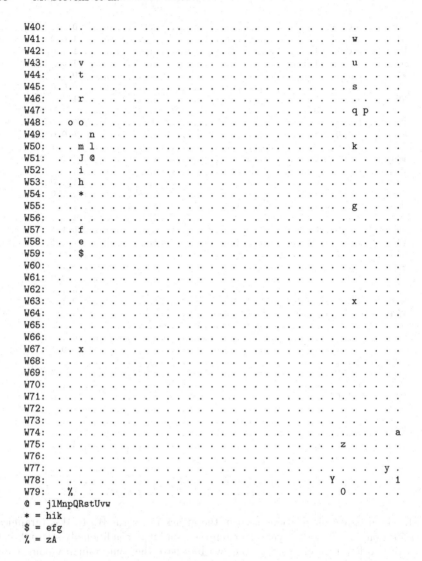

```
W14: ........ ........ ....xxxx x.x.....
W15: ........ .......x xxxxxxxx x..x....
W16: ........ .......x xxxxxxxx ........
W17: ........ ....xxxx xxxxxx.. ........
W18: ........ ........ xxxxxxxx xx.x....
W19: ........ ........ .x.xxxxx xx......
W20: ........ ........ x.xxx... .x......
```

Fig. 6. The 60 neutral bits. An "x" represents the presence of a neutral bit, and a "." the absence thereof. The LSB position is the rightmost one.

```
W10: ........ ........ ......BA ........
W11: ........ ........ .ba....D C.......
W12: ........ ........ ..dc.... ........
W13: ........ ........ ........ ........
W14: ........ ........ ........ .a......
W15: ........ ........ ........ ba......
W16: ........ ........ ........ .dc.....
```

Fig. 7. The local collision patterns for each of the four boomerangs. The position of the first difference to be introduced is highlighted with a capital letter; the correcting differences must then have a sign different from this one. Note that boomerang "A" uses one more difference than the others.

However, this source code does not directly enable the engineering of a SHA-1 collision attack. Any cryptanalytic tools needed for engineering a full SHA-1 collision attack will be released independently in a responsible manner.

Acknowledgements. We would like to express our gratitude to Orr Dunkelman for the use of his cluster with NVidia Tesla K10 cards. We also thank the anonymous reviewers for their helpful comments.

References

1. Biham, E., Chen, R.: Near-collisions of SHA-0. In: Franklin, M. (ed.) CRYPTO 2004. LNCS, vol. 3152, pp. 290–305. Springer, Heidelberg (2004)
2. Biham, E., Chen, R., Joux, A., Carribault, P., Lemuet, C., Jalby, W.: Collisions of SHA-0 and reduced SHA-1. In: Cramer [5], pp. 36–57
3. Brassard, G. (ed.): Advances in Cryptology - CRYPTO 1989. LNCS, vol. 435. Springer, Heidelberg (1990)
4. Chabaud, F., Joux, A.: Differential collisions in SHA-0. In: Krawczyk, H. (ed.) CRYPTO 1998. LNCS, vol. 1462, pp. 56–71. Springer, Heidelberg (1998)
5. Cramer, R. (ed.): Advances in Cryptology – EUROCRYPT 2005. LNCS, vol. 3494. Springer, Heidelberg (2005)
6. Damgård, I.B.: A design principle for hash functions. In: Brassard [3], pp. 416–427

7. De Cannière, C., Mendel, F., Rechberger, C.: Collisions for 70-Step SHA-1: on the full cost of collision search. In: Adams, C., Miri, A., Wiener, M. (eds.) SAC 2007. LNCS, vol. 4876, pp. 56–73. Springer, Heidelberg (2007)
8. De Cannière, C., Rechberger, C.: Finding SHA-1 characteristics: general results and applications. In: Lai, X., Chen, K. (eds.) ASIACRYPT 2006. LNCS, vol. 4284, pp. 1–20. Springer, Heidelberg (2006)
9. den Boer, B., Bosselaers, A.: An attack on the last two rounds of MD4. In: Feigenbaum, J. (ed.) CRYPTO 1991. LNCS, vol. 576, pp. 194–203. Springer, Heidelberg (1992)
10. den Boer, B., Bosselaers, A.: Collisions for the compression function of MD-5. In: Helleseth, T. (ed.) EUROCRYPT 1993. LNCS, vol. 765, pp. 293–304. Springer, Heidelberg (1994)
11. Dobbertin, H.: Cryptanalysis of MD4. In: Gollmann, D. (ed.) FSE 1996. LNCS, vol. 1039, pp. 53–69. Springer, Heidelberg (1996)
12. Forum, C.: Ballot 152 - Issuance of SHA-1 certificates through 2016. Cabforum mailing list (2015). https://cabforum.org/pipermail/public/2015-October/006048.html
13. Forum, C.: Ballot 152 - Issuance of SHA-1 certificates through 2016. Cabforum mailing list (2015). https://cabforum.org/pipermail/public/2015-October/006081.html
14. Grechnikov, E.A.: Collisions for 72-step and 73-step SHA-1: Improvements in the Method of Characteristics. IACR Cryptology ePrint Archive 2010, 413 (2010)
15. Grechnikov, E.A., Adinetz, A.V.: Collision for 75-step SHA-1: Intensive Parallelization with GPU. IACR Cryptology ePrint Archive 2011, 641 (2011)
16. Hashclash project webpage. https://marc-stevens.nl/p/hashclash/
17. Joux, A., Peyrin, T.: Hash functions and the (Amplified) boomerang attack. In: Menezes, A. (ed.) CRYPTO 2007. LNCS, vol. 4622, pp. 244–263. Springer, Heidelberg (2007)
18. Jutla, C.S., Patthak, A.C.: A matching lower bound on the minimum weight of sha-1 expansion code. Cryptology ePrint Archive, Report 2005/266 (2005)
19. Karpman, P., Peyrin, T., Stevens, M.: Practical free-start collision attacks on 76-step SHA-1. In: Gennaro, R., Robshaw, M. (eds.) Advances in Cryptology – CRYPTO 2015. LNCS, vol. 9215, pp. 623–642. Springer, Heidelberg (2015). http://dx.doi.org/10.1007/978-3-662-47989-6
20. Manuel, S.: Classification and generation of disturbance vectors for collision attacks against sha-1. Cryptology ePrint Archive, Report 2008/469 (2008)
21. Manuel, S.: Classification and generation of disturbance vectors for collision attacks against SHA-1. Des. Codes Cryptography **59**(1–3), 247–263 (2011)
22. Matusiewicz, K., Pieprzyk, J.: Finding good differential patterns for attacks on SHA-1. In: Ytrehus, Ø. (ed.) WCC 2005. LNCS, vol. 3969, pp. 164–177. Springer, Heidelberg (2006)
23. Mendel, F., Pramstaller, N., Rechberger, C., Rijmen, V.: The impact of carries on the complexity of collision attacks on SHA-1. In: Robshaw, M. (ed.) FSE 2006. LNCS, vol. 4047, pp. 278–292. Springer, Heidelberg (2006)
24. Merkle, R.C.: One way hash functions and DES. In: Brassard [3], pp. 428–446
25. Microsoft: SHA-1 Deprecation Update. Microsoft blog (2015)
26. Mozilla: Continuing to Phase Out SHA-1 Certificates. Mozilla Security Blog (2015)
27. National Institute of Standards and Technology: FIPS 180: Secure Hash Standard, May 1993
28. National Institute of Standards and Technology: FIPS 180–1: Secure Hash Standard, April 1995

29. National Institute of Standards and Technology: FIPS 180–2: Secure Hash Standard, August 2002
30. National Institute of Standards and Technology: Special Publication 800–57 - Recommendation for Key Management Part 1: General (Revision 3), July 2012
31. National Institute of Standards and Technology: FIPS 202: SHA-3 Standard: Permutation-Based Hash and Extendable-Output Functions, August 2015
32. Nvidia Corporation: Nvidia Geforce GTX 970 Specifications. http://www.geforce.com/hardware/desktop-gpus/geforce-gtx-970/specifications
33. Pramstaller, N., Rechberger, C., Rijmen, V.: Exploiting Coding Theory for Collision Attacks on SHA-1. In: Smart, N.P. (ed.) Cryptography and Coding 2005. LNCS, vol. 3796, pp. 78–95. Springer, Heidelberg (2005)
34. Rijmen, V., Oswald, E.: Update on SHA-1. In: Menezes, A. (ed.) CT-RSA 2005. LNCS, vol. 3376, pp. 58–71. Springer, Heidelberg (2005)
35. Rivest, R.L.: The MD4 message digest algorithm. In: Menezes, A., Vanstone, S.A. (eds.) CRYPTO 1990. LNCS, vol. 537, pp. 303–311. Springer, Heidelberg (1991)
36. Rivest, R.L.: RFC 1321: the MD5 message-digest algorithm, April 1992
37. Schneier, B.: When will we see collisions for sha-1? Schneier on Security (2012)
38. Services, A.W: Amazon EC2 - Virtual Server Hosting. https://aws.amazon.com, Retrieved Jan 2016
39. Shoup, V. (ed.): Advances in Cryptology – CRYPTO 2005. LNCS, vol. 3621. Springer, Heidelberg (2005)
40. Survey of the ssl implementation of the most popular web sites. TIM Trustworthy Internet Movement (2015). https://www.trustworthyinternet.org/ssl-pulse/
41. Stevens, M.: Attacks on Hash Functions and Applications. Ph.D. thesis, Leiden University, June 2012
42. Stevens, M.: Counter-cryptanalysis. In: Canetti, R., Garay, J.A. (eds.) CRYPTO 2013, Part I. LNCS, vol. 8042, pp. 129–146. Springer, Heidelberg (2013). http://dx.doi.org/10.1007/978-3-642-40041-4
43. Stevens, M.: New collision attacks on SHA-1 based on optimal joint local-collision analysis. In: Johansson, T., Nguyen, P.Q. (eds.) EUROCRYPT 2013. LNCS, vol. 7881, pp. 245–261. Springer, Heidelberg (2013). http://dx.doi.org/10.1007/978-3-642-38348-9
44. Stevens, M., Lenstra, A.K., de Weger, B.: Chosen-prefix collisions for MD5 and colliding X.509 certificates for different identities. In: Naor, M. (ed.) EUROCRYPT 2007. LNCS, vol. 4515, pp. 1–22. Springer, Heidelberg (2007)
45. Stevens, M., Sotirov, A., Appelbaum, J., Lenstra, A., Molnar, D., Osvik, D.A., de Weger, B.: Short chosen-prefix collisions for MD5 and the creation of a rogue CA certificate. In: Halevi, S. (ed.) CRYPTO 2009. LNCS, vol. 5677, pp. 55–69. Springer, Heidelberg (2009). http://dx.doi.org/10.1007/978-3-642-03356-8
46. Wang, X., Yin, Y.L., Yu, H.: Finding collisions in the full SHA-1. In: Shoup [38], pp. 17–36
47. Wang, X., Yu, H.: How to break MD5 and other hash functions. In: Cramer [5], pp. 19–35
48. Wang, X., Yu, H., Yin, Y.L.: Efficient collision search attacks on SHA-0. In: Shoup [38], pp. 1–16
49. Yajima, J., Iwasaki, T., Naito, Y., Sasaki, Y., Shimoyama, T., Kunihiro, N., Ohta, K.: A strict evaluation method on the number of conditions for the SHA-1 collision search. In: Abe, M., Gligor, V.D. (eds.) ASIACCS, pp. 10–20. ACM (2008)

New Attacks on the Concatenation and XOR Hash Combiners

Itai Dinur(✉)

Department of Computer Science, Ben-Gurion University, Beersheba, Israel
dinuri@cs.bgu.ac.il

Abstract. We study the security of the concatenation combiner $H_1(M)\|H_2(M)$ for two independent iterated hash functions with n-bit outputs that are built using the Merkle-Damgård construction. In 2004 Joux showed that the concatenation combiner of hash functions with an n-bit internal state does not offer better collision and preimage resistance compared to a single strong n-bit hash function. On the other hand, the problem of devising second preimage attacks faster than 2^n against this combiner has remained open since 2005 when Kelsey and Schneier showed that a single Merkle-Damgård hash function does not offer optimal second preimage resistance for long messages.

In this paper, we develop new algorithms for cryptanalysis of hash combiners and use them to devise the first second preimage attack on the concatenation combiner. The attack finds second preimages faster than 2^n for messages longer than $2^{2n/7}$ and has optimal complexity of $2^{3n/4}$. This shows that the concatenation of two Merkle-Damgård hash functions is not as strong a single ideal hash function.

Our methods are also applicable to other well-studied combiners, and we use them to devise a new preimage attack with complexity of $2^{2n/3}$ on the XOR combiner $H_1(M) \oplus H_2(M)$ of two Merkle-Damgård hash functions. This improves upon the attack by Leurent and Wang (presented at Eurocrypt 2015) whose complexity is $2^{5n/6}$ (but unlike our attack is also applicable to HAIFA hash functions).

Our algorithms exploit properties of random mappings generated by fixing the message block input to the compression functions of H_1 and H_2. Such random mappings have been widely used in cryptanalysis, but we exploit them in new ways to attack hash function combiners.

Keywords: Hash function · Cryptanalysis · Concatenation combiner · XOR combiner

1 Introduction

Hash functions are among the main building blocks of many cryptographic protocols used today. A hash function H takes as an input a message M of an arbitrary length, and maps it to an output $H(M)$ of a fixed length n. The main security properties expected from a hash function are:

M. Fischlin and J.-S. Coron (Eds.): EUROCRYPT 2016, Part I, LNCS 9665, pp. 484–508, 2016.
DOI: 10.1007/978-3-662-49890-3_19

1. **Collision Resistance:** It should be difficult to find a pair of different messages M and M' such that $H(M) = H(M')$.
2. **Preimage Resistance:** Given an arbitrary n-bit value V, it should be difficult to find any message M such that $H(M) = V$.
3. **Second Preimage Resistance:** Given a target message M, it should be difficult to find any different message M' such that $H(M) = H(M')$.

When the hash is function viewed as a "random oracle", it offers collision resistance up to a security level of $2^{n/2}$ due to the birthday paradox. The expected security level against preimage and second preimage attacks is 2^n.

In practice, widely deployed standards (such as MD5 and SHA-1) have failed to provide the expected security level [37,38]. As a result, researchers have proposed to combine the outputs of two hash functions to provide better security in case one (of even both) hash functions are weak. In fact, hash function combiners were also used in practice in previous versions of the SSL and TLS protocols [10,15].

The most well-known hash function combiner is the concatenation combiner that concatenates the outputs of two hash functions $H_1(M) \| H_2(M)$. This combiner was described already in 1993 [34] and has been subject to extensive analysis in various settings (e.g., [20,26]). From the theoretical side, researchers have studied the notion of a *robust combiner*, which is secure in case at least one of the combined hash functions is secure. Clearly, the concatenation combiner is secure with respect to the main security properties defined above, e.g., a collision $H_1(M) \| H_2(M) = H_1(M') \| H_2(M')$ implies $H_1(M) = H_1(M')$ and $H_2(M) = H_2(M')$. Lines of research regarding robust combiners include the study of advanced combiners in [13,14] and study of the minimal output length of robust combiners [5,33,35] (more recently works include [27,29]).

We are interested in this paper in the security of combiners of iterated hash functions that break the message M into blocks $m_1 \| m_2 \| \ldots \| m_L$ of fixed length and processes them by iterative applications of a compression function h (or several compression functions) that updates an internal state x_i using the previous state and the current message block $x_i = h(x_{i-1}, m_i)$. Hash functions whose internal state size is n-bits are known as "narrow-pipe" constructions and typically output the final state x_L. A very common way to build iterated hash functions is the Merkle-Damgård construction [8,28] which applies the same compression function h in all iterations, and adds a padding of the message length to final message block (known as Merkle-Damgård strengthening). Iterated hash function constructions (and in particular, the Merkle-Damgård construction) are very efficient and widely used in practice.

In a breakthrough paper published in 2004 [22], Joux showed that the collision and preimage resistance of the concatenation combiner of iterated narrow-pipe hash functions[1] is not much greater than that of a single n-bit hash function. The result of Joux was highly influential and resulted in numerous followup works in hash function analysis. The main technique introduced in [22] (known as Joux

[1] In fact, only one of the hash functions has to be iterated.

multi-collisions) generates a large multi-collision (many message that map to the same value) in a narrow-pipe hash function in time which is not much greater than $2^{n/2}$ (the time required to generate a single collision).

Joux's multi-collisions were used shortly afterwards by Kelsey and Schneier in order to show that the second preimage resistance of the Merkle-Damgård construction is less than the optimal 2^n for a long input target message M [23]. The idea is to compute the specific sequence of n-bit states $a_1, \ldots, a_L$ that is generated during the application of the compression function h on the message M. We then try to "connect" to some internal state a_i in this sequence with a different message prefix, and reuse the message suffix $m_{i+1} \| \ldots \| m_L$ to obtain the second preimage M'. However, this approach is foiled by the Merkle-Damgård strengthening if M and M' are not of the same length. The main idea of Kelsey and Schneier was use Joux's multi-collisions to construct an *expandable message* which essentially allows to expand the connected message prefix of M' to the desired length i, and overcome the length restriction.

Following the results of Joux (which showed that the concatenation combiner does not increase collision and preimage resistance), and the later results of Kelsey and Schneier (which showed that the second preimage resistance of the Merkle-Damgård construction is less than 2^n), a natural question is whether there exists a second preimage attack on the concatenation combiner of Merkle-Damgård hash functions that is faster than 2^n. Interestingly, the problem of devising such an attack remained open for a long time despite being explicitly mentioned in several papers including [12], and much more recently in [25]. In fact, although the works of [22,23] have attracted a significant amount of followup research on countermeasures against second preimage attacks (such as "hash twice" or "dithered hash") and attacks that break them [1–3], there has been no progress with respect to second preimage attacks on the basic concatenation combiner.

In this paper, we devise the first second preimage attack on the concatenation combiner of Merkle-Damgård hash functions which is faster than 2^n. As in related attacks (and in particular, [23]) we obtain a tradeoff between the complexity of the attack and the length of the target message. In particular, our second preimage attack is faster than 2^n only for input messages of length at least[2] $2^{2n/7}$. The optimal complexity[3] of our attack is $2^{3n/4}$, and is obtained for (very) long messages of length $2^{3n/4}$. Due to these constraints, the practical impact of our second preimage attack is limited and its main significance is theoretical. Namely, it shows that the concatenation of two Merkle-Damgård hash functions is not as strong a single ideal hash function.

The general framework of our attack is similar to the one of the long message second preimage attack of Kelsey and Schneier. Namely, we first compute the

[2] For example, for $n = 160$ and message block of length 512 bits (as in SHA-1), the attack is faster than 2^{160} only for messages containing at least 2^{46} blocks, or 2^{52} bytes.

[3] The complexity formulas do not take into account (small) constant factors, which are generally ignored throughout this paper.

sequences of internal states $a_1, \ldots, a_L$ and $b_1, \ldots, b_L$ by applying the compression functions h_1 and h_2 on the target message $M = m_1 \| \ldots \| m_L$. Our goal is then to "connect" to one of the state pairs (a_i, b_i) using a different message prefix of the same size. Once we manage to achieve this, we can reuse the same message suffix as in M and obtain a second preimage.

There are two main challenges in this approach, where the main challenge is connect to some state pair (a_i, b_i) generated by M from a different message. The secondary challenge is to ensure that the connected message prefixes are of the same length. We overcome the later challenge by building a (simultaneous) expandable message for two Merkle-Damgård hash functions with complexity that is not much larger than the complexity of building it for a single hash function [23]. The expandable message is built by using a cascaded form of Joux's multi-collisions, a technique which was used in previous works starting from Joux's original paper [22] and up to subsequent works such as [19,30]. A similar construction of an expandable message over two hash functions was proposed in the independent paper [21] by Jha and Nandi, which analyzes the zipper hash.

A much more difficult challenge is to actually connect to the target message on a state pair (a_i, b_i) from a different message of arbitrary (smaller) length. Indeed, the obvious approach of attempting to reach an arbitrary $2n$-bit state pair by trying random messages requires more than 2^n time, since the number of target state pairs is equal to the message length which is smaller than 2^n. A more promising approach is to use the *interchange structure*, which was recently introduced at Eurocrypt 2015 by Leurent and Wang [25]. Indeed, this structure is very useful in analysis of hash combiners as it breaks the dependency between the sequences of states computed during the computation of h_1 and h_2 on a common message. More specifically, the structure consists of an initial state pair (as, bs) and two sets of internal states A for H_1 and B for H_2 such that: for any value $a \in A$ and any value $b \in B$, it is possible to efficiently construct a message $M_{a,b}$ that sends (as, bs) to (a, b). Assume that there exists an index $i \in \{1, 2, \ldots, L\}$ such that $a_i \in A$ and $b_i \in B$, then we can connect to (a_i, b_i) using M_{a_i,b_i} as required. Unfortunately, this does not result in an efficient attack, essentially because the complexity of building an interchange structure for sufficiently large sets A and B is too high.

In this paper we use a different approach whose first step is to fix an arbitrary message block m, giving rise to functional graphs generated by the random n to n-bit mappings $f_1(x) = h_1(x, m)$ and $f_2(y) = h_2(y, m)$. Such random mappings have many interesting properties and have been extensively studied and used in cryptography in the classical works of Hellman [18] and van Oorschot and Wiener [36], and much more recently in [11,17,24,31,32]. However, in our case, the use of random mappings may seem quite artificial and unrelated to our goal of connecting to the arbitrary target message. Nevertheless, we will show how to exploit random mappings to search for a "special" state pair (a_p, b_p) whose states can be reached relatively easily from an arbitrary starting state pair (using the fixed message block m), thus connecting to the target message.

More specifically, we are particularly interested in "special" states of $a_1, \ldots, a_L$ and $b_1, \ldots, b_L$ that are located deep in the functional graphs defined by f_1 and f_2, i.e., these states can be reached after iterating f_1 and f_2 many times. Such special states (called *deep iterates*) are relatively easy to reach from arbitrary starting states. Our goal is to find a state pair (a_p, b_p) composed of two deep iterates in $f_1(x)$ and $f_2(y)$, respectively.[4] Once we find such a "special" state pair, we show how to simultaneously reach both of its states in an efficient manner from an arbitrary state pair. Combined with the expandable message, this gives the desired second preimage.

Interestingly, our algorithms for cryptanalysis of hash combiners are related to recent attacks on hash-based MACs [11,17,24,32], as in both settings the notion of distance of a node in the functional graph from a deep iterate plays an important role in the attack.

Our techniques are quite generic and can be applied to attack other Merkle-Damgård hash function combiners. In particular, they are applicable to another well-known combiner defined as $H_1(M) \oplus H_2(M)$ and known as the XOR combiner. At Eurocrypt 2015 [25], Leurent and Wang devised the first preimage attack against the XOR combiner (using the aforementioned interchange structure) with optimal complexity of $2^{5n/6}$. Here, we improve this attack for Merkle-Damgård hash functions and obtain an optimal complexity of $2^{2n/3}$. In practice, many concrete hash functions limit the maximal allowed message length L, and our attack on the XOR combiner gives a trade-off of $2^n \cdot L^{-2/3}$ (between L and the time complexity of attack). This improves the tradeoff of $2^n \cdot L^{-1/2}$, obtained by Leurent and Wang's attack. For a particular example mentioned in [25], we improve the complexity of finding a preimage of the XOR of two well-known Merkle-Damgård hash functions SHA-512 $\oplus$ WHIRLPOOL by a factor of about 2^{21} (from 2^{461} to 2^{440}). On the other hand, we stress that our techniques only apply in case that both hash functions combined use the Merkle-Damgård construction. In particular, our attacks are not applicable in case at least one combined hash function in built using the HAIFA mode [4] (that uses a block counter in every compression function call). In this case, the attack of Leurent and Wang remains the only published preimage attack on the XOR combiner that is faster than exhaustive search.

Finally, we point out that we can use the interchange structure [25] in order to optimize the complexity of our attacks on both the concatenation and XOR combiners. Although this does not lead to a very big improvement, it further demonstrates the wide applicability of this structure in cryptanalysis of hash function combiners.

The rest of the paper is organized as follows. In Sect. 2 we describe some preliminaries. In Sect. 3 we describe our second preimage attack on the concatenation combiner, while our preimage attack on the XOR combiner is described is Sect. 4. Finally, we conclude the paper in Sect. 5.

2 Preliminaries

In this section, we describe preliminaries that are used in the rest of the paper.

---

[4] The actual attack is slightly different, as it searches for deep iterates from which (a_p, b_p) can be reached with a common message block.

2.1 The Merkle-Damgård Construction [8,28]

The Merkle-Damgård construction [8,28] is a common way to build a hash function from a compression function $h : \{0,1\}^n \times \{0,1\}^k \rightarrow \{0,1\}^n$, where n denotes the size of the chaining value[5] and k denotes the block size.

The input message M is padded with its length (after additional padding which ensures that the final message length is a multiple of the block size k). The message is then divided into k-bit blocks $m_1\|m_2\|\ldots\|m_L$. The initial state of the hash function is an n-bit initial chaining value $x_0 = IV$. The compression function is then applied L times

$$x_i = h(x_{i-1}, m_i).$$

The hash value is set as $H(M) = x_L$. We extend the definition of h recursively to process an arbitrary number of k-bit message blocks, namely $h(x, m_1\|m_2\|\ldots\|m_L) = h(h(x, m_1), m_2\|\ldots\|m_L)$. Moreover, we denote by $|M|$ the length of M in blocks.

Given a state x and a message m such that $x' = h(x, m)$, we say that m *maps* the state x to the state x' (the compression function h used for this mapping will be clear from the context) and denote this by $x \xrightarrow{m} x'$. Throughout this paper we assume that the compression function is chosen uniformly at random from all $n+k$ to n-bit functions, which implies that our analysis applies to most (but not all) compression functions.

2.2 Hash Function Combiners

Given two hash functions H_1 and H_2, the concatenation combiner is defined as $H_1(M)\|H_2(M)$ while the XOR combiner is defined as $H_1(M) \oplus H_2(M)$. In this paper we are interested in the security of these combiners in the case that both H_1 and H_2 are based on the Merkle-Damgård construction with independent compression functions. We denote the IVs of H_1 and H_2 by IV_1 and IV_2 (respectively), their compression functions by h_1 and h_2, (respectively), and assume that both the chaining values and outputs are of size n (our techniques can also be extended to other cases). An additional technicality has to do with the message block sizes of H_1 and H_2, and we assume that both are equal to k. Once again, our techniques also apply in the case that the block sizes are different. A generic (although not always to most efficient) way to deal with this case is to align the block sizes by defining a "superblock" whose size is the least common multiple of the two block sizes.

We pair the (extended) compression functions h_1 and h_2 using the notation of $h_{1,2}$. Given two states x, y and a message m such that $x' = h_1(x, m)$ and $y' = h_2(y, m)$, we write $(x', y') = h_{1,2}((x, y), m)$. In this case, we say that m

[5] In this paper, we mainly consider "narrow-pipe" constructions in which the sizes of the chaining value and the hash function output are the same, but our techniques and analysis extend naturally (with additional cost) to "wide-pipe" constructions in which the chaining value size is larger.

maps (or sends) the state pair (x, y) to the state pair (x', y') (the compression functions h_1, h_2 are clear from the context) and denote this by $(x, y) \xrightarrow{m} (x', y')$.

2.3 Joux's Multi-collisions [22]

In a well-known result [22], Joux observed that finding a large multi-collision in an iterated hash function H is not much more expensive than finding a single collision. The algorithm starts from a state x_0 and evaluates the compression function for arbitrary message blocks until a collision is found. According to the birthday paradox, $2^{n/2}$ compression function evaluations are sufficient to find a collision $h(x_0, m_1) = h(x_0, m_1')$ for some m_1 and m_1'. Next, we set $x_1 = h(x_0, m_1)$ and continue the process iteratively t times. This gives 2^t t-block messages that reach x_t from x_0, as in the i'th block we can select either m_i or m_i'. Altogether, 2^t collisions are found with about $t \cdot 2^{n/2}$ compression function evaluations.

Joux's multi-collisions have numerous applications in cryptanalysis of hash functions. Next, we focus on one of the most relevant applications for this paper.

2.4 The Long Message Second Preimage Attack [23]

In [9], Dean devised a second preimage attack for long messages on specific Merkle-Damgård hash functions for which it is easy to find fixed points in their compression function. Given a target message $M = m_1 \| m_2 \| \ldots \| m_L$, the attacker computes the sequence of internal states $IV = a_0, a_1, \ldots, a_L$ generated during the invocation of the compression function on M. A simplified attack would now start from the state $IV = x_0$ and evaluate the compression function with arbitrary message blocks until a collision $h(x_0, m) = a_i$ is found for some message block m and index i. The attacker can now append the message suffix $m_{i+1} \| \ldots \| m_L$ to m, hoping to obtain the target hash value $H(M)$. However, this approach does not work due to the final padding of the message length which would be different if the message prefixes are of different lengths.

The solution of Dean was to compute an *expandable message* which consists of the initial state x_0 and another state $\hat{x}$ such that for each length κ (in some range) there is a message M_κ of κ blocks that maps x_0 to $\hat{x}$. Thus, the algorithm first finds a collision $h(\hat{x}, m) = a_i$, and the second preimage is computed as $M_{i-1} \| m \| m_{i+1} \| \ldots \| m_L$.

The assumption that it is easy to find fixed points in the compression function is used in [9] in efficient construction of the expandable message. In [23], Kelsey and Schneier described a more generic attack that uses multi-collisions of a special form to construct an expandable message, removing the restriction of Dean regarding fixed points. As in Joux's original algorithm, the multi-collisions are constructed iteratively in t steps. In the $i'th$ step, we find a collision between some m_i and m_i' such that $|m_i| = 1$ (it is a single block) and $|m_i'| = 2^{i-1} + 1$, namely $h(x_{i-1}, m_i) = h(x_{i-1}, m_i')$. This is done by picking an arbitrary prefix of size 2^{i-1} of m_i' denoted by m', computing $h(x_{i-1}, m') = x'$ and then looking for a collision $h(x_{i-1}, m_i) = h(x', m'')$ using a final block m'' (namely, $m_i' = m' \| m''$).

The construction of Kelsey and Schneier gives an expandable message that can be used to generate messages starting from x_0 and reaching $\hat{x} = x_t$ whose (integral) sizes range in the interval $[t, 2^t + t - 1]$ (it is denoted as a $(t, 2^t + t - 1)$-expandable message). A message of length $t \leq \kappa \leq 2^t + t - 1$ is generated by looking at the t-bit binary representation of $\kappa - t$. In iteration $i \in \{1, 2, \ldots, t\}$, we select the long message fragment m'_i if the $i'th$ LSB of $\kappa - t$ is set to 1 (otherwise, we select the single block m_i).

Given that the target message M is of length $L \leq 2^{n/2}$ blocks, the construction of the expandable message in the first phase of the attack requires less than $n \cdot 2^n$ computation, while obtaining the collision with one of the states computed during the computation of M requires about $1/L \cdot 2^n$ compression function evaluations according to the birthday paradox.

2.5 Functional Graph

In various phases of our attacks, we evaluate a compression function h with a fixed message input block m (e.g., the zero block), and simplify our notation by defining $f(x) = h(x, m)$. The mapping f gives rise a directed functional graph in which nodes are the n-bit states and an edge from node x to y is defined if and only if $f(x) = y$.

In this paper, we are particularly interested in nodes of f which are located deep in the functional graph. More specifically, x' is an iterate of depth i if there exists some ancestor node x' such that $x' = f^i(x)$. Deep iterates can be reached using *chains* evaluated from an arbitrary starting point x_0 by computing a sequence of nodes using the relation $x_{i+1} = f(x_i)$. We denote this sequence by $\overrightarrow{x}$.

A useful algorithm for expanding the functional graph of f is given below. This algorithm is not new and has been previously used (for example, in [17,32]). It takes an input parameter $g \geq n/2$ which determines the number of expanded nodes (and the running time of the algorithm).

1. Initialize $G = \emptyset$ as a data structure of evaluated nodes.
2. Until G contains 2^g nodes:
   (a) Pick an arbitrary starting point x_0 and evaluate the chain $x_{i+1} = f(x_i)$ until it cycles (there exists $x_i = x_j$ for $i \neq j$) or it hits a point in G. Add the points of the chain to G.

A simple but important observation that we exploit is that after we have executed the algorithm and developed 2^g nodes, then another chain from an arbitrary starting point is expected to collide with the evaluated graph at depth of roughly 2^{n-g}. This is a direct consequence of the birthday paradox. In particular, this observation implies that most chains developed by the algorithm will be extended to depth $\Omega(2^{n-g})$ (without colliding with G of cycling), and therefore a constant fraction of the developed nodes are iterates of depth 2^{n-g}. In total, the algorithm develops $\Theta(2^g)$ iterates of f of depth 2^{n-g} in 2^g time.

In this paper we will be interested in the probability of encountering a specific deep iterate at each stage of the evaluation of a chain from an arbitrary starting point.

Lemma 1. *Let f be an n-bit random mapping, and x'_0 an arbitrary point. Let $D \leq 2^{n/2}$ and define the chain $x'_i = f(x'_{i-1})$ for $i \in \{1,\ldots,D\}$ (namely, x'_D is an iterate of depth D). Let x_0 be a randomly chosen point, and define $x_d = f(x_{d-1})$. Then, for any $d \in \{1,\ldots,D\}$, $Pr[x_d = x'_D] = \Theta(d \cdot 2^{-n})$.*

Proof (Sketch). We can assume that the chains do not cycle (i.e., each chain contains distinct nodes), as $D \leq 2^{n/2}$. In order for $x_d = x'_D$ to occur then x_{d-i} should collide with x'_{D-i} for[6] some $0 \leq i \leq d$. For a fixed i, the probability for this collision is roughly[7] 2^{-n}, and summing over all $0 \leq i \leq d$ (are events are disjoint), we get that the probability is about $d \cdot 2^{-n}$.

Distinguished Points. The memory complexity of many algorithms that are based on functional graphs (e.g., parallel collision search [36]) can be reduced by utilizing the *distinguished points* method (which is attributed to Ron Rivest). Assume that our goal is to detect a collision of a chain (starting from an arbitrary node) with the nodes of G computed above, but without storing all the 2^g nodes in memory. The idea is to define a set of 2^g distinguished points (nodes) using a simple predicate (e.g. the $n-g$ LSBs of a node are zero). The nodes of G contain approximately $2^g \cdot 2^{g-n} = 2^{2g-n}$ distinguished points, and only they are stored in memory. A collision of an arbitrary chain with G is expected to occur at depth of about 2^{n-g}, and will be detected at the next distinguished point which is located (approximately) after traversing additional 2^{n-g} nodes. Consequently, we can detect the collision with a small overhead in time complexity, but a significant saving factor of 2^{n-g} in memory.

Interestingly, in the specific attack of Sect. 4, the distinguished points method is essential for reducing the time complexity of the algorithm.

3 A New Long Message Second Preimage Attack on the Concatenation Combiner

In this attack we are given a target message $M = m_1\|m_2\|\ldots\|m_L$ and our goal is to find another message M' such that $H_1(M')\|H_2(M') = H_1(M)\|H_2(M)$ (or equivalently $H_1(M') = H_1(M)$ and $H_2(M') = H_2(M)$). We denote the sequence of internal states computed during the invocation of h_1 (respectively, h_2) on M by $a_0, a_1, \ldots, a_L$ (respectively, $b_0, b_1, \ldots, b_L$). We start with a high-level overview of the attack and then give the technical details.

The attack is composed of three main phases. In the first phase, we build (a special form of) an expandable message, similarly to the second preimage attack

[6] A collision between x_{d-i} and x'_{D-i} occurs if $x_{d-i} = x'_{D-i}$ but $x_{d-i-1} \neq x'_{D-i-1}$.

[7] A more accurate analysis would take into account the event that the chains collide before x_{d-i}, but the probability for this is negligible.

on a single hash function [23]. This expandable message essentially consists of the initial state pair (IV_1, IV_2) and final state pair $(\hat{x}, \hat{y})$ such that for each length κ in some appropriate range (which is roughly $[n^2, L]$) there is a message M_κ of κ blocks that maps (IV_1, IV_2) to $(\hat{x}, \hat{y})$.

In the second phase our goal is to find a pair of states $(\bar{x}, \bar{y})$, a message block $\bar{m}$ and an index p such that $(\bar{x}, \bar{y}) \xrightarrow{\bar{m}} (a_p, b_p)$ (note that the state pair (a_p, b_p) is computed during the evaluation of the target message). Moreover, the state pair $(\bar{x}, \bar{y})$ should have a special property which is formally defined in the full description of the second phase.

In the third and final phase, we start from $(\hat{x}, \hat{y})$ and compute a message fragment $\hat{M}$ of length q (shorter than $p-1$) such that $(\hat{x}, \hat{y}) \xrightarrow{\hat{M}} (\bar{x}, \bar{y})$. This phase can be performed efficiently due to the special property of $(\bar{x}, \bar{y})$.

In order to compute the second preimage, we pick M_{p-q-1} using the expandable message, giving $(IV_0, IV_1) \xrightarrow{M_{p-q-1}} (\hat{x}, \hat{y})$, and concatenate $M_{p-q-1} \| \hat{M} \| \bar{m}$ in order to reach the state pair (a_p, b_p) from (IV_1, IV_2) with a message of appropriate length p. Indeed, we have

$$(IV_0, IV_1) \xrightarrow{M_{p-q-1}} (\hat{x}, \hat{y}) \xrightarrow{\hat{M}} (\bar{x}, \bar{y}) \xrightarrow{\bar{m}} (a_p, b_p).$$

Altogether, we obtain the second preimage

$$M' = M_{p-q-1} \| \hat{M} \| \bar{m} \| m_{p+1} \| \dots \| m_L.$$

This attack can be optimized using the interchange structure, as described in Appendix A.

Notation. For the sake of clarity, we summarize below the notation that is shared across the various phases. We note that each phase also introduces additional "internal" notation whose scope is only defined within the phase.

$M = m_1 \| \dots \| m_L$	: Target Message.
$a_0, \dots, a_L$ $(b_0, \dots, b_L)$	: Sequence of internal states computed during the invocation of h_1 (h_2) on M.
M'	: Computed second preimage.
$(\hat{x}, \hat{y})$	: Endpoint pair of expandable message (computed in Phase 1).
(a_p, b_p)	: State pair (in the sequences $a_0, \dots, a_L$ and $b_0, \dots, b_L$) on which the computation of M and M' coincides (computed in Phase 2).
$(\bar{x}, \bar{y})$, $\bar{m}$	: "Special" state pair and message block used to reach (a_p, b_p) (computed in Phase 2).
$\hat{M}$	: Message fragment that maps $(\hat{x}, \hat{y})$ to $(\bar{x}, \bar{y})$ (computed in Phase 3).
q	: Length of $\hat{M}$ (smaller than $p-1$)

Complexity Evaluation. Denote $L = 2^\ell$. For a parameter $g_1 \geq max(n/2, n-\ell)$, the complexity of the phases of the attack (as computed in their detail description) is given below (ignoring constant factors).

Phase 1: $L + n^2 \cdot 2^{n/2} = 2^\ell + n^2 \cdot 2^{n/2}$
Phase 2: $1/L \cdot 2^{2n-g_1} = 2^{2n-g_1-\ell}$
Phase 3: $2^{3g_1/2}$

We balance the second and third phases by setting $2n - g_1 - \ell = 3g_1/2$, or $g_1 = 2/5 \cdot (2n-\ell)$, giving time complexity of $2^{3/5 \cdot (2n-\ell)}$. This tradeoff holds as long as $2^\ell + n^2 \cdot 2^{n/2} \leq 2^{3/5(2n-\ell)}$, or $\ell \leq 3n/4$. The optimal complexity is $2^{3\ell/4}$, obtained for $\ell = 3n/4$. The attack is faster than 2^n (Joux's preimage attack) for[8] $\ell > n/3$. The message range for which the attack is faster than 2^n can be slightly improved to $L \geq 2^{2n/7}$ using the optimized attack, described in Appendix A.

3.1 Details of Phase 1: Constructing an Expandable Message

In this phase, we build a simultaneous expandable message for two Merkle-Damgård hash functions. This expandable message consists of the initial states (IV_1, IV_2) and final states $(\hat{x}, \hat{y})$ such that for each length κ in some appropriate range (determined below) there is a message M_κ of κ blocks that maps (IV_1, IV_2) to $(\hat{x}, \hat{y})$.

We set $C \approx n/2 + \log(n)$ as a constant. Our basic building block consists of two pairs of states (xa, ya) and (xb, yb) and two message fragments ms and ml that map the state pair (xa, ya) to (xb, yb). The message ms is the (shorter) message fragment of fixed size C, while ml is of size $i > C$. Below, we will show how to construct this building block for any state pair (xa, ya) and length $i > C$.

Given this building block and a positive parameter t, we build an expandable message in the range of $[C(C-1) + tC, C^2 - 1 + C(2^t + t - 1)]$. This is done by utilizing a sequence of $C - 1 + t$ basic building blocks. The first $C - 1$ building blocks are built with parameters $i \in \{C+1, C+2, \ldots, 2C-1\}$. It is easy to see that these structures give a $(C(C-1), C^2-1)$–expandable message by selecting at most one longer message fragment from the sequence, where the remaining $C-2$ (or $C-1$) fragments are of length C. The final t building blocks give a standard expandable message, but it is built in intervals of C. These t building blocks are constructed with parameters $i = C(2^{j-1} + 1)$ for $j \in \{1, \ldots, t\}$.

Given a length κ in the appropriate range of $[C(C-1) + tC, C^2 - 1 + C(2^t + t - 1)]$, we can construct a corresponding message by first computing κ (modulo C). We then select the length $\kappa' \in [C(C-1), C^2-1]$ such that $\kappa' \equiv \kappa$ (modulo C), defining the first $C-1$ message fragment choices. Finally, we compute $(\kappa - \kappa')/C$ which is an integer in the range of $[t, 2^t + t - 1]$ and select the final t message fragment choices as in a standard expandable message using the binary representation of $(\kappa - \kappa')/C$.

[8] Note that for $\ell > n/3$, $g_1 = 2/5 \cdot (2n - \ell) > 2n/3 > max(n/2, n - \ell)$, as required.

Construction of the Building Block. Given state pair (xa, ya) and length $i > C$, the algorithm for constructing the building block for the expandable message is based on multi-collisions as described below.

1. Pick an arbitrary prefix mp of size $i - C$ blocks and compute $xp = h_1(xa, mp)$.
2. Find a collision $h_1(xa, m_1) = h_1(xp, m_1') = x_2$ with single block messages m_1, m_1'.
3. Build a 2^{C-1} standard Joux multi-collision in h_1 starting from x_2 and denote its final endpoint by xb. Altogether we have a multi-collision in h_1 with 2^C messages that map xa to xb. Out of these messages, 2^{C-1} are of length C (obtained by first selecting m_1) and we denote this set of messages by S_1. The remaining 2^{C-1} messages are of length i (obtained by first selecting the $(i-C+1)$-block prefix $mp\|m_1'$), and we denote this set of messages by S_2.
4. Evaluate $yp = h_2(ya, mp)$ and store the result. Next, evaluate h_2 from ya on the two sets S_1 and S_2 (using the stored yp to avoid recomputing $h_2(ya, mp)$) and find a collision between them (such a collision is very likely to occur since $C-1 > n/2$). The collision gives the required $ms \in S_1$ and $ml \in S_2$ of appropriate sizes such that $yb \triangleq h_2(ya, ms) = h_2(ya, ml)$ and $xb \triangleq h_1(xa, ms) = h_1(xa, ml)$.

The complexity of Step 1 is less than i compression function evaluations. The complexity of Step 2 is about $2^{n/2}$, while the complexity of Step 3 is about $C \cdot 2^{n/2} \approx n \cdot 2^{n/2}$. The complexity of Step 4 is about $i + n \cdot 2^{n/2}$. In total, the complexity of constructing the basic building block is about $i + n \cdot 2^{n/2}$ (ignoring small factors).

Complexity Analysis of the Full Phase. The full expandable message requires computing $C - 1 + t$ building blocks whose sum of length parameters (dominated by the final building block) is about $C \cdot 2^t \approx n \cdot 2^t$. Assuming that $t < n$, we construct $C - 1 + t \approx n$ building blocks and the total time complexity of constructing the expandable message is about $n \cdot 2^t + n^2 \cdot 2^{n/2}$. Our attack requires the $(C(C-1)+tC, C^2-1+C(2^t+t-1))$-expandable message to extend up to length L, implying that $L \approx n \cdot 2^t$, and giving time complexity of about

$$L + n^2 \cdot 2^{n/2}.$$

3.2 Details of Phase 2: Finding a Target State Pair

In the second phase, we fix some message block m, giving rise to the functional graphs $f_1(x) = h_1(x, m)$ and $f_2(y) = h_1(y, m)$ and let $g_1 \geq n/2$ be a parameter (to be determined later). Our goal is to find a pair of states $(\bar{x}, \bar{y})$, a message block $\bar{m}$ and an index p such that the following two conditions hold:

1. The state $\bar{x}$ is an iterate of depth 2^{n-g_1} in the functional graph of $f_1(x)$ and $\bar{y}$ is an iterate of depth 2^{n-g_1} in the functional graph of $f_2(y)$.
2. The state pair $(\bar{x}, \bar{y})$ is mapped to (a_p, b_p) by $\bar{m}$, or $(\bar{x}, \bar{y}) \xrightarrow{\bar{m}} (a_p, b_p)$.

The algorithm of this phase is given below.

1. Fix an arbitrary single-block value m.
2. Expand the functional graph of f_1 using the procedure of Section 2.5 with parameter g_1. Store all encountered iterates of depth 2^{n-g_1} in a table T_1.
3. Similarly, expand the functional graph of f_2 using the procedure of Section 2.5 with parameter g_1. Store all encountered iterates of depth 2^{n-g_1} in a table T_2.
4. For single-block values $m' = 0, 1, \ldots$, perform the following steps:
   (a) For each node $x \in T_1$ evaluate $x' = h_1(x, m')$ and store the matches $x' = a_i$ with the[a] sequence $a_1, \ldots, a_L$ in a table T_1', sorted according to the index i of a_i.
   (b) Similarly, for each node $y \in T_2$ evaluate $y' = h_2(y, m')$ and look for matches $y' = b_j$ with the sequence $b_1, \ldots, b_L$. For each match with some b_j, search for the index j in the table T_1'. If a match $i = j$ is found, set $p \triangleq i$ (namely, $(a_p, b_p) \triangleq (x', y')$), $\bar{m} \triangleq m'$ and $(\bar{x}, \bar{y}) \triangleq (x, y)$. This gives $(\bar{x}, \bar{y}) \xrightarrow{\bar{m}} (a_p, b_p)$ as required. Otherwise (no match $i = j$ is found), go back to Step 4.

[a] More precisely, due to the minimal length restriction of the expandable message, matches $x' = a_i$ with i smaller than (approximately) $C^2 \approx n^2$ cannot be exploited in the full attack. Moreover, the maximal exploitable value of i is $L - 2$. However, the fraction of these nodes is very small and can be ignored in the complexity analysis.

The time complexity of steps 2 and 3 (which execute the algorithm of Sect. 2.5) is about 2^{g_1}. The time complexity of step 4.(a) and step 4.(b) is also bounded by 2^{g_1} (given that $a_1, \ldots, a_L$ and $b_1, \ldots, b_L$ are sorted in memory), as the size of T_1 and T_2 is at most 2^{g_1} and the number of matches found in each step can only be smaller.

We now calculate the expected number of executions of Step 4 until the required (a_p, b_p) is found. Using the analysis of Sect. 2.5, the expected size of T_1 and T_2 (that contain iterates of depth 2^{n-g_1}) is close to 2^{g_1}. According to the birthday paradox, the expected size of T_1' is about $L \cdot 2^{g_1-n}$. Similarly, the number of matches $y' = b_j$ is also about $L \cdot 2^{g_1-n}$. The probability of a match $i = j$ in Step 4.(b) is computed using a birthday paradox on the L possible indexes, namely, $1/L \cdot (L \cdot 2^{g_1-n})^2 = L \cdot 2^{2g_1-2n}$. As a result, Step 4 is executed about $1/L \cdot 2^{2n-2g_1}$ times until the required (a_p, b_p) is found (the executions with different blocks m' are essentially independent). Altogether, the total time complexity of this step is

$$2^{g_1} \cdot 1/L \cdot 2^{2n-2g_1} = 1/L \cdot 2^{2n-g_1}.$$

Since the index p is uniformly distributed in the interval $[1, L]$, we will assume that $p = \Theta(L)$.

3.3 Details of Phase 3: Hitting the Target State Pair

In the third and final phase, we start from $(\hat{x}, \hat{y})$ and compute a message fragment $\hat{M}$ of length $q < p - 1$ such that $(\hat{x}, \hat{y}) \xrightarrow{\hat{M}} (\bar{x}, \bar{y})$. We use in a strong way the fact that the state $\bar{x}$ (and $\bar{y}$) is a deep iterate (of depth 2^{n-g_1}) in the functional graph of $f_1(x)$ ($f_2(y)$).

This phase is carried out by picking an arbitrary starting message block ms, which gives points $x_0 = h_1(\hat{x}, ms)$ and $y_0 = h_2(\hat{y}, ms)$. We then continue to evaluate the chains $x_i = h_1(x_{i-1}, m)$ and $y_j = h_2(y_{j-1}, m)$ up to a maximal length L' (determined below). We hope to encounter $\bar{x}$ at some distance $q - 1$ from x_0 and to encounter $\bar{y}$ at the same distance $q - 1$ from y_0. Given that $q - 1 < p$, this will give the required $\hat{M} = ms\|[m]^{q-1}$ (where $[m]^{q-1}$ denotes the concatenation of $q - 1$ message blocks m), which is of length $q < p - 1$. In case $\bar{x}$ and $\bar{y}$ are encountered at different distances in the chains, or at least one of them is not encountered at all, we pick a different value for ms and start again.

The next question which we address is to what maximal length L' should we evaluate $\overrightarrow{x}$ and $\overrightarrow{y}$. As we wish to reach iterates $\bar{x}$ and $\bar{y}$ of depth 2^{n-g_1}, it can be shown that $L' = 2^{n-g_1}$ is optimal. Since the total chain length should be less than $p - 1$, this imposes the restriction $L' = 2^{n-g_1} < p - 1 < L$, or $2^{g_1} < 2^n/L$.

We now estimate the probability that $\bar{x}$ and $\bar{y}$ will be encountered at the same distance from the arbitrary starting points of the chains x_0 and y_0. This probability will allow us to compute the number of chains from different starting points that we need to evaluate in this phase of the attack, which is an important parameter in the complexity evaluation.

Since $\bar{x}$ is an iterate of depth 2^{n-g_1} in $f_1(x)$, it is an endpoint of a chain of states of length $L' = 2^{n-g_1}$ (such a chain was computed in Sect. 3.2). Let d be in the interval $[1, L'] = [1, 2^{n-g_1}]$, then according to Lemma 1, $Pr[x_d = \bar{x}] \approx d \cdot 2^{-n}$ (this is the probability that $\bar{x}$ will be encountered at distance d from x_0). Due to the independence of f_1 and f_2, $Pr[x_d = \bar{x} \bigwedge y_d = \bar{y}] \approx (d \cdot 2^{-n})^2$. Summing the probabilities of the (disjoint) events over all distances d in the interval $[1, 2^{n-g_1}]$, we conclude that the probability that $\bar{x}$ and $\bar{y}$ will be encountered at the same distance is about $(2^{n-g_1})^3 \cdot 2^{-2n} = 2^{n-3g_1}$.

The probability calculation seems to give rise to the conclusion that we need to compute about 2^{3g_1-n} chains from different starting points in this phase of the attack. This conclusion was verified experimentally, but its proof is incomplete since the various trials performed by selecting different starting points for the chains are dependent. More details can be found in Appendix B.

The Algorithm of Phase 3. The naive algorithm described above performs about 2^{3g_1-n} trials, where each trial evaluates chains of length $L' = 2^{n-g_1}$ from arbitrary points, giving a total time complexity of about $2^{3g_1-n+n-g_1} = 2^{2g_1}$. Since $g_1 \geq n/2$, the time complexity of the full algorithm is at least 2^n and it is not faster than Joux's preimage attack.

In order to optimize the algorithm, we further expand the graphs of f_1 and f_2. As a result, the evaluated chains are expected to collide with the graphs sooner (before they are evaluated to the full length of 2^{n-g_1}). Once a collision occurs, we use a lookahead procedure to calculate the distance of the chain's starting point from $\bar{x}$ (or $\bar{y}$). This lookahead procedure resembles the one used in recent attacks on hash-based MACs [17,32] (although the setting and actual algorithm in our case are obviously different).

More specifically, we pick a parameter $g_2 > g_1$ and execute the algorithm below (see Fig. 1 for illustration).

1. Develop 2^{g_2} nodes in the functional graphs of f_1 (and f_2) (as specified in Sect. 2.5) with the following modifications.
   - Store at each node its distance from $\bar{x}$ (or $\bar{y}$ in f_2) (the maximal stored distance is $L' = 2^{n-g_1}$): for each computed chain, once it hits a previously visited node in the graph, obtain its stored distance from $\bar{x}$ (or $\bar{y}$ in f_2) and update it in all the computed nodes in the current chain up to the maximal value $L' = 2^{n-g_1}$.
   - If a chain does not hit $\bar{x}$, then the distance of its nodes is undefined and stored as a special value $\perp$. Similarly, this special value is used for nodes whose distance from $\bar{x}$ is larger than L'.
   - The evaluated nodes for f_1 (f_2) are stored in the data structure G_1 (G_2).
2. For single-block values $ms = 0, 1, \ldots$, compute $x_0 = h_1(\hat{x}, ms)$ and $y_0 = h_2(\hat{y}, ms)$ and repeat the following step:
   (a) Compute the chains $\overrightarrow{x}$ and $\overrightarrow{y}$ up to maximal length $L' = 2^{n-g_1}$, or until they hit G_1 and G_2 (respectively).
      - If $\overrightarrow{x}$ (or $\overrightarrow{y}$) does not hit G_1 (G_2), return to Step 2.
      - Otherwise, once $\overrightarrow{x}$ ($\overrightarrow{y}$) hits G_1 (G_2), obtain the stored distance from $\bar{x}$ ($\bar{y}$) at the collision point. If the distance to $\bar{x}$ (or $\bar{y}$) is undefined, return to Step 2.
      - Compute the respective distances i and j of x_0 and y_0 from $\bar{x}$ and $\bar{y}$. If $i \neq j$, return to Step 2.
      - Otherwise ($i = j$), denote $q = i+1$. If $q \geq p-1$, return to Step 2.
      - Otherwise ($q < p - 1$), return the message $\hat{M} = ms\|[m]^i = ms\|[m]^{q-1}$ as output.

The time complexity of Step 1 is 2^{g_2}. As previously computed, in Step 2 we perform about 2^{3g_1-n} trials before encountering two starting points with the same distance to $\bar{x}$ and $\bar{y}$. According to the analysis of Sect. 2.5, each trial requires about 2^{n-g_2} computation (before hitting G_1 and G_2). Therefore, the total time complexity of this phase is $2^{g_2} + 2^{3g_1-n} \cdot 2^{n-g_2} = 2^{g_2} + 2^{3g_1-g_2}$. The complexity is minimized by setting $g_2 = 3g_1/2$ which balances the two terms and gives time complexity of

$$2^{3g_1/2}.$$

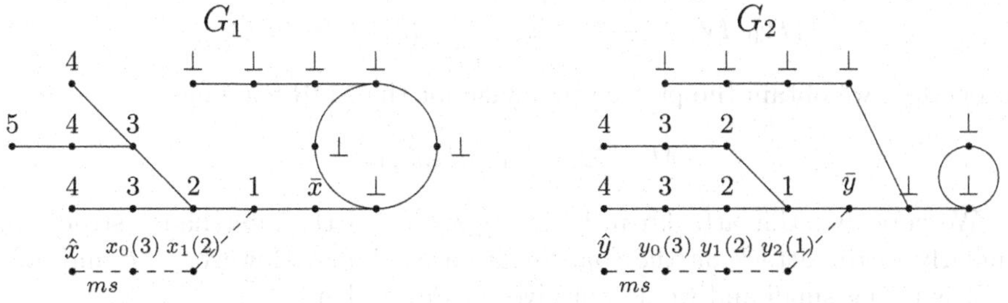

Fig. 1. Phase 3 of the attack

Finally, we note that the memory complexity of this algorithm can be optimized using distinguished points. A detailed way to achieve this will be presented in the closely related algorithm of Sect. 4.2.

4 A New Preimage Attack on the XOR Combiner

In this attack we are given a target n-bit preimage value V and our goal is to find a message M such that $H_1(M) \oplus H_2(M) = V$. Although the formal problem does not restrict M in any way, several concrete hash functions restrict the length of M. Therefore, we will also assume that the size of M is bounded by a parameter L. As in the previous attack, we start with a high-level overview and then give the technical details.

The attack is composed of three main phases which are similar to the second preimage attack on the concatenation combiner of Sect. 3. The first phase is identical to the first phase of the attack of Sect. 3. Namely, we build an expandable message that consists of the initial states (IV_1, IV_2) and final states $(\hat{x}, \hat{y})$ such that for each length κ in an appropriate range there is a message M_κ of κ blocks that maps (IV_1, IV_2) to $(\hat{x}, \hat{y})$. The description of this phase is given in Sect. 3.1 and is not repeated below.

In the second phase of the attack, we find a set S (of size 2^s) of tuples of the form $((x, y), w)$ such that w is a single block, $(x, y) \xrightarrow{w} (a, b)$, and $h_1(a, pad) \oplus h_2(b, pad) = V$, where pad is the final block of the (padded) preimage message of length L. Moreover, (x, y) has a special property that will be defined in the detailed description of this phase.

In the third and final phase, we start from $(\hat{x}, \hat{y})$ and compute a message fragment $\hat{M}$ of length q (shorter than $L - 2$) such that $(\hat{x}, \hat{y}) \xrightarrow{\hat{M}} (\bar{x}, \bar{y})$ for some $((\bar{x}, \bar{y}), \bar{m}) \in S$. For this tuple, denote $(\bar{a}, \bar{b}) \triangleq h_{1,2}((\bar{x}, \bar{y}), \bar{m})$.

Finally, we pick M_{L-q-2} using the expandable message, giving $(IV_0, IV_1) \xrightarrow{M_{L-q-2}} (\hat{x}, \hat{y})$, and concatenate $M_{L-q-2} \| \hat{M} \| \bar{m}$ in order to reach the state pair $(\bar{a}, \bar{b})$ from (IV_1, IV_2) with a message of appropriate length $L - 1$. Indeed, we have

$$(IV_0, IV_1) \xrightarrow{M_{L-q-2}} (\hat{x}, \hat{y}) \xrightarrow{\hat{M}} (\bar{x}, \bar{y}) \xrightarrow{\bar{m}} (\bar{a}, \bar{b}).$$

Altogether, we obtain the padded preimage for the XOR combiner

$$M = M_{L-q-2} \| \hat{M} \| \bar{m} \| pad.$$

We note that this attack can be optimized using the interchange structure, similarly to the attack on the concatenation combiner. However, the improvement is rather small and we do not give the details here.

Notation. We summarize below the notation that is shared across the various phases.

$\boldsymbol{V}$	: Target preimage.
$\boldsymbol{M}$	: Computed preimage.
$\boldsymbol{L}$	: Length of M.
$\boldsymbol{pad}$	: Final block of (the padded) M.
$(\hat{\boldsymbol{x}}, \hat{\boldsymbol{y}})$	: Endpoint pair of expandable message (computed in Phase 1).
$\boldsymbol{S}$:	Set containing tuples of the form $((x, y), w)$ such that w is a single block, $(x, y) \xrightarrow{w} (a, b)$, and $h_1(a, pad) \oplus h_2(b, pad) = V$ (computed in Phase 2).
$\boldsymbol{2^s}$	: Size of S.
$((\bar{\boldsymbol{x}}, \bar{\boldsymbol{y}}), \bar{\boldsymbol{m}})$	: State pair and message block in S used in M (computed in Phase 3).
$(\bar{\boldsymbol{a}}, \bar{\boldsymbol{b}})$	: State pair defined as $(\bar{a}, \bar{b}) \triangleq h_{1,2}((\bar{x}, \bar{y}), \bar{m})$ (computed in Phase 3).
$\hat{\boldsymbol{M}}$	: Message fragment used in M that maps $(\hat{x}, \hat{y})$ to $(\bar{x}, \bar{y})$ (computed in Phase 3).
$\boldsymbol{q}$	: The length of $\hat{M}$ (smaller than $L-2$).

Complexity Evaluation. Denote $L = 2^\ell$. For parameters $g_1 \geq max(n/2, n-\ell)$ and $s \geq 0$, the complexity of the phases of the attack (as computed in their detail description) is given below (ignoring constant factors).

Phase 1: $2^\ell + n^2 \cdot 2^{n/2}$
Phase 2: 2^{n+s-g_1}
Phase 3: $2^{3g_1/2-s/2} + L \cdot 2^{9g_1/2-2n-3s/2} + L \cdot 2^{2g_1-n} = 2^{3g_1/2-s/2} + 2^{\ell+9g_1/2-2n-3s/2} + 2^{\ell+2g_1-n}$

We balance the time complexities of the second phase and the first term in the expression of the third phase by setting $n + s - g_1 = 3g_1/2 - s/2$, or $s = 5g_1/3 - 2n/3$, giving a value of $2^{n/3+2g_1/3}$ for these terms. Furthermore, $\ell + 9g_1/2 - 2n - 3s/2 = \ell + 2g_1 - n$ and the time complexity expression of

Phase 3 is simplified to $2^{n/3+2g_1/3} + 2^{\ell+2g_1-n}$. Since g_1 is a positive factor in all the terms, we optimize the attack by picking the minimal value of g_1 under the restriction $g_1 \geq max(n/2, n-\ell)$. In case $\ell \leq n/2$, we set $g_1 = n-\ell$ and the total time complexity of the attack[9] is

$$2^{n/3+2(n-\ell)/3} = 2^{n-2\ell/3}.$$

The optimal complexity is $2^{2n/3}$, obtained for $\ell = n/2$ by setting $g_1 = n/2$.

4.1 Details of Phase 2: Finding a Set of Target State Pairs

In the second phase, we fix some message block m, giving rise to the functional graphs defined by the random mappings $f_1(x) = h_1(x, m)$ and $f_2(y) = h_1(y, m)$. Given parameters $g_1 \geq n/2$ and $s \geq 0$, our goal is to compute a set S (of size 2^s) of tuples of the form $((x, y), w)$ where w is a single block such that for each tuple:

1. The state x is an iterate of depth 2^{n-g_1} in the functional graph of $f_1(x)$ and y is an iterate of depth 2^{n-g_1} in the functional graph of $f_2(y)$.
2. $(x, y) \xrightarrow{w} (a, b)$ and $h_1(a, pad) \oplus h_2(b, pad) = V$, where pad is a final block of the (padded) preimage message of length L.

The algorithm of this phase is (obviously) somewhat different from the algorithm of Sect. 3.2 due to the fact that the goal of this attack and the actual combiner scheme attacked are different. This algorithm resembles the algorithm used in the final phase in Leurent and Wang's attack [25], as both look for state pairs (x, y) that give $h_1(x, w\|pad) \oplus h_2(y, w\|pad) = V$ (for some message block w). The difference is that in the attack of [25], (x, y) was an arbitrary endpoint pair of the interchange structure, while in our case, we look for x and y that are deep iterates.

1. Fix an arbitrary single-block value m.
2. Expand the functional graph of f_1 using the procedure of Sect. 2.5 with parameter g_1. Store all encountered iterates of depth 2^{n-g_1} in a table T_1.
3. Similarly, expand the functional graph of f_2 using the procedure of Sect. 2.5 with parameter g_1. Store all encountered iterates of depth 2^{n-g_1} in a table T_2.
4. Allocate a set $S = \emptyset$. For single-block values $w = 0, 1, \ldots,$ perform the following steps until S contains 2^s elements:
   (a) For each node $x \in T_1$ evaluate $h_1(x, w\|pad)$, and store the results in a table T_1', sorted according $h_1(x, w\|pad)$.
   (b) Similarly, for each node $y \in T_2$ evaluate $h_2(y, w\|pad) \oplus V$, and look for matches $h_2(y, w\|pad) \oplus V = h_1(x, w\|pad)$ with T_1'. For each match, add the tuple $((x, y), w)$ to S.

[9] Note that $\ell + 2g_1 - n = n - \ell < n - 2\ell/3$.

The time complexity of steps 2 and 3 is about 2^{g_1}. The time complexity of step 4.(a) and step 4.(b) is also bounded by 2^{g_1}. We now calculate the expected number of executions of Step 4 until 2^s matches are found and inserted into S.

According to the analysis of Sect. 2.5, the expected size of T_1 and T_2 (the number of deep iterates) is close to 2^{g_1}. Thus, for each execution of Step 4, the expected number of matches on n-bit values $h_2(y, w\|pad) \oplus V = h_1(x, w\|pad)$ is 2^{2g_1-n}. Consequently, Step 4 is executed 2^{s+n-2g_1} times in order to obtain 2^s matches. Altogether, the total time complexity of this step is

$$2^{n+s-2g_1+g_1} = 2^{n+s-g_1}.$$

4.2 Details of Phase 3: Hitting a Target State Pair

In the third and final phase, we start from $(\hat{x}, \hat{y})$ and compute a message $\hat{M}$ of length q (shorter than $L-2$) such that $(\hat{x}, \hat{y}) \xrightarrow{\hat{M}} (\bar{x}, \bar{y})$ for some $((\bar{x}, \bar{y}), \bar{m}) \in S$.

This phase is carried out by picking an arbitrary starting message block ms, which gives points $x_0 = h_1(\hat{x}, ms)$ and $y_0 = h_2(\hat{y}, ms)$. We then continue to evaluate the chains $x_{i+1} = h_1(x_i, m)$ and $y_{j+1} = h_2(y_j, m)$ up to length at most $L-3$. We hope to encounter $\bar{x}$ at some distance $q-1$ from x_0 and to encounter $\bar{y}$ at the same distance $q-1$ from y_0, where $((\bar{x}, \bar{y}), \bar{m}) \in S$ for some single block value $\bar{m}$. This gives the required $\hat{M} = ms\|[m]^{q-1}$.

The goal of this algorithm is very similar to one of the algorithm of Sect. 3.3, where the difference is the size of the set S, which essentially contained a single element[10] in Sect. 3.3, but can now have a larger size. This difference results in a complication that arises when the algorithm builds the functional graph of f_1 (and f_2), and has to keep track of distances of encountered nodes from all the 2^s nodes x (and y) that are in tuples of S (instead of merely keeping track of distances from a single node as in Sect. 3.3).

More formally, we define an S-*node* (for f_1) as a node x such that there exists a node y and a message block w such that $((x, y), w) \in S$. An S-node for f_2 in defined in a similar way. In order to avoid heavy update operations for the distances from all the S-nodes, we use distinguished points. Essentially, each computed chain is partitioned into intervals according to distinguished points, where each distinguished point stores only the distances to all the S-nodes that are contained in its interval up to the next distinguished point. Given a parameter $g_2 > g_1$, the algorithm for this phase is described below.

The time complexity of Step 1 is about 2^{g_1}, as described in Sect. 2.5 (note that we always perform a constant amount of work per developed node). Compared to the second step of the algorithm of Sect. 3.3, S contains 2^s elements (instead of 1), and this reduces by the same factor the expected number of trials we need to execute in order to reach some $((\bar{x}, \bar{y}), \bar{m}) \in S$ in Step 2. Reusing the analysis of Sect. 3.3, the expected number of trials (executions of Step 2) is reduced from 2^{3g_1-n} to 2^{3g_1-n-s}.

[10] One may ask why we did not compute a larger set S in Sect. 3.2. The reason for this is that it can be shown that in the previous case a set of size 1 is optimal.

1. Develop (about) 2^{g_2} nodes in the functional graphs of f_1 (and f_2) (as specified in Section 2.5) with the following modifications.
   - Store only distinguished points for which the $n - g_2$ LSBs are zero.
   - Once an S-node is encountered, update its distance in the previously encountered distinguished point (which is defined with high probability[a]).
   - Stop evaluating each chain once it hits a stored distinguished point.
   - The evaluated distinguished points for f_1 (f_2) are stored in the data structure G_1 (G_2).
2. For single-block values $ms = 0, 1, \ldots$, compute $x_0 = h_1(\hat{x}, ms)$ and $y_0 = h_2(\hat{y}, ms)$ and repeat the following step:
   (a) Compute chains $\overrightarrow{x}$ and $\overrightarrow{y}$ as specified below.
      - First, compute the chains in a standard way by evaluating the compression functions h_1 and h_2, until they hit stored distinguished points in G_1 and G_2 (respectively).
      - Then, allocate a table T_1 (T_2 for f_2) and continue traversing (only) the distinguished points of the chain (using the links in G_1 and G_2) up to depth $L-2$, while updating T_1 (T_2): for each visited distinguished point, add all its stored S-nodes to T_1 (T_2) with its distance from x_0 (y_0).
      - Once the maximal depth $L-2$ is reached, sort T_1 and T_2. Search for nodes $\bar{x}$ and $\bar{y}$ that were encountered at the same distance $q-1$ from x_0 and y_0 (respectively), such that $((\bar{x}, \bar{y}), \bar{m}) \in S$. If such $\bar{x} \in T_1$ and $\bar{y} \in T_2$ exist, return the message $\hat{M} = ms\|[m]^{q-1}$ and $\bar{m}$ (retrieved from S) as output. Otherwise (no such $\bar{x}$ and $\bar{y}$ were found), return to Step 2.

[a] Since $g_2 > g_1$, S-nodes are deeper iterates than distinguished points, and thus distinguished points are likely to be encountered in an arbitrary chain before an S-node.

The analysis of the complexity of Step 2.(a) is somewhat more involved compared to the corresponding step of Sect. 3.3. First, we estimate the expected number of nodes that we visit during the computation of a chain. Initially (as in Sect. 3.3), we compute about 2^{n-g_2} nodes until we hit stored distinguished points. Then, we continue by traversing (only) distinguished points up to depth of about L. The expected number of such points is $L \cdot 2^{g_2 - n}$. Therefore, we expect to visit about $2^{n-g_2} + L \cdot 2^{g_2-n}$ nodes while computing a chain.

Finally, we need to account for all the S-nodes encountered while traversing the chains of depth L. Basically, there are 2^s S-nodes which are iterates of depth 2^{n-g_1}, (essentially) randomly chosen in Phase 2 out of about 2^{g_1} such deep iterates. As a result, the probability of such a deep iterate to be an S-node is about 2^{s-g_1} (while other nodes have probability 0). Therefore, while traversing chains of depth L, we expect to encounter at most $L \cdot 2^{s-g_1}$ S-nodes (which is a bound on the sizes of T_1 and T_2). Altogether, the expected time complexity of a single execution of Step 2.(a) is at most $2^{n-g_2} + L \cdot 2^{g_2-n} + L \cdot 2^{s-g_1}$.

The total time complexity of this phase is $2^{g_2} + 2^{3g_1-n-s} \cdot (2^{n-g_2} + L \cdot 2^{g_2-n} + L \cdot 2^{s-g_1}) = 2^{g_2} + 2^{3g_1-g_2-s} + L \cdot 2^{3g_1+g_2-2n-s} + L \cdot 2^{2g_1-n}$. We set $g_2 = 3g_1/2 - s/2$ which balances the first two terms and gives time complexity of

$$2^{3g_1/2-s/2} + L \cdot 2^{9g_1/2-2n-3s/2} + L \cdot 2^{2g_1-n}.$$

The time complexity evaluation of the full attack at the beginning of this section shows that for the optimal parameters of this attack, the extra two terms $L \cdot 2^{9g_1/2-2n-3s/2} + L \cdot 2^{2g_1-n}$ are negligible compared to the other terms in the complexity equation. In other words, the distinguished points method allowed us to resolve with no overhead the complication of keeping track of distances from the S-nodes.

5 Conclusions and Open Problems

In this paper we devised the first second preimage attack on the concatenation combiner and improved the preimage attack on the XOR combiner (due to Leurent and Wang) in case both hash functions use the Merkle-Damgård construction. Since both of our second preimage and preimage attacks on the concatenation and XOR combiners have higher complexities than the lower bounds ($2^n/L$ and $2^{n/2}$, respectively), it would be interesting to further improve them, and it particular, extend the second preimage attack to shorter messages. There are many additional interesting future work items such as extending our algorithms to combine more than two hash functions. Indeed, while it is easy to extend the expandable message to more than two hash functions with small added complexity, extending the random mapping techniques is less obvious. Yet another open problem is to improve preimage attack of Leurent and Wang on the XOR combiner in case only one of the functions uses the Merkle-Damgård construction.

References

1. Andreeva, E., Bouillaguet, C., Dunkelman, O., Fouque, P.-A., Hoch, J., Kelsey, J., Shamir, A., Zimmer, S.: New second-preimage attacks on hash functions. J. Cryptol. 1–40. (to appear) (2015)
2. Andreeva, E., Bouillaguet, C., Dunkelman, O., Kelsey, J.: Herding, second preimage and trojan message attacks beyond Merkle-Damgård. In: Jacobson Jr., M.J., Rijmen, V., Safavi-Naini, R. (eds.) SAC 2009. LNCS, vol. 5867, pp. 393–414. Springer, Heidelberg (2009)
3. Andreeva, E., Bouillaguet, C., Fouque, P.-A., Hoch, J.J., Kelsey, J., Shamir, A., Zimmer, S.: Second preimage attacks on dithered hash functions. In: Smart, N.P. (ed.) EUROCRYPT 2008. LNCS, vol. 4965, pp. 270–288. Springer, Heidelberg (2008)
4. Biham, E., Dunkelman, O.: A framework for iterative hash functions- HAIFA. In: IACR Cryptology ePrint Archive (2007). http://eprint.iacr.org/2007/278

5. Boneh, D., Boyen, X.: On the impossibility of efficiently combining collision resistant hash functions. In: Dwork, C. (ed.) CRYPTO 2006. LNCS, vol. 4117, pp. 570–583. Springer, Heidelberg (2006)
6. Brassard, G. (ed.): CRYPTO 1989. LNCS, vol. 435. Springer, Heidelberg (1990)
7. Cramer, R. (ed.): EUROCRYPT 2005. LNCS, vol. 3494. Springer, Heidelberg (2005)
8. Damgård, I.: A design principle for hash functions. In: Brassard [6], pp. 416–427
9. Dean, R.D.: Formal Aspects of Mobile Code Security. Ph.D. thesis, Princeton University (1999)
10. Dierks, T., Rescorla, E.: The Transport Layer Security (TLS) Protocol Version 1.2. RFC 5246 (2008). https://www.ietf.org/rfc/rfc5246.txt
11. Dinur, I., Leurent, G.: Improved generic attacks against hash-based MACs and HAIFA. In: Garay, J.A., Gennaro, R. (eds.) [16], pp. 149–168
12. Dunkelman, O., Preneel, B.: Generalizing the herding attack to concatenated hashing schemes. In: ECRYPT Hash Workshopp (2007)
13. Fischlin, M., Lehmann, A.: Multi-property preserving combiners for hash functions. In: Canetti, R. (ed.) TCC 2008. LNCS, vol. 4948, pp. 375–392. Springer, Heidelberg (2008)
14. Fischlin, M., Lehmann, A., Pietrzak, K.: Robust multi-property combiners for hash functions. J. Cryptol. **27**(3), 397–428 (2014)
15. Freier, A.O., Karlton, P., Kocher, P.C.: The Secure Sockets Layer (SSL) Protocol Version 3.0.RFC 6101 (2011). http://www.ietf.org/rfc/rfc6101.txt
16. Garay, J.A., Gennaro, R. (eds.): CRYPTO 2014, Part I. LNCS, vol. 8616. Springer, Heidelberg (2014)
17. Guo, J., Peyrin, T., Sasaki, Y., Wang, L.: Updates on generic attacks against HMAC and NMAC. In: Garay, J.A., Gennaro, R. (eds.) [16], pp. 131–148
18. Hellman, M.E.: A cryptanalytic time-memory trade-off. IEEE Trans. Inf. Theory **26**(4), 401–406 (1980)
19. Hoch, J.J., Shamir, A.: Breaking the ICE - finding multicollisions in iterated concatenated and expanded (ICE) hash functions. In: Robshaw, M. (ed.) FSE 2006. LNCS, vol. 4047, pp. 179–194. Springer, Heidelberg (2006)
20. Hoch, J.J., Shamir, A.: On the strength of the concatenated hash combiner when all the hash functions are weak. In: Aceto, L., Damgård, I., Goldberg, L.A., Halldórsson, M.M., Ingólfsdóttir, A., Walukiewicz, I. (eds.) Automata, Languages and Programming. LNCS, vol. 5126, pp. 616–630. Springer, Heidelberg (2008)
21. Jha, A., Nandi, M.: Some Cryptanalytic Results on Zipper Hash and Concatenated Hash. IACR Cryptology ePrint Archive 2015:973 (2015)
22. Joux, A.: Multicollisions in iterated hash functions. Application to cascaded constructions. In: Franklin, M. (ed.) CRYPTO 2004. LNCS, vol. 3152, pp. 306–316. Springer, Heidelberg (2004)
23. Kelsey, J., Schneier, B.: Second preimages on n-Bit hash functions for much less than 2^n work. In: Cramer [7], pp. 474–490
24. Leurent, G., Peyrin, T., Wang, L.: New generic attacks against hash-based MACs. In: Sako, K., Sarkar, P. (eds.) ASIACRYPT 2013, Part II. LNCS, vol. 8270, pp. 1–20. Springer, Heidelberg (2013)
25. Leurent, G., Wang, L.: The sum can be weaker than each part. In: Oswald, E., Fischlin, M. (eds.) EUROCRYPT 2015. LNCS, vol. 9056, pp. 345–367. Springer, Heidelberg (2015)
26. Mendel, F., Rechberger, C., Schläffer, M.: MD5 is weaker than weak: attacks on concatenated combiners. In: Matsui, M. (ed.) ASIACRYPT 2009. LNCS, vol. 5912, pp. 144–161. Springer, Heidelberg (2009)

27. Mennink, B., Preneel, B.: Breaking and fixing cryptophia's short combiner. In: Gritzalis, D., Kiayias, A., Askoxylakis, I. (eds.) CANS 2014. LNCS, vol. 8813, pp. 50–63. Springer, Heidelberg (2014)
28. Merkle, R.C.: One way hash functions and DES. In: Brassard [6], pp. 428–446
29. Mittelbach, A.: Hash combiners for second pre-image resistance, target collision resistance and pre-image resistance have long output. In: Visconti, I., De Prisco, R. (eds.) SCN 2012. LNCS, vol. 7485, pp. 522–539. Springer, Heidelberg (2012)
30. Nandi, M., Stinson, D.R.: Multicollision attacks on some generalized sequential hash functions. IEEE Trans. Inf. Theory **53**(2), 759–767 (2007)
31. Perrin, L., Khovratovich, D.: Collision spectrum, entropy loss, T-Sponges, and cryptanalysis of GLUON-64. In: Cid, C., Rechberger, C. (eds.) FSE 2014. LNCS, vol. 8540, pp. 82–103. Springer, Heidelberg (2015)
32. Peyrin, T., Wang, L.: Generic universal forgery attack on iterative hash-based MACs. In: Nguyen, P.Q., Oswald, E. (eds.) EUROCRYPT 2014. LNCS, vol. 8441, pp. 147–164. Springer, Heidelberg (2014)
33. Pietrzak, K.: Non-trivial black-box combiners for collision-resistant hash-functions don't exist. In: Naor, M. (ed.) EUROCRYPT 2007. LNCS, vol. 4515, pp. 23–33. Springer, Heidelberg (2007)
34. Preneel, B.: Analysis and design of cryptographic hash functions. Ph.D. thesis, KU Leuven (1993)
35. Rjasko, M.: On existence of robust combiners for cryptographic hash functions. In: Vojtás, P. (ed.) Proceedings of the Conference onTheory and Practice of Information Technologies, ITAT 2009, Horskýhotel Kralova studna, Slovakia, September 25-29, 2009, volume 584 of CEUR Workshop Proceedings, pp. 71–76. CEUR-WS.org 2009
36. van Oorschot, P.C., Wiener, M.J.: Parallel collision search with cryptanalytic applications. J. Cryptol. **12**(1), 1–28 (1999)
37. Wang, X., Yin, Y.L., Yu, H.: Finding collisions in the full SHA-1. In: Shoup, V. (ed.) CRYPTO 2005. LNCS, vol. 3621, pp. 17–36. Springer, Heidelberg (2005)
38. Wang, X., Yu, H.: How to break MD5 and other hash functions. In: Cramer [7], pp. 19–35

A Optimizing the Second Preimage Attack Using the Interchange Structure

The interchange structure [25] is built with a parameter that we denote by r. It consists of a starting state pair (as, bs) and two sets of 2^r internal states A for H_1 and B for H_2 such that: for any value $a \in A$ and any value $b \in B$, it is possible to efficiently construct a message $M_{a,b}$ (of length 2^{2r}) such that $(as, bs) \xrightarrow{M_{a,b}} (a, b)$. We now describe how to use the interchange structure as a black box in order to optimize the second preimage attack of Sect. 3.

The idea is to insert the interchange structure after the expandable message in order to reach $(\bar{x}, \bar{y})$ more efficiently in the third phase of the attack. More specifically, consider Step 2 in the attack of Sect. 3.3. There, we start computing from the state pair $(\hat{x}, \hat{y})$ and evaluate chains independently for each single-block value $ms = 0, 1, \ldots$. In the optimized attack, we build the interchange structure with the starting state pair $(\hat{x}, \hat{y})$ and sets of 2^r states A, B. We pick some single-block value ms and compute two sets of 2^r chains starting from the states of A and B. Our goal is to find any pair of states $a \in A$ and $b \in B$ such that $(a, b) \xrightarrow{ms\|[m]^{q-1}} (\bar{x}, \bar{y})$ for some $q \le p - 2r - 1$. Therefore, we have $(\hat{x}, \hat{y}) \xrightarrow{M_{a,b}} (a, b) \xrightarrow{ms\|[m]^{q-1}} (\bar{x}, \bar{y})$, and we set $\hat{M} = M_{a,b}\|ms\|[m]^{q-1}$.

In the original attack, we evaluate and compare two chains for each execution of Step 2. In contrast, in the optimized attack, in each execution of modified Step 2 we evaluate $2 \cdot 2^r$ chains and compare them in $2^r \cdot 2^r = 2^{2r}$ pairs. Consequently, the time complexity of (modified) Step 2 is increased by 2^r, but we have to execute it 2^{2r} less times. The complexity evaluation of the attack is rather technical as we need to balance several parameters and account for the message length 2^{2r} of $M_{a,b}$.

We sketch the complexity evaluation for short messages, where the length of $M_{a,b}$ is roughly equal to L, i.e., we have $L = 2^{\ell} = 2^{2r}$ or $r = \ell/2$. According to Sect. 3.3, the complexity of Phase 3 in the original attack is $2^{g_2} + 2^{3g_1-n} \cdot 2^{n-g_2}$. Since building the interchange structure requires $2^{n/2+2r}$ time, the modified complexity is $2^{g_2} + 2^{3g_1-n-2r} . 2^{n-g_2+r} + 2^{n/2+2r} = 2^{g_2} + 2^{3g_1-g_2-\ell/2} + 2^{n/2+\ell}$ (setting $r = \ell/2$). We balance the first two terms and obtain $g_2 = 3g_1/2 - \ell/4$, giving time complexity of $2^{3g_1/2-\ell/4} + 2^{n/2+\ell}$. Recall from Sect. 3.2 that the complexity of Phase 2 is $2^{2n-g_1-\ell}$. We balance the second and third phases by setting $g_1 = 4n/5 - 3\ell/10$, which gives time complexity of $2^{6n/5-7\ell/10}$ for small values of ℓ in which the term $2^{n/2+\ell}$ is negligible. Therefore, we obtain an attack faster than 2^n for messages of length $L > 2^{2n/7}$ (which is a small improvement compared to $L \geq 2^{n/3}$, obtained without this optimization).

For larger values of ℓ we need to redo the computation and account for the term $2^{n/2+2r}$ in the complexity evaluation. However, since the improvement over the original attack in not very significant, we do not give the details here.

B On the Number of Required Chain Evaluations in Sect. 3.3

In Sect. 3.3 we concluded that the probability of encountering $\bar{x}$ and $\bar{y}$ at the same distance in chains (of f_1 and f_2) evaluated from arbitrary starting points x_0 and y_0 is about 2^{n-3g_1}. If the trials of chain evaluations were independent, this would have led to the conclusion that we need to compute about 2^{3g_1-n} chains from different starting points in Phase 3. However, the trials are dependent as explained below.

Reconsider Lemma 1 in case we select more than $2^n/D^2$ starting points x_0^i such that the chains (of length D) evaluated from them contain in total more than $D \cdot 2^n/D^2 = 2^n/D$ nodes. In this case, a new chain of length D (evaluated from x_0) is very likely to collide with a previously evaluated node before colliding with the original chain (evaluated from x_0') due to the birthday paradox. After a collision of the new chain with a previously evaluated node, the outcome of the trial is determined and cannot be analyzed probabilistically. Of course, this does not imply that Lemma 1 does not hold for more than $2^n/D^2$ trials, but it does imply that in a formal proof we need to account for the dependency of the trials when applying this lemma with more than $2^n/D^2$ trials.[11]

A potential way to handle this is to select more targets[12] of the form $(\bar{x}, \bar{y})$ in Phase 2, which reduces the number of trials that we need to perform in Phase 3 (as we need to reach only one target). This will enable us to complete the theoretical analysis of the attack, but it does result in a performance degradation (although the attack remains faster than 2^n for a modified set of parameters).

However, based on simulations (described below), we strongly believe that indeed about 2^{3g_1-n} trials are required in Sect. 3.3 in order to reach arbitrary iterates $\bar{x}$ and $\bar{y}$ of depth 2^{n-g_1} at the same distance. We note that assumptions of this type are not uncommon in analysis of random functions. A recent and related conjecture was made in [17].

Experimental Results. In our simulations we preformed hundreds of experiments on independent n-bit random mappings f_1 and f_2 for $n \in \{12, \ldots, 28\}$ with several[13] values of $n/3 \leq g_1 < n/2$. In the beginning of each experiment, we chose different mappings f_1 and f_2 and arbitrary deep iterates $\bar{x}, \bar{y}$ of depth 2^{n-g_1}. Each experiment was carried out by performing $2 \cdot 2^{3g_1-n}$ trials (with chains evaluated from arbitrary different starting points), trying to hit $\bar{x}$ and $\bar{y}$ at the same distance (up to 2^{n-g_1}). The success rate of the experiments was more than 50 % and did not drop as the value of n increased.

[11] Note that in our case $D = 2^{n-g_1}$ or $2^{g_1} = 2^n/D$, so $2^{3g_1-n} = 2^{2n}/D^3 > 2^n/D^2$.

[12] In a similar way to the algorithm of Sect. 4.

[13] Smaller values of n where chosen for smaller values of g_1, as these experiments are more expensive.

Cryptanalysis of the New CLT Multilinear Map over the Integers

Jung Hee Cheon[1(✉)], Pierre-Alain Fouque[2,3(✉)], Changmin Lee[1(✉)], Brice Minaud[2(✉)], and Hansol Ryu[1(✉)]

[1] Seoul National University, Seoul, Korea
{jhcheon,cocomi11,sol8586}@snu.ac.kr
[2] Université de Rennes 1, Rennes, France
pierre-alain.fouque@ens.fr, brice.minaud@gmail.com
[3] Institut Universitaire de France, Paris, France

Abstract. Multilinear maps serve as a basis for a wide range of cryptographic applications. The first candidate construction of multilinear maps was proposed by Garg, Gentry, and Halevi in 2013, and soon afterwards, another construction was suggested by Coron, Lepoint, and Tibouchi (CLT13), which works over the integers. However, both of these were found to be insecure in the face of so-called zeroizing attacks, by Hu and Jia, and by Cheon, Han, Lee, Ryu and Stehlé. To improve on CLT13, Coron, Lepoint, and Tibouchi proposed another candidate construction of multilinear maps over the integers at CRYPTO 2015 (CLT15).

This article presents two polynomial attacks on the CLT15 multilinear map, which share ideas similar to the cryptanalysis of CLT13. Our attacks allow recovery of all secret parameters in time polynomial in the security parameter, and lead to a full break of the CLT15 multilinear map for virtually all applications.

Keywords: Multilinear maps · Graded encoding schemes

1 Introduction

Cryptographic multilinear maps are a powerful and versatile tool to build cryptographic schemes, ranging from one-round multipartite Diffie-Hellman to witness encryption and general program obfuscation. The notion of cryptographic multilinear map was first introduced by Boneh and Silverberg in 2003, as a natural generalization of bilinear maps such as pairings on elliptic curves [BS03]. However it was not until 2013 that the first concrete instantiation over ideal lattices was realized by Garg, Gentry and Halevi [GGH13a], quickly inspiring another construction over the integers by Coron, Lepoint and Tibouchi [CLT13]. Alongside these first instantiations, a breakthrough result by Garg, Gentry, Halevi, Raykova, Sahai and Waters achieved (indistinguishability) obfuscation for all circuits from multilinear maps [GGH+13b]. From that point multilinear maps have garnered considerable interest in the cryptographic community, and a host of other applications have followed.

M. Fischlin and J.-S. Coron (Eds.): EUROCRYPT 2016, Part I, LNCS 9665, pp. 509–536, 2016.
DOI: 10.1007/978-3-662-49890-3_20

However this wealth of applications rests on the relatively fragile basis of only three constructions of multilinear maps to date: namely the original construction over ideal lattices [GGH13a], the construction over the integers [CLT13], and another recent construction over lattices [GGH15]. Moreover none of these constructions relies on standard hardness assumptions. In fact all three constructions have since been broken for their more "direct" applications such as one-round multipartite Diffie-Hellman [HJ15, CHL+15, Cor15]. Thus building candidate multilinear maps and assessing their security may be regarded as a challenging work in progress, and research in this area has been very active in recent years.

Following the attack by Cheon, Han, Lee, Ryu and Stehlé (CHLRS attack) on the [CLT13] multilinear map over the integers, several attempts to repair the scheme were published on ePrint, which hinged on hiding encodings of zero in some way; however these attempts were quickly proven insecure [CGH+15]. At CRYPTO 2015, Coron, Lepoint and Tibouchi set out to repair their scheme by following a different route [CLT15]: they essentially retained the structure of encodings from [CLT13], but added a new type of noise designed to thwart the CHLRS approach. Their construction was thus able to retain the attractive features of the original, namely conceptual simplicity, relative efficiency, and wide range of presumed hard problems on which applications could be built.

1.1 Our Contribution

In this paper we propose two polynomial attacks on the new multilinear map over the integers presented by Coron, Lepoint and Tibouchi at CRYPTO 2015 [CLT15]. These two attacks were originally published independently on ePrint by Cheon, Lee and Ryu [CLR15], and by Minaud and Fouque [MF15]. The present paper is a merge of the two results for publication at EUROCRYPT 2016.

The impact of both attacks is the same, and they both use the same starting point ("integer extraction"). The second half of the attacks is where they differ. In a nutshell, the attack by Cheon, Lee and Ryu looks into the exact expression of the value a in the term av_0 appearing in integer extractions. This makes it possible to uncover a matrix product similar to the CHLRS attack on CLT13, albeit a more complex one. As in the CHLRS attack, the secret parameters are then recovered as the eigenvalues of a certain matrix. For this reason we shall call this attack the *eigenvalue* attack.

By contrast the attack by Minaud and Fouque treats the value a in av_0 as a noise, which is removed by first recovering v_0 and taking equations modulo v_0. The secret parameter v_0 is recovered as (a divisor of) the determinant of a CHLRS-type matrix product. For this reason we shall call this attack the *determinant* attack. Once v_0 is recovered, CLT15 essentially collapses to CLT13 and can be broken by the CHLRS attack.

Both of the proposed attacks are polynomial in the security parameter. In addition, in the optimized version of the scheme where an exact multiple of x_0 is provided in the public parameters, the second attack is instant (as no determinant computation is actually required).

Moreover both attacks apply to virtually all possible applications of the CLT15 multilinear map. Indeed, while they do require low-level encodings of zero, these encodings are provided by the ladders given in the public parameters. In this respect CLT15 is weaker than CLT13. A closer look at the impact of our attacks is provided in Sect. 1.3.

We refer the reader to [MF15] for a third, probabilistic attack with similar properties.

1.2 Overview of the Attacks

We begin by briefly recalling the CLT15 multilinear map (more precisely, graded encoding scheme). The message space is $\mathbb{Z}_{g_1} \times \cdots \times \mathbb{Z}_{g_n}$ for some small primes $g_1, \ldots, g_n$, and $(m_1, \ldots, m_n)$ is encoded at some level $k \leq \kappa$ as:

$$\mathsf{CRT}_{(p_i)}\left(\frac{r_i g_i + m_i}{z^k}\right) + a x_0$$

where:

(p_i)	is a sequence of n large primes.
x_0	$= \prod p_i$.
$\mathsf{CRT}_{(p_i)}(x_i)$	is the unique integer in $[0, x_0)$ congruent to x_i modulo p_i.
z	is a fixed secret integer modulo x_0.
r_i	is a small noise.
a	is another noise.

Encodings at the same level can be added together, and the resulting encoding encodes the sum of the messages. Similarly encodings at levels i and j can be multiplied to yield an encoding at level $i + j$ of the coordinate-wise product of the encoded messages. This behavior holds as long as the values $r_i g_i + m_i$ do not go over p_i, *i.e.* reduction modulo p_i does not interfere. In order to prevent the size of encodings from increasing as a result of additions and multiplications, a *ladder* of encodings of zero of increasing size is published at each level. Encodings can then be reduced by subtracting elements of the ladder at the same level.

The power of the multilinear map comes from the zero-testing procedure, which allows users to test whether an encoding at the maximal level κ encodes zero. This is achieved by publishing a so-called zero-testing parameter denoted $\boldsymbol{p}_{zt} \in \mathbb{Z}$, together with a large prime $N \gg x_0$. An encoding at the maximal level κ may be written as:

$$e = \sum (r_i + m_i g_i^{-1} \bmod p_i) u_i + a x_0$$
$$\text{where } u_i \triangleq (g_i z^{-\kappa} (p_i^*)^{-1} \bmod p_i)\, p_i^* \quad \text{with } p_i^* = \prod_{j \neq i} p_j.$$

That is, some constants independent of the encoding have been folded with the CRT coefficients into u_i. Now $\boldsymbol{p}_{zt}$ is chosen such that $v_i \triangleq u_i \boldsymbol{p}_{zt} \bmod N$ and

$v_0 \triangleq x_0 \boldsymbol{p}_{zt} \bmod N$ satisfy $|v_i| \ll N$ and $|v_0| \ll N$. In this way, for any encoding e of zero at level κ, since $m_i = 0$, we have:

$$|e\boldsymbol{p}_{zt} \bmod N| = \left|\sum r_i v_i + a v_0\right| \ll N$$

provided the noises r_i and a are small enough. Thus, users can test whether e is an encoding of zero at level κ by checking whether $|e\boldsymbol{p}_{zt} \bmod N| \ll N$.

1.2.1 Integer Extraction (ϕ-value). Our attacks proceed in two steps. The first step is shared by both attacks and proceeds as follows. We define the integer extraction procedure $\phi : \mathbb{Z} \to \mathbb{Z}$. In short, ϕ computes $\sum_i r_i v_i + a v_0$ over the integers for any level-κ encoding e (of size up to the largest ladder element). Note that this value is viewed over the integers and not modulo N. If e is "small", then $\phi(e) = e\boldsymbol{p}_{zt} \bmod N$, *i.e.* ϕ matches the computation from the zero-testing procedure.

If e is "large" on the other hand, then e would need to be reduced by the ladder before zero-testing can be applied. However the crucial observation is that ϕ is $\mathbb{Z}$-linear as long as the values $r_i g_i + m_i$ associated with each encoding do not go over p_i. Thus e can be ladder-reduced into e', then $\phi(e') = e'\boldsymbol{p}_{zt} \bmod N$ is known, and $\phi(e)$ can be recovered from $\phi(e')$ by compensating the ladder reduction using $\mathbb{Z}$-linearity.

1.2.2 Eigenvalue Attack. The point of a CHLRS attack can be divided into two parts. The first is that, for a level-κ encoding of zero $e = \sum_{i=1}^n [\frac{r_i g_i}{z^\kappa}(\frac{x_0}{p_i})^{-1}]_{p_i} \frac{x_0}{p_i} + a x_0$,

$$[\boldsymbol{p}_{zt} \cdot e]_{x_0} = \sum_{i=1}^n r_i \hat{v}_i,$$

where $\hat{v}_i$ is common to all the encodings in CLT13, holds over the integers. The second point is that the zero-testing value of a product of two encodings is a quadratic form of some values related to each encoding. More precisely, for two encodings $e_1 = \sum_{i=1}^n [\frac{r_{i1} g_i}{z^t}(\frac{x_0}{p_i})^{-1}]_{p_i} \frac{x_0}{p_i} + a_1 x_0$ and $e_2 = \sum_{i=1}^n [\frac{r_{i2}}{z^{\kappa-t}}(\frac{x_0}{p_i})^{-1}]_{p_i} \frac{x_0}{p_i} + a_2 x_0$, the product is $e_1 e_2 \equiv \sum_{i=1}^n [\frac{r_{i1} r_{i2} g_i}{z^\kappa}(\frac{x_0}{p_i})^{-1}]_{p_i} \frac{x_0}{p_i} \bmod x_0$. Therefore, the zero-testing value of $e_1 e_2$ is

$$[\boldsymbol{p}_{zt} \cdot e_1 e_2]_{x_0} = \sum_{i=1}^n r_{i1} r_{i2} \hat{v}_i.$$

Let us look at CLT15 in these aspects. For a level-κ encoding of zero $e = \sum_{i=1}^n r_i u_{i\kappa} + a x_0$, the zero-testing value of x is written as

$$[\boldsymbol{p}_{zt} \cdot e]_N = \sum_{i=1}^n r_i v_i + a v_0,$$

for common v_i's, similar to CLT13. Let e_1 be a level-t encoding of zero, e_2 be a level-$(\kappa - t)$ encoding, and e be a product of e_1 and e_2. Then, these can be written as $e_1 = \sum_{i=1}^{n} r_{i1} u_{it} + a_1 x_0$, $e_2 = \sum_{i=1}^{n} r_{i2} u_{i\kappa - t} + a_2 x_0$, and $e = \sum_{i=1}^{n} r_{i1} r_{i2} u_{i\kappa} + a x_0$, for some integers $a, a_1, a_2, r_{i1}, r_{i2}, 1 \leq i \leq n$, where a is a quadratic form of $a_1, a_2, r_{i1}, r_{i2}, 1 \leq i \leq n$. Since the size of e is larger than that of x_0, we need to reduce the size of e to perform zero-testing. Let e' be a ladder-reduced encoding of e; then, it is of the form $e' = e - \sum_{j=0}^{M} b_j X_j = \sum_{i=1}^{n} (r_{i1} r_{i2} - \sum_{j=0}^{M} b_j s_{ij}) u_{i\kappa} + (a - \sum_{j=0}^{M} b_j q_j) x_0$, for some $b_0, \cdots, b_M \in \{0, 1\}$. In this case, the zero-testing value gives

$$
\begin{aligned}
[\boldsymbol{p}_{zt} \cdot e']_N &= \Big[\boldsymbol{p}_{zt} \cdot \big(e - \sum_{j=0}^{M} b_j X_j\big)\Big]_N \\
&= \sum_{i=1}^{n} \big(r_{i1} r_{i2} - \sum_{j=0}^{M} b_j s_{ij}\big) v_i + \big(a - \sum_{j=0}^{M} b_j q_j\big) v_0 \\
&= \sum_{i=1}^{n} \big(r_{i1} r_{i2}\big) v_i + a v_0 - \sum_{j=0}^{M} b_j \big(\sum_{i=1}^{n} s_{ij} v_i + q_j v_0\big).
\end{aligned}
$$

Therefore, if one has $\sum_{i=1}^{n} s_{ij} v_i + q_j v_0$ for all j, one can compute $\sum_{i=1}^{n} (r_{i1} r_{i2}) v_i + a v_0$ and follow a CHLRS attack strategy. For this purpose the integer extraction function ϕ provides exactly what we need.

By using $(n+1)$ level-t encodings of zero and $(n+1)$ level-$(\kappa - t)$ encodings, we constitute matrix equations that consist only of a product of matrices. As in [CHL+15], we have a matrix, the eigenvalues of which consist of the CRT components of an encoding. From these, we can recover all the secret parameters of the CLT15 scheme. Our attack needs only ladders and two level-0 encodings (which can be provided by ladder elements), and runs in polynomial time.

1.2.3 Determinant Attack. The determinant attack proceeds by first recovering x_0. Once x_0 is known, the original CHLRS attack can be applied by taking all values modulo v_0. We now explain how to recover x_0.

In the optimized variant of the scheme implemented in [CLT15], a small multiple $q x_0$ of x_0 is given in the public parameters. In that case $q x_0$ may be regarded as an encoding of zero at level κ, and $\phi(q x_0) = q v_0$. Since this holds over the integers, we can compute $q = \gcd(q x_0, q v_0)$ and then $x_0 = q x_0 / q$.

In the general case where no exact multiple of x_0 is given in the public parameters, pick $n+1$ encodings a_i at some level t, and $n+1$ encodings of zero b_i at level $\kappa - t$. Note that ladder elements provide encodings of zero even if the scheme itself does not. Then compute:

$$\omega_{i,j} \triangleq \phi(a_i b_j).$$

If we write $a_i \bmod v_0 = \mathsf{CRT}_{(p_j)}(a_{i,j}/z^t)$ and $b_i \bmod v_0 = \mathsf{CRT}_{(p_j)}(r_{i,j}g_j/z^{\kappa-t})$, then we get:

$$\omega_{i,j} \bmod v_0 = \sum_k a_{i,k} r_{j,k} v_k \bmod v_0.$$

Similar to the CHLRS attack on the CLT13 multilinear map, this equality can be viewed as a matrix product. Indeed, let Ω denote the $(n+1) \times (n+1)$ integer matrix with entries $\omega_{i,j}$, let A denote the $(n+1) \times n$ integer matrix with entries $a_{i,j}$, let R denote the $(n+1) \times n$ integer matrix with entries $r_{i,j}$, and finally let V denote the $n \times n$ diagonal matrix with diagonal entries v_i. If we embed everything into $\mathbb{Z}/v_0\mathbb{Z}$, then we have:

$$\Omega = A \cdot V \cdot R^{\mathrm{T}} \qquad \text{in } \mathbb{Z}/v_0\mathbb{Z}.$$

Since A and R are $(n+1) \times n$ matrices, this implies that Ω is not full-rank when embedded into $\mathbb{Z}/v_0\mathbb{Z}$. As a consequence v_0 divides $\det(\Omega)$. We can repeat this process with different choices of the families (a_i), (b_i) to build another matrix Ω' with the same property. Finally we recover v_0 as $v_0 = \gcd(\det(\Omega), \det(\Omega'))$, and $x_0 = v_0/\boldsymbol{p}_{zt} \bmod N$.

1.3 Impact of the Attacks

Two variants of the CLT15 multilinear map should be considered. Either a small multiple of x_0 is provided in the public parameters. In that case x_0 can be recovered instantly with the determinant attack, and the scheme becomes equivalent to CLT13 in terms of security (cf. Sect. 4.3.1). In particular it falls victim to the CHLRS attack when low-level encodings of zero are present, but it may still be secure for applications that do not require such encodings, such as obfuscation. However the scheme is strictly less efficient than CLT13 by construction, so there is no point in using CLT15 for those applications.

Otherwise, if no small multiple of x_0 is given out in the public parameters, then ladders of encodings of zero must be provided at levels below the maximal level. Thus we have access to numerous encodings of zero below the maximal level, even if the particular application of multilinear maps under consideration does not require them. As a result both the eigenvalue and the determinant attacks are applicable (cf. Sect. 4.3.3), and the secret parameters are still recovered in polynomial time, albeit less efficiently than the previous case.

In summary, the optimized version of CLT15 providing a small multiple of x_0 is no more secure than CLT13, and less efficient. On the other hand in the general non-optimized case, the scheme is broken for virtually all possible applications due to encodings of zero provided by the ladder. Thus overall the CLT15 scheme can be considered fully broken.

1.4 Organization of the Paper

For the sake of being self-contained, a presentation of multilinear maps and graded encoding schemes is provided in Appendix A. The CLT15 construction itself is described in Sect. 3. In Sect. 3.2 we recall the CHLRS attack on CLT13, as it shares similar ideas with our attacks. Readers already familiar with the CLT15 multilinear map can skip straight to Sect. 4 where we describe our main attacks.

2 Notation

The symbol $\triangleq$ denotes an equality by definition.

For n an integer, size(n) is the size of n in bits.

For a finite set S, we use $s \leftarrow S$ to denote the operation of uniformly choosing an element s from S.

Modular Arithmetic. The group $\mathbb{Z}/n\mathbb{Z}$ of integers modulo n is denoted by $\mathbb{Z}_n$. For $a, b, p \in \mathbb{Z}$, $a \equiv b \bmod p$ or $a \equiv_p b$ means that a is congruent to b modulo p. The notation "mod p" should be understood as having the lowest priority. For instance, the expression $a \cdot b \bmod p$ is equivalent to $(a \cdot b) \bmod p$.

We always view $a \bmod p$ (or $[a]_p$) as an integer in $\mathbb{Z}$. The representative closest to zero is always chosen, positive in case of tie. In other words $-p/2 < a \bmod p \leq p/2$.

Chinese Remainder Theorem. Given n prime numbers (p_i), we define p_i^* as in [Hal15a]:

$$p_i^* = \prod_{j \neq i} p_j.$$

For $(x_1, \ldots, x_n) \in \mathbb{Z}^n$, let $\mathsf{CRT}_{(p_i)}(x_i)$ denote the unique integer in $\mathbb{Z} \cap [0, \prod p_i)$ such that $\mathsf{CRT}_{(p_i)}(x_i) \bmod p_i = x_i \bmod p_i$, as per the Chinese Remainder Theorem.

It is useful to observe that for any $(x_1, \ldots, x_n) \in \mathbb{Z}^n$:

$$\mathsf{CRT}_{(p_i)}(x_i p_i^*) = \sum_i x_i p_i^* \bmod \prod_i p_i. \tag{1}$$

Matrix. For an $n \times n$ square matrix $\boldsymbol{H}$, we use (h_{ij}) to represent a matrix $\boldsymbol{H}$, the (i, j) component of which is h_{ij}. Similarly, for a vector $\boldsymbol{v} \in \mathbb{R}^n$, we define $(\boldsymbol{v})_j$ as the j-th component of $\boldsymbol{v}$. Let $\boldsymbol{H}^T$ be the transpose of $\boldsymbol{H}$ and $\|\boldsymbol{H}\|_\infty$ be the $\max_i \sum_{j=1}^n |h_{ij}|$. We denote by $\mathsf{diag}(d_1, \cdots, d_n)$ the diagonal matrix with diagonal coefficients equal to $d_1, \cdots, d_n$.

3 The CLT15 Multilinear Map and Its Cryptanalysis

In order to make our article self-contained, a short introduction to multilinear maps and graded encoding schemes is provided in Appendix A.

3.1 The CLT15 Multilinear Map over the Integers

Shortly after the multilinear map over ideal lattices by Garg, Gentry and Halevi [GGH13a], another construction over the integers was proposed by Coron, Lepoint and Tibouchi [CLT13]. However a devastating attack was published by Cheon, Han, Lee, Ryu and Stehlé at EUROCRYPT 2015 (on ePrint in late 2014). In the wake of this attack, a revised version of their multilinear map over the integers was presented by Coron, Lepoint and Tibouchi at CRYPTO 2015 [CLT15]. In the remainder of this article, we will refer to the original construction over the integers as CLT13, and to the new version from CRYPTO 2015 as CLT15.

In this section we recall the CLT15 construction. We omit aspects of the construction that are not relevant to our attack, and refer the reader to [CLT15] for more details. The message space is $R = \mathbb{Z}_{g_1} \times \cdots \times \mathbb{Z}_{g_n}$, for some (relatively small) primes $g_i \in \mathbb{N}$. An encoding of a message $(m_1, \ldots, m_n) \in \mathbb{Z}_{g_1} \times \cdots \times \mathbb{Z}_{g_n}$ at level $k \leq \kappa$ has the following form:

$$e = \mathsf{CRT}_{(p_i)}\left(\frac{r_i g_i + m_i}{z^k} \bmod p_i\right) + a x_0 \tag{2}$$

where:

- The p_i's are n large secret primes.
- The r_i's are random noise such that $|r_i g_i + m_i| \ll p_i$.
- $x_0 = \prod_{i \leq n} p_i$.
- z is a fixed secret integer modulo x_0.
- a is random noise.

The scheme relies on the following parameters:

λ : the security parameter.
κ : the multilinearity level.
n : the number of primes p_i.
η : the bit length of secret primes p_i.
$\gamma = n\eta$: the bit length of x_0.
α : the bit length of the g_i's.
ρ : the bit length of initial r_i's.
β : the bit size of matrix $\boldsymbol{H}$ used to zero-testing procedure.

Addition, negation and multiplication of encodings is exactly addition, negation and multiplication over the integers. Indeed, m_i is recovered from $e \cdot z^k$ as $m_i = (e \cdot z^k \bmod p_i) \bmod g_i$, and as long as $r_i g_i + m_i$ does not go over p_i, addition and multiplication will go through both moduli. Thus we have defined encodings and how to operate on them.

Regarding the sampling procedure from Appendix A.2, for our purpose, it suffices to know that it is realized by publishing a large number of level-0 encodings of random elements. Users can then sample a new random element as a subset sum of published elements. Likewise, the rerandomization procedure is achieved by publishing a large number of encodings of zero at each level, and an element is re-randomized by adding a random subset sum of encodings of zero at the same level. The encoding procedure is realized by publishing a single level-1 encoding y of 1 (by which we mean $(1, \ldots, 1) \in \mathbb{Z}_{g_1} \times \cdots \times \mathbb{Z}_{g_n}$): any encoding can then be promoted to an encoding of the same element at a higher level by multiplying by y.

Zero-testing in CLT13. We now move on to the crucial zero-testing procedure. This is where CLT13 and CLT15 differ. We begin by briefly recalling the CLT13 approach.

In CLT13, the product x_0 of the p_i's is public. In particular, every encoding can be reduced modulo x_0, and every value below should be regarded as being modulo x_0. Let $p_i^* = \prod_{j \neq i} p_j$. Using (1), define:

$$\boldsymbol{p}_{zt} \triangleq \sum_{i \leq n} \left(\frac{h_i z^\kappa}{g_i} \bmod p_i \right) p_i^* = \mathsf{CRT}_{(p_i)} \left(\frac{h_i z^\kappa}{g_i} p_i^* \bmod p_i \right) \qquad \bmod x_0.$$

where the h_i's are some relatively small numbers with $|h_i| \ll p_i$. Now take a level-κ encoding of zero:

$$e = \mathsf{CRT}_{(p_i)} \left(\frac{r_i g_i}{z^\kappa} \bmod p_i \right) \qquad \bmod x_0.$$

Since multiplication acts coordinate-wise on the CRT components, using (1) again, we have:

$$\omega \triangleq e \boldsymbol{p}_{zt} = \mathsf{CRT}_{(p_i)}(h_i r_i p_i^*) = \sum_i h_i r_i p_i^* \qquad \bmod x_0.$$

Since $p_i^* = x_0/p_i$, as long as we set our parameters so that $|h_i r_i| \ll p_i$, we have $|\omega| \ll x_0$.

Thus the zero-testing procedure is as follows: for a level-κ encoding e, compute $\omega = e\boldsymbol{p}_{zt} \bmod x_0$. Output 1, meaning we expect e to encode zero, iff the ν most significant bits of ω are zero, for an appropriately chosen ν. In [CLT13], multiple $\boldsymbol{p}_{zt}$'s can be defined in order to avoid false positives; we restrict our attention to a single $\boldsymbol{p}_{zt}$.

Zero-testing in CLT15. In CLT13, an encoding at some fixed level is entirely defined by its vector of associated values $c_i = r_i g_i + m_i$. Moreover, addition and multiplication of encodings act coordinate-wise on these values, and the value of the encoding itself is $\mathbb{Z}_{x_0}$-linear as a function of these values. Likewise, ω is $\mathbb{Z}_{x_0}$-linear as a function of the r_i's. This nice structure is an essential part of what makes the devastating attack, so called CHLRS attack [CHL+15] possible. In CLT15, the authors set out to break this structure by introducing a new noise component a.

For this purpose, the public parameters include a new prime number $N \gg x_0$, with size$(N) = \gamma + 2\eta + 1$. Meanwhile x_0 is kept secret, and no longer part of the public parameters. Encodings are thus no longer reduced modulo x_0, and take the general form given in (2), including a new noise value a. Equivalently, we can write an encoding e of (m_i) at level k as:

$$e = \sum_i \left(r_i + m_i (g_i^{-1} \bmod p_i)\right) u_i + a x_0 \tag{3}$$

$$\text{with } u_i \triangleq \left(g_i z^{-k} (p_i^*)^{-1} \bmod p_i\right) p_i^*.$$

That is, we fold the $g_i z^{-k}$ multiplier of r_i with the CRT coefficient into u_i.

The zero-testing parameter $\boldsymbol{p}_{zt}$ is now defined modulo N in such a way that:

$$\begin{array}{lll} & v_0 \triangleq x_0 \boldsymbol{p}_{zt} \bmod N & \forall i, v_i \triangleq u_i \boldsymbol{p}_{zt} \bmod N \\ \text{satisfy:} & |v_0| \ll N & |v_i| \ll N \end{array} \tag{4}$$

To give an idea of the sizes involved, size$(v_0) \approx \gamma$ and size$(v_i) \approx \gamma + \eta$ for $i > 0$. We refer the reader to [CLT15] for how to build such a $\boldsymbol{p}_{zt}$. The point is that if e is an encoding of zero at level κ, then we have:

$$\omega = e \boldsymbol{p}_{zt} \bmod N = \sum r_i v_i + a v_0 \bmod N.$$

In order for this quantity to be smaller than N, the size of a must be somehow controlled. Conversely as long as a is small enough and the noise satisfies $|r_i| \ll p_i$ then $|\omega| \ll N$. We state the useful lemma for an exact zero-testing, the so-called the zero-testing lemma, more precisely.

Lemma 1 (Zero-testing Lemma). *Let e be a level-κ encoding of zero with $e = \sum_{i=1}^n r_i u_i + a x_0, (r_1, \cdots, r_n, a \in \mathbb{Z})$. Then,*

$$[e\boldsymbol{p}_{zt}]_N = \sum_{i=1}^n r_i v_i + a v_0,$$

holds over the integers, if $|a| < 2^{2\eta - \beta - \log_2 n - 1}$ and $|r_i| < 2^{\eta - \beta - \log_2 n - 6}$ for $1 \le i \le n$.

Proof. By the construction of the zero-testing element, we have $e\boldsymbol{p}_{zt} \equiv \sum_{i=1}^n r_i v_i + a v_0 \bmod N$. It is sufficient to show that the right hand side is smaller than $N/2$. For $1 \le i \le n$,

$$v_i \equiv \sum_{j=1}^n h_j \alpha_j p_j^{-1} u_i \equiv h_i \beta_i + \sum_{j \neq i} h_j \alpha_j \left[\frac{g_i}{z^\kappa} \left(\frac{x_0}{p_i} \right)^{-1} \right]_{p_i} \frac{x_0}{p_i p_j} \bmod N,$$

and therefore, $|v_i| < 2^{\gamma + \eta + \beta + 4}$ for $1 \le i \le n$. Moreover, $v_0 = \sum_{j=1}^n h_j \alpha_j \frac{x_0}{p_j}$ and $|v_0| < n 2^{\gamma + \beta - 1}$. □

Thus the size of a must be controlled. The term ax_0 will be dominant in (3) in terms of size, so decreasing a is the same as decreasing the size of the encoding as a whole. The scheme requires a way to achieve this without altering the encoded value (and without publishing x_0).

For this purpose, inspired by [VDGHV10], a *ladder* $(X_i^{(k)})_{0 \le i \le \gamma'}$ of encodings of zero of increasing size is published for each level $k \le \kappa$, where $\gamma' = \gamma + \lfloor \log_2 \ell \rfloor$. The size of an encoding e at level k can then be reduced without altering the encoded value by recursively subtracting from e the largest ladder element smaller than e, until e is smaller than $X_0^{(\kappa)}$. More precisely we can choose $X_0^{(\kappa)}$ small enough that the previous zero-testing procedure goes through, and then choose $X_{\gamma'}^{(\kappa)}$ twice the size of $X_0^{(\kappa)}$, so that the product of any two encodings smaller than $X_0^{(\kappa)}$ can be reduced to an encoding smaller than $X_0^{(\kappa)}$. After each addition and multiplication, the size of the resulting encoding is reduced via the ladder.

In the end, the zero-testing procedure is very similar to CLT13: given a (ladder-reduced) level-κ encoding e, compute $\omega = e\boldsymbol{p}_{zt} \bmod N$. Then output 1, meaning we expect e to encode zero, iff the ν high-order bits of ω are zero.

Extraction. The extraction procedure simply outputs the ν high-order bits of ω, computed as above. For both CLT13 and CLT15, it can be checked that they only depend on the m_i's (as opposed to the noises a and the r_i's).

3.2 CHLRS Attack on CLT13

In this section we provide a short description of CHLRS attack on CLT13 [CHL+15], as elements of this attack appear in our own. We actually present (a close variant of) the slightly simpler version in [CGH+15].

Assume we have access to a level-0 encoding a of some random value, n level-1 encodings (b_i) of zero, and a level-1 encoding y of 1. This is the case for one-round multi-party Diffie-Hellman (see previous section). Let $a_i = a \bmod p_i$, *i.e.* a_i is the i-th value "$r_i g_i + m_i$" associated with a. For $i \le n$, define $r_{i,j} = b_i z / g_j \bmod p_j$, *i.e.* $r_{i,j}$ is the j-th value "r_j" associated with b_i (recall that b_i is an encoding of zero, so $m_j = 0$). Finally let $y_k = yz \bmod p_k$.

Now compute:

$$e_{i,j} = a \cdot b_i \cdot b_j \cdot y^{\kappa-2} \bmod x_0 \qquad \omega_{i,j} = e_{i,j}\boldsymbol{p}_{zt} \bmod x_0$$
$$e'_{i,j} = b_i \cdot b_j \cdot y^{\kappa-2} \bmod x_0 \qquad \omega'_{i,j} = e'_{i,j}\boldsymbol{p}_{zt} \bmod x_0$$

Note that:

$$\omega_{i,j} = \sum_k \left(a_k \frac{r_{i,k} g_k}{z} \frac{r_{j,k} g_k}{z} \frac{y_k^{\kappa-2}}{z^{\kappa-2}} \frac{h_k z^\kappa}{g_k} \bmod p_k \right) p_k^*$$
$$= \sum_k a_k r_{i,k} r_{j,k} c_k \qquad \text{with } c_k = g_k y_k^{\kappa-2} h_k p_k^*. \tag{5}$$

Crucially, in the second line, the modulo p_k disappears and the equation holds over the integers, because $e_{i,j}$ is a valid encoding of zero, so the correctness of the scheme requires $|e_{i,j} z^\kappa / g_k \bmod p_k| \ll p_k$.

Equation (5) may be seen as a matrix multiplication. Indeed, define Ω, resp. Ω', as the $n \times n$ matrix with entries $\omega_{i,j}$, resp. $\omega'_{i,j}$, and likewise R with entries $r_{i,j}$. Moreover let A, resp. C, be the diagonal matrix with diagonal entries a_i, resp. c_i. Then (5) may be rewritten:

$$\begin{aligned} \Omega &= R \cdot A \cdot C \cdot R^{\mathrm{T}} \\ \Omega' &= R \cdot C \cdot R^{\mathrm{T}} \\ \Omega \cdot (\Omega')^{-1} &= R \cdot A \cdot R^{-1}. \end{aligned}$$

Here matrices are viewed over $\mathbb{Q}$ for inversion (they are invertible whp).

Once $\Omega \cdot (\Omega')^{-1}$ has been computed, the (diagonal) entries of A can be recovered as its eigenvalues. In practice this can be achieved by computing the characteristic polynomial, and all computations can be performed modulo some prime p larger than the a_i's (which are size 2ρ).

Thus we recover the a_i's, and by definition $a_i = a \bmod p_i$, so p_i can be recovered as $p_i = \gcd(a - a_i, x_0)$. From there it is trivial to recover all other secret parameters of the scheme.

4 Main Attack

4.1 Integer Extraction (ϕ-Value)

Integer extraction essentially removes the extra noise induced by ladder reductions when performing computations on encodings. In addition, as we shall see in Sect. 4.3.2, this step is enough to recover x_0 when an exact multiple is known, as is the case in the optimized variant proposed and implemented in [CLT15].

In the remainder we say that an encoding at level k is small iff it is less than $X_0^{(k)}$ in absolute value. In particular, any ladder-reduced encoding is small.

Now, we describe our idea of attack. For a level-κ encoding of zero $e = \sum_{i=1}^n r_i u_i + a x_0$ of arbitrary size, if one can compute the integer value $\sum_{i=1}^n r_i v_i + a v_0$, which is not reduced modulus N, then a CHLRS attack can be applied similarly. Hence, we define the function ϕ such that it represents such a value and examine how to obtain the function values for a level-κ encoding of zero of arbitrary size.

When the size of e is small, by the zero-testing lemma, $[\boldsymbol{p}_{zt} \cdot e]_N$ gives the integer value $\sum_{i=1}^n r_i v_i + a v_0$. However, if the size of e is large, the zero-testing lemma does not hold and one cannot compute the integer value directly. To reach the goal, we use the ladder $X_j^{(\kappa)} = \sum_{i=1}^n r_{ij}^{(\kappa)} u_i + a_j^{(\kappa)}$. Let e be a level-κ encoding of zero. Then, we can compute the size-reduced encoding e' using the ladder and obtain the quantity (for short, we define γ' as $\gamma + \lfloor \log_2 \ell \rfloor$.)

$$\begin{aligned}[\boldsymbol{p}_{zt}\cdot e']_N &= \Big[\boldsymbol{p}_{zt}\cdot\Big(e-\sum_{j=0}^{\gamma'} b_j X_j^{(\kappa)}\Big)\Big]_N\\ &= \sum_{i=1}^{n}\Big(r_i-\sum_{j=0}^{\gamma'} b_j r_{ij}^{(\kappa)}\Big)v_i+\Big(a-\sum_{j=0}^{\gamma'} b_j a_j^{(\kappa)}\Big)v_0\\ &= \sum_{i=1}^{n} r_i v_i + a v_0 - \sum_{j=0}^{\gamma'} b_j\Big(\sum_{i=1}^{n} r_{ij}^{(\kappa)} v_i + a_j^{(\kappa)} v_0\Big).\end{aligned}$$

Therefore, if one can compute $\sum_{i=1}^{n} r_{ij}^{(\kappa)} v_i + a_j^{(\kappa)} v_0$ from $X_j^{(\kappa)}$, one can easily obtain $\sum_{i=1}^{n} r_i v_i + a v_0$.

To compute $\sum_{i=1}^{n} r_{ij}^{(\kappa)} v_i + a_j^{(\kappa)} v_0$ for all $j \in \{0,\cdots,\gamma+\lfloor\log_2\ell\rfloor\}$, we use an induction on j. When $j=0$, $[\boldsymbol{p}_{zt}\cdot X_0^{(\kappa)}]_N$ gives $\sum_{i=1}^{n} r_{i0}^{(\kappa)} v_i + a_0^{(\kappa)} v_0$, by the zero-testing lemma. Suppose we have $\sum_{i=1}^{n} r_{ij}^{(\kappa)} v_i + a_j^{(\kappa)} v_0$ for $j \in \{0,\cdots,t-1\}$; then, $[\boldsymbol{p}_{zt}\cdot X_t]_N = \sum_{i=1}^{n} r_{it}^{(\kappa)} v_i + a_t^{(\kappa)} v_0 - \sum_{j=0}^{t-1} b_j(\sum_{i=1}^{n} r_{ij}^{(\kappa)} v_i + a_j^{(\kappa)} v_0)$ for computable $b_i \in \{0,1\}$, where X_t is a size-reduced encoding of $X_t^{(\kappa)}$ using $\{X_0^{(\kappa)},\cdots,X_{t-1}^{(\kappa)}\}$. Since we know the latter terms, we can also compute $\sum_{i=1}^{n} r_{it}^{(\kappa)} v_i + a_t^{(\kappa)} v_0$. This idea can be extended to any level ladder.

Now, we give a precise description of function ϕ.

$$\begin{aligned}\phi:\quad & \mathbb{Z}\to\mathbb{Z}\\ & e \mapsto \sum_{i=1}^{n}\Big[e\cdot\frac{z^\kappa}{g_i}\Big]_{p_i} v_i + \frac{x-\sum_{i=1}^{n}[e\cdot\frac{z^\kappa}{g_i}]_{p_i} u_i}{x_0} v_0,\end{aligned}$$

where $v_i = [\boldsymbol{p}_{zt}\cdot u_i]_N (1 \le i \le n)$ and $v_0 = [\boldsymbol{p}_{zt}\cdot x_0]_N$. Note that ϕ is defined over the integers, and not modulo N. Indeed the v_i's are seen as integers: recall from Sect. 2 that throughout this paper $x \bmod N$ denotes an integer in $\mathbb{Z}\cap(-N/2, N/2]$.

Proposition 1. *Let e be an integer such that $e \equiv \frac{r_i\cdot g_i}{z^\kappa} \bmod p_i$ for $1 \le i \le n$. If $|r_i| < p_i/2$ for each i, then x can be uniquely expressed as $\sum_{i=1}^{n} r_i u_i + a x_0$ for some integer a, and $\phi(e) = \sum_{i=1}^{n} r_i v_i + a v_0$.*

Proof. We can see that $e \equiv \sum_{i=1}^{n} r_i u_i \bmod p_j$ for each j and thus there exists an integer a such that $e = \sum_{i=1}^{n} r_i u_i + a x_0$. For uniqueness, suppose e can be written as $e = \sum_{i=1}^{n} r_i' u_i + a' x_0$ for integers $r_1',\cdots,r_n',a'$ with $|r_i'| < p_i/2$. Then, $e \equiv r_i'[\frac{g_i}{z^\kappa}\left(\frac{x_0}{p_i}\right)^{-1}]_{p_i} \equiv \frac{r_i' g_i}{z^\kappa} \bmod p_i$, which implies $r_i \equiv r_i' \bmod p_i$. Since $|r_i - r_i'| < p_i$, we have $r_i' = r_i$ for each i and therefore $a' = a$, which proves the uniqueness. □

The point is that if e is a small encoding of zero at level κ, then $\phi(e) = e\boldsymbol{p}_{zt} \bmod N$. In that case $\phi(e)$ matches the extraction in the sense of the ext procedure of Appendix A.2 (more precisely ext returns the high-order bits of $\phi(e)$).

However we want to compute $\phi(e)$ even when e is larger. For this purpose, the crucial point is that ϕ is actually $\mathbb{Z}$-linear as long as for all encodings involved, the associated r_i's do not go over $p_i/2$, *i.e.* reduction modulo p_i does not interfere. More formally:

Proposition 2. *Let $e_1, \cdots, e_m$ be level-κ encodings of zero such that $e_j \equiv \frac{r_{ij} g_i}{z^\kappa} \bmod p_i$ and $|r_{ij}| < p_i/2$ for all $1 \le i \le n$, $1 \le j \le m$. Then, the equality*

$$\phi(\sum_{j=1}^{m} e_j) = \sum_{j=1}^{m} \phi(e_j),$$

holds if $\left|\sum_{j=1}^{m} r_{ij}\right| < \frac{p_i}{2}$, for all $1 \le i \le n$.

Proof. From Proposition 1, each e_j can be uniquely written as $e_j = \sum_{i=1}^{n} r_{ij} u_i + a_j x_0$ for some integer a_j, and $\phi(e_j) = \sum_{i=1}^{n} r_{ij} v_i + a_j v_0$. Then,

$$\begin{aligned}\sum_{j=1}^{m} \phi(e_j) &= \sum_{i=1}^{n} \Big(\sum_{j=1}^{m} r_{ij}\Big) \cdot v_i + \Big(\sum_{j=1}^{m} a_j\Big) \cdot v_0 \\ &= \phi\Big(\Big(\sum_{j=1}^{m} r_{ij}\Big) \cdot u_i + \Big(\sum_{j=1}^{m} a_j\Big) \cdot x_0\Big) = \phi\Big(\sum_{j=1}^{m} e_j\Big),\end{aligned}$$

where the source of the second equality is Proposition 1, since $\left|\sum_{j=1}^{m} r_{ij}\right| < p_i/2$. □

An important remark is that the conditions on the r_{ij}'s above are also required for the correctness of the scheme to hold. In other words, as long as we perform valid computations from the point of view of the multilinear map (*i.e.* there is no reduction of the r_{ij}'s modulo p_i, and correctness holds), then the $\mathbb{Z}$-linearity of ϕ also holds.

4.2 Eigenvalue Attack

Our strategy to attack CLT15 is similar to that in [CHL+15]. The goal is to construct a matrix equation over $\mathbb{Q}$ by computing the ϕ values of several products of level-$0, 1$, and $(\kappa - 1)$ encodings, fixed on level-0 encoding. We proceed using the following three steps.

- **(Step 1)** Compute the ϕ-value of level-κ ladder
- **(Step 2)** Compute the ϕ-value of level-κ encodings of large size
- **(Step 3)** Construct matrix equations over $\mathbb{Q}$.

Using the matrix equations in **Step 3**, we have a matrix, the eigenvalues of which are residue modulo p_i of level-0 encoding. From this, we deduce a secret modulus p_i.

4.2.1 Computing the ϕ-value of $X_j^{(\kappa)}$. To apply the zero-testing lemma to a level-κ encoding of zero $e = \sum_{i=1}^{n} r_i u_i + a x_0$, the size of r_i and a has to be bounded by some fixed values. By the parameter setting, η is larger than the maximum bit size of the noise r_i of a level-κ encoding obtained from the multiplication of lower level encodings. Hence, we need to reduce the size of e so that a satisfies the zero-testing lemma.

Let us consider a ladder of level-κ encodings of zero $\{X_j^{(\kappa)}\}$. This is provided to reduce the size of encodings to that of $2x_0$. More precisely, given a level-κ encoding of zero e of size smaller than $2^{2\gamma+\lfloor \log_2 \ell \rfloor}$, one can compute $e' = e - \sum_{j=0}^{\gamma'} b_j X_j^{(\kappa)}$ for $\gamma' = \gamma + \lfloor \log_2 \ell \rfloor$, which is an encoding of the same plaintext; its size is smaller than $X_0^{(\kappa)}$. As noted in [CLT15], the sizes of $X_j^{(\kappa)}$ are increasing and differ by only one bit, and therefore, $b_j \in \{0,1\}$, which implies the noise grows additively. We can reduce a to an integer much smaller than $2^{2\eta-\beta-1}/n$ so that the zero-testing lemma can be applied. We denote such e' as $[e]_{\boldsymbol{X}^{(\kappa)}}$. More generally, we use the notation

$$[e]_{\boldsymbol{X}^{(t)}} := [\cdots[[e]_{X_{\gamma'}^{(t)}}]_{X_{\gamma'-1}^{(t)}} \cdots]_{X_0^{(t)}} \quad \text{for } \boldsymbol{X}^{(t)} = (X_0^{(t)}, X_1^{(t)}, \ldots, X_{\gamma'}^{(t)}), 1 \le t \le \kappa.$$

Note that, if e satisfies the condition in Lemma 1, i.e., it is an encoding of zero of small size, then $\phi(e)$ is exactly the same as $[\boldsymbol{p}_{zt} \cdot e]_N$. However, if the size of e is large, it is congruent only to $[\boldsymbol{p}_{zt} \cdot e]_N$ modulo N. Now, we show how to compute the integer value $\phi(e)$ for an encoding e of zero, although e does not satisfy the condition in Lemma 1.

First, we adapt the size reduction process to a level-κ ladder itself. We can compute binary b_{ij} for each i, j, satisfying

$$[X_0^{(\kappa)}]_{\boldsymbol{X}^{(\kappa)}} = X_0^{(\kappa)}$$

$$[X_1^{(\kappa)}]_{\boldsymbol{X}^{(\kappa)}} = X_1^{(\kappa)} - b_{10} \cdot X_0^{(\kappa)}$$

$$[X_2^{(\kappa)}]_{\boldsymbol{X}^{(\kappa)}} = X_2^{(\kappa)} - \sum_{k=0}^{1} b_{2k} \cdot X_k^{(\kappa)}$$

$$\vdots$$

$$[X_j^{(\kappa)}]_{\boldsymbol{X}^{(\kappa)}} = X_j^{(\kappa)} - \sum_{k=0}^{j-1} b_{jk} \cdot X_k^{(\kappa)}.$$

Each $[X_j^{(\kappa)}]_{\boldsymbol{X}^{(\kappa)}}$ is an encoding of zero at level κ and therefore can be written as $[X_j^{(\kappa)}]_{\boldsymbol{X}^{(\kappa)}} = \sum_{i=1}^{n} r'_{ij} u_i + a'_j x_0$ for some integers r'_{ij} and a'_j. Moreover, its bit size is at most γ and therefore a'_j is small enough to satisfy the condition in Lemma 1. Therefore,

$$\phi([X_j^{(\kappa)}]_{\boldsymbol{X}^{(\kappa)}}) = [\boldsymbol{p}_{zt} \cdot [X_j^{(\kappa)}]_{\boldsymbol{X}^{(\kappa)}}]_N = \sum_{i=1}^{n} r'_{ij} v_i + a'_j v_0.$$

If we write $X_j^{(\kappa)} = \sum_{i=1}^n r_{ij}u_i + a_j x_0$ for some integer $r_{1j}, \ldots, r_{nj}, a_j$, we have $r'_{ij} = r_{ij} - \sum_{k=0}^{j-1} b_{jk} r_{ik}$ for each i and $a'_j = a_j - \sum_{k=0}^{j-1} b_{jk} a_k$, since all the coefficients of u_i are sufficiently smaller than p_i for each i. Therefore,

$$\sum_{i=1}^{n} r'_{ij} v_i + a'_j v_0 = \sum_{i=1}^{n} r_{ij} v_i + a_j v_0 - \sum_{k=0}^{j-1} b_{jk} \Big(\sum_{i=1}^{n} r_{ik} v_i + a_k v_0 \Big)$$

holds over the integers. Hence, we have the following inductive equations for $0 \le j \le \gamma'$.

$$\phi(X_j^{(\kappa)}) = \Big[\boldsymbol{p}_{zt} \cdot [X_j^{(\kappa)}]_{\boldsymbol{X}^{(\kappa)}} \Big]_N + \sum_{k=0}^{j-1} b_{jk} \cdot \phi\Big(X_k^{(\kappa)}\Big),$$

which gives all $\phi(X_0^{(\kappa)}), \phi(X_1^{(\kappa)}), \ldots, \phi(X_{\gamma'}^{(\kappa)})$, inductively. The computation consists of $(\gamma' + 1)$ zero-testing and $O(\gamma^2)$-times comparisons and subtractions of $(\gamma + \gamma')$-bit integers, and therefore, the total computation cost is $\widetilde{O}(\gamma^2)$ by using fast Fourier transform. Hence, we obtain the following lemma.

Lemma 2. *Given the public parameters of the CLT15 scheme, one can compute*

$$\phi(X_j^{(\kappa)}) = \Big[\boldsymbol{p}_{zt} \cdot [X_j^{(\kappa)}]_{\boldsymbol{X}^{(\kappa)}} \Big]_N + \sum_{k=0}^{j-1} b_{jk} \cdot \phi\left(X_k^{(\kappa)}\right)$$

in $\widetilde{O}(\gamma^2)$ bit computations.

4.2.2 Computing the ϕ-value of Level-κ Encodings of Large Size.

Using the ϕ values of the κ-level ladder, we can compute the ϕ value of any κ-level encoding of zero, the bit size of which is between γ and $\gamma + \gamma'$.

Lemma 3. *Let e be a level-κ encoding of zero, $e = \mathsf{CRT}_{(p_i)}\left(\frac{r_i g_i}{z^\kappa}\right) + q x_0 = \sum_{i=1}^n r_i u_i + a x_0$ for some integer $r_1, \ldots, r_n, a$ satisfying $|r_i| < 2^{\eta - \beta - \log_2 n - 7}$ for each i and $|a| < 2^{\gamma'}$. Given the public parameters of the CLT15 scheme, one can compute the value $\phi(e) = \sum_{i=1}^n r_i v_i + a v_0$ in $\widetilde{O}(\gamma^2)$ bit computations.*

Proof. Let e be a level-κ encoding of zero satisfying the above conditions. As in Sect. 4.2.1, we can find binary b_j's satisfying $[e]_{\boldsymbol{X}^{(\kappa)}} = e - \sum_{j=0}^{\gamma'} b_j \cdot X_j^{(\kappa)}$. Then, we have

$$\phi(e) = \phi([e]_{\boldsymbol{X}^{(\kappa)}}) + \sum_{j=0}^{\gamma'} b_j \cdot \phi(X_j^{(\kappa)}).$$

Since $[e]_{\boldsymbol{X}^{(\kappa)}}$ is a κ-level encoding of zero of at most γ-bit and the size of noise is bounded by $(\eta - \beta - \log_2 n - 6)$-bit, we can compute the value $\phi([e]_{\boldsymbol{X}^{(\kappa)}})$ via the zero-testing procedure. Finally, the ϕ values of the κ-level ladder and $\phi([e]_{\boldsymbol{X}^{(\kappa)}})$ give the value $\phi(e)$. The source of the complexity is Lemma 2. □

We apply Lemma 3 to obtain the ϕ value of a κ-level encoding of zero that is a product of two encodings of $(\gamma+\gamma')$-bit size.

Lemma 4. *Let X be a level-1 encoding and Y a level-$(\kappa-1)$ encoding of zero of bit size at most $\gamma+\gamma'$. Then, one can compute $\phi(XY)$ in $\widetilde{O}(\gamma^3)$ bit computations.*

Proof. We apply Lemma 3 to a product of two γ-bit encodings. From $[X_1^{(1)}]_{\boldsymbol{X}^{(1)}} = X_1^{(1)} - b \cdot X_0^{(1)}$ for some $b \in \{0,1\}$, we find $\phi(X_1^{(1)} \cdot X_0^{(\kappa-1)}) = \phi([X_1^{(1)}]_{\boldsymbol{X}^{(1)}} \cdot X_0^{(\kappa-1)}) + b \cdot \phi(X_0^{(1)} \cdot X_0^{(\kappa-1)})$, since $[X_1^{(1)}]_{\boldsymbol{X}^{(1)}}$ is γ-bit. Thus, we can obtain inductively all $\phi(X_j^{(1)} \cdot X_k^{(\kappa-1)})$ for each j,k from $\phi(X_{l_j}^{(1)} \cdot X_{l_k}^{(\kappa-1)})$, $0 \le l_j \le j, 0 \le l_k \le k, (l_j, l_k) \ne (j,k)$.

Let $[X]_{\boldsymbol{X}^{(1)}} = X - \sum_{j=0}^{\gamma'} b_j \cdot X_j^{(1)}$ and $[Y]_{\boldsymbol{X}^{(\kappa-1)}} = Y - \sum_{j=0}^{\gamma'} b'_j \cdot X_j^{(\kappa-1)}$. Then,

$$\begin{aligned}[X]_{\boldsymbol{X}^{(1)}} \cdot [Y]_{\boldsymbol{X}^{(\kappa-1)}} = XY &- \textstyle\sum_j b_j \cdot X_j^{(1)} \cdot Y \\ &- \textstyle\sum_j b'_j \cdot X_j^{(\kappa-1)} \cdot X + \sum_{j,k} b_j b'_k \cdot X_j^{(1)} \cdot X_k^{(\kappa-1)}.\end{aligned}$$

Note that the noise of $[[X]_{\boldsymbol{X}^{(1)}} \cdot [Y]_{\boldsymbol{X}^{(\kappa-1)}}]_{\boldsymbol{X}^{(\kappa)}}$ is bounded by $2\rho+\alpha+2\log_2(\gamma')+2$ and $\eta > \kappa(2\alpha+2\rho+\lambda+2\log_2 n+3)$, and therefore, we can adapt Proposition 2. Therefore, if we know the ϕ-value of each term, we can compute the ϕ-value of XY. Finally, Lemma 3 enables one to compute $\phi([X]_{\boldsymbol{X}^{(1)}} \cdot [Y]_{\boldsymbol{X}^{(\kappa-1)}})$. The second and third terms of the right hand side can be computed using $[X_j^{(1)}]_{\boldsymbol{X}^{(1)}}$, $[X_j^{(\kappa-1)}]_{\boldsymbol{X}^{(\kappa-1)}}$, and we know the ϕ-value of the last one. Since we perform zero-testings for $O(\gamma^2)$ encodings of zero, the complexity becomes $\widetilde{O}(\gamma^3)$. □

Note that the above Lemma can be applied to a level-t encoding X and a level-$(\kappa-t)$ encoding of zero Y. The proof is exactly the same, except for the indexes.

4.2.3 Constructing Matrix Equations over $\mathbb{Q}$. We reach the final stage. The following theorem is the result.

Theorem 1. *Given the public instances in [CLT15] and $\mathbf{p}_{zt}$, one can find all the secret parameters given in [CLT15] in $\widetilde{O}(\kappa^{\omega+4}\lambda^{2\omega+6})$ bit computations with $\omega \le 2.38$.*

Proof. We construct a matrix equation by collecting several ϕ-values of the product of level-0, 1 and $(\kappa-1)$ encodings. Let c, X, and Y be a level-0, 1, and $(\kappa-1)$ encoding, respectively, and additionally we assume Y is an encoding of zero. Let us express them as

$$\begin{aligned} c &= \mathsf{CRT}_{(p_i)}(c_i), \\ X &= \mathsf{CRT}_{(p_i)}\left(\frac{x_i}{z}\right) = x_i\left[z^{-1}\right]_{p_i} + q_i p_i, \\ Y &= \mathsf{CRT}_{(p_i)}\left(\frac{y_i g_i}{z^{\kappa-1}}\right) = \sum_{i=1}^{n} y_i \left[\frac{g_i}{z^{\kappa-1}}\left(p_i^*\right)^{-1}\right]_{p_i} \cdot p_i^* + a x_0. \end{aligned}$$

Assume that the size of each is less than $2x_0$. The product of c and X can be written as $cX = c_i x_i \left[z^{-1}\right]_{p_i} + q_i' p_i$ for some integer q_i'.

By multiplying cX and Y, we have

$$\begin{aligned}
&cXY \\
&= \sum_{i=1}^{n} \left(c_i x_i y_i \left[z^{-1}\right]_{p_i} \left[\frac{g_i}{z^{\kappa-1}} \left(\frac{x_0}{p_i}\right)^{-1}\right]_{p_i} \cdot \frac{x_0}{p_i} + y_i \left[\frac{g_i}{z^{\kappa-1}} \left(\frac{x_0}{p_i}\right)^{-1}\right]_{p_i} q_i' x_0 \right) + (cX)(a x_0) \\
&= \sum_{i=1}^{n} c_i x_i y_i u_i + \sum_{i=1}^{n} (c_i x_i y_i s_i + y_i \theta_i q_i') x_0 + acX x_0,
\end{aligned}$$

where $\theta_i = \left[\frac{g_i}{z^{\kappa-1}} \left(\frac{x_0}{p_i}\right)^{-1}\right]_{p_i}$, $\theta_i \left[z^{-1}\right]_{p_i} \frac{x_0}{p_i} = u_i + s_i x_0$ for some integer $s_i \in \mathbb{Z}$. Then, we can obtain $\phi(cXY) = \sum_{i=1}^{n} c_i x_i y_i v_i + \sum_{i=1}^{n} (c_i x_i y_i s_i + y_i \theta_i q_i') v_0 + acX v_0$ by Lemma 4.

By plugging $q_i' = \frac{1}{p_i}(cX - c_i x_i [z^{-1}]_{p_i})$ into the equation, we obtain

$$\begin{aligned}
\phi(cXY) &= \sum_{i=1}^{n} y_i \Big(v_i + s_i v_0 - \frac{\theta_i v_0}{p_i} [z^{-1}]_{p_i}\Big) c_i x_i + \sum_{i=1}^{n} y_i \frac{\theta_i v_0}{p_i} cX + a v_0 cX \\
&= \sum_{i=1}^{n} y_i w_i c_i x_i + \sum_{i=1}^{n} y_i w_i' cX + a v_0 cX,
\end{aligned}$$

where $w_i = v_i + s_i v_0 - \frac{\theta_i}{p_i} [z^{-1}]_{p_i} v_0$ and $w_i' = \frac{\theta_i v_0}{p_i}$. It can be written (over $\mathbb{Q}$) as

$$\phi(cXY) = \begin{pmatrix} y_1 \ y_2 \cdots y_n \ a \end{pmatrix} \begin{pmatrix} w_1 & & & 0 & w_1' \\ & w_2 & & & w_2' \\ & & \ddots & & \vdots \\ & & & w_n & w_n' \\ 0 & & & & v_0 \end{pmatrix} \begin{pmatrix} c_1 x_1 \\ c_2 x_2 \\ \vdots \\ c_n x_n \\ cX \end{pmatrix}. \tag{6}$$

Since $p_i w_i = p_i(v_i + s_i v_0) - \theta_i \left[z^{-1}\right]_{p_i} v_0 \equiv -\theta_i \left[z^{-1}\right]_{p_i} v_0 \not\equiv 0 \bmod p_i$, w_i is not equal to zero. Therefore, $v_0 \prod_{i=1}^{n} w_i \neq 0$ and thus the matrix in Eq. (6) is non singular. By applying Eq. (6) to various X, Y, taking for $0 \le j, k \le n$

$$\begin{aligned}
X &= [X_j^{(1)}]_{\boldsymbol{X}^{(1)}} = \mathsf{CRT}_{(p_i)} \left(\frac{x_{ij}}{z}\right), \\
Y &= [X_k^{(\kappa-1)}]_{\boldsymbol{X}^{(\kappa-1)}} = \sum_{i=1}^{n} y_{ik} \theta_i \frac{x_0}{p_i} + a_k x_0,
\end{aligned}$$

we finally obtain the matrix equation

$$\boldsymbol{W}_c = \begin{pmatrix} y_{10} & \cdots & y_{n0} & a_0 \\ & \ddots & & \vdots \\ y_{1n} & \cdots & y_{nn} & a_n \end{pmatrix} \begin{pmatrix} w_1 & & & 0 & w'_1 \\ & w_2 & & & w'_2 \\ & & \ddots & & \vdots \\ & & & w_n & w'_n \\ 0 & & & & v_0 \end{pmatrix} \begin{pmatrix} c_1 & & & & 0 \\ & c_2 & & & \\ & & \ddots & & \\ & & & c_n & \\ 0 & & & & c \end{pmatrix} \begin{pmatrix} x_{10} & \cdots & x_{1n} \\ & \ddots & \vdots \\ x_{n0} & & x_{nn} \\ X_0 & \cdots & X_n \end{pmatrix}$$
$$= \quad \boldsymbol{Y} \quad \boldsymbol{W} \quad \mathsf{diag}(c_1, \cdots, c_n, c) \quad \boldsymbol{X}.$$

We perform the same computation on $c = 1$, which is a level-0 encoding of $\mathbf{1} = (1, 1, \cdots, 1)$, and then, it implies

$$\boldsymbol{W}_1 = \boldsymbol{Y} \cdot \boldsymbol{W} \cdot \boldsymbol{I} \cdot \boldsymbol{X}.$$

From $\boldsymbol{W}_c$ and $\boldsymbol{W}_1$, we have a matrix that is similar to $\mathsf{diag}(c_1, \cdots, c_n, c)$:

$$\boldsymbol{W}_1^{-1} \cdot \boldsymbol{W}_c = \boldsymbol{X}^{-1} \cdot \mathsf{diag}(c_1, \cdots, c_n, c) \cdot \boldsymbol{X}.$$

Then, by computing the eigenvalues of $\boldsymbol{W}_1^{-1} \cdot \boldsymbol{W}_c$, we have $c_1, \cdots, c_n$, satisfying $p_i | (c - c_i)$ for each i. Using an additional level-0 encoding c', we obtain $\boldsymbol{W}_1^{-1} \cdot \boldsymbol{W}_{c'}$, and therefore, $c'_1, \cdots, c'_n$ with $p_i | (c' - c'_i)$ for each i. Computing $\gcd(c - c_i, c' - c'_i)$ gives the secret prime p_i.

Using $p_1, \cdots, p_n$, we can recover all the remaining parameters. By the definition of y and $X_j^{(1)}$, the equation $y/[X_j^{(1)}]_{x_0} \equiv (r_i g_i + 1)/(r_{ij}^{(1)} g_i) \bmod p_i$ is satisfied. Since $r_i g_i + 1$ and $r_{ij}^{(1)} g_i$ are smaller than $\sqrt{p_i}$ and are co-prime, one can recover them by rational reconstruction up to the sign. Therefore, we can obtain g_i by computing the gcd of $r_{i0}^{(1)} g_i, \cdots, r_{im}^{(1)} g_i$. Moreover, using $r_{ij}^{(1)} g_i$ and $[X_j^{(1)}]_{x_0}$, we can compute $[z]_{p_i}$ for each i and therefore z. Any other parameters are computed using z, g_i, and p_i.

Our attack consists of the following arithmetics: computing $\phi(X_j^{(\kappa)})$, $\phi(X_j^{(1)} \cdot X_k^{(\kappa-1)})$, constructing a matrix $\boldsymbol{W}_c$ and $\boldsymbol{W}_1$, matrix inversing and multiplying, and computing eigenvalues and the greatest common divisor. All of these are bounded by $\widetilde{O}(\gamma^3 + n^\omega \gamma) = \widetilde{O}(\kappa^6 \lambda^9)$ bit computations with $\omega \le 2.38$. For this algorithm to succeed, we need a property that $\boldsymbol{W}_1$ is non-singular. If we use the fact that the rank of a matrix $\boldsymbol{A} \in \mathbb{Z}^{(n+1)\times(n+1)}$ can be computed in time $\widetilde{\mathcal{O}}((n+1)^\omega \log \|\boldsymbol{A}\|_\infty)$ (see [Sto09]), we can find that $\boldsymbol{X}, \boldsymbol{Y} \cdot \boldsymbol{W} \in \mathbb{Q}^{(n+1)\times(n+1)}$ are non-singular in $\widetilde{\mathcal{O}}(2(\gamma + \log \ell)(n^\omega \log N)) = \widetilde{\mathcal{O}}(\kappa^{\omega+4} \lambda^{2\omega+6})$ by considering another $(n+1)$ subsets of $X_0^{(1)}, \cdots, X_{\gamma'}^{(1)}$ for X and also for Y. Therefore, the total complexity of our attack is $\widetilde{\mathcal{O}}(\kappa^{\omega+4} \lambda^{2\omega+6})$. □

4.3 Determinant Attack

4.3.1 On the Impact of Recovering x_0 . If x_0 is known, CLT15 essentially collapses to CLT13. In particular, all encodings can be reduced modulo x_0 so ladders are no longer needed. What is more, all $\omega_{i,j}$'s from the CHLRS attack can be reduced modulo $v_0 = x_0 \boldsymbol{p}_{zt} \bmod N$, which effectively removes the new noise a. As a direct consequence the CHLRS attack goes through and all secret parameters are recovered (cf. [CLT15, Sect. 3.3]). Moreover ladder elements reduced by x_0 provide low-level encodings of zero even if the scheme itself does not. Also note that the CHLRS attack is quite efficient as it can be performed modulo any prime larger than the values we are trying to recover, *i.e.* larger than $2^{2\rho}$.

4.3.2 Recovering x_0 when an Exact Multiple is Known. The authors of [CLT15] propose an optimized version of their scheme, where a multiple qx_0 of x_0 is provided in the public parameters. The size of q is chosen such that qx_0 is about the same size as N. Ladders at levels below κ are no longer necessary: every encoding can be reduced modulo qx_0 without altering encoded values or increasing any noise. The ladder at level κ is still needed as a preliminary to zero-testing, however it does not need to go beyond qx_0, which makes it much smaller. In the end this optimization greatly reduces the size of the public key and speeds up computations, making the scheme much more practical (cf. Sect. 4.3.4).

In this scenario, note that qx_0 may be regarded as an encoding of 0 at level κ (and indeed every level). Moreover by construction it is small enough to be reduced by the ladder at level κ with a valid computation (*i.e.* with low enough noise for every intermediate encoding involved that the scheme operates as desired and zero-extraction is correct). As a direct consequence we have:

$$\phi(qx_0) = qv_0$$

and so we can recover q as $q = \gcd(qx_0, \phi(qx_0))$, and get $x_0 = qx_0/q$. This attack has been verified on the reference implementation, and recovers x_0 instantly.

Remark. qv_0 is larger than N by design, so that it cannot be computed simply as $qx_0\boldsymbol{p}_{zt} \bmod N$ due to modular reductions (cf. [CLT15, Sect. 3.4]). The point is that our computation of ϕ is over the integers and not modulo N.

4.3.3 Recovering x_0 in the General Case. We now return to the non-optimized version of the scheme, where no exact multiple of x_0 is provided in the public parameters.

The second step of our attack recovers x_0 using a matrix product similar to the CHLRS attack (cf. Sect. 3.2), except we start with families of $n+1$ encodings rather than n. That is, assume that for some t we have $n+1$ level-t small encodings (a_i) of any value, and $n+1$ level-$(\kappa - t)$ small encodings (b_i) of zero. This is easily achievable for one-round multi-party Diffie-Hellman (cf. Sect. A.2), e.g. choose $t = 1$, then pick $(n+1)$ level-1 encodings (a_i) of zero from the public parameters, and let $b_i = a'_i y^{\kappa-2}$ for a'_i another family of $(n+1)$ level-1 encodings of zero and y any level-1 encoding, where the product is ladder-reduced at each level. In other applications of the multilinear map, observe that ladder elements provide plenty of small encodings of zero, as each ladder element can be reduced by the elements below it to form a small encoding of zero. Thus the necessary conditions to perform both our attack to recover x_0, and the follow-up CHLRS attack to recover other secret parameters once x_0 is known, are very lax. In this respect CLT15 is weaker than CLT13.

Let $a_{i,j} = a_i z \bmod p_j$, *i.e.* $a_{i,j}$ is the j-th value "$r_j g_j + m_j$" associated with a_i. Likewise for $i \leq n$, let $r_{i,j} = b_i z^{\kappa-1}/g_j \bmod p_j$, *i.e.* $r_{i,j}$ is the j-th value "r_j" associated with b_i (recall that b_i is an encoding of zero, so $m_j = 0$). Now compute:

$$\omega_{i,j} \triangleq \phi(a_i b_j).$$

If we look at the $\omega_{i,j}$'s modulo v_0 (which is unknown for now), everything behaves as in CLT13 since the new noise term av_0 disappears, and the ladder reduction at level κ is negated by the integer extraction procedure. Hence, similar to Sect. 3.2, we have:

$$\omega_{i,j} \bmod v_0 = \sum_k a_{i,k} r_{j,k} v_k \bmod v_0. \tag{7}$$

Again, Eq. (7) may be seen as a matrix product. Indeed, define Ω as the $(n+1) \times (n+1)$ integer matrix with entries $\omega_{i,j}$, let A be the $(n+1) \times n$ matrix with entries $a_{i,j}$, let R be the $(n+1) \times n$ matrix with entries $r_{i,j}$, and finally let V be the $n \times n$ diagonal matrix with diagonal entries v_i. Then (7) may be rewritten modulo v_0:

$$\Omega = A \cdot V \cdot R^{\mathrm{T}} \qquad \text{in } \mathbb{Z}_{v_0}.$$

Since A and R are $(n+1) \times n$ matrices, this implies that Ω is not full-rank when embedded into $\mathbb{Z}_{v_0}$. As a consequence v_0 divides $\det(\Omega)$, where the determinant is computed over the integers. Now we can build a new matrix Ω' in the same way using a different choice of b_i's, and recover v_0 as $v_0 = \gcd(\det(\Omega), \det(\Omega'))$. Finally we get $x_0 = v_0 / \boldsymbol{p}_{zt} \bmod N$ (note that $N \gg x_0$ by construction).

The attack has been verified on the reference implementation with reduced parameters.

Remark. As pointed out above, Ω cannot be full-rank when embedded into $\mathbb{Z}_{v_0}$. Our attack also requires that it *is* full-rank over $\mathbb{Q}$ (whp). This holds because while Ω can be nicely decomposed as a product when viewed modulo v_0, the "remaining" part of Ω, that is $\Omega - (\Omega \bmod v_0)$ is the matrix of the terms av_0 for each $\omega_{i,j}$, and the value a does have the nice structure of $\omega_{i,j} \bmod v_0$. This is by design, since the noise a was precisely added in CLT15 in order to defeat the matrix product structure of the CHLRS attack.

4.3.4 Attack Complexity. It is clear that the attack is polynomial, and asymptotically breaks the scheme. In this section we provide an estimate of its practical complexity. When an exact multiple of x_0 is known, the attack is instant as mentioned in Sect. 4.3.2, so we focus on the general case from Sect. 4.3.3.

In the general case, a ladder of encodings of size $\ell \approx \gamma$ is published at every level[1]. Using the scheme requires κ ladder reductions, *i.e.* $\kappa\ell$ additions of integers of size γ. Since there are κ users, this means the total computation incurred by using the scheme is close to $\kappa^2\gamma^2$. For the smallest 52-bit instance, this is already $\approx 2^{46}$. Thus using the scheme a hundred times is above the security parameter. This highlights the importance of the optimization based on publishing qx_0, which makes the scheme much more practical. More importantly for our current purpose, this makes it hard to propose an attack below the security parameters.

As a result, what we propose in terms of complexity evaluation is the following. For computations that compare directly to using the multilinear scheme, we will tally the complexity as the number of operations equivalent to using the scheme, in addition to the bit complexity. For unrelated operations, we will count the number of bit operations as usual.

There are two steps worth considering from a complexity point of view: computing Ω and computing its determinant. In practice both steps happen to have comparable complexity. Computing the final gcd is negligible in comparison using a subquadratic algorithm [Mol08], which is practical for our parameter size.

Computing Ω. As a precomputation, in order to compute ϕ, the integer extraction of ladder elements at level κ needs to be computed. This requires ℓ integer extractions, where $\ell \leq \gamma$. Computing Ω itself requires $(n+1)^2$ integer extractions of a single product. Each integer extraction requires 1 multiplication, and 2ℓ additions (as well as ℓ multiplications by small scalars). For comparison, using the multilinear scheme for one user requires 1 multiplication and ℓ additions on integers of similar size. Thus overall computing Ω costs about $\gamma + n^2$ times as much as simply *using* the multilinear scheme. For the 52-bit instance proposed in [CLT15] for instance, this means that if it is practical to use the scheme about a million times, then it is practical to compute Ω. Here by using the scheme we mean one (rather than κ^2) ladder reduction, so the bit complexity is $\mathcal{O}(\gamma^3 + n^2\gamma^2)$.

Computing the Determinant. Let n denote the size of a matrix Ω (it is $(n+1)$ in our case but we will disregard this), and β the number of bits of its largest entry. When computing the determinant of an integer matrix, one has to carefully control the size of the integers appearing in intermediate computations. It is generally possible to ensure that these integers do not grow past the size of the determinant. Using Hadamard's bound this size can be upper bounded as $\log(\det(\Omega)) \leq n(\beta + \frac{1}{2}\log n)$, which can be approximated to $n\beta$ in our case, since β is much larger than n.[2]

[1] As the level increases, it is possible to slightly reduce the size of the ladder. Indeed the acceptable level of noise increases with each level, up to ρ_f at level κ. As a consequence it is possible to leave a small gap between ladder elements as the level increases. For instance if the base level of noise is 2ρ for ladder elements, then at level κ it is possible to leave a gap of roughly $\rho_f - 2\rho - \log \ell$ bits between ladder elements. We disregard this effect, although it slighly improves our complexity.

[2] This situation is fairly unusual, and in the literature the opposite is commonly assumed; algorithms are often optimized for large n rather than large β.

As a result, computing the determinant using "naive" methods requires $\mathcal{O}(n^3)$ operations on integers of size up to $n\beta$, which results in a complexity $\widetilde{\mathcal{O}}(n^4\beta)$ using fast integer multiplication (but slow matrix multiplication). The asymptotic complexity is known to be $\widetilde{\mathcal{O}}(n^\omega\beta)$ [Sto05]; however we are interested in the complexity of practical algorithms. Computing the determinant can be reduced to solving the linear system associated with Ω with a random target vector: indeed the determinant can then be recovered as the least common denominator of the (rational) solution vector[3]. In this context the fastest algorithms use p-adic lifting [Dix82], and an up-to-date analysis using fast arithmetic in [MS04] gives a complexity $\mathcal{O}(n^3\beta\log^2\beta\log\log\beta)$ (with $\log n = o(\beta)$)[4].

For the concrete instantiations of one-round multipartite Diffie-Hellman implemented in [CLT15], this yields the following complexities:

Security parameter:	52	62	72	80
Building Ω:	2^{60}	2^{66}	2^{74}	2^{82}
Determinant:	2^{57}	2^{66}	2^{74}	2^{81}

Thus, beside being polynomial, the attack is actually coming very close to the security parameter as it increases to 80 bits.[5]

Acknowledgement. We would like to thank Damien Stehlé and the authors of CLT13 and CLT15 Jean-Sébastien Coron, Tancrède Lepoint, and Mehdi Tibouchi for fruitful discussions and remarks. The authors of the Seoul National University, Jung Hee Cheon, Changmin Lee, and Hansol Ryu, were supported by the National Research Foundation of Korea (NRF) grant funded by the Korea government (MSIP) (No. 2014R1A2A1A11050917).

References

[BF01] Boneh, D., Franklin, M.: Identity-based encryption from the weil pairing. In: Kilian, J. (ed.) CRYPTO 2001. LNCS, vol. 2139, pp. 213–229. Springer, Heidelberg (2001)

[BS03] Boneh, D., Silverberg, A.: Applications of multilinear forms to cryptography. Contemp. Math. **324**(1), 71–90 (2003)

[CGH+15] Coron, J.-S., Gentry, C., Halevi, S., Lepoint, T., Maji, H.K., Miles, E., Raykova, M., Sahai, A., Tibouchi, M.: Zeroizing without low-level zeroes: new attacks on multilinear maps and their limitations. In: Gennaro, R., Robshaw, M. (eds.) Advances in Cryptology– CRYPTO 2015. LNCS, vol. 9215, pp. 247–266. Springer, Heidelberg (2015)

[3] In general extra factors may appear, but this is not relevant for us.

[4] This assumes a multitape Turing machine model, which is somewhat less powerful than a real computer.

[5] We may note in passing that in a random-access or log-RAM computing model [Fur14], which is more realistic than the multitape model, the estimated determinant complexity would already be slightly lower than the security parameter.

[CHL+15] Cheon, J.H., Han, K., Lee, C., Ryu, H., Stehlé, D.: Cryptanalysis of the multilinear map over the integers. In: Oswald, E., Fischlin, M. (eds.) EUROCRYPT 2015. LNCS, vol. 9056, pp. 3–12. Springer, Heidelberg (2015)

[CLR15] Cheon, J.H., Lee, C., Ryu, H.: Cryptanalysis of the new CLT multilinear maps. Cryptology ePrint Archive, Report 2015/934 (2015). http://eprint.iacr.org/

[CLT13] Coron, J.-S., Lepoint, T., Tibouchi, M.: Practical multilinear maps over the integers. In: Canetti, R., Garay, J.A. (eds.) CRYPTO 2013, Part I. LNCS, vol. 8042, pp. 476–493. Springer, Heidelberg (2013)

[CLT15] Coron, J.-S., Lepoint, T., Tibouchi, M.: New multilinear maps over the integers. In: Gennaro, R., Robshaw, M. (eds.) Advances in Cryptology–CRYPTO 2015. LNCS, pp. 267–286. Springer, Heidelberg (2015)

[Cor15] Coron, J.-S.: Cryptanalysis of GGH15 multilinear maps. Cryptology ePrint Archive, Report 2015/1037 (2015). http://eprint.iacr.org/

[DH76] Diffie, W., Hellman, M.E.: Multiuser cryptographic techniques. In: Proceedings of the 7–10, June 1976, National Computer Conference and Exposition, pp. 109–112. ACM (1976)

[Dix82] Dixon, J.D.: Exact solution of linear equations using P-adic expansions. Nümer. Math. **40**(1), 137–141 (1982)

[Fur14] Fürer, M.: How fast can we multiply large integers on an actual computer? In: Pardo, A., Viola, A. (eds.) LATIN 2014: Theoretical Informatics. LNCS, pp. 660–670. Springer, Heidelberg (2014)

[GGH13a] Garg, S., Gentry, C., Halevi, S.: Candidate multilinear maps from ideal lattices. In: Johansson, T., Nguyen, P.Q. (eds.) EUROCRYPT 2013. LNCS, vol. 7881, pp. 1–17. Springer, Heidelberg (2013)

[GGH+13b] Garg, S., Gentry, C., Halevi, S., Raykova, M., Sahai, A., Waters, B.: Candidate indistinguishability obfuscation and functional encryption for all circuits. In: IEEE 54th Annual Symposium on Foundations of Computer Science (FOCS), pp. 40–49. IEEE (2013)

[GGH15] Gentry, C., Gorbunov, S., Halevi, S.: Graph-induced multilinear maps from lattices. In: Dodis, Y., Nielsen, J.B. (eds.) TCC 2015, Part II. LNCS, vol. 9015, pp. 498–527. Springer, Heidelberg (2015)

[Hal15a] Halevi, S.: Cryptographic graded-encoding schemes: Recent developments. TCS+ online seminar (2015). https://sites.google.com/site/plustcs/past-talks/20150318shaihaleviibmtjwatson

[Hal15b] Halevi, S.: Graded encoding, variations on a scheme. Technical report, Cryptology ePrint Archive, Report 2015/866 (2015). http://eprint.iacr.org

[HJ15] Hu, Y., Jia, H.: Cryptanalysis of GGH map. Technical report, Cryptology ePrint Archive, Report 2015/301 (2015)

[HSW13] Hohenberger, S., Sahai, A., Waters, B.: Full domain hash from (leveled) multilinear maps and identity-based aggregate signatures. In: Canetti, R., Garay, J.A. (eds.) CRYPTO 2013, Part I. LNCS, vol. 8042, pp. 494–512. Springer, Heidelberg (2013)

[Jou00] Joux, A.: A one round protocol for tripartite Diffie-Hellman. In: Bosma, A. (ed.) Algorithmic Number Theory. LNCS, vol. 1838, pp. 385–393. Springer, Heidelberg (2000)

[MF15] Minaud, B., Fouque, P.-A.: Cryptanalysis of the new multilinear map over the integers. Cryptology ePrint Archive, Report 2015/941 (2015). http://eprint.iacr.org

[Mol08] Möller, N.: On Schönhage's algorithm and subquadratic integer GCD computation. Math. Comput. **77**(261), 589–607 (2008)

[MS04] Mulders, T., Storjohann, A.: Certified dense linear system solving. J. Symbolic Comput. **37**(4), 485–510 (2004)

[Sha85] Shamir, A.: Identity-based cryptosystems and signature schemes. In: Blakely, G.R., Chaum, D. (eds.) CRYPTO 1984. LNCS, vol. 196, pp. 47–53. Springer, Heidelberg (1985)

[Sto05] Storjohann, A.: The shifted number system for fast linear algebra on integer matrices. J. Complex. **21**(4), 609–650 (2005). Festschrift for the 70th Birthday of Arnold Schonhage

[Sto09] Storjohann, A.: Integer matrix rank certification. In: Proceedings of the International Symposium on Symbolic and Algebraic Computation, pp. 333–340. ACM (2009)

[VDGHV10] van Dijk, M., Gentry, C., Halevi, S., Vaikuntanathan, V.: Fully homomorphic encryption over the integers. In: Gilbert, H. (ed.) EUROCRYPT 2010. LNCS, vol. 6110, pp. 24–43. Springer, Heidelberg (2010)

[Zim15] Zimmerman, J.: How to obfuscate programs directly. In: Oswald, E., Fischlin, M. (eds.) EUROCRYPT 2015. LNCS, vol. 9057, pp. 439–467. Springer, Heidelberg (2015)

A Short Introduction to Multilinear Maps

In this section we give a brief introduction to multilinear maps to make our article self-contained. In particular we only consider symmetric multilinear maps. We refer the interested reader to [GGH13a, Hal15b] for a more thorough presentation.

A.1 Multilinear Maps and Graded Encoding Schemes

Cryptographic multilinear maps were introduced by Boneh and Silverberg [BS03], as a natural generalization of bilinear maps stemming from pairings on elliptic curves, which had found striking new applications in cryptography [Jou00, BF01, ...]. A (symmetric) multilinear map is defined as follows.

Definition 1 (Multilinear Map [BS03]). *Given two groups $\mathbb{G}, \mathbb{G}_T$ of the same prime order, a map $e : \mathbb{G}^\kappa \to \mathbb{G}_T$ is a κ-*multilinear map *iff it satisfies the following two properties:*

1. *for all $a_1, \ldots, a_\kappa \in \mathbb{Z}$ and $x_1, \ldots, x_\kappa \in \mathbb{G}$,*

$$e(x_1^{a_1}, \ldots, x_\kappa^{a_\kappa}) = e(x_1, \ldots, x_\kappa)^{a_1 \cdots a_\kappa}$$

2. *if g is a generator of $\mathbb{G}$, then $e(g, \ldots, g)$ is a generator of $\mathbb{G}_T$.*

A natural special case are *leveled* multilinear maps:

Definition 2 (Leveled Multilinear Map [HSW13]). *Given $\kappa + 1$ groups $\mathbb{G}_1, \ldots, \mathbb{G}_\kappa, \mathbb{G}_T$ of the same prime order, and for each $i \leq \kappa$, a generator $g_i \in \mathbb{G}_i$, a* κ-leveled multilinear map *is a set of bilinear maps $\{e_{i,j} : \mathbb{G}_i \times \mathbb{G}_j \to G_{i+j} | i, j, i+j \leq \kappa\}$ such that for all i, j with $i + j \leq \kappa$, and all $a, b \in \mathbb{Z}$:*

$$e_{i,j}(g_i^a, g_j^b) = g_{i,j}^{ab}.$$

Similar to public-key encryption [DH76] and identity-based cryptosystems [Sha85], multilinear maps were originally introduced as a compelling target for cryptographic research, without a concrete instantiation [BS03]. The first multilinear map was built ten years later in the breakthrough construction of Garg, Gentry and Halevi [GGH13a]. More accurately, what the authors proposed was a *graded encoding scheme*, and to this day all known cryptographic multilinear maps constructions are actually variants of graded encoding schemes [Hal15b]. For this reason, and because both constructions have similar expressive power, the term "multilinear map" is used in the literature in place of "graded encoding scheme", and we follow suit in this article.

Graded encoding schemes are a relaxed definition of leveled multilinear map, where elements x_i^a for $x_i \in \mathbb{G}_i, a \in \mathbb{Z}$ are no longer required to lie in a group. Instead, they are regarded as "encodings" of a ring element a at level i, with no assumption about the underlying structure. Formally, encodings are thus defined as general binary strings in $\{0,1\}^*$. In the following definition, $S_i^{(\alpha)}$ should be regarded as the set of encodings of a ring element α at level i.

Definition 3 (Graded Encoding System [GGH13a]). *A* κ-graded encoding system *consists of a ring R and a system of sets $\mathcal{S} = \{S_i^{(\alpha)} \subset \{0,1\}^* | \alpha \in R, 0 \leq i \leq \kappa\}$, with the following properties:*

1. *For each fixed i, the sets $S_i^{(\alpha)}$ are pairwise disjoint as α spans R.*
2. *There is an associative binary operation '+' and a self-inverse unary operation '−' on $\{0,1\}^*$ such that for every $\alpha_1, \alpha_2 \in R$, every $i \leq \kappa$, and every $u_1 \in S_i^{(\alpha_1)}, u_2 \in S_i^{(\alpha_2)}$, it holds that:*

$$u_1 + u_2 \in S_i^{(\alpha_1+\alpha_2)} \quad \textit{and} \quad -u_1 \in S_i^{(-\alpha_1)}$$

   where $\alpha_1 + \alpha_2$ and $-\alpha_1$ are addition and negation in R.
3. *There is an associative binary operation '×' on $\{0,1\}^*$ such that for every $\alpha_1, \alpha_2 \in R$, every $i_1, i_2 \in \mathbb{N}$ such that $i_1 + i_2 \leq \kappa$, and every $u_1 \in S_{i_1}^{(\alpha_1)}, u_2 \in S_{i_2}^{(\alpha_2)}$, it holds that $u_1 \times u_2 \in S_{i_1+i_2}^{(\alpha_1 \cdot \alpha_2)}$. Here $\alpha_1 \cdot \alpha_2$ is the multiplication in R, and $i_1 + i_2$ is the integer addition.*

Observe that a leveled multilinear map is a graded encoding system where $R = \mathbb{Z}$ and, with the notation from the definitions, $S_i^{(\alpha)}$ contains the single element g_i^α. Also note that the behavior of addition and multiplication of encodings with respect to the levels i is the same as that of a graded ring, hence the *graded* qualifier.

All known constructions of graded encoding schemes do not fully realize the previous definition, insofar as they are "noisy"[6]. That is, all encodings have a certain amount of noise; each operation, and especially multiplication, increases this noise; and the correctness of the scheme breaks down if the noise goes above a certain threshold. The situation in this regard is similar to somewhat homomorphic encryption schemes.

A.2 Multilinear Map Procedures

The exact interface offered by a multilinear map, and called upon when it is used as a primitive in a cryptographic scheme, varies depending on the scheme. However the core elements are the same. Below we reproduce the procedures for manipulating encodings defined in [CLT15], which are a slight variation of [GGH13a].

In a nutshell, the scheme relies on a trusted third party that generates the instance (and is typically no longer needed afterwards). Users of the instance (that is, everyone but the generating trusted third party) cannot encode nor decode arbitrary encodings: they can only combine existing encodings using addition, negation and multiplication, and subject to the limitation that the level of an encoding cannot exceed κ. The power of the multilinear map comes from the zero-testing (resp. extraction) procedure, which allows users to test whether an encoding at level κ encodes zero (resp. roughly get a λ-bit "hash" of the value encoded by a level-κ encoding).

Here users are also given access to random level-0 encodings, and have the ability to re-randomize encodings, as well as promote any encoding to a higher-level encoding of the same element. These last functionalities are tailored towards the application of multilinear maps to one-round multi-party Diffie-Hellman. In general different applications of multilinear map require different subsets of the procedures below, and sometimes variants of them.

$\mathsf{instGen}(1^\lambda, 1^\kappa)$: the randomized instance procedure takes as input the security parameter λ, the multilinearity level κ, and outputs the public parameters $(\mathsf{pp}, \boldsymbol{p}_{zt})$, where pp is a description of a κ-graded encoding system as above, and $\boldsymbol{p}_{zt}$ is a zero-test parameter (see below).

$\mathsf{samp}(\mathsf{pp})$: the randomized sampling procedure takes as input the public parameters pp and outputs a level-0 encoding $u \in S_0^{(\alpha)}$ for a nearly uniform $\alpha \in R$.

$\mathsf{enc}(\mathsf{pp}, i, u)$: the possibly randomized encoding procedure takes as input the public parameters pp, a level $i \leq \kappa$, and a level-0 encoding $u \in S_0^\alpha$ for some $\alpha \in R$, and outputs a level-i encoding $u' \in S_i^{(\alpha)}$.

$\mathsf{reRand}(\mathsf{pp}, i, u)$: the randomized rerandomization procedure takes as input the public parameters pp, a level $i \leq \kappa$, and a level-i encoding $u \in S_i^\alpha$ for some $\alpha \in R$, and outputs another level-i encoding $u' \in S_i^{(\alpha)}$ of the same α, such

[6] In fact the question of achieving the functionality of multilinear maps without noise may be regarded as an important open problem [Zim15].

that for any $u_1, u_2 \in S_i^{(\alpha)}$, the output distributions of $\mathsf{reRand}(\mathsf{pp}, i, u_1)$ and $\mathsf{reRand}(\mathsf{pp}, i, u_2)$ are nearly the same.

$\mathsf{neg}(\mathsf{pp}, u)$: the negation procedure is deterministic and that takes as input the public parameters pp, and a level-i encoding $u \in S_i^{(\alpha)}$ for some $\alpha \in R$, and outputs a level-i encoding $u' \in S_i^{(-\alpha)}$.

$\mathsf{add}(\mathsf{pp}, u_1, u_2)$: the addition procedure is deterministic and takes as input the public parameters pp, two level-i encodings $u_1 \in S_i^{(\alpha_1)}, u_2 \in S_i^{(\alpha_2)}$ for some $\alpha_1, \alpha_2 \in R$, and outputs a level-i encoding $u' \in S_i^{(\alpha_1+\alpha_2)}$.

$\mathsf{mult}(\mathsf{pp}, u_1, u_2)$: the multiplication procedure is deterministic and takes as input the public parameters pp, two encodings $u_1 \in S_i^{(\alpha_1)}, u_2 \in S_j^{(\alpha_2)}$ of some $\alpha_1, \alpha_2 \in R$ at levels i and j such that $i + j \leq \kappa$, and outputs a level-$(i+j)$ encoding $u' \in S_{i+j}^{(\alpha_1 \cdot \alpha_2)}$.

$\mathsf{isZero}(\mathsf{pp}, u)$: the zero-testing procedure is deterministic and takes as input the public parameters pp, and an encoding $u \in S_\kappa^{(\alpha)}$ of some $\alpha \in R$ at the maximum level κ, and outputs 1 if $\alpha = 0$, 0 otherwise, with negligible probability of error (over the choice of $u \in S_\kappa^{(\alpha)}$).

$\mathsf{ext}(\mathsf{pp}, \boldsymbol{p}_{zt}, u)$: the extraction procedure is deterministic and takes as input the public parameters pp, the zero-test parameter $\boldsymbol{p}_{zt}$, and an encoding $u \in S_\kappa^{(\alpha)}$ of some $\alpha \in R$ at the maximum level κ, and outputs a λ-bit string s such that:

1. For $\alpha \in R$ and $u_1, u_2 \in S_\kappa^{(\alpha)}$, $\mathsf{ext}(\mathsf{pp}, \boldsymbol{p}_{zt}, u_1) = \mathsf{ext}(\mathsf{pp}, \boldsymbol{p}_{zt}, u_2)$.
2. The distribution $\{\mathsf{ext}(\mathsf{pp}, \boldsymbol{p}_{zt}, v) | \alpha \leftarrow R, v \in S_\kappa^{(\alpha)}\}$ is nearly uniform over $\{0,1\}^\lambda$.

Cryptanalysis of GGH Map

Yupu Hu(✉) and Huiwen Jia(✉)

ISN Laboratory, Xidian University, Xi'an 710071, China
yphu@mail.xidian.edu.cn, hwjia@stu.xidian.edu.cn

Abstract. Multilinear map is a novel primitive which has many cryptographic applications, and GGH map is a major candidate of K-linear maps for $K > 2$. GGH map has two classes of applications, which are applications with public tools for encoding and with hidden tools for encoding. In this paper, we show that applications of GGH map with public tools for encoding are not secure, and that one application of GGH map with hidden tools for encoding is not secure. On the basis of weak-DL attack presented by the authors themselves, we present several efficient attacks on GGH map, aiming at multipartite key exchange (MKE) and the instance of witness encryption (WE) based on the hardness of exact-3-cover (X3C) problem. First, we use special modular operations, which we call modified Encoding/zero-testing to drastically reduce the noise. Such reduction is enough to break MKE. Moreover, such reduction negates K-GMDDH assumption, which is a basic security assumption. The procedure involves mostly simple algebraic manipulations, and rarely needs to use any lattice-reduction tools. The key point is our special tools for modular operations. Second, under the condition of public tools for encoding, we break the instance of WE based on the hardness of X3C problem. To do so, we not only use modified Encoding/zero-testing, but also introduce and solve "combined X3C problem", which is a problem that is not difficult to solve. In contrast with the assumption that multilinear map cannot be divided back, this attack includes a division operation, that is, solving an equivalent secret from a linear equation modular some principal ideal. The quotient (the equivalent secret) is not small, so that modified Encoding/zero-testing is needed to reduce size. This attack is under an assumption that some two vectors are co-prime, which seems to be plausible. Third, for hidden tools for encoding, we break the instance of WE based on the hardness of X3C problem. To do so, we construct level-2 encodings of 0, which are used as alternative tools for encoding. Then, we break the scheme by applying modified Encoding/zero-testing and combined X3C, where the modified Encoding/zero-testing is an extended version. This attack is under two assumptions, which seem to be plausible. Finally, we present cryptanalysis of two simple revisions of GGH map, aiming at MKE. We show that MKE on these two revisions can be broken under the assumption that 2^K is polynomially large. To do so, we further extend our modified Encoding/zero-testing.

M. Fischlin and J.-S. Coron (Eds.): EUROCRYPT 2016, Part I, LNCS 9665, pp. 537–565, 2016.
DOI: 10.1007/978-3-662-49890-3_21

Keywords: Multilinear maps · Multipartite key exchange (MKE) · Witness encryption (WE) · Lattice based cryptography

1 Introduction

1.1 Background and Our Contributions

Multilinear map is a novel primitive. Mathematically speaking, multilinear map is a leveled encoding system. In other words, it is such a system that can multiply but cannot divide back, and goes further to let us recover some limited information. It is the solution of a long-standing open problem [1], and has many novel cryptographic applications, such as multipartite key exchange (MKE) [2], witness encryption (WE) [3–9], obfuscation [8–10], and so on. It also has several advantages in the traditional cryptographic area such as IBE, ABE [11], Broadcasting encryption, and so on. The first candidate of multilinear map is GGH map [2], and GGHLite map [12] is a special version of GGH map for the purpose of improving efficiency. Up until now, GGH map is a major candidate of K-linear maps for $K > 2$. It uses noisy encoding to obtain the trapdoor. The security of GGH map is not well-understood. In particular, hardness of lattice problems is necessary for its security, but it is not sufficient. GGH map has two classes of applications. The first class is applications with public tools for Encoding/zero-testing such as MKE [2], IBE, ABE, Broadcasting encryption, and so on. The second class contains applications with hidden tools for encoding such as GGHRSW obfuscation [8]. WE can be in the first and second classes. For the first class, WE tools for encoding are generated and published by the system, and can be used by any user. For the second class, WE tools for encoding are generated and hidden by a unique encrypter, and can only be used by him/herself. Besides, WE is another novel cryptographic notion and the instance of WE based on the hardness of exact-3-cover (X3C) problem is its first instance. Garg et al. provided in [2] a survey of relevant cryptanalysis techniques from the literature, and also described two new attacks on GGH map. In particular they presented the weak-DL attack, which indicated that GGH map makes division possible to some extent, and which is used in our attacks as well. We emphasize, however, that they did not show how to use that attack to break any of their proposed schemes.

In this paper, we show that applications of GGH map with public tools for encoding are not secure, and that one application of GGH map with hidden tools for encoding is not secure. We present several efficient attacks on GGH map, aiming at MKE and the instance of WE based on the hardness of X3C problem. In all of our attacks we begin by using the weak-DL attack from [2] to recover an "equivalent secret" which is equal to the original secret modulo some known ideal, but is not small. Then we proceed as follows.

First, we use special modular operations, which we call modified Encoding/zero-testing to drastically reduce the noise. Such reduction is enough to break MKE. Moreover, such reduction negates K-GMDDH assumption

(Assumption 5.1 of [11]), which is the security basis of the ABE scheme [11]. The procedure involves mostly simple algebraic manipulations, and rarely needs to use any lattice-reduction tools. The key point is our special tools for modular operations.

Second, under the condition of public tools for encoding, we break the instance of WE based on the hardness of X3C problem. To do so, we not only use modified Encoding/zero-testing, but also introduce and solve "combined X3C problem", which is a problem that is not difficult to solve. In contrast with the assumption that multilinear map cannot be divided back, this attack includes a division operation, that is, solving an equivalent secret from a linear equation modular some principal ideal. The quotient (the equivalent secret) is not small, so that modified Encoding/zero-testing is needed to reduce size. This attack is under an assumption that some two vectors are co-prime, which seems to be plausible.

Third, for hidden tools for encoding, we break the instance of WE based on the hardness of X3C problem. To do so, we construct level-2 encodings of 0, which are used as alternative tools for encoding. Then, we break the scheme by applying modified Encoding/zero-testing and combined X3C, where the modified Encoding/zero-testing is an extended version. This attack has several preparing works, including solving a new type of "equivalent secret". This attack is under two assumptions, which seem to be plausible.

Finally, we check whether GGH structure can be simply revised to avoid our attack. We present cryptanalysis of two simple revisions of GGH map, aiming at MKE. We show that MKE on these two revisions can be broken under the assumption that 2^K is polynomially large. To do so, we further extend our modified Encoding/zero-testing.

1.2 Principles and Main Techniques of Our Attack

Quite unlike the original DH maps and bilinear maps, all candidates of multilinear maps have a common security worry that zero-testing tools are public. This allows the adversary to zero-test messages freely. The adversary can choose those zero-tested messages that are small enough without protection of the modular operation. Such security worry has been used to break CLT map [13–17], which is another major candidate of multilinear maps, and which is simply over integers. Multilinear maps over the integer polynomials (GGH map [2] and GGHLite map [12]) haven't been broken because (1) (NTRU declaration) the product of a short polynomial and modular inverse of another short polynomial seems unable to be decomposed; and (2) the product of several short polynomials seems unable to be decomposed. However, the product of several short polynomials is a somewhat short polynomial. Although it cannot be decomposed, it can be used as a modulus to reduce the noise. On the other hand, breaking applications of GGH map with public tools for encoding does not mean solving the users' secrets. It only means solving "high-order bits of zero-test of the product of encodings of users' secrets", a weaker requirement. Therefore, by using our modified Encoding/zero-testing, we can easily migrate between modular operations and real number operations to find vulnerabilities which have not been found before. All of the above form the first principle of our attack. The second principle is that if one uses GGH map for constructing the instance of WE based on the hardness of X3C problem, special structure of GGH map allows us to transform the underlying X3C problem into a much easier combined X3C problem. Our main techniques are as follows.

Modified Encoding/zero-testing. For the secret of each user, we have an equivalent secret which is the sum of original secret and a noise. These equivalent secrets cannot be encoded, because they are not small. We compute the product of these equivalent secrets, rather than computing their modular product. Notice that the product is the sum of the product of original secrets and a noise. Then our modified Encoding/zero-testing is quite simple. It contains three simple operations, avoiding computing original secrets of users, and extracting same information. That is, it extracts same high-order bits of zero-tested message. Table 1 is a comparison between processing routines of GGH map and our work. It is a note of our claim that we can achieve the same purpose without knowing the secret of any user.

Table 1. Processing routines

GGH map	secrets → encodings → modular product → zero-testing → high-order bits
Our work	equivalent secrets → product → modified Encoding/zero-testing → high-order bits

Solving Combined Exact-3-cover (Combined X3C) Problem. The reason that X3C problem can be transformed into a combined X3C problem is that the special structure of GGH map sometimes makes division possible. We can solve combined X3C problem with non-negligible probability and break the instance of WE based on the hardness of X3C problem for public tools of encoding.

Finding Alternative Encoding Tools. When encoding tools are hidden, we can use redundant information to construct alternative encoding tools. For example, there are many redundant pieces beside X3C. Encodings of these redundant pieces can be composed into several level-2 encodings of 0. Only one level-2 encoding of 0 is enough to break the instance of WE based on the hardness of X3C problem for hidden tools of encoding. This technique can be adapted to other applications of GGH map, where although encoding tools are hidden, a large number of redundant information are needed to protect some secrets.

1.3 The Organization

In Subsect. 1.4 we review recent works related to multilinear map. In Sect. 2 we review GGH map and two applications, MKE and the instance of WE on X3C. In Sect. 3 we define special tools for our attack, which are special polynomials used for our modular operations. Also in this section, for the secret of each user, we generate an equivalent secret, which is not a short vector. Immediately, we obtain an "equivalent secret" of the product of the users' secrets, which is the product of the users' equivalent secrets. In Sect. 4 we present modified encoding/zero-testing. We show how "high-order bits of zero-test of the product of encodings of users' secrets" can be solved, so that MKE is broken. In Sect. 5 we show how to break the instance of WE on X3C problem with public tools for encoding. In this section, we first introduce and solve "combined X3C problem", then solve "high-order bits of zero-test of the product of encodings of users' secrets". In Sect. 6 we present an attack on the instance of WE based on the hardness of X3C problem with hidden tools for encoding. We show that this instance can be broken under several stronger assumptions. In Sect. 7 we present cryptanalysis of two simple revisions of GGH map, aiming at MKE. We show that MKE on these two revisions can be broken under the assumption that 2^K is polynomially large. Section 8 contains other results, some considerations, and poses several questions.

1.4 Related Works

Garg et al. presented in [2] three variants, which are "asymmetric encoding", "providing zero-test security" and "avoiding principal ideals". Arita and Handa [5] presented two applications of multilinear maps: MKE with smaller communication and an instance of WE. Their WE scheme (called AH scheme) has the security claim based on the hardness of Hamilton Cycle problem. The novelty is that they used an asymmetric multilinear map over integer matrices. Bellare and Hoang [6] presented adaptive witness encryption with stronger security than soundness security, named adaptive soundness security. Garg et al. [8] presented witness encryption by using indistinguishability obfuscation and Multilinear Jigsaw Puzzle, a simplified variant of multilinear maps. Extractable witness encryption was presented [7,9,10]. Gentry et al. designed multilinear maps based on graph [18]. Coron et al. presented efficient attack on CLT map for hidden tools for encoding [19]. Coron et al. designed CLT15 map [20]. Then Cheon et al. [21] and Minaud and Fouque [22] broke CLT15 respectively.

2 GGH Map and Two Applications

2.1 Notations and Definitions

We denote the rational numbers by $\mathbb{Q}$ and the integers by $\mathbb{Z}$. We specify that n-dimensional vectors of $\mathbb{Q}^n$ and $\mathbb{Z}^n$ are row vectors. We consider the $2n$'th cyclotomic polynomial ring $R = \mathbb{Z}[X]/(X^n + 1)$, and identify an element $u \in R$ with the coefficient vector of the degree-$(n-1)$ integer polynomial that represents u. In this way, R is identified with the integer lattice $\mathbb{Z}^n$. We also consider the ring $R_q = R/qR = \mathbb{Z}_q[X]/(X^n + 1)$ for a (large enough) integer q. Addition in these rings is done component-wise in their coefficients, and multiplication is polynomial multiplication modulo the ring polynomial $X^n + 1$. In some cases, we also consider the ring $\mathbb{K} = \mathbb{Q}[X]/(X^n + 1)$, which is likewise associated with the linear space $\mathbb{Q}^n$. We redefine the operation "mod q" as follows: if q is an odd, $a(\text{mod } q)$ is within $\{-(q-1)/2, -(q-3)/2, \dots, (q-1)/2\}$; if q is an even, $a(\text{mod } q)$ is within $\{-q/2, -(q-2)/2, \dots, (q-2)/2\}$. For $x \in R$, $\langle x \rangle = \{x \cdot u : u \in R\}$ is the principal ideal in R generated by x (alternatively, the sub-lattice of $\mathbb{Z}^n$ corresponding to this ideal). For $x \in R$, $y \in R$, $y(\text{mod } x)$ is such a vector: $y(\text{mod } x) = ax$, where each entry of a is within $[-0.5, 0.5)$, and $y - y(\text{mod } x) \in \langle x \rangle$. We refer the readers to Babai [23].

2.2 The GGH Construction

We secretly sample a short element $g \in R$. Let $\langle g \rangle$ be the principal ideal in R. g itself is kept secret, and no "good" description of $\langle g \rangle$ is made public. Another secret element $z \in R_q$ is chosen at random, and hence is not short.

An element y is called encoding parameter, or called level-1 encoding of 1, and is set in the following description. We secretly sample a short element $a \in R$, and let $y = (1 + ag)z^{-1}(\text{mod } q)$. The elements $\{x^{(i)}, i = 1, 2\}$ are called randomizers, or called level-1 encodings of 0, and are set as follows. We secretly sample a short element $b^{(i)} \in R$, and let $x^{(i)} = b^{(i)}gz^{-1}(\text{mod } q)$, $i = 1, 2$. The public element p_{zt} is called level-K zero-testing parameter, where $K \geq 3$ is an integer. p_{zt} is set as follows. We secretly sample a "somewhat small" element $h \in R$, and let $p_{zt} = (hz^K g^{-1})(\text{mod } q)$. Simply speaking, parameters y and $\{x^{(i)}, i = 1, 2\}$ are tools for encoding, while public parameter p_{zt} is tool of zero-test. $\{g, z, a, \{b^{(i)}, i = 1, 2\}, h\}$ are kept from all users. For MKE, y and $\{x^{(i)}, i = 1, 2\}$ are public. For WE, they can be either public or hidden.

Suppose a user has a secret $v \in R$, which is a short element. He secretly samples short elements $\{u^{(i)} \in R, i = 1, 2\}$. He computes noisy encoding $V = vy + (u^{(1)}x^{(1)} + u^{(2)}x^{(2)})(\text{mod } q)$, where $vy(\text{mod } q)$ and $(u^{(1)}x^{(1)} + u^{(2)}x^{(2)})(\text{mod } q)$ are respectively encoded secret and encoded noise. He publishes V. Then, GGH K-linear map includes $K, y, \{x^{(i)}, i = 1, 2\}, p_{zt}$, and all noisy encoding Vs for all users.

We call g grade 1 element, and denote σ as the standard deviation for sampling g. We call $\{a, \{b^{(i)}, i = 1, 2\}\}$ and $\{v, \{u^{(i)}, i = 1, 2\}\}$ grade 2 elements, and denote σ' as the standard deviation for sampling $\{a, \{b^{(i)}, i = 1, 2\}\}$ and $\{v, \{u^{(i)}, i = 1, 2\}\}$. Both σ and σ' are much smaller than $\sqrt{q}$, and GGH K-linear map [2] suggests $\sigma' = n\sigma$. Finally, we call h grade 3 element, and take $\sigma'' = \sqrt{q}$ as the standard deviation for sampling h. We say that g, $\{a, \{b^{(i)}, i = 1, 2\}\}$ and $\{v, \{u^{(i)}, i = 1, 2\}\}$ are "very small", and that h is "somewhat small". h cannot be "very small" for security reasons.

2.3 Application 1: MKE

Suppose that $K + 1$ users want to generate a commonly shared key by public discussion. To do so, each user k generates his secret $v^{(k)}$, and publishes the noisy encoding $V^{(k)}, k = 1, \dots, K + 1$. Then, each user can use his/her secret and other users' noisy encodings to compute KEY, the commonly shared key. KEY is high-order bits of any zero-tested message. For example, user k_0 first computes $v^{(k_0)} p_{zt} \prod_{k \neq k_0} V^{(k)} (\text{mod } q)$, then KEY is high-order bits of $v^{(k_0)} p_{zt} \prod_{k \neq k_0} V^{(k)} (\text{mod } q)$. That is, he/she first computes

$$v^{(k_0)} p_{zt} \prod_{k \neq k_0} V^{(k)} (\text{mod } q) =$$

$$h(1+ag)^K g^{-1} \prod_{k=1}^{K+1} v^{(k)} +$$

$$hv^{(k_0)} \sum_{\substack{S \subset \{1,\ldots,K+1\} \\ -\{k_0\}, |S| \geq 1}} (1+ag)^{K-|S|} g^{|S|-1} \prod_{\substack{k \in \{1,\ldots,K+1\} \\ -\{k_0\}-S}} (v^{(k)}) \prod_{t \in S} (u^{(t,1)} b^{(1)} + u^{(t,2)} b^{(2)}) (\text{mod } q).$$

It is the modular sum of two terms, zero-tested message and zero-tested noise. Zero-tested message is

$$h(1+ag)^K g^{-1} \prod_{k=1}^{K+1} v^{(k)} (\text{mod } q).$$

Zero-tested noise is

$$hv^{(k_0)} \sum_{\substack{S \subset \{1,\ldots,K+1\} \\ -\{k_0\}, |S| \geq 1}} (1+ag)^{K-|S|} g^{|S|-1} \prod_{\substack{k \in \{1,\ldots,K+1\} \\ -\{k_0\}-S}} (v^{(k)}) \prod_{t \in S} (u^{(t,1)} b^{(1)} + u^{(t,2)} b^{(2)}).$$

Notice that zero-tested noise is the sum of $3^K - 1$ terms. For example, $h(1+ag)^{K-1} b^{(1)} u^{(1,1)} \prod_{k=2}^{K+1} (v^{(k)})$ is a term of the zero-tested noise. Each term is the product of a "somewhat small" element and several "very small" elements. Therefore, zero-tested noise is "somewhat small", and it can be removed if we only extract high-order bits of $v^{(k_0)} p_{zt} \prod_{k \neq k_0} V^{(k)} (\text{mod } q)$. In other words, KEY is actually high-order bits of zero-tested message $h(1+ag)^K g^{-1} \prod_{k=1}^{K+1} v^{(k)} (\text{mod } q)$.

2.4 Application 2: The Instance of WE on Exact-3-cover

Definition 1. *A witness encryption scheme for an* **NP** *language L (with corresponding witness relation Rel) consists of the following two polynomial-time algorithms:*

Encryption. *The algorithm Encrypt$(1^\lambda, x, M)$ takes as input a security parameter 1^λ, a string x, and a message M, and outputs a ciphertext* CT.

Decryption. *The algorithm Decrypt(CT, w) takes as input a ciphertext* CT *and a string w, and outputs a message M if Rel$(w, x) = 1$ or the symbol $\perp$ otherwise.*

Exact-3-cover Problem [3,24]. If we are given a subset of $\{1, 2, \ldots, 3K\}$ containing 3 integers, we call it a piece. If we are given a collection of K pieces without intersection, we call it a X3C of $\{1, 2, \ldots, 3K\}$. The X3C problem is that for arbitrarily given $N(K)$ different pieces with a hidden X3C, find it.

It is clear that $1 \leq N(K) \leq C_{3K}^3$. Intuitively, the X3C problem is often not hard when $N(K) \leq O(K)$, because X3C is not hidden well. An extreme example is that if the number i is contained by only one piece $\{i, j, k\}$, then $\{i, j, k\}$ is certainly from X3C. Picking up $\{i, j, k\}$ and abandoning those pieces containing j or k, then other pieces form a reduced X3C problem on $\{1, 2, \ldots, 3K\} - \{i, j, k\}$. So that $N(K) \geq O(K^2)$ to avoid weak case. On the other hand, the larger $N(K)$ the easier our attack. So that in rest of this paper we will always take $N(K) = O(K^2)$.

Now we describe the WE based on the hardness of X3C problem from GGH structure.

Encryption. The encrypter samples short elements $v^{(1)}, v^{(2)}, \ldots, v^{(3K)} \in R$. He/she computes the encryption key as follows. He/she first computes $v^{(1)}v^{(2)} \ldots v^{(3K)}y^K\, p_{zt} (\text{mod } q)$, then takes $EKEY$ as its high-order bits. In fact, $EKEY$ is high-order bits of $v^{(1)}v^{(2)} \ldots v^{(3K)}(1 + ag)^K hg^{-1} (\text{mod } q)$. He/she can use $EKEY$ and an encryption algorithm to encrypt any plaintext. Then, he/she hides $EKEY$ into pieces as follows. He/she arbitrarily generates $N(K)$ different pieces of $\{1, 2, \ldots, 3K\}$, with a hidden X3C called XC. For each piece $\{i_1, i_2, i_3\}$, he/she computes noisy encoding of the product $v^{(i_1)}v^{(i_2)}v^{(i_3)}$, that is, secretly samples short elements $\{u^{(\{i_1,i_2,i_3\},i)} \in R, i = 1, 2\}$, then computes and publishes $V^{\{i_1,i_2,i_3\}} = v^{(i_1)}v^{(i_2)}v^{(i_3)}y + (u^{(\{i_1,i_2,i_3\},1)}x^{(1)} + u^{(\{i_1,i_2,i_3\},2)}x^{(2)}) (\text{mod } q)$.

Decryption. The one who knows XC computes the zero-test of $\prod_{\{i_1,i_2,i_3\} \in XC} V^{\{i_1,i_2,i_3\}} (\text{mod } q)$, that is, he/she computes $p_{zt} \prod_{\{i_1,i_2,i_3\} \in XC} V^{\{i_1,i_2,i_3\}} (\text{mod } q)$. Then, $EKEY$ is its high-order bits. In other words, $p_{zt} \prod_{\{i_1,i_2,i_3\} \in XC} V^{\{i_1,i_2,i_3\}} (\text{mod } q)$ is the modular sum of two terms, the first term is zero-tested message $v^{(1)}v^{(2)} \ldots v^{(3K)}(1 + ag)^K hg^{-1}$ $(\text{mod } q)$, while the second term is zero-tested noise which doesn't affect high-order bits of $p_{zt} \prod_{\{i_1,i_2,i_3\} \in XC} V^{\{i_1,i_2,i_3\}} (\text{mod } q)$.

3 Weak-DL Attack: Generating Equivalent Secrets

As the start of our attack, we will find equivalent secrets. The method is weak-DL attack [2].

3.1 Generating an Equivalent Secret for One User

We can obtain special elements $\{Y, X^{(i)}, i = 1, 2\}$, where

$$Y = y^{K-1}x^{(1)}p_{zt} (\text{mod } q) = h(1 + ag)^{K-1}b^{(1)},$$
$$X^{(i)} = y^{K-2}x^{(i)}x^{(1)}p_{zt} (\text{mod } q) = h(1 + ag)^{K-2}(b^{(i)}g)b^{(1)},$$
$$i = 1, 2.$$

Notice that the right sides of these equations have no operation "mod q". More precisely, each of $\{Y, X^{(i)}, i = 1, 2\}$ is a factor of a term of zero-tested noise.

For example, $Yu^{(1,1)}\prod_{k=2}^{K+1}(v^{(k)})$ is a term of the zero-tested noise. Therefore, each of $\{Y, X^{(i)}, i=1,2\}$ is far smaller than a term of the zero-tested noise. However, they are not small enough because of the existence of the factor h. We say they are "somewhat small", and take them as our tools.

Take the noisy encoding V (corresponding to the secret v and unknown $\{u^{(1)}, u^{(2)}\}$), and compute special element

$$W = Vy^{K-2}x^{(1)}p_{zt}(\text{mod } q) = vY + (u^{(1)}X^{(1)} + u^{(2)}X^{(2)}).$$

Notice that the right side of this equation has no operation "mod q". Then, compute

$$W(\text{mod } Y) = \left(u^{(1)}X^{(1)}(\text{mod } Y) + u^{(2)}X^{(2)}(\text{mod } Y)\right)(\text{mod } Y).$$

Step 1. By knowing $W(\text{mod } Y)$ and $\{X^{(1)}(\text{mod } Y), X^{(2)}(\text{mod } Y)\}$, we obtain $W' \in \langle X^{(i)}, i=1,2\rangle$ such that $W - W'(\text{mod } Y) = 0$. This is quite easy algebra, and we present the details in Appendix A. Notice that $W - W'$ is not a short vector. Denote $W' = u'^{(1)}X^{(1)} + u'^{(2)}X^{(2)}$.

Step 2. Compute $v^{(0)} = (W - W')/Y$ (division over real numbers with the quotient which is an integer vector). Then,

$$\begin{aligned} v^{(0)} &= v + ((u^{(1)}X^{(1)} + u^{(2)}X^{(2)}) - W')/Y \\ &= v + ((u^{(1)} - u'^{(1)})X^{(1)} + (u^{(2)} - u'^{(2)})X^{(2)})/Y \\ &= v + ((u^{(1)} - u'^{(1)})b^{(1)} + (u^{(2)} - u'^{(2)})b^{(2)})g/(1+ag). \end{aligned}$$

By considering another fact that g and $1+ag$ are co-prime, we have $v^{(0)} - v \in \langle g\rangle$. We call $v^{(0)}$ an equivalent secret of v, and call residual vector $v^{(0)} - v$ the noise. Notice that $v^{(0)}$ is not a short vector.

3.2 Generating an Equivalent Secret for the Product of Secrets

Suppose that each user k has his/her secret $v^{(k)}$ and we generate $v^{(0,k)}$, an equivalent secret of $v^{(k)}$, where $k = 1, \ldots, K+1$. For the product $\prod_{k=1}^{K+1} v^{(k)}$, we have an equivalent secret $\prod_{k=1}^{K+1} v^{(0,k)}$, where the noise is $\prod_{k=1}^{K+1} v^{(0,k)} - \prod_{k=1}^{K+1} v^{(k)} \in \langle g\rangle$. Notice that $\prod_{k=1}^{K+1} v^{(0,k)}$ is not a short vector.

4 Modified Encoding/zero-testing

In this section we transform $\prod_{k=1}^{K+1} v^{(0,k)}$ by our modified Encoding/zero-testing. Denote $\eta = \prod_{k=1}^{K+1} v^{(0,k)}$. The procedure has three steps, which are $\eta' = Y\eta$, $\eta'' = \eta'(\text{mod } X^{(1)})$, and $\eta''' = y(x^{(1)})^{-1}\eta''(\text{mod } q)$ (or $\eta''' = Y(X^{(1)})^{-1}\eta''(\text{mod } q)$). To help understanding their functions, we compare them with GGH processing procedure. The first operation is like a level-K encoding followed by a zero-testing, but there are three differences. Difference 1: The first operation doesn't use modular q. Difference 2: $\eta'(\text{mod } q)$ contains a modular q factor y^{K-1}, while zero-tested message contains a modular q factor y^K. In other words, $\eta'(\text{mod } q)$ lacks a y. Difference 3: $\eta'(\text{mod } q)$ contains a modular q factor $x^{(1)}$, while zero-tested message doesn't contain such modular q factor. In other words, $\eta'(\text{mod } q)$ has a surplus $x^{(1)}$. η'' is also like a level-K encoding followed by a zero-testing, and there are also three differences as above, but the size is reduced to "somewhat small". To obtain η''', we get rid of $x^{(1)}$ and put y in so that η''' is a level-K encoding followed by a zero-testing, and that we can guarantee zero-tested noise "somewhat small". Notice $\eta = \prod_{k=1}^{K+1} v^{(k)} + \xi g$, where $\xi \in R$.

Step 1. Compute $\eta' = Y\eta$. By noticing that Y is a multiple of $b^{(1)}$, we have a fact that $\eta' = Y \prod_{k=1}^{K+1} v^{(k)} + \xi' b^{(1)} g$, where $\xi' \in R$.

Step 2. Compute $\eta'' = \eta'(\text{mod } X^{(1)})$. There are 3 facts as follows.

(1) $\eta'' = Y \prod_{k=1}^{K+1} v^{(k)} + \xi'' b^{(1)} g$, where $\xi'' \in R$. Notice that η'' is the sum of η' and a multiple of $X^{(1)}$, and that $X^{(1)}$ is a multiple of $b^{(1)} g$.
(2) η'' has a similar size to that of $\sqrt{n} X^{(1)}$. In other words, η'' is smaller than one term of zero-tested noise. Notice standard deviations for sampling various variables.
(3) $Y \prod_{k=1}^{K+1} v^{(k)}$ has a similar size to that of one term of zero-tested noise.

The above 3 facts result in a new fact that $\xi'' b^{(1)} g = \eta'' - Y \prod_{k=1}^{K+1} v^{(k)}$ has a similar size to that of one term of zero-tested noise.

Step 3. Compute $\eta''' = y(x^{(1)})^{-1} \eta''(\text{mod } q)$. There are 3 facts as follows.

(1) $\eta''' = (h(1 + ag)^K g^{-1}) \prod_{k=1}^{K+1} v^{(k)} + \xi''(1 + ag)(\text{mod } q)$. Notice fact (1) of Step 2, and notice the definitions of Y and $X^{(1)}$.
(2) $\xi''(1 + ag)$ has a similar size to that of one term of zero-tested noise. In other words, $\xi''(1 + ag)$ is smaller than zero-tested noise. This fact is clear by noticing that $\xi'' b^{(1)} g$ has a similar size to that of one term of zero-tested noise, and by noticing that $1 + ag$ and $b^{(1)} g$ have a similar size.
(3) $(h(1+ag)^K g^{-1}) \prod_{k=1}^{K+1} v^{(k)}(\text{mod } q)$ is zero-tested message, therefore its high-order bits are what we want to obtain.

The above 3 facts result in a new fact that η''' is the modular sum of zero-tested message and a new zero-tested noise which is smaller than original zero-tested noise. Therefore, high-order bits of η''' are what we want to obtain. MKE has been broken. More important is that K-GMDDH assumption (Assumption 5.1 of [11]) is negated.

5 Breaking the Instance of WE Based on the Hardness of Exact-3-cover Problem with Public Tools for Encoding

Our modified Encoding/zero-testing cannot directly break the instance of WE based on the hardness of X3C problem, because the X3C is hidden. In this section we show that special structure of GGH map can simplify the X3C problem into a combined X3C problem, and then show how to use a combined exact cover to break the instance under the condition that low-level encodings of zero are made publicly available.

5.1 Combined Exact-3-cover Problem: Definition and Solution

Definition 2. *Suppose we are given $N(K) = O(K^2)$ different pieces of $\{1, 2, \ldots, 3K\}$. A subset $\{i_1, i_2, i_3\}$ of $\{1, 2, \ldots, 3K\}$ is called a combined piece, if*

(1) $\{i_1, i_2, i_3\}$ is not a piece;
(2) $\{i_1, i_2, i_3\} = \{j_1, j_2, j_3\} \cup \{k_1, k_2, k_3\} - \{l_1, l_2, l_3\}$;
(3) $\{j_1, j_2, j_3\}$, $\{k_1, k_2, k_3\}$ and $\{l_1, l_2, l_3\}$ are pieces;
(4) $\{j_1, j_2, j_3\}$ and $\{k_1, k_2, k_3\}$ don't intersect. (Then $\{j_1, j_2, j_3\} \cup \{k_1, k_2, k_3\} \supset \{l_1, l_2, l_3\}$).

Definition 3. *A subset $\{i_1, i_2, i_3\}$ of $\{1, 2, \ldots, 3K\}$ is called a second-order combined piece, if*

(1) $\{i_1, i_2, i_3\}$ is neither a piece nor a combined piece;
(2) $\{i_1, i_2, i_3\} = \{j_1, j_2, j_3\} \cup \{k_1, k_2, k_3\} - \{l_1, l_2, l_3\}$;
(3) $\{j_1, j_2, j_3\}$, $\{k_1, k_2, k_3\}$ and $\{l_1, l_2, l_3\}$ are pieces or combined pieces.
(4) $\{j_1, j_2, j_3\}$ and $\{k_1, k_2, k_3\}$ don't intersect. (Then $\{j_1, j_2, j_3\} \cup \{k_1, k_2, k_3\} \supset \{l_1, l_2, l_3\}$).

K pieces or combined pieces or second-order combined pieces without intersection are called a combined X3C of $\{1, 2, \ldots, 3K\}$. The combined X3C problem is that for arbitrarily given $N(K) = O(K^2)$ different pieces, find a combined X3C. We will show that the combined X3C problem is not difficult to solve. More specifically, suppose that $O(K^2)$ pieces are sufficiently randomly distributed, in them there is a hidden X3C, and the instance of X3C problem is assumed to be hard. Then we will prove that corresponding instance of combined X3C problem can be solved in polynomial time. Our proving procedure has two steps, which are obtaining combined pieces and obtaining second-order combined pieces.

Obtaining Combined Pieces. We take $P(E)$ as the probability of the event E, and $P(E|E')$ as the conditional probability of E under the condition E'. Arbitrarily take a subset $\{i_1, i_2, i_3\}$ which is not a piece. In Appendix B we show that $P(\{i_1, i_2, i_3\}$ is not a combined piece$) \approx exp\{-(O(K^2))^3/K^6\}$. For the sake of simple deduction, we temporarily assume $O(K^2) > K^2$, then this probability is smaller than e^{-1}. Now we construct all combined pieces from $O(K^2)$ pieces, and we have a result: there are more than $(1 - e^{-1})C_{3K}^3$ different subsets of $\{1, 2, \ldots, 3K\}$, each containing 3 elements, which are pieces or combined pieces.

Obtaining Second-Order Combined Pieces. There are less than $e^{-1}C_{3K}^3$ different subsets of $\{1, 2, \ldots, 3K\}$, each containing 3 elements, which are neither pieces nor combined pieces. Arbitrarily take one subset $\{i_1, i_2, i_3\}$ from them. By a deduction procedure similar to Appendix B, we can show that $P(\{i_1, i_2, i_3\}$ is not a second-order combined piece$)$ is negatively exponential in K. Now we construct all second-order combined pieces from more than $(1-e^{-1})C_{3K}^3$ pieces or combined pieces, and then we are almost sure to have a result: all C_{3K}^3 different subsets of $\{1, 2, \ldots, 3K\}$, each containing 3 elements, are pieces or combined pieces or second-order combined pieces. Therefore, the combined X3C problem is solved.

5.2 Positive/Negative Factors

Definition 4. *Take a fixed combined X3C. Take an element* $\{i_1, i_2, i_3\}$ *of this combined X3C.*

(1) If $\{i_1, i_2, i_3\}$ *is a piece, we count it as a positive factor.*
(2) If $\{i_1, i_2, i_3\}$ *is a combined piece,* $\{i_1, i_2, i_3\} = \{j_1, j_2, j_3\} \cup \{k_1, k_2, k_3\} - \{l_1, l_2, l_3\}$*, we count pieces* $\{j_1, j_2, j_3\}$ *and* $\{k_1, k_2, k_3\}$ *as positive factors, and count the piece* $\{l_1, l_2, l_3\}$ *as a negative factor.*
(3) Suppose $\{i_1, i_2, i_3\}$ *is a second-order combined piece,* $\{i_1, i_2, i_3\} = \{j_1, j_2, j_3\} \cup \{k_1, k_2, k_3\} - \{l_1, l_2, l_3\}$*, where* $\{j_1, j_2, j_3\}$*,* $\{k_1, k_2, k_3\}$ *and* $\{l_1, l_2, l_3\}$ *are pieces or combined pieces.*
  (3.1) If $\{j_1, j_2, j_3\}$ *is a piece, we count it as a positive factor; if* $\{j_1, j_2, j_3\}$ *is a combined piece, we count 2 positive factors corresponding to it as positive factors, and the negative factor corresponding to it as a negative factor.*
  (3.2) Similarly, if $\{k_1, k_2, k_3\}$ *is a piece, we count it as a positive factor; if* $\{k_1, k_2, k_3\}$ *is a combined piece, we count 2 positive factors corresponding to it as positive factors, and the negative factor corresponding to it as a negative factor.*
  (3.3) Oppositely, if $\{l_1, l_2, l_3\}$ *is a piece, we count it as a negative factor; if* $\{l_1, l_2, l_3\}$ *is a combined piece, we count 2 positive factors corresponding to it as negative factors, and the negative factor corresponding to it as a positive factor.*

Positive and negative factors are pieces. All positive factors form a collection, and all negative factors form another collection (notice that we use the terminology "collection" rather than "set", because it is possible that one piece is counted several times). Take CPF as the collection of positive factors, NPF as the number of positive factors. Take CNF as the collection of negative factors, NNF as the number of negative factors. Notice that some pieces may be counted repeatedly. It is easy to see that $NPF - NNF = K$. On the other hand, from C_{3K}^3 different subsets of $\{1, 2, \ldots, 3K\}$, there are $O(K^2)$ different pieces, more than $(1 - e^{-1})C_{3K}^3 - O(K^2)$ different combined pieces, and less than $e^{-1}C_{3K}^3$ different second-order combined pieces. Each piece is a positive factor, each combined piece is attached by 2 positive factors and a negative factor, each second-order combined piece is attached by at most 5 positive factors and 4 negative factors. Therefore, for a randomly chosen combined X3C, it is almost sure that $NPF \leq 3K$, resulting in $NNF \leq 2K$.

5.3 Our Construction

Randomly take a combined X3C. Obtain CPF, the collection of positive factors, and CNF, the collection of negative factors. For a positive factor $pf = \{i_1, i_2, i_3\}$, we denote $v^{(pf)} = v^{(i_1)}v^{(i_2)}v^{(i_3)}$ as the secret of pf, and $v'^{(pf)}$ as the equivalent secret of $v^{(pf)}$ obtained in Subsect. 3.1. Similarly we denote $v^{(nf)}$ and $v'^{(nf)}$ for a negative factor nf. Denote $PPF = \prod_{pf \in CPF} v'^{(pf)}$ as the product of equivalent secrets of all positive factors. Denote $PNF = \prod_{nf \in CNF} v'^{(nf)}$ as the product of equivalent secrets of all negative factors. Denote $PTS = \prod_{k=1}^{3K} v^{(k)}$ as the product of true secrets. The first clear equation is $\prod_{pf \in CPF} v^{(pf)} = PTS \times \prod_{nf \in CNF} v^{(nf)}$. Then, we have

Proposition 1.

(1) $PPF - \prod_{pf \in CPF} v^{(pf)} \in \langle g \rangle$.
(2) $PNF - \prod_{nf \in CNF} v^{(nf)} \in \langle g \rangle$.
(3) $PPF - PNF \times PTS \in \langle g \rangle$.

Proof. By considering Subsect. 3.1, we know that

(1) $PPF = \prod_{pf \in CPF} v^{(pf)} + \beta_{PF}$, where $\beta_{PF} \in \langle g \rangle$.
(2) $PNF = \prod_{nf \in CNF} v^{(nf)} + \beta_{NF}$, where $\beta_{NF} \in \langle g \rangle$.

On the other hand, (3) is true from

$$\prod_{pf \in CPF} v^{(pf)} = PTS \times \prod_{nf \in CNF} v^{(nf)}.$$

Proposition 1 is proven. □

Perhaps there is hope in solving PTS. However, we cannot filter off β_{PF} and β_{NF}, because no "good" description of $\langle g \rangle$ has been made public. Fortunately, we don't need to solve PTS for breaking the instance. We only need to find an equivalent secret of PTS, without caring about the size of the equivalent secret. Then, we can reduce zero-tested noise much smaller by our modified Encoding/zero-testing. Proposition 2 describes the shape of the equivalent secret of PTS under an assumption.

Proposition 2.

(1) If PTS' is an equivalent secret of PTS, then $PPF - PNF \times PTS' \in \langle g \rangle$.
(2) Assume that PNF and g are co-prime. If $PPF - PNF \times PTS' \in \langle g \rangle$, then PTS' is an equivalent secret of PTS.

Proof. (1) is clear by considering (3) of Proposition 1. If $PPF - PNF \times PTS' \in \langle g \rangle$, then $PNF \times (PTS' - PTS) \in \langle g \rangle$. According to our assumption, we have $(PTS' - PTS) \in \langle g \rangle$, hence (2) is proven. □

Now we want to find an equivalent secret of PTS. From viewpoint of multilinear map, this is a division operation: We "divide" PPF by PNF to obtain PTS'. Under our assumption, we only need to find a vector $PTS' \in R$ such that $PPF - PNF \times PTS' \in \langle g \rangle$ without caring about the size of PTS'. To do so we only need to obtain a "bad" description of $\langle g \rangle$. That is, we only need to obtain a public basis of the lattice $\langle g \rangle$; for example, the Hermite normal form. This is not a difficult task, and in Appendix C we will present our method for doing so. After obtaining a public basis G, the condition $PPF - PNF \times PTS' \in \langle g \rangle$ is transformed into an equivalent condition

$$PPF \times G^{-1} - PTS' \times \overline{PNF} \times G^{-1} \in R,$$

where G^{-1} is the inverse matrix of G, and

$$\overline{PNF} = \begin{bmatrix} PNF_0 & PNF_1 & \cdots PNF_{n-1} \\ -PNF_{n-1} & PNF_0 & \cdots PNF_{n-2} \\ \vdots & \vdots & \ddots \quad \vdots \\ -PNF_1 & -PNF_2 & \cdots \quad PNF_0 \end{bmatrix}.$$

Take each entry of $PPF \times G^{-1}$ and $\overline{PNF} \times G^{-1}$ as the form of reduced fraction, and take lcm as the least common multiple of all denominators, and then the condition is transformed into another equivalent condition

$$\begin{aligned}&(lcm \times PPF \times G^{-1}) (\text{mod } lcm)\\ &= PTS' \times (lcm \times \overline{PNF} \times G^{-1}) (\text{mod } lcm).\end{aligned}$$

This is a linear equation modular lcm, and it is easy to obtain a solution PTS'. After that we take our modified Encoding/zero-testing, exactly the same as in Sect. 4. Denote $\eta = PTS'$. Compute $\eta' = Y\eta$. Compute $\eta'' = \eta' (\text{mod } X^{(1)})$. Compute $\eta''' = y(x^{(1)})^{-1}\eta'' (\text{mod } q)$. Then, high-order bits of η''' are what we want to obtain. The instance has been broken.

We can explain that temporary assumption $O(K^2) > K^2$ is not needed for a successful attack. For smaller number of pieces, we can always generate combined pieces, second-order combined pieces, third-order combined pieces, ..., step by step, until we can easily obtain a combined X3C. From this combined X3C, each set is a piece or a combined piece or a second-order combined piece or a third-order combined piece or ..., rather than only a piece or a combined piece or a second-order combined piece. Then, we can obtain all positive and negative factors, which can be defined step by step. In other words, we can sequentially define positive/negative factors attached to a third-order combined piece, to a fourth-order combined piece, ..., and so on. Finally, we can break the instance by using the same procedure. The difference is merely a more complicated description. A question left is whether the assumption "PNF and g are co-prime" is a plausible case. It means that g and each factor of PNF are co-prime. The answer is seemingly yes. A test which we haven't run is that we take two different combined X3Cs, so that we obtain two different values of PNF. If they finally obtain the same high-order bits of η''', we can believe the assumption is true for two values of PNF.

6 Breaking the Instance of WE Based on the Hardness of Exact-3-cover Problem with Hidden Tools for Encoding

6.1 Preparing Work (1): Finding Level-2 Encodings of 0

Take two pieces $\{i_1, i_2, i_3\}$ and $\{j_1, j_2, j_3\}$ which do not intersect. From other pieces, randomly choose two pieces $\{k_1, k_2, k_3\}$ and $\{l_1, l_2, l_3\}$, then the probability that $\{k_1, k_2, k_3\} \cup \{l_1, l_2, l_3\} = \{i_1, i_2, i_3\} \cup \{j_1, j_2, j_3\}$ is about $\frac{1}{C_{3K}^6}$, which is polynomially small. From all of $N(K) = O(K^2)$ pieces, we construct all sets of 4 pieces, and we estimate the average number of such sets of 4 pieces $\{\{i_1, i_2, i_3\}, \{j_1, j_2, j_3\}, \{k_1, k_2, k_3\}, \{l_1, l_2, l_3\}\}$ that $\{i_1, i_2, i_3\}$ and $\{j_1, j_2, j_3\}$ do not intersect, and $\{k_1, k_2, k_3\} \cup \{l_1, l_2, l_3\} = \{i_1, i_2, i_3\} \cup \{j_1, j_2, j_3\}$. This number is of the order of magnitude $\frac{C_{O(K^2)}^4}{C_{3K}^6}$, meaning that we have "many" such sets. At least finding one such set is noticeable. Take one of such sets $\{\{i_1, i_2, i_3\}, \{j_1, j_2, j_3\}, \{k_1, k_2, k_3\}, \{l_1, l_2, l_3\}\}$ and corresponding encodings $\{V^{\{i_1,i_2,i_3\}}, V^{\{j_1,j_2,j_3\}}, V^{\{k_1,k_2,k_3\}}, V^{\{l_1,l_2,l_3\}}\}$, then

$$\left(V^{\{i_1,i_2,i_3\}} V^{\{j_1,j_2,j_3\}} - V^{\{k_1,k_2,k_3\}} V^{\{l_1,l_2,l_3\}}\right) (\text{mod } q) = ugz^{-2} (\text{mod } q),$$

where u is very small. We call it a level-2 encoding of 0. According to the statement above, we have "many" level-2 encodings of 0. Here we fix and remember one such encoding of 0, and call it V^*. Correspondingly, we fix and remember u.

6.2 Preparing Work (2): Supplement and Division

Take a combined X3C. Obtain CPF and CNF, collections of positive and negative factors. Suppose $NPF \leq 2K-2$ (therefore $NNF = NPF-K \leq K-2$. It is easy to see that this case is noticeable). Take a piece $\{i_1, i_2, i_3\}$ and supplement it $2K - NPF$ times into CPF, so that we have new $NPF = 2K$. Similarly, supplement such a piece $\{i_1, i_2, i_3\}$ $K - NNF = 2K - NPF$ times into CNF, so that we have a new $NNF = K$. We fix and remember the piece $\{i_1, i_2, i_3\}$.

Then, we divide the collection CPF into two subcollections, $CPF(1)$ and $CPF(2)$, where

(1) $\|CPF(1)\| = \|CPF(2)\| = K$. That is, $CPF(1)$ and $CPF(2)$ are of equal size.
(2) $CPF(2)$ contains $\{i_1, i_2, i_3\}$ at least twice.
(3) $CPF(1)$ contains two pieces $\{j_1, j_2, j_3\}$ and $\{k_1, k_2, k_3\}$ which do not intersect. We fix and remember these two pieces $\{j_1, j_2, j_3\}$ and $\{k_1, k_2, k_3\}$.

The purpose of such supplementation and division is the convenience for level-K zero-testing.

6.3 Preparing Work (3): Constructing the Equation

We have fixed and remembered five elements: V^* (a level-2 encoding of 0), u ($V^* = ugz^{-2}(\text{mod } q)$), $\{i_1, i_2, i_3\}$ (a piece contained by $CPF(2)$ at least twice), $\{j_1, j_2, j_3\}$ and $\{k_1, k_2, k_3\}$ (they are from $CPF(1)$, and do not intersect each other). Now we denote four elements as follows.

$$Dec(P(1)) = p_{zt}V^* \prod_{pf \in CPF(1) - \{\{j_1,j_2,j_3\},\{k_1,k_2,k_3\}\}} V^{(pf)}(\text{mod } q),$$

$$Dec(P(2)) = p_{zt}V^* \prod_{pf \in CPF(2) - \{\{i_1,i_2,i_3\},\{i_1,i_2,i_3\}\}} V^{(pf)}(\text{mod } q),$$

$$Dec(N) = p_{zt}V^* \prod_{nf \in CNF - \{\{i_1,i_2,i_3\},\{i_1,i_2,i_3\}\}} V^{(nf)}(\text{mod } q),$$

$$Dec(Original) = hV^*g^{-1}z^2 \prod_{k \in \{1,\ldots,3K\} - \{j_1,j_2,j_3,k_1,k_2,k_3\}} v^{(k)}(\text{mod } q).$$

We can rewrite $Dec(P(1))$, $Dec(P(2))$, $Dec(N)$, $Dec(Original)$, as follows.

$$Dec(P(1)) = hu \prod_{pf \in CPF(1) - \{\{j_1,j_2,j_3\},\{k_1,k_2,k_3\}\}} (v^{(pf)}(1 + ag) + u^{(pf,1)}b^{(1)}g + u^{(pf,2)}b^{(2)}g),$$

$$Dec(P(2)) = hu \prod_{pf \in CPF(2) - \{\{i_1,i_2,i_3\},\{i_1,i_2,i_3\}\}} (v^{(pf)}(1 + ag) + u^{(pf,1)}b^{(1)}g + u^{(pf,2)}b^{(2)}g),$$

$$Dec(N) = hu \prod_{nf \in CNF - \{\{i_1,i_2,i_3\},\{i_1,i_2,i_3\}\}} (v^{(nf)}(1 + ag) + u^{(nf,1)}b^{(1)}g + u^{(nf,2)}b^{(2)}g),$$

$$Dec(Original) = hu \prod_{k \in \{1,\ldots,3K\} - \{j_1,j_2,j_3,k_1,k_2,k_3\}} v^{(k)}.$$

Notice that $\{a, b^{(1)}, b^{(2)}\}$ has been fixed and remembered in Subsect. 2.2. Four facts about $\{Dec(P(1)), Dec(P(2)), Dec(N), Dec(Original)\}$ are as follows.

(1) They are all somewhat small.
(2) $Dec(P(1))$, $Dec(P(2))$, $Dec(N)$ can be obtained, while $Dec(Original)$ cannot.
(3) We have the equation

$$Dec(P(1)) \times Dec(P(2)) - Dec(N) \times Dec(Original) \in \langle (hu)^2 g \rangle \subset \langle hu^2 g \rangle.$$

This equation is clear by considering the encoding procedure and definitions of $\{Dec(P(1)), Dec(P(2)), Dec(N), Dec(Original)\}$.
(4) Conversely, suppose there is $D' \in R$ such that

$$Dec(P(1)) \times Dec(P(2)) - Dec(N) \times D' \in \langle hu^2 g \rangle.$$

Then, D' is the sum of $Dec(Original)$ and an element of $\langle ug \rangle$. Here we use a small assumption that $\frac{Dec(N)}{u}$ and (ug) are co-prime, which is noticeable. In other words, D' is a solution of the equation

$$Dec(P(1)) \times Dec(P(2)) \equiv Dec(N) \times D'(\text{mod } \langle hu^2 g \rangle),$$

if and only if D' is the sum of *Dec*(*Original*) and an element of $\langle ug \rangle$. Here "mod $\langle hu^2 g \rangle$" is general lattice modular operation by using a basis of the lattice $\langle hu^2 g \rangle$. We call D' "an equivalent secret" of *Dec*(*Original*). Notice that such new type of "equivalent secret" and original secret are congruent modular $\langle ug \rangle$ rather than modular $\langle g \rangle$.

6.4 Solving the Equation: Finding "An Equivalent Secret"

We want to obtain "an equivalent secret" of *Dec*(*Original*) without caring about the size. To do so we only need to obtain a basis of the lattice $\langle hu^2 g \rangle$ (the "bad" basis). If we can obtain many elements of $\langle hu^2 g \rangle$ which are somewhat small, obtaining a basis of $\langle hu^2 g \rangle$ is not hard work. Arbitrarily take $K-4$ pieces $\{piece(1), piece(2), \ldots, piece(K-4)\}$ without caring whether they are repeated. Then,

$$p_{zt}(V^*)^2 \prod_{k=1}^{K-4} V^{(piece(k))} (\text{mod } q) =$$

$$hu^2 g \prod_{k=1}^{K-4} (v^{(piece(k))}(1+ag) + u^{(piece(k),1)} b^{(1)} g + u^{(piece(k),2)} b^{(2)} g) \in \langle hu^2 g \rangle.$$

Thus, we can generate enough elements of $\langle hu^2 g \rangle$ which are somewhat small. This fact implies that finding a D' may be easy.

6.5 Reducing the Zero-Tested Noise Much Smaller

Suppose we have obtained D', "an equivalent secret" of *Dec*(*Original*). D' is the sum of *Dec*(*Original*) and an element of $\langle ug \rangle$, and D' is not a short vector. Arbitrarily take an element of $\langle hu^2 g \rangle$ which is somewhat small, and call it V^{**}. Compute $V^{***} = D' (\text{mod } V^{**})$. Two facts about V^{***} are as follows.

(1) $V^{***} = Dec(Original) + V^{****}$, where $V^{****} \in \langle ug \rangle$.
(2) Both V^{***} and *Dec*(*Original*) are somewhat small, so that V^{****} is somewhat small.

Then, compute

$$\begin{aligned} V^{\#} &= V^{***} V^{(j_1,j_2,j_3)} V^{(k_1,k_2,k_3)} (V^*)^{-1} (\text{mod } q) \\ &= \Big[\Big(Dec(Original) \times V^{(j_1,j_2,j_3)} V^{(k_1,k_2,k_3)} (V^*)^{-1} \Big) \\ &\quad + \Big(V^{****} \times V^{(j_1,j_2,j_3)} V^{(k_1,k_2,k_3)} (V^*)^{-1} \Big) \Big] (\text{mod } q). \end{aligned}$$

Two facts about $V^{\#}$ are as follows.

(1)

$$\left(Dec\,(Original) \times V^{(j_1,j_2,j_3)}V^{(k_1,k_2,k_3)}(V^*)^{-1}\right)(\text{mod } q)$$
$$= hg^{-1}V^{(j_1,j_2,j_3)}V^{(k_1,k_2,k_3)}z^2 \prod_{k\in\{1,\ldots,3K\}-\{j_1,j_2,j_3,k_1,k_2,k_3\}} v^{(k)}(\text{mod } q)$$
$$= hg^{-1}(v^{(j_1,j_2,j_3)}(1+ag)+u^{((j_1,j_2,j_3),1)}b^{(1)}g+u^{((j_1,j_2,j_3),2)}b^{(2)}g)$$
$$(v^{(k_1,k_2,k_3)}(1+ag)+u^{((k_1,k_2,k_3),1)}b^{(1)}g+u^{((k_1,k_2,k_3),2)}b^{(2)}g)$$
$$\prod_{k\in\{1,\ldots,3K\}-\{j_1,j_2,j_3,k_1,k_2,k_3\}} v^{(k)} \quad (\text{mod } q)$$

Therefore, its high-order bits are the secret key.

(2)

$$\left(V^{****} \times V^{(j_1,j_2,j_3)}V^{(k_1,k_2,k_3)}(V^*)^{-1}\right)(\text{mod } q)$$
$$= V^{****}(ug)^{-1}(v^{(j_1,j_2,j_3)}(1+ag)+u^{((j_1,j_2,j_3),1)}b^{(1)}g+u^{((j_1,j_2,j_3),2)}b^{(2)}g)$$
$$(v^{(k_1,k_2,k_3)}(1+ag)+u^{((k_1,k_2,k_3),1)}b^{(1)}g+u^{((k_1,k_2,k_3),2)}b^{(2)}g) \quad (\text{mod } q).$$

It is somewhat small because V^{****} is somewhat small, V^{****} is a multiple of (ug), and (ug) and

$$(v^{(j_1,j_2,j_3)}(1+ag)+u^{((j_1,j_2,j_3),1)}b^{(1)}g+u^{((j_1,j_2,j_3),2)}b^{(2)}g)\times$$
$$(v^{(k_1,k_2,k_3)}(1+ag)+u^{((k_1,k_2,k_3),1)}b^{(1)}g+u^{((k_1,k_2,k_3),2)}b^{(2)}g)$$

have same size.

These two facts mean that high-order bits of $V^{\#}$ are the secret key. The instance has been broken.

6.6 A Note

We have assumed that original $NPF \leq 2K-2$, and have supplemented pieces to make a new $NPF = 2K$. In fact, we can assume that original $NPF \leq 3K-2$, and supplement pieces to make a new $NPF = 3K$. In this case, we can still break the instance, but our attack will be a little bit more complicated.

7 Cryptanalysis of Two Simple Revisions of GGH Map

7.1 The First Simple Revision of GGH Map and Corresponding MKE

The first simple revision of GGH map is described as follows. All parameters of GGH map are reserved, except that we change encoding parameter y into encoding parameters $\{y^{(i)}, i = 1,2\}$, and accordingly we change Level-K zero-testing parameter p_{zt} into Level-K zero-testing parameters $\{p_{zt}^{(i)}, i = 1,2\}$.

Our encoding parameters are $\{y^{(i)}, i = 1, 2\}$, where $y^{(i)} = (y^{(0,i)} + a^{(i)}g)z^{-1} \pmod q$, $\{y^{(0,i)}, a^{(i)}, i = 1, 2\}$ are very small and are kept secret. We can see that $\{y^{(i)}, i = 1, 2\}$ are encodings of secret elements $\{y^{(0,i)}, i = 1, 2\}$, rather than encodings of 1. Accordingly, our level-K zero-testing parameters are $\{p_{zt}^{(i)}, i = 1, 2\}$, where $p_{zt}^{(i)} = hy^{(0,i)}z^K g^{-1} \pmod q$.

Suppose a user has a secret $(v^{(1)}, v^{(2)}) \in R^2$, where $v^{(1)}$ and $v^{(2)}$ are short elements. He/she secretly samples short elements $\{u^{(i)} \in R, i = 1, 2\}$. He/she computes noisy encoding $V = (v^{(1)}y^{(1)} + v^{(2)}y^{(2)}) + (u^{(1)}x^{(1)} + u^{(2)}x^{(2)}) \pmod q$. He/she publishes V. Then, the first revision of GGH map includes K, $\{y^{(i)}, i = 1, 2\}$, $\{x^{(i)}, i = 1, 2\}$, $\{p_{zt}^{(i)}, i = 1, 2\}$, and all noisy encoding V for all users. To guarantee our attack work, we assume that 2^K is polynomially large.

Suppose that $K + 1$ users want to generate KEY, a commonly shared key by public discussion. To do so, each user k generates his/her secret $(v^{(k,1)}, v^{(k,2)})$, and publishes the noisy encoding $V^{(k)}$, $k = 1, \ldots, K + 1$. Then, each user can use his/her secret and other users' noisy encodings to compute KEY, the commonly shared key. For example, user k_0 first computes $(v^{(k_0,1)}p_{zt}^{(1)} + v^{(k_0,2)}p_{zt}^{(2)}) \prod_{k \neq k_0} V^{(k)} \pmod q$, then takes KEY as its high-order bits. It is easy to see that

$$(v^{(k_0,1)}p_{zt}^{(1)} + v^{(k_0,2)}p_{zt}^{(2)}) \prod_{k \neq k_0} V^{(k)} \pmod q = (A + B^{(k_0)}) \pmod q,$$

such that

$$A = hg^{-1} \sum_{(j_1,\ldots,j_{K+1}) \in \{1,2\}^{K+1}} v^{(K+1,j_{K+1})} y^{(0,j_{K+1})} \prod_{k=1}^{K} v^{(k,j_k)}(y^{(0,j_k)} + a^{(j_k)}g) \pmod q,$$

which has no relation with user k_0; $B^{(k_0)}$ is the sum of several terms which are somewhat small. If related parameters are small enough, KEY is high-order bits of $A \pmod q$.

7.2 Generating "Equivalent Secret"

For the secret $(v^{(1)}, v^{(2)}) \in R^2$, we construct an "equivalent secret $(v'^{(1)}, v'^{(2)}) \in R^2$", such that

$$(v^{(1)}(y^{(0,1)}+a^{(1)}g)+v^{(2)}(y^{(0,2)}+a^{(2)}g))-(v'^{(1)}(y^{(0,1)}+a^{(1)}g)+v'^{(2)}(y^{(0,2)}+a^{(2)}g))$$

is a multiple of g. An equivalent requirement is that $(v^{(1)}y^{(0,1)} + v^{(2)}y^{(0,2)}) - (v'^{(1)}y^{(0,1)} + v'^{(2)}y^{(0,2)})$ is a multiple of g. That is enough, and we do not need $(v'^{(1)}, v'^{(2)})$ small. Take V, the noisy encoding of $(v^{(1)}, v^{(2)})$, we compute special element

$$\begin{aligned}W^* = V(y^{(1)})^{K-2}x^{(1)}p_{zt}^{(1)} \pmod q = hy^{(0,1)}\big[&v^{(1)}(y^{(0,1)} + a^{(1)}g)^{K-1}b^{(1)} \\ &+ v^{(2)}(y^{(0,2)} + a^{(2)}g)(y^{(0,1)} + a^{(1)}g)^{K-2}b^{(1)} \\ &+ u^{(1)}(b^{(1)}g)(y^{(0,1)} + a^{(1)}g)^{K-2}b^{(1)} \\ &+ u^{(2)}(b^{(2)}g)(y^{(0,1)} + a^{(1)}g)^{K-2}b^{(1)}\big].\end{aligned}$$

Notice that

(1) Right side of this equation has no operation "mod q", therefore W^* is somewhat small.
(2) Four vectors $hy^{(0,1)}(y^{(0,1)}+a^{(1)}g)^{K-1}b^{(1)}$, $hy^{(0,1)}(y^{(0,2)}+a^{(2)}g)(y^{(0,1)}+a^{(1)}g)^{K-2}b^{(1)}$, $hy^{(0,1)}(b^{(1)}g)(y^{(0,1)}+a^{(1)}g)^{K-2}b^{(1)}$ and $hy^{(0,1)}(b^{(2)}g)(y^{(0,1)}+a^{(1)}g)^{K-2}b^{(1)}$ can be obtained.

Now we start to find $(v'^{(1)}, v'^{(2)})$. First, compute $W^*(\text{mod } hy^{(0,1)}(y^{(0,1)}+a^{(1)}g)^{K-1}b^{(1)})$. Second, compute $\{v'^{(2)}, u'^{(1)}, u'^{(2)}\}$ such that

$$\begin{aligned}
&W^*(\text{mod } h\,y^{(0,1)}(y^{(0,1)}+a^{(1)}g)^{K-1}b^{(1)}) = \\
&\quad h\,y^{(0,1)}\big[v'^{(2)}(y^{(0,2)}+a^{(2)}g)(y^{(0,1)}+a^{(1)}g)^{K-2}b^{(1)}+ \\
&\quad u'^{(1)}(b^{(1)}g)(y^{(0,1)}+a^{(1)}g)^{K-2}b^{(1)}+ \\
&\quad u'^{(2)}(b^{(2)}g)(y^{(0,1)}+a^{(1)}g)^{K-2}b^{(1)}\big](\text{mod } hy^{(0,1)}(y^{(0,1)}+a^{(1)}g)^{K-1}b^{(1)}).
\end{aligned}$$

Solving this modular equation is quite easy algebra, as shown in Appendix A. Solutions are not unique, therefore $\{v'^{(2)}, u'^{(1)}, u'^{(2)}\} \neq \{v^{(2)}, u^{(1)}, u^{(2)}\}$. Third, compute $v'^{(1)}$ such that

$$\begin{aligned}
W^* = hy^{(0,1)}\big[&v'^{(1)}(y^{(0,1)}+a^{(1)}g)^{K-1}b^{(1)} \\
&+v'^{(2)}(y^{(0,2)}+a^{(2)}g)(y^{(0,1)}+a^{(1)}g)^{K-2}b^{(1)} \\
&+u'^{(1)}(b^{(1)}g)(y^{(0,1)}+a^{(1)}g)^{K-2}b^{(1)} \\
&+u'^{(2)}(b^{(2)}g)(y^{(0,1)}+a^{(1)}g)^{K-2}b^{(1)}\big],
\end{aligned}$$

which is another version of easy algebra. Finally, we obtain $(v'^{(1)}, v'^{(2)})$, and can easily check that $(v^{(1)}(y^{(0,1)}+a^{(1)}g)+v^{(2)}(y^{(0,2)}+a^{(2)}g))-(v'^{(1)}(y^{(0,1)}+a^{(1)}g)+v'^{(2)}(y^{(0,2)}+a^{(2)}g))$ is a multiple of g, although $v'^{(1)}$ and $v'^{(2)}$ are not short vectors.

7.3 Generalization of Modified Encoding/zero-testing: Our Attack on MKE

Suppose $K+1$ users hide $(v^{(k,1)}, v^{(k,2)})$ and publish $V^{(k)}, k=1,\ldots,K+1$, and for each user k we have obtained an equivalent secret $(v'^{(k,1)}, v'^{(k,2)})$. For each "$K+1$-dimensional boolean vector" $(j_1,\ldots,j_{K+1}) \in \{1,2\}^{K+1}$, we denote two products

$$v^{(j_1,\ldots,j_{K+1})} = \prod_{k=1}^{K+1} v^{(k,j_k)},$$

$$v'^{(j_1,\ldots,j_{K+1})} = \prod_{k=1}^{K+1} v'^{(k,j_k)}.$$

$v^{(j_1,\ldots,j_{K+1})}$ is clearly smaller than "somewhat small", because it does not include h. $v'^{(j_1,\ldots,j_{K+1})}$ is not a short vector. $v^{(j_1,\ldots,j_{K+1})}$ cannot be obtained, while $v'^{(j_1,\ldots,j_{K+1})}$ can. Suppose former K entries $\{j_1,\ldots,j_K\}$ include N_1 1s and N_2 2s, $N_1+N_2=K$. We denote the supporter $s^{(j_1,\ldots,j_{K+1})}$ as follows.

$$s^{(j_1,\ldots,j_{K+1})}=hy^{(0,j_{K+1})}(y^{(0,1)}+a^{(1)}g)^{N_1-1}(y^{(0,2)}+a^{(2)}g)^{N_2}b^{(1)} \quad \text{for } N_1\geq N_2,$$

$$s^{(j_1,\ldots,j_{K+1})}=hy^{(0,j_{K+1})}(y^{(0,1)}+a^{(1)}g)^{N_1}(y^{(0,2)}+a^{(2)}g)^{N_2-1}b^{(1)} \quad \text{for } N_1< N_2.$$

$s^{(j_1,\ldots,j_{K+1})}$ can be obtained. If $N_1\geq N_2$, $s^{(j_1,\ldots,j_{K+1})}=p_{zt}^{(j_{K+1})}(y^{(1)})^{N_1-1}(y^{(2)})^{N_2}x^{(1)}(\text{mod } q)$, and if $N_1<N_2$, $s^{(j_1,\ldots,j_{K+1})}=p_{zt}^{(j_{K+1})}(y^{(1)})^{N_1}(y^{(2)})^{N_2-1}x^{(1)}(\text{mod } q)$. $s^{(j_1,\ldots,j_{K+1})}$ is somewhat small. Then, we denote

$$V^{(N_1\geq N_2)}=\sum_{j_{K+1}=1}^{2}\sum_{N_1\geq N_2}v^{(j_1,\ldots,j_{K+1})}s^{(j_1,\ldots,j_{K+1})},$$

$$V^{(N_1<N_2)}=\sum_{j_{K+1}=1}^{2}\sum_{N_1<N_2}v^{(j_1,\ldots,j_{K+1})}s^{(j_1,\ldots,j_{K+1})},$$

$$V'^{(N_1\geq N_2)}=\sum_{j_{K+1}=1}^{2}\sum_{N_1\geq N_2}v'^{(j_1,\ldots,j_{K+1})}s^{(j_1,\ldots,j_{K+1})},$$

$$V'^{(N_1<N_2)}=\sum_{j_{K+1}=1}^{2}\sum_{N_1<N_2}v'^{(j_1,\ldots,j_{K+1})}s^{(j_1,\ldots,j_{K+1})}.$$

$V^{(N_1\geq N_2)}$ and $V^{(N_1<N_2)}$ are somewhat small, while $V'^{(N_1\geq N_2)}$ and $V'^{(N_1<N_2)}$ are not short vectors. $V^{(N_1\geq N_2)}$ and $V^{(N_1<N_2)}$ cannot be obtained, while $V'^{(N_1\geq N_2)}$ and $V'^{(N_1<N_2)}$ can be obtained, because $v'^{(j_1,\ldots,j_{K+1})}s^{(j_1,\ldots,j_{K+1})}$ can be obtained for each $(j_1,\ldots,j_{K+1})\in\{1,2\}^{K+1}$, and 2^K is polynomially large. Another fact is that ξ^* is a multiple of $b^{(1)}g$, where

$$\xi^*=(y^{(0,1)}+a^{(1)}g)(V'^{(N_1\geq N_2)}-V^{(N_1\geq N_2)})+(y^{(0,2)}+a^{(2)}g)(V'^{(N_1<N_2)}-V^{(N_1<N_2)}).$$

There are two reasons: (1) By considering the definitions of equivalent secrets, we know that ξ^* is a multiple of g. (2) By considering the definition of $s^{(j_1,\ldots,j_{K+1})}$, we know that ξ^* is a multiple of $b^{(1)}$. Here we use a small assumption that $b^{(1)}$ and g are co-prime. Notice that ξ^* is not a short vector, and that ξ^* cannot be obtained. Then, we compute a tool for the modular operations,

$$M=hy^{(0,1)}(b^{(1)})^K g^{K-1}=p_{zt}^{(1)}(x^{(1)})^K(\text{mod } q).$$

For the same reason, M is somewhat small. Then, we compute the modular operations

$$V''^{(N_1\geq N_2)}=V'^{(N_1\geq N_2)}(\text{mod } M),$$
$$V''^{(N_1<N_2)}=V'^{(N_1<N_2)}(\text{mod } M).$$

Both $V''^{(N_1 \geq N_2)}$ and $V''^{(N_1<N_2)}$ are somewhat small. Therefore, both $V''^{(N_1 \geq N_2)} - V^{(N_1 \geq N_2)}$ and $V''^{(N_1<N_2)} - V^{(N_1<N_2)}$ are somewhat small. Therefore, both $(y^{(0,1)} + a^{(1)}g)(V''^{(N_1 \geq N_2)} - V^{(N_1 \geq N_2)})$ and $(y^{(0,2)} + a^{(2)}g)(V''^{(N_1<N_2)} - V^{(N_1<N_2)})$ are somewhat small. Therefore,

$$\xi^{**} = (y^{(0,1)} + a^{(1)}g)(V''^{(N_1 \geq N_2)} - V^{(N_1 \geq N_2)}) + (y^{(0,2)} + a^{(2)}g)(V''^{(N_1<N_2)} - V^{(N_1<N_2)})$$

is somewhat small. On the other hand, ξ^{**} is a multiple of $b^{(1)}g$, because ξ^* is a multiple of $b^{(1)}g$. Therefore, $\xi^{**}/(b^{(1)}g)$ is somewhat small. Finally,

$$\begin{aligned}\frac{\xi^{**}}{(b^{(1)}g)} &= \xi^{**}(b^{(1)}g)^{-1} (\text{mod } q) \\ &= \left[\left((y^{(0,1)} + a^{(1)}g)V''^{(N_1 \geq N_2)} + (y^{(0,2)} + a^{(2)}g)V''^{(N_1<N_2)}\right)(b^{(1)}g)^{-1} - A\right] (\text{mod } q),\end{aligned}$$

which means that KEY is high-order bits of

$$\left[\left((y^{(0,1)} + a^{(1)}g)V''^{(N_1 \geq N_2)} + (y^{(0,2)} + a^{(2)}g)V''^{(N_1<N_2)}\right)(b^{(1)}g)^{-1}\right] (\text{mod } q),$$

which can be obtained, because $(y^{(0,1)} + a^{(1)}g)(b^{(1)}g)^{-1}(\text{mod } q)$ and $(y^{(0,2)} + a^{(2)}g)(b^{(1)}g)^{-1}(\text{mod } q)$ can be obtained.

7.4 The Second Simple Revision of GGH Map and Its Cryptanalysis

The second simple revision of GGH map is described as follows. All parameters of the first simple revision are reserved, except that we change K-order zero-testing parameters $\{p_{zt}^{(i)} = hy^{(0,i)} z^K g^{-1} (\text{mod } q), i = 1, 2\}$ into $\{p_{zt}^{(i)} = (y^{(0,i)} + h^{(i)}g) z^K g^{-1} (\text{mod } q), i = 1, 2\}$, where both $h^{(1)}$ and $h^{(2)}$ are somewhat small sampled with standard deviation $\sqrt{q}$. MKE is just the same procedure as the first simple revision, except for the different $\{p_{zt}^{(i)}, i = 1, 2\}$. Such a structure can be taken as a simplified version of Gu map-1 [25]. Our cryptanalysis obtains the same result: MKE can be broken under the assumption that 2^K is polynomially large. The deduction procedure is almost same, and we present it in Appendix D.

8 Some Considerations and Remaining Questions

There are many different variants of the GGH construction that one can consider, below we briefly discuss one of them. The variant which seems to defeat our attacks is using non-commutative operations (e.g., using matrices). However this greatly reduces the usability of this construction, for example the WE construction based on X3C requires commutativity. Other variants are under our study.

Trying to find extensions of these attacks and their limitations remains an interesting research direction. For example, we do not know whether the two simple revisions that we analyzed above can be used to construct a secure WE scheme based on X3C. It will also be very interesting to find a way to use our attacks against GGH-based obfuscation schemes.

Acknowledgments. We thank all the reviewers and editors for their valuable comments and works. We also appreciate Martin R. Albrecht for his implementation of our attack [27]. We are very grateful for help and suggestions from the authors of GGH map [2] and authors of the instance of WE based on the hardness of X3C problem [3]. We are very grateful to professor Dong Pyo Chi from UNIST, Korea, for pointing out mistakes in our work. This work was supported in part by the Natural Science Foundation of China under Grant 61173151 and 61472309.

References

1. Boneh, D., Silverberg, A.: Applications of multilinear forms to cryptography. Contemp. Math. **324**, 71–90 (2003)
2. Garg, S., Gentry, C., Halevi, S.: Candidate multilinear maps from ideal lattices. In: Johansson, T., Nguyen, P.Q. (eds.) EUROCRYPT 2013. LNCS, vol. 7881, pp. 181–184. Springer, Heidelberg (2013)
3. Garg, S., Gentry, C., Sahai, A., Waters, B.: Witness encryption and its applications. In: STOC (2013)
4. Gentry, C., Lewko, A., Waters, B.: Witness encryption from instance independent assumptions. In: Garay, J.A., Gennaro, R. (eds.) CRYPTO 2014, Part I. LNCS, vol. 8616, pp. 426–443. Springer, Heidelberg (2014)
5. Arita, S., Handa, S.: Two applications of multilinear maps: group key exchange and witness encryption. In: Proceedings of the 2nd ACM Workshop on ASIA Public-key Cryptography (ASIAPKC 2014), pp. 13–22. ACM, New York (2014)
6. Bellare, M., Hoang, V.T.: Adaptive witness encryption and asymmetric password-based cryptography. Cryptology ePrint Archive, Report 2013/704 (2013)
7. Goldwasser, S., Kalai, Y.T., Popa, R.A., Vaikuntanathan, V., Zeldovich, N.: How to run turing machines on encrypted data. In: Canetti, R., Garay, J.A. (eds.) CRYPTO 2013, Part II. LNCS, vol. 8043, pp. 536–553. Springer, Heidelberg (2013)
8. Garg, S., Gentry, C., Halevi, S., Raykova, M., Sahai, A., Waters, B.: Candidate indistinguishability obfuscation and functional encryption for all circuits. In: FOCS (2013)
9. Garg, S., Gentry, C., Halevi, S., Wichs, D.: On the implausibility of differing-inputs obfuscation and extractable witness encryption with auxiliary input. In: Garay, J.A., Gennaro, R. (eds.) CRYPTO 2014, Part I. LNCS, vol. 8616, pp. 518–535. Springer, Heidelberg (2014)
10. Boyle, E., Chung, K.-M., Pass, R.: On extractability obfuscation. In: Lindell, Y. (ed.) TCC 2014. LNCS, vol. 8349, pp. 52–73. Springer, Heidelberg (2014)
11. Garg, S., Gentry, C., Halevi, S., Sahai, A., Waters, B.: Attribute-based encryption for circuits from multilinear maps. In: Canetti, R., Garay, J.A. (eds.) CRYPTO 2013, Part II. LNCS, vol. 8043, pp. 479–499. Springer, Heidelberg (2013)
12. Langlois, A., Stehlé, D., Steinfeld, R.: GGHLite: more efficient multilinear maps from ideal lattices. In: Nguyen, P.Q., Oswald, E. (eds.) EUROCRYPT 2014. LNCS, vol. 8441, pp. 239–256. Springer, Heidelberg (2014)
13. Coron, J.-S., Lepoint, T., Tibouchi, M.: Practical multilinear maps over the integers. In: Canetti, R., Garay, J.A. (eds.) CRYPTO 2013, Part I. LNCS, vol. 8042, pp. 476–493. Springer, Heidelberg (2013)
14. Cheon, J.H., Han, K., Lee, C., Ryu, H., Stehlé, D.: Cryptanalysis of the multilinear map over the integers. In: Oswald, E., Fischlin, M. (eds.) EUROCRYPT 2015. LNCS, vol. 9056, pp. 3–12. Springer, Heidelberg (2015)

15. Gentry, C., Halevi, S., Maji, H.K., Sahai, A.: Zeroizing without zeroes: cryptanalyzing multilinear maps without encodings of zero. Cryptology ePrint Archive, Report 2014/929 (2014)
16. Boneh, D., Wu, D.J., Zimmerman, J.: Immunizing multilinear maps against zeroizing attacks. Cryptology ePrint Archive, Report 2014/930 (2014)
17. Coron, J.-S., Lepoint, T., Tibouchi, M.: Cryptanalysis of two candidate fixes of multilinear maps over the integers. Cryptology ePrint Archive, Report 2014/975 (2014)
18. Gentry, C., Gorbunov, S., Halevi, S.: Graph-induced multilinear maps from lattices. In: Dodis, Y., Nielsen, J.B. (eds.) TCC 2015, Part II. LNCS, vol. 9015, pp. 498–527. Springer, Heidelberg (2015)
19. Coron, J.-S., et al.: Zeroizing without low-level zeroes: new MMAP attacks and their limitations. In: Gennaro, R., Robshaw, M. (eds.) CRYPTO 2015, Part I. LNCS, vol. 9215, pp. 247–266. Springer, Heidelberg (2015)
20. Coron, J.-S., Lepoint, T., Tibouchi, M.: New multilinear maps over the integers. In: Gennaro, R., Robshaw, M. (eds.) CRYPTO 2015, Part I. LNCS, vol. 9215, pp. 267–286. Springer, Heidelberg (2015)
21. Cheon, J.H., Han, K., Lee, C., Ryu, H.: Cryptanalysis of the new CLT multilinear maps. Cryptology ePrint Archive, Report 2015/934 (2015)
22. Minaud, B., Fouque, P.-A.: Cryptanalysis of the new multilinear map over the integers. Cryptology ePrint Archive, Report 2015/941 (2015)
23. Babai, L.: On Lovász' lattice reduction and the nearest lattice point problem. Combinatorica **6**(1), 1–13 (1986)
24. Goldreich, O.: Computational Complexity: a Conceptual Perspective, 1st edn. Cambridge University Press, New York (2008)
25. Gu, C.: Multilinear maps using ideal lattices without encodings of zero. Cryptology ePrint Archive, Report 2015/023 (2015)
26. Nguyen, P.Q., Regev, O.: Learning a parallel piped: cryptanalysis of GGH and NTRU signatures. J. Crypt. **22**(2), 139–160 (2009)
27. Albrecht, M.R.: Sage code for GGH cryptanalysis by Hu and Jia. https://martinralbrecht.wordpress.com/2015/04/13/sage-code-for-ggh-cryptanalysis-by-hu-and-jia/

Appendix

A. Suppose $W(\bmod\ Y) = W'''Y$, $X^{(1)}(\bmod\ Y) = X'^{(1)}Y$, and $X^{(2)}(\bmod\ Y) = X'^{(2)}Y$. We want to obtain a solution $u'^{(i)} \in R$, $i = 1, 2$, such that $W'''Y = (u'^{(1)}X'^{(1)} + u'^{(2)}X'^{(2)})Y(\bmod\ Y)$. First, the equation has solution, because $\{u^{(i)} \in R, i = 1, 2\}$ is a solution. Second, the equation can be modified as an equivalent equation $W''' = (u'^{(1)}X'^{(1)} + u'^{(2)}X'^{(2)})(\bmod\ 1)$. Third, take each entry of W''', $X'^{(1)}$, and $X'^{(2)}$ as the form of reduced fraction, and take LCM as the least common multiple of all denominators, then the equation can be modified as an equivalent equation, which is a linear equation modular LCM:

$$(LCM)W''' = (u'^{(1)}((LCM)X'^{(1)}) + u'^{(2)}((LCM)X'^{(2)}))(\bmod\ (LCM)).$$

B. Arbitrarily take a subset $\{i_1, i_2, i_3\}$ which is not a piece. We will compute $P(\{i_1, i_2, i_3\}$ is not a combined piece). First, we take a random experiment: randomly choosing 3 subsets $\{j_1, j_2, j_3\}$, $\{k_1, k_2, k_3\}$, $\{l_1, l_2, l_3\}$ from $\{1, 2, \ldots, 3K\}$. Then, the probability of such event:

$$\{j_1, j_2, j_3\} \cup \{k_1, k_2, k_3\} \supset \{i_1, i_2, i_3\},$$

$$\{l_1, l_2, l_3\} = \{j_1, j_2, j_3\} \cup \{k_1, k_2, k_3\} - \{i_1, i_2, i_3\},$$

is

$$\frac{C_{3K}^3 C_6^3}{(C_{3K}^3)^3} \approx \frac{1}{K^6}.$$

Second, from $O(K^2)$ pieces we generate all 3-tuples of 3 pieces $\{\{j_1, j_2, j_3\}, \{k_1, k_2, k_3\}, \{l_1, l_2, l_3\}\}$. We know there are $O(K^2)(O(K^2) - 1)(O(K^2) - 2)$ 3-tuples. Then, the probability of such event: there is no a 3-tuples $\{\{j_1, j_2, j_3\}, \{k_1, k_2, k_3\}, \{l_1, l_2, l_3\}\}$, such that

$$\{j_1, j_2, j_3\} \cup \{k_1, k_2, k_3\} \supset \{i_1, i_2, i_3\},$$

$$\{l_1, l_2, l_3\} = \{j_1, j_2, j_3\} \cup \{k_1, k_2, k_3\} - \{i_1, i_2, i_3\},$$

is about

$$\left(1 - \frac{1}{K^6}\right)^{O(K^2)(O(K^2)-1)(O(K^2)-2)} \approx exp\Big\{ - \frac{(O(K^2))^3}{K^6}\Big\}.$$

C. We need to obtain Hermite normal form $G = \begin{bmatrix} G_0 & & & \\ G_1 & 1 & & \\ \vdots & & \ddots & \\ G_{n-1} & & & 1 \end{bmatrix}$, where each row of G is an element of $\langle g \rangle$, G_0 is the absolute value of the determinant of the matrix $\begin{bmatrix} g_0 & g_1 & \cdots & g_{n-1} \\ -g_{n-1} & g_0 & \cdots & g_{n-2} \\ \vdots & \vdots & \ddots & \vdots \\ -g_1 & -g_2 & \cdots & g_0 \end{bmatrix}$, and $G_i (\text{mod } G_0) = G_i$ for $i = 1, \ldots, n-1$.

For a principal ideal $\langle g' \rangle$, we call the determinant of $\begin{bmatrix} g'_0 & g'_1 & \cdots & g'_{n-1} \\ -g'_{n-1} & g'_0 & \cdots & g'_{n-2} \\ \vdots & \vdots & \ddots & \vdots \\ -g'_1 & -g'_2 & \cdots & g'_0 \end{bmatrix}$ corresponding determinant of $\langle g' \rangle$. We use the definition of parallel piped [26]. For a vector $\alpha \in R$, we call the set $PP(\alpha) = \{z \in R : z(\text{mod } \alpha) = z\}$ parallel piped of α.

Two Facts. We have $\{Y, X^{(i)}, i = 1, 2\}$, therefore we can obtain hermite normal forms of the principal ideals $\{\langle Y \rangle, \langle X^{(i)} \rangle, i = 1, 2\}$.

Suppose Hermite normal form of the principal ideal $\langle g' \rangle$ is $\begin{bmatrix} G'_0 & & & \\ G'_1 & 1 & & \\ \vdots & & \ddots & \\ G'_{n-1} & & & 1 \end{bmatrix}$,

$g \in R$ is a factor of g', and absolute value of corresponding determinant of $\langle g \rangle$ is G_0. Then, Hermite normal form of the principal ideal $\langle g \rangle$ is $\begin{bmatrix} G_0 & & & \\ G'_1 (\text{mod } G_0) & 1 & & \\ \vdots & & \ddots & \\ G'_{n-1} (\text{mod } G_0) & & & 1 \end{bmatrix}$.

Computing Hermite Normal Form of $\langle h(1+ag)^{K-2} b^{(1)} \rangle$. We take a trivial assumption that $1+ag$ and $b^{(1)} g$ are co-prime.

Step 1. By using $\{Y, (-Y_{n-1}, Y_0, \ldots, Y_{n-2}), \ldots, (-Y_1, \ldots, -Y_{n-1}, Y_0)\}$ as the basis, Gaussian sample Z, with sufficiently large deviation.

Step 2. Compute $Z' = Z (\text{mod } X^{(1)})$. Then, Z' is uniformly distributed over the intersection area $\langle h(1+ag)^{K-2} b^{(1)} \rangle \cap PP(X^{(1)})$. Algebra and Gaussian sampling theory have proven this result.

Step 3. Compute absolute value of corresponding determinant of $\langle Z' \rangle$.

Step 4. Repeat Step 1$\sim$3 polynomially many times, so that we obtain polynomially many absolute values of corresponding determinant.

Step 5. Compute the greatest common divisor of these polynomially many absolute values. Then, the greatest common divisor should be absolute value of corresponding determinant of $\langle h(1+ag)^{K-2} b^{(1)} \rangle$. By considering a fact stated in last subsection, we obtain Hermite normal form of $\langle h(1+ag)^{K-2} b^{(1)} \rangle$.

Computing Hermite Normal Form of $\langle h(1+ag)^{K-2} b^{(1)} g \rangle$. We take a trivial assumption that $b^{(1)}$ and $b^{(2)}$ are co-prime. The procedure is similar to last subsection.

Step 1. By using $\{X^{(2)}, (-X^{(2)}_{n-1}, X^{(2)}_0, \ldots, X^{(2)}_{n-2}), \ldots, (-X^{(2)}_1, \ldots, -X^{(2)}_{n-1}, X^{(2)}_0)\}$ as the basis, Gaussian sample Z, with sufficiently large deviation.

Step 2. Compute $Z' = Z (\text{mod } X^{(1)})$. Then, Z' is uniformly distributed over the intersection area $\langle h(1+ag)^{K-2} b^{(1)} g \rangle \cap PP(X^{(1)})$.

Step 3. Compute absolute value of corresponding determinant of $\langle Z' \rangle$.

Step 4. Repeat Step 1$\sim$3 polynomially many times, so that we obtain polynomially many absolute values of corresponding determinant.

Step 5. Compute the greatest common divisor of these polynomially many absolute values. Then, the greatest common divisor should be absolute value of corresponding determinant of $\langle h(1+ag)^{K-2} b^{(1)} g \rangle$, therefore, we obtain Hermite normal form of $\langle h(1+ag)^{K-2} b^{(1)} g \rangle$.

Obtaining Hermite Normal Form of $\langle g \rangle$. Divide absolute value of corresponding determinant of $\langle h(1+ag)^{K-2} b^{(1)} g \rangle$ by absolute value of corresponding determinant of $\langle h(1+ag)^{K-2} b^{(1)} \rangle$. Then, we obtain absolute value of corresponding determinant of $\langle g \rangle$, therefore we obtain Hermite normal form of $\langle g \rangle$.

D. Here we use several symbols which have been used for analyzing the first simple revision of GGH map. User k_0 first computes $(v^{(k_0,1)}p_{zt}^{(1)} + v^{(k_0,2)}p_{zt}^{(2)})\prod_{k\neq k_0} V^{(k)}(\text{mod } q)$, then takes KEY as its high-order bits. It is easy to see that

$$(v^{(k_0,1)}p_{zt}^{(1)} + v^{(k_0,2)}p_{zt}^{(2)}) \prod_{k\neq k_0} V^{(k)}(\text{mod } q) = (A + B^{(k_0)})(\text{mod } q),$$

such that

$$A = g^{-1} \sum_{(j_1,\dots,j_{K+1})\in\{1,2\}^{K+1}} v^{(K+1,j_{K+1})}(y^{(0,j_{K+1})} + h^{(j_{K+1})}g) \prod_{k=1}^{K} v^{(k,j_k)}(y^{(0,j_k)} + a^{(j_k)}g)(\text{mod } q),$$

which has no relation with user k_0; $B^{(k_0)}$ is the sum of several terms which are somewhat small. If related parameters are small enough, KEY is high-order bits of $A(\text{mod } q)$.

Generating "Equivalent Secret". For the secret $(v^{(1)}, v^{(2)}) \in R^2$, we construct an "equivalent secret $(v'^{(1)}, v'^{(2)}) \in R^2$", such that

$$(v^{(1)}(y^{(0,1)}+a^{(1)}g)+v^{(2)}(y^{(0,2)}+a^{(2)}g))-(v'^{(1)}(y^{(0,1)}+a^{(1)}g)+v'^{(2)}(y^{(0,2)}+a^{(2)}g))$$

is a multiple of g. One equivalent requirement is that $(v^{(1)}y^{(0,1)} + v^{(2)}y^{(0,2)}) - (v'^{(1)}y^{(0,1)} + v'^{(2)}y^{(0,2)})$ is a multiple of g. Another equivalent requirement is that

$$(v^{(1)}(y^{(0,1)}+h^{(1)}g)+v^{(2)}(y^{(0,2)}+h^{(2)}g))-(v'^{(1)}(y^{(0,1)}+h^{(1)}g)+v'^{(2)}(y^{(0,2)}+h^{(2)}g))$$

is a multiple of g. That is enough, and we do not need $(v'^{(1)}, v'^{(2)})$ small. Take V, the noisy encoding of $(v^{(1)}, v^{(2)})$, we compute special element

$$\begin{aligned} W^* = V(y^{(1)})^{K-2}x^{(1)}p_{zt}^{(1)}(\text{mod } q) = (y^{(0,1)} + h^{(1)}g)\big[&v^{(1)}(y^{(0,1)} + a^{(1)}g)^{K-1}b^{(1)} \\ &+ v^{(2)}(y^{(0,2)} + a^{(2)}g)(y^{(0,1)} + a^{(1)}g)^{K-2}b^{(1)} \\ &+ u^{(1)}(b^{(1)}g)(y^{(0,1)} + a^{(1)}g)^{K-2}b^{(1)} \\ &+ u^{(2)}(b^{(2)}g)(y^{(0,1)} + a^{(1)}g)^{K-2}b^{(1)}\big]. \end{aligned}$$

Notice that

(1) Right side of this equation has no operation "mod q", therefore W^* is somewhat small.
(2) Four vectors $(y^{(0,1)} + h^{(1)}g)(y^{(0,1)} + a^{(1)}g)^{K-1}b^{(1)}$, $(y^{(0,1)} + h^{(1)}g)(y^{(0,2)} + a^{(2)}g)(y^{(0,1)} + a^{(1)}g)^{K-2}b^{(1)}$, $(y^{(0,1)} + h^{(1)}g)(b^{(1)}g)(y^{(0,1)} + a^{(1)}g)^{K-2}b^{(1)}$ and $hy^{(0,1)}(b^{(2)}g)(y^{(0,1)} + a^{(1)}g)^{K-2}b^{(1)}$ can be obtained.

Now we start to find $(v'^{(1)}, v'^{(2)})$. First, compute $W^*(\text{mod } (y^{(0,1)} + h^{(1)}g)(y^{(0,1)} + a^{(1)}g)^{K-1}b^{(1)})$. Second, compute $\{v'^{(2)}, u'^{(1)}, u'^{(2)}\}$ such that

$$W^*(\mathrm{mod}\ (y^{(0,1)}+h^{(1)}g)(y^{(0,1)}+a^{(1)}g)^{K-1}b^{(1)}) = \\ (y^{(0,1)}+h^{(1)}g)\big[v'^{(2)}(y^{(0,2)}+a^{(2)}g)(y^{(0,1)}+a^{(1)}g)^{K-2}b^{(1)} \\ +u'^{(1)}(b^{(1)}g)(y^{(0,1)}+a^{(1)}g)^{K-2}b^{(1)} \\ +u'^{(2)}(b^{(2)}g)(y^{(0,1)}+a^{(1)}g)^{K-2}b^{(1)}\big](\mathrm{mod}\ (y^{(0,1)} \\ +h^{(1)}g)(y^{(0,1)}+a^{(1)}g)^{K-1}b^{(1)}).$$

Solving this modular equation is quite easy algebra, as in Appendix A. Solutions are not unique, therefore $\{v'^{(2)}, u'^{(1)}, u'^{(2)}\} \neq \{v^{(2)}, u^{(1)}, u^{(2)}\}$. Third, compute $v'^{(1)}$ such that

$$W^* = (y^{(0,1)}+h^{(1)}g)\big[v'^{(1)}(y^{(0,1)}+a^{(1)}g)^{K-1}b^{(1)} \\ +v'^{(2)}(y^{(0,2)}+a^{(2)}g)(y^{(0,1)}+a^{(1)}g)^{K-2}b^{(1)} \\ +u'^{(1)}(b^{(1)}g)(y^{(0,1)}+a^{(1)}g)^{K-2}b^{(1)} \\ +u'^{(2)}(b^{(2)}g)(y^{(0,1)}+a^{(1)}g)^{K-2}b^{(1)}\big],$$

which is another easy algebra. Finally, we obtain $(v'^{(1)}, v'^{(2)})$, and can easily check that $(v^{(1)}(y^{(0,1)}+a^{(1)}g)+v^{(2)}(y^{(0,2)}+a^{(2)}g)) - (v'^{(1)}(y^{(0,1)}+a^{(1)}g)+v'^{(2)}(y^{(0,2)}+a^{(2)}g))$ is a multiple of g, although $v'^{(1)}$ and $v'^{(2)}$ are not short vectors.

Generalization of Modified Encoding/zero-testing: Our Attack on MKE. Suppose $K+1$ users hide $(v^{(k,1)}, v^{(k,2)})$ and publish $V^{(k)}, k=1,\ldots,K+1$, and for each user k we have obtained equivalent secret $(v'^{(k,1)}, v'^{(k,2)})$. For each "$K+1$-dimensional boolean vector" $(j_1,\ldots,j_{K+1}) \in \{1,2\}^{K+1}$, we denote two products

$$v^{(j_1,\ldots,j_{K+1})} = \prod_{k=1}^{K+1} v^{(k,j_k)},$$

$$v'^{(j_1,\ldots,j_{K+1})} = \prod_{k=1}^{K+1} v'^{(k,j_k)}.$$

$v^{(j_1,\ldots,j_{K+1})}$ is clearly smaller than "somewhat small", because it does not contain $h^{(1)}$ and $h^{(2)}$. $v'^{(j_1,\ldots,j_{K+1})}$ is not a short vector. $v^{(j_1,\ldots,j_{K+1})}$ cannot be obtained, while $v'^{(j_1,\ldots,j_{K+1})}$ can. Suppose former K entries $\{j_1,\ldots,j_K\}$ include N_1 1s and N_2 2s, $N_1+N_2=K$. We denote the supporter $s^{(j_1,\ldots,j_{K+1})}$ as follows.

$$s^{(j_1,\ldots,j_{K+1})} = (y^{(0,j_{K+1})}+h^{(j_{K+1})}g)(y^{(0,1)}+a^{(1)}g)^{N_1-1}(y^{(0,2)}+a^{(2)}g)^{N_2}b^{(1)} \quad \text{for } N_1 \geq N_2,$$
$$s^{(j_1,\ldots,j_{K+1})} = (y^{(0,j_{K+1})}+h^{(j_{K+1})}g)(y^{(0,1)}+a^{(1)}g)^{N_1}(y^{(0,2)}+a^{(2)}g)^{N_2-1}b^{(1)} \quad \text{for } N_1 < N_2.$$

$s^{(j_1,\ldots,j_{K+1})}$ can be obtained. If $N_1 \geq N_2$, $s^{(j_1,\ldots,j_{K+1})} = p_{zt}^{(j_{K+1})}(y^{(1)})^{N_1-1}(y^{(2)})^{N_2}x^{(1)}(\mathrm{mod}\ q)$, and if $N_1 < N_2$, $s^{(j_1,\ldots,j_{K+1})} = p_{zt}^{(j_{K+1})}(y^{(1)})^{N_1}(y^{(2)})^{N_2-1}x^{(1)}(\mathrm{mod}\ q)$. $s^{(j_1,\ldots,j_{K+1})}$ is somewhat small. Then, we denote

$$V^{(N_1 \geq N_2)} = \sum_{j_{K+1}=1}^{2} \sum_{N_1 \geq N_2} v^{(j_1,\ldots,j_{K+1})} s^{(j_1,\ldots,j_{K+1})},$$

$$V^{(N_1<N_2)} = \sum_{j_{K+1}=1}^{2} \sum_{N_1<N_2} v^{(j_1,\ldots,j_{K+1})} s^{(j_1,\ldots,j_{K+1})},$$

$$V'^{(N_1\geq N_2)} = \sum_{j_{K+1}=1}^{2} \sum_{N_1\geq N_2} v'^{(j_1,\ldots,j_{K+1})} s^{(j_1,\ldots,j_{K+1})},$$

$$V'^{(N_1<N_2)} = \sum_{j_{K+1}=1}^{2} \sum_{N_1<N_2} v'^{(j_1,\ldots,j_{K+1})} s^{(j_1,\ldots,j_{K+1})}.$$

$V^{(N_1\geq N_2)}$ and $V^{(N_1<N_2)}$ are somewhat small, while $V'^{(N_1\geq N_2)}$ and $V'^{(N_1<N_2)}$ are not short vectors. $V^{(N_1\geq N_2)}$ and $V^{(N_1<N_2)}$ cannot be obtained, while $V'^{(N_1\geq N_2)}$ and $V'^{(N_1<N_2)}$ can be obtained, because $v'^{(j_1,\ldots,j_{K+1})} s^{(j_1,\ldots,j_{K+1})}$ can be obtained for each $(j_1,\ldots,j_{K+1}) \in \{1,2\}^{K+1}$, and 2^K is polynomially large. Another fact is that ξ^* is a multiple of $b^{(1)}g$, where

$$\xi^* = (y^{(0,1)} + a^{(1)}g)(V'^{(N_1\geq N_2)} - V^{(N_1\geq N_2)}) + (y^{(0,2)} + a^{(2)}g)(V'^{(N_1<N_2)} - V^{(N_1<N_2)}).$$

There are two reasons: (1) By considering the definitions of equivalent secrets, we know that ξ^* is a multiple of g. (2) By considering the definition of $s^{(j_1,\ldots,j_{K+1})}$, we know that ξ^* is a multiple of $b^{(1)}$. Here we use a small assumption that $b^{(1)}$ and g are co-prime. Notice that ξ^* is not a short vector, and that ξ^* cannot be obtained. Then, we compute a tool for modular operations,

$$M = (y^{(0,1)} + h^{(1)}g)(b^{(1)})^K g^{K-1} = p_{zt}^{(1)} (x^{(1)})^K (\text{mod } q).$$

For the same reason, M is somewhat small. Then, we compute the modular operations

$$V''^{(N_1\geq N_2)} = V'^{(N_1\geq N_2)} (\text{mod } M),$$
$$V''^{(N_1<N_2)} = V'^{(N_1<N_2)} (\text{mod } M).$$

Both $V''^{(N_1\geq N_2)}$ and $V''^{(N_1<N_2)}$ are somewhat small. Therefore, both $V''^{(N_1\geq N_2)} - V^{(N_1\geq N_2)}$ and $V''^{(N_1<N_2)} - V^{(N_1<N_2)}$ are somewhat small. Therefore, both $(y^{(0,1)}+a^{(1)}g)(V''^{(N_1\geq N_2)}-V^{(N_1\geq N_2)})$ and $(y^{(0,2)}+a^{(2)}g)(V''^{(N_1<N_2)} - V^{(N_1<N_2)})$ are somewhat small. Therefore,

$$\xi^{**} = (y^{(0,1)} + a^{(1)}g)(V''^{(N_1\geq N_2)} - V^{(N_1\geq N_2)}) + (y^{(0,2)} + a^{(2)}g)(V''^{(N_1<N_2)} - V^{(N_1<N_2)})$$

is somewhat small. On the other hand, ξ^{**} is a multiple of $b^{(1)}g$, because ξ^* is a multiple of $b^{(1)}g$. Therefore, $\xi^{**}/(b^{(1)}g)$ is somewhat small. Finally,

$$\begin{aligned}\frac{\xi^{**}}{(b^{(1)}g)} &= \xi^{**}(b^{(1)}g)^{-1}(\text{mod } q)\\ &= \left[\left((y^{(0,1)} + a^{(1)}g)V''^{(N_1\geq N_2)} + (y^{(0,2)} + a^{(2)}g)V''^{(N_1<N_2)}\right)(b^{(1)}g)^{-1} - A\right](\text{mod } q),\end{aligned}$$

which means that KEY is high-order bits of

$$\left[\left((y^{(0,1)} + a^{(1)}g)V''^{(N_1\geq N_2)} + (y^{(0,2)} + a^{(2)}g)V''^{(N_1<N_2)}\right)(b^{(1)}g)^{-1}\right](\text{mod } q).$$

Hash-Function Based PRFs: AMAC and Its Multi-User Security

Mihir Bellare[1(✉)], Daniel J. Bernstein[2,3], and Stefano Tessaro[4]

[1] Department of Computer Science and Engineering, University of California San Diego, San Diego, USA
mihir@eng.ucsd.edu
[2] University of Illinois at Chicago, Chicago, USA
[3] Technische Universiteit Eindhoven, Eindhoven, The Netherlands
[4] Department of Computer Science, University of California Santa Barbara, Santa Barbara, USA
http://cseweb.ucsd.edu/~mihir/
https://cr.yp.to/djb.html
http://www.cs.ucsb.edu/~tessaro/

Abstract. AMAC is a simple and fast candidate construction of a PRF from an MD-style hash function which applies the keyed hash function and then a cheap, un-keyed output transform such as truncation. Spurred by its use in the widely-deployed Ed25519 signature scheme, this paper investigates the provable PRF security of AMAC to deliver the following three-fold message: (1) First, we prove PRF security of AMAC. (2) Second, we show that AMAC has a quite unique and attractive feature, namely that its multi-user security is essentially as good as its single-user security and in particular superior in some settings to that of competitors. (3) Third, it is technically interesting, its security and analysis intrinsically linked to security of the compression function in the presence of leakage.

1 Introduction

This paper revisits a classical question, namely how can we turn a hash function into a PRF? The canonical answer is HMAC [4], which (1) first applies the keyed hash function to the message and then (2) re-applies, to the result, the hash function keyed with another key. We consider another, even simpler, candidate way, namely to change step (2) to apply a simple *un-keyed* output transform such as truncation. We call this AMAC, for augmented MAC. This paper investigates and establishes provable-security of AMAC, with good bounds, when the hash function is a classical MD-style one like SHA-512.

WHY? We were motivated to determine the security of AMAC by the following. *Usage.* AMAC with SHA-512 is used as a PRF in the Ed25519 signature scheme [8]. (AMAC under a key that is part of the signing key is applied to the hashed message to get coins for a Schnorr-like signature.) Ed25519 is widely deployed, including in SSH, Tor, OpenBSD and dozens of other places [10]. The security of AMAC for this

M. Fischlin and J.-S. Coron (Eds.): EUROCRYPT 2016, Part I, LNCS 9665, pp. 566–595, 2016.
DOI: 10.1007/978-3-662-49890-3_22

usage was questioned in `cfrg` forum debates on Ed25519 as a proposed standard. Analysis of AMAC is important to assess security of this usage and allow informed choices. *Speed.* AMAC is faster than HMAC, particularly on short messages. See [3]. *Context.* Sponge-based PRFs, where truncation is the final step due to its already being so for the hash function, have been proven secure [1,9,11,17,20]. Our work can be seen as stepping back to ask if truncation works in a similar way for classical MD-style hash functions.

FINDINGS IN A NUTSHELL. Briefly, the message of this paper is the following: (1) First, we are able to prove PRF security of AMAC. (2) Second, AMAC has a quite unique and attractive feature, namely that its multi-user security is essentially as good as its single-user security and in particular superior in some settings to that of competitors. (3) Third, it is technically interesting, its security and analysis intrinsically linked to security of the compression function in the presence of leakage, so that leakage becomes of interest for reasons entirely divorced from side-channel attacks. We now step back to provide some background and discuss our approach and results.

THE BASIC CASCADE. Let $\mathsf{h}\colon \{0,1\}^c \times \{0,1\}^b \to \{0,1\}^c$ represent a compression function taking a c-bit chaining variable and b-bit message block to return a c-bit output. The *basic cascade* of h is the function $\mathsf{h}^*\colon \{0,1\}^c \times (\{0,1\}^b)^+ \to \{0,1\}^c$ defined by

Basic Cascade $\mathsf{h}^*(K, \mathbf{X})$

$Y \leftarrow K$; For $i = 1, \ldots, n$ do $Y \leftarrow \mathsf{h}(Y, \mathbf{X}[i])$; Return Y

where $\mathbf{X}$ is a vector over $\{0,1\}^b$ whose length is denoted n and whose i-th component is denoted $\mathbf{X}[i]$. This construct is the heart of MD-style hash functions [13,21] like MD5, SHA-1, SHA-256 and SHA-512, which are obtained by setting K to a fixed, public value and then applying h^* to the padded message.

Now we want to key h^* to get PRFs. We regard h itself as a PRF on domain $\{0,1\}^b$, keyed by its c-bit chaining variable. Then h^* is the natural candidate for a PRF on the larger domain $(\{0,1\}^b)^+$. Problem is, h^* isn't secure as a PRF. This is due to the well-known *extension attack.* If I obtain $Y_1 = \mathsf{h}^*(K, X_1)$ for some $X_1 \in \{0,1\}^b$ of my choice, I can compute $Y_2 = \mathsf{h}^*(K, X_1X_2)$ for any $X_2 \in \{0,1\}^b$ of my choice *without knowing* K, via $Y_2 \leftarrow \mathsf{h}(Y_1, X_2)$. This clearly violates PRF security of h^*.

Although h^* is not a PRF, BCK2 [5] show that it is a prefix-free PRF. (A PRF as long as no input on which it is evaluated is a prefix of another. The two inputs X_1, X_1X_2 of the above attack violate this property.) When $b = 1$ and all inputs on which h^* is evaluated are of the same fixed length, the cascade h^* is the GGM construction of a PRF from a PRG [18].

To get a full-fledged PRF, NMAC applies h, under another key, to h^*. The augmented cascade $\mathsf{ACSC} = \mathsf{Out} \circ \mathsf{h}^*$ that we discuss next replaces NMAC's outer application of a keyed function with a simple un-keyed one.

AUGMENTED CASCADE. The augmented cascade is parameterized by some (keyless) function $\mathsf{Out}\colon \{0,1\}^c \to \mathsf{Out.R}$ that we call the output transform, and is obtained by simply applying this function to the output of the basic cascade:

$\underline{\text{Augmented Cascade } (\mathsf{Out} \circ \mathsf{h}^*)(K, \mathbf{X})}$
$Y \leftarrow \mathsf{h}^*(K, \mathbf{X})$; $Z \leftarrow \mathsf{Out}(Y)$; Return Z

AMAC is obtained from ACSC just as HMAC is obtained from NMAC, namely by putting the key in the input to the hash function rather than directly keying the cascade: $\mathsf{AMAC}(K, M) = \mathsf{Out}(H(K \| M))$. Just as NMAC is the technical core of HMAC, the augmented cascade is the technical core of AMAC, and our analysis will focus in it. We will be able to bridge to AMAC quite simply with the tools we develop.

The ACSC construction was suggested by cryptanalysts with the intuition that "good" choices of Out appear to allow $\mathsf{Out} \circ \mathsf{h}^*$ to evade the extension attack and thus possibly be a PRF. To understand this, first note that not all choices of Out are good. For example if Out is the identity function then the augmented cascade is the same as the basic one and the attack applies, or if Out is a constant function returning 0^r then $\mathsf{Out} \circ \mathsf{h}^*$ is obviously not a PRF over range $\{0,1\}^r$. Cryptanalysts have suggested some specific choices of Out, the most important being (1) truncation, where $\mathsf{Out}\colon \{0,1\}^c \rightarrow \{0,1\}^r$ returns, say, the first $r < c$ bits of its input, or (2) the mod function, as in Ed25519, where Out treats its input as an integer and returns the result modulo, say, a public r-bit prime number. Suppose r is sufficiently smaller than c (think $c = 512$ and $r = 256$). An adversary querying X_1 in the PRF game no longer gets back $Y_1 = \mathsf{h}^*(K, X_1)$ but rather $Z_1 = \mathsf{Out}(Y_1)$, and this does not allow the extension attack to proceed. On this basis, and for the choices of Out just named, the augmented cascade is already seeing extensive usage and is suggested for further usage and standardization.

This raises several questions. First, that $\mathsf{Out} \circ \mathsf{h}^*$ seems to evade the extension attack does not mean it is a PRF. There may be other attacks. The goal is to get a PRF, not to evade some specific attacks. Moreover we would like a proof that this goal is reached. Second, for which choices of Out does the construction work? We could try to analyze the PRF security of $\mathsf{Out} \circ \mathsf{h}^*$ in an ad hoc way for the specific choices of Out named above, but it would be more illuminating and useful to be able to establish security in a broad way, for all Out satisfying some conditions. These are the questions our work considers and resolves.

Connection to Leakage. If we want to prove PRF security of $\mathsf{Out} \circ \mathsf{h}^*$, a basic question to ask is, under what assumption on the compression function h? The natural one is that h is itself a PRF, the same assumption as for the proof of NMAC [2,16]. We observe that this is not enough. Consider an adversary who queries the one-block message X_1 to get back $Z_1 = \mathsf{Out}(Y_1)$ and then queries the two-block message $X_1 X_2$ to get back $Z_2 = \mathsf{Out}(Y_2)$ where by definition $Y_1 = \mathsf{h}^*(K, X_1) = \mathsf{h}(K, X_1)$ and $Y_2 = \mathsf{h}^*(K, X_1 X_2) = \mathsf{h}(Y_1, X_2)$. Note that Y_1 is being used as a key in applying h to X_2. But this key is not entirely unknown to the adversary because the latter knows $Z_1 = \mathsf{Out}(Y_1)$. If the application of h with key Y_1 is to provide security, it must be in the face of the fact that some information about this key, namely $\mathsf{Out}(Y_1)$, has been "leaked" to the adversary. As a PRF, h must thus be resilient to some leakage on its key, namely that represented by Out viewed as a leakage function.

APPROACH AND QUALITATIVE RESULTS. We first discuss our results at the qualitative level and then later at the (in our view, even more interesting) quantitative level. Theorems 3 and 4 show that if h is a PRF under Out-leakage then $\mathsf{Out} \circ \mathsf{h}^*$ is indistinguishable from the result of applying Out to a random function. (The compression function h being a PRF under Out-leakage means it retains PRF security under key K even if the adversary is given $\mathsf{Out}(K)$. The formal definition is in Sect. 4.) This result makes no assumptions about Out beyond that implicit in the assumption on h, meaning the result is true for *all* Out, and is in the standard model. As a corollary we establish PRF security of $\mathsf{Out} \circ \mathsf{h}^*$ for a large class of output functions Out, namely those that are close to regular. (This means that the distribution of $\mathsf{Out}(Y)$ for random Y is close to the uniform distribution on the range of Out.) In summary we have succeeded in providing conditions on Out, h under which $\mathsf{Out} \circ \mathsf{h}^*$ is proven to be PRF. Our conditions are effectively both necessary and sufficient and cover cases proposed for usage and standardization.

The above is a security proof for the augmented cascade $\mathsf{Out} \circ \mathsf{h}^*$ under the assumption that the compression function h is resistant to Out leakage. To assess the validity of this assumption, we analyze the security under leakage of an ideal compression function. Theorem 6 shows that an ideal compression function is resistant to Out-leakage as long as no range point of Out has too few pre-images. This property is in particular true if Out is close to regular. As a result, in the ideal model, we have a validation of our Out-leakage resilience assumption. Putting this together with the above we have a proof-based validation of the augmented cascade.

MULTI-USER SECURITY. The standard definition of PRF security of a function family F [18] is single user (su), represented by there being a single key K such that the adversary has access to an oracle FN that given x returns either $\mathsf{F}(K, x)$ or the result of a random function F on x. But in "real life" there are many users, each with their own key. If we look across the different entities and Internet connections active at any time, the number of users/keys is very large. The more appropriate model is thus a multi-user (mu) one, where, for a parameter u representing the number of users, there are u keys $K_1, \ldots, K_u$. Oracle FN now takes i, x with $1 \leq i \leq u$ and returns either $\mathsf{F}(K_i, x)$ or the result of a random function F_i on x. It is in this setting that we should address security.

Multi-user security is typically neglected because it makes no *qualitative* difference: BCK2 [5], who first formalized the notion, also showed by a hybrid argument that the advantage of an adversary relative to u users is not more than u times the advantage of an adversary of comparable resources relative to a single user. Our Lemma 1 is a generalization of this result. But this degradation in advantage is quite significant in practice, since u is large, and raises the important question of whether one can do quantitatively better. Clearly one cannot in general, but perhaps one can for specific, special function families F. If so, these function families are preferable in practice. This perspective is reflected in recent work like [22,25].

These special function families seem quite rare. But we show that the augmented cascade is one of them. In fact we show that mu security gives us a double benefit in this setting, one part coming from the cascade itself and the other from the security of the compression function under leakage, the end result being very good bounds for the mu security of the augmented cascade.

Theorem 3 establishes su security of the augmented cascade based not on the su, but on the mu security of the compression function under Out-leakage. The bound is very good, the advantage dropping only by a factor equal to the maximum length of a query. The interesting result is Theorem 4, establishing mu security of the augmented cascade under the same assumptions and with essentially the same bounds as Theorem 3 establishing its su security. In particular we do not lose a factor of the number of users u in the advantage. This is the first advance.

Now note that the assumption in both of the above-mentioned results is the mu (not su) security of the compression function under Out-leakage. Our final bound will thus depend on this. The second advance is that Theorem 6 shows mu security of the compression function under Out-leakage with bounds almost as good as for su security. This represents an interesting result of independent interest, namely that, under leakage, the mu security of an ideal compression function is almost as good as its su security. This is not true in the absence of leakage. The results are summarized via Fig. 4.

QUANTITATIVE RESULTS. We obtain good quantitative bounds on the mu prf security of the augmented cascade in the ideal compression function model by combining our aforementioned results on the mu prf security under leakage of an ideal compression function with our also aforementioned reduction of the security of the cascade to the security of the compression function under leakage. We illustrate these results for the case where the compression function is of form $\mathsf{h}\colon \{0,1\}^c \times \{0,1\}^b \to \{0,1\}^c$ and the output transform Out simply outputs the first r bits of its c-bit input, for $r \leq c$. We consider an attacker making at most q queries to a challenge oracle (that is either the augmented cascade or a random function), each query consisting of at most ℓ b-bit blocks, and q_{F} queries to the ideal compression function oracle. We show that such an attacker achieves distinguishing advantage at most

$$\frac{\ell^2 q^2 + \ell q q_{\mathrm{F}}}{2^c} + \frac{cr \cdot (\ell^2 q + \ell q_{\mathrm{F}})}{2^{c-r}}, \tag{1}$$

where we have intentionally omitted constant factors and lower order terms. Note that this bound holds *regardless of the number of users* u. Here c is large, like $c = 512$, so the first term is small. But $c-r$ is smaller, for example $c-r = 256$ with $r = 256$. The crucial merit of the bound of Eq. (1) is that the numerator in the second term does not contain quadratic terms like q^2 or $q \cdot q_{\mathrm{F}}$. In practice, q_{F} and q are the terms we should allow to be large, so this is significant. To illustrate, say for example $\ell = 2^{10}$ (meaning messages are about 128 KBytes if $b = 1024$) and $q_{\mathrm{F}} = 2^{100}$ and $q = 2^{90}$. The bound from Eq. (1) is about 2^{-128}, which is very good. But, had the second term been of the form $\ell^2(q_{\mathrm{F}}^2 + q^2)/2^{c-r}$ then the bound would be only 2^{-36}. See Sect. 8 for more information.

2-TIER CASCADE. We introduce and use an extension of the basic cascade h^*. Our 2-tier cascade is associated to two function families g, h. Under key K, it applies $\mathsf{g}(K, \cdot)$ to the first message block to get a sub-key K^* and the applies $\mathsf{h}^*(K^*, \cdot)$ to the rest of the message. The corresponding augmented cascade applies Out to the result. Our results about the augmented cascade above are in fact shown for the augmented 2-tier cascade. This generalization has both conceptual and analytical value. We briefly mention two instances. (1) First, we can visualize mu security of $\mathsf{Out} \circ \mathsf{h}^*$ as pre-pending the user identity to the message and then applying the 2-tier cascade with first tier a random function. This effectively reduces mu security to su security. With this strategy we prove Theorem 4 as a corollary of Theorem 3 and avoid a direct analysis of mu security. Beyond providing a modular proof this gives some insight into why the mu security is almost as good as the su security. (2) Second, just as NMAC is the technical core and HMAC the function used (because the latter makes blackbox use of the hash function), in our case the augmented cascade is the technical core but what will be used is AMAC, defined by $\mathsf{AMAC}(K, M) = \mathsf{Out}(H(K, M))$ where H is the hash function derived from compression function h: $\{0,1\}^c \times \{0,1\}^b \rightarrow \{0,1\}^c$ and K is a k-bit key. For the analysis we note (assuming $k = b$) that this is simply an augmented 2-tier cascade with the first tier being the dual of h, meaning the key and input roles are swapped. We thus directly get an analysis and proof for this case from our above-mentioned results. Obtaining HMAC from NMAC was more work [2,4] and required assumptions about PRF security of the dual function under related keys.

DAVIES-MEYER. Above we have assessed the PRF security under Out-leakage of the compression function by modeling the latter as ideal (random). But, following CDMP [12], one might say that the compression functions underlying MD-style hash functions are not un-structured enough to be treated as random because they are built from blockciphers via the Davies-Meyer (DM) construction. To address this we analyze the mu PRF security under Out-leakage of the DM construction in the ideal-cipher model. One's first thought may be that such an analysis would follow from our analysis for a random compression function and the indifferentiability [12,19] of DM from a random oracle, but the catch is that DM is *not* indifferentiable from a RO so a direct analysis is needed. The one we give in [3] shows mu security with good bounds. Similar analyses can be given for other PGV [24] compression functions.

2 Related Work

SPONGES. SHA-3 already internally incorporates a truncation output transform. The construction itself is a sponge. The suggested way to obtain a PRF is to simply key the hash function via its IV, so that the PRF is a keyed, truncated sponge. The security of this construct has been intensively analyzed [1,9,11,17,20] with Gaži, Pietrzak and Tessaro (GPT) [17] establishing PRF security with tight bounds. Our work can be seen as stepping back to ask whether the same truncation method would work for MD-style hash functions like SHA-512. Right now these older hash functions are much more widely deployed than SHA-3, and current standardization and deployment efforts continue to use them, making the analysis of constructions based on them important with regard to security in practice. The underlying construction in this case is the cascade, which is quite different from the sponge. The results and techniques of GPT [17] do not directly apply but were an important inspiration for our work.

We note that keyed sponges with truncation to an r-bit output from a c-bit state can easily be distinguished from a random function with advantage roughly $q^2/2^{c-r}$ or $qq_{\mathrm{F}}/2^{c-r}$, as shown for example in [17]. The bound of Eq. (1) is better, meaning the augmented cascade offers greater security. See [3] for more information.

CASCADE. BCK2 [5] show su security of the basic cascade (for prefix-free queries) in two steps. First, they show su security of the basic cascade (for prefix-free queries) assuming not su, but mu security of the compression function. Second, they apply the trivial bound mentioned above to conclude su security of the basic cascade for prefix-free queries assuming su security of the compression function. We follow their approach to establish su security of the augmented cascade, but there are differences as well: They have no output transform while we do, they assume prefix-free queries and we do not, we have leakage and they do not. They neither target nor show mu security of the basic cascade in any form, mu security arising in their work only as an intermediate technical step and only for the compression function, not for the cascade.

CHOP-MD. The chop-MD construction of CDMP [12] is the case of the augmented cascade in which the output transform is truncation. They claim this is indifferentiable from a RO when the compression function is ideal. This implies PRF security but their bound is $O(\ell^2(q+q_{\mathrm{F}})^2/2^{c-r})$ which as we have seen is significantly weaker than our bound of Eq. (1). Also, they have no standard-model proofs or analysis for this construction. In contrast our results in Sect. 5 establish standard-model security.

NMAC AND HMAC. NMAC takes keys $K_{\mathrm{in}}, K_{\mathrm{out}}$ and input $\mathbf{X}$ to return $\mathsf{h}(K_{\mathrm{out}}, \mathsf{h}^*(K_{\mathrm{in}}, \mathbf{X})\|\mathrm{pad})$ where pad is some $(b-c)$-bit constant and $b \geq c$. Through a series of intensive analyses, the PRF security of NMAC has been established based only on the assumed PRF security of the compression function h, and with tight bounds [2,4,16]. Note that NMAC is not a special case of the augmented cascade because Out is not keyed but the outer application of h in NMAC is keyed. In the model where the compression function is ideal, one can show bounds for NMAC that are somewhat better than for the augmented cascade. This is not surprising. Indeed, when attacking the augmented cascade, the adversary can learn far more information about the internal states of the hash computation. What is surprising (at least to us) is that the gap is actually quite small. See [3] for more information. We stress also that this is in the ideal model. In the standard model, there is no proof that NMAC has the type of good mu prf security we establish for the augmented cascade in Sect. 5.

AES AND OTHER MACs. Why consider new MACs? Why not just use an AES-based MAC like CMAC? The 128 bit key and block size limits security compared to $c = 512$ for SHA-512. A Schnorr signature takes the result of the PRF modulo a prime; the PRF output must have at least as many bits as the prime, and even more bits for most primes, to avoid the Bleichenbacher attack discussed in [23]. Also in that context a hash function is already being used to hash the message before signing so it is convenient to implement the PRF also with the same hash function. HMAC-SHA-512 will provide the desired security but AMAC has speed advantages, particularly on short messages, as discussed in [3], and is simpler. Finally, the question is in some sense moot since AMAC is already deployed and in widespread use via Ed25519 and we need to understand its security.

LEAKAGE. Leakage-resilience of a PRF studies the PRF security of a function h when the attacker can obtain the result of an *arbitrary* function, called the leakage function, applied to the key [14,15]. This is motivated by side-channel attacks. We are considering a much more restricted form of leakage where there is just one, very specific leakage function, namely Out. This arises naturally, as we have seen, in the PRF security of the augmented cascade. We are not considering side-channel attacks.

3 Notation

If $\mathbf{x}$ is a vector then $|\mathbf{x}|$ denotes its length and $\mathbf{x}[i]$ denotes its i-th coordinate. (For example if $\mathbf{x} = (10, 00, 1)$ then $|\mathbf{x}| = 3$ and $\mathbf{x}[2] = 00$.) We let ε denote the empty vector, which has length 0. If $0 \leq i \leq |\mathbf{x}|$ then we let $\mathbf{x}[1 \dots i] = (\mathbf{x}[1], \dots, \mathbf{x}[i])$, this being ε when $i = 0$. We let S^n denote the set of all length n vectors over the set S. We let S^+ denote the set of all vectors of positive length over the set S and $S^* = S^+ \cup \{\varepsilon\}$ the set of all finite-length vectors over the set S. As special cases, $\{0,1\}^n$ and $\{0,1\}^*$ denote vectors whose entries are bits, so that we are identifying strings with binary vectors and the empty string with the empty vector.

For sets A_1, A_2 we let $[\![A_1, A_2]\!]$ denote the set of all vectors $\mathbf{X}$ of length $|\mathbf{X}| \geq 1$ such that $\mathbf{X}[1] \in A_1$ and $\mathbf{X}[i] \in A_2$ for $2 \leq i \leq |\mathbf{X}|$.

We let $x \leftarrow\!\!\$\, X$ denote picking an element uniformly at random from a set X and assigning it to x. For infinite sets, it is assumed that a proper measure can be defined on X to make this meaningful. Algorithms may be randomized unless otherwise indicated. Running time is worst case. If A is an algorithm, we let $y \leftarrow A(x_1, \dots; r)$ denote running A with random coins r on inputs $x_1, \dots$ and assigning the output to y. We let $y \leftarrow\!\!\$\, A(x_1, \dots)$ be the result of picking r at random and letting $y \leftarrow A(x_1, \dots; r)$. We let $[A(x_1, \dots)]$ denote the set of all possible outputs of A when invoked with inputs $x_1, \dots$.

We use the code based game playing framework of [6]. (See Fig. 1 for an example.) By Pr[G] we denote the probability that game G returns true.

For an integer n we let $[1 \dots n] = \{1, \dots, n\}$.

4 Function-Family Distance Framework

We will be considering various generalizations and extensions of standard prf security. This includes measuring proximity not just to random functions but to some other family, multi-user security and leakage on the key. We also want to allow an easy later extension to a setting with ideal primitives. To enable all this in a unified way we introduce a general distance metric on function families and then derive notions of interest as special cases.

FUNCTION FAMILIES. A *function family* is a two-argument function $\mathsf{F}\colon \mathsf{F.K} \times \mathsf{F.D} \to \mathsf{F.R}$ that takes a key K in the key space $\mathsf{F.K}$ and an input x in the domain $\mathsf{F.D}$ to return an output $y \leftarrow \mathsf{F}(K, x)$ in the range $\mathsf{F.R}$. We let $f \leftarrow_{\$} \mathsf{F}$ be shorthand for $K \leftarrow_{\$} \mathsf{F.K}\,;\, f \leftarrow \mathsf{F}(K, \cdot)$, the operation of picking a function at random from family F.

An example of a function family that is important for us is the compression function underlying a hash function, in which case $\mathsf{F.K} = \mathsf{F.R} = \{0,1\}^c$ and $\mathsf{F.D} = \{0,1\}^b$ for integers c, b called the length of the chaining variable and the block length, respectively. Another example is a block cipher. However, families of functions do not have to be efficiently computable or have short keys. For sets D, R the *family* $\mathsf{A}\colon \mathsf{A.K} \times D \to R$ *of all functions from* D *to* R is defined simply as follows: let $\mathsf{A.K}$ be the set of all functions mapping D to R and let $\mathsf{A}(f, x) = f(x)$. (We can fix some representation of f as a key, for example the vector whose i-th component is the value f takes on the i-th input under some ordering of D. But this is not really necessary.) In this case $f \leftarrow_{\$} \mathsf{A}$ denotes picking at random a function mapping D to R.

Let $\mathsf{F}\colon \mathsf{F.K} \times \mathsf{F.D} \to \mathsf{F.R}$ be a function family and let $\mathsf{Out}\colon \mathsf{F.R} \to \mathsf{Out.R}$ be a function with domain the range of F and range $\mathsf{Out.R}$. Then the composition $\mathsf{Out} \circ \mathsf{F}\colon \mathsf{F.K} \times \mathsf{F.D} \to \mathsf{Out.R}$ is the function family defined by $(\mathsf{Out} \circ \mathsf{F})(K, x) = \mathsf{Out}(\mathsf{F}(K, x))$. We will use composition in some of our constructions.

BASIC DISTANCE METRIC. We define a general metric of distance between function families that will allow us to obtain other metrics of interest as special cases. Let $\mathsf{F}_0, \mathsf{F}_1$ be families of functions such that $\mathsf{F}_0.\mathsf{D} = \mathsf{F}_1.\mathsf{D}$. Consider game DIST on the left of Fig. 1 associated to $\mathsf{F}_0, \mathsf{F}_1$ and an adversary $\mathcal{A}$. Via oracle NEW, the adversary can create a new instance F_v drawn from F_c where c is the challenge bit. It can call this oracle multiple times, reflecting a multi-user setting. It can obtain $F_i(x)$ for any i, x of its choice with the restriction that $1 \le i \le v$ (instance i has been initialized) and $x \in \mathsf{F}_1.\mathsf{D}$. It wins if it guesses the challenge bit c. The advantage of adversary $\mathcal{A}$ is

$$\mathsf{Adv}^{\mathsf{dist}}_{\mathsf{F}_0,\mathsf{F}_1}(\mathcal{A}) = 2\Pr[\mathrm{DIST}_{\mathsf{F}_0,\mathsf{F}_1}(\mathcal{A})] - 1 \tag{2}$$

$$= \Pr[\mathrm{DIST}_{\mathsf{F}_0,\mathsf{F}_1}(\mathcal{A}) \mid c = 1] - (1 - \Pr[\mathrm{DIST}_{\mathsf{F}_0,\mathsf{F}_1}(\mathcal{A}) \mid c = 0])\,. \tag{3}$$

Equation (2) is the definition, while Eq. (3) is a standard alternative formulation that can be shown equal via a conditioning argument. We often use the second in proofs.

Game $\text{DIST}_{\mathsf{F}_0,\mathsf{F}_1}(\mathcal{A})$	Game $\text{DIST}_{\mathsf{F}_0,\mathsf{F}_1,\mathsf{Out}}(\mathcal{A})$
$v \leftarrow 0$	$v \leftarrow 0$
$c \leftarrow_{\$} \{0,1\}$; $c' \leftarrow_{\$} \mathcal{A}^{\text{NEW,FN}}$	$c \leftarrow_{\$} \{0,1\}$; $c' \leftarrow_{\$} \mathcal{A}^{\text{NEW,FN}}$
Return $(c = c')$	Return $(c = c')$
NEW()	NEW()
$v \leftarrow v+1$; $F_v \leftarrow_{\$} \mathsf{F}_c$	$v \leftarrow v+1$; $K_v \leftarrow_{\$} \mathsf{F}_1.\mathsf{K}$
FN(i,x)	If $(c=1)$ then $F_v \leftarrow \mathsf{F}_1(K_v,\cdot)$ else $F_v \leftarrow_{\$} \mathsf{F}_0$
Return $F_i(x)$	Return $\mathsf{Out}(K_v)$
	FN(i,x)
	Return $F_i(x)$

Fig. 1. Games defining distance metric between function families $\mathsf{F}_0, \mathsf{F}_1$. In the basic (left) case there is no leakage, while in the extended (right) case there is leakage represented by Out.

Let F be a function family and let A be the family of all functions from F.D to F.R. Let $\mathsf{Adv}^{\mathsf{prf}}_{\mathsf{F}}(\mathcal{A}) = \mathsf{Adv}^{\mathsf{dist}}_{\mathsf{F},\mathsf{A}}(\mathcal{A})$. This gives a metric of multi-user prf security. The standard (single user) prf metric is obtained by restricting attention to adversaries that make exactly one NEW query.

DISTANCE UNDER LEAKAGE. We extend the framework to allow leakage on the key. Let Out: $\mathsf{F}_1.\mathsf{K} \rightarrow \mathsf{Out.R}$ be a function with domain $\mathsf{F}_1.\mathsf{K}$ and range a set we denote Out.R. Consider game DIST on the right of Fig. 1, now associated not only to $\mathsf{F}_0, \mathsf{F}_1$ and an adversary $\mathcal{A}$ but also to Out. Oracle NEW picks a key K_v for F_1 and will return as leakage the result of Out on this key. The instance F_v is either $\mathsf{F}_1(K_v, \cdot)$ or a random function from F_0. Note that the leakage is on a key for a function from F_1 regardless of the challenge bit, meaning even if $c = 0$, we leak on the key K_v drawn from $\mathsf{F}_1.\mathsf{K}$. The second oracle is as before. The advantage of adversary $\mathcal{A}$ is

$$\mathsf{Adv}^{\mathsf{dist}}_{\mathsf{F}_0,\mathsf{F}_1,\mathsf{Out}}(\mathcal{A}) = 2\Pr[\text{DIST}_{\mathsf{F}_0,\mathsf{F}_1,\mathsf{Out}}(\mathcal{A})] - 1 \tag{4}$$

$$= \Pr[\,\text{DIST}_{\mathsf{F}_0,\mathsf{F}_1,\mathsf{Out}}(\mathcal{A}) \,|\, c = 1\,] - (1 - \Pr[\,\text{DIST}_{\mathsf{F}_0,\mathsf{F}_1,\mathsf{Out}}(\mathcal{A}) \,|\, c = 0\,])\,. \tag{5}$$

This generalizes the basic metric because $\mathsf{Adv}^{\mathsf{dist}}_{\mathsf{F}_0,\mathsf{F}_1}(\mathcal{A}) = \mathsf{Adv}^{\mathsf{dist}}_{\mathsf{F}_0,\mathsf{F}_1,\mathsf{Out}}(\mathcal{A})$ where Out is the function that returns ε on all inputs.

As a special case we get a metric of multi-user prf security under leakage. Let F be a function family and let A be the family of all functions from F.D to F.R. Let Out: $\mathsf{F.K} \rightarrow \mathsf{Out.R}$. Let $\mathsf{Adv}^{\mathsf{prf}}_{\mathsf{F},\mathsf{Out}}(\mathcal{A}) = \mathsf{Adv}^{\mathsf{dist}}_{\mathsf{F},\mathsf{A},\mathsf{Out}}(\mathcal{A})$.

NAIVE MU TO SU REDUCTION. Multi-user security for PRFs was first explicitly considered in [5]. They used a hybrid argument to show that the prf advantage of an adversary $\mathcal{A}$ against u users is at most u times the prf advantage of an adversary of comparable resources against a single user. The argument extends to the case where instead of prf advantage we consider distance and where leakage is present. This is summarized in Lemma 1 below.

We state this lemma to emphasize that mu security is not qualitatively different from su security, at least in this setting. The question is what is the quantitative difference. The lemma represents the naive bound, which always holds. The interesting element is that for the 2-tier augmented cascade, Theorem 4 shows that one can do better: the mu advantage is not a factor u less than the single-user advantage, but about the same. In the proof of the lemma in [3] we specify the adversary for the sake of making the reduction concrete but we omit the standard hybrid argument that establishes that this works.

Lemma 1. *Let* $\mathsf{F}_0, \mathsf{F}_1$ *be function families with* $\mathsf{F}_0.\mathsf{D} = \mathsf{F}_1.\mathsf{D}$ *and let* $\mathsf{Out}\colon \mathsf{F}_1.\mathsf{K} \to \mathsf{Out}.\mathsf{R}$ *be an output transform. Let* $\mathcal{A}$ *be an adversary making at most* u *queries to its* NEW *oracle and at most* q *queries to its* FN *oracle. The proof specifies an adversary* $\mathcal{A}_1$ *making one query to its* NEW *oracle and at most* q *queries to its* FN *oracle such that*

$$\mathsf{Adv}^{\mathsf{dist}}_{\mathsf{F}_0,\mathsf{F}_1,\mathsf{Out}}(\mathcal{A}) \leq u \cdot \mathsf{Adv}^{\mathsf{dist}}_{\mathsf{F}_0,\mathsf{F}_1,\mathsf{Out}}(\mathcal{A}_1). \tag{6}$$

The running time of $\mathcal{A}_1$ *is that of* $\mathcal{A}$ *plus the time for* u *computations of* F_0 *or* F_1. ■

5 The Augmented Cascade and Its Analysis

We first present a generalization of the basic cascade construction that we call the 2-tier cascade. We then present the augmented (2-tier) cascade construction and analyze its security.

2-TIER CASCADE CONSTRUCTION. Let $\mathcal{K}$ be a set. Let g, h be function families such that $\mathsf{g}\colon \mathsf{g}.\mathsf{K} \times \mathsf{g}.\mathsf{D} \to \mathcal{K}$ and $\mathsf{h}\colon \mathcal{K} \times \mathsf{h}.\mathsf{D} \to \mathcal{K}$. Thus, outputs of both g and h can be used as keys for h. This is the basis of our 2-tier version of the cascade. This is a function family $\mathbf{CSC}[\mathsf{g},\mathsf{h}]\colon \mathsf{g}.\mathsf{K} \times [\![\mathsf{g}.\mathsf{D}, \mathsf{h}.\mathsf{D}]\!] \to \mathcal{K}$. That is, a key is one for g. An input —as per the notation $[\![\cdot,\cdot]\!]$ defined in Sect. 3— is a vector $\mathbf{X}$ of length at least one whose first component is in $\mathsf{g}.\mathsf{D}$ and the rest of whose components are in $\mathsf{h}.\mathsf{D}$. Outputs are in $\mathcal{K}$. The function itself is defined as follows:

<u>Function $\mathbf{CSC}[\mathsf{g},\mathsf{h}](K,\mathbf{X})$</u>
$n \leftarrow |\mathbf{X}|$; $Y \leftarrow \mathsf{g}(K,\mathbf{X}[1])$
For $j = 2,\ldots,n$ do $Y \leftarrow \mathsf{h}(Y,\mathbf{X}[j])$
Return Y

We say that a function family G is a 2-tier cascade if $\mathsf{G} = \mathbf{CSC}[\mathsf{g},\mathsf{h}]$ for some g, h. If $\mathsf{f}\colon \mathcal{K}\times\mathsf{f}.\mathsf{D} \to \mathcal{K}$ then its basic cascade is recovered as $\mathbf{CSC}[\mathsf{f},\mathsf{f}]\colon \mathcal{K}\times\mathsf{f}.\mathsf{D}^{+} \to \mathcal{K}$. We will also denote this function family by f^*.

Recall that even if $\mathsf{f}\colon \{0,1\}^c\times\{0,1\}^b \to \{0,1\}^c$ is a PRF, f^* is not a PRF due to the extension attack. It is shown by BCK2 [5] to be a PRF when the adversary is restricted to prefix-free queries. When $b = 1$ and the adversary is restricted to queries of some fixed length ℓ, the cascade f^* is the GGM construction of a PRF from a PRG [18]. Bernstein [7] considers a generalization of the basic cascade in which the function applied depends on the block index and proves PRF security for any fixed number ℓ of blocks.

Our generalization to the 2-tier cascade has two motivations and corresponding payoffs. First, it will allow us to reduce mu security to su security in a simple, modular and tight way, the idea being that mu security of the basic cascade is su security of the 2-tier one for a certain choice of the 1st tier family. Second, it will allow us to analyze the blackbox AMAC construction in which the cascade is not keyed directly but rather the key is put in the input to the hash function.

THE AUGMENTED CASCADE. With $\mathcal{K}, \mathsf{g}, \mathsf{h}$ as above let $\mathsf{Out}\colon \mathcal{K} \to \mathsf{Out.R}$ be a function we call the output transform. The augmented (2-tier) cascade $\mathbf{ACSC}[\mathsf{g}, \mathsf{h}, \mathsf{Out}]\colon \mathsf{g.K} \times [\![\mathsf{g.D}, \mathsf{h.D}]\!] \to \mathsf{Out.R}$ is the composition of Out with $\mathbf{CSC}[\mathsf{g}, \mathsf{h}]$, namely $\mathbf{ACSC}[\mathsf{g}, \mathsf{h}, \mathsf{Out}] = \mathsf{Out} \circ \mathbf{CSC}[\mathsf{g}, \mathsf{h}]$, where composition was defined above. In code:

Function $\mathbf{ACSC}[\mathsf{g}, \mathsf{h}, \mathsf{Out}](K, \mathbf{X})$

$Y \leftarrow \mathbf{CSC}[\mathsf{g}, \mathsf{h}](K, \mathbf{X})$; $Z \leftarrow \mathsf{Out}(Y)$
Return Z

We say that a function family G^+ is an augmented (2-tier) cascade if $\mathsf{G}^+ = \mathbf{ACSC}[\mathsf{g}, \mathsf{h}, \mathsf{Out}]$ for some $\mathsf{g}, \mathsf{h}, \mathsf{Out}$.

The natural goal is that an augmented cascade G^+ be a PRF. This however is clearly not true for all Out. For example Out may be a constant function, or a highly irregular one. Rather than restrict Out at this point we target a general result that would hold for any Out. Namely we aim to show that $\mathbf{ACSC}[\mathsf{g}, \mathsf{h}, \mathsf{Out}]$ is close under our distance metric to the result of applying Out to a random function. Next we formalize and prove this.

SINGLE-USER SECURITY OF 2-TIER AUGMENTED CASCADE. Given $\mathsf{g}, \mathsf{h}, \mathsf{Out}$ defining the 2-tier augmented cascade $\mathsf{Out} \circ \mathbf{CSC}[\mathsf{g}, \mathsf{h}]$, we want to upper bound $\mathsf{Adv}^{\mathsf{dist}}_{\mathsf{Out} \circ \mathsf{A}, \mathsf{Out} \circ \mathbf{CSC}[\mathsf{g}, \mathsf{h}]}(\mathcal{A})$ for an adversary $\mathcal{A}$ making one NEW query, where A is the family of all functions with the same domain as $\mathbf{CSC}[\mathsf{g}, \mathsf{h}]$. We will do this in two steps. First, in Lemma 2, we will consider the case that the first tier is a random function, meaning $\mathsf{g} = \mathsf{r}$ is the family of all functions with the same domain and range as g. Then, in Theorem 3, we will use Lemma 2 to analyze the general case where g is a PRF. Most interestingly we will later use these single-user results to easily obtain, in Theorem 4, bounds for multi-user security that are essentially as good as for single-user security. This showcases a feature of the 2-tier cascade that is rare amongst PRFs. We now proceed to the above-mentioned lemma.

Lemma 2. *Let $\mathcal{K}, \mathcal{D}$ be non-empty sets. Let $\mathsf{h}\colon \mathcal{K} \times \mathsf{h.D} \to \mathcal{K}$ be a function family. Let r be the family of all functions with domain $\mathcal{D}$ and range $\mathcal{K}$. Let $\mathsf{Out}\colon \mathcal{K} \to \mathsf{Out.R}$ be an output transform. Let A be the family of all functions with domain $[\![\mathcal{D}, \mathsf{h.D}]\!]$ and range $\mathcal{K}$. Let $\mathcal{A}$ be an adversary making exactly one query to its* NEW *oracle followed by at most q queries to its* FN *oracle, the second* *argument of each of the queries in the latter case being a vector* $\mathbf{X} \in [\![\mathcal{D}, \mathsf{h.D}]\!]$ *with* $2 \leq |\mathbf{X}| \leq \ell + 1$. *Let adversary* $\mathcal{A}_\mathsf{h}$ *be as in Fig. 2. Then*

$$\mathsf{Adv}^{\mathrm{dist}}_{\mathsf{Out} \circ \mathsf{A}, \mathsf{Out} \circ \mathbf{CSC}[\mathsf{r},\mathsf{h}]}(\mathcal{A}) \leq \ell \cdot \mathsf{Adv}^{\mathrm{prf}}_{\mathsf{h},\mathsf{Out}}(\mathcal{A}_\mathsf{h}). \tag{7}$$

Adversary $\mathcal{A}_\mathsf{h}$ *makes at most* q *queries to its* New *oracle and at most* q *queries to its* Fn *oracle. Its running time is that of* $\mathcal{A}$ *plus the time for* $q\ell$ *computations of* h. ■

With the first tier being a random function, Lemma 2 is bounding the single-user ($\mathcal{A}$ makes one New query) distance of the augmented 2-tier cascade to the result of applying Out to a random function under our distance metric. The bound of Eq. (7) is in terms of the multi-user security of h as a PRF and grows linearly with one less than the maximum number of blocks in a query.

We note that we could apply Lemma 1 to obtain a bound in terms of the single-user PRF security of h, but this is not productive. Instead we will go the other way, later bounding the multi-user security of the 2-tier augmented cascade in terms of the multi-user PRF security of its component functions.

The proof below follows the basic paradigm of the proof of BCK2 [5], which is itself an extension of the classic proof of GGM [18]. However there are several differences: (1) The cascade in BCK2 is single-tier and non-augmented, meaning both the r component and Out are missing (2) BCK2 assume the adversary queries are prefix-free, meaning no query is a prefix of another, an assumption we do not make (3) BCK2 bounds prf security, while we bound the distance.

Proof (Lemma 2). Consider the hybrid games and adversaries in Fig. 2. The following chain of equalities establishes Eq. (7) and will be justified below:

$$\begin{aligned} \ell \cdot \mathsf{Adv}^{\mathrm{prf}}_{\mathsf{h},\mathsf{Out}}(\mathcal{A}_\mathsf{h}) &= \textstyle\sum_{g=1}^{\ell} \mathsf{Adv}^{\mathrm{prf}}_{\mathsf{h},\mathsf{Out}}(\mathcal{A}_g) & (8)\\ &= \textstyle\sum_{g=1}^{\ell} \Pr[\mathrm{H}_{g-1}] - \Pr[\mathrm{H}_g] & (9)\\ &= \Pr[\mathrm{H}_0] - \Pr[\mathrm{H}_\ell] & (10)\\ &= \mathsf{Adv}^{\mathrm{dist}}_{\mathsf{Out} \circ \mathsf{A}, \mathsf{Out} \circ \mathbf{CSC}[\mathsf{r},\mathsf{h}]}(\mathcal{A}) & (11) \end{aligned}$$

Adversary $\mathcal{A}_\mathsf{h}$ (bottom left of Fig. 2) picks g at random in the range $1, \ldots, \ell$ and runs adversary $\mathcal{A}_g$ (right of Fig. 2) so $\mathsf{Adv}^{\mathrm{prf}}_{\mathsf{h},\mathsf{Out}}(\mathcal{A}_\mathsf{h}) = (1/\ell) \cdot \sum_{g=1}^{\ell} \mathsf{Adv}^{\mathrm{prf}}_{\mathsf{h},\mathsf{Out}}(\mathcal{A}_g)$, which explains Eq. (8). For the rest we begin by trying to picture what is going on.

We imagine a tree of depth $\ell + 1$, meaning it has $\ell + 2$ levels. The levels are numbered $0, 1, \ldots, \ell + 1$, with 0 being the root. The root has $|\mathcal{D}|$ children while nodes at levels $1, \ldots, \ell$ have $|\mathsf{h.D}|$ children each. A query $\mathbf{X}$ of $\mathcal{A}$ in game $\mathrm{DIST}_{\mathsf{Out} \circ \mathsf{A}, \mathsf{Out} \circ \mathbf{CSC}[\mathsf{r},\mathsf{h}], \mathsf{Out}}(\mathcal{A})$ specifies a path in this tree starting at the root and terminating at a node at level $n = |\mathbf{X}|$. Both the path and the final node are viewed as named by $\mathbf{X}$. To a queried node $\mathbf{X}$ we associate two labels, an internal label $T_1[\mathbf{X}] \in \mathcal{K}$ and an external label $T_2[\mathbf{X}] = \mathsf{Out}(T_1[\mathbf{X}]) \in \mathsf{Out.R}$. The external label is the response to query $\mathbf{X}$. Since the first component of our 2-tier cascade is the family r of all functions from $\mathcal{D}$ to $\mathcal{K}$, we can view

Game H_s $(0 \le s \le \ell)$

$b' \leftarrow_{\$} \mathcal{A}^{\text{NEW}^*,\text{FN}^*}$
Return $(b' = 1)$

$\text{NEW}^*()$

$f \leftarrow_{\$} \mathsf{A}$

$\text{FN}^*(i, \mathbf{X})$

$n \leftarrow |\mathbf{X}|$
If $(n \le s)$ then $Y \leftarrow f(\mathbf{X})$
Else
  $Y \leftarrow f(\mathbf{X}[1..s+1])$
  For $j = s{+}2, \ldots, n$ do $Y \leftarrow \mathsf{h}(Y, \mathbf{X}[j])$
$T_1[\mathbf{X}] \leftarrow Y$; $T_2[\mathbf{X}] \leftarrow \mathsf{Out}(T_1[\mathbf{X}])$
Return $T_2[\mathbf{X}]$

Adversary $\mathcal{A}_{\mathsf{h}}^{\text{NEW},\text{FN}}$

$g \leftarrow_{\$} \{1, \ldots, \ell\}$; $b' \leftarrow_{\$} \mathcal{A}_g^{\text{NEW},\text{FN}}$
Return b'

Adversary $\mathcal{A}_g^{\text{NEW},\text{FN}}$ $(1 \le g \le \ell)$

$v \leftarrow 0$; $b' \leftarrow_{\$} \mathcal{A}^{\text{NEW}^*,\text{FN}^*}$; Return b'

$\text{NEW}^*()$

$\text{FN}^*(i, \mathbf{X})$

$n \leftarrow |\mathbf{X}|$
If $(n \le g-1)$ then
  If (not $T_1[\mathbf{X}]$) then
    $T_1[\mathbf{X}] \leftarrow_{\$} \mathcal{K}$; $T_2[\mathbf{X}] \leftarrow \mathsf{Out}(T_1[\mathbf{X}])$
If $(n \ge g)$ then
  If (not $U[\mathbf{X}[1..g]]$) then
    $v \leftarrow v+1$; $U[\mathbf{X}[1..g]] \leftarrow v$
    $T_2[\mathbf{X}[1..g]] \leftarrow \text{NEW}()$
If $(n \ge g+1)$ then
  $T_1[\mathbf{X}[1..g{+}1]] \leftarrow \text{FN}(U[\mathbf{X}[1..g]], \mathbf{X}[g{+}1])$
  For $j = g+2, \ldots, n$ do
    $T_1[\mathbf{X}[1..j]] \leftarrow \mathsf{h}(T_1[\mathbf{X}[1..j-1]], \mathbf{X}[j])$
  $T_2[\mathbf{X}] \leftarrow \mathsf{Out}(T_1[\mathbf{X}])$
Return $T_2[\mathbf{X}]$

Fig. 2. Games and adversaries for proof of Theorem 2.

$\mathrm{DIST}_{\mathsf{Out} \circ \mathsf{A}, \mathsf{Out} \circ \mathbf{CSC}[\mathsf{r},\mathsf{h}], \mathsf{Out}}(\mathcal{A})$ as picking $T_1[\mathbf{X}[1]]$ at random from $\mathcal{K}$ and then setting $T_1[\mathbf{X}] = \mathsf{h}^*(T_1[\mathbf{X}[1]], \mathbf{X}[2 \ldots n])$ for all queries $\mathbf{X}$ of $\mathcal{A}$.

Now we consider the hybrid games $\mathrm{H}_0, \ldots, \mathrm{H}_\ell$ of Fig. 2. They simulate $\mathcal{A}$'s NEW, FN oracles via procedures $\text{NEW}^*, \text{FN}^*$, respectively. By assumption $\mathcal{A}$ makes exactly one NEW^* query, and this will have to be its first. In response H_s picks at random a function f: $[\![\mathcal{D}, \mathcal{K}]\!] \rightarrow \mathcal{K}$. A query FN^* has the form $(i, \mathbf{X})$ but here i can only equal 1 and is ignored in responding. By assumption $2 \le |\mathbf{X}| \le \ell$. The game populates nodes at levels $2, \ldots, s$ of the tree with $T_1[\cdot]$ values that are obtained via f and thus are random elements of $\mathcal{K}$. For a node $\mathbf{X}$ at level $n \ge s+1$, the $T_1[\mathbf{X}[1 \ldots s+1]]$ value is obtained at random and then further values (if needed, meaning if $n \ge s+2$) are computed by applying the cascade h^* with key $T_1[\mathbf{X}[1 \ldots s+1]]$ to input $\mathbf{X}[s+2 \ldots n]$.

Consider game H_0, where $s = 0$. By assumption $n \ge 2$ so we will always be in the case $n \ge s+1$. In the Else statement, $Y \leftarrow f(\mathbf{X}[1])$ is initialized as a random element of $\mathcal{K}$. With this Y as the key, h^* is then applied to $\mathbf{X}[2 \ldots n]$ to get $T_1[\mathbf{X}]$. This means H_0 exactly mimics the $c = 1$ case of game $\mathrm{DIST}_{\mathsf{Out} \circ \mathsf{A}, \mathsf{Out} \circ \mathbf{CSC}[\mathsf{r},\mathsf{h}], \mathsf{Out}}(\mathcal{A})$, so that

$$\Pr[\mathrm{H}_0] = \Pr[\,\mathrm{DIST}_{\mathsf{Out} \circ \mathsf{A}, \mathsf{Out} \circ \mathbf{CSC}[\mathsf{r},\mathsf{h}]}(\mathcal{A}) \,|\, c = 1\,]. \tag{12}$$

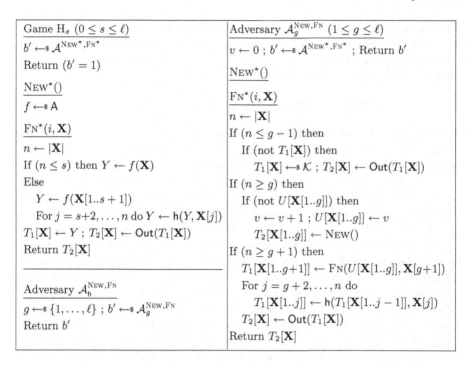

At the other extreme, consider game H_ℓ, where $s = \ell$. By assumption $n \le \ell+1$, yielding two cases. If $n \le \ell$ we are in the $n \le s$ case and the game, via f, the assigns $T_1[\mathbf{X}]$ a random value. If $n = \ell + 1$ we are in the $n \geq s + 1$ case, but the For loop does nothing so $T_1[\mathbf{X}]$ is again random. This means H_ℓ mimics the $c = 0$ case of game $\mathrm{DIST}_{\mathsf{Out} \circ \mathsf{A}, \mathsf{Out} \circ \mathbf{CSC}[\mathsf{r},\mathsf{h}], \mathsf{Out}}(\mathcal{A})$, except returning true exactly when the latter returns false. Thus

$$\Pr[\mathrm{H}_\ell] = 1 - \Pr[\,\mathrm{DIST}_{\mathsf{Out} \circ \mathsf{A}, \mathsf{Out} \circ \mathbf{CSC}[\mathsf{r},\mathsf{h}]}(\mathcal{A}) \,|\, c = 0\,]. \tag{13}$$

We will justify Eq. (9) in a bit but we can now dispense with the rest of the chain. Equation (10) is obvious because the sum "telescopes". Equation (11) follows from Eqs. (12) and (13) and the formulation of dist advantage of Eq. (5).

It remains to justify Eq. (9), for which we consider the adversaries $\mathcal{A}_1, \ldots, \mathcal{A}_\ell$ on the right side of Fig. 2. Adversary $\mathcal{A}_g$ is playing the PRF, formally game $\mathrm{DIST}_{\mathsf{B},\mathsf{h}}$ on the left of Fig. 1 in our notation, with B the family of all functions from $\mathsf{h}.\mathsf{D}$ to $\mathcal{K}$. It thus has oracles New, Fn. It will make crucial use of the assumed multi-user security of h, meaning its ability to query New many times, keeping track in variable u of the number of instances it creates. It simulates the oracles of $\mathcal{A}$ of the same names via procedures New*, Fn*, sampling functions lazily rather than directly as in the games. Arrays T_1, T_2, U are assumed initially to be everywhere $\perp$ and get populated as the adversary assigns values to entries. A test of the form "If (not $T_1[\mathbf{X}]$) ..." returns true if $T_1[\mathbf{X}] = \perp$, meaning has not yet been initialized. In response to the (single) New* query of $\mathcal{A}$, adversary $\mathcal{A}_g$ does nothing. Following that, its strategy is to have the $T_1[\cdot]$ values of level g nodes populated, not explicitly, but implicitly by the keys in game $\mathrm{DIST}_{\mathsf{B},\mathsf{h}}$ created by the adversary's own New queries, using array U to keep track of the user index associated to a node. $T_1[\cdot]$ values for nodes at levels $1, \ldots, g-1$ are random. At level $g+1$, the $T_1[\cdot]$ values are obtained via the adversary's Fn oracle, and from then on via direct application of the cascade h^*. One crucial point is that, if $\mathcal{A}_g$ does not know the $T_1[\cdot]$ values at level g, how does it respond to a length g query $\mathbf{X}$ with the right $T_2[\cdot]$ value? This is where the leakage enters, the response being the leakage provided by the New oracle. The result is that for every $g \in \{1, \ldots, \ell\}$ we have

$$\Pr[\,\mathrm{DIST}_{\mathsf{B},\mathsf{h}}(\mathcal{A}_g) \,|\, c = 1\,] = \Pr[\mathrm{H}_{g-1}] \tag{14}$$

$$1 - \Pr[\,\mathrm{DIST}_{\mathsf{B},\mathsf{h}}(\mathcal{A}_g) \,|\, c = 0\,] = \Pr[\mathrm{H}_g], \tag{15}$$

where c is the challenge bit in game $\mathrm{DIST}_{\mathsf{B},\mathsf{h}}$. Thus

$$\begin{aligned} \mathsf{Adv}^{\mathsf{prf}}_{\mathsf{h},\mathsf{Out}}(\mathcal{A}_g) &= \Pr[\,\mathrm{DIST}_{\mathsf{B},\mathsf{h}}(\mathcal{A}_g) \,|\, c = 1\,] - (1 - \Pr[\,\mathrm{DIST}_{\mathsf{B},\mathsf{h}}(\mathcal{A}_g) \,|\, c = 0\,]) \\ &= \Pr[\mathrm{H}_{g-1}] - \Pr[\mathrm{H}_g]. \end{aligned} \tag{16}$$

This justifies Eq. (9). ∎

We now extend the above to the case where the first tier g of the 2-tier cascade is a PRF rather than a random function. We will exploit PRF security of g to reduce this to the prior case. Since the proof uses standard methods, it is relegated to [3].

Theorem 3 *Let $\mathcal{K}$ be a non-empty set. Let* $\mathsf{g}\colon \mathsf{g.K} \times \mathsf{g.D} \to \mathcal{K}$ *and* $\mathsf{h}\colon \mathcal{K} \times \mathsf{h.D} \to \mathcal{K}$ *be function families. Let* $\mathsf{Out}\colon \mathcal{K} \to \mathsf{Out.R}$ *be an output transform. Let* A *be the family of all functions with domain* $[\![\mathsf{g.D}, \mathsf{h.D}]\!]$ *and range* $\mathcal{K}$. *Let* $\mathcal{A}$ *be an adversary making exactly one query to its* New *oracle followed by at most* q *queries to its* Fn *oracle, the second argument of each of the queries in the latter case being a vector* $\mathbf{X} \in [\![\mathsf{g.D}, \mathsf{h.D}]\!]$ *with* $2 \leq |\mathbf{X}| \leq \ell + 1$. *The proof shows how to construct adversaries* $\mathcal{A}_{\mathsf{h}}, \mathcal{A}_{\mathsf{g}}$ *such that*

$$\mathsf{Adv}^{\mathsf{dist}}_{\mathsf{Out} \circ \mathsf{A}, \mathsf{Out} \circ \mathbf{CSC}[\mathsf{g},\mathsf{h}]}(\mathcal{A}) \leq \ell \cdot \mathsf{Adv}^{\mathsf{prf}}_{\mathsf{h},\mathsf{Out}}(\mathcal{A}_{\mathsf{h}}) + 2\,\mathsf{Adv}^{\mathsf{prf}}_{\mathsf{g}}(\mathcal{A}_{\mathsf{g}}). \tag{17}$$

Adversary $\mathcal{A}_{\mathsf{h}}$ *makes at most* q *queries to its* New *oracle and at most* q *queries to its* Fn *oracle. Adversary* $\mathcal{A}_{\mathsf{g}}$ *makes one query to its* New *oracle and at most* q *queries to its* Fn *oracle. The running time of both constructed adversaries is about that of* $\mathcal{A}$ *plus the time for* $q\ell$ *computations of* h. ■

Multi-user security of 2-tier augmented cascade. We now want to assess the multi-user security of a 2-tier augmented cascade. This means we want to bound $\mathsf{Adv}^{\mathsf{dist}}_{\mathsf{Out} \circ \mathsf{A}, \mathsf{Out} \circ \mathbf{CSC}[\mathsf{g},\mathsf{h}]}(\mathcal{A})$ with everything as in Theorem 3 above except that $\mathcal{A}$ can now make any number u of New queries rather than just one. We could do this easily by applying Lemma 1 to Theorem 3, resulting in a bound that is u times the bound of Eq. (17). We consider Theorem 4 below the most interesting result of this section. It says one can do much better, and in fact the bound for the multi-user case is not much different from that for the single-user case.

Theorem 4 *Let $\mathcal{K}$ be a non-empty set. Let* $\mathsf{g}\colon \mathsf{g.K} \times \mathsf{g.D} \to \mathcal{K}$ *and* $\mathsf{h}\colon \mathcal{K} \times \mathsf{h.D} \to \mathcal{K}$ *be function families. Let* $\mathsf{Out}\colon \mathcal{K} \to \mathsf{Out.R}$ *be an output transform. Let* A *be the family of all functions with domain* $[\![\mathsf{g.D}, \mathsf{h.D}]\!]$ *and range* $\mathcal{K}$. *Let* $\mathcal{A}$ *be an adversary making at most* u *queries to its* New *oracle and at most* q *queries to its* Fn *oracle, the second argument of each of the queries in the latter case being a vector* $\mathbf{X} \in [\![\mathsf{g.D}, \mathsf{h.D}]\!]$ *with* $2 \leq |\mathbf{X}| \leq \ell + 1$. *The proof shows how to construct adversaries* $\mathcal{A}_{\mathsf{h}}, \mathcal{A}_{\mathsf{g}}$ *such that*

$$\mathsf{Adv}^{\mathsf{dist}}_{\mathsf{Out} \circ \mathsf{A}, \mathsf{Out} \circ \mathbf{CSC}[\mathsf{g},\mathsf{h}]}(\mathcal{A}) \leq \ell \cdot \mathsf{Adv}^{\mathsf{prf}}_{\mathsf{h},\mathsf{Out}}(\mathcal{A}_{\mathsf{h}}) + 2\,\mathsf{Adv}^{\mathsf{prf}}_{\mathsf{g}}(\mathcal{A}_{\mathsf{g}}). \tag{18}$$

Adversary $\mathcal{A}_{\mathsf{h}}$ *makes at most* q *queries to its* New *oracle and at most* q *queries to its* Fn *oracle. Adversary* $\mathcal{A}_{\mathsf{g}}$ *makes* u *queries to its* New *oracle and at most* q *queries to its* Fn *oracle. The running time of both constructed adversaries is about that of* $\mathcal{A}$ *plus the time for* $q\ell$ *computations of* h. ■

A comparison of Theorems 3 and 4 shows that the bound of Eq. (18) is the same as that of Eq. (17). So where are we paying for u now not being one? It is reflected only in the resources of adversary $\mathcal{A}_{\mathsf{g}}$, the latter in Theorem 4 making u queries to its New oracle rather than just one in Theorem 3.

The proof below showcases one of the advantages of the 2-tier cascade over the basic single-tier one. Namely, by appropriate choice of instantiation of the first tier, we can reduce multi-user security to single-user security in a modular way. In this way we avoid re-entering the proofs above. Indeed, the ability to do this is one of the main reasons we introduced the 2-tier cascade.

Proof (Theorem 4). Let $\mathcal{D} = [1 \ldots u]$. Let $\overline{\mathsf{r}}$ be the family of all functions with domain $\mathcal{D}$ and range $\mathsf{g.K}$. Let function family $\overline{\mathsf{g}}\colon \overline{\mathsf{r}}.\mathsf{K} \times (\mathcal{D} \times \mathsf{g.D}) \to \mathcal{K}$ be defined by $\overline{\mathsf{g}}(f,(i,x)) = \mathsf{g}(f(i),x)$. Let B be the family of all functions with domain $[\![\mathcal{D} \times \mathsf{g.D}, \mathsf{h.D}]\!]$ and range $\mathcal{K}$. The main observation is as follows. Suppose $i \in \mathcal{D}$ and $\mathbf{X} \in [\![\mathsf{g.D}, \mathsf{h.D}]\!]$. Let $\mathbf{Y} \in [\![\mathcal{D} \times \mathsf{g.D}, \mathsf{h.D}]\!]$ be defined by $\mathbf{Y}[1] = (i, \mathbf{X}[1])$ and $\mathbf{Y}[j] = \mathbf{X}[j]$ for $2 \le j \le |\mathbf{X}|$. Let $f\colon \mathcal{D} \to \mathsf{g.K}$ be a key for $\overline{\mathsf{g}}$. Then $f(i) \in \mathsf{g.K}$ is a key for g, and

$$\mathbf{CSC}[\overline{\mathsf{g}},\mathsf{h}](f,\mathbf{Y}) = \mathbf{CSC}[\mathsf{g},\mathsf{h}](f(i),\mathbf{X}). \tag{19}$$

Think of $f(i)$ as the key for instance i. Then Eq. (19) allows us to obtain values of $\mathbf{CSC}[\mathsf{g},\mathsf{h}]$ for different instances $i \in \mathcal{D}$ via values of $\mathbf{CSC}[\overline{\mathsf{g}},\mathsf{h}]$ on a single instance with key f. This will allow us to reduce the multi-user security of $\mathbf{CSC}[\mathsf{g},\mathsf{h}]$ to the single-user security of $\mathbf{CSC}[\overline{\mathsf{g}},\mathsf{h}]$. Theorem 3 will allow us to measure the latter in terms of the prf security of h under leakage and the (plain) prf security of $\overline{\mathsf{g}}$. The final step will be to measure the prf security of $\overline{\mathsf{g}}$ in terms of that of g.

Proceeding to the details, let adversary $\mathcal{B}$ be as follows:

Adversary $\mathcal{B}^{\text{New,Fn}}$	$\text{Fn}^*(i,\mathbf{X})$
New()	$\mathbf{Y}[1] \leftarrow (i,\mathbf{X}[1])$
$b' \leftarrow_{\$} \mathcal{A}^{\text{New}^*,\text{Fn}^*}$; Return b'	For $j = 2,\ldots,\lvert\mathbf{X}\rvert$ do $\mathbf{Y}[j] \leftarrow \mathbf{X}[j]$
New*()	$Z \leftarrow \text{Fn}(1,\mathbf{Y})$; Return Z

Then we have

$$\mathsf{Adv}^{\mathsf{dist}}_{\mathsf{Out}\circ\mathsf{A},\mathsf{Out}\circ\mathbf{CSC}[\mathsf{g},\mathsf{h}]}(\mathcal{A}) = \mathsf{Adv}^{\mathsf{dist}}_{\mathsf{Out}\circ\mathsf{B},\mathsf{Out}\circ\mathbf{CSC}[\overline{\mathsf{g}},\mathsf{h}]}(\mathcal{B}) \tag{20}$$

$$\le \ell \cdot \mathsf{Adv}^{\mathsf{prf}}_{\mathsf{h},\mathsf{Out}}(\mathcal{A}_{\mathsf{h}}) + 2\,\mathsf{Adv}^{\mathsf{prf}}_{\overline{\mathsf{g}}}(\mathcal{A}_{\overline{\mathsf{g}}}) \tag{21}$$

Adversary $\mathcal{B}$ is allowed only one New query, and begins by making it so as to initialize instance 1 in its game. It answers queries of $\mathcal{A}$ to its New oracle via procedure New*. Adversary $\mathcal{A}$ can make up to u queries to New*, but, as the absence of code for New* indicates, this procedure does nothing, meaning no action is taken when $\mathcal{A}$ makes a New* query. When $\mathcal{A}$ queries its Fn oracle, $\mathcal{B}$ answers via procedure Fn*. The query consists of an instance index i with $1 \le i \le u$ and a vector $\mathbf{X}$. Adversary $\mathcal{B}$ creates $\mathbf{Y}$ from $\mathbf{X}$ as described above. Namely it modifies the first component of $\mathbf{X}$ to pre-pend i, so that $\mathbf{Y}[1] \in \mathcal{D}\times\mathsf{g.D}$ is in the domain of $\overline{\mathsf{g}}$. It leaves the rest of the components unchanged, and then calls its own Fn oracle on vector $\mathbf{Y} \in [\![\mathcal{D} \times \mathsf{g.D}, \mathsf{h.D}]\!]$. The instance used is 1, regardless of i, since $\mathcal{B}$ has only one instance active. The result Z of Fn is returned to $\mathcal{A}$ as the answer to its query. Eq. (20) is now justified by Eq. (19), thinking of $f(i)$ as the key K_i chosen in game $\text{DIST}_{\mathsf{Out}\circ\mathsf{A},\mathsf{Out}\circ\mathbf{CSC}[\mathsf{g},\mathsf{h}]}(\mathcal{A})$ where f is the (single) key chosen in game $\text{DIST}_{\mathsf{Out}\circ\mathsf{B},\mathsf{Out}\circ\mathbf{CSC}[\overline{\mathsf{g}},\mathsf{h}]}(\mathcal{B})$. Theorem 3 applied to $\overline{\mathsf{g}},\mathsf{h}$ and adversary $\mathcal{B}$ provides the adversaries $\mathcal{A}_{\mathsf{h}}, \mathcal{A}_{\overline{\mathsf{g}}}$ of Eq. (21).

Now consider adversary $\mathcal{A}_{\mathsf{g}}$ defined as follows:

Adversary $\mathcal{A}_{\mathsf{g}}^{\text{New},\text{Fn}}$	$\text{Fn}^*(j, X)$
For $i = 1, \ldots, u$ do New()	$(i, x) \leftarrow X$; $Y \leftarrow \text{Fn}(i, x)$
$b' \leftarrow_{\$} \mathcal{A}_{\overline{\mathsf{g}}}^{\text{New}^*,\text{Fn}^*}$; Return b'	Return Y
$\text{New}^*()$	

Adversary $\mathcal{A}_{\mathsf{g}}$ begins by calling its New oracle u times to initialize u instances. It then runs $\mathcal{A}_{\overline{\mathsf{g}}}$, answering the latter's oracle queries via procedures $\text{New}^*, \text{Fn}^*$. By Theorem 3 we know that $\mathcal{A}_{\overline{\mathsf{g}}}$ makes only one New^* query. In response the procedure New^* above does nothing. When $\mathcal{A}_{\overline{\mathsf{g}}}$ makes query j, X to Fn^* we know that $j = 1$ and $X \in \mathcal{D} \times \mathsf{g.D}$. Procedure Fn^* parses X as (i, x). It then invokes its own Fn oracle with instance i and input x and returns the result Y to $\mathcal{A}_{\overline{\mathsf{g}}}$. We have

$$\mathsf{Adv}_{\mathsf{g}}^{\mathsf{prf}}(\mathcal{A}_{\mathsf{g}}) = \mathsf{Adv}_{\overline{\mathsf{g}}}^{\mathsf{prf}}(\mathcal{A}_{\overline{\mathsf{g}}}). \tag{22}$$

Equations (21) and (22) imply Eq. (18). ■

One might ask why prove Theorem 4 for a 2-tier augmented cascade $\mathsf{Out} \circ \mathbf{CSC}[\mathsf{g}, \mathsf{h}]$ instead of a single tier one $\mathsf{Out} \circ \mathbf{CSC}[\mathsf{h}, \mathsf{h}]$. Isn't the latter the one of ultimate interest in usage? We establish a more general result in Theorem 4 because it allows us to analyze AMAC itself by setting g to the dual of h [2], and also for consistency with Theorem 3.

6 Framework for Ideal-Model Cryptography

In Sect. 5 we reduced the (mu) security of the augmented cascade tightly to the assumed mu prf security of the compression function under leakage. To complete the story, we will, in Sect. 7, bound the mu prf security of an ideal compression function under leakage and thence obtain concrete bounds for the mu security of the augmented cascade in the same model. Additionally, we will consider the same questions when the compression function is not directly ideal but obtained via the Davies-Meyer transform on an ideal blockcipher, reflecting the design in popular hash functions. If we gave separate, ad hoc definitions for all these different constructions in different ideal models for different goals, it would be a lot of definitions. Accordingly we introduce a general definition of an ideal primitive (that may be of independent interest) and give a general definition of PRF security of a function family with access to an instance of an ideal primitive, both for the basic setting and the setting with leakage. A reader interested in our results on the mu prf security of ideal primitives can jump ahead to Sect. 7 and refer back here as necessary.

Idealized cryptography. We define an *ideal primitive* to simply be a function family $\mathbf{P}\colon \mathbf{P}.\mathsf{K} \times \mathbf{P}.\mathsf{D} \to \mathbf{P}.\mathsf{R}$. Below we will provide some examples but first let us show how to lift security notions to idealized models using this definition by considering the cases of interest to us, namely PRFs and PRFs under leakage.

Game $\mathrm{PRF}_{\mathsf{F},\mathbf{P}}(\mathcal{A})$	Game $\mathrm{PRF}_{\mathsf{F},\mathsf{Out},\mathbf{P}}(\mathcal{A})$
$v \leftarrow 0$	$v \leftarrow 0$
$c \leftarrow\!\$ \{0,1\}$; $\mathsf{P} \leftarrow\!\$ \mathbf{P}$; $c' \leftarrow\!\$ \mathcal{A}^{\mathrm{New,Fn,Prim}}$	$c \leftarrow\!\$ \{0,1\}$; $\mathsf{P} \leftarrow\!\$ \mathbf{P}$; $c' \leftarrow\!\$ \mathcal{A}^{\mathrm{New,Fn,Prim}}$
Return $(c = c')$	Return $(c = c')$
<u>New()</u>	<u>New()</u>
$v \leftarrow v + 1$	$v \leftarrow v + 1$; $K_v \leftarrow\!\$ \mathsf{F.K}$
If $(c = 1)$ then $F_v \leftarrow\!\$ \mathsf{F}^{\mathrm{Prim}}$	If $(c = 1)$ then $F_v \leftarrow \mathsf{F}^{\mathrm{Prim}}(K_v, \cdot)$
Else $F_v \leftarrow\!\$ \mathsf{A}$	Else $F_v \leftarrow\!\$ \mathsf{A}$
	Return $\mathsf{Out}(K_v)$
<u>Fn(i, x)</u>	<u>Fn(i, x)</u>
Return $F_i(x)$	Return $F_i(x)$
<u>Prim(x)</u>	<u>Prim(x)</u>
$y \leftarrow \mathsf{P}(x)$; Return y	$y \leftarrow \mathsf{P}(x)$; Return y

Fig. 3. Games defining prf security of function family F in the presence of an ideal primitive P. In the basic (left) case there is no leakage, while in the extended (right) case there is leakage represented by Out.

An *oracle function family* F specifies for each function P in its *oracle space* F.O a function family F^{P}: $\mathsf{F.K} \times \mathsf{F.D} \to \mathsf{F.R}$. We say F and ideal primitive $\mathbf{P}$ are *compatible* if $\{\, \mathbf{P}(\mathsf{KK}, \cdot) \ : \ \mathsf{KK} \in \mathbf{P}.\mathsf{K} \,\} \subseteq \mathsf{F.O}$, meaning instances of $\mathbf{P}$ are legitimate oracles for F. These represent constructs whose security we want to measure in an idealized model represented by $\mathbf{P}$.

We associate to $\mathsf{F}, \mathbf{P}$ and adversary $\mathcal{A}$ the game PRF in the left of Fig. 3. In this game, A is the family of all functions with domain F.D and range F.R. The game begins by picking an instance P: $\mathbf{P}.\mathsf{D} \to \mathbf{P}.\mathsf{R}$ of $\mathbf{P}$ at random. The function P is provided as oracle to F and to $\mathcal{A}$ via procedure Prim. The game is in the multi-user setting, and when $c = 1$ it selects a new instance F_v at random from the function family F^{P}. Otherwise it selects F_v to be a random function from F.D to F.R. As usual a query i, x to Fn must satisfy $1 \leq i \leq v$ and $x \in \mathsf{F.D}$. A query to Prim must be in the set $\mathbf{P}.\mathsf{D}$. We let $\mathsf{Adv}^{\mathsf{prf}}_{\mathsf{F},\mathbf{P}}(\mathcal{A}) = 2\Pr[\mathrm{PRF}_{\mathsf{F},\mathbf{P}}(\mathcal{A})] - 1$ be the advantage of $\mathcal{A}$.

We now extend this to allow leakage on the key. Let Out: $\mathsf{F.K} \to \mathsf{Out.R}$ be a function with domain F.K and range Out.R. Game PRF on the right of Fig. 3 is now associated not only to $\mathsf{F}, \mathbf{P}$ and an adversary $\mathcal{A}$ but also to Out. The advantage of $\mathcal{A}$ is $\mathsf{Adv}^{\mathsf{prf}}_{\mathsf{F},\mathsf{Out},\mathbf{P}}(\mathcal{A}) = 2\Pr[\mathrm{PRF}_{\mathsf{F},\mathsf{Out},\mathbf{P}}(\mathcal{A})] - 1$.

Capturing particular ideal models. The above framework allows us to capture the random oracle model, ideal cipher model and many others as different choices of the ideal primitive $\mathbf{P}$. Not all of these are relevant to our paper but we discuss them to illustrate how the framework captures known settings.

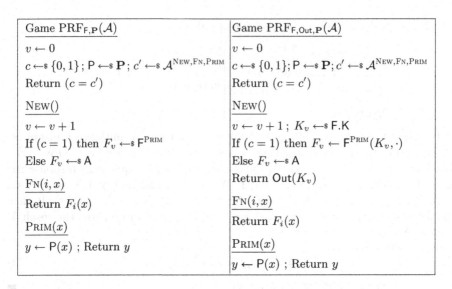

Let $\mathcal{Y}$ be a non-empty set. Let $\mathbf{P}.\mathsf{K}$ be the set of all functions $\mathsf{P}\colon \{0,1\}^* \to \mathcal{Y}$. (Each function is represented in some canonical way, in this case for example as a vector over $\mathcal{Y}$ of infinite length.) Let $\mathbf{P}\colon \mathbf{P}.\mathsf{K} \times \{0,1\}^* \to \mathcal{Y}$ be defined by $\mathbf{P}(\mathsf{P}, x) = \mathsf{P}(x)$. Then $\mathsf{P} \leftarrow\!\!\$\, \mathbf{P}$ is a random oracle with domain $\{0,1\}^*$ and range $\mathcal{Y}$. In this case, an oracle function family compatible with $\mathbf{P}$ is simply a function family in the random oracle model, and its prf security in the random oracle model is measured by $\mathsf{Adv}^{\mathsf{prf}}_{\mathsf{F},\mathbf{P}}(\mathcal{A})$.

Similarly let $\mathbf{P}.\mathsf{K}$ be the set of all functions $\mathsf{P}\colon \{0,1\}^* \times \mathbb{N} \to \{0,1\}^*$ with the property that $|\mathsf{P}(x,l)| = l$ for all $(x,l) \in \{0,1\}^* \times \mathbb{N}$. Let $\mathbf{P}\colon \mathbf{P}.\mathsf{K} \times (\{0,1\}^* \times \mathbb{N}) \to \{0,1\}^*$ be defined by $\mathbf{P}(\mathsf{P},(x,l)) = \mathsf{P}(x,l)$. Then $\mathsf{P} \leftarrow\!\!\$\, \mathbf{P}$ is a variable output length random oracle with domain $\{0,1\}^*$ and range $\{0,1\}^*$.

Let $\mathcal{D}$ be a non-empty set. To capture the single random permutation model, let $\mathbf{P}.\mathsf{K}$ be the set of all permutations $\pi\colon \mathcal{D} \to \mathcal{D}$. Let $\mathbf{P}.\mathsf{D} = \mathcal{D} \times \{+,-\}$. Let $\mathbf{P}.\mathsf{R} = \mathcal{D}$. Define $\mathbf{P}(\pi,(x,+)) = \pi(x)$ and $\mathbf{P}(\pi,(y,-)) = \pi^{-1}(y)$ for all $\pi \in \mathbf{P}.\mathsf{K}$ and all $x, y \in \mathcal{D}$. An oracle for an instance $\mathsf{P} = \mathbf{P}(\pi,\cdot)$ of $\mathbf{P}$ thus allows evaluation of both π and π^{-1} on inputs of the caller's choice.

Finally we show how to capture the ideal cipher model. If $\mathcal{K}, \mathcal{D}$ are non-empty sets, a function family $E\colon \mathcal{K} \times \mathcal{D} \to \mathcal{D}$ is a blockcipher if $E(K,\cdot)$ is a permutation on $\mathcal{D}$ for every $K \in \mathcal{K}$, in which case $E^{-1}\colon \mathcal{K} \times \mathcal{D} \to \mathcal{D}$ denotes the blockcipher in which $E^{-1}(K,\cdot)$ is the inverse of the permutation $E(K,\cdot)$ for all $K \in \mathcal{K}$. Let $\mathbf{P}.\mathsf{K}$ be the set of all block ciphers $E\colon \mathcal{K} \times \mathcal{D} \to \mathcal{D}$. Let $\mathbf{P}.\mathsf{D} = \mathcal{K} \times \mathcal{D} \times \{+,-\}$. Let $\mathbf{P}.\mathsf{R} = \mathcal{D}$. Define $\mathbf{P}(E,(K,X,+)) = E(K,X)$ and $\mathbf{P}(E,(K,Y,-)) = E^{-1}(K,Y)$ for all $E \in \mathbf{P}.\mathsf{K}$ and all $X, Y \in \mathcal{D}$. An oracle for an instance $\mathsf{P} = \mathbf{P}(E,\cdot)$ of $\mathbf{P}$ thus allows evaluation of both E and E^{-1} on inputs of the caller's choice.

7 Security of the Compression Function Under Leakage

In Sect. 5 we reduced the (multi-user) security of the augmented cascade tightly to the assumed multi-user prf security of the compression function under leakage. To complete the story, we now study (bound) the multi-user prf security of the compression function under leakage. This will be done assuming the compression function is ideal. Combining these results with those of Sect. 5 we will get concrete bounds for the security of the augmented cascade for use in applications, discussed in [3].

In the (leak-free) multi-user setting, it is well known that prf security of a compression function decreases linearly in the number of users. We will show that this is an extreme case, and as the amount of leakage increases, the multi-user prf security degrades far more gracefully in the number of users (Theorem 6). This (perhaps counterintuitive) phenomenon will turn out to be essential to obtain good bounds on augmented cascades. We begin below with an informal overview of the bounds and why this phenomenon occurs.

Overview of bounds. The setting of an ideal compression function mapping $\mathcal{K} \times \mathcal{X} \to \mathcal{D}$ is formally captured, in the framework of Sect. 6, by the ideal primitive $\mathbf{F}\colon \mathbf{F}.\mathsf{K} \times (\mathcal{K} \times \mathcal{X}) \to \mathcal{K}$ defined as follows. Let $\mathbf{F}.\mathsf{K}$ be the set of all functions mapping $\mathcal{K} \times \mathcal{X} \to \mathcal{K}$ and let $\mathbf{F}(\mathsf{f},(K,X)) = \mathsf{f}(K,X)$. Now,

	$\mathsf{Adv}^{\mathsf{prf}}_{\mathsf{CF},\mathbf{F}}(\mathcal{B})$	$\mathsf{Adv}^{\mathsf{prf}}_{\mathsf{CF},\mathsf{Out},\mathbf{F}}(\mathcal{B})$
su	$\frac{q_{\mathrm{F}}}{2^c}$	$\frac{q_{\mathrm{F}}}{2^{c-r}}$
mu, trivial	$\frac{u(q+q_{\mathrm{F}})}{2^c}$	$\frac{u(q+q_{\mathrm{F}})}{2^{c-r}}$
mu, dedicated	$\frac{u^2+2uq_{\mathrm{F}}}{2^{c+1}}$	$\frac{u^2+2uq_{\mathrm{F}}+1}{2^c}+\frac{3crq_{\mathrm{F}}}{2^{c-r}}$

Fig. 4. Upper bounds on prf advantage of an adversary $\mathcal{B}$ attacking an ideal compression function mapping $\{0,1\}^c \times \mathcal{X}$ to $\{0,1\}^c$. Left: Basic case, without leakage. **Right:** With leakage Out being the truncation function that returns the first $r \le c$ bits of its output. **First row:** Single user security, q_{F} is the number of queries to the ideal compression function. **Second row:** Multi-user security as obtained trivially by applying Lemma 1 to the su bound, u is the number of users. **Third row:** Multi-user security as obtained by a dedicated analysis, with the bound in the leakage case being from Theorem 6.

the construction we are interested in is the simplest possible, namely the compression function itself. Formally, again as per Sect. 6, this means we consider the oracle function family CF whose oracle space CF.O consists of all functions $\mathsf{f}\colon \mathcal{K} \times \mathcal{X} \to \mathcal{K}$, and with $\mathsf{CF}^{\mathsf{f}} = \mathsf{f}$.

For this overview we let $\mathcal{K} = \{0,1\}^c$. We contrast the prf security of an ideal compression function along two dimensions: (1) Number of users, meaning su or mu, and (2) basic (no leakage) or with leakage. The bounds are summarized in Fig. 4 and discussed below. When we say the (i,j) table entry we mean the row i, column j entry of the table of Fig. 4.

First consider the basic (no leakage) case. We want to upper bound $\mathsf{Adv}^{\mathsf{prf}}_{\mathsf{CF},\mathbf{F}}(\mathcal{B})$ for an adversary $\mathcal{B}$ making q_{F} queries to the ideal compression function (oracle Prim) and q queries to oracle Fn. In the su setting (one New query) it is easy to see that the bound is the $(1,1)$ table entry. This is because a fairly standard argument bounds the advantage by the probability that $\mathcal{B}$ makes a Prim query containing the actual secret key K used to answer Fn queries. We refer to issuing such a query as *guessing the secret key* K. Note that this probability is actually independent of the number q of Fn queries and q does not figure in the bound. Now move to the mu setting, and let $\mathcal{B}$ make u queries to its New oracle. Entry (2,1) of the table is the trivial bound obtained via Lemma 1 applied with F_1 being our ideal compression function and F_0 a family of all functions, but one has to be careful in applying the lemma. The subtle point is that adversary $\mathcal{A}_1$ built in Lemma 1 runs $\mathcal{B}$ but makes an additional q queries to Prim to compute the function F_1, so its advantage is the $(1,1)$ table entry with q_{F} replaced by $q_{\mathrm{F}}+q$. This term gets multiplied by u according to Eq. (6), resulting in our $(1,2)$ table entry. A closer look shows one can do a tad better: the bound of the $(1,1)$

table entry extends with the caveat that a collisions between two different keys also allows the adversary to distinguish. In other words, the advantage is now bounded by the probability that $\mathcal{B}$ guesses *any* of the u keys $K_1, \ldots, K_u$, or that any two of these keys collide. This yields the $(1,3)$ entry of the table. Either way, the (well known) salient point here is that the advantage in the mu case is effectively u times the one in the su case.

We show that the growth of the advantage as a function of the number of users becomes far more favorable when the adversary obtains some leakage about the secret key under some function Out. For concreteness we take the leakage function to be truncation to r bits, meaning $\mathsf{Out} = \mathsf{TRUNC}_r$ is the function that returns the first $r \leq c$ bits of its input. (Theorem 6 will consider a general Out.) Now we seek to bound $\mathsf{Adv}^{\mathsf{prf}}_{\mathsf{CF},\mathsf{Out},\mathbf{F}}(\mathcal{B})$. Now, given only $\mathsf{TRUNC}_r(K)$ for a secret key K, then there are only 2^{c-r} candidate secret keys consistent with this leakage, thus increasing the probability that the adversary can guess the secret key. Consequently, the leakage-free bound from of the (1,1) entry generalizes to the bound of the (2,1) entry. Moving to multiple users, the (2,2) entry represents the naive bound obtained by applying Lemma 1. It is perhaps natural to expect that this is best possible as in the no-leakage case. We however observe that this is overly pessimistic. To this end, we exploit the following simple fact: *Every* PRIM *query* (K, X) *made by* $\mathcal{B}$ *to the ideal compression function can only help in guessing a key* K_i *such that* $\mathsf{Out}(K) = \mathsf{Out}(K_i)$. In particular, every PRIM query (K, X) has only roughly $m \cdot 2^{-(c-r)}$ chance of guessing one of the u keys, where m is the number of generated keys K_i such that $\mathsf{Out}(K_i) = K$. A standard balls-into-bins arguments (Lemma 5) can be used to infer that except with small probability (e.g., 2^{-c}), we always have $m \leq 2u/2^r + 3cr$ for any K. Combining these two facts yields our bound, which is the (3,2) entry of the table. Theorem 6 gives a more general result and the full proof. Note that if $r = 0$, i.e., nothing is leaked, this is close to the bound of the (1,3) entry and the bound does grow linearly with the number of users, but as r grows, the $3crq_{\mathrm{F}} \cdot 2^{-(c-r)}$ term becomes the leading one, and does *not* grow with u. We now proceed to the detailed proof of the (3,2) entry.

COMBINATORIAL PRELIMINARIES. Our statements below will depend on an appropriate multi-collision probability of the output function Out: $\mathsf{Out.D} \to \mathsf{Out.R}$. In particular, for any $X_1, \ldots, X_u \in \mathsf{Out.R}$, we first define

$$\mu(X_1, \ldots, X_u) = \max_{Y \in \mathsf{Out.R}} |\{ i \,:\, X_i = Y \}| ,$$

i.e., the number of occurrences of the most frequent value amongst $X_1, \ldots, X_u$. In particular, this is an integer between 1 and u, and $\mu(X_1, \ldots, X_u) = 1$ if all elements are distinct, whereas $\mu(X_1, \ldots, X_u) = u$ if they are all equal. (Note when $u = 1$ the function has value 1.) Then, the m-collision probability of Out for u users is defined as

$$\mathsf{P}^{\mathsf{coll}}_{\mathsf{Out}}(u, m) = \Pr_{K_1, \ldots, K_u \twoheadleftarrow \mathsf{Out.D}} [\mu(\mathsf{Out}(K_1), \ldots, \mathsf{Out}(K_u)) \geq m] . \quad (23)$$

We provide a bound on $\mathsf{P}^{\mathsf{coll}}_{\mathsf{Out}}(u, m)$ for the case where $\mathsf{Out}(K)$, for a random K, is close enough to uniform. (We stress that a combinatorial restriction on Out is necessary for this probability to be small – it would be one if Out is the contant function, for example.) To this end, denote

$$\delta(\mathsf{Out}) = \mathbf{SD}(\mathsf{Out}(\mathrm{K}), \mathrm{R}) = \frac{1}{2} \sum_{y \in \mathsf{Out.R}} \left| \Pr[\mathsf{Out}(\mathrm{K}) = y] - \frac{1}{|\mathsf{Out.R}|} \right|, \tag{24}$$

i.e., the statistical distance between Out(K), where K is uniform on Out.D, and a random variable R uniform on Out.R.

We will use the following lemma, which we prove using standard balls-into-bins techniques. The proof is deferred to [3].

Lemma 5 (Multi-collision probability). *Let* $\mathsf{Out} : \mathsf{Out.D} \to \mathsf{Out.R}$, $u \geq 1$, *and* $\lambda \geq 0$. *Then, for any* $m \leq u$ *such that*

$$m \geq \frac{2u}{|\mathsf{Out.R}|} + \lambda \ln |\mathsf{Out.R}|, \tag{25}$$

we have

$$\mathsf{P}^{\mathsf{coll}}_{\mathsf{Out}}(u, m) \leq u \cdot \delta(\mathsf{Out}) + \exp(-\lambda/3).$$ ■

We stress that the factor 2 in Eq. (25) can be omitted (one can use an additive Chernoff bound when u is sufficiently large in the proof given below, rather than a multiplicative one) at the cost of a less compact statement. As this factor will not be crucial in the following, we keep this simpler variant.

For the analysis below, we also need to use a lower bound the number of potential preimages of a given output. To this end, given Out: $\mathsf{Out.D} \to \mathsf{Out.R}$, we define

$$\rho(\mathsf{Out}) = \min_{y \in \mathsf{Out.R}} \left| \mathsf{Out}^{-1}(y) \right|.$$

Security of ideal compression functions. The following theorem establishes the multi-user security under key-leakage of a random compression function. We stress that the bound here does *not* depend on the number of queries the adversary $\mathcal{B}$ makes to oracle Fn. Also, the parameter m can be set arbitrarily in the theorem statement for better flexibility, even though our applications below will mostly use the parameters from Lemma 5.

Theorem 6. *Let* Out: $\mathcal{K} \to \mathsf{Out.R}$. *Then, for all* $m \geq 1$, *and all adversaries* $\mathcal{B}$ *making* u *queries to* New, *and* q_{F} *queries to* Prim,

$$\mathsf{Adv}^{\mathsf{prf}}_{\mathsf{CF},\mathsf{Out},\mathbf{F}}(\mathcal{B}) \leq \frac{u^2}{2|\mathcal{K}|} + \mathsf{P}^{\mathsf{coll}}_{\mathsf{Out}}(u, m) + \frac{(m-1) \cdot q_{\mathrm{F}}}{\rho(\mathsf{Out})}.$$ ■

The statement could be rendered useless whenever $\rho(\mathsf{Out}) = 1$ because a single point has a single pre-image. We note here that Theorem 6 can easily be generalized to use a "soft" version of $\rho(\mathsf{Out})$ guaranteeing that the number of preimages of a point is bounded from below by $\rho(\mathsf{Out})$, except with some small probability ϵ, at the cost of an extra additive term $u \cdot \epsilon$. This more general version will not be necessary for our applications. We also note that it is unclear how to use the *average* number of preimages of Out(K) in our proof.

Game G_0, $\boxed{G_1}$

$v \leftarrow 0$

$c' \leftarrow\!\!\$\ \mathcal{B}^{\text{NEW},\text{FN},\text{PRIM}}$

Return $(c' = 1)$

NEW()

$v \leftarrow v + 1\,;\ K_v \leftarrow\!\!\$\ \mathcal{K}$

Return $\mathsf{Out}(K_v)$

PRIM(k, x)

if $T_F[k, x] = \bot$ then
  $T_F[k, x] \leftarrow\!\!\$\ \mathcal{K}$
  If $\exists j : k = K_j$ and $T_{FN}[j, x] \neq \bot$ then
    $\mathsf{bad}_1 \leftarrow \mathsf{true}$
    $\boxed{T_F[k, x] \leftarrow T_{FN}[j, x]}$
Return $T_F[k, x]$

FN(i, x)

If $T_{FN}[i, x] = \bot$ then
  $T_{FN}[i, x] \leftarrow\!\!\$\ \mathcal{K}$
  If $T_F[K_i, x] \neq \bot$ then
    $\mathsf{bad}_1 \leftarrow \mathsf{true}$
    $\boxed{T_{FN}[i, x] \leftarrow T_F[K_i, x]}$
  else if $\exists j \neq i$: $K_j = K_i$
      and $T_{FN}[j, x] \neq \bot$ then
    $\mathsf{bad}_2 \leftarrow \mathsf{true}$
    $\boxed{T_{FN}[i, x] \leftarrow T_{FN}[j, x]}$
Return $T_{FN}[i, x]$

Fig. 5. Games G_0 and G_1 in the proof of Theorem 6. The boxed assignment statements are only executed in Game G_1, but not in Game G_0.

Proof (Theorem 6*).* The first step of the proof involves two games, G_0 and G_1, given in Fig. 5. Game G_1 is semantically equivalent to $\mathrm{PRF}_{\mathsf{CF},\mathsf{Out},\mathbf{F}}$ with challenge bit $c = 1$, except that we have modified the concrete syntax of the oracles. In particular, the randomly sampled function $\mathsf{f} \leftarrow\!\!\$\ \mathbf{F}$ is now implemented via lazy sampling, and the table entry $T_F[k, x]$ contains the value of $\mathsf{f}(k, x)$ if it has been queried. Otherwise, T_F is $\bot$ on all entries which have not been set. Also, the game keeps another table T_{FN} such that $T_{FN}[i, x]$ contains the value returned upon a query FN(i, x). Note that the game enforces that any point in time, if $T_{FN}[i, x]$ and $T_F[K_i, x]$ are both set (i.e., they are not equal $\bot$), then we also have $T_{FN}[i, x] = T_F[K_i, x]$ and that, moreover, if $K_i = K_j$, then $T_{FN}[i, x] = T_{FN}[j, x]$ whenever both are not $\bot$. Finally, whenever any of these entries is set for the first time, then it is set to a fresh random value from $\mathcal{K}$. This guarantees that the combined behavior of the FN and the PRIM oracles are the same as in $\mathrm{PRF}_{\mathsf{CF},\mathsf{Out},\mathbf{F}}$ for the case $c = 1$. Thus,

$$\Pr[\,G_1\,] = \Pr[\,\mathrm{PRF}_{\mathsf{CF},\mathsf{Out},\mathbf{F}} \mid c = 1\,].$$

It is easier to see that in game G_0, in contrast, the PRIM and FN oracles always return random values, and thus, since we are checking whether c' equals 1, rather than c, we get $\Pr[\,G_0\,] = 1 - \Pr[\,\mathrm{PRF}_{\mathsf{CF},\mathsf{Out},\mathbf{F}} \mid c = 0\,]$, and consequently,

$$\mathsf{Adv}^{\mathsf{prf}}_{\mathsf{CF},\mathsf{Out},\mathbf{F}}(\mathcal{B}) = \Pr[\,G_1\,] - \Pr[\,G_0\,].$$

Both games G_0 and G_1 also include two flags bad_1 and bad_2, initially false, which can be set to true when specific events occur. In particular, bad_1 is set

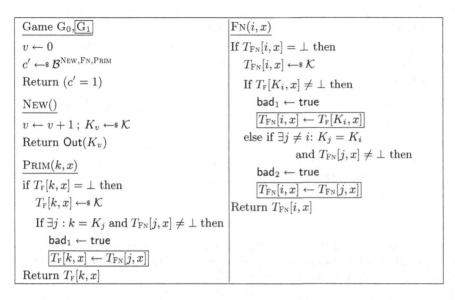

Game H_0	$\text{New}()$
$v \leftarrow 0$	$v \leftarrow v{+}1;\ K_v \leftarrow\!\!\$\ \mathcal{K};\ Y_v \leftarrow \mathsf{Out}(K_v)$
$c' \leftarrow\!\!\$\ \mathcal{B}^{\text{New},\text{Fn},\text{Prim}}$	Return Y_v
Return $(\exists j, x\colon T_{\text{F}}[K_j, x] \neq \bot)$	$\text{Prim}(k, x)$
Game H_1	if $T_{\text{F}}[k, x] = \bot$ then $T_{\text{F}}[k, x] \leftarrow\!\!\$\ \mathcal{K}$
$v \leftarrow 0$	Return $T_{\text{F}}[k, x]$
$c' \leftarrow\!\!\$\ \mathcal{B}^{\text{New},\text{Fn},\text{Prim}}$	$\text{Fn}(i, x)$
for $i = 0$ to $v - 1$ do	If $T_{\text{FN}}[i, x] = \bot$ then $T_{\text{FN}}[i, x] \leftarrow\!\!\$\ \mathcal{K}$
$\quad K'_i \leftarrow\!\!\$\ \{\, k' : \mathsf{Out}(k') = Y_i \,\}$	Return $T_{\text{FN}}[i, x]$
Return $(\exists j, x\colon T_{\text{F}}[K'_j, x] \neq \bot)$	

Fig. 6. Games H_0 and H_1 in the proof of Theorem 6. Both games share the same New, Prim, and Fn oracles, the only difference being the additional re-sampling of the secret keys K'_i in the main procedure of H_1.

whenever one of the following two events happens: Either $\mathcal{B}$ queries $\text{Fn}(i, x)$ after querying $\text{Prim}(K_i, x)$, or $\mathcal{B}$ queries $\text{Prim}(K_i, x)$ after querying $\text{Fn}(i, x)$. Moreover, bad_2 is set whenever $\mathcal{B}$ queries $\text{Fn}(i, x)$ after $\text{Fn}(j, x)$, $K_i = K_j$, and $\text{Prim}(K_i, x) = \text{Prim}(K_j, x)$ was not queried earlier. (Note that if the latter condition is not true, then bad_1 has been set already.) It is immediate to see that G_0 and G_1 are identical until $\mathsf{bad}_1 \vee \mathsf{bad}_2$ is set. Therefore, by the fundamental lemma of game playing [6],

$$\mathsf{Adv}^{\mathsf{prf}}_{\mathsf{CF},\mathsf{Out},\mathbf{F}}(\mathcal{B}) = \Pr[\, G_1 \,] - \Pr[\, G_0 \,] \leq \Pr[\, G_0 \text{ sets } \mathsf{bad}_1 \,] + \Pr[\, G_0 \text{ sets } \mathsf{bad}_2 \,]. \tag{26}$$

We immediately note that in order for bad_2 to be set in G_0, we *must* have $K_i = K_j$ for distinct $i \neq j$, i.e., two keys must collide. Since we know that at most u calls are made to New, a simple Birthday bound yields

$$\Pr[\, G_0 \text{ sets } \mathsf{bad}_2 \,] \leq \frac{u^2}{2 \cdot |\mathcal{K}|}. \tag{27}$$

The rest of the proof thus deals with the more difficult problem of bounding $\Pr[\, G_0 \text{ sets } \mathsf{bad}_1 \,]$. To simplify this task, we first introduce a new game, called H_0 (cf. Fig. 6), which behaves as G_0, except that it only checks at the end of the game whether the bad event triggering bad_1 has occurred during the interaction, in which case the game outputs true. Note that we are relaxing this check a bit further compared with G_0, allowing it to succeed as long as a query to Prim of form (K_j, x) for some j and some x was made, even if $\text{Fn}(j, x)$ was never queried before. Therefore,

$$\Pr[\, G_0 \text{ sets } \mathsf{bad}_1 \,] \leq \Pr[\, H_0 \,]. \tag{28}$$

Note that in H_0, the replies to all oracle calls made by $\mathcal{B}$ do not depend on the keys $K_1, K_2, \ldots$ anymore, *except* for the leaked values $\mathsf{Out}(K_1), \mathsf{Out}(K_2), \ldots$ returned by calls to NEW. We introduce a new and final game H_1 which modifies H_0 by pushing the sampling of the actual key values as far as possible in the game: That is, we first only gives values to $\mathcal{B}$ with the correct leakage *distribution*, and in the final phase of H_1, when computing the game output, we sample keys that are consistent with this leakage. In other words, in the final check we replace the keys $K_1, K_2, \ldots$ with *freshly* sampled key $K'_1, K'_2, \ldots$, which are uniform, under the condition that $\mathsf{Out}(K_i) = \mathsf{Out}(K'_i) = Y_i$.

It is not hard to see that $\Pr[H_0] = \Pr[H_1]$. This follows from two observations: First, for every i, the joint distribution of $(K_i, Y_i = \mathsf{Out}(K_i))$ is identical to that of $(K'_i, Y_i = \mathsf{Out}(K_i))$, since given Y_i, both K_i and K'_i are uniformly distributed over the set of pre-images of Y_i. Second, the behavior of both H_0 and H_1, before the final check to decide their outputs, only depends on values $Y_i = \mathsf{Out}(K_i)$, and *not* on the K_i's. The actual keys K_i are only used for the final check, and since the probability distributions of K_i and K'_i conditioned on $\mathsf{Out}(Y_i)$ are identical, then so are the probabilities of outputting true in games H_0 and H_1.

Thus, combining Eqs. (26), (27), and (28), we have

$$\mathsf{Adv}^{\mathsf{prf}}_{\mathsf{CF},\mathsf{Out},\mathbf{F}}(\mathcal{B}) \leq \frac{u^2}{2 \cdot |\mathcal{K}|} + \Pr[H_1]. \tag{29}$$

We are left with computing an upper bound on $\Pr[H_1]$. For this purpose, denote by $\mathcal{S}$ the set of pairs (k, x) on which $T_{\mathrm{F}}[k, x] \neq \perp$ after $\mathcal{B}$ outputs its bit c' in H_1. Also, let $\mathcal{Y}$ be the multi-set $\{Y_0, Y_1, \ldots, Y_{u-1}\}$ of values output by NEW to $\mathcal{B}$, and denote $\overline{\mathcal{Y}}$ the resulting set obtained by removing repetitions. Note that $|\mathcal{S}| \leq q_{\mathrm{F}}$ and $|\overline{\mathcal{Y}}| \leq |\mathcal{Y}| \leq u$, and the first inequality may be strict, since some elements can be repeated due to collisions $\mathsf{Out}(K_i) = \mathsf{Out}(K_j)$.

Asume that now $\mathcal{S}$ and $\mathcal{Y}$ are given and fixed. We proceed to compute the probability that H_1 outputs true conditioned on the event that $\mathcal{S}$ and $\mathcal{Y}$ have been generated. For notational help, for every $y \in \overline{\mathcal{Y}}$, also denote

$$\mathcal{S}_y = \{ (k, x) \in \mathcal{S} \;:\; \mathsf{Out}(k) = y \},$$

and let $q_y = |\mathcal{S}_y|$. Also, let n_y be the number of occurrence of $y \in \overline{\mathcal{Y}}$ in $\mathcal{Y}$. Note that except with probability $\mathsf{P}^{\mathsf{coll}}(u, m)$, we have $n_y \leq m - 1$ for all $y \in \overline{\mathcal{Y}}$, and thus

$$\begin{aligned} \Pr[H_1] &\leq \Pr\left[\exists y \in \overline{\mathcal{Y}} \;:\; n_y \geq m\right] + \Pr[H_1 \mid \forall y \in \overline{\mathcal{Y}} \;:\; n_y < m] \\ &= \mathsf{P}^{\mathsf{coll}}_{\mathsf{Out}}(u, m) + \Pr[H_1 \mid \forall y \in \overline{\mathcal{Y}} \;:\; n_y < m]. \end{aligned} \tag{30}$$

Therefore, let us assume we are given $\mathcal{S}$ and $\mathcal{Y}$ sich that $n_y \leq m-1$ for all $y \in \overline{\mathcal{Y}}$. Denote by $\Pr[H_1 \mid \mathcal{S}, \mathcal{Y}]$ the probability that H_1 outputs true conditioned on the fact that this $\mathcal{S}$ and $\mathcal{Y}$ has been generated. Using the fact that the keys

$K_0', K_1', \ldots K_{u-1}'$ are sampled independently of $\mathcal{S}$, we compute

$$\Pr[\mathrm{H}_1|\mathcal{S},\mathcal{Y}] = \Pr\left[\exists j, x : (K_j', x) \in \mathcal{S}\right] \leq \sum_{y\in\mathcal{Y}} \frac{q_y \cdot n_y}{|\mathsf{Out}^{-1}(y)|}$$

$$\leq (m-1) \cdot \sum_{y\in\mathcal{Y}} \frac{q_y}{|\mathsf{Out}^{-1}(y)|} \leq \frac{m-1}{\rho(\mathsf{Out})} \sum_{y\in\mathcal{Y}} q_y \leq \frac{(m-1)q_{\mathrm{F}}}{\rho(\mathsf{Out})}.$$

Since the bound holds for all such $\mathcal{S}$ and $\mathcal{Y}$, we also have

$$\Pr[\,\mathrm{H}_1 \mid \forall y \in \overline{\mathcal{Y}} \,:\, n_y < m\,] \leq \frac{(m-1)q_{\mathrm{F}}}{\rho(\mathsf{Out})}. \tag{31}$$

The final bound follows by combining Eqs. (29), (30), and (31). ■

Security of the Davies-Meyer construction. One might object that practical compression functions are not un-structured enough to be treated as random because they are built from blockciphers via the Davies-Meyer construction. Accordingly, in [3], we study the mu PRF security under leakage of the Davies-Meyer construction with an ideal blockcipher and show that bounds of the quality we have seen for a random compression function continue to hold.

8 Quantitative Bounds for Augmented Cascades and AMAC

We consider two instantiations of augmented cascades, one using bit truncation, the other using modular reduction. We give concrete bounds on the mu prf security of these constructions in the ideal compression function model, combining results from above. This will give us good guidelines for a comparison with existing constructions – such as NMAC and sponges – in [3].

Bit truncation. Let $\mathcal{K} = \{0,1\}^c$, and $\mathsf{Out} = \mathsf{TRUNC}_r : \{0,1\}^c \to \{0,1\}^r$, for $r \leq c$, outputs the first r bits of its inputs, i.e., $\mathsf{TRUNC}_r(X) = X[1\ldots r]$. Note that $\delta(\mathsf{TRUNC}_r) = 0$, since omitting $c-r$ bits does not affect uniformity, and $\rho(\mathsf{TRUNC}_r) = 2^{c-r}$, since every r-bit strings has 2^{c-r} preimages. Then, combining Lemma 5 with Theorem 6, using $m = 2u/2^r + 3cr$, we obtain the following corollary, denoting with $\mathbf{F}_c$ the ideal compression function for $\mathcal{K} = \{0,1\}^c$. (We do not specify $\mathcal{X}$ further, as it does not influence the statement.)

Corollary 7. *For any $c \leq r$, and all adversaries $\mathcal{B}$ making u queries to* New *and q_{F} queries to* Prim*,*

$$\mathsf{Adv}^{\mathsf{prf}}_{\mathsf{CF},\mathsf{TRUNC}_r,\mathbf{F}_c}(\mathcal{B}) \leq \frac{u^2}{2^{c+1}} + \frac{2u \cdot q_{\mathrm{F}}}{2^c} + \frac{3cr \cdot q_{\mathrm{F}}}{2^{c-r}} + \exp(-c).$$ ■

We can then use this result to obtain our bounds for the augmented cascade **ACSC**[CF, CF, TRUNC_r] when using an ideal compression function $\{0,1\}^c \times \mathcal{X} \to \{0,1\}^c$. The proof is in [3].

Theorem 8 (mu prf security for r-bit truncation). *For any $r \leq n$, and all adversaries $\mathcal{A}$ making q queries to* Fn *consisting of vectors from $\mathcal{X}^*$ of length at most ℓ, q_{F} queries to* Prim, *and $u \leq q$ queries to* New,

$$\mathsf{Adv}^{\mathsf{prf}}_{\mathbf{ACSC}[\mathsf{CF},\mathsf{CF},\mathsf{TRUNC}_r],\mathbf{F}_c}(\mathcal{A}) \leq \frac{5\ell^2q^2 + 3\ell q q_{\mathrm{F}}}{2^c} + \frac{3cr(\ell^2 q + \ell q_{\mathrm{F}})}{2^{c-r}} + \ell \exp(-c), \quad \blacksquare$$

Modular reduction. Our second example becomes particularly important for the application to the Ed25519 signature scheme.

Here, we let $\mathcal{K} = \mathbb{Z}_N$, and consider the output function $\mathsf{Out} = \mathsf{MOD}_M : \mathbb{Z}_N \to \mathbb{Z}_M$ for $M \leq N$ is such that $\mathsf{MOD}_M(X) = X \mod M$. (Note that as a special case, we think of $\mathcal{K} = \{0,1\}^c$ here as $\mathbb{Z}_{2^c}$.) We need the following two properties of MOD_M, proved in [3].

Lemma 9. *For all $M \leq N$: (1) $\rho(\mathsf{MOD}_M) \geq \frac{N}{M} - 1$, (2) $\delta(\mathsf{MOD}_M) \leq M/N$.*

Then, combining Lemmas 5 and 9 with Theorem 6, using $m = 2u/M + 3 \ln N \ln M$, we obtain the following corollary, denoting with $\mathbf{F}_N$ the ideal compression function with $\mathcal{K} = Z_N$. (As above, we do not specify $\mathcal{X}$ further, as it does not influence the statement.)

Corollary 10. *For any $M \leq N/2$, and all adversaries $\mathcal{B}$ making u queries to* New *and q_{F} queries to* Prim,

$$\mathsf{Adv}^{\mathsf{prf}}_{\mathsf{CF},\mathsf{MOD}_M,\mathbf{F}_N}(\mathcal{B}) \leq \frac{u^2}{2N} + \frac{uM}{N} + \frac{4u \cdot q_{\mathrm{F}}}{N} + \frac{6M \ln N \ln M \cdot q_{\mathrm{F}}}{N} + \frac{1}{N}. \quad \blacksquare$$

This can once again be used to obtain the final analysis of the augmented cascade using modular reduction. The proof is similar to that of Theorem 8 and is deferred to [3].

Theorem 11 (mu prf security for modular reduction). *For any $M \leq N/2$, and all adversaries $\mathcal{A}$ making q queries to* Fn *consisting of vectors from $\mathcal{X}^*$ of length at most ℓ, q_{F} queries to* Prim, *and $u \leq q$ queries to* New,

$$\mathsf{Adv}^{\mathsf{prf}}_{\mathbf{ACSC}[\mathsf{CF},\mathsf{CF},\mathsf{MOD}_M],\mathbf{F}_N}(\mathcal{A}) \leq \frac{5\ell^2q^2 + 3\ell q q_{\mathrm{F}}}{N} + \frac{7M \ln N \ln M(\ell^2 q + \ell q_{\mathrm{F}})}{N} + \frac{\ell}{N}. \quad \blacksquare$$

Bounds for AMAC. The above bounds are for augmented cascades, but they can easily be adapted to AMAC, at the cost of adding an extra additive term, which we now discuss. Recall that $\mathsf{AMAC}(K, M) = \mathsf{Out}(H(K \| M))$, where the iterated hash function H is derived from a compression function h. We only consider here the special case where the key K is completely handled by the first compression function call of H (and is exactly a random element of $\mathcal{X}$), and the message is processed from the second call onwards. In other words, AMAC is the 2-tier cascade with the first tier being the dual of h, meaning the key and input roles are swapped. In particular, we can use Theorem 4, which would give us a modified version of the above bounds with an additional additive term, accounting $2\,\mathsf{Adv}^{\mathsf{prf}}_{\mathrm{g}}(\mathcal{A}_{\mathrm{g}})$ for $\mathcal{A}_{\mathrm{g}}$ as given in the reduction. This can easily be upper bounded (using the dedicated mu bound from Fig. 4) as

$$2 \cdot \mathsf{Adv}^{\mathsf{prf}}_{\mathrm{g}}(\mathcal{A}_{\mathrm{g}}) \leq \frac{u^2 + u(q_{\mathrm{F}} + q\ell)}{|\mathcal{X}|} \leq \frac{q^2 + q(q_{\mathrm{F}} + q\ell)}{|\mathcal{X}|}.$$

Acknowledgments. Bellare was supported in part by NSF grants CNS-1526801 and CNS-1228890, ERC Project ERCC FP7/615074 and a gift from Microsoft. Bernstein was supported in part by NSF grant CNS-1314919 and NWO grant 639.073.005. Tessaro was supported in part by NSF grant CNS-1423566. This work was done in part while Bellare and Tessaro were visiting the Simons Institute for the Theory of Computing, supported by the Simons Foundation and by the DIMACS/Simons Collaboration in Cryptography through NSF grant CNS-1523467. We thank the Eurocrypt 2016 reviewers for their comments.

References

1. Andreeva, E., Daemen, J., Mennink, B., Van Assche, G.: Security of keyed sponge constructions using a modular proof approach. In: Leander, G. (ed.) FSE 2015. LNCS, vol. 9054, pp. 364–384. Springer, Heidelberg (2015)
2. Bellare, M.: New proofs for NMAC and HMAC: security without collision-resistance. In: Dwork, C. (ed.) CRYPTO 2006. LNCS, vol. 4117, pp. 602–619. Springer, Heidelberg (2006)
3. Bellare, M., Bernstein, D.J., Tessaro, S.: Hash-function based PRFs: AMAC and its multi-user security. Cryptology ePrint Archive, Report 2016/142 (2016). https://eprint.iacr.org/
4. Bellare, M., Canetti, R., Krawczyk, H.: Keying hash functions for message authentication. In: Koblitz, N. (ed.) CRYPTO 1996. LNCS, vol. 1109, pp. 1–15. Springer, Heidelberg (1996)
5. Bellare, M., Canetti, R., Krawczyk, H.: Pseudorandom functions revisited: the cascade construction and its concrete security. In: 37th FOCS, pp. 514–523. IEEE Computer Society Press, October 1996
6. Bellare, M., Rogaway, P.: The security of triple encryption and a framework for code-based game-playing proofs. In: Vaudenay, S. (ed.) EUROCRYPT 2006. LNCS, vol. 4004, pp. 409–426. Springer, Heidelberg (2006)
7. Bernstein, D.J.: Extending the Salsa20 nonce. In: Symmetric key encryption workshop (SKEW). https://cr.yp.to/papers.html#xsalsa
8. Bernstein, D.J., Duif, N., Lange, T., Schwabe, P., Yang, B.-Y.: High-speed high-security signatures. In: Preneel, B., Takagi, T. (eds.) CHES 2011. LNCS, vol. 6917, pp. 124–142. Springer, Heidelberg (2011)
9. Bertoni, G., Daemen, J., Peeters, M., Assche, G.: On the security of the keyed sponge construction. In: Symmetric key encryption workshop (SKEW), February 2011
10. Brown, N.: Things that use Ed25519. http://ianix.com/pub/ed25519-deployment.html

11. Chang, D., Dworkin, M., Hong, S., Kelsey, J., Nandi, M.: A keyed sponge construction with pseudorandomness in the standard model. In: The Third SHA-3 Candidate Conference (March 2012) (2012)
12. Coron, J.-S., Dodis, Y., Malinaud, C., Puniya, P.: Merkle-damgård revisited: how to construct a hash function. In: Shoup, V. (ed.) CRYPTO 2005. LNCS, vol. 3621, pp. 430–448. Springer, Heidelberg (2005)
13. Damgård, I.B.: A design principle for hash functions. In: Brassard, G. (ed.) CRYPTO 1989. LNCS, vol. 435, pp. 416–427. Springer, Heidelberg (1990)
14. Dodis, Y., Pietrzak, K.: Leakage-resilient pseudorandom functions and side-channel attacks on feistel networks. In: Rabin, T. (ed.) CRYPTO 2010. LNCS, vol. 6223, pp. 21–40. Springer, Heidelberg (2010)
15. Dziembowski, S., Pietrzak, K.: Leakage-resilient cryptography. In: 49th FOCS, pp. 293–302. IEEE Computer Society Press, October 2008
16. Gaži, P., Pietrzak, K., Rybár, M.: The exact PRF-security of NMAC and HMAC. In: Garay, J.A., Gennaro, R. (eds.) CRYPTO 2014, Part I. LNCS, vol. 8616, pp. 113–130. Springer, Heidelberg (2014)
17. Gazi, P., Pietrzak, K., Tessaro, S.: The exact PRF security of truncation: tight bounds for keyed sponges and truncated CBC. In: Gennaro, R., Robshaw, M.J.B. (eds.) CRYPTO 2015. LNCS, vol. 9215, pp. 368–387. Springer, Heidelberg (2015)
18. Goldreich, O., Goldwasser, S., Micali, S.: How to construct random functions. J. ACM **33**(4), 792–807 (1986)
19. Maurer, U.M., Renner, R.S., Holenstein, C.: Indifferentiability, impossibility results on reductions, and applications to the random oracle methodology. In: Naor, M. (ed.) TCC 2004. LNCS, vol. 2951, pp. 21–39. Springer, Heidelberg (2004)
20. Mennink, B., Reyhanitabar, R., Vizár, D.: Security of full-state keyed spongeand duplex: applications to authenticated encryption. In: Iwata, T., Cheon, J.H. (eds.) ASIACRYPT 2015. LNCS, vol. 9453, pp. 465–489. Springe, Heidelberg (2015)
21. Merkle, R.C.: One way hash functions and DES. In: Brassard, G. (ed.) CRYPTO 1989. LNCS, vol. 435, pp. 428–446. Springer, Heidelberg (1990)
22. Mouha, N., Luykx, A.: Multi-key security: the even-mansour construction revisited. In: Gennaro, R., Robshaw, M.J.B. (eds.) CRYPTO 2015. LNCS, vol. 9215, pp. 209–223. Springer, Heidelberg (2015)
23. De Mulder, E., Hutter, M., Marson, M.E., Pearson, P.: Using Bleichenbacher's solution to the hidden number problem to attack nonce leaks in 384-bit ECDSA. In: Bertoni, G., Coron, J.-S. (eds.) CHES 2013. LNCS, vol. 8086, pp. 435–452. Springer, Heidelberg (2013)
24. Preneel, B., Govaerts, R., Vandewalle, J.: Hash functions based on block ciphers: a synthetic approach. In: Stinson, D.R. (ed.) CRYPTO 1993. LNCS, vol. 773, pp. 368–378. Springer, Heidelberg (1994)
25. Tessaro, S.: Optimally secure block ciphers from ideal primitives. In: Iwata, T., Cheon, J.H. (eds.) ASIACRYPT 2015. LNCS, vol. 9453, pp. 437–462. Springer, Heidelberg (2015)

On the Influence of Message Length in PMAC's Security Bounds

Atul Luykx[1,2,3(✉)], Bart Preneel[1,2], Alan Szepieniec[1,2], and Kan Yasuda[1,3]

[1] Department of Electrical Engineering, ESAT/COSIC, KU Leuven, Leuven, Belgium
{atul.luykx,bart.preneel}@esat.kuleuven.be
[2] iMinds, Ghent, Belgium
[3] NTT Secure Platform Laboratories, NTT Corporation, Tokyo, Japan

Abstract. Many MAC (Message Authentication Code) algorithms have security bounds which degrade linearly with the message length. Often there are attacks that confirm the linear dependence on the message length, yet PMAC has remained without attacks. Our results show that PMAC's message length dependence in security bounds is non-trivial. We start by studying a generalization of PMAC in order to focus on PMAC's basic structure. By abstracting away details, we are able to show that there are two possibilities: either there are infinitely many instantiations of generic PMAC with security bounds independent of the message length, or finding an attack against generic PMAC which establishes message length dependence is computationally hard. The latter statement relies on a conjecture on the difficulty of finding subsets of a finite field summing to zero or satisfying a binary quadratic form. Using the insights gained from studying PMAC's basic structure, we then shift our attention to the original instantiation of PMAC, namely, with Gray codes. Despite the initial results on generic PMAC, we show that PMAC with Gray codes is one of the more insecure instantiations of PMAC, by illustrating an attack which roughly establishes a linear dependence on the message length.

Keywords: Unforgeability · Integrity · Verification · Birthday bound · Tag · PMAC · Message length

1 Introduction

When searching for optimal cryptographic schemes, security bounds provide an important tool for selecting the right parameters. Security bounds, as formalized by Bellare et al. [1], capture the concept of explicitly measuring the effect of an adversary's resources on its success probability in breaking the scheme. They enable one to determine how intensively a scheme can be used in a session. Therefore, provably reducing the impact of an adversary's resources from, say, a quadratic to a linear term, can mean an order of magnitude increase in a scheme's lifetime. Conversely, finding attacks which confirm an adversary's success rate, relative to its allotted resources, prove claims of security bound optimality.

M. Fischlin and J.-S. Coron (Eds.): EUROCRYPT 2016, Part I, LNCS 9665, pp. 596–621, 2016.
DOI: 10.1007/978-3-662-49890-3_23

MAC algorithms provide a good example of schemes which have been studied extensively to determine optimal bounds. A MAC's longevity is defined as the number of times the MAC can be used under a single key: it can be measured as a function of the number of tagging queries, q, and the largest message length, ℓ, used before a first forgery attempt is successful. The impact of an adversary's resources, q and ℓ, on its success probability in breaking a MAC is then described via an upper bound of the form $f(q, \ell) \cdot \epsilon$, where f is a function, often a polynomial, and ϵ is a quantity dependent on the MAC's parameters. The maximum number of queries $q_{\max}$ with length $\ell_{\max}$ one can make under a key is computed by determining when $f(q_{\max}, \ell_{\max}) \cdot \epsilon$ is less than some threshold success probability. For example, if one is comfortable with adversaries which have a one in a million chance of breaking the scheme, but no more, then one would determine $q_{\max}$ and $\ell_{\max}$ via

$$f(q_{\max}, \ell_{\max}) \cdot \epsilon \leq 10^{-6}. \tag{1}$$

Given that $q_{\max}$ and $\ell_{\max}$ depend only on f, it becomes important to find the f which establishes the tightest upper bound on the success probability.

The optimality of f depends on the environment in which the MAC operates, or in other words, the assumptions made on the MAC. For instance, stateful MACs, such as the Wegman-Carter construction [21], can achieve bounds independent of q and ℓ. In this case, an adversary's success remains negligible regardless of q and ℓ, as long as the construction receives nonces, that is, additional unique input. Therefore, determining $q_{\max}$ and $\ell_{\max}$ for Wegman-Carter MACs amounts to solving $\epsilon \ll 1$, which is true under the assumption that nonces are unique. Similarly, XOR MAC [3] with nonces achieves a security upper bound of $\epsilon = 1/2^{\tau}$, with τ the tag length in bits, which is the optimal bound for any MAC. Randomized, but stateless MACs can achieve bounds similar to stateful MACs, as shown by Minematsu [14].

In contrast, deterministic and stateless MACs necessarily have a lower bound of $q^2/2^n$, where n is the inner state size, due to a generic attack by Preneel and van Oorschot [18]. This means that for any f,

$$f(q, \ell) \cdot \epsilon \geq \frac{q^2}{2^n}, \tag{2}$$

hence any deterministic, stateless MAC must use fewer than $2^{n/2}$ tagging queries per key.

Given this lower limit on f, one would perhaps expect to find schemes for which the proven upper bound is $q^2/2^n$. Yet many deterministic, stateless MACs have upper bounds including an ℓ-factor. Block cipher based MACs, such as CBC-MAC [4], OMAC [12], and PMAC [7], were originally proven with an upper bound on the order of $q^2\ell^2/2^n$, growing quadratically as a function of ℓ. Much effort has been placed in improving the bounds to a linear dependence on ℓ, resulting in bounds of the form $q^2\ell/2^n$ [5,11,15,16].

For certain deterministic, stateless schemes the dependence on ℓ has been proven to be necessary. Dodis and Pietrzak [9] point out that this is the case for polynomial based MACs, and try to avoid the dependence by introducing randomness. Pietrzak [17] notes that the EMAC bound must depend on ℓ. Gazi, Pietrzak, and Rybár [10] give an attack on NMAC showing its dependence on ℓ. Nevertheless, there are no known generic attacks establishing a lower bound of the form $\ell^{\epsilon}/2^n$ for any $\epsilon > 0$.

PMAC, introduced by Black and Rogaway [7], stands out as a construction for which little analysis has been performed showing the necessity of ℓ in the bound. It significantly differs in structure from other MACs (see Fig. 1 and Definition 3), which gives it many advantages:

1. it is efficient, since nearly all block cipher calls can be made in parallel,
2. it is simple, which in turn enables simple analysis,
3. and its basic structure lends itself to high-security extensions, such as PMAC-Plus [22], PMAC-with-Parity [23], and PMACX [24].

The disadvantage of having such a different structure is that no known attacks can help to establish ℓ-dependency.

Contributions. We start by abstracting away some details of PMAC in order to focus on its basic structure. We do so by considering *generic* PMAC, which is a generalized version of PMAC accepting an arbitrary block cipher and constants, and with an additional independent key. We prove that one of the following two statements is true:

1. either there are infinitely many instances of generic PMAC for which there are no attacks with success probability greater than $2q^2/2^n$,
2. or finding an attack against generic PMAC with success probability greater than $2q^2/2^n$ is computationally hard.

The second statement relies on a conjecture which we explain below.

Then we focus on an instantiation of generic PMAC, namely PMAC with Gray codes, introduced by Black and Rogaway [7]. We show that PMAC with Gray codes is an instantiation which does not meet the optimal bound of $2q^2/2^n$, by finding an attack with success probability $(2^{k-1} - 1)/2^n$ with $\ell = 2^k$, establishing a dependence on ℓ for every power of two.

Approach. Proving the above results requires viewing the inputs to PMAC's block cipher calls in a novel way: as a set of points P lying in a finite affine plane. Keys are identified as slopes of lines in the affine plane. A collision is guaranteed to occur under a specific key w if and only if each line with slope w covers an even number of points in P; in this case we say that w *evenly covers* P.

Maximizing the collision probability means finding a set of points P for which there is large set of slopes W evenly covering P. But finding such a set W is non-trivial: the x-coordinates of the points in P must either contain a subset summing to zero, or satisfying some quadratic form.

Finding a subset summing to zero is the *subset sum* (SS) problem, which is known to be **NP**-complete. The second problem we call the *binary quadratic form*

(BQF) problem (see Definition 9), and there is reason to believe this problem is **NP**-complete as well (see Appendix B). As a result, we conjecture that finding solutions to the union of the two problems is computationally hard.

By reducing SS and the BQF problem to finding slopes W evenly covering points P, we establish our results.

Related Work. Rogaway [19] has shown that the dependence on ℓ disappears if you consider a version of PMAC with an ideal tweakable block cipher. PMAC's basic structure has also been used to design schemes where the impact of ℓ is reduced by construction: Yasuda's PMAC-with-Parity [23] and Zhang's PMACX [24] get bounds of the form $q^2\ell^2/2^{2n}$.

For EMAC, Pietrzak [17] proved that if $\ell \leq 2^{n/8}$ and $q \geq \ell^2$, then the bound's order of growth is independent of ℓ. The proven bound is

$$128 \cdot \frac{q^2\ell^8}{2^{2n}} + 16 \cdot \frac{q^2}{2^n} + \frac{q(q-1)}{2^{n+1}} \,. \tag{3}$$

Note that the condition on ℓ means that EMAC's bound is not truly independent of ℓ. An example of a construction which has a bound which is truly independent of ℓ is a variant of PMAC described by Yasuda [23, Sect. 1]. This construction achieves a bound that does *not* grow as a function of ℓ, with the limitation that $\ell \leq 2^{n/2}$ and at a rate of two block cipher calls per block of message. The construction works by splitting the message into half blocks, and then appending a counter to each half-block, to create a full block. Each full block is input into a block cipher, and all the block cipher outputs are XORed together, and finally input into a last, independent block cipher.

2 Preliminaries

2.1 Notation

If X is a set then $\overline{\mathsf{X}}$ is its complement, X^q is the Cartesian product of q copies of X, $\mathsf{X}^{\leq \ell} = \bigcup_{i=1}^{\ell} \mathsf{X}^i$, and $\mathsf{X}^+ = \bigcup_{i=1}^{\infty} \mathsf{X}^i$. If $\boldsymbol{x} \in \mathsf{X}^q$, then its coordinates are $(x_1, x_2, \ldots, x_q)$. If $f : \mathsf{X} \to \mathsf{Y}$ then define $\widetilde{f} : \mathsf{X}^+ \to \mathsf{Y}^+$ to be the mapping

$$\widetilde{f}(x_1, \ldots, x_q) = (f(x_1), \ldots, f(x_q)) \,. \tag{4}$$

If $\boldsymbol{a} \in \mathsf{X}^{\ell_1}$ and $\boldsymbol{b} \in \mathsf{X}^{\ell_2}$, then $\boldsymbol{a} \| \boldsymbol{b}$ is the concatenation of $\boldsymbol{a}$ and $\boldsymbol{b}$, that is,

$$\boldsymbol{a} \| \boldsymbol{b} := (a_1, a_2, \ldots, a_{\ell_1}, b_1, b_2, \ldots, b_{\ell_2}) \in \mathsf{X}^{\ell_1 + \ell_2} \,. \tag{5}$$

If $\boldsymbol{a} \in \mathsf{X}^{\ell}$ and $\mu \leq \ell$, then $\boldsymbol{a}_{\leq \mu} := (a_1, a_2, \ldots, a_{\mu})$. If X is a field, then for $\boldsymbol{a} \in \mathsf{X}^{\ell}$, $\mathbf{1} \cdot \boldsymbol{a} = \sum_{i=1}^{\ell} a_i$. Furthermore, when considering elements (x, y) of X^2, we call the left coordinate of the pair the x-coordinate, and the other the y-coordinate.

2.2 Primitives

A *uniformly distributed random function* (URF) from M to T is a uniformly distributed random variable over the set of all functions from M to T. A *uniformly distributed random permutation* (URP) on X is a uniformly distributed random variable over the set of all permutations on X.

A *pseudo-random function* (PRF) is a function $\Phi : \mathsf{K} \times \mathsf{M} \to \mathsf{T}$ defined on a set of keys K and messages M with output in T. We write $\Phi_k(m)$ for $\Phi(k, m)$. The *PRF-advantage* of an adversary A against the PRF Φ is the probability that A distinguishes Φ_k from \$, where k is a uniformly distributed random variable over K, and \$ is a URF. More formally, the advantage of A can be described as

$$\left|\mathbf{Pr}\left[A^{\Phi_k} = 1\right] - \mathbf{Pr}\left[A^{\$} = 1\right]\right| , \tag{6}$$

where $A^O = 1$ is the event that A outputs 1 given access to oracle O.

A *pseudorandom permutation* (PRP) is a function $E : \mathsf{K} \times \mathsf{X} \to \mathsf{X}$ defined on a set of keys K, where $E(k, \cdot)$ is a permutation for each $k \in \mathsf{K}$. As with PRFs, we write $E_k(x)$ for $E(k, x)$. The *PRP-advantage* of an adversary A versus E is defined similarly to the PRF-advantage, and can be described as follows:

$$\left|\mathbf{Pr}\left[A^{E_k} = 1\right] - \mathbf{Pr}\left[A^{\pi} = 1\right]\right| , \tag{7}$$

where k is uniformly distributed over K, and π is a URP.

2.3 Message Authentication

A MAC consists of a tagging and a verification algorithm. The tagging algorithm accepts messages from some message set M and produces tags from a tag set T. The verification algorithm receives message-tag pairs (m, t) as input, and outputs 1 if the pair (m, t) is valid, and 0 otherwise. The insecurity of a MAC is measured as follows.

Definition 1. *Let A be an adversary with access to a MAC. The advantage of A in breaking the MAC is the probability that A is able to produce a message-tag pair (m, t) for which the verification algorithm outputs 1, where m has not been previously queried to the tagging algorithm.*

PRF-based MACs use a PRF $\Phi : \mathsf{K} \times \mathsf{M} \to \mathsf{T}$ to define the tagging algorithm. The verification algorithm outputs 1 if $\Phi_k(m) = t$, and 0 otherwise. As shown by the following theorem, the insecurity of a PRF-based MAC can be reduced to the insecurity of the PRF, allowing us to focus on Φ.

Theorem 1 ([2]). *Let α denote the advantage of adversary A in breaking a PRF-based MAC with underlying PRF Φ. Say that A makes q tagging queries and v verification queries. Then there exists a PRF-adversary B making $q + v$ PRF queries such that*

$$\alpha \le \frac{v}{|\mathsf{T}|} + \beta , \tag{8}$$

where β is the advantage of B.

Some PRFs are constructed using a smaller PRP $E_k : \mathsf{K} \times \mathsf{X} \to \mathsf{X}$. If Φ^{E_k} denotes a PRF using E_k, then one can reduce the PRF-advantage of an adversary against Φ^{E_k} to the PRF-advantage of an adversary against Φ^{π}, where π is a URP over X. The result is well-known, and used, for example, to prove the security of PMAC [7].

Theorem 2. *Let α denote the PRF-advantage of adversary A against Φ^{E_k}. Say that A makes q queries to the PRF. Then there exists a PRF-adversary B against Φ^{π} making q queries and a PRP-adversary C against E such that*

$$\alpha \leq \beta + \gamma, \tag{9}$$

where β is the advantage of B and γ is the advantage of C.

The above theorem lets us focus on PRFs built with URPs instead of PRPs.

3 PMAC

PMAC is a PRF-based MAC, which means we can focus on the underlying PRF. Throughout this paper we identify PMAC with its PRF. Furthermore, we focus on PMAC defined with a URP.

The original PMAC specifications [7,19] have as message space the set of arbitrary length strings. Although our results focus on the dependency of PMAC on message length, it will suffice to consider strings with length a multiple of some block size in order to illustrate how the security bounds evolve as a function of message length. With this in mind, we define PHASH, first introduced by Minematsu and Matsushima [15]. Figure 1 depicts a diagram of PHASH.

Definition 2 (PHASH). *Let X be a finite field of characteristic two with N elements. Let $\mathsf{M} := \mathsf{X}^{\leq N}$ and let $\boldsymbol{c} \in \mathsf{X}^N$ be a sequence containing all elements of X. Let π be a URP over X. Let $\omega = \pi(0)$, then $PHASH : \mathsf{M} \to \mathsf{X}$ is defined to be*

$$PHASH(\boldsymbol{m}) := \mathbf{1} \cdot \overline{\pi}\left(\boldsymbol{m} + \omega \boldsymbol{c}_{\leq \ell}\right), \tag{10}$$

where $\boldsymbol{m}$ has length ℓ.

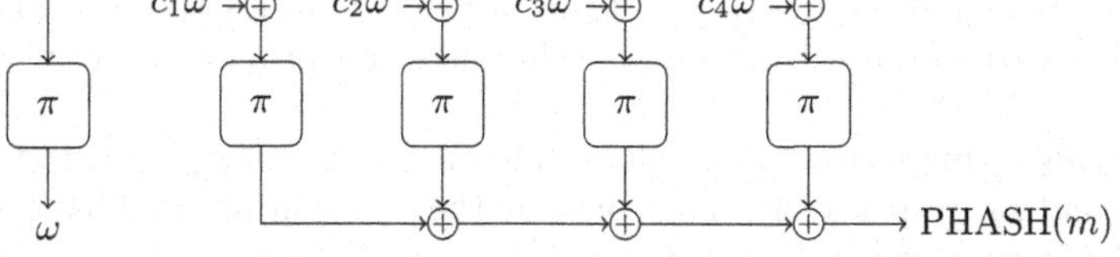

Fig. 1. PHASH evaluated on a message $m = (m_1, m_2, m_3, m_4)$.

PHASH maps messages to a single block. PMAC sends this block through a last transformation, whose output will be the tag. We describe two different generic versions of PMAC, one in which the last transformation is independent of PHASH, and one in which it is not.

Definition 3 (PMAC). *Consider PHASH* : $\mathsf{M} \to \mathsf{X}$ *with URP* π *and let* c^* *denote the last element of* $\boldsymbol{c}$. *If* y *is the output of PHASH under message* $\boldsymbol{m}$, *PMAC evaluated on* $\boldsymbol{m}$ *is* $\pi(y + c^*\omega)$.

Definition 4 (PMAC*). *Consider PHASH* : $\mathsf{M} \to \mathsf{X}$ *with URP* π. *Let* $\phi : \mathsf{X} \to \mathsf{X}$ *be an independent URF. Then PMAC* is the composition of PHASH with* ϕ.

Although PMAC* is defined with an independent outer URF instead of a URP, all the results in the paper hold with slight modifications to the bounds if a URP is used.

The two specifications of PMAC define the sequence $\boldsymbol{c}$ differently. Our attack against PMAC applies to the specification with Gray codes [7], which we will define in Sect. 6.4. As pointed out by Nandi and Mandal [16], in order to get a PRF-advantage upper bound of the form $q^2\ell/N$, the only requirement on $\boldsymbol{c}$ is that each of its components are distinct.

4 PHASH Collision Probability

Definition 5. *The collision probability of PHASH is*

$$\max_{\boldsymbol{m}^1, \boldsymbol{m}^2 \in \mathsf{M}, \boldsymbol{m}^1 \neq \boldsymbol{m}^2} \mathbf{Pr}\left[PHASH(\boldsymbol{m}^1) = PHASH(\boldsymbol{m}^2)\right]. \tag{11}$$

PHASH's collision probability is closely linked with the security of PMAC and PMAC*. In particular, if an adversary finds a collision in PHASH, then it is able to distinguish PMAC and PMAC* from a URF. The converse is true for PMAC*, which is a well-known result; see for example Dodis and Pietrzak [9]. Concluding that a distinguishing attack against PMAC results in a collision found for PHASH has not been proven and is outside of the scope of the paper, although we conjecture that the statement holds. In either case, understanding the effect of the message length on PHASH's collision probability will give us a good understanding of PMAC's message length dependence.

In this section we compute bounds on the collision probability for PHASH. Minematsu and Matsushima [15] prove an upper bound for the collision probability of PHASH. We use their proof techniques and provide a lower bound as well.

Throughout this section we fix two different messages $\boldsymbol{m}^1$ and $\boldsymbol{m}^2$ in M of length ℓ_1 and ℓ_2, respectively, and consider the collision probability over these messages. Let $\boldsymbol{m} = \boldsymbol{m}^1 \| \boldsymbol{m}^2$ and $\boldsymbol{d} = \boldsymbol{c}_{\leq \ell_1} \| \boldsymbol{c}_{\leq \ell_2}$.

If there exists i such that $m_i^1 = m_i^2$, then these blocks will cancel each other out in Eq. (11) and will not affect the collision probability, hence we remove them. Let $i_1, i_2, \ldots, i_k$ denote the indices of the blocks for which $\boldsymbol{m}^1$ equals $\boldsymbol{m}^2$, then define $\boldsymbol{m}^*$ to be $\boldsymbol{m}$ with the entries indexed by $i_1, i_2, \ldots, i_k$ and $i_1+\ell_1, i_2+\ell_1, \ldots, i_k+\ell_1$ removed; $\boldsymbol{d}^*$ is defined similarly and ℓ^* denotes the length of $\boldsymbol{m}^*$ and $\boldsymbol{d}^*$.

Let $\boldsymbol{x}^w := \boldsymbol{m}^* + w\boldsymbol{d}^*$ for $w \in \mathsf{X}$. The vector $\boldsymbol{x}^w$ represents the inputs to the permutation π when $\pi(0)$ equals w, meaning the equality $\text{PHASH}(\boldsymbol{m}^1) = \text{PHASH}(\boldsymbol{m}^2)$ can be written as

$$\mathbf{1} \cdot \widetilde{\pi}\left(\boldsymbol{x}^w\right) = 0\,, \tag{12}$$

given that $\pi(0) = w$. If there is a component of $\boldsymbol{x}^w$ which does not equal any of the other components, then Eq. (12) will contain a π-output which is roughly independent of the other outputs, thereby making a collision unlikely when $\pi(0) = w$. For example, say that $\boldsymbol{x}^w = (a, b, c, b)$, then Eq. (12) becomes $\pi(a)+\pi(b)+\pi(c)+\pi(b) = \pi(a)+\pi(c)$, which equals 0 with negligible probability.

Similarly, if there are an odd number of components of $\boldsymbol{x}^w$ which equal each other, but do not equal any other components, then they will not cancel out, resulting again in an unlikely collision. For example, if $\boldsymbol{x}^w = (a, a, a, b, b)$, then Eq. (12) becomes $\pi(a)$. In fact, a collision is only guaranteed under a given key w when each component of $\boldsymbol{x}^w$ is paired with another component so that each pair cancels each other out in Eq. (12). Bounding the collision probability in Eq. (11) amounts to determining how many keys w there are for which each component of $\boldsymbol{x}^w$ is paired.

We formalize these "equality classes" of components of $\boldsymbol{x}^w$ as follows. Define I to be the set of integers from 1 to ℓ^*, $\{1, \ldots, \ell^*\}$, then the components of $\boldsymbol{x}^w = (x_1^w, x_2^w, \ldots, x_{\ell^*}^w)$, induce the following equivalence relation on I: i is equivalent to j if and only if $x_i^w = x_j^w$. For $i \in I$, let $[i]$ denote i's equivalence class, and $\#[i]$ the number of elements in $[i]$. Let R^w denote the set of equivalence class representatives where each representative is the smallest element of its class. Let R_e^w be those $i \in R^w$ such that $\#[i]$ is even, and R_o^w the complement of R_e^w in R^w. Taking the example $\boldsymbol{x}^w = (c, c, c, b, b, b, b, a)$, then R^w would equal $\{1, 4, 8\}$ and R_e^w is $\{4\}$.

Define $\mathbf{W}$ to be the set of $w \in \mathsf{X}$ such that R_o^w is empty. In other words, the set $\mathbf{W}$ is the set of keys w for which $\boldsymbol{m}^1$ and $\boldsymbol{m}^2$ are guaranteed to collide.

Proposition 1. *Let $F = PHASH$, then*

$$\frac{|\mathbf{W}|}{N} \leq \mathbf{Pr}\left[F(\boldsymbol{m}^1) = F(\boldsymbol{m}^2)\right] \leq \frac{|\mathbf{W}|}{N} + \frac{1}{N - \ell^* + 1}\,. \tag{13}$$

Proof. Let Π be the set of permutations on X. Let δ_w be the number of distinct components in $0\|\boldsymbol{x}^w$ and let S_w be the set of $\boldsymbol{y}$ such that $\mathbf{1} \cdot \boldsymbol{y} = 0$ and $w\|\boldsymbol{y}$ *matches* $0\|\boldsymbol{x}^w$, where two sequences $\boldsymbol{a}$ and $\boldsymbol{b}$ of the same length match if $a_i = a_j$ if and only if $b_i = b_j$, for all i, j. We have that

$$\mathbf{Pr}\left[F(\boldsymbol{m}^1) + F(\boldsymbol{m}^2) = 0\right] = \mathbf{Pr}\left[\mathbf{1} \cdot \widetilde{\pi}(\boldsymbol{x}^\omega) = 0\right] \tag{14}$$

$$= \frac{1}{N!} \cdot \left|\left\{p \in \Pi \mid \mathbf{1} \cdot \widetilde{p}\left(\boldsymbol{x}^{p(0)}\right) = 0\right\}\right| \tag{15}$$

$$= \frac{1}{N!} \cdot \sum_{w \in \mathsf{X}} \sum_{\boldsymbol{y} \in S_w} |\{p \in \Pi \mid \widetilde{p}(0\|\boldsymbol{x}^w) = w\|\boldsymbol{y}\}| \ . \tag{16}$$

Note that for all w and $\boldsymbol{y} \in S_w$,

$$|\{p \in \Pi \mid \widetilde{p}(0\|\boldsymbol{x}^w) = w\|\boldsymbol{y}\}| = (N - \delta_w)! \ , \tag{17}$$

hence we get

$$\mathbf{Pr}\left[F(\boldsymbol{m}^1) = F(\boldsymbol{m}^2)\right] = \frac{1}{N!} \cdot \sum_{w \in \mathsf{X}} (N - \delta_w)! \cdot |S_w| \ . \tag{18}$$

Let $\boldsymbol{y}$ be such that $w\|\boldsymbol{y}$ matches $0\|\boldsymbol{x}^w$. Note that $y_i = y_j$ if and only if i is equivalent to j, and for any $i \in R^w$,

$$\sum_{j \in [i]} y_j = \begin{cases} 0 & \text{if } \#[i] \text{ is even} \\ y_i & \text{otherwise}. \end{cases} \tag{19}$$

Then $\boldsymbol{y} \in S_w$ if and only if $w\|\boldsymbol{y}$ matches $0\|\boldsymbol{x}^w$ and $\sum_{i \in R_o^w} y_i = 0$.

Let w be such that $x_i^w \neq 0$ for all i. The number of $\boldsymbol{y}$ such that $w\|\boldsymbol{y}$ matches $0\|\boldsymbol{x}^w$ and $\sum_{i \in R_o^w} y_i = 0$ can be counted as follows. Consider $\boldsymbol{y} = (y_1, \ldots, y_{\ell^*})$ satisfying the requirements, and enumerate the values in R_e^w: $i_1, i_2, \ldots, i_k$. By fixing $y_{i_1}, y_{i_2}, \ldots, y_{i_k}$, we determine all components of $\boldsymbol{y}$ contained in the equivalence classes of R_e^w. Since $y_{i_1}, y_{i_2}, \ldots, y_{i_k}$ is a sequence of k distinct values, all different from w, there are $(N-1)!/(N-k-1)!$ possibilities for $y_{i_1}, y_{i_2}, \ldots, y_{i_k}$. If $R_o^w \neq \emptyset$, then we enumerate the elements of R_o^w: $j_1, j_2, \ldots, j_l$. Similar to R_e^w, by determining $y_{j_1}, y_{j_2}, \ldots, y_{j_l}$ we will determine the remaining components of $\boldsymbol{y}$. The sequence $y_{j_1}, y_{j_2}, \ldots, y_{j_l}$ contains l distinct values, all different from $y_{i_1}, y_{i_2}, \ldots, y_{i_k}$ and w, and such that $y_{j_1} + y_{j_2} + \cdots + y_{j_l} = 0$, resulting in at most $(N-k-1)!/(N-k-l)!$ possibilities. Putting this together, and observing that $k + l = |R_e^w| + |R_o^w| = \delta_w - 1$, we get $|S_w| \leq \frac{(N-1)!}{(N-\delta_w+1)!}$ when $R_o^w \neq \emptyset$ and $x_i^w \neq 0$ for all i. If $R_o^w = \emptyset$, then $|S_w| = \frac{(N-1)!}{(N-\delta_w)!}$.

By following similar reasoning, we get that if w is such that there exists $x_i^w = 0$, $|S_w| \leq \frac{(N-1)!}{(N-\delta_w+1)!}$ when $R_o^w \neq \emptyset$, and $|S_w| = \frac{(N-1)!}{(N-\delta_w)!}$ otherwise.

Putting the above together, we have

$$\mathbf{Pr}\left[F(\boldsymbol{m}^1) = F(\boldsymbol{m}^2)\right] \leq \frac{|\mathbf{W}|}{N} + \frac{1}{N} \sum_{w \in \overline{\mathbf{W}}} \frac{1}{N - \delta_w + 1}, \tag{20}$$

and since the computation of $|S_w|$ is exact when $R_o^w = \emptyset$, we get

$$\frac{|\mathbf{W}|}{N} \leq \mathbf{Pr}\left[F(\boldsymbol{m}^1) = F(\boldsymbol{m}^2)\right] . \tag{21}$$

□

5 Necessary Conditions for a Collision

This section provides a geometric interpretation of the set $\mathbf{W}$ which facilitates finding necessary conditions for $\mathbf{W}$ to contain more than two elements.

5.1 Evenly Covered Sets

Recall that an element w of X is in $\mathbf{W}$ only if $R_o^w = \emptyset$, meaning $\#[i]$ is even for all $i \in R^w$. Two components x_i^w and x_j^w of $\boldsymbol{x}^w$ are equal if and only if

$$w = \frac{m_i^* - m_j^*}{d_j^* - d_i^*} , \tag{22}$$

since the points such that $(d_i, m_i) = (d_j, m_j)$ were removed earlier when forming $\boldsymbol{m}^*$ from $\boldsymbol{m}$. In particular, Eq. (22) says that x_i^w equals x_j^w if and only if the points (d_i^*, m_i^*) and (d_j^*, m_j^*) lie on a line with slope w. Since $\#[i]$ is even, we know that there are an even number of points on the line through (d_i^*, m_i^*) with slope w, which motivates the following definition.

Definition 6. *Let* $\mathsf{P} \subset \mathsf{X}^2$ *be a set of points. A line* evenly covers P *if it contains an even number of points from* P. *A slope* $w \in \mathsf{X}$ *evenly covers* P *if all lines with slope* w *evenly cover* P. *A subset of* X *evenly covers* P *if all slopes in the subset evenly cover* P.

We let $\mathbf{P}$ denote the set of points (d_i, m_i) for $1 \leq i \leq \ell$. Applying the above definition together with Eq. (22), we get the following proposition.

Proposition 2. *An element* $w \in \mathsf{X}$ *is in* $\mathbf{W}$ *if and only if* w *evenly covers* $\mathbf{P}$.

Using this geometric interpretation, we obtain the upper bound proved by Minematsu and Matsushima [15] for the collision probability of PHASH.

Proposition 3.

$$|\mathbf{W}| \leq \ell^* - 1 \tag{23}$$

Proof. Given a point $p_0 \in \mathbf{P}$, all possible slopes connecting p_0 to another point in $\mathbf{P}$ can be generated from the lines connecting the points. This results in at most $|\mathbf{P}| - 1$ different slopes covering $\mathbf{P}$, hence an upper bound for $|\mathbf{W}|$ is $|\mathbf{P}| - 1 = \ell^* - 1$. □

It is easy to construct sets evenly covered by two slopes. Consider $\mathsf{P} := \{(x_1, 0), (x_1, 1), (x_2, 0), (x_2, 1)\}$, depicted in Fig. 2. The possible slopes are 0 and $(x_1 + x_2)^{-1}$. Throughout the paper we do not consider ∞ to be a slope, since such a slope would only be possible if $d_i^* = d_j^*$ in Eq. (22), which happens only if $m_i^* = m_j^*$. The lines with slope 0, from $(x_1, 0)$ to $(x_2, 0)$ and from $(x_1, 1)$ to $(x_2, 1)$, evenly cover P. Similarly, the lines with slope $(x_1 + x_2)^{-1}$, from $(x_1, 0)$

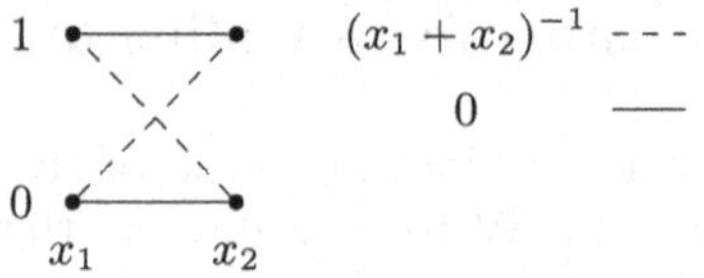

Fig. 2. A set of four points evenly covered by the slopes 0 and $(x_1 + x_2)^{-1}$. The x-coordinates of the points are x_1 and x_2, and the y-coordinates are 0 and 1.

to $(x_2, 1)$ and from $(x_1, 1)$ to $(x_2, 0)$, also evenly cover P. Therefore P is evenly covered by $\{0, (x_1 + x_2)^{-1}\}$.

The above set can be converted into two messages: $\boldsymbol{m}_1 = (0, 0)$ and $\boldsymbol{m}_2 = (1, 1)$. Setting $x_1 = c_1$ and $x_2 = c_2$, then we know that the collision probability of $\boldsymbol{m}_1$ and $\boldsymbol{m}_2$ is at least $2/N$.

Proposition 4. *There exist messages* $\boldsymbol{m}_1$ *and* $\boldsymbol{m}_2$ *such that* $|\mathbf{W}| \geq 2$.

Note that **P** constructed from $\boldsymbol{m}^*$ contains at most two points per x-coordinate.

5.2 Properties of Evenly Covered Sets

Although Proposition 3 gives a good upper bound for the collision probability of PHASH, it does not use any of the structure of evenly covered sets. In this section we explore various properties of evenly covered sets, allowing us to relate their discovery to **NP**-hard problems in Sect. 5.3.

The following lemma shows that removing an evenly covered subset from an evenly covered set results in an evenly covered set.

Lemma 1. *Let* $\mathsf{P} \subset \mathsf{X}^2$ *and let* $\mathsf{W} \subset \mathsf{X}$ *be a set evenly covering* P. *Say that* P *contains a subset* P' *evenly covered by* W *as well, then* $\mathsf{P} \setminus \mathsf{P}'$ *is evenly covered by* W.

Proof. Let $\mathsf{Q} := \mathsf{P} \setminus \mathsf{P}'$. The set W evenly covers Q if and only if every line with slope $w \in \mathsf{W}$ contains an even number of points in Q. Let $p \in \mathsf{Q}$ and $w \in \mathsf{W}$ and consider the line λ with slope w through point p. By hypothesis, λ evenly covers P and P'. By removing P' from P, an even number of points are removed from λ, resulting in λ evenly covering Q. □

If a set P is evenly covered by at least two slopes u and v, then all the points in the set lie in a *loop*.

Definition 7. *Let* $\mathsf{P} \subset \mathsf{X}^2$ *be evenly covered by* $\mathsf{W} \subset \mathsf{X}$. *A* (u, v)*-loop in* (W, P) *is a sequence of points* $(p_1, p_2, \ldots, p_k)$ *with two different slopes* $u, v \in \mathsf{W}$ *such that* p_i *and* $p_{i+1 \pmod k}$ *lie on a line with slope* u *for* i *odd, and on a line with slope* v *otherwise.*

The set from Fig. 2 contains $(0,(x_1+x_2)^{-1})$-loops. In fact, there are always at least four points in any (u,v)-loop. Note that there must be at least three points since there are two distinct slopes. If there are only three points then p_1 is connected to p_2 via u, p_2 is connected to p_3 via v, and p_3 must be connected to p_1 via u, resulting in all three lying on the same line with slope u, but also p_2 lying on a line with slope v with p_3, resulting in a contradiction. Figure 3 shows a set with more complicated loops, including two which loop over all points in the set.

Lemma 2. *Let $\mathsf{P} \subset \mathsf{X}^2$ be evenly covered by $\mathsf{W} \subset \mathsf{X}$. Let $u, v \in \mathsf{W}$, then every point in P is in a (u,v)-loop starting with slope u and ending with slope v.*

Proof. Let $p_0 \in \mathsf{P}$, then by hypothesis there is another point p_1 in P lying on a line with slope u connecting to p_0. Similarly, there is a point p_2 different from p_0 and p_1 lying on a line with slope v connected to p_1. Continuing like this, we can create a sequence of points $p_0, p_1, \ldots, p_k$ until $p_{k+1} = p_i$ for some $i \leq k$, with the property that adjacent points in the sequence are connected by lines alternating with slope u and v.

If $i = 0$, then we are done. Otherwise, consider p_{i-1}, p_i, p_{i+1}, and p_k. Say that p_{i-1} is connected to p_i via a line with slope u, so that p_i is connected to p_{i+1} via a line with slope v. If p_k is connected to p_i via a line with slope v, then there are three points on the same line with slope v: p_i, p_{i+1}, and p_k. This means there is a fourth point p^* on the same line. Since p_k is connected to p_{i+1} via v, the sequence $p_{i+1}, p_{i+2}, \ldots, p_k$ forms a (u,v)-loop. We remove the (u,v)-loop from P, which is evenly covered by u and v, resulting in a set evenly covered by u and v, and we continue by induction. Similar reasoning can be applied when p_k is connected to p_i via u. □

Proposition 5. *The sum of the x-coordinates in a (u,v)-loop must be zero.*

Proof. Say that $(x_1,y_1),(x_2,y_2),\ldots,(x_k,y_k)$ are the points in the loop. Then

$$y_i + y_{i+1} = \delta_i(x_i + x_{i+1 \,(\mathrm{mod}\ k)}), \tag{24}$$

where δ_i is u if i is odd, and v otherwise. Since

$$(y_1+y_2)+(y_2+y_3)+\cdots+(y_{k-1}+y_k)+(y_k+y_1)=0, \tag{25}$$

we have that

$$u(x_1+x_2)+v(x_2+x_3)+u(x_3+x_4)+\cdots \\ +u(x_{k-1}+x_k)+v(x_k+x_1)=0, \tag{26}$$

therefore

$$(u+v)(x_1+x_2+\cdots+x_k)=0. \tag{27}$$

Since $u \neq v$, it must be the case that $x_1+x_2+\cdots+x_k=0$. □

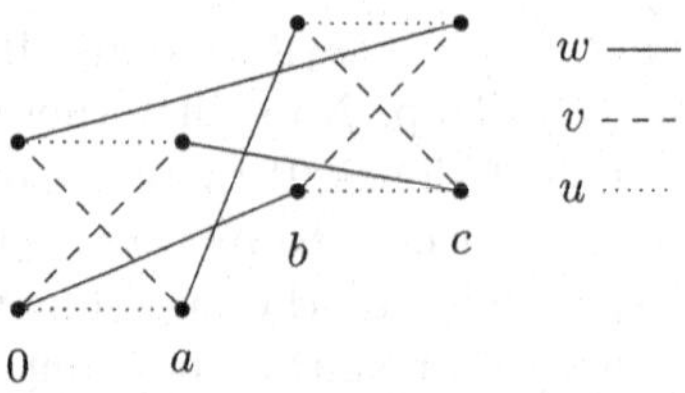

Fig. 3. A set of points evenly covered by the slopes u, v, and w. Each point is accompanied by another point with the same x-coordinate. The x-coordinates of the pairs are indicated below the lower points.

Adversaries can only construct sets P where there are at most two points per x-coordinate. Therefore, either all loops only contain points (x, y) for which there is exactly one other point (x, y') with the same x-coordinate, or there exists a loop with a point which is the only one with that x-coordinate. For example, Figs. 2 and 3 depict evenly covered sets where every loop always contains all x-coordinate pairs. If we consider the only loop in Fig. 2, then we get

$$0 \cdot (x_1 + x_2) + (x_1 + x_2)^{-1}(x_2 + x_1) + 0 \cdot (x_1 + x_2) + (x_1 + x_2)^{-1}(x_2 + x_1), \quad (28)$$

which trivially equals zero. All loops in Fig. 3 also trivially sum to zero.

In contrast, Fig. 4 depicts an evenly covered set in which we get a non-trivial sum of the x-coordinates:

$$u \cdot a + v(a + c) + u(c + b) + v \cdot b = (u + v)(a + b + c) = 0, \quad (29)$$

hence such a set only exists if $a + b + c = 0$.

Therefore, Proposition 5 only poses a non-trivial restriction on the x-coordinates if there is a loop which contains a point without another point sharing its x-coordinate. If the loop contains all pairs of points with the same x-coordinates, then the x-coordinates will trivially sum to zero. This is why in the case of Fig. 2 there are no restrictions on the x-coordinates, other than the fact that they must be distinct, resulting in the existence of sets evenly covered by two slopes.

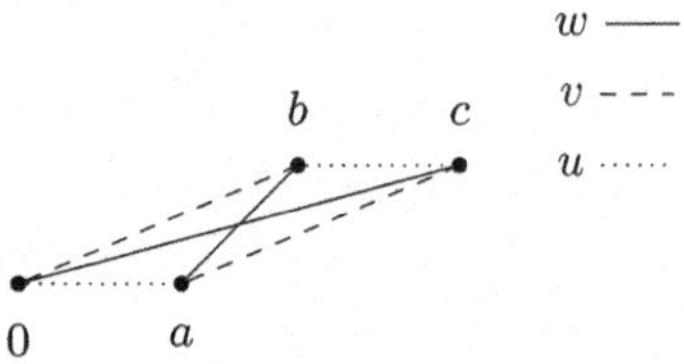

Fig. 4. A set of points evenly covered by the slopes u, v, and w. None of the points are accompanied by another point with the same x-coordinate. The points are labelled by their x-coordinates.

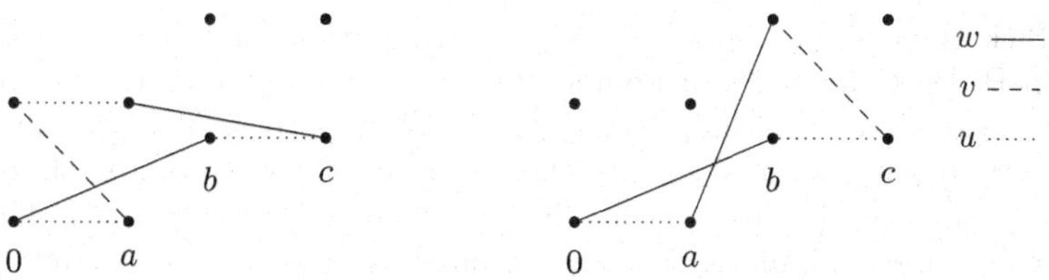

Fig. 5. Illustration of loops with three slopes.

In the case of Fig. 3 however, there are additional restrictions on the x-coordinates. Consider the two points at x-coordinate 0. Then there is part of a (u, v)-loop connecting them, and part of a (u, w)-loop connecting them, and combining both parts we get a full loop using all three slopes; see the left hand side of Fig. 5. A similar loop involving all three slopes can be constructed around the points with x-coordinate b. Using these two loops, we get the following equations. From the left hand side of Fig. 5 we have

$$ua + va = wb + u(b + c) + w(a + c) + ua \tag{30}$$

$$(u + v)a = (w + u)(a + b + c) . \tag{31}$$

From the right hand side of Fig. 5 we have

$$(u + v)(b + c) = wb + ua + w(a + b) \tag{32}$$

$$(u + v)(b + c) = (w + u)a . \tag{33}$$

Combining both, we get the following:

$$\frac{a + b + c}{a} = \frac{a}{b + c} \tag{34}$$

$$a^2 + b^2 + c^2 + ab + ac = 0 . \tag{35}$$

The last equation above can be described as a so-called *quadratic form*. A quadratic form over X is a homogeneous multivariate polynomial of degree two. In our case, the quadratic form can be written as $\boldsymbol{x}^T Q \boldsymbol{x}$, where $\boldsymbol{x} \in \mathsf{X}^n$ is the list of variables, and $Q \in \{0, 1\}^{n \times n}$ is a matrix with entries in $\{0, 1\}$. We say that $\boldsymbol{x}_*$ is a *solution* to Q if $\boldsymbol{x}_*^T Q \boldsymbol{x}_* = 0$, and the quadratic form Q is *non-trivial* if there exists $\boldsymbol{x} \neq 0$ such that $\boldsymbol{x}^T Q \boldsymbol{x} \neq 0$.

So the evenly covered set from Fig. 3 only exists if the x-coordinates satisfy some non-trivial quadratic form. The same is true for any evenly covered set where all loops always contain pairs of points with the same x-coordinate.

Proposition 6. *Let $\mathsf{P} \subset \mathsf{X}^2$ be evenly covered by $\mathsf{W} \subset \mathsf{X}$ with $\mathsf{W} \geq 3$. Say that all loops in P contain only pairs of points with the same x-coordinates. Then there exists a subset S of k x-coordinates, and a non-trivial quadratic form described by a matrix $Q \in \{0, 1\}^{k \times k}$ over k variables, such that when the k elements of S are placed in a vector $\boldsymbol{x}_* \in \mathsf{X}^k$, $\boldsymbol{x}_*^T Q \boldsymbol{x}_* = 0$.*

Proof. Pick three slopes, u, v, w in W. We know that there are at least four points in P. Pick two pairs of points with the same x-coordinates: (p, p') and (q, q'). Consider the (u, v)-loop starting at p. By hypothesis it must contain p'. We let $\boldsymbol{a} = (a_1, a_2, \ldots, a_{k_a})$ denote the sequence of x-coordinates of the part of the (u, v)-loop from p to p'. Note that a_1 equals a_{k_a} since p and p' have the same x-coordinates. Similarly, the (u, v)-loop starting at q must contain q', and we denote the sequence of x-coordinates of the part of the (u, v)-loop from q to q' by $\boldsymbol{b} = (b_1, b_2, \ldots, b_{k_b})$. The same holds for the (v, w)-loops containing p and q, and we define the x-coordinate sequences $\boldsymbol{e}$ and $\boldsymbol{f}$ similarly.

Let y denote the difference in the y-coordinates of p and p'. For $\boldsymbol{a}$ we have the following:

$$u(a_1 + a_2) + v(a_2 + a_3) + \cdots + \delta(u, v)_{k_a}(a_{k_a-1} + a_{k_a}) = y\,, \tag{36}$$

where $\delta(u, v)_{k_a}$ is u if k_a is even and v otherwise. Collecting the terms, if k_a is even, we get

$$u(a_1 + a_2 + \cdots + a_{k_a-1} + a_{k_a}) + v(a_2 + \cdots + a_{k_a-1}) = y\,, \tag{37}$$

and since $a_1 = a_{k_a}$, we know that

$$(u + v)(a_2 + \cdots + a_{k_a-1}) = y\,. \tag{38}$$

If k_a is odd, then we get

$$(u + v)(a_1 + a_2 + \cdots + a_{k_a-1}) = y\,. \tag{39}$$

Note that it cannot be the case that $\sum a_i = 0$, since $y \neq 0$.

Similar reasoning applied to $\boldsymbol{b}$ gives

$$\begin{aligned} (v + w)(b_2 + \cdots + b_{k_b-1}) &= y \;\text{ if } k_b \text{ is even} \\ (v + w)(b_1 + \cdots + b_{k_b-1}) &= y \;\text{ otherwise}\,. \end{aligned} \tag{40}$$

Regardless of k_a and k_b's parities, setting both equations equal to each other results in the following equation:

$$\frac{u + v}{v + w} = \frac{\sum b_i}{\sum a_i}\,. \tag{41}$$

Applying the same result to $\boldsymbol{e}$ and $\boldsymbol{f}$, we get

$$\frac{u + v}{v + w} = \frac{\sum f_i}{\sum e_i}\,. \tag{42}$$

As a result, we have

$$\left(\sum b_i\right)\left(\sum e_i\right) + \left(\sum a_i\right)\left(\sum f_i\right) = 0\,, \tag{43}$$

which is the solution to a quadratic form. □

5.3 Computational Hardness

As shown in Propositions 5 and 6, either there is a loop where the x-coordinates non-trivally sum to zero, or there is a subset of the x-coordinates which form the solution to some non-trivial quadratic form. The former is Subset Sum (SS), whereas the latter we name the binary quadratic form (BQF) problem.

Definition 8 (Subset Sum Problem (SS)). *Given a finite field* X *of characteristic two and a subset* $S \subset \mathsf{X}$, *determine whether there is a subset* $S_0 \subset S$ *such that* $\sum_{x \in S_0} x = 0$.

Definition 9 (Binary Quadratic Form Problem (BQF)). *Given a finite field* X *of characteristic two and a subset* $S \subset \mathsf{X}$, *determine whether there is a non-trivial quadratic form* $Q \in \{0,1\}^{k \times k}$ *with a solution* $\boldsymbol{x}_*$ *made up of distinct components from* S.

SS is know to be **NP**-complete. In Appendix B we show that BQF-t, a generalization of BQF, is **NP**-complete as well. The problem of finding either a subset summing to zero or a non-trivial quadratic form we call the *SS-or-BQF* problem.

Conjecture 1. There do not exist polynomial time algorithms solving SS-or-BQF.

Definition 10 (PHASH Problem). *Given a finite field* X *of characteristic two and a sequence of masks* $\mathbf{c}$, *determine whether there is a collision in PHASH with probability greater than* $2/N$, *where* $N = |\mathsf{X}|$.

Given a collision in PHASH one can easily find a solution to SS-or-BQF. The converse does not necessarily hold, which means SS-or-BQF cannot be reduced to the PHASH problem in general, although we can conclude the following.

Theorem 3. *One of the following two statements holds.*

1. *There are infinitely many input sizes for which the PHASH problem does not have a solution, but SS-or-BQF does.*
2. *For sufficiently large input sizes, SS-or-BQF can be reduced to the PHASH problem.*

Proof. Both the PHASH and SS-or-BQF problems are decision problems, so the output of the algorithms solving the problems is a yes or a no, indicating whether the problems have a solution or not. Note that the inputs to both problems are identical. The reductions consist of simply converting the input to one problem into the input of the other, and then directly using the output of the algorithm solving the problem.

We proved that a yes instance for PHASH becomes a yes instance for SS-or-BQF: if you have an instance of SS-or-BQF, then you can convert it into a PHASH problem, and if you are able to determine that PHASH has a collision with sufficient probability, then SS-or-BQF has a solution. Similarly, a no instance for SS-or-BQF means a no instance for PHASH.

The issue is when there exists a no instance for PHASH and a yes instance for SS-or-BQF for a particular input size. If there are finitely many input sizes for which there is a no instance for PHASH and a yes instance for SS-or-BQF simultaneously, then there exists an r such that for all input sizes greater than r a no instance for PHASH occurs if and only if a no instance for SS-or-BQF occurs, and a yes instance for PHASH occurs if and only if a yes instance for SS-or-BQF occurs. Therefore, an algorithm which receives a no instance for PHASH can say that the corresponding SS-or-BQF problem is a no instance, and similarly for the yes instances, which is our reduction. Otherwise there are infinitely many input sizes for which PHASH is a no instance, and SS-or-BQF is a yes instance. □

If statement 1 holds, then there are infinitely many candidates for an instantiation of PMAC* with security bound independent of the message length. If statement 2 holds, and we assume that SS-or-BQF is hard to solve, then finding a collision for generic PHASH is computationally hard.

6 Finding Evenly Covered Sets

The previous section focused on determining necessary conditions for the existence of evenly covered sets, illustrating the difficulty with which such sets are found. Nevertheless, finding evenly covered sets becomes feasible in certain situations. In this section we provide an alternative description of evenly covered sets in order to find sufficient conditions for their existence.

6.1 Distance Matrices

Let $(x_1, y_1), (x_2, y_2), \ldots, (x_n, y_n)$ be an enumeration of the elements of $\mathsf{P} \subset \mathsf{X}^2$. If $w \in \mathsf{X}$ covers P evenly, then the line with equation $y = w(x - x_1) + y_1$ must meet P in an even number of points. In particular, there must be an even number of x_i values for which $w(x_i - x_1) + y_1 = y_i$, or in other words, the vector

$$w \cdot (x_1 - x_1, x_2 - x_1, \ldots, x_n - x_1) \tag{44}$$

must equal

$$(y_1 - y_1, y_2 - y_1, \ldots, y_n - y_1) \tag{45}$$

in an even number of coordinates. The same must hold for the lines starting from all other points in P.

Let Δ^x be the matrix with (i, j) entry equal to $x_i - x_j$ and Δ^y the matrix with (i, j) entry equal to $y_i - y_j$. We write $A \sim B$ if matrix $A \in \mathsf{X}^{n \times n}$ equals matrix $B \in \mathsf{X}^{n \times n}$ in an even number of entries in each row. Then, following the reasoning from above, we have that $w \in \mathsf{X}$ covers P evenly only if $\Delta^y \sim w\Delta^x$.

The matrices Δ^x and Δ^y are so-called *distance* matrices, that is, symmetric matrices with zero diagonal. Entry (i, j) in these distance matrices represents the "distance" between x_i and x_j, or y_i and y_j. In fact, starting from distance matrices M and D such that $M \sim wD$ we can also recover a set P evenly covered by w: interpret the matrices M and D as the distances between the points in the set P. This proves the following lemma.

Lemma 3. *Let $k \leq n-1$ and let $\mathsf{W} \subset \mathsf{X}$ be a set of size k. There exist n by n distance matrices M and D such that $M \sim wD$ for all $w \in \mathsf{W}$ if and only if there exists P with $|\mathsf{P}| = n$ and W evenly covers P.*

From the above lemma we can conclude that the existence of $\mathsf{P} \subset \mathsf{X}^2$ evenly covered by $\mathsf{W} \subset \mathsf{X}$ is not affected by the following transformations:

1. Translating the set P by any vector in X^2. This also preserves the set W.
2. Subtracting any element $w_0 \in \mathsf{W}$ from the set W.
3. Scaling the set P in either x or y-direction by a non-zero scalar in X.
4. Scaling the set W by any non-zero element of X.

6.2 Connection with Graphs

Let $\mathsf{P} \subset \mathsf{X}^2$ be evenly covered by $\mathsf{W} \subset \mathsf{P}$. The pair (P, W) has a natural graph structure with vertices P and an edge connecting two vertices p_1 and p_2 if and only if the line connecting them has slope in W. Figures 2 and 3 provide diagrams which can also be viewed as examples of the natural graph structure. In this section we connect the existence of evenly covered sets with so-called *factorizations* of a graph. See Appendix A for a review of the basic graph theoretic definitions used in this section.

Each vertex in the natural graph has at least $|\mathsf{W}|$ neighbours, and if there are two points per line in P, then the graph is $|\mathsf{W}|$-regular. Vertices have more than $|\mathsf{W}|$ neighbours only if they are on a line with more than two points. Since we are not interested in the redundancy from connecting a point with all points on the same line, we only consider graphs without the additional edges.

Definition 11. *A graph associated to (P, W) is a $|\mathsf{W}|$-regular graph G with P as its set of vertices and an edge between two vertices p_1 and p_2 only if the line connecting p_1 with p_2 has slope in W.*

Any graph associated to (P, W) is a subgraph of the natural graph structure described above, and there could be multiple associated graphs, depending upon what edges are chosen to connect multiple points lying on the same line. For example, Fig. 6 depicts an evenly covered set with twelve points, six of which

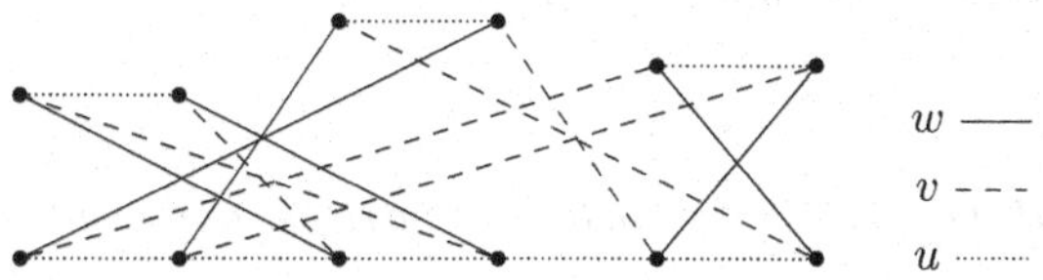

Fig. 6. Non-trivial example of a set with 12 points evenly covered by three slopes. Horizontal points lie on the same y-coordinate, and vertical points on the same x-coordinate. Since there are six points on a line with slope u, the natural graph is not regular.

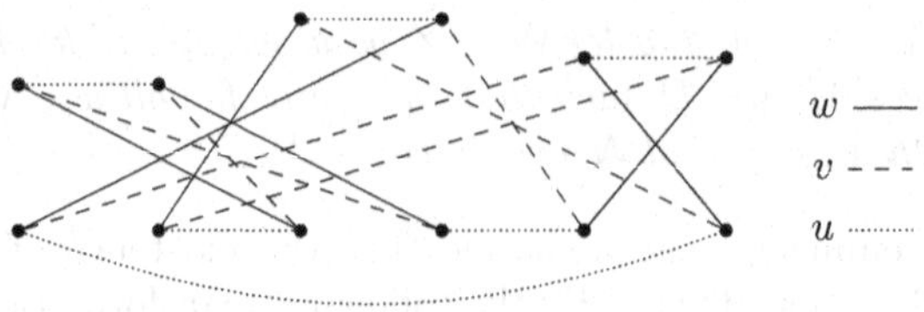

Fig. 7. The diagram from Fig. 6 converted into an associated graph. The slopes u, v, and w induce a natural 1-factorization of the graph.

lie on the same line. As depicted in Fig. 7, it can easily be converted into an associated graph.

The following definition allows us to describe another property that associated graphs have.

Definition 12. *A k-factor of a graph G is a k-regular subgraph with the same vertex set as G. A k-factorization partitions the edges of a graph in disjoint k-factors.*

Associated graphs have a 1-factorization induced by W, where each 1-factor is composed of the edges associated to the same slope in W. See Fig. 7 for an example.

We know that every pair (P, W) has an associated $|\mathsf{W}|$-regular graph with 1-factorization. In order to determine the existence of evenly covered sets we need to consider when a k-regular graph with 1-factorization describes the structure of some pair (P, W) with $|\mathsf{W}| = k$. By first fixing a graph with a 1-factorization, it is possible to set up a system of equations to determine the existence of distance matrices M and D, and slopes W such that $M \sim wD$ for all $w \in \mathsf{W}$. Then, by applying Lemma 3, we will have our desired pair (P, W).

Definition 13. *Let G be a regular graph with vertices $(v_1, \ldots, v_n)$ and a 1-factorization, and let $\mathsf{X}^{n\times n}$ denote the set of matrices over X. Define $\mathbf{S}_G \subset \mathsf{X}^{n\times n}$ to be the matrices where entry (i, j) equals entry (k, l) if and only if the edges (v_i, v_j) and (v_k, v_l) are in the same 1-factor of G.*

Proposition 7. *There exists a set $\mathsf{P} \subset \mathsf{X}^2$ with n elements evenly covered by $\mathsf{W} \subset \mathsf{X}$ with $|\mathsf{W}| = k$ if and only if there exists a k-regular graph G of order n with a 1-factorization such that there is a solution to*

$$M = S \circ D, \tag{46}$$

where $S \in \mathbf{S}_G$, $M, D \in \mathsf{X}^{n\times n}$ are distance matrices, and $\circ$ denotes elementwise multiplication.

Therefore by picking a regular graph with a 1-factorization and solving a system of equations, we can determine the existence of pairs (P, W) for various sizes, in order to determine a lower bound for PHASH's collision probability.

6.3 Latin Squares and Abelian Subgroups

In this section we consider what happens when we solve Eq. (46) with a 1-factorization of the complete graph of order n. Since we look at complete graphs, finding a solution would imply the existence of sets with n points evenly covered by $n-1$ slopes, the optimal number as shown by Proposition 3. We describe a necessary and sufficient condition on the matrix D from Eq. (46), which in turn becomes a condition on the x-coordinates of the evenly covered sets.

As described by Laywine and Mullen [13, Sect. 7.3], 1-factorizations of a complete graph G of order n, with n even, are in one-to-one correspondence with reduced, symmetric, and unipotent Latin squares, that is, n by n matrices with entries in $\mathbb{N}$ such that

1. the first row enumerates the numbers from 1 to n,
2. the matrix is symmetric, that is, entry (i, j) equals entry (j, i),
3. the diagonal consists of just ones,
4. and each natural number from 1 to n appears just once in every row and column.

The correspondence between 1-factorizations of complete graphs and Latin squares works by identifying row i and column i with a vertex in the graph, labelling the 1-factor containing edge $(1, i)$ with i, and then setting entry (i, j) equal to the label of the 1-factor containing edge (i, j). This is exactly the structure of the matrices in $\mathbf{S}_G$.

Let n be a power of two. The *abelian 2-group of order* n is a commutative group in which every element has order two, that is, $a + a = 0$ for all elements a in the group. The Cayley table of the abelian 2-group of order n can be written as a reduced, symmetric, and unipotent Latin square.

Definition 14. *The* (i, j) *entry of the Cayley table of the abelian 2-group with* ℓ *elements is denoted* $\gamma(i, j)$.

Lemma 4. $\gamma(i, \gamma(i, j)) = j$.

Proposition 8. *Let* G *denote the complete graph of order* n*, where* n *is a power of two, with 1-factorization induced by the Cayley table of the abelian 2-group of order* n*. Then Eq.* (46) *has a solution if and only if the first row of* D *forms an additive subgroup of* X *of order* n*.*

The above proposition shows that the graph structure corresponding to the abelian 2-group induces the same additive structure on the x-coordinates of the evenly covered set. This transfer of structure only works with this particular 1-factorization of the complete graph. In general, reduced, symmetric, and unipotent Latin squares do not even correspond to the Cayley table of some group: associativity is not guaranteed. Furthermore, 1-factorizations of non-complete graphs do not necessarily even form Latin squares; see for example Fig. 6.

Proof. Denote the first row of S by $s_1, s_2, \ldots, s_n$, and the first row of D by $d_1, \ldots, d_n$. Note that D is entirely determined by its first row, since the (i,j) entry of D is $d_i + d_j$, and since S follows the form of γ, it is entirely determined by its first row as well. In particular, the (i,j) entry of S is $s_{\gamma(i,j)}$, where $\gamma(i,j)$ is the (i,j) entry of the Cayley table.

We need to determine the conditions under which $S \circ D$ is a distance matrix, as a function of $s_1, \ldots, s_n$ and $d_1, \ldots, d_n$. This happens if and only if the (i,j) entry of $S \circ D$ is equal to $s_i d_i + s_j d_j$:

$$s_i d_i + s_j d_j = s_{\gamma(i,j)}(d_i + d_j)\,. \tag{47}$$

Furthermore, it must be the case that

$$s_i d_i + s_{\gamma(i,j)} d_{\gamma(i,j)} = s_j(d_i + d_{\gamma(i,j)})\,, \tag{48}$$

since $\gamma(i, \gamma(i,j)) = j$. Therefore

$$s_j d_j + s_{\gamma(i,j)} d_{\gamma(i,j)} = s_{\gamma(i,j)}(d_i + d_j) + s_j(d_i + d_{\gamma(i,j)}) \tag{49}$$

$$(s_j + s_{\gamma(i,j)})(d_i + d_j + d_{\gamma(i,j)}) = 0\,. \tag{50}$$

Since S must follow the Latin square structure, the first row of S must consist of n distinct entries, hence $s_j \neq s_{\gamma(i,j)}$ and so $d_i + d_j + d_{\gamma(i,j)} = 0$. Therefore, $d_1, \ldots, d_n$ satisfies the equations of the Cayley table, hence they form an additive subgroup of X.

Continuing, we have the following equations:

$$s_i d_i + s_j d_j + s_{\gamma(i,j)} d_{\gamma(i,j)} = 0\,. \tag{51}$$

In order for these equations to be satisfied, $s_1 d_1, \ldots, s_n d_n$ must form an additive subgroup of X as well. In particular, there must exist an isomorphism ϕ mapping d_i to $s_i d_i$, which can be written as $d_i^{-1}\phi(d_i) = s_i$ for $i > 1$. The only requirement for the existence of such an isomorphism is that $x^{-1}\phi(x)$ must map to distinct values. Picking $x \mapsto x^2$ as the isomorphism, we have our desired result. Note that the d_i must be distinct, otherwise the s_i are not distinct, contradicting the fact that S follows the Latin square structure. □

6.4 Application to PMAC

Before we present an attack, we first need the following lemma.

Lemma 5. *Let* P *and* P' *be disjoint subsets of* X^2 *evenly covered by* $\mathsf{W} \subset \mathsf{X}$. *Then* $\mathsf{P} \cup \mathsf{P}'$ *is evenly covered by* W.

A collision in PHASH with probability $(\ell - 1)/N$ can be found as follows. Take $\mathbf{c}$ and let k be the smallest index such that $\mathbf{c}_{\leq k}$ contains a subsequence $\mathbf{c}'$ of length ℓ such that the elements $\{c'_1 + c'_1, c'_1 + c'_2, \ldots, c'_1 + c'_\ell\}$ form an additive subgroup of X. Let μ be the mapping which maps indices of $\boldsymbol{c}'$ onto indices of $\boldsymbol{c}$, so that $c'_i = c_{\mu(i)}$.

Let D be a distance matrix in $\mathsf{X}^{\ell \times \ell}$ such that its first row is equal to $(c'_1 + c'_1, c'_1 + c'_2, \ldots, c'_1 + c'_\ell)$; recall that a distance matrix is completely determined by its first row. Let G be the complete graph of order ℓ with 1-factorization determined by the abelian 2-group of order ℓ. Solve Eq. (46), that is, find a distance matrix M such that there exists $S \in \mathbf{S}_G$ where

$$M = S \circ D. \tag{52}$$

Let $\boldsymbol{m}^1$ denote the first row of M, and let W denote the elements making up the first row of S, without the first row element. Then the set $\mathsf{P} := \{(c'_1, m^1_1), \ldots, (c'_\ell, m^1_\ell)\}$ is evenly covered by W, which contains $\ell - 1$ slopes.

By translating P vertically by some constant, say 1, construct the disjoint set P', which is also evenly covered by W. Therefore, by Lemma 5, the union of P and P' is evenly covered by W. Let $\boldsymbol{m}^2$ denote the y-coordinates of P'.

Define $\overline{\boldsymbol{m}^1}$ to be the vector of length k where for all $i \leq \ell$, $\overline{m^1}_{\mu(i)} = m^1_i$, and for all i not in the range of μ, $\overline{m^1}_i = 0$. Define $\overline{\boldsymbol{m}^2}$ similarly. Then $\overline{\boldsymbol{m}^1}$ and $\overline{\boldsymbol{m}^2}$ collide with probability $(\ell - 1)/N$.

For sufficiently large k, $\boldsymbol{c}_{\leq k}$ will always contain additive subgroups. In particular, one can find such subgroups in PMAC with Gray codes [7], where $\boldsymbol{c}$ is defined as follows. In this case $\mathsf{X} := \{0,1\}^\nu$ is the set of ν-bit strings, identified in some way with a finite field of size 2^ν. We define the following sequence of vectors λ^ν:

$$\lambda^1 = (0, 1) \tag{53}$$

$$\lambda^{\nu+1} = (0\|\lambda^\nu_1, 0\|\lambda^\nu_2, \ldots, 0\|\lambda^\nu_{2^\nu}, 1\|\lambda^\nu_{2^\nu}, \ldots, 1\|\lambda^\nu_2, 1\|\lambda^\nu_1)\,. \tag{54}$$

Note that λ^ν contains all strings in X. Then $\boldsymbol{c}$ is λ^ν without the first component, meaning $\boldsymbol{c}$ contains all strings in X without the zero string. Similarly, the sequence $(c_1, \ldots, c_{2^\kappa})$ contains all strings starting with $\nu - \kappa$ zeros, i.e. $0^{\nu-\kappa}\| \{0,1\}^\kappa$, excluding the zero string. Note that $c_1 = 0^{\nu-1}1$. The sequence $(c_1 + c_1, c_1 + c_2, \ldots, c_1 + c_{2^\kappa})$ contains all strings in $0^{\nu-\kappa}\| \{0,1\}^\kappa$ except for c_1, meaning it contains an additive subgroup of order $2^{\kappa-1}$. This results in an attack using messages of length $k = 2^\kappa$ with success probability $(2^\kappa - 1)/2^\nu$.

Acknowledgments. We would like to thank Tomer Ashur, Bart Mennink, and the reviewers for providing useful comments, and also Kazumaro Aoki for his help in exploring subset sums in finite fields. This work was supported in part by the Research Council KU Leuven: GOA TENSE (GOA/11/007). In addition, this work was supported by the Research Fund KU Leuven, OT/13/071. Atul Luykx and Alan Szepieniec are supported by Ph.D. Fellowships from the Institute for the Promotion of Innovation through Science and Technology in Flanders (IWT-Vlaanderen).

A Basic Graph Theoretic Definitions

1. A *neighbour* of a vertex v in a graph G is a vertex with an edge connecting it to v.
2. A graph G is said to be k-*regular* if every vertex of G has exactly k neighbours.
3. A *subgraph* of a graph G is a graph with vertex set and edge set subsets of G's vertex and edge sets, respectively.
4. A *complete graph* is a graph in which every vertex is connected to every other vertex via an edge.

B BQF-t is NP-complete

Definition 15 (BQF-t). *Given a finite field* X *with characteristic 2 and a vector* $\boldsymbol{x}_* \in \mathsf{X}^k$ *and a target element* $t \in \mathsf{X}$, *determine if there is a non-trivial binary quadratic form* $Q \in \{0,1\}^{k\times k}$ *such that* $\boldsymbol{x}_*^T Q \boldsymbol{x}_* = t$.

Note. The word 'binary' in our use of the term 'binary quadratic form' refers to the coefficients of the quadratic form matrix Q and not to the number of variables.

Proposition 9. *BQF-t* $\in$ **NP**

Proof. Given a BQF-t yes-instance $(\mathsf{X}, \boldsymbol{x}_*, t)$ of $(k+2) \times \ell$ bits, there exists a certificate of $k^2 \times \ell$ bits that proves it is a yes-instance, namely the matrix Q such that $\boldsymbol{x}_*^T Q \boldsymbol{x}_* = t$. Moreover, the validity of this certificate can be verified by computing $\boldsymbol{x}_*^T Q \boldsymbol{x}_*$ and testing if it is indeed equal to t. This evaluation requires $(n+1) \times n$ multiplications and the same number of additions in the finite field X. After testing equality, the non-triviality of Q is verified by testing whether $Q^T + Q \neq 0$, costing another n^2 finite field additions and as many equality tests. Thus, for every yes-instance of BQF-t, there exists a polynomial-size certificate whose validity is verifiable in polynomial time. Hence, BQF-t $\in$ **NP**. □

Proposition 10. *BQF-t is* **NP**-*hard.*

Proof. We show that BQF-t is **NP**-hard by reducing the subset-sum problem SS, another **NP**-hard problem, to it. In particular, we show that SS$\leq$ BQF-t under deterministic polynomial-time Karp reductions.

Given an instance (X, S) of SS, the goal is to find a subset $S_0 \subset S$ such that $\sum_{x\in S_0} x = 1$. Note the target of SS can be changed without loss of generality. We transform this problem instance to an instance $(\mathsf{X}', \boldsymbol{x}_*, t)$ of BQF-t as follows.

Let $k = \#S$, the number of elements in S and let each unique element s_i of S be indexed by $i \in \{1, \ldots, k\}$. Choose a degree $2k+1$ irreducible polynomial $\psi(z) \in \mathsf{X}[z]$ and define the extension field $\mathsf{X}' = \mathsf{X}[z]/\langle\psi(z)\rangle$. Then define the vector $\boldsymbol{x}_*$ as follows:

$$\boldsymbol{x}_* = \begin{pmatrix} z^1 s_1 \\ z^2 s_2 \\ \vdots \\ z^k s_k \\ z^{-1} \\ z^{-2} \\ \vdots \\ z^{-k} \end{pmatrix}.$$

The BQF-t instance is $(\mathsf{X}', \boldsymbol{x}_*, 1)$. It now remains to be shown that (1) this transformation is computable in polynomial time; (2) if the SS problem instance is a yes-instance, then the BQF-t problem instance is yes-instance; (3) conversely, if the SS problem instance is a no-instance, then the BQF-t problem instance is a no-instance.

1. It is known to be possible to deterministically select an irreducible polynomial over a finite field of small characteristic in polynomial time [20]. After selecting the polynomials, the inverse of z is computed using the polynomial-time extended GCD algorithm and all the necessary powers of z and z^{-1} are found after two times k multiplications. Lastly, the proper powers of z are combined with the s_i elements using k multiplications for the construction of the first half of the vector $\boldsymbol{x}_*$; the second half of this vector has already been computed. So since this transformation consists of a polynomial-number of polynomial-time steps, its total running time is also polynomial.
2. If the SS instance is a yes-instance, then there exist k binary weights $w_i \in \{0, 1\}$ for all $i \in \{1, \ldots, k\}$ such that $\sum_{i=1}^{k} w_i s_i = 1$. The existence of these weights imply the existence of the matrix Q, as defined below. This matrix consists of four $k \times k$ submatrices and only the diagonal of the upper right submatrix is nonzero. In fact, this diagonal is where the weights w_i appear.

$$Q = \left(\begin{array}{c|c} & \begin{matrix} w_1 & & \\ & \ddots & \\ & & w_k \end{matrix} \\ \hline & \end{array}\right) \tag{55}$$

Indeed, the BQF-t instance is guaranteed to be a yes-instance as

$$\boldsymbol{x}_*^T Q \boldsymbol{x}_* = \sum_{i=1}^{k} z^i s_i w_i z^{-i} = 1$$

if and only if

$$\sum_{i=1}^{k} w_i s_i = 1 \ ,$$

which is the solution to the SS problem. Also, Q is non-trivial if there exists at least one nonzero weight w_i.

3. If the SS instance is a no-instance, then no set of weights w_i such that $\sum_{i=1}^{k} w_i s_i = 1$ exists. Consequently, no Q satisfying $\boldsymbol{x}_*^T Q \boldsymbol{x}_* = 1$ can exist. The reason is that all the elements of the Q-matrix except for the upper right diagonal are multiplied with higher or lower powers of z, which make them linearly independent from 1. Hence, neither the upper right diagonal nor any other set of nonzero elements in Q can make the total quadratic form equal to one. □

Corollary 1. *BQF-t is* **NP***-complete.*

References

1. Bellare, M., Desai, A., Jokipii, E., Rogaway, P.: A concrete security treatment of symmetric encryption. In: FOCS, pp. 394–403. IEEE Computer Society (1997)
2. Bellare, M., Goldreich, O., Mityagin, A.: The power of verification queries in message authentication and authenticated encryption. IACR Cryptology ePrint Archive 2004, 309 (2004)
3. Bellare, M., Guérin, R., Rogaway, P.: XOR MACs: new methods for message authentication using finite pseudorandom functions. In: Coppersmith [8], pp. 15–28. http://dx.doi.org/10.1007/3-540-44750-4_2
4. Bellare, M., Kilian, J., Rogaway, P.: The security of cipher block chaining. In: Desmedt, Y.G. (ed.) CRYPTO 1994. LNCS, vol. 839, pp. 341–358. Springer, Heidelberg (1994). http://dx.doi.org/10.1007/3-540-48658-5_32
5. Bellare, M., Pietrzak, K., Rogaway, P.: Improved security analyses for CBC MACs. In: Shoup, V. (ed.) CRYPTO 2005. LNCS, vol. 3621, pp. 527–545. Springer, Heidelberg (2005). http://dx.doi.org/10.1007/11535218_32
6. Biryukov, A. (ed.): Fast Software Encryption. LNCS, vol. 4593. Springer, Heidelberg (2007)
7. Black, J.A., Rogaway, P.: A block-cipher mode of operation for parallelizable message authentication. In: Knudsen, L.R. (ed.) EUROCRYPT 2002. LNCS, vol. 2332, pp. 384–397. Springer, Heidelberg (2002). http://dx.doi.org/10.1007/3-540-46035-7_25
8. Coppersmith, D. (ed.): Advances in Cryptology – CRYPTO 1995. LNCS, vol. 963. Springer, Heidelberg (1995)
9. Dodis, Y., Pietrzak, K.: Improving the security of MACs via randomized message preprocessing. In: Biryukov [6], pp. 414–433. http://dx.doi.org/10.1007/978-3-540-74619-5_26
10. Gaži, P., Pietrzak, K., Rybár, M.: The exact PRF-security of NMAC and HMAC. In: Garay, J.A., Gennaro, R. (eds.) CRYPTO 2014, Part I. LNCS, vol. 8616, pp. 113–130. Springer, Heidelberg (2014). http://dx.doi.org/10.1007/978-3-662-44371-2_7

11. Gazi, P., Pietrzak, K., Tessaro, S.: The exact PRF security of truncation: tight bounds for keyed sponges and truncated CBC. In: Gennaro, R., Robshaw, M. (eds.) Advances in Cryptology - CRYPTO 2015. LNCS, vol. 9215, pp. 368–387. Springer, Heidelberg (2015). http://dx.doi.org/10.1007/978-3-662-47989-6_18
12. Iwata, T., Kurosawa, K.: Stronger security bounds for OMAC, TMAC, and XCBC. In: Johansson, T., Maitra, S. (eds.) INDOCRYPT 2003. LNCS, vol. 2904, pp. 402–415. Springer, Heidelberg (2003). http://dx.doi.org/10.1007/978-3-540-24582-7_30
13. Laywine, C.F., Mullen, G.L.: Discrete Mathematics Using Latin Squares, vol. 49. Wiley, New York (1998)
14. Minematsu, K.: How to Thwart birthday attacks against MACs via small randomness. In: Hong, S., Iwata, T. (eds.) FSE 2010. LNCS, vol. 6147, pp. 230–249. Springer, Heidelberg (2010). http://dx.doi.org/10.1007/978-3-642-13858-4_13
15. Minematsu, K., Matsushima, T.: New bounds for PMAC, TMAC, and XCBC. In: Biryukov [6], pp. 434–451
16. Nandi, M., Mandal, A.: Improved security analysis of PMAC. J. Math. Cryptology **2**(2), 149–162 (2008)
17. Pietrzak, K.: A tight bound for EMAC. In: Bugliesi, M., Preneel, B., Sassone, V., Wegener, I. (eds.) ICALP 2006. LNCS, vol. 4052, pp. 168–179. Springer, Heidelberg (2006). http://dx.doi.org/10.1007/11787006_15
18. Preneel, B., van Oorschot, P.C.: MDx-MAC and building fast MACs from hash functions. In: Coppersmith [8], pp. 1–14
19. Rogaway, P.: Efficient instantiations of tweakable blockciphers and refinements to modes OCB and PMAC. In: Lee, P.J. (ed.) ASIACRYPT 2004. LNCS, vol. 3329, pp. 16–31. Springer, Heidelberg (2004)
20. Shoup, V.: New algorithms for finding irreducible polynomials over finite fields. In: 29th Annual Symposium on Foundations of Computer Science, White Plains, New York, USA, 24–26 October 1988, pp. 283–290. IEEE Computer Society (1988). http://dx.doi.org/10.1109/SFCS.1988.21944
21. Wegman, M.N., Carter, L.: New hash functions and their use in authentication and set equality. J. Comput. Syst. Sci. **22**(3), 265–279 (1981). http://dx.doi.org/10.1016/0022-0000(81)90033-7
22. Yasuda, K.: A new variant of PMAC: beyond the birthday bound. In: Rogaway, P. (ed.) CRYPTO 2011. LNCS, vol. 6841, pp. 596–609. Springer, Heidelberg (2011). http://dx.doi.org/10.1007/978-3-642-22792-9_34
23. Yasuda, K.: PMAC with parity: minimizing the query-length influence. In: Dunkelman, O. (ed.) CT-RSA 2012. LNCS, vol. 7178, pp. 203–214. Springer, Heidelberg (2012). http://dx.doi.org/10.1007/978-3-642-27954-6_13
24. Zhang, Y.: Using an error-correction code for fast, beyond-birthday-bound authentication. In: Nyberg, K. (ed.) CT-RSA 2015. LNCS, vol. 9048, pp. 291–307. Springer, Heidelberg (2015). http://dx.doi.org/10.1007/978-3-319-16715-2_16

Lucky Microseconds: A Timing Attack on Amazon's *s2n* Implementation of TLS

Martin R. Albrecht(✉) and Kenneth G. Paterson(✉)

Information Security Group, Royal Holloway, University of London, Egham, Surrey TW20 0EX, UK
{martin.albrecht,kenny.paterson}@rhul.ac.uk

Abstract. *s2n* is an implementation of the TLS protocol that was released in late June 2015 by Amazon. It is implemented in around 6,000 lines of C99 code. By comparison, OpenSSL needs around 70,000 lines of code to implement the protocol. At the time of its release, Amazon announced that *s2n* had undergone three external security evaluations and penetration tests. We show that, despite this, *s2n* — as initially released — was vulnerable to a timing attack in the case of CBC-mode ciphersuites, which could be extended to complete plaintext recovery in some settings. Our attack has two components. The first part is a novel variant of the Lucky 13 attack that works even though protections against Lucky 13 were implemented in *s2n*. The second part deals with the randomised delays that were put in place in *s2n* as an *additional* countermeasure to Lucky 13. Our work highlights the challenges of protecting implementations against sophisticated timing attacks. It also illustrates that standard code audits are insufficient to uncover all cryptographic attack vectors.

Keywords: TLS · CBC-mode encryption · Timing attack · Plaintext recovery · Lucky 13 · s2n

1 Introduction

In late June 2015, Amazon announced a new implementation of TLS (and SSLv3), called *s2n* [Lab15,Sch15]. A particular feature of *s2n* is its small codebase: while *s2n* relies on OpenSSL or any of its forks for low-level cryptographic processing the core of the TLS protocol implementation is written in around 6,000 lines of C99. This is intended to make *s2n* easier to audit. Indeed, Amazon also announced that *s2n* had undergone three external security evaluations and penetration tests prior to release. No details of these audits appear to be in the

M.R. Albrecht—This author's research supported by EPSRC grant EP/L018543/1.
K.G. Paterson—This author's research supported by EPSRC grants EP/L018543/1 and EP/M013472/1, and a research programme funded by Huawei Technologies and delivered through the Institute for Cyber Security Innovation at Royal Holloway, University of London.

M. Fischlin and J.-S. Coron (Eds.): EUROCRYPT 2016, Part I, LNCS 9665, pp. 622–643, 2016.
DOI: 10.1007/978-3-662-49890-3_24

public domain at the time of writing. Given the recent travails of SSL/TLS in general and the OpenSSL implementation in particular, *s2n* generated significant interest in the security community and technical press.[1]

We show that *s2n* — as initially released — was vulnerable to a timing attack on its implementation of CBC-mode ciphersuites. Specifically, we show that the two levels of protection offered against the Lucky 13 attack [AP13] in *s2n* at the time of first release were imperfect, and that a novel variant of the Lucky 13 attack could be mounted against *s2n*.

The attack is particularly powerful in the web setting, where an attack involving malicious client-side Javascript (as per BEAST, POODLE [MDK14] and Lucky 13) results in the complete recovery of HTTP session cookies, and user credentials such as BasicAuth passwords. In this setting, an adversary runs malicious JavaScript in a victim's browser and additionally performs a Person-in-the-Middle attack. We note, though, that many modern browsers prefer TLS 1.2 AEAD cipher suites avoiding CBC-mode, making them immune to the attack described in this work if the sever also supports TLS 1.2 cipher suites as *s2n* does. The issues identified in this work have since been addressed in *s2n*, partly in response to this work, and current versions are no longer vulnerable to the attacks described in this work.

We stress that the problem we identify in *s2n* does not arise from reusing OpenSSL's crypto code, but rather from *s2n*'s own attempt to protect itself against the Lucky 13 attack when processing incoming TLS records. It does this in two steps: (1) using additional cryptographic operations, to equalise the running time of the record processing; and (2) introducing random waiting periods in case of an error such as a MAC failure.

Step (1) involves calls to a function `s2n_hmac_update`, which in turn makes hash compression function calls to, for example, OpenSSL or LibreSSL. The designers of *s2n* chose to draw a line above which to start their implementation, roughly aligned at the boundary between low-level crypto functions and the protocol itself. The first part of our attack is focused at the lowest level above that line. Specifically, we show that the desired additional cryptographic operations may not be carried out as anticipated: while *s2n* always fed the same number of *bytes* to `s2n_hmac_update`, to defeat timing attacks, this need not result in the same number of *compression function calls* of the underlying hash function. Indeed this latter number may vary depending on the padding length byte which controls after how many bytes `s2n_hmac_digest` is called, this call producing a digest over all data submitted so far. We can also arrange that subsequent calls to `s2n_hmac_update` do not trigger any compression function calls at all. This has the effect of removing the timing equalisation and reopening the window for an attack in the style of Lucky 13.

The second part of our attack is focussed on step (2), the random waiting periods introduced in *s2n* as an additional protection against timing attacks.

[1] See for example http://www.theregister.co.uk/2015/07/01/amazon_s2n_tls_library/, http://www.securityweek.com/amazon-releases-new-open-source-implementation-tls-protocol.

The authors of [AP13] showed that adding random delays as a countermeasure to Lucky 13 would be ineffective if the maximum delay was too small. The *s2n* code had a maximum waiting period that is enormous relative to the processing time for a TLS record, 10s compared to around 1 µs, putting the attack techniques of [AP13] well out of contention. However, the initial release of *s2n* used timing delays generated by calls to `sleep` and `usleep`, giving them a granularity much greater than the timing differences arising from the failure to equalise the running time in step (1). Consequently, at a high level, we were able to bypass step (2) by "mod-ing out" the timing delays provided by `sleep` and `usleep`. However, the reality is slightly more complex than this simple description would suggest, because those functions do not provide delays that are exact multiples of 1 µs but instead themselves have distributions that need to be taken into account in our statistical analysis. Weaknesses in random delays as countermeasures to timing side-channels have been point out before, cf. [CK10]. In contrast to previous work, though, here the source of timing differences was not close enough to uniform, allowing our analysis of the low-level code to "leak through" the random timing delays, despite them being very large.

Our attack illustrates that protecting TLS's CBC construction against attacks in the style of Lucky 13 is hard (cf. [AIES15]). It also shows that standard code audits may be insufficient to uncover all cryptographic attack vectors.

Our attack can be prevented by more carefully implementing countermeasures to the Lucky 13 attack that were presented in [AP13]. A fully constant time/constant memory access patch can be found in the OpenSSL implementation; its complexity is such that around 500 lines of new code were required to implement it, and it is arguable whether the code would be understandable by all but a few crypto-expert developers. It is worth noting that the countermeasure against Lucky 13 in OpenSSL does not respect the separation adopted in the *s2n* design, i.e. it avoids higher-level interfaces to HMAC but makes hash compression function calls directly on manually constructed blocks.[2] The *s2n* code was patched to prevent our attacks using a different strategy, (mostly) maintaining the above-mentioned separation. At a high-level, the first step of our attacks exploits that *s2n* counted bytes submitted to HMAC instead of compression function calls. In response, *s2n* now counts the number of compression function calls. Furthermore, the second *s2n* countermeasure was strengthened by switching from using `usleep` to using `nanosleep`.

1.1 Disclosure and Remediation

We notified Amazon of the issue in step (1) of their countermeasures, in the function `s2n_verify_cbc` in *s2n* on 5th July 2015. Subsequently and in response, this function was revised to address the issue reported. This issue in itself does not constitute a successful attack because *s2n* also implemented step (2), the randomised waiting period, as was pointed out to us by the developers of *s2n*. This countermeasure has since been strengthened by switching to the use of

[2] See [Lan13] for a detailed description of the patch.

nanosleep to implement randomised wait periods. This transition was already planned by the developers of *s2n* prior to learning about our work, but the change was accelerated in response to it. Our work shows that the switch to using nanosleep was a good decision because this step prevents the attacks described in this work.[3]

1.2 Lucky 13 Remedies in Other Libraries

As mentioned above OpenSSL prevents the Lucky 13 attack in 500 lines of code which achieves fully constant time/memory access [Lan13]. GnuTLS does not completely eliminate all potential sources of timing differences, but makes sure the number of compression function calls is constant and other major sources of timing differences are eliminated. As reported in [Mav13] this results in timing differences in the tens of nanonseconds, likely too small to be exploited in practice. In contrast, GoTLS as of now does not implement any countermeasure to Lucky 13. However, a patch is currently under review to equalise the number of compression function calls regardless of padding value [VF15]. This fix does not promise constant time/memory access. Botan does not implement any countermeasure to Lucky 13.[4] WolfSSL implements the recommended countermeasures to Lucky 13 from [AP13].[5]

2 The TLS Record Protocol and S2n

The main component of TLS of interest here is the Record Protocol, which uses symmetric key cryptography (block ciphers, stream ciphers and MAC algorithms) in combination with sequence numbers to build a secure channel for transporting application-layer data. In SSL and versions of TLS prior to TLS 1.2, the only encryption option uses a MAC-Encode-Encrypt (MEE) construction. Here, the plaintext data to be transported is first passed through a MAC algorithm (along with a group of 13 header bytes) to create a MAC tag. The supported MAC algorithms are all HMAC-based, with MD5, SHA-1 and SHA-256 being typical hash algorithms. Then an encoding step takes place. For the RC4 stream cipher, this just involves concatenation of the plaintext and the MAC tag, while for CBC-mode encryption (the other possible option), the plaintext, MAC tag, and some encryption padding of a specified format are concatenated. In the encryption step, the encoded plaintext is encrypted with the selected cipher. In the case where CBC-mode is selected, the block cipher is DES, 3DES

[3] We also note that the first fix was still vulnerable to a timing attack in step (1), as reported in [ABBD15]. This further highlights the delicacy of protecting against timing side-channel attacks and that the move towards using nanosleep was a good decision.

[4] https://github.com/randombit/botan/blob/master/src/lib/tls/tls_record.cpp#L398.

[5] http://www.yassl.com/forums/topic328-wolfssl-releases-protocol-fix-for-lucky-thirteen-attack.html.

or AES (with DES being deprecated in TLS 1.2). The *s2n* implementation supports 3DES and AES. Following [PRS11], we refer to this MEE construction as MEE-TLS-CBC.

The MEE construction used in the TLS has been the source of many security issues and attacks [Vau02, CHVV03, Moe04, PRS11, AP12, AP13]. These all stem from how the padding that is required in MEE-TLS-CBC is handled during decryption, specifically the fact that the padding is added *after* the MAC has been computed and so forms unauthenticated data in the encoded plaintext. This long sequence of attacks shows that handling padding arising during decryption processing is a delicate and complex issue for MEE-TLS-CBC. It, along with the attacks on RC4 in TLS [ABP+13], has been an important spur in the TLS community's push to using TLS 1.2 and its Authenticated Encryption modes. AES-GCM is now widely supported in implementations. However, the MEE construction is still in widespread use, as highlighted by the fact that Amazon chose to support it in its minimal TLS implementation *s2n*.

2.1 MEE-TLS-CBC

We now explain the core encryption process for MEE-TLS-CBC in more detail.

Data to be protected by TLS is received from the application and may be fragmented and compressed before further processing. An individual record R (viewed as a byte sequence of length at least zero) is then processed as follows. The sender maintains an 8-byte sequence number $\texttt{SQN}$ which is incremented for each record sent, and forms a 5-byte field $\texttt{HDR}$ consisting of a 2-byte version field, a 1-byte type field, and a 2-byte length field. The sender then calculates a MAC over the bytes $\texttt{SQN}||\texttt{HDR}||R$; let T denote the resulting MAC tag. Note that exactly 13 bytes of data are prepended to the record R here before the MAC is computed. The size of the MAC tag is 16 bytes (HMAC-MD5), 20 bytes (HMAC-SHA-1), or 32 bytes (HMAC-SHA-256). We let t denote this size in bytes.

The record is then encoded to create the plaintext P by setting $P = R||T||\mathsf{pad}$. Here pad is a sequence of padding bytes chosen such that the length of P in bytes is a multiple of b, where b is the block-size of the selected block cipher (so $b = 8$ for 3DES and $b = 16$ for AES). In all versions of TLS, the padding must consist of $p + 1$ copies of some byte value p, where $0 \leq p \leq 255$. In particular, at least one byte of padding must always be added. The padding may extend over multiple blocks, and receivers must support the removal of such extended padding. In SSL the padding format is not so strictly specified: it is only required that the last byte of padding must indicate the total number of additional padding bytes. The attack on *s2n* that we present works irrespective of whether the padding format follows the SSL or the TLS specification.

In the encryption step, the encoded record P is encrypted using CBC-mode of the selected block cipher. TLS 1.1 and 1.2 mandate an explicit IV, which should be randomly generated. TLS 1.0 and SSL use a chained IV; our attack works for either option. Thus, the ciphertext blocks are computed as:

$$C_j = E_{K_e}(P_j \oplus C_{j-1})$$

where P_i are the blocks of P, C_0 is the IV, and K_e is the key for the block cipher E. For TLS (and SSL), the ciphertext data transmitted over the wire then has the form:

$$\mathtt{HDR}||C$$

where C is the concatenation of the blocks C_i (including or excluding the IV depending on the particular SSL or TLS version). Note that the sequence number is not transmitted as part of the message.

Simplistically, the decryption process reverses this sequence of steps: first the ciphertext is decrypted block by block to recover the plaintext blocks:

$$P_j = D_{K_e}(C_j) \oplus C_{j-1},$$

where D denotes the decryption algorithm of the block cipher. Then the padding is removed, and finally, the MAC is checked, with the check including the header information and a version of the sequence number that is maintained at the receiver.

However, in order to avoid a variety of known attacks, these operations must be performed without leaking any information about what the composition of the plaintext blocks is in terms of message, MAC field and padding, and indeed whether the format is even valid. The difficulties and dangers inherent in this are explained at length in [AP13].

For TLS, any error arising during decryption should be treated as fatal, meaning an encrypted error message is sent to the sender and the session terminated with all keys and other cryptographic material being disposed of.

2.2 Details of HMAC

As mentioned above, TLS exclusively uses the HMAC algorithm [KBC97], with HMAC-MD5, HMAC-SHA-1, and HMAC-SHA-256 being supported in TLS 1.2.[6] To compute the MAC tag T for a message M with key K_a, HMAC applies the specified hash algorithm H twice, in an iterated fashion:

$$T = H((K_a \oplus \mathtt{opad})||H((K_a \oplus \mathtt{ipad})||M)).$$

Here `opad` and `ipad` are specific 64-byte values, and the key K_a is zero-padded to bring it up to 64 bytes before the XOR operations are performed. All the hash functions H used in TLS have an iterated structure, processing messages in chunks of 64 bytes (512 bits) using a compression function, with the output of each compression step being chained into the next step. Also, for all relevant hash functions used in TLS, an 8-byte length field followed by padding of a specified byte format are appended to the message M to be hashed. The padding is at least 1 byte in length and extends the data to a (56 mod 64)-byte boundary.

[6] TLS ciphersuites using HMAC with SHA-384 are specified in RFC 5289 (ECC cipher suites for SHA256/SHA384) and RFC 5487 (Pre-Shared Keys SHA384/AES) but we do not consider the SHA-384 algorithm further here.

In combination, these features mean that HMAC implementations for MD5, SHA-1 and SHA-256 have a distinctive timing profile. Messages M of length up to 55 bytes can be encoded into a single 64-byte block, meaning that the first, inner hash operation in HMAC is done in 2 compression function evaluations, with 2 more being required for the outer hash operation, for a total of 4 compression function evaluations. Messages M containing from 56 up to $64 + 55 = 119$ bytes can be encoded in two 64-byte blocks, meaning that the inner hash is done in 3 compression function evaluations, with 2 more being required for the outer operation, for a total of 5. In general, an extra compression function evaluation is needed for each additional 64 bytes of message data. A single compression function evaluation takes typically a few hundred clock cycles.[7]

Implementations typically implement HMAC via an "IUF" interface, meaning that the computation is first initialised (I), then the computation is updated (U) as many times as are needed with each update involving the buffering and/or hashing of further message bytes. When the complete message has been processed, a finalisation (F) step is performed. In *s2n*, OpenSSL or any of its forks is used to implement HMAC. The initialisation step `s2n_hmac_init` carries out a compression function call on the 64-byte string $K_a \oplus$ `ipad`. The update step `s2n_hmac_update` involves buffering of message bytes and calls to the compression function on buffered 64-byte chunks of message. Note that no compression function call will be made until at least 64 bytes have been buffered. The finalisation step `s2n_hmac_digest` consists of adding the length encoding and padding, performing final compression function calls to compute the inner hash and then performing the outer hash operation (itself involving 2 compression function evaluations).

2.3 HMAC Computations After Decryption in *s2n*

The *s2n* implementation uses the code in Fig. 1 to check the MAC on a record in the function `s2n_verify_cbc`. This code is followed by a constant-time padding check that need not concern us here (except to note that the fact that it is constant time helps our attack, since it enables us to isolate timing differences coming from this code fragment). In Fig. 1, the content of buffer `decrypted->data` is the plaintext after CBC-mode decryption. The header `SQN||HDR` of 13 bytes is dealt with by the calling function.

Notice how the code first computes, using the last byte of plaintext, a value for `padding_length`, the presumed length of padding that should be removed (excluding the pad length byte). Arithmetic is then performed to find `payload_length`, the presumed length of the remaining payload over which the HMAC computation is to be done. The actual HMAC computation is performed via an initialise call (not shown), and then the code in line 78 (update via the function `s2n_hmac_update`) and line 84 (finalise via the function `s2n_hmac_digest`). Line 86 compares the computed HMAC value with that

[7] For example, SHA-256 takes about 550 cycles per block on one of our test systems, an Intel Core i7–4850HQ CPU @ 2.30 GHz, whereas SHA-1 takes about 300 cycles.

contained in the plaintext, and sets a flag **mismatches** if they do not match as expected.

Line 79 copies the HMAC state to a dummy state, so that line 89 can perform a dummy **s2n_hmac_update** computation on data from the plaintext buffer. This attempts to ensure that the number of hash computations carried out is the same, irrespective of the amount of padding that should be removed. This is in an effort to remove the timing channel exploited in the Lucky 13 attack. The number of bytes over which the update is performed is equal to **decrypted->size - payload_length - mac_digest_size - 1**, which is one less than the number of bytes in the plaintext buffer excluding the 13 bytes of SQN||HDR, the message, and the MAC value. Recall, however, that this update operation may not actually result in any compression function computations being carried out. What happens depends on exactly how many bytes are already sitting unprocessed in the internal buffer and how many are added to it in the call.

2.4 Randomised Waiting Period

In order to additionally protect against attacks exploiting timing side-channels, *s2n* implements the following countermeasure: whenever an error occurs, the

```
67    int payload_and_padding_size = decrypted->size - mac_digest_size;
68
69    /* Determine what the padding length is */
70    uint8_t padding_length = decrypted->data[decrypted->size - 1];
71
72    int payload_length = payload_and_padding_size - padding_length \
        - 1;
73    if (payload_length < 0) {
74        payload_length = 0;
75    }
76
77    /* Update the MAC */
78    GUARD(s2n_hmac_update(hmac, decrypted->data, payload_length));
79    GUARD(s2n_hmac_copy(&copy, hmac));
80
81    /* Check the MAC */
82    uint8_t check_digest[S2N_MAX_DIGEST_LEN];
83    lte_check(mac_digest_size, sizeof(check_digest));
84    GUARD(s2n_hmac_digest(hmac, check_digest, mac_digest_size));
85
86    int mismatches = s2n_constant_time_equals(decrypted->data +
                                              payload_length,
                                              check_digest,
                                              mac_digest_size) ^ 1;
87
88    /* Compute a MAC on the rest of the data so that we perform
         the same number of hash operations */
89    GUARD(s2n_hmac_update(&copy, decrypted->data + payload_length +
                            mac_digest_size,
                            decrypted->size - payload_length -
                            mac_digest_size - 1));
```

Fig. 1. Excerpt from **s2n_verify_cbc**, *s2n*'s code for checking the MAC on a TLS record implementation waits for a random period of time before sending an error message. We reproduce the relevant code excerpts in Fig. 2; at a high level, when a MAC failure occurs, the following steps are taken:

- All available data is erased. Depending on the amount of buffered data, the time this takes may vary.
- All connection data is wiped, which may also introduce a timing difference.
- A random integer x between 1,000 and 10,001,000 is requested. Since rejection sampling is used to generate x, this might also introduce some timing variation.
- This random integer is then fed to `usleep` and `sleep` calls (after the appropriate scaling), causing a random delay of at least x μs.

```
s2n_record_read.c
int s2n_record_parse(struct s2n_connection *conn)
...
    /* Padding */
    if (cipher_suite->cipher->type == S2N_CBC) {
        if (s2n_verify_cbc(conn, mac, &en) < 0) {
            GUARD(s2n_stuffer_wipe(&conn->in));
            S2N_ERROR(S2N_ERR_BAD_MESSAGE);
            return -1;
        }

s2n_recv.c
int s2n_read_full_record(struct s2n_connection *conn, \
                         uint8_t *record_type, int *isSSLv2)
    /* Decrypt and parse the record */
    if (s2n_record_parse(conn) < 0) {
        GUARD(s2n_connection_wipe(conn));
        if (conn->blinding == S2N_BUILT_IN_BLINDING) {
            int delay;
            GUARD(delay = s2n_connection_get_delay(conn));
            GUARD(sleep(delay / 1000000));
            GUARD(usleep(delay % 1000000));
        }
        return -1;
    }
```

Fig. 2. Excerpts from `s2n_record_read.c` and `s2n_recv.c`, *s2n*'s code for adding a random waiting period

We note that this countermeasure, which is activated by default, is designed as an API mode which can in principle be disabled. This is to support implementations which provide their own timing channel countermeasures. If the variable `blinding` is not equal to `S2N_BUILT_IN_BLINDING` then none of the countermeasure code is run.[8] Since this countermeasure introduces a delay of up to 10 s in

[8] However, we note that a bug in the version of *s2n* that we studied prevented this from ever happening, because the call to wipe the connection data erased this configuration flag as well.

case of an error, it might be tempting for some application developers to disable it. However, note that the *s2n* documentation strongly advises against disabling this counter measure without replacing it by an equivalent one on the application level.

3 The Attack Without the Random Waiting Period Countermeasure

We first describe our variant of the Lucky 13 attack against *s2n* assuming the random waiting period countermeasure is not present. We show how to deal with this additional countermeasure in Sect. 4.

For simplicity of presentation, in what follows, we assume the CBC-mode IVs are explicit (as in TLS 1.1 and 1.2). We also assume that $b = 16$ (so our block cipher is AES). It is easy to construct variants of our attacks for implicit IVs and for $b = 8$. The MAC algorithm is HMAC-H where H is either MD5, SHA-1 or SHA-256. We focus at first on the case where the MAC algorithm is HMAC-SHA-256, so that $t = 32$. We explain below how the attack can be adapted to $t = 16$ and $t = 20$ (HMAC-MD5 and HMAC-SHA-1, respectively).

Let C^* be any ciphertext block whose corresponding plaintext P^* the attacker wishes to recover. Let C' denote the ciphertext block preceding C^*. Note that C' may be the IV or the last block of the preceding ciphertext if C^* is the first block of a ciphertext. We have:

$$P^* = D_{K_e}(C^*) \oplus C'.$$

Let Δ be an arbitrary block of 16 bytes and consider the decryption of a ciphertext $C^{\text{att}}(\Delta)$ of the form

$$C^{\text{att}}(\Delta) = \texttt{HDR}||C_0||C_1||C_2||C_3||C' \oplus \Delta||C^*$$

consisting of a header field `HDR` containing an appropriate value in the length field, an IV block, and 5 non-IV blocks. The IV block and the first 3 non-IV blocks are arbitrary, the penultimate block $C_4 = C' \oplus \Delta$ is an XOR-masked version of C' and the last block is $C_5 = C^*$. The corresponding 80-byte plaintext is $P = P_1||P_2||P_3||P_4||P_5$ in which

$$\begin{aligned} P_5 &= D_{K_e}(C^*) \oplus (C' \oplus \Delta) \\ &= P^* \oplus \Delta. \end{aligned}$$

Notice that P_5 is closely related to the unknown, target plaintext block P^*. Notice also that, via line 67 of the code in Fig. 1, the variable `payload_and_padding_size` is set to $80 - 32 = 48$ (recall that the 13-byte string `SQN`||`HDR` was fed to HMAC by the calling function and is buffered but otherwise unprocessed at this point). We now consider 2 distinct cases:

1. Suppose P_5 ends with a byte value from the set {0x00,...,0x04}. In this case, the code sets padding_length to be at most 4 and then, at line 72, payload_length is set to a value that is at least $48 - 4 - 1 = 43$ (and at most 47). This means that when the HMAC computation is performed in lines 78 (update) and 84 (finalise), the internal buffer contains at least 56 bytes (because 13 bytes were already buffered by the calling function) and exactly 5 calls to the compression function will be made, including one call that initialises HMAC and 2 that finalises it. The time equalising code at line 89 adds between 0 and 4 bytes to the internal buffer, which still holds the previous message bytes. However, because of the short length of our chosen ciphertext, the buffer ends up being exactly 60 bytes in size. This number is obtained by considering the 13 bytes of SQN||HDR, the payload_length bytes added to the buffer at line 78 and the decrypted->size - payload_length - mac_digest_size - 1 bytes added to the buffer at line 89. Combining these, one arrives at there being 12 + decrypted->size - mac_digest_size bytes in the buffer. This evaluates to 60 for the particular values in the attack. Notably, this number is independent of payload_length and padding_length. The call at line 89 is to the update function rather than the finalise function, so at least 64 bytes would be needed in the buffer to cause any compression function computations to be performed at this point. Thus no compression function call is made as a consequence of the call to s2n_hmac_update at line 89.
2. Suppose P_5 ends with a byte value from the set {0x05,...,0xff}. In this case, the code sets padding_length to be at least 5 and then, at line 72, payload_length is set to a value that is at most $48 - 5 - 1 = 42$ (and at least 0). This means that when the HMAC computation is performed in lines 78 (update) and 84 (finalise), the internal buffer contains at most 55 bytes and exactly 4 calls to the compression function will be made (again, including the initialisation and finalisation calls). The time equalising code at line 89 will again result in no additional calls to the compression function being made, as the internal buffer is again too small at exactly 60 bytes in size (recall that the buffer size is independent of payload_length and padding_length).

Based on this case analysis, a timing difference will arise in HMAC processing of the attack ciphertext $C^{\text{att}}(\Delta)$, according to whether the last byte of $P_5 = P^* \oplus \Delta$ is from the set {0x00,...,0x04} or not. The difference is equal to that taken by one compression function call. This timing difference becomes evident on the network in the form of a difference in the arrival time of an error message at the man-in-the-middle attacker who injects the attack ciphertext. The difference is of the same size as that observed in the plaintext recovery attack presented in [AP13], a few hundred clock cycles on a modern processor. Of course, as in [AP13], this time difference would be affected by noise arising from network jitter, but it is sufficiently big to enable it to be detected. Furthermore, if the attacker can arrange to be co-resident with the victim in a cloud environment, a realistic prospect as shown by a line of work culminating in [VZRS15], the attacker can perform a Person-in-the-Middle attack and observe the usage of resources on the server by being co-resident.

As was the case in [AP13], the attack can be iterated as often as is desired and with different values of Δ, provided the same plaintext is repeated at a predictable location across multiple sessions. The attack as presented already takes care of the complication that each trial will involve a different key in a different TLS session; only P^* needs to be constant for it to work.

By carefully exploring the timing behaviour for different values in the last byte of Δ (each value being tried sufficiently often so as to minimise the effect of noise), the attacker can deduce the value of the last byte of P^*. For example, the attacker can try every value in the 6 most significant bits in the last byte of Δ to identify a value Δ^* for which the time taken is relatively high. This indicates that the last byte of $P^* \oplus \Delta^*$ is in the set $\{\texttt{0x00}, \ldots, \texttt{0x04}\}$; a more refined analysis can then be carried out on the 3 least significant bits of the last byte of Δ^* to identify the exact value of the last byte of P^*. The worst case cost of this version of the attack is $64 + 8 = 72$ trials (multiplied by a factor corresponding to the number of trials per Δ needed to remove noise).

The attack cost can be reduced further by using initially longer ciphertexts, because the peculiar characteristics of the *s2n* code mean that this choice results in there being a greater number of values for (the last byte of) Δ that result in a higher processing time; the precise value of the last byte of P^* can then be pinned down by using progressively shorter ciphertexts. We omit the details of this enhancement.

3.1 Extending to Full Plaintext Recovery

In the web setting, with HTTP session cookies as the target, the attack extends in a straightforward manner to full plaintext recovery using by-now-standard techniques involving malicious client-side Javascript and careful HTTP message padding. A good explanation of how this is achieved can be found in [MDK14] describing the POODLE attack on TLS. `BasicAuth` passwords also form a good target; see [GPdM15] for details.

3.2 Variants for HMAC-MD5 and HMAC-SHA-1

Assume $b = 16$ (as in AES) and consider the case of HMAC-MD5. Then, because $t = 16$ in this case, and t is still a multiple of b, the attack described above works perfectly, except that we need to use a ciphertext having 4 non-IV blocks instead of 5. The attack also works for $b = 8$ for both HMAC-MD5 and HMAC-SHA-256 by doubling the number of non-IV blocks used.

For HMAC-SHA-1, we have $t = 20$. Assume $b = 16$ (AES). Then a similar case analysis as above shows that using a ciphertext with 4 blocks result in a slow execution time if and only if the last plaintext block P_4 ends with `0x00`. This leads to a plaintext recovery attack requiring, in the worst case, 256 trials per byte. The attack adapts to the $b = 8$ case by again doubling the number of non-IV blocks used.

4 Defeating the Random Wait Period Countermeasure

As described in Sect. 2.4, *s2n* implemented a second countermeasure against attacks exploiting timing channels. In this section, we show how it could be defeated.

4.1 Characterising the Timing Delays

To start off, we notice that at the price of increasing the number of samples by a factor of roughly ten, we can assume that `sleep` at line 103 in the code in Fig. 2 is called with parameter zero, by rejecting in an attack any sample where the overall time is more than 1s. This removes one potential source of randomness. As shown in Fig. 3, calling `sleep(0)` has a rather stable timing profile.

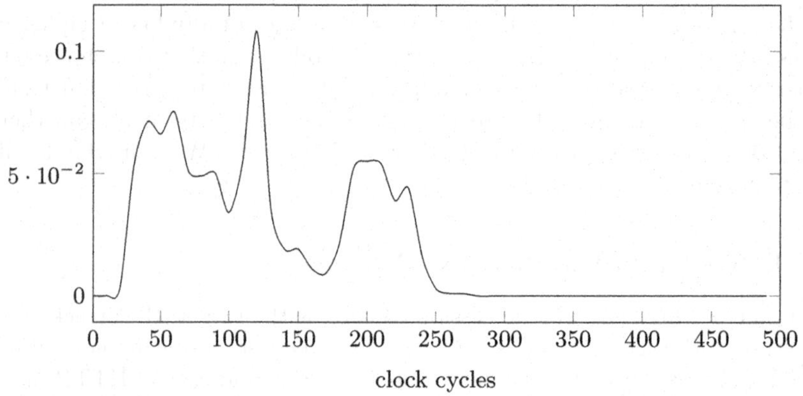

Fig. 3. Distribution of clock ticks for calling `sleep(0)` on Intel(R) Xeon(R) CPU E5-2667 v2 @ 3.30 GHz.

Next, we consider calls to `usleep` with a random delay as a source of timing randomness. For this, note that `usleep` has a granularity of 1μs. On our main test machine, which is clocked at 3.3 GHz, this translates to 3,300 clock cycles.[9] From this, we might expect that if we take our timings modulo the clock ticks per μs (namely, 3,300 on our test machine), we could filter out all the additional noise contributed by the `usleep(delay)` call. However, `usleep(delay)` does not guarantee to return after *exactly* `delay` μs, or even to return after an exact number of μs. Instead, it merely guarantees that it will return after *at least* `delay` μs have elapsed. Indeed, on a typical UNIX system, waking up a process from sleep can take an unpredictable amount of time depending on global the state of the OS.

[9] We note, however, that modern CPUs reclock their CPUs dynamically both below the base operating frequency and above it (e.g. Intel Turbo Boost). This must be taken into account when measuring time delays in elapsed clock cycles.

However, despite this, **usleep** does show exploitable non-uniform behaviour on the systems we tested. Figures 4 and 5 illustrate this behaviour. Figure 4 shows raw timings (in clock cycles) for **usleep**(d), normalised to remove the minimum possible delay, namely $3,300 \cdot d$ clock cycles. Figure 5 shows the distribution of timings (in clock cycles) for **usleep(delay)** with **delay** uniformly random in an

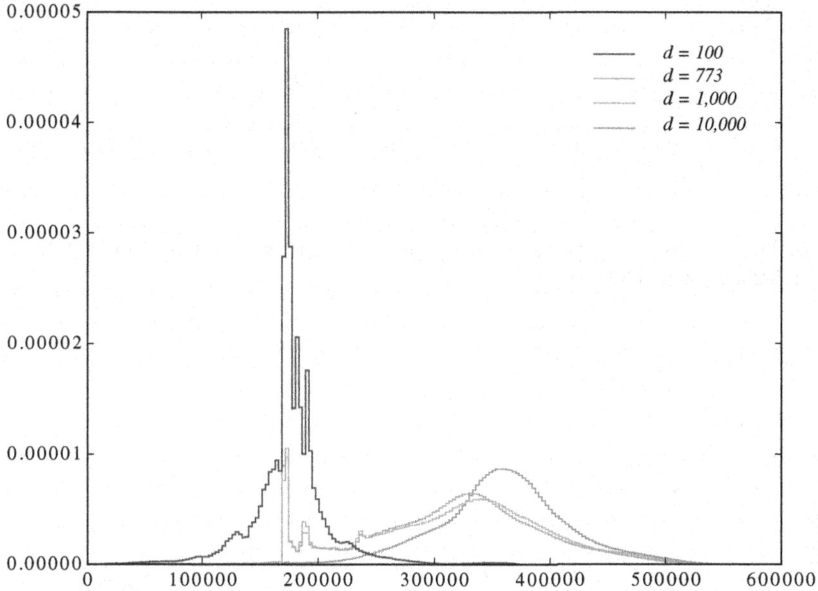

Fig. 4. Distribution of **usleep**(d)$-3,300 \cdot d$ (in clock cycles) on Intel(R) Xeon(R) CPU E5-2667 v2 @ 3.30 GHz. Probability on the y axis.

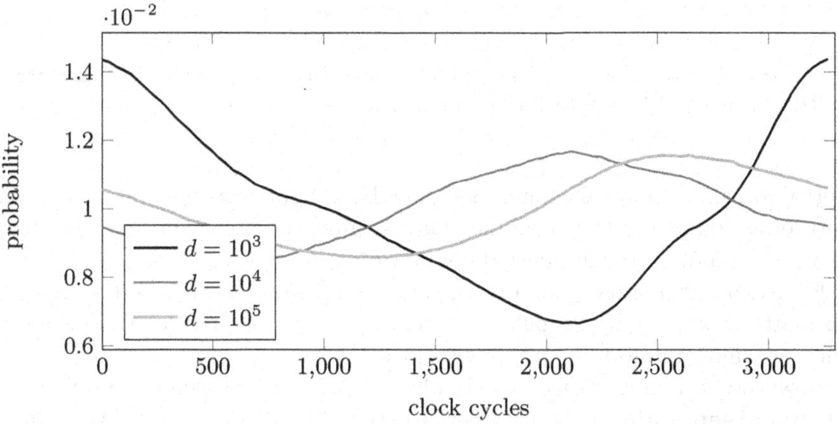

Fig. 5. Distribution of clock ticks modulo 3,300 for **usleep(delay)** with **delay** uniformly random in $[0, d)$, on Intel(R) Xeon(R) CPU E5-2667 v2 @ 3.30 GHz.

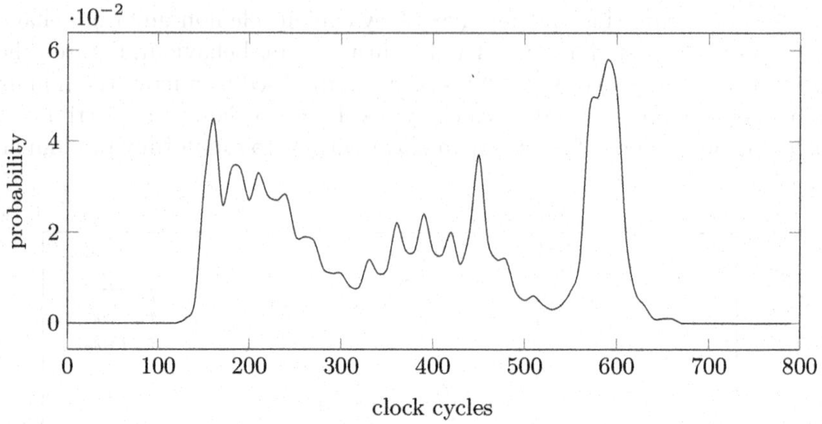

Fig. 6. Distribution of clock ticks for calling **s2n_stuffer_wipe** on Intel(R) Xeon(R) CPU E5-2667 v2 @ 3.30 GHz.

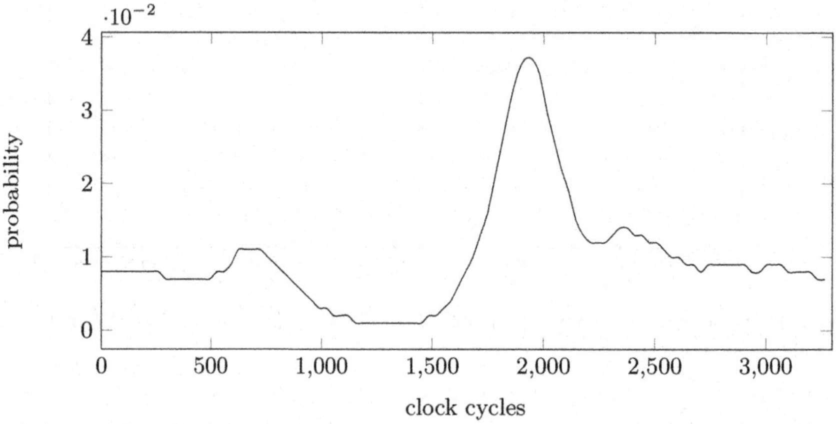

Fig. 7. Distribution of clock ticks modulo 3300 for calling **s2n_public_random** on Intel(R) Xeon(R) CPU E5-2667 v2 @ 3.30 GHz.

interval $[0, d)$, but now taken modulo 3,300. Both figures are generated from data captured on our main test machine. They exhibit the non-uniformity needed to bypass the random waiting period countermeasure in *s2n*.

Figures 6 and 7 show that, like the call to `usleep`, the calls to the functions `s2n_stuffer_wipe` and `s2n_public_random` also do not produce timing profiles which are uniform modulo 1µs (3,300 clock cycles).

However, it is not enough to simply characterise the timing profile of the calls to `usleep`; rather it is necessary to study the distribution of the running time of the entire random timing delay code in Fig. 2, in combination with the code for checking the MAC on a TLS record in Fig. 1, for different values of the mask Δ in the attack in Sect. 3. Figure 8 brings different sources of timing difference together and shows that the timing distributions (modulo 3,300) that are obtained for different mask values are indeed still rather easily distinguishable. The figure is for samples with the maximum delay restricted to 100,000 µs instead of 10 s. We stress that this is a synthetic benchmark for studying the behaviour of the various sources of timing randomness and does not necessarily represent actual behaviour. See Sect. 5 for experiments with the actual *s2n* implementation of these countermeasures.

4.2 Distinguishing Attack

Having characterised the timing behaviour of the *s2n* code, as exemplified in Fig. 8, we are now in a position to describe a statistical attack recovering plaintext bytes and its performance. In fact, the approach is completely standard: given the preceding analysis, we expect the timing distributions modulo 1µs for ciphertexts in the attack of Sect. 3 to fall into two classes depending on the value of the last byte of $P^* \oplus \Delta$, one class $H = \{\texttt{0x00}, \ldots, \texttt{0x04}\}$, the other class $L = \{\texttt{0x05}, \ldots, \texttt{0xff}\}$; if the observed distributions for all values in L (resp. H) are close to each other but the Kullback-Leibler (KL) divergence between distributions from L and H is large (and equal to D, say), then, applying standard statistical machinery, we know that we will require about $1/D$ samples to distinguish samples from the two distributions. As Tables 1 and 2 demonstrate, the requirements on KL divergence for values in L and H are indeed satisfied, even for relatively large values for the maximum delay.

For example, assuming for the sake of argument that no additional noise is introduced by network jitter or other sources, we would be able to distinguish the value 0x00 from 0xc8 in the last byte of $P^* \oplus \Delta$ with $1/(3.6/1,000) \approx 280$ TLS sessions if the maximum delay were restricted to 100,000 μs. Using rejection sampling, i.e. discarding all samples with a delay greater than 100,000 μs from

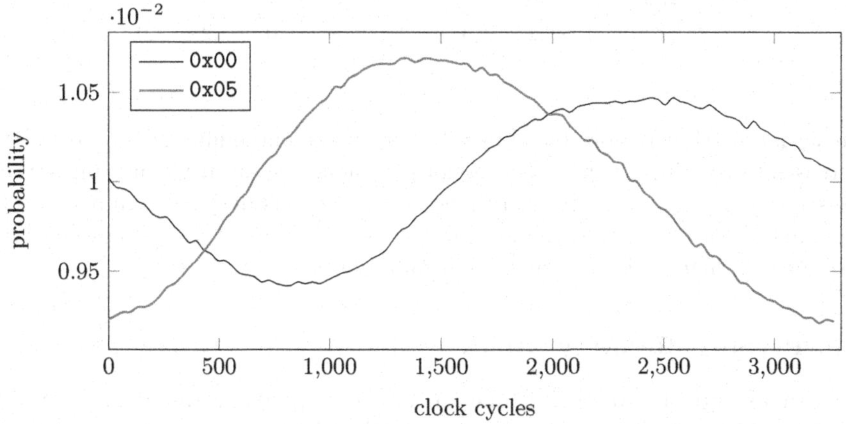

Fig. 8. Distribution of clock ticks modulo 3,300 for timing signals on Intel(R) Xeon(R) CPU E5-2667 v2 @ 3.30 GHz with the maximum delay restricted to $d = 100,000$.

Table 1. KL divergence multiplied by 1,000 of time distributions in clock cycles modulo 3,300 with the maximum delay limited to 1,000μs on Intel(R) Xeon(R) CPU E5-2667 v2 @ 3.30 GHz.

	0x00	0x04	0x05	0x10	0x20	0x30	0x40	0x64	0xc8
0x00	.0	.7	14.1	15.1	17.7	13.2	18.4	17.4	17.6
0x04	.7	.0	15.4	16.8	19.5	15.3	20.0	18.9	19.3
0x05	14.0	15.3	.0	.1	.2	.3	.3	.2	.2
0x10	15.0	16.6	.1	.0	.1	.2	.2	.1	.1
0x20	17.4	19.2	.2	.1	.0	.5	.0	.0	.0
0x30	13.0	15.1	.3	.2	.5	.0	.7	.5	.5
0x40	18.2	19.7	.3	.2	.0	.7	.0	.0	.0
0x64	17.2	18.7	.2	.1	.0	.5	.0	.0	.0
0xc8	17.4	19.0	.2	.1	.0	.5	.0	.0	.0

Table 2. KL divergence (scaled by 1,000 for readability) of time distributions in clock cycles modulo 3,300 with the maximum delay limited to 100,000 μs on Intel(R) Xeon(R) CPU E5-2667 v2 @ 3.30 GHz.

	0x00	0x04	0x05	0x10	0x20	0x30	0x40	0x64	0xc8
0x00	.0	.0	2.4	1.9	2.3	2.0	2.8	2.1	3.6
0x04	.0	.0	2.3	1.8	2.1	2.0	2.6	1.9	3.3
0x05	2.4	2.3	.0	.0	.0	.1	.0	.0	.2
0x10	1.9	1.8	.0	.0	.1	.1	.1	.0	.3
0x20	2.3	2.1	.0	.1	.0	.2	.0	.0	.1
0x30	2.0	2.0	.1	.1	.2	.0	.3	.2	.5
0x40	2.8	2.7	.0	.1	.0	.3	.0	.1	.0
0x64	2.1	1.9	.0	.0	.0	.2	.1	.0	.2
0xc8	3.6	3.4	.2	.3	.1	.5	.0	.2	.0

the actual distribution produced by *s2n* (where the maximum delay is 10 s), this increases to roughly 28,000 TLS sessions for a successful distinguishing attack. We stress that this estimate is optimistic because it is derived from a synthetic benchmark not the actual implementation and because the surrounding code and network jitter will introduce additional noise.

4.3 Plaintext Recovery Attack

We can extend this distinguishing attack to a plaintext recovery attack in the following (standard) way. We assume that in a characterisation step, we have obtained, for possible value x of the last byte in block P_5, a histogram of the timing distribution modulo 1 μs for ciphertexts $C^{\text{att}}(\Delta)$ of the form used in

Table 3. Timing of function `s2n_verify_cbc` (in cycles) with H = SHA-256 for different values of last byte in the `decrypted` buffer, each cycle count averaged over 2^8 trials.

Byte value	Cycles	Byte value	Cycles	Byte value	Cycles
0x00	2251.96	0x05	1746.49	...	...
0x01	2354.57	0x06	1747.65	0xfc	1640.79
0x02	2252.07	0x07	1705.62	0xfd	1634.61
0x03	2135.11	0x08	1808.73	0xfe	1648.70
0x04	2130.02	0x09	1806.50	0xff	1634.64

the attack. We assume these timings are distributed into B equal-sized bins, and so the empirical probability of each bin $p_{x,b}$ for $0 \leq b < B$ can be calculated. (In fact, since we expect that timing behaviours for the classes H and L are similar, it is sufficient to sample for two values x, one from each class.)

Now, in the actual attack, for each value δ of the last byte of Δ, we obtain N samples for ciphertexts $C^{\text{att}}(\Delta)$ for which the timing delay is at most 100,000 μs. This then requires a total of about $256 \cdot 100 \cdot N$ TLS sessions. We bin these into B bins as above, letting $n_{\delta,b}$ denote the number of values in bin b for last byte value δ. Now for each candidate value y for the last byte of P^*, we compute the log likelihood for the candidate, using the formula:

$$LL(y) = \sum_{\delta \in \{\texttt{0x00},\ldots,\texttt{0xFF}\}} n_{\delta,b} \cdot \log(p_{\delta \oplus y,b}) \,.$$

We then output as the preferred candidate for the last plaintext byte the value y^* having the highest value of $LL(y)$ amongst all candidates.

We omit the detailed analysis of the performance of this attack, pausing only to note that it will require more samples than the distinguishing attack because the underlying statistical problem is to now separate one correct candidate from 255 wrong candidates, and this is more demanding than the basic distinguishing problem.

To wrap up, we note that `nanosleep`, which is now used in *s2n* to add a random time delay, has a granularity of nanoseconds, does not show this behaviour, and therefore thwarts the attacks described in this work.

5 Proof of Concept

We confirmed that *s2n* does indeed behave as expected using the following two experiments.

For the first experiment, we setup a `s2n_blob` buffer of length 93 and filled it with random data. Then, we assigned all possible padding length values `0x00` to `0xff` by overwriting the last byte of the buffer and timed how long the function `s2n_verify_cbc` took to return. As expected, the padding length values between `0x00` and `0x04` resulted in timings about 500–550 cycles longer than all other values. The timing difference was clear and stable. Some sample data is shown in Tables 3 and 4. We note that at present we cannot explain the variation within the second and third columns of those tables.

Table 4. Timing of function `s2n_verify_cbc` (in cycles) with H = SHA-1 for different values of last byte in the `decrypted` buffer, each cycle count averaged over 2^{10} trials.

Byte value	Cycles	Byte value	Cycles	Byte value	Cycles
0x00	1333.99	0x05	1095.01	...	...
0x01	1174.29	0x06	1092.68	0xfc	1062.37
0x02	1178.52	0x07	1065.08	0xfd	1035.48
0x03	1156.56	0x08	1102.31	0xfe	1035.15
0x04	1140.14	0x09	1101.04	0xff	1036.02

For the second experiment, we ran the attack against the actual *s2n* implementation instead of running a synthetic benchmark. That is, we timed the execution of `s2n_recv` under the attack described in Sect. 3. However, to speed up execution we patched *s2n* to only sample random delays up to 10,000 μs. As highlighted in Table 5, this, too, shows marked non-uniform timing behaviour modulo 1 μs.

Table 5. KL divergence observed the full attack against actual *s2n* implementation (scaled by 10^5 for readability) using 2^{24} samples on Intel(R) Xeon(R) CPU E5-2667 v2 @ 3.30 GHz.

	0x00	0x01	0x02	0x03	0x04	0x05	0x0a	0x10	0x20
0x00	.0	.4	.2	.1	.4	1.7	1.6	1.9	2.2
0x01	.4	.0	.4	.3	.3	2.6	2.6	2.8	3.2
0x02	.2	.4	.0	.1	.2	2.3	2.2	2.6	2.8
0x03	.1	.3	.1	.0	.3	2.1	1.9	2.3	2.7
0x04	.4	.3	.2	.3	.0	2.6	2.6	2.9	3.2
0x05	1.7	2.6	2.3	2.1	2.6	.0	.1	.2	.3
0x0a	1.6	2.6	2.2	1.9	2.6	.1	.0	.2	.3
0x10	1.9	2.8	2.6	2.3	2.9	.2	.2	.0	.2
0x20	2.2	3.2	2.8	2.7	3.2	.3	.3	.2	.0

We did not adjust our proof-of-concept code to realise a full plaintext recovery attack, because (a) *s2n* has since been patched in response to this work and because (b) the cost is somewhat dependent on the target machine and operating system. We note, though, that an attack can establish the characteristics of a target machine by establishing genuine TLS sessions (where, hence, padding bytes are known) but with some random bits flipped.

The complete source codes for our experiments (which borrow heavily from the *s2n* test suite) are available at https://bitbucket.org/malb/research-snippets.

6 Discussion

Our attack successfully overcomes both levels of defence against timing attacks that were instituted in *s2n*, the first level being the inclusion of extra cryptographic operations in an attempt to equalise the code's running time and the second level being the use of a random wait interval in the event of an error such as a MAC failure.

Fundamentally, the first level could be bypassed because *s2n* counted bytes going into `s2n_hmac_update` instead of computing the number of compression function calls that need to be performed as suggested in [AP13]. A call to `s2n_hmac_update` in itself will not necessarily trigger a compression function call if insufficient data for such a call is provided. A call to `s2n_hmac_digest`, however, will pad the data and trigger several compression function calls, the number also depending on the data already submitted at the time of the call. We note that in OpenSSL this issue is avoided by effectively re-implementing HMAC in the function `ssl3_cbc_digest_record`, i.e. by performing lower-level cryptographic operations within the protocol layer. In contrast, *s2n* is specifically aimed at separating those layers. In response to this work, *s2n* now sensibly counts the number of compression function calls performed, somewhat maintaining this separation.

The second level could be bypassed because, while the randomised wait periods were large, they were not sufficiently random to completely mask the timing signal remaining from the first step of our attack. Note that the analysis in [AP13] of the effectiveness of random delays in preventing the Lucky 13 attack assumed the delays were uniformly distributed; under this assumption, their analysis shows that the count measure is not effective unless the maximum delay is rather large. What the second step of our attack shows is that, even if the maximum delay is very large, non-uniformity in the distribution of the delay can be exploited. In short, it is vital to carefully study any source of timing delay to ensure it is of an appropriate quality when using it for this kind of protection.

Our experiments indicate that the distribution of `nanonsleep` as implemented on Linux is sufficiently close to uniform to thwart the attack described in this work. We note, however, that this puts a high security burden on this function which is not designed for this purpose. In particular, `nanosleep(2)` states (emphasis added): "nanosleep() suspends the execution of the calling thread until either *at least* the time specified in *req has elapsed, or the delivery of a signal that triggers the invocation of a handler in the calling thread or that terminates the process".

Finally, since randomised waiting can also have a significant performance impact, this work further highlights that MAC-then-Encrypt constructions such as MEE-TLS should be avoided where possible.

Acknowledgement. We would like to thank Colm MacCarthaigh and the rest of the *s2n* development team for pointing out the randomised waiting countermeasure and for helpful discussions on an earlier draft of this work.

References

[ABBD15] Almeida, J.B., Barbosa, M., Barthe, G., Dupressoir, F.: Verifiable side-channel security of cryptographic implementations: constant-time MEE-CBC. IACR Cryptology ePrint Archive, 2015:1241 (2015)

[ABP+13] AlFardan, N.J., Bernstein, D.J., Paterson, K.G., Poettering, B., Schuldt, J.C.N.: On the security of RC4 in TLS. In: King, S.T. (ed.) Proceedings of the 22nd USENIX Security Symposium, Washington D.C., USA, pp. 305–320. USENIX, August 2013

[AIES15] Apecechea, G.I., Inci, M.S., Eisenbarth, T., Sunar, B.: Lucky 13 strikes back. In: Bao, F., Miller, S., Zhou, J., Ahn, G.-J. (eds.) Proceedings of the 10th ACM Symposium on Information, Computer and Communications Security, ASIA CCS 2015, Singapore, April 14–17, pp. 85–96. ACM (2015)

[AP12] AlFardan, N., Paterson, K.G.: Plaintext-recovery attacks against datagram TLS. In: Network and Distributed System Security Symposium (NDSS 2012) (2012)

[AP13] AlFardan, N., Paterson, K.G.: Lucky thirteen: breaking the TLS and DTLS record protocols. In: Sommer, R. (ed.) Proceedings of the 2013 IEEE Symposium on Security and Privacy (S&P 2013), San Diego, CA, USA, pp. 526–540. IEEE Press, May 2013

[CHVV03] Canvel, B., Hiltgen, A.P., Vaudenay, S., Vuagnoux, M.: Password interception in a SSL/TLS channel. In: Boneh, D. (ed.) CRYPTO 2003. LNCS, vol. 2729, pp. 583–599. Springer, Heidelberg (2003)

[CK10] Coron, J.-S., Kizhvatov, I.: Analysis and improvement of the random delay countermeasure of CHES 2009. In: Mangard, S., Standaert, F.-X. (eds.) CHES 2010. LNCS, vol. 6225, pp. 95–109. Springer, Heidelberg (2010)

[GPdM15] Garman, C., Paterson, K.G., Van der Merwe, T.: Attacks only get better: Password recovery attacks against RC4 in TLS. In: Jung, J., Holz, T. (eds.) 24th USENIX Security Symposium, USENIX Security 15, Washington, D.C., USA, August 12–14, pp. 113–128. USENIX Association (2015)

[KBC97] Krawczyk, H., Bellare, M., Canetti, R.: HMAC: Keyed-Hashing for Message Authentication. RFC 2104 (Informational), February 1997

[Lab15] Amazon Web Services Labs. s2n: an implementation of the TLS/SSL protocols (2015). https://github.com/awslabs/s2n

[Lan13] Langley, A.: Lucky thirteen attack on TLS CBC, February 2013. https://www.imperialviolet.org/2013/02/04/luckythirteen.html

[Mav13] Mavrogiannopoulos, N.: Time is money (in CBC ciphersuites), February 2013. http://nmav.gnutls.org/2013/02/time-is-money-for-cbc-ciphersuites.html

[MDK14] Möller, B., Duong, T., Kotowicz, K.: This POODLE bites: Exploiting the SSL 3.0 fallback, September 2014

[Moe04] Moeller, B.: Security of CBC ciphersuites in SSL/TLS: Problems and countermeasures. Unpublished manuscript, May 2004. http://www.openssl.org/~bodo/tls-cbc.txt

[PRS11] Paterson, K.G., Ristenpart, T., Shrimpton, T.: Tag size *Does* matter: attacks and proofs for the TLS record protocol. In: Wang, X., Lee, D.H. (eds.) ASIACRYPT 2011. LNCS, vol. 7073, pp. 372–389. Springer, Heidelberg (2011)

[Sch15] Schmidt, S.: Introducing s2n, a new open source TLS implementation, June 2015. https://blogs.aws.amazon.com/security/post/TxCKZM94ST1S6Y/Introducing-s2n-a-New-Open-Source-TLS-Implementation

[Vau02] Vaudenay, S.: Security flaws induced by CBC padding - applications to SSL, IPSEC, WTLS. In: Knudsen, L.R. (ed.) EUROCRYPT 2002. LNCS, vol. 2332, pp. 534–546. Springer, Heidelberg (2002)

[VF15] Valsorda, F., Fitzpatrick, B.: crypto/tls: implement countermeasures against CBC padding oracles, December 2015. https://go-review.googlesource.com/#/c/18130/

[VZRS15] Varadarajan, V., Zhang, Y., Ristenpart, T., Swift, M.M.: A placement vulnerability study in multi-tenant public clouds. In: Jung, J., Holz, T. (eds.) 24th USENIX Security Symposium, USENIX Security 15, Washington, D.C., USA, August 12–14, pp. 913–928. USENIX Association (2015)

An Analysis of OpenSSL's Random Number Generator

Falko Strenzke(✉)

cryptosource GmbH, Darmstadt, Germany
fstrenzke@cryptosource.de

Abstract. In this work we demonstrate various weaknesses of the random number generator (RNG) in the OpenSSL cryptographic library. We show how OpenSSL's RNG, knowingly in a low entropy state, potentially leaks low entropy secrets in its output, which were never intentionally fed to the RNG by client code, thus posing vulnerabilities even when in the given usage scenario the low entropy state is respected by the client application. Turning to the core cryptographic functionality of the RNG, we show how OpenSSL's functionality for adding entropy to the RNG state fails to be effectively a mixing function. If an initial low entropy state of the RNG was falsely presumed to have 256 bits of entropy based on wrong entropy estimations, this causes attempts to recover from this state to succeed only in long term but to fail in short term. As a result, the entropy level of generated cryptographic keys can be limited to 80 bits, even though thousands of bits of entropy might have been fed to the RNG state previously. In the same scenario, we demonstrate an attack recovering the RNG state from later output with an off-line effort between 2^{82} and 2^{84} hash evaluations, for seeds with an entropy level n above 160 bits. We also show that seed data with an entropy of 160 bits, fed into the RNG, under certain circumstances, might be recovered from its output with an effort of 2^{82} hash evaluations. These results are highly relevant for embedded systems that fail to provide sufficient entropy through their operating system RNG at boot time and rely on subsequent reseeding of the OpenSSL RNG. Furthermore, we identify a design flaw that limits the entropy of the RNG's output to 240 bits in the general case even for an initially correctly seeded RNG, despite the fact that a security level of 256 bits is intended.

1 Introduction

The ability to generate high entropy random numbers is crucial to the generation of secret keys, initialization vectors, and other values that the security of cryptographic operations depends on. Thus, random number generators (RNGs) are the backbone of basically any cryptographic architecture. Numerous works have already dealt with the security of RNGs of operating systems [1–4]. In [5], the predictability of OpenSSL's [6] RNG on the Android [7] operating system is investigated. That work reveals the problem of a too low entropy level of

M. Fischlin and J.-S. Coron (Eds.): EUROCRYPT 2016, Part I, LNCS 9665, pp. 644–669, 2016.
DOI: 10.1007/978-3-662-49890-3_25

the OpenSSL RNG output as a consequence of its weak seeding through the operating system entropy sources at boot time.

In contrast, in the present work, we analyse the security features of the OpenSSL RNG itself. Specifically, we analyse the behaviour of the RNG in high and low entropy states. In a low entropy state, i.e. before the RNG has been properly seeded, we certainly know it to be unable to produce cryptographically secure random numbers – but we also expect it not to do any damage if we respect this condition by refraining from using it in cryptographic algorithms. Furthermore, in such a situation, we expect the RNG to produce cryptographically strong output after it has been reseeded with fresh high entropy seed data. As we shall see, neither property is fulfilled by the OpenSSL RNG.

We wish to point out that we are not addressing the RNG recovery problem by continuous entropy collection, which is for instance the subject of [8]. The OpenSSL RNG does not even attempt this, but solely relies on reseeding by the client application. The problems discussed here are in the context of the explicit invocation of these methods for reseeding the RNG from the client application.

As a very fundamental result, we prove that even when seeded initially with a 256 bit entropy seed, the RNG output may only have an entropy level of 240 bits for up to several hundreds of output bytes. The remainder of our findings is concerned with the behaviour of the RNG when it is initially in a low entropy level. We show that in this state, various functions of OpenSSL silently feed data to the RNG that is potentially secret and of low entropy. As a consequence, the RNG's function for outputting low entropy random numbers, which is available before the complete seeding of the RNG, is prone to leak these low entropy secrets. The potentially leaked values we have identified are keys of weak ciphers such as DES and the previous contents of buffers overwritten with random bytes. The latter is problematic in that wiping secret data by overwriting them with (pseudo) random data is an established practice.

Furthermore, we analyse the recovery ability of the RNG from a low entropy state. There are two scenarios in which this becomes relevant: first, the RNG might falsely presume a high entropy state based on false entropy estimations by the client application feeding it with seed data, or during the automatic seeding from the operating system RNG, which OpenSSL performs for instance on Linux systems. This makes our findings relevant for embedded system that feature only a small entropy level in their operating system RNG at boot time, for instance if they rely on reseeding the OpenSSL RNG by using a seed-file after the initial automatic seed from the operating system RNG. The second scenario is that though the RNG was seeded with correctly estimated high entropy data, its current state is revealed to an attacker through a break-in into the system. The latter scenario mainly applies to standard platforms such as servers or personal computers. In either scenario, it is vital that through the addition of further high entropy seeding data to the RNG state an immediate recovery is possible in the sense that subsequently generated output will also be of high entropy. We find that this recovery essentially fails: when the RNG is in a state with low entropy level close or equal to zero, we show that despite the feeding of a arbitrarily high amount of entropy to the RNG, various attacks are possible that allow to predict previous and future output of the RNG with a computational effort of around 2^{80} hash evaluations.

The adversarial model underlying all the weaknesses presented in this work is that of a passive adversary who receives output from the attacked RNG and wishes to predict previous or future RNG output, or in some cases even the seed values. Most of the weaknesses identified in this work, in order to be actually exploited, demand a computational effort beyond what is believed be practical today even for computationally strong adversaries but might easily become feasible within a decade. Due to the entirely passive nature of the attacks, it would be possible for an adversary to record the RNG output and carry out the computational part of the attack at a later point in time when he has gained sufficient computational power.

All our results are based on the analysis of OpenSSL version 1.0.2a, but to the best of our knowledge apply to all versions, as the RNG is legacy feature of the library.

2 API, Life Cycle of OpenSSL's RNG, and the Associated Vulnerabilities

In this section we describe the RNG's API functions that are relevant to our analysis and how they relate to its life cycle states. The purpose of this description is to give a high level understanding of RNG's operation as it is necessary to understand the low entropy secret leakage issues discussed in Sect. 3 and introduce our formal life cycle states that are relevant for the remaining issues. Furthermore, we give an overview of the vulnerabilities presented in this work and relate them to the respective life cycle states where they are manifest. The core cryptographic operation of the RNG will be introduced later in Sect. 4.

The implementation of the RNG is found in the file `md_rand.c`. It defines the default RNG to be used in OpenSSL, though in principle the framework allows for switching to different RNG implementations provided by the user. As any purely software-based RNG it is based on a pseudo random number generator (PRNG). The API of functions related to OpenSSL's RNG are described in the respective manual page [9]. The functions relevant to our problem domain are described in the following.

`void RAND_add(const void *buf, int num, double entropy)` "adds" the entropy contained in `buf` of length `num` to the RNG's internal state, where `entropy` shall be an estimate of the actual entropy contained in `buf`. The newly fed data modifies the RNG's state such that any subsequently generated random output will be affected by it. In this work we show that the claimed "addition" of entropy suffers from a severe weakness. However, for the issue of low entropy secret leakage, this feature is irrelevant.

`void RAND_seed(const void *buf, int num)` is a wrapper for `RAND_add`. It calls that function with `entropy` = `num`, i.e. it expects the seed data to have maximal entropy.

`int RAND_poll()` draws entropy from the operating system's randomness sources, e.g. `/dev/random` on Linux, and feeds it to the RNG.

`int RAND_load_file(const char *filename, long max_bytes)` just loads a file and feeds up to `max_bytes` from that file to the RNG using `RAND_add`. Depending on the operating system, it feeds some further data to the RNG, but this is irrelevant to the issues discussed in this work.

`int RAND_bytes(unsigned char *buf, int num)` outputs `num` random bytes into `buf`. If the entropy level of the RNG, computed as the sum of the estimates provided by the calls to `RAND_add`, is less than the specified minimum of 256 bits, this function returns with an error code.

`int RAND_pseudo_bytes(unsigned char *buf, int num)`
performs the same operation for the random output generation as `RAND_bytes`, except that it also generates output if the minimum entropy level has not been reached.

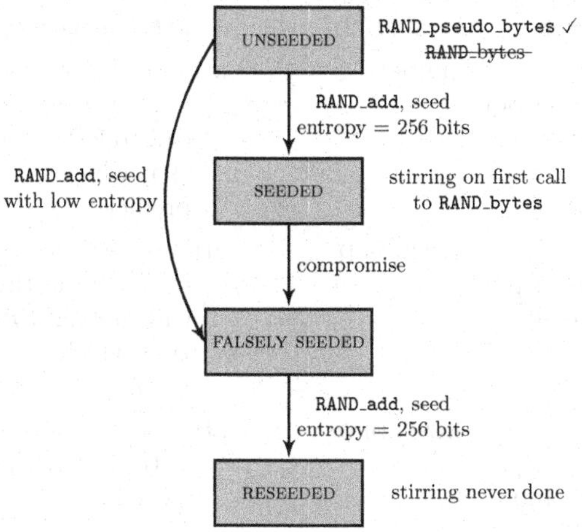

Fig. 1. Depiction of the life cycle states of OpenSSL's RNG.

Figure 1 shows the formalized life cycle states of the RNG. It always starts in the state UNSEEDED. In this state, it has zero entropy. If on the system a random device such as `/dev/random` on Unix is available, a call to `RAND_bytes` or `RAND_pseudo_bytes` will transfer the RNG automatically into the state SEEDED or FALSELY SEEDED by drawing 32 bytes of randomness from that device with a presumed entropy of 256 bits and feeding them to the RNG through a call to `RAND_add`. The distinction between SEEDED and FALSELY SEEDED is not reflected by the RNG state directly, but only implicitly through the quality of the seed. In case of a seed with considerably lower entropy than 256 bits drawn from the random device, we identify the resulting state as FALSELY SEEDED. Another possibility of entering this state is through a compromise of the SEEDED state through a break-in into the system. Recovery from the FALSELY SEEDED state is attempted by feeding the RNG with a high entropy seed.

The resulting state is referred to as RESEEDED. It is distinguished from the SEEDED state by the fact that the so-called "stirring" operation, which distributes the entropy within the RNG state, is never carried out in this state. This operation is executed in the first call to `RAND_bytes` in the SEEDED state and never again after that. But, as we shall see, it is essential that it is carried out after a call to `RAND_add` for the distribution of the entropy in the RNG state and its safety. The details of these considerations will be given in the later sections when we turn to the core cryptographic design of the RNG and also explain the exact effect of the stirring operation.

Table 1. Overview of the identified weaknesses.

Issue	State	Condition	Section
LESLI: low entropy secret leakage in output of `RAND_pseudo_bytes`	UNSEEDED, FALSELY SEEDED	attacker has access to output of `RAND_pseudo_bytes`	3
ELO-240: entropy limitation of the output of `RAND_bytes` to 240 bits	SEEDED	attacker has access to some output from the same call to `RAND_bytes` as that which he wishes to predict	5
ELO-80: entropy limitation of the output of `RAND_bytes` to 80 bits	RESEEDED	attacker has access to some output from the same call to `RAND_bytes` as that which he wishes to predict	6
ELO-160: entropy limitation of the output of `RAND_bytes` to 160 bits	RESEEDED	attacker has access to output after the reseeding	6
DEJA-SEED: recovery of the seed data of entropy of 160 bits and the resulting RNG state with an effort of about 2^{82} hash evaluations given that the seed is prepended with a known value of a specific length	RESEEDED	attacker has access to output after the reseeding at a specific offset	7.1
DEJA-STATE: for instance recovery of the RNG state after a 320-bit entropy reseed with an effort of 2^{84} hash evaluations	RESEEDED	attacker has access to output after the reseeding at a specific offset	7.2

We want to point out how easy it is to get into the RESEEDED state on a device with low boot time entropy provided by the operating system RNG. It is for instance achieved through the following sequence:

```
RAND_pseudo_bytes()
RAND_load_file().
```

The first call triggers the automatic initial seeding of the RNG. The second call attempts to seed the RNG using a high entropy seed file.

Table 1 shows an overview of our contributions. In the first column, following a short identifier for easier referencing of the respective issue, a brief description is given. In the next column the state in which the issue arises is listed. The remaining two columns specify the condition under which the issue arises and the number of the section within this work that explains the issue. LESLI is the abbreviation of "low entropy secret leakage issue". The label ELO... is based on the abbreviation of "entropy limitation of output". The ELO-issues apply to the roughly 1kB of output generated after the reseeding. The last two issues allow an attacker to recover the RNG state, if he gets output at a specific "position" after the reseeding. Here, "position" refers to an offset determined by the sum of length parameters to later calls to either `RAND_add` or `RAND_bytes`.

3 Low Entropy Secret Leakage in Low Entropy States of the RNG

A low entropy state of an RNG certainly makes it impossible to generate secure keys or to carry out cryptographic operations safely that depend on generation of random values. But there is no indication for application developers that are using a cryptographic library through its API to assume that it is generally unsafe either to use functions of the library appearing totally disjoint from the RNG functionality or to make use of the RNG for purposes where the low entropy state would not be a problem from a cryptographic perspective. Note that OpenSSL's function `RAND_pseudo_bytes` explicitly has the purpose of generating output before the RNG is sufficiently seeded. In the following sections we learn that OpenSSL violates the above assumptions, potentially resulting in the leakage of various secrets through the RNG output.

3.1 The General Problem

The basis of the low entropy secret leakage problems we investigate in the following, which we refer to as LESLI, is a fundamental one: given an RNG in a low entropy state, any further seed data fed to the RNG to increase its entropy, is leaked through the RNG output if the resulting state still fails to have a sufficient entropy level. The attack is simply carried out by iterating through all the possible seed inputs, generating the resulting outputs in the attacker's own instance of the RNG, and comparing these outputs to those of the attacked device. If they are equal, the seed values used in that attack iteration are the actual values used to seed the attacked RNG. Thus, any secret value that was part of the seed data is recovered. A requirement for this attack to work is that the number of output bits from RNG is approximately at least as high as the number of entropy bits in its state from the attacker's point of view. In general, the expectation value for the number of collisions, i.e. the number of wrong input values that map to the output value identified as correct, is

$$\mathrm{e} = \frac{2^n - 1}{2^l}, \tag{1}$$

where n is the entropy of the input in bits and l is the size of the output in bits. This relation holds under the assumption that the mapping of RNG input to its output is a random mapping. Assuming $n = l$, we find that on average there will be one collision.

On the basis of these considerations, it is rather doubtful that OpenSSL's manual pages suggest the feeding of low entropy secrets such as user-entered passwords through the function `RAND_add()` to increase the RNG's entropy level. They state: "RAND_add() may be called with sensitive data such as user entered passwords. The seed values cannot be recovered from the PRNG output" [10]. This is, as we have seen above, only true if the resulting entropy level of the RNG is sufficiently high[1]. From this analysis we learn that the feeding of low entropy secrets to RNGs such as that of OpenSSL is a risky and doubtful approach. It is only safe in situations where the RNG already has an entropy level that is secure with respect to brute force state recovery attacks – and in these situations it is needed the least.

However, to avoid the same-state problem, the feeding of low entropy data is indeed useful. Given that two systems share the same but otherwise high entropic RNG state, the feeding of a single bit with value zero to the first RNG and one bit with value one to the second, both RNGs will be in a secure state – as long as they don't appear as adversaries to one another. OpenSSL uses this approach to be secure with respect to the well known process-forking problem of RNGs by feeding the process-ID to the RNG before generating output [11].

But even forearmed with this knowledge, not following OpenSSL's manual pages' encouragement to feed low entropy secrets to the RNG, we run into problems, as various OpenSSL API functions silently feed secret data to RNG, as we explain in the following.

3.2 Leakage of Secrets Overwritten with Random Data

The first leakage problem we present occurs in the following scenario: Assume an application is developed for an embedded system using OpenSSL as the cryptographic library. The system designers are aware of the fact that they might have a low entropy state problem in OpenSSL's RNG. However, they only use OpenSSL for the following purposes (possibly they might later seed it with a

[1] Generally, an entropy of 80 bits is regarded as the minimum to achieve at least short term security.

fresh high-entropy seed and use it also for different purposes): They overwrite short PIN numbers the system temporarily stores during user interactions. They follow a common security advice to overwrite these critical secrets with random numbers using OpenSSL's `RAND_pseudo_bytes` function. This measure to wipe secret data is not necessarily useful in all application contexts. However, it is useful when one cannot exclude the possibility of working on memory-mapped files [12]. Furthermore it can be useful to prevent compiler optimizations from removing the wiping procedure [13][2]. Another reason for this measure is side-channel security: using a fixed byte value such as zero to overwrite the secret bytes could result in leakage of their Hamming weights through power consumption or electromagnetic emission.

Furthermore, in our scenario, we assume that the application uses the RNG to generate weak random numbers with calls to `RAND_pseudo_bytes`, which are output by the device, for instance as nonces for cryptographic purposes.

Neither usage of the potentially predictable random numbers is a problem from a cryptographic point of view: given that there is at least some minimal entropy in the RNG state, side-channel attacks will be severely complicated and also the other two purposes of the random overwriting will not be impeded. Nonce values only need to have the property to be non-repetitive, a property that is not affected by the low entropy state problem even if the RNG state was completely known to the attacker – at least until a reboot that might incur the same-state problem.

However, in the described usage scenario, the secret PIN is leaked through the random numbers output by the device. This is due to the fact that the `RAND_add` function uses the initial content of the memory area to be overwritten as an additional seed value. For the detailed description of function of `RAND_add`, which shows how exactly the initial buffer contents affect the PRNG state, refer to Sect. 4. The RNG's state becomes dependent on the PIN number, and the attacker simply has to execute a brute force search on the joint input space of all possible RNG states before the call to `RAND_add` and all possible PIN values as additional seeds and match the resulting RNG output to the nonces recorded from the device under attack[3].

Furthermore, despite the realistic scenario where the `RAND_pseudo_bytes` function is used to wipe secrets from RAM, there is certainly also a potential leakage problem when previously uninitialized buffers are overwritten with random bytes. Uninitialized buffers on the heap or stack may also by chance contain sensitive low entropy data from previous operations of the application. Which memory locations are reused in parts of the program on the heap or stack is often determin-

[2] In that reference randomizing the target buffer is not suggested, however, since calls to an RNG function have a side effect (on the RNG state), it is almost impossible for the compiler to remove that call. OpenSSL itself uses a similar approach internally, though not by using an actual RNG.

[3] Note that the usage of uninitialized memory for the purpose of random number generation can lead to an even greater threat, namely the compiler's decision to remove subsequent operations on variables that become "tainted" by the uninitialized data [14]. However, this does not seem to apply to OpenSSL's implementation [15].

istic or under the influence of the attacker. In addition, a program may implement buffer reuse at the source code level as an optimization technique.

3.3 Leakage of des Keys in PKCS#8 Conversion

In the file `evp_pkey.c`, in a function used for converting private keys to PKCS#8 format, `RAND_add` is called with the key data as a seed. Given that DES keys, that might otherwise be used in a secure cryptographic construction, can be brute force attacked, they are at risk to be leaked through `RAND_pseudo_bytes`.

4 Detailed Description of OpenSSL's RNG

In this section we give a complete description of the OpenSSL's RNG, since this is necessary to explain the further vulnerabilities presented in this work. However, we omit some details such as the seeding of the RNG with the process-ID (PID) and certain counter values. The PID generally does not feature any entropy since PIDs are predictable on Linux [16], and the counters only depend on the number of bytes provided in the calls to `RAND_add` and `RAND_bytes` and thus can be assumed to be known from the application program's source code and the sequence of high level operations. For a description involving these counters, see [5]. Furthermore, management operations such as checking and updating the level of the estimated entropy are ignored. Algorithm 1 and 2 provide the algorithmic descriptions – simplified in this sense – of the functions `RAND_add` and `RAND_bytes`. First, we explain the symbols used in the algorithmic description. The RNG state is comprised of four elements: md_0 refers to a message digest of 20 bytes length. The state bytes are an array of 1023 bytes represented by s. Furthermore, the index of the current state byte is labelled p (corresponds to

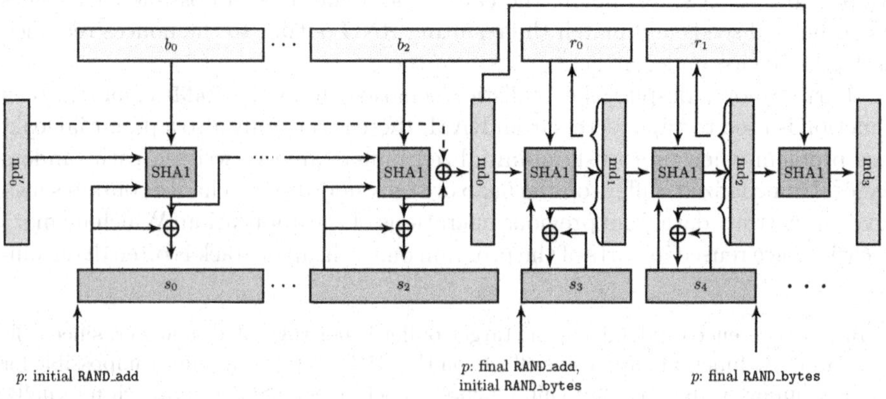

Fig. 2. Depiction of actions induced by a call to function `RAND_add` and a subsequent call to `RAND_bytes`. The blocks b_i, s_0 to s_2 have a length of 20 bytes each, r_i, s_3 and s_4 have a length of 10 bytes each.

Algorithm 1. Simplified algorithmic description of RAND_add

Input: md_0, s, p, q, $b = (b_0||b_1||\ldots||b_{n-1})$ where each b_i is 20 bytes long except for the final one which is potentially shorter

Output: md_0, s, p, q

for $i = 1$ to n **do**
  $t = \text{size}(b_{i-1}) - 1$
  $\text{md}_i = \text{SHA1}(\text{md}_{i-1}||s[p : p + t \mod 1023]||b_{i-1})$
  $s[p : p + t \mod 1023] = s[p : p + t \mod 1023] \oplus \text{md}_i[0 : t]$
  $p = p + t + 1 \mod 1023$
end for
$\text{md}_0 = \text{md}_i \oplus \text{md}_0$
$q = \min(q + \text{size}(b), 1023)$
return md_0, s, p, q

Algorithm 2. Simplified algorithmic description of RAND_bytes

Input: md_0, s, p, q, $r = (r_0||r_1||\ldots||r_{n-1})$ where each r_i is 10 bytes long except for the final one which is potentially shorter

Output: md_0, s, p, $r = (r_0||r_1||\ldots||r_{n-1})$

for $i = 1$ to n **do**
  $\text{md}_i = \text{SHA1}(\text{md}_{i-1}||r_{i-1}||s[p : p + 9 \mod q]$
  $r_{i-1} = \text{md}_i[10 : 10 + \text{size}(r_{i-1}) - 1]$
  $s[p : p + 9 \mod q] = s[p : p + 9 \mod q] \oplus \text{md}_i[0 : 9]$
  $p = p + 10 \mod q$
end for
$\text{md}_0 = \text{SHA1}(\text{md}_i||\text{md}_0)$
return md_0, s, p, r

Algorithm 3. The stirring operation executed within RAND_bytes

$i = 0$
$c = c[0 : 19]$ // constant
while $i < 1023$ **do**
  call RAND_add with $r = c$
  $i = i + 20$
end while

the variable `state_index` in the source code) and q is the sum of the number of state bytes updated by RAND_add (corresponds to the code variable `state_num`). Both algorithms update all four state elements, with the exception that RAND_-bytes does not update q. Both s and md_0 have arbitrary starting values as they use uninitialized memory. However, this can in general not be viewed as reliable source of entropy. The initial value of p and q is zero. q is increased up to 1023 in each call to RAND_add by the number of input bytes. For most of the analyses we conduct in the following sections q is always equal to 1023, since in the state RESEEDED, which will be the starting point for all remaining issues except for $ELO - 240$, the previous stirring operation has already increased q to 1023.

Algorithm 1 specifies the cryptographic operations carried out by a call to RAND_add. Apart from the four state elements, it has an additional input b which represents the seed data. It is partitioned into blocks b_i having a size of 20 bytes each, except for the last block, which has a potentially smaller size. The "size()" operation returns the size of the respective block in bytes. In the loop, iteratively, new values of md_i are computed as the SHA1 hash value of the specified elements. Here "$x[y]$" indicates the y-th byte of x and "$||$" denotes concatenation. $x[y:z]$ is the block formed by the bytes $x[y]||x[y+1]||\ldots||x[z]$, where the index z may also indicate a byte position before y, in which case the wraparound is performed at the highest byte position of x. During each iteration, up to 20 state bytes are updated with the iteratively computed md_i. At the end of the operation md_0 is updated as the XOR of the previous value of md_0 and the final md_i.

The action of the function RAND_bytes is given in Algorithm 2. Additionally to the state elements, the memory area r to be filled with random bytes is an input and output value. Here, r is partitioned into blocks r_i, each having a size of 10 bytes except for the last block, which again has a potentially smaller size. After the computation of the value md_i in each iteration of the loop, one half of this byte string is XORed back to the state bytes that were used to feed the hash function in that iteration, and the other half is written to the 10 byte output block r_{i-1}. If the final output block is shorter than 10 bytes, a correspondingly smaller number of bytes is written to it. Finally, the value md_0 as part of the RNG state is updated.

Note that both Algorithms 1 and 2 implement a seamless wraparound when reaching the end of the state bytes s, so that the beginning and end of this array do not have any special properties. Algorithm 3 specifies the stirring operation. As already explained in Sect. 2, this algorithm is carried out on the first call to RAND_bytes after the RNG state has reached an entropy level of 256 bits by its own accounting based on the entropy measures provided in the calls to RAND_add. Here, c is a 20 byte constant value. By feeding a total of 1040 bytes to the RNG, a certain distribution of the entropy in the state s is achieved.

Figure 2 depicts the operations carried out by a call to RAND_add and a subsequent call to RAND_bytes. On the bottom, the respective values of p are indicated: its initial value at the start of RAND_add, its value after the execution of that function, which corresponds to the initial value in the subsequent call to RAND_bytes, and its final value after termination of the latter function. In the figure, for better readability, the state bytes have been arranged as blocks s_i. Note that this partitioning of the state bytes is done dynamically within both functions starting from the current position indicated by p when they are called.

5 Restriction of the RNG's Output to an Entropy of 240 Bits

The first design flaw of OpenSSL's core RNG, which we refer to as ELO-240, leads to the possibility that generated keys are limited to 240 bits of security whereas it tries to achieve 256 bit – however, due to lack of documentation, the intended security level can be only inferred from the source code. In contrast to further weaknesses reported in this work, this issue arises even when the RNG is initially seeded with correctly estimated high entropy data before any output is generated. From a practical point of view, this problem is meaningless, since it does not allow any practical attacks, not even in the foreseeable distant future, but it shows us a design flaw of the RNG, which will turn out to be relevant for the further issues reported in this work. From a strictly formal point of view, we find that it is commonly agreed that keys with 256 bit security shall be used for applications where long term security matters, as it is reflected by the standardized key sizes for AES and elliptic curve keys, and that thus an RNG should produce output with the corresponding entropy level.

In order to understand how this limitation arises, we consider the following example. The RNG, in its initial state, is seeded with a 256 bit entropy seed, the length of which is rather irrelevant as long as it remains considerably shorter than the state length of 1023 bytes. To simplify things, we assume that the length of the seed data is 256 bits. As a result of this seeding operation, the first 32 bytes of the state contain high entropy data, the remaining state bytes contain zero entropy, and p points to the byte with index 32 (counted from zero). With a subsequent call to `RAND_bytes` the stirring operation is induced. With a sequence of calls to `RAND_add`, the stirring operation, Algorithm 3, completely cycles over all state bytes once, except for the first 17 bytes pointed to by p before the operation, which are processed twice. During this operation, from the entropy added into the first 32 bytes, only 160 bits flow into the remaining state bytes through md_0. Accordingly, an attacker could theoretically enumerate all possible values of the state bytes from position 32 to 1022 by 2^{160} guesses. We now assume that a call to `RAND_bytes` is made to draw 42 random bytes from the RNG. The first ten bytes are generated on the basis of current state block indicated by p. During the output generation, the new value of md_i is calculated as $\text{md}_1 = \text{SHA1}(\text{md}_0||r_0^{\text{init}}||s_0)$, with r_0^{init} being the initial value of the output buffer, which we assume to carry no entropy. Half of md_1, i.e. 80 bits, are output as r_0. Then the following 10-byte blocks and the 2-byte final block are computed iteratively in the same manner. Now assume that the first generated 10-byte block, r_0, is output to the attacker. This means that the attacker learns 80 bits of md_1. Accordingly, the entropy of md_1 is reduced to 80 bits from his point of view. Since all state bytes that flow into the output generation of the further output blocks r_1, r_2, and r_3 have a total entropy of 160 bits as we have seen above, and that 80 bits of entropy flow into the generation from md_1, the remaining generated 32 bytes not output to the attacker have an entropy of 240 bits.

Note that this theoretical attack does not apply when the attacker receives output bytes from one call to `RAND_bytes` and the output value he wishes to predict from a different call. This is due to the fact that at the end of `RAND_bytes`, as given in Algorithm 2, md_0 is updated as $\text{md}_0 = \text{SHA1}(\text{md}_i||\text{md}_0)$, where md_0 on the right hand side is another source of entropy, which foils his knowledge gained through the output r_0. Thus, we conclude that this weakness can be seen as due to a design fault which causes the security of the RNG output to depend on the call sequence. In reality it may be well possible that a client application generates multiple symmetric keys for different users or a key along with the first CBC initialisation vector in a single call to `RAND_bytes`. In the theoretic view of RNGs, implementation details such as concrete call sequences to draw random bytes find no regard.

6 Low Entropy Recovery Failure

We now investigate what happens when OpenSSL's RNG executed the stirring operation in a low entropy state and the client application subsequently adds high entropy seed data to the RNG before generating 256 bit cryptographic keys. The two issues identified in this scenario are ELO-80 and ELO-160.

A principally useful notion for a function to feed entropy to an RNG is that of a mixing function, defined as follows [2]: A mixing function f guarantees that

$$H(f(I,S)) \geq H(S) \text{ and } H(f(I,S) \geq H(I),$$

where $H()$ denotes the Shannon entropy, I the input seed data and S the state of the RNG. A function adhering to this notion guarantees that it does not reduce the entropy of the RNG state and that after its operation the state will have at least the same entropy as the input seed data.

From a purely formal perspective, OpenSSL's RNG fulfils both requirements. However, the definition of the mixing function as provided in the reference and shown above turns out to be of limited usefulness: Given an RNG that uses only a part of its internal state for the production of output, such as it is the case with OpenSSL' RNG, the definition should refer to the entropy of newly generated output instead of that of the state S. With respect to this adjusted definition of a mixing function, we will learn that `RAND_add` fulfils the first requirement, but not the second: when the RNG is in a low entropy state, even after adding high entropy seed data, the RNG will effectively remain in a low entropy state in short term, i.e. generate low entropy output.

We develop the following scenario. After an initial entropy update of the RNG of 32 bytes with a believed entropy of 256 bits, but an actual entropy of $x < 256$ bits, the stirring operation is carried out. This causes, following the analysis from Sect. 5, the whole state $s[0:1023]$ to contain x bits of entropy. Now we assume that a second call to `RAND_add` is made, this time with a 32 byte string with the full entropy of 256 bits. After the stirring operation, p pointed to position $32 + 17 = 59$. Thus the high entropy update affects the state bytes $s[59]$ to $s[90]$, and p afterwards points to $s[91]$. Now a call is made to `RAND_bytes` for the output comprised of the 10 byte blocks r_0, r_1, ..., r_y. The only limitations of the length of the output is that it may only cause `RAND_bytes` to process low entropy state bytes, i.e. not to reach the high entropy block starting at position 59 again, and thus can be of a length of several hundreds of bytes. The first three blocks r_0, r_1, and r_2 are output to the attacker. We now show that he can fully recover the remaining output r_3, r_4, ..., r_y with a complexity of 2^{80+x} hash evaluations. The analysis follows that of Sect. 5.

Through the stirring operation carried out after the initial addition of the x bit entropy seed, x bits of entropy were distributed to all the state blocks $s[i]$. After the second high entropy update, only the state bytes $s[59]$ to $s[90]$ hold 256 bits of entropy. In the subsequent call to **RAND_bytes**, the only source of entropy are md_0, carrying 160 bits, and the processed state bytes starting from $s[91]$. After having seen output r_0, the entropy of md_1 is reduced to 80 bits for the attacker. He now iterates through all the possible values of the unknown 80 bits from md_1 and x bits of the initial entropy seed, i.e. a total of 2^{80+x} possibilities. Each guess for the 2^x initial states implies a value for state bytes $s[91]$ through $s[1022]$. With access to the values of r_0, r_1 and r_2 he can reliably identify the correct guess by comparing his simulated RNG output to the actual output. He now has completely determined the state of the RNG except for the high entropy block spanning from $s[59]$ to $s[90]$ and the value of md_0 before the call to **RAND_add**. He can thus now predict as many output bytes from the same call as can be generated before again processing the high entropy state bytes, which is a little less than one 1 kB. After that call, he loses information about the new value of md_0, according to the update $\mathrm{md}_0 = \mathrm{SHA1}(\mathrm{md}_i || \mathrm{md}_0)$ at the end of **RAND_bytes**. This is the manifestation of ELO-80.

If the attacker has no access to output from the same call, we find the ELO-160 issue, described in the following. After the high entropy reseed without the stirring operation carried out, output bytes of the RNG while processing the low entropy state bytes $s[91]$ through $s[1022]$ have an entropy of $160 + x$ bits. However, if he sees enough output bytes from a single call to **RAND_bytes** to reliably determine the initial seed with entropy level x and thus the values of all the state bytes $s[91]$ through $s[1022]$ as described above, then the entropy of output produced with further calls to **RAND_bytes** when processing these state bytes is only 160 bits (stemming only from md_0).

7 State Recovery Attacks in the RESEEDED State

We now investigate the possibility of another class of attacks against the RESEEDED state. We label these attacks after the deja-vu effect since they exploit the "re-entering" of the state bytes where the high entropy seed was added by exploiting the wraparound at the end of the state bytes in the RNG. In the following sections, we develop two different attacks in the same scenario: we assume that in the FALSELY SEEDED state with zero entropy a high entropy bit string v is fed to RNG. The first attack, DEJA-SEED, explained in Sect. 7.1, recovers the seed v and the RNG state after the reseed. The second attack, subject of Sect. 7.2, is named DEJA-STATE and only recovers the RNG state after the feeding of v. Both attacks recover any secret random values generated after the reseeding.

For the development of both attacks in the following sections, we assume an initial seeding with an entropy of zero. Section 7.3 explains how the attacks can be adjusted if that is not the case and how this affects their complexity.

7.1 Recovery of RNG State and Seed Data

The DEJA-SEED attack presumes the following scenario: The RNG is in the state FALSELY SEEDED with zero entropy. Then, a 160-bit entropy seed v is fed to the RNG using `RAND_add`. Afterwards, a 160-bit key (for instance for HMAC) is generated. To simplify the discussion, we use again fixed positions of the state bytes. Let the 160-bit entropy seed data be fed to the RNG when $p = 40$. Assume the added seed data v is of the following form: a publicly known 10-byte constant part followed by the 20 byte seed data with full entropy. After the seed has been added, p points to s[70]. Now the 160-bit key k is generated (from the analysis of Sect. 6 it follows that it achieves the full 160-bit entropy level, provided that no RNG output from the same call to `RAND_bytes` is accessible to the attacker). Afterwards, either through further additions of seed data or through output generation, p reaches the value 0 again. In this situation, a 90 byte output is generated with a single call to `RAND_bytes`, which is accessible to the attacker. The attacker uses this output to recover the seed v and subsequently the 160-bit key k as follows.

Figure 3 depicts the attack. The state bytes shown at the top represent the RNG's state after the initial zero entropy seed and before the high entropy seed with v. As indicated, the feeding of v alters the state bytes $s[40:69]$. The values of md entering the respective computations are shown above the state bytes as md'_i, with md'_0 being the value before the call to `RAND_add`. Here, the state bytes prior to the feeding are labelled as s' where they differ from those of the state s after the feeding of v and the key generation. In the lower half of the figure, the execution of the call to `RAND_bytes` is depicted that produces the

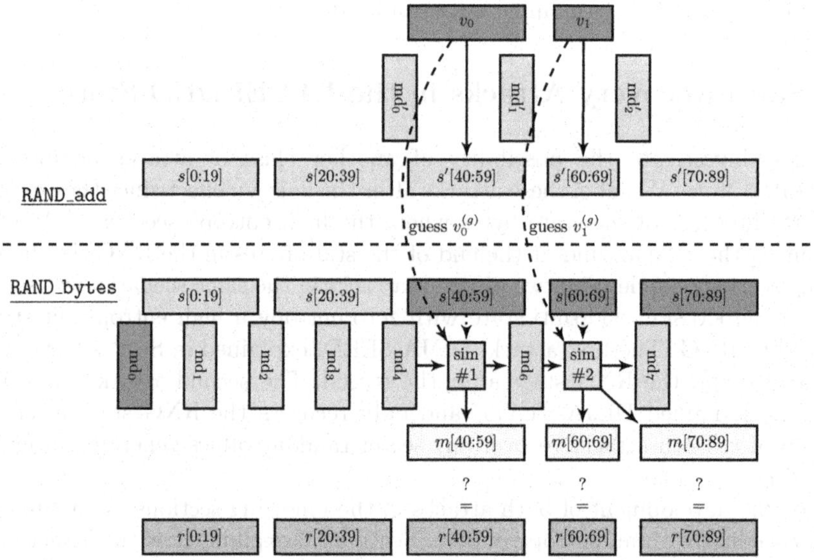

Fig. 3. Depiction of the DEJA-SEED attack.

output the attacker uses for the attack. Below the state bytes s after the feeding of v, the values of md_i belonging to the respective RNG state are shown. The attacker uses $r[0:39]$ to recover md_1 with an effort of 2^{80} simulations of output generation: With the knowledge of $r[0:9]$ he has to guess 2^{80} possibilities of the other half of md_1. The identification of the correct guess is reliable since he can use $r[10:39]$ with a total size of 240 bits for the matching.

For the output generation of a single 20-byte block, a SHA-1 computation with an input of 38 bytes has to be carried out. Since in the vast majority of the guesses, already $r[10:19]$ differs from the simulated output, the cost for the generation of further simulated output values for the few cases where $r[20:39]$ has to be checked, can be ignored. Accordingly, the cost for this step is 2^{80} computations of SHA-1 with a 38-byte input.

Note that the state block $s[0:39]$ is still completely known to the attacker and thus the 80-bit second half of md_1 is the only value unknown to him that serves as an input to the output generation of $r[10:39]$. The attacker knows the value of md_i for as long as known state bytes are processed, i.e. until the start of the processing of $s[40]$. From Algorithm 1, Steps 3 and 4, we find the relation

$$s[40:59] = s'[40:59] \oplus \mathrm{SHA1}(\mathrm{md}'_0||s'[40:59]||v_0) = f(v_0), \tag{2}$$

where md'_0 is the value of md_0 at the beginning of Algorithm 1 during the feeding of v and $s'[40:59]$ indicates the respective value of the variable prior to the feeding of v and which is thus known to the attacker. The result of the hash function on the right hand side amounts to md'_1. Accordingly, we view $s[40:59]$ as a function $f()$ of v_0, the only unknown value for the attacker in the update of $s[40:59]$. The seed bytes of v_0, which flow into that computation, contain 80 bits of entropy: their first half is the fixed 10 byte value, their second half has full 80 bit entropy. The attacker recovers the value of v_0 as follows: He iterates through all the 2^{80} possible values of v_0. For each guessed value $v_0^{(g)}$, he computes the resulting value of $s[40:59]$ according to (2). This procedure is indicated in Fig. 3 as "sim#1". For each guess of $v_0^{(g)}$, this procedure outputs a value $m[40:59]^{(g)}$ where, according to Algorithm 2, Steps 2 and 3,

$$m[40:49]^{(g)} = \mathrm{md}_5[10:19] = \mathrm{SHA1}\left(\mathrm{md}_4||r[40:49]^{\mathrm{init}}||f(v_0^{(g)})[0:9]\right)[10:19],$$

$m[50:59]^{(g)}$ is computed accordingly as

$$m[50:59]^{(g)} = \mathrm{md}_6[10:19] = \mathrm{SHA1}\left(\mathrm{md}_5||r[50:59]^{\mathrm{init}}||f(v_0^{(g)})[10:19]\right)[10:19],$$

and r_i^{init} indicates the initial contents of the output buffer, which in our model does not contain any entropy. If the attacker finds $m[40:59]^{(g)} = r[40:59]$, then he concludes that $v_0 = v_0^{(g)}$ and he has determined the first seed block.

Since here the output space ($r[40:59]$ with 160 bits) used for the verification has the double size of the input space (v_0 with 80 bits), the chance for a collision is overwhelmingly small and the attacker can determine $v_0^{(g)}$ with certainty.

After having recovered v_0, he applies the same brute force recovery to v_1. This procedure is denoted as "sim#2" in the figure. The only difference to the attack on v_0 is that for each guess of $v_1^{(g)}$, also the subsequent state bytes starting from $s[70]$ are determined. Using the 320 bits of $r[60:69]$ and $r[70:89]$ to match the guess values $m[60:69]^{(g)}$ and $m[70:89]^{(g)}$, he can reliably verify his 80-bit guess of v_1. At this point, the attacker has completely recovered the state of the RNG after the feeding of v and v itself. If no further entropy was added after the feeding of v, he can predict any future output of the RNG. In any case he recovers the key k directly generated after the feeding of v.

In this example, the attacker needs an effort of $3 \cdot 2^{80} \approx 2^{82}$ hash evaluations (recovery of md with 56-byte hash input; of v_0, and v_1 each with 38-byte hash inputs, and with an average effort of 2^{80} hash evaluations for each of the three) to recover a 160 bit seed. With less prepended constant data, and thus a greater entropy in v_0, the attack complexity increases accordingly.

7.2 Recovery of only the RNG State

In this section we present the DEJA-STATE attack, also applicable in the RNG state RESEEDED. It is similar to the DEJA-SEED attack from Sect. 7.1, the difference being that not the high entropy seed value is recovered, but only the RNG state. Furthermore, there is no requirement on the seed data v to be prepended with constant data, since here we attack the state bytes in blocks of 80 bits, as they are processed by `RAND_bytes`, anyway. This attack allows the prediction of all output of the RNG after the high entropy reseeding like in the DEJA-SEED attack.

We assume that a 160-bit seed with full entropy was fed to the RNG through a call to `RAND_add` when p pointed to $s[40]$. Accordingly, the state bytes $s[40:59]$ are affected by this update. The attack starts in the same way as the DEJA-SEED attack from Sect. 7.1, including the recovery of md_1 through the first output blocks, up to the point where in the DEJA-SEED attack the values for v_0 and v_1 would be guessed. From this point on, the DEJA-STATE attack proceeds as follows.

The state of the attack is that md_4, the value entering the output generation of $r[40:49]$ based on $s[40:49]$, is known. Figure 4 depicts this analogously to the previous attack. Now the attacker performs the following steps:

1. He iterates through all the 2^{80} possible values of $s[40:49]$. For each guess $s[40:49]^{(g)}$, he simulates the attacked RNG's output generation, creating a match value

$$m[40:49]^{(g)} = \text{md}_5[10:19] = \text{SHA1}\left(\text{md}_4 || r[40:49]^{\text{init}} || s[40:49]^{(g)}\right)[10:19].$$

Here we again assume all values r^{init} to be known. He determines the correct guess as the one where $m[40:49]^{(g)} = r[40:49]$. This allows him to verify his input space of 2^{80} on an output space of 2^{80}, giving him an expectation value for the number of collisions of one according to (1), since also a hash

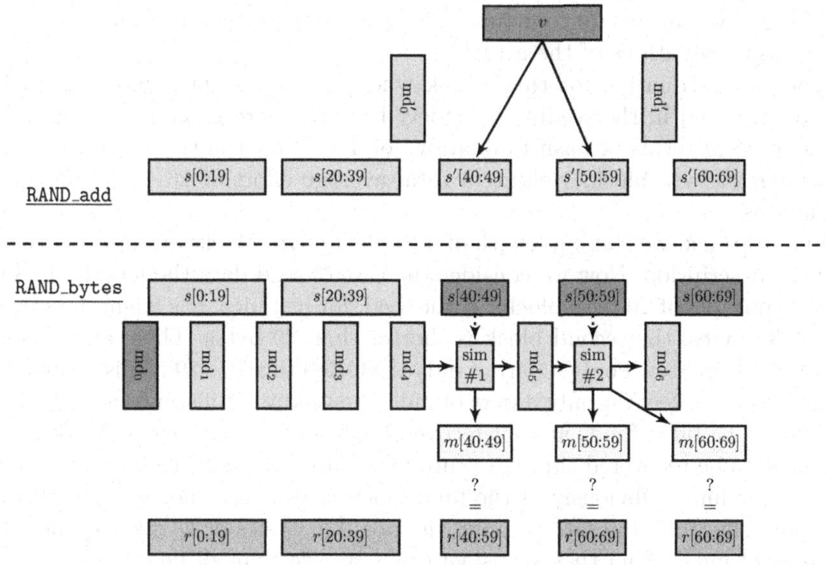

Fig. 4. Depiction of the DEJA-STATE attack.

function with truncated output can be viewed as a random mapping. In the following, we proceed with the description of the attack as though there were no collisions and take them into account when we calculate the complexity of the attack. This procedure is indicated as "sim#1" in the figure.

2. He applies the same procedure to recover $s[50:59]$, where he uses the updated value of md that enters this iteration as md_5, thus recovering the value of md_6. This procedure is indicated as "sim#2" in the figure, which also includes the following two items.
3. Now he knows $s[40:59]$ and also md_6. Since he also knows $s'[40:59]$, the state before the feeding of v, he can calculate md'_1, the value of md after the processing of v during `RAND_add`, as $\mathrm{md}'_1 = s'[40:59] \oplus s[40:59]$, according to Algorithm 1, Step 4.
4. He computes the final updated value of md at the end of `RAND_add`, during the feeding of v, as $\mathrm{md}''_0 = \mathrm{md}'_1 \oplus \mathrm{md}'_0$, according to Algorithm 1, Step 7. Again, md'_0 is known to him as a value from the state prior to the feeding of v. With $s[40:59]$ and the updated value md''_0, he has recovered the complete state after the reseeding with v. This allows him to identify the correct guess for $s[40:49]$ and $s[50:59]$ with respect to the occurring collisions based on further output from $r[60]$ on.

We now estimate the average complexity of the attack: The attacker iteratively applies the single block recovery procedure of a complexity of 2^{80} hash evaluations to the initial recovery of md_0 and to each of the two blocks $r[40:49]$ and $r[50:59]$. Since for each block he has on average one collision, on average he has to process $r[50:59]$ two times. This means he has to go through an effort of 2^{80} hash evaluations for four times on average, thus yielding an average effort of 2^{82} hash evaluations of the attack.

The cost estimation for this attack when two full 20-byte blocks with full entropy are used in the seeding is achieved easily: there is on average an additional effort in terms of hash evaluations of $4 \cdot 2^{80}$ for the third and $8 \cdot 2^{80}$ for the fourth 10-byte block, yielding a total average effort of $16 \cdot 2^{80} = 2^{84}$ hash evaluations.

In our attack, we chose a length of a single 20 byte blocks on purpose to simplify the description. Now we consider the case of seed data the length of which is not a multiple of 20-byte blocks. If the seed, for instance, has a length between 31 and 39 bytes, the second block is shorter than 20 bytes. Then, according to Algorithm 1, in Step 4 only a part of md_2' is XORed to $s[70:79]$. This means that the attacker can recover only a part of md_2'. Assuming for instance a seed length of 32 bytes, 64 bits of md_2' are not recoverable from the state bytes. Accordingly, they have to be recovered through brute force effort, using further output blocks for the matching. Obviously, if the final block v_l becomes shorter than 80 bits, it is more efficient to iterate through the possible values of v_l than to guess the lost part of md_2'. From these considerations we see that all possible seed value lengths of maximally 40 bytes can be attacked with an extra average effort 2^{83}, where we also account for the 8 expected collisions when processing $r[70:79]$.

Like in the DEJA-SEED attack from Sect. 7.1, the attacker recovers completely the state of the RNG after the feeding of v. Thus he can predict all RNG output from that point on without any further effort.

7.3 Dealing with Non-Zero Initial Entropy

In Sects. 7.1 and 7.2 we have assumed an initial entropy of zero for the RNG state before the reseed. Given that the seed used for the initial seeding of the RNG had a low but non-zero entropy of x bits, two approaches can be considered dealing with this entropy in the DEJA-SEED and DEJA-STATE attacks.

If the attacker has access to RNG output before the reseeding, he can recover the initial state with an effort of 2^x hash evaluations. This effort will be negligible compared to that of the presented attacks for actual low entropy states. If the attacker has no access to output prior to the reseeding, he has to recover the initial state parallel to the recovery of md_1 using the first output blocks. This means that additionally to the 2^{80} input space of the unknown half of md_1, he has also to iterate over the 2^x input space for the initial entropy (each guess for the initial seed implies a guess for the whole state s), yielding a total effort of 2^{80+x} hash evaluations. Thus, depending on the initial entropy level, this can become the dominating cost for DEJA-SEED and DEJA-STATE attacks.

8 Conclusion

In this section we discuss the impact, the possibilities for removing of the discovered issues and summarize the theoretical conclusions of our findings.

8.1 Theoretical Aspects of Our Findings

Our central result is that under worst conditions, OpenSSL's RNG only achieves a security level of 80 bits. This sounds devastating, but when we discuss the impact of our results, we will find that from a realistic perspective the majority of real-world systems will be affected to a lesser extend, if at all.

The common notions of security applied to RNGs [4,17] are the well established *forward* and *backward security*, i.e. the security of an RNG under the assumption of the disclosure of its state at a point forward or backward in time, respectively; as well as the notion of *resilience*. The latter would be violated if an attacker can reduce the entropy of the RNG's output by feeding specially prepared seed data to the RNG.

From these notions, backward security is not even attempted by the RNG itself, but when it is attempted by the client application, then it suffers from the DEJA-SEED and DEJA-STATE attacks that allow the disclosure of the RNG state based on the generated output after a reseeding. Accordingly, the backward security of the RNG is impaired. The RNG's forward security in the state RESEEDED is affected by the same attacks since they allow the recovery of previous output – which certainly is also possible with access to the RNG state instead of its output.

Our findings do not suggest that the RNG's resilience is defective in any way. As it seems, the transformations leading to updated RNG states during the addition of seed data are sound in this respect.

Moreover, we find that the addition of seed data to the RNG is not optimal in the sense that if the RNG is in a low entropy state, then the added seed data remains recoverable from the RNG state with a much lower complexity than that corresponding to their combined entropy. Since, to our knowledge, the security of the seed data in the RNG state is not covered by any of the existing notions, we propose the *forward security of seed data* as a new security notion for RNGs.

We also had to point out a shortcoming in the existing notion of a mixing function: using the entropy of the RNG state in its definition, its application leads to meaningless results if the RNG does not use its complete RNG state in a symmetric way for the output production. Accordingly, we propose the notion of an *effective mixing function*, wherein the role of the RNG state is replaced by the subsequent output of the RNG.

What remains from our findings is the low entropy secret leakage issues (LESLI). It is a rather trivial requirement for an RNG not to produce output before sufficiently seeded, however, to our knowledge, this has so far only been viewed from the angle of the low-entropy-recovery problem, which imposes different restrictions than the prevention of seed data leakage. Accordingly, also this problem seems to require formal treatment in RNG security models.

8.2 Impact

Estimating the impact of our findings, we conclude that, excluding the possibility of system break-ins revealing the RNG state to an attacker for the time being, any application running on a system that features a sufficiently high boot time entropy and automatic seeding of the OpenSSL RNG will be safe, since then the initial seed will be of high entropy. The same applies to systems that perform the high entropy reseeding before ever having generated any random numbers. Thus, the only issue such systems suffer from will be ELO-240, meaning that the RNG will produce about 1KB of 240-bit entropy output before it runs at full 256-bit security. Though a clear error in the cryptographic design, this vulnerability can be seen as purely cosmetic problem of the RNG, as generally no system building on ordinary software implementations will be able to achieve the corresponding security level with respect to other aspects such as a general assured security features and physically secure key storage.

However, for any system where the potentially automatic seeding from the operating system RNG delivers a low entropy level, or where this feature is absent, and the client application for instance relies on the loading of a seed-file, the issues ELO-80, ELO-160, DEJA-SEED and DEJA-STATE come up as a threat. The first two are a comparatively minor threat, since ELO-80 depends on the attacker receiving some bytes from the same call to `RAND_bytes` as the one he wishes to predict output from, which is presumably not possible in most designs. And ELO-160 at least maintains a reasonable security level of 160 bits which will most likely remain impossible to break even for "nation-strength" adversaries in the long term. However, especially DEJA-STATE is applicable in general scenarios and has such a small complexity that it forms a concrete threat, since an entropy level of around 80 bits is generally assumed to provide only short term security. So-called "lightweight" cryptographic algorithms such as PRESENT [18] provide 80 bit security. Even though today such a computational effort must be assumed to still be impossible to realize, this can quickly change in the near future. Since all the attacks presented in this work are entirely passive procedures, the RNG output values can be recorded and the computational part of the attack be carried out once the necessary computational resources become available to the attacker. From this perspective, the OpenSSL RNG must be considered as broken in the RESEEDED state, i.e. a scenario where the stirring operation is ceased due to a falsely believed high entropy level. In order to assess the security of an application on a potentially vulnerable system, one must assure that before the high entropy seeding, no other low entropy seeding with subsequent output generation was performed. This is an undesirable situation, since such an analysis is inherently non-local, making it a complex and error prone task.

Good news is that RSA key generation remains safe if it is performed directly after the reseeding, i.e. without any other output generation in between. First of all, there are no RSA keys in use with a security level of 256 bits (this would correspond to a modulus of 15360 bits), so that ELO-240 has no effect, but much more importantly, in an RSA key generation so much output is drawn from the RNG that none of the issues apply any more after such an operation. This is due to the fact that also the RNG's output generation has a "stirring" effect on the state. However, this helpful effect of RSA key generation certainly only applies if it is performed before the generation of any other random output.

LESLI remains an issue for any system that temporarily operates in the UNSEEDED or FALSELY SEEDED state. Here, the leakage of intentionally overwritten or uninitialized/reused memory is a potential threat that could affect real world systems. But the number of actually exploitable applications must be deemed to be small, since in general the use of the library with the RNG being in a low entropy state will be unintentional and is thus likely to incur greater problems than low entropy secret leakage.

The main category of systems potentially affected by the any of the issues identified in this work must be assumed to be embedded systems and mobile platforms. As it is well known, such platforms often feature insufficient entropy levels of their operating system RNGs at least at boot time, without this being detectable by the RNG implementation. Accordingly, the use of a seed file, which stores entropy for applications across application restarts or system reboots, is a common mitigating measure under these circumstances. As we have learned, this approach is at risk to be affected by the issues reported in this work.

8.3 Repair of OpenSSL's RNG

From our analysis of the individual issues it becomes clear what the two main problems of OpenSSL's RNG are: the cessation the stirring operation after having entered the SEEDED state, and subversion of its remaining backbone, the running md, by leaking 80 bits of it through the output. We want to point out that this problem must have been clear to some extent to the designer, since a source code comment in `md_rand.c` just above the code implementing the stirring operation states:

```
/*
 * In the output function only half of 'md' remains secret, so we
 * better make sure that the required entropy gets 'evenly
 * distributed' through 'state', our randomness pool. The input
 * function (ssleay_rand_add) chains all of 'md', which makes it more
 * suitable for this purpose.
 */
```

However, it seems that neither the exact nature of the entropy distribution through the stirring operation nor the necessity of being able to recover from a compromised state at any point during the RNG's life cycle was seen. In any case, if the RNG was intentionally designed to be only one-time seedable, then at least this would have to be stated in the documentation.

The straightforward repair is given by two measures. First, make the updated md in `RAND_bytes` dependent (through hashing or XOR) on the previous value of md after the generation of each block, as it is the case when the vulnerable implementation is used only with calls generating 10 bytes or less of output. Second, the stirring operation must be carried out after each call to `RAND_add` before the generation of new output.

This brings us to a further point, already discussed in the previous section, namely the forward security of the RNG with respect to the recovery of potentially low entropy seed data fed to it. With the current approach of executing the stirring function from `RAND_bytes`, even in the case of correct entropy estimation during the initial seeding, the seed values would remain recoverable through an attack in the style of DEJA-SEED directly on the state bytes s in the period between their feeding through `RAND_add` and the next call to `RAND_bytes`. The correct approach would be to implement `RAND_add` in such a way that a brute force attack on the state s would have the same complexity as the total entropy of all the data fed to it so far.

Concerning LESLI, the only solution is to let `RAND_pseudo_bytes` generate output using a different RNG state than that used for `RAND_bytes`. This would also remove another subtle issue concerning the security with respect to process forking and the RNG's entropy calculation, which is reported in a source code comment of `md_rand.c` itself.

The above described measures will remove all vulnerabilities discovered in this work. However, with respect to efficiency aspects, a standard construction for instance using a CTR mode generator would be much more favourable than a repair of the current design. An AES-based RNG will generally be able to achieve a higher efficiency due to the wide-spread hardware support for this cipher, whereas hardware support of hash functions is very rare. As it becomes evident from our results, the employment of such a large state as used by the OpenSSL RNG does not have any positive effects on security. It remains completely unclear what was the goal of this design choice. A substantial reduction of its size could also help to increase the RNG's performance.

8.4 Countermeasures in Client Code

In order for users of vulnerable versions of OpenSSL to be able to use the RNG without being affected by the issues ELO-240, ELO-80, ELO-160, DEJA-SEED, and DEJA-STATE, in Appendix A, we provide secure wrapper functions for OpenSSL's RNG functionality. The LESLI issue is based on a fundamental design problem that cannot be repaired by wrapper functions. Accordingly, we advise not to use the function `RAND_pseudo_bytes` at all. All calls to that function should be replaced by calls to `RAND_bytes` instead. Note that it is important to check the return value of `RAND_bytes`: if this function fails due to an insufficiently seeded RNG, although returning an error value, it still outputs random bytes. Our secure version of `RAND_bytes` retains this behaviour. Failing to check the error value means, among other problems, that the LESLI issue remains even when abstaining from using `RAND_pseudo_bytes`.

Calls to `RAND_add` shall be replaced with `RAND_add_secure_240bits` or `RAND_add_secure_256bits`. Which of the respective version, "240bits" or "256bits" shall be used depends on whether the user deems it sufficient to have 240 bit entropy output or needs the full 256 bit security. These functions are implemented as follows: after adding the seed through `RAND_add`, the 240 bit version executes a single stirring operation, the 256 bit version repeats this operation once more. The secure functions to replace `RAND_seed`, `RAND_poll` and `RAND_load_file` follow the same principle and naming convention.

Calls to `RAND_bytes` shall be replaced with `RAND_bytes_secure`. This function sets the target buffer to all zeroes before calling `RAND_bytes`, thus avoiding the leakage of its previous contents in future RNG output under all conditions. If this behaviour is not desired, the function can be modified correspondingly. Furthermore, the function makes calls to `RAND_bytes` block-wise with a block length of maximally 10 bytes. This avoids the issues that rely on the learning of half of md through the RNG output, which are ELO-240, ELO-80, DEJA-SEED, and DEJA-STATE.

For users which cannot dispense with `RAND_pseudo_bytes`, we also provide the function `RAND_pseudo_bytes_secure`, which at least prevents leakage of the previous contents of the target buffer.

References

1. Gutterman, Z., Pinkas, B., Reinman, T.: Analysis of the linux random number generator. In: Proceedings of the 2006 IEEE Symposium on Security and Privacy. SP 2006, Washington, DC, USA, pp. 371–385. IEEE Computer Society (2006)
2. Lacharme, P., Röck, A., Strubel, V., Videau, M.: The Linux Pseudorandom Number Generator Revisited (2012). http://eprint.iacr.org/2012/251.pdf
3. Kaplan, D., Kedmi, S., Hay, R., Dayan, A.: Attacking the linux PRNG on android: Weaknesses in seeding of entropic pools and low boot-time entropy. In: Proceedings of the 8th USENIX Conference on Offensive Technologies. WOOT 2014, Berkeley, CA, USA, p. 14. USENIX Association (2014)
4. Dodis, Y., Pointcheval, D., Ruhault, S., Vergniaud, D., Wichs, D.: Security analysis of pseudo-random number generators with Input: /Dev/Random is not robust. In: Proceedings of the 2013 ACM SIGSAC Conference on Computer & Communications Security. CCS 2013, pp. 647–658. ACM, New York (2013)
5. Kim, S.H., Han, D., Lee, D.H.: Predictability of android openSSL's pseudo random number generator. In: Proceedings of the 2013 ACM SIGSAC Conference on Computer –Communications Security. CCS 2013, pp. 659–668. ACM, New York, NY, USA (2013)
6. The OpenSSL Library. http://www.openssl.org
7. The Android operating system. https://www.android.com/
8. Dodis, Y., Shamir, A., Stephens-Davidowitz, N., Wichs, D.: How to eat your entropy and have it too – optimal recovery strategies for compromised RNGs. In: Garay, J.A., Gennaro, R. (eds.) CRYPTO 2014, Part II. LNCS, vol. 8617, pp. 37–54. Springer, Heidelberg (2014)
9. OpenSSL: Manual page of RAND. https://www.openssl.org/docs/crypto/RAND.html
10. OpenSSL: Manual page of RAND_add(). https://www.openssl.org/docs/crypto/RAND_add.html
11. Viega, J.: Practical random number generation in software. In: Proceedings of the 19th Annual Computer Security Applications Conference. ACSAC 2003, Washington,DC, USA, IEEE Computer Society (2003)
12. Ferguson, N., Schneier, B., Kohno, T.: Cryptography Engineering. John Wiley & Sons Inc, New York (2010)
13. Software Engineering Institute. https://buildsecurityin.us-cert.gov/bsi-rules/home/g1/771-BSI.html

14. Wang, X.: More randomness or less (2012). http://kqueue.org/blog/2012/06/25/more-randomness-or-less/
15. Strenzke, F.: uninitialized RAM and random number generation: Threats from compiler optimizations (2015). http://cryptosource.de/posts/uninit_data_rng_en.html
16. Michael Brooks: stackoverflow: How does Linux determine the next PID? http://stackoverflow.com/questions/3446727/how-does-linux-determine-the-next-pid
17. Barak, B., Halevi, S.: A model and architecture for pseudo-random generation with applications to /dev/random. In: Proceedings of the 12th ACM Conference on Computer and Communications Security. CCS 2005, pp. 203–212. ACM (2005)
18. Bogdanov, A.A., Knudsen, L.R., Leander, G., Paar, C., Poschmann, A., Robshaw, M., Seurin, Y., Vikkelsoe, C.: PRESENT: An ultra-lightweight block cipher. In: Paillier, P., Verbauwhede, I. (eds.) CHES 2007. LNCS, vol. 4727, pp. 450–466. Springer, Heidelberg (2007)

A Secure Wrapper Functions for the OpenSSL RNG

In this section we give secure wrapper functions for the OpenSSL RNG functionality. Please refer to Sect. 8.4 for an explanation of the countermeasures used in these functions.

```
void RAND_add_secure_240bits( const void* buf , int num, double
entropy) {
  int n = 1023;
  const unsigned char dummy_seed[20] = { 0 };
  if(buf)
  {
    RAND_add(buf, num, entropy);
  }
  while (n > 0)
  {
    RAND_add(dummy_seed, sizeof(dummy_seed), 0.0);
    n -= sizeof(dummy_seed);
  }
}


void RAND_add_secure_256bits(const void* buf , int num, double
entropy) {
  RAND_add_secure_240bits( buf, num, entropy);
  RAND_add_secure_240bits( NULL, 0, 0.0);
}


void RAND_seed_secure_240bits(const void *buf, int num) {
  RAND_add_secure_240bits(buf, num, (double)num);
}


void RAND_seed_secure_256bits(const void *buf, int num) {
  RAND_add_secure_256bits(buf, num, (double)num);
}


int RAND_poll_secure_240bits() {
  int result = RAND_poll();
  RAND_add_secure_240bits( NULL, 0, 0.0);
  return result;
}
```

```
int RAND_poll_secure_256bits() {
  int result = RAND_poll();
  RAND_add_secure_256bits( NULL, 0, 0.0);
  return result;
}

int RAND_load_file_secure_240bits(const char *file, long max_bytes)
{
 int ret = RAND_load_file(file, max_bytes);
 RAND_add_secure_240bits(NULL, 0, 0.0);
 return ret;
}

int RAND_load_file_secure_256bits(const char *file, long max_bytes)
{
 int ret = RAND_load_file(file, max_bytes);
 RAND_add_secure_256bits(NULL, 0, 0.0);
 return ret;
}

int RAND_bytes_secure(unsigned char *buf, int num) {
  memset(buf, 0, num);
  int final_ret = 1;
  while(num)
  {
    int ret;
    int this_round = num > 10 ? 10 : num;
    ret = RAND_bytes(buf, this_round);
    if(ret != 1)
    {
      final_ret = ret;
    }
    buf += this_round;
    num -= this_round;
  }
  return final_ret;
}

int RAND_pseudo_bytes_secure(unsigned char *buf, int num) {
  memset(buf, 0, num);
  return RAND_pseudo_bytes(buf, num);
}
```

Safely Exporting Keys from Secure Channels
On the Security of EAP-TLS and TLS Key Exporters

Christina Brzuska[1], Håkon Jacobsen[2(✉)], and Douglas Stebila[3,4]

[1] Hamburg University of Technology, Hamburg, Germany
brzuska@tuhh.de
[2] Norwegian University of Science and Technology, Trondheim, Norway
hakoja@item.ntnu.no
[3] Queensland University of Technology, Brisbane, Australia
[4] McMaster University, Hamilton, ON, Canada
douglas@stebila.ca

Abstract. We investigate how to safely export additional cryptographic keys from secure channel protocols, modelled with the *authenticated and confidential channel establishment (ACCE)* security notion. For example, the EAP-TLS protocol uses the Transport Layer Security (TLS) handshake to output an additional shared secret which can be used for purposes outside of TLS, and the RFC 5705 standard specifies a general mechanism for exporting keying material from TLS. We show that, for a class of ACCE protocols we call "TLS-like" protocols, the EAP-TLS transformation can be used to export an additional key, and that the result is a secure AKE protocol in the Bellare–Rogaway model. Interestingly, we are able to carry out the proof without looking at the specifics of the TLS protocol itself (beyond the notion that it is "TLS-like"), but rather are able to use the ACCE property in a semi black-box way. To facilitate our modular proof, we develop a novel technique, notably an encryption-based key checking mechanism that is used by the security reduction. Our results imply that EAP-TLS using secure TLS 1.2 ciphersuites is a secure authenticated key exchange protocol.

1 Introduction

Secure channel protocols are widely used in practice to allow two parties to authenticate each other and securely transmit data. A common design paradigm is to use an *authenticated key exchange (AKE)* protocol to authenticate parties based on public key certificates and to establish a session key, and then use a *stateful authenticated encryption* scheme to encrypt and authenticate the transmission of application data. Real-world secure channel protocols such as TLS, SSH, IPsec, Google's QUIC, the EMV chip-and-pin system, and IEEE 802.11i all follow this paradigm.

Håkon Jacobsen was supported by a STSM Grant from COST Action IC1306.
Douglas Stebila was supported by Australian Research Council (ARC) Discovery Project grant DP130104304.

M. Fischlin and J.-S. Coron (Eds.): EUROCRYPT 2016, Part I, LNCS 9665, pp. 670–698, 2016.
DOI: 10.1007/978-3-662-49890-3_26

For theoreticians, this paradigm is desirable because it allows for modular proofs via composability. A classic result by Canetti and Krawczyk [11] shows how to provably construct a secure channel by running a key exchange protocol that satisfies standard key indistinguishability notions, and then using the key output by the AKE protocol as the symmetric key in authenticated encryption.

For practitioners, this paradigm is desirable because it is efficient and allows to use and combine simple software and hardware components in a variety of ways to form the overall system.

Despite the merits of modularity, most real-world designs are not as clean. In TLS versions up to 1.2, a key exchange protocol, the so-called handshake protocol, is used to establish a premaster secret, which is then used to derive a master secret, which is then used to derive session keys. The final messages of the handshake protocol are encrypted using the session keys, and then application data can be sent, encrypted using the same session keys. SSH has a similar design. In this design, the session keys do *not* satisfy the standard key indistinguishability notion for key exchange security: an adversary can decide whether they have been given the real session key or a random one simply by trial decrypting the encrypted handshake messages.

Early work on proving the security of TLS avoided this problem by showing that a modified version of the TLS handshake yields indistinguishable session keys [29], but this is unsatisfactory since it does not consider the TLS protocol as used in practice. In 2012, Jager, Kohlar, Schäge, and Schwenk (JKSS) [20] introduced the *authenticated and confidential channel establishment* (ACCE) security notion, which treats the key exchange and authenticated encryption as a single monolithic object, allowing them to prove security of the signed Diffie–Hellman ciphersuites in the unmodified TLS 1.2 protocol. ACCE has been applied or adapted to prove security of most other TLS ciphersuites [21,24,26], as well as SSH [4], QUIC [15,27], and the EMV chip-and-pin system [10].

The ACCE notion is not necessarily ideal to cryptographers; its monolithic nature can make modular analysis more difficult, and in particular individual components of ACCE-secure protocols cannot necessarily be used independently. For example, although TLS 1.2's signed Diffie–Hellman ciphersuite is ACCE-secure, one has no security assurance that the session key satisfies any independent security notion: we only have the assurance that the session key is safe to use with the corresponding authenticated encryption scheme in the manner described by the protocol.

Moreover, practitioners seem to like to use the TLS handshake in order to establish keying material for their own purposes. A prominent example is the EAP-TLS protocol [33], which uses the TLS handshake to derive a session key between two peers in the Extensible Authentication Protocol (EAP) [1]. More generally, the practice of *exporting* additional keys from the master secret in the TLS handshake has been formalized in the proposed IETF standard RFC 5705 on TLS key material exporters [31].

However, is it actually safe to use keys exported from the master secret in the TLS handshake? Solely assuming ACCE security of TLS does not at first sight seem to say anything about the *internal* variables of TLS, such as the master secret. However, interestingly, inspired by Morrissey, Smart, and Warinschi (MSW) [29] we can show that the ACCE security of TLS implies that the master secret is *unpredictable.* If the master secret were predictable, then we would be able to break the security of the ACCE channel. This intuition lies at the heart of our proof which uses the ACCE property of TLS in a (semi-)black-box way.

Our Contributions. In this paper we analyze the security of key exporters from ACCE protocols in the provable security setting. Concretely, for TLS we show that if one derives an additional exported key from the TLS master secret—independently of the other handshake messages—then TLS (outputting this additional exported key as the session key) constitutes a secure AKE protocol in the sense of Bellare and Rogaway [2]. However, while our starting point is the TLS protocol, our result is in fact more general, pertaining to a wider class of protocols which we call *TLS-like ACCE protocols.* Roughly speaking, these are protocols which satisfy the ACCE security notion and, like TLS, establish a master secret during the handshake, and from the master secret derive both the channel encryption key and the additional exported key. Apart from this requirement, our result has no other dependencies on the specifics of the protocol. In other words, our main result is a general theorem showing that the transformation specified by EAP-TLS as a key exporter turns any ACCE protocol which has a concept of a master secret into an AKE protocol.

An immediate application of our result is a proof of security in the Bellare-Rogaway model for TLS Key Material Exporters [31] and EAP-TLS [33]. The former has never been subject to a formal security analysis, while the latter has only been analyzed in the symbolic model by He et al. [17] who gave a proof in the context of IEEE 802.11.

Motivation for Our Approach. MSW [29] proved that a modified version of the TLS handshake yields indistinguishable session keys. Specifically, they considered a variant of TLS were the final messages are sent unencrypted. As an intermediate step in their analysis, they showed that the TLS master secret is unpredictable, i.e., that no adversary is able to output the full master secret of a fresh target session. They modeled the key derivation function (KDF) in TLS as a random oracle, and as the inputs to the random oracle are unpredictable, the session keys derived from the master secret are indistinguishable from random.

Similar to MSW, we want to use the fact that the master secret is unpredictable to show that *export* keys are indistinguishable from random. This should be possible even for the *unmodified* TLS protocol, because exported keys are not used to encrypt messages during the handshake phase. One obvious approach would be to reuse one of the existing security proofs which shows TLS to be ACCE secure. Specifically, in these proofs the master secret of a particular session is typically swapped out with a completely random value, allowing the rest of the proof to continue on the assumption that the master secret is completely hidden from the adversary. Due to the unpredictability of the master secret, the adversary will not be able to detect the switch. Using this truly random master secret, we could extend the proof with one additional step where we derive the export key through a random oracle query. It would then follow that the derived export key is indistinguishable from random.

However, such a result could not be re-used across different TLS ciphersuites, nor hold for future versions of TLS. Instead, for every variant of TLS, one would have to copy-paste the corresponding security proof and augment it accordingly to account for the extra export key. This approach is of course inherently non-modular since it is tied to the innards of each particular proof. Still, it seems likely that most of these proofs would be fairly similar in terms of technique, and also reasonably independent of the specific details of the TLS protocol itself.

The question is whether we can isolate exactly those properties of the TLS protocol that these proofs rely on. If so, we could extract a generic proof of TLS key exporters that works across different versions unmodified. Moreover, it would be even better if we could have a result that is not tied to TLS at all, but rather one that targets an appropriate abstract security notion.

Essentially, this is what we do in this paper. We identify some features of the TLS protocol which, when added to a generic ACCE protocol, are sufficient to establish the indistinguishability of the export keys derived from the protocol. Note that, apart from the features that we identify, the result is completely independent of the internals of TLS. Below we describe these features.

Technical Overview of Our Result. Surprisingly, the number of additional features we require in addition to a generic ACCE protocol is rather minimal and consists of the following three requirements (which we make more precise in Sect. 3). We call an ACCE protocol that satisfies these requirements *TLS-like.*

(i) The handshake includes a random *nonce* from each party.
(ii) Each party maintains a value called the *master secret* during the handshake.
(iii) The session key is derived from the master secret, the nonces, and possibly some other public information.

Our result can now be more precisely formulated as follows: starting from an ACCE secure TLS-like protocol Π, we create an AKE secure protocol Π^+, where Π^+ consists of first running protocol Π until a session accepts (according to Π), then deriving one additional key from the master secret and nonces of Π. This key—which is distinct from the session key in the underlying protocol Π—becomes the session key of Π^+. In our security proof the key derivation step will be modeled using a random oracle. The construction of Π^+ from Π precisely captures the definition used in TLS key exporters [31] and EAP-TLS [33].

Note that while we put no security requirements on the master secret of a TLS-like protocol, it is pivotal in our proof to relate the indistinguishability of the session keys in Π^+ to the ACCE security of Π. As mentioned previously, we build on the idea used by MSW [29] to show that unless the adversary queries the random oracle on the exact master secret of a party, it has no advantage in distinguishing the corresponding exported session key in Π^+. MSW proved that an application key agreement protocol (having indistinguishable session keys) could be built out of a master key agreement protocol (having unpredictable master secrets). In their security reduction the simulator could simulate the application key agreement protocol since it had access to a *perfect* key-checking oracle, allowing it to test the validity of master secrets supplied to the random oracle. Our proof is complicated by the fact that we do a reduction to a (TLS-like) ACCE protocol for which there is no key-checking oracle available. The main technical novelty of our proof is to show that we can still create an approximation of the key-checking oracle as long as we allow a (small) one-sided error probability. This emulated key-checking oracle suffices to simulate the AKE experiment of protocol Π^+ in our reduction to the ACCE security of Π.

To give some intuition for our key-checking oracle within the ACCE setting, suppose we want to test whether the value ms' is the master secret of some session π. First, we use ms', the nonces π accepted with, and the KDF of Π (all available due to the TLS-like requirement on Π) to derive a *guess* on π's session key *in* Π. Next, we obtain a ciphertext C of a random message under π's *actual* session key in Π, using our access to a left-or-right encryption oracle in the ACCE game. Finally, we *locally* decrypt C using the guessed session key of Π, i.e., we do not use the decrypt oracle of the ACCE game. If this decryption gives back the random message we started with, we guess that ms' was the correct master secret of π; otherwise, we guess that it was incorrect.

In the above we tacitly assumed that different master secrets derive different session keys (using the same nonces). Normally, this would follow directly from the pseudorandomness of the KDF used in Π. However, since we do not require the master secrets to be independent and uniformly distributed, we cannot invoke this property of the KDF. Instead, we have to explicitly assume that different master secrets do not "collide" to the same session key. We expect this property to hold for most real-world KDFs. Concretely, we show in Theorem 2 (Appendix A) that the HMAC-based KDF used in TLS has this property.

Alternatives to using the ACCE Security Notion? The main reason for using the ACCE security notion in our analysis is that is has proved to be a very useful model for studying the security of two-stage channel establishment protocols. As already mentioned, it has been used repeatedly to analyze real-world protocols such as TLS, SSH, and QUIC. Since our result applies to *any* ACCE protocol that is TLS-like, it can be applied to all these protocols in a near black-box manner. In particular, we can plug in any existing ACCE result without having to re-do any of the steps carried out in the (ACCE) proof of the protocol itself. For example, our result applies unmodified to every ciphersuite version of TLS for which there exist an ACCE proof. Moreover, we can even apply our theorem to future versions of TLS, as long as these continue to be TLS-like and derive their channel keys using a collision resistant KDF.

Still, in the specific case of TLS, one might ask whether another approach could give a simpler, yet equally modular proof of the same result, namely that EAP-TLS (and more generally, TLS key exporters) constitutes a secure AKE.

Krawczyk, Paterson, and Wee (KPW) [24] showed that all the major handshake variants of TLS satisfy a security notion on its key encapsulation mechanism (KEM) called IND-CCCA [18]. If we could reduce the AKE security of

EAP-TLS to the IND-CCCA security of the TLS-KEM, then the results of KPW would give us all the major TLS ciphersuites "for free".

Unfortunately, it is not obvious how such a result can be obtained in a black-box manner from the KEM in [24]. Technically, in order to reduce the AKE security of EAP-TLS to the IND-CCCA security of the TLS-KEM, we need to be able to simulate the key derivation step in the AKE game of EAP-TLS. This requires knowledge about the sessions' master secrets. However, the KEM defined by KPW does not contain the TLS master secret. This means that an adversary against the TLS-KEM in the IND-CCCA game cannot simulate the Test-challenge for some adversary playing in the AKE game against EAP-TLS. Moreover, as remarked by KPW [24, Remark 4], if the KEM key actually *was* defined to be the TLS master secret, then the resulting scheme would be insecure for TLS-RSA provided that RSA PKCS#1v1.5 is re-randomizable[1].

Other Modular Approaches to Analyzing TLS. Canetti and Krawczyk [11] presented a model that allows to analyze protocols in modular way. Unfortunately, since TLS does not meet the stringent requirement of key indistinguishability, it cannot be analyzed within their framework. Küsters and Thuengerthal [25] analyzed the core of TLS in their simulation-based universal composability model called IITM. Unlike some other UC models, the IITM model has the appealing feature that it does not rely on pre-established session identifiers. Brzuska et al. [8] introduced a framework that uses so-called key-independent reductions and allows to analyze protocols such as TLS. Their analysis is in a game-based setting and, up to some small technical differences between models, implies ACCE security of TLS. Kohlweiss et al. [22] recently used the abstract cryptography framework by Maurer and Renner [28] for a modular analysis of TLS.

2 Protocol Definitions

2.1 Execution Environment

Parties. A two-party protocol is carried out by a set of parties $\mathcal{P} = \{P_1, \ldots, P_{n_\mathcal{P}}\}$. Each party P_i has an associated long-term key pair (sk_i, pk_i). We presuppose the existence of a public key infrastructure (PKI) by assuming that every party has an authenticated copy of all the other parties' public keys pk_i. For simplicity we restrict to the setting of mutual authentication, but our results apply equally to the server-only authenticated setting.

Sessions. Each party can take part in multiple executions of the protocol, both concurrently and subsequently. Each run of the protocol is called a *session.* Let n_π denote the maximum number of sessions per party; for party P_i's sth session, we associate an oracle π_i^s which embodies this (local) session's execution of the

[1] On the other hand, Bhargavan et al. [6] conjecture that re-randomizing RSA PKCS#1v1.5 is infeasible, allowing the master secret to be used as the KEM key in TLS-RSA too. We forgo the issue by not reducing to the KEM-security of TLS.

Table 1. State variables for session oracle π_i^s.

Variable	Description
ρ	the role $\rho \in \{\mathsf{init}, \mathsf{resp}\}$ of the session in the protocol execution, being either the *initiator* or the *responder*
pid	the identity $\mathsf{pid} \in \mathcal{P}$ of the intended communication partner of π_i^s
pk	the public key of $\pi_i^s.\mathsf{pid}$
α	the state $\alpha \in \{\mathtt{accepted}, \mathtt{rejected}, \mathtt{running}\}$ of the session oracle
T	the ordered transcript of all messages sent and received by π_i^s
k	the symmetric session-key $k \in \mathcal{K}$ derived by π_i^s
γ	the status $\gamma \in \{\perp, \mathtt{revealed}\}$ of the session key $\pi_i^s.k$
sid	a session identifier $\mathsf{sid} \in \{0,1\}^*$ locally computable by π_i^s
b	a random bit $b \in \{0,1\}$ sampled at the initialization of π_i^s
st	additional auxiliary state that might be needed by the protocol

protocol, maintains the state specific to this session (as described in Table 1), and has access to the long-term secret key sk_i of the party. We put the following correctness requirements on the variables α, k, sid and pid:

$$\pi_i^s.\alpha = \mathtt{accepted} \implies \pi_i^s.k \neq \perp \wedge \pi_i^s.\mathsf{sid} \neq \perp, \tag{1}$$

$$\pi_i^s.\alpha = \pi_j^t.\alpha = \mathtt{accepted} \wedge \pi_i^s.\mathsf{sid} = \pi_j^t.\mathsf{sid} \implies \begin{cases} \pi_i^s.k = \pi_j^t.k \\ \pi_i^s.\mathsf{pid} = P_j \\ \pi_j^t.\mathsf{pid} = P_i \end{cases}. \tag{2}$$

Adverserial Queries. The adversary is assumed to control the network, and interacts with the oracles by issuing queries to them. Below we describe the admissible queries.

- $\mathsf{NewSession}(P_i, \rho, \mathsf{pid})$: This query creates a new session π_i^s with at party P_i, having role ρ and intended partner pid. Based on pid, π_i^s sets the variable pk correspondingly. The session's state is set to $\pi_i^s.\alpha = \mathtt{running}$ and, if $\rho = \mathsf{init}$, it also produces the first message of the protocol which is returned to the adversary.
- $\mathsf{Send}(\pi_i^s, m)$: This query allows the adversary to send any message m to the session oracle π_i^s. If $\pi_i^s.\alpha \neq \mathtt{running}$ return $\perp$. Otherwise, the oracle responds according to the protocol specification, which depends on its role and current internal state.
- $\mathsf{Corrupt}(P_i)$: Return the private key $P_i.sk$ held by party P_i. If $\mathsf{Corrupt}(P_i)$ was the τ-th query issued by $\mathcal{A}$, then we say that P_i is *τ-corrupted.* For uncorrupted parties we define $\tau := \infty$.
- $\mathsf{Reveal}(\pi_i^s)$: This query returns the session key $\pi_i^s.k$ and sets $\pi_i^s.\gamma = \mathtt{revealed}$.

2.2 AKE Protocols

An *authenticated key exchange protocol* (AKE) is a two-party protocol satisfying the syntactical requirement of (1) and (2), and where the security is defined in terms of an AKE security experiment played between a challenger and an adversary. This experiment uses the execution environment described in Sect. 2.1, but has one additional query:

- $\mathsf{Test}(\pi_i^s)$: This query may be asked only once during the course of the game. If $\pi_i^s.\alpha \neq \texttt{accepted}$, then the oracle returns $\perp$. Otherwise, based on $b = \pi_i^s.b$, it returns k_b, where $k_0 \leftarrow \mathcal{K}$ is an independent uniformly sampled key and $k_1 := \pi_i^s.k$. The key k_b is called the Test-*challenge.*

The adversary can win in the AKE experiment in one of two ways: (i) by making a session accept maliciously or (ii) by guessing the secret bit of the Test-session. We formalize these winning conditions below. We simultaneously consider AKE protocols with and without *perfect forward secrecy* (PFS) [13].

Definition 1. *Two sessions π_i^s and π_j^t are* partners *if $\pi_i^s.\mathsf{sid} = \pi_j^t.\mathsf{sid}$.*

Definition 2. *A session π_i^s is said to be* fresh *(resp.* PFS-fresh*), with intended partner P_j, if*

(a) $\pi_i^s.\alpha = \texttt{accepted}$ and $\pi_i^s.\mathsf{pid} = P_j$ when $\mathcal{A}$ issued its τ_0-th query,
(b) $\pi_i^s.\gamma \neq \texttt{revealed}$ and P_i is uncorrupted (resp. τ-corrupted with $\tau_0 < \tau$)[2],
(c) for any partner oracle π_j^t of π_i^s, we have that $\pi_j^t.\gamma \neq \texttt{revealed}$ and P_j is uncorrupted (resp. τ'-corrupted with $\tau_0 < \tau'$).

Definition 3 (Entity Authentication). *A session π_i^s is said to have* accepted maliciously *(resp.* accepted maliciously with PFS*) in the AKE security experiment with intended partner P_j, if*

(a) $\pi_i^s.\alpha = \texttt{accepted}$ and $\pi_i^s.\mathsf{pid} = P_j$ when $\mathcal{A}$ issued its τ_0-th query,
(b) P_i and P_j are uncorrupted (resp. τ- and τ'-corrupted with $\tau_0 < \tau, \tau'$), and
(c) there is no unique session π_j^t such that π_i^s and π_j^t are partners.

We let $\mathbf{Adv}_\Pi^{\mathsf{auth}}(\mathcal{A})$ (resp. $\mathbf{Adv}_\Pi^{\mathsf{auth\text{-}PFS}}(\mathcal{A})$) denote the probability that an adversary $\mathcal{A}$ gets a session to accept maliciously (resp. accepts maliciously with PFS) during the AKE security experiment.

Definition 4 (Key Indistinguishability). *An adversary $\mathcal{A}$ that issued its* Test*-query to session π_i^s during the AKE security experiment,* answers the Test-challenge correctly *(resp.* answers the Test-challenge correctly with PFS*) if it terminates with output b', such that*

(a) π_i^s is fresh (resp. PFS-fresh) with some intended partner P_j, and
(b) $\pi_i^s.b = b'$.

[2] For simplicity we do not model *key-compromise impersonation attacks* in this paper, which should allow P_i itself to be τ-corrupted, with $\tau < \tau_0$.

We assign the following advantage measure to the event that $\mathcal{A}$ answers the Test*-challenge correctly (resp. answers the* Test*-challenge correctly with PFS):*

$$\mathbf{Adv}_{\Pi}^{\mathsf{key\text{-}ind(\text{-}PFS)}}(\mathcal{A}) := \left|\Pr[\pi_i^s.b = b'] - \frac{1}{2}\right|. \quad (3)$$

Definition 5 (AKE Security). *An adversary $\mathcal{A}$* wins *(resp.* wins with PFS*) in the AKE security experiment if a session to accept maliciously (resp. accept maliciously with PFS) or it answers the* Test*-challenge correctly (resp. answers the* Test*-challenge correctly with PFS). We assign the following advantage measure to the event that $\mathcal{A}$ wins (resp. wins with PFS):*

$$\mathbf{Adv}_{\Pi}^{\mathsf{AKE(\text{-}PFS)}}(\mathcal{A}) := \mathbf{Adv}_{\Pi}^{\mathsf{auth(\text{-}PFS)}}(\mathcal{A}) + \mathbf{Adv}_{\Pi}^{\mathsf{key\text{-}ind(\text{-}PFS)}}(\mathcal{A}). \quad (4)$$

2.3 ACCE Protocols

Jager et al. [20] introduced the notion of *authenticated and confidential channel establishment (ACCE)* protocols in order to model TLS. An ACCE protocol is a two-party protocol satisfying the syntactical requirement of Eqs. (1) and (2) and where the session key k is used to key a *stateful length-hiding authenticated encryption scheme* (sLHAE) $\mathsf{stE} = (\mathsf{st.Gen}, \mathsf{stE.Init}, \mathsf{stE.Enc}, \mathsf{stE.Dec})$ (following the definition in [24]). For correctness, we require that if the *deterministic* algorithm $\mathsf{st.Init}$ produced initial states st_E^0, st_D^0, and the ACCE session key k was used to produce a sequence of encryptions $(C_i, st_E^{i+1}) \leftarrow \mathsf{stE.Enc}(k, \ell, m_i, H_i, st_E^i)$ where no C_i equal $\perp$, then the sequence of decryptions $(m_i', st_D^{i+1}) \leftarrow \mathsf{stE.Dec}(k, C_i, H_i, st_D^i)$ is such that $m_i' = m_i$ for each $i \geq 0$. For security, we define an ACCE security experiment based on the execution environment described in Sect. 2.1 that has the following two additional queries (note that there is no Test query).

- $\mathsf{Encrypt}(\pi_i^s, \ell, m_0, m_1, H)$: This query takes as input a ciphertext length specification ℓ, two messages m_0, m_1, and a header H. If $\pi_i^s.\alpha \neq \mathtt{accepted}$, the query returns $\perp$. Otherwise, π_i^s has (by assumption) computed its session key k and run the $\mathsf{stE.Init}$ algorithm of a sLHAE scheme st_{E} to initiate states $\pi_i^s.st_E$ and $\pi_i^s.st_D$. Depending on the bit $\pi_i^s.b$, this call returns the encryption of either m_0 or m_1 using stE. For details, see Fig. 1.
- $\mathsf{Decrypt}(\pi_i^s, C, H)$: This query takes as input a ciphertext C and a header H. If $\pi_i^s.\alpha \neq \mathtt{accepted}$, then the query returns $\perp$. Otherwise, it (statefully) decrypts (C, H) using the underlying sLHAE scheme stE. For details, see Fig. 1.

The adversary can win in the ACCE experiment in one of two ways: (i) by making a session accept maliciously according to Definition 3 (as in the AKE security experiment), or (ii) by breaking one of the sLHAE channels through guessing the corresponding session's secret bit, (we formally define this condition below). Partnering and freshness in the ACCE experiment are defined exactly like in the AKE experiment, i.e., according to Definitions 1 and 2, respectively.

Definition 6 (Channel Security). *An adversary $\mathcal{A}$* breaks the channel *(resp.* breaks the channel with PFS*) in the ACCE security experiment if it terminates with output (π_i^s, b'), such that*

$\underline{\mathsf{Encrypt}(\pi_i^s, \ell, m_0, m_1, H)\text{:}}$

if $\pi_i^s.\alpha \neq$ **accepted:**
  return $\bot$:
$u \leftarrow u + 1$;
$(C^{(0)}, st_{\mathrm{E}}^{(0)}) \leftarrow \mathsf{stE.Enc}(k, \ell, m_0, H, st_{\mathrm{E}})$;
$(C^{(1)}, st_{\mathrm{E}}^{(1)}) \leftarrow \mathsf{stE.Enc}(k, \ell, m_1, H, st_{\mathrm{E}})$;
if $C^{(0)} = \bot$ or $C^{(1)} = \bot$:
  return $\bot$;
$(\mathsf{C}[u], \mathsf{H}[u], st_{\mathrm{E}}) := (C^{(b)}, H, st_{\mathrm{E}}^{(b)})$
return $\mathsf{C}[u]$;

$\underline{\mathsf{Decrypt}(\pi_i^s, C, H)\text{:}}$

if $\pi_i^s.\alpha \neq$ **accepted:**
  return $\bot$:
if $b = 0$:
  return $\bot$;
$\pi_j^t \leftarrow \pi_i^s$'s partner or $\bot$;
$v \leftarrow v + 1$;
$(m, st_{\mathrm{D}}) \leftarrow \mathsf{stE.Dec}(k, C, H, st_{\mathrm{D}})$;
if $v > \pi_j^t.u$ or $C \neq \pi_j^t.\mathsf{C}[v]$ or $H \neq \pi_j^t.\mathsf{H}[v]$:
  in-sync $\leftarrow$ **false**;
if in-sync $=$ **false:**
  return m;
return $\bot$;

Fig. 1. The Encrypt and Decrypt queries of the ACCE security experiment. The variables $k, b, st_{\mathrm{E}}, st_{\mathrm{D}}, \mathsf{C}, \mathsf{H}, u$ and v all belong to the internal state of π_i^s. The variables C and H are lists, initially empty. The counters u and v are initialized to 0, and `in-sync` is set to `true` at the beginning of every session π_i^s. In case π_i^s does not have a partner when answering a Decrypt query, then `in-sync = false`.

(a) π_i^s *is fresh (resp. PFS-fresh) with some intended partner* P_j*, and*
(b) $\pi_i^s.b = b'$.

We assign the following advantage measure to the event that $\mathcal{A}$ *breaks the channel (resp. breaks the channel with PFS):*

$$\mathbf{Adv}_{\Pi}^{\mathsf{chan(\text{-}PFS)}}(\mathcal{A}) := \left| \Pr[\pi_i^s.b = b'] - \frac{1}{2} \right|. \tag{5}$$

Definition 7 (ACCE Security). *An adversary* $\mathcal{A}$ wins *(resp.* wins with PFS*) in the ACCE security experiment if it either gets a session to accept maliciously (resp. accept maliciously with PFS) or breaks the channel (resp. breaks the challenge with PFS). We assign the following advantage measure to the event that* $\mathcal{A}$ *wins (resp. wins with PFS) in the ACCE experiment:*

$$\mathbf{Adv}_{\Pi}^{\mathsf{ACCE(\text{-}PFS)}}(\mathcal{A}) := \mathbf{Adv}_{\Pi}^{\mathsf{auth(\text{-}PFS)}}(\mathcal{A}) + \mathbf{Adv}_{\Pi}^{\mathsf{chan(\text{-}PFS)}}(\mathcal{A}). \tag{6}$$

3 TLS-Like Protocols

Definition 8. *An ACCE protocol* Π *is said to be* TLS-like *if*

(i) each session uniformly at random generates and transmits a distinguished nonce value $n \xleftarrow{\$} \{0,1\}^\lambda$ *during its run of the protocol,*
(ii) each session holds a variable $\pi_i^s.ms \in \{0,1\}^\lambda \cup \{\bot\}$*, called the* master secret*,*
(iii) if n_1, n_2 *are the two nonces on a session's transcript* T*, then the* session key *is derived as*

$$k \leftarrow \mathsf{Kdf}(ms, n_1 \| n_2, F_\Pi(T)), \tag{7}$$

where $\mathsf{Kdf}\colon \{0,1\}^\lambda \times \{0,1\}^{2\lambda} \times \{0,1\}^* \to \{0,1\}^\lambda$ *and* $F_\Pi\colon \{0,1\}^* \to \{0,1\}^*$ *are deterministic functions.*

Remark 1. The function F_Π is protocol specific and meant to capture any additional input that might be used to derive the session keys. In TLS, $F_\Pi(T)$ is the empty string, while for example in IPSec (IKEv2), $F_\Pi(T)$ is the Security Parameter Index (SPI) of the initiator and responder.

Remark 2. Clearly TLS is TLS-like, but most other real-world protocols, like SSH, IPSec and QUIC, belong to this class as well.

4 Constructing an AKE Protocol from a TLS-Like ACCE Protocol

4.1 Construction

Let Π be a TLS-like ACCE protocol with key derivation function Kdf and let $G\colon \{0,1\}^\lambda \times \{0,1\}^{2\lambda} \times \{0,1\}^* \to \{0,1\}^\lambda$ be a random oracle. From Π and G we create an AKE protocol Π^+ as follows. Protocol Π^+ consists of first running protocol Π as usual until a session accepts, then it derives an additional key $ek \leftarrow G(ms, n_C \| n_S, aux)$, where ms is the master secret of Π, n_C and n_S are the nonces, and $aux \in \{0,1\}^*$ is an (optional) string containing selected values from the session's transcript T. The key ek becomes the *session key* in protocol Π^+. The session identifier in Π^+ is inherited from Π.

By construction, a session in Π^+ derives (at least) two keys: its "true" session key in the sense of the AKE-model, i.e., the key ek derived from G, and the channel encryption key derived in the underlying protocol Π using $\Pi.\mathsf{Kdf}$. To avoid confusion, we will call the former key the *export key*; while we will call the latter key the *channel key* and denote it ck. In particular, in the formal AKE security experiment the session key variable $\pi_i^s.k$ will store the export key ek, while the channel key ck will simply be part of π_i^s's internal state, written $\pi_i^s.ck$.

4.2 Main Result

Informally, our main result shows that the construction described above transforms a TLS-like ACCE protocol Π into an AKE protocol Π^+. However, in our proof we need to rely on two additional assumptions besides the ACCE-notion: (1) the key derivation function $\Pi.\mathsf{Kdf}$ used to derive the channel keys in Π^+ is *collision resistant* in a particular sense (Definition 9) and (2) the session identifier allows for *public session matching* (Definition 10) and contains the sessions' nonces and $F_\Pi(T)$ value (q.v. Eq. (7)).

Definition 9 (KDF Collision Resistance). *Let* KDF *be an oracle implementing the key derivation function of a TLS-like ACCE protocol* Π. *Define the following advantage measure for an adversary* $\mathcal{A}$:

$$\mathbf{Adv}^{\mathsf{KDFcoll}}_{\Pi.\mathsf{Kdf}}(\mathcal{A}) := \Pr\left[((ms, ms'), n, s) \leftarrow \mathcal{A}^{\mathsf{KDF}} : \begin{array}{c} \mathsf{KDF}(ms, n, s) = \mathsf{KDF}(ms', n, s) \\ ms \neq ms' \end{array}\right]. \tag{8}$$

A triple $((ms, ms'), n, s)$ *satisfying the criteria in (8) is called a* (KDF) collision *for* $\Pi.\mathsf{Kdf}$.

Remark 3. Definition 9 is a variant of the more common notion of collision resistant *hash functions.* The difference is that KDF collision resistance is about collisions in the *keys*, not the messages.

Definition 10 (Public Session Matching). *A session identifier* sid *allows for* public session matching *in security experiment* E, *if there exists an efficient algorithm* $\mathcal{M}$*—having access to all the queries/responses exchanged between* $\mathcal{A}$ *and the challenger—that can always answer whether or not two accepted sessions are partners during the execution of* E, *i.e.:*

$$\forall k \in \mathbb{N}, \text{ and } \forall \pi_i^s, \pi_j^t \text{ having accepted before } \mathcal{A}\text{'s } \tau_k\text{-th query}:$$
$$\mathcal{M}(\pi_i^s, \pi_j^t) := \begin{cases} 1 & \text{if } \pi_i^s.\mathsf{sid} = \pi_j^t.\mathsf{sid}, \\ 0 & \text{otherwise.} \end{cases} \tag{9}$$

Theorem 1. *Let* Π *be a TLS-like ACCE protocol having a session identifier that allows for public session matching and contains the sessions' nonces and* $F_\Pi(T)$ *values. Let* Π^+ *be the protocol derived from* Π *and random oracle* G, *using the construction described in Sect. 4.1. Then for any adversary* $\mathcal{A}$ *in the AKE security experiment against* Π^+

$$\mathbf{Adv}_{\Pi^+}^{\mathsf{AKE(\text{-}PFS)}}(\mathcal{A}) \leq 6 \cdot \mathbf{Adv}_{\Pi}^{\mathsf{ACCE(\text{-}PFS)}}(\mathcal{B}) + 3 \cdot \mathbf{Adv}_{\Pi.\mathsf{Kdf}}^{\mathsf{KDFcoll}}(\mathcal{C}) + \frac{6qn_{\mathcal{P}}n_\pi}{2^{c\lambda}} + \frac{(n_{\mathcal{P}}n_\pi)^2}{2^{\lambda+1}}, \tag{10}$$

where λ *is the security parameter,* $n_{\mathcal{P}}$ *is the number of parties,* n_π *is the number of sessions at each party,* q *is* $\mathcal{A}$*'s number of random oracle queries, and* $c \in \mathbb{N}$ *is an arbitrary constant.*

The main idea behind the proof of Theorem 1 is to relate the security of the derived export keys to the security of the channel keys in the underlying ACCE protocol Π. Roughly speaking, by using the property that TLS-like protocols derive their channel keys from the master secret and nonces, we establish that two sessions derive the same export key if and only if they derive the same channel key (barring certain bad events which we bound). This fact will make it possible to derive the sessions' export keys in Π^+ independently of their master secrets, and still fully simulate the random oracle G.

4.3 Proof of Theorem 1

Let $\mathcal{A}$ be the adversary in an AKE security experiment against protocol Π^+. From $\mathcal{A}$ we construct an algorithm $\mathcal{B}$ against the ACCE security of the underlying protocol Π. Our proof proceeds through a *sequence of games* ([3,32]), where each consecutive game aims to reduce the challenger's dependency on the sessions' master secrets and the random oracle G, in order to derive the export keys in protocol Π^+. Eventually, in the final game, the random oracle G will have been completely replaced by a local list L_G, and the Π^+ export keys are derived independently of the sessions' master secrets. Thus, at this point, algorithm $\mathcal{B}$ will be able to simulate the game.

Game 0. This is the original AKE security experiment for protocol Π^+:

$$\mathbf{Adv}_{\Pi^+}^{\mathsf{AKE(\text{-}PFS)}}(\mathcal{A}) = \mathbf{Adv}_{\Pi^+}^{\mathsf{G}_0}(\mathcal{A}). \tag{11}$$

Game 1. Game 1 proceeds like in Game 0, but aborts if two sessions generate the same nonce value. Since there are $n_\mathcal{P} \cdot n_\pi$ generated nonces, the probability of there being at least one collision is bounded by $(n_\mathcal{P} n_\pi)^2 \cdot 2^{-(\lambda+1)}$. By the Difference Lemma ([32]) we have

$$\mathbf{Adv}_{\Pi^+}^{\mathsf{G}_0}(\mathcal{A}) \leq \mathbf{Adv}_{\Pi^+}^{\mathsf{G}_1}(\mathcal{A}) + \frac{(n_\mathcal{P} n_\pi)^2}{2^{\lambda+1}}. \tag{12}$$

The remaining games are aimed at removing the challenger's dependency on the random oracle and enabling it to derive the Π^+ export keys without knowing the sessions' master secrets. To this end, the challenger will begin to maintain a list L_G which it will use to simulate the random oracle G and derive the sessions' export keys. The entries of L_G are tuples of the form $(ms, n, aux, ek, [*])$, where $ms \in \{0,1\}^\lambda \cup \{\perp\}$, $n \in \{0,1\}^{2\lambda}$, $ek \in \{0,1\}^\lambda$, $aux \in \{0,1\}^*$, and $[*]$ denotes a list that contains zero or more session oracles. Specifically, we use the notation "$[\,]$" to denote an empty list, "$[\pi_i^s]$" for a list containing exactly π_i^s, "$[\pi_i^s, *]$" for a list containing π_i^s plus zero or more (unspecified) sessions, and "$[*]$" for a list containing zero or more (unspecified) sessions. L_G is initially empty and is filled out either in response to $\mathcal{A}$'s random oracle queries or when a session reaches the `accepted` state.

All the remaining games either change the way export keys are derived for newly accepted sessions (which we call the "Send-code"), or how they answer random oracle calls (which we call the "G-code"). The evolution of the Send-code in Game 2 through Game 6 is shown in Fig. 2, while the corresponding G-code is shown in Fig. 3. We annotate the changes made to a game relative to the previous one using red boxes. Note that some games make changes to both the Send-code and G-code simultaneously.

Game 2. This game introduces the list L_G. When a session π_i^s accepts with master secret ms, nonces $n = n_C \| n_S$, and auxiliary data aux, the challenger uses the Send-code shown in the panel labeled "Game 2" in Fig. 2 to derive its export key. It uses the G-code shown in the panel labeled "Game 2' in Fig. 3 to answer the adversary's random oracle queries. We claim that

$$\mathbf{Adv}_{\Pi^+}^{\mathsf{G}_1}(\mathcal{A}) = \mathbf{Adv}_{\Pi^+}^{\mathsf{G}_2}(\mathcal{A}). \tag{13}$$

Since the challenger considers all of the input values to the random oracle when answering from L_G in this game—in particular, it explicitly looks at the master secrets of the sessions—and because a random oracle always returns the same value when given the same input twice, the answers in Game 2 are distributed exactly like in Game 1.

In the remaining games, we define $ck\text{-}\mathsf{coll}_i$ to be the event that during the run of Game i, the challenger calls the key derivation function $\Pi.\mathsf{Kdf}$ on two different

```
▷ Game 2
  // look at the master secret
if ∃(ms, n, aux, ek, [*]) ∈ L_G:
    π_i^s.k ← ek;
    update (ms, n, aux, ek, [*]) to (ms, n, aux, ek, [*, π_i^s]);
  // no match found – derive new key
else
    ek ← G(ms, n, aux);
    π_i^s.k ← ek;
    L_G ← L_G ∪ {(ms, n, aux, ek, [π_i^s])};
```

```
▷ Game 3
  // match partner sessions
if ∃(*, n, aux, ek, [π_j^t]) ∈ L_G ∧ π_i^s.sid = π_j^t.sid:
    π_i^s.k ← ek;
    update (*, n, aux, ek, [π_j^t]) to (*, n, aux, ek, [π_j^t, π_i^s]);
  // look at the master secret
else if ∃(ms, n, aux, ek, [*]) ∈ L_G:
    π_i^s.k ← ek;
    update (ms, n, aux, ek, [*]) to (ms, n, aux, ek, [*, π_i^s]);
  // no match found – derive new key
else
    ek ← G(ms, n, aux);
    π_i^s.k ← ek;
    L_G ← L_G ∪ {(ms, n, aux, ek, [π_i^s])};
```

```
▷ Game 4
  // match partner sessions
if ∃(*, n, aux, ek, [π_j^t]) ∈ L_G ∧ π_i^s.sid = π_j^t.sid:
    π_i^s.k ← ek;
    update (*, n, aux, ek, [π_j^t]) to (*, n, aux, ek, [π_j^t, π_i^s]);
  // match non-fresh sessions based on their channel keys
else if ∃(*, n, aux, ek, [π_j^t]) ∈ L_G ∧ π_i^s, π_j^t non-fresh ∧ π_i^s.ck = π_j^t.ck:
    π_i^s.k ← ek;
    update (*, n, aux, ek, [π_j^t]) to (*, n, aux, ek, [π_j^t, π_i^s]);
  // look at the master secret
else if ∃(ms, n, aux, ek, [*]) ∈ L_G:
    π_i^s.k ← ek;
    update (ms, n, aux, ek, [*]) to (ms, n, aux, ek, [*, π_i^s]);
  // no match found – derive new key
else
    ek ← G(ms, n, aux);
    π_i^s.k ← ek;
    L_G ← L_G ∪ {(ms, n, aux, ek, [π_i^s])};
```

```
▷ Game 5
  // match partner sessions
if ∃(*, n, aux, ek, [π_j^t]) ∈ L_G ∧ π_i^s.sid = π_j^t.sid:
    π_i^s.k ← ek;
    update (*, n, aux, ek, [π_j^t]) to (*, n, aux, ek, [π_j^t, π_i^s]);
  // match non-fresh sessions based on their channel keys
else if ∃(*, n, aux, ek, [π_j^t]) ∈ L_G ∧ π_i^s, π_j^t non-fresh ∧ π_i^s.ck = π_j^t.ck:
    π_i^s.k ← ek;
    update (*, n, aux, ek, [π_j^t]) to (*, n, aux, ek, [π_j^t, π_i^s]);
  // look at the master secret
else if ∃(ms, n, aux, ek, [*]) ∈ L_G:
    π_i^s.k ← ek;
    update (ms, n, aux, ek, [*]) to (ms, n, aux, ek, [*, π_i^s]);
  // no match found – derive new key
else
    ek ←$ {0,1}^λ;
    π_i^s.k ← ek;
    L_G ← L_G ∪ {(ms, n, aux, ek, [π_i^s])};
```

```
▷ Game 6
  // match partner sessions
if ∃(*, n, aux, ek, [π_j^t]) ∈ L_G ∧ π_i^s.sid = π_j^t.sid:
    π_i^s.k ← ek;
    update (*, n, aux, ek, [π_j^t]) to (*, n, aux, ek, [π_j^t, π_i^s]);
  // match non-fresh sessions based on their channel keys
else if ∃(*, n, aux, ek, [π_j^t]) ∈ L_G ∧ π_i^s, π_j^t non-fresh ∧ π_i^s.ck = π_j^t.ck:
    π_i^s.k ← ek;
    update (*, n, aux, ek, [π_j^t]) to (*, n, aux, ek, [π_j^t, π_i^s]);
  // has G been queried on a master secret ms' valid for π_i^s?
else if ∃(ms', n, aux, ek, []) ∈ L_G ∧ KO(π_i^s, ms') = true:
    π_i^s.k ← ek;
    update (ms', n, aux, ek, []) to (ms', n, aux, ek, [π_i^s]);
  // no match found – derive new key
else
    ek ←$ {0,1}^λ;
    π_i^s.k ← ek;
    L_G ← L_G ∪ {(⊥, n, aux, ek, [π_i^s])};
```

```
▷ B's simulation
  // match partner sessions
if ∃(*, n, aux, ek, [π_j^t]) ∈ L_G ∧ M(π_i^s, π_j^t) = 1:
    π_i^s.k ← ek;
    update (*, n, aux, ek, [π_j^t]) to (*, n, aux, ek, [π_j^t, π_i^s]);
  // match non-fresh sessions based on their channel keys
else if ∃(*, n, aux, ek, [π_j^t]) ∈ L_G ∧ π_i^s, π_j^t non-fresh ∧ π_i^s.ck = π_j^t.ck:
    π_i^s.k ← ek;
    update (*, n, aux, ek, [π_j^t]) to (*, n, aux, ek, [π_j^t, π_i^s]);
  // has G been queried on a master secret ms' valid for π_i^s?
else if ∃(ms', n, aux, ek, []) ∈ L_G ∧ CheckKey(π_i^s, ms') = true:
    π_i^s.k ← ek;
    update (ms', n, aux, ek, []) to (ms', n, aux, ek, [π_i^s]);
  // no match found – derive new key
else
    ek ←$ {0,1}^λ;
    π_i^s.k ← ek;
    L_G ← L_G ∪ {(⊥, n, aux, ek, [π_i^s])};
```

Fig. 2. How to derive the export key ek of a session π_i^s that accepted with master secret ms, nonces $n = n_C \| n_S$, and auxiliary data aux, in Game 2 to Game 6, and in $\mathcal{B}$'s simulation. Variables with underscores denote those that are "pattern matched" against π_i^s's variables. For example, π_i^s is "matched" to $(a, \underline{b}, \underline{c}, ek, [*]) \in L_G$ only if $n_C \| n_S = b$, and $aux = c$. In particular, ms could be different from a.

master secrets $ms \neq ms'$, but with the same nonces $n = n_C \| n_S$ and additional input $F_\Pi(T)$, such that $\Pi.\mathsf{Kdf}(ms, n, F_\Pi(T)) = \Pi.\mathsf{Kdf}(ms', n, F_\Pi(T))$. We call event ck-$\mathsf{coll_i}$ a *channel key collision.*

Game 3. In this game the Send-code is modified so that when a session accepts, the challenger first checks whether the session's partner is present in a tuple on L_G before deriving its export key (see the panel labeled "Game 3" in Fig. 2).

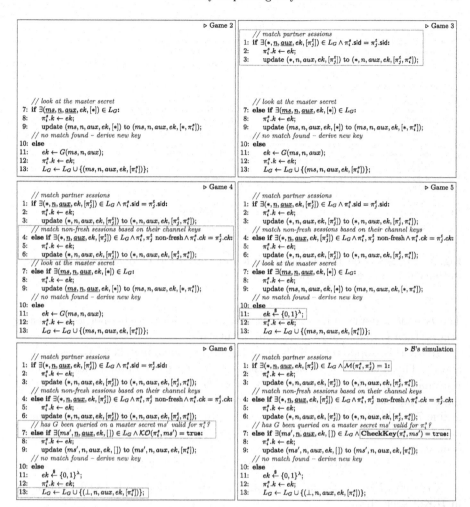

▷ Game 2
```
   // look for previous G queries on the same values
1: if ∃(ms, n, aux, ek, [*]) ∈ L_G:
2:     return ek;

   // no match found - derive new key
5: else
6:     ek ← G(ms, n, aux);
7:     L_G ← L_G ∪ {(ms, n, aux, ek, [])};
8:     return ek;
```

▷ Game 5
```
   // look for previous G queries on the same values
1: if ∃(ms, n, aux, ek, [*]) ∈ L_G:
2:     return ek;

   // no match found - derive new key
5: else
6:     ek ←$ {0,1}^λ;
7:     L_G ← L_G ∪ {(ms, n, aux, ek, [])};
8:     return ek;
```

▷ Game 6
```
   // look for previous G queries on the same values
1: if ∃(ms, n, aux, ek, [*]) ∈ L_G:
2:     return ek;
   // test if ms matches any already accepted sessions
3: else if ∃(⊥, n, aux, ek, [π_i^s, *]) ∈ L_G ∧ KO(π_i^s, ms) = true:
4:     return ek;
   // no match found - derive new key
5: else
6:     ek ←$ {0,1}^λ;
7:     L_G ← L_G ∪ {(ms, n, aux, ek, [])};
8:     return ek;
```

▷ B's simulation
```
   // look for previous G queries on the same values
1: if ∃(ms, n, aux, ek, [*]) ∈ L_G:
2:     return ek;
   // test if ms matches any already accepted sessions
3: else if ∃(⊥, n, aux, ek, [π_i^s, *]) ∈ L_G ∧ CheckKey(π_i^s, ms) = true:
4:     return ek;
   // no match found - derive new key
5: else
6:     ek ←$ {0,1}^λ;
7:     L_G ← L_G ∪ {(ms, n, aux, ek, [])};
8:     return ek;
```

Fig. 3. How $\mathcal{A}$'s G queries, being of the form $G(ms, n, aux)$, are answered in Game 2 to Game 6, and in $\mathcal{B}$'s simulation.

The G-code remains unchanged. We claim that unless a channel key collision occurs, then Game 3 and Game 2 are identical.

To see this, suppose the if-check at line 1 of Game 3 matched two sessions π_i^s and π_j^t. This means that $\pi_i^s.\mathsf{sid} = \pi_j^t.\mathsf{sid}$, which by Eq. (2), implies that they have the same channel key. Then our assumption that no key collision occurs further implies that they must also have the same master secret. Hence, the else-if check at line 7 would also have matched π_i^s and π_j^t in Game 2. This shows that Game 2 and Game 3 matches exactly the same sessions when no channel key collision occurs, hence

$$\mathbf{Adv}_{\Pi+}^{\mathsf{G}_2}(\mathcal{A}) \leq \mathbf{Adv}_{\Pi+}^{\mathsf{G}_3}(\mathcal{A}) + \Pr[ck\text{-}\mathsf{coll}_3]. \tag{14}$$

To bound $\Pr[ck\text{-}\mathsf{coll}_3]$ we create an algorithm $\mathcal{C}'$ that finds (KDF) collisions in $\Pi.\mathsf{Kdf}$ such that

$$\Pr[ck\text{-}\mathsf{coll}_3] \leq \mathbf{Adv}_{\Pi.\mathsf{Kdf}}^{\mathsf{KDFcoll}}(\mathcal{C}'). \tag{15}$$

Algorithm $\mathcal{C}'$ emulates adversary $\mathcal{A}$ and the challenger in an execution of Game 3 by instantiating all the parties' long-term keys and running all the sessions according to the specification of the game. If event ck-coll_1 happened during this run, say due to calls $\Pi.\mathsf{Kdf}(ms, n, F_\Pi(T))$ and $\Pi.\mathsf{Kdf}(ms', n, F_\Pi(T))$, then algorithm $\mathcal{C}'$ outputs $((ms, ms'), n, F_\Pi(T))$ as its collision for $\Pi.\mathsf{Kdf}$.

Since $\mathcal{C}'$ holds all the keys, it can simulate Game 3 perfectly. In particular, it can correctly simulate the random oracle G in those places where it is called inside of Game 3 (i.e., line 11 of the Send-code, and line 6 of the G-code). Thus, the probability that $\mathcal{C}'$ finds a collision in $\Pi.\mathsf{Kdf}$ is exactly the probability that event ck-coll_1 occurs during its simulation of Game 3 for $\mathcal{A}$.

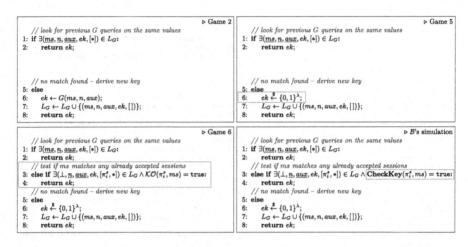

Remark 4. The reason we have to condition on there being no channel key collision in Game 3 is because we do not assume that equal session identifiers

implies equal master secrets (cf. Eq. (2)). It is conceivable that two partner sessions might end up with the same channel key (and export key) even if their master secrets differ. This would lead to a discrepancy in how G queries are answered in Game 2 and Game 3.

Game 4. In this game the Send-code is augmented by matching non-fresh sessions based on their channel keys (see Fig. 2). That is, if two non-fresh sessions are found to have the same channel key (and the same nonces and auxiliary data), then they are given the same export key too. Again, as long as a channel key collision does not occur (event $ck\text{-}\mathsf{coll}_1$), then Game 4 and Game 3 are identical. Similarly, to bound $\Pr[ck\text{-}\mathsf{coll}_4]$ we build an algorithm $\mathcal{C}''$ against the collision resistance of $\Pi.\mathsf{Kdf}$ just like $\mathcal{C}'$ in Game 3. Thus

$$\mathbf{Adv}^{\mathsf{G}_3}_{\Pi+}(\mathcal{A}) - \mathbf{Adv}^{\mathsf{G}_4}_{\Pi+}(\mathcal{A}) \leq \Pr[ck\text{-}\mathsf{coll}_4] \leq \mathbf{Adv}^{\mathsf{KDFcoll}}_{\Pi.\mathsf{Kdf}}(\mathcal{C}''). \tag{16}$$

Game 5. In this game the challenger replaces the calls to the random oracle (both in the Send-code and in the G-code) with strings drawn uniformly at random. We claim that this change does not affect $\mathcal{A}$'s view compared to Game 4 in any way, hence

$$\mathbf{Adv}^{\mathsf{G}_4}_{\Pi+}(\mathcal{A}) = \mathbf{Adv}^{\mathsf{G}_5}_{\Pi+}(\mathcal{A}). \tag{17}$$

To prove (17) we show that the challenger in Game 4 never repeats a call to the random oracle on the same input. Thus, replacing these calls with uniformly drawn strings in Game 5 yields exactly the same distribution on the export keys.

Suppose at some point during Game 4 the challenger made the random oracle call $G(ms, n, aux)$ for the first time (either due to a session accepting, or because $\mathcal{A}$ made this exact G query). Suppose the random oracle responded with ek, and let $t = (ms, n, aux, ek, [*])$ be the tuple that was added to L_G in response to this call.

If the adversary later makes a G query on the same values, i.e. a query of the form $G(ms, n, aux)$, then line 1 of the G-code will be used to answer the query. Thus, the random oracle call on line 6 of the G-code would never be made on the same values twice in Game 4.

Likewise, if a session π_i^s accepts with the same values, i.e., master secret ms, nonces $n = n_C \| n_S$, and auxiliary data aux, *after* the initial G query was made, then the else-if check on line 7 of the Send-code would match π_i^s to t. Thus, the random oracle call on line 11 of the Send-code would not be made on the same values twice in Game 4 either.

In the final game the challenger will derive the sessions' export keys independently of their master secrets. To do this, it will use a probabilistic *key-checking oracle* $\mathcal{KO}$ to test whether the adversary ever queried the random oracle at the correct master secret of a session. Oracle $\mathcal{KO}$ is defined as follows:

$$\mathcal{KO}(\pi_i^s, ms') := \begin{cases} \texttt{true} & \text{with probability 1 when } \pi_i^s.ms = ms', \\ \texttt{false} & \text{with probability } (1-\epsilon) \text{ when } \pi_i^s.ms \neq ms' \ . \end{cases} \tag{18}$$

Specifically, $\mathcal{KO}$ has a *one-sided error probability* since it can potentially return `true` (with probability ϵ) when $\pi_i^s.ms \neq ms$. Based on $\mathcal{KO}$ we define the following event, which will be important in our later analysis:

$$Q : \mathcal{KO} \text{ returns } \mathtt{true} \text{ when called on a fresh session.} \tag{19}$$

We will later show that $\mathcal{A}$ has zero advantage in guessing the Test-challenge correctly unless Q happens (Lemma 1). Note that, if event Q happened, say due to a call $\mathcal{KO}(\pi_i^s, ms')$, then this does not necessarily imply that $\pi_i^s.ms = ms'$; namely, event Q also includes those cases where $\mathcal{KO}$ erroneously returns `true`.

Game 6. Game 6 modifies the else-if clause at line 7 of the Send-code in Game 5 to use the key-checking oracle $\mathcal{KO}$ instead of explicitly looking at a session's master secret. In addition, if a session accepts without a match on L_G, then Game 6 omits its master secret from the tuple that gets added to L_G (line 13). The G-code of Game 6 is also changed to use $\mathcal{KO}$, as shown in Fig. 3.

We claim that as long as $\mathcal{KO}$ does not make a mistake, then Game 6 and Game 5 are identical:

$$\mathbf{Adv}_{\Pi^+}^{\mathsf{G}_5}(\mathcal{A}) \leq \mathbf{Adv}_{\Pi^+}^{\mathsf{G}_6}(\mathcal{A}) + \epsilon. \tag{20}$$

Let $t^{\perp}$ denote the tuple derived from $t = (ms, n, aux, ek, [*]) \in L_G$ by setting $ms = \perp$. To show (20) we prove the following three invariants.

(i) A session π_i^s accepts with master secret ms, nonces $n = n_C \| n_S$ and auxiliary data aux in Game 5 if and only if it accepts with the same master secret, nonces and auxiliary data, and at the same time instance, in Game 6.
(ii) A session π_i^s gets matched to a tuple $t \in L_G$ by one of the if/else-if clauses in the Send-code of Game 6 if and only if π_i^s gets matched to t or $t^{\perp}$ by the corresponding else/if-else clause in Game 6.
(iii) A G query is answered using tuple $t = (ms, n, aux, ek, [*]) \in L_G$ at line 1 of the G-code in Game 5 if and only if it is answered by $t \in L_G$ at line 1, or $t^{\perp} \in L_G$ at line 3, in Game 6.

We only show that (i) holds for the *first* accepting session since (ii) and (iii) implies that it also holds for all subsequent sessions.

(i) Fix a tape of random coins and some adversary $\mathcal{A}$, and consider a run of $\mathcal{A}$ in Game 5 and Game 6 using this tape as the source of randomness (both for the adversary and the challenger). Suppose π_i^s was the first session that accepted in this run of Game 5, say with values ms, $n = n_C \| n_S$, aux. If $\mathcal{A}$ made no G queries before π_i^s accepted, then π_i^s would have accepted with the same values (and at the same time) in the corresponding run in Game 6 too, since there are no differences between the two games up until this point. On the other hand, if $\mathcal{A}$ first made, say q_0, G queries before π_i^s accepted, then these queries would have been answered identically by the G-code in both Game 5 and Game 6 (in particular, by the else-clause at line 5). Hence, π_i^s would have accepted identically in both games also in the case where $\mathcal{A}$ made prior G queries.

(ii) Note that the first two if/else-if clauses in the Send-code do not look at the master secret explicitly (as indicated by the "$*$"). Thus, these two checks behave identically in Game 5 and Game 6.

Next, if π_i^s got matched to $t = (ms, n, aux, ek, [*])$ at line 7 in Game 5, then we claim that $[*] = [\,]$. To see this, suppose $[*] = [\pi_j^t, *]$. Clearly $\pi_i^s.\mathsf{sid} \neq \pi_j^t.\mathsf{sid}$, since otherwise the if-check at line 1 would already have matched π_i^s and t. Furthermore, since we can assume that π_i^s has not accepted maliciously (otherwise the game would already have ended), both π_i^s and π_j^t must be non-fresh by the assumption that the nonces are part of the session identifiers and are unique (Game 1). But then the else-if check at line 4 would have matched π_i^s and t, contradicting our assumption that π_i^s got matched to t at line 7. Hence $[*] = [\,]$. It follows that π_i^s would also have gotten matched to t at line 7 in Game 6 (by assumption, $\pi_i^s.ms = ms$, so $\mathcal{KO}$ is guaranteed to return `true`).

Conversely, if π_i^s got matched to $t = (ms', n, aux, ek, [\,])$ at line 7 in Game 6, it means that $\mathcal{KO}(\pi_i^s, ms') =$ `true`. Since we have conditioned on $\mathcal{KO}$ not making a mistake, $\pi_i^s.ms = ms'$. Moreover, since line 7 is the only check that considers tuples having $[*] = [\,]$ in Game 5, it follows that π_i^s would have gotten matched to t at this line in Game 5 too.

(iii) Line 1 of the G-code ensures that the answers to G queries are consistent with respect to repeated queries in both Game 5 and Game 6, so we only consider non-repeated G-queries.

Suppose $t = (ms, n, aux, ek, [*]) \in L_G$ was used to answer a G query of the form $G(ms, n, aux)$ in Game 5. Note that if $[*] = [\,]$, then this was a repeated G query, so we assume $[*] = [\pi_i^s, *]$. By (i) and (ii), $t^\perp$ must have been on L_G prior to the G query being made in Game 6, and consequently line 3 would have been used to answer it in this game ($\mathcal{KO}$ is guaranteed to return `true` since $\pi_i^s.ms = ms$).

Conversely, if $t^\perp = (\perp, n, aux, [\pi_i^s, *]) \in L_G$ was used at line 3 to answer the query $G(ms', n, aux)$ in Game 6, then $\mathcal{KO}(\pi_i^s, ms') =$ `true`. Since we have conditioned on $\mathcal{KO}$ not making a mistake, it follows that $\pi_i^s.ms = ms'$. Thus, when $\mathcal{A}$ makes the G query in Game 5, t would already be on L_G by (i) and (ii), yielding the right answer at line 1.

This establishes (20). We now turn to the analysis of Game 6.

Analyzing Game 6. It remains to bound the right-hand terms in Eq. (20). First we show that unless $\mathcal{A}$ manages to get event Q to happen (q.v. Eq. (19)), then it has zero advantage in guessing the Test-challenge correctly.

Lemma 1. *Suppose $\mathcal{A}$ issued its* Test*-query against session π_i^s during Game 6, and that it output b' as its answer to the* Test*-challenge. Then*

$$\Pr[\pi_i^s.b = b' \mid \overline{Q}] = \frac{1}{2}, \tag{21}$$

i.e. $\mathcal{A}$ has zero advantage in answering the Test*-challenge correctly if event Q did not happen during Game 6.*

Proof. That event Q did not happen means that $\mathcal{KO}$ never returned `true` for any fresh session during Game 6. Since $\mathcal{KO}$ is always correct when rejecting a key, i.e. when outputting `false`, this implies that $\mathcal{A}$ never queried the random oracle on the correct master secret of any fresh session. In particular, this means that the derived export key of the Test-session in Game 6 is distributed exactly like that of a random key. Thus, the hidden bit of the Test-session is independent of the derived export key from $\mathcal{A}$'s point of view. □

Lemma 1 implies that it is sufficient to bound the probability of event Q and the probability of a session accepting maliciously in order to bound $\mathcal{A}$'s advantage in Game 6. To this end, we construct an ACCE adversary $\mathcal{B}$ against the underlying protocol Π, which instantiates the key-checking oracle $\mathcal{KO}$ of Game 6 with a concrete procedure called **CheckKey**, such that

$$\mathbf{Adv}_{\Pi+}^{\mathsf{G_6\text{-}auth}}(\mathcal{A}) \leq \mathbf{Adv}_{\Pi}^{\mathsf{auth\text{-}(PFS)}}(\mathcal{B}), \tag{22}$$

$$\Pr[Q] \leq 2 \cdot \mathbf{Adv}_{\Pi}^{\mathsf{chan\text{-}(PFS)}}(\mathcal{B}) + \frac{2 \cdot q n_P n_\pi}{2^{c\lambda}}. \tag{23}$$

Moreover, the **CheckKey** procedure will allow us to put a concrete bound on the failure probability ϵ in Eq. (20), specifically

$$\epsilon \leq 2 \cdot \Pr[Q] + \mathbf{Adv}_{\Pi.\mathsf{Kdf}}^{\mathsf{KDFcoll}}(\mathcal{C}'''). \tag{24}$$

We prove (22), (23), and (24), in Lemmas 3, 4, and 2, respectively.

Description of Algorithm $\mathcal{B}$. Algorithm $\mathcal{B}$ plays in an ACCE security experiment against protocol Π and will use adversary $\mathcal{A}$ of Game 6 to win. Roughly speaking, algorithm $\mathcal{B}$ will simulate Game 6 for $\mathcal{A}$ by "embedding" the sessions in its own ACCE experiment into Game 6 and outfitting them with export keys. To derive these export keys, $\mathcal{B}$ maintains the list L_G which it fills out, and answers from, according to the Send and G-code shown in the last panels of Figs. 2 and 3, respectively (both labeled "$\mathcal{B}$'s simulation"). The difference between Game 6 and $\mathcal{B}$'s simulation is that $\mathcal{B}$ has to "implement" the key-checking oracle $\mathcal{KO}$ and also be able to correctly match partnered sessions.

To match partnered sessions, $\mathcal{B}$ uses one of the public session matching algorithms $\mathcal{M}$ guaranteed to exist for sid (since Π is TLS-like).

To instantiate the $\mathcal{KO}$ oracle, $\mathcal{B}$ uses the aforementioned procedure called **CheckKey**, which is formally defined in Algorithm 1 below. We will later show that **CheckKey** has the same properties as the key-checking oracle $\mathcal{KO}$ (as defined in Eq. (18)), but first we describe $\mathcal{B}$'s simulation in detail.

At the beginning of its ACCE security experiment, $\mathcal{B}$ receives the public keys of all the parties from its challenger E which it forwards to $\mathcal{A}$. Then $\mathcal{B}$ initializes L_G to an empty list and runs $\mathcal{A}$, answering its queries as follows:

- NewSession$(P_i, \rho, \mathsf{pid})$: $\mathcal{B}$ forwards the query to its own ACCE challenger and, if $\rho =$ `init`, returns the corresponding response back to $\mathcal{A}$.

- Send(π_i^s, m): $\mathcal{B}$ forwards the query to its own challenger and returns its response back to $\mathcal{A}$. Additionally, if m caused π_i^s to accept then $\mathcal{B}$ derives its export key by running the Send-code shown in the last panel of Fig. 2.
- Corrupt(P_i): $\mathcal{B}$ issues Corrupt(P_i) to its own challenger to obtain the secret key of P_i which it returns back to $\mathcal{A}$.
- $G(ms, n, aux)$: $\mathcal{B}$ answers this query by running the G-code shown in the last panel of Fig. 3.
- Reveal(π_i^s)/Test(π_i^s): If $\pi_i^s.\alpha \neq$ `accepted`, then $\mathcal{B}$ returns $\perp$. Otherwise, there will be an entry $(*, n, aux, ek, [\pi_i^s, *]) \in L_G$, and $\mathcal{B}$ returns ek.

In addition to the above, $\mathcal{B}$ stops and outputs a guess (π_i^s, b') to its ACCE challenger if one of the following events happen.

- *Two sessions generated the same nonce*: select π_i^s arbitrarily among the fresh sessions and draw b' randomly.
- *Event Q happened due to a call to* **CheckKey**(π_i^s, ms): if the ciphertext C decrypted to m_0 at line 20 of Algorithm 1, output $(\pi_i^s, 0)$, otherwise, if it decrypted to m_1, output $(\pi_i^s, 1)$.
- *$\mathcal{A}$ outputs a guess for the* Test-*challenge*: select π_i^s arbitrarily among the fresh sessions and draw b' randomly[3].

This ends the description of algorithm $\mathcal{B}$. Note that the only thing that differs between $\mathcal{B}$'s simulation and Game 6 is $\mathcal{B}$'s usage of the **CheckKey** procedure and the algorithm $\mathcal{M}$ for matching sessions. By definition, the latter is always correct, so $\mathcal{B}$'s simulation is sound given that **CheckKey** correctly implements the $\mathcal{KO}$ oracle.

Analysis of CheckKey. We need to show that **CheckKey** has the same properties as the key-checking oracle $\mathcal{KO}$ used in Game 6, i.e. that it always returns `true` if called on the right master secret of a session and returns `false` (with high probability) when not. The idea of **CheckKey** is to derive from the supplied master secret a guess on the session's channel key and then compare this to the channel key actually held by the session.

For non-fresh sessions this is straightforward since $\mathcal{B}$ can just make a Reveal query in order to obtain their channel keys and make the comparison directly (line 9 in Algorithm 1). On the other hand, issuing a Reveal query to a fresh session would "destroy" its status as a valid target in the ACCE game, preventing $\mathcal{B}$ from capitalizing on the event where $\mathcal{A}$ queries the random oracle on the master secret of a fresh session.

For fresh sessions **CheckKey** instead tests the validity of a derived channel key *indirectly* by trying to (locally) decrypt a ciphertext that was legitimately created with the actual channel key of the session. To obtain this ciphertext,

[3] By Lemma 1 it is immaterial whether $\mathcal{B}$ outputs a random bit or uses $\mathcal{A}$'s guess on the Test-challenge, since Q did not happen.

Algorithm 1. CheckKey(π_i^s, ms)

Note: The procedure is parameterized by $c \in \mathbb{N}$. Calls on the same input always return the same value, i.e. **CheckKey** records its results for every input combination. To simplify the presentation, we leave out the code that deals with this below.

Precondition: $(C_1, H_1), (C_2, H_2), \ldots, (C_k, H_k)$, are the encrypted handshake messages (if any) output by π_i^s during the run of Π^+, together with the corresponding additional data.

1: $x, y \xleftarrow{\$} \{0,1\}^\lambda$;
2: $(m_0, m_1) := (0\|x, 1\|y)$;
3:
4: // n_C, n_S *are the nonces, and* T *the transcript,* π_i^s *accepted with.*
5: $ck' \leftarrow \Pi.\mathsf{Kdf}(ms, n_C\|n_S, F_\Pi(T))$;
6:
7: **if** π_i^s is non-fresh **then**
8: $\quad ck \leftarrow \mathsf{Reveal}(\pi_i^s)$;
9: $\quad$ **return** $ck \stackrel{?}{=} ck'$;
10: **else**
11: $\quad C \leftarrow \mathsf{Encrypt}(\pi_i^s, \ell, m_0, m_1, H)$; ▷ *obtain an encryption of* $m_{\pi_i^s.b}$ *under* $\pi_i^s.ck$.
12:
13: $\quad$ // *"recreate" a decrypt state* st'_D *matching the encrypt state used to create* C.
14: $\quad (*, st'_D) \leftarrow \mathsf{stE.Init}$;
15: $\quad$ **for all** (C_r, H_r) **do**
16: $\quad\quad (*, st'_D) \leftarrow \mathsf{stE.Dec}(ck', C_r, H_r, st'_D)$;
17: $\quad$ **for all** C's from previous calls to **CheckKey**($\pi_i^s, *$) **do**
18: $\quad\quad (*, st'_D) \leftarrow \mathsf{stE.Dec}(ck', C, H, st'_D)$;
19:
20: $\quad (m', *) \leftarrow \mathsf{stE.Dec}(ck', C, H, st'_D)$; ▷ *locally decrypt* C *using* ck' *and* st'_D.
21: $\quad$ **return** $m' \stackrel{?}{\in} \{m_0, m_1\}$;

CheckKey exploits $\mathcal{B}$'s access to a left-or-right encryption oracle for every session in the ACCE game (i.e., the Encrypt query). However, **CheckKey** is complicated by the statefulness of the sLHAE scheme. That is, before attempting to (locally) decrypt the ciphertext at line 20 of Algorithm 1, **CheckKey** first needs to "recreate" a valid decryption state. This is done as follows: starting from the initial state of the sLHAE scheme, **CheckKey** chronologically decrypts each encrypted message output by the session during the handshake (if any). Then it decrypts all ciphertext messages created in prior calls to **CheckKey** (because these advance the session's encrypt state st_E). Finally, it attempts the decryption of C. If the correct channel key was used, then this process is guaranteed to generate a decryption state st'_D that "matches"[4] the encrypt state st_E which was used to create the ciphertext C (due to the correctness of the sLHAE scheme).

[4] The recreated state st'_D does not necessarily have to be *equal* to the decryption state held by π_i^s — it only needs to yield a valid decryption.

Since Π.Kdf is deterministic, the above shows that **CheckKey**(π_i^s, ms) will always return `true` if ms is equal to the master secret of π_i^s, since the derived channel key ck' will then equal $\pi_i^s.ck$.

Conversely, if **CheckKey** is called on a wrong master secret, then it does indeed have a one-sided error probability. In particular, let fresh (resp. non-fresh) denote that **CheckKey** was called on a fresh (resp. non-fresh) session, and let CKerror denote that a call to **CheckKey** erroneously returned `true`. Note that for a fresh session π_i^s and master secret ms', this requires that the decryption of C at line 20 of Algorithm 1 returned one of the two messages (m_0, m_1) associated to the pair (π_i^s, ms'). Letting $b = \pi_i^s.b$, we write correctDec for the event that C decrypted to m_b, and wrongDec for the event that it decrypted to $m_{\bar{b}}$[5]. Consequently, CKerror can be partitioned as follows, depending on whether the session was fresh or not.

$$\begin{aligned} \mathsf{CKerror} &= (\mathsf{CKerror} \cap \mathsf{fresh}) \cup (\mathsf{CKerror} \cap \textsf{non-fresh}) && (25)\\ &= (\mathsf{CKerror} \cap (\mathsf{correctDec} \cup \mathsf{wrongDec})) \cup (\mathsf{CKerror} \cap \textsf{non-fresh}). && (26) \end{aligned}$$

By the above we have shown that **CheckKey** correctly implements the key-checking oracle $\mathcal{KO}$. Moreover, we can now provide concrete bounds on the error probability ϵ in (20) by bounding CKerror.

Lemma 2.

$$\begin{aligned} \Pr[\mathsf{CKerror}] &= \Pr[\mathsf{CKerror} \cap \mathsf{fresh}] + \Pr[\mathsf{CKerror} \cap \textsf{non-fresh}] && (27)\\ &\le \Pr[\mathsf{correctDec}] + \Pr[\mathsf{wrongDec}] + \Pr[ck\textsf{-coll}_6] && (28)\\ &\le 2 \cdot \Pr[Q] + \mathbf{Adv}_{\Pi.\mathsf{Kdf}}^{ck\textsf{-coll}}(\mathcal{C}'''). && (29) \end{aligned}$$

Proof. For CKerror $\cap$ fresh, note that correctDec and wrongDec are mutually exclusive since $\mathcal{B}$ aborts as soon as one of them happens. Also, in the context of **CheckKey**, they are both sub-events of Q. Thus, $\Pr[\mathsf{CKerror} \cap \mathsf{fresh}] = \Pr[\mathsf{correctDec}] + \Pr[\mathsf{wrongDec}] \le 2 \cdot \Pr[Q]$.

If CKerror $\cap$ non-fresh happens in Game 6, then event ck-coll$_6$ must by definition have happened too. Hence $\Pr[\mathsf{CKerror} \cap \textsf{non-fresh}] \le \Pr[ck\textsf{-coll}_6]$. Furthermore, the bound $\Pr[ck\textsf{-coll}_6] \le \mathbf{Adv}_{\Pi.\mathsf{Kdf}}^{ck\textsf{-coll}}(\mathcal{C}''')$ follows from the same strategy used in the game hop from Game 2 to 3, and from Game 3 to 4. That is, we construct an algorithm $\mathcal{C}'''$ that plays the challenger in Game 6; once ck-coll$_6$ occurs in this game, then $\mathcal{C}'''$ has found a collision in Π.Kdf. □

Analysis of $\mathcal{B}$. Having shown that $\mathcal{B}$'s simulation of Game 6 is sound, we now turn to bounding $\mathcal{A}$'s advantage in Game 6 in terms of $\mathcal{B}$'s advantage in the ACCE security experiment.

Lemma 3.

$$\mathbf{Adv}_{\Pi+}^{\mathsf{G_6\text{-}auth}}(\mathcal{A}) \le \mathbf{Adv}_{\Pi}^{\mathsf{auth\text{-}(PFS)}}(\mathcal{B}). \qquad (30)$$

[5] Event correctDec can either happen legitimately ($\pi_i^s.ms = ms'$), or because of an error ($\pi_i^s.ms \ne ms'$). On the other hand, event wrongDec can only happen due to an error.

Proof. Since $\mathcal{B}$'s simulation of Game 6 is sound, and because the protocols Π^+ and Π have the same session identifier, it follows that $\mathcal{A}$ gets a session to accept maliciously in Game 6, if and only if the session accepts maliciously in the underlying ACCE security experiment in $\mathcal{B}$'s simulation. □

Lemma 4.

$$\mathbf{Adv}_{\Pi^+}^{\mathsf{G_6\text{-}chan}}(\mathcal{A}) \le \Pr[Q] \le 2 \cdot \mathbf{Adv}_{\Pi}^{\mathsf{chan\text{-}(PFS)}}(\mathcal{B}) + \frac{2qn_{\mathcal{P}}n_{\pi}}{2^{c\lambda}}. \tag{31}$$

Proof. The first inequality follows from Lemma 1. The proof of the second inequality amounts to a direct calculation based on conditional probabilities. Suppose $\mathcal{B}$ halted with output (π_i^s, b') in its ACCE security experiment, where π_i^s is some fresh session. By conditioning on whether event Q happened or not during $\mathcal{B}$'s simulation of Game 6 for $\mathcal{A}$, we get that $\mathcal{B}$'s probability of breaking the ACCE channel is:

$$\Pr[\pi_i^s.b = b'] = \Pr[\pi_i^s.b = b' \mid Q] \cdot \Pr[Q] + \Pr[\pi_i^s.b = b' \mid \overline{Q}] \cdot \Pr[\overline{Q}] \tag{32}$$

$$\overset{(a)}{=} \Pr[\pi_i^s.b = b' \mid Q] \cdot \Pr[Q] + \frac{1}{2}(1 - \Pr[Q]) \tag{33}$$

$$\overset{(b)}{=} \Big(\overbrace{\Pr[\pi_i^s.b = b' \mid Q \cap \mathsf{correctDec}]}^{=1} \cdot \Pr[\mathsf{correctDec} \mid Q] + \overbrace{\Pr[\pi_i^s.b = b' \mid Q \cap \mathsf{wrongDec}]}^{=0} \cdot \Pr[\mathsf{wrongDec} \mid Q]\Big) \cdot \Pr[Q] \tag{34}$$

$$+ \frac{1}{2}(1 - \Pr[Q])$$

$$= \Pr[\mathsf{correctDec} \mid Q] \cdot \Pr[Q] + \frac{1}{2}(1 - \Pr[Q]) \tag{35}$$

$$= \Pr[\mathsf{correctDec} \cap Q] - \frac{1}{2} \cdot \Pr[Q] + \frac{1}{2} \tag{36}$$

$$\overset{(c)}{=} (\Pr[Q] - \Pr[\mathsf{wrongDec} \cap Q]) - \frac{1}{2}\Pr[Q] + \frac{1}{2} \tag{37}$$

$$= \frac{1}{2}\Pr[Q] - \Pr[\mathsf{wrongDec} \cap Q] + \frac{1}{2} \tag{38}$$

$$\overset{(d)}{=} \frac{1}{2}\Pr[Q] - \Pr[\mathsf{wrongDec}] + \frac{1}{2} \tag{39}$$

$$\ge \frac{1}{2}\Pr[Q] - \frac{qn_{\mathcal{P}}n_{\pi}}{2^{c\lambda}} + \frac{1}{2}. \tag{40}$$

In (a) we used the fact that $\mathcal{B}$ outputs a random bit when Q does not happen, (b) and (c) used that $Q = \mathsf{correctDec} \cup \mathsf{wrongDec}$ and $\mathsf{correctDec} \cap \mathsf{wrongDec} = \emptyset$, and (d) used that $\mathsf{wrongDec} \subseteq Q$. We prove the final inequality as follows.

Let $\overline{b} = 1 - \pi_i^s.b$ and let (m_0, m_1) be the two messages associated to the pair (π_i^s, ms) in **CheckKey**. Since $m_{\overline{b}}$ is independent of the ciphertext C produced at line 11 of Algorithm 1, the probability that C decrypts to $m_{\overline{b}}$ at line 20 is statistically bounded by $2^{-c\lambda}$ for any key k. By taking the union bound over all parties, the number of sessions per party, and the number of random oracle calls, we get that $\Pr[\mathsf{wrongDec}] \leq qn_{\mathcal{P}}n_{\pi}/2^{c\lambda}$.

Solving (40) for $\Pr[Q]$ yields the second inequality in Lemma 4. □

Concluding the Proof of Theorem 1. Applying Lemmas 2, 3, and 4, we get that the right-hand side of Eq. (20) is bounded by

$$\mathbf{Adv}_{\Pi}^{\mathsf{auth\text{-}(PFS)}}(\mathcal{B}) + 6 \cdot \mathbf{Adv}_{\Pi}^{\mathsf{chan\text{-}(PFS)}}(\mathcal{B}) + \frac{6qn_{\mathcal{P}}n_{\pi}}{2^{c\lambda}} + \mathbf{Adv}_{\Pi.\mathsf{Kdf}}^{\mathsf{KDFcoll}}(\mathcal{C}'''). \tag{41}$$

By collecting all the probabilities from Game 0 to Game 6, and letting $\mathcal{C} = \max_{\mathbf{Adv}_{\Pi.\mathsf{Kdf}}^{\mathsf{KDFcoll}}}\{\mathcal{C}', \mathcal{C}'', \mathcal{C}'''\}$, the theorem follows.

4.4 Application to EAP-TLS and TLS Key Material Exporters

EAP [1] is a widely used authentication framework which defines a set of generic message formats and message flows. EAP is not a specific authentication mechanism on its own, but is instead used to encapsulate another concrete authentication protocol, like TLS, IKEv2 or IEEE 802.1X, known as a *method.* Each EAP method can additionally specify a way of generating keying material, known as *export keys*, both for internal and external use. For example, in EAP-TLS [33] the export key ek is derived as follows:

$$ek := \mathsf{tls.PRF}(ms, \text{“}\texttt{client EAP encryption}\text{”}, n_C \| n_S), \tag{42}$$

where ms is the master secret and n_C, n_S the nonces established during the TLS handshake. How export keys should be derived from the TLS handshake in settings outside of EAP is defined in RFC 5705: “Keying Material Exporters for Transport Layer Security (TLS)” [31]. Besides a different constant label string, RFC 5705 defines ek almost exactly as in (42). The only difference is that it also allows an extra context value aux to be added into the key derivation together with the nonces. For both EAP-TLS and RFC 5705 the security requirement on ek is that it be indistinguishable from random.

In order to apply Theorem 1 to EAP-TLS, we have to show that TLS is in fact a TLS-like ACCE protocol, using a session identifier that satisfies the requirements of the theorem. Since several works have already proven TLS to be ACCE secure, it only remains to demonstrate that the session identifier used in these prior analyses allowed for public session matching and contained the sessions' nonces.

As an example, in their analysis of TLS, Krawczyk, Paterson, and Wee [24] defined their session identifier to consist of the two first flows between the client and the server, in addition to the client's KEM-value (either a Diffie-Hellman share or the pre-master secret encrypted with the server's public RSA key). This session identifier includes the parties' nonces, and allows for public session matching since it only consists of public values. Thus, using the TLS analysis of Krawczyk et al. [24], we can apply Theorem 1 with Π = TLS, and Π^+ = EAP-TLS, in order to get the following result.

Corollary 1 (AKE Security of EAP-TLS).

$$\mathbf{Adv}^{\mathsf{AKE}}_{\mathsf{EAP\text{-}TLS}}(\mathcal{A}) \le 6 \cdot \mathbf{Adv}^{\mathsf{ACCE}}_{\mathsf{TLS}}(\mathcal{B}) + 3 \cdot \mathbf{Adv}^{\mathsf{KDFcoll}}_{\mathsf{tls.PRF}}(\mathcal{C}) + \frac{6qn_{\mathcal{P}}n_{\pi}}{2^{c\lambda}} + \frac{(n_{\mathcal{P}}n_{\pi})^2}{2^{\lambda+1}}, \quad (43)$$

where $\Pi.\mathsf{Kdf} = \mathsf{tls.PRF}$, *and all other quantities are defined as stated in Theorem 1.*

Remark 5. The KDF used in TLS is based on HMAC [23], and its KDF collision resistance follows from the (hash function) collision resistance of the underlying hash function used in HMAC (see Theorem 2, Appendix A).

Remark 6. JKSS [20] used *matching conversations* as their partnering mechanism in their analysis of TLS. Since matching conversations contain the parties' nonces and trivially allow for public session matching, it would seem like JKSS's analysis could also be used with Theorem 1 in order to establish Corollary 1.

However, there is a subtle technical difference between the ACCE model as defined in this paper and the ACCE model as defined by JKSS, stemming from the difference in choice of partnering mechanism. Specifically, in JKSS's definition of ACCE [19, Definition 11] one must forbid the adversary from issuing a Reveal query towards the server after it sent out its last message, but *before* the client to which it has a matching conversation received it. This is to avoid a trivial attack whereby the adversary re-encrypts the final message towards the client, getting it to accept maliciously (see [19, Remark 6] for further details)[6].

In contrast, the definition of ACCE used in this paper (in particular, Definition 3) allows all Reveal queries. It should be noted that the trivial attack in JKSS's model does not imply any actual weakness in TLS, but rather highlights a peculiarity of using matching conversations as the partnering mechanism when defining ACCE.

Remark 7. Brzuska et al. [8] defined their session identifier to consist of the parties' nonces and identities, together with the TLS *pre-master secret.* Unfortunately, basing the session identifier upon secret values does not in general allow for public session matching. For instance, if the KEM used in the TLS handshake was a *re-randomizable* encryption scheme [12, 30], then the choice of Brzuska et al. [8] would not allow for public session matching (see also [9] for further details).

Remark 8. Bhargavan et al. [5] showed that the *full* TLS protocol, including resumption and renegotiation, is vulnerable to an *unknown key-share* attack [7]. The attack allows an adversary to synchronize the master secret and nonces of two non-partnered sessions, leading them to derive the same channel key. While the attack carries over to EAP-TLS, it does not invalidate Corollary 1, since our model does not consider resumption nor renegotiation. However, it should be noted that this has been done for the sake of simplicity, not because of an

[6] This extra requirement was not included in the original published version [20], but was later added to the online version [19].

essential limitation in our analysis. Our result can be extended to incorporate features like renegotiation, resumption or ciphersuite and version negotiation, either by using the *multi-phase* ACCE model of Giesen et al. [16] or the *multi-ciphersuite* ACCE model of Bergsma et al. [4]. The former has been used to prove results on TLS with renegotiation [16], while the latter has been used to prove results on SSH and TLS with ciphersuite and version negotiation [4,14]. Since our proof uses the underlying ACCE protocol in an almost black-box way, by adopting one of the above models we would essentially "inherit" their corresponding results for EAP-TLS as well.

Acknowledgments. We would like to thank Colin Boyd and Britta Hale for helpful comments and discussions. Part of this work was done while Christina Brzuska was working for Microsoft Research, Cambridge, UK. Christina Brzuska is grateful to NXP Semiconductors for supporting her chair for IT Security Analysis. Håkon Jacobsen was hosted by Microsoft Research, Cambridge, UK, for parts of this work. Some of this work performed while Douglas Stebila was hosted by the Norwegian University of Science and Technology.

A KDF Collision Resistance of the TLS KDF

Let H be a hash function, and let $\overline{H}$ denote the HMAC function using H as its underlying hash function, namely

$$\overline{H}(k, m) \stackrel{\text{def}}{=} H\left(k \oplus \mathtt{opad} \| H(k \oplus \mathtt{ipad} \| m)\right), \tag{44}$$

where $\mathtt{ipad}$ and $\mathtt{opad}$ are distinct constants.

The TLS 1.2 KDF is defined as follows, where the variable t depends on how much keying material is needed.

$$\mathsf{tls.PRF}(ms, L, n) \stackrel{\text{def}}{=} \Big\|_{i=1}^{t} \overline{H}(ms, A(i) \| L \| n) \tag{45}$$

$$A(1) = \overline{H}(ms, n) \tag{46}$$

$$A(i) = \overline{H}(ms, A(i-1)) \tag{47}$$

In TLS 1.2, $L =$ "$\mathtt{key\ expansion}$" and $n = n_C \| n_S$, where and n_C, n_S are the client and server nonces, respectively. For simplicity, we write $S = L \| n$.

Theorem 2. *A KDF collision (Definition 9) in* $\mathsf{tls.PRF}$ *implies a collision in* H.

Proof. Suppose $\mathsf{tls.PRF}(ms, L, n) = \mathsf{tls.PRF}(ms', L, n)$, with $ms \neq ms'$. By (45) we specifically have that

$$\overline{H}(ms, A(1) \| S) = \overline{H}(ms', A'(1) \| S), \tag{48}$$

where $A'(1) = \overline{H}(ms', n)$. Expanding (48) using (44) we get:

$$\begin{aligned} H\left(ms \oplus \texttt{opad} \| H\left(ms \oplus \texttt{ipad} \| A(1) \| S\right)\right) \\ = \\ H\left(ms' \oplus \texttt{opad} \| H\left(ms' \oplus \texttt{ipad} \| A'(1) \| S\right)\right). \end{aligned} \tag{49}$$

Letting $X = H\left(ms \oplus \texttt{ipad} \| A(1) \| S\right)$ and $Y = H\left(ms' \oplus \texttt{ipad} \| A'(1) \| S\right)$ denote the "inner" hash function values, (49) becomes:

$$H(ms \oplus \texttt{opad} \| X) = H(ms' \oplus \texttt{opad} \| Y). \tag{50}$$

Since $ms \oplus \texttt{opad} \neq ms' \oplus \texttt{opad}$, it follows that $ms \oplus \texttt{opad} \| X$ and $ms' \oplus \texttt{opad} \| Y$ constitute a collision in H. □

Remark 9. The construction of tls.PRF in TLS 1.0/1.1 is different from that in TLS 1.2 (shown in Eq. (45)). In versions prior to TLS 1.2, tls.PRF is defined as $P_{\text{MD5}} \oplus P_{\text{SHA1}}$, where P_{MD5} and P_{SHA1} are equal to the right-hand side of Eq. (45) with $\overline{H}$ using MD5 and SHA1, respectively. Theorem 2 only applies to the construction used in TLS 1.2.

References

1. Aboba, B., Blunk, L.J., Vollbrecht, J.R., Carlson, J., Levkowetz, H.: Extensible authentication protocol. RFC 3748, RFC Editor, June 2004. https://tools.ietf.org/html/rfc3748
2. Bellare, M., Rogaway, P.: Entity authentication and key distribution. In: Stinson, D.R. (ed.) CRYPTO 1993. LNCS, vol. 773, pp. 232–249. Springer, Heidelberg (1994)
3. Bellare, M., Rogaway, P.: The security of triple encryption and a framework for code-based game-playing proofs. In: Vaudenay, S. (ed.) EUROCRYPT 2006. LNCS, vol. 4004, pp. 409–426. Springer, Heidelberg (2006)
4. Bergsma, F., Dowling, B., Kohlar, F., Schwenk, J., Stebila, D.: Multi-ciphersuite security of the Secure Shell (SSH) Protocol. In: Proceedings of the 2014 ACM SIGSAC Conference on Computer and Communications Security, CCS 2014, pp. 369–381. ACM, NY, USA, New York (2014)
5. Bhargavan, K., Delignat-Lavaud, A., Fournet, C., Pironti, A., Strub, P.Y.: Triple handshakes and cookie cutters: Breaking and fixing authentication over TLS. In: 2014 IEEE Symposium on Security and Privacy (SP), pp. 98–113, May 2014
6. Bhargavan, K., Fournet, C., Kohlweiss, M., Pironti, A., Strub, P.-Y., Zanella-Béguelin, S.: Proving the TLS handshake secure (As It Is). In: Garay, J.A., Gennaro, R. (eds.) CRYPTO 2014, Part II. LNCS, vol. 8617, pp. 235–255. Springer, Heidelberg (2014)
7. Blake-Wilson, S., Menezes, A.: Unknown key-share attacks on the station-to-station (STS) protocol. In: Imai, H., Zheng, Y. (eds.) PKC 1999. LNCS, vol. 1560, p. 154. Springer, Heidelberg (1999)
8. Brzuska, C., Fischlin, M., Smart, N.P., Warinschi, B., Williams, S.C.: Less is more: relaxed yet composable security notions for key exchange. Int. J. Inf. Secur. **12**(4), 267–297 (2013)

9. Brzuska, C., Fischlin, M., Warinschi, B., Williams, S.C.: Composability of Bellare-Rogaway key exchange protocols. In: Proceedings of the 18th ACM Conference on Computer and Communications Security, CCS 2011, pp. 51–62. ACM, NY, USA, New York (2011)
10. Brzuska, C., Smart, N.P., Warinschi, B., Watson, G.J.: An analysis of the EMV channel establishment protocol. In: Sadeghi, A., Gligor, V.D., Yung, M. (eds.) 2013 ACM SIGSAC Conference on Computer and Communications Security, CCS 2013, Berlin, Germany, 4–8 November 2013, pp. 373–386. ACM (2013)
11. Canetti, R., Krawczyk, H.: Analysis of key-exchange protocols and their use for building secure channels. In: Pfitzmann, B. (ed.) EUROCRYPT 2001. LNCS, vol. 2045, p. 453. Springer, Heidelberg (2001)
12. Canetti, R., Krawczyk, H., Nielsen, J.B.: Relaxing chosen-ciphertext security. In: Boneh, D. (ed.) CRYPTO 2003. LNCS, vol. 2729, pp. 565–582. Springer, Heidelberg (2003)
13. Diffie, W., Van Oorschot, P.C., Wiener, M.J.: Authentication and authenticated key exchanges. Des. Codes Crypt. **2**(2), 107–125 (1992)
14. Dowling, B., Stebila, D.: Modelling ciphersuite and version negotiation in the TLS protocol. In: Foo, E., Stebila, D. (eds.) ACISP 2015. LNCS, vol. 9144, pp. 270–288. Springer, Heidelberg (2015)
15. Fischlin, M., Günther, F.: Multi-stage key exchange and the case of Google's QUIC protocol. In: Ahn, G., Yung, M., Li, N. (eds.) Proceedings of the 2014 ACM SIGSAC Conference on Computer and Communications Security, Scottsdale, AZ, USA, 3–7 November 2014, pp. 1193–1204. ACM (2014)
16. Giesen, F., Kohlar, F., Stebila, D.: On the security of TLS renegotiation. In: Sadeghi, A., Gligor, V.D., Yung, M. (eds.) 2013 ACM SIGSAC Conference on Computer and Communications Security, CCS 2013, Berlin, Germany, 4–8 November 2013, pp. 387–398. ACM (2013)
17. He, C., Sundararajan, M., Datta, A., Derek, A., Mitchell, J.C.: A modular correctness proof of IEEE 802.11i and TLS. In: Proceedings of the 12th ACM Conference on Computer and Communications Security, CCS 2005, pp. 2–15. ACM, NY, USA, New York (2005)
18. Hofheinz, D., Kiltz, E.: Secure hybrid encryption from weakened key encapsulation. In: Menezes, A. (ed.) Advances in Cryptology - CRYPTO 2007. LNCS, vol. 4622, pp. 553–571. Springer, Berlin, Heidelberg (2007)
19. Jager, T., Kohlar, F., Schäge, S., Schwenk, J.: On the security of TLS-DHE in the standard model. Cryptology ePrint Archive, report 2011/219(2011). https://eprint.iacr.org/2011/219
20. Jager, T., Kohlar, F., Schäge, S., Schwenk, J.: On the security of TLS-DHE in the standard model. In: Safavi-Naini, R., Canetti, R. (eds.) CRYPTO 2012. LNCS, vol. 7417, pp. 273–293. Springer, Heidelberg (2012)
21. Kohlar, F., Schäge, S., Schwenk, J.: On the security of TLS-DH and TLS-RSA in the standard model. Cryptology ePrint Archive, report 2013/367 (2013). https://eprint.iacr.org/2013/367
22. Kohlweiss, M., Maurer, U., Onete, C., Tackmann, B., Venturi, D.: (De-) constructing TLS. Cryptology ePrint Archive, report 2014/020 (2014). https://eprint.iacr.org/2014/020
23. Krawczyk, H., Bellare, M., Canetti, R.: HMAC: Keyed-Hashing for Message Authentication. RFC 2104 (Informational), February 1997. https://tools.ietf.org/html/rfc2104

24. Krawczyk, H., Paterson, K.G., Wee, H.: On the security of the TLS protocol: A systematic analysis. In: Canetti, R., Garay, J. (eds.) Advances in Cryptology - CRYPTO 2013. LNCS, vol. 8042, pp. 429–448. Springer, Berlin, Heidelberg (2013)
25. Küsters, R., Tuengerthal, M.: Composition theorems without pre-established session identifiers. In: Chen, Y., Danezis, G., Shmatikov, V. (eds.) Proceedings of the 18th ACM Conference on Computer and Communications Security, CCS 2011, Chicago, Illinois, USA, 17–21 October 2011, pp. 41–50. ACM (2011)
26. Li, Y., Schäge, S., Yang, Z., Kohlar, F., Schwenk, J.: On the security of the pre-shared key ciphersuites of TLS. In: Krawczyk, H. (ed.) PKC 2014. LNCS, vol. 8383, pp. 669–684. Springer, Heidelberg (2014)
27. Lychev, R., Jero, S., Boldyreva, A., Nita-Rotaru, C.: How secure and quick is quic? provable security and performance analyses. In: 2015 IEEE Symposium on Security and Privacy, SP 2015, San Jose, CA, USA, 17–21 May 2015, pp. 214–231. IEEE Computer Society (2015)
28. Maurer, U., Renner, R.: Abstract cryptography. In: Chazelle, B. (ed.) Proceedings of Innovations in Computer Science - ICS 2010, Tsinghua University, Beijing, China, 7–9 January 2011, pp. 1–21. Tsinghua University Press (2011)
29. Morrissey, P., Smart, N.P., Warinschi, B.: A modular security analysis of the TLS handshake protocol. In: Pieprzyk, J. (ed.) Advances in Cryptology - ASIACRYPT 2008. LNCS, vol. 5350, pp. 55–73. Springer, Berlin Heidelberg (2008)
30. Prabhakaran, M., Rosulek, M.: Rerandomizable RCCA encryption. In: Menezes, A. (ed.) CRYPTO 2007. LNCS, vol. 4622, pp. 517–534. Springer, Heidelberg (2007)
31. Rescorla, E.: Keying material exporters for transport layer security (TLS). RFC 5705, RFC Editor, March 2010. https://tools.ietf.org/html/rfc5705
32. Shoup, V.: Sequences of games: a tool for taming complexity in security proofs. Cryptology ePrint Archive, report 2004/332 (2004). https://eprint.iacr.org/2004/332
33. Simon, D., Aboba, B., Hurst, R.: The EAP-TLS Authentication Protocol. RFC 5216, RFC Editor, March 2008

Valiant's Universal Circuit is Practical

Ágnes Kiss(✉) and Thomas Schneider

TU Darmstadt, Darmstadt, Germany
{agnes.kiss,thomas.schneider}@crisp-da.de

Abstract. Universal circuits (UCs) can be programmed to evaluate any circuit of a given size k. They provide elegant solutions in various application scenarios, e.g. for private function evaluation (PFE) and for improving the flexibility of attribute-based encryption (ABE) schemes. The optimal size of a universal circuit is proven to be $\Omega(k \log k)$. Valiant (STOC'76) proposed a size-optimized UC construction, which has not been put in practice ever since. The only implementation of universal circuits was provided by Kolesnikov and Schneider (FC'08), with size $\mathcal{O}(k \log^2 k)$.

In this paper, we refine the size of Valiant's UC and further improve the construction by (at least) $2k$. We show that due to recent optimizations and our improvements, it is the best solution to apply in the case for circuits with a constant number of inputs and outputs. When the number of inputs or outputs is linear in the number of gates, we propose a more efficient hybrid solution based on the two existing constructions. We validate the practicality of Valiant's UC, by giving an example implementation for PFE using these size-optimized UCs.

Keywords: Universal circuit · Size-optimization · Private function evaluation

1 Introduction

Any computable function $f(x)$ can be represented as a Boolean circuit with input bits $x = (x_1, \ldots, x_u)$. Universal circuits (UCs) are programmable circuits, which means that beyond the true u inputs, they receive $p = (p_1, \ldots, p_m)$ program bits as further inputs. By means of these program bits, the universal circuit is programmed to evaluate the function, such that $UC(x,p) = f(x)$. The advantage of universal circuits in general is that one can apply the same UC for computing different functions of the same size. An analogy between universal circuits and a universal Turing machine allows to turn any function into data in the form of a program description. Thus, the size-depth problem of UCs can be related to the time-space problem for Turing machines [Val76].

Efficient constructions considering both the size and the depth of the UC were proposed. The first approach was the optimization of the size by Valiant [Val76], resulting in a construction with asymptotically optimal size $\mathcal{O}(k \log k)$ and depth $\mathcal{O}(k)$, where k denotes the size of the simulated circuits. The second optimization

M. Fischlin and J.-S. Coron (Eds.): EUROCRYPT 2016, Part I, LNCS 9665, pp. 699–728, 2016.
DOI: 10.1007/978-3-662-49890-3_27

was proposed with respect to the UC depth in [CH85], where a construction with linear depth $\mathcal{O}(d)$ in the simulated circuit depth d and size $\mathcal{O}(\frac{k^3 d}{\log k})$ was designed. In this paper, due to the applications that we revisit in Sect. 1.2, e.g., diagnostic programs, blinded policies and database queries, we concentrate on the existing size-optimized UCs and note, that the asymptotically optimal size is $\Omega(k \log k)$ [Val76, Weg87].

The most prominent application of universal circuits is the evaluation of private functions based on *secure function evaluation* (SFE) or *secure two-party computation.* SFE enables two parties P_1 and P_2 to evaluate a publicly known function $f(x, y)$ on their private inputs x and y, ensuring that none of the participants learns anything about the other participant's input. SFE ensures that both P_1 and P_2 learn the correct result of the evaluation. Many secure computation protocols use Boolean circuits for representing the desired functionality, such as Yao's garbled circuit protocol [Yao86, LP09a] and the GMW protocol [GMW87]. In some applications the function itself should be kept secret. This setting is called *private function evaluation* (PFE), where we assume that only one of the parties P_1 knows the function $f(x)$, whereas the other party P_2 provides the input to the private function. P_2 learns no information about f besides the size of the circuit defining the function and the number of inputs and outputs.

PFE can be reduced to SFE [AF90, SYY99, Pin02, KS08b] by securely evaluating a UC that is programmed by P_1 to evaluate the function f on P_2's input x. Thus, P_1 provides the program bits for the UC and P_2 provides his private input x into an SFE protocol that computes a UC. The complexity of PFE in this case is determined mainly by the complexity of the UC construction. The security follows from that of the SFE protocol that is used to evaluate the UC. If the SFE protocol is secure against semi-honest, covert or malicious adversaries, then the PFE protocol is secure in the same adversarial setting.

1.1 Related Work on Universal Circuits and Private Function Evaluation

Universal Circuits. Valiant presented an asymptotically optimal universal circuit construction with size $\approx 4.75(u+v+k^*)\log_2(u+v+k^*)$ [Val76], relying on edge-universal graphs. u, k and v denote the respective number of inputs, gates and outputs in the simulated circuit, and k^* is the number of gates in the equivalent fanout-2 circuit, with $k \leq k^* \leq 2k+v$. Valiant's size-optimized UC construction was recapitulated in [Weg87, Sect. 4.8]. However, Valiant's construction has been considered to be mostly a proof of existence of a universal circuit, whereas details needed for the practical realization, e.g., how to derive the program for the UC are left open. Kolesnikov and Schneider proposed a UC construction with size $\approx 0.75k\log_2^2 k + 2.25k\log_2 k + k\log_2 u + (0.5k+0.5v)\log_2 v$ [KS08b, Sch08]. They present the first implementation of PFE using UCs by extending the Fairplay secure computation framework [MNPS04]. Some building blocks of this construction are of interest, but due to its asymptotically non-optimal size, we show in Sect. 3.2 that Valiant's UC construction results in smaller UCs for circuits in the most general case. The UC constructions from [Val76, KS08b] were generalized for circuits consisting of gates with more than two inputs in [SS08]. In this paper, we show the practicality of Valiant's UC construction.

In concurrent and independent work [LMS16], Lipmaa et al. also bring the same UC construction to practice. They detail a k-way recursive construction for UCs, instantiate it for $k \in \{2, 4\}$ as in [Val76], and descrease its *total* number of gates compared to that of Valiant's construction. However, in contrast to our optimizations, their number of AND gates is exactly the same and therefore their improvement does not affect PFE with UC, when XOR gates are evaluated for free [KS08a]. Currently their implementation for generating and programming UCs supports the 2-way recursive construction, the same construction that we study and realize in practice in this work.

Private Function Evaluation. In [KM11], Katz and Malka presented an approach for PFE that does not rely on UCs. They use (singly) homomorphic public-key encryption as well as a symmetric-key encryption scheme and achieve constant-round PFE with linear communication complexity. However, the number of public-key operations is linear in the circuit size and due to the gap between the efficiency of public-key and symmetric-key operations, this results in a less efficient protocol for circuits with reasonable size. Their protocol is secure against semi-honest adversaries and uses Yao's garbled circuit technique [Yao86]. Mohassel and Sadeghian consider PFE with semi-honest adversaries in [MS13]. Their generic PFE framework can be instantiated with different secure computation protocols. The first version uses homomorphic encryption with which they achieve linear complexity in the circuit size and the second alternative relies solely on oblivious transfers (OT), that results in a method with $\mathcal{O}(k \log k)$ symmetric-key operations, where k denotes the circuit size. The OT-based construction is more desirable in practice, since using OT extension, the number of expensive public-key operations can significantly be reduced, s.t. it is independent of the number of OTs [IKNP03, ALSZ13]. The asymptotical complexity of the OT-based construction of [MS13] and Valiant's UCs for PFE is the same, and therefore we compare these solutions for PFE in more detail in Sect. 4.2. Mohassel et al. extend the framework from [MS13] to malicious adversaries in [MSS14] and show that an actively secure PFE framework with linear complexity $\mathcal{O}(k)$ is feasible, using singly homomorphic encryption.

1.2 Applications of Universal Circuits

Universal circuits have several applications, which we summarize in this section.

Private Function Evaluation. As mentioned before, UCs can be used to securely evaluate a private function using a generic secure computation protocol. [CCKM00] shows an application for secure computation, where evaluating UCs or other PFE protocols would ensure privacy: when *autonomous mobile agents* migrate between several distrusting hosts, the privacy of the inputs of the hosts is achieved using SFE, while privacy of the mobile agent's code can be guaranteed with PFE. Privacy-preserving *credit checking* using garbled circuits is described in [FAZ05]. Their original scheme cannot represent any policy, though by evaluating a UC, their scheme can be extended to more complicated credit checking policies. [OI05] show a method to *filter remote streaming data* obliviously, using secret keywords and their combinations. Their scheme can additionally preserve data privacy by using PFE to search the matching data with a private search function. Privacy-preserving evaluation of *diagnostic programs* was considered in [BPSW07], where the owner of the program does not want to reveal the diagnostic method and the user does not want to reveal his data. Example applications for such programs include medical systems [BFK+09] and remote software fault diagnosis, where in both cases the function and the user's input are desired to be handled privately. In the protocol presented in [BPSW07], the diagnostic programs are represented as binary decision trees or branching programs which can easily be converted into a Boolean circuit representation and evaluated using PFE based on universal circuits. Besides, PFE can be applied to create *blinded policy evaluation protocols* [FAL06,FLA06]. [FAL06] utilizes UCs for so-called oblivious circuit policies and [DDKZ13] for hiding the circuit topology in order to create one-time programs. Since PFE using UCs utilizes general secure computation protocols, it is possible to outsource the function and the data to two or multiple servers (using XOR secret sharing) and then run private queries on these. This is not directly possible with other PFE protocols, e.g., with the protocol presented in [KM11] or the homomorphic encryption-based protocols from [MS13,MSS14].

Beyond Private Function Evaluation. Besides being used for PFE, UCs can be applied in various other scenarios. Efficient *verifiabile computation* on encrypted data was studied in [FGP14]. A verifiable computation scheme was proposed for arbitrary computations and a UC is required to hide the function. [GGPR13] make use of universal circuits for reducing the verifier's preprocessing step. In [GHV10], a *multi-hop homomorphic encryption* scheme is proposed that also uses a universal circuit evaluator to achieve the privacy of the function. When the common reference string is dependent on a function that the verifier is interested in outsourcing, then the function description can be provided as input to a UC of appropriate size. In [PKV+14,FVK+15], universal circuits are used for hiding *queries in database management systems* (DBMSs). The Blind Seer DBMS was improved in [PKV+14] by making use of a simpler UC for evaluating queries, which does not hide the circuit topology. The authors mention that in case the topology of the SQL formula and the circuit have to be kept private, a UC can be utilized. As described in [Att14], the *Attribute-Based Encryption* (ABE) schemes for some polynomial-size circuits can be turned into ciphertext-policy ABE by using universal circuits. The ABE scheme of [GGHZ14] also uses UCs.

1.3 Outline and Our Contributions

In Sect. 2, we revisit the two existing size-optimized UC constructions of [Val76,KS08b]. We put an emphasis on the asymptotically size-optimal method proposed by Valiant in [Val76]. This complex construction makes use of an internal graph representation and programs a so-called edge-universal graph. However, the algorithm for programming a universal circuit is not explicitly described and in the presence of the included optimizations is not straightforwardly applicable. In Sect. 2.1, we recapitulate Valiant's recursive edge-universal graph construction and describe how the construction of UCs can be reduced to this problem. In Sect. 2.2, we briefly summarize the main building blocks of the UC construction of [KS08b].

Optimized Size and Depth of Valiant's UC Construction: In Sect. 3, we elaborate on the concrete size of Valiant's UC construction. We refine upper and lower bounds for the size of the edge-universal graph and approximate a closed formula with $\leq 2\,\%$ deviation from the actual size in Sect. 3.1. We include two optimizations detailed in Sect. 3.2, achieving altogether a linear improvement of at least $4u+4v+2k$. We give hybrid constructions for cases with many inputs and outputs in the same section. In Sect. 3.2, we compare the refined concrete size and the depth of Valiant's construction with that of [KS08b] and conclude the advantage of Valiant's method (potentially using building blocks from [KS08b]).

Valiant's Size-Optimized UC Construction in Practice: In Sect. 4, we detail the steps of our algorithm for a practical realization of Valiant's UC construction and provide an example application for PFE. We describe the internal representations and the algorithms in our UC compiler in Sect. 4.1, along with detailed implementations of universal gates and switches. We compare our resulting PFE with the OT-based protocol from [MS13] in Sect. 4.2. We show concrete example circuits and elaborate on the number of symmetric-key operations and the performance of our UC compiler.

2 Existing Universal Circuit Constructions

In this section, we summarize the two size-optimized universal circuit constructions: of [Val76] in Sect. 2.1 and of [KS08b] in Sect. 2.2.

2.1 Valiant's Universal Circuit Construction

In this section, we describe Valiant's edge-universal graph construction for graphs for which all nodes have at most one incoming and at most one outgoing edge and detail how two such graphs can be used for constructing universal circuits [Val76].

Edge-Universal Graphs. $G = (V, E)$ is a directed graph with the set of nodes $V = \{1, \ldots, n\}$ and the set of edges $E \subseteq V \times V$. A directed graph has *fanin* or *fanout* ℓ if each of its nodes has at most ℓ edges directed into or out of it, respectively. $\Gamma_\ell(n)$ denotes the set of all acyclic directed graphs with n nodes and fanin and fanout ℓ. Further on, we require a labelling of the nodes in a topological order, i.e., $i > j$ implies that there is no directed path from i to j. In a graph in $\Gamma_\ell(n)$, a topological ordering can always be found with computational complexity $\mathcal{O}(n + \ell n)$.

An *edge-embedding* of graph $G = (V, E)$ into $G' = (V', E')$ is a mapping that maps V into V' one-to-one, with possible additional nodes in V', and E into directed paths in E', such that they are pairwise edge-disjoint, i.e., an edge can be used only in one path. A graph G' is *edge-universal* for $\Gamma_\ell(n)$ if it has distinguished poles $\{p_1, \ldots, p_n\} \subseteq V'$ and every graph $G \in \Gamma_\ell(n)$ with node set $V = \{1, \ldots, n\}$ can be edge-embedded into G' by a mapping φ^G such that $\varphi^G : i \mapsto p_i$ and $\varphi^G : (i, j) \mapsto \{\text{path from pole } p_i \text{ to pole } p_j\}$ for each $i, j \in V$.

Here, we recapitulate Valiant's construction for acyclic edge-universal graph for $\Gamma_1(n)$, denoted by U_n, that has fewer than $2.5n \log_2 n$ nodes, fanin and fanout 2 and poles with fanin and fanout 1. Valiant presents another edge-universal graph construction with a lower multiplicative constant $2.375n \log_2 n$. We omit that version of the algorithm for two reasons: firstly, our aim is to show the practicality of Valiant's approach and secondly, including all the optimizations even in the simpler construction is a challenging task in practice. The more efficient algorithm uses four subgraphs instead of two at each recursion and utilizes a skeleton with a more complex structure. For more details on this improved algorithm, the reader is referred to [Val76, LMS16]. We leave showing the practicality of the improved method as future work.

Valiant's Edge-Universal Graph Construction for $\Gamma_1(n)$ Graphs: The edge-universal graph for $\Gamma_1(n)$, denoted by U_n, is constructed with poles $\{p_1, \ldots, p_n\}$ with fanin and fanout 1, which are connected according to the skeleton shown in Figs. 1a and b. The poles are emphasized as special nodes with squares, and the additional nodes are shown as circles. The recursive construction works as follows: the nodes denoted by $\{q_1, \ldots, q_{\lceil \frac{n-2}{2} \rceil}\}$ and $\{r_1, \ldots, r_{\lfloor \frac{n-2}{2} \rfloor}\}$ are considered as the poles of two smaller edge-universal graphs called subgraphs $Q_{\lceil \frac{n-2}{2} \rceil}$ and $R_{\lfloor \frac{n-2}{2} \rfloor}$, respectively, that are otherwise not shown. Since they are poles of the two subgraphs with such a skeleton but not of U_n, they will have at most the allowed fanin and fanout 2: they inherit one incoming and one outgoing edge from the outer skeleton, and at most one incoming and one outgoing edge from the subgraph. $Q_{\lceil \frac{n-2}{2} \rceil}$ (and $R_{\lfloor \frac{n-2}{2} \rfloor}$) is then constructed similarly: the skeleton is completed and two smaller graphs with sizes $\lceil \frac{\lceil \frac{n-2}{2} \rceil - 2}{2} \rceil$ and $\lfloor \frac{\lceil \frac{n-2}{2} \rceil - 2}{2} \rfloor$ (and sizes $\lceil \frac{\lfloor \frac{n-2}{2} \rfloor - 2}{2} \rceil$ and $\lfloor \frac{\lfloor \frac{n-2}{2} \rfloor - 2}{2} \rfloor$) are constructed. For starting off the recursion, U_1 is a graph with a single pole while U_2 and U_3 are graphs with two and three connected poles, respectively. Valiant gives special constructions for U_4, U_5 and U_6 and shows that it is possible to obtain the respective edge-universal graphs with altogether 3, 7 and 9 additional nodes, respectively, as shown in Figs. 1c, d, and e.

We recapitulate the proof from [Val76] that U_n is edge-universal for $\Gamma_1(n)$, such that any graph with n nodes and fanin and fanout 1 can be edge-embedded into U_n. According to the definition of edge-embedding, it has to be shown that given any $\Gamma_1(n)$ graph G with set of edges E, for any $(i, j) \in E$ and $(k, l) \in E$

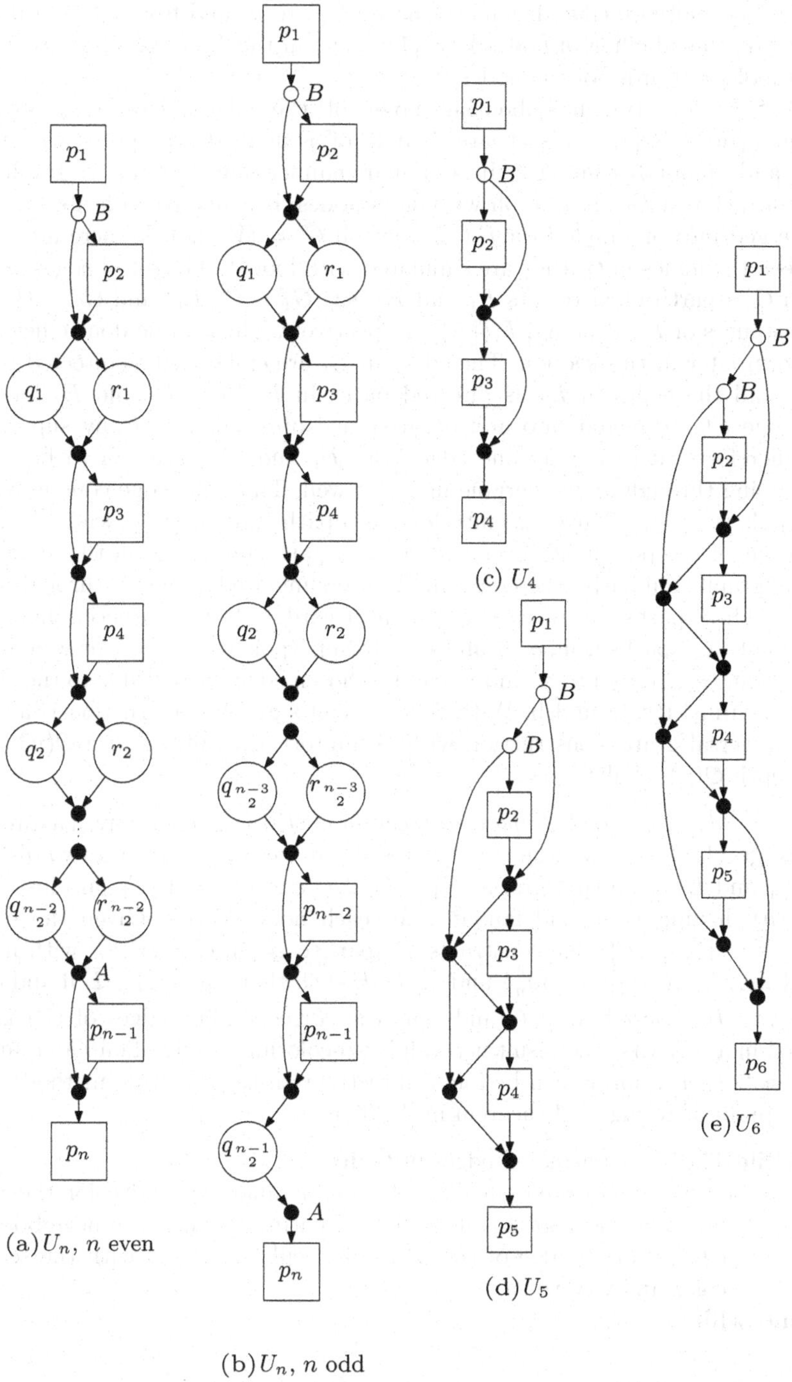

Fig. 1. Skeleton of Valiant's edge-universal graph and optimized cases.

we can find pairwise edge-disjoint paths from p_i to p_j and from p_k to p_l in U_n. As before, the labelling of nodes $V = \{1, \dots, n\}$ in the $\Gamma_1(n)$ graph is according to a topological order of the nodes.

Firstly, each two neighbouring poles of the edge-universal graph, p_{2s} and p_{2s+1} for $s \in \{1, \dots, \lceil \frac{n}{2} \rceil\}$, are thought of as merged superpoles, with their fanin and fanout becoming 2. In a similar manner, any $G \in \Gamma_1(n)$ graph can be regarded as a $\Gamma_2(\lceil \frac{n}{2} \rceil)$ graph with supernodes, i.e. each pair $(2s, 2s+1)$ will be merged into one node in a $\Gamma_2(\lceil \frac{n}{2} \rceil)$ graph $G' = (V', E')$. If there are edges between the nodes in G, they are simulated with loops[1]. The set of edges of this graph G is partitioned to sets E_1 and E_2, s.t. $G_1 = (V, E_1)$ and $G_2 = (V, E_2)$ are instances of $\Gamma_1(\lceil \frac{n}{2} \rceil)$ and $\Gamma_1(\lfloor \frac{n}{2} \rfloor)$, respectively. This can be done efficiently, as shown later in this section. The edges in E_1 are embedded as directed paths in Q, and the edges in E_2 as directed paths in R. Both E_1 and E_2 have at most one edge directed into and at most one directed out of any supernode and therefore, there is only one edge from E_1 and one from E_2 to be simulated going through any superpole in U_n as well. Thus, the edge coming into a superpole (p_{2s}, p_{2s+1}) in E_1 is embedded as a path through q_{s-1}, while the edge going out of the pole in E_1 is embedded as a path through q_s in the appropriate subgraph. Similarly, the edges in E_2 are simulated as edges through r_{s-1} and r_s. These paths can be chosen disjoint according to the induction hypothesis. Finally, the paths from q_{s-1} and r_{s-1} to superpole (p_{2s-1}, p_{2s}) as well as the paths from (p_{2s-1}, p_{2s}) to q_s and r_s can be chosen edge-disjoint due to the skeleton shown in Figs. 1a and b. With this, Valiant's graph construction is a valid edge-universal graph construction with asymptotically optimal size $\mathcal{O}(n \log n)$, and depth $\mathcal{O}(n)$ [Val76].

Valiant's Edge-Universal Graph Construction for $\Gamma_2(n)$ Graphs: Given a directed acyclic graph $G \in \Gamma_2(n)$, the set of edges E can be separated into two distinct sets E_1 and E_2, such that graphs $G_1 = (V, E_1)$ and $G_2 = (V, E_2)$ are instances of $\Gamma_1(n)$, having fanin and fanout 1 for each node [Val76]. Given the set of nodes $V = \{1, \dots, n\}$, one constructs a bipartite graph $\overline{G} = (\overline{V}, \overline{E})$ with nodes $\overline{V} = \{m_1, \dots, m_n, m'_1, \dots, m'_n\}$ and edges $\overline{E}$ such that $(m_i, m'_j) \in \overline{E}$ if and only if $(i, j) \in E$. The edges of $\overline{G}$ and thus the corresponding edges of G can be colored in a way that the result is a valid two-coloring. Having fanin and fanout at most 2, such coloring can be found directly with the following method, used in the proof of Kőnig-Hall theorem in [LP09b]:

1: **while** There are uncolored edges in $\overline{G}$ **do**
2: Choose an uncolored edge $e = (m_i, m'_j)$ randomly and color the path or cycle that contains it in an alternating manner: the neighbouring edge(s) of an edge of the first color will be colored with the second color and vice versa.
3: **end while**

[1] We note that these G' graphs are constructed from the original $\Gamma_1(n)$ graph G in order to define the correct embedding. Therefore, they are not required to be acyclic.

This coloring can be performed in $\mathcal{O}(n)$ steps and it defines the edges in E_1 and E_2, s.t. E_1 contains the edges colored with color one and E_2 the ones with color two and $G_1 = (V, E_1)$ and $G_2 = (V, E_2)$ (cf. full version [KS16]).

With this method, the problem of constructing edge-universal graphs for $\Gamma_2(n)$ can be reduced to the $\Gamma_1(n)$ construction. After constructing two edge-universal graphs for $\Gamma_1(n)$ (i.e. $U_{n,1}$ and $U_{n,2}$), their poles are merged and an edge-universal graph for $\Gamma_2(n)$ is obtained. The merged poles now have fanin and fanout 2, since the poles of $U_{n,1}$ and $U_{n,2}$ previously had fanin and fanout 1. E_1 can then be edge-embedded using the edges of $U_{n,1}$ and E_2 using the edges of $U_{n,2}$.

Universal Circuits. We now describe how to construct UCs by means of Valiant's edge-universal graph construction for $\Gamma_2(n)$ graphs [Val76]. Our goal is to obtain an acyclic circuit built from special gates that simulate any acyclic Boolean circuit with u inputs, v outputs and k gates. In the circuit, the inputs of the gates are either connected to an input variable, to the output of another gate or are assigned a fixed constant. Due to the nature of Valiant's edge-universal graph construction, we have two restrictions on the original circuit. Firstly, all the gates must have at most two inputs and secondly, the fanout of inputs and gates must be at most 2, i.e., each input of the circuit and each output of any gate can only be the input of at most two later gates. This is necessary in order to guarantee that the graph of the original circuit has fanin and fanout 2. We note that the first restriction was present in case of the construction in [KS08b] as well, but the output of any input or any gate could be used multiple times. However, it was proven in [Val76] that the general case, where the fanout of the circuit can be any integer $m \geq 2$, can be transformed to the special case when $m \leq 2$ by introducing copy gates, where the resulting circuit will have k^* gates with $k \leq k^* \leq 2k + v$, where k denotes the number of gates and v the number of outputs in the circuit. We detail how this can be done in Sect. 4.1.

After this transformation, given a circuit C with u inputs, v outputs and k^* gates with fanin and fanout 2, the graph of C, denoted by G^C consists of a node for each gate, input and output variable and thus is in $\Gamma_2(u + v + k^*)$. The wires of circuit C are represented by edges in G^C. A *topological ordering* of the gates is chosen, which ensures that gate g_i has no inputs that are outputs of a later gate g_j, where $j > i$. The inputs and the outputs can be ordered arbitrarily within themselves as long as the inputs are kept before the topologically ordered gates and the outputs after them. Even though the output nodes cause an overhead in Valiant's UC, they are required to fully hide the topology of the circuit in the corresponding universal circuit. If, one can observe which gates provide the output of the computation, it might reveal information about the structure of the circuit, e.g. how many times is the result of an output gate used after being calculated. We ensure by adding nodes corresponding to the outputs that the last v nodes in U_{u+v+k^*} are the ones providing the outputs. We note that our understanding of universal circuits here slightly differs from Valiant's, since he constructs U_{u+k^*} [Val76].

Therefore, after obtaining G^C a Γ_2 edge-universal graph U_{u+v+k^*} is constructed, into which G^C is edge-embedded. Valiant shows in [Val76] how to obtain the universal circuit corresponding to U_{u+v+k^*} and how to program it according to the edge-embedding of G^C. Firstly, the first u poles become inputs, the next k^* poles are so-called universal gates, and the last v poles are outputs in the universal circuit. A *universal gate* denoted by $U(\mathrm{in}_1, \mathrm{in}_2; c_0, c_1, c_2, c_3)$, can compute any function with two inputs in_1 and in_2 and four control bits c_0, c_1, c_2 and c_3 as in Eq. 1.

$$\mathrm{out}_1 = c_0\overline{\mathrm{in}_1\mathrm{in}_2} \oplus c_1\overline{\mathrm{in}_1}\mathrm{in}_2 \oplus c_2\mathrm{in}_1\overline{\mathrm{in}_2} \oplus c_3\mathrm{in}_1\mathrm{in}_2. \tag{1}$$

The rest of the nodes of the edge-universal graph are translated into *universal switches* or *X gates*, denoted by $(\mathrm{out}_1, \mathrm{out}_2) = X(\mathrm{in}_1, \mathrm{in}_2; c)$ that are defined by one control bit c and return the two input values either in the same or in reversed order as in Eq. 2.

$$\mathrm{out}_1 = \overline{c}\,\mathrm{in}_1 \oplus c\,\mathrm{in}_2, \qquad \mathrm{out}_2 = c\,\mathrm{in}_1 \oplus \overline{c}\,\mathrm{in}_2. \tag{2}$$

The programming of the universal circuit means specifying the control bit of each universal switch and the four control bits of each universal gate. The universal gates are programmed according to the simulated gates in C and the universal switches according to the paths defined by the edge-embedding of the graph of the circuit G^C in the edge-universal graph U_{u+v+k^*}. Depending on if the path takes the same direction during the embedding (e.g. arrives from the left and continues on the left) or changes its direction at a given node (e.g. arrives from the left and continues on the right), the control bit of the universal switch can be programmed accordingly. In Sect. 4.1, we detail our concrete method for programming the universal circuit and discuss efficient implementations of universal gates and switches.

2.2 Universal Circuit Construction from [KS08b]

The universal circuit construction from [KS08b] is built from three main building blocks (cf. full version [KS16]) that we summarize in this section. The construction uses efficient building blocks for hiding the wiring of the u inputs and v outputs, using the fact that the maximum number of inputs to a circuit with k gates is $2k$ and the maximum number of outputs is k. A recursive building block with size $\mathcal{O}(k \log^2 k)$ is constructed for hiding the wiring between the gates.

For hiding the input wiring, a *selection block* $S^u_{2k \geq u}$ is used, i.e., a programmable block that selects for $2k$ outputs one of $u \leq 2k$ inputs. This means that with the u inputs of circuit C, it can be programmed to assign the output wires according to the original structure of C and assign duplicates to the rest of the wires. The authors show an efficient implementation of selection blocks with size $\mathcal{O}(k \log k)$ and depth $\mathcal{O}(k)$ with a small constant factor [KS08b].

For hiding the output wiring, the authors use a smaller selection block. We note that the usage of their so-called *truncated permutation block* is enough to program the output wires according to the original topology of C as no duplicates can occur. This truncated permutation block $TP_v^{k \geq v}$ permutes a subset of the maximal k inputs to the $v \leq k$ outputs. An efficient construction of size $\mathcal{O}(k \log v)$ and depth $\mathcal{O}(\log k)$ is given in [KS08b].

A *universal block* UB_k is placed between the input selection block and the output permutation block. It takes care of the simulation of the gates using universal gates and ensures that each possible wiring can be implemented in the UC. The universal block construction is recursive, makes use of two universal blocks of smaller size with a selection block and a mixing block (essentially a layer of universal switches with one output) in between them. The $\mathcal{O}(k \log^2 k)$ size of this universal block is asymptotically not optimal and its $\mathcal{O}(k \log k)$ depth is also a factor of $\log k$ larger than Valiant's UC's. Thus, despite the efficiency of the other two building blocks, the construction from [KS08b] yields larger circuits than Valiant's UC in most cases. However, we note that using some of its building blocks can be beneficial in some scenarios (cf. Sect. 3.2).

3 The Size and the Depth of Valiant's Construction

In this section, we obtain new formulae for the size and the depth of Valiant's construction: the Γ_1 edge-universal graph construction is described in Sect. 3.1 and the universal circuit construction in Sect. 3.2. The *size of the edge-universal graph* is the number of nodes, counting all the poles and nodes created while using Valiant's construction. The *depth of the edge-universal graph* is the number of nodes on the longest path between any two nodes. When considering UCs and the PFE application, since XOR gates can be evaluated for free in secure computation [KS08a], the *ANDsize of the universal circuit* is the number of AND gates that are needed to realize the UC in total. The *ANDdepth of the universal circuit* in this scenario is the maximum number of AND gates between any input and output. For the sake of generality, we give the *total size* and *depth* of Valiant's UC construction with respect to both the AND and XOR gates that are used. Our implementation of universal gates and switches is optimized for PFE (cf. Sect. 4.1) and therefore uses the fewest AND gates possible. However, the total size and depth can be relevant when optimizing for other applications, in which case our implementation gives an upper bound that can be improved. For instance, when XOR and AND gates have the same costs, one needs to minimize the total number of gates instead of the number of AND gates as in [LMS16].

3.1 The Size and the Depth of the Γ_1 Edge-Universal Graph

In the skeleton, node A in Fig. 1a is redundant, since one can choose to embed the edge $(y, n-1)$ as (p_y, p_{n-1}) through Q, and (z, n) as (p_z, p_n) through R for any y and z nodes [Val76]. Thus, the number of nodes other than poles $\textsc{Exact}(n)$, for even n becomes

$$\textsc{Exact}(n) = 2 \cdot \textsc{Exact}\left(\frac{n-2}{2}\right) + 5 \cdot \frac{n-2}{2}. \tag{3}$$

For odd n, the construction makes use of $\frac{n-1}{2}$ poles in Q and $\frac{n-3}{2}$ poles in R. Then, edge (y, n) is embedded as (p_y, p_n) through Q for any y node, and node A is again redundant. Thus,

$$\textsc{Exact}(n) = \textsc{Exact}\left(\frac{n-1}{2}\right) + \textsc{Exact}\left(\frac{n-3}{2}\right) + 5 \cdot \frac{n-3}{2} + 3. \quad (4)$$

Using these recursive formulae, given the value n, it is possible to obtain the exact number of nodes other than poles in U_n. Valiant includes optimizations for starting off the recursion: for $1, 2, 3, 4, 5$ and 6 nodes; the respective number of additional nodes are $0, 0, 0, 3, 7$ and 9 (cf. Figs. 1c, d and e). Thus, a simple algorithm using dynamic programming based on the recursion relations of Eqs. 3 and 4 yields the exact number of nodes other than the original n poles that are created during the edge-universal graph construction. It depends on the parity of the input n at each iteration and unfortunately does not yield a closed formula for the size of Valiant's edge-universal graph construction, which is $n + \textsc{Exact}(n)$.

Valiant states that using his method, an edge-universal graph for $\Gamma_1(n)$ can be found "*with fewer than* $\frac{19}{8} n \log_2 n$ *nodes, and fanin and fanout 2*" [Val76]. As mentioned in Sect. 2.1, we consider the more detailed algorithm that yields the result with a slightly larger prefactor of $2.5n \log_2 n$ instead of $2.375n \log_2 n$. In this section, we sharpen this bound and give an approximate closed formula for the size of the construction. We first give upper and lower bounds, and then derive an approximation for a closed formula. For our lower bound, we consider the case when only the formula for even numbers, i.e., Eq. 3, is considered. This yields our lower bound of

$$n+5\left(\sum_{i=0}^{\log_2 n-1} 2^i\left(\frac{n}{2^{i+1}} - \frac{2(2^{i+1}-1)}{2^{i+1}}\right)\right) = 2.5n\log_2 n - 9n + 5\log_2 n + 10. \quad (5)$$

The upper bound can be obtained similarly, considering the case when only the formula for odd numbers with $5 \cdot (\frac{n-1}{2})$ is considered

$$n+5\left(\sum_{i=0}^{\log_2 n-1} 2^i\left(\frac{n}{2^{i+1}} - \frac{2^{i+1}-1}{2^{i+1}}\right)\right) = 2.5n\log_2 n - 4n + 2.5\log_2 n + 5. \quad (6)$$

Figure 2 depicts our upper and lower bounds along with Valiant's upper bound on the same construction for up to 100 000 nodes. We observe that the mean of our bounds is very close to the exact number of nodes. Figure 3 shows that already after a couple of hundreds of poles, it only slightly deviates from the exact number of nodes $\textsc{Exact}(n)$. Thus, we accept

$$\text{size}(U_n) \approx 2.5n\log_2 n - 6.5n + 3.75\log_2 n + 7.5 \quad (7)$$

as a good approximation of the closed formula for the size of the construction, noting that an estimated deviation of at most 2 % compared to the exact number of nodes, i.e., $\varepsilon \leq 0.02 \cdot \text{size}(U_n)$ may occur.

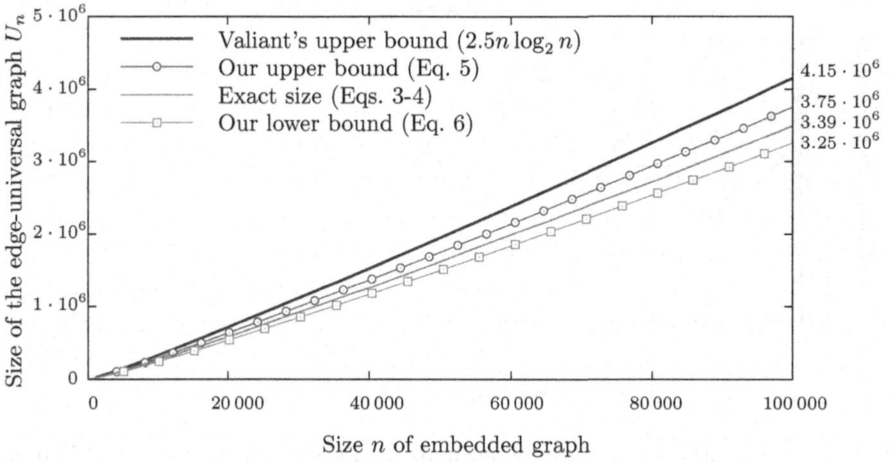

Fig. 2. Our upper and lower bounds for the size of Valiant's edge-universal graph construction for $\Gamma_1(n)$ graphs, along with Valiant's upper bound on the same construction and the exact size $\text{EXACT}(n)$, considering the size of the embedded graph $n \in \{1, \ldots, 100\,000\}$ (Color figure online).

The depth of the edge-universal graph, i.e., the maximum number of nodes between any two nodes is defined by the number of nodes between p_1 and p_n in the skeleton (cf. Figs. 1a and b). Thus, $\text{depth}(U_n) = 3n - 3$ for even n and $\text{depth}(U_n) = 3n - 2$ for odd n.

3.2 The Size and the Depth of Valiant's Universal Circuit

As described in Sect. 2.1, a universal circuit is constructed by means of an edge-universal graph for graphs with fanin and fanout 2, which is in turn constructed from two Γ_1 edge-universal graphs with poles merged together and thus taken only once into consideration. When constructing a UC, the number of inputs u, the number of outputs v and the number of gates k is public. We set k^* as the number of gates in the equivalent fanout-2 circuit, where $k \leq k^* \leq 2k + v$, in order to be able to later fairly compare with the UC construction of [KS08b]. We consider k^* as the public parameter instead of k, since without the knowledge of the original number of simulated gates, it does not reveal information about the simulated circuit. If the original k is public, one can hide k^* by setting it to its maximal value $2k + v$. Thus, using Valiant's UC construction, a Γ_2 edge-universal graph with $u + v + k^*$ poles is constructed and thus, our approximative formula for the size of the Γ_2 edge-universal graph corresponding to the graph of the circuit would become $2 \cdot \text{size}(U_{u+v+k^*}) - (u + v + k^*)$ and the exact number would be $u + v + k^* + 2 \cdot \text{EXACT}(u + v + k^*)$, i.e., the $u + v + k^*$ merged poles of the two edge-universal graphs plus the exact number of nodes other than poles. Therefore, the size of Valiant's UC is

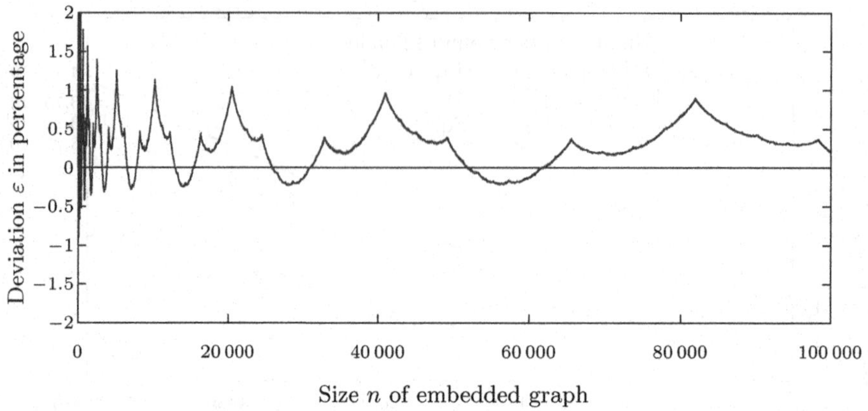

Fig. 3. The deviation of the mean of our upper and lower bounds (Eqs. 5 and 6) from the exact size of the edge-universal graph $\text{Exact}(n) + n$, considering the size of the embedded graph $n \in \{1, \ldots, 100\,000\}$.

$$\begin{aligned}\text{size}(UC^{\text{Valiant}}_{u,v,k^*}) \approx &[5(u+v+k^*)\log_2(u+v+k^*) - 15(u+v+k^*) \\ &+ 7.5\log_2(u+v+k^*) + 15] \cdot \text{size}(X) + k^* \cdot \text{size}(U)\end{aligned} \tag{8}$$

and the depth stays

$$\text{depth}(UC^{\text{Valiant}}_{u,v,k^*}) \approx [2(u+v+k^*) - 2] \cdot \text{depth}(X) + k^* \cdot \text{depth}(U). \tag{9}$$

When transforming the Γ_2 edge-universal graph into a UC, the first u poles are associated with inputs, the last v poles with outputs, and the k^* poles between are realized with universal gates (cf. Eq. 1) and their programming is defined by the corresponding gates in the simulated circuit. The rest of the nodes of the edge-universal graph are translated into universal switches (cf. Eq. 2), whose programming is defined by the edge-embedding of the graph of the circuit into the Γ_2 edge-universal graph. Thus, the size and depth of Valiant's UC can be directly derived from the size of the Γ_2 edge-universal graph. However, we include two optimizations to obtain a smaller size of the UC. The first optimization improves already the size of the edge-universal graph and the second optimization is applied when translating the edge-universal graph into a UC description (cf. Sect. 4.1).

1. **Optimization for Input and Output Nodes:** We observe that obviously circuit inputs need no ingoing edges and circuit outputs need no outgoing edges. Therefore, since u, v and k^* are publicly known, we optimize by deleting nodes that become redundant while canceling the edges going to the first u (input) and coming from the last v (output) nodes. Depending on the parity of u, v and $u+v+k^*$, the number of redundant switching nodes is $u+v-3\pm1$ in both Γ_1 edge-universal graphs that build up the graph of the UC. Therefore, we have, on average, $2(u+v-3)$ redundant nodes, which number we use in our calculations further on. This optimization also affects the depth by, on average, $u+v-3$.

2. **Optimization for Fanin-1 Nodes:** We observe that in the skeleton of the Γ_1 edge-universal graph construction there is a fanin-1 node (denoted with B in Figs. 1a and b). Such fanin-1 nodes exist in the base-cases for a small number of poles as well (cf. Figs. 1c, d and e). These nodes are important to achieve fanin and fanout 2 of each nodes in the graph, but can be ignored and replaced with wires when translated into a circuit description, essentially resulting in the same UC. According to Valiant's construction, these gates would translate into universal switches with one real input (and an other arbitrary one). Instead, we translate each of them into two wires and therefore set the second input to the same as the first one. Since at least one such node can be ignored in each subgraph when nodes are translated into gates, this results in altogether around

$$2 \cdot \left(\sum_{i=0}^{\log_2(u+v+k^*)-1} 2^i \right) - 1 = 2(u+v+k^*) - 3 \tag{10}$$

less gates for the two Γ_1 edge-universal graphs. This improvement has no effect on the depth of the construction.

Since both the size and the depth are dependent on the underlying representation of the circuit building blocks (of the universal gate U and of the universal switch or X gate), and the secure computation protocol, we express the size of the universal circuit with the size and depth of U and of X as parameters. Including the above optimizations of altogether $4(u+v)+2k^*-9$, the approximate formula for the size of Valiant's optimized UC construction becomes

$$\begin{aligned}\text{size}(UC^{\text{opt}}_{u,v,k^*}) \approx & [5(u+v+k^*)\log_2(u+v+k^*) - 17k^* - 19(u+v) \\ & + 7.5\log_2(u+v+k^*) + 24] \cdot \text{size}(X) + k^* \cdot \text{size}(U).\end{aligned} \tag{11}$$

To obtain the exact size of the UC, we use the recursive relations depicted in Eqs. 3 and 4 and include our optimizations. Thus, we obtain

$$\begin{aligned}\text{size}_{\text{exact}}(UC^{\text{opt}}_{u,v,k^*}) = & [2 \cdot \textsc{Exact}(u+v+k^*) \\ & - 4(u+v) - 2k^* + 9] \cdot \text{size}(X) + k^* \cdot \text{size}(U).\end{aligned} \tag{12}$$

From the depth of the edge-universal graph, the depth of the UC becomes

$$\text{depth}(UC^{\text{opt}}_{u,v,k^*}) \approx [u+v+2k^*+3] \cdot \text{depth}(X) + k^* \cdot \text{depth}(U). \tag{13}$$

Depending on the application, size(X) and size(U) as well as depth(X) and depth(U) can be optimized. Due to the PFE application, where XOR gates can be evaluated for free, we assess the ANDsize and ANDdepth of our AND-optimized implementations of universal gates and switches (cf. Sect. 4.1). In general, a universal gate can be realized with 3 AND gates (and 6 XOR gates), and ANDdepth of 2 (total depth of 6). Universal switches can be realized with only one AND gate (and 3 XOR gates), and ANDdepth of 1 (total depth of 3) [KS08a].

For private function evaluation, the size and the depth of U can be further optimized depending on the underlying secure computation protocol. In case the SFE implementation uses Yao's garbled circuit protocol [Yao86], both ANDsize(U) and ANDdepth(U) can be minimized to 1, due to the fact that in some garbling schemes the evaluator does not learn the type of the evaluated gate such as in case of garbled 3-row-reduction [NPS99]. Therefore, a universal gate can be implemented with one 2-input non-XOR gate [PSS09].

Optimized Hybrid Universal Circuit Construction: We investigate if hybrid methods utilizing building blocks of both UC constructions, i.e., of both [Val76] summarized in Sect. 2.1 and [KS08b] in Sect. 2.2, could yield better size. The simulation of the k gates of the original circuit is asymptotically more efficient using Valiant's UC construction due to the logarithmic factor, despite the overhead caused by taking the equivalent fanout-2 circuit with k^* gates, where $k \leq k^* \leq 2k + v$. However, we calculate if the modular approach of [KS08b] using a selection block $S^u_{m \geq u}$ for selecting the input variables or a truncated permutation block $TP^{k^* \geq v}_v$ for the output variables results in a smaller size.

Placing a selection block on top of Valiant's UC with m universal gates would imply a selection block $S^u_{m \geq u}$ which is then programmed to direct the u inputs of the circuit to the proper inputs of the m universal gates. Depending on how the output nodes are represented, m is either $2(k^* + v)$ for the case when including the outputs in Valiant's construction or $2k^*$ for the construction with a truncated permutation block. In the latter case, $TP^{k^* \geq v}_v$ takes care of permuting a subset of the outputs of the k^* gates, resulting in the v outputs of the UC. A selection block $S^u_{m \geq u}$ has size $\frac{u+m}{2} \log_2 u + m \log_2 m - u + 1$ and depth $2 \log_2 u + 2 \log_2 m + m - 2$, and a truncated permutation block $TP^{k^* \geq v}_v$ has size $\frac{k^*+v}{2} \log_2 v - 2v + k^* + 1$ and depth $\log_2 k^* + \log_2 v - 1$ [KS08b] (cf. full version [KS16]).

Let us take three scenarios into consideration, depending on the number of inputs u and the number of outputs v. The number of gates in the circuit to be simulated is k and the number of gates in the equivalent fanout-2 circuit is k^* with $k \leq k^* \leq 2k + v$.

1. **Constant I/O Case: $u = c_1$ constant, $v = c_2$ constant:** If both u and v are constant values c_1 and c_2 respectively, as is the case in many applications that compute a non-trivial function with relatively few inputs and outputs, the size of the selection block becomes $\approx 2k^* \log_2 k^* + (2 + \log_2 c_1)k^*$ and the size of the truncated permutation block is $\approx (0.5 \log_2 c_2 + 1)\, k^*$. With Valiant's UC construction, the overhead caused by a constant number of inputs and outputs is around $5(c_1 + c_2) \log_2 k^*$. The depth of Valiant's UC is only affected with constant overhead, while the depth of the selection and permutation blocks are $\approx 2k^* + 2 \log_2 k^*$ and $\approx \log_2 k$, respectively. Thus, both for the inputs and the outputs, Valiant's UC is an asymptotically better solution in the case with a constant number of inputs and outputs.

2. **Many Inputs:** $u \sim k$, $v = c$ **constant:** For many inputs where u is around the number of gates k and we have a constant number of c outputs, we include these c nodes in Valiant's UC instead of using a truncated permutation block due to the same reasoning as in the previous case. However, a selection block can be constructed to direct k inputs to $k^* + c$ universal gates. Thus, its size becomes $\approx 2k^* \log_2 k^* + k^* \log_2 k + 0.5k \log_2 k + 2k^* - k + 3c \log_2 k^*$ and its depth $\approx 2k^* + 2\log_2 k^* + 2\log_2 k$. In case of Valiant's UC construction, k inputs result in an overhead of $\approx 5k \log_2 k - 9k + 5c \log_2 k$ for the size and $\approx k$ for the depth, since a large part (up to a half) of the circuit is built in order to hide the input wiring. Therefore, in this scenario it is often worth to use a hybrid method, utilizing the selection block from [KS08b] for input selection. Our *many inputs hybrid* construction places a selection block on top of a UC with $k^* + c$ universal gates and has approximate size when $u \sim k$ and v is constant c

$$\begin{aligned} \text{size}(UC_{k,c,k^*}^{\text{many I}}) \approx [&7k^* \log_2 k^* + k^* \log_2 k + 0.5k \log_2 k - k - 15k^* \\ &+ (7.5 + 5c) \log_2 k^* + 3c \log_2 k^* + \mathcal{O}(1)] \cdot \text{size}(X) + k^* \cdot \text{size}(U) \end{aligned} \tag{14}$$

and approximate depth

$$\begin{aligned} \text{depth}(UC_{k,c,k^*}^{\text{many I}}) \approx [4k^* + 2\log_2 k^* + 2\log_2 k + \mathcal{O}(1)] \cdot \text{depth}(X) \\ + k^* \cdot \text{depth}(U). \end{aligned} \tag{15}$$

3. **Maximal I/O Case:** $u \sim 2k$, $v \sim k$: For circuits with $u \sim 2k$ inputs and $v \sim k$ outputs, we discuss the possibility of using both an input selection block and an output permutation block. The size of the selection block is $\approx 2k^* \log_2 k^* + k^* \log_2 k + k \log_2 k + 3k^* - k$ and its depth becomes $\approx 2k^* + 2\log_2 k^* + 2\log_2 k$, which is more beneficial (when it comes to the size) than the $\approx 10k \log_2 k - 12k$ size overhead and $\approx 2k$ depth overhead in Valiant's construction caused by $2k$ inputs (up to half of the UC is constructed for inputs only). The truncated permutation block has size $\approx 0.5k^* \log_2 k + 0.5k \log_2 k + k^* - 2k$ and depth $\approx \log_2 k^* + \log_2 k$, while the same amount of outputs in Valiant's construction introduces at least $5k \log_2 k - 9k$ new switches with depth of $\approx k$. Thus, for the case when the maximal $2k$ inputs and k outputs are considered, we conclude that it is advantageous to use our *maximal I/O hybrid* construction, utilizing Valiant's graph construction for the k^* gates [Val76], a selection block for the inputs and a truncated permutation block for the outputs [KS08b]. This yields an approximate size when $u \sim 2k$ and $v \sim k$

$$\begin{aligned} \text{size}(UC_{2k,k^*,k}^{\max \text{I/O}}) \approx [&7k^* \log_2 k^* + 1.5k^* \log_2 k + 1.5k \log_2 k - 13k^* - 3k \\ &+ 7.5 \log_2 k^* + \mathcal{O}(1)] \cdot \text{size}(X) + k^* \cdot \text{size}(U) \end{aligned} \tag{16}$$

and an approximate depth

$$\begin{aligned} \text{depth}(UC_{2k,k^*,k}^{\max \text{I/O}}) \approx [4k^* + 3\log_2 k^* + 3\log_2 k + \mathcal{O}(1)] \cdot \text{depth}(X) \\ + k^* \cdot \text{depth}(U). \end{aligned} \tag{17}$$

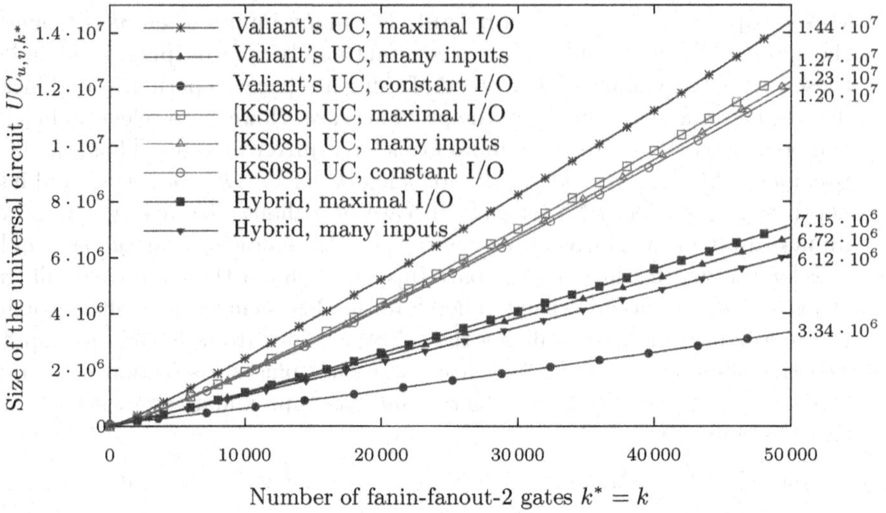

Fig. 4. Comparison between the sizes of the universal circuit constructions for $k^* = k \in \{0, \ldots, 50\,000\}$ gates, considering the three scenarios: *constant I/O* with constant number of inputs and outputs, *many inputs* with $\sim k$ inputs and constant outputs and *maximal I/O* with $\sim 2k$ inputs and $\sim k$ outputs (Color figure online).

We conclude that in case of a large number of inputs and outputs it is beneficial to construct a hybrid UC, making use of both existing constructions (cf. Sects. 2.1 and 2.2). Most practical applications have input and output with constant size and only some specific applications use input size linear in the number of gates (e.g. simple computations on large databases). Thus, we consider Valiant's construction as the most beneficial for general purposes, however we have shown, that one can optimize the construction for many inputs or outputs by adding selection or truncated permutation blocks from [KS08b].

Comparison with the Universal Circuit Construction from [KS08b]. In [KS08b], a universal circuit construction was proposed with approximate size $1.5k\log_2^2 k + 2.5k\log_2 k$. This was calculated with the doubled size of the universal switches, not yet considering the free-XOR optimizations of [KS08a]. We recalculated the size of the construction with our additional optimization for the outputs described in Sect. 2.2. We give our detailed calculations in the full version [KS16] and summarize its exact size here as

$$\text{size}(UC_{u,v,k}^{[\text{KS08b}]}) = [0.75k\log_2^2 k + 2.25k\log_2 k + (0.5+k)\log u + (0.5k+0.5v)\log v + 5k - u - 2v]\cdot \text{size}(X) + k\cdot\text{size}(U), \tag{18}$$

and from [KS08b] we know that its depth is

$$\text{depth}(UC_{u,v,k}^{[\text{KS08b}]}) = [k\log_2 k + 2k + 7\log_2 k + 2\log_2 u + \log_2 v - 14]\cdot k\cdot\text{depth}(U). \tag{19}$$

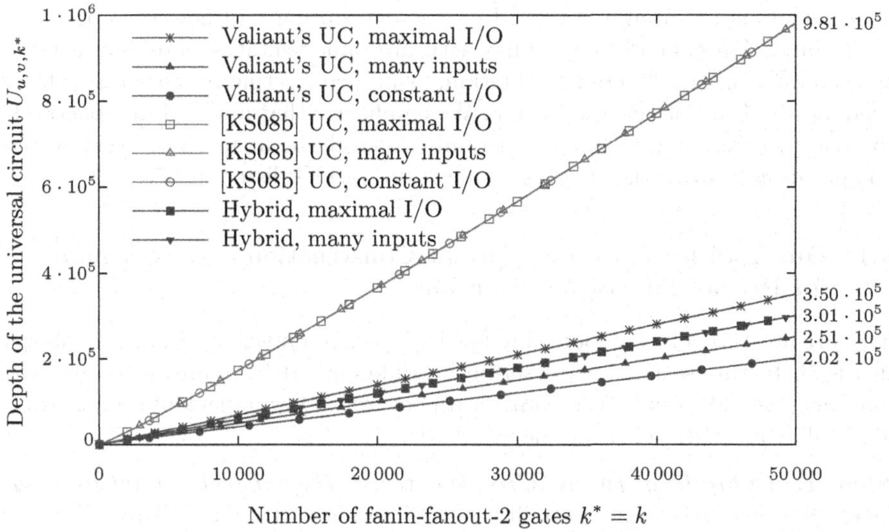

Fig. 5. Comparison between the depths of the universal circuit constructions for $k^* = k \in \{0, \ldots, 50\,000\}$ gates, considering the three scenarios: *constant I/O* with constant number of inputs and outputs, *many inputs* with $\sim k$ inputs and constant outputs and *maximal I/O* with $\sim 2k$ inputs and $\sim k$ outputs (Color figure online).

It was concluded in [KS08b] that this construction outperforms Valiant's construction for circuits with up to 5 000 gates. However, this was achieved using the assumption that Valiant's UC has size $\approx 9.5(u + 2v + 2k)\log_2(u + 2v + 2k)$, which can vary between two to four times its actual size. On the one hand, a factor of two of this difference is due to the free-XOR optimizations in [KS08a]. On the other hand, [KS08b] used the maximal $k^* = 2k+v$ in their approximation. In Sect. 4.2, we show on concrete example circuits that k^* stays significantly below this upper bound. The construction described in detail in Sect. 2.1 has a larger constant factor 5, but due to the logarithmic factor it outperforms the construction from [KS08b] (Sect. 2.2) already for a few hundred gates in the constant I/O case. Figures 4 and 5 compare the sizes and depth of the different UC constructions, respectively in the three scenarios described above, with the lowest possible gate number $k^* = k$. When considering the hybrid approach, the method corresponding to the given scenario is indeed always the most efficient construction for many inputs and/or outputs. We give a comparison for the upper bound case $k^* = 2k+v$, for the sizes of all universal circuit constructions for well-known circuits from [TS15] and compare their structure in the full version [KS16].

4 Implementing Valiant's Universal Circuit in Practice

In this section, we detail the challenges that we faced while demonstrating the practicality of Valiant's universal circuit construction. We show how to construct a universal circuit from a standard circuit description and how to program it accordingly. We validate our results with an implementation, creating a novel toolchain for private function evaluation, using two existing frameworks as frontend and backend of our application. We emphasize that our tool for constructing and programming UC is generic and can easily be adapted to other secure computation frameworks or other applications of UCs listed in Sect. 1.2.

4.1 Our Tool for Universal Circuit Construction and Toolchain for Private Function Evaluation

The architecture of our toolchain for PFE using universal circuits is shown in Fig. 6. In this section, we describe its different artifacts and its use of the Fairplay [MNPS04] and ABY [DSZ15] frameworks. Our implementation is available online at http://encrypto.de/code/UC.

Step 1. Compiling Input Circuits from High-Level Functionality: Due to its easy adoptability, we decided to use the Fairplay compiler [MNPS04,BNP08] with the FairplayPF extension [KS08b] to translate the functionality described in the high-level SFDL format to the Fairplay circuit description called Secure Hardware Definition Language (SHDL). The FairplayPF extension already converts circuits with gates of an arbitrary fanin into gates with at most two inputs, which is required for Valiant's construction as well. However, in case of Valiant's UC construction, there is another restriction on the input circuit. It has to have fanout 2, i.e., the outputs of all the gates and inputs can only be used as the input of at most two later gates.

In case the input circuit does not follow this restriction, an algorithm places a binary tree in place of each gate with fanout larger than 2, following Valiant's proposition: *"Any gate with fanout $x+2$ can be replaced by a binary fanout tree with $x+1$ gates"* [Val76, Corollary 3.1]. This is done using so-called *copy gates*, i.e., identity gates, each of them eliminating one from the extra fanout of the original gate. An upper bound can be given on the number of copy gates. The class of Boolean functions with u inputs and v outputs that can be realized by acyclic circuits with k gates and arbitrary fanout, can also be realized with an acyclic fanout-2 circuit with $k \leq k^* \leq 2k+v$ gates [Val76, Corollary 3.1]. We give concrete examples in Sect. 4.2 on how this conversion changes the input circuit size for practical circuits and show that in most cases, the resulting number of gates remains significantly below the upper bound $2k+v$.

Step 2. Obtaining the Γ_2 Graph of the Circuit: From the SHDL description of a C circuit with fanin and fanout 2, the Γ_2 graph G^C of the circuit C can be directly generated as described in Sect. 2.1: with the number of inputs u, the number of outputs v and the number of gates k^* in circuit C, G^C has $u+v+k^*$ nodes and the wires are represented as edges in the graph. Then, the first u nodes in the topological order correspond to the inputs, the last v nodes to the outputs and the nodes in between them to the k^* ordered gates. We note that since C had fanin and fanout 2, the resulting G^C graph is in $\Gamma_2(u+v+k^*)$.

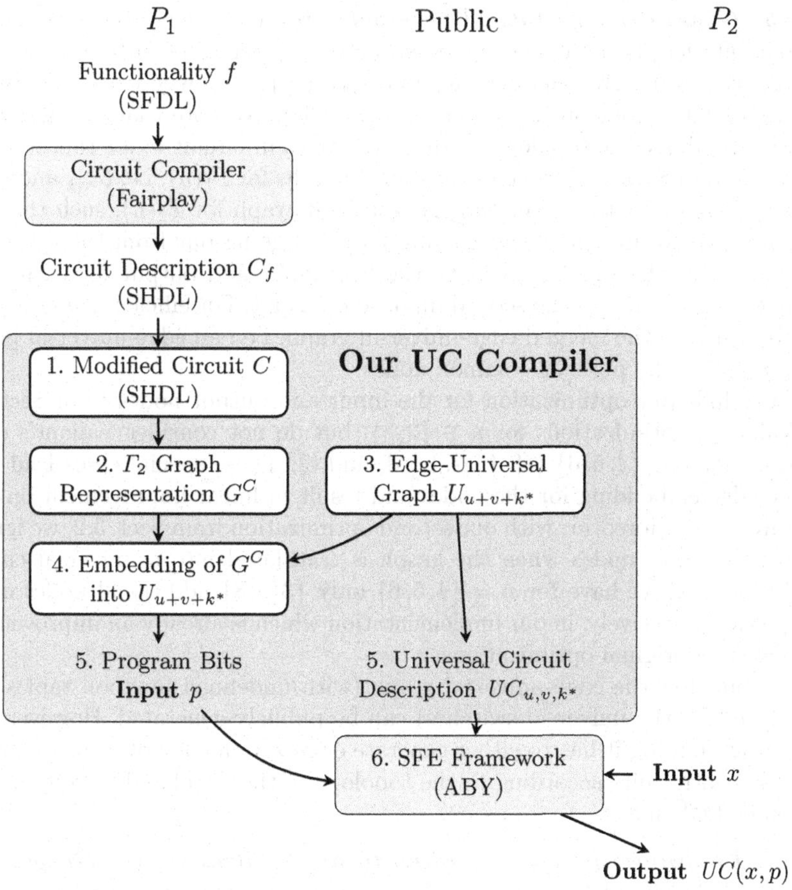

Fig. 6. Our toolchain for universal circuits and private function evaluation.

Therefore in G^C, each node can have at most two incoming edges, one defined to be the first and the other the second. It is possible in the modified SHDL circuit description that an internal value becomes two times the first or two times the second input of gates. This is due to the fact that in the original SHDL circuit with arbitrary fanout, a value could be the input of arbitrary number of later gates. Transforming the circuit to a fanout-2 circuit by adding copy gates allows a value to be an input only two times, but the order of the inputs is fixed. Therefore, in such a case when a value is the second time the same input to a gate (i.e., first or second), besides the two inputs, the two middle bits of the function table of the gate must be reversed as well (i.e., to compute $f(\text{in}_1, \text{in}_2)$ instead of $f(\text{in}_2, \text{in}_1)$) for the correct programming of the universal circuit in Step 5.

***Step 3. Generating Γ_2 Edge-Universal Graph U_n*:** Knowing the number of input bits u, the number of gates k^* and the number of output bits v one can construct the corresponding edge-universal graph U_n, where $n = u + v + k^*$, with out input-output optimization from Sect. 3.2. We note that no knowledge is necessary about the topology or the gate tables in circuit C for this step. As we described in Sect. 2.1, two edge-universal graphs for $\Gamma_1(n)$, i.e. $U_{n,1}$ and $U_{n,2}$, are merged in order to obtain an edge-universal graph for $\Gamma_2(n)$, such that the poles are merged and the edges coming into and going out from them become as follows: the edges in $U_{n,1}$ will be the first input and output for each pole, the edges in $U_{n,2}$ will be the second input and output. For efficiency reasons, we directly generate the merged edge-universal graph, i.e., an edge-universal graph for $\Gamma_2(n)$, with the poles as common nodes.

We include our optimization for the input and output nodes from Sect. 3.2 and Valiant's optimizations for $n \in \{2, 3\}$, but do not consider Valiant's optimizations for $n \in \{4, 5, 6\}$ (cf. Figs. 1c, d, and e). These special cases lead to a specific edge-embedding for the nodes and result in linear improvement only in very rare cases. Moreover, with our second optimization from Sect. 3.2, we ignore most of the extra nodes when the graph is translated into a universal circuit description, i.e., we have for $n = \{4, 5, 6\}$ only $\{3, 5, 8\}$ additional nodes other than poles, respectively, in our implementation which is already an improvement over Valiant's original optimizations.

We note that the edge-universal graph (with undefined function tables and control bits for the universal switches) can be publicly generated. However, the party programming it has to either generate or receive a copy of it for programming the control bits according to the topology of the simulated circuit (i.e., to edge-embed G^C into U_n).

Step 4. Programming U_n According to an Arbitrary $\Gamma_2(n)$ Graph: The Γ_2 graph of the circuit G^C with n nodes is partitioned into two $\Gamma_1(n)$ graphs G_1^C and G_2^C which are embedded into the two edge-universal graphs for $\Gamma_1(n)$ that build up U_n. Valiant proved in [Val76] that for any topologically ordered $\Gamma_1(n)$ graph, for any $(i, j) \in E$ and $(k, l) \in E$ edges there exist edge-disjoint paths in U_n between the ith and the jth poles and between the kth and the lth poles. We described Valiant's method in Sect. 2.1 and here we show the algorithm that uniquely defines these paths in U_n.

For the description of our algorithm, we first define a $\Gamma_1(n)$ supergraph, which is a $\Gamma_1(n)$ graph with additionally a binary tree of Γ_1 graphs of decreasing size. These Γ_1 graphs uniquely define the embedding of the edges into U_n. When embedding an edge (i, j) of the topologically ordered graph G into the edge-universal graph, one needs to construct the supergraph of G as described in Algorithm 1 and then look at the binary tree in the supergraph. The path of the edge (i, j) defines the edge-embedding uniquely. This means that if edge $(\lceil \frac{i}{2} \rceil, \lceil \frac{j}{2} \rceil - 1)$ is in the left subgraph of G, then it can be embedded through subgraph Q in U_n, otherwise it is in the right subgraph of G and can be embedded through subgraph R in U_n. The unique embedding happens through $\{r_{\lceil \frac{i}{2} \rceil}, r_{\lceil \frac{j}{2} \rceil - 1}\}$ or through $\{q_{\lceil \frac{i}{2} \rceil}, q_{\lceil \frac{j}{2} \rceil - 1}\}$, utilizing the unique shortest path between them, through subpoles further identified by smaller subgraphs of G.

Algorithm 1. SUPERGRAPH(G)

Input: $\Gamma_1(n)$ graph G with set of nodes $V = \{1, \ldots, n\}$
Output: $\Gamma_1(n)$ supergraph

Create a graph H with $\lceil \frac{n}{2} \rceil - 1$ nodes ▷ H Γ_2 graph (with possible loops)
if there exist an edge (i, j) in G **and** $\lceil \frac{j}{2} \rceil - 1 \geq \lceil \frac{i}{2} \rceil$ **then**
  Add edge $\left(\lceil \frac{i}{2} \rceil, \lceil \frac{j}{2} \rceil - 1\right)$ in H ▷ each pair of nodes in G is one node in H
end if
Partition H into two Γ_1 graphs G_1 of size $\lceil \frac{n}{2} \rceil - 1$ and G_2 of size $\lfloor \frac{n}{2} \rfloor - 1$ using König's theorem as in §2.1
  ▷ in odd case, the $(e, \lceil \frac{n}{2} \rceil - 1)$ edge in H for arbitrary e will be added in G_1
if size(G_1) $\neq 0$ **then**
  SUPERGRAPH(G_1)
  Store G_1 as the left subgraph of G
end if
if size(G_2) $\neq 0$ **then**
  SUPERGRAPH(G_2)
  Store G_2 as the right subgraph of G
end if
delete H
return G

When the embedding is done (cf. full version [KS16]), for defining the control bits, each node x has at most two nodes that have ingoing edges to x, one is represented as the left parent and one as the right parent of x in the edge-universal graph. The two consecutive nodes are also saved as left and right children of x. Now, when x is a switching node and we take edges (v, x) and (x, w) in the path, we save for x if parent v and child w are on the same or on the opposite side in the edge-universal graph. This defines the control bit of each universal switch in the translated universal circuit, where left and right parent and child translate to first and second input and output, respectively. We note that in order to program U_n correctly, we require that if x is the left (right) parent of v in the edge-universal graph, then v is the left (right) child of x.

Step 5. Generating the Output Circuit Description and the Programming of the Universal Circuit: After embedding the graph of the simulated circuit into the edge-universal graph U_n, we write the resulting circuit in a file using our own circuit description. In the edge-universal graph, each node stores the program bit resulting from the edge-embedding (control bit c of the corresponding universal switch in Eq. 2) and each pole stores four bits corresponding to the simulated circuit (the four control bits of the function table, c_0, c_1, c_2, c_3 in Eq. 1, their order possibly changed in Step 2). Thus, after topologically ordering U_n, one can directly write out the gate identifiers into a circuit file and the program bits to a programming file.

Our circuit description format starts with enumerating the inputs and ends with enumerating the outputs. We have universal gates denoted by U, universal switches denoted by X or Y depending on the number of outputs (X with two outputs and Y with one). We note that we replace any gates that have only one input by wires in the UC, thus achieving our fanin-1 node optimization from Sect. 3.2. The wires are represented in the following manner:

$$\begin{array}{lllll} U & \text{in}_1 & \text{in}_2 & \text{out}_1 & \\ X & \text{in}_1 & \text{in}_2 & \text{out}_1 & \text{out}_2 \\ Y & \text{in}_1 & \text{in}_2 & \text{out}_1 & \end{array} \quad (20)$$

denotes that wire out_1 (and possibly out_2) is coming from a gate with input wires in_1 and in_2. The program bits are not represented in the circuit format, but in a separate file, for each universal gate we save a four-bit number representing the control bits and for each universal switch we store the control bit. The output nodes are outputs of Y universal switches and are marked in the end of the file as $O \quad \text{o}_1 \quad \text{o}_2 \quad \ldots \quad \text{o}_v$. The circuit and its programming are given in plain text files.

Step 6. Evaluating Universal Circuits for PFE in ABY: As an example application of UCs, we implement PFE using SFE of a universal circuit. We adapted the ABY secure two-party computation framework [DSZ15] for this purpose. Firstly, since ABY uses the free-XOR optimization from [KS08a], we construct universal gates and switches with low ANDsize and ANDdepth given in Sect. 3.2. With the cost metric we consider, X and Y gates have the same AND complexity, optimized in [KS08a] and are obtained as

$$\begin{aligned} \text{out}_1 &= Y(\text{in}_1, \text{in}_2; c) = (\text{in}_1 \oplus \text{in}_2)c \oplus \text{in}_1 \\ (\text{out}_1, \text{out}_2) &= X(\text{in}_1, \text{in}_2; c) = (e \oplus \text{in}_1, e \oplus \text{in}_2) \text{ with } e = (\text{in}_1 \oplus \text{in}_2)c \end{aligned} \quad (21)$$

with ANDsize and ANDdepth of 1 for both universal switches. X gates are realized with one additional XOR gate compared to Y gates.

Our efficient implementation of generic universal gates uses Y gates yielding

$$\text{out}_1 = U(\text{in}_1, \text{in}_2; c_0, c_1, c_2, c_3) = Y[Y(c_0, c_1; \text{in}_2), Y(c_2, c_3; \text{in}_2); \text{in}_1] \quad (22)$$

with ANDsize$(U) = 3$ and ANDdepth$(U) = 2$. This universal gate implementation is generic and works in all secure computation protocols. However, for Yao's garbled circuits protocol, one can further optimize it to ANDsize$(U) =$ ANDdepth$(U) = 1$, as in some garbling schemes such as the garbled 3-row-reduction [NPS99] the gate being evaluated remains oblivious to the evaluator.

After constructing the efficient building blocks, the output circuit file of our UC compiler is parsed, a circuit is generated accordingly and programmed with the input program bits. We conclude that our toolchain is the first implementation of Valiant's size-optimized universal circuit that supports efficient private function evaluation.

4.2 Comparison of Our PFE-Toolchain with Other PFE Protocols

Mohassel et al. in [MS13] design a generic framework for PFE and apply it to three different scenarios: to the m-party GMW protocol [GMW87], to Yao's

Table 1. The number of symmetric-key operations using different PFE protocols: Valiant's UC with SFE, the universal circuit construction from [KS08b] or Mohassel et al.'s OT-based method from [MS13]. u, v and k denote the number of inputs, outputs and gates in the simulated circuit, and k^* denotes the number of gates in the equivalent fanout-2 circuit.

Circuit	u	k	v	$k^* - k$ ($\frac{k^*}{k}$)	Valiant	[KS08b]	OT-based [MS13]
`AES-non-exp`	256	31 924	128	14 539 (1.46)	$1.150 \cdot 10^7$	$2.797 \cdot 10^7$	$6.243 \cdot 10^6$
`AES-expanded`	1 536	25 765	128	11 089 (1.43)	$9.211 \cdot 10^6$	$2.206 \cdot 10^7$	$4.943 \cdot 10^6$
`DES-non-exp`	128	19 464	64	12 290 (1.63)	$7.502 \cdot 10^6$	$1.560 \cdot 10^7$	$3.639 \cdot 10^6$
`md5`	512	43 234	128	22 623 (1.52)	$1.700 \cdot 10^7$	$3.995 \cdot 10^7$	$8.681 \cdot 10^6$
`add_32`	64	187	33	58 (1.31)	35 512	55 341	19 939
`comp_32`	64	150	1	1 (1.01)	19 384	40 222	15 424
`mult_32x32`	64	6 995	64	5 079 (1.73)	$2.522 \cdot 10^6$	$4.647 \cdot 10^6$	$1.184 \cdot 10^6$
`Branching_18`	72	121	4	3 (1.02)	17 312	30 994	11 994
`CreditCheck`	25	50	1	6 (1.12)	5 056	9 348	4 198
`MobileCode`	80	64	16	0 (1.00)	12 528	13 727	5 644

garbled circuits [Yao86] and to arithmetic circuits using homomorphic encryption [CDN01]. Both the two-party version of their framework with the GMW protocol and the solution with Yao's garbled circuit protocol has two alternatives: using homomorphic encryption they achieve linear complexity $\mathcal{O}(k)$ in the circuit size k and when using a solution solely based on oblivious transfers (OTs), they obtain a construction with $\mathcal{O}(k \log k)$ symmetric-key operations. The OT-based construction in both cases is more desirable in practice, since using OT extension the number of public-key operations can be reduced significantly [IKNP03, ALSZ13].

Since the asymptotical complexity of this construction and using Valiant's UC for PFE is the same, we compare these methods for PFE. We revisit the formulas provided in [MS13] for the PFE protocol based on Yao's garbled circuits and elaborate on the number of symmetric-key operations when the different PFE protocols are used. Mohassel et al. show that the total number of switches in their framework is $4k \log_2(2k) + 1$ that are evaluated using OT extension, for which they calculate $8k \log_2(2k) + 8$ symmetric-key operations together with $5k$ operations for evaluating the universal gates with Yao's protocol. We count only the work of the party that performs most of the work, i.e., $4k$ symmetric-key operations for creating a garbled circuit with k gates and 3 symmetric-key operations (two calls to a hash function and one call to a pseudorandom function (PRF)) for each OT using today's most efficient OT extension of [ALSZ13]. Hence, according to our estimations, the protocol of [MS13] requires $12 \log_2(2k) + 4k + 12$ symmetric-key operations.

In the same way, we assume that in our case, for evaluating both the universal gates and switches, the garbler needs $4k$ symmetric-key operations. Thus, for a fair comparison, we essentially update Table 4 from the full version of [MS13, Appendix J.1], where Valiant's UC size was calculated with assumed $k^* = 2k+v$, without calculating 4 operations for the garbling.

We took our example circuit files of varying size in Table 1 from two different sources and elaborate on the resulting number of symmetric-key operations using the different constructions. The first 7 circuits we obtained from the function set of [TS15] and the last three from the FairplayPF extension of the Fairplay compiler [MNPS04, KS08b]. The example circuits that we took from [TS15] had to be converted to our desired SHDL format, which was a necessary step in order to be able to elaborate on the performance of these more complicated circuits as well. We included the INV gates in the function table of the consecutive gate and therefore, resulted in smaller gate numbers k for the equivalent SHDL circuits with arbitrary fanout. Then, these SHDL circuits were considered as input circuits for our tool.

We now compare the size of the three two-party PFE protocols: the two UC-based PFE with secure computation and the OT-based method of [MS13]. We assess our findings in Table 1. We note that our numbers are estimations, i.e., we do not consider that [MS13] works with circuits made up solely of NAND gates. Since Valiant's UC construction depends also on the number of gates with fanout more than 2 in the original circuit, we include the number of copy gates, $(k^* - k)$ in the table. We emphasize the ratio between the new number of gates k^* and the original number of gates k and conclude that in general circuits, it is well below the maximal $\frac{k^*}{k} \sim 2$. The size of the UC construction from [KS08b] obviously makes their method less efficient, in our examples using more than twice as many symmetric-key operations as the method with Valiant's UC and four times as many as Mohassel et al.'s efficient OT-based method [MS13]. We conclude that universal circuits are not the most efficient solution to perform PFE, however, we show the feasibility of generating and evaluating UCs simulating large circuits. We emphasize that even though the PFE-specific protocol from [MS13] achieves better results for PFE, universal circuits are generic and can be applied for various other scenarios (cf. Sect. 1.2), and the most efficient UC construction is Valiant's construction.

Our Experimental Results. We validated the practicality of Valiant's universal circuit construction with an efficient implementation. We ran our experiments on two Desktop PCs, each equipped with an Intel Haswell i7-4770K CPU with 3.5 GHz and 16 GB RAM, that are connected via Gigabit-LAN and give our benchmarks in Table 2. We are able to generate UCs up to around 300 000 gates of the simulated circuit, i.e., which results in billions of gates in the UC. Until now, the only implementation of universal circuits was given in [KS08b], which is outperformed by Valiant's construction already for a couple of hundred gates (cf. Figs. 4 and 5) due to its asymptotically larger complexity. We show the real practicality of UCs through experimental results proving the efficiency of our

Table 2. Running time and communication for our UC-based PFE implementation with ABY. We include the compile time, the I/O time of the UC compiler, and the evaluation time (in milliseconds) and the total communication (in Kilobytes) between the parties in GMW as well as in Yao sharing.

Circuit	UC Compile	UC I/O	GMW		Yao	
	Time (ms)	Time (ms)	Time (ms)	Communic. (KB)	Time (ms)	Communic. (KB)
AES-non-exp	2 909.2	6 331.2	5 522.08	137 561.13	2 349.35	88 417.61
AES-expanded	2 103.7	5 063.6	4 136.72	109 033.79	1 878.75	70 097.48
DES-non-exp	1 596.2	4 173.5	2 695.51	76 644.38	1 310.52	48 180.69
md5	4 043.5	8 785.4	7 041.12	169 558.83	3 547.68	110 043.59
add_32	11.4	63.8	31.97	457.77	26.49	224.77
comp_32	5.8	34.1	29.94	340.23	8.90	159.73
mult_32x32	328.9	1 443.2	1 092.46	31 053.53	539.98	18741.85
Branching_18	4.8	31.4	26.23	307.77	17.34	145.87
CreditCheck	1.2	11.4	26.25	113.35	5.67	45.15
MobileCode	3.2	26.3	25.71	202.50	28.16	103.45

implementation of PFE with the ABY framework [DSZ15]. Furthermore, due to its asymptotically smaller depth, we are also able to evaluate our generated UCs with the GMW protocol [GMW87], whereas the construction from [KS08b] was only evaluated with Yao's garbled circuit protocol. We do not directly compare our runtimes with the method of [MS13], since to the best of our knowledge, their framework has not yet been implemented.

Converting from circuit descriptions and writing into and reading out from files slows down the program significantly, but it still achieves good performance for practical circuits such as AES and DES. Our implementation in ABY can evaluate most of the circuits in both the GMW and Yao's protocols, but for some examples it runs out of memory (e.g. SHA-256). However, improvements on SFE protocols imply improvements on UC-based PFE frameworks as well. As can be seen in Table 2, the evaluation time and the communication in case of Yao's garbled cirucit protocol is about a factor of two smaller than that of the GMW protocol. This difference is due to the more efficient universal gate construction with only one gate for the case of Yao's protocol in contrast to the universal gates used in the GMW protocol with ANDsize = 3 and ANDdepth = 2.

Acknowledgements. This work has been co-funded by the European Union's 7th Framework Program (FP7/2007–2013) under grant agreement n. 609611 (PRACTICE), by the German Federal Ministry of Education and Research (BMBF) within CRISP, by the DFG as part of project E3 within the CRC 1119 CROSSING, and by the Hessian LOEWE excellence initiative within CASED. We thank Michael Zohner and Daniel Demmler for helping with the implementation in ABY, and the anonymous reviewers of Eurocrypt 2016 for their helpful comments on our paper.

References

[AF90] Abadi, M., Feigenbaum, J.: Secure circuit evaluation. J. Cryptol. **2**(1), 1–12 (1990)

[ALSZ13] Asharov, G., Lindell, Y., Schneider, T., Zohner, M.: More efficient oblivious transfer and extensions for faster secure computation. In: ACM CCS 2013, pp. 535–548. ACM (2013)

[Att14] Attrapadung, N.: Fully secure and succinct attribute based encryption for circuits from multi-linear maps. IACR Cryptology ePrint Archive 2014:772 (2014)

[BFK+09] Barni, M., Failla, P., Kolesnikov, V., Lazzeretti, R., Sadeghi, A.-R., Schneider, T.: Secure evaluation of private linear branching programs with medical applications. In: Backes, M., Ning, P. (eds.) ESORICS 2009. LNCS, vol. 5789, pp. 424–439. Springer, Heidelberg (2009)

[BNP08] Ben-David, A., Nisan, N., Pinkas, B.: FairplayMP: a system for secure multi-party computation. In: ACM CCS 2008, pp. 257–266. ACM (2008)

[BPSW07] Brickell, J., Porter, D.E., Shmatikov, V., Witchel, E.: Privacy-preserving remote diagnostics. In: ACM CCS 2007, pp. 498–507. ACM (2007)

[CCKM00] Cachin, C., Camenisch, J.L., Kilian, J., Müller, J.: One-round secure computation and secure autonomous mobile agents. In: Welzl, E., Montanari, U., Rolim, J.D.P. (eds.) ICALP 2000. LNCS, vol. 1853, p. 512. Springer, Heidelberg (2000)

[CDN01] Cramer, R., Damgård, I.B., Nielsen, J.B.: Multiparty computation from threshold homomorphic encryption. In: Pfitzmann, B. (ed.) EUROCRYPT 2001. LNCS, vol. 2045, pp. 280–299. Springer, Heidelberg (2001)

[CH85] Cook, S.A., Hoover, H.J.: A depth-universal circuit. SIAM J. Comput. **14**(4), 833–839 (1985)

[DDKZ13] Durnoga, K., Dziembowski, S., Kazana, T., Zajac, M.: One-time programs with limited memory. In: Lin, D., Xu, S., Yung, M. (eds.) Inscrypt 2013. LNCS, vol. 8567, pp. 377–394. Springer, Heidelberg (2013)

[DSZ15] Demmler, D., Schneider, T., Zohner, M.: ABY - a framework for efficient mixed-protocol secure two-party computation. In: Network and Distributed System Security (NDSS 2015). The Internet Society (2015). http://encrypto.de/code/ABY

[FAL06] Frikken, K.B., Atallah, M.J., Li, J.: Attribute-based access control with hidden policies and hidden credentials. IEEE Trans. Comput. **55**(10), 1259–1270 (2006)

[FAZ05] Frikken, K.B., Atallah, M.J., Zhang, C.: Privacy-preserving credit checking. In: ACM Electronic Commerce (EC 2005), pp. 147–154. ACM (2005)

[FGP14] Fiore, D., Gennaro, R., Pastro, V.: Efficiently verifiable computation on encrypted data. In: ACM CCS 2014, pp. 844–855. ACM (2014)

[FLA06] Frikken, K.B., Li, J., Atallah, M.J., Trust negotiation with hidden credentials, hidden policies, and policy cycles. In: Network and Distributed System Security (NDSS 2006), pp. 157–172. The Internet Society (2006)

[FVK+15] Fisch, B., Vo, B., Krell, F., Kumarasubramanian, A., Kolesnikov, V., Malkin, T., Bellovin, S.M.: Malicious-client security in Blind Seer: a scalable private DBMS. In: IEEE Symposium on Security and Privacy (S&P 2015), pp. 395–410. IEEE (2015)

[GGHZ14] Garg, S., Gentry, C., Halevi, S., Zhandry, M.: Fully secure attribute based encryption from multilinear maps. IACR Cryptology ePrint Archive 2014:622 (2014)

[GGPR13] Gennaro, R., Gentry, C., Parno, B., Raykova, M.: Quadratic span programs and succinct NIZKs without PCPs. In: Johansson, T., Nguyen, P.Q. (eds.) EUROCRYPT 2013. LNCS, vol. 7881, pp. 626–645. Springer, Heidelberg (2013)

[GHV10] Gentry, C., Halevi, S., Vaikuntanathan, V.: i-hop homomorphic encryption and rerandomizable Yao circuits. In: Rabin, T. (ed.) CRYPTO 2010. LNCS, vol. 6223, pp. 155–172. Springer, Heidelberg (2010)

[GMW87] Goldreich, O., Micali, S., Wigderson, A.: How to play any mental game or a completeness theorem for protocols with honest majority. In: ACM Symposium on Theory of Computing (STOC 1987), pp. 218–229. ACM (1987)

[IKNP03] Ishai, Y., Kilian, J., Nissim, K., Petrank, E.: Extending oblivious transfers efficiently. In: Boneh, D. (ed.) CRYPTO 2003. LNCS, vol. 2729, pp. 145–161. Springer, Heidelberg (2003)

[KM11] Katz, J., Malka, L.: Constant-round private function evaluation with linear complexity. In: Lee, D.H., Wang, X. (eds.) ASIACRYPT 2011. LNCS, vol. 7073, pp. 556–571. Springer, Heidelberg (2011)

[KS08a] Kolesnikov, V., Schneider, T.: Improved garbled circuit: free XOR gates and applications. In: Aceto, L., Damgård, I., Goldberg, L.A., Halldórsson, M.M., Ingólfsdóttir, A., Walukiewicz, I. (eds.) ICALP 2008, Part II. LNCS, vol. 5126, pp. 486–498. Springer, Heidelberg (2008)

[KS08b] Kolesnikov, V., Schneider, T.: A practical universal circuit construction and secure evaluation of private functions. In: Tsudik, G. (ed.) FC 2008. LNCS, vol. 5143, pp. 83–97. Springer, Heidelberg (2008)

[KS16] Kiss, Á., Schneider, T.: Valiant's universal circuit is practical. Cryptology ePrint Archive, Report 2016/093 (2016). http://eprint.iacr.org/2016/093

[LMS16] Lipmaa, H., Mohassel, P., Sadeghian, S.: Valiant's universal circuit: improvements, implementation, and applications. Cryptology ePrint Archive, Report 2016/017 (2016). http://ia.cr/2016/017

[LP09a] Lindell, Y., Pinkas, B.: A proof of security of Yao's protocol for two-party computation. J. Cryptol. **22**(2), 161–188 (2009)

[LP09b] Lovász, L., Plummer, M.D.: Matching Theory. AMS Chelsea Publishing Series. American Mathematical Soc., Providence (2009)

[MNPS04] Malkhi, D., Nisan, N., Pinkas, B., Sella, Y.: Fairplay - secure two-party computation system. In: USENIX Security Symposium, USENIX 2004, pp. 287–302 (2004)

[MS13] Mohassel, P., Sadeghian, S.: How to hide circuits in MPC an efficient framework for private function evaluation. In: Johansson, T., Nguyen, P.Q. (eds.) EUROCRYPT 2013. LNCS, vol. 7881, pp. 557–574. Springer, Heidelberg (2013)

[MSS14] Mohassel, P., Sadeghian, S., Smart, N.P.: Actively secure private function evaluation. In: Sarkar, P., Iwata, T. (eds.) ASIACRYPT 2014, Part II. LNCS, vol. 8874, pp. 486–505. Springer, Heidelberg (2014)

[NPS99] Naor, M., Pinkas, B., Sumner, R.: Privacy preserving auctions and mechanism design. In: ACM Electronic Commerce (EC 1999), pp. 129–139 (1999)

[OI05] Ostrovsky, R., Skeith III, W.E.: Private searching on streaming data. In: Shoup, V. (ed.) CRYPTO 2005. LNCS, vol. 3621, pp. 223–240. Springer, Heidelberg (2005)

[Pin02] Pinkas, B.: Cryptographic techniques for privacy-preserving data mining. SIGKDD Explor. **4**(2), 12–19 (2002)

[PKV+14] Pappas, V., Krell, F., Vo, B., Kolesnikov, V., Malkin, T., Choi, S.G., George, W., Keromytis, A.D., Bellovin, S., Seer, B.: A scalable private DBMS. In: IEEE Symposium on Security and Privacy (S&P 2014), pp. 359–374. IEEE (2014)
[PSS09] Paus, A., Sadeghi, A.-R., Schneider, T.: Practical secure evaluation of semi-private functions. In: Abdalla, M., Pointcheval, D., Fouque, P.-A., Vergnaud, D. (eds.) ACNS 2009. LNCS, vol. 5536, pp. 89–106. Springer, Heidelberg (2009)
[Sch08] Schneider, T.: Practical secure function evaluation. Master's thesis, University Erlangen-Nürnberg, Germany, 27 February 2008
[SS08] Sadeghi, A.-R., Schneider, T.: Generalized universal circuits for secure evaluation of private functions with application to data classification. In: Lee, P.J., Cheon, J.H. (eds.) ICISC 2008. LNCS, vol. 5461, pp. 336–353. Springer, Heidelberg (2009)
[SYY99] Sander, T., Young, A.L., Yung, M.: Non-interactive cryptocomputing for NC^1. In: Foundations of Computer Science (FOCS 1999), pp. 554–567. IEEE (1999)
[TS15] Tillich, S., Smart, N.: Circuits of basic functions suitable for MPC and FHE (2015). http://www.cs.bris.ac.uk/Research/CryptographySecurity/MPC/
[Val76] Valiant, L.G.: Universal circuits (preliminary report). In: ACM Symposium on Theory of Computing (STOC 1976), pp. 196–203. ACM (1976)
[Weg87] Wegener, I.: The Complexity of Boolean Functions. Wiley, New York (1987)
[Yao86] Yao, A.C.-C.: How to generate and exchange secrets (extended abstract). In: Foundations of Computer Science (FOCS 1986), pp. 162–167. IEEE (1986)

Nonce-Based Cryptography: Retaining Security When Randomness Fails

Mihir Bellare and Björn Tackmann(✉)

Department of Computer Science and Engineering,
University of California, San Diego, USA
{mihir,btackmann}@eng.ucsd.edu
http://cseweb.ucsd.edu/~mihir/
http://cseweb.ucsd.edu/~btackmann/

Abstract. We take nonce-based cryptography beyond symmetric encryption, developing it as a broad and practical way to mitigate damage caused by failures in randomness, whether inadvertent (bugs) or malicious (subversion). We focus on definitions and constructions for nonce-based public-key encryption and briefly treat nonce-based signatures. We introduce and construct hedged extractors as a general tool in this domain. Our nonce-based PKE scheme guarantees that if the adversary wants to violate IND-CCA security then it must do *both* of the following: (1) fully compromise the RNG (2) penetrate the sender system to exfiltrate a seed used by the sender.

1 Introduction

An old security adage says there is no point putting strong locks on the door if you leave the window open. The lock here is modern public-key encryption, proven to meet the strong IND-CCA goal. The window is the assumption made in these proofs that, at every encryption, the encryptor has access to perfect, fresh (independent from prior) randomness. To allow encryption to fulfill in practice the promise it makes in theory, we must close the window. This paper develops nonce-based public-key encryption as a practical way to do this. It goes on to develop nonce-based digital signatures.

Randomness. That randomness failures occur and lead to cryptographic failures is by now very well known and does not need to be belabored. The news of interest is perhaps that it is getting worse. Let us explain. There are two sources of randomness failures. The first, which has been with us a while and is not going away, is *bugs*. A good example is the Debian Linux vulnerability present from September 2006 to May 2008 where a programmer removed some lines of code from the OpenSSL source, resulting in there being only 15 bits of entropy in the seed for the PRNG [1]. HDWH [19] finds cryptographic vulnerabilities in a significant fraction of TLS and SSH servers due to what they call "malfunctioning RNGs." And the list goes on. The second source of randomness failures, which may have been with us for a while but of which we have learned only recently due

M. Fischlin and J.-S. Coron (Eds.): EUROCRYPT 2016, Part I, LNCS 9665, pp. 729–757, 2016.
DOI: 10.1007/978-3-662-49890-3_28

to the mass-surveillance revelations, is *subversion*, the deliberate and targeted attempt to weaken randomness. At the South by South West (SXSW) 2014 conference, Snowden said "we know that the encryption algorithms we are using today work... it is the random number generators that are attacked as opposed to the encryption algorithms themselves." The prime example here is Dual EC, a RNG the NSA designed to have a backdoor and then pushed into standards and adoption. The ability to compromise security in practice via the backdoor has been demonstrated in [14].

PRIOR WORK. The basic definitions of security for public-key encryption (PKE), namely IND-CPA [18] and IND-CCA [5,15], provide no guarantees if the randomness is bad. There is a long line of work giving new definitions of security that do provide such guarantees, and building PKE schemes to meet these definitions.

The simplest way to avoid vulnerabilities due to poor randomness is to not use randomness at all. Deterministic PKE [3,6,12,17] however only provides security when the messages have high min-entropy. This limits utility (for example, we may want to encrypt votes, which have low min-entropy) and, although in a different context, CGPR [13] show that in practice the entropy of "real" data is often quite low. Hedged PKE [4,7,27] extends Deterministic PKE to provide privacy as long as the message and randomness together have sufficient min-entropy. This is a significant benefit and we recommend that one use Hedged PKE whenever possible. But the limitations remain. Since messages regularly do not in fact have entropy [13], and the "randomness" can be entirely predictable (this happened both with the Debian Linux bug and the Dual EC subversion), the message and randomness together still may not have enough entropy for hedged encryption to provide security. A further limitation of both Deterministic and Hedged PKE is that security is only provided for messages that do not depend on the public key. (This second limitation can be partially addressed but at some cost [26].) Yilek [32] defines and achieves security against randomness-reset attacks, where the randomness is perfect but the adversary can force its re-use across different encryptions. This is useful in the context of virtual machine resets but not more broadly. PSS [24] introduce related-randomness attacks, where encryption is under adversary-specified functions of some initial uniform randomness. However, negative results they provide show that for many functions one cannot achieve security.

In summary, all these notions have some limitations and the practical benefit they provide is not clear. Most importantly, these were all designed in the older mindset of RNG failures due to bugs and can break down severely when the RNG is subverted. The latter is the new reality against which we need to defend.

NONCES. Rogaway [28,29] introduced nonce-based symmetric-key encryption, where the encryption algorithm is deterministic, taking the shared key, message and a quantity called a nonce. Security is provided as long as the nonce does not repeat. The notion was strengthened by Rogaway and Shrimpton (RS) [30]. Rogaway suggests that packet sequence numbers may play the role of the nonce. Motivations he provides include reducing implementor error and achieving stronger notions of security. We suggest that nonces can be used much more broadly and are a good defense against poor randomness. The goal we pursue in depth is nonce-based PKE.

THE OBVIOUS EXTENSION FAILS. Towards this we begin by noting that a direct extension of nonce-based symmetric-key encryption as defined in [28–30] to the public-key setting does not work. Such an extension would have the encryption algorithm $\mathcal{E}$ be deterministic, taking input the encryption key *ek*, message m and a nonce n to return the ciphertext $C = \mathcal{E}(ek, m, n)$. The privacy game would give the adversary an oracle that takes messages m_0, m_1 and a nonce n (in the definitions of [28–30], nonces are adversarially chosen subject to not repeating) to return $C = \mathcal{E}(ek, m_b, n)$, where b is a challenge bit chosen at random by the game. Security would require that the adversary has little advantage in guessing b. But such security would not be achieved because in the public-key setting the adversary has *ek* and can itself encrypt. Thus it could query its oracle with any m_0, m_1, n of its choice to get back ciphertext $C = \mathcal{E}(ek, m_b, n)$, and itself compute $C_0 = \mathcal{E}(ek, m_0, n)$. If $C = C_0$ it knows that the challenge b is 0, else it is 1.

NONCE-BASED PKE. We define a nonce-based PKE scheme NPE as follows. The receiver runs key-generation algorithm $\mathsf{NPE.Kg}$ as usual to get an encryption key *ek* and decryption key *dk*. Not as usual, the sender begins by locally running a *seed generation* algorithm $\mathsf{NPE.sKg}$ to get a *seed xk*. The encryption algorithm $\mathsf{NPE.Enc}$ is *deterministic*, taking in addition to the usual *ek* and message m, two new inputs, a nonce n and the *seed xk*, and returning ciphertext $C = \mathsf{NPE.Enc}(ek, xk, m, n)$. Decryption is unchanged, taking *dk* and a ciphertext C to return a message. The receiver does *not* need to know the seed, and the keys and seed are entirely independent. The sender can either re-use the same seed across multiple encryptions, or generate a fresh one at every encryption, or anything in between, and the receiver will be oblivious to all of this.

Security is captured via two games and corresponding requirements. Nonce-based privacy One (NBP1) asks that IND-CCA privacy be maintained as long as message-nonce pairs do not repeat. That is, the only way security fails is if, for the same message, a nonce is re-used. This is a very strong guarantee. However there is one caveat, namely this holds when the seed is kept private from the adversary. Nonce-based privacy Two (NBP2) addresses the possibility that the adversary compromises the sender's system and obtains the seed. Even in this case, it guarantees IND-CCA privacy as long as nonces are unpredictable to the adversary. The formalizations are in terms of a stateful *nonce generator* NG that takes an adversary specified input η to return the next nonce, so that nonces are (indirectly) under adversary control.

In practice we would expect a combination of a variety of things to be used as the nonce, for example the current time (this does not repeat) and prior ciphertexts, but, also, randomness from the system RNG, since, for NBP2, nonce unpredictability is required. (This is a departure from the symmetric setting.) However guarantees in the face of poor randomness are much better than before, as we now explain.

WHAT THIS BUYS US. Intuitively our definitions are saying the following. Consider two cases. The expected and good case is that the sender seed stays private. In this case we get IND-CCA privacy *regardless of the quality of the randomness*, the only requirement being that message-nonce pairs do not repeat. The latter is a mild condition, such repetition being unlikely with reasonable nonces, even simply using the date and time as the nonce. The other case is that the sender's system is compromised and the seed is exposed. In this case, we are effectively in the setting of standard PKE and we cannot deterministically provide IND-CCA. We guarantee that we do no worse than standard PKE, meaning we provide IND-CCA as long as the randomness (here part of the nonce) is good. But in fact we do better, since the requirement on nonces is only unpredictability. This means that we are safe even if the outputs of the RNG are correlated and structured, as long as they remain unpredictable.

Put another way, if a subvertor wanted to compromise privacy, it would not suffice to compromise the RNG. They would have to also break in to the sender's system, find the seed, and exfiltrate it. Frequent rotation of seeds (which has effectively no cost) makes this even harder. This ups the ante. Now, it is true that with the NSA's capabilities in malware, we should not under-estimate their ability to penetrate a target sender. But this would have to be done on per-sender basis, making mass surveillance harder.

HEDGED EXTRACTORS AND OUR SCHEME. It is easy to achieve either of NBP1 or NBP2 in isolation. We can get an NBP1 nonce-based PKE scheme by encrypting under a conventional (randomized) IND-CCA PKE scheme with the coins set to the result of a PRF keyed by the sender seed and applied to the message and nonce, but there is no reason this scheme would also be NBP2 secure. We can get an NBP2 nonce-based PKE scheme by encrypting under the conventional IND-CCA PKE scheme with the coins set to the result of an extractor keyed by the seed and applied to the (message and) nonce, but there is no reason this scheme would also be NBP1 secure. To simultaneously get both properties, we introduce and use *hedged extractors*.

A hedged extractor HE takes a seed (also called a key) xk, a message m and a nonce n to deterministically return a string $r = \mathsf{HE}(xk, (m, n))$. It has two properties: (1) It is a PRF, meaning if xk is random and hidden then the outputs look random even to an adversary that picks m, n, and (2) it is an extractor, meaning if xk is random but known, then r looks random if (m, n) is unpredictable (meaning, has enough min-entropy). Again, achieving either goal in isolation is trivial. The task is to achieve them simultaneously, in the same construction. We give two solutions, one in the ROM, the other in the standard model. The first is easy but practical, and likely to be what we would use, namely to simply apply the RO to $xk, (m, n)$. The second combines a PRF with a strong randomness extractor via XOR. The ROM solution delivers optimal security, the standard-model one a bit less due to the inherent limitations of strong randomness extractors, namely that they are only guaranteed to work for seed-independent inputs that retain min-entropy even conditioned on prior inputs.

Our nonce-based PKE scheme is then simply defined via the same paradigm as for the in-isolation cases, namely we encrypt under a conventional (randomized) IND-CCA PKE scheme with the coins set to the result of a hedged extractor keyed by the sender seed and applied to the message and nonce. Both NBP1 and NBP2 security are proven for this scheme assuming IND-CCA of the conventional scheme and security of the hedged extractor.

DISCUSSION AND PRAGMATICS. We can view nonce-based cryptography as moving the traditional abstraction boundary between cryptography and system RNGs closer to the cryptography. The view is that, in the presence of bad RNGs, a safer and better architecture is that the cryptography take on as large a share of the burden of providing security as possible, in other words, rely on its environment as little as possible. Our suggestion here is that the environment is relied on only to produce nonces with relatively weak requirements. This view is in some ways the opposite of that represented by work that aims to strengthen RNGs against failure or subversion [16]. In practice the two can co-exist and their combined presence will increase security.

Our nonce-based encryption scheme is simple and modular, a way to transform any given conventional IND-CCA scheme into an NBP1+NBP2 secure nonce-based scheme. With a practical choice of hedged extractor such as our ROM one, we retain the efficiency attributes of the initial PKE scheme. In our scheme, decryption is unchanged. The decryptor does not need to change its software or even know that nonces are being used. These attributes make it easier to deploy nonce-based PKE as a practical defense against poor randomness.

In the above-discussed prior work aimed at increasing resistance of PKE to randomness failures, the *model* was unchanged in the sense that the object whose security was being considered continued in syntax to be a classical public-key encryption scheme as per [18]. Nonce-based encryption is a new model (because the sender has a seed) and a new syntax (there is a seed generation algorithm and the encryption algorithm is different). It is these changes, and in particular not just the nonce, but the combination of nonce and seed, that are a game changer and result in significantly better guarantees against poor randomness compared to prior work.

Picking a seed, like picking a key, does require (good) randomness. The viewpoint here, as in all the prior work discussed above, is that there is a difference between static and dynamic randomness usage. We assume good randomness for key generation because effort can be invested in it. Current key-generation software often has the user generate coins by waving their mouse around. Good seed generation would require similar effort, but one would expect to use a seed for some time so this effort is not frequent. This is flexible. If a seed is lost due to a system reboot, or compromised, the user can elect to make the effort to pick a new one. If you are encrypting from multiple systems (your desktop, laptop and phone) each can have its own, independently chosen seed.

NONCE-BASED SIGNATURES AND BEYOND. We define nonce-based signatures, where the signing algorithm is deterministic, and takes not only the signing key and message, but also a seed and nonce. We require that (1) if the seed remains hidden then we have regular security (unforgeability) regardless of how nonces are generated, and (2) if the seed is exposed, then we have security as long as the nonces were unpredictable. Section 5 formalizes this and shows how to convert any signature scheme into a nonce-based one with these security properties using a hedged extractor.

Due to their speed and short signature sizes, the most attractive signature schemes for practice are elliptic-curve versions of DSA, El Gamal and Schnorr [31]. However, they are randomized, and fail spectacularly when the randomness is bad. Discussions on the `cfrg` forum show overwhelming support for making these schemes deterministic. This is easily done, by deriving the coins either as a PRF, keyed by a seed that is part of the secret key and applied to the message, or as a RO applied to the secret key and message [9,21,23,25], and the popular Ed25519 signature scheme of [11] already embodies this. Making a scheme nonce-based complements this traditional de-randomization, retaining the benefits of deterministic signing while adding further ones. See Sect. 5 for more extensive background and discussion.

Nonces in combination with seeds can similarly be used in many other areas of cryptography to provide resilience in the face of poor randomness or even provide other gains. Our work aims to be illustrative rather than exhaustive.

RELATED WORK. BKS [8] introduce stateful PKE. Here also the sender can maintain a seed. They show that this leads to significant efficiency gains. Their schemes are however randomized, and there are no nonces. An interesting direction for future work is to combine their methods with ours to get similar efficiency gains for nonce-based PKE.

Rogaway [29] discusses nonces as "surfacing the IV." As motivation, he says that when IVs are implicit, implementors and even books get things wrong. He says that often nonces are readily available, for example packet sequence numbers. He does not seem to explicitly mention robustness in the face of randomness failure as a goal in the symmetric case. Intriguingly, in the final section of the paper, he goes on to say: " ... it makes just as much sense to consider nonce-based public-key encryption schemes as it does to consider nonce-based symmetric encryption schemes. This provides an approach to effectively weakening the requirement for randomness on the sender." Our work has pursued this suggestion. It is surprising that this waited 12 years.

2 Notation and Standard Definitions

NOTATION. We let ε denote the empty string. If X is a finite set, we let $x \leftarrow\!\!\$\ X$ denote picking an element of X uniformly at random and assigning it to x. Algorithms may be randomized unless otherwise indicated. Running time is worst case. If A is an algorithm, we let $y \leftarrow A(x_1, \ldots; r)$ denote running A with random coins r on inputs $x_1, \ldots$ and assigning the output to y. We let $y \leftarrow\!\!\$\ A(x_1, \ldots)$ be the result of picking r at random and letting $y \leftarrow A(x_1, \ldots; r)$. We let $[A(x_1, \ldots)]$ denote the set of all possible outputs of A when invoked with inputs $x_1, \ldots$. We use the code based game playing framework of [10].

Game $\mathbf{G}^{\mathrm{prf}}_{\mathsf{F}}(\mathcal{A})$	Game $\mathbf{G}^{\mathrm{ind}}_{\mathsf{PE}}(\mathcal{A})$	Game $\mathbf{G}^{\mathrm{uf}}_{\mathsf{DS}}(\mathcal{A})$

$fk \leftarrow\$ \mathsf{F.Keys}$
$c \leftarrow\$ \{0,1\}$
$c' \leftarrow\$ \mathcal{A}^{\mathrm{FN,RO}}$
Return $(c = c')$
 	
FN(x)
If $(c = 1)$ then
  $S[x] \leftarrow \mathsf{F}^{\mathrm{RO}}(fk, x)$
Else
  If $S[x] = \bot$ then
    $S[x] \leftarrow\$ \mathsf{F.Rng}$
Return $S[x]$
 	
RO(x, l)
If $T[x, l] = \bot$ then
  $T[x, l] \leftarrow\$ \{0,1\}^l$
Return $T[x, l]$
	$(ek, dk) \leftarrow\$ \mathsf{PE.Kg}$
	$b \leftarrow\$ \{0,1\}$; $C \leftarrow \emptyset$
	$b' \leftarrow\$ \mathcal{A}^{\mathrm{ENC,DEC}}(ek)$
	Return $(b = b')$
	
	ENC(m_0, m_1)
	If $(\lvert m_0 \rvert \neq \lvert m_1 \rvert)$ then return $\bot$
	$c \leftarrow\$ \mathsf{PE.Enc}(ek, m_b)$
	$C \leftarrow C \cup \{c\}$
	Return c
	
	DEC(c)
	If $(c \in C)$ then return $\bot$
	$m \leftarrow \mathsf{PE.Dec}(ek, dk, c)$
	Return m
		$(sk, vk) \leftarrow\$ \mathsf{DS.Kg}$
		$M \leftarrow \emptyset$
		$(m, s) \leftarrow\$ \mathcal{A}^{\mathrm{SIG}}(vk)$
		$v \leftarrow (\mathsf{DS.Ver}(vk, m, s) = 1)$
		Return $(v \wedge (m \notin M))$
		
		SIG(m)
		$s \leftarrow \mathsf{DS.Sig}(sk, m)$
		$M \leftarrow M \cup \{m\}$
		Return s

Fig. 1. Games for defining PRF security of a function family F, standard IND-CCA security of a standard PKE scheme PE and EUF-CMA security of a signature scheme DS.

(See Fig. 1 for an example.) By Pr[G] we denote the event that the execution of game G results in the game returning true. Random oracles are variable output length, represented by a game procedure RO that takes x, l and returns a random string of length l. The min-entropy of a random variable X over $\mathcal{X}$ is defined as $\mathrm{H}_\infty(\mathsf{X}) = -\log(\max_{x \in \mathcal{X}}(\Pr[\mathsf{X} = x]))$.

FUNCTION FAMILIES. A family of functions $\mathsf{F}\colon \mathsf{F.Keys} \times \mathsf{F.Dom} \to \mathsf{F.Rng}$ is a two-argument function that takes a key K in the key space $\mathsf{F.Keys}$, an input x in the domain $\mathsf{F.Dom}$ and returns an output $\mathsf{F}(K, x)$ in the range $\mathsf{F.Rng}$. In the ROM, F takes an oracle RO.

PSEUDO-RANDOM FUNCTIONS. The security of F as a PRF is defined via game $\mathbf{G}^{\mathrm{prf}}_{\mathsf{F}}(\mathcal{A})$ that is associated to adversary $\mathcal{A}$ and shown in Fig. 1. Here F could have access to a RO and thus the game is in the ROM. Tables S, T are assumed initially $\bot$ everywhere. The advantage of $\mathcal{A}$ is defined as $\mathbf{Adv}^{\mathsf{prf}}_{\mathsf{F}}(\mathcal{A}) = 2\Pr[\mathbf{G}^{\mathrm{prf}}_{\mathsf{F}}(\mathcal{A})] - 1$.

PUBLIC-KEY ENCRYPTION. A *public-key encryption scheme* PE specifies the following. Receiver key-generation algorithm $\mathsf{PE.Kg}$ returns an encryption key ek and associated decryption key dk. Encryption algorithm $\mathsf{PE.Enc}$ takes ek and message $m \in \{0,1\}^*$ to return a ciphertext c. Deterministic decryption algorithm $\mathsf{PE.Dec}$ takes ek, dk and ciphertext c to return a value in $\{0,1\}^* \cup \{\bot\}$, and we require standard decryption correctness. The advantage of an adversary $\mathcal{A}$ in breaking the IND-CCA security of PE is defined as $\mathbf{Adv}^{\mathsf{ind}}_{\mathsf{PE}}(\mathcal{A}) =$

$2\Pr[\mathbf{G}^{\mathsf{ind}}_{\mathsf{PE}}(\mathcal{A}) - 1$ for the game $\mathbf{G}^{\mathsf{ind}}_{\mathsf{PE}}(\mathcal{A})$ described in Fig. 1. This represents a conventional (not nonce-based scheme), and thus PE.Enc is randomized. We will use such schemes as base schemes and we will need to surface their coins, writing $c \leftarrow \mathsf{PE.Enc}(ek, m; r)$ to mean that PE.Enc is run with coins r to deterministically return c. The length of the coins is denoted PE.rl.

DIGITAL SIGNATURES. A digital signature scheme DS specifies the following. Signer key-generation algorithm DS.Kg returns a signature key sk and a verification key vk. Signing algorithm DS.Sig takes sk and message $m \in \{0,1\}^*$ to return a signature $s \in \{0,1\}^{\mathsf{DS.ol}}$. Verification algorithm DS.Ver takes vk, message $m \in \{0,1\}^*$, and signature $s \in \{0,1\}^{\mathsf{DS.ol}}$, to return a bit $b \in \{0,1\}$. The advantage of an adversary $\mathcal{A}$ in breaking the EUF-CMA security of DS is defined as $\mathbf{Adv}^{\mathsf{uf}}_{\mathsf{DS}}(\mathcal{A}) = \Pr[\mathbf{G}^{\mathsf{uf}}_{\mathsf{DS}}(\mathcal{A})]$ for the game $\mathbf{G}^{\mathsf{uf}}_{\mathsf{DS}}(\mathcal{A})$ described in Fig. 1. Again we may need to surface the coins $r \in \{0,1\}^{\mathsf{DS.rl}}$ of DS.Sig, writing $s \leftarrow \mathsf{DS.Sig}(sk, m; r)$.

3 Hedged Extractors

Our nonce-based schemes work simply by supplying coins to a base scheme via a *hedged extractor* keyed by the sender seed. This primitive, that we introduce and build here, is a function family that has two security properties. The first is that it is a PRF. The second, which we define and call ror (real or random) security, formalizes randomness of outputs when the key (seed) is known. Clearly this can only be achieved with some restrictions, and the type of ror security achieved will vary across constructions, from the "best possible" achieved by our ROM construction to a weaker, but we think still meaningful, version for our standard model construction. To make the goals precise we first introduce the notion of a nonce generator.

NONCE GENERATORS. A nonce generator is an algorithm NG that, on input a *nonce selector* η and a current state St, returns a nonce n, belonging to the range set $\mathsf{NG.Rng} \subseteq \{0,1\}^*$ of NG, together with an updated state, written $(n, St) \leftarrow\!\!\$\ \mathsf{NG}(\eta, St)$. We say the generator has nonce length $\mathsf{NG.nl} \in \mathbb{N}$ if $\mathsf{NG.Rng} = \{0,1\}^{\mathsf{NG.nl}}$. Let $\mathcal{P}$ be an adversary called a predictor and consider game $\mathbf{G}^{\mathsf{pred}}_{\mathsf{NG}}(\mathcal{P})$ of Fig. 2. Let

$$\mathbf{Adv}^{\mathsf{pred}}_{\mathsf{NG}}(\mathcal{P}) = \Pr[\mathbf{G}^{\mathsf{pred}}_{\mathsf{NG}}(\mathcal{P})] \quad \text{and} \quad \mathbf{Adv}^{\mathsf{pred}}_{\mathsf{NG}}(q_1, q_2) = \max_{\mathcal{P}} \mathbf{Adv}^{\mathsf{pred}}_{\mathsf{NG}}(\mathcal{P}) \,,$$

where the maximum is over all $\mathcal{P}$ making at most $q_1 \in \mathbb{N}$ queries to GEN and $q_2 \in \{0,1\}$ queries to EXPOSE. Now let us explain. A call to GEN generates the next nonce in the sequence, returning nothing to the adversary, The adversary can influence the choice of nonces through its choice of the selector η. The EXPOSE oracle allows to additionally get access to the state of the nonce generator. To win, the adversary needs to guess some generated nonce or create a collision between generated nonces.

Game $\mathbf{G}^{\text{pred}}_{\mathsf{NG}}(\mathcal{P})$	Game $\mathbf{G}^{\text{ror}}_{\mathsf{HE},\mathsf{NG}}(\mathcal{G})$
$St \leftarrow \varepsilon$; $s \leftarrow 0$; $N \leftarrow \emptyset$	$St \leftarrow \varepsilon$; $xk \leftarrow\!\!\$\ \mathsf{HE.Keys}$; $c \leftarrow\!\!\$\ \{0,1\}$; $s \leftarrow 0$
$n \leftarrow\!\!\$\ \mathcal{P}^{\text{GEN,EXPOSE}}$	$c' \leftarrow\!\!\$\ \mathcal{G}^{\text{RoR,EXPOSE,RO}}(xk)$
Return $((n \in N)$ OR coll)	Return $(c = c')$
GEN(η)	RoR(m, η)
If $s = 1$ then return $\bot$	If $s = 1$ then return $\bot$
$(n, St) \leftarrow\!\!\$\ \mathsf{NG}(\eta, St)$	$(n, St) \leftarrow\!\!\$\ \mathsf{NG}(\eta, St)$
If $(n \in N)$ then coll $\leftarrow$ true	If $(c = 1)$ then $r \leftarrow \mathsf{HE}^{\text{RO}}(xk, (m, n))$
$N \leftarrow N \cup \{n\}$	Else $r \leftarrow\!\!\$\ \mathsf{HE.Rng}$
EXPOSE	Return r
$s \leftarrow 1$	EXPOSE
Return St	$s \leftarrow 1$; Return St
	RO(x, l)
	If $T[x, l] = \bot$ then $T[x, l] \leftarrow\!\!\$\ \{0,1\}^l$
	Return $T[x, l]$

Fig. 2. Games for defining predictability of the nonce generator NG and real-or-random security of function family HE.

Nonce generators represent another departure from nonce-based symmetric encryption. In the latter the adversary picks the nonce, but we saw in Sect. 1 that this does not work in the public-key setting. Instead, we model the process of a sender picking a nonce via a nonce generator.

In discussions, we refer to NG as weakly unpredictable if it is unpredictable for adversaries making no EXPOSE query, meaning $\mathbf{Adv}^{\text{pred}}_{\mathsf{NG}}(q_1, 0)$ is "small" for "practical" values of q_1, and strongly unpredictable if it is unpredictable even for adversaries making an EXPOSE query, meaning $\mathbf{Adv}^{\text{pred}}_{\mathsf{NG}}(q_1, 1)$ is "small" for "practical" values of q_1. If NG is strongly unpredictable it is also weakly unpredictable, but not necessarily vice versa. That is, the class of weakly unpredictable nonce generators is larger than the class of strongly unpredictable nonce generators.

Real or random security. Let $\mathsf{HE}\colon \mathsf{HE.Keys} \times \mathsf{HE.Dom} \rightarrow \mathsf{HE.Rng}$ be an oracle family of functions (this means it may have access to a random oracle). The first input is referred to as the "key" or the "seed." The domain has the form $\mathsf{HE.Dom} = \{0,1\}^* \times \mathsf{HE.NS}$, so that an input is a pair of strings, the first referred to as the "message" and the second as the "nonce," the latter drawn from a nonce space HE.NS associated to HE. Consider game $\mathbf{G}^{\text{ror}}_{\mathsf{HE},\mathsf{NG}}(\mathcal{G})$ of Fig. 2 associated to HE, nonce generator NG and an adversary $\mathcal{G}$. The number of queries to EXPOSE is either 0 or 1, and the number to other oracles is arbitrary. Let

$$\mathbf{Adv}^{\text{ror}}_{\mathsf{HE},\mathsf{NG}}(\mathcal{G}) = 2\Pr[\mathbf{G}^{\text{ror}}_{\mathsf{HE},\mathsf{NG}}(\mathcal{G})] - 1 .$$

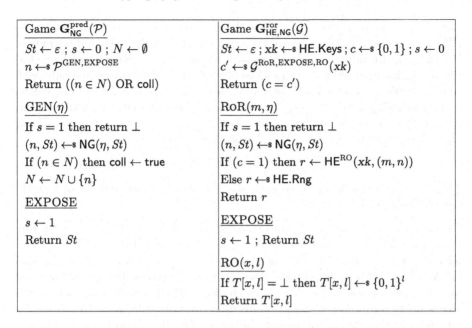

Note that here the adversary is given the key (seed) *xk* as input, unlike in the PRF notion, modeling exposure of the sender seed. Security asks that outputs of $\mathsf{HE}(xk, (\cdot, \cdot))$, for adversary-chosen messages and nonces from the nonce generator, are indistinguishable from random. Clearly, this will be possible only with certain restrictions, which will emerge when we discuss our constructions below.

In game $\mathbf{G}^{\mathrm{ror}}_{\mathsf{HE},\mathsf{NG}}(\mathcal{G})$, we say that adversary $\mathcal{G}$ is *agnostic* if its RoR queries do not depend on the seed *xk*. More formally, there exists a pair $(\mathcal{G}_1, \mathcal{G}_2)$ of algorithms such that $\mathcal{G}^{\mathrm{RoR,EXPOSE,RO}}(xk)$ does the following:

$$St \leftarrow\$\ \mathcal{G}_1^{\mathrm{RoR,EXPOSE,RO}}(\varepsilon)\ ;\ c' \leftarrow \mathcal{G}_2^{\mathrm{EXPOSE,RO}}(xk, St)\ ;\ \text{Return } c'\ .$$

This represents one of the restrictions we will impose to achieve ror security in the standard model.

HEDGED EXTRACTORS. A *hedged extractor* $\mathsf{HE}\colon \mathsf{HE.Keys} \times \mathsf{HE.Dom} \to \mathsf{HE.Rng}$ is an oracle family of functions as above where the goal is that (1) HE is a PRF, meaning $\mathbf{Adv}^{\mathrm{prf}}_{\mathsf{HE}}(\mathcal{A})$ is low for any adversary $\mathcal{A}$ of practical resources, and also (2) $\mathbf{Adv}^{\mathrm{ror}}_{\mathsf{HE},\mathsf{NG}}(\mathcal{G})$ is small for some class of nonce generators NG and some class of ror adversaries $\mathcal{G}$, both specified via results for individual hedged extractors. We give a ROM construction and standard model one. Both achieve PRF security, but differ in the type of ror security achieved. The ROM construction achieves ror security for unpredictable nonce generators (both weak and strong) and for all ror adversaries. This is "best possible" because the unpredictability assumption is easily seen to be necessary. The standard model construction achieves ror security for strongly unpredictable generators and agnostic ror adversaries. These restrictions reflect limitations of the randomness extractors that are our underlying tool. The restriction to agnostic adversaries reflects that randomness extractors only work on seed-independent distributions, and the strong unpredictability requirement on the generator reflects that when extracting from a sequence of inputs, one needs not only that each has some min-entropy, but that it does even given the others.

ROM HEDGED EXTRACTOR. We start by giving a simple and efficient construction **HE1** of a hedged extractor in the ROM. Let ℓ be a desired number of output bits for the extractor, and k a desired seed (key) length. Associated to ℓ, k is the hedged extractor $\mathsf{HE} = \mathbf{HE1}[\ell, k]\colon\ \{0,1\}^k \times (\{0,1\}^* \times \{0,1\}^*) \to \{0,1\}^\ell$ defined by

$$\mathsf{HE}^{\mathrm{RO}}(xk, x) = \mathrm{RO}((xk, x), \ell)\ .$$

Here $\mathsf{HE.Keys} = \{0,1\}^k$, $\mathsf{HE.Dom} = \{0,1\}^* \times \mathsf{HE.NS}$ with $\mathsf{HE.NS} = \{0,1\}^*$, and $\mathsf{HE.Rng} = \{0,1\}^\ell$.

The following lemma states that this construction achieves PRF security and also achieves real or random security assuming only that the nonce generator is unpredictable. Note the latter only requires each nonce to individually be unpredictable, but nonces may be arbitrarily correlated, and it could be that given n_1 one can easily predict n_2. But the extractor works nonetheless.

Lemma 1. *Let* $\ell, k \geq 1$ *be integers and let* $\mathsf{HE} = \mathbf{HE1}[\ell, k]$ *be the ROM function family associated to* ℓ *and* k *as above.*

1. *If $\mathcal{A}$ is an adversary making q_2 queries to its* RO *oracle, then*

$$\mathbf{Adv}^{\mathsf{prf}}_{\mathsf{HE}}(\mathcal{A}) \le \frac{q_2}{2^k} \,. \tag{1}$$

2. *Let* NG *be a nonce generator. If $\mathcal{G}$ is an adversary making q_1 queries to its* RoR *oracle, q_2 queries to its* RO *oracle, and $q_3 \in \{0,1\}$ queries to its* EXPOSE *oracle, then*

$$\mathbf{Adv}^{\mathsf{ror}}_{\mathsf{HE},\mathsf{NG}}(\mathcal{G}) \le q_2 \cdot \mathbf{Adv}^{\mathsf{pred}}_{\mathsf{NG}}(q_1, q_3) \,. \tag{2}$$

Note that in the 2nd case, the reduction preserves the number of EXPOSE queries, meaning the number reflected by q_3 is the number made by $\mathcal{G}$. This is the best one could hope for.

Proof (Lemma 1). For the proof of Eq. 1, consider the games $\mathrm{G}_0, \mathrm{G}_1$ of Fig. 3, where G_1 contains the boxed code and G_0 does not. Letting c denote the challenge bit in game $\mathbf{G}^{\mathrm{prf}}_{\mathsf{HE}}(\mathcal{A})$, the following, justified below, establishes Eq. 1:

$$\begin{aligned}
\mathbf{Adv}^{\mathsf{prf}}_{\mathsf{HE}}(\mathcal{A}) &= \Pr[\,\mathbf{G}^{\mathrm{prf}}_{\mathsf{HE}}(\mathcal{A}) \,|\, c = 1\,] - \left(1 - \Pr[\,\mathbf{G}^{\mathrm{prf}}_{\mathsf{HE}}(\mathcal{A}) \,|\, c = 0\,]\right) &(3)\\
&= \Pr[\mathrm{G}_1] - \Pr[\mathrm{G}_0] &(4)\\
&\le \Pr[\mathrm{G}_0 \text{ sets } \mathsf{bad}] &(5)\\
&\le \frac{q_2}{2^k} \,. &(6)
\end{aligned}$$

Equation 3 is a standard re-formulation of the definition of the advantage. In game G_0, replies to queries to the FN and RO oracles are independently distributed, so that it is equivalent to the $c = 0$ case of game $\mathbf{G}^{\mathrm{prf}}_{\mathsf{HE}}(\mathcal{A})$, up to the flipping of the outcomes from true to false. In game G_1, the replies to FN queries are given by $\mathsf{HE}^{\mathrm{RO}}$, making it equivalent to the $c = 1$ case of game $\mathbf{G}^{\mathrm{prf}}_{\mathsf{HE}}(\mathcal{A})$. This justifies Eq. 4. Games $\mathrm{G}_0, \mathrm{G}_1$ are identical until bad (differ only in statements following the setting of bad to true), so the Fundamental Lemma of Game Playing [10] justifies Eq. 5. In game G_0, replies to all oracle queries are random and independent of xk so the probability that the latter is queried as part of an RO query is at most the quantity of Eq. 6.

For Eq. 2, consider the games $\mathrm{G}_2, \mathrm{G}_3$ of Fig. 3, where G_3 contains the boxed code and G_2 does not. Let predictor adversary $\mathcal{P}$ be as specified in Fig. 3. Letting c denote the challenge bit in game $\mathbf{G}^{\mathrm{ror}}_{\mathsf{HE},\mathsf{NG}}(\mathcal{G})$, the following, justified below, establishes Eq. 2:

$$\begin{aligned}
\mathbf{Adv}^{\mathsf{ror}}_{\mathsf{HE},\mathsf{NG}}(\mathcal{G}) &= \Pr[\,\mathbf{G}^{\mathrm{ror}}_{\mathsf{HE},\mathsf{NG}}(\mathcal{G}) \,|\, c = 1\,] - (1 - \Pr[\,\mathbf{G}^{\mathrm{ror}}_{\mathsf{HE},\mathsf{NG}}(\mathcal{G}) \,|\, c = 0\,]) &(7)\\
&= \Pr[\mathrm{G}_3] - \Pr[\mathrm{G}_2] &(8)\\
&\le \Pr[\mathrm{G}_2 \text{ sets } \mathsf{bad}] &(9)\\
&\le q_2 \cdot \mathbf{Adv}^{\mathsf{pred}}_{\mathsf{NG}}(\mathcal{G}) &(10)\\
&\le q_2 \cdot \mathbf{Adv}^{\mathsf{pred}}_{\mathsf{NG}}(q_1, q_3) \,. &(11)
\end{aligned}$$

Game G_0 / $\boxed{\mathrm{G}_1}$

$xk \leftarrow\!\!\$\ \{0,1\}^k$; $c' \leftarrow\!\!\$\ \mathcal{A}^{\mathrm{FN},\mathrm{RO}}$; Return $(c' = 1)$

<u>FN(w)</u>

If $S[w] = \bot$ then $S[w] \leftarrow\!\!\$\ \{0,1\}^\ell$; Return $S[w]$

<u>RO(w, l)</u>

If $T[w,l] = \bot$ then
  $T[w,l] \leftarrow\!\!\$\ \{0,1\}^l$; $(u, x) \leftarrow w$
  If $((u = xk)$ and $(l = \ell))$ then
    If $(S[w] = \bot)$ then $S[w] \leftarrow\!\!\$\ \{0,1\}^\ell$; bad $\leftarrow$ true ; $\boxed{T[w,l] \leftarrow S[w]}$
Return $T[w,l]$

Game G_2 / $\boxed{\mathrm{G}_3}$

$St \leftarrow \varepsilon$; $xk \leftarrow\!\!\$\ \{0,1\}^k$ $S \leftarrow \emptyset$
$c' \leftarrow\!\!\$\ \mathcal{G}^{\mathrm{RoR},\mathrm{EXPOSE},\mathrm{RO}}(xk)$; Return $(c' = 1)$

<u>RoR(m, η)</u>

If $s = 1$ then return $\bot$
$(n, St) \leftarrow\!\!\$\ \mathsf{NG}(\eta, St)$
$r \leftarrow R[m,n]$; $R[m,n] \leftarrow\!\!\$\ \{0,1\}^\ell$
If $((m,n) \in S)$ then bad $\leftarrow$ true ; $\boxed{R[m,n] \leftarrow r}$
If $(T[(xk,(m,n)),\ell] \neq \bot)$ then
  bad $\leftarrow$ true ; $\boxed{R[m,n] \leftarrow T[(xk,(m,n)),\ell]}$
$S \leftarrow S \cup \{(m,n)\}$
Return $R[m,n]$

<u>EXPOSE</u>

$s \leftarrow 1$
Return St

<u>RO(w, l)</u>

If $(T[w,l] = \bot)$ then
  $T[w,l] \leftarrow\!\!\$\ \{0,1\}^l$; $(u,x) \leftarrow w$; $(m,n) \leftarrow x$
  If $((u = xk)$ and $(l = \ell)$ and $((m,n) \in S))$ then
    bad $\leftarrow$ true ; $\boxed{T[w,l] \leftarrow R[m,n]}$
Return $T[w,l]$

Adversary $\mathcal{P}^{\mathrm{GEN}}$

$xk \leftarrow\!\!\$\ \{0,1\}^k$; $i \leftarrow 0$
$g \leftarrow\!\!\$\ \{1, \ldots, q_2\}$
$c' \leftarrow\!\!\$\ \mathcal{G}^{\mathrm{RoR},\mathrm{EXPOSE},\mathrm{RO}}(xk)$
Return n_g

<u>RoR(m, η)</u>

If $s = 1$ then return $\bot$
GEN(η)
$r \leftarrow\!\!\$\ \{0,1\}^\ell$
Return r

<u>EXPOSE</u>

$s \leftarrow 1$
$St \leftarrow$ EXPOSE
Return St

<u>RO(w, l)</u>

If $(T[w,l] = \bot)$ then
  $T[w,l] \leftarrow\!\!\$\ \{0,1\}^l$
  $(u,x) \leftarrow w$; $(m,n) \leftarrow x$
  $i \leftarrow i + 1$; $n_i \leftarrow n$
Return $T[w,l]$

Fig. 3. Games and adversary for proof of Lemma 1.

Equation 7 is a standard re-formulation of the definition of the advantage. In game G_2, replies to queries to the RoR and RO oracles are independently distributed, so that it is equivalent to the $c = 0$ case of game $\mathbf{G}^{\mathrm{ror}}_{\mathsf{HE},\mathsf{NG}}(\mathcal{G})$, up to the flipping of the outcomes from true to false. In game G_3, the replies to RoR queries are given by $\mathsf{HE}^{\mathrm{RO}}$, making it equivalent to the $c = 1$ case of game $\mathbf{G}^{\mathrm{ror}}_{\mathsf{HE},\mathsf{NG}}(\mathcal{G})$. This justifies Eq. 8. Games G_2, G_3 are identical until bad, so the Fundamental Lemma of Game Playing [10] justifies Eq. 9. In game G_2, replies to all oracle queries are random and independent, so adversary $\mathcal{P}$ can simulate the oracles of adversary $\mathcal{G}$ directly. Its output is a random one of the nonces in a RO-query of $\mathcal{G}$, whence Eq. 10. Equation 11 is because $\mathcal{P}$ makes q_1 calls to its GEN oracle and q_3 queries to its EXPOSE oracle. □

Standard-model hedged extractor. Next we give a standard-model hedged extractor based on a PRF F and an almost-XOR-universal hash function H. We use the latter essentially as a strong extractor. The construction is simple: the PRF is evaluated on the message and nonce, and the hash function is evaluated only on the nonce. The results are combined via a simple XOR operation. The intuition behind this is that as long as at least one of the outputs generated by the two schemes is random, then the result is also random. PRF security of the hedged extractor is proved assuming only on the assumed PRF security of F. Real-or-random security of the hedged extractor is shown for a restricted class of nonce generators NG and adversaries $\mathcal{G}$. Namely NG must retain unpredictability even in the presence of an EXPOSE query revealing the state, and $\mathcal{G}$'s RoR queries must not depend on the seed. These restrictions reflect inherent limitations of strong extractors.

We start with some definitions. For $\epsilon \in [0,1]$, function family H is *ϵ-almost XOR-universal* [22] if $\mathsf{H.Rng} = \{0,1\}^\ell$ for some $\ell \in \mathbb{N}$ and, for all distinct $x, y \in \mathsf{H.Dom}$ and all $s \in \{0,1\}^\ell$, we have

$$\Pr[\mathsf{H}(hk,x) \oplus \mathsf{H}(hk,y) = s \;:\; hk \leftarrow_{\$} \mathsf{H.Keys}] \leq \epsilon.$$

Our standard model construction is as follows. Let ℓ be a desired number of output bits for the extractor. Let $\mathsf{F}\colon \mathsf{F.Keys} \times (\{0,1\}^* \times \{0,1\}^*) \to \{0,1\}^\ell$ be a function family assumed to be a PRF, and let $\mathsf{H}\colon \mathsf{H.Keys} \times \mathsf{H.Dom} \to \{0,1\}^\ell$ be an almost-XOR-universal hash function with $\mathsf{H.Dom} \subseteq \{0,1\}^*$. We associate to $\ell, \mathsf{F}, \mathsf{H}$ the standard-model hedged extractor $\mathsf{HE} = \mathbf{HE2}[\mathsf{F},\mathsf{H}]\colon (\mathsf{F.Keys} \times \mathsf{H.Keys}) \times (\{0,1\}^* \times \mathsf{H.Dom}) \to \{0,1\}^\ell$ defined by

<u>Algorithm $\mathsf{HE}(xk, (x, y))$</u>

$(hk, fk) \leftarrow xk$; $z_1 \leftarrow \mathsf{H}(hk, y)$; $z_2 \leftarrow \mathsf{F}(fk, (x, y))$; Return $z_1 \oplus z_2$

Here $\mathsf{HE.Keys} = \mathsf{F.Keys} \times \mathsf{H.Keys}$, $\mathsf{HE.Dom} = \{0,1\}^* \times \mathsf{HE.NS}$ with $\mathsf{HE.NS} = \mathsf{H.Dom}$, and $\mathsf{HE.Rng} = \{0,1\}^\ell$. The following says this hedged extractor achieves PRF security and restricted real-or-random security.

Lemma 2. *Let $\ell \geq 1$ be an integer. Let $\mathsf{F}\colon \mathsf{F.Keys} \times (\{0,1\}^* \times \{0,1\}^*) \to \{0,1\}^\ell$ be a function family. Let $\mathsf{H}\colon \mathsf{H.Keys} \times \mathsf{H.Dom} \to \{0,1\}^\ell$ be a $(1+\gamma) \cdot 2^{-\ell}$-almost-XOR-universal hash function. Let $\mathsf{HE} = \mathbf{HE2}[\mathsf{F},\mathsf{H}]$ be the function family associated to ℓ, F and H as above.*

1. *If $\mathcal{A}$ is an adversary making q queries to its* FN *oracle then there is an adversary $\mathcal{B}$ (described in the proof) such that*

$$\mathbf{Adv}^{\mathsf{prf}}_{\mathsf{HE}}(\mathcal{A}) \leq \mathbf{Adv}^{\mathsf{prf}}_{\mathsf{F}}(\mathcal{B}) . \tag{12}$$

Adversary $\mathcal{B}$ also makes q queries to its FN *oracle and has running time that of $\mathcal{A}$ plus the time for q computations of* H.

2. *Let* NG *be a nonce generator that produces outputs in the set* H.Dom. *If $\mathcal{G}$ is an agnostic adversary making q queries to its* RoR *oracle and $\mathbf{Adv}^{\mathsf{pred}}_{\mathsf{NG}}(q,1) \leq 2^{-m}$, then*

$$\mathbf{Adv}^{\mathsf{ror}}_{\mathsf{HE},\mathsf{NG}}(\mathcal{G}) \leq \frac{q}{2}\sqrt{\gamma + \frac{2^{\ell}}{2^m}} . \tag{13}$$

To prove this we first need some more definitions. Recall that for $\epsilon \in [0,1]$, a function family H is *ϵ-almost universal* if $\Pr[\mathsf{H}(hk,x) = \mathsf{H}(hk,y) : hk \leftarrow\!\!\$\ \mathsf{H.Keys}] \leq \epsilon$ for all distinct $x, y \in \mathsf{H.Dom}$. For a function family H with $\mathsf{H.Dom} = \{0,1\}^* \times \{0,1\}^*$, we say that H is *$\epsilon$-almost universal in the second input component* if for all $x_1, y_1 \in \{0,1\}^*$ and all $x_2, y_2 \in \{0,1\}^*$ with $x_2 \neq y_2$, $\Pr[\mathsf{H}(hk,(x_1,x_2)) = \mathsf{H}(hk,(y_1,y_2)) : hk \leftarrow\!\!\$\ \mathsf{H.Keys}] \leq \epsilon$. For $k \in \mathbb{N}$ and $\epsilon \in [0,1]$, a (k,ϵ)-strong extractor SE is a (standard model) family of functions such that $\mathbf{Adv}^{\mathsf{ror}}_{\mathsf{E},\mathsf{NG}}(1,0,1) \leq \epsilon$ for all NG with $\mathbf{Adv}^{\mathsf{pred}}_{\mathsf{NG}}(1,1) \leq 2^{-k}$. This is a re-formulation of the standard requirement that is clearly equivalent to it. The Leftover Hash Lemma, a celebrated result by Impagliazzo, Levin, and Luby [20], states that an almost universal hash function is a strong extractor:

Lemma 3 (Leftover Hash Lemma). *Let $\gamma, k > 0$. Let* H *be an $(1+\gamma)/|\mathsf{H.Rng}|$-almost universal hash function family. Then* H *is an (k,ϵ)-strong extractor where*

$$\epsilon = \frac{1}{2}\sqrt{\gamma + \frac{|\mathsf{H.Rng}|}{2^k}} .$$

The original result is stated in a different formalism and a slightly more restricted form, but it generalizes to the form stated here. (The formulation of the bounds comes from [2].)

Since we use the nonce that is used in the hash function H also outside of it, namely in the evaluation of the PRF F, we cannot immediately apply the leftover-hash lemma. However, we show that if the function H is almost XOR-universal, then the hedged extractor HE obtained by XORing the output of F to the output of H is almost universal and therefore serves as a strong extractor.

Lemma 4. *Let $\mathsf{H} : \mathsf{H.Keys} \times \mathsf{H.Dom} \to \{0,1\}^{\ell}$ be an ϵ-almost-XOR-universal hash function. Let $\mathsf{F} : \mathsf{F.Keys} \times (\{0,1\}^* \times \{0,1\}^*) \to \{0,1\}^{\ell}$ a function family and $\mathsf{H.Dom} \subseteq \{0,1\}^*$. Then* HE *as defined above is ϵ-almost-universal in the second input component.*

Proof. For all $m, m' \in \{0,1\}^*$ and $n, n' \in \mathsf{H.Dom}$ with $n \neq n'$:

$$\begin{aligned}
&\Pr\left[\mathsf{HE}(xk,(m,n)) = \mathsf{HE}(xk,(m',n')) \;:\; xk \leftarrow\!\!\$\; \mathsf{HE.Keys}\right] \\
&= \Pr\Big[\mathsf{H}(hk,n) \oplus \mathsf{F}(fk,(m,n)) = \mathsf{H}(hk,n') \oplus \mathsf{F}(fk,(m',n')) \\
&\qquad\qquad : (hk,fk) \leftarrow\!\!\$\; \mathsf{H.Keys} \times \mathsf{F.Keys}\Big] \\
&\leq \max_{fk \in \mathsf{F.Keys}} \Pr\Big[\mathsf{H}(hk,n) \oplus \mathsf{H}(hk,n') = \mathsf{F}(fk,(m,n)) \oplus \mathsf{F}(fk,(m',n')) \\
&\qquad\qquad : hk \leftarrow\!\!\$\; \mathsf{H.Keys}\Big] \leq \epsilon,
\end{aligned}$$

since for each $fk \in \mathsf{F.Keys}$ the term $\mathsf{F}(fk,(m,n)) \oplus \mathsf{F}(fk,(m',n'))$ describes a fixed value $s \in \{0,1\}^\ell$ and H is ϵ-almost XOR-universal. □

A strong extractor will only guarantee the outputs to be random and independent of the seed if the inputs to the extractor, that is the nonces, do not depend on the seed. Once the seed is exposed to the adversary, no guarantee on further outputs can be given. Therefore, for the game $\mathbf{G}^{\mathrm{ror}}_{\mathsf{HE},\mathsf{NG}}(\mathcal{G})$ we restrict our attention to agnostic adversaries $\mathcal{G}$.

Proof (Lemma 2). We first prove Eq. 12. Adversary $\mathcal{B}$ starts by choosing a seed $hk \leftarrow\!\!\$\; \mathsf{H.Keys}$ uniformly at random. It then uses the assumed adversary $\mathcal{A}$, and whenever $\mathcal{A}$ makes a query $\textsc{Fn}(x,y)$, then $\mathcal{B}$ first computes $z_1 \leftarrow \mathsf{H}(hk,y)$ and then makes a query $z_2 \leftarrow \textsc{Fn}(x,y)$, and returns $z_1 \oplus z_2$. Finally, $\mathcal{B}$ provides the same output as $\mathcal{A}$. The view of $\mathcal{A}$ has the same distribution in $\mathbf{G}^{\mathrm{prf}}_{\mathsf{HE}}(\mathcal{A})$ and in $\mathbf{G}^{\mathrm{prf}}_{\mathsf{F}}(\mathcal{B})$, and therefore the distribution of the output is also the same. This implies Eq. 12.

We prove Eq. 13 using a hybrid argument similarly to Zuckerman [33, Lemma 6]. The hybrid argument involves a game G as specified in Fig. 4 and adversaries $\mathcal{G}_1, \ldots, \mathcal{G}_q$ that each query the oracle RoR only once. This is achieved by having $\mathcal{G}_i$ answer the ith query of $\mathcal{G}$ by using the RoR oracle, by using for all previous queries uniformly random values and computing all subsequent values by evaluating HE. In more detail, for all queries $j = 1, \ldots, i-1$, the adversary $\mathcal{G}_i$ will call its $\mathrm{GEN}(\eta)$ oracle and then sample a value from $\mathsf{HE.Rng} = \{0,1\}^\ell$ uniformly at random. For the ith query, the adversary $\mathcal{G}_i$ will call $\textsc{RoR}(m,\eta)$ and obtain the output r, as well as the nonce generator's state St through a subsequent EXPOSE query. The answer to $\mathcal{G}$ will be r in this case. For all queries $j = i+1, \ldots, q$, adversary $\mathcal{G}_i$ will compute the nonce via $(n_j, St) \leftarrow \mathsf{NG}(\eta, St)$ and the output as $\mathsf{HE}(xk, m, n_j)$. This is possible because it uses the seed xk only after the challenge query. We can therefore specify the adversary as a pair $\mathcal{G}_i = (\mathcal{G}_{i,\mathsf{prep}}, \mathcal{G}_{i,\mathsf{chal}})$ as described in Fig. 4.

To conclude the hybrid argument, we first observe that the view of $\mathcal{G}$ in $\mathbf{G}^{\mathrm{ror}}_{\mathsf{HE},\mathsf{NG}}(\mathcal{G})$ with $c = 1$ is the same as its view in $\mathrm{G}(\mathcal{G}_1)$ with $c = 1$. Subsequently, for each $i = 1, \ldots, q-1$, the view of $\mathcal{G}$ in $\mathrm{G}(\mathcal{G}_i)$ with $c = 0$ is the same as in $\mathrm{G}(\mathcal{G}_{i+1})$ with $c = 1$. Finally, the view of $\mathcal{G}$ in $\mathrm{G}(\mathcal{G}_q)$ with $c = 0$ is the same as in $\mathbf{G}^{\mathrm{ror}}_{\mathsf{HE},\mathsf{NG}}(\mathcal{G})$ with $c = 0$.

Game $\mathrm{G}(\mathcal{G})$
$St \leftarrow \varepsilon$; $xk \leftarrow\$ \mathsf{E.Keys}$; $c \leftarrow\$ \{0,1\}$; $s \leftarrow 0$
$St_{\mathcal{G}} \leftarrow\$ \mathcal{G}_{\mathsf{prep}}^{\mathrm{GEN}}$
$c' \leftarrow\$ \mathcal{G}_{\mathsf{chal}}^{\mathrm{RoR,EXPOSE}}(St_{\mathcal{G}}, xk)$
Return $(c = c')$

$\mathrm{RoR}(m, \eta)$
If $s = 1$ then return $\bot$
$(n, St) \leftarrow\$ \mathsf{NG}(\eta, St)$
If $(c = 1)$ then $r \leftarrow \mathsf{E}^{\mathrm{RO}}(xk, m, n)$
Else $r \leftarrow\$ \mathsf{E.Rng}$
Return r

$\mathrm{GEN}(\eta)$
$(n, St) \leftarrow\$ \mathsf{NG}(\eta, St)$

EXPOSE
$s \leftarrow 1$
Return St

Fig. 4. Hybrid game for E, as needed in the proof of Lemma 2.

We still have to bound the advantage of adversary $\mathcal{G}_i$. By Lemma 4, we know that HE is $(1+\gamma)/2^\ell$-almost universal in its second input and therefore by a slight generalization of Lemma 3 an (m,ϵ)-strong extractor for $\epsilon = 1/2\sqrt{\gamma + 2^{\ell-m}}$. Fix i, then $\mathcal{G}_{i,\mathsf{prep}}$ makes $i-1$ queries to its GEN oracle. We can therefore define a nonce generator $\mathsf{RG}(\mathcal{G}_{i,\mathsf{guess}}, \mathsf{NG})$ as follows: Set $St \leftarrow \varepsilon$ and execute $(n, St_{\mathcal{G}}) \leftarrow\$ \mathcal{G}_{i,\mathsf{guess}}^{\mathrm{GEN}}$ with the oracle as defined in the game. Compute a random x via $(x, St) \leftarrow\$ \mathsf{NG}(\eta, St)$ and output a pair $(x, (St, St_{\mathcal{G}}))$.

This implies that $\mathbf{Adv}^{\mathsf{pred}}_{\mathsf{RG}(\mathcal{G}_{i,\mathsf{prep}},\mathsf{NG})}(1,1) \le \mathbf{Adv}^{\mathsf{pred}}_{\mathsf{NG}}(i,1)$, since otherwise $\mathcal{G}_{i,\mathsf{prep}}$ could be used in the statement against the assumption on NG. But now we can view $\mathcal{G}_{i,\mathsf{chal}}$ as an adversary against HE in the original $\mathbf{G}^{\mathrm{ror}}$ game, and $2\Pr[\mathrm{HYB}_{\mathsf{HE},\mathsf{NG}}(\mathcal{G}_i)] - 1 \le \mathbf{Adv}^{\mathsf{ror}}_{\mathsf{SE},\mathsf{RG}(\mathcal{G}_{i,\mathsf{prep}},\mathsf{NG})}(\mathcal{G}_{i,\mathsf{chal}}) \le \epsilon$. We obtain a factor q through the hybrid argument; this completes the proof. □

4 Nonce-Based Public-Key Encryption

In this section we define nonce-based public-key encryption, giving first a syntax and then two security goals, NBP1 and NPB2. We give a construction, simultaneously meeting both goals, based on a hedged extractor. Instantiating the latter via our constructions of Sect. 3 yields concrete nonce-based PKE schemes, one in the ROM and the other in the standard model.

SYNTAX. A *nonce-based public-key encryption scheme* NPE specifies the following. Receiver key-generation algorithm NPE.Kg returns an encryption key *ek* and associated decryption key *dk*. Sender seed-generation algorithm NPE.sKg returns a seed *xk*. Encryption algorithm NPE.Enc takes *ek*, *xk*, message $m \in \{0,1\}^*$ and nonce n from nonce set NPE.NS to return a ciphertext c. This algorithm is *deterministic*. Decryption algorithm NPE.Dec (also deterministic) takes ek, dk and ciphertext c to return a value in $\{0,1\}^* \cup \{\bot\}$. The scheme NPE is correct if for all $(ek, dk) \in [\mathsf{NPE.Kg}]$, all $xk \in [\mathsf{NPE.sKg}]$, all $m \in \{0,1\}^*$ and all $n \in \mathsf{NPE.NS}$ we have $\mathsf{NPE.Dec}(ek, dk, \mathsf{NPE.Enc}(ek, xk, m, n)) = m$.

The receiver key-generation algorithm has the same role as in a randomized PKE scheme. The sender seed-generation algorithm is a new element of nonce-based PKE schemes. Encryption is changed to take a nonce rather than randomness and, importantly, is now deterministic. Decryption is as in a standard PKE scheme. Unlike symmetric nonce-based encryption, the decryption algorithm is not given the nonce.

Nonce-based encryption is a sender-side hardening and can be added to an existing encryption scheme in such a way that the receiver is oblivious to its presence and the receiver implementation needs no changes.

Game $\mathbf{G}^{\mathrm{nbp1}}_{\mathsf{NPE,NG}}(\mathcal{A})$	Game $\mathbf{G}^{\mathrm{nbp2}}_{\mathsf{NPE,NG}}(\mathcal{A})$
$(ek, dk) \leftarrow\$ \mathsf{NPE.Kg}$; $xk \leftarrow\$ \mathsf{NPE.sKg}$ $b \leftarrow\$ \{0,1\}$; $St \leftarrow \varepsilon$; $C, S_0, S_1 \leftarrow \emptyset$ $b' \leftarrow\$ \mathcal{A}^{\mathrm{ENC,DEC,RO}}(ek)$ Return $(b = b')$ $\mathrm{ENC}(m_0, m_1, \eta)$ If $(\|m_0\| \neq \|m_1\|)$ then return $\bot$ $(n, St) \leftarrow\$ \mathsf{NG}(\eta, St)$ If $((m_0, n) \in S_0)$ then return $\bot$ If $((m_1, n) \in S_1)$ then return $\bot$ $S_0 \leftarrow S_0 \cup \{(m_0, n)\}$ $S_1 \leftarrow S_1 \cup \{(m_1, n)\}$ $c \leftarrow \mathsf{NPE.Enc}^{\mathrm{RO}}(ek, xk, m_b, n)$ $C \leftarrow C \cup \{c\}$ Return c $\mathrm{DEC}(c)$ If $(c \in C)$ then $m \leftarrow \bot$ Else $m \leftarrow \mathsf{NPE.Dec}^{\mathrm{RO}}(ek, dk, c)$ Return m $\mathrm{RO}(x, l)$ If $T[x, l] = \bot$ then $T[x, l] \leftarrow\$ \{0,1\}^l$ Return $T[x, l]$	$(ek, dk) \leftarrow\$ \mathsf{NPE.Kg}$ $xk \leftarrow\$ \mathsf{NPE.sKg}$ $b \leftarrow\$ \{0,1\}$; $St \leftarrow \varepsilon$; $C \leftarrow \emptyset$ $b' \leftarrow\$ \mathcal{A}^{\mathrm{ENC,DEC,RO}}(ek, xk)$ Return $(b = b')$ $\mathrm{ENC}(m_0, m_1, \eta)$ If $(\|m_0\| \neq \|m_1\|)$ then return $\bot$ $(n, St) \leftarrow\$ \mathsf{NG}(\eta, St)$ $c \leftarrow \mathsf{NPE.Enc}^{\mathrm{RO}}(ek, xk, m_b, n)$ $C \leftarrow C \cup \{c\}$ Return c $\mathrm{DEC}(c)$ If $(c \in C)$ then return $\bot$ $m \leftarrow \mathsf{NPE.Dec}^{\mathrm{RO}}(ek, dk, c)$ Return m $\mathrm{RO}(x, l)$ If $T[x, l] = \bot$ then $T[x, l] \leftarrow\$ \{0,1\}^l$ Return $T[x, l]$

Fig. 5. Games for security goals for nonce-based asymmetric encryption scheme NPE. Both are relative to nonce generator NG.

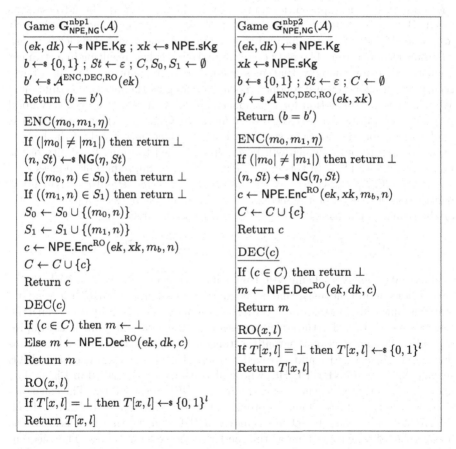

SECURITY DEFINITIONS. Let NPE be a nonce-based PKE scheme. Let NG be a nonce generator returning nonces in NPE.NS. We associate to NPE, NG and adversary $\mathcal{A}$ the games of Fig. 5. We define the advantages of $\mathcal{A}$ in these games via

$$\mathbf{Adv}^{\mathsf{nbp1}}_{\mathsf{NPE,NG}}(\mathcal{A}) = 2\Pr[\mathbf{G}^{\mathrm{nbp1}}_{\mathsf{NPE,NG}}(\mathcal{A})] - 1$$

and

$$\mathbf{Adv}^{\mathsf{nbp2}}_{\mathsf{NPE,NG}}(\mathcal{A}) = 2\Pr[\mathbf{G}^{\mathrm{nbp2}}_{\mathsf{NPE,NG}}(\mathcal{A})] - 1 ,$$

respectively. The games are described in the ROM; standard-model definitions are derived by considering only schemes and adversaries that do not query the RO oracle. The notation "nbp" stands for "nonce-based privacy." We proceed to discuss the definitions.

Game $\mathbf{G}^{\mathrm{nbp1}}_{\mathsf{NPE,NG}}(\mathcal{A})$ formalizes security in the case where the sender's seed is not exposed, captured in the formalism by the fact that the adversary is not given the seed as input, while game $\mathbf{G}^{\mathrm{nbp2}}_{\mathsf{NPE,NG}}(\mathcal{A})$ formalizes security in the case where the sender's seed is exposed, formalized by its being given to the adversary as input. Both ask for indistinguishability-style security under a chosen-ciphertext attack. In $\mathbf{G}^{\mathrm{nbp1}}_{\mathsf{NPE,NG}}(\mathcal{A})$, the natural restriction that one would consider is to ask that nonces not repeat. The restriction we make is weaker, resulting in a stronger security condition, namely that message-nonce pairs may not repeat. (Thus, security is provided even if a nonce repeats, as long as the message is different.) We will achieve this notion for *any* nonce generator, meaning we get very good privacy with minimal restrictions on nonces. In $\mathbf{G}^{\mathrm{nbp2}}_{\mathsf{NPE,NG}}(\mathcal{A})$, no restriction is made, so a priori the notion is stronger, but we will achieve it only if the hedged extractor is ror-secure, which will be further reduced to unpredictability of the nonce generator. Thus, in this case, security requires unpredictable nonces.

In game $\mathbf{G}^{\mathrm{nbp2}}_{\mathsf{NPE,NG}}(\mathcal{A})$, we say that adversary $\mathcal{A}$ is *agnostic* if its ENC, DEC queries do not depend on the seed xk. More formally, there exists a pair $(\mathcal{A}_1, \mathcal{A}_2)$ of algorithms such that $\mathcal{A}^{\mathrm{ENC,DEC,RO}}(ek, xk)$ does the following:

$$St \leftarrow\!\!\$\; \mathcal{A}_1^{\mathrm{ENC,DEC,RO}}(ek)\,;\; c' \leftarrow \mathcal{A}_2^{\mathrm{RO}}(xk, St)\,;\;\; \text{Return } c' \,.$$

SCHEME. We specify a scheme NPE that achieves both the NPB1 and NPB2 security notions simultaneously, that is, it guarantees security if either the sender's state remains secret and as long as the message-nonce pairs are unique, or even if the sender's state is leaked to the adversary as long as the nonces are sufficiently unpredictable. The construction is actually a transform **R2NPE** that takes a base, randomized PKE scheme PE and a hedged extractor HE to return a nonce-based PKE scheme NPE $=$ **R2NPE**[PE, HE] whose algorithms are described in Fig. 6. The nonce space is that of the hedged extractor, i.e. NPE.NS $=$ HE.NS. The construction requires that the randomness provided by HE is sufficient, i.e., PE.rl $=$ HE.ol.

We first prove that the scheme achieves NBP1-security, that is, it is secure as a nonce-based scheme as long as the sender's seed remains secret. The theorem bounds the adversaries advantage by advantages of other, related adversaries against the underlying probabilistic public-key scheme and the PRF-property of the hedged extractor HE. For the constructions described in Sect. 3, the latter

Algorithm $\mathsf{NPE.Kg}$	Algorithm $\mathsf{NPE.Enc}^{\mathrm{RO}}(ek, xk, m, n)$
$(ek, dk) \leftarrow\!\!\$\ \mathsf{PE.Kg}$ Return (ek, dk)	$r \leftarrow \mathsf{HE}^{\mathrm{RO}}(xk, (m, n))$ $c \leftarrow \mathsf{PE.Enc}(ek, m; r)$ Return c
Algorithm $\mathsf{NPE.sKg}$	**Algorithm $\mathsf{NPE.Dec}(ek, dk, c)$**
$xk \leftarrow\!\!\$\ \mathsf{HE.Keys}$ Return xk	$m \leftarrow \mathsf{PE.Dec}(ek, dk, c)$ Return m

Fig. 6. The nonce-based public-key encryption scheme NPE, based on probabilistic public-key encryption scheme PE and hedged extractor HE. The nonce space is the same as for the hedged extractor.

advantage is then bounded by Lemmas 1 (for the ROM construction) and 2 (for the standard-model construction), respectively.

Theorem 5. *Let* PE *be a (standard, randomized) public-key encryption scheme. Let* HE *be a hedged extractor. Let nonce-based public-key encryption scheme* NPE = **R2NPE**[PE, HE] *be associated to them as above. Let* NG *be a nonce generator. Let* $\mathcal{A}$ *be an adversary making at most* q_1 *queries to its* ENC *oracle,* q_2 *queries to its* DEC *oracle, and* q_3 *queries to its* RO *oracle. Then the proof specifies adversaries* $\mathcal{B}$ *and* $\mathcal{G}$ *such that*

$$\mathbf{Adv}^{\mathsf{nbp1}}_{\mathsf{NPE},\mathsf{NG}}(\mathcal{A}) \leq 2\mathbf{Adv}^{\mathsf{prf}}_{\mathsf{HE}}(\mathcal{G}) + \mathbf{Adv}^{\mathsf{ind}}_{\mathsf{PE}}(\mathcal{B}) , \tag{14}$$

where adversary $\mathcal{B}$ *makes at most* q_1 *queries to its* ENC *oracle and* q_2 *queries to its* DEC *oracle; in terms of computation it evaluates* NG *for* q_1 *times and manages an array of generated ciphertexts. Adversary* $\mathcal{G}$ *makes at most* q_1 *queries to its* FN *oracle and* q_3 *queries to its* RO *oracle; it generates keys for* PE *and compute* q_1 *encryptions and* q_2 *decryptions in addition to the computation of* $\mathcal{A}$.

Proof. Adversary $\mathcal{B}$ obtains ek and starts by setting $St \leftarrow \varepsilon$, $S_0 \leftarrow \emptyset$, and $S_1 \leftarrow \emptyset$. Adversary $\mathcal{B}$ then runs the adversary $\mathcal{A}$ internally on input ek. For queries $\mathrm{ENC}(m_0, m_1, \eta)$, adversary $\mathcal{B}$ computes $(n, St) \leftarrow\!\!\$\ \mathsf{NG}(\eta, St)$, checks whether $(m_0, n) \in S_0$ or $(m_1, n) \in S_1$, and returns $\perp$ in that case. Otherwise, it queries its own oracle $\mathrm{ENC}(m_0, m_1)$ and returns the result to $\mathcal{A}$. Queries $\mathrm{DEC}(c)$ are answered using the respective oracles in the game, and potential queries $\mathrm{RO}(x, l)$ are answered by emulating a random oracle via lazy sampling as also described in the game.

For each $d \in \{0, 1\}$, adversary $\mathcal{G}_d$ against the PRF initializes $St \leftarrow \varepsilon$, $S_0 \leftarrow \emptyset$, and $S_1 \leftarrow \emptyset$ as $\mathcal{B}$ and generates a key pair $(ek, dk) \leftarrow\!\!\$\ \mathsf{PE.Kg}$. Adversary $\mathcal{G}_d$ then runs adversary $\mathcal{A}$ on input ek. Upon a query $\mathrm{ENC}(m_0, m_1, \eta)$ from $\mathcal{A}$, adversary $\mathcal{G}_d$ computes $(n, St) \leftarrow\!\!\$\ \mathsf{NG}(\eta, St)$, checks whether $(m_0, n) \in S_0$ or $(m_1, nonce) \in S_1$, and returns $\perp$ in that case. $\mathcal{G}_d$ gets r from its oracle $\mathrm{FN}(m_d, n)$, computes $c \leftarrow \mathsf{PE.Enc}(ek, m_d; r)$ and returns c. Queries $\mathrm{DEC}(c)$ are answered by computing $m \leftarrow \mathsf{PE.Dec}(ek, dk, c)$ and returning m. Potential queries $\mathrm{RO}(x, l)$

are referred to the random oracle provided in the game. When $\mathcal{A}$ provides its output b', then $\mathcal{G}_d$ outputs $b' \oplus d$.

We define a hybrid game G in which all computations are performed as before, except that in the encryption with a new pair of message m_b and nonce n a fresh random string $r \leftarrow \{0,1\}^\ell$ is used. The difference with the game $\mathbf{G}^{\mathrm{nbp1}}_{\mathsf{NPE,NG}}(\mathcal{A})$ can be highlighted as described in game G in Fig. 7.

Game G	Game H
$\vdots$	$\vdots$
$S_1 \leftarrow S_1 \cup \{(m_1, n)\}$	$(n, St) \leftarrow\$\ \mathsf{NG}(\eta, St)$
$c \leftarrow \mathsf{NPE.Enc}^{\mathrm{RO}}(ek, xk, m_b, n)$	$c \leftarrow \mathsf{NPE.Enc}^{\mathrm{RO}}(ek, xk, m_b, n)$
$r \leftarrow\$\ \{0,1\}^\ell$	$r \leftarrow\$\ \{0,1\}^\ell$
$c \leftarrow \mathsf{PE.Enc}(ek, m_b; r)$	$c \leftarrow \mathsf{PE.Enc}(ek, m_b; r)$
$C \leftarrow C \cup \{c\}$	$C \leftarrow C \cup \{c\}$
Return c	Return c
$\vdots$	$\vdots$

Fig. 7. Intermediate games used in the proofs of Theorem 5 (left) and Theorem 6 (right).

The view of $\mathcal{A}$ in G is exactly the same as in $\mathbf{G}^{\mathrm{ind}}_{\mathsf{PE}}(\mathcal{B})$. Furthermore, we observe that

$$2\Pr[\mathbf{G}^{\mathrm{prf}}_{\mathsf{HE}}(\mathcal{G}_0)] = \Pr[\mathbf{G}^{\mathrm{nbp1}}_{\mathsf{NPE,NG}}(\mathcal{A})|b=0] - \Pr[\mathrm{G}|b=0] + 1,$$

the reason is that if the bit $b = 0$ is chosen in $\mathbf{G}^{\mathrm{prf}}_{\mathsf{HE}}(\mathcal{G}_0)$, then the view of $\mathcal{A}$ is exactly as in G with $b = 0$; all ciphertexts are encryptions of m_0 with fresh randomness. Analogously, if $b = 1$ is chosen in $\mathbf{G}^{\mathrm{prf}}_{\mathsf{HE}}(\mathcal{G}_0)$, then the view of $\mathcal{A}$ is exactly as in $\mathbf{G}^{\mathrm{nbp1}}_{\mathsf{NPE,NG}}(\mathcal{A})$ with $b = 0$; all ciphertexts are encryptions of m_0 with randomness computed via NG and HE from the message m_0 and the input η, but in this case $\mathcal{G}_0$ outputs the "wrong" bit. In the same sense,

$$2\Pr[\mathbf{G}^{\mathrm{prf}}_{\mathsf{HE}}(\mathcal{G}_1)] = \Pr[\mathbf{G}^{\mathrm{nbp1}}_{\mathsf{NPE,NG}}(\mathcal{A})|b=1] - \Pr[\mathrm{G}|b=1] + 1,$$

since if the bit b is chosen as $b = 0$ in $\mathbf{G}^{\mathrm{prf}}_{\mathsf{HE}}(\mathcal{G}_0)$, then the view of $\mathcal{A}$ is exactly as in G with $b = 1$; all ciphertexts are encryptions of m_1 with fresh randomness. This is the reason for $\mathcal{G}_1$ to invert the output of $\mathcal{A}$. Overall, we obtain

$$\begin{aligned}
\mathbf{Adv}^{\mathrm{nbp1}}_{\mathsf{NPE,NG}}(\mathcal{A}) &= \Pr[\mathbf{G}^{\mathrm{nbp1}}_{\mathsf{NPE,NG}}(\mathcal{A})|b=0] + \Pr[\mathbf{G}^{\mathrm{nbp1}}_{\mathsf{NPE,NG}}(\mathcal{A})|b=1] - 1 \\
&= \Pr[\mathbf{G}^{\mathrm{nbp1}}_{\mathsf{NPE,NG}}(\mathcal{A})|b=0] - \Pr[\mathrm{G}|b=0] \\
&\quad + \Pr[\mathbf{G}^{\mathrm{nbp1}}_{\mathsf{NPE,NG}}(\mathcal{A})|b=1] - \Pr[\mathrm{G}|b=1] + 2\Pr[\mathrm{G}] - 1 \\
&\leq 2\Pr[\mathbf{G}^{\mathrm{prf}}_{\mathsf{HE}}(\mathcal{G}_0)] - 1 + 2\Pr[\mathbf{G}^{\mathrm{prf}}_{\mathsf{HE}}(\mathcal{G}_1)] - 1 + \mathbf{Adv}^{\mathsf{nbp1}}_{\mathsf{PE}}(\mathcal{B}) \\
&= \mathbf{Adv}^{\mathsf{prf}}_{\mathsf{HE}}(\mathcal{G}_0) + \mathbf{Adv}^{\mathsf{prf}}_{\mathsf{HE}}(\mathcal{G}_1) + \mathbf{Adv}^{\mathsf{nbp1}}_{\mathsf{PE}}(\mathcal{B}).
\end{aligned}$$

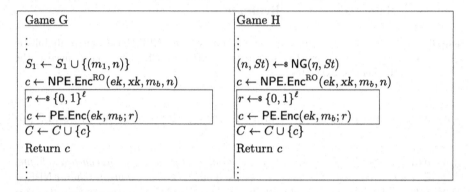

The proof concludes by defining $\mathcal{G}$ as choosing $\mathcal{G}_0$ or $\mathcal{G}_2$ with probability 1/2 each. □

The property of a hedged extractor to serve simultaneously as extractor and as a PRF implies that the scheme described above is also secure even if the sender's seed leaks, as long as the nonces are sufficiently unpredictable. The reduction in the theorem below preserves agnosticity, meaning if $\mathcal{A}$ is agnostic, so is $\mathcal{G}$. This allows us to draw conclusions based on Lemma 2 in the case of the standard-model hedged extractor.

Theorem 6. *Let* PE *be a (standard, randomized) public-key encryption scheme. Let* HE *be a hedged extractor. Let nonce-based public-key encryption scheme* NPE = **R2NPE**[PE, HE] *be associated to them as above. Let* NG *be a nonce generator. Let $\mathcal{A}$ be an adversary making at most q_1 queries to its* ENC *oracle, q_2 queries to its* DEC *oracle, and q_3 queries to its* RO *oracle. Then the proof specifies adversaries $\mathcal{B}$ and $\mathcal{G}$ such that*

$$\mathbf{Adv}^{\mathsf{nbp2}}_{\mathsf{NPE,NG}}(\mathcal{A}) \leq 2\mathbf{Adv}^{\mathsf{ror}}_{\mathsf{HE,NG}}(\mathcal{G}) + \mathbf{Adv}^{\mathsf{ind}}_{\mathsf{PE}}(\mathcal{B}) \ , \qquad (15)$$

where adversary $\mathcal{B}$ makes at most q_1 queries to its ENC *oracle and q_2 queries to its* DEC *oracle, and emulates a random oracle for q_3 queries. Adversary $\mathcal{G}$ makes at most q_1 queries to its* RoR *oracle and q_3 queries to its* RO *oracle; it generates keys for* PE *and computes q_1 encryptions and q_2 decryptions in addition to the computation of $\mathcal{A}$. Furthermore, if $\mathcal{A}$ is agnostic, then $\mathcal{G}$ is also agnostic.*

Proof. The proof follows the same ideas as the one for Theorem 5. Adversary $\mathcal{B}$ against the underlying public-key encryption scheme PE behaves as follows. It obtains an encryption key ek as an input and generates a seed $xk \leftarrow_{\$} \mathsf{HE.Keys}$, and then runs adversary $\mathcal{A}$ on input (ek, xk). Upon a query $\mathrm{ENC}(m_0, m_1, \eta)$ by $\mathcal{A}$, adversary $\mathcal{B}$ queries its oracle $\mathrm{ENC}(m_0, m_1)$ and returns the result to $\mathcal{A}$. Queries $\mathrm{DEC}(c)$ are answered by making the same query to its own oracle; the queries $\mathrm{RO}(x, l)$ are answered by emulating a random oracle via lazy sampling. Adversary $\mathcal{B}$ outputs the same output bit as $\mathcal{A}$.

For each $d \in \{0, 1\}$, adversary $\mathcal{G}_d$ against the extractor behaves as follows. It obtains an extractor seed $xk \in \mathsf{HE.Keys}$, generates a key pair $(ek, dk) \leftarrow_{\$} \mathsf{PE.Kg}$, and then runs adversary $\mathcal{A}$ on input (ek, xk). Upon a query $\mathrm{ENC}(m_0, m_1, \eta)$ from $\mathcal{A}$, adversary $\mathcal{G}_d$ calls $\mathrm{RoR}(m_d, \eta)$ to obtain a random string r. It then computes $c \leftarrow \mathsf{PE.Enc}(ek, m_d; r)$ and returns c. Queries $\mathrm{DEC}(c)$ are answered by computing $m \leftarrow \mathsf{PE.Dec}(ek, dk, c)$ and returning m. Potential queries $\mathrm{RO}(x, l)$ are referred to the random oracle provided in the game. When $\mathcal{A}$ provides its output b', then $\mathcal{G}_d$ outputs $b' \oplus d$. Note that $\mathcal{G}_d$ uses the seed only through $\mathcal{A}$, and if $\mathcal{A}$ uses it only after all encryption queries, then $\mathcal{G}_d$ uses it only after all RoR queries. Thus, if $\mathcal{A}$ is agnostic, then $\mathcal{G}_d$ is also agnostic.

We define a hybrid game H in which all computations are performed as before, except that in the encryption a fresh random string $r \leftarrow \{0, 1\}^\ell$ is used; the difference with the game $\mathbf{G}^{\mathrm{nbp2}}_{\mathsf{NPE,NG}}(\mathcal{A})$ is highlighted by the boxed code in Fig. 7.

The view of $\mathcal{A}$ in H is exactly the same as in $\mathbf{G}^{\mathrm{nbp2}}_{\mathsf{PE}}(\mathcal{B})$. Furthermore, we observe that

$$2\Pr[\mathbf{G}^{\mathrm{ror}}_{\mathsf{HE},\mathsf{NG}}(\mathcal{G}_0)] \;=\; \Pr[\mathbf{G}^{\mathrm{nbp2}}_{\mathsf{NPE},\mathsf{NG}}(\mathcal{A})|b=0] - \Pr[\mathrm{H}|b=0] + 1,$$

the reason is that if the bit $b = 0$ is chosen in $\mathbf{G}^{\mathrm{ror}}_{\mathsf{HE},\mathsf{NG}}(\mathcal{G}_0)$, then the view of $\mathcal{A}$ is exactly as in H with $b = 0$; all ciphertexts are encryptions of m_0 with fresh randomness. Analogously, if $b = 1$ is chosen in $\mathbf{G}^{\mathrm{ror}}_{\mathsf{HE},\mathsf{NG}}(\mathcal{G}_0)$, then the view of $\mathcal{A}$ is exactly as in $\mathbf{G}^{\mathrm{nbp2}}_{\mathsf{NPE},\mathsf{NG}}(\mathcal{A})$ with $b = 0$; all ciphertexts are encryptions of m_0 with randomness computed via HE and NG from the message m_0 and the input η, but in this case $\mathcal{G}_0$ outputs the "wrong" bit. In the same sense,

$$2\Pr[\mathbf{G}^{\mathrm{ror}}_{\mathsf{HE},\mathsf{NG}}(\mathcal{G}_1)] \;=\; \Pr[\mathbf{G}^{\mathrm{nbp2}}_{\mathsf{NPE},\mathsf{NG}}(\mathcal{A})|b=1] - \Pr[\mathrm{H}|b=1] + 1,$$

since if the bit b is chosen as $b = 0$ in $\mathbf{G}^{\mathrm{ror}}_{\mathsf{HE},\mathsf{NG}}(\mathcal{G}_0)$, then the view of $\mathcal{A}$ is exactly as in H with $b = 1$; all ciphertexts are encryptions of m_1 with fresh randomness. This is the reason for $\mathcal{G}_1$ to invert the output of $\mathcal{A}$. The final computation follows exactly as in Theorem 5. □

Settings with multiple senders and multiple receivers. The security properties defined above take into account only a single sender and a single receiver. Realistic settings, however, involve multiple senders and multiple receivers. This means that, on the one hand, encryptions toward the same receivers will be made with respect to different sender seeds. On the other hand, senders will use the same seed to generate randomness for encryptions toward different receivers. To achieve security in these settings, we extend the games $\mathbf{G}^{\mathrm{nbp2}}_{\mathsf{NPE},\mathsf{NG}}(\mathcal{A})$ and $\mathbf{G}^{\mathrm{nbp1}}_{\mathsf{NPE},\mathsf{NG}}(\mathcal{A})$ by the following oracle:

$\underline{\mathrm{ENC2}(ek, m, \eta)}$
$(n, St) \leftarrow\!\!\$\; \mathsf{NG}(\eta, St)$
$c \leftarrow \mathsf{NPE.Enc}(ek, xk, m, n)$
Return c

The scheme discussed above can easily be shown to achieve the correspondingly modified games; the hedged extractor provides uniform and independent randomness for each encryption.

Extending this game to multiple senders and multiple receives is done by generating multiple sets of keys and seeds in the game and extending the oracles with arguments to select the desired sender and/or receiver of the messages to be processed. The proof then follows by two hybrid arguments, one for reducing the number of senders, and one for reducing the number of receivers. The oracle ENC2 is required in the step to reduce the number of receivers to simulate encryptions toward the receivers not captured in the game.

5 Nonce-Based Signatures

In this section we define and construct nonce-based digital signature schemes.

BACKGROUND. Eliminating randomness in signing is not new and is easily done. A simple way to convert a given randomized EUF-CMA digital signature scheme DS into a deterministic one is as follows. Let F be a PRF. The key-generation algorithm lets $(sk, vk) \leftarrow\$ \mathsf{DS.Kg}$ and $fk \leftarrow\$ \mathsf{F.Keys}$, and stores the pair (fk, sk) as the secret signing key of the new scheme. A signature on a message $m \in \{0,1\}^*$ is then computed by first evaluating $r \leftarrow \mathsf{F}(fk, m)$ and then $s \leftarrow \mathsf{DS.Sig}(sk, m; r)$. This method goes back to MNPV [23] and it is easy to show that it works, meaning the constructed, deterministic signature scheme retains the EUF-CMA security of the starting randomized one assuming F is a PRF.

The above solution, however, changes the secret key, which is not always desirable. For example it may be a problem to retrofit deployed schemes with the modification, or if the same signature key is used by multiple applications and the format cannot easily be changed. A folklore solution is to leave the keys unchanged and obtain the coins r via a random oracle applied to the existing secret key sk and the message. In the case that DS is ECDSA, this was proven to work by KM [21]. It was proven to work in general (meaning, for any base EUF-CMA scheme) by BPS [9]. Such de-randomization is used in the Ed25519 signature scheme [11] and is specified for DSA and ECDSA in an RFC by Pornin [25].

NONCE-BASED SIGNATURES. In our model, the signer has a secret key as well as a seed. Signing uses both these and a nonce, and is deterministic. If the seed (and secret key) are kept private, we get the usual EUF-CMA level of security, regardless of how nonces are generated. So far this is providing the same security as deterministic signature schemes. The added condition is that if the seed is exposed (but the secret key isn't) then we still retain security as long as the nonces are unpredictable.

The secret key and seed are held by the same entity, namely the signer. So one may ask how it could be that the seed is exposed but the secret key isn't. That is, either the system is secure, in which case both are secure, or not, in which case both are exposed. If so, indeed, nonce-based signatures do not provide anything over and above classical deterministic signatures. However, we can imagine settings where the level of security for the secret key and seed are different. For example the secret key may be already stored in hardware, and the seed not. The seed may be stored at a different place than the signature scheme's secret key, and it may be re-generated at any frequency that seems appropriate for the application (ranging from never, at fixed time intervals, at every system reboot, or at every signature operation). Being a signer-only modification of the signature generation, a user may also use different seeds on different machines, or use the modified scheme with the standard probabilistic one.

The scheme we propose is again based on hedged extractors.

DEFINITIONS. A *nonce-based signature scheme* NDS specifies the following. Signer key-generation algorithm NDS.Kg returns a signature key sk, a verification key vk, and a seed xk. *Deterministic* signature algorithm NDS.sign takes sk, xk, message $m \in \{0,1\}^*$, and nonce n from nonce set NDS.NS, to return a signature $s \in \{0,1\}^{\mathsf{NDS.ol}}$. Deterministic verification algorithm NDS.vrf takes

Game $\mathbf{G}^{\mathrm{nbuf1}}_{\mathsf{NDS},\mathsf{NG}}(\mathcal{A})$ / Game $\boxed{\mathbf{G}^{\mathrm{nbuf2}}_{\mathsf{NDS},\mathsf{NG}}(\mathcal{A})}$

$(sk, vk, xk) \leftarrow\$ \mathsf{NDS.Kg}$
$St \leftarrow \varepsilon$; $M \leftarrow \emptyset$
$(m, s) \leftarrow\$ \mathcal{A}^{\mathrm{SIG},\mathrm{RO}}(vk \boxed{, xk})$
$v \leftarrow \mathsf{NDS.vrf}(vk, m, s)$
Return $((v = 1) \wedge (m \notin M))$

SIG(m, η)

$(n, St) \leftarrow\$ \mathsf{NG}(\eta, St)$
$s \leftarrow \mathsf{NDS.sign}(sk, xk, m, n)$
$M \leftarrow M \cup \{m\}$
Return s

RO(x, l)

If $T[x, l] = \bot$ then $T[x, l] \leftarrow\$ \{0,1\}^l$
Return $T[x, l]$

Algorithm $\mathsf{NDS.Kg}$

$xk \leftarrow\$ \mathsf{HE.Keys}$
$(sk, vk) \leftarrow\$ \mathsf{DS.Kg}$
Return (sk, vk, xk)

Algorithm $\mathsf{NDS.sign}^{\mathrm{RO}}(sk, xk, m, n)$

$r \leftarrow \mathsf{HE}^{\mathrm{RO}}(xk, (m, n))$
$c \leftarrow \mathsf{DS.Sig}(sk, m; r)$
Return c

Fig. 8. Games for security goals for nonce-based digital scheme NDS relative to nonce-generator NG, and our nonce-based digital signature scheme.

vk, message $m \in \{0,1\}^*$, and candidate signature $s \in \{0,1\}^{\mathsf{NDS.ol}}$, to return a bit $b \in \{0,1\}$. The scheme NDS is correct if for all $(sk, vk, xk) \in [\mathsf{NDS.Kg}]$, all $m \in \{0,1\}^*$ and $n \in \mathsf{NDS.NS}$, the verification of true signatures succeeds: $\mathsf{NDS.vrf}(vk, m, \mathsf{NDS.sign}(sk, xk, m, n)) = 1$.

To formalize security, we consider the games $\mathbf{G}^{\mathrm{nbuf1}}_{\mathsf{NDS},\mathsf{NG}}(\mathcal{A})$ and $\mathbf{G}^{\mathrm{nbuf2}}_{\mathsf{NDS},\mathsf{NG}}(\mathcal{A})$ in Fig. 8 associated to nonce-based signature scheme NDS, nonce generator NG returning nonces in NDS.NS, and adversary $\mathcal{A}$, where the second game includes the boxed code and the first does not. We let

$$\mathbf{Adv}^{\mathrm{nbuf1}}_{\mathsf{NDS},\mathsf{NG}}(\mathcal{A}) = \Pr[\mathbf{G}^{\mathrm{nbuf1}}_{\mathsf{NDS},\mathsf{NG}}(\mathcal{A})] \text{ , and}$$
$$\mathbf{Adv}^{\mathrm{nbuf2}}_{\mathsf{NDS},\mathsf{NG}}(\mathcal{A}) = \Pr[\mathbf{G}^{\mathrm{nbuf2}}_{\mathsf{NDS},\mathsf{NG}}(\mathcal{A})] .$$

As usual the games are described in the ROM, with standard-model definitions are derived by considering only schemes and adversaries that do not query the RO oracle. The difference between the games is tiny, and in just one line of the code, namely that in the second game, the adversary gets the seed xk as an additional input. The first game captures the case that the seed is not exposed, and we will guarantee security for any nonce generator. The second game captures the case that the seed is exposed, in which case we will provide security for unpredictable nonce generators.

Regular signature schemes can be viewed as a special case of nonce-based ones where the seed xk is defined to be the empty string and security is measured relative only to the nonce generator that always returns a uniformly random string.

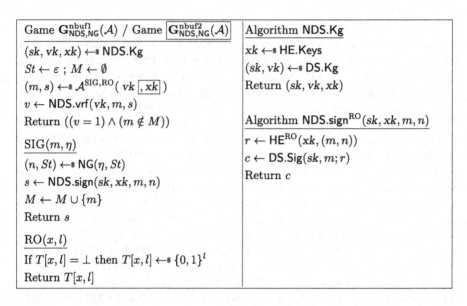

In game $\mathbf{Adv}^{\mathsf{nbuf2}}_{\mathsf{NDS},\mathsf{NG}}(\mathcal{A})$, we say that adversary $\mathcal{A}$ is *agnostic* if its SIG queries do not depend on the seed xk. More formally, there exists a pair $(\mathcal{A}_1, \mathcal{A}_2)$ of algorithms such that $\mathcal{A}^{\mathrm{SIG},\mathrm{RO}}(vk, xk)$ does the following:

$$St \leftarrow\!\!\$\; \mathcal{A}_1^{\mathrm{SIG},\mathrm{RO}}(vk)\;;\;(m,s) \leftarrow \mathcal{A}_2^{\mathrm{RO}}(xk, St)\;;\;\text{ Return } (m,s)\;.$$

SCHEME. We specify a transform **R2NDS** that takes a (standard, randomized) signature scheme DS and a hedged extractor HE and returns the nonce-based signature scheme NDS = **R2NDS**[DS, HE] whose algorithms are described in Fig. 8. The nonce space is the same as for the hedged extractor, i.e. NDS.NS = HE.NS. The length of the signatures is preserved, NDS.ol = DS.ol. The construction requires that HE provides sufficient randomness, i.e., DS.rl = HE.ol.

We first show that the described scheme NDS is indeed secure according to the game $\mathbf{G}^{\mathsf{nbuf1}}_{\mathsf{NDS},\mathsf{NG}}(\mathcal{A})$, that is, in case the seed is not exposed. In this case and if the hedged extractor is a good pseudo-random function, we achieve the same security guarantees as achieved by the original scheme if proper randomness is used.

Theorem 7. *Let* DS *be a (standard, randomized) digital-signature scheme. Let* HE *be a hedged extractor. Let nonce-based digital signature scheme* NDS = **R2NDS**[DS, HE] *be associated to them as above. Let* NG *be a nonce generator. Let $\mathcal{A}$ be an adversary making at most q_1 queries to its* SIG *oracle and q_2 queries to its* RO *oracle. Then the proof specifies adversaries $\mathcal{B}$ and $\mathcal{G}$ such that*

$$\mathbf{Adv}^{\mathsf{nbuf1}}_{\mathsf{NDS},\mathsf{NG}}(\mathcal{A}) \leq \mathbf{Adv}^{\mathsf{prf}}_{\mathsf{HE}}(\mathcal{G}) + \mathbf{Adv}^{\mathsf{uf}}_{\mathsf{DS}}(\mathcal{B})\;, \tag{16}$$

where adversary $\mathcal{B}$ makes at most q_1 queries to its SIG *oracle; besides emulating a Random Oracle it performs almost the same computation as $\mathcal{A}$. Adversary $\mathcal{G}$ makes at most q_1 queries to its* RoR *oracle and q_2 queries to its* RO *oracle; in terms of computation it generates a key pair for* DS *and computes q_1 signatures.*

Proof. Adversary $\mathcal{B}$ against DS behaves as follows. When started with input vk, it executes $\mathcal{A}(vk)$. Upon a query $\mathrm{SIG}(m, \eta)$ from $\mathcal{A}$, adversary $\mathcal{B}$ queries $\mathrm{SIG}(m)$, obtaining a signature s, and returns s to $\mathcal{A}$. Queries to the oracle RO are exactly as in the game, that is, by lazy sampling of a random function.

Adversary $\mathcal{G}$ against the hedged extractor behaves as follows. It generates a key pair $(sk, vk) \leftarrow\!\!\$\; \mathsf{DS.Kg}$, and then runs adversary $\mathcal{A}$ on input vk. Upon a query $\mathrm{SIG}(m, \eta)$ from $\mathcal{A}$, adversary $\mathcal{G}$ calls $\mathrm{FN}(m, \eta)$ to obtain a random string r. It then computes $s \leftarrow \mathsf{DS.Sig}(sk, m; r)$ and returns s. Potential queries $\mathrm{RO}(x, l)$ are referred to the random oracle provided in the game. When $\mathcal{A}$ provides its output (m, s), then $\mathcal{G}$ outputs the result of $\mathsf{DS.Ver}(vk, m, s) = 1$.

We define a hybrid game G that is defined almost identically with $\mathbf{G}^{\mathsf{nbuf1}}_{\mathsf{NDS},\mathsf{NG}}(\mathcal{A})$, with the only difference that the randomness used in SIG queries is uniformly random instead of derived via the hedged extractor. The difference with the game $\mathbf{G}^{\mathsf{nbuf1}}_{\mathsf{NDS},\mathsf{NG}}(\mathcal{A})$ is highlighted in Fig. 9.

The view of $\mathcal{A}$ in G is the same as in $\mathbf{G}^{\mathsf{uf}}_{\mathsf{DS}}(\mathcal{B})$; in both cases the signatures are computed with fresh randomness.

Game G	Game H
⋮	⋮
$\underline{\mathrm{SIG}(m,\eta)}$	$\underline{\mathrm{SIG}(m,\eta)}$
$(n, St) \leftarrow_{\$} \mathsf{NG}(\eta, St)$	$(n, St) \leftarrow_{\$} \mathsf{NG}(\eta, St)$
$s \leftarrow \mathsf{NDS.sign}(sk, xk, m, n)$	$s \leftarrow \mathsf{NDS.sign}(sk, xk, m, n)$
$\boxed{r \leftarrow_{\$} \{0,1\}^{\ell}}$	$\boxed{r \leftarrow_{\$} \{0,1\}^{\ell}}$
$\boxed{s \leftarrow \mathsf{DS.Sig}(sk, m; r)}$	$\boxed{s \leftarrow \mathsf{DS.Sig}(sk, m; r)}$
$M \leftarrow M \cup \{m\}$	$M \leftarrow M \cup \{m\}$
Return s	Return s
⋮	⋮

Fig. 9. Intermediate games used in the proofs of Theorem 7 (left) and Theorem 8 (right).

We observe that

$$\begin{aligned} 2\Pr[\mathbf{G}^{\mathrm{prf}}_{\mathsf{HE}}(\mathcal{G})] &= \Pr[\mathbf{G}^{\mathrm{prf}}_{\mathsf{HE}}(\mathcal{G})|b=1] + \Pr[\mathbf{G}^{\mathrm{prf}}_{\mathsf{HE}}(\mathcal{G})|b=0] \\ &= \Pr[\mathbf{G}^{\mathrm{prf}}_{\mathsf{HE}}(\mathcal{G})|b=1] + 1 - \Pr[\neg\mathbf{G}^{\mathrm{prf}}_{\mathsf{HE}}(\mathcal{G})|b=0] \\ &= \Pr[\mathbf{G}^{\mathrm{nbuf1}}_{\mathsf{NDS,NG}}(\mathcal{A})] + 1 - \Pr[\mathbf{G}^{\mathrm{uf}}_{\mathsf{DS}}(\mathcal{B})], \end{aligned}$$

the reason is that if the bit b is chosen as $b = 0$ in $\mathbf{G}^{\mathrm{prf}}_{\mathsf{HE}}(\mathcal{G})$, then the view of $\mathcal{A}$ is exactly as in $\mathbf{G}^{\mathrm{uf}}_{\mathsf{DS}}(\mathcal{B})$; all signatures are generated using fresh randomness. Analogously, if b is chosen as $b = 1$ in $\mathbf{G}^{\mathrm{prf}}_{\mathsf{HE}}(\mathcal{G})$, then the view of $\mathcal{A}$ is exactly as in $\mathbf{G}^{\mathrm{nbuf1}}_{\mathsf{NDS,NG}}(\mathcal{A})$ with $b = 0$; all signatures are generated with randomness computed via HE and NG from the message m and the input η. For $b = 1$ the probability of $\mathcal{G}$ guessing correctly is the same as the probability of $\mathcal{A}$ forging the signature, but for $b = 0$ the probability of $\mathcal{G}$ guessing correctly is the same as the probability of $\mathcal{A}$ *not* forging a signature. The above equation implies

$$\mathbf{Adv}^{\mathrm{prf}}_{\mathsf{HE}}(\mathcal{G}) = \mathbf{Adv}^{\mathrm{nbuf1}}_{\mathsf{NDS,NG}}(\mathcal{A}) - \mathbf{Adv}^{\mathrm{uf}}_{\mathsf{DS}}(\mathcal{B})$$

and therefore the inequality claimed in the theorem statement. □

The second statement concerns the security in the sense of $\mathbf{G}^{\mathrm{nbuf2}}_{\mathsf{NDS,NG}}(\mathcal{A})$, that is, the signatures are indeed secure even if the seed is exposed, as long as the nonces contain a sufficient amount of min-entropy. For the scheme based on our standard-model hedged extractor, we again restrict the statement to agnostic adversaries $\mathcal{A}$. The scheme based on our ROM-based hedged extractor is again secure against all (i.e., not necessarily agnostic) adversaries.

Theorem 8. *Let* DS *be a (standard, randomized) digital-signature scheme. Let* HE *be a hedged extractor. Let nonce-based digital signature scheme* NDS = **R2NDS**[DS, HE] *be associated to them as above. Let* NG *be a nonce generator.*

Let $\mathcal{A}$ be an adversary making at most q_1 queries to its SIG *oracle and q_2 queries to its* RO *oracle. Then the proof specifies adversaries $\mathcal{B}$ and $\mathcal{G}$ such that*

$$\mathbf{Adv}^{\mathsf{nbuf2}}_{\mathsf{NDS,NG}}(\mathcal{A}) \leq \mathbf{Adv}^{\mathsf{ror}}_{\mathsf{HE,NG}}(\mathcal{G}) + \mathbf{Adv}^{\mathsf{uf}}_{\mathsf{DS}}(\mathcal{B}) \;, \tag{17}$$

where adversary $\mathcal{B}$ makes at most q_1 queries to its SIG *oracle; besides emulating a Random Oracle it performs almost the same computation as $\mathcal{A}$. Adversary $\mathcal{G}$ makes at most q_1 queries to its* RoR *oracle and q_2 queries to its* RO *oracle; in terms of computation it generates a key pair for* DS *and computes q_1 signatures. Furthermore, if $\mathcal{A}$ is agnostic, then $\mathcal{G}$ is also agnostic.*

Proof. Adversary $\mathcal{B}$ against DS behaves as follows. When started with input vk, it samples $xk \leftarrow\$ \mathsf{HE.Keys}$ and executes $\mathcal{A}(vk, xk)$. Upon a query $\mathrm{SIG}(m, \eta)$ from $\mathcal{A}$, adversary $\mathcal{B}$ queries $\mathrm{SIG}(m)$, obtaining a signature s, and returns s to $\mathcal{A}$. Queries to the oracle RO are exactly as in the game, that is, by lazy sampling of a random function.

Adversary $\mathcal{G}$ against the hedged extractor behaves as follows. It obtains a seed $xk \in \mathsf{HE.Keys}$ and generates a key pair $(sk, vk) \leftarrow\$ \mathsf{DS.Kg}$, and then runs adversary $\mathcal{A}$ on input (vk, xk). Upon a query $\mathrm{SIG}(m, \eta)$ from $\mathcal{A}$, adversary $\mathcal{G}$ calls $\mathrm{RoR}(m, \eta)$ to obtain a random string r. It then computes $s \leftarrow \mathsf{DS.Sig}(sk, m; r)$ and returns s. Potential queries $\mathrm{RO}(x, l)$ are referred to the random oracle provided in the game. When $\mathcal{A}$ provides its output (m, s), then $\mathcal{G}$ outputs the result of $\mathsf{DS.Ver}(vk, m, s) = 1$.

We define a hybrid game H that is defined almost identically with $\mathbf{G}^{\mathsf{nbuf2}}_{\mathsf{NDS,NG}}(\mathcal{A})$, with the only difference that the randomness used in SIG queries is uniformly random instead of derived via the hedged extractor. The difference with the game $\mathbf{G}^{\mathsf{nbuf2}}_{\mathsf{NDS,NG}}(\mathcal{A})$ is highlighted in Fig. 9.

The view of $\mathcal{A}$ in H is the same as in $\mathbf{G}^{\mathsf{uf}}_{\mathsf{DS}}(\mathcal{B})$; in both cases the signatures are computed with fresh randomness. The remainder of the proof follows almost exactly as in Theorem 7. □

Acknowledgments. Bellare was supported in part by NSF grants CNS-1526801 and CNS-1228890, ERC Project ERCC FP7/615074 and a gift from Microsoft. Tackmann was supported in part by the Swiss National Science Foundation (SNF) via Fellowship No. P2EZP2_155566 and by NSF grant CNS-1228890. We thank the anonymous reviewers for their helpful and detailed comments.

References

1. OpenSSL predictable random number generator. Debian security advisory, May 2008. http://www.debian.org/security/2008/dsa-1571
2. Barak, B., Dodis, Y., Krawczyk, H., Pereira, O., Pietrzak, K., Standaert, F.-X., Yu, Y.: Leftover hash lemma, revisited. In: Rogaway, P. (ed.) CRYPTO 2011. LNCS, vol. 6841, pp. 1–20. Springer, Heidelberg (2011)
3. Bellare, M., Boldyreva, A., O'Neill, A.: Deterministic and efficiently searchable encryption. In: Menezes, A. (ed.) CRYPTO 2007. LNCS, vol. 4622, pp. 535–552. Springer, Heidelberg (2007)

4. Bellare, M., Brakerski, Z., Naor, M., Ristenpart, T., Segev, G., Shacham, H., Yilek, S.: Hedged public-key encryption: how to protect against bad randomness. In: Matsui, M. (ed.) ASIACRYPT 2009. LNCS, vol. 5912, pp. 232–249. Springer, Heidelberg (2009)
5. Bellare, M., Desai, A., Pointcheval, D., Rogaway, P.: Relations among notions of security for public-key encryption schemes. In: Krawczyk, H. (ed.) CRYPTO 1998. LNCS, vol. 1462, pp. 26–45. Springer, Heidelberg (1998)
6. Bellare, M., Fischlin, M., O'Neill, A., Ristenpart, T.: Deterministic encryption: definitional equivalences and constructions without random oracles. In: Wagner, D. (ed.) CRYPTO 2008. LNCS, vol. 5157, pp. 360–378. Springer, Heidelberg (2008)
7. Bellare, M., Hoang, V.T.: Resisting randomness subversion: fast deterministic and hedged public-key encryption in the standard model. In: Oswald, E., Fischlin, M. (eds.) EUROCRYPT 2015. LNCS, vol. 9057, pp. 627–656. Springer, Heidelberg (2015)
8. Bellare, M., Kohno, T., Shoup, V.: Stateful public-key cryptosystems: How toencrypt with one 160-bit exponentiation. In: Juels, A., Wright, R.N., Vimercati, S. (eds.) ACM CCS 2006, pp. 380–389. ACM Press, October/November 2006
9. Bellare, M., Poettering, B., Stebila, D.: From identification to signatures, tightly: A framework and generic transforms. Cryptology ePrint Archive, report 2015/1157 (2015). http://eprint.iacr.org/2015/1157
10. Bellare, M., Rogaway, P.: The security of triple encryption and a framework for code-based game-playing proofs. In: Vaudenay, S. (ed.) EUROCRYPT 2006. LNCS, vol. 4004, pp. 409–426. Springer, Heidelberg (2006)
11. Bernstein, D.J., Duif, N., Lange, T., Schwabe, P., Yang, B.-Y.: High-speed high-security signatures. In: Preneel, B., Takagi, T. (eds.) CHES 2011. LNCS, vol. 6917, pp. 124–142. Springer, Heidelberg (2011)
12. Boldyreva, A., Fehr, S., O'Neill, A.: On notions of security for deterministic encryption, and efficient constructions without random oracles. In: Wagner, D. (ed.) CRYPTO 2008. LNCS, vol. 5157, pp. 335–359. Springer, Heidelberg (2008)
13. Cash, D., Grubbs, P., Perry, J., Ristenpart, T.: Leakage-abuse attacks against searchable encryption. In: Ray, N., Kruegel. (eds.), ACM CCS 2015, pp. 668–679. ACM Press, October 2015
14. Checkoway, S., Fredrikson, Niederhagen, R., Everspaugh, M., Green, A., Lange, T., Ristenpart, T., Bernstein, D.J., Maskiewicz, J., Shacham, H.: On the practical exploitability of dual EC in TLS implementations. In: USENIX Security, vol. 1 (2014)
15. Cramer, R., Shoup, V.: A practical public key cryptosystem provably secure against adaptive chosen ciphertext attack. In: Krawczyk, H. (ed.) CRYPTO 1998. LNCS, vol. 1462, pp. 13–25. Springer, Heidelberg (1998)
16. Dodis, Y., Ganesh, C., Golovnev, A., Juels, A., Ristenpart, T.: A formal treatment of backdoored pseudorandom generators. In: Oswald, E., Fischlin, M. (eds.) EUROCRYPT 2015. LNCS, vol. 9056, pp. 101–126. Springer, Heidelberg (2015)
17. Fuller, B., O'Neill, A., Reyzin, L.: A unified approach to deterministic encryption: new constructions and a connection to computational entropy. In: Cramer, R. (ed.) TCC 2012. LNCS, vol. 7194, pp. 582–599. Springer, Heidelberg (2012)
18. Goldwasser, S., Micali, S.: Probabilistic encryption. J. Comput. Syst. Sci. **28**(2), 270–299 (1984)
19. Heninger, N., Durumeric, Z., Wustrow, E.: Halderman. Mining your PS and QS: Detection of widespread weak keys in network devices. In: USENIX Security Symposium, pp. 205–220 (2012)

20. Impagliazzo, R., Levin, L.A., Luby, M.: Pseudo-random generation from oneway functions (extended abstracts). In: 21st ACM STOC, pp. 12–24. ACM Press, May 1989
21. Koblitz, N., Menezes, A.: The random oracle model: a twenty-year retrospective. Cryptology ePrint Archive, report 2015/140 (2015). http://eprint.iacr.org/2015/140
22. Krawczyk, H.: LFSR-based hashing and authentication. In: Desmedt, Y.G. (ed.) CRYPTO 1994. LNCS, vol. 839, pp. 129–139. Springer, Heidelberg (1994)
23. Vitek, J., Naccache, D., Pointcheval, D., Vaudenay, S.: Computational alternatives to random number generators. In: Tavares, S., Meijer, H. (eds.) SAC 1998. LNCS, vol. 1556, pp. 72–80. Springer, Heidelberg (1999)
24. Paterson, K.G., Schuldt, J.C.N., Sibborn, D.L.: Related randomness attacks for public key encryption. In: Krawczyk, H. (ed.) PKC 2014. LNCS, vol. 8383, pp. 465–482. Springer, Heidelberg (2014)
25. Pornin, T.: Deterministic usage of the digital signature algorithm (DSA) and elliptic curve digital signature algorithm (ECDSA). IETF RFC 6979 (2013)
26. Raghunathan, A., Segev, G., Vadhan, S.: Deterministic public-key encryption for adaptively chosen plaintext distributions. In: Johansson, T., Nguyen, P.Q. (eds.) EUROCRYPT 2013. LNCS, vol. 7881, pp. 93–110. Springer, Heidelberg (2013)
27. Ristenpart, T., Yilek, S.: When good randomness goes bad: Virtual machine resetvulnerabilities and hedging deployed cryptography. In: NDSS 2010. The InternetSociety, February/March 2010
28. Rogaway, P.: Authenticated-encryption with associated-data. In: Atluri, V. (ed.) ACM CCS 2002, pp. 98–107. ACM Press, November 2002
29. Rogaway, P.: Nonce-based symmetric encryption. In: Roy, B., Meier, W. (eds.) FSE 2004. LNCS, vol. 3017, pp. 348–359. Springer, Heidelberg (2004)
30. Rogaway, P., Shrimpton, T.: A provable-security treatment of the key-wrap problem. In: Vaudenay, S. (ed.) EUROCRYPT 2006. LNCS, vol. 4004, pp. 373–390. Springer, Heidelberg (2006)
31. Schnorr, C.-P.: Efficient signature generation by smart cards. J. Cryptol. **4**(3), 161–174 (1991)
32. Yilek, S.: Resettable public-key encryption: how to encrypt on a virtual machine. In: Pieprzyk, J. (ed.) CT-RSA 2010. LNCS, vol. 5985, pp. 41–56. Springer, Heidelberg (2010)
33. Zuckerman, D.: Simulating Bpp. using a general weak random source. In: 32nd FOCS, pp. 79–89. IEEE Computer Society Press, October 1991

Honey Encryption Beyond Message Recovery Security

Joseph Jaeger[1], Thomas Ristenpart[2], and Qiang Tang[3](✉)

[1] University of California, San Diego, USA
jsjaeger@eng.ucsd.edu
[2] Cornell Tech, New York, USA
ristenpart@cornell.edu
[3] Cornell University, Ithaca, USA
qt44@cornell.edu

Abstract. Juels and Ristenpart introduced honey encryption (HE) and showed how to achieve message recovery security even in the face of attacks that can exhaustively try all likely keys. This is important in contexts like password-based encryption where keys are very low entropy, and HE schemes based on the JR construction were subsequently proposed for use in password management systems and even long-term protection of genetic data. But message recovery security is in this setting, like previous ones, a relatively weak property, and in particular does not prohibit an attacker from learning partial information about plaintexts or from usefully mauling ciphertexts.

We show that one can build HE schemes that can hide partial information about plaintexts and that prevent mauling even in the face of exhaustive brute force attacks. To do so, we introduce target-distribution semantic-security and target-distribution non-malleability security notions. We prove that a slight variant of the JR HE construction can meet them. The proofs require new balls-and-bins type analyses significantly different from those used in prior work. Finally, we provide a formal proof of the folklore result that an unbounded adversary which obtains a limited number of encryptions of known plaintexts can always succeed at message recovery.

1 Introduction

Password-based encryption (PBE) suffers from the threat of brute-force attacks. People pick poor, easy-to-predict passwords and so an attacker, given a ciphertext, can try decrypting it with the most likely password, the next most likely, and so on. It is easy to determine when the right password is found, and so as long as the password falls in this list the attacker wins, recovering the password and the full plaintext. Unfortunately, studies indicate that the most common password is typically selected by almost 1 % of users [10], meaning that passwords have less than $\mu = 7$ bits of min-entropy. The straightforward attack succeeds with probability a bit more than $q/2^{\mu}$ where q is the number of decryption attempts. Bellare, Ristenpart, and Tessaro [7] proved a closely matching upper bound, perhaps suggesting that the case was closed and that, for PBE, one cannot do better.

M. Fischlin and J.-S. Coron (Eds.): EUROCRYPT 2016, Part I, LNCS 9665, pp. 758–788, 2016.
DOI: 10.1007/978-3-662-49890-3_29

Honey Encryption. Juels and Ristenpart (JR) [23], however, showed how one might provably achieve security for relatively low-entropy keys—even when attackers can try decrypting a ciphertext with all possible keys. Intuitively, their approach makes attacks unsuccessful by ensuring that all plaintexts generated during a brute-force attack look plausible. This approach was used previously for the special case of uniformly random plaintexts by Kausik and Hoover [26]. JR proposed a more general cryptographic primitive that they called *honey encryption* (HE). An HE scheme is tailored to an estimate of the (possibly non-uniform) distribution of messages for which it will be employed. We refer to this distribution as the target distribution. Decrypting an HE ciphertext with an incorrect key yields a decoy (or honey) message that appears, to the attacker, to be a fresh sample from the target distribution. An attacker that knows no further information about the true message will be unable to pick it out from the decoys.

JR gave a framework for building HE schemes that composes a distribution-transforming encoder (DTE) with an encryption scheme. A DTE is a kind of randomized encoding scheme tailored to the target distribution. They propose that HE schemes should achieve security in two distinct settings, what we will call the high-entropy key setting and the low-entropy key setting. The former is the conventional setting in which security rests on the adversary being unable to do work proportional to 2^{μ}. Here they show that DTE-then-Encrypt can use standard mechanisms to provably achieve the conventional goals of [7].

The novelty lies in the low-entropy setting, where we assume that keys have some entropy μ but that adversaries can nevertheless do work much greater than 2^{μ}. For simplicity here one most often just assumes unbounded attackers. In this context, JR formalized a message recovery security goal. They then proved that in some useful cases DTE-then-Encrypt constructions can achieve close to optimal message recovery security: for a (relatively high-entropy) message encrypted under a key whose maximum probability of taking on any particular value is at most $1/2^{\mu}$, then an unbounded adversary's ability to guess the correct message, even given the ciphertext, is at most $1/2^{\mu}$ plus a negligible amount. Given that an attacker can always output the decryption of the challenge ciphertext under the most likely key, the JR result is essentially tight.

The DTE-then-Encrypt construction provides a recipe for building HE for particular applications, as one need only build a custom DTE for the setting by way of some estimate of the message distribution. Chatterjee et al. [11] showed how to do so for messages that are themselves lists of human-chosen passwords and built a prototype password vault system based on HE. Huang et al. [21] showed how to construct DTEs for messages that describe a person's genetic information. The application was for building a secure, long-term genetic information store. In both contexts they rely on JR's goal of MR security.

But MR security has several deficiencies from the viewpoint of modern security goals for conventional symmetric encryption (SE), and even for the applications for which researchers have explored use of HE. For SE one strives for authenticated encryption security [6], or its robust variants [19,29]. These notions allow chosen message and ciphertext attacks. Informally speaking, they demand that not even a single bit of information about plaintexts can be learned by an adversary and that ciphertexts cannot be forged. We are therefore left with a significant gap between the JR results and what we might like in terms of security. In the genetic store application, for example, it could be that using an only MR-secure HE scheme would leak most of your genome. All this begs the question of whether there exist stronger security goals for HE and constructions that meet them.

Our Contributions. In this work, we provide a systematic study of stronger notions of security for HE schemes in the low-entropy key setting. The bad news first: we formally rule out the ability to strengthen the JR security notions to allow known-message attacks when attackers can exhaust the key space. While this result seems intuitively obvious, and was taken for granted in [23], showing it formally for arbitrary HE schemes required a surprising amount of care. Having done so, we return to unknown message attack settings, but here provide good news in the way of stronger security goals and proofs that simple constructions meet them. First, we give a semantic security-style notion suitable for unknown message attacks and, second, a notion of target-distribution non-malleability. We show how the JR construction meets the first, and a new construction that achieves both. In the remainder of the introduction we provide more overview of these results.

Impossibility of Known-message Attack Security. The JR security message recovery (MR) definition works as follows. A challenge message is drawn from the target distribution, encrypted under a key, and the resulting ciphertext is given to the adversary. It wins if it can output the challenge message. While the adversary knows the target distribution and the distribution from which keys are drawn, it does not get access to any known message, ciphertext pairs under the key. We extend the notion to additionally give the adversary an oracle from which it can obtain message-ciphertext pairs more messages drawn from the target distribution, yielding a known-message attack variant. We denote this notion by MR-KMA.

Intuitively MR-KMAshould be unachievable when the adversary can exhaustively search the key space. The adversary simply queries the oracle on several different messages, runs a brute-force attack to find the key that is consistent with all the message-ciphertext pairs, and uses that to decrypt the challenge ciphertext. While this attack might seem to work against all schemes, in fact there exist many for which there will be a large set of consistent keys. In the most extreme case, all keys will be consistent after any number of queries when encryption is the identity function for each key. One approach to deal with this is to make assumptions about the underlying scheme that allow one to show that after sufficiently many queries the consistent set will shrink to one. For example, if the encryption scheme has "sufficiently random" mappings for distinct keys. But we would like to make no assumptions about the HE scheme.

Our attacker instead simply embraces that there may be a large set of consistent keys, and just uses one of them at random to decrypt the challenge ciphertext. We then have to lower bound the probability that a random key from the consistent set decrypts the challenge ciphertext to the target plaintext. In fact we do not know how to (or whether one can) prove this for an adversary that makes a fixed number of queries. Rather we show that there exists some number of queries between zero and κ, where 2^κ is the size of the key space, for which an adversary will achieve advantage at least $1/2\kappa$.

In the end, our result rules out security against known-message attacks. We also note that the proof techniques here already apply to (non-stateful) symmetric encryption as they do not take advantage of any properties specific to HE. We are, in fact, unaware of any previous general lower bound on message recovery for exhaustive key search attacks against conventional symmetric encryption schemes. Finally, the proof technique can generalize as well to message authentication goals, such as unforgeability under chosen-message attack.

Protecting Partial Information. We now return to unknown message attacks, but seek to strengthen the security goals along two dimensions. First, we consider partial information leakage. MR security is potentially adequate in settings for which the encrypted message is, say, an authentication credential which must be supplied in full elsewhere (the original motivating settings in [23]). It is likely to prove insufficient more generally. Schemes meeting MR might trivially leak a significant amount of information about messages. The seminal work of Goldwasser and Micali [17] argued (in the context of public-key encryption) that one should instead prefer encryption to hide all partial information about plaintexts. This stronger goal, called semantic security, was subsequently adapted to (at least) the settings of symmetric encryption [3], deterministic symmetric encryption [15,30], and deterministic and hedged public-key encryption [1,2,4].

Unfortunately the traditional symmetric encryption semantic security notion (denoted SS below) [3], along with its variants of indistinguishability under chosen plaintext attack [3], are unachievable when keys are low entropy. (This is a corollary of our negative results about MR-CPA.)

We therefore introduce a new semantic-security style notion suitable for the low-entropy key setting. We call it target-distribution semantic security (TDSS). In it, an adversary is given the encryption of a message drawn from a target distribution and must predict a boolean function applied to the plaintext. It needs to do this better than is possible when predicting the predicate without the ciphertext. The key difference from SS is that it is asked to hold only for a specific message distribution, the target, and not for all message distributions. Interestingly we could find no meaningful indistinguishability-style variant of TDSS (unlike in the conventional setting, where we have the notion of IND-CPA and, moreover, an equivalence between it and SS [3]).

We relate the MR and TDSS notions, in particular using a result from Dodis and Smith [15] (see also [4]) that straightforwardly adapts to our setting. We use it as an intermediate step to show that predicting predicates implies predicting functions for TDSS. Since MR security is equivalent to predicting the identity function, we obtain that TDSS implies MR security. There exists a simple separation showing that MR does not imply TDSS.

We go on to analyze the DTE-then-Encrypt scheme due to JR, showing via a new balls-and-bins analysis an upper bound of about $2\omega_k^{\frac{7}{16}} + 2e^{-\frac{1}{3\omega_k^{1/8}}}$ on the advantage of unbounded TDSS attackers. Like the MR proof by JR, ours is in the random oracle model [8]. Because TDSS focuses on predicates, the new balls-and-bins analysis necessarily focuses on the trickier setting of having many more balls (representing keys here) than the two bins (the possible predicate outputs). Our proof crucially relies, as did JR's, on a majorization lemma due to Berenbrink et al. [9] to transition the balls-and-bins analysis from non-uniform keys to uniform ones. In comparison to MR security our new bound is quantitatively weaker: JR showed MR advantage upper bounded by ω_k (when message distribution entropy is sufficiently large). Here we instead lose about half the entropy of the key. Nevertheless our result may be close to optimal (see Remark 1).

Non-malleability. The JR message recovery security goal, as well as the TDSS goal above, do not rule out active attackers manipulating ciphertexts. Indeed, DTE-then-Encrypt instantiations used in [11,21,23] are trivially malleable as they encrypt the DTE output by XOR'ing it with a pad derived from a hash of the key. An attacker can flip particular bits of the ciphertext and know that the resulting ciphertext will be decrypted to a plaintext related in a predictable way to the original. This is true regardless of the unpredictability of either the key or message.

Complicating matters, achieving MR or TDSS security seems to rule out preventing manipulation by including in an HE scheme typical mechanisms such as authentication tags or redundancy. Intuitively, this is because they would seem to always help the attacker rule out incorrect keys. We therefore turn to weaker notions like non-malleability [16,25], which again are unachievable in the low-entropy key setting (by our negative results above) but may be adaptable to unknown message settings because their goals do not seem to inherently conflict with confidentiality goals like MR and TDSS.

We introduce a target-distribution non-malleability (TDNM) notion for HE schemes when used with low-entropy keys. Informally, an attacker should not be able maul a ciphertext C to produce a new ciphertext $\tilde{C}$ in a way that some fixed relation R over the associated plaintexts is met with probability higher than one can achieve without access to C. All this holds for C being the encryption of a message taken from the target distribution.

We propose a simple construction that we call DTE-then-Encipher. It composes a DTE with a block cipher with sufficiently large domain. Modeling the cipher as ideal allows us to prove both TDSS and TDNM security. The TDNM proof shares some similarity to the TDSS proof of DTE-then-Encrypt, but requires additional techniques. In particular, the balls-in-bins analysis here cannot use the majorization lemma of [9], and so we perform a new majorization-style analysis that exploits Schur convexity [22].

Further Related Work. Entropic security was considered in [15,30] as a statistical analogue of semantic security, and like HE they can also resist unbounded attackers. They show security against can be achieved when $\mu_k + \mu_m \geq n$, where μ_k, μ_m are the min-entropy of the key and message distribution, respectively, and n is the message length in bits. They show one can do no better in their setting, which requires security to hold over all distributions with the indicated min-entropy. HE low-entropy key security instead relaxes this to focus on specific target distributions, thereby skirting their lower bounds on required entropy, and providing meaningful security even when $\mu_k + \mu_m < n$.

2 Notation and Definitions

Notation. If n is an integer we let $\mathbb{Z}_n$ be the set $\{0, \ldots, n-1\}$. We use $y \leftarrow\!\!\$\ A(x)$ to denote running randomized algorithm A on input x and setting y equal to its output. If instead A is deterministic we write $y \leftarrow A(x)$. If G is a game we let $\Pr[\mathrm{G} \Rightarrow \mathsf{true}]$ denote the probability that G outputs true.

Let $\mathcal{S}$ be a set. A distribution on $\mathcal{S}$ is a function $p : \mathcal{S} \rightarrow [0,1]$ such that $\sum_{s \in \mathcal{S}} p(s) = 1$. The maximum probability ω of a distribution p is defined to be $\omega = \max_{s \in \mathcal{S}} p(s)$. The min-entropy μ of p is defined to be $\mu = -\log \omega$. When referencing min-entropy and maximum probability the distribution will always be clear from context. By $s \leftarrow_p \mathcal{S}$ we denote sampling an element $s \in \mathcal{S}$ according to the distribution p. That is, each $s \in \mathcal{S}$ is chosen with probability $p(s)$. For $B \subseteq \mathcal{S}$ we overload notation and let $p(B) = \sum_{s \in B} p(s)$.

Hash Functions. A hash function H is a function $\mathsf{H} : \{0,1\}^* \rightarrow \{0,1\}^n$ which maps strings of arbitrary length to strings of some fixed length n. The length n will always be clear from context. In this work, we model hash functions as random oracles.

Symmetric Encryption. A symmetric encryption scheme $\mathsf{SE} = (\mathtt{Enc}, \mathtt{Dec})$ is a pair of algorithms defined relative to a key space $\mathcal{K}$ and message space $\mathcal{M}$. The randomized encryption algorithm $\mathtt{Enc}$ takes as input a key $K \in \mathcal{K}$ and a message $M \in \mathcal{M}$ and outputs a ciphertext $C \in \mathcal{C}$. The deterministic decryption algorithm $\mathtt{Dec}$ takes as input a key $K \in \mathcal{K}$ and a ciphertext $C \in \mathcal{C}$ and outputs a message $M \in \mathcal{M}$. We require that a symmetric encryption scheme must be correct, meaning that for all $K \in \mathcal{K}$ and all $M \in \mathcal{M}$, $\Pr[\mathtt{Dec}(K, \mathtt{Enc}(K, M)) = M] = 1$.

Majorization. We say $\bar{p}$ majorizes $\bar{q}$ (denoted as $\bar{p} \succ \bar{q}$), if the two vectors $\bar{p} = \langle p_1, p_2, \ldots, p_n \rangle$, and $\bar{q} = \langle q_1, q_2, \ldots, q_n \rangle$ (written in descending order such that for all $i \in [1, n-1]$ that $p_i \geq p_{i+1}$, and $q_i \geq q_{i+1}$) satisfy $\sum_{i=1}^{k} p_i \geq \sum_{i=1}^{k} q_i$ for all $k \in \{1, \ldots, n\}$. When $\bar{p}$ denotes the probabilities of a distribution with support size n, it is easy to see that $\bar{q} \succ \bar{p}$, if $\bar{q}$ is defined as $q_i = p_1$ for $1 \leq i \leq \lceil 1/p_1 \rceil$ and $q_i = 0$ for $\lceil 1/p_1 \rceil + 1 \leq i \leq n$.

$\mathrm{MR}^{\mathcal{A}}_{\mathsf{HE},p_m,p_k}$	$\mathrm{SAMP1}^{\mathcal{D}}_{\mathsf{DTE},p_m}$	$\mathrm{SAMP0}^{\mathcal{D}}_{\mathsf{DTE}}$
$K^* \leftarrow_{p_k} \mathcal{K}$	$M \leftarrow_{p_m} \mathcal{M}$	$S \leftarrow_{\$} \mathcal{S}$
$M^* \leftarrow_{p_m} \mathcal{M}$	$S \leftarrow_{\$} \mathsf{encode}(M)$	$b \leftarrow_{\$} \mathcal{D}(S)$
$C^* \leftarrow_{\$} \mathtt{HEnc}(K^*, M^*)$	$b \leftarrow_{\$} \mathcal{D}(S)$	Return $(b = 1)$
$M \leftarrow \mathcal{A}(C^*)$	Return $(b = 1)$	
Return $(M = M^*)$		

Fig. 1. Left: Game defining message recovery security. **Middle and Right:** Games defining security of a DTE.

3 Background on Honey Encryption

Honey Encryption Schemes. An HE scheme $\mathsf{HE} = (\mathtt{HEnc}, \mathtt{HDec})$ is a symmetric encryption scheme for some key space $\mathcal{K}$ and message space $\mathcal{M}$. Typically $\mathcal{K}$ will be strings representating human-chosen passwords, but HE can be applied in other settings as well. HE schemes should meet conventional security goals for password-based symmetric encryption [7,24]. Differentiating HE schemes from conventional ones, however, is that they are designed relative to a specific (estimated) distribution over $\mathcal{M}$. This allows schemes that achieve a level of security even when the keys are relatively predictable, or have low min-entropy, from an attacker's perspective. Again, human-chosen passwords are the canonical example of such keys.

We let p_m represent the message distribution on the message space $\mathcal{M}$ and μ_m, ω_m denote its min-entropy and maximum probability respectively. Similarly we let p_k represent the key distribution on the key space $\mathcal{K}$ and let μ_k, ω_k denote its min-entropy and maximum probability respectively. In the low-entropy settings we focus on, we assume that ω_k is large enough that an attacker can easily perform work proportional to 2^{μ_k}. For simplicity in fact we will in our treatment simply assume adversaries can run in unbounded time. Our results extend to this setting, but also can be translated to computationally bounded settings in a straightforward manner.

MR Decurity. Juels and Ristenpart [23] formalized and built schemes to achieve message recovery (MR) security. Their MR security game is defined in Fig. 1 for a scheme HE and distributions p_m, p_k. An MR adversary $\mathcal{A}$ takes as input a ciphertext encrypting a challenge message chosen according to p_m and outputs a message $M \in \mathcal{M}$. The adversary wins if it outputs the challenge message. More precisely, we measure the advantage of a (computationally unbounded) MR adversary $\mathcal{A}$ against scheme HE and distributions p_m and p_k by

$$\mathbf{Adv}^{\mathrm{mr}}_{\mathsf{HE},p_m,p_k}(\mathcal{A}) = \Pr\left[\mathrm{MR}^{\mathcal{A}}_{\mathsf{HE},p_m,p_k} \Rightarrow \mathsf{true}\right].$$

Distribution-transforming Encoders. A distribution-tranforming encoder (let us use DTE for short) is a pair of algorithms $\mathsf{DTE} = (\mathsf{encode}, \mathsf{decode})$

defined relative to a message space $\mathcal{M}$ and a set $\mathcal{S}$ called the seed space. Via $S \leftarrow\!\!\$\ \mathsf{encode}(M)$ the randomized encoding algorithm encode taking a message $M \in \mathcal{M}$ as input and outputs a seed $S \in \mathcal{S}$. A DTE must satisfy correctness, that for any message $M \in \mathcal{M}$, $\Pr[\mathsf{decode}(\mathsf{encode}(M)) = M] = 1$. Like HE schemes, a DTE is designed for a specific message distribution p_m.

Following [23], the security property desired for a DTE is that it is hard for an adversary to distinguish between $S \in \mathcal{S}$ chosen uniformly at random and chosen by first picking a message according to p_m and then applying encode. This property is formalized by two of the games shown in Fig. 1. We measure the advantage of an adversary $\mathcal{D}$ against DTE and distribution p_m by

$$\mathbf{Adv}^{\rm dte}_{\mathsf{DTE},p_m}(\mathcal{D}) = \Pr[\mathrm{SAMP1}^{\mathcal{D}}_{\mathsf{DTE},p_m}] - \Pr[\mathrm{SAMP0}^{\mathcal{D}}_{\mathsf{DTE}}]\ ,$$

and the DTE-goodness is defined by $\mathbf{Adv}^{\rm dte}_{\mathsf{DTE},p_m} = \max_{\mathcal{D}} \mathbf{Adv}^{\rm dte}_{\mathsf{DTE},p_m}(\mathcal{D})$ and where the the maximization is over all, even computationally unbounded, adversaries $\mathcal{D}$. When the DTE in question is clear we let p_d represent the distribution induced on $\mathcal{M}$ by sampling a random seed from $\mathcal{S}$ and applying decode. Formally,

$$p_d(M) = \Pr[M' = M : S \leftarrow\!\!\$\ \mathcal{S}; M' \leftarrow \mathsf{decode}(S)]\ .$$

DTE-then-Encrypt. JR introduced a framework of constructing HE schemes for a target distribution p_m from a symmetric encryption scheme SE and a distribution-transforming encoder DTE. More specifically, the DTE-then-Encrypt framework encrypts a message by applying the DTE encoding first and then encrypting the encoding using SE. Security requires some easy-to-meet properties of SE, such as that it does not pad out inputs.

In more detail, let H be a hash function and r an integer representing the number of random bits to be used by encryption. Then the scheme which we will denote by HE[DTE, H] is shown in Fig. 2.

Note that as written, this scheme does not achieve the password-based encryption security goals of [7] for the high-entropy key setting. It is easy to modify the scheme to do so: simply replace the hash function with an appropriate password-based key derivation function (PBKDF). One can also deal with using fixed-output-length hash functions with large seed spaces by appropriate use of a mode of operation. See [23] for more detailed discussion.

$\mathtt{HEnc}(K, M)$	$\mathtt{HDec}(K, C)$
$S \leftarrow\!\!\$\ \mathsf{encode}(M)$	$(R, C_2) \leftarrow C$
$R \leftarrow\!\!\$\ \{0,1\}^{rl}$	$S \leftarrow \mathsf{H}(R\|\|K) \oplus C_2$
$C_2 \leftarrow \mathsf{H}(R\|\|K) \oplus S$	$M \leftarrow \mathsf{decode}(S)$
Return (R, C_2)	Return M

Fig. 2. The DTE-then-Encrypt construction HE[DTE, H], using hash function H and DTE DTE = (encode, decode).

4 Impossibility of KMA Security with Low-Entropy Keys

Recall that the MR security notion is a relatively weak goal in various ways. One such weakness is that it is only an unknown-message attack and provides adversaries with no plaintext-ciphertext examples. In this section we show that one cannot hope to achieve security in the low-entropy key setting when given a relatively small number of plaintext-ciphertext examples in a known-message attack. Making this claim formal required a surprising amount of care.

MR-KMA Security Definition. Let game MR-KMA be defined as in Fig. 3 for scheme HE and distributions p_m, p_k. This game is exactly the same as the MR security game except the adversary additionally has access to an encryption oracle which samples a message M according to p_m and returns an encryption of M under the secret key. We measure the advantage of a (computationally unbounded) adversary $\mathcal{A}$ against HE with distributions p_m and p_k by

$$\mathbf{Adv}^{\text{mr-kma}}_{\mathsf{HE},p_m,p_k}(\mathcal{A}) = \Pr[\text{MR-KMA}^{\mathcal{A}}_{\mathsf{HE},p_m,p_k} \Rightarrow \mathsf{true}].$$

The "obvious" Attack Strategy. A straightforward strategy for an MR-KMA adversary is to use its encryption oracle to receive q distinct message-ciphertext pairs. Then, use test decryption under all keys to find those keys that correctly decrypt all the ciphertexts correctly. We refer to such a key as being consistent. The intuition is that even for small q the set of consistent keys will be a singleton and that, necessarily, it is the key chosen by the experiment. This intuition stems from the fact that for a "reasonable" scheme, the probability that the wrong key decrypts all the ciphertexts correctly is low.

Of course formally this logic is meaningless as it makes unspecified assumptions on the scheme. Indeed there are many examples of schemes for which the set of consistent keys will be large, no matter how large q gets. In the most egregious case, where HEnc and HDec implement the identity function for all keys,

$\text{MR-KMA}^{\mathcal{A}}_{\mathsf{HE},p_m,p_k}$
$K^* \leftarrow_{p_k} \mathcal{K}$
$M^* \leftarrow_{p_m} \mathcal{M}$
$C^* \leftarrow_\$ \mathtt{HEnc}(K^*, M^*)$
$M \leftarrow_\$ \mathcal{A}^{\mathbf{Enc}}(C^*)$
Return $(M = M^*)$

Enc()
$M \leftarrow_{p_m} \mathcal{M}$
$C \leftarrow_\$ \mathtt{HEnc}(K^*, M)$
Return (M, C)

Fig. 3. Game defining message recovery security under a known message attack.

then the set of consistent keys will always be $\mathcal{K}$. Clearly this scheme is not MR secure, but the point is that when giving a proof that holds for all schemes we must handle such degenerate cases. This issue (and the particular degenerate example just given) is related to the well-known fact that key recovery security does not imply message recovery security for all schemes. Nevertheless, we are unaware of any proofs showing that no SE scheme can resist message recovery attacks that exhaustively search the key space.

A Lower Bound on MR-KMA Security. Given the example that ruling out keys may not work very well, we give a slightly different adversary. Our adversary, shown in Fig. 4, runs the attack as described above, but simply finishes by decrypting the challenge using a uniformly chosen key from the set of consistent keys. It is clear, for example, that in the trivial identity-function scheme mentioned above all keys will be consistent with the challenge and this attack achieves advantage one.

We must lower-bound the success probability for any scheme. Doing so requires showing that with high probability the uniformly selected consistent key must be consistent also with the challenge ciphertext. Due to technical difficulties relating to our proof, we cannot give an exact number of oracle queries for which this attack has a high advantage. Instead we show that for some number of queries which is at most $\kappa = \lceil \log |\mathcal{K}| \rceil$ this attack has a high success probability of at least $1/(2\kappa)$. For concreteness we then say that the advantage of an adversary who picks the number of queries at random from $0, \ldots, \kappa$ will have advantage at least $1/(2\kappa^2)$. These results give us the following theorem.

```
Adversary A(C*)
q ←$ Z_κ; S_q ← ∅
For i = 1, ..., q do
  (M_i, C_i) ← Enc(M_i)
For K ∈ K do
  If (∀i HDec(K, C_i) = M_i)
    S_q ← S_q ∪ {K}
K ←$ S_q
Return HDec(K, C*)
```

Fig. 4. Adversary for MR-KMA making at most $\kappa = \lceil \log |\mathcal{K}| \rceil$ encryption queries.

Theorem 1. *Let HE be an encryption scheme and $\kappa = \lceil \log |\mathcal{K}| \rceil$. Then for any p_m, p_k the adversary $\mathcal{A}$ shown in Fig. 4 makes at most $\kappa - 1$ oracle queries and has advantage*

$$\mathbf{Adv}^{\text{mr-kma}}_{HE,p_m,p_k}(\mathcal{A}) \geq \frac{1}{2\kappa^2}. \tag{1}$$

The idea of the proof is to note that the advantage of the adversary $\mathcal{A}$ for a particular value of q is equal to the probability that a randomly chosen key that is consistent with q message-ciphertext pairs is also consistent with a $(q+1)$-th pair (the challenge message and ciphertext). Then letting S_q denote the set of keys consistent after q pairs. We have that the advantage of $\mathcal{A}$ for a particular value of q is $\mathbb{E}[|S_{q+1}|/|S_q|]$, where the expectation is taken over the appropriate experiment (defined below). Intuitively, this ratio can only be really small for a small number of q's because each S_q must contain between 1 and 2^κ keys.

Before presenting the full proof we formalize the above intuition with the following lemma about random variables.

Lemma 1. *If $s_0, \ldots, s_\kappa$ are positive integer-valued random variables such that $s_0 \le 2^\kappa$ and $s_{q+1} \le s_q$ for $q \in \mathbb{Z}_\kappa$, then $\max_{q \in \mathbb{Z}_\kappa} \mathbb{E}[s_{q+1}/s_q] \ge \frac{1}{2\kappa}$.*

Proof. Let $\epsilon = \max_{q \in \mathbb{Z}_\kappa} \mathbb{E}[s_{q+1}/s_q]$. We will use an inductive argument to prove that $\Pr[s_q \ge 2^{\kappa-q}] \le 2q\epsilon$ for $1 \le q \le \kappa$. Then considering when q is κ and noting that $s_\kappa \ge 1$ always we have $1 = \Pr[s_\kappa \ge 1] \le 2\kappa\epsilon$. Solving for ϵ gives the desired bound.

We now give the inductive argument. First, Markov's inequality can be used to bound the probability that s_{q+1} is at least half s_q by $\Pr[s_{q+1}/s_q \ge 1/2] \le 2\mathbb{E}[s_{q+1}/s_q]$. Rewriting and bounding $\mathbb{E}[s_{q+1}/s_q]$ by ϵ we get

$$\Pr[s_{q+1} \ge (1/2)s_q] \le 2\epsilon \tag{2}$$

for all $q \in \mathbb{Z}_\kappa$.

Recalling that $s_0 \le 2^\kappa$, the base case is easily derived by $\Pr[s_1 \ge 2^{\kappa-1}] \le \Pr[s_1 \ge (1/2)s_0] \le 2\epsilon$.

Now suppose $1 < q \le \kappa$ and $\Pr[s_{q-1} \ge 2^{\kappa-(q-1)}] \le 2(q-1)\epsilon$. By definition we have,

$$\begin{aligned}\Pr[s_q \ge 2^{\kappa-q}] = {} & \Pr[s_q \ge 2^{\kappa-q} | s_{q-1} < 2^{\kappa-(q-1)}] \Pr[s_{q-1} < 2^{\kappa-(q-1)}] \\ & + \Pr[s_q \ge 2^{\kappa-q} | s_{q-1} \ge 2^{\kappa-(q-1)}] \Pr[s_{q-1} \ge 2^{\kappa-(q-1)}];\end{aligned}$$

The first part of the equation can be bounded using our inductive assumption:

$$\begin{aligned}\Pr[s_q \ge 2^{\kappa-q} | s_{q-1} \ge 2^{\kappa-(q-1)}] \Pr[s_{q-1} \ge 2^{\kappa-(q-1)}] & \le \Pr[s_{q-1} \ge 2^{\kappa-(q-1)}] \\ & \le 2(q-1)\epsilon;\end{aligned}$$

To bound the second part note that conditioned on the fact that s_{q-1} is less than $2^{\kappa-(q-1)}$, it can only hold that s_q is greater than $2^{\kappa-q}$ if s_q is greater than $(1/2)s_{q-1}$. This gives us

$$\Pr[s_q \ge 2^{\kappa-q} | s_{q-1} < 2^{\kappa-(q-1)}] \le \Pr[s_q \ge (1/2)s_{q-1} | s_{q-1} < 2^{\kappa-(q-1)}] \,.$$

Then from the definition of conditional probability and using (2) we get that

$$\begin{aligned}\Pr[s_q \ge (1/2)s_{q-1} | s_{q-1} < 2^{\kappa-(q-1)}] \Pr[s_{q-1} < 2^{\kappa-(q-1)}] & \le \Pr[s_q \ge (1/2)s_{q-1}] \\ & \le 2\epsilon;\end{aligned}$$

Putting the above equations together we get $\Pr[s_q \ge 2^{\kappa-q}] \le 2q\epsilon$, completing the proof. □

We now use the above result to prove Theorem 1. The proof proceeds by showing that the advantage of adversary $\mathcal{A}$ for a particular q is $\mathbb{E}[|S_{q+1}|/|S_q|]$ where S_q is the set of consistent keys after q message-ciphertext pairs and then noting that the size of these sets fulfill the conditions of the lemma above.

Game G	Game H	Experiment E
$K^* \leftarrow_{p_k} \mathcal{K}$; $M^* \leftarrow_{p_m} \mathcal{M}$	$K^* \leftarrow_{p_k} \mathcal{K}$; $q \leftarrow\!\$\ \mathbb{Z}_\kappa$	$S_0 \leftarrow \mathcal{K}$; $S_1, \ldots, S_\kappa \leftarrow \emptyset$
$C^* \leftarrow\!\$\ \mathtt{HEnc}(K^*, M^*)$	$S_0 \leftarrow \mathcal{K}$; $S_1, \ldots, S_{q+1} \leftarrow \emptyset$	$K^* \leftarrow_{p_k} \mathcal{K}$
$q \leftarrow\!\$\ \mathbb{Z}_\kappa$; $S_q \leftarrow \emptyset$	For $i = 1, \ldots, q$ do	For $i = 1, \ldots, \kappa$ do
For $i = 1, \ldots, q$ do	$M_i \leftarrow_{p_m} \mathcal{M}$	$M_i \leftarrow_{p_m} \mathcal{M}$
$M_i \leftarrow_{p_m} \mathcal{M}$	$C_i \leftarrow\!\$\ \mathtt{HEnc}(K^*, M_i)$	$C_i \leftarrow\!\$\ \mathtt{HEnc}(K^*, M_i)$
$C_i \leftarrow\!\$\ \mathtt{HEnc}(K^*, M_i)$	For $K \in S_{i-1}$ do	For $K \in S_{i-1}$ do
For $K \in \mathcal{K}$ do	If $(\mathtt{HDec}(K, C_i) = M_i)$	If $(\mathtt{HDec}(K, C_i) = M_i)$
If $(\forall i\ \mathtt{HDec}(K, C_i) = M_i)$	$S_i \leftarrow\!\$\ S_i \cup \{K\}$	$S_i \leftarrow\!\$\ S_i \cup \{K\}$
$S_q \leftarrow S_q \cup \{K\}$	$M^* \leftarrow_{p_m} \mathcal{M}$	
$K \leftarrow\!\$\ S_q$	$C^* \leftarrow\!\$\ \mathtt{HEnc}(K^*, M^*)$	
$M \leftarrow \mathtt{HDec}(K, C^*)$	For $K \in S_q$ do	
Return $(M = M^*)$	If $(\mathtt{HDec}(K, C^*) = M^*)$	
	$S_{q+1} \leftarrow S_{q+1} \cup \{K\}$	
	$K \leftarrow\!\$\ S_q$	
	Return $(K \in S_{q+1})$	

Fig. 5. Left and Middle: Games used in MR-KMA proof. **Right:** Experiment used in MR-KMA proof.

Proof (of Theorem 1). First note that $\mathbf{Adv}^{\text{mr-kma}}_{\mathsf{HE},p_m,p_k}(\mathcal{A}) = \Pr[\mathrm{G} \Rightarrow \mathsf{true}]$, where game G is defined on the left side of Fig. 5. This is clear because G is simply the game $\text{MR-KMA}^{\mathcal{A}}_{\mathsf{HE},p_m,p_k}$ with the code of $\mathcal{A}$ inserted.

Now consider game H shown in the middle of Fig. 5. Game H is obtained from G via a few simple transforms. In it S_q is computed iteratively one (M, C) pair at a time, the choice of M^* and C^* is deferred until they are used, and instead of checking whether $M = M^*$ the game equivalently checks whether the randomly chosen K falls in the subset of S_q that decrypts C^* to M^* which is called S_{q+1}. It is thus clear that $\Pr[\mathrm{H} \Rightarrow \mathsf{true}] = \Pr[\mathrm{G} \Rightarrow \mathsf{true}]$.

Noting that $S_{q+1} \subseteq S_q$ holds for every q, it is clear from the last two lines of H that once q, S_q, and S_{q+1} are chosen, the probability that H will output true is $\mathbb{E}[|S_{q+1}|/|S_q|]$. Thus we have that $\Pr[\mathrm{H} \Rightarrow \mathsf{true}] = \sum_{q=0}^{\kappa}(1/\kappa)\mathbb{E}[|S_{q+1}|/|S_q|]$.

Next we transition our analysis to considering the experiment E shown in Fig. 5. Note that the distribution of S_{q+1} and S_q for any $q \in \mathbb{Z}_\kappa$ in E is identical to the distribution in H. For $0 \leq q \leq \kappa$, let s_q be the random variable representing $|S_q|$ in E and ϵ be $\max_{q \in \mathbb{Z}_\kappa} \mathbb{E}[s_{q+1}/s_q]$ where the expectation is taken in experiment E.

Since all S_q always contains at least K^*, each s_q must be positive. Thus $s_0, \ldots, s_\kappa$ are positive integer-valued random variables which fulfill the conditions of Lemma 1 so we have $\epsilon \geq \frac{1}{2\kappa}$. Then the following sequence of inequalities exhibits (1):

$$\mathbf{Adv}^{\text{mr-kma}}_{\mathsf{HE},p_m,p_k}(\mathcal{A}) = \Pr[\mathrm{H} \Rightarrow \mathsf{true}] = \sum_{q=0}^{\kappa} \frac{1}{\kappa}\mathbb{E}[s_{q+1}/s_q] \geq \frac{1}{\kappa} \cdot \epsilon \geq \frac{1}{2\kappa^2}. \qquad \square$$

Extensions. While we focused above on known-message attacks, our proof techniques carry over to the more typical setting of chosen-plaintext attacks. Here the adversary has access instead to an encryption oracle that takes as input an adversarially chosen message, encrypts it using the secret key, and returns the ciphertext.

Furthermore, the ideas behind our proof can be extended to cover unforgeability under chosen-message attacks for, e.g., message authentication codes [5,18]. Here an adversary with access to a tagging oracle tries to come up with a valid message-tag pair for a message it has not queried yet. The adversary used to prove its impossibility would use a fixed sequence of messages $M_1 \ldots, M_\kappa$ and use a random key consistent with the first q messages to sign the next message (a fixed sequence of messages is used here to avoid the problem of the adversary trying to tag a message it has already been given the correct tag for). Then essentially the exact same analysis shows that this adversary will succeed with high probability. We omit the details for brevity.

5 Stronger Message Privacy for HE Schemes

Given the impossibility result of the last section, we turn to explore achievable but still meaningful security notions that capture the goal of hiding partial informations about the messages encrypted by an HE scheme. In this section, we propose a semantic security-style definition tailored to the low-entropy key setting. We call it targeted-distribution semantic security (TDSS). We will also investigate its relationship with MR security. We then go on to show that the DTE-then-Encrypt construction meets this stronger notion of security, though with concrete security bounds slightly worse than what could be proved in the MR case.

5.1 TDSS Security and its Relation to MR Security

Recall that semantic security style notions ask, roughly, that an attacker given the encryption of a message cannot predict a predicate on it with probability better than is possible without the encryption. In the symmetric encryption setting, semantic security was first formalized by Bellare et al. [3] where they give the adversary a chosen-message encryption oracle. By our impossibility results in the last section, we cannot do so, and instead return to an unknown-message only attack setting for the target message distribution. We refer to this as target-distribution semantic security (TDSS).

Let $\mathcal{M}$ be a message space and p_m be an associated target distribution. Let HE be an HE scheme for $\mathcal{M}$. We let $f : \mathcal{M} \to \{0,1\}$ be a predicate on messages. Let $p_f(b) = \Pr[f(M) = b \mid M \leftarrow_{p_m} \mathcal{M}]$ and let $\omega_f = \max(p_f(0), p_f(1))$.

The TDSS security games are shown in Fig. 6. In game $\text{TDSS1}^{\mathcal{A},f}_{\mathsf{HE},p_m,p_k}$ an adversary $\mathcal{A}$ is charged with predicting $f(M)$ given an encryption of it. In game $\text{TDSS0}^{\mathcal{A}_s,f}_{p_m}$ an adversary $\mathcal{A}_s$, called the simulator, which attempts to guess $f(M)$ without access to a ciphertext. The optimal simulator $\mathcal{A}_s$ for any p_m, f pair simply outputs most likely value of $f(M)$ given the message distribution and predicate f. This forces $\Pr[\text{TDSS0}^{\mathcal{A}_s,f}_{p_m} \Rightarrow \mathsf{true}] = \omega_f$. We therefore define the advantage of adversary $\mathcal{A}$ against the TDSS security of an HE scheme HE with respect to distributions p_m, p_k and predicate f by

$$\mathbf{Adv}^{\mathrm{tdss}}_{\mathsf{HE},p_m,p_k}(\mathcal{A}, f) = \Pr\left[\text{TDSS1}^{\mathcal{A},f}_{\mathsf{HE},p_m,p_k} \Rightarrow \mathsf{true}\right] - \omega_f\ .$$

When working with a random oracle H, the game allows the adversary and the encryption algorithm to query H but f must be independent of H.

$\text{TDSS1}^{\mathcal{A},f}_{\mathsf{HE},p_m,p_k}$	$\text{TDSS0}^{\mathcal{A}_s,f}_{\mathsf{HE},p_m,p_k}$
$K \leftarrow_{p_k} \mathcal{K}$	$M \leftarrow_{p_m} \mathcal{M}$
$M \leftarrow_{p_m} \mathcal{M}$	$b \leftarrow_\$ \mathcal{A}_s$
$C \leftarrow_\$ \mathtt{HEnc}(K, M)$	Return $(b = f(M))$
$b \leftarrow_\$ \mathcal{A}(C)$	
Return $(b = f(M))$	

Fig. 6. Games defining TDSS security.

And the TDSS security of HE is measured by

$$\mathbf{Adv}^{\mathrm{tdss}}_{\mathsf{HE},p_m,p_k} = \max_{\mathcal{A},f} \mathbf{Adv}^{\mathrm{tdss}}_{\mathsf{HE},p_m,p_k}(\mathcal{A}, f)\ .$$

The maximization is over all, even unbounded adversaries $\mathcal{A}$ and arbitrary predicates f. It is easy to derive ways of measuring restricted versions of TDSS security, such as by placing computational limits on $\mathcal{A}$ or restricting the class of predicates. We will not consider such weaker notions further.

TDSS and MR. We will now consider the relation between MR security and TDSS security. It is not hard to see that MR security does not imply TDSS security. We can easily construction an HE scheme such that one bit of the message is revealed completely but the rest is secure using a good HE scheme, thus making the resulting scheme secure in the MR sense but not in the TDSS sense.

Intuitively, TDSS should imply MR, but proving this is not as easy as the other direction. Consider the trivial reduction in which a TDSS adversary $\mathcal{B}$ runs an MR adversary $\mathcal{A}$ and then computes the predicate on the message returned by $\mathcal{A}$. It's clear that $\Pr[\text{TDSS0}^{\mathcal{B},f}_{\mathsf{HE},p_m,p_k} \Rightarrow \mathsf{true}] \geq \mathbf{Adv}^{\mathrm{mr}}_{\mathsf{HE},p_m,p_k}(\mathcal{A})$, but this might be *smaller* than ω_f even if $\mathcal{A}$ is a very good MR adversary.

Fortunately, Dodis and Smith [15] showed that in the information theoretic setting, a good predictor for a function can be turned into a good predictor for a boolean predicate. Viewing a MR adversary as a predictor for the identity function, we can use this to convert a good MR adversary into a good TDSS adversary. We defer the proof of the following theorem to the full version.

Theorem 2. *Let HE be a honey encryption scheme for message distribution* p_m. *(i) If* $\mathbf{Adv}^{\mathrm{mr}}_{HE,p_m,p_k} \geq \omega_m + \omega_m^{2/3}$, *then* $\mathbf{Adv}^{\mathrm{mr}}_{HE,p_m,p_k} \leq \omega_m + 4 \cdot \mathbf{Adv}^{\mathrm{tdss}}_{HE,p_m,p_k}$. *(ii) There exists message distribution* p'_m, *honey encryption scheme* HE', *predicate* f, *and TDSS adversary* $\mathcal{A}$ *such that for any* p_k, HE' *satisfies* $\mathbf{Adv}^{\mathrm{mr}}_{HE',p'_m,p_k} = \mathbf{Adv}^{\mathrm{mr}}_{HE,p_m,p_k}$ *and* $\mathbf{Adv}^{\mathrm{tdss}}_{HE',p'_m,p_k}(\mathcal{A}, f) = \frac{1}{2}$.

5.2 TDSS Security of DTE-then-Encrypt

We turn to showing that the DTE-then-Encrypt construction (refer back to Fig. 2 in Sect. 3) achieves TDSS security in the random oracle model.

Our analysis proceeds in a modular fashion similar to the JR proof of MR security for this construction, but with important differences. First we use DTE security to transition to a game in which the ciphertext is chosen uniformly at random and the challenge key is not sampled until after the adversary has run (one might look ahead to Fig. 8 for the games). In this game, we can show that the advantage of any adversary is no better than the advantage of an adversary $\mathcal{A}^*$ that decrypts the ciphertext using all possible keys, computes the predicate value on the resulting plaintext, and outputs the bit which has the higher cumulative mass of keys that resulted in this bit.

One can then view the game measuring this optimal adversary's success equivalently as a balls-and-bins experiment. The detailed experiment is shown in Fig. 7. Here the balls represent keys and each ball has weight indicated by p_k. There are two bins B_0 and B_1, and throwing a ball into a bin corresponds to seeing the predicate value arrived at by decrypting the fixed ciphertext under the key associated to that ball. Ball throws are independent because H is modeled as a RO.

To our knowledge, in the case that the number of balls is much larger than the number of bins, existing analyses of balls-and-bins experiments only provide an asymptotic bound [28] and in the case that bins are chosen uniformly. We instead analyze the maximum load in the case of non-uniform bin selection and uniformly weighted balls (with the same weights). We can finally then apply a majorization lemma [9] to get a concrete upper bound in the general case of non-uniform balls. We break down the analysis into a series of lemmas, and give the final theorem at the end of this section.

The following lemma captures the first part of our analysis, reducing the security of $\mathsf{HE}[\mathsf{DTE}, \mathsf{H}]$ to the security of DTE, the expected maximum load $\mathbb{E}[\mathrm{L}^{\mathsf{H},\mathsf{DTE},f}_{p_k}]$ in experiment $\mathrm{E}^{\mathsf{H},\mathsf{DTE},f}_{p_k}$, and the bias ω_f of the predicate f on p_m.

Experiment $\mathrm{E}_{p_k}^{\mathsf{H},\mathsf{DTE},f}$

$R \leftarrow\!\!\$ \{0,1\}^r$; $C_2 \leftarrow\!\!\$ \mathcal{S}$
For $K \in \mathcal{K}$ do
  $S \leftarrow \mathsf{H}(R\|K) \oplus C_2$
  $M \leftarrow \mathsf{decode}(S)$
  $b \leftarrow f(M)$
  $B_b \leftarrow B_b \cup \{K\}$
$\mathrm{L}_{p_k}^{\mathsf{H},\mathsf{DTE},f} \leftarrow \max_{b\in\{0,1\}} p_k(B_b)$

Fig. 7. Balls-into-bins experiment used to analyze the security of $\mathsf{HE}[\mathsf{DTE},\mathsf{H}]$.

Lemma 2. *Let* HE *be* $\mathsf{HE}[\mathsf{DTE},\mathsf{H}]$ *as defined in Sect. 3 for distributions* p_m, p_k. *Let* f *be a predicate on* $\mathcal{M}$, $\mathcal{A}$ *be any adversary, then we have:*

$$\mathbf{Adv}^{\mathrm{tdss}}_{\mathsf{HE},p_m,p_k}(\mathcal{A},f) \leq \mathbf{Adv}^{\mathrm{dte}}_{\mathsf{DTE},p_m} + \mathbb{E}[\mathrm{L}_{p_k}^{\mathsf{H},\mathsf{DTE},f}] - \omega_f.$$

Proof. We will use the sequence of games shown in Fig. 8 to transition to a game in which the optimal strategy is clearly $\mathcal{A}^*$ (shown in Fig. 9) which outputs the bit most likely to be the output of f applied to the decryption of the challenge ciphertext under a randomly chosen key.

First note that $\Pr[\mathrm{G}_0 \Rightarrow \mathsf{true}] = \Pr[\mathrm{TDSS0}^{\mathcal{A},f,0}_{\mathsf{HE},p_m,p_k} \Rightarrow \mathsf{true}]$, which is clear because game G_0 is simply the TDSS0 game with the code of HE inserted. Thus,

$$\mathbf{Adv}^{\mathrm{tdss}}_{\mathsf{HE},p_m,p_k}(\mathcal{A},f) = \Pr\left[\,\mathrm{G}_0 \Rightarrow \mathsf{true}\,\right] - \omega_f\ .$$

We can then use the security of DTE to transition to game G_1 because G_1 is identical to G_0 except instead of a random message being sampled and then encoded, a random seed is sampled and then decoded. Consider the adversary $\mathcal{D}$ against the security of DTE shown on the left side of Fig. 9. Adversary $\mathcal{D}$ uses its input S to simulate the view of $\mathcal{A}$, returning 1 if $\mathcal{A}$ selects the correct bit and 0 otherwise. It's easy to verify that $\Pr[\mathrm{G}_0 \Rightarrow \mathsf{true}] = \Pr[\mathrm{SAMP1}^{\mathcal{D}}_{\mathsf{DTE},p_m} \Rightarrow \mathsf{true}]$ and $\Pr[\mathrm{G}_1 \Rightarrow \mathsf{true}] = \Pr[\mathrm{SAMP0}^{\mathcal{D}}_{\mathsf{DTE}} \Rightarrow \mathsf{true}]$. This gives us:

$$\Pr\left[\,\mathrm{G}_0 \Rightarrow \mathsf{true}\,\right] \leq \mathbf{Adv}^{\mathrm{dte}}_{\mathsf{DTE},p_m} + \Pr\left[\,\mathrm{G}_1 \Rightarrow \mathsf{true}\,\right]\ .$$

Next we will see that G_2 is equivalent to G_1. In game G_2 the ciphertext C is sampled uniformly at random while the sampling of K and computation of M are delayed until the adversary has already executed. Note that in G_1 because S was a uniformly chosen element of $\mathcal{S}$, C_2 was also a uniform element of $\mathcal{S}$ independent of the choice of K. Thus we can instead select C_2 at random and defer the choice of K (and thus S and M) until after $\mathcal{A}$ is executed. Consequently,

$$\Pr\left[\,\mathrm{G}_1 \Rightarrow \mathsf{true}\right] = \Pr[\mathrm{G}_2 \Rightarrow \mathsf{true}\,]\ .$$

Now we argue that $\mathcal{A}^*$ is the best possible adversary in game G_2. To see this note that in G_2 the choice of the challenge key K is independent of the input to

$\mathcal{A}$, so the values L_0 and L_1 calculated by $\mathcal{A}^*$ are exactly the probabilities that 0 and 1 will be the correct output, respectively. Thus it's clear that the b^* output by $\mathcal{A}^*$ is the optimal output. Letting $\Pr[G_2^* \Rightarrow \mathsf{true}]$ denote the probability that $\mathcal{A}^*$ succeeds in G_2 we have $\Pr[G_2 \Rightarrow \mathsf{true}] \leq \Pr[G_2^* \Rightarrow \mathsf{true}]$.

Finally we note that the weight L^* of the maximally loaded bin in the balls-in-bins experiment $\mathrm{L}_{p_k}^{\mathsf{H},\mathsf{DTE},f}$ is identical to the probability that the output of $\mathcal{A}^*$ is correct for the chosen (R, C_2). So we have $\Pr[G_2^* \Rightarrow \mathsf{true}] = \mathbb{E}[\mathrm{L}_{p_k}^{\mathsf{H},\mathsf{DTE},f}]$.

Putting everything together gives the desired theorem. □

Game G_0	Game G_1	Game G_2
$K \leftarrow_{p_k} \mathcal{K}$	$K \leftarrow_{p_k} \mathcal{K}$	$R \leftarrow_\$ \{0,1\}^r$
$M \leftarrow_{p_m} \mathcal{M}$	$S \leftarrow_\$ \mathcal{S}$	$C_2 \leftarrow_\$ \mathcal{S}$
$S \leftarrow_\$ \mathsf{encode}(M)$	$M \leftarrow \mathsf{decode}(S)$	$C \leftarrow (R, C_2)$
$R \leftarrow_\$ \{0,1\}^r$	$R \leftarrow_\$ \{0,1\}^r$	$b \leftarrow_\$ \mathcal{A}(C)$
$C_2 \leftarrow \mathsf{H}(R \| K) \oplus S$	$C_2 \leftarrow \mathsf{H}(R \| K) \oplus S$	$K \leftarrow_{p_k} \mathcal{K}$
$C \leftarrow (R, C_2)$	$C \leftarrow (R, C_2)$	$S \leftarrow \mathsf{H}(R \| K) \oplus C_2$
$b \leftarrow_\$ \mathcal{A}(C)$	$b \leftarrow_\$ \mathcal{A}(C)$	$M \leftarrow \mathsf{decode}(S)$
Return $(b = f(M))$	Return $(b = f(M))$	Return $(b = f(M))$

Fig. 8. Games used in proof of Theorem 2.

Adversary $\mathcal{D}(S)$	Adversary $\mathcal{A}^*(C)$
$K \leftarrow_{p_k} \mathcal{K}$	$(R, C_2) \leftarrow C$
$M \leftarrow \mathsf{decode}(S)$	For $K \in \mathcal{K}$ do
$R \leftarrow_\$ \{0,1\}^r$	$\quad S \leftarrow \mathsf{H}(R \| K) \oplus C_2$
$C_2 \leftarrow \mathsf{H}(R \| K) \oplus S$	$\quad M \leftarrow \mathsf{decode}(S)$
$C \leftarrow (R, C_2)$	$\quad L_{f(M)} \leftarrow L_{f(M)} + p_k(K)$
$b \leftarrow_\$ \mathcal{A}(C)$	$b^* \leftarrow \mathrm{argmax}_{b \in \{0,1\}} L_b$
If $(b = f(M))$	Return b^*
$\quad$ Return 1	
Return 0	

Fig. 9. Adversaries used in proof of Theorem 2.

Before we move onto the next step, we first simplify notation. Recall that p_d is the distribution on $\mathcal{M}$ given by applying decode to uniform samples from $\mathcal{S}$. When H is a random oracle each of its outputs is a uniform and independent sample S from $\mathcal{S}$. Thus we can view each message M as being independently sampled according to p_d. Now we can see that experiment $\mathrm{E}_{p_k}^{\mathsf{H},\mathsf{DTE},f}$ is equivalent

Experiment $\mathrm{E}_{p_k}^{p_d,f}$
For $K \in \mathcal{K}$ do
$\quad M \leftarrow_{p_d} \mathcal{M}$
$\quad b_K \leftarrow f(M)$
$\quad B_{b_K} \leftarrow B_{b_K} \cup \{K\}$
$\mathrm{L}_{p_k}^{p_d,f} \leftarrow \max_{b\in\{0,1\}} p_k(B_b)$

Fig. 10. Simplified balls-into-bin experiment.

to a new experiment $\mathrm{E}_{p_k}^{p_d,f}$ in Fig. 10, which is more intuitive. Thus $\mathbb{E}[\mathrm{L}_{p_k}^{\mathsf{H},\mathsf{DTE},f}] = \mathbb{E}[\mathrm{L}_{p_k}^{p_d,f}]$.

Next, we will recall a majorization lemma so that we can transition to a balls and bins experiment with uniform ball weights. Let $K_1, \ldots, K_{|\mathcal{K}|}$ denote an ordering of $\mathcal{K}$ according to weight, that is, for all $i \in \{1, \ldots, |\mathcal{K}| - 1\}$ we have $p_k(K_i) \geq p_k(K_{i+1})$. Then we let p'_k be defined such that for $i \leq \lceil 1/\omega_k \rceil$ we have $p'_k(K_i) = \omega_k$ and $p'_k(K_i) = 0$ otherwise. (Note that p'_k may no longer define a distribution because it's elements may sum to more than one, but this is not important for our analysis below.) Recalling the notion of majorization defined in Sect. 2, we see that p'_k majorizes p_k. The following is a special case of a lemma from [9].

Lemma 3 (BFHM08). *For all p_d, f, and weight vectors p'_k, p_k for which p'_k majorizes p_k it holds that $\mathbb{E}[\mathrm{L}_{p_k}^{p_d,f}] \leq \mathbb{E}[\mathrm{L}_{p'_k}^{p_d,f}]$.*

We can now concentrate on establishing an upper-bound on $\mathbb{E}[\mathrm{L}_{p'_k}^{p_d,f}]$, where p'_k consists of $a = \lceil 1/\omega_k \rceil$ weights all equal to ω_k. Note that we have here ignored the keys of weight zero, but this is clearly without loss of generality since they have no influence on bin loads. The following lemma is gives a bound on the expected maximum load.

Lemma 4. *Let f be a predicate, p_d be a distribution, and p_t be the distribution over $\{0,1\}$ defined by sampling from p_d and applying f. Let p'_k be a weight vector with $a = \lceil 1/\omega_k \rceil$ values each equal to $\omega_k \leq 1$. Then for all s satisfying $a^{s-1} \leq \omega_t$*

$$\mathbb{E}[\mathrm{L}_{p'_k}^{p_d,f}] \leq (1 + \omega_k)(\omega_t + a^{s-1} + 2e^{\frac{-a^{2s-1}}{3}}).$$

Proof. As per the lemma statement, we have that p_t is defined by $p_t(b) = \Pr[f(M) = b : M \leftarrow_{p_d} \mathcal{M}]$ and that ω_t is the associated probability of the most probable value in p_t. That is, let $b^* = \mathrm{argmax}_{b\in\{0,1\}}\, p_t(b)$ and then $\omega_t = p_t(b^*)$. For simplicity we will assume without loss of generality that $b^* = 1$.

Referring to experiment $\mathrm{E}_{p'_k}^{p_d,f}$ (Fig. 7 with p_k replaced by p'_k), b_K is a random variable which equals 1 if K_i is thrown into B_1 and 0 otherwise. Noting then that $|B_1| = \sum_{K\in\mathcal{K}} b_K$ it is easy to see that $\mathbb{E}[|B_1|] = a\omega_t$. Let $B = (1/\omega_k) \cdot \mathrm{L}_{p'_k}^{p_d,f}$ be the random variable corresponding to the number of balls that fall into the maximally loaded bin at the end of the experiment. Then we can see that for all n, $\Pr[B \geq n] \leq 2 \cdot \Pr[|B_1| \geq n]$ because if $B \geq n$ then either B_0 or B_1 must have size at least n and from our assumption that $b^* = 1$, $\Pr[|B_0| \geq n]$ is clearly less than $\Pr[|B_1| \geq n]$.

To complete the proofs we carefully chose a value of n, so that we can bound the probability that B is greater than n and obtain the desired result by pessimistically assuming that B is a whenever it is greater than n and n otherwise.

Recall that Chernoff's bound tells us that for any $0 \leq \delta \leq 1$, $\Pr[|B_1| \geq (1+\delta)\mathbb{E}[|B_{b^*}|]] \leq e^{\frac{-\delta^2 \mathbb{E}[|B_1|]}{3}}$. Then setting $\delta = a^{s-1}/\omega_t$ (which is less than 1 from our choice s) and let we get $\Pr[|B_1| \geq a\omega_t + a^s] \leq e^{\frac{-a^{2s-1}}{3}}$. Then we get the following sequence of inequalities.

$$
\begin{aligned}
\mathbb{E}[B] &= \sum_{i=1}^{a} i \cdot \Pr[B = i] \\
&\leq (a\omega_t + a^s)\Pr[B < a\omega_t + a^s] + a \cdot \Pr[B \geq a\omega_t + a^s] \\
&\leq (a\omega_t + a^s) + 2a \cdot \Pr[|B_1| \geq a\omega_t + a^s] \\
&\leq a(\omega_t + a^{s-1} + 2e^{\frac{-a^{2s-1}}{3}})
\end{aligned}
$$

Multiplying both sides of the inequality by ω and noting that $\lceil 1/\omega_k \rceil \cdot \omega_k \leq 1 + \omega_k$ gives the bound on $\mathbb{E}[\mathrm{L}_{p_k}^{p_d,f}]$. □

At this point we can combine Lemmas 2, 3 and 4 in turn to derive the following sequence of inequalities:

$$
\begin{aligned}
\mathbf{Adv}^{\mathrm{tdss}}_{\mathsf{HE},p_m,p_k}(\mathcal{A}, p_m) &\leq \mathbf{Adv}^{\mathrm{dte}}_{\mathsf{DTE},p_m} + \mathbb{E}[\mathrm{L}_{p_k}^{\mathsf{H},\mathsf{DTE},f}] - \omega_f \\
&\leq \mathbf{Adv}^{\mathrm{dte}}_{\mathsf{DTE},p_m} + \mathbb{E}[\mathrm{L}_{p_k}^{p_d,f}] - \omega_f \\
&\leq \mathbf{Adv}^{\mathrm{dte}}_{\mathsf{DTE},p_m} + \mathbb{E}[\mathrm{L}_{p'_k}^{p_d,f}] - \omega_f \\
&\leq \mathbf{Adv}^{\mathrm{dte}}_{\mathsf{DTE},p_m} + (1+\omega_k)(\omega_t + a^{s-1} + 2e^{\frac{-a^{2s-1}}{3}}) - \omega_f
\end{aligned}
$$

along with the restriction that $a^{s-1} \leq \omega_t$ needed for the last transition. To proceed we note that we can apply again the security of our DTE to transition ω_t to ω_f. Consider the adversary $\mathcal{D}_f$ who just decodes its input S and outputs f applied to the resulting message. It is easy to verify that $\Pr[\mathrm{SAMP1}^{\mathcal{D}_f}_{\mathsf{DTE},p_m}] = p_f(1)$ and $\Pr[\mathrm{SAMP0}^{\mathcal{D}_f}_{\mathsf{DTE}}] = p_t(1)$ which gives us:

$$
|\omega_f - \omega_t| \leq \mathbf{Adv}^{\mathrm{dte}}_{\mathsf{DTE},p_m}(\mathcal{D}_f) \leq \mathbf{Adv}^{\mathrm{dte}}_{\mathsf{DTE},p_m}.
$$

One can apply this to the last inequality above and rearrange, as well as to the restriction on a^{s-1}. Together all this, combined with maximizing over $\mathcal{A}, f$ yields a proof of the following theorem.

Theorem 3. *Let* HE *be* HE[DTE, H] *as defined in Sect. 3 for distributions* p_m, p_k *and with* H *modeled as a random oracle. Then for any* $\mathcal{A}$*, and any* s *satisfying* $a^{s-1} \leq 1/2 - \mathbf{Adv}^{\mathsf{dte}}_{\mathsf{DTE},p_m}$,

$$\mathbf{Adv}^{\mathsf{tdss}}_{\mathsf{HE},p_m,p_k} \leq \omega_k + \lceil 1/\omega_k \rceil^{s-1} + 2\mathbf{Adv}^{\mathsf{dte}}_{\mathsf{DTE},p_m} + 2\exp\left(\frac{-\lceil 1/\omega_k \rceil^{2s-1}}{3}\right).$$

Remark 1. (1) This bound does not seem too far from optimal. From existing results about the estimation of the number of balls in the uniform bin case (see full version for details.) we can see that when $\omega_f = \frac{1}{2}$, choosing $\alpha = \frac{3}{4}$, $\mathbb{E}[L^{p_d,f}_{p'_k}] \geq \Pr[X > k_\alpha] \cdot k_\alpha \omega_k \geq (1 - o(1))(\frac{1}{2} + \sqrt{\frac{\omega_k}{3}})$, thus the advantage of the TDSS adversary is at least at the order of $\omega_k^{\frac{1}{2}}$. (2). As long as p_k has more than 3 bits entropy, we can easily find s such that $\lceil 1/\omega_k \rceil^{s-1} \leq \frac{1}{2}$. (3). If we choose $s = \frac{9}{16}$, $\mathbf{Adv}^{\mathsf{tdss}}_{\mathsf{HE},p_m,p_k} \leq 2\omega_k^{\frac{7}{16}} + 2e^{-\frac{1}{3\omega_k^{1/8}}} + 2\epsilon$. When $\omega_k = 2^{-30}$, the bound is close to 2^{-13}, for which we lose about half of the entropy in the key. If we don't mind losing more entropy (choosing larger s), we can tolerate even smaller ω_k. (3). The condition that $a^{s-1} \leq 1/2 - \mathbf{Adv}^{\mathsf{dte}}_{\mathsf{DTE},p_m}$, simply comes from the condition that $a^{s-1} \leq \omega_f$ for all f.

6 Non-malleability for HE Schemes

The TDSS security goal provides stronger confidentiality properties than MR. But it still does not speak to the threat of attackers mauling ciphertexts. In particular, the DTE-then-Encrypt construction as instantiated with a hash function is trivially malleable: an attacker can, without any knowledge of the key or plaintext, flip bits in a ciphertext so that when it is later decrypted the resulting plaintext is different from the original in a predictable way. Unfortunately the negative results in Sect. 4 suggest that we cannot meet traditional non-malleability security goals [16,25], let alone ciphertext integrity notions [6], when attackers can exhaustively search the key space.

Analogously to the last section, we therefore provide a notion of target-distribution non-malleability (TDNM) for HE schemes for use in the low-entropy key setting. TDNM, like TDSS, is an unknown-message attack setting, and intuitively demands that even if the key space is searchable, the ability of an attacker to successfully maul a ciphertext is not improved by having access to the ciphertext. We then give a construction called DTE-then-Encipher and show it enjoys both TDSS and TDNM security in the ideal cipher model.

6.1 TDNM Security

We adjust the standard non-malleability notion for symmemtric encryption [25] to consider only messages from a target distribution. Informally, TDNM security requires that given a ciphertext, it is difficult to come up with a new ciphertext so that the underling messages satisfy some relation. This is formalized by the two games shown in Fig. 11.

Both games are defined with respect to a binary relation $R\colon \mathcal{M} \times \mathcal{M} \to \{0,1\}$. To simplify notation we sometimes say $R(M, M')$ is true if $R(M, M') = 1$ and false otherwise. We let $p_R = \Pr[R(M, M') = 1 \mid M, M' \leftarrow_{p_m} \mathcal{M}]$. The first game has an adversary $\mathcal{A}$ attempt to maul a ciphertext C so as to satisfy R. The second game has another adversary $\mathcal{A}_s$, called the simulator, attempt to do so without access to C.

$\text{TDNM1}^{\mathcal{A},R}_{\mathsf{HE},p_m,p_k}$	$\text{TDNM0}^{\mathcal{A}_s,R}_{\mathsf{HE},p_m,p_k}$
$K \leftarrow_{p_k} \mathcal{K}; M \leftarrow_{p_m} \mathcal{M}$	$K \leftarrow_{p_k} \mathcal{K}; M \leftarrow_{p_m} \mathcal{M}$
$C \leftarrow\!\!\$\ \texttt{HEnc}(K, M)$	$\tilde{C} \leftarrow\!\!\$\ \mathcal{A}_s$
$\tilde{C} \leftarrow\!\!\$\ \mathcal{A}(C)$	$\tilde{M} \leftarrow \texttt{HDec}(K, \tilde{C})$
If $(\tilde{C} = C)$	Return $R(M, \tilde{M})$
Return false	
$\tilde{M} \leftarrow \texttt{HDec}(K, \tilde{C})$	
Return $R(M, \tilde{M})$	

Fig. 11. Games defining TDNM security.

The TDNM advantage of an adversary $\mathcal{A}$ with respect to a binary relation R, HE scheme HE, and distributions p_m, p_k is defined by

$$\mathbf{Adv}^{\mathrm{tdnm}}_{\mathsf{HE},p_m,p_k}(\mathcal{A}, R) = \Pr\left[\text{TDNM1}^{\mathcal{A},R}_{\mathsf{HE},p_m,p_k} \Rightarrow \mathsf{true}\right] - \max_{\mathcal{A}_s} \Pr\left[\text{TDNM0}^{\mathcal{A}_s,R}_{\mathsf{HE},p_m,p_k} \Rightarrow \mathsf{true}\right]$$

We can then define the TDNM advantage of an HE with distributions p_m, p_k by

$$\mathbf{Adv}^{\mathrm{tdnm}}_{\mathsf{HE},p_m,p_k} = \max_{\mathcal{A},R} \mathbf{Adv}^{\mathrm{tdnm}}_{\mathsf{HE},p_m,p_k}(\mathcal{A}, R).$$

Remark 2. There are two points we would like to note.

1. TDNM with n-ary relations: For simplicity we choose the simpler binary form of the TDNM definition. Of course, we may generalize it to an n-ary relations, but one must be careful about concrete security with respect to n. Imagine that n equals the size of the key space. Then TDNM can be broken for the relation that returns true if at least one of $M_1, \ldots, M_n$ is the challenge message. An adversary that generates each C_i by decrypting the challenge ciphertext using a different key and re-encrypting the message with the same key, sill succeed with probability 1.
2. Relationship between TDNM and TDSS (or MR): It is easy obvious that MR and TDSS do not imply TDNM, the encode-then-encrypt construction serves as an example. However, the other directions are less clear. Intuitively, an MR adversary can be used as an imperfect decryption oracle, this property may be explored for the TDNM adversary to compute a new ciphertext

encrypting the same message. We conjecture that TDNM implies MR (and TDSS), at least under certain conditions. On the other hand, proving unconditional implication results would require new observations. The straightforward transformation of an MR adversary to find the secret key would likely incur a large reduction loss which we can not afford. In all those notions, the bound is already quite small.

Note that this implication is in contrast with the classical notions when adversary has no access to the encryption oracle. A trivial scheme than outputs the plaintext directly together with a MAC is unforgeable, but has no message security.

It would be an interesting open problem to give a complete characterization of the security notions of the honey encryption schemes.

6.2 The DTE-then-Encipher Construction

Intuitively, to achieve non-malleability we would like a scheme for which modifying any portion of a ciphertext would yield a ciphertext that will be decrypted to an independent message. Revisiting the DTE-then-Encrypt construction, a natural route to achieving this property is to replace the (malleable) encryption with one that is non-malleable. A good block cipher has the property that changing any bit of a ciphertext will randomize the decrypted plaintext. In our low-key setting standard security properties like being a pseudorandom permutation are insufficient, and we will instead turn to the ideal cipher model. Here we model a deterministic, length-preserving encryption scheme ($\mathtt{Enc}, \mathtt{Dec}$) as a family of $|\mathcal{K}|$ random permutations, one for each key. The resulting DTE-then-Encipher construction is shown in Fig. 12. We denote it by HE-NM.

$\mathtt{HEnc}(K, M)$:	$\mathtt{HDec}(K, C)$
$S \leftarrow \mathtt{encode}(M)$	$S \leftarrow \mathtt{Dec}(K, C)$
$C \leftarrow \mathtt{Enc}(K, S)$	$M \leftarrow \mathtt{decode}(S)$
return C	return M

Fig. 12. The TDNM construction HE-NM

To instantiate $\mathtt{Enc}, \mathtt{Dec}$ one could use a standard block cipher such as AES, but this will only work when the seed space of the DTE used is exactly the set of bit strings of length equal to the block size of the cipher, e.g., 128. One can turn to constructions that are proven indifferntiable [27] from an ideal cipher of the appropriate domain size. Coron et al. [12] show how to build form an ideal cipher $E\colon \{0,1\}^{2k} \times \{0,1\}^n$ an ideal cipher with domain $2n$ and key length k. One could repeatedly use this construction to extend the domain sufficiently. In theory one could also build large-domain ciphers using a hash function (modeled as a RO) within the Feistel-like constructions analyzed in [13,14,20], but the bounds are too loose to be of practical use. We leave as an open question finding more efficient constructions of TDNM constructions, and focus in the remainder on analysis assuming a suitable ideal cipher.

The TDSS security for HE[DTE, H] can be adjusted to apply to HE-NM in a straightforward manner. We focus below on establishing TDNM security.

Proof Intuition. Intuitively, in an DTE-then-Encipher construction, any two different ciphertext would be decrypted to a pair of (nearly) uniform encoded strings, and they will thus decoded to two randomly sampled messages. However, to formally demonstrate the analysis for TDNM, it still requires us to show that the maximum probability that an (unbounded) adversary can generate a ciphertext that is correlated with a given ciphertext is not much better than without having it. In particular, the adversary can even enumerate all possible ciphertexts that are not equal to the given ciphertext C, and try decrypting each of them using all possible keys. Based on the decrypted message pairs, she may choose one to try to maximize the chance of success.

The nontrivial part of the analysis concentrates on bounding the maximal possible success probability in $\mathrm{TDNM1}^{\mathcal{A},R}_{\mathsf{HE\text{-}NM},p_m,p_k}$. Again we first do game changes so that the adversary would output the modified ciphertext before the key is selected. In this case, for each pair of ciphertext, we can define clearly a set of "preferable" keys for which the decrypted messages resulting from decrypting using these keys satisfy the relation. After exhausting searching all possible ciphertexts, the maximum probability that an adversaries can win with is achieved by outputting the ciphertext $\tilde{C}$ which defines the set of "preferable" keys which has the maximum accumulated probability among all those sets, i.e., the largest possible probability that a randomly selected key will fall into a preferable set. Bounding the accumulated probability can again be transformed into bounding the maximum weight of balls in a bin. The difference compared to the TDSS analysis is that now in every experiment, we will throw $|\mathcal{K}|$ balls into two bins, but once for every single ciphertext. Letting N be the number of possible ciphertexts in the range of the scheme, we therefore analyze N experiments and find the maximum load among all N experiments.

It is not hard to see the expected load in one experiment would be p_R, however, directly bounding the load using, e.g., a Chernoff bound would not be very effective since the expected value is small but the "bad" event (load with significant deviation compared to the expected value) happens with a significant probability. To proceed forward in a similar way as the TDSS analysis, we would reduce the bound to the flat distribution. Unfortunately, in this case, the expected maximum load defined in the TDNM analysis is over several independent balls-into-bins experiments instead of a single such experiment we can not directly apply the majorization lemma from [9].

We turn to a more general majorization technique that uses the property of Schur convexity [22] (to be defined below). A Schur convex function preserves the order under majorization, i.e., if $\bar{p} \succ \bar{q}$, for a Schur convex function f, it holds that $f(\bar{p}) \geq f(\bar{q})$. To bound the expected maximum load in N experiments, we then proceed in two steps. First we argue the expected value of maximum weight across all bins in the N experiments as a function over the key distribution p_k is indeed Schur convex. Since the flat weight vector ($\lceil 1/\omega_k \rceil$ keys each with probability at most ω_k) majorizes the key distribution, we then bound the expected maximum load in N experiments for the flat weight vector which can be done by counting the maximum number of balls falling into a bin in N experiments.

First we introduce some notion we will use for the generalized majorization technique.

Schur Convex Functions. A function $g : \mathbb{R}^n \to \mathbb{R}$ is called Schur-convex if for any $p, q \in \mathbb{R}^n$, if $p \succ q$ then $g(p) \geq g(q)$. A useful result from Schur [22], tells us that any function satisfying two properties known as convex and symmetric must be Schur-convex.

We say g is convex if for any $t \in [0,1]$ and $p, q \in \mathbb{R}^n$ we have $g(tp+(1-t)q)] \leq tg(p) + (1-t)g(q)$. Finally, g is symmetric if the value of the function does not change if the input vector is permuted, that is if $\phi : n \to n$ is a permutation and $p^\phi \in \mathbb{R}^n$ is define by $p^\phi(i) = p(\phi(i))$ for all $1 \leq i \leq n$ then $g(p) = g(p^\phi)$.

Lemma 5 (Schur23). *If a function $g : \mathbb{R}^n \to \mathbb{R}$ is symmetric and convex, then g is Schur-convex.*

6.3 Security Proofs for DTE-then-Encipher

We are now in position to formalize the proof intuition above. The main theorem of this section, given below, establishes an upper bound on the TDNM security of the DTE-then-Encipher construction.

Theorem 4. *Let HE-NM be defied as in Fig. 12 for distributions p_m, p_k, where (Enc, Dec) is an ideal cipher with Enc : $\mathcal{K} \times \{0,1\}^\ell \to \{0,1\}^\ell$. Let $\epsilon = \mathbf{Adv}^{\mathrm{dte}}_{\mathsf{DTE},p_m}$. Then for any s satisfying $\lceil 1/\omega_k \rceil^{s-1} \leq p_R - \epsilon - 2^\ell$ we have*

$$\mathbf{Adv}^{\mathrm{tdnm}}_{\mathsf{HE\text{-}NM},p_m,p_k} \leq \omega_k(1 + \lceil 1/\omega_k \rceil^s) + 2^{1-\ell} + 2^{\ell - \frac{\lceil 1/\omega_k \rceil^{2s-1}}{3}} + 2\epsilon$$

Remark 3. The given bound will typically be quantitatively similar to that of the TDSS advantage, since the message length ℓ is quite small comparing to $\lceil 1/\omega_k \rceil^{2s-1}$. When we take $\omega_k = 2^{-40}, \ell = 128, s = \frac{5}{8}$, we can get $\mathbf{Adv}^{\mathrm{tdnm}}_{\mathsf{HE\text{-}NM},p_m,p_k}$ is around 2^{-15}.

Proof. First we give a lower bound for $\max_{\mathcal{B}} \Pr[\mathrm{TDNM0}^{\mathcal{B},R}_{\mathsf{HE},p_m,p_k} \Rightarrow \mathsf{true}]$. Consider the simulator $\mathcal{A}_s$ that simply outputs a ciphertext $\tilde{C}$ randomly sampled from $\mathcal{C}$. It's easy to verify that the probability $\mathcal{A}_s$ succeeds is equal to the probability that a random sample according to p_m and a random sample according to p_d satisfy the relation, i.e., $\Pr[R(M, \tilde{M}) = 1 : M \leftarrow_{p_m} \mathcal{M}, \tilde{M} \leftarrow_{p_d} \mathcal{M}]$.

Denoting this quantity by p_R^d, let $\mathcal{D}_R$ be the adversary against the security of DTE which simply samples a random M according to p_m and decodes its input S to obtain $\tilde{M}$, then outputs 1 if M and $\tilde{M}$ satisfy R. It is easy to verify that $\Pr[\mathrm{SAMP1}^{\mathcal{D}_R}_{\mathsf{DTE},p_m}] = p_R$ and $\Pr[\mathrm{SAMP0}^{\mathcal{D}_R}_{\mathsf{DTE}}] = p_R^d$ which gives us:

$$\max_{\mathcal{A}_s} \Pr[\mathrm{TDNM0}^{\mathcal{B},R}_{\mathsf{HE},p_m,p_k} \Rightarrow \mathsf{true}] \geq p_R - \mathbf{Adv}^{\mathrm{dte}}_{\mathsf{DTE},p_m}.$$

Transitioning $\mathrm{TDNM1}^{\mathcal{A},R}_{\mathsf{HE},p_m,p_k}$. Now we analyze the maximum winning probability of an adversary $\mathcal{A}$ in $\mathrm{TDNM1}^{\mathcal{A},R}_{\mathsf{HE\text{-}NM},p_m,p_k}$. Consider the sequence of game shown in Fig. 13. Game G_0 is the simply $\mathrm{TDNM1}^{\mathcal{A},R}_{\mathsf{HE\text{-}NM},p_m,p_k}$ with the encryption code of HE-NM inserted. Thus,

$$\Pr[\mathrm{G}_0 \Rightarrow \mathsf{true}] = \Pr[\mathrm{TDNM1}^{\mathcal{A},R}_{\mathsf{HE},p_m,p_k} \Rightarrow \mathsf{true}].$$

We can then use the security of DTE to transition to game G_1 because G_1 is identical to G_0 except instead of a random message being sampled and then encoded, a random seed is sampled and then decoded. Consider the adversary $\mathcal{D}$ against the security of DTE shown in the center of Fig. 14. Adversary $\mathcal{D}$ uses its input to simulate the view of $\mathcal{A}$ returning 1 is $\mathcal{A}$ wins and 0 otherwise. It is easy to check that when in SAMP1, $\mathcal{D}$ perfectly simulated game G_0 for $\mathcal{A}$ and when in SAMP0 it perfect simulates game G_1. It is then clear that

$$\Pr[\mathrm{G}_1 \Rightarrow \mathsf{true}] \le \Pr[\mathrm{G}_0 \Rightarrow \mathsf{true}] + \mathbf{Adv}^{\mathsf{dte}}_{\mathsf{DTE},p_m}.$$

Finally game G_2 is simply a rewriting of G_1 so that the sampling of K is delayed until after $\mathcal{A}$ has already executed. It is clear that,

$$\Pr[\mathrm{G}_2 \Rightarrow \mathsf{true}] = \Pr[\mathrm{G}_1 \Rightarrow \mathsf{true}].$$

Game G_0	Game G_1	Game G_2
$K \leftarrow_{p_k} \mathcal{K}$; $M \leftarrow_{p_m} \mathcal{M}$	$K \leftarrow_{p_k} \mathcal{K}$; $S \leftarrow\!\!\$\ \mathcal{S}$	$C \leftarrow\!\!\$\ \mathcal{C}$ $\tilde{C} \leftarrow\!\!\$\ \mathcal{A}(C)$
$S \leftarrow\!\!\$\ \mathsf{encode}(M)$	$C \leftarrow \mathtt{Enc}(K,S)$	If $(\tilde{C} = C)$
$C \leftarrow \mathtt{Enc}(K,S)$	$\tilde{C} \leftarrow\!\!\$\ \mathcal{A}(C)$	Return false
$\tilde{C} \leftarrow\!\!\$\ \mathcal{A}(C)$	If $(\tilde{C} = C)$	$K \leftarrow_{p_k} \mathcal{K}$
If $(\tilde{C} = C)$	Return false	$M \leftarrow \mathtt{HDec}(K,C)$
Return false	$M \leftarrow \mathsf{decode}(S)$	$\tilde{M} \leftarrow \mathtt{HDec}(K,\tilde{C})$
$\tilde{M} \leftarrow \mathtt{HDec}(K,\tilde{C})$	$\tilde{M} \leftarrow \mathtt{HDec}(K,\tilde{C})$	Return $R(M,\tilde{M})$
Return $R(M,\tilde{M})$	Return $R(M,\tilde{M})$	

Fig. 13. Game transition for TDNM analysis

Next, we will focus on bounding the winning probability of $\mathcal{A}$ in game G_2. Consider the attacking strategy described on the right side of Fig. 14. The adversary $\mathcal{A}^*$ takes a ciphertext C as input. For each other $C' \in \mathcal{C}$, adversary $\mathcal{A}^*$ tries decrypting both C and C' using all possible keys and defines a set $\mathcal{K}_{C'}$ consisting of all the keys for which the corresponding decrypted messages M and M' satisfy the relation R. Then $\mathcal{A}^*$ defines a quantity $p_{C'}$ as the probability that a key sampled according to p_k will fall into $\mathcal{K}_{C'}$.

Recall that in G_2, the key is selected after the adversary outputs the ciphertext $\tilde{C}$, thus we can see that the winning probability of an adversary $\mathcal{A}$ in G_2 will be exactly the value $p_{C'}$ calculated by $\mathcal{A}^*$ corresponding to the output C'.

Adversary $\mathcal{D}(S)$
$K \leftarrow_{p_k} \mathcal{K}$
$C \leftarrow \mathtt{Enc}(K, S)$
$\tilde{C} \leftarrow_{\$} \mathcal{A}(C)$
If $(\tilde{C} = C)$
Return 0
$M \leftarrow \mathsf{decode}(S)$
$\tilde{M} \leftarrow \mathtt{HDec}(K, \tilde{C})$
Return $R(M, \tilde{M})$

Adversary $\mathcal{A}^*(C)$
$p_C \leftarrow 0$
For $C' \in \mathcal{C} \setminus \{C\}$
$\mathcal{K}'_C \leftarrow \emptyset$
For $K \in \mathcal{K}$
$M \leftarrow \mathtt{HDec}(K, C)$
$M' \leftarrow \mathtt{HDec}(K, C')$
If $R(M, M')$
$\mathcal{K}_{C'} \leftarrow \mathcal{K}_{C'} \cup \{K\}$
$p_{C'} \leftarrow p_k(\mathcal{K}_{C'})$
$\tilde{C} \leftarrow \text{argmax}_{C' \in \mathcal{C}}\, p_{C'}$
Return $\tilde{C}$

Fig. 14. Adversaries used in proof of Theorem 4.

Thus because $\mathcal{A}^*$ outputs $\tilde{C}$ maximizing this value, it is clear that $\mathcal{A}^*$ is an optimal adversary for G_2. Thus letting G_2^* denote the game G_2 when run with $\mathcal{A}^*$ it is clear that:

$$\Pr[\mathrm{G}_2^* \Rightarrow \mathsf{true}] \geq \Pr[\mathrm{G}_2 \Rightarrow \mathsf{true}].$$

Furthermore we can clear see that $\Pr[\mathrm{G}_2^* \Rightarrow \mathsf{true}]$ will be exactly the expected value of $\max_{C \in \mathcal{C}} p_C$, denoted by $\mathbb{E}[\max_{C \in \mathcal{C}} p_C]$.

Schur convexity of $\mathbb{E}[\max_{C \in \mathcal{C}} p_C]$. To apply the majorization technique, we will argue the Schur convexity of the quantity we want to bound. We will argue $\mathbb{E}[\max_{C \in \mathcal{C}} p_C]$ is symmetric and convex. then following Lemma 5, it is Schur convex.

For a given key distribution p_k let P_{p_k} be a random variable denoting the value $\max_{C' \in \mathcal{C}} p'_C$ when $\mathcal{A}^*$ uses distribution p_k as the key distribution.

It is clear that $\mathbb{E}[P_p]$ is symmetric because the key are only used for the ideal cipher (Enc, Dec) whose a priori behavior of the key used as input. Thus permuting the corresponding probabilities of the keys will does not change the expected value of P.

To see that $\mathbb{E}[P_p]$ is convex let $p, q \in \mathbb{R}^{|\mathcal{K}|}$, $t \in \mathbb{R}$, and set $r = tp + (1-t)q$. We would like to show that $\mathbb{E}[P_r] \leq t \cdot \mathbb{E}[P_p] + (1-t) \cdot \mathbb{E}[P_q]$. Note that the corresponding executions of $\mathcal{A}^*$ differ only in the weights assigned to the keys, so the distributions of which keys are included in the various sets $\mathcal{K}_C$ are the same between them.

For a fixed choice of random coins, let C_r, C_p, C_q denote the respective output of $\mathcal{A}^*$ in the different experiments. Then from the definition of r and the fact that the ciphertexts are chosen to maximize the weights of the corresponding $\mathcal{K}_C$ we get:

$$\begin{aligned} P_r &= r(\mathcal{K}_{C_r}) \\ &= t \cdot p(\mathcal{K}_{C_r}) + (1-t)q(C_r) \\ &\leq t \cdot p(\mathcal{K}_{C_p}) + (1-t)q(C_q) \\ &= tP_p + (1-t)P_q. \end{aligned}$$

Because the above holds for every choice of random coins in the corresponding experiments it is clear that $\mathbb{E}[P_r] \leq t \cdot \mathbb{E}[P_p] + (1-t) \cdot \mathbb{E}[P_q]$, so $\mathbb{E}[P_p]$ is convex.

Having now shown that $\mathbb{E}[P_p]$ is symmetric and convex, Lemma 5 tells us it is Schur-convex.

Bounding $\mathbb{E}[P_{p'_k}]$ for flat distribution p'_k. Now as in our TDSS analysis let p'_k be defined such that for $i \leq \lceil 1/\omega_k \rceil$ we have $p'_k(K_i) = \omega_k$ and $p'_k(K_i) = 0$ otherwise, and note that p'_k majorizes p_k. Since $\mathbb{E}[P_p]$ is Schur convex, and p'_k majorizes p_k, we have:

$$\mathbb{E}[P_{p_k}] \leq \mathbb{E}[P_{p'_k}]$$

Next, we will focus on bounding $\mathbb{E}[P_{p'_k}]$.

Experiment $\mathrm{E}^{p'_d,R}_{p'_k}$

For $K \in \mathcal{K}$ do
  $(M, M') \leftarrow_{p'_d} \mathcal{M} \times \mathcal{M}$
  $b \leftarrow R(M, M')$
  $B_b \leftarrow B_b \cup \{K\}$
$\mathrm{L}^{p'_d,R}_{p_k} \leftarrow p_k(B_1)$

Fig. 15. Ball-into-bins experiments used to analyze the security of HE-NM.

Let us rephrase the quantity $p_{C'}$ using the terminology of a balls-into-bins game. Letting the challenge ciphertext C be fixed we can think of each $K \in \mathcal{K}$ as a ball into bins B_0 and B_1 according to the value of $R(M, M')$ where M and M' are obtained by decrypting C and C' respectively with K. Because each decryption uses the ideal cipher it is clear that each key is thrown independently.

Because decrypting applying `Dec` to ciphertext results in a uniformly random S we would like to say that we can view M and M' as both being drawn independently according to p_d. However, this is not quite true because there is a small dependence between the samples because `Dec` applied to C and C' results in uniformly chosen S and S' with the restriction that $S \neq S'$. Let p'_d denote the distribution on $\mathcal{M} \times \mathcal{M}$ obtained by applying decode to two uniformly chosen seeds with the restriction that the seeds are not equal. Then we can view the values of M and M' for each K in the balls-into-bins experiment as being independent samples from p'_d.

Putting this together we can thing of the quantity $p_{C'}$ as the load $\mathrm{L}_{p'_k}^{p'_d,R}$ in the balls-into-bins experiment $\mathrm{E}_{p'_k}^{p'_d,R}$ shown in Fig. 15.

Let $p'_R = \Pr[R(M,M') | (M,M') \leftarrow_{p'_d} \mathcal{M} \times \mathcal{M}]$ denote the probability that (M,M') sampled according to p'_d satisfies R and $a = \lceil 1/\omega_k \rceil$ denote the number of balls thrown in experiment $\mathrm{E}_{p'_k}^{p'_d,R}$. Then it is clear that the expected number of balls that fall into bin B_1 is a ap'_R.

Then letting X denote the number of balls thrown into B_1 and $\delta = a^{s-1}/p'_R$ (which is less than 1 from our choice of s) we can apply Chernoff's bound to get:

$$\Pr[X \geq ap'_R + a^s] \leq e^{\frac{-a^{2s-1}}{3}}.$$

Now we can complete the proof by using this to bound the expected value of $\max_{C \in \mathcal{C}} p'_C$ for $\mathcal{A}^*$. For this to be greater than $\omega_k(ap'_R + a^s)$ it must be the case that for some $C \in \mathcal{C}$, p'_C is greater than $\omega_k(ap'_R + a^s)$. Then from the union bound we get

$$\begin{aligned}
\Pr\left[\max_{C \in \mathcal{C}} p'_C \geq \omega_k(ap'_R + a^s)\right] &\leq \sum_{C \in \mathcal{C}} \Pr[p'_C \geq \omega_k(ap'_R + a^s)] \\
&\leq \sum_{C \in \mathcal{C}} \Pr[\mathrm{L}_{p'_k}^{p'_d,R} \geq \omega_k(ap'_R + a^s)] \\
&= (|\mathcal{C}| - 1) \cdot \Pr[X \geq ap'_R + a^s] \\
&\leq (|\mathcal{C}| - 1) e^{\frac{-a^{2s-1}}{3}}.
\end{aligned}$$

Note that applying the union bound in this manner allows us to ignore the dependence that exists between difference p'_C for different C.

Finally we can bound the expected value of $\max_{C \in \mathcal{C}} p'_C$ by pessimistically assuming it is always 1 whenever it is greater than $\omega_k(ap'_R + a^s)$ and it is $\omega_k(ap'_R + a^s)$ otherwise. Recalling that $|\mathcal{C}| = 2^\ell$ and letting $P = \max_{C \in \mathcal{C}} p'_C$, this gives us the following sequence of inequalities:

$$\begin{aligned}
\mathbb{E}[P] &\leq \omega_k(ap'_R + a^s) \Pr[P \leq \omega_k(ap'_R + a^s)] + 1 \cdot \Pr[P \geq \omega_k(ap'_R + a^s)] \\
&\leq \omega_k(ap'_R + a^s) + (2^\ell - 1) e^{\frac{-a^{2s-1}}{3}} \\
&\leq \omega_k(ap'_R + a^s) + 2^{\ell - \frac{a^{2s-1}}{3}}
\end{aligned}$$

From the definition of p'_R it is clear that $p'_R \leq p_R + 1/|\mathcal{S}| = p_R + 2^{-\ell}$. Putting everything together we get the final bound of

$$\begin{aligned}
\mathbf{Adv}^{\mathrm{tdnm}}_{\mathsf{HE\text{-}NM}, p_m, p_k} &\leq \omega_k(\lceil 1/\omega_k \rceil (p_R + 2^{-\ell}) + \lceil 1/\omega_k \rceil^s) + 2^{\ell - \frac{\lceil 1/\omega_k \rceil^{2s-1}}{3}} - p_R + 2\epsilon \\
&\leq (1 + \omega_k)(p_R + 2^{-\ell}) + \omega_k \lceil 1/\omega_k \rceil^s + 2^{\ell - \frac{\lceil 1/\omega_k \rceil^{2s-1}}{3}} - p_R + 2\epsilon \\
&\leq \omega_k(1 + \lceil 1/\omega_k \rceil^s) + 2^{1-\ell} + 2^{\ell - \frac{\lceil 1/\omega_k \rceil^{2s-1}}{3}} + 2\epsilon.
\end{aligned}$$

This completes the proof. □

7 Conclusions and Open Problems

In this work, we initiated the study of security notions for honey encryption schemes stronger than the previously proposed goal of resistance to message recovery attacks. We, first, proved that message recovery is always possible with a known-message attack. Formally proving this folklore result was more nuanced than expected. We then defined semantic security and non-malleability for honey encryption schemes with respect to targeted message distributions, and we showed that the simple constructions of encode-then-encrypt and encode-then-encipher achieve targeted distribution semantic security and targeted distribution non-malleability, respectively. The general technique for balls-into-bins type of analysis using Schur convexity may be of independent interest.

Security notions for symmetric key encryption with low-entropy keys are still not yet fully understood. For honey encryption schemes, completely characterizing the relations among various security notions remains an open problem whose solution would expand on our results. Also, although replacing a random oracle with a k-wise independent hash function to get a standard model construction for TDSS seems intuitive, formally analyzing its security requires more delicate balls-into-bins analysis than we have provided here. Last, our TDNM construction relies on an ideal cipher with large block size. Obtaining a construction provably TDNM secure in the random oracle model therefore represents an important open question.

Acknowledgements. We thank the anonymous reviewers for valuable comments. Joseph Jaeger was supported in part by NSF grants CNS-1526801 and CNS-1228890, ERC Project ERCC FP7/615074 and a gift from Microsoft. Thomas Ristenpart was supported by NSF grants CNS-1514163, CNS-1546033, CNS-1065134, CNS-1330308 and a gift from Microsoft. Qiang Tang was supported by NSF grants CNS-1518765 and CNS-1514261.

References

1. Bellare, M., Boldyreva, A., O'Neill, A.: Deterministic and efficiently searchable encryption. In: Menezes, A. (ed.) CRYPTO 2007. LNCS, vol. 4622, pp. 535–552. Springer, Heidelberg (2007)
2. Bellare, M., Brakerski, Z., Naor, M., Ristenpart, T., Segev, G., Shacham, H., Yilek, S.: Hedged public-key encryption: how to protect against bad randomness. In: Matsui, M. (ed.) ASIACRYPT 2009. LNCS, vol. 5912, pp. 232–249. Springer, Heidelberg (2009)
3. Bellare, M., Desai, A., Jokipii, E., Rogaway, P.: A concrete security treatment of symmetric encryption. In: 38th Annual Symposium on Foundations of Computer Science, FOCS 1997, Miami Beach, Florida, USA, pp. 394–403, 19–22 October 1997
4. Bellare, M., Fischlin, M., O'Neill, A., Ristenpart, T.: Deterministic encryption: definitional equivalences and constructions without random oracles. In: Wagner, D. (ed.) CRYPTO 2008. LNCS, vol. 5157, pp. 360–378. Springer, Heidelberg (2008)

5. Bellare, M., Kilian, J., Rogaway, P.: The security of the cipher block chaining message authentication code. J. Comput. Syst. Sci. **61**(3), 362–399 (2000)
6. Bellare, M., Namprempre, C.: Authenticated encryption: relations among notions and analysis of the generic composition paradigm. In: Okamoto, T. (ed.) ASIACRYPT 2000. LNCS, vol. 1976, pp. 531–545. Springer, Heidelberg (2000)
7. Bellare, M., Ristenpart, T., Tessaro, S.: Multi-instance security and its application to password-based cryptography. In: Safavi-Naini, R., Canetti, R. (eds.) CRYPTO 2012. LNCS, vol. 7417, pp. 312–329. Springer, Heidelberg (2012)
8. Bellare, M., Rogaway, P.: Random oracles are practical: a paradigm for designing efficient protocols. In: ACM Conference on Computer and Communications Security, pp. 62–73 (1993)
9. Berenbrink, P., Friedetzky, T., Hu, Z., Martin, R.: On weighted balls-into-bins games. J. Comput. Syst. Sci. **409**(3), 511–520 (2008)
10. Bonneau, J.: The science of guessing: analyzing an anonymized corpus of 70 million passwords. In: Proceedings of the 2012 IEEE Symposium on Security and Privacy, SP 2012, pp. 538–552 (2012)
11. Chatterjee, R., Bonneau, J., Juels, A., Ristenpart, T.: Cracking-resistant password vaults using natural language encoders. In: 2015 IEEE Symposium on Security and Privacy, SP 2015, San Jose, CA, USA, pp. 481–498, 17–21 May 2015
12. Coron, J.-S., Dodis, Y., Mandal, A., Seurin, Y.: A domain extender for the ideal cipher. In: Micciancio, D. (ed.) TCC 2010. LNCS, vol. 5978, pp. 273–289. Springer, Heidelberg (2010)
13. Dachman-Soled, D., Katz, J., Thiruvengadam, A.: 10-round feistel is indifferentiable from an ideal cipher. Cryptology ePrint Archive, report 2015/876 (2015). http://eprint.iacr.org/
14. Dai, Y., Steinberger, J.: Feistel networks: indifferentiability at 10 rounds. Cryptology ePrint Archive, report 2015/874 (2015). http://eprint.iacr.org/
15. Dodis, Y., Smith, A.: Entropic security and the encryption of high entropy messages. In: Kilian, J. (ed.) TCC 2005. LNCS, vol. 3378, pp. 556–577. Springer, Heidelberg (2005)
16. Dolev, D., Dwork, C., Naor, M.: Nonmalleable cryptography. SIAM J. Comput. **30**(2), 391–437 (2000)
17. Goldwasser, S., Micali, S.: Probabilistic encryption and how to play mental poker keeping secret all partial information. In: Proceedings of the 14th Annual ACM Symposium on Theory of Computing, San Francisco, California, USA, pp. 365–377, 5–7 May 1982
18. Goldwasser, S., Micali, S., Rivest, R.: A digital signature scheme secure against adaptive chosen-message attacks. SIAM J. Comput. **17**(2), 281–308 (1988)
19. Hoang, V.T., Krovetz, T., Rogaway, P.: Robust authenticated-encryption AEZ and the problem that it solves. In: Oswald, E., Fischlin, M. (eds.) EUROCRYPT 2015. LNCS, vol. 9056, pp. 15–44. Springer, Heidelberg (2015)
20. Holenstein, T., Künzler, R., Tessaro, S.: The equivalence of the random oracle model and the ideal cipher model, revisited. In: Proceedings of the Forty-third Annual ACM Symposium on Theory of Computing, pp. 89–98. ACM (2011)
21. Huang, Z., Ayday, E., Fellay, J., Hubaux, J., Juels, A.: Genoguard: Protecting genomic data against brute-force attacks. In: IEEE Symposium on Security and Privacy, SP 2015, San Jose, CA, USA, May 17–21, 2015, pp. 447–462 (2015)
22. Schur, I.: Über eine klassevon mittelbildungen mit anwendungen die determinanten. Theorie Sitzungsber. Berlin. Math. Gesellschaft **22**, 9–20 (1923)

23. Juels, A., Ristenpart, T.: Honey encryption: Security beyond the brute-force bound. In: 2014 Proceedings of Advances in Cryptology - EUROCRYPT -33rd Annual International Conference on the Theory and Applications of Cryptographic Techniques, Copenhagen, Denmark, May 11–15, pp. 293–310 (2014)
24. Kaliski, B.: PKCS #5: Password-Based Cryptography Specification Version 2.0. RFC 2898 (Informational), September 2000
25. Katz, J., Yung, M.: Complete characterization of security notions for probabilistic private-key encryption. In: Proceedings of the Thirty-Second Annual ACM Symposium on Theory of Computing, 21–23 May, Portland, OR, USA, pp. 245–254 (2000)
26. Kausik, B.: Method and apparatus for cryptographically camouflaged cryptographic key storage, certification and use, January 2. US Patent 6,170,058 (2001)
27. Maurer, U.M., Renner, R.S., Holenstein, C.: Indifferentiability, impossibility results on reductions, and applications to the random oracle methodology. In: Naor, M. (ed.) TCC 2004. LNCS, vol. 2951, pp. 21–39. Springer, Heidelberg (2004)
28. Raab, M., Steger, A.: "Balls into bins"– a simple and tight analysis. In: Rolim, J.D.P., Serna, M., Luby, M. (eds.) RANDOM 1998. LNCS, vol. 1518, pp. 159–170. Springer, Heidelberg (1998)
29. Rogaway, P., Shrimpton, T.: A provable-security treatment of the key-wrap problem. In: Vaudenay, S. (ed.) EUROCRYPT 2006. LNCS, vol. 4004, pp. 373–390. Springer, Heidelberg (2006)
30. Russell, A., Wang, H.: How to fool an unbounded adversary with a short key. In: 2002 Proceedings od Advances in Cryptology - EUROCRYPT, International Conference on the Theory and Applications of Cryptographic Techniques, Amsterdam, The Netherlands, April 28 - May 2, pp. 133–148 (2002)

Improved Progressive BKZ Algorithms and Their Precise Cost Estimation by Sharp Simulator

Yoshinori Aono[1(✉)], Yuntao Wang[2], Takuya Hayashi[1], and Tsuyoshi Takagi[3,4]

[1] National Institute of Information and Communications Technology, Tokyo, Japan
aono@nict.go.jp

[2] Graduate School of Mathematics, Kyushu University, Fukuoka, Japan

[3] Institute of Mathematics for Industry, Kyushu University, Fukuoka, Japan

[4] CREST, Japan Science and Technology Agency, Kyoto, Japan

Abstract. In this paper, we investigate a variant of the BKZ algorithm, called progressive BKZ, which performs BKZ reductions by starting with a small blocksize and gradually switching to larger blocks as the process continues. We discuss techniques to accelerate the speed of the progressive BKZ algorithm by optimizing the following parameters: blocksize, searching radius and probability for pruning of the local enumeration algorithm, and the constant in the geometric series assumption (GSA). We then propose a simulator for predicting the length of the Gram-Schmidt basis obtained from the BKZ reduction. We also present a model for estimating the computational cost of the proposed progressive BKZ by considering the efficient implementation of the local enumeration algorithm and the LLL algorithm. Finally, we compare the cost of the proposed progressive BKZ with that of other algorithms using instances from the Darmstadt SVP Challenge. The proposed algorithm is approximately 50 times faster than BKZ 2.0 (proposed by Chen-Nguyen) for solving the SVP Challenge up to 160 dimensions.

Keywords: Lattice basis reduction · Progressive BKZ · Gram-Schmidt orthogonal basis · Geometric series assumption

1 Introduction

Lattices in cryptography have been actively used as the foundation for constructing efficient or high-functional cryptosystems such as public-key encryptions [17,26,41], fully homomorphic encryptions [10,22], and multilinear maps [21]. The security of lattice-based cryptography is based on the hardness of solving the (approximate) shortest vector problems (SVP) in the underlying lattice [15,32,35,36]. In order to put lattice-based cryptography into practical use, we must precisely estimate the secure parameters in theory and practice by analyzing the previously known efficient algorithms for solving the SVP.

Y. Aono—This work was supported by JSPS KAKENHI Grant Number 26730069.

M. Fischlin and J.-S. Coron (Eds.): EUROCRYPT 2016, Part I, LNCS 9665, pp. 789–819, 2016.
DOI: 10.1007/978-3-662-49890-3_30

Currently the most efficient algorithms for solving the SVP are perhaps a series of BKZ algorithms [13,14,46,47]. Numerous efforts have been made to estimate the security of lattice-based cryptography by analyzing the BKZ algorithms. Lindner and Peikert [32] gave an estimation of secure key sizes by connecting the computational cost of BKZ algorithm with the root Hermite factor from their experiment using the NTL-BKZ [49]. Furthermore, van de Pol and Smart [51] estimated the key sizes of fully homomorphic encryptions using a simulator based on Chen-Nguyen's BKZ 2.0 [13]. Lepoint and Naehrig [31] gave a more precise estimation using the parameters of the full-version of BKZ 2.0 paper [14]. On the other hand, Liu and Nguyen [33] estimated the secure key sizes of some LWE-based cryptosystems by considering the BDD in the associated q-ary lattice. Aono et al. [7] gave another security estimation for LWE-based cryptosystems by considering the challenge data from the Darmstadt Lattice Challenge [50]. Recently, Albrecht et al. presented a comprehensive survey on the state-of-the-art of hardness estimation for the LWE problem [5].

The above analyzing algorithms are usually called "lattice-based attacks", which have a generic framework consisting of two parts:

(1) Lattice reduction: This step aims to decrease the norm of vectors in the basis by performing a lattice reduction algorithm such as the LLL or BKZ algorithm.

(2) Point search: This step finds a short vector in the lattice with the reduced basis by performing the enumeration algorithm.

In order to obtain concrete and practical security parameters for lattice-based cryptosystems, it is necessary to investigate the trade-offs between the computational cost of a lattice reduction and that of a lattice point search.

For our total cost estimation, we further limit the lattice-based attack model by (1) using our improved progressive BKZ algorithm for lattice reduction, and (2) using the standard (sometimes randomized) lattice vector enumeration algorithm with sound pruning [20]. To predict the computational cost under this model, we propose a simulation method to generate the computing time of lattice reduction and the lengths of the Gram-Schmidt vectors of the basis to be computed.

BKZ Algorithms: Let $B = (\mathbf{b}_1, \ldots, \mathbf{b}_n)$ be the basis of the lattice. The BKZ algorithms perform the following local point search and update process from index $i = 1$ to $n - 1$. The local point search algorithm, which is essentially the same as the algorithm used in the second part of the lattice-based attacks, finds a short vector in the local block $B_i = \pi_i(\mathbf{b}_i, \ldots, \mathbf{b}_{i+\beta-1})$ of the fixed blocksize β (the blocksize shrinks to $n - i + 1$ for large $i \geq n - \beta + 1$). Here, the lengths of vectors are measured under the projection π_i which is defined in Sect. 2.1. Then, the update process applies lattice reduction for the degenerated basis $(\mathbf{b}_1, \ldots, \mathbf{b}_{i-1}, \mathbf{v}, \mathbf{b}_i, \ldots, \mathbf{b}_n)$ after inserting vector $\mathbf{v}$ at i-th index.

The point search subroutine finds a short vector in some searching radius $\alpha \cdot \mathrm{GH}(B_i)$ with some probability which is defined over random local blocks of the fixed dimension. Here, $\mathrm{GH}(B_i)$ is an approximation of the length of the shortest vector in the sublattice generated by B_i.

Table 1. Technical comparison from BKZ 2.0

Technique	BKZ 2.0 [13]	Our algorithm
Enumeration setting		
randomizing basis [20]	yes	no
optimal pruning [20]	yes	yes
blocksize β	fixed	iteratively increasing (Sect. 6.1)
search radius $\alpha \cdot \mathrm{GH}(B_i)$	$\sqrt{1.1} \cdot \mathrm{GH}(B_i)$	} optimized by GSA (Sect. 4)
probability p	optimized by simulator	
Preprocessing local block	optimal BKZ strategy	progressive BKZ
Terminating BKZ strategy	simulator based (fixed)	FEC based (adaptive, Sect. 5)
Predicting $\|\mathbf{b}_i^*\|$	simulator based	simulator based (Sect. 5.1)

In the classical BKZ algorithms [46,47], the local point search calls a single execution of a lattice vector enumeration algorithm with a reasonable pruning for searching tree. The BKZ 2.0 algorithm proposed by Chen and Nguyen [13] uses the extreme pruning technique [20], which performs the lattice enumeration with success probability p for $\lfloor 1/p \rceil$ different bases $G_1, \ldots, G_{\lfloor 1/p \rceil}$ obtained by randomizing the local basis B_i. They use the fixed searching radius as $\sqrt{1.1} \cdot \mathrm{GH}(B_i)$. We stress that BKZ 2.0 is practically the fastest algorithm for solving the approximate SVP of large dimensions. Indeed, many top-records in the Darmstadt Lattice Challenge [50] have been solved by BKZ 2.0 (Table 1).

Our Contributions: In this paper we revisit progressive BKZ algorithms, which have been mentioned in several studies; these include [13,19,25,45,48]. The main idea of progressive BKZ is that performing BKZ iteratively starting with a small blocksize is practically faster than the direct execution of BKZ with a larger blocksize. The method used to increase the blocksize β strongly affects the overall computational cost of progressive BKZ. The research goal here is to find an optimal method of increasing the blocksize β according to the other parameters in the BKZ algorithms.

One major difference between BKZ 2.0 and our algorithm is the usage of randomized enumeration in local blocks. To find a very short vector in each local block efficiently, BKZ 2.0 uses the randomizing technique in [20]. Then, it reduces each block to decrease the cost of lattice enumeration. Although it is significantly faster than the enumeration without pruning, it introduces overhead because the bases are not good in practice after they have been randomized. To avoid this overhead, we adopted the algorithm with a single enumeration with a low probability.

Moreover, BKZ of a large blocksize with large pruning (i.e., a low probability) is generally better in both speed and quality of basis than that of a small blocksize with few pruning (i.e., a high probability), as a rule of thumb. We pursue this idea and add the freedom to choose the radius $\alpha \cdot \mathrm{GH}(L)$ of the enumeration of the local block; this value is fixed in BKZ 2.0 as $\sqrt{1.1} \cdot \mathrm{GH}(L)$.

To optimize the algorithm, we first discuss techniques for optimizing the BKZ parameters of enumeration subroutine, including the blocksize β, success probability p of enumeration, and α to set the searching radius of enumeration as $\alpha \cdot \mathrm{GH}(B_i)$. We then show the parameter relationship that minimizes the computational cost for enumeration of a BKZ-β-reduced basis. Next, we introduce the new usage of *full enumeration cost* (FEC), derived from Gama-Nguyen-Regev's cost estimation [20] with a Gaussian heuristic radius and without pruning, to define the quality of the basis and to predict the cost after BKZ-β is performed. Using this metric, we can determine the timing for increasing blocksize β that provides an optimized strategy; in previous works, the timing was often heuristic.

Furthermore, we propose a new BKZ simulator to predict the Gram-Schmidt lengths $\|\mathbf{b}_i^*\|$ after BKZ-β. Some previous works aimed to find a short vector as fast as possible, and did not consider other quantities. However, additional information is needed to analyze the security of lattice-based cryptosystems. In literatures, a series of works on lattice basis reduction [13,14,19,44] have attempted to predict the Gram-Schmidt lengths $\|\mathbf{b}_i^*\|$ after lattice reduction. In particular, Schnorr's GSA is the first simulator of Gram-Schmidt lengths and the information it provides is used to analyze the random sampling algorithm. We follow this idea, i.e., predicting Gram-Schmidt lengths to analyze other algorithms.

Our simulator is based on the Gaussian heuristic with some modifications, and is computable directly from the lattice dimension and the blocksize. On the other hand, Chen-Nguyen's simulator must compute the values sequentially; it has an inherent problem of accumulative error, if we use the strategy that changes blocksize many times. We also investigate the computational cost of our implementation of the new progressive BKZ, and show our estimation for solving challenge problems in the Darmstadt SVP Challenge and Ideal Lattice Challenge [50]. Our cost estimation is derived by setting the computation model and by curve fitting based on results from computer experiments. Using our improved progressive BKZ, we solved Ideal Lattice Challenge of 600 and 652 dimensions in the exact expected times of $2^{20.7}$ and $2^{24.0}$ s, respectively, on a standard PC.

Finally, we compare our algorithm with several previous algorithms. In particular, compared with Chen-Nguyen's BKZ 2.0 algorithm [13,14] and Schnorr's blocksize doubling strategy [48], our algorithm is significantly faster. For example, to find a vector shorter than $1.05 \cdot \mathrm{GH}(L)$, which is required by the SVP Challenge [50], our algorithm is approximately 50 times faster than BKZ 2.0 in a simulator-based comparison up to 160 dimensions.

Roadmap: In Sect. 2 we introduce the basic facts on lattices. In Sect. 3 we give an overview of BKZ algorithms, including Chen-Nguyen's BKZ 2.0 [13] and its cost estimation; we also state some heuristic assumptions. In Sect. 4, we propose the optimized BKZ parameters under the Schnorr's geometric series assumption (GSA). In Sect. 5, we explain the basic variant of the proposed progressive BKZ algorithm and its simulator for the cost estimation. In Sect. 6, we discuss the optimized block strategy that improved the speed of the proposed progressive BKZ algorithm. In Sect. 7, we show the cost estimation for processing local blocks based on our implementation. Due to the spacing limitation, we omit the

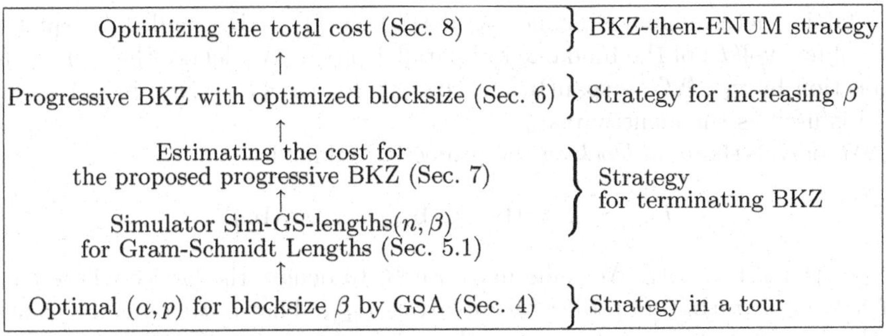

Fig. 1. Roadmap of this paper: optimizing parameters from local to global

details of our implementation (See the full version [8] for the details). We then discuss an extended strategy using many random reduced bases [20] besides our progressive BKZ in Sect. 8. Finally, Sect. 9 gives the results of our simulation to solve the SVP Challenge problems and compares these results with previous works (Fig. 1).

2 Lattice and Shortest Vector

A lattice L is generated by a basis B which is a set of linearly independent vectors $\mathbf{b}_1, \ldots, \mathbf{b}_n$ in $\mathbb{R}^m$. We will refer to it as $L(\mathbf{b}_1, \ldots, \mathbf{b}_n) = \{\sum_{i=1}^{n} x_i \mathbf{b}_i, x_i \in \mathbb{Z}\}$. Throughout this paper, we assume $m = O(n)$ to analyze the computational cost, though it is not essential. The length of $\mathbf{v} \in \mathbb{R}^m$ is the standard Euclidean norm $\|\mathbf{v}\| := \sqrt{\mathbf{v} \cdot \mathbf{v}}$, where the dot product of any two lattice vectors $\mathbf{v} = (v_1, \ldots, v_m)$ and $\mathbf{w} = (w_1, \ldots, w_m)$ is defined as $\mathbf{v} \cdot \mathbf{w} = \sum_{i=1}^{m} v_i w_i$. For natural numbers i and j with $i < j$, $[i : j]$ is the set of integers $\{i, i+1, \ldots, j\}$. Particularly, $[1 : j]$ is denoted by $[j]$.

The gamma function $\Gamma(s)$ is defined for $s > 0$ by $\Gamma(s) = \int_0^\infty t^{s-1} \cdot e^{-t} dt$. The beta function is $\mathrm{B}(x, y) = \int_0^1 t^{x-1}(1-t)^{y-1} dt$. We denote by $\mathrm{Ball}_n(R)$ the n-dimensional Euclidean ball of radius R, and then its volume $V_n(R) = R^n \cdot \frac{\pi^{n/2}}{\Gamma(n/2+1)}$. Stirling's approximation yields $\Gamma(n/2+1) \approx \sqrt{\pi n}(n/2)^{n/2} e^{-n/2}$ and $V_n(1)^{-1/n} \approx \sqrt{n/(2\pi e)} \approx \sqrt{n/17}$.

2.1 Gram-Schmidt Basis and Projective Sublattice

For a given lattice basis $B = (\mathbf{b}_1, \ldots, \mathbf{b}_n)$, we define its *Gram-Schmidt orthogonal basis* $B^* = (\mathbf{b}_1^*, \ldots, \mathbf{b}_n^*)$ by $\mathbf{b}_i^* = \mathbf{b}_i - \sum_{j=1}^{i-1} \mu_{ij} \mathbf{b}_j^*$ for $1 \le j < i \le n$, where $\mu_{ij} = (\mathbf{b}_i \cdot \mathbf{b}_j^*)/\|\mathbf{b}_j^*\|^2$ are the Gram-Schmidt coefficients (abbreviated as *GS-coefficients*). We sometimes refer to $\|\mathbf{b}_i^*\|$ as the *Gram-Schmidt lengths* (abbreviated as *GS-lengths*). We also use the Gram-Schmidt variables (abbreviated as *GS-variables*) to denote the set of GS-coefficients μ_{ij} and lengths $\|\mathbf{b}_i^*\|$.

The lattice determinant is defined as $\det(L) := \prod_{i=1}^{n} \|\mathbf{b}_i^*\|$ and it is equal to the volume $\mathrm{vol}(L)$ of the fundamental parallelepiped. We denote the orthogonal projection by $\pi_i : \mathbb{R}^m \mapsto \mathrm{span}(\mathbf{b}_1, \ldots, \mathbf{b}_{i-1})^\perp$ for $i \in \{1, \ldots, n\}$. In particular, $\pi_1(\cdot)$ is used as the identity map.

We denote the *local block* by the projective sublattice

$$L_{[i:j]} := L(\pi_i(\mathbf{b}_i), \pi_i(\mathbf{b}_{i+1}), \ldots, \pi_i(\mathbf{b}_j))$$

for $j \in \{i, i+1, \ldots, n\}$. We sometimes use B_i to denote the lattice whose basis is $(\pi_i(\mathbf{b}_i), \ldots, \pi_i(\mathbf{b}_j))$ of projective sublattice $L_{[i:j]}$. That is, we omit the change of blocksize $\beta = j - i + 1$ if it is clear by context.

2.2 Shortest Vector and Gaussian Heuristic

A non-zero vector in a lattice L that has the minimum norm is called the *shortest vector*. We use $\lambda_1(L)$ to denote the norm of the shortest vector. The notion is also defined for a projective sublattice as $\lambda_1(L_{[i:j]})$ (we occasionally refer to this as $\lambda_1(B_i)$ in this paper).

The *shortest vector problem* (SVP) is the problem of finding a vector of length $\lambda_1(L)$. For a function $\gamma(n)$ of a lattice dimension n, the standard definition of γ-approximate SVP is the problem of finding a vector shorter than $\gamma(n) \cdot \lambda_1(L)$.

An n-dimensional lattice L and a continuous (usually convex and symmetric) set $S \subset \mathbb{R}^m$ are given. Then the *Gaussian heuristic* says that the number of points in $S \cap L$ is approximately $\mathrm{vol}(S)/\mathrm{vol}(L)$.

In particular, taking S as the origin-centered ball of radius R, the number of lattice points is approximately $V_n(R)/\mathrm{vol}(L)$, which derives the length of shortest vector λ_1 by R so that the volume of the ball is equal to that of the lattice:

$$\lambda_1(L) \approx \det(L)^{1/n}/V_n(1)^{1/n} = \frac{(\Gamma(n/2+1)\det(L))^{1/n}}{\sqrt{\pi}}$$

This is usually called the *Gaussian heuristic of a lattice*, and we denote it by $\mathrm{GH}(L) = \det(L)^{1/n}/V_n(1)^{1/n}$.

For our analysis, we use the following lemma on the randomly generated points.

Lemma 1. *Let $x_1, \ldots, x_K$ be K points uniformly sampled from the n-dimensional unit ball. Then, the expected value of the shortest length of vectors from origin to these points is*

$$\mathbf{E}\Big[\min_{i\in[K]} \|x_i\|\Big] = K \cdot B\Big(K, \frac{n+1}{n}\Big) := K \cdot \int_0^1 t^{1/n}(1-t)^{K-1}dt.$$

In particular, letting $K = 1$, the expected value is $n/(n+1)$.

Proof. Since the cumulative distribution function of each $\|x_i\|$ is $F_i(r) = r^n$, the cumulative function of the shortest length of the vectors is

$F_{\min}(r) = 1 - (1 - F_i(r))^K = 1 - (1 - r^n)^K$. Its probability density function is $P_{\min}(r) = \frac{dF}{dr} = Kn \cdot r^{n-1}(1 - r^n)^{K-1}$. Therefore, the expected value of the shortest length of the vectors is

$$\int_0^1 r P_{\min}(r) dr = K \cdot \int_0^1 t^{1/n}(1-t)^{K-1} dt.$$

□

2.3 Enumeration Algorithm [20,28,46]

We explain the enumeration algorithm for finding a short vector in the lattice. The pseudo code of the enumeration algorithm is given in [20,46]. For given lattice basis $(\mathbf{b}_1, \ldots, \mathbf{b}_n)$, and its Gram-Schmidt basis $(\mathbf{b}_1^*, \ldots, \mathbf{b}_n^*)$, the enumeration algorithm considers a search tree whose nodes are labeled by vectors. The root of the search tree is the zero vector; for each node labeled by $\mathbf{v} \in L$ at depth $k \in [n]$, its children have labels $\mathbf{v} + a_{n-k} \cdot \mathbf{b}_{n-k}$ ($a_{n-k} \in \mathbb{Z}$) whose projective length $\|\pi_{n-k}(\sum_{i=n-k}^{n} a_i \cdot \mathbf{b}_i)\|$ is smaller than a bounding value $R_{k+1} \in (0, \|\mathbf{b}_1\|]$. After searching all possible nodes, the enumeration algorithm finds a lattice vector shorter than R_n at a leaf of depth n, or its projective length is somehow short at a node of depth $k < n$. It is clear that by taking $R_k = \|\mathbf{b}_1\|$ for all $k \in [n]$, the enumeration algorithm always finds the shortest vector $\mathbf{v}_1$ in the lattice, namely $\|\mathbf{v}_1\| = \lambda_1(L)$.

Because $\|\mathbf{b}_1\|$ is often larger than $\lambda_1(L)$, we can set a better searching radius $R_n = \mathrm{GH}(L)$ to decrease the computational cost. We call this *the full enumeration algorithm* and define the full enumeration cost $\mathrm{FEC}(B)$ as the cost of the algorithm for this basis. With the same argument in [20], we can evaluate $\mathrm{FEC}(B)$ using the following equation.

$$\mathrm{FEC}(B) = \sum_{k=1}^{n} \frac{V_k(\mathrm{GH}(L))}{\prod_{i=n-k+1}^{n} \|\mathbf{b}_i^*\|}.$$

Because full enumeration is a cost-intensive algorithm, several improvements have been proposed by considering the trade-offs between running time, searching radius, and success probability [20,47]. Gama-Nguyen-Regev [20] proposed a cost estimation model of the lattice enumeration algorithm to optimize the bounding functions of $R_1, \ldots, R_n$, which were mentioned above. The success probability p of finding a single vector within a radius c is given by

$$p = \Pr_{(x_1,\ldots,x_n) \leftarrow c \cdot S_n}\left[\sum_{i=1}^{\ell} x_i^2 < R_\ell^2 \text{ for } \forall\, \ell \in [n]\right],$$

where S_n is the surface of the n-dimensional unit ball. Then, the cost of the enumeration algorithm can be estimated by the number of processed nodes, i.e.,

$$N = \frac{1}{2} \sum_{k=1}^{n} \frac{\mathrm{vol}\{(x_1, \ldots, x_k) \in \mathbb{R}^k : \sum_{i=1}^{\ell} x_i^2 < R_\ell^2 \text{ for } \forall\, \ell \in [k]\}}{\prod_{i=n-k+1}^{n} \|\mathbf{b}_i^*\|}. \quad (1)$$

Note that the factor 1/2 is based on the symmetry. Using the methodology in [20], Chen-Nguyen proposed a method to find the optimal bounding functions of $R_1, \ldots, R_n$ that minimizes N subject to p.

In this paper, we use the *lattice enumeration cost*, abbreviated as *ENUM cost*, to denote the number N in Eq. (1). For a lattice L defined by a basis B and parameters $\alpha > 0$ and $p \in [0,1]$, we use $\text{ENUMCost}(B; \alpha, p)$ to denote the minimized cost N of lattice enumeration with radius $c = \alpha \cdot \text{GH}(L)$ subject to the success probability p. This notion is also defined for a projective sublattice.

3 Lattice Reduction Algorithms

Lattice reduction algorithms transform a given lattice basis $(\mathbf{b}_1, \ldots, \mathbf{b}_n)$ to another basis whose Gram-Schmidt lengths are relatively shorter.

LLL Algorithm [30]: The LLL algorithm transforms the basis $(\mathbf{b}_1, \ldots, \mathbf{b}_n)$ using the following two operations: size reduction $\mathbf{b}_i \leftarrow \mathbf{b}_i - \lfloor \mu_{ji} \rceil \mathbf{b}_j$ for $j \in [i-1]$, and neighborhood swaps between $\mathbf{b}_i$ and $\mathbf{b}_{i+1}$ if $\|\mathbf{b}_{i+1}^*\|^2 \leq 1/2 \|\mathbf{b}_i^*\|^2$ until no update occurs.

BKZ Algorithms [46,47]. For a given lattice basis and a fixed blocksize β, the BKZ algorithm processes the following operation in the local block B_i, i.e., the projected sublattice $L_{[i,i+\beta-1]}$ of blocksize β, starting from the first index $i = 1$ to $i = n-1$. Note that the blocksize β reluctantly shrinks to $n-i+1$ for large $i > n-\beta+1$, and thus we sometimes use β' to denote the dimension of B_i, i.e. $\beta' = \min(\beta, n-i+1)$.

At index i, the standard implementation of the BKZ algorithm calls the enumeration algorithm for the local block B_i. Let $\mathbf{v}$ be a shorter vector found by the enumeration algorithm. Then the BKZ algorithm inserts $\mathbf{v}$ into $\mathbf{b}_{i-1}$ and $\mathbf{b}_i$, and constructs the degenerated basis $(\mathbf{b}_1, \ldots, \mathbf{b}_{i-1}, \mathbf{v}, \mathbf{b}_i, \ldots, \mathbf{b}_{\min(i+\beta-1,n)})$. For this basis, we apply the LLL algorithm (or BKZ with a smaller blocksize) so that the basis of shorter independent vectors can be obtained. One set of these procedures from $i = 1$ to $n-1$ is usually called a *tour*. The original version of the BKZ algorithm stops when no update occurs during $n-1$ iterations. In this paper, we refer to the BKZ algorithm with blocksize β as the BKZ-β.

HKZ Reduced Basis: The lattice basis $(\mathbf{b}_1, \ldots, \mathbf{b}_n)$ is called Hermite-Korkine-Zolotarev (HKZ) reduced [38, Chapter 2] if it is size-reduced $|\mu_{ji}| \leq 1/2$ for all i and j, and $\pi_i(\mathbf{b}_i)$ is the shortest vector in the projective sublattice $L_{[i:n]}$ for all i. We can estimate the Gram-Schmidt length of the HKZ-reduced basis by using the Gaussian heuristic as $\|\mathbf{b}_i^*\| = \text{GH}(L_{[i:n]})$. Since the HKZ-reduced basis is completely reduced in this sense, we will use this to discuss the lower bound of computing time in Sect. 8.2.

3.1 Some Heuristic Assumptions in BKZ

Gaussian Heuristic in Small Dimensions: Chen and Nguyen observed that the length $\lambda_1(B_i)$ of the shortest vector in the local block B_i is usually larger

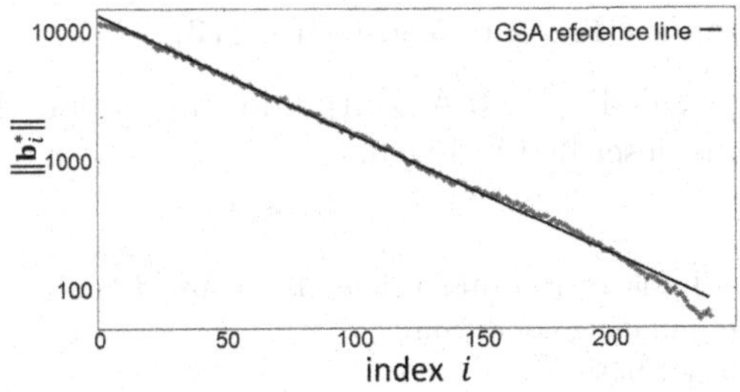

Fig. 2. Semi-log graph of $\|\mathbf{b}_i^*\|$ of a 240-dimensional highly reduced basis

than $\mathrm{GH}(B_i)$ in small dimensions i.e., small β' [13]. They gave the averaged values of $\|\mathbf{b}_i^*\|/\det(L)^{1/n}$ for the last indexes of highly reduced bases to modify their BKZ simulator, see [13, Appendix C]. For their 50 simulated values for $\|\mathbf{b}_{n-49}^*\|, \ldots, \|\mathbf{b}_n^*\|$, we define the modified Gaussian heuristic constant by

$$\tau_i := \frac{\lambda_1(\pi_{n-i+1}(L))}{\mathrm{GH}(\pi_{n-i+1}(L))} = \frac{\|\mathbf{b}_{n-i+1}^*\|}{V_i(1)^{-1/i} \cdot \prod_{j=n-i+1}^{n} \|\mathbf{b}_j^*\|^{1/i}}. \tag{2}$$

We will use τ_i for $i \leq 50$ to denote these modifying constants; for $i > 50$ we define $\tau_i = 1$ following Chen-Nguyen's simulator [13].

In the rest of this paper, we assume that the shortest vector lengths of β-dimensional local blocks B_i of reduced bases satisfies

$$\lambda_1(B_i) \approx \begin{cases} \tau_\beta \cdot \mathrm{GH}(B_i) & (\beta \leq 50) \\ \mathrm{GH}(B_i) & (\beta > 50) \end{cases}$$

on average.

We note that there exists a mathematical theory to guarantee $\tau_i \to 1$ for random lattices when the dimension goes to infinity [42]. Though it does not give the theoretical guarantee $\tau_i = 1$ for BKZ local blocks, they are very close in our preliminary experiments.

Geometric Series Assumption (GSA): Schnorr [44] introduced *geometric series assumption* (GSA), which says that the Gram-Schmidt lengths $\|\mathbf{b}_i^*\|$ in the BKZ-reduced basis decay geometrically with quotient r for $i = 1, \ldots, n$, namely, $\|\mathbf{b}_i^*\|^2/\|\mathbf{b}_1\|^2 = r^{i-1}$, for some $r \in [3/4, 1)$. Here r is called *the GSA constant.* Figure 2 shows the Gram-Schmidt lengths of a 240-dimensional reduced basis after processing BKZ-100 using our algorithm and parameters.

It is known that GSA does not hold exactly in the first and last indexes [11]. Several previous works [3,11,44] aimed to modify the reduction algorithm that outputs the reduced basis satisfying GSA. However, it seems difficult to obtain such a reduced basis in practice. In this paper, we aim to modify the parameters in the first and last indexes so that the proposed simulator performs with optimal efficiency (See Sect. 5.1).

3.2 Chen-Nguyen's BKZ 2.0 Algorithm [13]

We recall Chen-Nguyen's BKZ 2.0 Algorithm in this section. The outline of the BKZ 2.0 algorithm is described in Fig. 3.

Input: A lattice basis B of n dimensions, blocksize β, and some terminating condition.
Output: A reduced basis B.
1: $B \leftarrow \mathrm{LLL}(B)$;
2: **for** $i = 1$ to $n - 1$
3: Set probability p for local block B_i of fixed blocksize $\beta_i' = \min(\beta, n-i+1)$ and let $M = \lfloor 1/p \rceil$;
4: Generate randomized local blocks $G_1, \ldots, G_M$ from local block B_i, and preprocess $G_1, \ldots, G_M$ (reduction by LLL and small blocksize BKZ);
5: Find a vector $\mathbf{v}$ using lattice enumeration with radius $c = \min\{||\mathbf{b}_i^*||, \sqrt{1.1} \cdot \mathrm{GH}(B_i)\}$ for $G_1, \ldots, G_M$ with probability p;
6: **if** $\mathbf{v}$ satisfies $\|\mathbf{v}\| < \|\mathbf{b}_i^*\|$ **then** update basis B by $\mathbf{v}$;
7: **end-for**
8: **if** terminating condition is satisfied **then** return B **else** goto Step 2;

Fig. 3. Outline of BKZ 2.0

Speed-Up Techniques for BKZ 2.0: BKZ 2.0 employs four major speed-up techniques that differentiate it from the original BKZ:

1. BKZ 2.0 employs the extreme pruning technique [20], which attempts to find shorter vectors in the local blocks B_i with low probability p by randomizing basis B_i to more blocks $G_1, \ldots, G_M$ where $M = \lfloor 1/p \rceil$.
2. For the search radius $\min\{||\mathbf{b}_i^*||, \alpha \cdot \mathrm{GH}(B_i)\}$ in the enumeration algorithm of the local block B_i, Chen and Nguyen set the value as $\alpha = \sqrt{1.1}$ from their experiments, while the previous works set the radius as $\|\mathbf{b}_i^*\|$.
3. In order to reduce the cost of the enumeration algorithm, BKZ 2.0 preprocesses the local blocks by executing the sequence of BKZ algorithm, e.g., 3 tours of BKZ-50 and then 5 tours of BKZ-60, and so on. The parameters blocksize, number of rounds and number of randomized bases, are precomputed to minimize the total enumeration cost.
4. BKZ 2.0 uses the terminating condition introduced in [23], which aborts BKZ within small number of tours. It can find a short vector faster than the full execution of BKZ.

Chen-Nguyen's BKZ 2.0 Simulator: In order to predict the computational cost and the quality of the output basis, they also propose the simulating procedure of the BKZ 2.0 algorithm. Let $(\ell_1, \ldots, \ell_n)$ be the simulated values of the GS-lengths $\|\mathbf{b}_i^*\|$ for $i = 1, \ldots, n$. Then, the simulated values of the determinant and the Gaussian heuristic are represented by $\prod_{j=1}^{n} \ell_j$ and $\mathrm{GH}(B_i) = V_{\beta'}(1)^{-1/\beta'} \prod_{j=i}^{i+\beta'-1} \ell_i$ where $\beta' = \min\{\beta, n-i+1\}$, respectively.

Input: A lattice basis B of n dimensions, blocksize β
Output: A reduced basis B.
1: $B \leftarrow \mathrm{LLL}(B)$;
2: $flag = 1$ // set $flag = 0$ when the basis is updated.
3: **for** $i = 1$ to $n - 1$
4: Set (α, p) for local block B_i of fixed blocksize $\beta'_i = \min(\beta, n - i + 1)$;
5: Execute lattice enumeration with probability p and radius $\alpha \cdot \mathrm{GH}(B_i)$;
6: **if** $\mathbf{v}$ satisfies $\|\mathbf{v}\| < \alpha \cdot \mathrm{GH}(B_i)$, **then** update basis B by $\mathbf{v}$ and $flag = 0$;
7: **end-for**
8: **if** $flag = 1$ **then** return B **else** goto Step 2;

Fig. 4. Plain BKZ algorithm

They simulate a BKZ tour of blocksize β assuming that each enumeration procedure finds a vector of projective length $\mathrm{GH}(B_i)$. Roughly speaking, their simulator updates (ℓ_i, ℓ_{i+1}) to (ℓ'_i, ℓ'_{i+1}) for $i = 1, \ldots, n-1$, where $\ell'_i = \mathrm{GH}(B_i)$ and $\ell'_{i+1} = \ell_{i+1} \cdot (\ell_i / \ell'_i)$. Here, the last 50 GS-lengths are modified using an HKZ reduced basis. The details of their simulator are given in [13, Algorithm 3].

They also present the upper and lower bounds for the number of processed nodes during the lattice enumeration of blocksize β. From [14, Table 4], we extrapolate the costs as

$$\log_2(\mathrm{Cost}_\beta) = 0.000784314\beta^2 + 0.366078\beta - 6.125 \quad (3)$$

Then, the total enumeration cost of performing the BKZ 2.0 algorithm using blocksize β and t tours is given by

$$t \cdot \sum_{i=1}^{n-1} \mathrm{Cost}_{\min\{\beta, n-i+1\}}. \quad (4)$$

To convert the number of nodes into single-threaded time in seconds, we use the rational constant $4 \cdot 10^9 / 200 = 2 \cdot 10^7$, because they assumed that processing one node requires 200 clock cycles in a standard CPU, and we assume it can work at 4.0GHz.

We note that there are several models to extrapolate $\log_2(\mathrm{Cost}_\beta)$. Indeed, Lepoint and Naehrig [31] consider two models by a quadratic interpolation and a linear interpolation from the table. Albrecht et al. [5] showed another BKZ 2.0 cost estimation that uses an interpolation using the cost model $\log_2(\mathrm{Cost}_\beta) = O(n \log n)$. It is a highly non-trivial task to find a proper interpolation that estimates a precise cost of the BKZ 2.0 algorithm.

We further mention that the upper bound of the simulator is somewhat debatable, because they use the enumeration radius $c = \min\{\sqrt{1.1} \cdot \mathrm{GH}(B_i), \|\mathbf{b}^*_i\|\}$ for $i < n - 30$ in their experiment whereas they assume $c = \mathrm{GH}(B_i)$ for the cost estimation in their upper bound simulation. Thus, the actual cost of BKZ 2.0 could differ by a factor of $1.1^{O(\beta)}$.

4 Optimizing Parameters in Plain BKZ

In this section we consider the plain BKZ algorithm described in Fig. 4, and roughly predict the GS-lengths of the output basis, which were computed by the GSA constant r. Using this analysis, we can obtain the optimal settings for parameters (α, p) in Step 4 of the plain BKZ algorithm of blocksize β.

4.1 Relationship of Parameters α, P, β, R

We fix the values of parameters (β, α) and assume that the lattice dimension n is sufficiently large.

Suppose that we found a vector $\mathbf{v}$ of $\|\mathbf{v}\| < \alpha \cdot \mathrm{GH}(B_i)$ in the local block B_i. We update the basis B_i by inserting $\mathbf{v}$ at i-th index, and perform LLL or small blocksize BKZ on the updated basis.

When the lattice dimension is large, Rogers' theorem [42] says that approximately $\alpha^n/2$ vector pairs $(\mathbf{v}, -\mathbf{v})$ exist within the ball of radius $c = \alpha \cdot \mathrm{GH}(L)$. Since the pruning probability is defined for a single vector pair, we expect the actual probability that the enumeration algorithm finds at least one vector shorter than c is roughly

$$1 - (1-p)^{\alpha^n/2} \approx p \cdot \frac{\alpha^n}{2}. \tag{5}$$

From relation (5), there may exist one lattice vector in the searching space by setting parameter p as

$$p = \frac{2}{\alpha^\beta}. \tag{6}$$

Remark 1. The probability setting of Eq. (6) is an optimal choice under our assumption. If p is smaller, the enumeration algorithm finds no short vector with high probability and basis updating at i-th index does not occur, which is a waste of time. On the other hand, if we take a larger p so that there exist $p \cdot \alpha^\beta/2 > 1$ vector pairs, the computational time of the enumeration algorithm increases more than $p \cdot \alpha^\beta/2$ times [20]. Although it can find shorter vectors, this is also a waste of time from the viewpoint of basis updating.

Assume that one vector is found using the enumeration, and also assume that the distribution of it is the same as the random point in the β-dimensional ball of radius $\alpha \cdot \mathrm{GH}(B_i)$. Then, the expected value of $\|\mathbf{v}\|$ is $\frac{\beta}{\beta+1}\alpha \cdot \mathrm{GH}(B_i)$ by letting $K = 1$ in Lemma 1. Thus, we can expect that this is the value $\|\mathbf{b}_i^*\|$ after update.

Therefore, after executing a sufficient number of BKZ tours, we can expect that all the lengths $\|\mathbf{b}_i^*\|$ of the Gram-Schmidt basis satisfy

$$\|\mathbf{b}_i^*\| = \frac{\beta}{\beta+1}\alpha \cdot \mathrm{GH}(B_i) \tag{7}$$

on average. Hence, under Schnorr's GSA, we have the relation

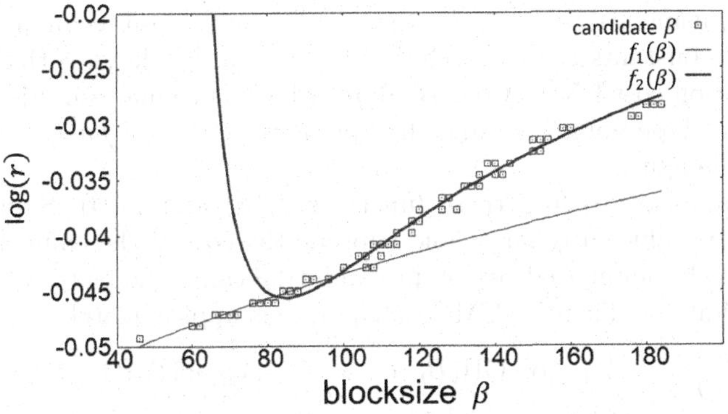

Fig. 5. Relation between β and r that minimizes the computational cost

$$\|\mathbf{b}_i^*\| = \frac{\alpha\beta}{\beta+1} \cdot V_\beta(1)^{-1/\beta}\|\mathbf{b}_i^*\| \prod_{j=1}^{\beta} r^{(j-1)/2\beta}, \tag{8}$$

and the GSA constant is

$$r = \left(\frac{\beta+1}{\alpha\beta}\right)^{\frac{4}{\beta-1}} \cdot V_\beta(1)^{\frac{4}{\beta(\beta-1)}}. \tag{9}$$

Therefore, by fixing (α, β), we can set the probability p and obtain r as a rough prediction of the quality of the output lattice basis. We will use the relations (6) and (9) to set our parameters. Note that any two of β, α, p and r are determined from the other two values.

Remark 2. Our estimation is somehow underestimate, i.e., in our experiments, the found vectors during BKZ algorithm are often shorter than the estimation in Eq. (7). This gap is mainly from the estimation in (5), which can be explained as follows. Let $(R_1, \ldots, R_\beta)$ be a bounding function of probability p for a vector of length $\|v\|$. Then, the probability p' for a vector of length $\|v'\|$ of a shorter vector is the same as the scaled bounding function $(R'_1, \ldots, R'_\beta)$ where $R'_i = \min\{1.0, R_i \cdot \|v\|/\|v'\|\}$. Here, p' is clearly larger than p due to $R'_i \geq R_i$ for $i \in [\beta]$. Therefore, when the above parameters are used, the quality of the output basis is better than that derived from Eq. (9) if we perform a sufficient number of tours. Hence, within a few tours, our algorithm can output a basis which has a good quality predicted by our estimation in this section.

4.2 Optimizing Parameters

Now for a fixed parameter pair (β, r), the cost ENUMCost$(B_i; \alpha, p)$ of the enumeration algorithm in local block B_i satisfying GSA is computable. Concretely, we compute α using the relation (9), fix p by (6), and simulate the Gram-Schmidt lengths of B_i using $\|\mathbf{b}_i^*\| = r^{(i-1)/2}$. Using the computation technique in [20], for several GSA constants r, we search for the optimal blocksize β that minimizes the enumeration cost ENUMCost$(B_i; \alpha, p)$. The small squares in Fig. 5 show the results. From these points, we find the functions $f_1(\beta)$ and $f_2(\beta)$, whose graphs are also in the figure.

We explain how to find these functions $f_1(\beta)$ and $f_2(\beta)$. Suppose lattice dimension n is sufficiently large, and suppose the cost of the enumeration algorithm is roughly dominated by the probability p times the factor at $k = n/2$ in the summation (1). Then ENUMCost$(B_i; \alpha, p)$ is approximately

$$D = p \cdot \frac{V_{\beta/2}(\alpha \cdot \mathrm{GH}(B_r))}{\prod_{i=\beta/2+1}^{\beta} \|\mathbf{b}_i^*\|} = 2\alpha^{-\beta/2} \frac{V_{\beta/2}(1) V_\beta(1)^{-1/2}}{r^{\beta^2/16}},$$

where from Eq. (9) we have obtained

$$D \approx Const. \times r^{(\beta^2-2\beta)/16} \cdot \left(\frac{\beta}{e\pi}\right)^{\beta/4}, \text{ and } \frac{\partial \log D}{\partial \beta} \approx \frac{\beta-1}{8} \log r + \frac{1}{4} + \frac{1}{4} \log \frac{\beta}{e\pi}.$$

In order to minimize D, we roughly need the above derivative to be zero; thus, we use the following function of β for our cost estimation with constants c_i

$$\log(r) = 2 \cdot (\log \beta + 1 - \log(e\pi))/(\beta - 1) = \frac{\log c_1 \beta}{c_2 \beta + c_3}.$$

From this observation, we fix the fitting function model as $f(\beta) = \frac{\log(c_1\beta + c_2)}{c_3\beta + c_4}$.

By using the least squares method implemented in `gnuplot`, we find the coefficients c_i so that $f(\beta)$ is a good approximation of the pairs $(\beta_i, \log(r_i))$. In our curve fitting, we separate the range of β into the interval $[40, 100]$, and the larger one. This is needed for converging to $\log(r) = 0$ when β is sufficiently large; however, our curve fitting using a single natural function did not achieve it. Curves $f_1(\beta)$ and $f_2(\beta)$ in Fig. 5 are the results of our curve fitting for the range $[40, 100]$ and the larger one, respectively.

For the range of $\beta \in [40, 100]$, we have obtained

$$\log(r) = f_1(\beta) := -18.2139/(\beta + 318.978) \tag{10}$$

and for the larger blocksize $\beta > 100$,

$$\log(r) = f_2(\beta) := (-1.06889/(\beta - 31.0345)) \cdot \log(0.417419\beta - 25.4889). \tag{11}$$

Note that we will use the relation (10) when the blocksize is smaller than 40.

Moreover, we obtain pairs of β and minimize ENUMCost$(B_i; \alpha, p)$, in accordance with the above experiments. Using the curve fitting that minimizes $\sum_\beta |f(\beta) - \log_2 \text{ENUMCost}(B_i; \alpha, p)|^2$ using `gnuplot`, we find the extrapolating formula

$$\log_2 \text{MINCost}(\beta) := \begin{cases} 0.1375\beta + 7.153 & (\beta \in [60, 105]) \\ 0.000898\beta^2 + 0.270\beta - 16.97 & (\beta > 105) \end{cases} \tag{12}$$

to $\log_2 \text{ENUMCost}(B_i; \alpha, p)$. We will use this as the standard of the enumeration cost of blocksize β.

Remark 3. Our estimation from the real experiments is $0.25\beta \cdot \text{ENUMCost}(B_i; \alpha, p)$ (See, Sect. 7.1 of the full-version [8]), which crosses over the estimation of BKZ 2.0 simulator (3) at $\beta = 873$. Thus, the performance of BKZ 2.0 might be better in some extremely high block sizes, while our algorithm has a better performance in the realizable block sizes < 200.

4.3 Parameter Settings in Step 4 in Fig. 4

Using the above arguments, we can fix the optimized pair (α, p) for each blocksize β. That is, to process a local block of blocksize β in Step 4 of the plain BKZ algorithm in Fig. 4, we compute the corresponding r by Eqs. (10) and (11), and additionally obtain the parameters α by Eq. (9) and p by Eq. (6). These are our basic parameter settings.

Modifying Blocksize at First Indexes: We sometimes encounter the phenomenon in which the actual $\text{ENUMCost}(B_i; \alpha, p)$ in small indexes is much smaller than that in middle indexes. This is because $||\mathbf{b}_i^*||$ is smaller than $\text{GH}(B_i)$ in small indexes. In other words, $\mathbf{b}_i$ is hard to update using the enumeration of blocksize β. To speed up the lattice reduction, we use a heuristic method that enlarges the blocksizes as follows.

From the discussion in the above subsection, we know the theoretical value of the enumeration cost at blocksize β. On the other hand, in the actual computing of BKZ algorithms, the enumeration cost is increased because the sequence $(||\mathbf{b}_i^*||, \ldots, ||\mathbf{b}_{i+\beta-1}^*||)$, which mainly affects the computing cost, does not follow the GSA of slope r exactly. Thus, we define the expected enumeration cost in blocksize β as $\beta \cdot \text{MINCost}(\beta)$. With this expectation, we reset the blocksize as the minimum β satisfying $\text{ENUMCost}(B_{[i:i+\beta-1]}; \alpha, p) > \beta \cdot \text{MINCost}(\beta)$.

Modifying (α, p) at Last Indexes: For large indexes such as $i > n - \beta$, the blocksize of a local block shrinks to $\beta' = \min(\beta, n-i+1)$. In our implementation, we increase the success probability to a new p', while $\text{ENUMCost}(B_i; \alpha', p')$ is smaller than $\beta \cdot \text{MINCost}(\beta)$. We also reset the radius as $\alpha' = (2/p')^{1/\beta}$ from Eq. (6).

5 Our Proposed Progressive BKZ: Basic Variant

In this section, we explain the basic variant of our proposed progressive BKZ algorithm.

In general, if the blocksize of the BKZ algorithm increases, a shorter vector $\mathbf{b}_1$ can be computed; however, the running cost will eventually increase. The progressive BKZ algorithm starts a BKZ algorithm with a relatively small blocksize β_{start} and increases the blocksize to β_{end} by some criteria. The idea of the progressive BKZ algorithm has been mentioned in several literatures, for example, [13,25,45,48]. The research challenge in the progressive BKZ algorithm is to find an effective criteria for increasing blocksizes that minimizes the total running time.

```
Input: A lattice basis B of n dimensions, starting blocksize β_start,
       and ending blocksize β_end.
Output: A reduced basis B.
B ← LLL(B);
for β = β_start to β_end do
  while FEC(B) > Sim-FEC(n, β) do
    for i = 1 to n − 1
      Set (α, p) for local block B_i of blocksize β' = min(β, n − i + 1)
       using the setting in Section 4.3;
      Preprocess the basis by the progressive BKZ;
      Execute lattice enumeration with probability p and radius α · GH(B_i);
      if v satisfies ||v|| < α · GH(B_i) then update basis B by v;
    end-for
  end-while
end-for
```

Fig. 6. Our progressive BKZ algorithm (basic variant)

In this paper we employ the *full enumeration cost* (FEC) in Sect. 2.3, in order to evaluate the quality of the basis for finding the increasing criteria. Recall that the FEC of basis $B = (\mathbf{b}_1, \ldots, \mathbf{b}_n)$ of n-dimensional lattice L is defined by $\mathrm{FEC}(B) = \sum_{k=1}^{n} \frac{V_k(\mathrm{GH}(L))}{\prod_{i=n-k+1}^{n} \|\mathbf{b}_i^*\|}$, where $\|\mathbf{b}_i^*\|$ represents the GS-lengths. Note that $\mathrm{FEC}(B)$ eventually decreases after performing several tours of the BKZ algorithm using the fixed blocksize β.

Moreover, we construct a simulator that evaluates the GS-lengths by the optimized parameters α, p, β, r for the BKZ algorithm described in the local block discussion in Sect. 4.3. The simulator for an n-dimensional lattice only depends on the blocksize β of the local block; we denote by Sim-GS-lengths(n, β) the simulated GS-lengths $(\ell_1, \ldots, \ell_n)$. The construction of simulator will be presented in Sect. 5.1.

For this purpose, we define some functions defined on the simulated GS-lengths $(\ell_1, \ldots, \ell_n)$. $\text{Sim-GH}(\ell_1, \ldots, \ell_n) = V_n(1)^{-1/n} \prod_{j=1}^{n} \ell_j^{1/n}$ is the simulated Gaussian heuristic. The simulated value of full enumeration cost is

$$\text{Sim-FEC}(\ell_1, \ldots, \ell_n) := \sum_{k=1}^{n} \frac{V_k(\text{Sim-GH}(\ell_1, \ldots, \ell_n))}{\prod_{i=n-k+1}^{n} \ell_i}.$$

Further, for $(\ell_1, \ldots, \ell_n) = \text{Sim-GS-lengths}(n, \beta)$, we use the notation $\text{Sim-FEC}(n, \beta) := \text{Sim-FEC}(\ell_1, \ldots, \ell_n)$ in particular. The simulated enumeration cost $\text{Sim-ENUMCost}(\ell_1, \ldots, \ell_\beta; \alpha, p)$ is defined by $\mathrm{ENUMCost}(B; \alpha, p)$ for a lattice basis B that has GS-lengths $\|\mathbf{b}_i^*\| = \ell_i$ for $i \in [\beta]$.

The key point of our proposed progressive BKZ algorithm is to increase the blocksize β if $\mathrm{FEC}(B)$ becomes smaller than $\text{Sim-FEC}(n, \beta)$. In other words, we perform the BKZ tours of blocksize β while $\mathrm{FEC}(B) > \text{Sim-FEC}(n, \beta)$. We describe the proposed progressive BKZ in Fig. 6.

Remark 4. In the basic variant of our progressive BKZ described in Sect. 6.1, we increase the blocksize β in increments of one in Step 2. However, we will present an optimal strategy for increasing the blocksize in Sect. 5.

5.1 Sim-GS-lengths(n, β): Predicting Gram-Schmidt Lengths

In the following, we construct a simulator for predicting the Gram-Schmidt lengths $\|\mathbf{b}_i^*\|$ obtained from the plain BKZ algorithm of blocksize β.

Our simulator consists of two phases. First, we generate approximated GS-lengths using Gaussian heuristics; we then modify it for the first and last indexes of GSA in Sect. 3.1. We will explain how to compute $(\ell_1, \ldots, \ell_n)$ as the output of Sim-GS-lengths(n, β).

First Phase: Our simulator computes the initial value of $(\ell_1, \ldots, \ell_n)$.

We start from the last index by setting $\ell_n = 1$, and compute ℓ_i backwards. From Eqs. (2) and (7) we are able to simulate the GS-lengths ℓ_i by solving the following equation of ℓ_i:

$$\ell_i = \max\left\{\frac{\beta'}{\beta'+1}\alpha, \tau_{\beta'}\right\} \cdot \mathrm{GH}(\ell_i, \ldots, \ell_{i+\beta'-1}), \text{ where } \beta' = \min(\beta, n-i+1). \tag{13}$$

Here, α is the optimized radius parameter in Sect. 4.3 and $\tau_{\beta'}$ is the coefficient of the modified Gaussian heuristic.

This simple simulation in the first phase is sufficient for smaller blocksizes ($\beta < 30$). However, for simulating larger blocksizes, we must modify the GS-lengths of the first and last indexes in Sect. 3.1.

Second Phase: To modify the results of the simple simulation, we consider our two modifying methods described in Sect. 4.3. We recall that MINCost(β) is the standard value of the enumeration cost of blocksize β.

We first consider the modification for the last indexes $i > n - \beta + 1$, i.e., a situation in which the blocksize is smaller than β. We select the modified probability p_i at index i so that Sim-ENUMCost$(\ell_i, \ldots, \ell_n; \alpha_i, p_i)$ = MINCost(β), where $\ell_i, \ldots, \ell_n$ is the result of the first simulation, and we use $\alpha_i = (2/p_i)^{n-i+1}$. After all (α_i, p_i) for $n - \beta + 1 \le i \le n$ are fixed, we modify the GS-lengths by solving the following equation of ℓ_i again:

$$\ell_i = \max\left\{\frac{\beta'}{\beta'+1}\alpha_i, \tau_{\beta'}\right\} \cdot \mathrm{GH}(\ell_i, \ldots, \ell_n) \text{ where } \beta' = n - i + 1.$$

Next, using the modified $(\ell_1, \ldots, \ell_n)$, we again modify the first indexes as follows. We determine the integer parameter $b > 0$ for the size of enlargement. For $b = 1, 2, \ldots$, we reset the blocksize at index i as $\beta_i := \beta + \max\{(b - i + 1)/2, b - 2(i-1)\}$ for $i \in \{1, \ldots, b\}$. Using these blocksizes, we recompute the GS-lengths by solving Eq. (13) from $i = \beta_i$ to 1. Then, we compute Sim-ENUMCost$(\ell_1, \ldots, \ell_{\beta+b}; \alpha, p)$. We select the maximum b such that this simulated enumeration cost is smaller than $2 \cdot$ MINCost(β).

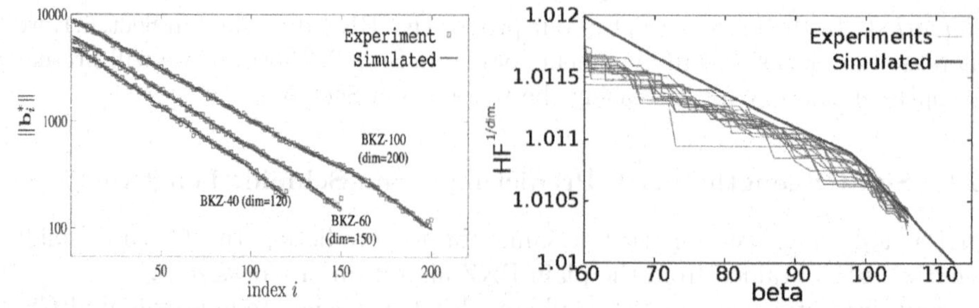

Fig. 7. Left figure: Semi-log graph of $\|\mathbf{b}_i^*\|$ of reduced random lattices from the SVP Challenge problem generator: Simulation (bold lines) vs. Experiment (small squares). **Right figure**: The root Hermite factor of reduced random 300-dimensional bases after BKZ-β. Simulation (bold red lines) vs. Experiment (thin blue lines) (Color figure online).

Experimental Result of Our GS-lengths Simulator: We performed some experiments on the GS-lengths for some random lattices from the Darmstadt SVP Challenge [50]. We computed the GS-lengths for 120, 150 and 200 dimensions using the proposed progressive BKZ algorithm, with ending blocksizes of 40, 60, and 100, respectively (Note that the starting blocksize is irrelevant to the quality of the GS-lengths). The simulated result is shown in Fig. 7. Almost all small squares of the computed GS-lengths are plotted on the bold line obtained by our above simulation. Our simulator can precisely predict the GS-lengths of these lattices. The progress of the first vector, which uses 300-dimensional lattices, is also shown in the figure.

5.2 Expected Number of BKZ Tours at Step 3

At Step 3 in the proposed algorithm (Fig. 6) we iterate the BKZ tour with blocksize β as long as the full enumeration cost FEC(B) is larger than the simulated cost Sim-FEC(n, β). In the following we estimate the expected number of BKZ tours (we denote it as $\sharp tours$) at blocksize β.

In order to estimate $\sharp tours$, we first compute $(\ell_1, \ldots, \ell_n)$ and the output of Sim-GS-lengths($n, \beta - 1$), and update it by using the modified Chen-Nguyen's BKZ 2.0 simulator described in Sect. 3.2, until Sim-FEC$(\ell_1, \ldots, \ell_n)$ is smaller than Sim-FEC(n, β). We simulate a BKZ tour by updating the pair (ℓ_i, ℓ_{i+1}) to (ℓ'_i, ℓ'_{i+1}) for $i = 1, \ldots, n-1$ according to the following rule:

$$\ell'_i = \max\left\{\frac{\beta}{\beta+1}\alpha, \tau_\beta\right\} \cdot \mathrm{GH}(\ell_i, \ldots, \ell_{\min(n,i+\beta-1)})$$
$$\text{and } \ell'_{i+1} = \ell_{i+1} \cdot (\ell_i/\ell'_i).$$

At the simulation of t-th BKZ tour, write the input GS-lengths $(\ell'_1, \ldots, \ell'_n)$; i.e., the output of the $(t-1)$-th BKZ tour. We further denote the output of t-th BKZ tour as $(\ell_1, \ldots, \ell_n)$. Suppose they satisfy

$$\text{Sim-FEC}(\ell'_1, \ldots, \ell'_n) > \text{Sim-FEC}(n, \beta) > \text{Sim-FEC}(\ell_1, \ldots, \ell_n).$$

Then, our estimation of $\sharp tours$ is the interpolated value:

$$\sharp tours = (t-1) + \frac{\text{Sim-FEC}(\ell'_1,\ldots,\ell'_n) - \text{Sim-FEC}(n,\beta)}{\text{Sim-FEC}(\ell'_1,\ldots,\ell'_n) - \text{Sim-FEC}(\ell_1,\ldots,\ell_n)}. \quad (14)$$

Note that we can use this estimation for other BKZ strategies, although we estimate the number of BKZ tours from BKZ-$(\beta-1)$ basis to BKZ-β basis, using BKZ-β algorithm. We will estimate the tours for other combinations of starting and ending blocksizes, and use them in the algorithm.

6 Our Progressive BKZ: Optimizing Blocksize Strategy

We propose how to optimally increase the blocksize β in the proposed progressive BKZ algorithm. Several heuristic strategies for increasing the blocksizes have been proposed. The following sequences of blocksizes after LLL-reduction have been used in the previous literatures:

$20 \to 21 \to 22 \to 23 \to 24 \to \cdots$	Gama and Nguyen [19]
$2 \to 4 \to 8 \to 16 \to 32 \to \cdots$	Schnorr and Shevchenko [48],
$2 \to 4 \to 6 \to 8 \to 10 \to \cdots$	Haque, Rahman, and Pieprzyk [25],
$50 \to 60 \to 70 \to 80 \to 90 \to \cdots$	Chen and Nguyen [13, 14]

The timings for changing to the next blocksize were not explicitly given. They sometimes continue the BKZ tour until no update occurs as the original BKZ. In this section we try to find the sequence of the blocksizes that minimizes the total cost of the progressive BKZ to find a BKZ-β reduced basis. To find this strategy, we consider all the possible combinations of blocksizes used in our BKZ algorithm and the timing to increase the blocksizes.

Notations on Blocksize Strategy: We say a lattice basis B of dimension n is β-reduced when FEC(B) is smaller than Sim-FEC(n,β). For a tuple of blocksizes $(\beta^{alg},\beta^{start},\beta^{goal})$ satisfying $2 \le \beta^{start} < \beta^{goal} \le \beta^{alg}$, the notation

$$\beta^{start} \xrightarrow{\beta^{alg}} \beta^{goal}$$

is the process of the BKZ following algorithm. The input is a β^{start}-reduced basis B, and the algorithm updates B using the tours of BKZ-β^{alg} algorithm with parameters in Sect. 4.3. It stops when FEC$(B) <$ Sim-FEC(n,β^{goal}).

TimeBKZ$(n, \beta^{start} \xrightarrow{\beta^{alg}} \beta^{goal})$ is the computing time in seconds of this algorithm. We provide a concrete simulating procedure in this and the next sections. We assume that TimeBKZ is a function of $n, \beta^{alg}, \beta^{start}$ and β^{goal}.

To obtain a BKZ-β reduced basis from an LLL reduced basis, many blocksize strategies are considered as follows:

$$\beta_0^{goal} = \text{LLL} \xrightarrow{\beta_1^{alg}} \beta_1^{goal} \xrightarrow{\beta_2^{alg}} \beta_2^{goal} \xrightarrow{\beta_3^{alg}} \cdots \xrightarrow{\beta_D^{alg}} \beta_D^{goal} (=\beta). \quad (15)$$

We denote this sequence as $\{(\beta_j^{alg},\beta_j^{goal})\}_{j=1,\ldots,D}$, and regard it as the progressive BKZ given in Fig. 8.

Input: A lattice basis B of n dimensions,
Blocksize strategy $\{(\beta_j^{alg}, \beta_j^{goal})\}_{j=1,\ldots,D}$
Output: A β_D^{goal}-reduced basis B.
1: $B \leftarrow \mathrm{LLL}(B)$;
2: **for** $j = 1$ **to** D **do**
3: **while** $\mathrm{FEC}(B) > \text{Sim-FEC}(n, \beta_j^{goal})$ **do**
4-9: The same as Step 4-9 in Figure 6 with blocksize β_j^{alg}
10: **end-while**
11: **end-for**

Fig. 8. Our progressive BKZ algorithm with blocksize strategy

6.1 Optimizing Blocksize Strategies

Our goal in this section is to find the optimal sequence that minimizes the total computing time

$$\sum_{i=1}^{D} \mathrm{TimeBKZ}(n, \beta_{i-1}^{goal} \overset{\beta_i^{alg}}{\rightarrow} \beta_i^{goal}) \tag{16}$$

of the progressive BKZ algorithm to find a BKZ-β_D^{goal} basis.

Based on our experimental results, which are given in Sect. 7, we can estimate the computing time of the BKZ algorithm:

$$\begin{aligned} &\mathrm{TimeBKZ}(n, \beta^{start} \overset{\beta^{alg}}{\rightarrow} \beta^{goal}) \text{ [sec.]} \\ &= \sum_{t=1}^{\sharp tours} \left[1.5 \cdot 10^{-10} \cdot (\beta^{alg})^2 n^3 + 1.5 \cdot 10^{-8} \cdot \beta^{alg} \sum_{i=1}^{n-1} \mathrm{ENUMCost}(B_i; \alpha, p)\right] \end{aligned} \tag{17}$$

when dimension n is small ($n < 400$), and

$$\begin{aligned} &\mathrm{TimeBKZ}(n, \beta^{start} \overset{\beta^{alg}}{\rightarrow} \beta^{goal}) \text{ [sec.]} \\ &= \sum_{t=1}^{\sharp tours} \left[2.5 \cdot 10^{-4} \cdot \frac{n - \beta^{alg}}{250 - \beta^{alg}} \cdot n^2. + 3.0 \cdot 10^{-8} \cdot \beta^{alg} \sum_{i=1}^{n-1} \mathrm{ENUMCost}(B_i; \alpha, p)\right] \end{aligned} \tag{18}$$

when dimension n is large ($n \geq 400$). The difference is caused by the difference in the types to compute Gram-Schmidt variables in implementation. The former and latter implementation employ `quad_float` and `RR` (320 bits) respectively, where `RR` is the arbitrary precision floating point type in the NTL library [49]. To compute $\sharp tours$ we use the procedure in Sect. 5.2. The input of the ENUMCost function is from Sim-GS-lengths(n, β^{start}) at the first tour. From the second tour, we use the updated GS-lengths by the Chen-Nguyen's simulator with blocksize β^{alg}.

Using these computing time estimations, we discuss how to find the optimal blocksize strategy (15) that minimizes the total computing time. In this optimizing procedure, the input consists of n and β, the lattice dimension and the goal blocksize. We denote TimeBKZ(n, β^{goal}) to be the minimized time in seconds to find a β-reduced basis from an LLL reduced basis, that is, the minimum of (16) from among the possible blocksize strategies. By definition, we have

$$\text{TimeBKZ}(n, \beta^{goal}) = \min_{\beta', \beta^{alg}} \left\{ \text{TimeBKZ}(n, \beta') + \text{TimeBKZ}(n, \beta' \xrightarrow{\beta^{alg}} \beta^{goal}) \right\}$$

where we take the minimum over the pair of blocksizes (β', β^{alg}) satisfying $\beta' < \beta^{goal} \le \beta^{alg}$.

For the given (n, β), our optimizing algorithm computes TimeBKZ$(n, \bar{\beta})$ from small $\bar{\beta}$ to the target $\bar{\beta} = \beta$. As the base case, we define that TimeBKZ$(n, 20)$ represents the time to compute a BKZ-20 reduced basis using a fixed blocksize, starting from an LLL reduced basis:

$$\text{TimeBKZ}(n, 20) := \min_{\beta^{alg}} \left\{ \text{TimeBKZ}(n, \text{LLL} \xrightarrow{\beta^{alg}} 20) \right\}.$$

6.2 Simulating Time to Find Short Vectors in Random Lattices

In this section, we give our simulating result of finding short vectors for random lattices. For the given lattice dimension n and the target length, we simulate the necessary BKZ blocksize β so that ℓ_1 of Sim-GS-lengths(n, β) is smaller than the target length. Then, we simulate TimeBKZ(n, β) by using the method in Sect. 6.1.

As an example, in Table 2, we show the optimized blocksize strategy and computing time to find a 102-reduced basis in $n = 600$ dimension. We estimate blocksize 102 is necessary to find a vector shorter than $n \cdot \det(L)^{1/n}$, which is the condition to enter the Hall of Fame in the Approximate Ideal Lattice Challenge [50].

Table 2. The optimized blocksize strategy and computational time in seconds in 600-dimensional lattice.

$\xrightarrow{\beta^{alg}} \beta^{goal}$	LLL	$\xrightarrow{32} 21$	$\xrightarrow{50} 36$	$\xrightarrow{58} 46$	$\xrightarrow{65} 55$	$\xrightarrow{71} 61$	$\xrightarrow{75} 70$	$\xrightarrow{81} 76$	$\xrightarrow{85} 84$
$\log_2$(Time [sec.])		15.61	15.86	16.04	16.21	16.31	16.51	16.70	17.07
$\xrightarrow{\beta^{alg}} \beta^{goal}$		$\xrightarrow{89} 88$	$\xrightarrow{91} 90$	$\xrightarrow{93} 92$	$\xrightarrow{99} 98$	$\xrightarrow{101} 100$	$\xrightarrow{103} 102$		
$\log_2$(Time [sec.])		17.42	17.67	17.97	18.89	19.49	20.09		

Table 3 shows the blocksize and predicted total computing time in seconds to find a vector shorter than $n \cdot \text{GH}(L)$ (this corresponds to the n-approximate SVP from the learning with errors problem [41].), $n \cdot \det(L)^{1/n}$ (from the Approximate Ideal Lattice Challenge published in Darmstadt [50]), and $\sqrt{n} \cdot \text{GH}(L)$. For comparison, the simulating result of BKZ 2.0 is given to find $n \cdot \det(L)^{1/n}$. Recall that their estimated cost in seconds is given by $\sharp\text{ENUM}/2 \cdot 10^7$. From Table 3, our algorithm is asymptotically faster than BKZ 2.0.

Table 3. Simulated $\log_2$(Time [sec.]) of our algorithm and BKZ 2.0 for large dimensions to find short vectors. The time is after LLL-reduced basis. Because the estimate for BKZ 2.0 is only the cost for enumeration, our algorithm appears to be slow in small blocksizes.

Goal	$n \cdot \mathrm{GH}(L)$		$n \cdot \det(L)^{1/n}$			$\sqrt{n} \cdot \mathrm{GH}(L)$	
n	β	$\log_2$(Ours)	β	$\log_2$(Ours)	$\log_2$(BKZ 2.0)	β	$\log_2$(Ours)
600	35	15.8	102	20.1	16.0	145	38.4
650	45	16.6	114	24.3	21.9	157	51.0
700	59	17.3	124	28.3	28.2	169	60.4
800	100	20.8	144	38.6	41.3	193	82.1

6.3 Comparing with Other Heuristic Blocksize Strategies

In this section, we compare the blocksize strategy of our progressive BKZ in Fig. 8. Using a random 256-dimensional basis, we experimented and simulated the progressive BKZ to find a BKZ-128 reduced basis with the three following strategies:

$$2 \overset{4}{\rightarrow} 4 \overset{8}{\rightarrow} 8 \overset{16}{\rightarrow} 16 \overset{32}{\rightarrow} 32 \overset{64}{\rightarrow} 64 \overset{128}{\rightarrow} 128$$

(Schnorr-Shevchenko's doubling strategy [48])

$$2 \overset{20}{\rightarrow} 20 \overset{21}{\rightarrow} 21 \overset{22}{\rightarrow} 22 \overset{23}{\rightarrow} 23 \overset{24}{\rightarrow} 24 \overset{25}{\rightarrow} \cdots \overset{128}{\rightarrow} 128$$

(Simplest step-by-step in Fig. 6)

$$2 \overset{30}{\rightarrow} 20 \overset{35}{\rightarrow} 25 \overset{39}{\rightarrow} 29 \overset{43}{\rightarrow} 33 \overset{47}{\rightarrow} 37 \overset{48}{\rightarrow} \cdots \overset{128}{\rightarrow} 128$$

(Optimized blocksize strategy in Fig. 8)

In experiment, our simple and optimized strategy takes about 27.1 min and about 11.5 min respectively to achieve BKZ-64 basis after the LLL reduction. On the other hand, Schnorr-Schevchenko's doubling strategy takes about 21 min.

After then, the doubling strategy switches to BKZ-128 and takes about 14 single-core days to process the first one index, while our strategies comfortably continues the execution of progressive BKZ.

Our simulator predicts that it takes about $2^{25.3}$, $2^{25.1}$ and $2^{37.3}$ s to finish BKZ-128 by our simple, optimized, and Schnorr-Schevchenko's doubling strategy, respectively. Our strategy is about 5000 times faster than the doubling strategy.

Interestingly, we find that the computing time of simple blocksize strategy is close to that of optimized strategy in many simulations when the blocksize is larger than about 100. Hence, the simple blocksize strategy would be better than the optimizing blocksize strategy in practice, because the latter needs a heavy precomputing as in Sect. 6.1.

7 Our Implementation and Cost Estimation for Processing Local Blocks

In this section we describe how to derive the estimation of the computing times of Eqs. (17) and (18) of Step 3–10 in Fig. 6. Remark that due to the page limitation, we omit almost of detailed description from the full-version [8].

The total computing time is the sum of times to process local blocks (corresponds to Step 5–8 in Fig. 6):

$$\text{TimeBKZ}(n, \beta^{start} \overset{\beta^{alg}}{\rightarrow} \beta^{goal}) = \sum_{t=1}^{\sharp tours} \sum_{i=1}^{n-1} \left[\text{Time of processing local block } B_i \text{with parameters } (\alpha, p) \right]. \quad (19)$$

We constructed our model of computing time for small dimensional lattices ($dim < 400$) as follows.

$$Time_{Sim\text{-}small}(dim, \beta, A_1, W_1) = \sum_{\beta^{start}}^{\beta^{goal}} \sum_{t=1}^{\sharp tours} \left[A_1 \cdot \beta^2 n^3 + W_1 \cdot \beta \sum_{i=1}^{n-1} \text{ENUMCost}(B_i; \alpha, p) \right] [\text{sec.}]. \quad (20)$$

And for the large dimensions as

$$Time_{Sim\text{-}large}(dim, \beta, A_2, W_2) = \sum_{\beta_{start}}^{\beta_{goal}} \sum_{t=1}^{\sharp tours} \left[A_2 \cdot \frac{n-\beta}{H-\beta} \cdot Hn^2 + W_2 \cdot \beta \sum_{i=1}^{n-1} \text{ENUMCost}(B_i; \alpha, p) \right] [\text{sec.}]. \quad (21)$$

In this section, we conduct the computer experiments with the simple blocksize strategy:

$$2 \overset{20}{\rightarrow} 20 \overset{21}{\rightarrow} 21 \overset{22}{\rightarrow} 22 \overset{23}{\rightarrow} 23 \overset{24}{\rightarrow} 24 \overset{25}{\rightarrow} \cdots$$

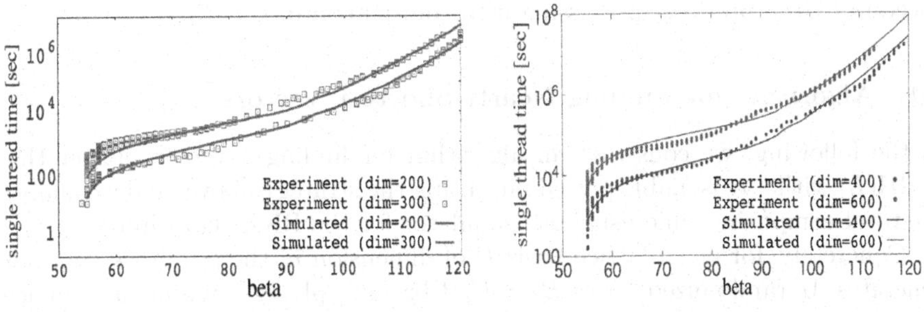

Fig. 9. Result of our parameter fitting for cost estimation. Left Figure: implementation for small dimensional lattices. Right Figure: implementation for large dimensional lattices. In both graphs, experimental results are plotted by small squares and the simulating results are drawn in bold lines.

using a lattice generated by the SVP challenge problem generator, and then we estimate the undefined variables W_1, W_2, A_1 and A_2 by the experimental computing time after BKZ-55, i.e., $\beta^{start} = 55$.

We find the suitable coefficients (A_1, W_1) by using the standard curve fitting method in semi-log scale, which minimize

$$\sum_{dim \in \{200,300\}} \sum_{\beta=55} \left| \log\left(T(dim, \beta, A_1, W_1)\right) - \log\left(Time_{Exp}(dim, \beta)\right) \right|^2,$$

where $T(dim, \beta, A_1, W_1) = Time_{Sim\text{-}large}(dim, \beta, A_1, W_1)$ in the small dimensional situation. For the large dimensional situation, we use the result of $dim \in \{600, 800\}$ to fix A_2 and W_2.

We find suitable coefficients

$$\begin{array}{ll} A_1 = 1.5 \cdot 10^{-10} & \text{and } W_1 = 1.5 \cdot 10^{-8} \\ A_2 = 10^{-6} & \text{and } W_2 = 3.0 \cdot 10^{-8}. \end{array} \tag{22}$$

The fitting results are given in Fig. 9. Using the Eqs. (20) and (21) with the above coefficients (22), we can estimate the computing times of our progressive BKZ algorithm.

8 Pre/Post-Processing the Entire Basis

In this section, we consider an extended strategy that enhances the speed of our progressive BKZ by pre/post-precessing the entire basis.

In pre-processing we first generate a number of randomized bases for input basis. Each basis is then reduced by using the proposed progressive BKZ algorithm. Finally we perform the enumeration algorithm for each reduced basis with some low probability in the post-processing. This strategy is essentially the same as the extreme pruning technique [20]. However, it is important to note that we do not generate a randomized basis inside the progressive BKZ. Our simulator for the proposed progressive BKZ is so precise that we can also estimate the speedup by the pre/post-precessing using our simulator.

8.1 Algorithm for Finding Nearly Shortest Vectors

In the following, we construct an algorithm for finding a vector shorter than $\gamma \cdot \mathrm{GH}(L)$ with a reasonable probability using the strategy above, and we analyze the total computing time using our simulator for the BKZ algorithm.

Concretely, for given lattice basis B of dimension n, the pre-processing part generates M randomized bases $B_i = U_i B$ by multiplying unimodular matrices U_i for $i = 1, \ldots, M$. Next, we apply our progressive BKZ for finding the BKZ-β reduced basis. The cost to obtain the randomized reduced bases is estimated by $M \cdot (\mathrm{TimeRandomize}(n) + \mathrm{TimeBKZ}(n, \beta))$. Here, TimeRandomize includes the cost of generating a random unimodular matrix and matrix multiplication,

Table 4. The cost of solving SVP Challenge using our optimal blocksize strategy

dim	M	$\log_2$(Time [sec.])	Optimal blocksize strategy in Sect. 6
100	5	10.8	$LLL \stackrel{30}{\to} 20 \stackrel{35}{\to} 25 \stackrel{36}{\to} 26 \stackrel{40}{\to} 30 \stackrel{42}{\to} 32 \stackrel{44}{\to} 34 \stackrel{48}{\to} 38 \stackrel{49}{\to} 39 \stackrel{50}{\to} 40 \stackrel{51}{\to} 41 \stackrel{52}{\to} 42 \stackrel{54}{\to} 44 \stackrel{58}{\to} 48 \stackrel{60}{\to} 50 \stackrel{62}{\to} 52 \stackrel{64}{\to} 54 \stackrel{67}{\to} 57 \stackrel{71}{\to} 70 \stackrel{81}{\to} 79$
120	35	20.3	$LLL \stackrel{30}{\to} 20 \stackrel{35}{\to} 24 \stackrel{36}{\to} 26 \stackrel{40}{\to} 30 \stackrel{42}{\to} 32 \stackrel{44}{\to} 34 \stackrel{48}{\to} 38 \stackrel{49}{\to} 39 \stackrel{50}{\to} 40 \stackrel{51}{\to} 41 \stackrel{52}{\to} 42 \stackrel{54}{\to} 44 \stackrel{58}{\to} 48 \stackrel{60}{\to} 50 \stackrel{62}{\to} 52 \stackrel{64}{\to} 54 \stackrel{66}{\to} 56 \stackrel{68}{\to} 58 \stackrel{69}{\to} 59 \stackrel{72}{\to} 71 \stackrel{84}{\to} 84 \stackrel{93}{\to} 93 \stackrel{99}{\to} 99 \stackrel{101}{\to} 100 \stackrel{105}{\to} 105$
140	164	30.3	$LLL \stackrel{30}{\to} 20 \stackrel{35}{\to} 24 \stackrel{36}{\to} 26 \stackrel{40}{\to} 30 \stackrel{42}{\to} 32 \stackrel{44}{\to} 34 \stackrel{48}{\to} 38 \stackrel{49}{\to} 39 \stackrel{50}{\to} 40 \stackrel{52}{\to} 42 \stackrel{54}{\to} 44 \stackrel{58}{\to} 48 \stackrel{62}{\to} 52 \stackrel{64}{\to} 54 \stackrel{66}{\to} 56 \stackrel{68}{\to} 58 \stackrel{70}{\to} 60 \stackrel{71}{\to} 61 \stackrel{72}{\to} 70 \stackrel{80}{\to} 79 \stackrel{87}{\to} 87 \stackrel{93}{\to} 93 \stackrel{99}{\to} 99 \stackrel{103}{\to} 103 \stackrel{107}{\to} 107 \stackrel{111}{\to} 111 \stackrel{115}{\to} 115 \stackrel{118}{\to} 118 \stackrel{123}{\to} 122$
160	49	41.2	$LLL \stackrel{30}{\to} 20 \stackrel{35}{\to} 24 \stackrel{36}{\to} 26 \stackrel{40}{\to} 30 \stackrel{42}{\to} 32 \stackrel{44}{\to} 34 \stackrel{48}{\to} 38 \stackrel{50}{\to} 40 \stackrel{52}{\to} 42 \stackrel{54}{\to} 44 \stackrel{58}{\to} 48 \stackrel{62}{\to} 52 \stackrel{64}{\to} 54 \stackrel{66}{\to} 56 \stackrel{68}{\to} 58 \stackrel{70}{\to} 60 \stackrel{72}{\to} 62 \stackrel{74}{\to} 70 \stackrel{76}{\to} 72 \stackrel{82}{\to} 80 \stackrel{87}{\to} 86 \stackrel{92}{\to} 92 \stackrel{99}{\to} 99 \stackrel{100}{\to} 100 \stackrel{103}{\to} 103 \stackrel{104}{\to} 104 \stackrel{107}{\to} 107 \stackrel{108}{\to} 108 \stackrel{110}{\to} 110 \stackrel{112}{\to} 112 \stackrel{114}{\to} 114 \stackrel{116}{\to} 116 \stackrel{119}{\to} 119 \stackrel{121}{\to} 121 \stackrel{125}{\to} 125 \stackrel{129}{\to} 129 \stackrel{135}{\to} 134 \stackrel{138}{\to} 136 \stackrel{141}{\to} 139 \stackrel{144}{\to} 143$
180	2148	52.4	$LLL \stackrel{30}{\to} 20 \stackrel{35}{\to} 24 \stackrel{36}{\to} 26 \stackrel{40}{\to} 30 \stackrel{44}{\to} 34 \stackrel{48}{\to} 38 \stackrel{50}{\to} 40 \stackrel{52}{\to} 42 \stackrel{54}{\to} 44 \stackrel{58}{\to} 48 \stackrel{62}{\to} 52 \stackrel{64}{\to} 54 \stackrel{66}{\to} 56 \stackrel{68}{\to} 58 \stackrel{70}{\to} 59 \stackrel{72}{\to} 70 \stackrel{80}{\to} 76 \stackrel{85}{\to} 82 \stackrel{87}{\to} 86 \stackrel{93}{\to} 92 \stackrel{99}{\to} 98 \stackrel{103}{\to} 103 \stackrel{104}{\to} 104 \stackrel{107}{\to} 107 \stackrel{108}{\to} 108 \stackrel{110}{\to} 110 \stackrel{113}{\to} 113 \stackrel{114}{\to} 114 \stackrel{116}{\to} 116 \stackrel{118}{\to} 118 \stackrel{120}{\to} 120 \stackrel{122}{\to} 122 \stackrel{125}{\to} 125 \stackrel{127}{\to} 127 \stackrel{130}{\to} 130 \stackrel{133}{\to} 133 \stackrel{136}{\to} 136 \stackrel{139}{\to} 138 \stackrel{141}{\to} 141 \stackrel{144}{\to} 144 \stackrel{148}{\to} 147 \stackrel{149}{\to} 148 \stackrel{152}{\to} 149 \stackrel{154}{\to} 154 \stackrel{155}{\to} 155 \stackrel{157}{\to} 157 \stackrel{160}{\to} 158$

which is negligibly smaller than TimeBKZ in general. Thus we assume the computational cost for lattice reduction is $M \cdot \mathrm{TimeBKZ}(n, \beta)$.

Finally, in the post-processing part, we execute the standard enumeration algorithm with the searching radius parameter $\alpha = \gamma$ and probability parameter $p = 2 \cdot \gamma^{-n}/M$. As with the similar argument in Sect. 4.1, there exist about $\gamma^n/2$ short vector pairs in $\mathrm{Ball}_n(\gamma \cdot \mathrm{GH}(L))$. Therefore, the probability that one enumeration finds the desired vector is about $(\gamma^n/2) \cdot (2 \cdot \gamma^{-n}/M) = 1/M$ and the total probability of success is $1 - (1 - 1/M)^M \approx 0.632$.

Consequently, the total computing cost in our model is

$$M \cdot \left(\mathrm{TimeBKZ}(n, \beta) + \frac{\mathrm{ENUMCost}(B; \gamma, p = 2 \cdot \gamma^{-n}/M)}{6 \cdot 10^7} \right) \text{ [sec.]}, \qquad (23)$$

where $\mathrm{TimeBKZ}(n, \beta)$ and $\mathrm{ENUMCost}(B; \gamma, p)$ are defined by Sect. 6.1 and Sect. 2.3, respectively. We can optimize this total cost by finding the minimum of formula (23) over parameter (β, M). Here, note that the constant $6 \cdot 10^7$ comes from our best benchmarking record of lattice enumeration. In Table 4, we provide the detailed simulating result with setting $\gamma = 1.05$ to analyze the hardness of the Darmstadt SVP Challenge in several dimensions. A comparison with previous works are given in Sect. 9 (See the line **C** in Fig. 10).

8.2 Lower Bound of the Cost by an Idealized Algorithm

Here we discuss the lower bound of the total computing cost of the proposed progressive BKZ algorithm (or other reduction algorithm) with the pre/post-processing.

The total cost is estimated by the sum of the computational time for the randomization, the progressive BKZ algorithm, and the enumeration algorithm by the following extremely idealized situations. Note that we believe that they are beyond the most powerful cryptanalysis which we can achieve in the future, and thus we say that this is the lower bound in our model.

(a) The cost for the randomization becomes negligibly small. The algorithm for randomizing the basis would not only be the method of multiplying random unimodular bases, and we could find an ideal randomization at a negligibly small cost. Thus, TimeRandomize(n) = 0.

(b) The cost for the progressive BKZ algorithm does not become lower than that of computing the Gram-Schmidt lengths. Even though the progressive BKZ algorithm ideally improved, we always need the Gram-Schmidt basis computation used for the enumeration algorithm or the LLL algorithm. The computation of the Gram-Schmidt basis (even though the computation is performed in an approximation using floating-point operations with a sufficient precision) includes $\Theta(n^3)$ floating point arithmetic operations via the Cholesky factorization algorithm (See, for example [38, Chapter 5]). A modern CPU can perform a floating point operation in one clock cycle, and it can work at about 4.0 GHz. Thus, we assume that the lower bound of the time in seconds is $(4.0 \cdot 10^9)^{-1} \cdot n^3$.

(c) The reduced basis obtained by the progressive BKZ (or other reduction algorithm) becomes ideally reduced. We define *the simulated γ-approximate HKZ basis $B_{\gamma\text{-}HKZ}$* by a basis satisfying

$$||\mathbf{b}_i^*|| = \tau_{n-i+1}\mathrm{GH}(L_{[i:n]}) \text{ for } i = 2, \ldots, n \text{ and } ||\mathbf{b}_1|| = \gamma \cdot \mathrm{GH}(L).$$

For any fixed γ and p, we assume this basis minimizes the cost for enumeration over any basis satisfying $||\mathbf{b}_1|| \geq \gamma \cdot \mathrm{GH}(L)$.

Therefore, the lower bound of the total cost of the idealized algorithm in seconds is given by

$$\min_{M \in \mathbb{N}} M \cdot \left((4.0 \cdot 10^9)^{-1} \cdot n^3 + \frac{ENUMCost(B_{\gamma\text{-}HKZ}; \alpha, p/M)}{6 \cdot 10^7} \right). \tag{24}$$

Setting $\gamma = 1.05$, we analyze the lower bound cost to enter the SVP Challenge. (See the line **D** in Fig. 10).

9 Simulation Results for SVP Challenge and Comparison

In this section, we give our simulation results using our proposed progressive BKZ algorithm together with the pre/post-processing strategy in Sect. 8.1 for solving the Darmstadt SVP Challenge [50], which tries to find a vector shorter than $1.05 \cdot \mathrm{GH}(L)$ in the random lattice L of dimension n.

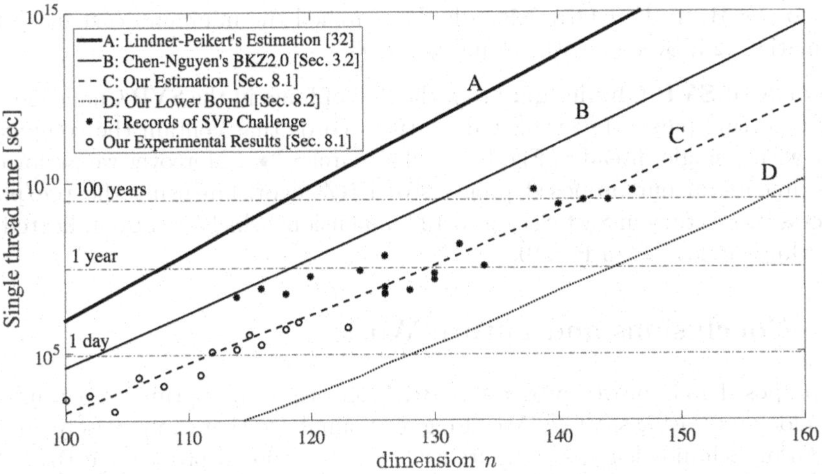

Fig. 10. Comparing cost in seconds. **A**: Lindner-Peikert estimation, **B**: Chen-Nguyen's BKZ 2.0 simulation, **C**: Simulating estimation of our randomized BKZ-then-ENUM algorithm, **D**: Lower bound in the randomized BKZ-then-ENUM strategy. Records in the SVP Challenge are indicated by the black circles "•", and our experimental results are indicated by the white circles "o".

We also simulate the cost estimation of Lindner and Peikert [32] and that of Chen and Nguyen [13] in the same model. The summery of our simulation results and the latest records published in the SVP Challenge are given in Fig. 10. The outlines of our estimations **A** to **D** in Fig. 10 are given below.

From our simulation, the proposed progressive BKZ algorithm is about 50 times faster than BKZ 2.0 and about 100 times slower than the idealized algorithm that achieves the lower bound in our model of Sect. 8.2.

A: Lindner-Peikert's Estimation [32]**:** From the experiments using the BKZ implementation in the NTL library [49], they estimated that the BKZ algorithm can find a short vector of length $\delta^n \det(L)^{1/n}$ in $2^{1.8/\log_2(\delta)-110}$ [sec.] in the n-dimensional lattice. The computing time of Lindner-Peikert's model becomes

$$Time_{\mathrm{LP}} = 2^{1.8/\log_2(\delta)-110} \text{ with } \delta = 1.05^{1/n} \cdot V_n(1)^{-1/n^2},$$

because this δ attains $1.05 \cdot \mathrm{GH}(L) = \delta^n \det(L)^{1/n}$.

B: Chen-Nguyen's BKZ 2.0 [13,14]**:** We estimated the cost of BKZ 2.0 using the simulator in Sect. 3.2. Following the original paper [13], we assume that a blocksize is fixed and the estimation is the minimum of (4) over all possible pairs of the blocksize β and the number t of tours. Again we convert the number of nodes into the single-threaded time, we divide the number by $2 \cdot 10^7$.

C: Our Estimation: We searched the minimum cost using the estimation (23) over M and β with setting $\gamma = 1.05$.

D: Lower Bound in Our Model: We searched the minimum cost using the estimation (24) over M with setting $\gamma = 1.05$.

Records of SVP Challenge: From the hall of fame in the SVP Challenge [50] and reporting paper [18], we listed up the records that contain the computing time with a single thread in Fig. 10, as black circles "•". Moreover we performed experiments on our proposed progressive BKZ algorithm using the pre/post-processing strategy in Sect. 8.1 up to 123 dimensions which are also indicated by the white circles "○" in Fig. 10.

10 Conclusions and Future Work

We proposed an improved progressive BKZ algorithm with optimized parameters and block-increasing strategy. We also gave a simulator that can precisely predict the Gram-Schmidt lengths computed using the proposed progressive BKZ. We also presented the efficient implementation of the enumeration algorithm and LLL algorithm, and the total cost of the proposed progressive BKZ algorithm was precisely evaluated by the sharp simulator.

Moreover, we showed a comparison with other algorithms by simulating the cost of solving the instances from the Darmstadt SVP Challenge. Our progressive BKZ algorithm is about 50 times faster than the BKZ 2.0 proposed by Chen and Nguyen for solving the SVP Challenges up to 160 dimensions. Finally, we discussed a computational lower bound of the proposed progressive BKZ algorithm under certain ideal assumptions. These simulation results contribute to the estimation of the secure parameter sizes used in lattice based cryptography.

We outline some future works: (1) constructing a BKZ simulator without using our ENUMCost, (2) adopting our simulator with other strategies such as BKZ-then-Sieve strategy for computing a short vector more efficiently, and (3) estimating the secure key length of lattice-based cryptosystems using the lower bound of the proposed progressive BKZ.

References

1. Aggarwal, D., Dadush, D., Regev, O., Stephens-Davidowitz, N.: Solving the shortest vector problem in 2^n time using discrete Gaussian sampling: extended abstract. In: STOC 2015, pp. 733–742 (2015)
2. Ajtai, M.: The shortest vector problem in L_2 is NP-hard for randomized reductions. In: STOC, pp. 10–19 (1998)
3. Ajtai, M.: The worst-case behavior of Schnorr's algorithm approximating the shortest nonzero vector in a lattice. In: STOC, pp. 396–406 (2003)
4. Albrecht, M.R., Fitzpatrick, R., Göpfert, F.: On the efficacy of solving LWE by reduction to Unique-SVP. In: Lee, H.-S., Han, D.-G. (eds.) ICISC 2013. LNCS, vol. 8565, pp. 293–310. Springer, Heidelberg (2014)
5. Albrecht, M.R., Player, R., Scott, S.: On the concrete hardness of learning with errors. J. Math. Cryptology **9**(3), 169–203 (2015)
6. Aono, Y.: A faster method for computing Gama-Nguyen-Regev's extreme pruning coefficients (2014). arXiv:1406.0342

7. Aono, Y., Boyen, X., Phong, L.T., Wang, L.: Key-private proxy re-encryption under LWE. In: Paul, G., Vaudenay, S. (eds.) INDOCRYPT 2013. LNCS, vol. 8250, pp. 1–18. Springer, Heidelberg (2013)
8. Aono, Y., Wang, Y., Hayashi, T., Takagi, T.: Improved progressive BKZ algorithms and their precise cost estimation by sharp simulator. In: IACR Cryptology ePrint Archive 2016: 146 (2016)
9. Bai, S., Galbraith, S.D.: Lattice decoding attacks on binary LWE. In: Susilo, W., Mu, Y. (eds.) ACISP 2014. LNCS, vol. 8544, pp. 322–337. Springer, Heidelberg (2014)
10. Brakerski, Z., Vaikuntanathan, V.: Fully homomorphic encryption from Ring-LWE and security for key dependent messages. In: Rogaway, P. (ed.) CRYPTO 2011. LNCS, vol. 6841, pp. 505–524. Springer, Heidelberg (2011)
11. Buchmann, J., Ludwig, C.: Practical lattice basis sampling reduction. In: Hess, F., Pauli, S., Pohst, M. (eds.) ANTS 2006. LNCS, vol. 4076, pp. 222–237. Springer, Heidelberg (2006)
12. Chen, Y.: Réduction de réseau et sécurité concrète du chiffrement complètement homomorphe, Doctoral dissertation (2013)
13. Chen, Y., Nguyen, P.Q.: BKZ 2.0: better lattice security estimates. In: Lee, D.H., Wang, X. (eds.) ASIACRYPT 2011. LNCS, vol. 7073, pp. 1–20. Springer, Heidelberg (2011)
14. Chen, Y., Nguyen, P.Q.: BKZ 2.0: better lattice security estimates, the full version. http://www.di.ens.fr/~ychen/research/Full_BKZ.pdf
15. Coppersmith, D., Shamir, A.: Lattice attacks on NTRU. In: Fumy, W. (ed.) EUROCRYPT 1997. LNCS, vol. 1233, pp. 52–61. Springer, Heidelberg (1997)
16. Fincke, U., Pohst, M.: Improved methods for calculating vectors of short length in a lattice, including a complexity analysis. Math. Comp. **44**, 463–471 (1985)
17. Fischlin, R., Seifert, J.-P.: Tensor-based trapdoors for CVP and their application to public key cryptography. In: Walker, M. (ed.) Cryptography and Coding 1999. LNCS, vol. 1746, pp. 244–257. Springer, Heidelberg (1999)
18. Fukase, M., Kashiwabara, K.: An accelerated algorithm for solving SVP based on statistical analysis. J. Inf. Process. **23**(1), 67–80 (2015)
19. Gama, N., Nguyen, P.Q.: Predicting lattice reduction. In: Smart, N.P. (ed.) EUROCRYPT 2008. LNCS, vol. 4965, pp. 31–51. Springer, Heidelberg (2008)
20. Gama, N., Nguyen, P.Q., Regev, O.: Lattice enumeration using extreme pruning. In: Gilbert, H. (ed.) EUROCRYPT 2010. LNCS, vol. 6110, pp. 257–278. Springer, Heidelberg (2010)
21. Garg, S., Gentry, C., Halevi, S.: Candidate multilinear maps from ideal lattices. In: Johansson, T., Nguyen, P.Q. (eds.) EUROCRYPT 2013. LNCS, vol. 7881, pp. 1–17. Springer, Heidelberg (2013)
22. Gentry, C.: Fully homomorphic encryption using ideal lattices. In: STOC 2009, pp. 169–178 (2009)
23. Hanrot, G., Pujol, X., Stehlé, D.: Analyzing blockwise lattice algorithms using dynamical systems. In: Rogaway, P. (ed.) CRYPTO 2011. LNCS, vol. 6841, pp. 447–464. Springer, Heidelberg (2011)
24. Hanrot, G., Stehlé, D.: Improved Analysis of Kannan's shortest lattice vector algorithm. In: Menezes, A. (ed.) CRYPTO 2007. LNCS, vol. 4622, pp. 170–186. Springer, Heidelberg (2007)
25. Haque, M., Rahman, M.O., Pieprzyk, J.: Analysing progressive-BKZ lattice reduction algorithm. In: NCICIT 2013, pp. 73–80 (2013)
26. Hoffstein, J., Pipher, J., Silverman, J.H.: An Introduction to Mathematical Cryptography. Springer-Verlag New York, New York (2008)

27. Ishiguro, T., Kiyomoto, S., Miyake, Y., Takagi, T.: Parallel Gauss Sieve algorithm: solving the SVP challenge over a 128-dimensional ideal lattice. In: Krawczyk, H. (ed.) PKC 2014. LNCS, vol. 8383, pp. 411–428. Springer, Heidelberg (2014)
28. Kannan, R.: Improved algorithms for integer programming and related lattice problems. In: STOC, pp. 193–206 (1983)
29. Korkine, A., Zolotareff, G.: Sur les formes quadratiques. Math. Ann. **6**(3), 366–389 (1873)
30. Lenstra, A.K., Lenstra Jr., H.W., Lovász, L.: Factoring polynomials with rational coefficients. Math. Ann. **261**(4), 515–534 (1982)
31. Lepoint, T., Naehrig, M.: A comparison of the homomorphic encryption schemes FV and YASHE. In: Pointcheval, D., Vergnaud, D. (eds.) AFRICACRYPT 2014. LNCS, vol. 8469, pp. 318–335. Springer, Heidelberg (2014)
32. Lindner, R., Peikert, C.: Better key sizes (and attacks) for LWE-based encryption. In: Kiayias, A. (ed.) CT-RSA 2011. LNCS, vol. 6558, pp. 319–339. Springer, Heidelberg (2011)
33. Liu, M., Nguyen, P.Q.: Solving BDD by enumeration: an update. In: Dawson, E. (ed.) CT-RSA 2013. LNCS, vol. 7779, pp. 293–309. Springer, Heidelberg (2013)
34. Micciancio, D.: The shortest vector problem is NP-hard to approximate to within some constant. In: FOCS, pp. 92–98 (1998)
35. Nguyên, P.Q.: Cryptanalysis of the Goldreich-Goldwasser-Halevi cryptosystem from Crypto'97. In: Wiener, M. (ed.) CRYPTO 1999. LNCS, vol. 1666, pp. 288–304. Springer, Heidelberg (1999)
36. Nguyên, P.Q., Regev, O.: Learning a parallelepiped: cryptanalysis of GGH and NTRU signatures. In: Vaudenay, S. (ed.) EUROCRYPT 2006. LNCS, vol. 4004, pp. 271–288. Springer, Heidelberg (2006)
37. Nguyên, P.Q., Stehlé, D.: LLL on the average. In: Hess, F., Pauli, S., Pohst, M. (eds.) ANTS 2006. LNCS, vol. 4076, pp. 238–256. Springer, Heidelberg (2006)
38. Nguyen, P.Q., Vallée, B.: The LLL Algorithm: Survey and Applications. Springer-Verlag, Heidelberg (2009)
39. Plantard, T., Schneider, M.: Creating a challenge for ideal lattices. In: IACR Cryptology ePrint Archive 2013: 039 (2013)
40. Plantard, T., Susilo, W., Zhang, Z.: Adaptive precision floating point LLL. In: Boyd, C., Simpson, L. (eds.) ACISP 2013. LNCS, vol. 7959, pp. 104–117. Springer, Heidelberg (2013)
41. Regev, O.: On lattices, learning with errors, random linear codes, and cryptography. J. ACM, **56**(6), Article no. 34 (2009)
42. Rogers, C.A.: The number of lattice points in a set. Proc. London Math. Soc. **3**(6), 305–320 (1956)
43. Schnorr, C.P.: A hierarchy of polynomial time lattice basis reduction algorithms. Theor. Comput. Sci. **53**(2–3), 201–224 (1987)
44. Schnorr, C.P.: Lattice reduction by random sampling and birthday methods. In: Alt, H., Habib, M. (eds.) STACS 2003. LNCS, vol. 2607, pp. 145–156. Springer, Heidelberg (2003)
45. Schnorr, C.P.: Accelerated and Improved Slide- and LLL-Reduction. ECCC TR11-050 (2011)
46. Schnorr, C.P., Euchner, M.: Lattice basis reduction: improved practical algorithms and solving subset sum problems. Math. Program. **66**(1–3), 181–199 (1994)
47. Schnorr, C.-P., Hörner, H.H.: Attacking the Chor-Rivest cryptosystem by improved lattice reduction. In: Guillou, L.C., Quisquater, J.-J. (eds.) EUROCRYPT 1995. LNCS, vol. 921, pp. 1–12. Springer, Heidelberg (1995)

48. Schnorr, C.P., Shevchenko, T.: Solving subset sum problems of density close to 1 by "randomized" BKZ-reduction. In: IACR Cryptology ePrint Archive 2012: 620 (2012)
49. Shoup, V.: NTL: a library for ng number theory. http://www.shoup.net/ntl/
50. TU Darmstadt Lattice Challenge. http://www.latticechallenge.org/
51. van de Pol, J., Smart, N.P.: Estimating key sizes for high dimensional lattice-based systems. In: Stam, M. (ed.) IMACC 2013. LNCS, vol. 8308, pp. 290–303. Springer, Heidelberg (2013)

Practical, Predictable Lattice Basis Reduction

Daniele Micciancio and Michael Walter(✉)

University of California, San Diego, USA
{daniele,miwalter}@eng.ucsd.edu

Abstract. Lattice reduction algorithms are notoriously hard to predict, both in terms of running time and output quality, which poses a major problem for cryptanalysis. While easy to analyze algorithms with good worst-case behavior exist, previous experimental evidence suggests that they are outperformed in practice by algorithms whose behavior is still not well understood, despite more than 30 years of intensive research. This has lead to a situation where a rather complex simulation procedure seems to be the most common way to predict the result of their application to an instance. In this work we present new algorithmic ideas towards bridging this gap between theory and practice. We report on an extensive experimental study of several lattice reduction algorithms, both novel and from the literature, that shows that theoretical algorithms are in fact surprisingly practical and competitive. In light of our results we come to the conclusion that in order to predict lattice reduction, simulation is superfluous and can be replaced by a closed formula using weaker assumptions.

One key technique to achieving this goal is a novel algorithm to solve the Shortest Vector Problem (SVP) in the dual without computing the dual basis. Our algorithm enjoys the same practical efficiency as the corresponding primal algorithm and can be easily added to an existing implementation of it.

1 Introduction

Lattice basis reduction is a fundamental tool in cryptanalysis and it has been used to successfully attack many cryptosystems, based on both lattices, and other mathematical problems. (See for example [9,23,39,44–47,61,62,66].) The success of lattice techniques in cryptanalysis is due to a large extent to the fact that reduction algorithms perform much better in practice than predicted by their theoretical worst-case analysis. Basis reduction algorithms have been investigated in many papers over the past 30 years [3,6,8,10,12–16,18,20,21, 26,28,32,36,40–42,45,48,50,51,54–56,58–60,63,65,67–69], but the gap between theoretical analysis and practical performance is still largely unexplained. This gap hinders our ability to estimate the security of lattice based cryptographic

Research supported in part by the DARPA SafeWare program and NSF grant CNS-1117936. Opinions, findings and conclusions or recommendations expressed in this material are those of the author(s) and do not necessarily reflect the views of DARPA or NSF.

M. Fischlin and J.-S. Coron (Eds.): EUROCRYPT 2016, Part I, LNCS 9665, pp. 820–849, 2016.
DOI: 10.1007/978-3-662-49890-3_31

functions, and it has been widely recognized as one of the main obstacles to the use of lattice cryptography in practice. In this work, we make some modest progress towards this challenging goal.

By and large, the current state of the art in lattice basis reduction (in theory and in practice) is represented by two algorithms:

- the eminently practical Block-Korkine-Zolotarev (BKZ) algorithm of Schnorr and Euchner [54,60], in its modern BKZ 2.0 incarnation [8] incorporating pruning, recursive preprocessing and early termination strategies [14,18],
- the Slide reduction algorithm of Gama and Nguyen [15], an elegant generalization of LLL [27,40] which provably approximates short lattice vectors within factors related to Mordell's inequality.

Both algorithms make use of a Shortest Vector Problem (SVP) oracle for lower dimensional lattices, and are parameterized by a bound k (called the "block size") on the dimension of these lattices. The Slide reduction algorithm has many attractive features: it makes only a polynomial number of calls to the SVP oracle, all SVP calls are to projected sub-lattices in exactly the same dimension k, and it achieves the best known worst-case upper bound on the length of its shortest output vector: $\gamma_k^{(n-1)/(2(k-1))} \det(L)^{1/n}$, where $\gamma_k = \Theta(k)$ is the Hermite constant, and $\det(L)$ is the determinant of the lattice. Unfortunately, it has been reported [15,16] that in experiments the Slide reduction algorithm is outperformed by BKZ, which produces much shorter vectors for comparable block size. In fact, [15] remarks that even BKZ with block size $k = 20$ produces better reduced bases than Slide reduction with block size $k = 50$. As a consequence, the Slide reduction algorithm is never used in practice, and it has not been implemented and experimentally tested beyond the brief claims made in the initial work [15,16].[1]

On the other hand, while surprisingly practical in experimental evaluations, the BKZ algorithm has its own shortcomings too. In its original form, BKZ is not even known to terminate after a polynomial number of calls to the SVP oracle, and its observed running time has been reported [16] to grow superpolynomially in the lattice dimension, even when the block size is fixed to some relatively small value $k \approx 30$. Even upon termination, the best provable bounds on the output quality of BKZ are worse than Slide reduction by at least a polynomial factor [15].[2] In practice, in order to address running time issues, BKZ is often employed with an "early termination" strategy [8] that tries to determine heuristically when no more progress is expected from running the algorithm. Theoretical bounds on the quality of the output after a polynomial number of iterations

[1] While we are referencing two separate works, both refer to the same experimental study.

[2] We remark that polynomial approximation factors, while being asymptotically insignificant, can make a substantial difference in practice, as lattice-based cryptography relies on the hardness of approximating lattice problems within factors that are super-linear in the lattice dimension. In fact, much effort has been put on minimizing such factors in the design of cryptographic constructions [1,2,31,33,35,49,52,53].

have been proved [18], but they are worse than Slide reduction by even larger polynomial factors. Another fact that complicates the analysis (both in theory and in practice) of the output quality of BKZ is the fact that the algorithm makes SVP calls in all dimensions up to the block size. In theory, this results in a formula that depends on all worst-case (Hermite) constants γ_i for $i \leq k$. In practice, the output quality and running time is evaluated by a simulator [8] that initially attempts to numerically estimate the performance of the SVP oracle on random lattices in all possible dimensions up to k.

Our Contribution. We introduce new algorithmic techniques that can be used to design improved lattice basis reduction algorithms, analyze their theoretical performance, implement them, and report on their practical behavior through a detailed set of experiments with block size as high as 75, and several data points per dimension for (still preliminary, but already meaningful) statistical estimation.

One of our main findings is that the Slide reduction algorithm is much more practical than originally thought, and as the dimension increases, it performs almost as well as BKZ, while at the same time, offering a simple closed-formula to evaluate its output quality. This provides a simple and effective method to evaluate the impact of lattice basis reduction attacks on lattice cryptography, without the need to run simulators or other computer programs [8,68]. Key to our findings, is a new procedure to enumerate shortest lattice vectors in *dual* lattices, without the need to explicitly compute a dual basis. Interestingly, our dual enumeration procedure is almost identical (syntactically) to the standard enumeration procedure to find short vectors in a (primal) lattice, and, as expected, it is just as efficient in practice. Using our new procedure, we are able to conduct experiments using Slide reduction with significantly larger block size than previously reported, and observe that the gap between theoretical (more predicable) algorithms and practical heuristics gets pretty narrow already for moderate block size and dimension.

For small block sizes (say, up to 40), there is still a substantial gap between the output quality of Slide reduction and BKZ in practice. For this setting, we design a new variant of BKZ, based on lattice duality and a new notion of block reduced basis. Our new DBKZ algorithm can be efficiently implemented using our dual enumeration procedure, achieving running times comparable to BKZ, and matching its experimental output quality for small block size almost exactly. At the same time, our algorithm has various advantages over BKZ, that make it a better target for theoretical analysis: it only makes calls to an SVP oracle in fixed dimension k, and it is self dual, in the sense that it performs essentially the same operations when run on a basis or its dual. The fact that all SVP calls on projected sublattices are in the same fixed dimension k has several important implications. First, it results in a simpler bound on the length of the shortest output vector, which can be expressed as a function of just γ_k. More importantly, this allows to get a practical estimate on the output quality simply by replacing γ_k with the value predicted by the Gaussian Heuristic $GH(k)$, commonly used in lattice cryptanalysis. We remark that the $GH(k)$ formula has been validated for moderately large values of k, where it gives fairly accurate estimates on the shortest vector length in k-dimensional sublattices. However, early work on predicting lattice reduction [16] has also shown that for small k (say, up to $k \leq 25$), BKZ sublattices do not follow the Gaussian Heuristic. As a result, while the BKZ 2.0 simulator of [8] makes extensive use of $GH(k)$ for large values of k, it also needs to resort to cumbersome experimental estimations for predicting the result of SVP calls in dimension lower than k. By making only SVP calls on k-dimensional sublattices, our algorithm obviates the need for any such experimental estimates, and allows to predict the output quality (under the same, or weaker heuristic assumptions than the BKZ 2.0 simulator) just using the GH(k) formula. We stress that this is not only true for the length of the shortest vector found by our algorithm, but one can estimate many more properties of the resulting basis. This is important in many cryptanalytic settings, where lattice reduction is used as a preprocessing for other attacks. In particular, using the Gaussian Heuristic we are able to show that a large part of the basis output by our algorithm can be expected to follow the Geometric Series Assumption [57], an assumption often made about the output of lattice reduction, but so far never proven. (See Sect. 5 for details.) One last potential advantage of only making SVP calls in fixed dimension k (and, consequently, the ability to use the Gaussian Heuristic for all of them) is that it opens up the possibility of even more accurate stochastic simulations (or analytic solutions) where the GH(k) deterministic formula is replaced by a probability distribution (following the length of the shortest vector in a random k-dimensional lattice). We leave the investigation of such a stochastic simulator to future work.

Technical Ideas. Enumeration algorithms (as typically used within block basis reduction) find short vectors in a lattice by examining all possible coordinates $x_1, \ldots, x_n$ of candidate short lattice vectors $\sum_i \mathbf{b}_i \cdot x_i$ with respect to the given lattice basis, and using the length of the projected lattice vector to prune the search. Our dual lattice enumeration algorithm works similarly, but without explicitly computing a basis for the dual lattice. The key technical idea is that one can enumerate over the scalar products $y_i = \langle \mathbf{b}_i, \mathbf{v} \rangle$ of the candidate short dual vectors $\mathbf{v}$ and the primal basis vectors $\mathbf{b}_i$.[3] Perhaps surprisingly, one can also compute the length of the projections of the dual lattice vector $\mathbf{v}$ (required to prune the enumeration tree), without explicitly computing $\mathbf{v}$ or a dual basis. The simplicity of the algorithm is best illustrated just by looking at the pseudo code, and comparing it side-to-side to the pseudo code of standard (primal) lattice enumeration. (See Algorithms 2 and 3 in Sect. 7.) The two programs are almost identical, leading to a dual enumeration procedure that is just as efficient as primal enumeration, and allowing the application of all standard optimizations (e.g., all various forms of pruning) that have been developed for enumerating in primal lattices.

[3] By definition of dual lattice, all these products y_i are integers, and, in fact, they are the coordinates of $\mathbf{v}$ with respect to the standard *dual basis* of $\mathbf{b}_1, \ldots, \mathbf{b}_n$.

On the basis reduction front, our DBKZ algorithm is based on a new notion of block-reduced basis. Just as for BKZ, DBKZ-reduction is best described as a recursive definition. In fact, the recursive condition is essentially the same for both algorithms: given a basis $\mathbf{B}$, if $\mathbf{b}$ is a shortest vector in the sublattice generated by the first k basis vectors $\mathbf{B}_{[1,k]}$, we require the projection of $\mathbf{B}$ orthogonal to $\mathbf{b}$ to satisfy the recursive reduction property. The difference between BKZ and DBKZ is that, while BKZ requires $\mathbf{B}_{[1,k]}$ to start with a shortest lattice vector $\mathbf{b} = \mathbf{b}_1$, in DBKZ we require it to *end* with a shortest *dual vector*.[4] This simple twist in the definition of reduced basis leads to a much simpler bound on the length of $\mathbf{b}$, improving the best known bound for BKZ reduction, and matching the theoretical quality of Slide reduction.

Experiments. To the best of our knowledge, we provide the first experimental study of lattice reduction with large block size parameter beyond BKZ. Even for BKZ we improve on the currently only study involving large block sizes [8] by collecting multiple data points per block size parameter. This allows us to apply standard statistical methods to try to get a sense of the main statistical parameters of the output distribution. Clearly, learning more about the output distribution of these algorithms is highly desirable for cryptanalysis, as an adversary is drawing samples from that distribution and will utilize the most convenient sample, rather than a sample close to the average.

Finally, in contrast to previous experimental work [8,16], we contribute to the community by making our code[5] and data[6] publicly available. To the best of our knowledge, this includes the first publicly available implementation of dual SVP reduction and Slide reduction. At the time of publication of this work, a modified version of our implementation of dual SVP reduction has been integrated into the main branch of `fpLLL` [4]. We hope that this will spur more research into the predictability of lattice reduction algorithms.

2 Preliminaries

Notation. Numbers and reals are denoted by lower case letters. For $n \in \mathbb{Z}_+$ we denote the set $\{0, \ldots, n\}$ by $[n]$. For vectors we use bold lower case letters and the i-th entry of a vector $\mathbf{v}$ is denoted by v_i. Let $\langle \mathbf{v}, \mathbf{w} \rangle = \sum_i v_i \cdot w_i$ be the scalar product of two vectors. If $p \geq 1$ we define the p norm of a vector $\mathbf{v}$ to be $\|\mathbf{v}\|_p = \left(\sum |v_i|^p\right)^{1/p}$. We will only be concerned with the norms given by $p = 1$, 2, and ∞. Whenever we omit the subscript p, we mean the standard Euclidean norm, i.e. $p = 2$. We define the projection of a vector $\mathbf{b}$ orthogonal to a vector $\mathbf{v}$ as $\pi_{\mathbf{v}}(\mathbf{b}) = \mathbf{b} - \frac{\langle \mathbf{b}, \mathbf{v} \rangle}{\|\mathbf{v}\|^2}\mathbf{v}$. Matrices are denoted by bold upper case letters. The i-th column of a matrix $\mathbf{B}$ is denoted by $\mathbf{b}_i$. Furthermore, we denote the submatrix comprising the columns from the i-th to the j-th column

[4] To be precise, we require $\mathbf{b}_k^*/\|\mathbf{b}_k^*\|^2$ to be a shortest vector in the dual lattice of $\mathbf{B}_{[1,k]}$. See Sect. 3 for details.

[5] http://cseweb.ucsd.edu/~miwalter/src/fplll-dual_enum/fplll-dual_enum.zip.

[6] http://cseweb.ucsd.edu/~miwalter/src/fplll-dual_enum/results.zip.

(inclusive) as $\mathbf{B}_{[i,j]}$ and the horizontal concatenation of two matrices $\mathbf{B}_1$ and $\mathbf{B}_2$ by $[\mathbf{B}_1|\mathbf{B}_2]$. For any matrix $\mathbf{B}$ and $p \geq 1$ we define the induced norm to be $\|\mathbf{B}\|_p = \max_{\|\mathbf{x}\|_p=1}(\|\mathbf{Bx}\|_p)$. For $p = 1$ (resp. ∞) this is often denoted by the column (row) sum norm; for $p = 2$ this is also known as the spectral norm. It is a classical fact that $\|\mathbf{B}\|_2 \leq \sqrt{\|\mathbf{B}\|_1 \|\mathbf{B}\|_\infty}$. Finally, we extend the projection operator to matrices, where $\pi_{\mathbf{V}}(\mathbf{B})$ is the matrix obtained by applying $\pi_{\mathbf{V}}$ to every column $\mathbf{b}_i$ of $\mathbf{B}$ and $\pi_{\mathbf{V}}(\mathbf{b}_i) = \pi_{\mathbf{v}_k}(\cdots(\pi_{\mathbf{v}_1}(\mathbf{b}_i))\cdots)$.

2.1 Lattices

A *lattice* Λ is a discrete subgroup of $\mathbb{R}^m$ and is generated by a matrix $\mathbf{B} \in \mathbb{R}^{m\times n}$, i.e. $\Lambda = \mathcal{L}(\mathbf{B}) = \{\mathbf{Bx} \colon \mathbf{x} \in \mathbb{Z}^n\}$. If $\mathbf{B}$ has full column rank, it is called a *basis* of Λ and $\dim(\Lambda) = n$ is the dimension (or rank) of Λ. A lattice has infinitely many bases, which are related to each other by right-multiplication with unimodular matrices. With each matrix $\mathbf{B}$ we associate its *Gram-Schmidt-Orthogonalization* (GSO) $\mathbf{B}^*$, where the i-th column $\mathbf{b}_i^*$ of $\mathbf{B}^*$ is defined as $\mathbf{b}_i^* = \pi_{\mathbf{B}^*_{[1,i-1]}}(\mathbf{b}_i) = \mathbf{b}_i - \sum_{j<i} \mu_{i,j}\mathbf{b}_j^*$ and $\mu_{i,j} = \langle \mathbf{b}_i, \mathbf{b}_j^* \rangle / \|\mathbf{b}_j^*\|^2$ (and $\mathbf{b}_1^* = \mathbf{b}_1$). For every lattice basis there are infinitely many bases that have the same GSO vectors $\mathbf{b}_i^*$, among which there is a (not necessarily unique) basis that minimizes $\|\mathbf{b}_i\|$ for all i. Transforming a basis into this form is commonly known as *size reduction* and is easily and efficiently done using a slight modification of the Gram-Schmidt process. In this work we will implicitly assume all bases to be size reduced. The reader can simply assume that any basis operation described in this work is followed by a size reduction. For a fixed matrix $\mathbf{B}$ we extend the projection operation to indices: $\pi_i(\cdot) = \pi_{\mathbf{B}^*_{[1,i-1]}}(\cdot)$, so $\pi_1(\mathbf{B}) = \mathbf{B}$. Whenever we refer to the *shape* of a basis $\mathbf{B}$, we mean the vector $(\|\mathbf{b}_i^*\|)_{i\in[n]}$. We define $\mathbf{D}^\dagger$ to be the GSO of $\mathbf{D}$ in reverse order.

For every lattice Λ there are a few invariants associated to it. One of them is its determinant $\det(\mathcal{L}(\mathbf{B})) = \prod_i \|\mathbf{b}_i^*\|$ for any basis $\mathbf{B}$. Even though the basis of a lattice is not uniquely defined, the determinant is and it is efficiently computable given a basis. Furthermore, for every lattice Λ we denote the length of its shortest non-zero vector (also known as the *first minimum*) by $\lambda_1(\Lambda)$, which is always well defined. We use the short-hand notations $\det(\mathbf{B}) = \det(\mathcal{L}(\mathbf{B}))$ and $\lambda_1(\mathbf{B}) = \lambda_1(\mathcal{L}(\mathbf{B}))$. Minkowski's theorem is a classic result that relates the first minimum to the determinant of a lattice. It states that $\lambda_1(\Lambda) \leq \sqrt{\gamma_n}\det(\Lambda)^{1/n}$, for any Λ with $\dim(\Lambda) = n$, where $\Omega(n) \leq \gamma_n \leq n$ is Hermite's constant. Finding a (even approximate) shortest nonzero vector in a lattice, commonly known as the *Shortest Vector Problem* (SVP), is NP-hard under randomized reductions [25,34].

For every lattice Λ, its *dual* is defined as $\hat{\Lambda} = \{\mathbf{w} \in \mathrm{span}(\Lambda) | \langle \mathbf{w}, \mathbf{v}\rangle \in \mathbb{Z} \text{ for all } \mathbf{v} \in \Lambda\}$. It is a classical fact that $\det(\hat{\Lambda}) = \det(\Lambda)^{-1}$. For a lattice basis $\mathbf{B}$, let $\mathbf{D}$ be the unique matrix that satisfies $\mathrm{span}(\mathbf{B}) = \mathrm{span}(\mathbf{D})$ and $\mathbf{B}^T\mathbf{D} = \mathbf{D}^T\mathbf{B} = \mathbf{I}$. Then $\widehat{\mathcal{L}(\mathbf{B})} = \mathcal{L}(\mathbf{D})$ and we denote $\mathbf{D}$ as the *dual basis* of $\mathbf{B}$. It follows that for any vector $\mathbf{w} = \mathbf{Dx}$ we have that $\mathbf{B}^T\mathbf{w} = \mathbf{x}$, i.e. we can recover the coefficients $\mathbf{x}$ of $\mathbf{w}$ with respect to the dual basis $\mathbf{D}$ by multiplication with the transpose of the primal basis $\mathbf{B}^T$. Given a lattice basis, its dual basis is computable in polynomial time, but requires at least $\Omega(n^3)$ bit operations using matrix inversion. Finally, if $\mathbf{D}$ is the dual basis of $\mathbf{B}$, their GSOs are related by $\|\mathbf{b}_i^*\| = 1/\|\mathbf{d}_i^\dagger\|$.

In this work we will often modify a lattice basis $\mathbf{B}$ such that its first vector satisfies $\alpha\|\mathbf{b}_1\| \leq \lambda_1(\mathbf{B})$ for some $\alpha \leq 1$. We will call this process *SVP reduction* of $\mathbf{B}$. Given an SVP oracle, it can be accomplished by using the oracle to find the shortest vector in $\mathcal{L}(\mathbf{B})$, prepending it to the basis, and running LLL (cf. Sect. 2.3) on the resulting generating system. Furthermore, we will modify a basis $\mathbf{B}$ such that its dual $\mathbf{D}$ satisfies $\alpha\|\mathbf{d}_n\| \leq \lambda_1(\widehat{\mathcal{L}(\mathbf{B})})$, i.e. its reversed dual basis is SVP reduced. This process is called *dual SVP reduction*. Note that if $\mathbf{B}$ is dual SVP reduced, then $\|\mathbf{b}_n^*\|$ is maximal among all bases of $\mathcal{L}(\mathbf{B})$. The obvious way to achieve dual SVP reduction is to compute the dual of the basis, SVP reduce it as described above, and compute the primal basis. We present an alternative way to achieve this in Sect. 7. In the context of reduction algorithms, the relaxation factor α is usually needed for proofs of termination or running time and only impacts the analysis of the output quality in lower order terms. In this work, we will sweep it under the rug and take it implicitly to be a constant close to 1. Finally, we will apply SVP and dual SVP reduction to projected blocks of a basis $\mathbf{B}$, for example we will (dual) SVP reduce the block $\pi_i(\mathbf{B}_{[i,i+k]})$. By that we mean that we will modify $\mathbf{B}$ in such a way that $\pi_i(\mathbf{B}_{[i,i+k]})$ is (dual) SVP reduced. This can easily be achieved by applying the transformations to the original basis vectors instead of their projections.

2.2 Enumeration Algorithms

In order to solve SVP in practice, enumeration algorithms are usually employed, since these are the most efficient algorithms for currently realistic dimensions. The standard enumeration procedure, usually attributed to Fincke, Pohst [11], and Kannan [24] can be described as a recursive algorithm: given as input a basis $\mathbf{B} \in \mathbb{Z}^{m\times n}$ and a radius r, it first recursively finds all vectors $\mathbf{v}' \in \mathcal{L}(\pi_2(\mathbf{B}))$ with $\|\mathbf{v}'\| \leq r$, and then for each of them finds all $\mathbf{v} \in \mathcal{L}(\mathbf{B})$, s.t. $\pi_2(\mathbf{v}) = \mathbf{v}'$ and $\|\mathbf{v}\| \leq r$, using $\mathbf{b}_1$. This essentially corresponds to a breadth first search on a large tree, where layers correspond to basis vectors and the nodes to the respective coefficients. While it is conceptually simpler to think of enumeration as a BFS, implementations usually employ a depth first search for performance reasons. Pseudo code can be found in Algorithm 3 in Sect. 7.

There are several practical improvements of this algorithm collectively known as *SchnorrEuchner enumeration* [60]: First, due to the symmetry of lattices, we can reduce the search space by ensuring that the last non zero coefficient is always positive. Furthermore, if we find a vector shorter than the bound r, we can update the latter. And finally, we can enumerate the coefficients of a basis vector in order of the length of the resulting (projected) vector and thus increase the chance of finding some short vector early, which will update the bound r and keep the search space smaller.

It has also been demonstrated [14] that reducing the search space (and thus the success probability) – a technique known as pruning – can speed up enumeration by exponential factors. For more details on recent improvements we refer to [14,19,20,36,69].

2.3 Lattice Reduction

As opposed to exact SVP algorithms, lattice reductions approximate the shortest vector. The quality of their output is usually measured in the length of the shortest vector they are able to find with respect to the root determinant of the lattice. This quantity is denoted by the *Hermite factor* $\bar{\delta} = \|\mathbf{b}_1\|/\det(\mathbf{B})^{1/n}$. The Hermite factor depends on the lattice dimension n, but the experiments of [16] suggest that the *root Hermite factor* $\delta = \bar{\delta}^{1/n}$ converges to a constant as n increases for popular reduction algorithms. During our experiments we found that to be true at least for large enough dimensions ($n \geq 140$).

The *LLL* algorithm [27] is a polynomial time basis reduction algorithm. A basis $\mathbf{B} \in \mathbb{Z}^{m \times n}$ can be defined to be LLL reduced if $\mathbf{B}_{[1,2]}$ is SVP reduced and $\pi_2(\mathbf{B})$ is LLL reduced. From this it is straight forward to prove that LLL reduction achieves a root Hermite factor of at most $\delta \leq \gamma_2^{1/4} \approx 1.0746$. However, LLL has been reported to behave much better in practice [16,43].

BKZ [54] is a generalization of LLL to larger block size. A basis $\mathbf{B}$ is BKZ reduced with block size k (denoted by BKZ-k) if $\mathbf{B}_{[1,\min(k,n)]}$ is SVP reduced and $\pi_2(\mathbf{B})$ is BKZ-k reduced. BKZ achieves this by simply scanning the basis from left to right and SVP reducing each projected block of size k (or smaller once it reaches the end) by utilizing a SVP oracle for all dimensions $\leq k$. It iterates this process (which is usually called a *tour*) until no more change occurs. When $k = n$, this is usually referred to as *HKZ* reduction and is essentially equivalent to solving SVP. The following bound for the Hermite factor holds for $\mathbf{b}_1$ of a BKZ-k reduced basis [18]:

$$\|\mathbf{b}_1\| \leq 2\gamma_k^{\frac{n-1}{2(k-1)}+\frac{3}{2}} \det(\mathbf{B})^{1/n} \tag{1}$$

Equation (1) shows that the root Hermite factor achieved by BKZ-k is at most $\lesssim \gamma_k^{\frac{1}{2(k-1)}}$. Furthermore, while there is no polynomial bound on the number of calls BKZ makes to the SVP oracle, Hanrot, Pujol, and Stehlé showed in [18] that one can terminate BKZ after a polynomial number of calls to the SVP oracle and still provably achieve the bound (1). Finally, BKZ has been repeatedly reported to behave very well in practice [8,16]. For these reasons, BKZ is very popular in practice and implementations are readily available in different libraries, e.g. in `NTL` [64] or `fpLLL` [4].

In [15], Gama and Nguyen introduced a different block reduction algorithm, namely *Slide reduction*. It is also parameterized by a block size k, which is required to divide the lattice dimension n, but uses a SVP oracle only in dimension k.[7] A basis $\mathbf{B}$ is defined to be slide reduced, if $\mathbf{B}_{[1,k]}$ is SVP reduced, $\pi_2(\mathbf{B}_{[2,k+1]})$ is dual SVP reduced (if $k > n$), and $\pi_{k+1}(\mathbf{B}_{[k+1,n]})$ is slide reduced. Slide reduction, as described in [15], reduces a basis by first alternately SVP reducing all blocks $\pi_{ik+1}(\mathbf{B}_{[ik+1,(i+1)k]})$ and running LLL on $\mathbf{B}$. Once no more changes occur, the blocks $\pi_{ik+2}(\mathbf{B}_{[ik+2,(i+1)k+1]})$ are dual SVP reduced. This entire process is iterated until no more changes occur. Upon termination, the basis is guaranteed to satisfy

$$\|\mathbf{b}_1\| \leq \gamma_k^{\frac{n-1}{2(k-1)}} \det(\mathbf{B})^{1/n} \tag{2}$$

This is slightly better than Eq. (1), but the achieved root Hermite factor is also only guaranteed to be less than $\gamma_k^{\frac{1}{2(k-1)}}$. Slide reduction has the desirable properties of only making a polynomial number of calls to the SVP oracle and that all calls are in dimension k (and not in lower dimensions). The latter allows for a cleaner analysis, for example when combined with the Gaussian Heuristic (cf. Sect. 2.4). Unfortunately, Slide reduction has been reported to be greatly inferior to BKZ in experiments [16], so it is rarely used in practice and we are not aware of any publicly available implementation.

2.4 The Gaussian Heuristic

The Gaussian Heuristic gives an approximation of the number of lattice points in a "nice" subset of $\mathbb{R}^n$. More specifically, it says that for a given set S and a lattice Λ, we have $|S \cap \Lambda| \approx \text{vol}(S)/\det(\Lambda)$. The heuristic has been proved to be very useful in the average case analysis of lattice algorithms. For example, it can be used to estimate the complexity of enumeration algorithms [14,19] or the output quality of lattice reduction algorithms [8]. For the latter, note that reduction algorithms work by repeatedly computing the shortest vector in some lattice and inserting this vector in a certain position of the basis. To estimate the effect such a step has on the basis, it is useful to be able to predict how long such a vector might be. This is where the Gaussian Heuristic comes in: using the above formula, one can estimate how large the radius of an n-dimensional ball (this is the "nice" set) needs to be such that we can expect it to contain a non-zero lattice point (where $n = \dim(\Lambda)$). Using the volume formula for the n-dimensional ball, we get an estimate for the shortest non-zero vector in a lattice Λ:

$$GH(\Lambda) = \frac{(\Gamma(n/2+1) \cdot \det(\Lambda))^{1/n}}{\sqrt{\pi}} \tag{3}$$

If k is an integer, we define $GH(k)$ to be the Gaussian Heuristic (i.e. Eq. (3)) for k-dimensional lattices with unit determinant. The heuristic has been tested

[7] Strictly speaking, the algorithm as described in [15] uses HKZ reduction and thus requires an SVP oracle in lower dimensions as well. However, the entire analysis in [15] only relies on the SVP reducedness of the projected blocks and thus the HKZ reduction can be replaced by SVP reduction, which we do in the following.

experimentally [14], also in the context of lattice reduction [8,16], and been found to be too rough in small dimensions, but to be quite accurate starting in dimension > 45. In fact, for a precise definition of random lattices (which we are not concerned with in this work) it can be shown that the expected value of the first minimum of the lattice (over the choice of the lattice) converges to Eq. (3) as the lattice dimension tends to infinity.[8]

Heuristic 1 [Gaussian Heuristic]. *For a given lattice Λ, $\lambda_1(\Lambda) = GH(\Lambda)$.*

Invoking Heuristic 1 for all projected sublattices that the SVP oracle is called on during the process, the root Hermite factor achieved by lattice reduction (usually with regards to BKZ) is commonly estimated to be [5]

$$\delta \approx GH(k)^{\frac{1}{k-1}}. \tag{4}$$

However, since the Gaussian Heuristic only seems to hold in large enough dimensions and BKZ makes calls to SVP oracles in all dimensions up to the block size k, it is not immediately clear how justified this estimation is. While there is a proof by Chen [7] that under the Gaussian Heuristic, Eq. (4) is accurate for BKZ, this is only true as the lattice dimension tends to infinity. It might be reasonable to assume that this also holds in practice as long as the lattice dimension is large enough compared to the block size, but in practice and cryptanalytic settings this is often not the case. In fact, in order to achieve an approximation good enough to break a cryptosystem, a block size at least linear in the lattice dimension is often required. As another approach to predicting the output of BKZ, Chen and Nguyen proposed a simulation routine [8]. Unfortunately, the simulator approach has several drawbacks. Obviously, it requires more effort to apply than a closed formula like (4), since it needs to be implemented and "typical" inputs need to be generated or synthesized (among others, the shape of a "typical" HKZ reduced basis in dimension 45). On top of that, the accuracy of the simulator is based on several additional heuristic assumptions, the validity of which has not been independently verified.

To the best of our knowledge there have been no attempts to make similar predictions for Slide reduction, as it is believed to be inferior to BKZ and thus usually not considered for cryptanalysis.

3 Self-Dual BKZ

In this section we describe our new reduction algorithm. Like BKZ it is parameterized by a block size k and a SVP oracle in dimension k, and acts on the input basis $\mathbf{B} \in \mathbb{Z}^{m \times n}$ by iterating tours. The beginning of every tour is exactly like a BKZ tour, i.e. SVP reducing every block $\pi_i(\mathbf{B}_{[i,i+k-1]})$ from $i = 1$ to $n - k + 1$.

[8] One can also formulate Heuristic 1 for a given lattice by assuming it "behaves like a random lattice". Depending on the exact definition of what it means for a lattice to "behave like a random lattice", this version is either stronger as or equivalent to Heuristic 1.

We will call this part a *forward tour*. For the last block, which BKZ simply HKZ reduces and where most of the problems for meaningful predictions stem from, we do something different. Instead, we dual SVP the last block and proceed by dual SVP reducing all blocks of size k backwards (which is a *backward tour*). After iterating this process (which we call a *tour* of Self-Dual BKZ) the algorithm terminates when no more progress is made. The algorithm is formally described in Algorithm 1.

Algorithm 1. Self-Dual BKZ

`procedure` DBKZ ($\mathbf{B}$, k, SVP_k)
Input: A lattice basis $\mathbf{B} \in \mathbb{Z}^{m \times n}$, a block size k, a SVP oracle in dimension k
Output: A k-reduced basis $\mathbf{B}'$ (See Definition 1 for a formal definition.)

1 **do**
2   **for** $i = 1 \ldots n - k$
3   SVP reduce $\pi_i(\mathbf{B}_{[i,i+k-1]})$ using SVP_k
4   **for** $i = n - k + 1 \ldots 1$
5   dual SVP reduce $\pi_i(\mathbf{B}_{[i,i+k-1]})$ using SVP_k
6 **while** *progress is made*
7 **return** $\mathbf{B}$

Note that, like BKZ, Self-Dual BKZ (DBKZ) is a proper block generalization of the LLL algorithm, which corresponds to the case $k = 2$.

The terminating condition in Line 6 is left ambiguous at this point on purpose as there are several sensible ways to approach this as we will see in the next section. One has to be careful to, on the one hand guarantee termination, while on the other hand achieving a meaningful reducedness definition.

3.1 Analysis

The output of Algorithm 1 satisfies the following reducedness definition upon termination:

Definition 1. *A basis* $\mathbf{B} = [\mathbf{b}_1, \ldots, \mathbf{b}_n]$ *is k-reduced if either* $n < k$, *or it satisfies the following conditions:*

- $\|\mathbf{b}_k^*\|^{-1} = \lambda_1(\widehat{\mathcal{L}(\mathbf{B}_{[1,k]})})$, *and*
- *for some SVP reduced basis* $\tilde{\mathbf{B}}$ *of* $\mathcal{L}(\mathbf{B}_{[1,k]})$, $\pi_2([\tilde{\mathbf{B}}|\mathbf{B}_{[k+1,n]}])$ *is k-reduced.*

We first prove that Algorithm 1 indeed achieves Definition 1 when used with a specific terminating condition:

Lemma 1. *Let* $\mathbf{B}$ *be an n-dimensional basis. If* $\pi_{k+1}(\mathbf{B})$ *is the same before and after one loop of Algorithm 1, then* $\mathbf{B}$ *is k-reduced.*

Proof. The proof is inductive: for $n = k$ the result is trivially true. So, assume $n > k$, and that the result already holds for $n-1$. At the end of each iteration, the first block $\mathbf{B}_{[1,k]}$ is dual-SVP reduced by construction. So, we only need to verify that for some $\tilde{\mathbf{B}}$ an SVP reduced basis for $\mathcal{L}(\mathbf{B}_{[1,k})$, the projection $\pi_2([\tilde{\mathbf{B}}|\mathbf{B}_{[k+1,n]}])$ is also k-reduced. Let $\tilde{\mathbf{B}}$ be the SVP reduced basis produced in the first step. Note that the first and last operation in the loop do not change $\mathcal{L}(\mathbf{B}_{[1,k]})$ and $\mathbf{B}_{[k+1,n]}$. It follows that $\pi_{k+1}(\mathbf{B})$ is the same before and after the partial tour (the tour without the first and the last step) on the projected basis $\pi_2([\tilde{\mathbf{B}}|\mathbf{B}_{[k+1,n]}])$, and so $\pi_{k+2}(\mathbf{B})$ is the same before and after the partial tour. By induction hypothesis, $\pi_2([\tilde{\mathbf{B}}|\mathbf{B}_{[k+1,n]}])$ is k-reduced. □

Lemma 1 gives a terminating condition which ensures that the basis is reduced. We remark that it is even possible to adapt the proof such that it is sufficient to check that the shape of the projected basis $\pi_{k+1}(\mathbf{B})$ is the same before and after the tour, which is much closer to what one would do in practice to check if *progress was made* (cf. Line 6). However, this requires to relax the definition of SVP-reduction slightly, such that the first vector is not necessarily a shortest vector, but merely a short vector achieving Minkowski's bound. Since this is the only property of SVP reduced bases we need for the analysis below, this does not affect the worst case output quality. Finally, we are aware that it is not obvious that either of these conditions are ever met, e.g. (the shape of) $\pi_{k+1}(\mathbf{B})$ might loop indefinitely. However, in Sect. 4 we show that one can put a polynomial upper bound on the number of loops without sacrificing worst case output quality.

To show that the output quality of Self-Dual BKZ in the worst case is at least as good as BKZ's worst case behavior, we analyze the Hermite factor it achieves:

Theorem 1. *If* $\mathbf{B}$ *is k-reduced, then* $\lambda_1(\mathbf{B}_{[1,k]}) \leq \sqrt{\gamma_k}^{\frac{n-1}{k-1}} \cdot \det(\mathbf{B})^{1/n}$.

Proof. Assume without loss of generality that $\mathcal{L}(\mathbf{B})$ has determinant 1, and let Δ be the determinant of $\mathcal{L}(\mathbf{B}_{[1,k]})$. Let $\lambda \leq \sqrt{\gamma_k}\Delta^{1/k}$ and $\hat{\lambda} \leq \sqrt{\gamma_k}\Delta^{-1/k}$ be the lengths of the shortest nonzero primal and dual vectors of $\mathcal{L}(\mathbf{B}_{[1,k]})$. We need to prove that $\lambda \leq \sqrt{\gamma_k}^{\frac{n-1}{k-1}}$.

We first show, by induction on n, that the determinant Δ_1 of the first $k-1$ vectors is at most $\sqrt{\gamma_k}^{n-k+1}\det(\mathbf{B})^{(k-1)/n} = \sqrt{\gamma_k}^{n-k+1}$. Since $\mathbf{B}$ is k-reduced, this determinant equals $\Delta_1 = \hat{\lambda} \cdot \Delta \leq \sqrt{\gamma_k}\Delta^{1-1/k}$. (This alone already proves the base case of the induction for $n = k$.) Now, let $\tilde{\mathbf{B}}$ be a SVP reduced basis of $\mathcal{L}(\mathbf{B}_{[1,k]})$ satisfying the k-reduction definition, and consider the determinant $\Delta_2 = \Delta/\lambda$ of $\pi_2(\tilde{\mathbf{B}})$. Since $\pi_2([\tilde{\mathbf{B}}|\mathbf{B}_{[k+1,n]}])$ has determinant $1/\|\tilde{\mathbf{b}}_1\| = 1/\lambda$, by induction hypothesis we have $\Delta_2 \leq \sqrt{\gamma_k}^{n-k}(1/\lambda)^{(k-1)/(n-1)}$.

$$\Delta = \lambda\Delta_2 \leq \sqrt{\gamma_k}^{n-k}\lambda^{\frac{n-k}{n-1}} \leq \sqrt{\gamma_k}^{n-k}(\sqrt{\gamma_k}\Delta^{\frac{1}{k}})^{\frac{n-k}{n-1}} = \sqrt{\gamma_k}^{\frac{(n-k)n}{n-1}}\Delta^{\frac{n-k}{k(n-1)}}.$$

Rising both sides to the power $(n-1)/n$ we get $\Delta^{1-\frac{1}{n}} \leq \sqrt{\gamma_n}^{n-k} \Delta^{\frac{1}{k}-\frac{1}{n}}$, or, equivalently, $\Delta^{1-\frac{1}{k}} \leq \sqrt{\gamma_k}^{n-k}$. It follows that $\Delta_1 = \hat{\lambda}\Delta \leq \sqrt{\gamma_k}\Delta^{1-\frac{1}{k}} \leq \sqrt{\gamma_k}^{n-k+1}$, concluding the proof by induction.

We can now prove the main theorem statement. Recall from the inductive proof that $\Delta \leq \sqrt{\gamma_k}^{n-k} \lambda^{\frac{n-k}{n-1}}$. Therefore, $\lambda \leq \sqrt{\gamma_k}\Delta^{1/k} \leq \sqrt{\gamma_k}^{\frac{n}{k}} \lambda^{\frac{n-k}{k(n-1)}}$. Solving for λ, proves the theorem. □

4 Dynamical System

Proving a good running time on DBKZ directly seems just as hard as for BKZ, so in this section we analyze the DBKZ algorithm using the dynamical system technique from [18].

Let $\mathbf{B} = [\mathbf{b}_1, \ldots, \mathbf{b}_n]$ be an input basis to DBKZ, and assume without loss of generality that $\det(\mathbf{B}) = 1$. During a forward tour, our algorithm computes a sequence of lattice vectors $\mathbf{B}' = [\mathbf{b}'_1, \ldots, \mathbf{b}'_{n-k}]$ where each $\mathbf{b}'_i$ is set to a shortest vector in the projection of $[\mathbf{b}_i, \ldots, \mathbf{b}_{i+k-1}]$ orthogonal to $[\mathbf{b}'_1, \ldots, \mathbf{b}'_{i-1}]$. This set of vectors can be extended to a basis $\mathbf{B}'' = [\mathbf{b}''_1, \ldots, \mathbf{b}''_n]$ for the original lattice. Since $[\mathbf{b}'_1, \ldots, \mathbf{b}'_{i-1}]$ generates a primitive sublattice of $[\mathbf{b}_i, \ldots, \mathbf{b}_{i+k-1}]$, the projected sublattice has determinant $\det(\mathcal{L}(\mathbf{b}_1, \ldots, \mathbf{b}_{i+k-1}))/\det(\mathcal{L}(\mathbf{b}'_1, \ldots, \mathbf{b}'_{i-1}))$, and the length of its shortest vector is

$$\|(\mathbf{b}'_i)^*\| \leq \sqrt{\gamma_k} \left(\frac{\det(\mathcal{L}(\mathbf{b}_1, \ldots, \mathbf{b}_{i+k-1}))}{\det(\mathcal{L}(\mathbf{b}'_1, \ldots, \mathbf{b}'_{i-1}))} \right)^{1/k}. \tag{5}$$

At this point, simulations based on the Gaussian Heuristics typically assume that (5) holds with equality. In order to get a rigorous analysis without heuristic assumptions, we employ the amortization technique of [18,19]. For every $i = 1, \ldots, n-k$, let $x_i = \log\det(\mathbf{b}_1, \ldots, \mathbf{b}_{k+i-1})$ and $x'_i = \log\det(\mathbf{b}'_1, \ldots, \mathbf{b}'_i)$. Using (5), we get for all $i = 1, \ldots, n-k$,

$$\begin{aligned} x'_i &= x'_{i-1} + \log\|(\mathbf{b}'_i)^*\| \\ &\leq x'_{i-1} + \alpha + \frac{x_i - x'_{i-1}}{k} \\ &= \omega x'_{i-1} + \alpha + (1-\omega)x_i \end{aligned}$$

where $\omega = (1 - 1/k)$, $\alpha = \frac{1}{2}\log\gamma_k$ and $x'_0 = 0$. By induction on i,

$$x'_i \leq \alpha\frac{1-\omega^i}{1-\omega} + (1-\omega)\sum_{j=1}^{i} \omega^{i-j} x_j,$$

or, in matrix notation $\mathbf{x}' \leq \mathbf{b} + \mathbf{A}\mathbf{x}$ where

$$\mathbf{b} = \alpha k \begin{bmatrix} 1-\omega \\ \vdots \\ 1-\omega^{n-k} \end{bmatrix} \qquad \mathbf{A} = \frac{1}{k} \begin{bmatrix} 1 & & & \\ \omega & 1 & & \\ \vdots & & \ddots & \\ \omega^{n-k-1} & \cdots & \omega & 1 \end{bmatrix}.$$

Since all the entries of $\mathbf{A}$ are positive, we also see that if $X_i \geq x_i$ are upper bounds on the initial values x_i for all i, then the vector $X' = \mathbf{A}X + \mathbf{b}$ gives upper bounds on the output values $x'_i \leq X'_i$.

The vector $\mathbf{x}'$ describes the shape of the basis matrix before the execution of a backward tour. Using lattice duality, the backward tour can be equivalently formulated by the following steps:

1. Compute the reversed dual basis $\mathbf{D}$ of $\mathbf{B}'$
2. Apply a forward tour to $\mathbf{D}$ to obtain a new dual basis $\mathbf{D}'$
3. Compute the reversed dual basis of $\mathbf{D}'$

The reversed dual basis computation yields a basis $\mathbf{D}$ such that, for all $i = 1, \ldots, n-k$,

$$\begin{aligned} y_i &= \log\det(\mathbf{d}_1, \ldots, \mathbf{d}_{k+i-1}) \\ &= -\log(\det(\mathbf{B}')/\det([\mathbf{b}'_1, \ldots, \mathbf{b}'_{n-k+1-i}])) \\ &= \log\det([\mathbf{b}'_1, \ldots, \mathbf{b}'_{n-k+1-i}]) = x'_{n-k+1-i}. \end{aligned}$$

So, the vector $\mathbf{y}$ describing the shape of the dual basis at the beginning of the backward tour is just the reverse of $\mathbf{x}'$. It follows that applying a full (forward and backward) DBKZ tour produces a basis such that if X are upper bounds on the log determinants $\mathbf{x}$ of the input matrix, then the log determinants of the output matrix are bounded from above by

$$\mathbf{R}(\mathbf{A}\mathbf{R}(\mathbf{A}X + \mathbf{b}) + \mathbf{b}) = (\mathbf{R}\mathbf{A})^2 X + (\mathbf{R}\mathbf{A} + \mathbf{I})\mathbf{R}\mathbf{b}$$

where $\mathbf{R}$ is the coordinate reversal permutation matrix. This leads to the study of the discrete time affine dynamical system

$$X \mapsto (\mathbf{R}\mathbf{A})^2 X + (\mathbf{R}\mathbf{A} + \mathbf{I})\mathbf{R}\mathbf{b}. \tag{6}$$

4.1 Output Quality

We first prove that this system has at most one fixed point.

Claim. The dynamical system (6) has at most one fixed point.

Proof. Any fixed point is a solution to the linear system $((\mathbf{R}\mathbf{A})^2 - \mathbf{I})X + (\mathbf{R}\mathbf{A} + \mathbf{I})\mathbf{R}\mathbf{b} = \mathbf{0}$. To prove uniqueness, we show that the matrix $((\mathbf{R}\mathbf{A})^2 - \mathbf{I})$ is non-singular, i.e., if $(\mathbf{R}\mathbf{A})^2\mathbf{x} = \mathbf{x}$ then $\mathbf{x} = \mathbf{0}$. Notice that the matrix $\mathbf{R}\mathbf{A}$ is symmetric, so we have $(\mathbf{R}\mathbf{A})^2 = (\mathbf{R}\mathbf{A})^T\mathbf{R}\mathbf{A} = \mathbf{A}^T\mathbf{A}$. So proving $((\mathbf{R}\mathbf{A})^2 - \mathbf{I})$ is non-singular is equivalent to showing that 1 is not an eigenvalue of $\mathbf{A}^T\mathbf{A}$. We have $\rho(\mathbf{A}^T\mathbf{A}) = \|\mathbf{A}\|_2^2 \leq \|\mathbf{A}\|_1 \|\mathbf{A}\|_\infty$, where $\rho(\cdot)$ denotes the spectral radius of the given matrix (i.e. the largest eigenvalue in absolute value). But we also have

$$\|\mathbf{A}\|_\infty = \|\mathbf{A}\|_1 = \frac{1}{k}\sum_{i=0}^{n-k-1} \omega^i = \frac{1 - \omega^{n-k}}{k(1-\omega)} = 1 - \omega^{n-k} < 1 \tag{7}$$

which shows that the absolute value of any eigenvalue of $\mathbf{A}^T\mathbf{A}$ is strictly smaller than 1. □

We need to find a fixed point for (6). We have proved that $(\mathbf{RA})^2 - \mathbf{I}$ is a non-singular matrix. Since $(\mathbf{RA})^2 - \mathbf{I} = (\mathbf{RA}+\mathbf{I})(\mathbf{RA}-\mathbf{I})$, it follows that $(\mathbf{RA} \pm \mathbf{I})$ are also non singular. So, we can factor $(\mathbf{RA}+\mathbf{I})$ out of the fixed point equation $((\mathbf{RA})^2 - \mathbf{I})\mathbf{x} + (\mathbf{RA}+\mathbf{I})\mathbf{Rb} = \mathbf{0}$, and obtain $(\mathbf{RA}-\mathbf{I})\mathbf{x} + \mathbf{Rb} = 0$. This shows that the only fixed point of the full dynamical system (if it exists) must also be a fixed point of a forward tour $\mathbf{x} \mapsto \mathbf{R}(\mathbf{Ax}+\mathbf{b})$.

Claim. The fixed point of the dynamical system $\mathbf{x} \mapsto \mathbf{R}(\mathbf{Ax}+\mathbf{b})$ is given by

$$x_i = \frac{(n-k-i+1)(k+i-1)}{k-1}\alpha. \tag{8}$$

Proof. The unique fixed point of the system is given by the solution to the linear system $(\mathbf{R}-\mathbf{A})\mathbf{x} = \mathbf{b}$. We prove that (8) is a solution to the system by induction on the rows. For the first row, the system yields

$$x_{n-k} - x_1/k = \alpha. \tag{9}$$

From (8) we get that $x_{n-k} = \frac{n-1}{k-1}\alpha$ and $x_1 = \frac{k(n-k)}{k-1}\alpha$. Substituting these into (9), the validity is easily verified.

The r-th row of the system is given by

$$x_{n-k-r+1} - \frac{1}{k}\left(\sum_{j=1}^{r} \omega^{r-j} x_j\right) = \frac{1-\omega^r}{1-\omega}\alpha \tag{10}$$

which is equivalent to

$$x_{n-k-r+1} + \omega\left(x_{n-k-r+2} - \frac{1}{k}\left(\sum_{j=1}^{r-1} \omega^{r-1-j} x_j\right)\right) - \frac{x_r}{k} - \omega x_{n-k-r+2} = \frac{1-\omega^r}{1-\omega}\alpha. \tag{11}$$

By induction hypothesis, this is equivalent to

$$\omega\left(\frac{1-\omega^{r-1}}{1-\omega}\right)\alpha + x_{n-k-r+1} - \frac{x_r}{k} - \omega x_{n-k-r+2} = \frac{1-\omega^r}{1-\omega}\alpha. \tag{12}$$

Substituting (8) in for $i = n-k-r+1$, r, and $n-k-r+2$, we get

$$x_{n-k-r+1} - \frac{x_r}{k} - \omega x_{n-k-r+2} = \frac{kr(n-r) - (n-r-k+1)(r+k-1) - (k-1)(r-1)(n-r+1)}{k(k-1)}\alpha$$

which, after some tedious, but straight forward, calculation can be shown to be equal to α (i.e. the fraction simplifies to 1). This in turn shows that the left hand side of (12) is equivalent to

$$\omega\left(\frac{1-\omega^{r-1}}{1-\omega}\right)\alpha + \alpha$$

which is equal to its right hand side. □

Note that since x_1 corresponds to the log determinant of the first block, applying Minkowski's theorem results in the same worst case Hermite factor as proved in Theorem 1.

4.2 Convergence

Consider any input vector $\mathbf{v}$ and write it as $\mathbf{v} = \mathbf{x} + \mathbf{e}$, where $\mathbf{x}$ is the fixed point of the dynamical system as in (8). The system sends $\mathbf{v}$ to $\mathbf{v} \mapsto \mathbf{RAv} + \mathbf{b} = \mathbf{RAx} + \mathbf{RAe} + \mathbf{b} = \mathbf{x} + \mathbf{RAe}$, so the difference $\mathbf{e}$ to the fixed point is mapped to $\mathbf{RAe}$ in each iteration. In order to analyze the convergence of the algorithm, we consider the induced norm of the matrix $\|\mathbf{RA}\|_p = \|\mathbf{A}\|_p$, since after t iterations the difference is $(\mathbf{RA})^t\mathbf{e}$ and so its norm is bounded by $\|(\mathbf{RA})^t\mathbf{e}\|_p \leq \|(\mathbf{RA})^t\|_p\|\mathbf{e}\|_p \leq \|\mathbf{RA}\|_p^t\|\mathbf{e}\|_p$. So if the induced norm of $\mathbf{A}$ is strictly smaller than 1, the corresponding norm of the error vector follows an exponential decay. While the spectral norm of $\mathbf{A}$ seems hard to bound, the 1 and the infinity norm are straight forward to analyze. In particular, we saw in (7) that $\|\mathbf{A}\|_\infty = 1 - \omega^{n-k}$. This proves that the algorithm converges. Furthermore, let the input be a basis $\mathbf{B}$ (with $\det(\mathbf{B}) = 1$), the corresponding vector $\mathbf{v} = (\log\det(\mathbf{b}_1, \ldots, \mathbf{b}_{k+i-1}))_{1\leq i\leq n}$ and write $\mathbf{v} = \mathbf{x} + \mathbf{e}$. Then we have $\|\mathbf{e}\|_\infty = \|\mathbf{v} - \mathbf{x}\|_\infty \leq \|\mathbf{v}\|_\infty + \|\mathbf{x}\|_\infty \leq \text{poly}(n, \text{size}(\mathbf{B}))$. This implies that for

$$t = \text{polylog}(n, \text{size}(\mathbf{B}))/\omega^{n-k} \approx O(e^{(n-k)/k})\text{polylog}(n, \text{size}(\mathbf{B})) \qquad (13)$$

we have that $\|(\mathbf{RA})^t\mathbf{e}\| \leq c$ for constant c. Eq. (13) already shows that for $k = \Omega(n)$, the algorithm converges in a number of tours polylogarithmic in the lattice dimension n, i.e. makes at most $\tilde{O}(n)$ SVP calls. In the initial version of this work, proving polynomial convergence for arbitrary k was left as an open problem. Recently, Neumaier filled this gap [38]. We reformulate his proof using our notation in the full version of this paper [37].

5 Heuristic Analysis

In the context of cryptanalysis, we are more interested in the average case behavior of algorithms. For this we can use a very simple observation to predict the Hermite factor achieved by DBKZ. Note that the proof of Theorem 1 is based solely on Minkowski's bound $\lambda_1(\mathbf{B}) \leq \sqrt{\gamma_n}\det(\mathbf{B})^{1/n}$. Replacing it with Heuristic 1 yields the following corollary.

Corollary 1. *Applying Heuristic 1 to every lattice that is passed to the SVP oracle during the execution of Algorithm 1, if* $\mathbf{B}$ *is* k*-reduced, then* $\lambda_1(\mathbf{B}_{1,k}) = GH(k)^{\frac{n-1}{k-1}}\det(\mathbf{B})^{1/n}$.

As the Hermite factor is the most relevant quantity in many cryptanalytic settings, Corollary 1 is already sufficient for many intended applications in terms of output quality. We remark that the proof of achieved worst-case output quality of Slide reduction also only relies on Minkowski's bound. This means the same observation can be used to predict the average case behavior of Slide reduction and yields the same estimate as Corollary 1. In fact, from the recursive definition of Slide reduction it is clear that this yields even more information about the returned basis: we can use Corollary 1 to predict the norm of $\|\mathbf{b}_{ik+1}\|$ for all $i \in [n/k]$. A short calculation shows that these vectors follow a geometric series, supporting a frequently assumed behavior of lattice reduction, namely the *Geometric Series Assumption* [57].

However, many attacks [30,45] require to estimate the average case output much more precisely. Fortunately, applying a similar trick as in Corollary 1 to the dynamical systems analysis in Sect. 4 allows us to obtain much more information about the basis. For this, note that again we can replace Minkowski's theorem in the analysis by Heuristic 1. This transformation changes the dynamical system in (6) only slightly, the only difference being that $\alpha = \frac{1}{2}\log GH(k)$. As the analysis is independent of the constant α, we can translate the fixed point in (8) to information about the shape of the basis that DBKZ is likely to return.

Corollary 2. *Applying Heuristic 1 to every lattice that is passed to the SVP oracle during the execution of Algorithm 1, the fixed point of the heuristic dynamical system, i.e. (6) with $\alpha = \frac{1}{2}\log GH(k)$, is (8) with the same α and implies that after one more forward tour, the basis satisfies*

$$\|\mathbf{b}_i^*\| = GH(k)^{\frac{n+1-2i}{2(k-1)}} \det(\mathcal{L}(\mathbf{B}))^{\frac{1}{n}} \tag{14}$$

for all $i \leq n - k$.

Proof. According to (8), upon termination of Algorithm 1 the output basis satisfies

$$\log(\det([\mathbf{b}_1, \ldots, \mathbf{b}_i])) = \frac{(n-k-i+1)(k+i-1)}{k-1}\alpha$$

By Heuristic 1 we have $\log\|\mathbf{b}_1\| = \alpha + x_1/k$, from which Eq. (14) easily follows for $i = 1$. Now assume (14) holds for all $j < i$. Then we have, again by Heuristic 1, $\log\|\mathbf{b}_i^*\| = \alpha + (x_i - \sum_{j<i}\log\|\mathbf{b}_j^*\|)/k$. Invoking the induction hypothesis, Eq. (14) easily follows for all $i \leq n - k$. □

Corollary 2 shows that the output of the DBKZ algorithm, if terminated after a forward tour, can be expected to closely follow the GSA, at least for all $i \leq n-k$ and can be computed using simple closed formulas. It is noteworthy that the self-dual properties of DBKZ imply that if terminated after a backward tour, the GSA holds for all $i \geq k$. This means, depending on the application one can choose which part of the output basis to predict. Moreover, we see that DBKZ allows to predict a much larger part of the basis than Slide reduction solely based on the Gaussian Heuristic. If one is willing to make additional assumptions, i.e. assumptions about the shape of a k-dimensional HKZ reduced basis, the BKZ simulator allows to predict the shape of the entire basis output by BKZ. Obviously, the same assumptions can be used to estimate the remaining parts of the shape of the basis in the case of Slide reduction and DBKZ, since a final application of a HKZ reduction to individual blocks of size k only requires

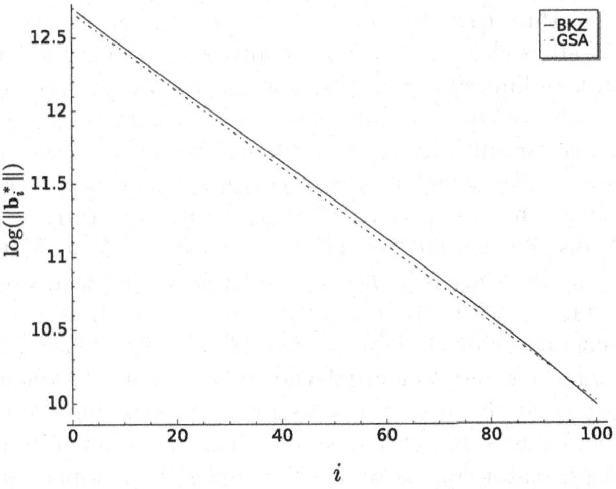

Fig. 1. Expected shape of the first 100 basis vectors in dimension $n = 200$ after BKZ compared to the GSA. Note that the latter corresponds exactly to the expected shape of the first 100 basis vectors after DBKZ (cf. 2).

negligible amount of time compared to the running time of the entire algorithm. Furthermore, since the estimation of the known part of the shape (from Corollary 2 and 1) do not depend on these additional assumptions, the estimation for Slide reduction and DBKZ is much less sensitive to the (in-)correctness of these assumptions, while errors propagate during the BKZ simulation.

To compare the expected output of BKZ, DBKZ, and Slide reduction, we generated a Goldstein-Mayer lattice [17] in dimension $n = 200$ with numbers of bit size 2000, applied LLL to it, and simulated the execution of BKZ with block size $k = 100$ until no more progress was made. The output in terms of the logarithm of the shape of the basis for the first 100 basis vectors is shown in Fig. 1 and compared to the GSA. Recall that the latter represents the expected output of DBKZ and, to some degree, Slide reduction. Under the assumption that Heuristic 1 and the BKZ simulator are accurate, one would expect BKZ to behave a little worse than the other two algorithms in terms of output quality.

6 Experiments

For an experimental comparison, we implemented DBKZ and Slide reduction in `fpLLL`. SVP reduction in `fplll` is implemented in the standard way as described in Sect. 2.1. For dual SVP reduction we used the algorithm explained in the Sect. 7.

6.1 Methodology

In the context of cryptanalysis we are usually interested in the root Hermite factor achievable using lattice reduction in order to choose parameters for cryptosystems, as this often determines the success probability and/or complexity of an attack. It is clear that merely reporting on the average root Hermite factor achieved is of limited use for this. Instead we will view the resulting root Hermite factor achieved by a certain reduction algorithm (with certain parameters) as a random variable and try to estimate the main statistical parameters of its distribution. We believe this will eventually allow for more meaningful security estimates. The only previous experimental work studying properties of the underlying distribution of the root Hermite factor [16] suggests that it is a Gaussian-like but the study is limited to relatively small block sizes.

Since experiments with lattice reduction are rather time consuming, it is infeasible to generate as much data as desirable to estimate statistical parameters like the mean value and standard deviation accurately. A standard statistical technique to overcome this is to use bootstrapping to compute confidence intervals for these parameters. Roughly speaking, in order to compute the confidence interval for an estimator from a set of N samples, we sample l sets of size N with replacement from the original samples and compute the estimator for each of them. Intuitively, this should give a sense of the variability of the estimator computed on the samples. Our confidence interval with confidence parameter α, according to the bootstrap percentile interval method, is simply the $\alpha/2$ and $1 - \alpha/2$ quantiles. For further discussion we refer to [70]. Throughout this work we use $\alpha = .05$ and $l = 100$. The complete confidence intervals for mean value and standard deviation are listed in Appendix A. Whenever we refer to the standard deviation of a distribution resulting from the application of a reduction algorithm and computing the root Hermite factor achieved, we mean the maximum of the corresponding confidence interval.

It is folklore that the output quality of lattice reduction algorithms measured in the root Hermite factor depends mostly on the block size parameter rather than on properties of the input lattice, like the dimension or bit size of the numbers, at least when the lattice dimension and size of the numbers is large enough. A natural approach to comparing the different algorithms would be to fix a number of lattices of certain dimension and bit size and run the different algorithms with varying block size on them. Unfortunately, Slide reduction requires the block size to divide the dimension.[9] To circumvent this we select the dimension of the input lattices depending on the block sizes we want to test, i.e. $n = t \cdot k$, where k is the block size and t is a small integer. This is justified as most lattice attacks involve choosing a suitable sublattice to attack, where such a requirement can easily be taken into account. Since for very small dimensions block reduction performs a little better then in larger dimensions, we need to deal with a trade-off here: on the one hand we need to ensure that the lattice dimension n is large enough, even for small block sizes, so that the result is not biased positively for small block sizes due to the small dimension. On the other hand, if the lattice dimension grows very large we would have to

[9] While it is trivial to generalize Slide reduction to other block sizes, the performance in terms of the achieved output quality of the basis deteriorates somewhat in this case compared to other reduction algorithms [29].

increase the precision of the GSO computation significantly which would result in an artificial slow down and thus limit the data we are able to collect. Our experiments and previous work [16] suggest that the bias for small dimensions weakens sufficiently as soon as the lattice dimension is larger than 140, so for the lattice dimension n we select the smallest multiple t of the block size k such that $t \cdot k \geq 140$.

For each block size we generated 10 different subset sum lattices in dimension n in the sense of [19] and we fix the bit size of the numbers to $10 \cdot n$ following previous work [19,36]. Experimental studies [43] have shown that this notion of random lattices is suitable in this context as lattice reduction behaves similarly on them as on "random" lattices in a mathematically more precise sense [17].[10] Then we ran each of the three reduction algorithms with corresponding block size on each of those lattices. For BKZ and DBKZ we used the same terminating condition: the algorithms terminate when the slope of the shape of the basis does not improve during 5 loop iterations in a row (this is the default terminating condition in `fpLLL`'s BKZ routine with auto abort option set). Finally, for sufficiently large block sizes ($k > 45$), we preprocessed the local blocks with BKZ-($k/2$) before calling the SVP oracle, since this has been shown to achieve good asymptotic running time [69] and also seemed a good choice in practice in our experiments.

6.2 Results

Figure 2 shows the average output quality including the confidence interval produced by each of the three algorithms in comparison with the prediction based

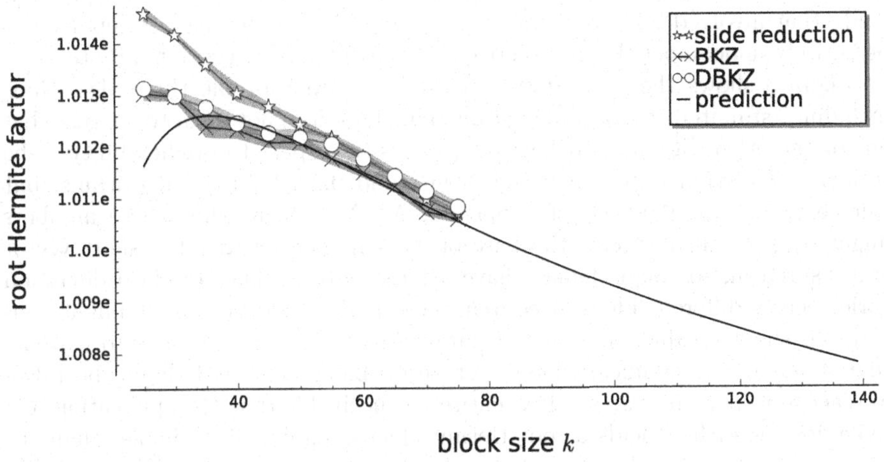

Fig. 2. Confidence interval of average root Hermite factor for random bases as computed by different reduction algorithms and the prediction given by Eq. (4).

[10] In fact, subset sum lattices are extremely similar to the random lattices of [17].

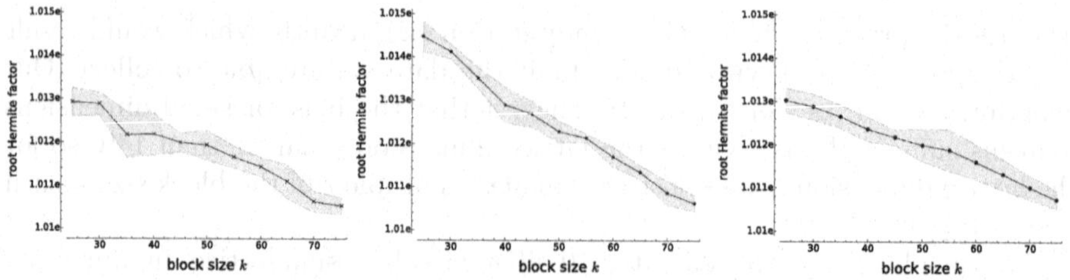

(a) BKZ (b) Slide Reduction (c) DBKZ

Fig. 3. Same as Fig. 2 with estimated standard deviation

on the Gaussian Heuristic (cf. Eq. (4)). It demonstrates that BKZ and DBKZ have comparable performance in terms of output quality and clearly outperform Slide reduction for small block sizes (< 50), which confirms previous reports [16]. For some of the small block sizes (e.g. $k = 35$) BKZ seems to perform unexpectedly well in our experiments. To see if this is indeed inherent to the algorithms or a statistical outlier owed to the relatively small number of data points, we ran some more experiments with small block sizes. We report on the results in the full version [37], where we show that the performance of BKZ and DBKZ are actually extremely close for these parameters.

Furthermore, Fig. 2 shows that all three algorithms tend towards the prediction given by Eq. (4) in larger block sizes, supporting the conjecture, and Slide reduction becomes quite competitive. Even though BKZ still seems to have a slight edge for block size 75, note that the confidence intervals for Slide reduction and BKZ are heavily overlapping here. This is in contrast to the only previous study that involved Slide reduction [16], where Slide reduction was reported to be entirely noncompetitive in practice and thus mainly of theoretical interest.

Figure 3 shows the same data separately for each of the three algorithms including estimated standard deviation. The data does not seem to suggest that one or the other algorithm behaves "nicer" with respect to predictability – the standard deviation ranges between 0.0002 and 0.0004 for all algorithms, but can be as high as 0.00054 (cf. Appendix A). Note that while these numbers might seem small, it affects the base of the exponential that the short vector is measured in, so small changes have a large impact. The standard deviation varies across different block sizes, but there is no evidence that it might converge to smaller values or even 0 in larger block sizes. So we have to assume, that it remains a significant factor for larger block sizes and should be taken into account in cryptanalysis. It is entirely conceivable that the application of a reduction algorithm yields a root Hermite factor significantly smaller than the corresponding mean value.

In order to compare the runtime of the algorithms we ran separate experiments, because due to the way we selected the dimension, the data would exhibit a somewhat strange "zigzag" behavior. For each block size $50 \leq k \leq 75$ we generated again 10 random subset sum lattices with dimension $n = 2k$ and the bit size of the numbers was fixed to 1400. Figure 4 shows the average runtime for each of the algorithms and block size in log scale. It shows that the runtime of all three algorithms follows a close to single exponential (in the block size) curve. This supports the intuition that the runtime mainly depends on the complexity of the SVP oracle, since we are using an implementation that preprocesses the local blocks before enumeration with large block size. This has been shown to achieve an almost single exponential complexity (up to logarithmic factors in the exponent) [69].

The data also shows that in terms of runtime, Slide reduction outperforms both, BKZ and DBKZ. But again, with increasing block size the runtime of the different algorithms seem to converge to each other. Combined with the data from Fig. 2 this suggests that all three algorithms offer a similar trade-off between runtime and achieved Hermite factor for large block sizes. This shows that Slide reduction is not only theoretically interesting with its cleaner and tighter analysis of both, output quality and runtime, but also quite competitive in practice. It should be noted that we analyzed Slide reduction as described in [15]. While significant research effort has been spent on improving BKZ, essentially nothing along these lines has been done with regards to Slide reduction. We hope that the results reported here will initiate more research into improvements, both in practice and theory, of Slide reduction.

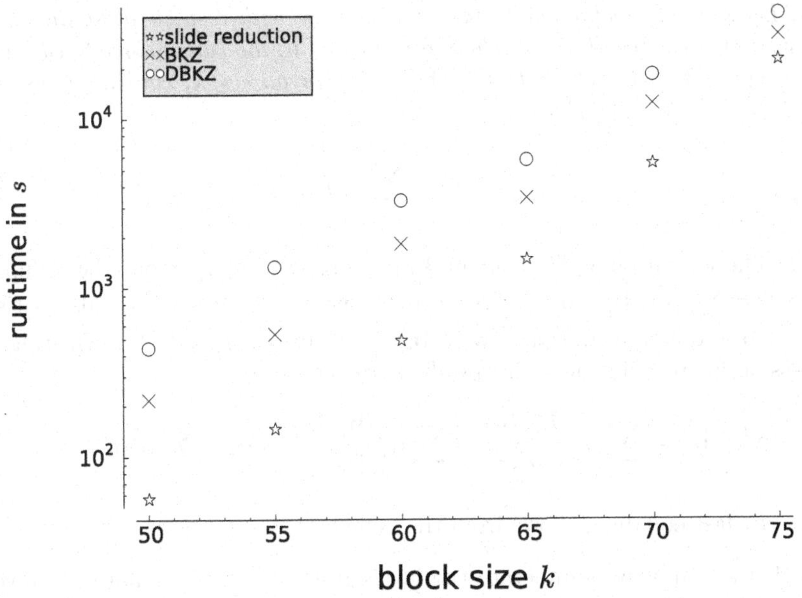

Fig. 4. Average runtime in seconds for random bases in dimension $n = 2k$ for different reduction algorithms (in log scale).

7 Dual Enumeration

Similar to Slide reduction, DBKZ makes intensive use of dual SVP reduction of projected blocks. The obvious way to achieve this reduction is to compute the dual basis for the projected block, run the primal SVP reduction on it, and finally compute the primal basis of the block. While the transition between primal and dual basis is a polynomial time computation and is thus dominated by the enumeration step, it does involve matrix inversion, which can be quite time consuming in practice. To address this issue, Gama and Nguyen [15] proposed a different strategy. Note that SVP reduction, as performed by enumeration, consists of two steps: (1) the coordinates of a shortest vector in the given basis are computed, and (2) this vector is inserted into the basis. Gama and Nguyen observe that for dual SVP reduction, (2) can be achieved using the coordinates obtained during the dual enumeration by solely operating on the primal basis. Furthermore, note that the enumeration procedure (step (1)) only operates on the GSO of the basis so it is sufficient for (1) to invert the GSO matrices of the projected block, which is considerably easier since they consist of a diagonal and an upper triangular matrix. However, this still incurs a computational overhead of $\Omega(n^3)$.

We now introduce a way to find the coordinates of a shortest vector in the dual lattice without computing the dual basis or dual GSO.

Lemma 2. *Let* $\mathbf{B}$ *be a lattice basis and* $\mathbf{w}$ *an arbitrary vector in the linear span of* $\mathbf{B}$. *Let* $\mathbf{x}$ *be the coefficient vector expressing* $\mathbf{w}$ *with respect to the dual basis, i.e.,* $x_i = \langle \mathbf{w}, \mathbf{b}_i \rangle$ *for all* $i \leq n$. *Then, for any* $k \leq n$, *the (uniquely defined) vector* $\mathbf{w}^{(k)} \in \operatorname{span}(\mathbf{B}_{[1,k]})$ *such that* $\langle \mathbf{w}^{(k)}, \mathbf{b}_i \rangle = x_i$ *for all* $i \leq k$, *can be expressed as* $\mathbf{w}^{(k)} = \sum_{i \leq k} \alpha_i \mathbf{b}_i^* / \|\mathbf{b}_i^*\|^2$ *where*

$$\alpha_i = x_i - \sum_{j<i} \mu_{i,j} \alpha_j. \tag{15}$$

Proof. The condition $\mathbf{w}^{(k)} \in \operatorname{span}(\mathbf{B}_{[1,k]})$ directly follows from the definition of $\mathbf{w}^{(k)} = \sum_{i \leq k} \alpha_i \mathbf{b}_i^* / \|\mathbf{b}_i^*\|^2$. We need to show that this vector also satisfies the scalar product conditions $\langle \mathbf{w}^{(k)}, \mathbf{b}_i \rangle = x_i$ for all $i \leq k$. Substituting the expression for $\mathbf{w}^{(k)}$ in the scalar product we get

$$\langle \mathbf{w}^{(k)}, \mathbf{b}_i \rangle = \sum_{j \leq k} \alpha_j \frac{\langle \mathbf{b}_j^*, \mathbf{b}_i \rangle}{\|\mathbf{b}_j^*\|^2} = \sum_{j \leq i} \alpha_j \frac{\langle \mathbf{b}_j^*, \mathbf{b}_i \rangle}{\|\mathbf{b}_j^*\|^2} = \alpha_i + \sum_{j<i} \alpha_j \mu_{i,j} = x_i$$

where the last equality follows from the definition of α_i. □

This shows that if we enumerate the levels from $k = 1$ to n (note the reverse order as opposed to primal enumeration) we can easily compute α_k from all the given or previously computed quantities in $O(n)$. The length of $\mathbf{w}^{(k)}$ is given by

$$\|\mathbf{w}^{(k)}\|^2 = \sum_{i \leq k} \alpha_i^2 / \|\mathbf{b}_i^*\|^2 = \|\mathbf{w}^{(k-1)}\|^2 + \alpha_k^2 / \|\mathbf{b}_k^*\|^2. \tag{16}$$

To obtain an algorithm that is practically as efficient as primal enumeration, it is necessary to apply the same standard optimizations known as SchnorrEuchner enumeration to the dual enumeration. It is obvious that we can exploit lattice symmetry and dynamic radius updates in the same fashion as in the primal enumeration. The only optimization that is not entirely obvious is enumerating the values for x_k in order of increasing length of the resulting partial solution. However, from Eqs. (15) and (16) it is clear that we can start by selecting $x_k = \lfloor \sum_{j<k} \mu_{k,j}\alpha_j \rceil$ in order to minimize the first value of α_k, and then proceed by alternating around this first value just as in the SchnorrEuchner primal enumeration algorithm.

It is also noteworthy that being able to compute partial solutions even allows us to apply pruning [14] directly. In summary this shows that dual SVP enumeration should be just as efficient as primal enumeration. To illustrate this, Algorithms 2 and 3 show the SchnorrEuchner variant of the two enumeration procedures.[11]

Algorithm 2. Dual Enumeration

procedure DualEnum(μ, $(\|\mathbf{b}_i^*\|^2)_{i\in[n]}$, A)
Input: The GSO of a lattice μ and $(\|\mathbf{b}_i^*\|^2)_{i\in[n]}$ and an upper bound A to the squared length of a shortest dual vector
Output: The coordinates of a shortest dual vector in the dual basis $\mathbf{D}$

1 $k \leftarrow 1$
2 **while** $k \geq 1$
3 $\quad \alpha_k \leftarrow x_k - \sum_{j<k} \mu_{k,j}\alpha_j$
4 $\quad l_k \leftarrow l_{k-1} + \alpha_k^2/\|\mathbf{b}_k^*\|^2$
5 $\quad$ **if** $l_k \leq A$ and $k = n$ **then**
6 $\quad\quad \mathbf{s} \leftarrow \mathbf{x}$, $A \leftarrow l_k$
7 $\quad$ **if** $l_k \leq A$ and $k < n$ **then**
8 $\quad\quad k \leftarrow k+1$, $x_k \leftarrow \lfloor \sum_{j<k} \mu_{k,j}\alpha_j \rceil$
9 $\quad$ **else**
10 $\quad\quad k \leftarrow k-1$, $x_k \leftarrow \text{nextX}(k)$
11 **return s**

Algorithm 3. Primal Enumeration

procedure PrimalEnum(μ, $(\|\mathbf{b}_i^*\|^2)_{i\in[n]}$, A)
Input: The GSO of a lattice μ and $(\|\mathbf{b}_i^*\|^2)_{i\in[n]}$ and an upper bound A to the squared length of a shortest vector
Output: The coordinates of a shortest vector in the basis $\mathbf{B}$

1 $k \leftarrow n$
2 **while** $k \leq n$
3 $\quad \alpha_k \leftarrow x_k + \sum_{j>k} \mu_{j,k}x_j$
4 $\quad l_k \leftarrow l_{k+1} + \alpha_k^2\|\mathbf{b}_k^*\|^2$
5 $\quad$ **if** $l_k \leq A$ and $k = 1$ **then**
6 $\quad\quad \mathbf{s} \leftarrow \mathbf{x}$, $A \leftarrow l_k$
7 $\quad$ **if** $l_k \leq A$ and $k > 1$ **then**
8 $\quad\quad k \leftarrow k-1$, $x_k \leftarrow \lfloor -\sum_{j>k} \mu_{j,k}x_j \rceil$
9 $\quad$ **else**
10 $\quad\quad k \leftarrow k+1$, $x_k \leftarrow \text{nextX}(k)$
11 **return s**

Implementation Notes. To give some experimental evidence that the dual enumeration is just as efficient as primal enumeration, we implemented it in `fpLLL`.[12] Note that Algorithm 2 can be easily added to an implementation of Algorithm 3 by using special cases in data accesses and a few operations. Furthermore, in order to avoid the division in line 4 we precomputed the values $1/\|\mathbf{b}_k^*\|^2$ for all k.

[11] The function `nextX` simply selects the next value for a specific variable in order to alternate correctly around the center of the interval of valid values. We omit details here since it works identical in both algorithms and requires auxiliary variables that would clutter the code unnecessarily.

[12] At the point of publication of this work, a modified version of this implementation is now included in the main branch of `fplll`.

We compared the implementation with the primal enumeration on 10 random bases (in the same sense as in Sect. 6) in dimension $35 \leq n \leq 50$. As expected, the rate of enumeration was close to equal in both cases – around $3.2 \cdot 10^7$ nodes per second (cf. Table 1), which corresponds to slightly more than 100 cycles per node on our 3.4 Ghz test machine. The slight discrepancies (and the low rate for $n = 35$) can be explained by the variable number of nodes that were enumerated and thus certain setup costs are amortized over a different number of nodes.

Table 1. Rate of enumeration (in 10^7 nodes per s) in primal and dual enumeration

n	35	40	45	50
primal	2.73	3.16	3.13	3.17
dual	2.77	3.19	3.18	3.27

8 Conclusion and Future Work

While our experimental study of lattice reduction confirms that the *average* root Hermite factor achieved by lattice reduction is indeed, as conjectured, given by Eq. (4), the standard deviation is large enough that it is conceivable that a *single* instance finds a much shorter vector. This means that cryptanalytic estimates should take this into account.

It is clear that we need to learn more about the underlying distribution in order to aid parameter selection. For example, using more data one could try to verify experimentally if the distribution follows a (possibly truncated) Gaussian as already suspected in [16] for small block sizes, which would allow for much tighter bounds and meaningful estimates. A brief inspection of our data suggests that this might be true even for larger block sizes, but 10 data points per experiment is not sufficient to allow for any further conclusions about the distribution. In any case, we believe our results show that simply relying on the average of a handful of data points is not very meaningful and we hope that this work can serve as a starting point for more sophisticated approaches to selecting parameters secure against attacks involving lattice reduction.

With our new dual enumeration algorithm we provide another tool to practically examine different reduction algorithms. This should facilitate experimental research into reduction algorithms that make use of dual SVP reduction, like variants of Slide reduction. Future lines of research could explore if, for example, the block Rankin reduction algorithm of [28] can be efficiently implemented by using it to apply the densest sublattice algorithm of [10] to the dual lattice. This could be used to achieve potentially stronger notions of reduction with better output quality.

Acknowledgment. We thank Arnold Neumaier for completing the analysis of the convergence of DBKZ. Furthermore, we thank Arnold and the anonymous reviewers of the Eurocrypt 2016 committee for many helpful comments on a previous version of this work. We also thank Florian Göpfert for helpful discussions in the early stages of the development of the dual enumeration algorithm.

References

1. Ajtai, M.: Generating hard instances of lattice problems. Complex. Comput. Proofs Quaderni di Matematica **13**, 1–32 (2004). Preliminary version in STOC 1996
2. M. Ajtai and C. Dwork. A public-key cryptosystem with worst-case/average-case equivalence. In: Proceedings of STOC 1997, pp. 284–293. ACM, May 1997
3. Akhavi, A., Stehlé, D.: Speeding-up lattice reduction with random projections (extended abstract). In: Laber, E.S., Bornstein, C., Nogueira, L.T., Faria, L. (eds.) LATIN 2008. LNCS, vol. 4957, pp. 293–305. Springer, Heidelberg (2008)
4. Albrecht, M., Cadé, D., Pujol, X., Stehlé, D.: fplll-4.0, a floating-point LLL implementation. http://perso.ens-lyon.fr/damien.stehle
5. Albrecht, M.R., Player, R., Scott, S.: On the concrete hardness of learning with errors. Cryptology ePrint Archive, Report 2015/046 (2015). http://eprint.iacr.org/
6. Bachem, A., Kannan, R.: Lattices and basis reduction algorithm. Technical Report 84–006, Mathematisches Institut, Universität zu Köln (1984)
7. Chen, Y.: Réduction de réseau et sécurité concrète du chiffrement complètement homomorphe. Ph.D. thesis, ENS, Paris, Thse de doctorat dirige par Nguyen, Phong-Quang Informatique Paris 7 (2013)
8. Chen, Y., Nguyen, P.Q.: BKZ 2.0: Better lattice security estimates. In: Lee, D.H., Wang, X. (eds.) ASIACRYPT 2011. LNCS, vol. 7073, pp. 1–20. Springer, Heidelberg (2011)
9. Coupé, C., Nguyên, P.Q., Stern, J.: The effectiveness of lattice attacks against low-exponent RSA. In: Imai, H., Zheng, Y. (eds.) PKC 1999. LNCS, vol. 1560, pp. 204–218. Springer, Heidelberg (1999)
10. Dadush, D., Micciancio, D.: Algorithms for the densest sub-lattice problem. In: Proceedings of the Twenty-Fourth Annual ACM-SIAM Symposium on Discrete Algorithms, SODA 2013, pp. 1103–1122. SIAM (2013)
11. Fincke, U., Pohst, M.: Improved methods for calculating vectors of short length in a lattice, including a complexity analysis. Math. Comput. **44**, 463–471 (1985)
12. Gama, N., Howgrave-Graham, N., Koy, H., Nguyên, P.Q.: Rankin's constant and blockwise lattice reduction. In: Dwork, C. (ed.) CRYPTO 2006. LNCS, vol. 4117, pp. 112–130. Springer, Heidelberg (2006)
13. Gama, N., Howgrave-Graham, N., Nguyên, P.Q.: Symplectic lattice reduction and NTRU. In: Vaudenay, S. (ed.) EUROCRYPT 2006. LNCS, vol. 4004, pp. 233–253. Springer, Heidelberg (2006)
14. Gama, N., Nguyen, P.Q., Regev, O.: Lattice enumeration using extreme pruning. In: Gilbert, H. (ed.) EUROCRYPT 2010. LNCS, vol. 6110, pp. 257–278. Springer, Heidelberg (2010)
15. Gama, N., Nguyen, P.Q.: Finding short lattice vectors within Mordell's inequality. In: Proceedings of STOC, pp. 207–216. ACM, May 2008
16. Gama, N., Nguyen, P.Q.: Predicting lattice reduction. In: Smart, N.P. (ed.) EUROCRYPT 2008. LNCS, vol. 4965, pp. 31–51. Springer, Heidelberg (2008)
17. Goldstein, D., Mayer, A.: On the equidistribution of Hecke points. Forum Mathematicum **15**(2), 165–189 (2003)

18. Hanrot, G., Pujol, X., Stehlé, D.: Analyzing blockwise lattice algorithms using dynamical systems. In: Rogaway, P. (ed.) CRYPTO 2011. LNCS, vol. 6841, pp. 447–464. Springer, Heidelberg (2011)
19. Hanrot, G., Stehlé, D.: Improved analysis of Kannan's shortest lattice vector algorithm. In: Proceedings of Crypto [22], pp. 170–186
20. Helfrich, B.: Algorithms to construct minkowski reduced and hermite reduced lattice bases. Theoret. Comput. Sci. **41**(2–3), 125–139 (1985)
21. Howgrave-Graham, N.: A hybrid lattice-reduction and meet-in-the-middle attack against NTRU. In: Proceedings of Crypto [22], pp. 150–169
22. Menezes, A. (ed.): CRYPTO 2007. LNCS, vol. 4622. Springer, Heidelberg (2007)
23. Joux, A., Stern, J.: Lattice reduction: A toolbox for the cryptanalyst. J. Cryptol. **11**(3), 161–185 (1998)
24. Kannan, R.: Improved algorithms for integer programming, related lattice problems. In: Proceedings of the Fifteenth Annual ACM Symposium on Theory of Computing - STOC 1983, pp. 193–206. ACM, April 1983. Journal version in Math. of Operation Research **12**(3), 415–440 (1987)
25. Khot, S.: Hardness of approximating the shortest vector problem in lattices. J. ACM **52**(5), 789–808 (2005). Preliminary version in FOCS 2004
26. Koy, H., Schnorr, C.-P.: Segment LLL-reduction of lattice bases. In: Silverman, J.H. (ed.) CaLC 2001. LNCS, vol. 2146, pp. 67–80. Springer, Heidelberg (2001)
27. Lenstra, A.K., Lenstra, Jr., H.W., Lovász, L.: Factoring polynomials with rational coefficients. Math. Ann. **261**, 513–534 (1982)
28. Li, J., Nguyen, P.Q.: Approximating the densest sublattice from Rankins inequality. LMS J. Comput. Math. **17**, 92–111 (2014)
29. Li, J., Wei, W.: Slide reduction, successive minima, several applications. Bull. Aust. Math. Soc. **88**, 390–406 (2013)
30. Lindner, R., Peikert, C.: Better key sizes (and attacks) for LWE-based encryption. In: Kiayias, A. (ed.) CT-RSA 2011. LNCS, vol. 6558, pp. 319–339. Springer, Heidelberg (2011)
31. Lyubashevsky, V., Micciancio, D.: On bounded distance decoding, unique shortest vectors, and the minimum distance problem. In: Halevi, S. (ed.) CRYPTO 2009. LNCS, vol. 5677, pp. 577–594. Springer, Heidelberg (2009)
32. Mehrotra, S., Li, Z.: Segment LLL reduction of lattice bases using modular arithmetic. Algorithms **3**(3), 224–243 (2010)
33. Micciancio, D.: Almost perfect lattices, the covering radius problem, applications to Ajtai's connection factor. SIAM J. Comput. **34**(1), 118–169 (2004). Preliminary version in STOC 2002
34. Micciancio, D.: Inapproximability of the shortest vector problem: Toward a deterministic reduction. Theory Comput. **8**(1), 487–512 (2012)
35. Micciancio, D., Regev, O.: Worst-case to average-case reductions based on Gaussian measure. SIAM J. Comput. **37**(1), 267–302 (2007). Preliminary version in FOCS 2004
36. Micciancio, D., Walter, M.: Fast lattice point enumeration with minimal overhead. In: Indyk, P. (ed.) Proceedings of the Twenty-Sixth Annual ACM-SIAM Symposium on Discrete Algorithms, SODA, San Diego, CA, USA, January 4–6, 2015, pp. 276–294. SIAM (2015)
37. Micciancio, D., Walter, M.: Practical, predictable lattice basis reduction. Cryptology ePrint Archive, Report 2015/1123 (2015). http://eprint.iacr.org/
38. Neumaier, A.: Bounding basis reduction properties. Cryptology ePrint Archive, Report 2016/004 (2016). http://eprint.iacr.org/

39. Nguyên, P.Q.: Cryptanalysis of the goldreich-goldwasser-halevi cryptosystem from Crypto'97. In: Wiener, M. (ed.) CRYPTO 1999. LNCS, vol. 1666, pp. 288–304. Springer, Heidelberg (1999)
40. Nguyen, P.: Hermite's constant and lattice algorithms. In: Nguyen, P.Q., Vallée, B. (eds.) The LLL Algorithm, pp. 19–69. Springer, Heidelberg (2010)
41. Nguyen, P.Q.: Lattice reduction algorithms: theory and practice. In: Paterson, K.G. (ed.) EUROCRYPT 2011. LNCS, vol. 6632, pp. 2–6. Springer, Heidelberg (2011)
42. Nguyên, P.Q., Stehlé, D.: Low-dimensional lattice basis reduction revisited. In: Buell, D.A. (ed.) ANTS 2004. LNCS, vol. 3076, pp. 338–357. Springer, Heidelberg (2004). Journal version in ACM Trans. on Algorithms
43. Nguyên, P.Q., Stehlé, D.: LLL on the average. In: Hess, F., Pauli, S., Pohst, M. (eds.) ANTS 2006. LNCS, vol. 4076, pp. 238–256. Springer, Heidelberg (2006)
44. Nguyên, P.Q., Stern, J.: Cryptanalysis of the Ajtai-Dwork cryptosystem. In: Krawczyk, H. (ed.) CRYPTO 1998. LNCS, vol. 1462, pp. 223–242. Springer, Heidelberg (1998)
45. Nguyen, P., Stern, J.: Lattice reduction in cryptology: an update. In: Bosma, W. (ed.) ANTS 2000. LNCS, vol. 1838, pp. 85–112. Springer, Heidelberg (2000)
46. Nguyên, P.Q., Stern, J.: The two faces of lattices in cryptology. In: Silverman, J.H. (ed.) CaLC 2001. LNCS, vol. 2146, pp. 146–180. Springer, Heidelberg (2001)
47. Nguyen, P.Q., Shparlinski, I.E.: The insecurity of the elliptic curve digital signature algorithm with partially known nonces. Des. Codes Crypt. **30**(2), 201–217 (2003)
48. Nguyen, P.Q., Stehlé, D.: Low-dimensional lattice basis reduction revisited. ACM Trans. Algorithms **5**(4), 46 (2009). Preliminary version in ANTS 2004
49. Peikert, C.: Public-key cryptosystems from the worst-case shortest vector problem. In: Proceedings of STOC, pp. 333–342. ACM (2009)
50. Plantard, T., Susilo, W.: Recursive lattice reduction. In: Garay, J.A., De Prisco, R. (eds.) SCN 2010. LNCS, vol. 6280, pp. 329–344. Springer, Heidelberg (2010)
51. Pohst, M.: A modification of the LLL-reduction algorithm. J. Symbolic Comput. **4**(1), 123–127 (1987)
52. Regev, O.: New lattice based cryptographic constructions. J. ACM **51**(6), 899–942 (2004). Preliminary version in STOC 2003
53. Regev, O.: On lattices, learning with errors, random linear codes, cryptography. J. ACM **56**(6), 34 (2009). Preliminary version in STOC 2005
54. Schnorr, C.-P.: A hierarchy of polynomial time lattice basis reduction algorithms. Theoret. Comput. Sci. **53**(2–3), 201–224 (1987)
55. Schnorr, C.-P.: A more efficient algorithm for lattice basis reduction. J. Algorithms **9**(1), 47–62 (1988)
56. Schnorr, C.-P.: Block reduced lattice bases and successive minima. Comb. Probab. Comput. **3**, 507–522 (1994)
57. Schnorr, C.-P.: Lattice reduction by random sampling and birthday methods. In: Alt, H., Habib, M. (eds.) STACS 2003. LNCS, vol. 2607, pp. 145–156. Springer, Heidelberg (2003)
58. Schnorr, C.P.: Fast LLL-type lattice reduction. Inf. Comput. **204**(1), 1–25 (2006)
59. Schnorr, C.P.: Progress on LLL and lattice reduction. In: Nguyen, P.Q., Vallée, B. (eds.) The LLL Algorithm, pp. 145–178. Springer, Heidelberg (2010)
60. Schnorr, C.-P., Euchner, M.: Lattice basis reduction: Improved practical algorithms and solving subset sum problems. Math. Program. **66**(1–3), 181–199 (1991). Preliminary version in FCT 1994
61. Schnorr, C.-P., Fischlin, M., Koy, H., May, A.: Lattice attacks on GGH cryptosystem. Rump session of Crypto 1997 (1997)

62. Schnorr, C.-P., Hörner, H.H.: Attacking the chor-rivest cryptosystem by improved lattice reduction. In: Guillou, L.C., Quisquater, J.-J. (eds.) EUROCRYPT 1995. LNCS, vol. 921, pp. 1–12. Springer, Heidelberg (1995)
63. Seysen, M.: Simultaneous reduction of a lattice basis and its reciprocal basis. Combinatorica **13**(3), 363–376 (1993)
64. Shoup, V.: NTL: a library for ng number theory. http://www.shoup.net/ntl/index.html
65. Storjohann, A.: Faster algorithms for integer lattice basis reduction. Technical Report 249, Swiss Federal Institute of Technology, ETH-Zurich, Department of Computer Science, Zurich, Switzerland, July 1996
66. Vallée, B., Girault, M., Toffin, P.: How to break okamoto's cryptosystem by reducing lattice bases. In: Günther, C.G. (ed.) EUROCRYPT 1988. LNCS, vol. 330, pp. 281–291. Springer, Heidelberg (1988)
67. Vallée, B., Vera, A.: Probabilistic analyses of lattice reduction algorithms. In: Nguyen, P.Q., Vallée, B. (eds.) The LLL Algorithm, pp. 71–143. Springer, Heidelberg (2010)
68. van de Pol, J., Smart, N.P.: Estimating key sizes for high dimensional lattice based systems. Cryptology ePrint Archive, Report 2013/630 (2013). http://eprint.iacr.org/
69. Walter, M.: Lattice point enumeration on block reduced bases. In: Lehmann, A., Wolf, S. (eds.) Information Theoretic Security. LNCS, vol. 9063, pp. 269–282. Springer, Heidelberg (2015)
70. Wasserman, L.: All of Nonparametric Statistics. Springer Texts in Statistics. Springer, New York (2006)

A Experimental Data

See Table 2.

Table 2. Confidence intervals for mean value μ and standard deviation σ of root Hermite factor achieved by lattice reduction with block size k

k	$\mu_{[min]}$	$\mu_{[max]}$	$\sigma_{[min]}$	$\sigma_{[min]}$
BKZ				
25	1.0129	1.0132	0.00014910	0.00031542
30	1.0129	1.0131	0.00013775	0.00028044
35	1.0122	1.0127	0.00030272	0.00053605
40	1.0122	1.0125	0.00020367	0.00045143
45	1.0120	1.0122	0.00011779	0.00022000
50	1.0119	1.0122	0.00019697	0.00035792
55	1.0117	1.0119	0.00012984	0.00025347
60	1.0114	1.0116	0.00009925	0.00022226
65	1.0111	1.0114	0.00012052	0.00038869
70	1.0106	1.0109	0.00012487	0.00026133
75	1.0105	1.0107	0.00009756	0.00020439
Slide reduction				
25	1.0145	1.0148	0.00016524	0.00034219
30	1.0141	1.0142	0.00005426	0.00017119
35	1.0135	1.0137	0.00013333	0.00022903
40	1.0129	1.0133	0.00018023	0.00041554
45	1.0127	1.0129	0.00013605	0.00027860
50	1.0123	1.0125	0.00014877	0.00026298
55	1.0121	1.0122	0.00005830	0.00009498
60	1.0117	1.0119	0.00013659	0.00022897
65	1.0114	1.0115	0.00009455	0.00017193
70	1.0109	1.0111	0.00012178	0.00023823
75	1.0106	1.0108	0.00010597	0.00019067
Self-Dual BKZ				
25	1.0130	1.0133	0.00012817	0.00029479
30	1.0129	1.0131	0.00017812	0.00027150
35	1.0127	1.0129	0.00016756	0.00025963
40	1.0123	1.0126	0.00013635	0.00028876
45	1.0122	1.0123	0.00010143	0.00018625
50	1.0120	1.0123	0.00018334	0.00038216
55	1.0119	1.0123	0.00026051	0.00046222
60	1.0116	1.0120	0.00018311	0.00040919
65	1.0113	1.0116	0.00014412	0.00037256
70	1.0110	1.0112	0.00013096	0.00030097
75	1.0107	1.0109	0.00011095	0.00021169

Author Index

Printed in the United States
By Bookmasters

Printed in the United States
By Bookmasters